## About the Cover

The cover image illustrates a sphidron, which is a circular disc formed by spiral arms. Sphidrons are a variety of spidrons, which were invented by Daniel Erdely in 1979 as a homework assignment for Erno Rubik, inventor of the Rubik's Cube.

▶ **Meet the Artist**

The cover art was generated by Paul Nylander, a mechanical engineer with strong programming and mathematical skills that he uses to design complex engineering systems. Nylander says that he always enjoyed science and art as a hobby. When he was in high school, he had some aptitude for math, but programming was difficult for him. However, he became much more interested in programming when he began studying computer graphics. Most of Nylander's artwork is created in Mathematica, POV-Ray, and C++.

You can find a short bio and description of his work at:
http://virtualmathmuseum.org/mathart/ArtGalleryNylander/Nylanderindex.html.

Cover: Sphidron. By Daniel Erdely and Paul Nylander (bugman123.com)

GLENCOE

# ALGEBRA 1

McGraw Hill

mheducation.com/prek-12

**McGraw Hill**

Copyright © 2018 McGraw-Hill Education

All rights reserved. No part of this publication may be reproduced or distributed in any form or by any means, or stored in a database or retrieval system, without the prior written consent of McGraw-Hill Education, including, but not limited to, network storage or transmission, or broadcast for distance learning.

Send all inquiries to:
McGraw-Hill Education
8787 Orion Place
Columbus, OH 43240

ISBN: 978-0-07-898515-7 (Teacher Edition, Vol. 1)
MHID: 0-07-898515-3 (Teacher Edition, Vol. 1)
ISBN: 978-0-07-898518-8 (Teacher Edition, Vol. 2)
MHID: 0-07-898518-8 (Teacher Edition, Vol. 2)
ISBN: 978-0-07-903989-7 (Student Edition)
MHID: 0-07-903989-8 (Student Edition)

Printed in the United States of America.

8 9 10 MER 24 23 22 21

Understanding by Design® is a registered trademark of the Association for Supervision and Curriculum Development ("ASCD").

**STEM**

McGraw-Hill is committed to providing instructional materials in Science, Technology, Engineering, and Mathematics (STEM) that give all students a solid foundation, one that prepares them for college and careers in the 21st century.

# Contents in Brief

**Chapter 0**
Preparing for Algebra

**Chapter 1**
Expressions and Functions

**Chapter 2**
Linear Equations

**Chapter 3**
Linear and Nonlinear Functions

**Chapter 4**
Equations of Linear Functions

**Chapter 5**
Linear Inequalities

**Chapter 6**
Systems of Linear Equations and Inequalities

**Chapter 7**
Exponents and Exponential Functions

**Chapter 8**
Polynomials

**Chapter 9**
Quadratic Functions and Equations

**Chapter 10**
Statistics

## SUGGESTED PACING GUIDE

Each chapter includes multiple days for review and assessment.

| | |
|---|---|
| Chapter 0 | 6 days |
| Chapter 1 | 14 days |
| Chapter 2 | 12 days |
| Chapter 3 | 14 days |
| Chapter 4 | 11 days |
| Chapter 5 | 11 days |
| Chapter 6 | 12 days |
| Chapter 7 | 15 days |
| Chapter 8 | 13 days |
| Chapter 9 | 16 days |
| Chapter 10 | 9 days |
| **Total** | **133 days** |

# Authors

Our lead authors ensure that the Macmillan/McGraw-Hill and Glencoe/McGraw-Hill mathematics programs are truly vertically aligned by beginning with the end in mind — success in Algebra 1 and beyond. By "backmapping" the content from the high school programs, all of our mathematics programs are well articulated in their scope and sequence.

## LEAD AUTHORS

### John A. Carter, Ph.D.
**Mathematics Teacher**
WINNETKA, ILLINOIS

**Areas of Expertise:**
Using technology and manipulatives to visualize concepts; mathematics achievement of English-language learners

### Gilbert J. Cuevas, Ph.D.
**Professor of Mathematics Education, Texas State University—San Marcos**
SAN MARCOS, TEXAS

**Areas of Expertise:**
Applying concepts and skills in mathematically rich contexts; mathematical representations; use of technology in the development of geometric thinking

### Roger Day, Ph.D., NBCT
**Mathematics Department Chairperson, Pontiac Township High School**
PONTIAC, ILLINOIS

**Areas of Expertise:**
Understanding and applying probability and statistics; mathematics teacher education

### In Memoriam
### Carol Malloy, Ph.D.

Dr. Carol Malloy was a fervent supporter of mathematics education. She was a Professor at the University of North Carolina, Chapel Hill, NCTM Board of Directors member, President of the Benjamin Banneker Association (BBA), and 2013 BBA Lifetime Achievement Award for Mathematics winner. She joined McGraw-Hill in 1996. Her influence significantly improved our programs' focus on real-world problem solving and equity. We will miss her inspiration and passion for education.

## PROGRAM AUTHORS

### Berchie Holliday, Ed.D.
**National Mathematics Consultant**
SILVER SPRING, MARYLAND

**Areas of Expertise:**
Using mathematics to model and understand real-world data; the effect of graphics on mathematical understanding

### Beatrice Moore Luchin
**Mathematics Consultant**
HOUSTON, TEXAS

**Areas of Expertise:**
Mathematics literacy; working with English language learners

## CONTRIBUTING AUTHORS

### Dinah Zike FOLDABLES
**Educational Consultant**
Dinah-Might Activities, Inc.
SAN ANTONIO, TEXAS

### Jay McTighe
**Educational Author and Consultant**
COLUMBIA, MARYLAND

# Consultants and Reviewers

These professionals were instrumental in providing valuable input and suggestions for improving the effectiveness of the mathematics instruction.

## LEAD CONSULTANT

**Viken Hovsepian**
Professor of Mathematics
Rio Hondo College
WHITTIER, CALIFORNIA

## CONSULTANTS

### MATHEMATICAL CONTENT

**Grant A. Fraser, Ph.D.**
Professor of Mathematics
California State University, Los Angeles
LOS ANGELES, CALIFORNIA

**Arthur K. Wayman, Ph.D.**
Professor of Mathematics Emeritus
California State University, Long Beach
LONG BEACH, CALIFORNIA

### GIFTED AND TALENTED

**Shelbi K. Cole**
Research Assistant
University of Connecticut
STORRS, CONNECTICUT

### COLLEGE READINESS

**Robert Lee Kimball, Jr.**
Department Head, Math and Physics
Wake Technical Community College
RALEIGH, NORTH CAROLINA

### DIFFERENTIATION FOR ENGLISH-LANGUAGE LEARNERS

**Susana Davidenko**
State University of New York
CORTLAND, NEW YORK

**Alfredo Gómez**
Mathematics/ESL Teacher
George W. Fowler High School
SYRACUSE, NEW YORK

### GRAPHING CALCULATOR

**Ruth M. Casey**
$T^3$ National Instructor
FRANKFORT, KENTUCKY

**Jerry Cummins**
Former President
National Council of Supervisors of Mathematics
WESTERN SPRINGS, ILLINOIS

### MATHEMATICAL FLUENCY

**Robert M. Capraro**
Associate Professor
Texas A&M University
COLLEGE STATION, TEXAS

### PRE-AP

**Dixie Ross**
Lead Teacher for Advanced Placement Mathematics
Pflugerville High School
PFLUGERVILLE, TEXAS

### READING AND WRITING

**ReLeah Cossett Lent**
Author and Educational Consultant
MORGANTON, GEORGIA

**Lynn T. Havens**
Director of Project CRISS
KALISPELL, MONTANA

## REVIEWERS

**Sherri Abel**
Mathematics Teacher
Eastside High School
TAYLORS, SOUTH CAROLINA

**Kelli Ball, NBCT**
Mathematics Teacher
Owasso 7th Grade Center
OWASSO, OKLAHOMA

**Cynthia A. Burke**
Mathematics Teacher
Sherrard Junior High School
WHEELING, WEST VIRGINIA

**Patrick M. Cain, Sr.**
Assistant Principal
Stanhope Elmore High School
MILLBROOK, ALABAMA

**Robert D. Cherry**
Mathematics Instructor
Wheaton Warrenville South High School
WHEATON, ILLINOIS

**Tammy Cisco**
8th Grade Mathematics/ Algebra Teacher
Celina Middle School
CELINA, OHIO

**Amber L. Contrano**
High School Teacher
Naperville Central High School
NAPERVILLE, ILLINOIS

**Catherine Creteau**
Mathematics Department
Delaware Valley Regional High School
FRENCHTOWN, NEW JERSEY

**Glenna L. Crockett**
Mathematics Department Chair
Fairland High School
FAIRLAND, OKLAHOMA

**Jami L. Cullen**
Mathematics Teacher/Leader
Hilltonia Middle School
COLUMBUS, OHIO

**Franco DiPasqua**
Director of K-12 Mathematics
West Seneca Central Schools
WEST SENECA, NEW YORK

**Kendrick Fearson**
Mathematics Department Chair
Amos P. Godby High School
TALLAHASSEE, FLORIDA

**Lisa K. Gleason**
Mathematics Teacher
Gaylord High School
GAYLORD, MICHIGAN

**Debra Harley**
Director of Math & Science
East Meadow School District
WESTBURY, NEW YORK

**Tracie A. Harwood**
Mathematics Teacher
Braden River High School
BRADENTON, FLORIDA

**Bonnie C. Hill**
Mathematics Department Chair
Triad High School
TROY, ILLINOIS

**Clayton Hutsler**
Teacher
Goodwyn Junior High School
MONTGOMERY, ALABAMA

**Gureet Kaur**
7th Grade Mathematics Teacher
Quail Hollow Middle School
CHARLOTTE, NORTH CAROLINA

**Rima Seals Kelley, NBCT**
Mathematics Teacher/ Department Chair
Deerlake Middle School
TALLAHASSEE, FLORIDA

**Holly W. Loftis**
8th Grade Mathematics Teacher
Greer Middle School
LYMAN, SOUTH CAROLINA

**Katherine Lohrman**
Teacher, Math Specialist, New Teacher Mentor
John Marshall High School
ROCHESTER, NEW YORK

**Carol Y. Lumpkin**
Mathematics Educator
Crayton Middle School
COLUMBIA, SOUTH CAROLINA

**Ron Mezzadri**
Supervisor of Mathematics K-12
Fair Lawn Public Schools
FAIR LAWN, NEW JERSEY

**Bonnye C. Newton**
SOL Resource Specialist
Amherst County Public Schools
AMHERST, VIRGINIA

**Kevin Olsen**
Mathematics Teacher
River Ridge High School
NEW PORT RICHEY, FLORIDA

**Kara Painter**
Mathematics Teacher
Downers Grove South High School
DOWNERS GROVE, ILLINOIS

**Sheila L. Ruddle, NBCT**
Mathematics Teacher, Grades 7 and 8
Pendleton County Middle/ High School
FRANKLIN, WEST VIRGINIA

**Angela H. Slate**
Mathematics Teacher/Grade 7, Pre-Algebra, Algebra
LeRoy Martin Middle School
RALEIGH, NORTH CAROLINA

**Cathy Stellern**
Mathematics Teacher
West High School
KNOXVILLE, TENNESSEE

**Dr. Maria J. Vlahos**
Mathematics Division Head for Grades 6-12
Barrington High School
BARRINGTON, ILLINOIS

**Susan S. Wesson**
Mathematics Consultant/ Teacher (Retired)
Pilot Butte Middle School
BEND, OREGON

**Mary Beth Zinn**
High School Mathematics Teacher
Chippewa Valley High Schools
CLINTON TOWNSHIP, MICHIGAN

# Solutions That Work

**Glencoe High School Math Series is about connecting math content, rigor, and adaptive instruction for student success.**

**Glencoe High School Math Series** provides you the flexibility to meet your classroom needs. As you build teacher-student relationships and a classroom environment that encourages learning, this program provides a flexible print and digital solution that allows you to:

- **Personalize learning** for all students with differentiated instruction and math content that meets rigorous standards.

- **Engage students** in a variety of ways with projects, activities, and resources that make math come to life and enable students to connect math to their world.

- **Drive success** on state assessments with dynamic digital tools that support and inform instructional decisions.

"Great…easy textbook to use, good online resources, and many examples, including great exit ticket ideas and warm up questions."

--*High School Teacher*

connectED.mcgraw-hill.com

# Rigor in Instruction

**Glencoe High School Math Series** is about connecting math content and rigor. Rigor is built-in throughout the entire program with a strong focus on promoting conceptual understanding and encouraging students to think critically.

## Performance Tasks

A Performance Task is available for each chapter that enables students to practice persevering through a rich, multi-step task. The Performance Tasks are previewed at the beginning of the chapter. Students will learn concepts and skills as they work through each lesson that will help them to finish the task at the end of the chapter.

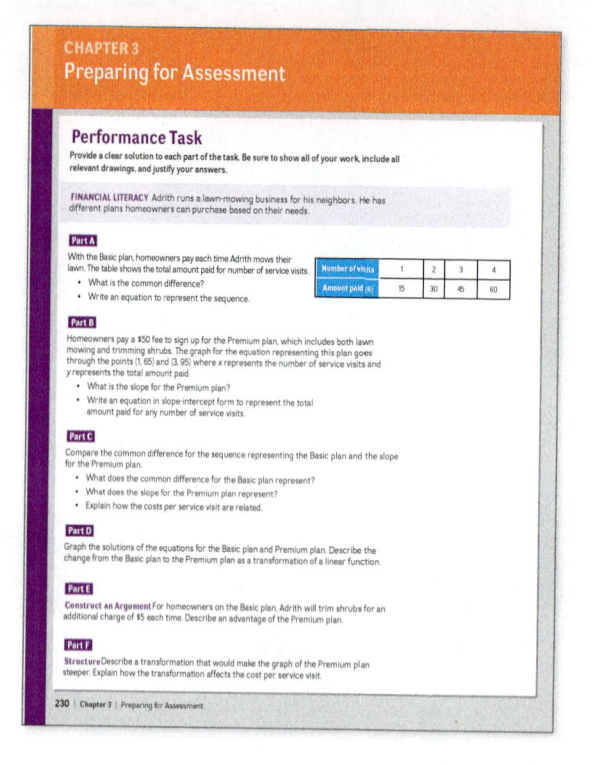

## Multi-Step Questions

Multi-step questions allow students to practice reasoning, modeling, and looking for structure in different situations. These questions are incorporated into the Preparing for Assessment page at the end of each lesson and can also be found online in eAssessment.

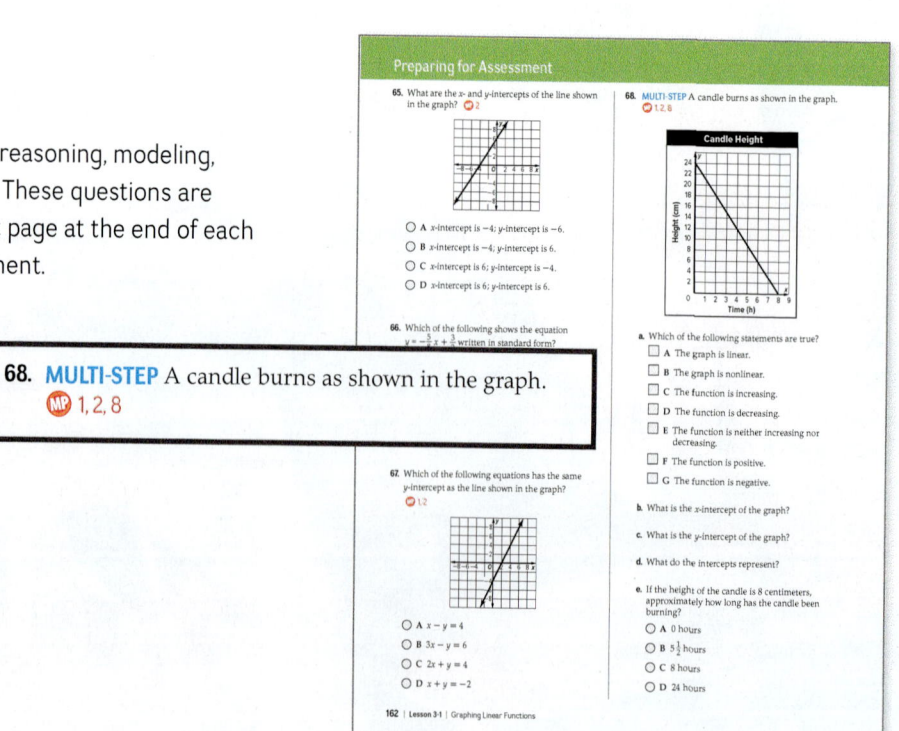

> 68. **MULTI-STEP** A candle burns as shown in the graph.
> MP 1, 2, 8

# Standards for Mathematical Practice

The Standards for Mathematical Practice describe how students should approach mathematics. The goal of the practice standards is to instill in all students the abilities to be mathematically literate and to create a positive disposition for the importance of using math effectively.

**You can use the following tools to incorporate the practices into your teaching.**

> Mathematical Practice Study Tips located in the margins of the student edition.

> Look for the MP icon **MP** to see which practice is being taught.

> Mathematical Practice Strategies show you how to incorporate the practices in each chapter.

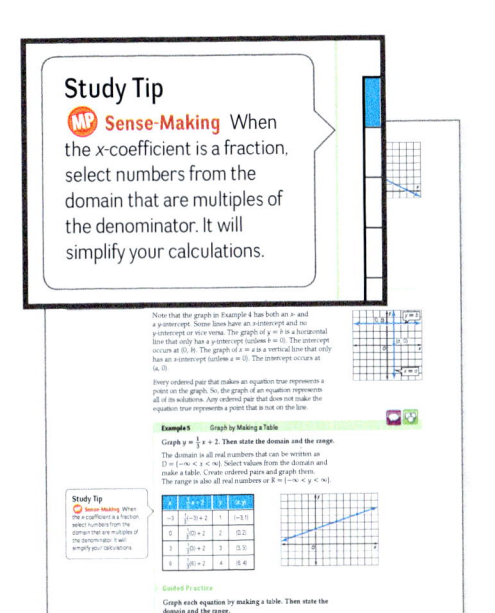

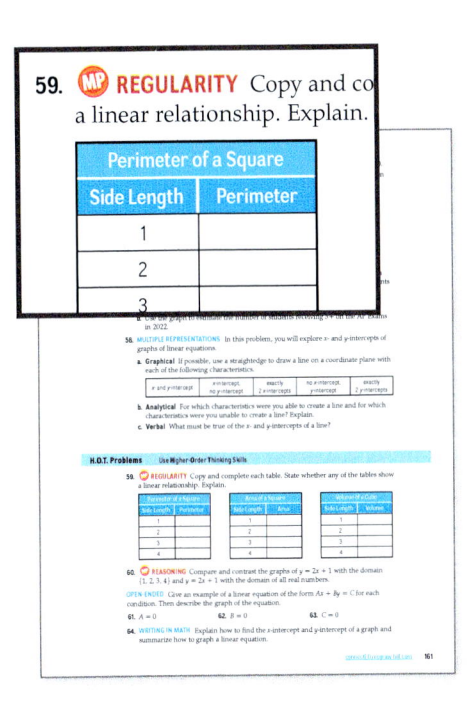

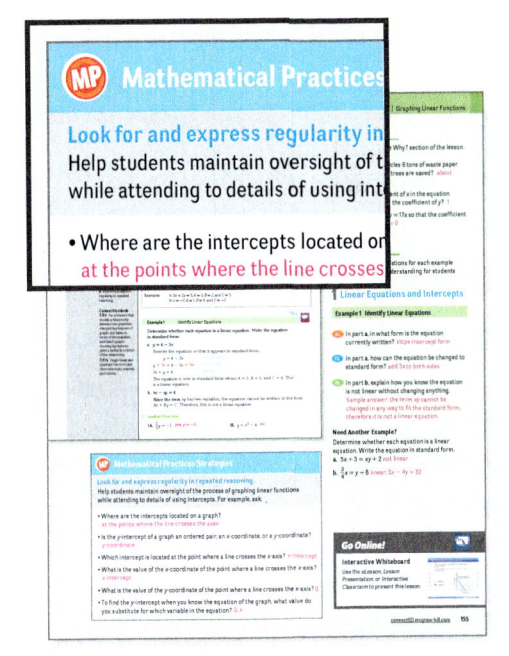

# Interactive Student Guide

Working together, the Student Edition and *Interactive Student Guide* (ISG) promote a deep understanding of the content to ensure student success. Topics from the ISG correspond to the lessons within the Student Edition and point-of-use references are located in the margins of the Teacher Edition.

connectED.mcgraw-hill.com    T11

# Understanding by Design

What should students know and be able to do? Understanding by Design can be used to help teachers identify learning goals, develop revealing assessments of student understanding, and plan effective and engaging learning activities.

## Backward Design

Understanding by Design (UbD) is a framework that uses backward design to create a coherent curriculum by considering the desired results first and then planning instruction.

The backward design process guided the development of Glencoe Algebra 1, Glencoe Geometry, and Glencoe Algebra 2.

### Identifying Desired Results

The first step in developing an effective curriculum using the UbD framework is to consider the goals. What should students know and be able to do?

**Glencoe High School Math Series** addresses the big ideas of algebra and focuses student attention on Essential Questions, which are located within each chapter of the Student and Teacher Editions.

An Essential Question is provided at the beginning of each chapter. These thought provoking questions can be used as:

> a discussion starter for your class; throughout the discussion identify what the students already know and what they would like to know about the topic. Revisit these notes throughout the chapter.

> a benchmark of understanding; post these questions in a prominent place and have students expand upon their initial response as their understanding of the subject material grows.

Follow-up Essential Questions can be found throughout each chapter. These questions challenge students to apply specific knowledge to a broader context, thus deepening their understanding.

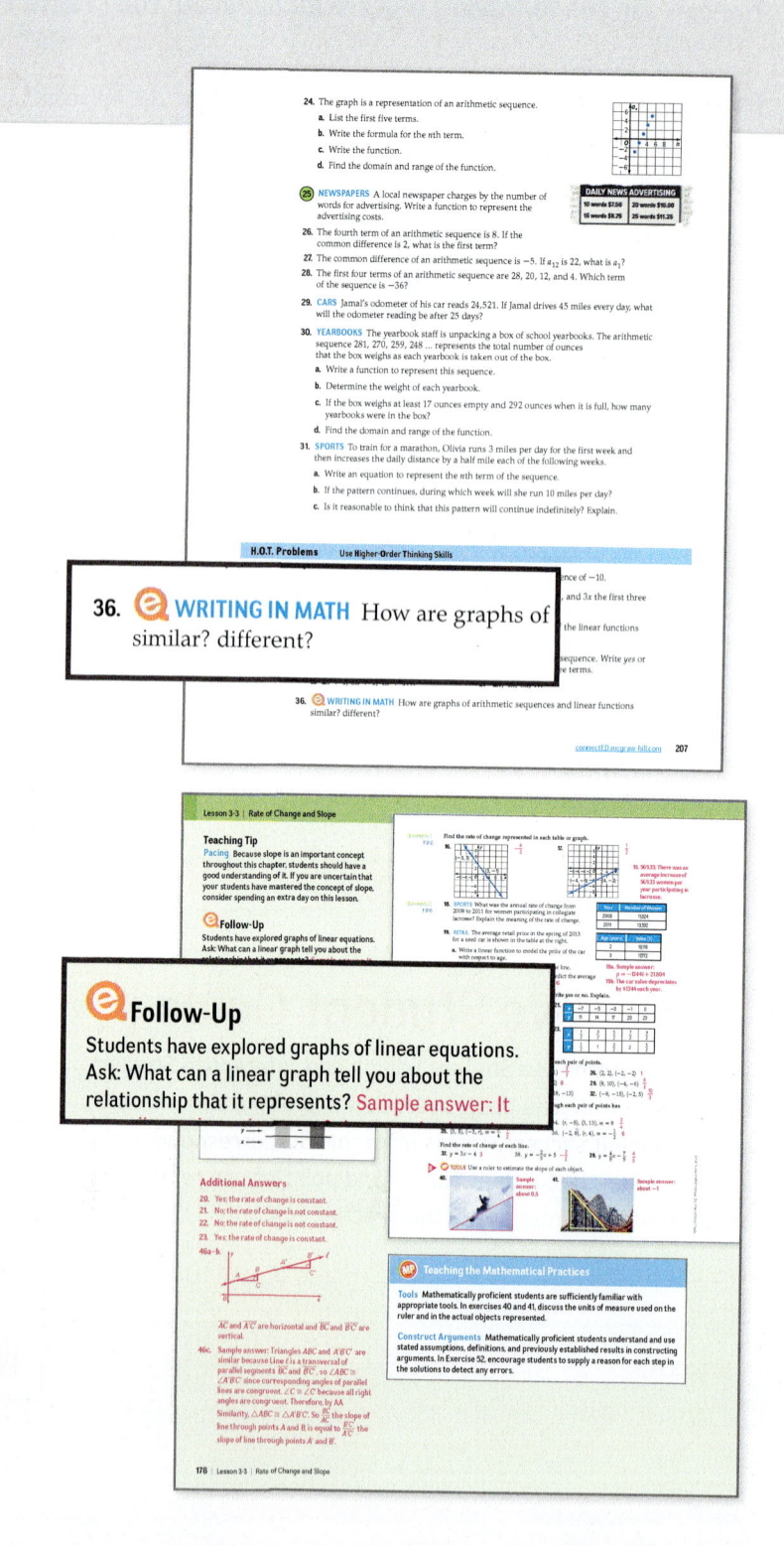

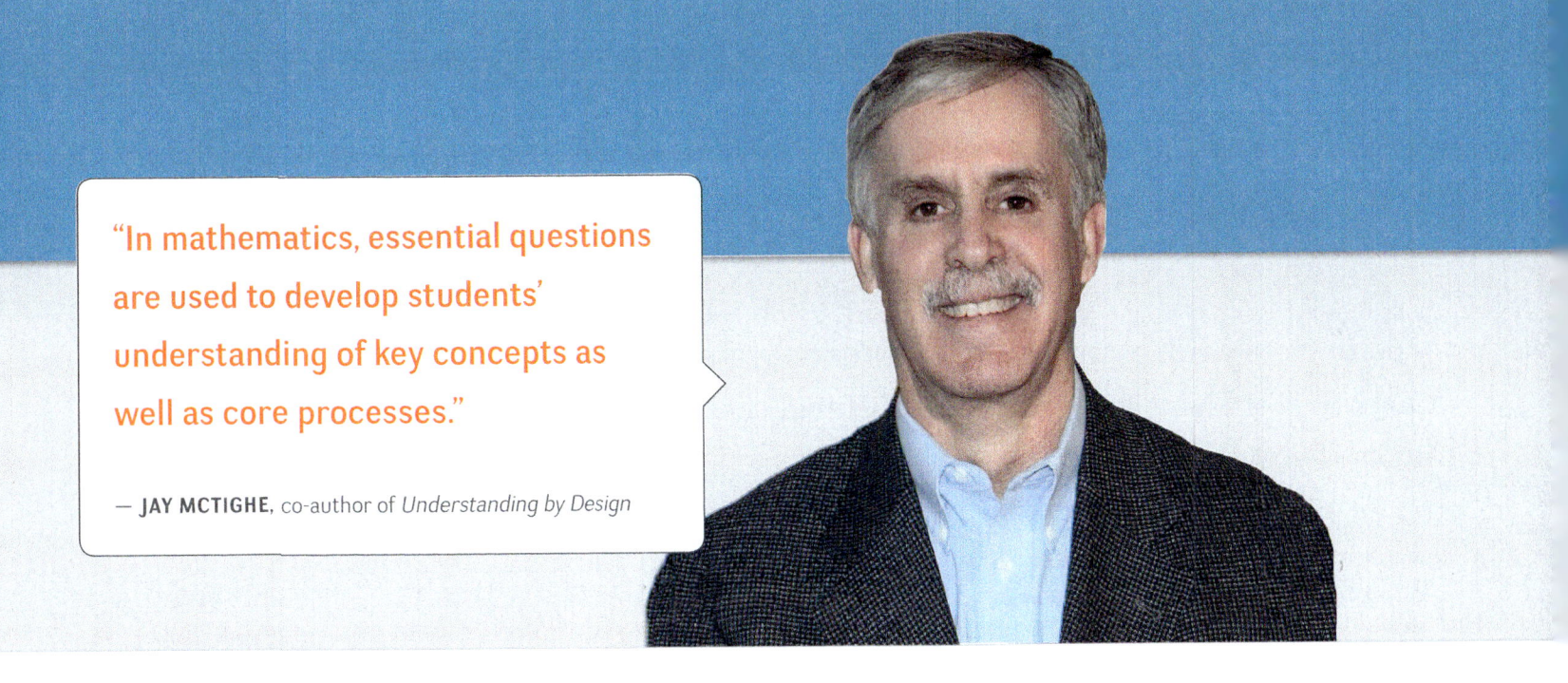

*"In mathematics, essential questions are used to develop students' understanding of key concepts as well as core processes."*

— **JAY MCTIGHE**, co-author of *Understanding by Design*

## Determine Acceptable Evidence

A variety of assessment opportunities are available that enable students to show evidence of their understanding.

> Practice and Problem Solving and H.O.T. Problems allow students to explain, interpret, and apply mathematical concepts.

> Mid-Chapter Quizzes and Chapter Tests offer more traditional methods of assessment.

> McGraw-Hill eAssessment can also be used to customize and create assessments.

## Plan Learning Experiences and Instruction

There are numerous performance activities available throughout the program, including:

> Algebra Labs that offer students hands-on learning experiences, and

> Graphing Technology Labs that use graphing calculators to aid student understanding.

You can also visit connectED.mcgraw-hill.com to choose from an extensive collection of resources to use when planning instruction, such as presentation tools, chapter projects, editable worksheets, digital animations, and eTools.

**40. WRITING IN MATH** Compare and contr[ast] functions with the graphs of linear fu[nctions] and minima.

# Built-In Differentiated Instruction

Approximately 43% of teachers feel their classes are so mixed in terms of students' learning abilities that they can't teach them effectively.

The **Glencoe High School Math Series** fully supports the 3-tier RtI model with print and digital resources to diagnose students, identify areas of need, and conduct short, frequent assessments for accurate data-driven decision making. Every lesson provides easy-to-use resources that consider the needs of all students.

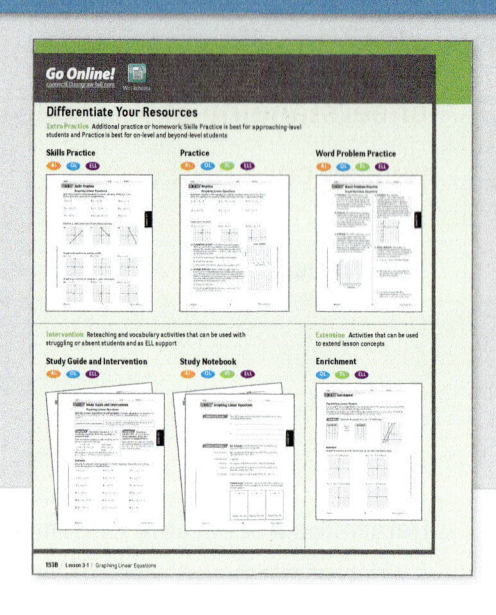

## RtI: Response to Intervention

### TIER 1: Daily Intervention

#### OL On grade level

Core instruction targets on-level students. Comprehensive instructional materials help you personalize instruction for every student.

- Diagnostic Teaching
- Options for Differentiated Instruction
- Leveled Exercise Sets, Resources, and Technology
- Data-Driven Decision Making

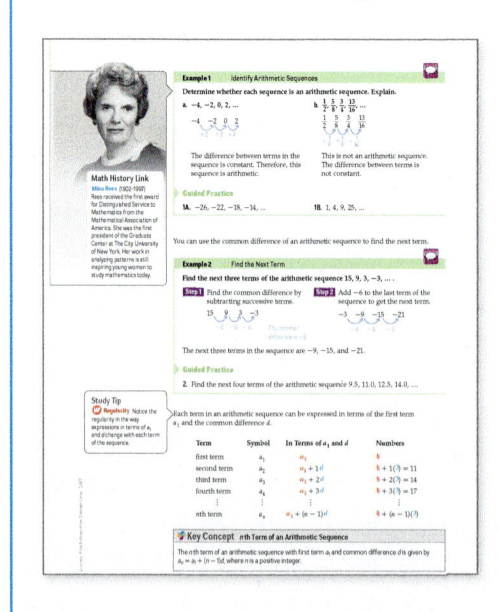

#### BL Beyond grade level

At every step, resources and assignments are available for advanced learners.

- Higher-Order Thinking Questions
- Differentiated Homework Options
- Enrichment Masters
- Differentiated Instruction: Extension

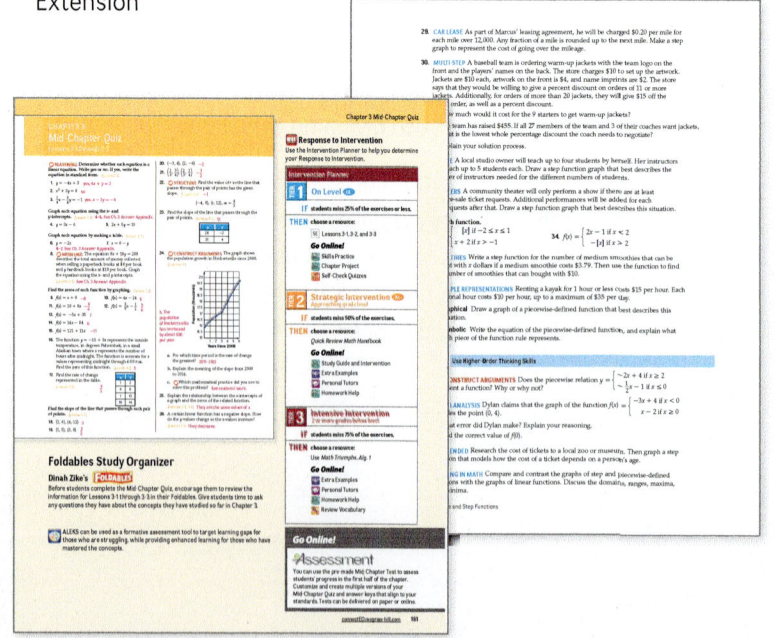

T14

**ELL** English Language Learners

Comprehensive resources are found throughout the program.

- Teacher Edition with strategies to modify activities and lesson content
- Multilingual eGlossary with definitions for each vocabulary word in 13 languages

## TIER 2: Strategic Intervention

**AL** Approaching grade level

You can choose from a myriad of intervention tips and ancillary materials to support struggling learners.

- Using Manipulatives
- Alternate Teaching Strategies
- Online resources, including animations, examples, and Personal Tutors

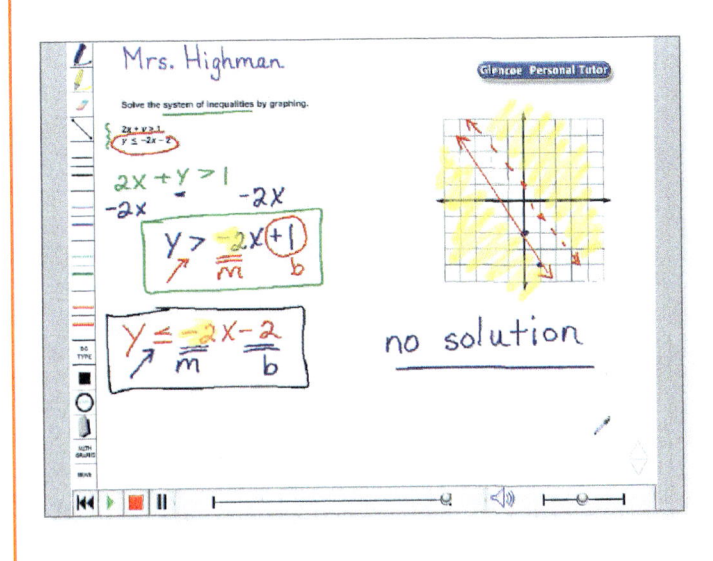

## TIER 3: Intensive Intervention

**AL** Significantly below grade level

For students who are far below grade level, **Math Triumphs** provides step-by-step instruction, vocabulary support, and meaningful practice.

connectED.mcgraw-hill.com    T15

# Assessment Tools that Inform Instruction

A comprehensive variety of print and online assessment options are built into the **Glencoe High School Math Series**. Using traditional assessment combined with our online test generator and reporting system, you can use data to make on–the-spot instructional decisions for the entire class or individual students.

## Chapter Assessment Resources

Prebuilt assessment resources are provided with every chapter to quickly print and deliver on demand. As an added feature, most are available in editable Word documents to allow for maximum flexibility and customization:

- Chapter Quizzes
- Mid-Chapter Tests
- Vocabulary Tests
- Extended Response Tests
- Standardized Tests

> The personalized study resources your students need today – to master state assessments tomorrow.

## eAssessment

With a simple, intuitive interface that enables teachers to create customized assessments and homework, web-based eAssessment means easy, anytime access via a secure online test center. Manage your content, create and assign quizzes and tests that can be assigned online or in traditional format. The reporting system enables you to see performance at both a class and a student level which allows you to modify instruction or provide targeted support to improve student achievement.

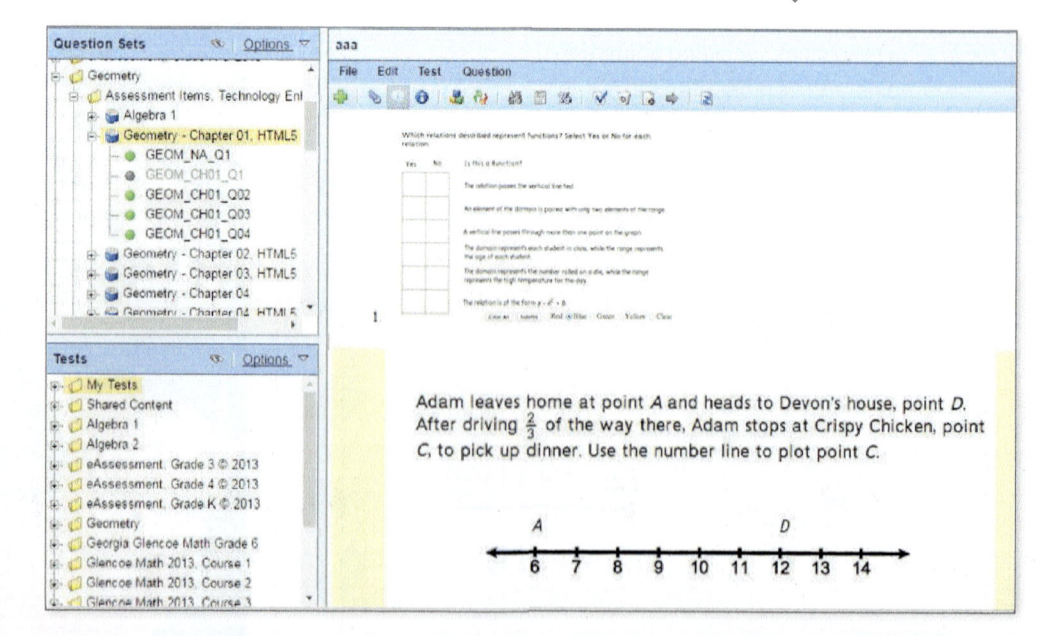

> Help students learn faster, study more efficiently, and retain more knowledge.

# LEARNSMART®

Using this robust, online personalized study resource, students have a modernized study partner to practice for high-stakes testing. By following a personalized study plan, each student will see the course topics they have mastered or which they need to revisit. Learning resources are tagged to questions to ensure students have multiple opportunities to refresh their content knowledge. Advanced reporting enables you to maximize your instructional time leading up to end of course or end of year assessments.

# ALEKS®

Differentiating instruction can be a challenge. It's **ALEKS**® to the rescue! This online resource uses adaptive questioning to provide each student with a personalized learning path.

### Support each student's unique needs

- Knowledge checks identify what a student knows and where there are gaps
- Personalized instruction allows students to progress at their own pace
- Robust Reporting tools provide instructionally-actionable data
- Targeted instruction on Ready to Learn topics build learning momentum and confidence
- Open-response environment ensures students demonstrate understanding—no guessing!

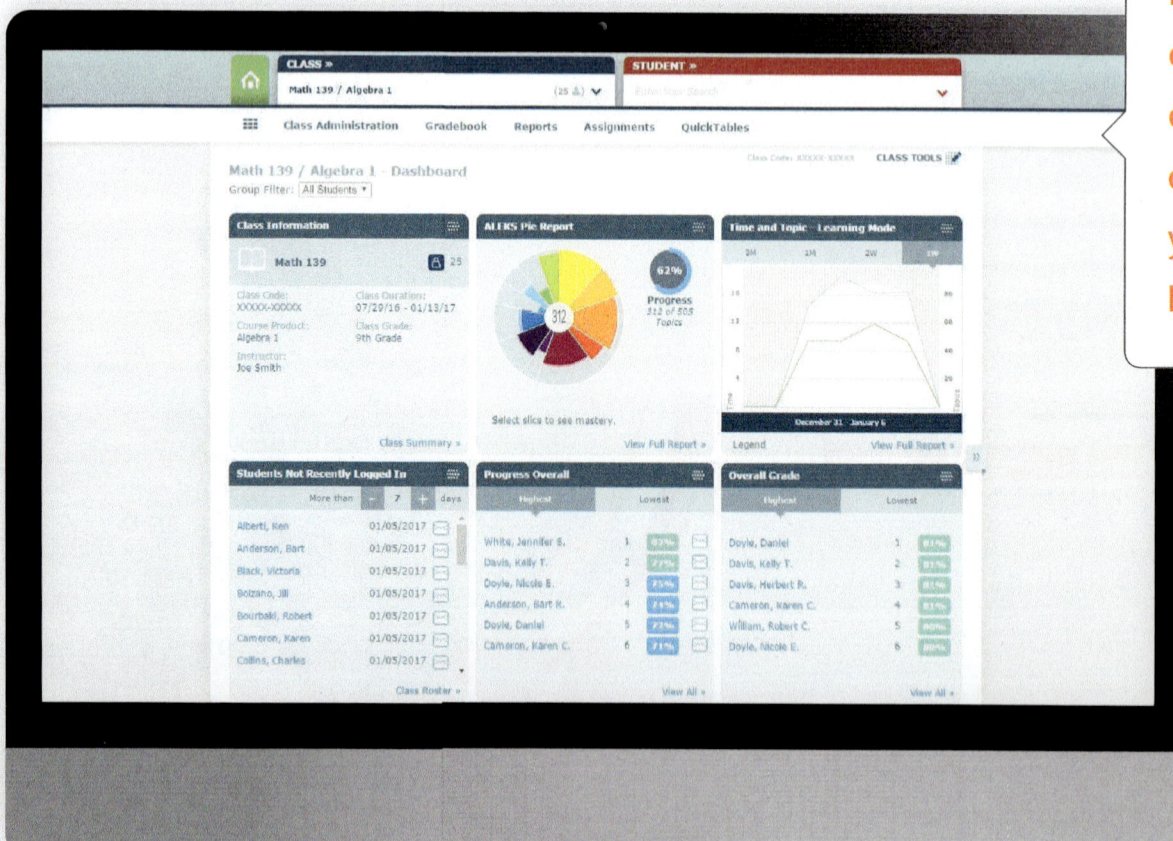

> Learn how **ALEKS** can enhance your current math curriculum. Contact your McGraw-Hill Education rep today.

# Use ConnectED as Your Portal

Nearly 84% of teachers say that they only spend half of their time teaching, as opposed to disciplining or doing administrative work.

ConnectED is a time-saving online portal that has all of your digital program resources in one place. Through this portal, you can access the complete Student Edition and Teacher Edition, digital animations and tutorials, editable worksheets, presentation tools, and assessment resources.

With ConnectED Mobile you can browse through your course content on the go. The app includes a powerful eBook engine where you can download, view, and interact with your books.

## connectED allows you to:

- build lesson plans with easy-to-find print and digital resources
- search for activities to meet a variety of learning modalities
- teach with technology by providing virtual manipulatives, lesson animations, whole-class presentations, and more
- personalize instruction with print and digital resources
- provide students with anywhere, anytime access to student resources and tools, including eBooks, tutorials, animations, and the eGlossary
- access eAssessment, which allows you to assign online assessments, track student progress, generate reports, and differentiate instruction

connectED.mcgraw-hill.com

# Digital Tools to Enhance Learning Opportunities

Today's students have an unprecedented access to and appetite for technology and new media. They perceive technology as their friend and rely on it to study, work, play, relax, and communicate.

Your students are accustomed to the role that computers play in today's world. They may well be the first generation whose primary educational tool is a computer or a cell phone. The eStudentEdition allows your students to access their math curriculum anytime, anywhere. Along with your teacher edition, **Glencoe Algebra 1** provides a blended instructional solution for your next-generation students.

## Investigate

**Sketchpad** Discover concepts using The Geometer's Sketchpad®.

**Vocabulary** tools include fun Vocabulary Review Games.

**Tools** enhance understanding through exploration.

## Go Online!

Look for the *Go Online!* feature in your Teacher Edition to easily find digital tools to engage students and personalize instruction.

> Each chapter includes a curated list of differentiated resources as well as featured interactive whiteboard resources with suggested pacing.
> Each lesson includes a list of digital resources with implementation tips and suggested pacing.

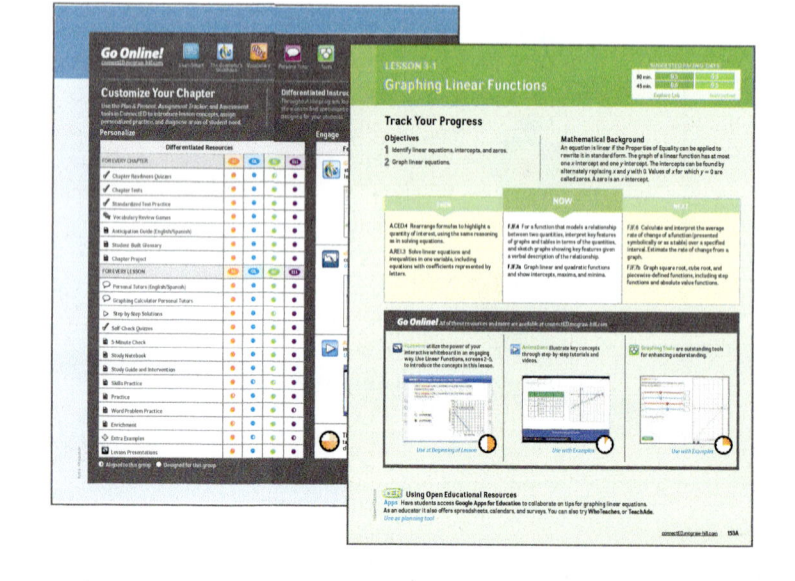

## The Geometer's Sketchpad

*The Geometer's Sketchpad* is proven to increase engagement, understanding, and achievement. This premier digital learning tool makes math come alive by giving students a tangible, visual way to see the math in action through dynamic model manipulation of lines, shapes, and functions.

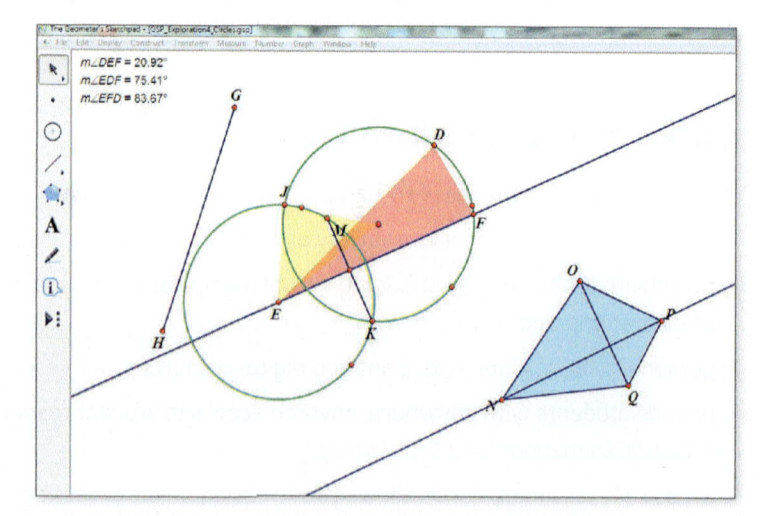

## Learn

**LearnSmart**
Topic based online assessment

**Animations**
illustrate key concepts through step-by-step tutorials and videos.

**Tutors**
See and hear a teacher explain how to solve problems.

**Calculator Resources**
provides other calculator keystrokes for each Graphing Technology Lab.

## Practice

**Self-Check Practice**
allows students to check their understanding and send results to their teacher.

**eBook**
Interactive learning experience with links directly to assets

## eToolkit

*eToolkit* enables students to use virtual manipulatives to extend learning beyond the classroom by modifying concrete models in a real-time, interactive format focused on problem-based learning.

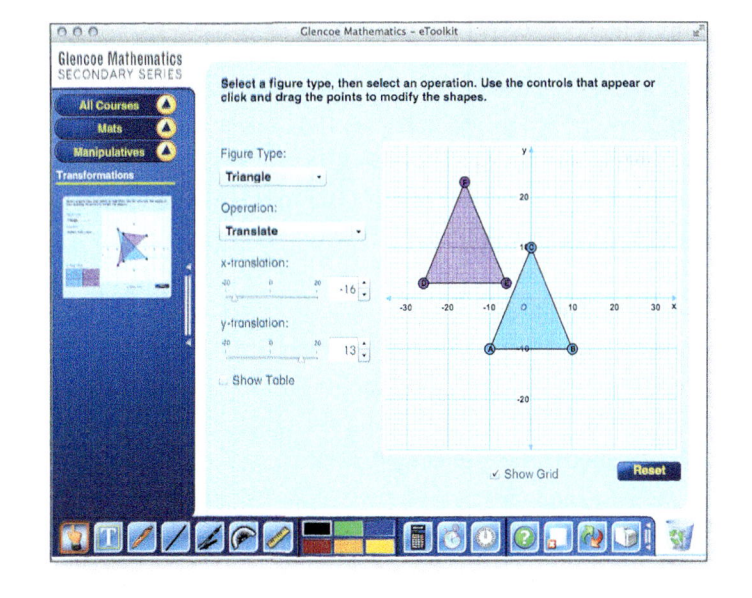

## Interactive Student Guide

*Interactive Student Guide* is a dynamic digital resource to help you and your students meet the increasing demands for rigor.

You can be empowered to teach confidently knowing every lesson includes standards-aligned content and emphasizes the Standards for Mathematical Practice. This guide works together with the student edition to ensure that students can reflect on comprehension and application, use mathematics in real-world applications, and internalize concepts to develop "second nature" recall.

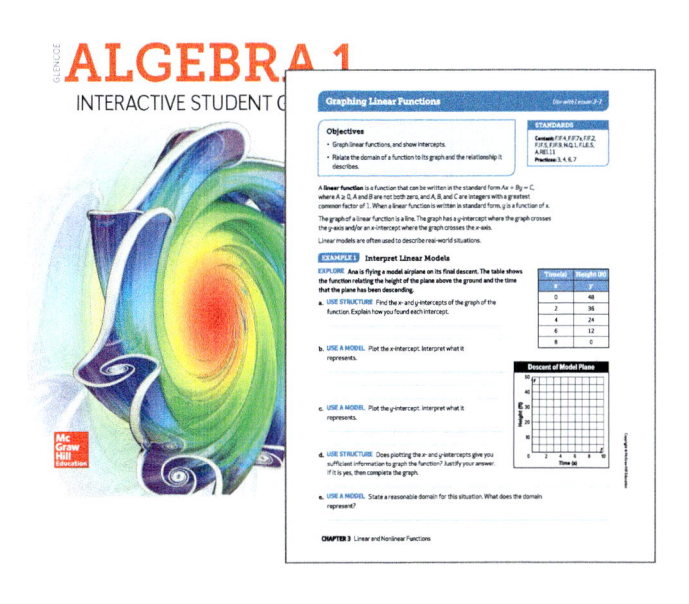

connectED.mcgraw-hill.com   T21

# 21st Century Skills

The U.S. Department of Labor estimates that 90% of 21st-century skilled workforce jobs will require post-secondary education. To be competitive in tomorrow's global workforce, American students must be prepared to succeed in college.

A strong high school curriculum is a good predictor of college readiness (Adelman, 2006). Students who take at least three years of college-preparatory mathematics using programs like **Glencoe Algebra 1**, **Glencoe Geometry**, and **Glencoe Algebra 2** are less likely to need remedial courses in college than students who do not (Abraham & Creech, 2002).

## College Readiness

David Conley at the University of Oregon developed the following criteria for college readiness.

### Key Content Knowledge

**Glencoe High School Math Series** is aligned to rigorous state and national standards, including the *NCTM Principles & Standards for School Mathematics*, the College Board Standards for College Success, and the American Diploma Project's Benchmarks. Correlations to these standards can be found at **connectED.mcgraw-hill.com**.

### Habits of Mind

These include critical thinking skills such as analysis, interpretation, problem solving, and reasoning. Students can hone critical higher-order thinking skills through the use of **H.O.T. (Higher Order Thinking) Problems**.

### Contextual Skills

These are practical skills like understanding the admissions process and financial aid, placement testing, and communicating with professors. Throughout each Glencoe mathematics program, students are required to write, explain, justify, prove, and analyze.

### Academic Behaviors

These include general skills such as reading comprehension, time management, note-taking, and metacognition. Reading Math tips and Vocabulary Links help students with reading comprehension. Study Notebooks and Anticipation Guides help students build note-taking skills and aid with metacognition.

## Developing STEM Careers

With **Glencoe High School Math Series**, you can unleash your students' curiosity about the world around them and prepare them for exciting STEM (**S**cience, **T**echnology, **E**ngineering, and **M**ath) careers.

### Examples
Examples are relevant, connecting in-class experiences to the world beyond the classroom.

### Real-World Careers
Real-World Careers are engaging, providing information on exciting careers.

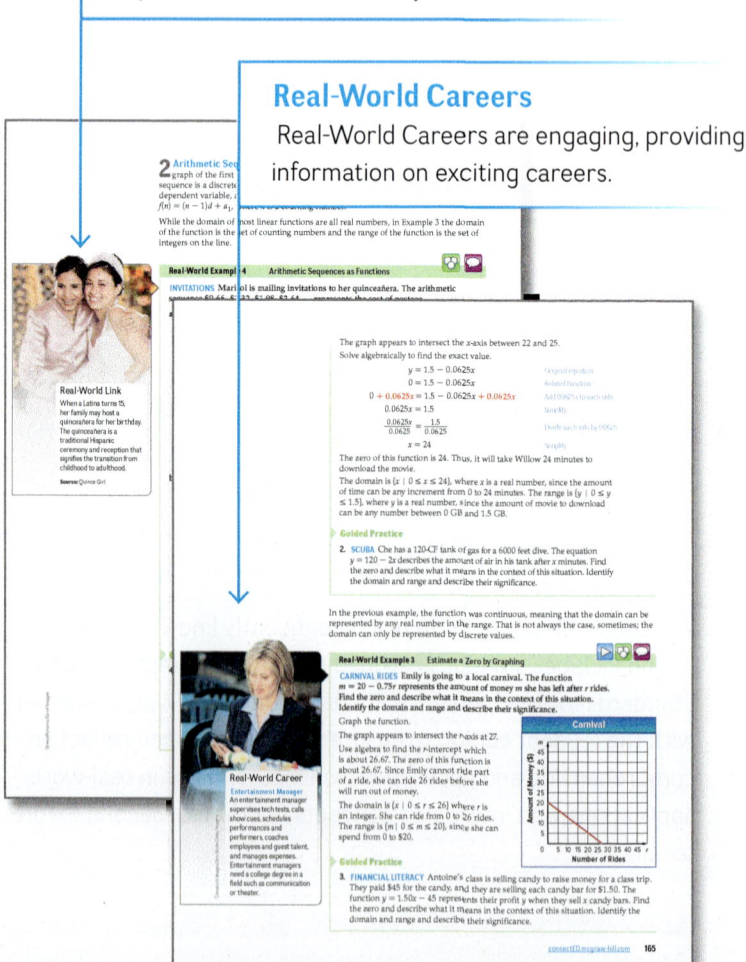

"The current and future health of America's 21st Century Economy depends directly on how broadly and deeply Americans reach a new level of literacy—'21st Century Literacy'—that includes strong academic skills, thinking, reasoning, teamwork skills, and proficiency in using technology."

–21st Century Workforce Commission National Alliance of Business

# Developing 21st Century Skills

The Partnership for 21st Century Skills identifies the following key student elements of a 21st century education.

## Core Subjects and 21st Century Themes

**Glencoe High School Math Series** has been aligned to rigorous state and national standards, *NCTM Principles & Standards for School Mathematics*, and the American Diploma Project's Benchmarks. Throughout each program, students solve problems that incorporate 21st century themes, such as financial literacy.

## Learning and Innovation Skills

Students who are prepared for increasingly complex life and work environments are creative and innovative critical thinkers, problem solvers, effective communicators, and know how to work collaboratively. Throughout each **Glencoe High School Math Series**, students are required to write, explain, justify, prove, and analyze. Students use critical thinking skills through the use of H.O.T. (Higher Order Thinking) Problems and are encouraged to work collaboratively in labs.

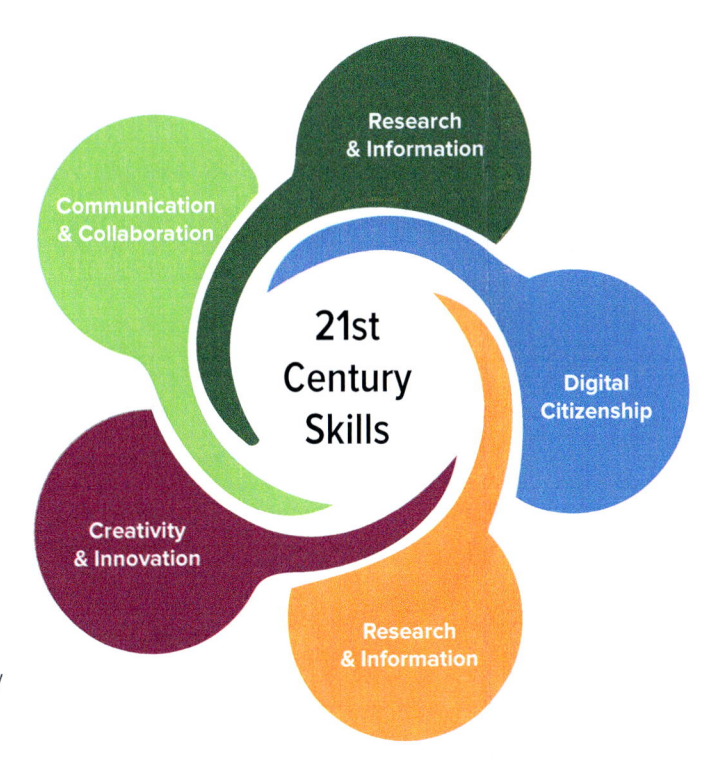

## Information, Media, and Technology Skills

Students use technology, including graphing calculators and the Internet, to develop 21st century mathematics knowledge and skills throughout each program.

## 21st Century Assessments

**Glencoe High School Math Series** offers a variety of frequent and meaningful assessments built right into the curriculum structure and teacher support materials. These programs include both traditional and nontraditional methods of assessment, including quizzes and tests, performance tasks, and open-ended assessments. Digital assessment solutions offer additional options for creating, customizing, administering, and instantly grading assessments.

# Professional Development

McGraw-Hill Education recognizes that learning is a lifelong endeavor. To ensure student and teacher success, we have built in resources to the Glencoe High School Math Series featuring best practices, implementation support, alternative teaching practices, and much more.

# Correlations

## Common Core State Standards, Traditional Algebra I Pathway, Correlated to *Glencoe Algebra 1*

Lessons in which the standard is the primary focus are indicated in **bold**.

| Standards | Student Edition Lesson(s) | Student Edition Page(s) |
|---|---|---|
| **Number and Quantity** | | |
| **The Real Number System N-RN** | | |
| Extend the properties of exponents to rational exponents. | 7-3 | 410–417 |
| 1. Explain how the definition of the meaning of rational exponents follows from extending the properties of integer exponents to those values, allowing for a notation for radicals in terms of rational exponents. | | |
| 2. Rewrite expressions involving radicals and rational exponents using the properties of exponents. | 7-3, 7-4 | 410–417, 419–427 |
| Use properties of rational and irrational numbers. | **Extend 1-4, Extend 7-4** | 30–31, 428–429 |
| 3. Explain why the sum or product of two rational numbers is rational; that the sum of a rational number and an irrational number is irrational; and that the product of a nonzero rational number and an irrational number is irrational. | | |
| **Quantities ★ N-Q** | | |
| Reason quantitatively and use units to solve problems. | Throughout the text; for example, 2-6, 2-7, Extend 3-2, 4-4, 7-5, 10-3 | 115–121, 122–127, 157–158, 247–253, 665–671 |
| 1. Use units as a way to understand problems and to guide the solution of multi-step problems; choose and interpret units consistently in formulas; choose and interpret the scale and the origin in graphs and data displays. | | |
| 2. Define appropriate quantities for the purpose of descriptive modeling. | 1-5 | 33–41 |
| 3. Choose a level of accuracy appropriate to limitations on measurement when reporting quantities. | 1-5 | 33–41 |
| **Algebra** | | |
| **Seeing Structure in Expressions A-SSE** | | |
| Interpret the structure of expressions | 1-1, 1-4, 8-1, 9-1 | 5–9, 23–29, 491–497, 559–569 |
| 1. Interpret expressions that represent a quantity in terms of its context. ★ | | |
| a. Interpret parts of an expression, such as terms, factors, and coefficients. | | |
| b. Interpret complicated expressions by viewing one or more of their parts as a single entity. | 1-2, 1-3, 3-7 | 10–15, 16–22, 197–203 |
| 2. Use the structure of an expression to identify ways to rewrite it. | 1-1, 1-2, 1-3, 1-4, 7-1, 7-2, 7-3, Explore 8-5, 8-5, Explore 8-6, 8-6, 8-7 | 5–9, 10–15, 16–22, 23–29, 395–401, 402–409, 410–417, 519, 520–526, 529–530, 531–538, 539–545 |

★ Mathematical Modeling Standards

# Correlations

| Standards | Student Edition Lesson(s) | Student Edition Page(s) |
|---|---|---|
| Write expressions in equivalent forms to solve problems.<br><br>3. Choose and produce an equivalent form of an expression to reveal and explain properties of the quantity represented by the expression. ★<br><br>   a. Factor a quadratic expression to reveal the zeros of the function it defines. | 8-5, 8-6, 8-7, 9-4 | 520–526, 531–538, 539–545, 588–595 |
|    b. Complete the square in a quadratic expression to reveal the maximum or minimum value of the function it defines. | 9-5, Extend 9-5 | 596–602, 603-604 |
|    c. Use the properties of exponents to transform expressions for exponential functions. | 7-8 | 458-461 |
| **Arithmetic with Polynomials and Rational Expressions A-APR** | | |
| Perform arithmetic operations on polynomials.<br><br>1. Understand that polynomials form a system analogous to the integers, namely, they are closed under the operations of addition, subtraction, and multiplication; add, subtract, and multiply polynomials. | Explore 8-1, 8-1, 8-2, Explore 8-3, 8-3, 8-4 | 489–490, 491–497, 498–503, 504–505, 506–511, 512–517 |
| **Creating Equations ★ A-CED** | | |
| Create equations that describe numbers or relationships.<br><br>1. Create equations and inequalities in one variable and use them to solve problems. | 2-1, 2-2, 2-3, 2-4, 2-5, 5-1, 5-2, 5-3, 5-4, 5-5, 7-7, 9-3, 9-5, 9-6 | 79–84, 87–93, 95–100, 101–106, 107–113, 289–294, 296–301, 302–307, 309–314, 316–320, 448–455, 580–586, 596–602, 606–613 |
| 2. Create equations in two or more variables to represent relationships between quantities; graph equations on coordinate axes with labels and scales. | Extend 1-7, 3-1, Extend 3-2, Explore 3-4, 3-4, Extend 3-7, 4-1, 4-2, 4-3, 4-4, 4-6, 4-7, 6-1, Extend 6-1, 6-2, 6-3, 6-4, 6-5, 6-6, 7-7, 8-6, 8-7, 9-7 | 57, 143–150, 157–158, 170, 171–178, 204, 225–231, 232–238, 239–245, 247–253, 259–266, 267–274, 339–345, 346–347, 348–353, 354–360, 361–366, 368–373, 376–380, 448–455, 531–538, 539–545, 614–621 |
| 3. Represent constraints by equations or inequalities, and by systems of equations and/or inequalities, and interpret solutions as viable or nonviable options in a modeling context. | 4-1, 5-4, 5-6, 6-1, 6-2, 6-6 | 225–231, 309–314, 321–326, 339–345, 348–353, 376–380 |
| 4. Rearrange formulas to highlight a quantity of interest, using the same reasoning as in solving equations. | 2-7, 3-4 | 122–127, 171–178 |
| **Reasoning with Equations and Inequalities A-REI** | | |
| Understand solving equations as a process of reasoning and explain the reasoning.<br><br>1. Explain each step in solving a simple equation as following from the equality of numbers asserted at the previous step, starting from the assumption that the original equation has a solution. Construct a viable argument to justify a solution method. | 2-2, 2-3, 2-4, 2-5, 2-6, 2-7, 9-4 | 87–93, 95–100, 101–106, 107–113, 115–121, 122–127, 588–595 |
| Solve equations and inequalities in one variable.<br><br>3. Solve linear equations and inequalities in one variable, including equations with coefficients represented by letters. | Explore 2-2, 2-2, Explore 2-3, 2-3, 2-4, 2-5, 2-6, 2-7, 5-1, Explore 5-2, 5-2, 5-3, 5-4, 5-5 | 85–86, 87–93, 94, 95–100, 101–106, 107–113, 115–121, 122–127, 289–294, 295, 296–301, 302–307, 309–314, 316–320 |

★ Mathematical Modeling Standards

| Standards | Student Edition Lesson(s) | Student Edition Page(s) |
|---|---|---|
| 4. Solve quadratic equations in one variable. | 9-5, 9-6 | 596–602, 606–613 |
| a. Use the method of completing the square to transform any quadratic equation in $x$ into an equation of the form $(x - p)^2 = q$ that has the same solutions. Derive the quadratic formula from this form. | | |
| b. Solve quadratic equations by inspection (e.g., for $x^2 = 49$), taking square roots, completing the square, the quadratic formula and factoring, as appropriate to the initial form of the equation. Recognize when the quadratic formula gives complex solutions and write them as $a \pm bi$ for real numbers $a$ and $b$. | 9-3, 9-4, 9-5, 9-6 | 580–586, 588–595, 596–602, 606–613 |
| Solve systems of equations. | Extend 8-5 | 527–528 |
| 5. Prove that, given a system of two equations in two variables, replacing one equation by the sum of that equation and a multiple of the other produces a system with the same solutions. | | |
| 6. Solve systems of linear equations exactly and approximately (e.g., with graphs), focusing on pairs of linear equations in two variables. | 6-1, Extend 6-1, 6-2, 6-3, 6-4, 6-5, Extend 6-5 | 339–345, 346–347, 348–353, 354–360, 361–366, 368–373, 374–375 |
| 7. Solve a simple system consisting of a linear equation and a quadratic equation in two variables algebraically and graphically. | 9-7 | 614–621 |
| Represent and solve equations and inequalities graphically. | 1-6, 1-7, 3-1, 3-2, 7-5, 9-1 | 42–48, 49–56, 143–150, 151–156, 430–435, 559–569 |
| 10. Understand that the graph of an equation in two variables is the set of all its solutions plotted in the coordinate plane, often forming a curve (which could be a line). | | |
| 11. Explain why the $x$-coordinates of the points where the graphs of the equations $y = f(x)$ and $y = g(x)$ intersect are the solutions of the equation $f(x) = g(x)$; find the solutions approximately, e.g., using technology to graph the functions, make tables of values, or find successive approximations. Include cases where $f(x)$ and/or $g(x)$ are linear, polynomial, rational, absolute value, exponential, and logarithmic functions. ★ | 3-7, Extend 6-1, Extend 7-5, 9-7 | 197–203, 346–347, 436–437, 614–621 |
| 12. Graph the solutions to a linear inequality in two variables as a halfplane (excluding the boundary in the case of a strict inequality), and graph the solution set to a system of linear inequalities in two variables as the intersection of the corresponding half-planes. | 5-6, Extend 5-6, 6-6, Extend 6-6 | 321–326, 327, 376–380, 381 |

## Functions

### Interpreting Functions F-IF

| | | |
|---|---|---|
| Understand the concept of a function and use function notation. | 1-6, 1-7 | 42–48, 49–56 |
| 1. Understand that a function from one set (called the domain) to another set (called the range) assigns to each element of the domain exactly one element of the range. If $f$ is a function and $x$ is an element of its domain, then $f(x)$ denotes the output of $f$ corresponding to the input $x$. The graph of $f$ is the graph of the equation $y = f(x)$. | | |
| 2. Use function notation, evaluate functions for inputs in their domains, and interpret statements that use function notation in terms of a context. | 1-7, 4-2, 7-5, 7-7, 9-1 | 49–56, 232–238, 430–435, 448–455, 559–569 |
| 3. Recognize that sequences are functions, sometimes defined recursively, whose domain is a subset of the integers. | 3-6, 7-9, 7-10 | 190–196, 462–467, 469–474 |

★ Mathematical Modeling Standards

# Correlations

| Standards | Student Edition Lesson(s) | Student Edition Page(s) |
|---|---|---|
| Interpret functions that arise in applications in terms of the context.<br><br>4. For a function that models a relationship between two quantities, interpret key features of graphs and tables in terms of the quantities, and sketch graphs showing key features given a verbal description of the relationship. ★ | 1-8, Explore 3-1, 3-1, 3-5, 3-7, 7-5, 9-1 | 58–64, 141–142, 143–150, 181–189, 197–203, 430–435, 559–569 |
| 5. Relate the domain of a function to its graph and, where applicable, to the quantitative relationship it describes. | 1-7, 1-8, 3-7, 7-5, 9-1 | 49–56, 58–64, 197–203, 430–435, 559–569 |
| 6. Calculate and interpret the average rate of change of a function (presented symbolically or as a table) over a specified interval. Estimate the rate of change from a graph. ★ | Explore 3-3, 3-3, Extend 7-9, Extend 9-1, 9-8 | 159, 160–168, 468, 570, 622–627 |
| Analyze functions using different representations.<br><br>7. Graph functions expressed symbolically and show key features of the graph, by hand in simple cases and using technology for more complicated cases. ★<br><br>   a. Graph linear and quadratic functions and show intercepts, maxima, and minima. | 3-1, 3-2, Extend 3-2, Explore 3-4, 3-4, 3-5, 9-1, 9-2, 9-3, 9-7 | 143–150, 151–156, 157–158, 170, 171–178, 181–189, 559–569, 571–579, 580–586, 614–621 |
|    b. Graph square root, cube root, and piecewise-defined functions, including step functions and absolute value functions. | 3-7, Extend 3-7, 3-8 | 197–203, 204, 205–211 |
|    e. Graph exponential and logarithmic functions, showing intercepts and end behavior, and trigonometric functions, showing period, midline, and amplitude. | 7-5, 7-6 | 430–435, 438–447 |
| 8. Write a function defined by an expression in different but equivalent forms to reveal and explain different properties of the function.<br><br>   a. Use the process of factoring and completing the square in a quadratic function to show zeros, extreme values, and symmetry of the graph, and interpret these in terms of a context. | 9-4, 9-5, Extend 9-5 | 588–595, 596–602, 603–604 |
|    b. Use the properties of exponents to interpret expressions for exponential functions. | 7-1, 7-2, 7-5, 7-7, 7-8 | 395–401, 402–409, 430–435, 448–455, 458–461 |
| 9. Compare properties of two functions each represented in a different way (algebraically, graphically, numerically in tables, or by verbal descriptions). | 1-7, 4-2, 7-10, 9-1, 9-2 | 49–56, 232–238, 469–474, 559–569, 571–579 |
| **Building Functions F-BF** | | |
| Build a function that models a relationship between two quantities.<br><br>1. Write a function that describes a relationship between two quantities. ★<br><br>   a. Determine an explicit expression, a recursive process, or steps for calculation from a context. | 1-7, 3-1, 3-4, 3-6, 4-1, 4-2, 4-3, 4-4, 4-6, 4-7, 7-7, 7-10 | 49–56, 143–150, 171–178, 190–196, 225–231, 232–238, 239–245, 247–253, 259–266, 267–274, 448–455, 469–474 |
|    b. Combine standard function types using arithmetic operations. | 9-9 | 630–636 |
| 2. Write arithmetic and geometric sequences both recursively and with an explicit formula, use them to model situations, and translate between the two forms. ★ | 3-6, 7-9, 7-10 | 190–196, 464–469, 469–474 |

★ Mathematical Modeling Standards

| Standards | Student Edition Lesson(s) | Student Edition Page(s) |
|---|---|---|
| Build new functions from existing functions.<br><br>3. Identify the effect on the graph of replacing $f(x)$ by $f(x) + k$, $k\,f(x)$, $f(kx)$, and $f(x + k)$ for specific values of $k$ (both positive and negative); find the value of $k$ given the graphs. Experiment with cases and illustrate an explanation of the effects on the graph using technology. | 3-5, 3-8, 7-6, 9-2 | 181–189, 205–211, 438–447, 571–579 |
| 4. Find inverse functions.<br><br>a. Solve an equation of the form $f(x) = c$ for a simple function $f$ that has an inverse and write an expression for the inverse. | 4-7, Extend 4-7 | 267–274, 275 |

### Linear, Quadratic, and Exponential Models F-LE

| | | |
|---|---|---|
| Construct and compare linear, quadratic, and exponential models and solve problems.<br><br>1. Distinguish between situations that can be modeled with linear functions and with exponential functions.<br><br>a. Prove that linear functions grow by equal differences over equal intervals, and that exponential functions grow by equal factors over equal intervals. | Explore 3-3, 3-3, Extend 3-4, Extend 7-5, 9-8 | 159, 160–168, 171–178, 436–437, 622–627 |
| b. Recognize situations in which one quantity changes at a constant rate per unit interval relative to another. | 9-8 | 622–627 |
| c. Recognize situations in which a quantity grows or decays by a constant percent rate per unit interval relative to another. | 9-8 | 622–627 |
| 2. Construct linear and exponential functions, including arithmetic and geometric sequences, given a graph, a description of a relationship, or two input-output pairs (include reading these from a table). | 3-6, 4-1, 4-2, 4-3, 4-4, 4-6, 7-7, 7-9, 9-8, Extend 9-8 | 190–196, 225–231, 232–238, 239–245, 247–253, 259–266, 448–455, 462–467, 622–627, 628–629 |
| 3. Observe using graphs and tables that a quantity increasing exponentially eventually exceeds a quantity increasing linearly, quadratically, or (more generally) as a polynomial function. | 9-8, Extend 9-8 | 622–627, 628–629 |
| Interpret expressions for functions in terms of the situation they model.<br><br>5. Interpret the parameters in a linear or exponential function in terms of a context. | Explore 3-4, 3-4, 3-5, 4-1, 4-4, 7-5, 7-6, 7-7, 9-8 | 170, 171–178, 181–189, 225–231, 247–253, 430–435, 438–447, 448–455, 622–627 |

## Statistics and Probability

### Interpreting Categorical and Quantitative Data S-ID

| | | |
|---|---|---|
| Summarize, represent, and interpret data on a single count or measurement variable.<br><br>1. Represent data with plots on the real number line (dot plots, histograms, and box plots). | 10-2, 10-3, 10-4, 10-5 | 657–664, 665–671, 673–679, 680–687 |
| 2. Use statistics appropriate to the shape of the data distribution to compare center (median, mean) and spread (interquartile range, standard deviation) of two or more different data sets. | 10-5 | 680–687 |
| 3. Interpret differences in shape, center, and spread in the context of the data sets, accounting for possible effects of extreme data points (outliers). | 10-4, 10-5 | 673–679, 680–687 |

# Correlations

| Standards | Student Edition Lesson(s) | Student Edition Page(s) |
|---|---|---|
| Summarize, represent, and interpret data on two categorical and quantitative variables.<br><br>5. Summarize categorical data for two categories in two-way frequency tables. Interpret relative frequencies in the context of the data (including joint, marginal, and conditional relative frequencies). Recognize possible associations and trends in the data. | 10-6 | 688–697 |
| 6. Represent data on two quantitative variables on a scatter plot, and describe how the variables are related.<br><br>  a. Fit a function to the data; use functions fitted to data to solve problems in the context of the data. Use given functions or choose a function suggested by the context. Emphasize linear, quadratic, and exponential models. | 4-4, 4-6, Extend 9-8 | 247–253, 259–266, 628–629 |
|   b. Informally assess the fit of a function by plotting and analyzing residuals. | 4-6 | 259–266 |
|   c. Fit a linear function for a scatter plot that suggests a linear association. | 4-4, 4-6 | 247–253, 259–266 |
| Interpret linear models.<br><br>7. Interpret the slope (rate of change) and the intercept (constant term) of a linear model in the context of the data. | 4-3 | 239–245 |
| 8. Compute (using technology) and interpret the correlation coefficient of a linear fit. | 4-6 | 259–266 |
| 9. Distinguish between correlation and causation. | 4-5 | 254–258 |

# Standards for Mathematical Practice

*Glencoe Algebra 1* exhibits these practices throughout the entire program. All of the Standards for Mathematical Practice are covered in each chapter. The MP icon notes specific areas of coverage.

| Mathematical Practices | Student Edition Lessons |
|---|---|
| 1. Make sense of problems and persevere in solving them. | Throughout the text, for example: 1-4, 1-6, 1-8, 2-2, 2-4, 2-5, 2-6, 3-8, 4-2, 4-4, 5-2, 5-4, 6-4, 7-5, 7-7, 8-7, 9-2, 9-4, 9-5, 10-2 |
| 2. Reason abstractly and quantitatively. | Throughout the text, for example: 1-3, Extend 1-4, 2-1, 2-2, 3-3, 3-4, 3-5, 4-5, 5-1, Explore 5-5, 6-2, 6-5, 7-2, Extend 7-4, 7-6, 8-5, 9-1, Explore 9-3, 9-5, 9-7, 10-2, 10-3, 10-5 |
| 3. Construct viable arguments and critique the reasoning of others. | Throughout the text, for example: 1-3, 1-7, 2-4, 2-5, 2-6, 2-7, 3-5, 4-1, 4-5, 5-5, 6-1, 7-10, 8-1, 8-2, Explore 8-3, Explore 8-5, Explore 8-6, 9-3, 10-1 |
| 4. Model with mathematics. | Throughout the text, for example: 1-1, 1-5, 2-4, 2-5, 2-6, 2-7, 3-2, 3-5, 3-7, 4-2, 4-4, 4-6, 5-1, 6-5, 7-6, 7-8, Extend 7-9, 8-7, Extend 9-1, Explore 9-3, 9-5, 9-7, 9-9, 10-1, 10-6 |
| 5. Use appropriate tools strategically. | Throughout the text, for example: Extend 1-7, 2-2, Extend 3-2, Explore 3-4, Extend 3-7, 4-3, Extend 4-7, 5-6, Extend 5-6, Extend 6-1, Extend 6-6, 7-3, Extend 7-7, Explore 8-1, 9-3, Extend 9-8, 10-4 |
| 6. Attend to precision. | Throughout the text, for example: 1-5, 2-7, 4-1, 5-2, 6-6, 7-4, 8-3, Explore 9-3, 9-4, 9-6, 10-3 |
| 7. Look for and make use of structure. | Throughout the text, for example: 1-2, Explore 2-3, 2-4, 2-5, 2-6, 2-7, Explore 3-1, 3-8, 4-2, 4-7, Explore 5-2, 5-3, 5-5, 6-3, Extend 6-5, 6-6, 7-4, 7-8, 7-9, Extend 8-5, 8-7, 9-5, 9-8, 9-9, 10-1, 10-6 |
| 8. Look for and express regularity in repeated reasoning. | Throughout the text, for example: 1-4, Explore 2-2, 2-3, 3-1, Explore 3-3, 3-4, Extend 3-4, 3-6, 4-5, 5-4, 6-1, 7-1, 7-4, Extend 7-5, 8-2, Explore 8-3, 8-3, 8-4, 8-6, 8-7, Extend 9-1, 9-2, Extend 9-5, 10-6 |

connectED.mcgraw-hill.com

# CHAPTER 0
# Preparing for Algebra

| | | |
|---|---|---|
| | **Get Started on Chapter 0** | **P1** |
| | Pretest | **P3** |
| 0-1 | **Plan for Problem Solving** | **P5** |
| 0-2 | **Real Numbers** | **P7** |
| 0-3 | **Operations with Integers** | **P11** |
| 0-4 | **Adding and Subtracting Rational Numbers** | **P13** |
| 0-5 | **Multiplying and Dividing Rational Numbers** | **P17** |
| 0-6 | **The Percent Proportion** | **P20** |
| 0-7 | **Perimeter** | **P23** |
| 0-8 | **Area** | **P26** |
| 0-9 | **Volume** | **P29** |
| 0-10 | **Surface Area** | **P31** |
| 0-11 | **Simple Probability and Odds** | **P33** |
| | Posttest | **P37** |

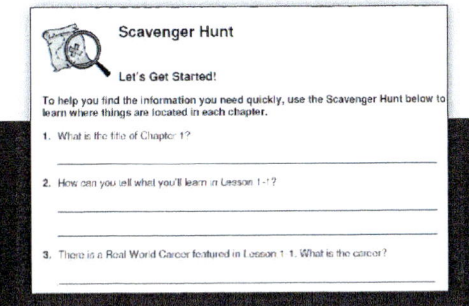

**Worksheets** help to explain key concepts, let you practice your skills, and offer opportunities for extending the lessons. Find them in the Resources in ConnectED.

**Go Online!**
connectED.mcgraw-hill.com

# CHAPTER 1
## Expressions and Functions

| | | | |
|---|---|---|---|
| | Get Ready for Chapter 1 | 2 | |
| 1-1 | Variables and Expressions | 5 | A.SSE.1a, A.SSE.2 |
| 1-2 | Order of Operations | 10 | A.SSE.1b, A.SSE.2 |
| 1-3 | Properties of Numbers | 16 | A.SSE.1b, A.SSE.2 |
| 1-4 | Distributive Property | 23 | A.SSE.1a, A.SSE.2 |
| | Extend 1-4 Algebra Lab: Operations with Rational Numbers | 30 | N.RN.3 |
| | Mid-Chapter Quiz | 32 | |
| 1-5 | Descriptive Modeling and Accuracy | 33 | N.Q.2, N.Q.3 |
| 1-6 | Relations | 42 | A.REI.10, F.IF.1 |
| 1-7 | Functions | 49 | F.IF.1, F.IF.2, F.IF.5 |
| | Extend 1-7 Graphing Technology Lab: Representing Functions | 57 | A.CED.2 |
| 1-8 | Interpreting Graphs of Functions | 58 | F.IF.4, F.IF.5, F.IF.9 |

### ASSESSMENT

| | | |
|---|---|---|
| Study Guide and Review | 65 | |
| Practice Test | 71 | |
| Performance Task | 72 | |
| Preparing for Assessment, Test-Taking Strategy | 73 | |
| Preparing for Assessment, Cumulative Review | 74 | |

**Go Online!**
connectED.mcgraw-hill.com

**Personal Tutors** let you hear a real teacher discuss each step to solving a problem. There is a Personal Tutor for each example, just a click away in ConnectED.

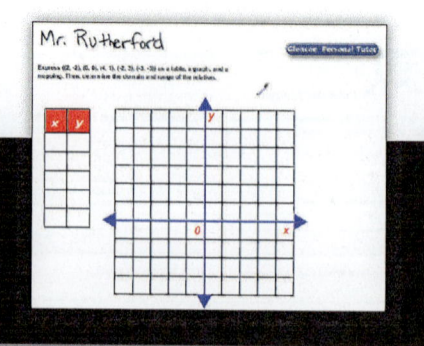

# CHAPTER 2
# Linear Equations

| | | | |
|---|---|---|---|
| | Get Ready for Chapter 2 | 76 | |
| 2-1 | Writing Equations | 79 | A.CED.1 |
| | Explore 2-2 Algebra Lab: Solving Equations | 85 | A.REI.3 |
| 2-2 | Solving One-Step Equations | 87 | A.REI.1, A.REI.3 |
| | Explore 2-3 Algebra Lab: Solving Multi-Step Equations | 94 | A.REI.3 |
| 2-3 | Solving Multi-Step Equations | 95 | A.REI.1, A.REI.3 |
| 2-4 | Solving Equations with the Variable on Each Side | 101 | A.REI.1, A.REI.3 |
| 2-5 | Solving Equations Involving Absolute Value | 107 | A.REI.1, A.REI.3 |
| | Mid-Chapter Quiz | 114 | |
| 2-6 | Ratios and Proportions | 115 | A.REI.1, A.REI.3 |
| 2-7 | Literal Equations and Dimensional Analysis | 122 | A.CED.4, A.REI.3 |

## ASSESSMENT

| | | |
|---|---|---|
| Study Guide and Review | 128 | |
| Practice Test | 133 | |
| Performance Task | 134 | |
| Preparing for Assessment, Test-Taking Strategy | 135 | |
| Preparing for Assessment, Cumulative Review | 136 | |

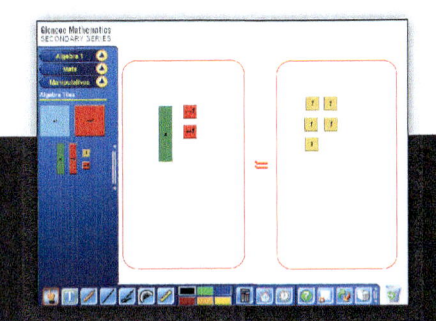

Using the algebra tiles in the **Virtual Manipulatives** can be helpful as you study this chapter. Find them in the eToolkit in ConnectED.

**Go Online!**
connectED.mcgraw-hill.com

# CHAPTER 3
## Linear and Nonlinear Functions

| | | |
|---|---|---|
| **Get Ready for Chapter 3** | **138** | |
| Explore 3-1 Algebra Lab: Analyzing Linear Graphs | 141 | F.IF.4 |
| **3-1 Graphing Linear Functions** | **143** | F.IF.4, F.IF.7a |
| **3-2 Zeros of Linear Functions** | **151** | A.REI.10, F.IF.7a |
| Extend 3-2 Graphing Technology Lab: Graphing Linear Functions | 157 | N.Q.1, F.IF.7a |
| Explore 3-3 Algebra Lab: Rate of Change of a Linear Function | 159 | F.IF.6, F.LE.1a |
| **3-3 Rate of Change and Slope** | **160** | F.IF.6, F.LE.1a |
| **Mid-Chapter Quiz** | **169** | |
| Explore 3-4 Graphing Technology Lab: Investigating Slope-Intercept Form | 170 | A.CED.2, F.IF.7a |
| **3-4 Slope-Intercept Form** | **171** | A.CED.2, F.IF.7a |
| Extend 3-4 Algebra Lab: Linear Growth Patterns | 179 | F.LE.1a |
| **3-5 Transformations of Linear Functions** | **181** | F.IF.7a, F.BF.3 |
| **3-6 Arithmetic Sequences as Linear Functions** | **190** | F.BF.2, F.LE.2 |
| **3-7 Piecewise and Step Functions** | **197** | F.IF.4, F.IF.7b |
| Extend 3-7 Graphing Technology Lab: Piecewise-Linear Functions | 204 | F.IF.7b |
| **3-8 Absolute Value Functions** | **205** | F.IF.7b, F.BF.3 |

### ASSESSMENT

| | |
|---|---|
| **Study Guide and Review** | 212 |
| **Practice Test** | 217 |
| **Performance Task** | 218 |
| **Preparing for Assessment, Test-Taking Strategy** | 219 |
| **Preparing for Assessment, Cumulative Review** | 220 |

## Go Online!
connectED.mcgraw-hill.com

With the **Graphing Tools** in ConnectED, you can explore how changing the values affects the graph of a linear function.

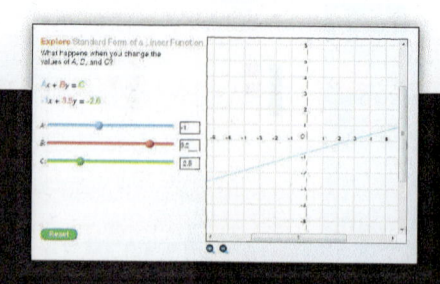

# CHAPTER 4
# Equations of Linear Functions

| | | | |
|---|---|---|---|
| | Get Ready for Chapter 4 | 222 | |
| 4-1 | Writing Equations in Slope-Intercept Form | 225 | F.LE.5 |
| 4-2 | Writing Equations in Standard and Point-Slope Forms | 232 | F.IF.2 |
| 4-3 | Parallel and Perpendicular Lines | 239 | S.ID.7 |
| | Mid-Chapter Quiz | 246 | |
| 4-4 | Scatter Plots and Lines of Fit | 247 | S.ID.6a, S.ID.6c |
| 4-5 | Correlation and Causation | 254 | S.ID.9 |
| 4-6 | Regression and Median-Fit Lines | 259 | S.ID.6, S.ID.8 |
| 4-7 | Inverses of Linear Functions | 267 | A.CED.2, F.BF.4a |
| | Extend 4-7 Algebra Lab: Drawing Inverses | 275 | F.BF.4a |

## ASSESSMENT

| | | |
|---|---|---|
| Study Guide and Review | 276 | |
| Practice Test | 281 | |
| Performance Task | 282 | |
| Preparing for Assessment, Test-Taking Strategy | 283 | |
| Preparing for Assessment, Cumulative Review | 284 | |

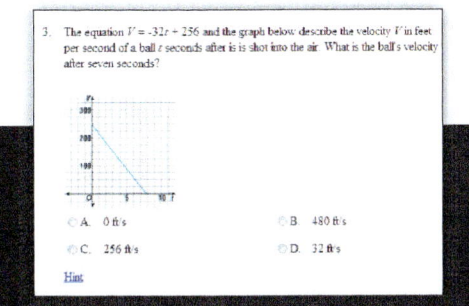

# CHAPTER 5
# Linear Inequalities

| | | |
|---|---|---|
| **Get Ready for Chapter 5** | 286 | |
| **5-1 Solving Inequalities by Addition and Subtraction** | 289 | A.CED.1, A.REI.3 |
| Explore 5-2 Algebra Lab: Solving Inequalities | 295 | A.REI.3 |
| **5-2 Solving Inequalities by Multiplication and Division** | 296 | A.CED.1, A.REI.3 |
| **5-3 Solving Multi-Step Inequalities** | 302 | A.CED.1, A.REI.3 |
| **Mid-Chapter Quiz** | 308 | |
| Explore 5-4 Algebra Lab: Reading Compound Statements | 309 | |
| **5-4 Solving Compound Inequalities** | 310 | A.CED.1, A.CED.3 |
| **5-5 Solving Inequalities Involving Absolute Value** | 316 | A.CED.1, A.REI.3 |
| **5-6 Graphing Inequalities in Two Variables** | 321 | A.CED.3, A.REI.12 |
| Extend 5-6 Graphing Technology Lab: Graphing Inequalities | 327 | A.REI.12 |

## ASSESSMENT

| | |
|---|---|
| **Study Guide and Review** | 328 |
| **Practice Test** | 331 |
| **Performance Task** | 332 |
| **Preparing for Assessment, Test-Taking Strategy** | 333 |
| **Preparing for Assessment, Cumulative Review** | 334 |

**Go Online!**
connectED.mcgraw-hill.com

**Animations** demonstrate Key Concepts and topics from the chapter. Click to watch animations in ConnectED.

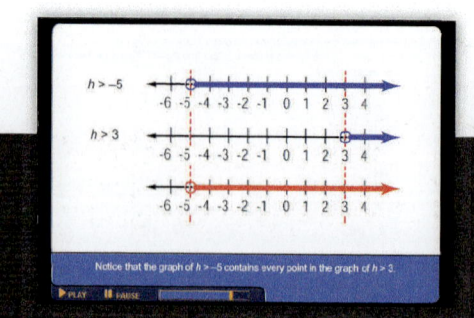

# CHAPTER 6
## Systems of Linear Equations and Inequalities

| | | | |
|---|---|---|---|
| | **Get Ready for Chapter 6** | 336 | |
| 6-1 | **Graphing Systems of Equations** | 339 | A.CED.3, A.REI.6 |
| | Extend 6-1 Graphing Technology Lab: Systems of Equations | 346 | A.REI.6, A.REI.11 |
| 6-2 | **Substitution** | 348 | A.CED.3, A.REI.6 |
| 6-3 | **Elimination Using Addition and Subtraction** | 354 | A.CED.2, A.REI.6 |
| 6-4 | **Elimination Using Multiplication** | 361 | A.REI.5, A.REI.6 |
| | Mid-Chapter Quiz | 367 | |
| 6-5 | **Applying Systems of Linear Equations** | 368 | A.REI.6 |
| | Extend 6-5 Algebra Lab: Using Matrices to Solve Systems of Equations | 374 | A.REI.6 |
| 6-6 | **Systems of Inequalities** | 376 | A.CED.3, A.REI.12 |
| | Extend 6-6 Graphing Technology Lab: Systems of Inequalities | 381 | A.REI.12 |

### ASSESSMENT

| | |
|---|---|
| **Study Guide and Review** | 382 |
| **Practice Test** | 387 |
| **Performance Task** | 388 |
| **Preparing for Assessment, Test-Taking Strategy** | 389 |
| **Preparing for Assessment, Cumulative Review** | 390 |

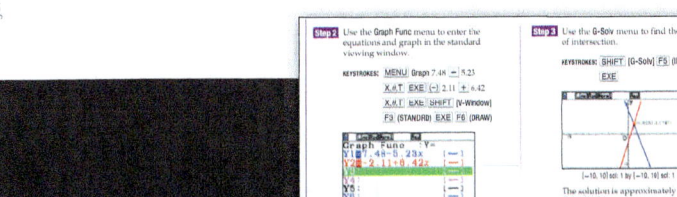

**Graphing Calculator Keystrokes** help you make the most of your graphing calculator. Find the keystrokes for your calculator in the Resources in ConnectED.

**Go Online!**
connectED.mcgraw-hill.com

# CHAPTER 7
# Exponents and Exponential Functions

| | | | |
|---|---|---|---|
| | **Get Ready for Chapter 7** | 392 | |
| 7-1 | **Multiplication Properties of Exponents** | 395 | A.SSE.2, F.IF.8b |
| 7-2 | **Division Properties of Exponents** | 402 | A.SSE.2, F.IF.8b |
| 7-3 | **Rational Exponents** | 410 | N.RN.1, N.RN.2 |
| | **Mid-Chapter Quiz** | 418 | |
| 7-4 | **Radical Expressions** | 419 | N.RN.2 |
| | Extend 7-4 Algebra Lab: Sums and Products of Rational and Irrational Numbers | 428 | N.RN.3 |
| 7-5 | **Exponential Functions** | 430 | F.IF.7e, F.LE.5 |
| | Extend 7-5 Algebra Lab: Exponential Growth Patterns | 436 | F.LE.1a |
| 7-6 | **Transformations of Exponential Functions** | 438 | F.IF.7e, F.BF.3, F.LE.5 |
| 7-7 | **Writing Exponential Functions** | 448 | A.CED.2, F.LE.2, F.LE.5 |
| | Extend 7-7 Graphing Technology Lab: Solving Exponential Equations and Inequalities | 455 | A.REI.11 |
| 7-8 | **Transforming Exponential Expressions** | 457 | A.SSE.3c, F.IF.8b |
| 7-9 | **Geometric Sequences as Exponential Functions** | 461 | F.BF.2, F.LE.2 |
| | Extend 7-9 Algebra Lab: Average Rate of Change of Exponential Functions | 467 | F.IF.6 |
| 7-10 | **Recursive Formulas** | 468 | F.IF.3, F.BF.2 |

## ASSESSMENT

| | | |
|---|---|---|
| **Study Guide and Review** | 474 | |
| **Practice Test** | 481 | |
| **Performance Task** | 482 | |
| **Preparing for Assessment, Test-Taking Strategy** | 483 | |
| **Preparing for Assessment, Cumulative Review** | 484 | |

**Go Online!** connectED.mcgraw-hill.com

**Geometer's Sketchpad®** gives you a tangible, visual way to learn mathematics. Investigate exponents and exponential functions with sketches in ConnectED.

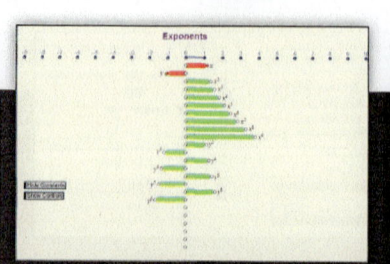

# CHAPTER 8
# Polynomials

| | | | |
|---|---|---|---|
| | **Get Ready for Chapter 8** | 486 | |
| | Explore 8-1 Algebra Lab: Adding and Subtracting Polynomials | 489 | A.APR.1 |
| 8-1 | **Adding and Subtracting Polynomials** | 491 | A.SSE.1a, A.APR.1 |
| 8-2 | **Multiplying a Polynomial by a Monomial** | 498 | A.APR.1 |
| | Explore 8-3 Algebra Lab: Multiplying Polynomials | 504 | A.APR.1 |
| 8-3 | **Multiplying Polynomials** | 506 | A.APR.1 |
| 8-4 | **Special Products** | 512 | A.APR.1 |
| | **Mid-Chapter Quiz** | 518 | |
| | Explore 8-5 Algebra Lab: Factoring Using the Distributive Property | 519 | A.SSE.2 |
| 8-5 | **Using the Distributive Property** | 520 | A.SSE.2, A.SSE.3a |
| | Extend 8-5 Algebra Lab: Proving the Elimination Method | 527 | A.REI.5 |
| | Explore 8-6 Algebra Lab: Factoring Trinomials | 529 | A.SSE.2 |
| 8-6 | **Factoring Quadratic Trinomials** | 531 | A.SSE.2, A.SSE.3a |
| 8-7 | **Factoring Special Products** | 539 | A.SSE.2, A.SSE.3a |

## ASSESSMENT

| | | |
|---|---|---|
| **Study Guide and Review** | 546 | |
| **Practice Test** | 551 | |
| **Performance Task** | 552 | |
| **Preparing for Assessment, Test-Taking Strategy** | 553 | |
| **Preparing for Assessment, Cumulative Review** | 554 | |

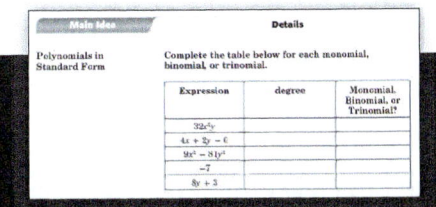

# CHAPTER 9
# Quadratic Functions and Equations

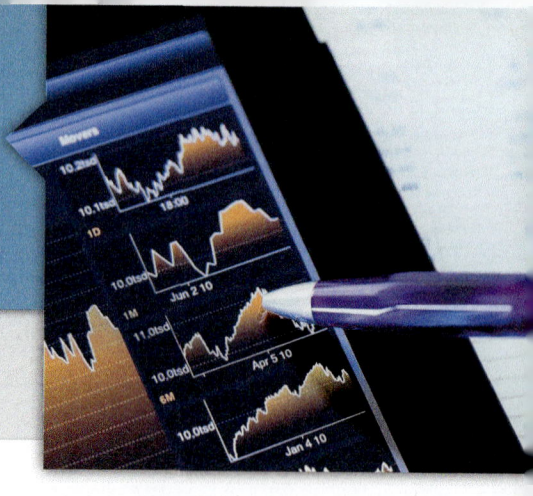

| | | | |
|---|---|---|---|
| | **Get Ready for Chapter 9** | 556 | |
| 9-1 | **Graphing Quadratic Functions** | 559 | F.IF.4, F.IF.7a |
| | Extend 9-1 Algebra Lab: Rate of Change of a Quadratic Function | 570 | F.IF.6 |
| 9-2 | **Transformations of Quadratic Functions** | 571 | F.IF.7a, F.BF.3 |
| 9-3 | **Solving Quadratic Equations by Graphing** | 580 | A.REI.4, F.IF.7a |
| | Extend 9-3 Graphing Technology Lab: Quadratic Inequalities | 587 | |
| 9-4 | **Solving Quadratic Equations by Factoring** | 588 | A.SSE.3a, A.REI.4b |
| 9-5 | **Solving Quadratic Equations by Completing the Square** | 596 | A.SSE.3b, A.REI.4, F.IF.8a |
| | Extend 9-5 Algebra Lab: Finding the Maximum or Minimum Value | 603 | A.SSE.3b, F.IF.8a |
| | **Mid-Chapter Quiz** | 605 | |
| 9-6 | **Solving Quadratic Equations by Using the Quadratic Formula** | 606 | A.CED.1, A.REI.4b |
| 9-7 | **Solving Systems of Linear and Quadratic Equations** | 614 | A.CED.2, A.REI.7 |
| 9-8 | **Analyzing Functions with Successive Differences** | 622 | F.IF.6, F.LE.1, F.LE.2, F.LE.3 |
| | Extend 9-8 Graphing Technology Lab: Curve Fitting | 628 | F.LE.2, S.ID.6a |
| 9-9 | **Combining Functions** | 630 | F.BF.1b |

## ASSESSMENT

| | | |
|---|---|---|
| **Study Guide and Review** | 637 | |
| **Practice Test** | 643 | |
| **Performance Task** | 644 | |
| **Preparing for Assessment, Test-Taking Strategy** | 645 | |
| **Preparing for Assessment, Cumulative Review** | 646 | |

# Go Online!
connectED.mcgraw-hill.com

With **Graphing Calculator** Easy Files™, your graphing calculator can do more than graph! Practice vocabulary in English or Spanish with Lesson Vocabulary Review Files. Or review with a 5-Minute Check. Ask your teacher to assign them to you in ConnectED.

# CHAPTER 10
# Statistics

| | | | |
|---|---|---|---|
| | Get Ready for Chapter 10 | 648 | |
| 10-1 | Measures of Center | 651 | Prep for S.ID.2 |
| 10-2 | Representing Data | 657 | S.ID.1, N.Q.1 |
| 10-3 | Measures of Spread | 665 | S.ID.2 |
| | Mid-Chapter Quiz | 672 | |
| 10-4 | Distributions of Data | 673 | S.ID.1, S.ID.3 |
| 10-5 | Comparing Sets of Data | 680 | S.ID.2, S.ID.3 |
| 10-6 | Summarizing Categorical Data | 688 | S.ID.5 |

## ASSESSMENT

| | | |
|---|---|---|
| Study Guide and Review | 698 | |
| Practice Test | 703 | |
| Performance Task | 704 | |
| Preparing for Assessment, Test-Taking Strategy | 705 | |
| Preparing for Assessment, Cumulative Review | 706 | |

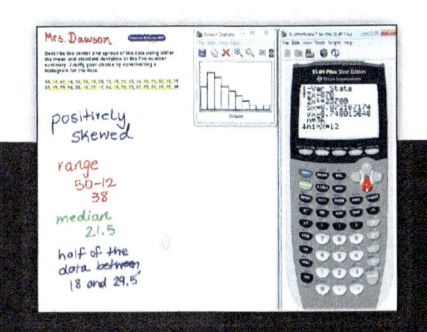

**Personal Tutors** that use graphing calculator technology show you every step to solving problems with this powerful tool. Find them in the Resources in ConnectED.

**Go Online!**
connectED.mcgraw-hill.com

# Student Handbook

**Built-In Workbook**

Extra Practice ........................................................... R1

**Reference**

Selected Answers and Solutions ............................. R11

Glossary/Glosario .................................................... R82

Index ....................................................................... R100

Formulas and Measures ........................... Inside Back Cover

Symbols and Properties ........................... Inside Back Cover

# Standards for Mathematical Practice

*Glencoe Algebra 1* exhibits these practices throughout the entire program. All of the Standards for Mathematical Practice will be covered in each chapter. The MP icon notes specific areas of coverage.

| Mathematical Practices | What does it mean? |
|---|---|
| 1. Make sense of problems and persevere in solving them. | Solving a mathematical problem takes time. Use a logical process to make sense of problems, understand that there may be more than one way to solve a problem, and alter the process if needed. |
| 2. Reason abstractly and quantitatively. | You can start with a concrete or real-world context and then represent it with abstract numbers or symbols (decontextualize), find a solution, then refer back to the context to check that the solution makes sense (contextualize). |
| 3. Construct viable arguments and critique the reasoning of others. | Sound mathematical arguments require a logical progression of statements and reasons. Mathematically proficient students can clearly communicate their thoughts and defend them. |
| 4. Model with mathematics. | Modeling links classroom mathematics and statistics to everyday life, work, and decision-making. High school students at this level are expected to apply key takeaways from earlier grades to high-school level problems. |
| 5. Use appropriate tools strategically. | Certain tools, including estimation and virtual tools are more appropriate than others. You should understand the benefits and limitations of each tool. |
| 6. Attend to precision. | Precision in mathematics is more than accurate calculations. It is also the ability to communicate with the language of mathematics. In high school mathematics, precise language makes for effective communication and serves as a tool for understanding and solving problems. |
| 7. Look for and make use of structure. | Mathematics is based on a well-defined structure. Mathematically proficient students look for that structure to find easier ways to solve problems. |
| 8. Look for and express regularity in repeated reasoning. | Mathematics has been described as the study of patterns. Recognizing a pattern can lead to results more quickly and efficiently. |

# FOLDABLES® by Dinah Zike

## Folding Instructions

The following pages offer step-by-step instructions to make the Foldables® study guides.

### Layered-Look Book

1. Collect three sheets of paper and layer them about 1 cm apart vertically. Keep the edges level.
2. Fold up the bottom edges of the paper to form 6 equal tabs.
3. Fold the papers and crease well to hold the tabs in place. Staple along the fold. Label each tab.

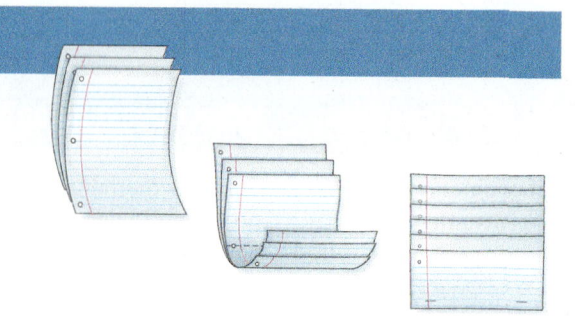

### Shutter-Fold and Four-Door Books

1. Find the middle of a horizontal sheet of paper. Fold both edges to the middle and crease the folds. Stop here if making a shutter-fold book. For a four-door book, complete the steps below.
2. Fold the folded paper in half, from top to bottom.
3. Unfold and cut along the fold lines to make four tabs. Label each tab.

### Concept-Map Book

1. Fold a horizontal sheet of paper from top to bottom. Make the top edge about 2 cm shorter than the bottom edge.
2. Fold width-wise into thirds.
3. Unfold and cut only the top layer along both folds to make three tabs. Label the top and each tab.

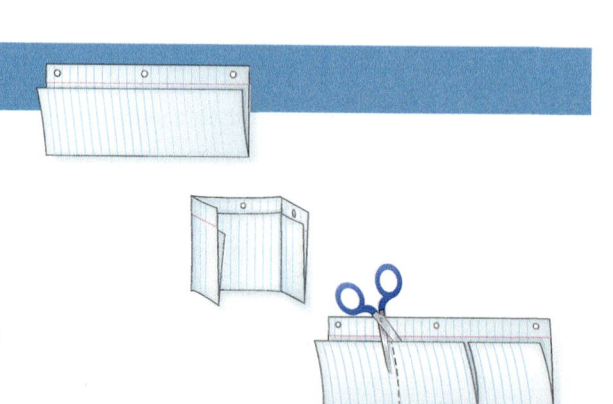

### Vocabulary Book

1. Fold a vertical sheet of notebook paper in half.
2. Cut along every third line of only the top layer to form tabs. Label each tab.

## Pocket Book

1. Fold the bottom of a horizontal sheet of paper up about 3 cm.
2. If making a two-pocket book, fold in half. If making a three-pocket book, fold in thirds.
3. Unfold once and dot with glue or staple to make pockets. Label each pocket.

## Bound Book

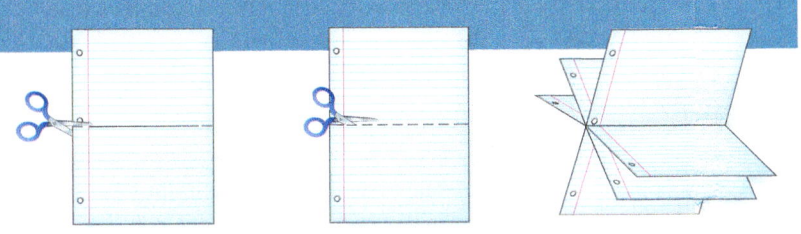

1. Fold several sheets of paper in half to find the middle. Hold all but one sheet together and make a 3-cm cut at the fold line on each side of the paper.
2. On the final page, cut along the fold line to within 3-cm of each edge.
3. Slip the first few sheets through the cut in the final sheet to make a multi-page book.

## Top-Tab Book

1. Layer multiple sheets of paper so that about 2–3 cm of each can be seen.
2. Make a 2–3-cm horizontal cut through all pages a short distance (3 cm) from the top edge of the top sheet.
3. Make a vertical cut up from the bottom to meet the horizontal cut.
4. Place the sheets on top of an uncut sheet and align the tops and sides of all sheets. Label each tab.

## Accordion Book

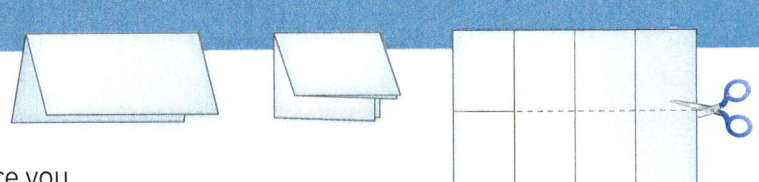

1. Fold a sheet of paper in half. Fold in half and in half again to form eight sections.
2. Cut along the long fold line, stopping before you reach the last two sections.
3. Refold the paper into an accordion book. You may want to glue the double pages together.

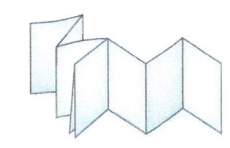

# CHAPTER 6
# Systems of Linear Equations and Inequalities

**SUGGESTED PACING (DAYS)**
90 min. 5 | 1
45 min. 10 | 2
Instruction | Review & Assess

## Track Your Progress

This chapter focuses on content from the Creating Equations and Reasoning with Equations & Inequalities domains.

### THEN

**A.CED.1** Create equations and inequalities in one variable and use them to solve problems.

**A.REI.3** Solve linear equations and inequalities in one variable, including equations with coefficients represented by letters.

**A.REI.12** Graph the solutions to a linear inequality in two variables as a half-plane (excluding the boundary in the case of a strict inequality), and graph the solution set to a system of linear inequalities in two variables as the intersection of the corresponding half-planes.

### NOW

**A.CED.2** Create equations in two or more variables to represent relationships between quantities; graph equations on coordinate axes with labels and scales.

**A.CED.3** Represent constraints by equations or inequalities, and by systems of equations and/or inequalities, and interpret solutions as viable or nonviable options in a modeling context.

**A.REI.6** Solve systems of linear equations exactly and approximately (e.g., with graphs), focusing on pairs of linear equations in two variables.

### NEXT

**A.SSE.2** Use the structure of an expression to identify ways to rewrite it.

**F.BF.2** Write arithmetic and geometric sequences both recursively and with an explicit formula, use them to model situations, and translate between the two forms.

**F.IF.8b** Graph square root, cube root, and piecewise-defined functions, including step functions and absolute value functions.

**N.RN.2** Rewrite expressions involving radicals and rational exponents using the properties of exponents.

## Standards for Mathematical Practice

All of the Standards for Mathematical Practice will be covered in this chapter. The MP icon notes specific areas of coverage.

### MP Teaching the Mathematical Practices
Help students develop the mathematical practices by asking questions like these.

**Questioning Strategies** As students approach problems in this chapter, help them develop mathematical practices by asking:

**Sense-Making**
- How do you determine the number of solutions in a system of linear equations?
- In what situations can you apply systems of linear equations?

**Reasoning**
- Why is it possible to eliminate parts of an equation by adding or subtracting the same value on each side of an equation?
- How do you determine the best method to use to solve systems of equations?

**Construct Arguments**
- How can systems of equations be solved using substitution?
- How can systems of equations be solved using elimination with multiplication?

**Using Tools**
- How can a graph be used to represent a system of equations?
- How can a graph be used in systems of inequalities?

**Precision**
- Why is it important to ensure that what is done to one side of an equation is also done to the other?

### Go Online!

**StudySync: SMP Modeling Videos**

These demonstrate how to apply the Standards for Mathematical Practice to collaborate, discuss, and solve real-world math problems.

336A | Chapter 6 | Systems of Linear Equations and Inequalities

# Go Online!
connectED.mcgraw-hill.com

LearnSmart | The Geometer's Sketchpad | Vocabulary | Personal Tutor | Tools | Calculator Resources | Self-Check Practice | Animations

## Customize Your Chapter
Use the *Plan & Present*, *Assignment Tracker*, and *Assessment* tools in ConnectED to introduce lesson concepts, assign personalized practice, and diagnose areas of student need.

### Differentiated Instruction
Throughout the program, look for the icons to find specialized content designed for your students.

- **AL** Approaching Level
- **OL** On Level
- **BL** Beyond Level
- **ELL** English Language Learners

## Personalize

| Differentiated Resources | | | | |
|---|---|---|---|---|
| FOR EVERY CHAPTER | AL | OL | BL | ELL |
| ✓ Chapter Readiness Quizzes | ● | ● | ◐ | ● |
| ✓ Chapter Tests | ● | ● | ● | ● |
| ✓ Standardized Test Practice | ● | ● | ● | ● |
| Vocabulary Review Games | ● | ● | ● | ● |
| Anticipation Guide (English/Spanish) | ● | ● | ● | ● |
| Student-Built Glossary | ● | ● | ● | ● |
| Chapter Project | ● | ● | ● | ● |
| FOR EVERY LESSON | AL | OL | BL | ELL |
| Personal Tutors (English/Spanish) | ● | ● | ● | ● |
| Graphing Calculator Personal Tutors | ● | ● | ● | ● |
| Step-by-Step Solutions | ● | ● | ◐ | ● |
| Self-Check Quizzes | ● | ● | ● | ● |
| 5-Minute Check | ● | ● | ● | ● |
| Study Notebook | ● | ● | ● | ● |
| Study Guide and Intervention | ● | ● | ◐ | ● |
| Skills Practice | ● | ◐ | ◐ | ● |
| Practice | ◐ | ● | ● | ● |
| Word Problem Practice | ● | ● | ● | ◐ |
| Enrichment | ◐ | ● | ● | ● |
| Extra Examples | ● | ◐ | ◐ | ◐ |
| Lesson Presentations | ● | ● | ● | ● |

◐ Aligned to this group   ● Designed for this group

## Engage

### Featured IWB Resources

**Geometer's Sketchpad** provides students with a tangible, visual way to learn. *Use with Lessons 6-1 and 6-6.*

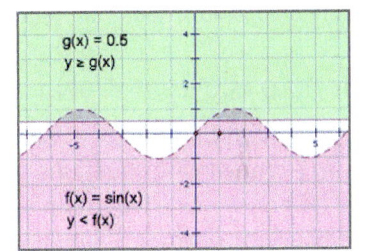

**eLessons** engage students and help build conceptual understanding of big ideas. *Use with Lessons 6-1 through 6-5.*

**Animations** help students make important connections through motion. *Use with Lessons 6-1 and 6-6.*

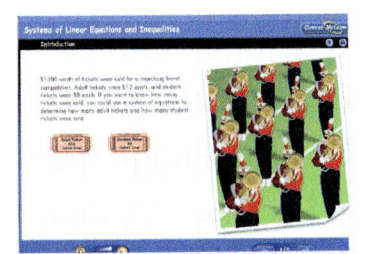

**Time Management** How long will it take to use these resources? Look for the clock in each lesson interleaf.

Chapter 6 | Systems of Linear Equations and Inequalities

# Introduce the Chapter

## Mathematical Background
A system of equations is a set of equations with the same variables. Systems of equations can be solved by graphing the equations on the same coordinate plane or by using algebraic methods, depending on the degree of precision needed for the solution. Systems of inequalities are solved by graphing the inequalities and identifying the set of all points that satisfy both inequalities.

## e Essential Question
At the end of this chapter, students should be able to answer the Essential Question.

How can you find the solution to a math problem?
Sample answers: Use a graph; analyze a table of values; solve an equation; guess and check.

## Apply Math to the Real World
**MUSIC** In this activity, students will use what they already know about writing equations to determine the amount of money made from admission fees to a marching band competition. Have students work individually or in small groups to complete this activity. MP 1

## Go Online!

### Chapter Project
**Save the Date** Students can use what they have learned about systems of equations to plan a fundraising dinner in support of a cause that is important to them. This project addresses global awareness, as well as several specific skills identified as being essential to success by the Framework for 21st Century Learning. MP 1, 3, 4, 6

---

# CHAPTER 6
# Systems of Linear Equations and Inequalities

**THEN**
You solved linear equations in one variable.

**NOW**
In this chapter you will:
- Solve systems of linear equations by graphing, substitution, and elimination.
- Solve systems of linear inequalities by graphing.

## MP WHY

**MUSIC** Marching band competitions are a source of funds for the bands that host them. They use these funds to buy new uniforms, instruments, and to help with travel expenses.

**Use the Mathematical Practices to complete the activity.**

1. **Use Tools** Use the Internet to find information about a local band competition or find out if your school hosts a competition. What are the admission fees for students and adults for this event?

2. **Reasoning** If you knew how many adult and student tickets were sold, what could you determine? Could you use an equation to determine how much money the band made?

3. **Apply Math** Using the admission fees you found, write an equation showing the total amount of band income if 200 adults and 100 students attend the competition.

4. **Model with Mathematics** Use the Equation Chart tool in ConnectED to represent the necessary equation.

5. **Discuss** Could you calculate how many student tickets and how many adult tickets were sold if you knew only the price of each and the total amount of income?

---

## ALEKS

**Your Student Success Tool** ALEKS is an adaptive, personalized learning environment that identifies precisely what each student knows and is ready to learn—ensuring student success at all levels.

- **Formative Assessment:** Dynamic, detailed reports monitor students' progress toward standards mastery.
- **Automatic Differentiation:** Strengthen prerequisite skills and target individual learning gaps.
- **Personalized Instruction:** Supplement in-class instruction with personalized assessment and learning opportunities.

Chapter 6 | Systems of Linear Equations and Inequalities

## Go Online to Guide Your Learning

### Explore & Explain

**The Geometer's Sketchpad**
Visualize and model slope using the Slope of a Line sketch in Lesson 6-1.

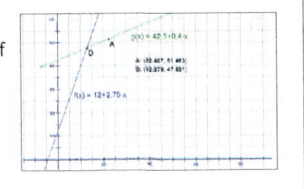

**eToolkit**
Investigate using the **Equation Chart Mat** and the **Algebra Tiles** tool to model substitution.

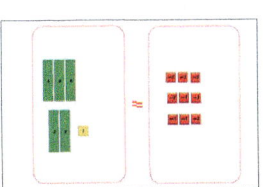

**eBook**
**Interactive Student Guide**
Before starting the chapter, answer the **Chapter Focus** preview questions. Check your answers as you complete each lesson. At the end of the chapter, try the **Performance Task**.

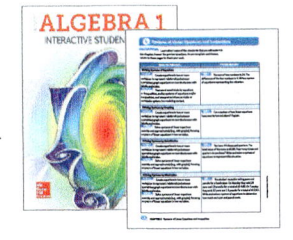

### Organize

**Foldables**
**Systems of Linear Equations and Inequalities** Make this Foldable to help you organize your notes about solving systems of equations and inequalities. Begin with a sheet of notebook paper folded lengthwise. Cut and label six tabs using the lesson titles.

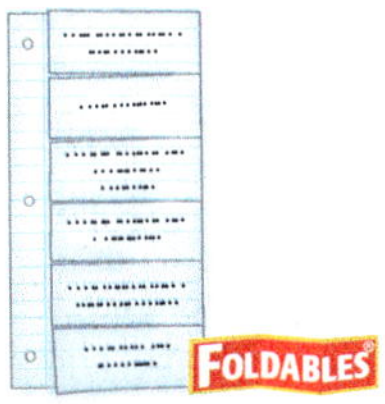

### Collaborate

**Chapter Project**
In the **Save the Date** project, you will use what you have learned about systems of equations to complete a project.

### Focus

**LEARNSMART**
Need help studying? Complete the **Linear and Exponential Relationships** domain in LearnSmart to review for the chapter test.

**ALEKS**
You can use the **Systems** and **Linear Inequalities** topics in ALEKS to find out what you know about linear functions and what you are ready to learn.*

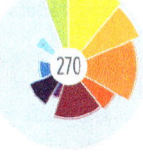

*Ask your teacher if this is part of your program.

---

### Dinah Zike's FOLDABLES

**Focus** Students write notes about solving systems of linear equations and inequalities as they are presented in the lessons of this chapter.

**Teach** Have students make and label their Foldables as illustrated. Have students write a word or concept from the lesson on the back of each lesson's tab. Under the word or concept, ask students to include a definition and an example.

**When to Use It** Encourage students to add to their Foldables as they work through the chapter and to use them to review for the chapter test.

---

### Go Online!

**Independent and Notebook Foldables**
Not all Foldables are created equal! In this video, you will learn how to create *independent* and *notebook* Foldables.

connectED.mcgraw-hill.com    337

# Get Ready for the Chapter

## RtI Response to Intervention
Use the Concept Check results and the Intervention Planner chart to help you determine your Response to Intervention.

### Intervention Planner

**TIER 1 — On Level (OL)**

**IF** students miss 25% of the exercises or less,

**THEN** choose a resource:

*Go Online!*
- Skills Practice, Chapter 1
- Skills Practice, Chapter 2
- Chapter Project

**TIER 2 — Approaching Level (AL)**

**IF** students miss 50% of the exercises,

**THEN** choose a resource:

*Go Online!*
- Study Guide and Intervention, Chapter 1
- Extra Examples
- Personal Tutors
- Homework Help

*Quick Review Math Handbook*

**TIER 3 — Intensive Intervention**

**IF** students miss 75% of the exercises,

**THEN** use *Math Triumphs, Alg. 1*

*Go Online!*
- Extra Examples
- Personal Tutors
- Homework Help
- Review Vocabulary

---

## Get Ready for the Chapter

### Connecting Concepts

**Concept Check**
Review the concepts used in this chapter by answering each question below.

1. Is the expression $6^2$ the same as $(6)^2$?
   **Yes, these expressions are the same.**
2. Is the expression $3 + 6^2$ the same as $(3 + 6)^2$?
   **No, these expressions are different.**
3. What would be the first step in evaluating the expression $\left(\frac{7}{11}\right)2$? **Rewrite the expression as $\frac{7}{11} \cdot \frac{7}{11}$**
4. The formula for the area of a triangle is $A = \frac{1}{2}bh$, where $A$ represents the area, $b$ is the base, and $h$ is the height. Solve the formula for the base of the triangle. $b = \frac{2A}{h}$
5. Explain what step you would do first to solve $2x + 4y = 12$ for $x$.
6. Given $3x = 36 - 12y$, what would you do to simplify the equation? **Divide each side by 3.**
7. Given $5x - 10y = 40$, what would the equation be if you solve for $y$? $y = \frac{x-8}{2}$
8. What do you call two or more inequalities with the same variables? **a system of inequalities**

5. **Subtract 4y from each side to isolate the term with x on one side of the equation.**

**Performance Task Preview**
You can use the concepts and skills in the chapter to solve problems in a real-world setting. Understanding systems of equations will help you finish the Performance Task at the end of the chapter.

**In this Performance Task you will:**
- make sense of problems
- reason abstractly and quantitatively
- construct viable arguments
- model with mathematics
- look for and make use of structure

### New Vocabulary

*Go Online! for Vocabulary Review Games and key vocabulary in 13 languages.*

| English | | Español |
|---|---|---|
| system of equations | p. 339 | sistema de ecuaciones |
| consistent | p. 339 | consistente |
| independent | p. 339 | independiente |
| dependent | p. 339 | dependiente |
| inconsistent | p. 339 | inconsistente |
| substitution | p. 348 | sustitución |
| elimination | p. 354 | eliminación |
| matrix | p. 374 | matriz |
| element | p. 374 | elemento |
| dimension | p. 374 | dimensión |
| augmented matrix | p. 374 | matriz ampliada |
| row reduction | p. 375 | reducción de fila |
| identity matrix | p. 375 | matriz |
| system of inequalities | p. 376 | sistema de desigualdades |

### Review Vocabulary

**domain** *dominio* the set of the first numbers of the ordered pairs in a relation

**intersection** *intersección* the graph of a compound inequality containing *and*; the solution is the set of elements common to both graphs

**Proportion**

**proportion** *proporción* an equation stating that two ratios are equal

$$\frac{24}{30} \overset{\div 6}{\underset{\div 6}{=}} \frac{4}{5}$$

## Key Vocabulary (ELL)

Introduce the key vocabulary in the chapter using the routine below.

**Define** A system of equations is a set of equations with the same variables.

**Example** $y = 34.2 - 14.9x$ and $y = 3.3 + 4.7x$

**Ask** What variables do both equations have in common? *x* and *y*

**LESSON 6-1**

# Graphing Systems of Equations

**SUGGESTED PACING (DAYS)**

| | | |
|---|---|---|
| 90 min. | 0.75 | .25 |
| 45 min. | 1.5 | 0.5 |
| | Instruction | Extend Lab |

## Track Your Progress

### Objectives

1. Determine the number of solutions a system of linear equations has.
2. Solve systems of linear equations by graphing.

### Mathematical Background

A system of equations can be solved by graphing the equations on the same coordinate plane. The graphs of the equations can intersect at one point (exactly one solution), be parallel (no solution), or be the same line (infinite number of solutions).

### THEN

**A.CED.1** Create equations and inequalities in one variable and use them to solve problems.

### NOW

**A.CED.3** Represent constraints by equations or inequalities, and by systems of equations and/or inequalities, and interpret solutions as viable or nonviable options in a modeling context.

**A.REI.6** Solve systems of linear equations exactly and approximately (e.g., with graphs), focusing on pairs of linear equations in two variables.

### NEXT

**A.REI.11** Explain why the $x$-coordinates of the points where the graphs of the equations $y = f(x)$ and $y = g(x)$ intersect are the solutions of the equation $f(x) = g(x)$; find the solutions approximately, e.g., using technology to graph the functions, make tables of values, or find successive approximations. Include cases where $f(x)$ and/or $g(x)$ are linear, polynomial, rational, absolute value, exponential, and logarithmic functions.

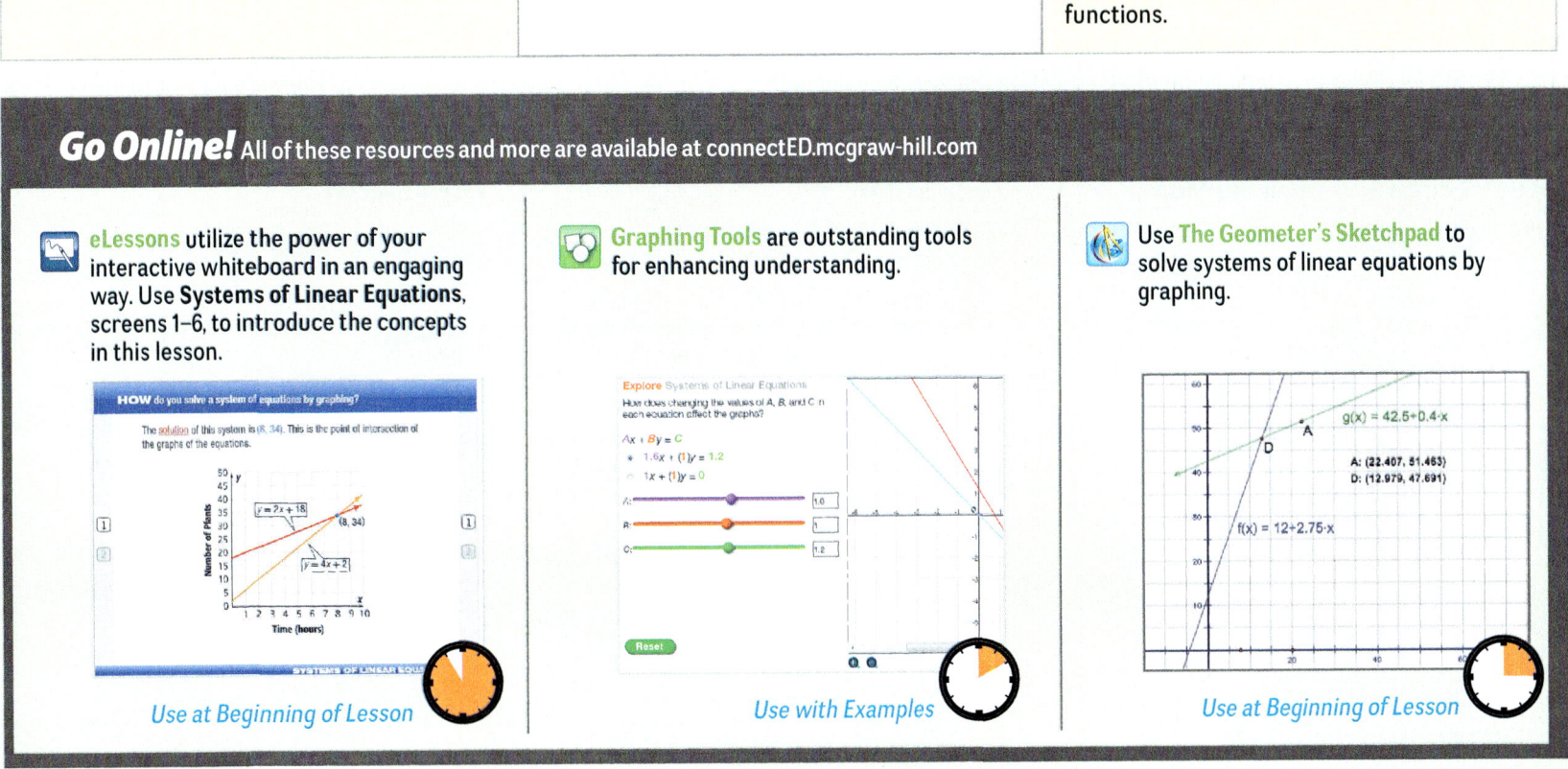

## Go Online! All of these resources and more are available at connectED.mcgraw-hill.com

**eLessons** utilize the power of your interactive whiteboard in an engaging way. Use **Systems of Linear Equations**, screens 1–6, to introduce the concepts in this lesson.

*Use at Beginning of Lesson*

**Graphing Tools** are outstanding tools for enhancing understanding.

*Use with Examples*

Use **The Geometer's Sketchpad** to solve systems of linear equations by graphing.

*Use at Beginning of Lesson*

## OER Using Open Educational Resources

**Tutorial** Have students go to **onlinemathlearning.com** and search solving systems of equations by graphing. On this site, they can watch tutorials and see worked out examples. *Use as flipped learning or review*

connectED.mcgraw-hill.com    339A

# Go Online!
connectED.mcgraw-hill.com
Worksheets

# Differentiate Your Resources

**Extra Practice** Additional practice or homework; Skills Practice is best for approaching-level students and Practice is best for on-level and beyond-level students

## Skills Practice
AL  OL  ELL

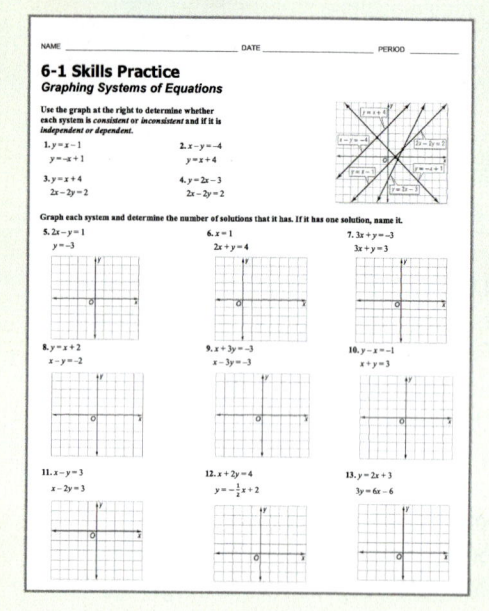

## Practice
AL  OL  BL  ELL

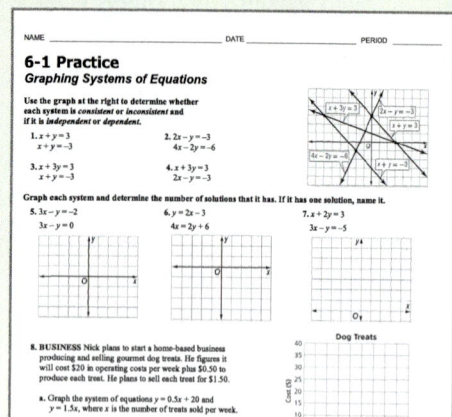

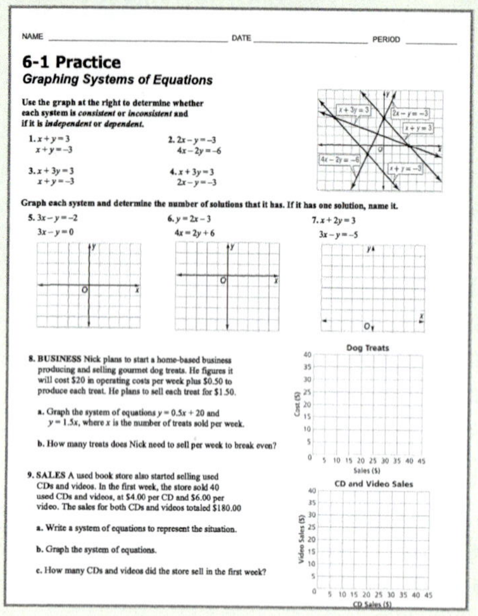

## Word Problem Practice
AL  OL  BL  ELL

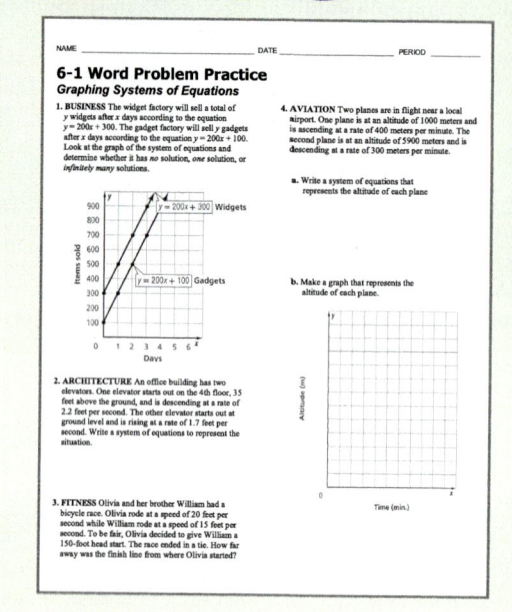

**Intervention** Reteaching and vocabulary activities that can be used with struggling or absent students and as ELL support

## Study Guide and Intervention
AL  OL  ELL

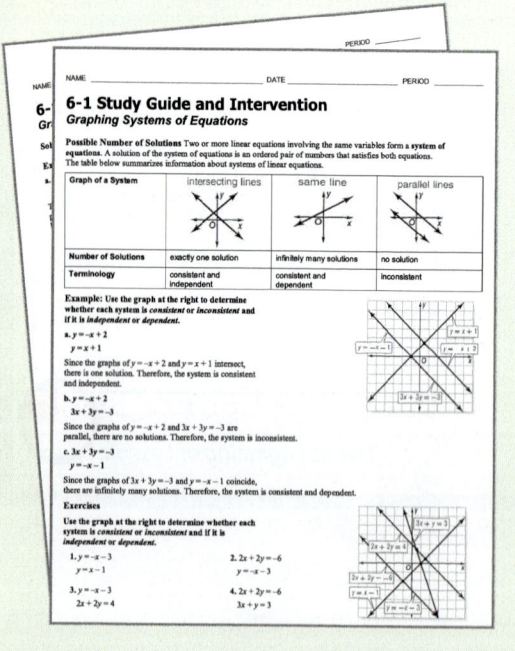

## Study Notebook
AL  OL  BL  ELL

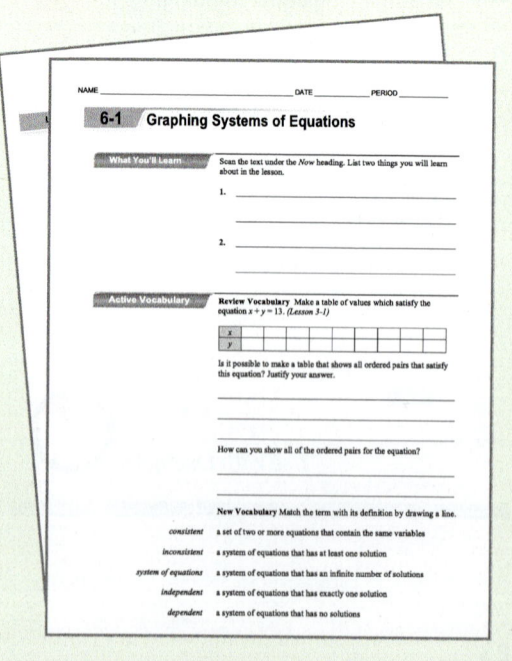

**Extension** Activities that can be used to extend lesson concepts

## Enrichment
OL  BL  ELL

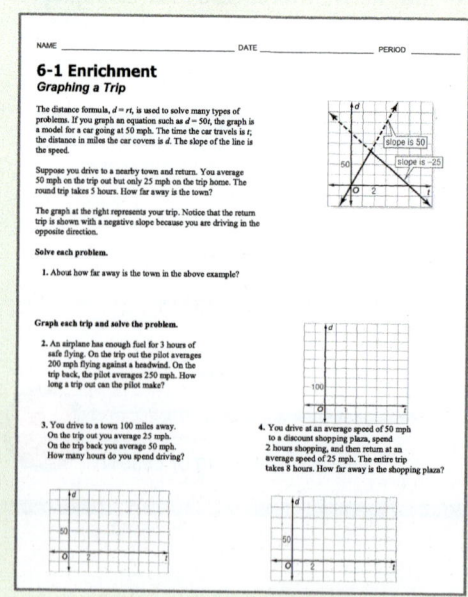

339B | Lesson 6-1 | Graphing Systems of Equations

# LESSON 1
# Graphing Systems of Equations

| ::Then | ::Now | ::Why? |
|---|---|---|
| • You graphed linear equations. | 1 Determine the number of solutions a system of linear equations has.<br>2 Solve systems of linear equations by graphing. | A volleyball team is selling T-shirts. There is a $600 set-up fee and then each T-shirt costs $3 to print. They plan to sell the T-shirts for $15 each. The volleyball team wants to know how many shirts they will need to sell in order to make a profit.<br><br>Graphing a system can show when a profit is made. The cost of producing the T-shirts can be modeled by the equation $y = 3x + 600$, where $y$ represents the cost of production and $x$ is the number of T-shirts produced. |

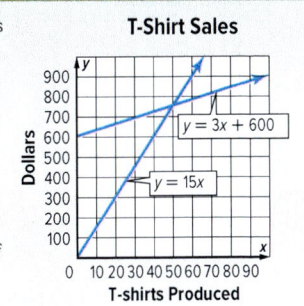

**T-Shirt Sales** graph showing $y = 3x + 600$ and $y = 15x$.

### New Vocabulary
system of equations
consistent
independent
dependent
inconsistent

### Mathematical Practices
3 Construct viable arguments and critique the reasoning of others.
8 Look for and express regularity in repeated reasoning.

### Content Standards
**A.CED.3** Represent constraints by equations or inequalities, and by systems of equations and/or inequalities, and interpret solutions as viable or nonviable options in a modeling context.
**A.REI.6** Solve systems of linear equations exactly and approximately (e.g. with graphs), focusing on pairs of linear equations in two variables.

**1 Possible Number of Solutions** The income from the T-shirts sold can be modeled by the equation $y = 15x$, where $y$ represents the total income of selling the T-shirts, and $x$ is the number of T-shirts sold.

If we graph these equations, we can see at which point the volleyball team begins making a profit. The point where the two graphs intersect is where the volleyball team breaks even. At this point, the income equals the cost. This happens when the volleyball team sells 50 T-shirts. If the volleyball team sells more than 50 T-shirts, they will make a profit.

The two equations, $y = 3x + 600$ and $y = 15x$, form a **system of equations**. The ordered pair that is a solution of both equations is the solution of the system. A system of two linear equations can have one solution, an infinite number of solutions, or no solution.

- If a system has at least one solution, it is said to be **consistent**. The graphs intersect at one point or are the same line.
- If a consistent system has exactly one solution, it is said to be **independent**. If it has an infinite number of solutions, it is said to be **dependent**. This means that there are unlimited solutions that satisfy both equations.
- If a system has no solution, it is said to be **inconsistent**. The graphs are parallel.

### Concept Summary  Possible Solutions

| Number of Solutions | exactly one | infinite | no solution |
|---|---|---|---|
| Terminology | consistent and independent | consistent and dependent | inconsistent |
| Graph | (intersecting lines) | (same line) | (parallel lines) |

---

### MP Mathematical Practices Strategies

**Make sense of problems and persevere.** Remind students that systems of linear equations can vary in the number of solutions they have.

- If two lines have the same slope, but different $y$-intercepts, how many times do they intersect? **0**
- If two lines have different slopes, how many times do they intersect? **1**
- If two lines have the same slope and same intercept, how many times do they intersect? **infinitely many**

---

**Lesson 6-1 | Graphing Systems of Equations**

## Launch

Have students read the Why? section of the lesson. Ask:

- What does it mean to *break even* in this situation? **The income from T-shirt sales and the cost of producing the T-shirts are equal.**
- What is the break-even dollar amount? **$750**
- How many solutions does the system have? Explain. **One; there can be only one break-even point, and that is where the graphs intersect.**

## Teach

Ask the scaffolded questions for each example to build conceptual understanding for students at all levels.

### 1 Possible Number of Solutions

#### Example 1  Number of Solutions

**AL** In part **a**, are the functions linear? How do you know? **Yes; the equations are written in slope-intercept form.**

**OL** In part **b**, explain why there is no solution of the system. **Sample answer: the lines are parallel, which means the lines will never intersect. If they never intersect, then there cannot be a solution.**

**BL** In part **b**, describe a way to determine that there is no solution to the system without graphing. **Sample answer: the equations have the same slope but different $y$-intercepts, therefore, they are parallel and will never intersect.**

*(continued on the next page)*

### Go Online!

**Interactive Whiteboard**
Use the *eLesson* or *Lesson Presentation* to present this lesson.

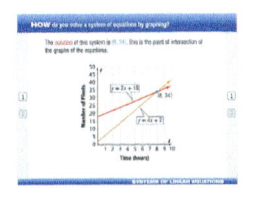

## Lesson 6-1 | Graphing Systems of Equations

**Need Another Example?**

Use the graph to determine whether each system is *consistent* or *inconsistent* and whether it is *independent* or *dependent*.

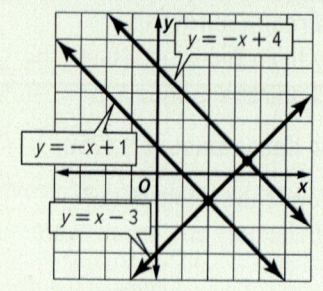

a. $y = -x + 1$
   $y = -x + 4$
   inconsistent

b. $y = x - 3$
   $y = -x + 1$
   consistent and independent

### Teaching Tip

**Interactive Whiteboard** Create a table with columns: *no solutions*, *1 solution*, and *infinitely many solutions*. Write several systems of equations on the board. Have students drag each system to the correct column.

## 2 Solve By Graphing

**Example 2 Solve by Graphing**

**AL** In part b, what must be done before graphing?
Write the equations in slope-intercept form.

**OL** In part b, what do you notice about the equations once they are written in slope-intercept form? They have the same slope, but different y-intercepts.

**BL** In part a, explain how to check to make sure the solution is correct for both equations.
Sample answer: Substitute 3 for x in both equations and substitute 1 for y in both equations. Simplify, and make sure the equations are equivalent on both sides of the equals signs.

---

**Study Tip**

**Reasoning** When both equations are of the form $y = mx + b$, the values of $m$ and $b$ can determine the number of solutions.

| Compare m and b | Number of Solutions |
|---|---|
| different m values | one |
| same m value, but different b values | none |
| same m value, and same b value | infinite |

**Example 1** Number of Solutions    A.REI.6

Use the graph at the right to determine whether each system is *consistent* or *inconsistent* and if it is *independent* or *dependent*.

a. $y = -2x + 3$
   $y = x - 5$

Since the graphs of these two lines intersect at one point, there is exactly one solution. Therefore, the system is consistent and independent.

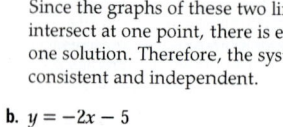

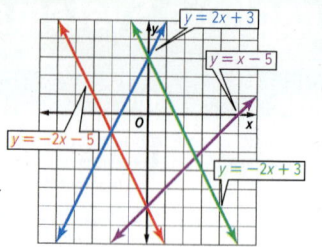

b. $y = -2x - 5$
   $y = -2x + 3$

Since the graphs of these two lines are parallel, there is no solution of the system. Therefore, the system is inconsistent.

**Guided Practice**

1A. $y = 2x + 3$   consistent and
    $y = -2x - 5$  independent

1B. $y = x - 5$   consistent and
    $y = -2x - 5$  independent

**2 Solve by Graphing** One method of solving a system of equations is to graph the equations carefully on the same coordinate grid and find their point of intersection. This point is the solution of the system.

**Example 2** Solve by Graphing    A.REI.6

Graph each system and determine the number of solutions that it has. If it has one solution, name it.

a. $y = -3x + 10$
   $y = x - 2$

The graphs appear to intersect at the point (3, 1). You can check this by substituting 3 for x and 1 for y.

CHECK  $y = -3x + 10$         Original equation
       $1 \stackrel{?}{=} -3(3) + 10$   Substitution
       $1 \stackrel{?}{=} -9 + 10$     Multiply.
       $1 = 1$ ✓

       $y = x - 2$             Original equation
       $1 \stackrel{?}{=} 3 - 2$        Substitution
       $1 = 1$ ✓                Multiply.

The solution is (3, 1).

b. $2x - y = -1$
   $4x - 2y = 6$

**Review Vocabulary**
parallel lines  never intersect and have the same slope

The lines have the same slope but different y-intercepts, so the lines are parallel. Since they do not intersect, there is no solution of this system. The system is inconsistent.

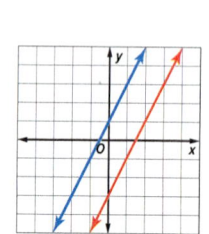

---

**Differentiated Instruction** AL OL ELL

**Interpersonal Learners** Have students work in pairs or groups to check the solutions for Exercises 16–24 and 27–38. Suggest that they use the Study Tip (p. 340) on comparing $m$ and $b$ when both equations are in the form of $y = mx + b$. Have students write equations in slope-intercept form, if necessary.

Lesson 6-1 | Graphing Systems of Equations

### Guided Practice

Graph each system and determine the number of solutions that it has. If it has one solution, name it. **2A–2B. See margin for graphs.**

**2A.** $x - y = 2$
$3y + 2x = 9$ **1 solution; (3, 1)**

**2B.** $y = -2x - 3$
$6x + 3y = -9$ **infinitely many**

We can use what we know about systems of equations to solve many real-world problems involving constraints that are modeled by two or more different functions.

A.CED.3, A.REI.6

**Real-World Example 3** Write and Solve a System of Equations

The number of girls participating in high school soccer and track and field has steadily increased over the past few years. Use the information in the table to predict the approximate year when the number of girls participating in these two sports will be the same.

| High School Sport | Number of Girls Participating in 2007 (thousands) | Number of Girls Participating in 2012 (thousands) |
|---|---|---|
| soccer | 345 | 371 |
| track and field | 458 | 467 |

**Source:** National Federation of State High School Associations

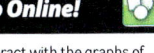

**Step 1** Find the average rate of increase for both sports.

Soccer: $\frac{371 - 345}{2012 - 2007} = 5.2$    Track and field: $\frac{467 - 458}{2012 - 2007} = 1.8$

**Step 2** Write equations describing the number of girls participating in each sport if the average rate of increase stays the same.

Let $y$ = the number of girls participating. Let $x$ = the number of years after 2012.

Soccer: $y = 5.2x + 371$    Track and field: $y = 1.8x + 467$

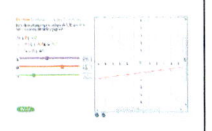

**Step 3** Graph both equations. The graph appears to intersect at (28, 517).

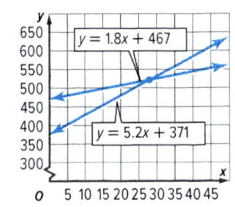

Use substitution to check this answer.

$y = 5.2x + 371$        $y = 1.8x + 467$
$517 \stackrel{?}{=} 5.2(28) + 371$    $517 \stackrel{?}{=} 1.8(28) + 467$
$517 \approx 516.6$ ✓     $517 \approx 517.4$ ✓

The solution means that approximately 28 years after 2012—or in 2040—the number of girls participating in high school soccer and track and field will be the same, about 517,000.

### Guided Practice

**3. VIDEO GAMES** Joe and Josh each want to buy a video game. Joe has $14 and saves $10 a week. Josh has $26 and saves $7 a week. In how many weeks will they have the same amount? **4 weeks**

**Real-World Link**
In 2015, 33 million girls participated in high school sports. This was an all-time high for female participation.
**Source:** National Federation of State High School Associations

**Go Online!**
Interact with the graphs of systems of linear equations by using the Graphing Tools in ConnectED.

## Additional Answers (Guided Practice)

**2A.**

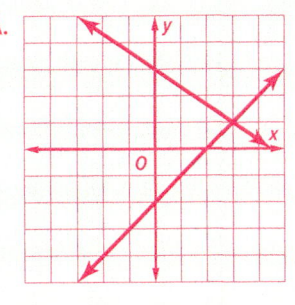

**2B.**

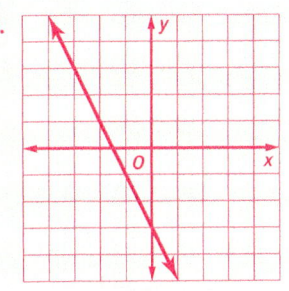

---

### Need Another Example?

Graph each system and determine the number of solutions that it has. If it has one solution, name it.

**a.** $y = 2x + 3$
$8x - 4y = -12$ **infinitely many solutions**

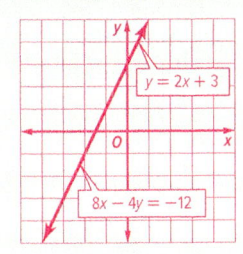

**b.** $x - 2y = 4$
$x - 2y = -2$ **no solution**

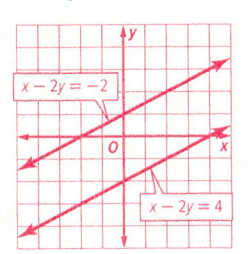

### Example 3 Write and Solve a System of Equations

**AL** What information must be found to solve the problem? **the solution to the system of equations**

**OL** What must be done before graphing? **For both sports, find the average rate of increase and use it to write an equation, in slope-intercept form.**

**BL** Interpret the solution to the system of equations in the context of the problem. **The solution, (28, 517), means that 28 years after 2012 (2040) the number of participants in each sport will be about the same, 517,000.**

### Need Another Example?

**Bicycling** Naresh rode 20 miles last week and plans to ride 35 miles per week. Diego rode 50 miles last week and plans to ride 25 miles per week. Predict the week in which Naresh and Diego will have ridden the same number of miles. **week 3**

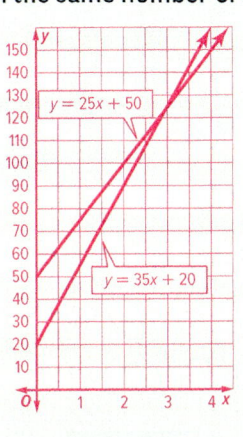

connectED.mcgraw-hill.com    341

### Lesson 6-1 | Graphing Systems of Equations

## Practice

**Formative Assessment** Use Exercises 1–9 to assess students' understanding of the concepts in this lesson.

The Practice and Problem Solving exercises assess the content taught in the lesson. The Preparing for Assessment page is meant to be used as preparation for end-of-course assessments.

## Extra Practice

See page R6 for extra exercises for students who are approaching level or for on-level students who need additional reinforcement.

## Additional Answers

7. 1 solution, (−4, 0)

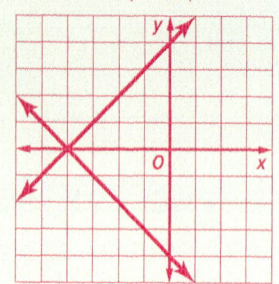

8. 1 solution, (−1, 2)

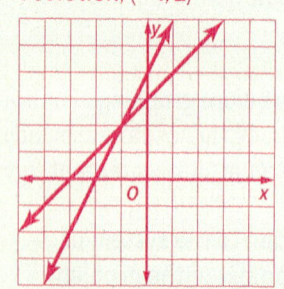

9b.

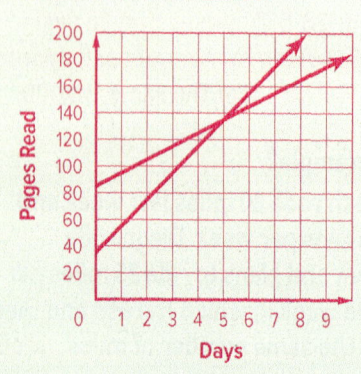

---

**Check Your Understanding**  ○ = Step-by-Step Solutions begin on page R11.

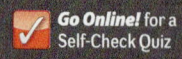

**Go Online!** for a Self-Check Quiz

**Example 1**
A.REI.6

Use the graph at the right to determine whether each system is *consistent* or *inconsistent* and whether it is *independent* or *dependent*.

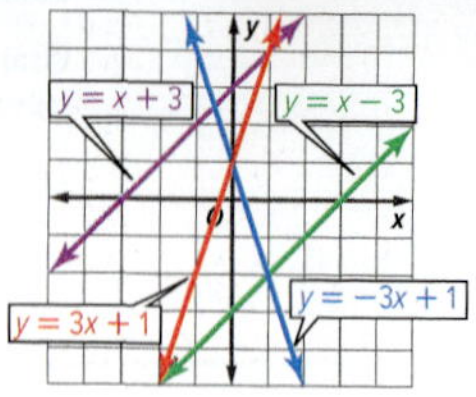

1. $y = -3x + 1$
   $y = 3x + 1$   consistent and independent

2. $y = 3x + 1$
   $y = x - 3$   consistent and independent

3. $y = x - 3$
   $y = x + 3$   inconsistent

4. $y = x + 3$
   $x - y = -3$   consistent and dependent

5. $x - y = -3$
   $y = -3x + 1$   consistent and independent

6. $y = -3x + 1$
   $y = x - 3$   consistent and independent

**Example 2**
A.REI.6

Graph each system and determine the number of solutions that it has. If it has one solution, name it.  **7–8. See margin.**

7. $y = x + 4$
   $y = -x - 4$

8. $y = x + 3$
   $y = 2x + 4$

**Example 3**
A.CED.3

9. **MODELING** Alberto and Ashanti are reading a graphic novel.  **9a. Alberto: $y = 20x + 35$; Ashanti: $y = 10x + 85$**

   a. Write an equation to represent the pages each boy has read.

   b. Graph each equation. **See margin.**

   c. How long will it be before Alberto has read more pages than Ashanti? Check and interpret your solution. **(5, 135); Alberto will have read more after 5 days.**

Alberto: 35 pages read; 20 pages each day
Ashanti: 85 pages read; 10 pages each day

---

**Practice and Problem Solving**  Extra Practice is on page R6.

**Example 1**
A.REI.6

Use the graph at the right to determine whether each system is *consistent* or *inconsistent* and whether it is *independent* or *dependent*.

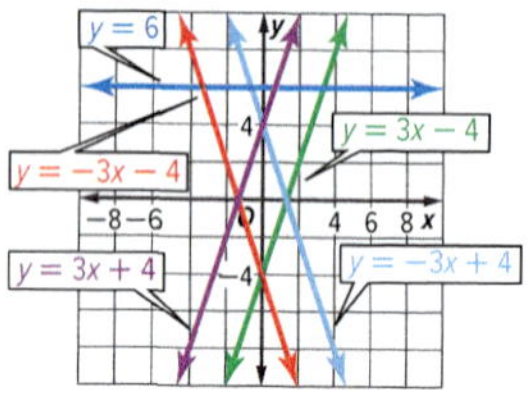

10. $y = 6$
    $y = 3x + 4$

11. $y = 3x + 4$
    $y = -3x + 4$

12. $y = -3x + 4$
    $y = -3x - 4$

13. $y = -3x - 4$
    $y = 3x - 4$

14. $3x - y = -4$
    $y = 3x + 4$

15. $3x - y = 4$
    $3x + y = 4$

**Example 2**
A.REI.6

Graph each system and determine the number of solutions that it has. If it has one solution, name it.  **16–24. See Ch. 6 Answer Appendix.**

16. $y = -3$
    $y = x - 3$

17. $y = 4x + 2$
    $y = -2x - 4$

18. $y = x - 6$
    $y = x + 2$

19. $x + y = 4$
    $3x + 3y = 12$

20. $x - y = -2$
    $-x + y = 2$

21. $x + 2y = 3$
    $x = 5$

22. $2x + 3y = 12$
    $2x - y = 4$

23. $2x + y = -4$
    $y + 2x = 3$

24. $2x + 2y = 6$
    $5y + 5x = 15$

10. consistent and independent
11. consistent and independent
12. inconsistent
13. consistent and independent
14. consistent and dependent
15. consistent and independent

---

### Differentiated Homework Options

| Levels | AL Basic | OL Core | BL Advanced |
|---|---|---|---|
| Exercises | 10–26, 48–59 | 11–25 odd, 27–37 odd, 40, 48–59 | 27–52, (optional: 53–59) |
| 2-Day Option | 11–25 odd, 53–59 | 10–26 | |
| | 10–26 even, 48–52 | 27–38, 40, 48–59 | |

You can use ALEKS to provide additional remediation support with personalized instruction and practice.

---

**Go Online!**  eBook

**Interactive Student Guide**
Use the *Interactive Student Guide* to deepen conceptual understanding.
· Writing Systems of Equations
· Solving Systems by Graphing

**Example 3**
A.CED.3,
A.REI.6

**25. SCHOOL DANCE** Akira and Jen are competing to see who can sell the most tickets for the Winter Dance. On Monday, Akira sold 22 and then sold 30 per day after that. Jen sold 53 on Monday and then sold 20 per day after that.

a. Write equations for the number of tickets each person has sold.
Akira: $y = 30x + 22$; Jen: $y = 20x + 53$
b. Graph each equation. **See margin.**
c. Solve the system of equations. Check and interpret your solution.
**(3.1, 115); After about 3 days, Akira will have sold more tickets.**

**26. MODELING** If $x$ is the number of years since 2013 and $y$ is the percent of people paying bills, the following equations represent the percent of people writing checks to pay their bills and the percent of people paying their bills online.

Writing checks: $y = -2x + 24$    Online: $y = 6x + 55$

a. Graph the system of equations. **See margin.**
b. Estimate the year check writing and the online payments were used equally. **2009**

**B** Graph each system and determine the number of solutions that it has. If it has one solution, name it. **27–38. See Ch. 6 Answer Appendix for graphs.**

**27.** $y = \frac{1}{2}x$  **1 solution;**
$y = x + 2$  **(−4, −2)**

**28.** $y = 6x + 6$  **1 solution;**
$y = 3x + 6$  **(0, 6)**

**29.** $y = 2x - 17$  **1 solution;**
$y = x - 10$  **(7, −3)**

**30.** $8x - 4y = 16$  **1 solution;**
$-5x - 5y = 5$  **(1, −2)**

**31.** $3x + 5y = 30$  **1 solution;**
$3x + y = 18$  **(5, 3)**

**32.** $-3x + 4y = 24$  **1 solution;**
$4x - y = 7$  **(4, 9)**

**33.** $2x - 8y = 6$  **infinitely**
$x - 4y = 3$  **many**

**34.** $4x - 6y = 12$  **infinitely**
$-2x + 3y = -6$  **many**

**35.** $2x + 3y = 10$  **no solution**
$4x + 6y = 12$

**36.** $3x + 2y = 10$  **1 solution;**
$2x + 3y = 10$  **(2, 2)**

**37.** $3y - x = -2$  **no solution**
$y - \frac{1}{3}x = 2$

**38.** $\frac{8}{5}y = \frac{2}{5}x + 1$  **infinitely many**
$\frac{2}{5}y = \frac{1}{10}x + \frac{1}{4}$

**39. MULTI-STEP** In 2012, website A had 512 million visitors, and website B had 131 million visitors. In 2016, website A had 476 million visitors and website B had 251 million visitors. **a–c. See margin.**

a. What conclusions can you make?
b. Describe your solution process in making your conclusions.
c. What assumptions did you make?

**40. ELECTRONICS** Suppose $x$ represents the number of years since 2012 and $y$ represents the number of units sold. Then the number of tablets sold each year since 2012, in millions, can be modeled by the equation $y = 12.5x + 10.9$. The number of laptops sold each year since 2012, in millions, can be modeled by the equation $y = -9.1x + 78.8$.

a. Graph each equation and find its domain and range. **See Ch. 6 Answer Appendix for graph.**
b. In which year did tablet sales surpass laptop sales? **2015**

**40a.** The domain of each of these equations is the set of real numbers greater than or equal to 0. The range of each of these equations is the set of real numbers greater than or equal to 0.

**C** Graph each system and determine the number of solutions that it has. If it has one solution, name it. **41–42. See margin for graphs.**

**41.** $2y = 1.2x - 10$  **no solution**
$4y = 2.4x$

**42.** $x = 6 - \frac{3}{8}y$  **infinitely many**
$4 = \frac{2}{3}x + \frac{1}{4}y$

---

**Lesson 6-1 | Graphing Systems of Equations**

**MP Teaching the Mathematical Practices**

**Modeling** Mathematically proficient students apply the mathematics they know to solve problems arising in everyday life. In Exercise 9, point out that Ashanti starts with more pages read but since Alberto is reading faster he will catch up. In Exercise 26, ask students what the slope in each equation represents in the context of the problem.

**Levels of Complexity Chart**

The levels of the exercises progress from 1 to 3, with Level 1 indicating the lowest level of complexity.

| Exercises | 10–26 | 27–40, 53–59 | 41–52 |
|---|---|---|---|
| **C** Level 3 |  |  | ● |
| **B** Level 2 |  | ● |  |
| Level 1 | ● |  |  |

## Additional Answers

**25b.** [Graph showing Tickets sold vs. Number of Days, with two lines intersecting]

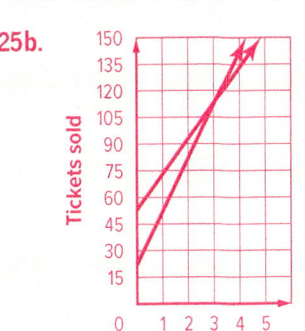

**26a.** [Graph showing $y = 6x + 55$ and $y = -2x + 24$]

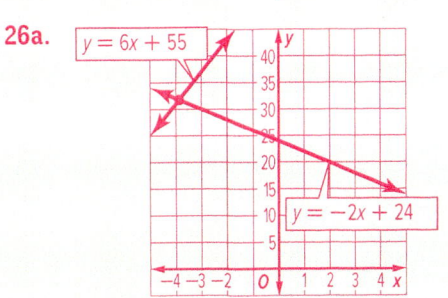

**39a.** Sample answers: At this rate, website A will have no visitors by 2069. Both websites will have the same number of visitors sometime during 2021. Website B will have 1 billion visitors by 2039.

**39b.** Sample answer: First, I found the average change in visitors per year for both websites: $\frac{476 - 512}{4}$ or $-9$, $\frac{251 - 131}{4}$ or 30. Next, I wrote equations to represent the situation: $y = -9x + 512$ and $y = 30x + 131$. Then, I graphed the equations and saw that they intersected between 9 and 10. So the number of visitors to the websites is the same sometime during the ninth year after 2012, or during 2021. I also saw from the graph when website A would reach 0 and website B would reach 1 billion.

**39c.** I assumed that these trends happened at a constant rate throughout the years, instead of the number of visitors going up and down over the years.

**41.** [Graph showing two parallel lines]

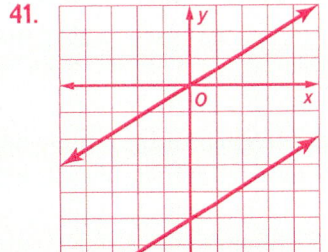

**42.** [Graph showing two intersecting lines]

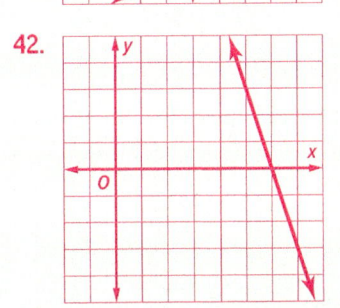

# Lesson 6-1 | Graphing Systems of Equations

## Watch Out!

**Error Analysis** For Exercise 48, suggest that students try several prices at both discounts, such as $50, $100, and $150, and then decide whether Francisca or Alan is correct.

## Teaching the Mathematical Practices

**Construct Arguments** Mathematically proficient students make conjectures and build a logical progression of statements to explore the truth of their conjectures. In Exercise 49, tell students to think about how many solutions are possible for a system of two linear equations.

## Additional Answers

**46b.** (6, 6)

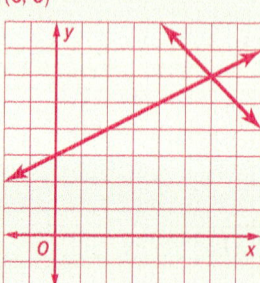

**47.**

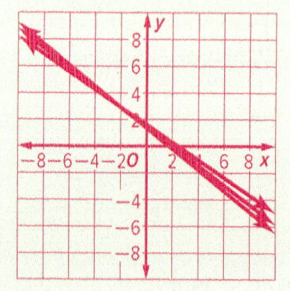

**52.** Graphing clearly shows whether a system of equations has one solution, no solution, or infinitely many solutions. However, finding the exact values of $x$ and $y$ from a graph can be difficult.

## Go Online!

**eSolutions Manual** Create worksheets, answer keys, and solutions handouts for your assignments.

344 | Lesson 6-1 | Graphing Systems of Equations

---

**46c.** Both sides of the equation in part a are set equal to $y$ in the system of linear equations in part b.

**46d.** You can find the solution from the $x$-coordinate of the intersection of the two lines in the system of equation.

Write a system of linear equations for each graph given. Then name the solution, if it has one.

**43.** $y = 3x - 3$, $y = 3x + 4$; no solution

**44.** $y = 3x - 4$, $y = -3x + 2$; (1, −1)

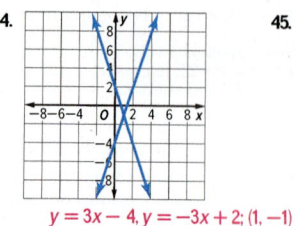

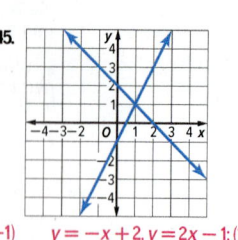

**45.** $y = -x + 2$, $y = 2x - 1$; (1, 1)

**46. MULTIPLE REPRESENTATIONS** In this problem, you will explore different methods for finding the intersection of the graphs of two linear equations.

a. **Algebraic** Use algebra to solve the equation $\frac{1}{2}x + 3 = -x + 12$. **6**

b. **Graphical** Use a graph to solve $y = \frac{1}{2}x + 3$ and $y = -x + 12$. **See margin.**

c. **Analytical** How is the equation in part **a** related to the system in part **b**?

d. **Verbal** Explain how to use the graph in part **b** to solve the equation in part **a**.

A.CED.3, A.REI.6

### H.O.T. Problems — Use Higher-Order Thinking Skills

**47. CHALLENGE** Use graphing to find the solution of the system of equations $2x + 3y = 5$, $3x + 4y = 6$, and $4x + 5y = 7$. **(−2, 3); See margin for graph.**

**49.** Always; if the equations are linear and have more than one common solution, they must be consistent and dependent, which means that they have an infinite number of solutions in common.

**48. ERROR ANALYSIS** Store A is offering a 10% discount on the purchase of all electronics in their store. Store B is offering $10 off all the electronics in their store. Francisca and Alan are deciding which offer will save them more money. Is either of them correct? Explain your reasoning.

Francisca: ............

Alan: Store A has the better offer because 10% of the sale price is a greater discount than $10.

**48.** Francisca; if the item is less than $100, then $10 off is better. If the item is more than $100, then the 10% is better.

**49. CONSTRUCT ARGUMENTS** Determine whether a system of two linear equations with (0, 0) and (2, 2) as solutions *sometimes*, *always*, or *never* has other solutions. Explain.

**50.** Sample answer: $4x + 2y = 14$, $12x + 6y = 18$; This system is inconsistent while the others are consistent and independent.

**50. WHICH ONE DOESN'T BELONG?** Which one of the following systems of equations doesn't belong with the other three? Explain your reasoning.

| $4x - y = 5$ | $-x + 4y = 8$ | $4x + 2y = 14$ | $3x - 2y = 1$ |
| $-2x + y = -1$ | $3x - 6y = 6$ | $12x + 6y = 18$ | $2x + 3y = 18$ |

**51. OPEN-ENDED** Write three equations such that they form three systems of equations with $y = 5x - 3$. The three systems should be inconsistent, consistent and independent, and consistent and dependent, respectively.

Sample answers: $y = 5x + 3$; $y = -5x - 3$; $2y = 10x - 6$

**52. WRITING IN MATH** Describe the advantages and disadvantages to solving systems of equations by graphing. **See margin.**

---

## Differentiated Instruction OL BL

**Extension** Write a system of three equations in two variables on the board. Have the students determine if the system has *one* solution, *no* solution, or *infinitely many* solutions. If it has one solution, name it. For example,

$$x + y = 2$$
$$x - y = 0$$
$$y = -2$$

has *no* solution because the three lines do not intersect at one point.

Lesson 6-1 | Graphing Systems of Equations

## Preparing for Assessment

53. Jenny is selling T-shirts and sweatshirts to raise money for the pep squad. She sold 12 shirts total. The number of T-shirts Jenny sold was 2 more than 4 times the number of sweatshirts she sold. How many of each type of shirt did Jenny sell? MP 4
A.REI.6  A
- A  2 sweatshirts; 10 T-shirts
- B  10 sweatshirts; 2 T-shirts
- C  10 sweatshirts; 50 T-shirts
- D  50 sweatshirts; 10 T-shirts

54. Which system of equations could have (1, 5) as a solution, if $b < 0$ and $c > 0$? MP 2  A.REI.6  C
- A  $y = -x + 6$
     $y = -x + b$
- B  $y = -x + 6$
     $y = cx + 6$
- C  $y = 2x + 3$
     $y = bx + c$
- D  $y = 2x + 3$
     $y = 2x + b$

55. Some values for two linear equations are shown in the tables below.

Equation 1

| x | y |
|---|---|
| −2 | 8 |
| 0 | 6 |
| 5 | 1 |
| 10 | −4 |

Equation 2

| x | y |
|---|---|
| −4 | −8 |
| −1 | −2 |
| 2 | 4 |
| 5 | 10 |

What is the solution to the system of equations represented by these tables? MP 8  A.REI.6  B
- A  (0, 6)
- B  (2, 4)
- C  (4, 2)
- D  (6, 0)

56. A quarter, some dimes, and some nickels are worth $1. If there are 12 coins altogether, how many dimes are there? MP 2, 4  A.REI.6  4

57. **MULTI-STEP** In 2017, the population of the town of Smithfield was 7200 and growing by 100 residents per year. The population of the town of Plymouth was 8850 in 2017 and decreasing by 50 residents per year. MP 4  A.REI.6
$y = 7200 + 100x$
$y = 8850 - 50x$
a. Write equations to represent the population of each town.
b. In what year will the towns be expected to have the same population? 2028
c. What is the population of both cities when they have the same population? 8300 people
d. Describe your solution process.
e. What assumptions did you make?
Sample answer: I assumed the populations continued to change at the same rate.

58. What system of equations is given by the lines in the graph? MP 1  A.REI.6  D

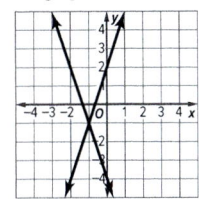

57. d Sample answer: Create a system of equations, graph the equations, and find the point of intersection.

- A  $y = 3x + 4$
     $y = -3x + 2$
- B  $y = -x + 1$
     $y = -2x - 2$
- C  $y = 2x + 3$
     $y = -4x - 3$
- D  $y = 3x + 2$
     $y = -3x - 4$

59. On a computer game, Fermin's score was 12 points less than twice Lisa's score. The total of both scores was 18 points. How many points did each person score? MP 4  A.REI.6  C
- A  Lisa: 2 points, Fermin: 16 points
- B  Lisa: 8 points, Fermin: 10 points
- C  Lisa: 10 points, Fermin: 8 points
- D  Lisa: 15 points, Fermin: 3 points

### MP Standards for Mathematical Practices

| Emphasis On | Exercises |
|---|---|
| 1 Make sense of problems and persevere in solving them. | 58 |
| 2 Reason abstractly and quantitatively. | 54, 56 |
| 3 Construct viable arguments and critique the reasoning of others. | 49 |
| 4 Model with mathematics. | 9, 25, 26, 40, 53, 57, 59 |

59.
| A | Wrote and graphed the equation $y = 2x + 12$ |
|---|---|
| B | Reversed the x- and y-coordinates |
| C | CORRECT |
| D | Wrote and graphed the equation $y = x - 12$ |

## Assess

**Ticket Out the Door** Give students a small piece of grid paper. Have them draw a graph that represents a system of equations that is consistent and dependent.

## Preparing for Assessment

Exercises 53–59 require students to use the skills they will need on standardized assessments. Each exercise is dual-coded with content standards and mathematical practice standards.

### Dual Coding

| Items | Content Standards | MP Mathematical Practices |
|---|---|---|
| 53 | A.REI.6 | 4 |
| 54 | A.REI.6 | 2 |
| 55 | A.REI.6 | 8 |
| 56 | A.REI.6 | 2, 4 |
| 57 | A.REI.6 | 4 |
| 58 | A.REI.6 | 1 |
| 59 | A.REI.6 | 4 |

## Diagnose Student Errors

Survey student responses for each item. Class trends may indicate common errors and misconceptions.

55.
| A | Interpreted the y-intercept of one equation as the common solution |
|---|---|
| B | CORRECT |
| C | Reversed the x- and y-values |
| D | Interpreted the x-intercept of one equation as the common solution |

## Go Online!

### Quizzes
Students can use *Self-Check Quizzes* to check their understanding of this lesson.

connectED.mcgraw-hill.com  345

Extend 6-1

# Launch

**Objective** Use a graphing calculator to solve a system of equations.

## Materials
- TI–83/84 Plus or other graphing calculator

## Teaching Tip
Remind students that the equations must be solved for *y* before they are entered in the calculator.

# Teach

**Working in Cooperative Groups** Put students in groups of two or three, mixing abilities. Have groups complete Activities 1 and 2. **ELL**

## Activity 1

- In Step 2, remind students to clear all previous equations from the **Y=** list. Have students graph each system using the standard viewing window. If the intersection is not visible, have them adjust the window to an area suggested by the directions of the lines.

- In Step 3, point out that the GUESS feature that appears after the second [ENTER] gives students an opportunity to use the arrow keys to estimate the solution to the system and then check their estimates by pressing [ENTER] the third time.

**Practice** Have students complete Exercises 1–10.

## Go Online!

### Graphic Calculators
Students can use the Graphing Calculator Personal Tutors to review the use of the graphing calculator to represent functions. They can also use the Other Calculator Keystrokes, which cover lab content for students with calculators other than the TI-84 Plus.

---

# EXTEND 6-1
## Graphing Technology Lab
## Systems of Equations

You can use a graphing calculator to graph and solve a system of equations.

**Mathematical Practices**
5 Use appropriate tools strategically.

**Content Standards**
**A.REI.6** Solve systems of linear equations exactly and approximately (e.g., with graphs), focusing on pairs of linear equations in two variables.

**A.REI.11** Explain why the *x*-coordinates of the points where the graphs of the equations $y = f(x)$ and $y = g(x)$ intersect are the solution of the equation $f(x) = g(x)$; find the solutions approximately, e.g., using technology to graph the functions, make tables of values, or find successive approximations. Include cases where $f(x)$ and/or $g(x)$ are linear, polynomial, rational, absolute value, exponential, and logarithmic functions.

### Activity 1  Solve a System of Equations

Work cooperatively. Solve the system of equations. Round to the nearest hundredth.

$5.23x + y = 7.48$
$6.42x - y = 2.11$

**Step 1** Solve each equation for *y* to enter them into the calculator.

| | |
|---|---|
| $5.23x + y = 7.48$ | First equation |
| $5.23x + y - 5.23x = 7.48 - 5.23x$ | Subtract 5.23x from each side. |
| $y = 7.48 - 5.23x$ | Simplify. |
| $6.42x - y = 2.11$ | Second equation |
| $6.42x - y - 6.42x = 2.11 - 6.42x$ | Subtract 6.42x from each side. |
| $-y = 2.11 - 6.42x$ | Simplify. |
| $(-1)(-y) = (-1)(2.11 - 6.42x)$ | Multiply each side by −1. |
| $y = -2.11 + 6.42x$ | Simplify. |

**Step 2** Enter these equations in the **Y=** list and graph in the standard viewing window.

**KEYSTROKES:** [Y=] 7.48 [−] 5.23 [X,T,θ,n] [ENTER] [(−)] 2.11 [+] 6.42 [X,T,θ,n] [ZOOM] 6

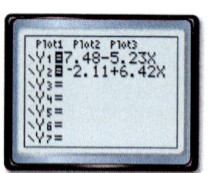

**Step 3** Use the **CALC** menu to find the point of intersection.

**KEYSTROKES:** [2nd] [CALC] 5 [ENTER] [ENTER] [ENTER]

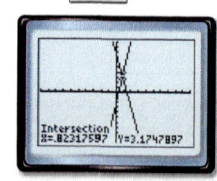

[−10, 10] scl: 1 by [−10, 10] scl: 1

The solution is approximately (0.82, 3.17).

When you solve a system of equations with $y = f(x)$ and $y = g(x)$, the solution is an ordered pair that satisfies both equations. The solution always occurs when $f(x) = g(x)$. Thus, the *x*-coordinate of the solution is the value of *x* where $f(x) = g(x)$.

One method you can use to solve an equation with one variable is by graphing and solving a system of equations based on the equation. To do this, write a system using both sides of the equation. Then use a graphing calculator to solve the system.

## Follow-Up

Students have solved systems of equations by graphing with and without graphing calculators.
Ask: What are the advantages of using technology to solve systems of equations?
**Sample answer:** The graphing calculator makes it easy to find the exact solution using the **intersection** option from the **CALC** menu.

# Extend 6-1

## Activity 2  Use a System to Solve a Linear Equation

Work cooperatively. Use a system of equations to solve $5x + 6 = -4$.

**Step 1** Write a system of equations. Set each side of the equation equal to $y$.

$y = 5x + 6$  First equation
$y = -4$  Second equation

**Step 2** Enter these equations in the **Y=** list and graph.

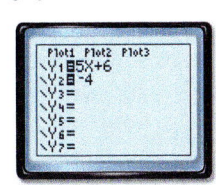

**Step 3** Use the **CALC** menu to find the point of intersection.

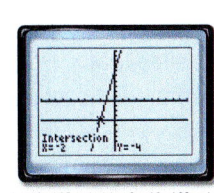

[−10, 10] scl: 1 by [−10, 10] scl: 1

The solution is $-2$.

### Exercises

Work cooperatively. Use a graphing calculator to solve each system of equations. Write decimal solutions to the nearest hundredth.

1. $y = 2x - 3$
   $y = -0.4x + 5$  **(3.33, 3.67)**

2. $y = 6x + 1$
   $y = -3.2x - 4$  **(−0.54, −2.26)**

3. $x + y = 9.35$
   $5x - y = 8.75$  **(3.02, 6.33)**

4. $2.32x - y = 6.12$
   $4.5x + y = -6.05$  **(0.01, −6.10)**

5. $5.2x - y = 4.1$
   $1.5x + y = 6.7$  **(1.61, 4.28)**

6. $1.8 = 5.4x - y$
   $y = -3.8 - 6.2x$  **(−0.17, −2.73)**

7. $7x - 2y = 16$
   $11x + 6y = 32.3$  **(2.51, 0.78)**

8. $3x + 2y = 16$
   $5x + y = 9$  **(0.29, 7.57)**

9. $0.62x + 0.35y = 1.60$
   $-1.38x + y = 8.24$  **(−1.16, 6.63)**

10. $75x - 100y = 400$
    $33x - 10y = 70$  **(1.18, −3.12)**

Use a graphing calculator to solve each equation. Write decimal solutions to the nearest hundredth.

11. $4x - 2 = -6$  **−1**

12. $3 = 1 + \frac{x}{2}$  **4**

13. $\frac{x+4}{-2} = -1$  **−2**

14. $\frac{3}{2}x + \frac{1}{2} = 2x - 3$  **7**

15. $4x - 9 = 7 + 7x$  **−5.33**

16. $-2 + 10x = 8x - 1$  **0.5**

17. **WRITING IN MATH** Explain why you can solve an equation like $r = ax + b$ by solving the system of equations $y = r$ and $y = ax + b$.  **At the intersection of the graphs of $y = r$ and $y = ax + b$, the y-values are equal. Therefore, at that point $r = ax + b$.**

## Activity 2

- For Activity 2, students will need to clear the **Y=** list to begin the activity. Remind students to press ⊖ to enter $-4$ rather than the key for the subtraction symbol.
- In Step 3, students should use the same keystrokes used in Activity 1, Step 3.

**Ask:**
- What does the solution in Step 3 mean? For the equation $y = 5x + 6$, $x = -2$ when $y = -4$.
- How can you check your solution? Substitute $-2$ for $x$ in the original equation to check whether the equation is true.

**Practice** Have students complete Exercises 11–17.

## Assess

### Formative Assessment

- Use Exercise 7 to assess whether students can use a graphing calculator to solve a system of equations.
- Use Exercise 15 to assess whether students can graph a system to solve an equation with one variable.

### From Concrete to Abstract

Exercise 17 asks students to give an algebraic justification for using a system of equations to solve a linear equation with one variable.

### Extending the Concept

**Ask:**
- When would a system of two linear equations not have a point of intersection? Sample answer: Two linear equations will not have a point of intersection if their graphs are parallel.

## Go Online!

The most up-to-date resources available for your program can be found at **connectED.mcgraw-hill.com**.

# LESSON 6-2
# Substitution

**SUGGESTED PACING (DAYS)**
90 min. — 0.5
45 min. — 1
Instruction

## Track Your Progress

### Objectives
1. Solve systems of equations by using substitution.
2. Solve real-world problems involving systems of equations by using substitution.

### Mathematical Background
The substitution method uses the Substitution Property of Equality to solve systems of equations. In the system $y = x + 5$ and $x + y = -11$, I can *substitute* $x + 5$ for $y$ in the second equation to solve the system. Then I can *substitute* the result, $x = -8$ into either equation to find a value for $y$.

### THEN
**A.REI.3** Solve linear equations and inequalities in one variable, including equations with coefficients represented by letters.

### NOW
**A.CED.3** Represent constraints by equations or inequalities, and by systems of equations and/or inequalities, and interpret solutions as viable or nonviable options in a modeling context.

**A.REI.6** Solve systems of linear equations exactly and approximately (e.g., with graphs), focusing on pairs of linear equations in two variables.

### NEXT
**A.CED.2** Create equations in two or more variables to represent relationships between quantities; graph equations on coordinate axes with labels and scales.

## Go Online! All of these resources and more are available at connectED.mcgraw-hill.com

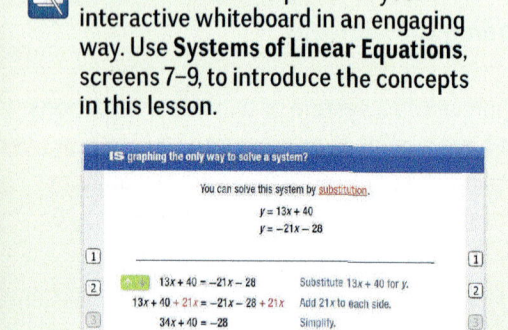

**eLessons** utilize the power of your interactive whiteboard in an engaging way. Use **Systems of Linear Equations**, screens 7–9, to introduce the concepts in this lesson.

*Use at Beginning of Lesson*

**Personal Tutors** (for every example) let students hear real teachers solve problems. Students can pause and repeat as many times as necessary.

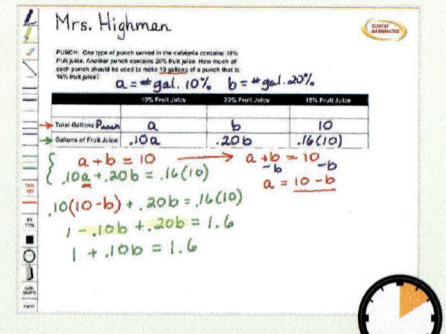

*Use with Examples*

**eToolkit** allows students to explore and enhance their understanding of math concepts. Use the Equation Chart Mat and Algebra Tiles tool to model substitution.

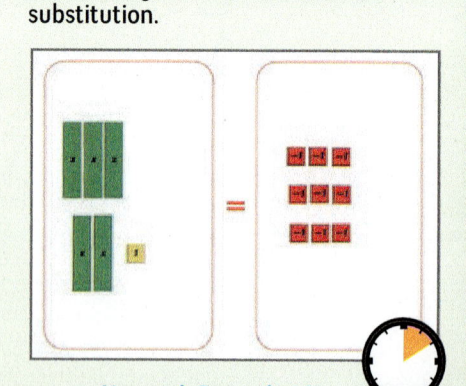

*Use with Example 1 Step 2*

## OER Using Open Educational Resources
**Video Sharing** Have students work in groups to record and upload songs on **YouTube** about solving systems of equations by substitution to check their understanding of the process. If you are unable to access YouTube, try **KidsTube**, **MathATube**, **SchoolTube**, or **TeacherTube**. *Use as homework*

# Go Online!
connectED.mcgraw-hill.com — Worksheets

# Differentiate Your Resources

**Extra Practice** Additional practice or homework; Skills Practice is best for approaching-level students and Practice is best for on-level and beyond-level students

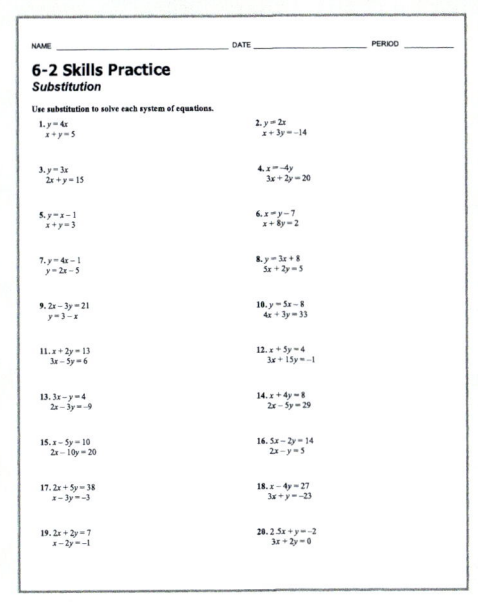

## Skills Practice
AL  OL  ELL

## Practice
AL  OL  BL  ELL

## Word Problem Practice
AL  OL  BL  ELL

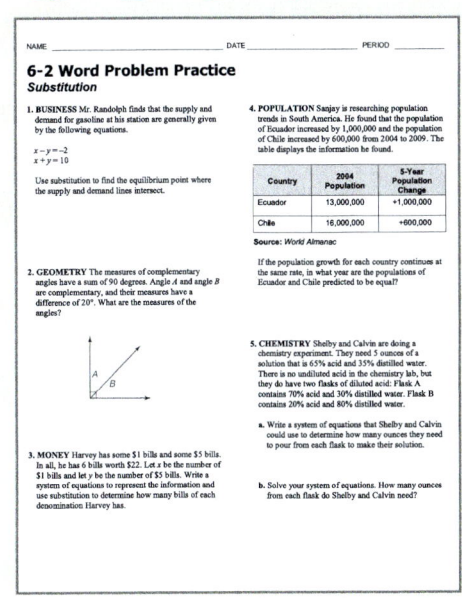

**Intervention** Reteaching and vocabulary activities that can be used with struggling or absent students and as ELL support

## Study Guide and Intervention
AL  OL  ELL

## Study Notebook
AL  OL  BL  ELL

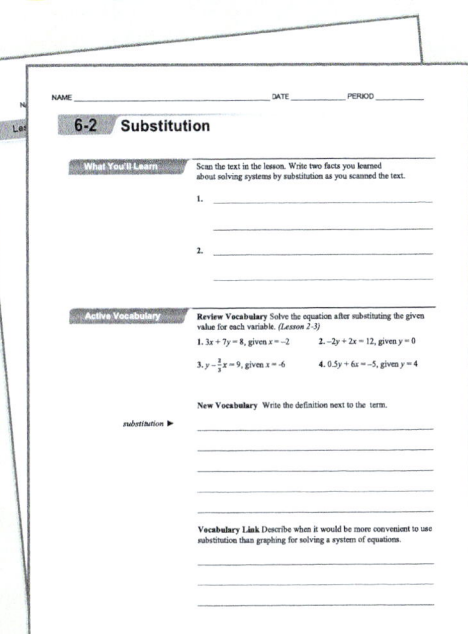

**Extension** Activities that can be used to extend lesson concepts

## Enrichment
OL  BL  ELL

connectED.mcgraw-hill.com     348B

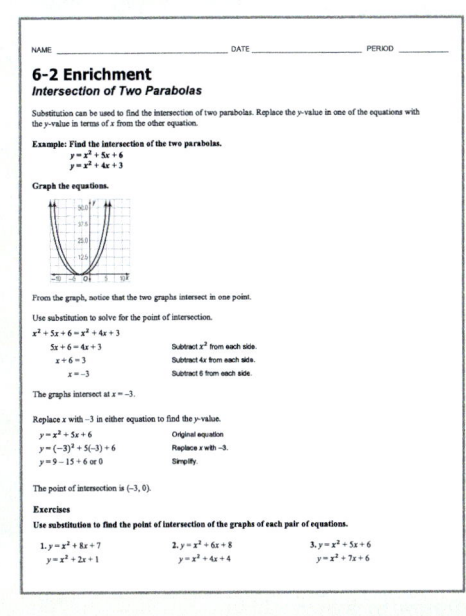

## Lesson 6-2 | Substitution

# Launch

Have students read the Why? section of the lesson. Ask:

- What are the rates of decrease for movies A and B? **$16 million per week; $10 million per week**
- If $x$ = the number of weeks from the opening week, and $y$ = total earnings, what equation represents the earnings in week $x$ for movie A? **$y = -16x + 31$** for movie B? **$y = -10x + 21$**
- Why might solving the system of equations using substitution be better than graphing the equations to determine when the movies have the same earnings? **Sample answer: Using substitution might give a more exact answer, because the graphs might intersect between weeks.**

# Teach

Ask the scaffolded questions for each example to build conceptual understanding for students at all levels.

## 1 Solve by Substitution

**Example 1 Solve a System by Substitution**

**AL** What does it mean to substitute? **Sample answer: to plug in an equivalent value or quantity for another value**

**OL** What is the first step to solving the system of equations? **Since one of the equations is already solved for $y$, we can substitute the quantity $2x + 1$ for $y$ in the second equation and solve for $x$.**

**BL** What is an advantage of solving a system of equations by substitution instead of by graphing? **Substitution sometimes gives you a more exact answer than graphing.**

## Go Online!

**Interactive Whiteboard**
Use the *eLesson* or *Lesson Presentation* to present this lesson.

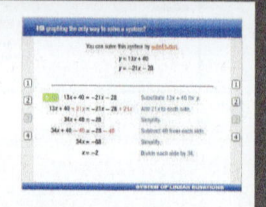

---

# LESSON 2
# Substitution

**::Then** · You solved systems of equations by graphing.

**::Now** 
1. Solve systems of equations by using substitution.
2. Solve real-world problems involving systems of equations by using substitution.

**::Why?** · Two movies were released at the same time. Movie A earned $31 million in its opening week, but fell to $15 million the following week. Movie B opened earning $21 million and fell to $11 million the following week. If the earnings for each movie continue to decrease at the same rate, when will they earn the same amount?

**New Vocabulary**
substitution

**Mathematical Practices**
2 Reason abstractly and quantitatively.

**Content Standards**
A.CED.3 Represent constraints by equations or inequalities, and by systems of equations and/or inequalities, and interpret solutions as viable or nonviable options in a modeling context.

A.REI.6 Solve systems of linear equations exactly and approximately (e.g. with graphs), focusing on pairs of linear equations in two variables.

**1 Solve by Substitution** You can use a system of equations to find when the movie earnings are the same. One method of finding an exact solution of a system of equations is called **substitution**.

**Key Concept** Solving by Substitution

**Step 1** When necessary, solve at least one equation for one variable.
**Step 2** Substitute the resulting expression from Step 1 into the other equation to replace the variable. Then solve the equation.
**Step 3** Substitute the value from Step 2 into either equation, and solve for the other variable. Write the solution as an ordered pair.

A.REI.6

**Example 1** Solve a System by Substitution

Use substitution to solve the system of equations.
$y = 2x + 1$ ← **Step 1** The first equation is already solved for $y$.
$3x + y = -9$

**Step 2** Substitute $2x + 1$ for $y$ in the second equation.

$3x + y = -9$    Second equation
$3x + 2x + 1 = -9$    Substitute $2x + 1$ for $y$.
$5x + 1 = -9$    Combine like terms.
$5x = -10$    Subtract 1 from each side.
$x = -2$    Divide each side by 5.

**Step 3** Substitute $-2$ for $x$ in either equation to find $y$.
$y = 2x + 1$    First equation
$\phantom{y} = 2(-2) + 1$    Substitute $-2$ for $x$.
$\phantom{y} = -3$    Simplify.

The solution is $(-2, -3)$.
**CHECK** You can check your solution by graphing.

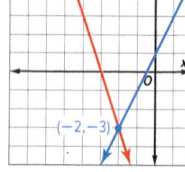

▶ **Guided Practice**

**1A.** $y = 4x - 6$
$5x + 3y = -1$   **(1, -2)**

**1B.** $2x + 5y = -1$
$y = 3x + 10$   **(-3, 1)**

---

**MP Mathematical Practices Strategies**

**Reason abstractly and quantitatively.**
Reinforce that there are different methods to solving systems of equations and determining the best one to use is an important skill.

- How can you check a solution of a system of equations. **by graphing**
- If the result of solving by substitution is an identity, then how many solutions are there? **an infinite number**
- If the result of solving by substitution results in a false statement, then how many solutions are there? **0**

## Lesson 6-2 | Substitution

**Study Tip**

**MP Perseverance** If both equations are in the form $y = mx + b$, they can simply be set equal to each other and then solved for $x$. The solution for $x$ can then be used to find the value of $y$.

If a variable is not isolated in one of the equations in a system, solve an equation for a variable first. Then you can use substitution to solve the system.

A.REI.6

### Example 2  Solve and then Substitute

Use substitution to solve the system of equations.
$x + 2y = 6$
$3x - 4y = 28$

**Step 1** Solve the first equation for $x$ since the coefficient is 1.

| $x + 2y = 6$ | First equation |
| $x + 2y - 2y = 6 - 2y$ | Subtract 2y from each side. |
| $x = 6 - 2y$ | Simplify. |

**Step 2** Substitute $6 - 2y$ for $x$ in the second equation to find the value of $y$.

| $3x - 4y = 28$ | Second equation |
| $3(6 - 2y) - 4y = 28$ | Substitute 6 − 2y for x. |
| $18 - 6y - 4y = 28$ | Distributive Property |
| $18 - 10y = 28$ | Combine like terms. |
| $18 - 10y - 18 = 28 - 18$ | Subtract 18 from each side. |
| $-10y = 10$ | Simplify. |
| $y = -1$ | Divide each side by −10. |

**Step 3** Find the value of $x$.

| $x + 2y = 6$ | First equation |
| $x + 2(-1) = 6$ | Substitute −1 for y. |
| $x - 2 = 6$ | Simplify. |
| $x = 8$ | Add 2 to each side. |

▶ **Guided Practice**

**2A.** $4x + 5y = 11$
$y - 3x = -13$  (4, −1)

**2B.** $x - 3y = -9$
$5x - 2y = 7$  (3, 4)

Generally, if you solve a system of equations and the result is a false statement such as $3 = -2$, there is no solution. If the result is an identity, such as $3 = 3$, then there are an infinite number of solutions.

A.REI.6

### Example 3  No Solution or Infinitely Many Solutions

Use substitution to solve the system of equations.
$y = 2x - 4$
$-6x + 3y = -12$

Substitute $2x - 4$ for $y$ in the second equation.

| $-6x + 3y = -12$ | Second equation |
| $-6x + 3(2x - 4) = -12$ | Substitute 2x − 4 for y. |
| $-6x + 6x - 12 = -12$ | Distributive Property |
| $-12 = -12$ | Combine like terms. |

This statement is an identity. Thus, there are an infinite number of solutions.

**Study Tip**

**Dependent Systems** There are infinitely many solutions of the system in Example 3 because the equations in slope-intercept form are equivalent, and they have the same graph.

---

### Differentiated Instruction AL OL

**IF** students have difficulty solving a system using substitution,

**THEN** suggest that they use the Study Tip to write both equations in slope-intercept form, set the expressions in $x$ equal to each other, solve for $x$, and then use the value for $x$ to find $y$. Have students compare the methods and decide which method they prefer.

---

**Need Another Example?**

Use substitution to solve the system of equations.
$y = -4x + 12$
$2x + y = 2$  (5, −8)

### Example 2  Solve and then Substitute

**AL** Which variable in which equation will be the easiest to solve for? *x in the first equation*

**OL** When the quantity $6 - 2y$ is substituted for $x$, why is it necessary to use parentheses? *Sample answer: the entire quantity needs to be multiplied by 3 so if parentheses are not used, only the 6 will be multiplied by 3.*

**BL** What can be determined from the $x$- and $y$-values of the system of equations? *Sample answer: The point where the functions intersect is (8, −1).*

**Need Another Example?**

Use substitution to solve the system of equations.
$x - 2y = -3$
$3x + 5y = 24$  (3, 3)

**Watch Out!**

**Preventing Errors** Point out that if neither of the equations gives one variable in terms of the other, you must solve for one variable first. The easiest choice in Example 2 is to solve the first equation for $x$ by subtracting $2y$ from both sides.

### Example 3  No Solution or Infinitely Many Solutions

**AL** What is the first step to solving the system? *Substitute the quantity $2x - 4$ into the second equation for $y$.*

**OL** What would a solution like $12 = 7$ mean? *Sample answer: This would mean that there is no solution.* What would it look like on a graph? *The lines would be parallel.*

**BL** What is another way to determine that this is an identity? *Sample answer: Solve the second equation for $y$, to get $y = 2x - 4$. This equation is the same as the first equation. That means that any value of $x$ will be a solution.*

*(continued on the next page)*

connectED.mcgraw-hill.com  349

## Lesson 6-2 | Substitution

**Need Another Example?**
Use substitution to solve the system of equations.
$2x + 2y = 8$
$x + y = -2$ **no solution**

## 2 Solve Real-World Problems

**Example 4 Write and Solve a System of Equations**

**AL** How can terms for the total cost of each item be written? **Sample answer: The price given is per unit, so multiply the per unit cost by the total number of units sold.**

**OL** What steps are needed to solve the system? **Solve for one of the variables in the first equation, then substitute the quantity into the second equation and solve for the variable. Finally, substitute the value back into the first equation and solve for the other variable.**

**BL** Interpret the solution to the system of equations in the context of the problem. **Sample answer: $t = 72$ is the number of speakers sold; $c = 53$ is the number of car stereo systems sold.**

**Need Another Example?**
**Nature Center** A nature center charges $35.25 for a yearly membership and $6.25 for a single admission. Last week it sold a combined total of 50 yearly memberships and single admissions for $660.50. How many memberships and how many single admissions were sold? **12 memberships and 38 single admissions**

## Practice

**Formative Assessment** Use Exercises 1–7 to assess students' understanding of the concepts in this lesson.

The Practice and Problem Solving exercises assess the content taught in the lesson. The Preparing for Assessment page is meant to be used as preparation for end-of-course assessments.

---

**Guided Practice** Use substitution to solve each system of equations.

**3A.** $2x - y = 8$
$y = 2x - 3$ **no solution**

**3B.** $4x - 3y = 1$
$6y - 8x = -2$ **infinite number**

**2 Solve Real-World Problems** You can use substitution to find the solution of a real-world problem involving constraints modeled by a system of equations.
A.CED.3, A.REI.6

**Real-World Example 4** Write and Solve a System of Equations

**MUSIC** A store sold a total of 125 car stereo systems and speakers in one week. The stereo systems sold for $104.95, and the speakers sold for $18.95. The sales from these two items totaled $6926.75. How many of each item were sold?

| Number of Units Sold | c | t | 125 |
|---|---|---|---|
| Sales ($) | 104.95c | 18.95t | 6926.75 |

Let $c$ = the number of car stereo systems sold, and let $t$ = the number of speakers sold.

So, the two equations are $c + t = 125$ and $104.95c + 18.95t = 6926.75$.
Notice that $c + t = 125$ represents combinations of car stereo systems and speakers with a sum of 125. The equation $104.95c + 18.95t = 6926.75$ represents the combinations of car stereo systems and speakers with a sales of $6926.75. The solution of the system of equations represents the option that meets both of the constraints.

**Step 1** Solve the first equation for $c$.

$c + t = 125$     First equation
$c + t - t = 125 - t$     Subtract $t$ from each side.
$c = 125 - t$     Simplify.

**Step 2** Substitute $125 - t$ for $c$ in the second equation.

$104.95c + 18.95t = 6926.75$     Second equation
$104.95(125 - t) + 18.95t = 6926.75$     Substitute $125 - t$ for $c$.
$13{,}118.75 - 104.95t + 18.95t = 6926.75$     Distributive Property
$13{,}118.75 - 86t = 6926.75$     Combine like terms.
$-86t = -6192$     Subtract 13,118.75 from each side.
$t = 72$     Divide each side by $-86$.

**Step 3** Substitute 72 for $t$ in either equation to find the value of $c$.

$c + t = 125$     First equation
$c + 72 = 125$     Substitute 72 for $t$.
$c = 53$     Subtract 72 from each side.

The store sold 53 car stereo systems and 72 speakers.

**Guided Practice**

**4. BASEBALL** As of 2016, the New York Yankees and the Cincinnati Reds together had won a total of 32 World Series. The Yankees had won 5.4 times as many as the Reds. How many World Series had each team won? **Cincinnati, 5; New York, 27**

---

**Real-World Link**

**Sound Engineering Technician** Sound engineering technicians record, synchronize, mix, and reproduce music, voices, and sound effects in recording studios, sporting arenas, and theater, movie, or video productions. They need to have at least a 2-year associate's degree in electronics.

**Go Online!**

In *Personal Tutor* videos for this lesson, teachers describe how to solve systems of equations using substitution. Watch with a partner. Then try describing how to solve a problem for them. Have them ask questions to help your understanding. **ELL**

---

**Differentiated Instruction** AL OL BL ELL

**Beginning** Ask questions about the lesson content to elicit yes/no answers: "Look at Example 1. Is one of the equations solved for one of the variables?" *yes* "Is the first step of solving the system complete?" *yes*

**Intermediate/Advanced** Ask questions about the lesson content to elicit short answers: "Look at Example 1. Which equation is solved for a variable?" *the first* "What should be substituted for $y$ in the second equation?" *$2x + 1$*

**Advanced High** Ask questions about lesson content to elicit complete sentences: "How does Example 1 compare to Example 2?" *In Example 1, Step 1 is already completed.* "How would you choose the variable to solve for to solve Example 2 by substitution?" *Solving the first equation for $x$ makes sense because its coefficient is 1.*

Lesson 6-2 | Substitution

### Check Your Understanding
= Step-by-Step Solutions begin on page R11. **Go Online!** for a Self-Check Quiz

**Examples 1–3** Use substitution to solve each system of equations.
A.REI.6

1. $y = x + 5$
   $3x + y = 25$ **(5, 10)**

2. $x = y - 2$
   $4x + y = 2$ **(0, 2)**

3. $3x + y = 6$
   $4x + 2y = 8$ **(2, 0)**

4. $2x + 3y = 4$
   $4x + 6y = 9$ **no solution**

5. $x - y = 1$
   $3x = 3y + 3$ **infinitely many**

6. $2x - y = 6$
   $-3y = -6x + 18$ **infinitely many**

**Example 4**
A.CED.3, A.REI.6

7. **GEOMETRY** The sum of the measures of angles X and Y is 180°. The measure of angle X is 24° greater than the measure of angle Y.
   a. Define the variables, and write equations for this situation. $x = m\angle X, y = m\angle Y;$ $x + y = 180, x = 24 + y$
   b. Find the measure of each angle. $x = 102°, y = 78°$

### Practice and Problem Solving
Extra Practice is on page R6.

**Examples 1–3** Use substitution to solve each system of equations.
A.REI.6

8. $y = 5x + 1$
   $4x + y = 10$ **(1, 6)**

9. $y = 4x + 5$
   $2x + y = 17$ **(2, 13)**

10. $y = 3x - 34$
    $y = 2x - 5$ **(29, 53)**

11. $y = 3x - 2$
    $y = 2x - 5$ **(−3, −11)**

12. $2x + y = 3$
    $4x + 4y = 8$ **(1, 1)**

13. $3x + 4y = -3$
    $x + 2y = -1$ **(−1, 0)**

14. $y = -3x + 4$
    $-6x - 2y = -8$ **infinitely many**

15. $-1 = 2x - y$
    $8x - 4y = -4$ **infinitely many**

16. $x = y - 1$
    $-x + y = -1$ **no solution**

17. $y = -4x + 11$
    $3x + y = 9$ **(2, 3)**

18. $y = -3x + 1$
    $2x + y = 1$ **(0, 1)**

19. $3x + y = -5$
    $6x + 2y = 10$ **no solution**

20. $5x - y = 5$
    $-x + 3y = 13$ **(2, 5)**

21. $2x + y = 4$
    $-2x + y = -4$ **(2, 0)**

22. $-5x + 4y = 20$
    $10x - 8y = -40$ **infinitely many**

**Example 4**
A.CED.3, A.REI.6

23. **ECONOMICS** In 2000, the demand for nurses was 2,000,000, while the supply was only 1,890,000. The projected demand for nurses in 2020 is 2,810,414, while the supply is only projected to be 2,001,998. **23a.** Let $d$ = demand for nurses; $s$ = supply of nurses; $t$ = number of years; $d = 40,521t + 2,000,000$; $s = 5,600t + 1,890,000$
    a. Define the variables, and write equations to represent these situations.
    b. Use substitution to determine during which year the supply of nurses was equal to the demand. **during 1996**

24. **REASONING** The table shows the approximate number of tourists in two areas of the world during a recent year and the average rates of change in tourism.

| Destination | Number of Tourists | Average Rates of Change in Tourists (millions per year) |
|---|---|---|
| South America and the Caribbean | 40.3 million | increase of 0.8 |
| Middle East | 17.0 million | increase of 1.8 |

a. Define the variables, and write an equation for each region's tourism rate. **See margin.**
b. If the trends continue, in how many years would you expect the number of tourists in the regions to be equal? **in 23.3 yr, or about 23 yr 4 mo**

### Differentiated Homework Options

| Levels | **AL** Basic | **OL** Core | **BL** Advanced |
|---|---|---|---|
| Exercises | 8–23, 27–39 | 9–23 odd, 24–39 | 26–31, 32–39 (optional) |
| 2-Day Option | 9–23 odd, 32–39 | 8–23 | |
| | 8–22 even, 27–31 | 24–39 | |

You can use ALEKS to provide additional remediation support with personalized instruction and practice.

### Additional Answers

**24a.** Let $x$ = number of years since base year, and $y$ = number of tourists in millions; South America and Caribbean, $y = 0.8x + 40.3$; Middle East, $y = 1.8x + 17.0$

---

### Teaching the Mathematical Practices

**Reasoning** Mathematically proficient students attend to the meaning of quantities, not just how to compute them. In Exercise 24, point out that the table headings tell the meanings of the values in the table.

### Extra Practice

See page R6 for extra exercises for students who are approaching level or for on-level students who need additional reinforcement.

### Watch Out!

**Error Analysis** In Exercise 27, remind students that both Guillermo and Cara could be wrong, and that errors can be made in interpreting a solution as well as finding it.

### Levels of Complexity Chart

The levels of the exercises progress from 1 to 3, with Level 1 indicating the lowest level of complexity.

| Exercises | 8–23 | 24–25, 32–39 | 26–31 |
|---|---|---|---|
| Level 3 | | | ● |
| Level 2 | | ● | |
| Level 1 | ● | | |

### Standards for Mathematical Practice

| Emphasis On | Exercises |
|---|---|
| 1 Make sense of problems and persevere in solving them. | 28 |
| 2 Reason abstractly and quantitatively. | 24, 29, 34, 37, 38 |
| 3 Construct viable arguments and critique the reasoning of others. | 27 |
| 4 Model with mathematics. | 25, 26, 35, 36, 39 |
| 8 Look for and express regularity in repeated reasoning. | 32, 33 |

### Go Online! eBook

**Interactive Student Guide**
Use the *Interactive Student Guide* to deepen conceptual understanding.
· Solving Systems by Substitution

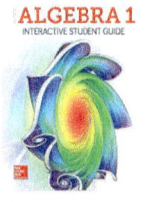

## Teaching the Mathematical Practices

**Sense-Making** Mathematically proficient students look for entry points to the solution of a problem. In Exercise 28, encourage students to write a few ratios equivalent to 7:5. Then ask them to write an expression for all the ratios in terms of a variable.

## Additional Answers

**26a.** Let $x$ = number of tickets purchased and let $y$ = the cost; $y = 65x + 10$, $y = 69x + 13.60$

**26b.**

**Booker's Concert Tickets**

| Number of Tickets | Cost($) |
|---|---|
| 1 | 75 |
| 2 | 140 |
| 3 | 205 |
| 4 | 270 |
| 5 | 335 |

**Paula's Concert Tickets**

| Number of Tickets | Cost($) |
|---|---|
| 1 | 82.60 |
| 2 | 151.60 |
| 3 | 220.60 |
| 4 | 289.60 |
| 5 | 358.60 |

**26d.** Since the graphs do not intersect in the first quadrant, Paula is always paying more for the tickets than Booker.

**30.** Sample answer: $a + b = 25$, $24a + 16b = 464$; Let $a$ = the number of tops Allison bought, and let $b$ = the number of tops Beth bought; together, Allison and Beth bought 25 tops. Allison spent $24 per top, and Beth spent $16 per top. Together they spent $464. How many tops did each girl buy? Allison bought 8 tops, and Beth bought 17 tops.

**Go Online!**

**eSolutions Manual**

Create worksheets, answer keys, and solutions handouts for your assignments.

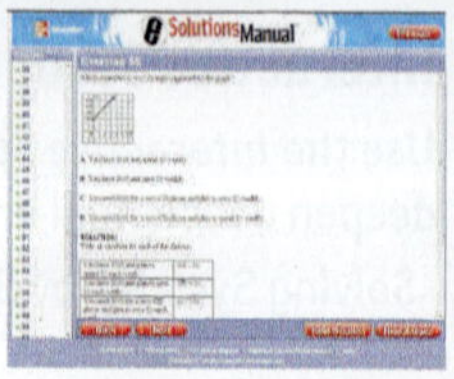

---

**25.** **SPORTS** The table shows the winning times for the Ironman World Championship.

| Year | Men's | Women's |
|---|---|---|
| 2000 | 8:21:00 | 9:26:16 |
| 2012 | 8:18:37 | 9:15:54 |

**a.** The times are in hours, minutes, and seconds. Rewrite the times rounded to the nearest minute. men: 501, 499; women: 566, 556

**b.** Let the year 2000 be 0. Assume that the rate of change remains the same for years after 2000. Write an equation to represent each of the men's and women's winning times $y$ in any year $x$. $y = -\frac{1}{6}x + 501$; $y = -\frac{5}{6}x + 566$

**c.** If the trend continues, when would you expect the men's and women's winning times to be the same? Explain your reasoning. 2097; the graphs intersect around (97, 485)

**26.** **CONCERT TICKETS** Booker is buying tickets online for a concert. He finds tickets for himself and his friends for $65 each plus a one-time fee of $10. Paula is looking for tickets to the same concert. She finds them at another website for $69 and a one-time fee of $13.60. d. See margin.

**a.** Define the variables, and write equations to represent this situation. See margin.

**b.** Create a table of values for 1 to 5 tickets for each person's purchase. See margin.

**c.** Graph each of these equations. See Ch. 6 Answer Appendix.

**d.** Use the graph to determine who received the better deal. Explain why. See margin.

27. Neither; Guillermo substituted incorrectly for $b$. Cara solved correctly for $b$, but misinterpreted the pounds of apples bought.

A.CED.3, A.REI.6

### H.O.T. Problems Use Higher-Order Thinking Skills

29. Sample answer: The solutions found by each of these methods should be the same. However, it may be necessary to estimate using a graph. So, when a precise solution is needed, you should use substitution.

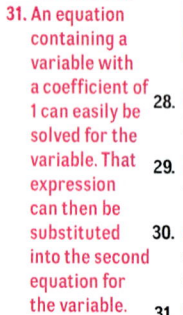

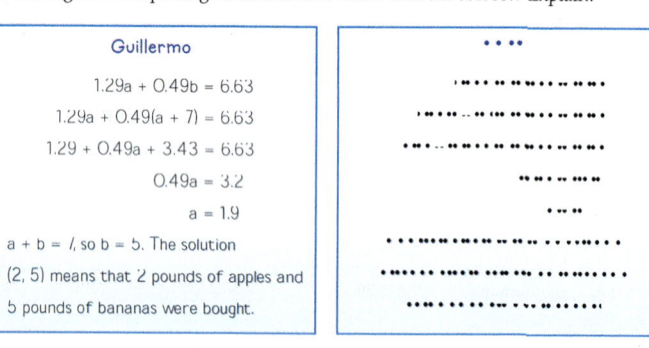

31. An equation containing a variable with a coefficient of 1 can easily be solved for the variable. That expression can then be substituted into the second equation for the variable.

**27.** **ERROR ANALYSIS** In the system $a + b = 7$ and $1.29a + 0.49b = 6.63$, $a$ represents pounds of apples and $b$ represents pounds of bananas. Guillermo and Cara are finding and interpreting the solution. Is either of them correct? Explain.

Guillermo

$1.29a + 0.49b = 6.63$
$1.29a + 0.49(a + 7) = 6.63$
$1.29 + 0.49a + 3.43 = 6.63$
$0.49a = 3.2$
$a = 1.9$
$a + b = 7$, so $b = 5$. The solution (2, 5) means that 2 pounds of apples and 5 pounds of bananas were bought.

[illegible]

**28.** **SENSE-MAKING** A local charity has 60 volunteers. The ratio of boys to girls is 7:5. Find the number of boy volunteers and the number of girl volunteers. 25 girls and 35 boys

**29.** **REASONING** Compare and contrast the solution of a system found by graphing and the solution of the same system found by substitution.

**30.** **OPEN-ENDED** Create a system of equations that has one solution. Illustrate how the system could represent a real-world situation and describe the significance of the solution in the context of the situation. See margin.

**31.** **WRITING IN MATH** Explain how to determine what to substitute when using the substitution method of solving systems of equations.

## Preparing for Assessment

**32.** The perimeter of a rectangle is 72 inches. Its length is 6 inches greater than twice its width. What is the length of the rectangle? **MP 8 A.REI.6  C**

- A  10 inches
- B  11 inches
- C  26 inches
- D  28 inches

**33.** Solve the system of equations. **MP 8 A.REI.6  C**

$$3x + 2y = -8$$
$$-6x - 4y = 12$$

- A  $(0, -2)$
- B  $(-2, 0)$
- C  no solution
- D  infinitely many solutions

**34.** For which value of $c$ does the system of equations have infinitely many solutions? **MP 2 A.CED.3  D**

$$9x = -3y + c$$
$$y + 3x = 5$$

- A  3
- B  5
- C  10
- D  15

**35.** A collection of nickels and dimes has a value of $1.85. The value of the dimes is $0.10 less than twice the value of the nickels. Which system of equations can be used to find the number of nickels and the number of dimes? **MP 4 A.CED.3  D**

- A  $n + d = 185; n - d = 1$
- B  $n + d = 185; 2n - d = 1$
- C  $5n + 10d = 185; d = 2n - 1$
- D  $5n + 10d = 185; n - d = 1$

**36.** Given a temperature in degrees Celsius C, you can find the temperature in degrees Fahrenheit using the following formula.

$$F = \frac{9}{5}C + 32$$

Which of these equations could be used to create a system of equations to find the temperature in degrees Fahrenheit that is double the temperature in degrees Celsius? **MP 4 A.CED.3  B**

- A  $2F = C$
- B  $2C = F$
- C  $F = 2 + C$
- D  $C = 2 + F$

**37. MULTI-STEP** The sum of two numbers is three times the difference of the two numbers. The greater number is five more than the lesser number. **MP 2 A.CED.3**

a. Write a system of equations to represent this situation. $x + y = 3(x - y); x = y + 5$

b. What are the values of the lesser and greater numbers? **5, 10**

c. Could you have substituted for $x$ or $y$ to solve the system of equations? Explain.
**Sample answer: The system can be solved by substituting either $y + 5$ for $x$ or $x - 5$ for $y$.**

**38.** How many solutions are possible for the following system of equations? **MP 2 A.REI.6  B**

$$x + 2y = 0$$
$$2x = y$$

- A  0
- B  1
- C  2
- D  infinitely many

**39.** A school paid $248 for new basketball and soccer jerseys. The number of basketball jerseys purchased was 6 less than 3 times the number of soccer ball jerseys purchased. Basketball jerseys cost $28 each and soccer ball jerseys cost $20 each. How many total jerseys did the school purchase? **MP 4 A.REI.6  C**

- A  4 jerseys
- B  6 jerseys
- C  10 jerseys
- D  12 jerseys

---

### Differentiated Instruction OL BL

**Extension** Ask students to explain how you can use the graph of $y = -2x + 6$ to solve the inequality $-2x + 6 < 0$. **Sample explanation: The $x$-intercept ($x = 3$) is the value of $x$ where $-2x + 6 = 0$. All values of $x$ that are more than the $x$-intercept are solutions of $-2x + 6 < 0$. So, $x > 3$.**

**39.**

| | |
|---|---|
| A | Solved for the number of soccer jerseys |
| B | Solved for the number of basketball jerseys |
| C | CORRECT |
| D | Substitution error |

---

Lesson 6-2 | Substitution

## Assess

**Name the Math** Write a system of equations on the board. Have students pair up and tell each other how they would solve the system. **ELL**

## Preparing for Assessment

Exercises 32–39 require students to use the skills they will need on standardized assessments. Each exercise is dual-coded with content standards and mathematical practice standards.

### Dual Coding

| Items | Content Standards | MP Mathematical Practices |
|---|---|---|
| 32 | A.REI.6 | 8 |
| 33 | A.REI.6 | 8 |
| 34 | A.CED.3 | 2 |
| 35 | A.CED.3 | 4 |
| 36 | A.CED.3 | 4 |
| 37 | A.CED.3 | 2 |
| 38 | A.REI.6 | 2 |
| 39 | A.REI.6 | 4 |

## Diagnose Student Errors

Survey student responses for each item. Class trends may indicate common errors and misconceptions.

**35.**

| | |
|---|---|
| A | Confused the number of coins with the value of each coin |
| B | Wrote an incorrect expression and confused the number of coins with the value of each coin |
| C | Wrote an incorrect expression and confused the number of coins with the value of each coin |
| D | CORRECT |

## Go Online!

### Quizzes

Students can use *Self-Check Quizzes* to check their understanding of this lesson. You can also give *Quiz 1*, which covers the content in Lessons 6-1 and 6-2.

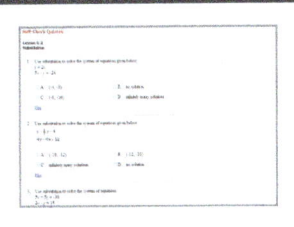

# LESSON 6-3
## Elimination Using Addition and Subtraction

**SUGGESTED PACING (DAYS)**

| | |
|---|---|
| 90 min. | 1 |
| 45 min. | 2 |

Instruction

## Track Your Progress

### Objectives

1. Solve systems of equations by using elimination with addition.
2. Solve systems of equations by using elimination with subtraction.

### Mathematical Background

Elimination using addition or subtraction uses the Addition and Subtraction Properties of Equality to combine two equations in a system of equations and eliminate one variable. After eliminating one variable, solve for the remaining variable. Then, substitute this value back into either original equation to find the value of the eliminated variable.

### THEN

**A.CED.1** Create equations and inequalities in one variable and use them to solve problems.

**A.REI.3** Solve linear equations and inequalities in one variable.

### NOW

**A.CED.2** Create equations in two or more variables to represent relationships between quantities; graph equations on coordinate axes with labels and scales.

**A.REI.6** Solve systems of linear equations exactly, focusing on pairs of linear equations in two variables.

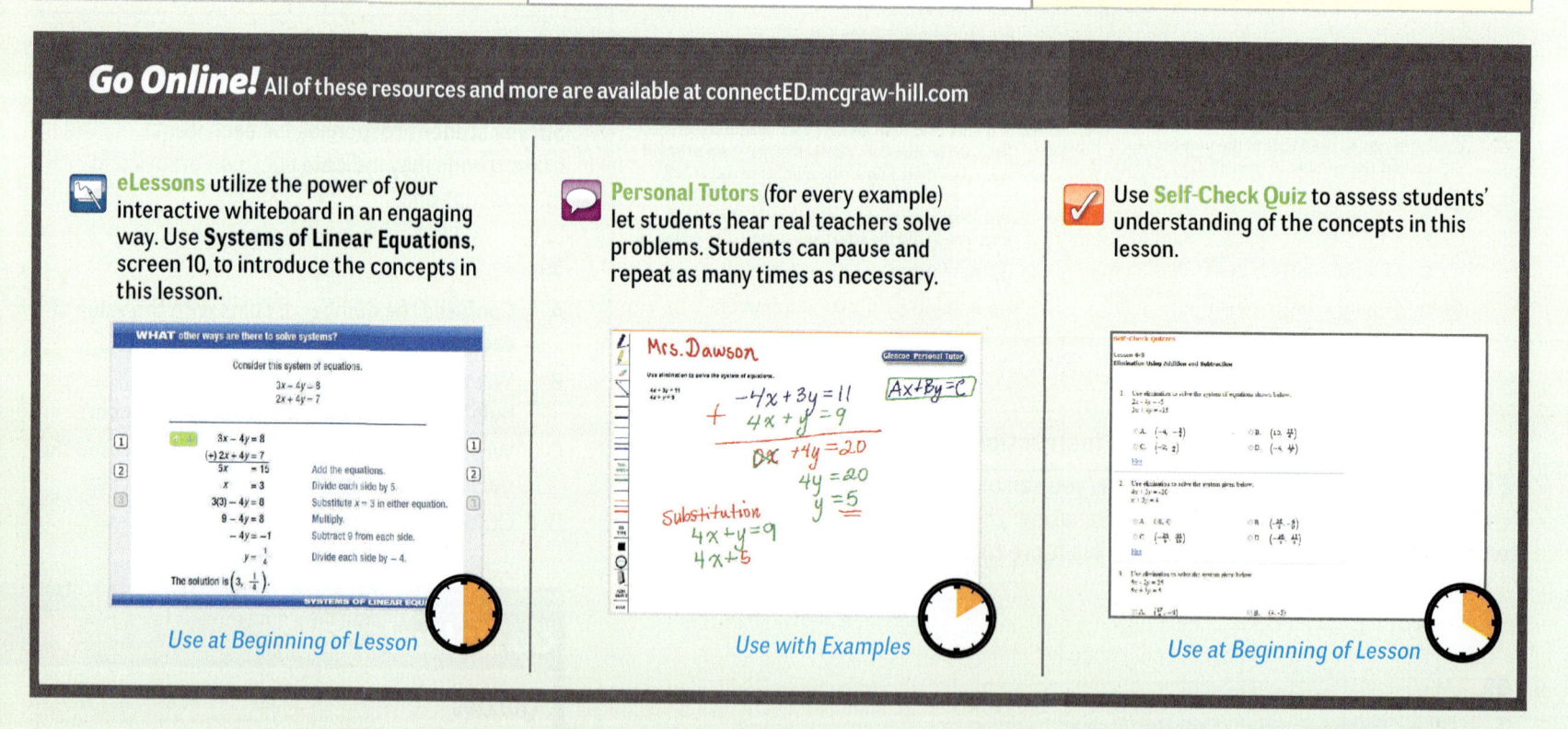

### NEXT

**A.REI.5** Prove that, given a system of two equations in two variables, replacing one equation by the sum of that equation and a multiple of the other produces a system with the same solutions.

## Go Online! All of these resources and more are available at connectED.mcgraw-hill.com

**eLessons** utilize the power of your interactive whiteboard in an engaging way. Use **Systems of Linear Equations**, screen 10, to introduce the concepts in this lesson.

*Use at Beginning of Lesson*

**Personal Tutors** (for every example) let students hear real teachers solve problems. Students can pause and repeat as many times as necessary.

*Use with Examples*

Use **Self-Check Quiz** to assess students' understanding of the concepts in this lesson.

*Use at Beginning of Lesson*

## OER Using Open Educational Resources

**Lesson Sharing** You can access already made slide presentations on **slideshare.net** about solving systems of equations by elimination using addition and subtraction. You can use these slides as ideas to add to your lesson or present them via your interactive white board. The comment section is a great place to learn from other educators who have already test drove this presentation in their own classrooms. *Use as flipped learning or as an instructional aide*

354A | Lesson 6-3 | Elimination Using Addition and Subtraction

## Go Online!
connectED.mcgraw-hill.com — Worksheets

# Differentiate Your Resources

**Extra Practice** Additional practice or homework; Skills Practice is best for approaching-level students and Practice is best for on-level and beyond-level students

### Skills Practice
AL  OL  ELL

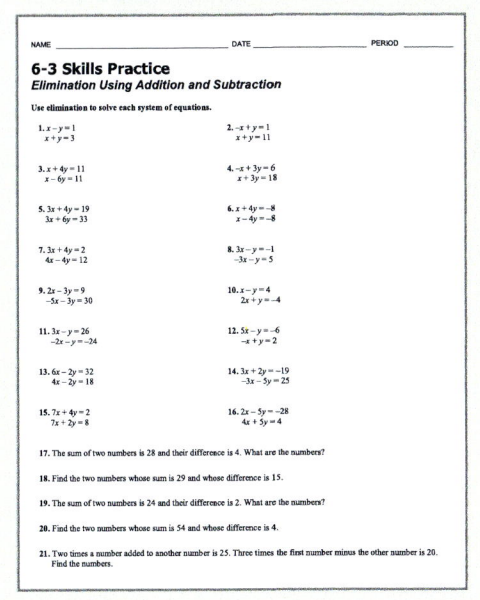

### Practice
AL  OL  BL  ELL

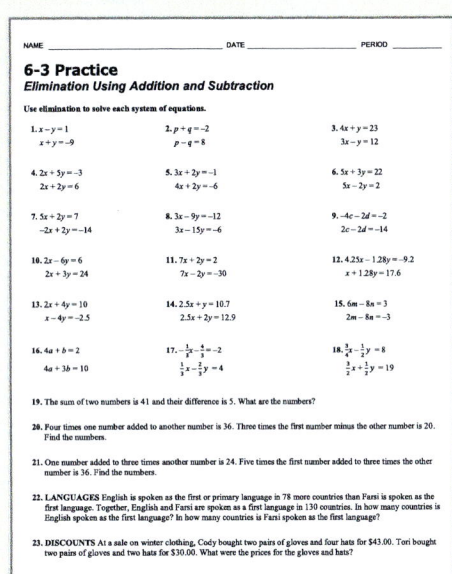

### Word Problem Practice
AL  OL  BL  ELL

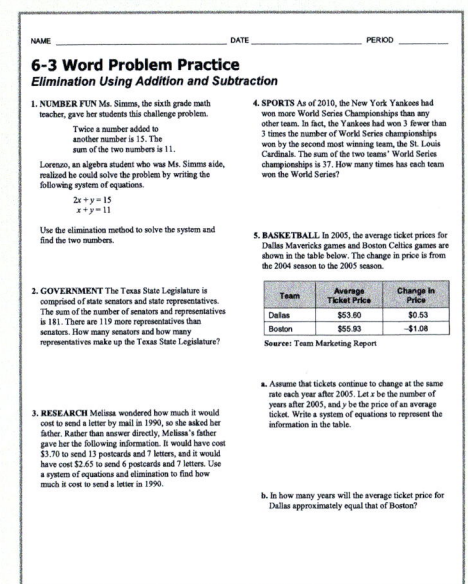

---

**Intervention** Reteaching and vocabulary activities that can be used with struggling or absent students and as ELL support

### Study Guide and Intervention
AL  OL  ELL

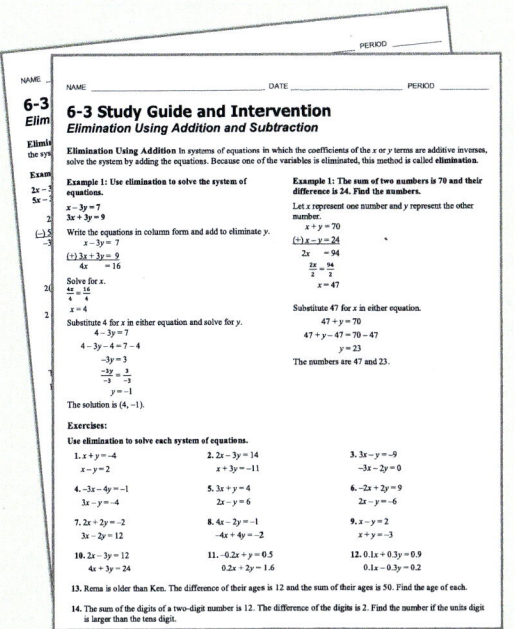

### Study Notebook
AL  OL  BL  ELL

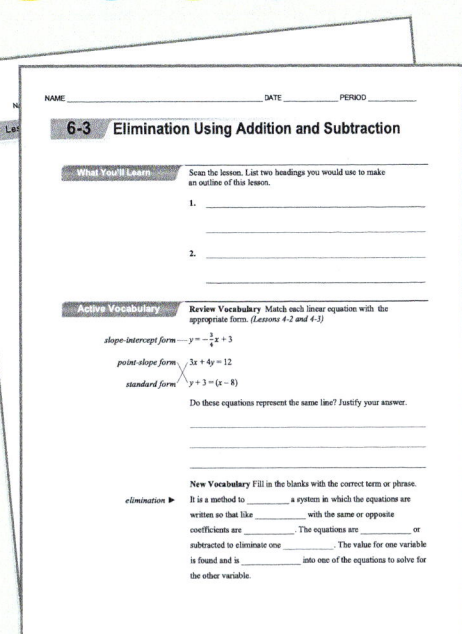

**Extension** Activities that can be used to extend lesson concepts

### Enrichment
OL  BL  ELL

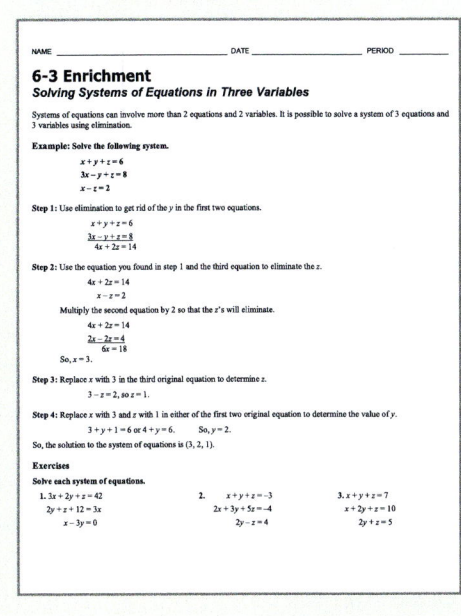

connectED.mcgraw-hill.com  354B

Lesson 6-3 | Elimination Using Addition and Subtraction

# Launch

Have students read the Why? section of the lesson. Ask:

- Why does the system $a + b = 12$ and $a - b = 2$ represent the situation? **There are 12 months in a year and the difference between $a$ and $b$ is 2.**

- Why is $b$ eliminated when you add $a + b = 12$ and $a - b = 2$? **$b$ is eliminated because $b + (-b) = 0$.**

- If you add the equations, then $2a = 14$. How many months will the mean high temperature be below 70° F? **7 months**

- If you know the value of $a$, how can you find the value of $b$? **Substitute the value of $a$ for $a$ in one of the equations and solve for $b$.**

- When might you use elimination with addition to solve a system of equations? **when the coefficients of one variable are opposites**

# Teach

Ask the scaffolded questions for each example to build conceptual understanding for students at all levels.

## 1 Elimination Using Addition

**Example 1 Elimination Using Addition**

**AL** What is being eliminated? **one of the variables**

**OL** Why should elimination by addition be used? **The $y$-terms in the equations are opposites. If they are added together, they will cancel each other out and the $y$ will be eliminated.**

**BL** Why in this case might it be easier to use elimination instead of substitution? **Sample answer: By using elimination, the system can be solved for a variable in fewer steps.**

## Go Online!

**Interactive Whiteboard**
Use the *eLesson* or *Lesson Presentation* to present this lesson.

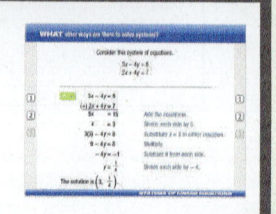

---

# LESSON 3
# Elimination Using Addition and Subtraction

| :Then | :Now | :Why? |
|---|---|---|
| • You solved systems of equations by using substitution. | 1 Solve systems of equations by using elimination with addition. <br><br> 2 Solve systems of equations by using elimination with subtraction. | • In Chicago, Illinois, there are two more months $a$ when the mean high temperature is below 70°F than there are months $b$ when it is above 70°F. The system of equations $a + b = 12$ and $a - b = 2$ represents this situation. |

**New Vocabulary**
elimination

**Mathematical Practices**
7 Look for and make use of structure.

**Content Standards**
A.CED.2 Create equations in two or more variables to represent relationships between quantities; graph equations on coordinate axes with labels and scales.

A.REI.6 Solve systems of linear equations exactly and approximately (e.g., with graphs), focusing on pairs of linear equations in two variables.

**1 Elimination Using Addition** If you add these equations, the variable $y$ will be eliminated. Using addition or subtraction to solve a system is called **elimination**.

### Key Concept  Solving by Elimination

**Step 1** Write the system so like terms with the same or opposite coefficients are aligned.

**Step 2** Add or subtract the equations, eliminating one variable. Then solve the equation.

**Step 3** Substitute the value from Step 2 into one of the equations and solve for the other variable. Write the solution as an ordered pair.

A.REI.6

**Example 1   Elimination Using Addition**

Use elimination to solve the system of equations.
$4x + 6y = 32$
$3x - 6y = 3$  ← **Step 1** $6y$ and $-6y$ have opposite coefficients.

**Step 2** Add the equations.

$\phantom{(+)\ }4x + 6y = 32$
$(+)\ 3x - 6y = \phantom{0}3$
$\phantom{(+)\ }7x \phantom{+ 6y\ } = 35$    The variable $y$ is eliminated.
$\phantom{(+)\ }\frac{7x}{7} = \frac{35}{7}$    Divide each side by 7.
$\phantom{(+)\ }x = 5$    Simplify.

**Step 3** Substitute 5 for $x$ in either equation to find the value of $y$.

$4x + 6y = 32$    First equation
$4(5) + 6y = 32$    Replace $x$ with 5.
$20 + 6y = 32$    Multiply.
$20 + 6y - 20 = 32 - 20$    Subtract 20 from each side.
$6y = 12$    Simplify.
$\frac{6y}{6} = \frac{12}{6}$    Divide each side by 6.
$y = 2$    Simplify.

The solution is (5, 2).

### MP  Mathematical Practices Strategies

**Look for and make use of structure.**

Help students recognize the structure of a pair of equations to eliminate a variable. For example, ask:

- What is an additive inverse? What is the additive inverse of $5y$? **The additive inverse of a number $n$ is the number that, when added to $n$, yields zero; $-5y$ is the additive inverse of $5y$ because $5y + (-5y) = 0$.**

- Can you solve the system $\begin{cases} mx + ny = p \\ mx - ny = q \end{cases}$ by elimination? Explain how you can tell. **Yes; because $ny + (-ny) = 0$, the $y$-variable will be eliminated when the pair of equations are added together**

- Can you solve the system $\begin{cases} mx + ny = p \\ rx + ny = q \end{cases}$ by elimination? Explain how you can tell. **Yes; because $ny = ny$, you can subtract the two equations; then the $y$-variable will be eliminated**

354 | Lesson 6-3 | Elimination Using Addition and Subtraction

## Lesson 6-3 | Elimination Using Addition and Subtraction

▶ **Guided Practice**

**1A.** $-4x + 3y = -3$
$4x - 5y = 5$ (0, −1)

**1B.** $4y + 3x = 22$
$3x - 4y = 14$ (6, 1)

We can use elimination to find specific numbers that are described as being related to each other.

A.CED.2, A.REI.6

**Example 2** Write and Solve a System of Equations

Negative three times one number plus five times another number is −11. Three times the first number plus seven times the other number is −1. Find the numbers.

| Negative three times one number | plus | five times another number | is | −11 |
|---|---|---|---|---|
| $-3x$ | $+$ | $5y$ | $=$ | $-11$ |
| Three times the first number | plus | seven times the other number | is | −1 |
| $3x$ | $+$ | $7y$ | $=$ | $-1$ |

**Steps 1 and 2** Write the equations vertically and add.

$-3x + 5y = -11$
$(+) \; 3x + 7y = -1$
$\phantom{(+) \;} 12y = -12$    The variable x is eliminated.
$\dfrac{12y}{12} = \dfrac{-12}{12}$    Divide each side by 12.
$y = -1$    Simplify.

**Step 3** Substitute −1 for y in either equation to find the value of x.

$3x + 7y = -1$    Second equation
$3x + 7(-1) = -1$    Replace y with −1.
$3x + (-7) = -1$    Simplify.
$3x + (-7) + 7 = -1 + 7$    Add 7 to each side.
$3x = 6$    Simplify.
$\dfrac{3x}{3} = \dfrac{6}{3}$    Divide each side by 3.
$x = 2$    Simplify.

The numbers are 2 and −1.

**CHECK**
$-3x + 5y = -11$    First equation
$-3(2) + 5(-1) \stackrel{?}{=} -11$    Substitute 2 for x and −1 for y.
$-11 = -11$ ✓    Simplify.
$3x + 7y = -1$    Second equation
$3(2) + 7(-1) \stackrel{?}{=} -1$    Substitute 2 for x and −1 for y.
$-1 = -1$ ✓    Simplify.

**Study Tip**
**Coefficients** When the coefficients of a variable are the same, subtracting the equations will eliminate the variable. When the coefficients are opposites, adding the equations will eliminate the variable.

**Problem-Solving Tip**
**MP Perseverance**
Checking your answers in both equations of a system helps ensure there are no calculation errors.

▶ **Guided Practice**

**2.** The sum of two numbers is −10. Negative three times the first number minus the second number equals 2. Find the numbers.   4, −14

---

**Need Another Example?**
Use elimination to solve the system of equations.
$-3x + 4y = 12$
$3x - 6y = 18$ (−24, −15)

**Example 2** Write and Solve a System of Equations

**AL** What words tell me the operations to perform in the equations? **times and plus**

**OL** What terms will be eliminated when using addition? **the x terms**

**BL** Does it matter which equation we substitute the −1 back into? Explain. **no; Sample answer: we will substitute for x and solve for y and get the same answer no matter which equation we use.**

**Need Another Example?**
Four times one number minus three times another number is 12. Two times the first number added to three times the second number is 6. Find the numbers. **The numbers are 3 and 0.**

**MP Teaching the Mathematical Practices**

**Perseverance** Mathematically proficient students check their answers. Encourage students to check their answers in both equations, stressing that they should use the original equations.

---

## Differentiated Instruction AL OL ELL

**Kinesthetic Learners** Students may benefit from using concrete models to solve systems of equations with elimination. Have students write the terms of the equations on pieces of paper or use algebra tiles or other models to represent the equations. When they eliminate a variable, have them remove the model for that variable. The act of removing the terms should help them remember eliminating the variable.

---

**Go Online!**

The most up-to-date resources available for your program can be found at **connectED.mcgraw-hill.com**.

Lesson 6-3 | Elimination Using Addition and Subtraction

## 2 Elimination Using Subtraction

**Example 3 Solve a System of Equations**

**AL** What methods do we have to solve the system? solve by graphing, solve by substitution, elimination by addition, elimination by subtraction

**OL** What is the best method to use to solve the system? Explain. Since both equations contain $2t$, we should use elimination by subtraction.

**BL** If someone was unsure of what method to use to solve the system, what is a test taking strategy they could try? Explain. Sample answer: they could test each of the multiple choice answers until they find one that works. They would start by substituting $-7$ for $r$ and 15 for $t$ into both of the equations.

**Need Another Example?**
Use elimination to solve the system of equations.
$4x + 2y = 28$
$4x - 3y = 18$ (6, 2)

**Example 4 Write and Solve a System of Equations**

**AL** What is another way to say *find the slope*? find the rate of change

**OL** How do you write terms for the total amount of money each girl earns per week? Sample answer: the amount of money given to us is per hour, so we have to multiply the per hour earnings by the total number of hours worked. We don't know the number of hours worked, so we can use variables to represent hours.

**BL** Interpret the solution to the system of equations in the context of the situation. Sample answer: Since $j = 15$, that tells us that in a typical week, Jackie works 15 hours. Since $c = 22$, we know that Cheryl works 22 hours in a typical week. We also can then determine that the week Jackie doubled her hours, she worked 30 hours.

---

## 2 Elimination Using Subtraction
Sometimes we can eliminate a variable by subtracting one equation from another.

A.REI.6

**Example 3** Solve a System of Equations

Solve the system of equations. $2t + 5r = 6$
$9r + 2t = 22$

A $(-7, 15)$  B $\left(7, \frac{8}{9}\right)$  C $(4, -7)$  D $\left(4, -\frac{2}{5}\right)$

**Read the Item**

Since both equations contain $2t$, use elimination by subtraction.

**Solve the Item**

**Step 1** Subtract the equations.

$\phantom{(-)}5r + 2t = \phantom{0}6$    Write the system so like terms are aligned.
$(-)\, 9r + 2t = \phantom{0}22$
$\overline{\phantom{(-)}-4r \phantom{+ 2t} = -16}$    The variable $t$ is eliminated.
$\phantom{(-)-4}r = 4$    Simplify.

**Step 2** Substitute 4 for $r$ in either equation to find the value of $t$.

$5r + 2t = 6$    First equation
$5(4) + 2t = 6$    $r = 4$
$20 + 2t = 6$    Simplify.
$20 + 2t - 20 = 6 - 20$    Subtract 20 from each side.
$2t = -14$    Simplify.
$t = -7$    Simplify.

The solution is $(4, -7)$. The correct answer is C.

**Guided Practice**

3. Solve the system of equations. G    $8b + 3c = 11$
                                                    $8b + 7c = 7$

F $(1.5, -1)$    G $(1.75, -1)$    H $(1.75, 1)$    J $(1.5, 1)$

A.CED.2, A.REI.6

**Real-World Example 4** Write and Solve a System of Equations

**JOBS** Cheryl and Jackie work at an ice cream shop. Cheryl earns $9.25 per hour and Jackie earns $8.75 per hour. During a typical week, Cheryl and Jackie earn $334.75 together. One week, Jackie doubles her work hours, and the girls earn $466. How many hours does each girl work during a typical week?

**Understand** You know how much Cheryl and Jackie each earn per hour and how much they earned together.

**Plan** Let $c$ = Cheryl's hours and $j$ = Jackie's hours.

| Cheryl's pay | plus | Jackie's pay | equals | $334.75 |
|---|---|---|---|---|
| $9.25c$ | + | $8.75j$ | = | 334.75 |
| Cheryl's pay | plus | Jackie's pay | equals | $466 |
| $9.25c$ | + | $8.75(2)j$ | = | 466 |

**Real-World Link**
The five most dangerous jobs for teens are: agriculture, construction, traveling sales, landscaping, and driving.

**Source:** National Consumers League

---

**Watch Out!**

**Preventing Errors** When using elimination with subtraction to solve systems of equations, many students forget to distribute the negative sign over every term of the equation that is subtracted. Since subtraction is the same as adding the inverse, you might suggest that students change the signs of the terms and then add to eliminate the variable.

356 | Lesson 6-3 | Elimination Using Addition and Subtraction

**Study Tip**

**Another Method** Instead of subtracting the equations, you could also multiply one equation by −1 and then add the equations.

**Solve** Subtract the equations to eliminate one of the variables. Then solve for the other variable.

$9.25c + 8.75j = 334.75$     Write the equations vertically.
$(-)\ 9.25c + 8.75(2)j = 466$

$\overline{\quad\quad\quad -8.75j = -131.25}$     Subtract. The variable $c$ is eliminated.

$\dfrac{-8.75j}{-8.75} = \dfrac{-131.25}{-8.75}$     Divide each side by −8.75.

$j = 15$     Simplify.

Now substitute 15 for $j$ in either equation to find the value of $c$.

$9.25c + 8.75j = 334.75$     First equation
$9.25c + 8.75(15) = 334.75$     Substitute 15 for $j$.
$9.25c + 131.25 = 334.75$     Simplify.
$9.25c = 203.50$     Subtract 131.25 from each side.
$c = 22$     Divide each side by 9.25.

**Check** Substitute both values into the other equation to see if the equation holds true. If $c = 22$ and $j = 15$, then $9.25(22) + 17.50(15)$ or 466.

Cheryl works 22 hours, while Jackie works 15 hours during a typical week.

Systems of equations can be solved using any of the solution methods you have learned. By thinking about the system before beginning the solution process, you reduce the work required to solve the problem.

▶ **Guided Practice**

4. **PARTIES** Tamera and Adelina are throwing a birthday party for their friend. Tamera invited 5 fewer friends than Adelina. Together they invited 47 guests. How many guests did each girl invite? **Tamera, 21; Adelina, 26**

---

**Check Your Understanding**    ◯ = Step-by-Step Solutions begin on page R11.

✓ **Go Online!** for a Self-Check Quiz

Examples 1, 3   Use elimination to solve each system of equations.
A.REI.6

1. $5m - p = 7$
   $7m - p = 11$   **(2, 3)**

2. $8x + 5y = 38$
   $-8x + 2y = 4$   **(1, 6)**

③ $7f + 3g = -6$
   $7f - 2g = -31$   **(−3, 5)**

4. $6a - 3b = 27$
   $2a - 3b = 11$   **(4, −1)**

Example 2
A.CED.2,
A.REI.6

5. **MP REASONING** The sum of two numbers is 24. Five times the first number minus the second number is 12. What are the two numbers? **6, 18**

Example 4
A.CED.2,
A.REI.6

6. **RECYCLING** The recycling and reuse industry employs approximately 1,025,000 more workers than the waste management industry. Together they provide 1,275,000 jobs. How many jobs does each industry provide? **recycling and reuse, 1,150,000; waste management, 125,000**

---

### Differentiated Homework Options

| Levels | **AL** Basic | **OL** Core | **BL** Advanced |
|---|---|---|---|
| Exercises | 7–23, 34–37, 39–46 | 7–29 odd, 30–37, 39–46 | 33–39, 40–46 (optional) |
| 2-Day Option | 7–23 odd, 40–46 | 7–23 | |
| | 8–22 even, 34–37, 39 | 24–37, 39–46 | |

You can use ALEKS to provide additional remediation support with personalized instruction and practice.

---

**Lesson 6-3 | Elimination Using Addition and Subtraction**

**Need Another Example?**

**Rentals** A hardware store earned $956.50 from renting ladders and power tools last week. The store charged customers for a total of 36 days for ladders and 85 days for power tools. This week the store charged 36 days for ladders, 70 days for power tools, and earned $829. How much does the store charge per day for ladders and for power tools? **$6.50 per day for ladders and $8.50 per day for power tools**

## Practice

**Formative Assessment** Use Exercises 1–6 to assess students' understanding of the concepts in this lesson.

The Practice and Problem Solving exercises assess the content taught in the lesson. The Preparing for Assessment page is meant to be used as preparation for end-of-course assessments.

## Extra Practice

See page R6 for extra exercises for students who are approaching level or for on-level students who need additional reinforcement.

**MP Teaching the Mathematical Practices**

**Reasoning** Mathematically proficient students create a coherent representation of the problem at hand. In Exercise 5, make sure students use different variables for the two numbers.

### Levels of Complexity Chart

The levels of the exercises progress from 1 to 3, with Level 1 indicating the lowest level of complexity.

| Exercises | 7–23 | 24–32, 40–46 | 33–39 |
|---|---|---|---|
| Level 3 | | | ● |
| Level 2 | | ● | |
| Level 1 | ● | | |

## Go Online!   eBook

**Interactive Student Guide**
Use the *Interactive Student Guide* to deepen conceptual understanding.
· Solving Systems by Elimination

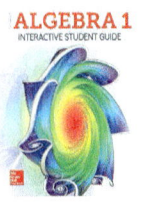

connectED.mcgraw-hill.com    357

# Lesson 6-3 | Elimination Using Addition and Subtraction

**MP Teaching the Mathematical Practices**

**Sense-Making** Mathematically proficient students can draw diagrams of important features and relationships. In Exercise 30, encourage students to draw a diagram of the statue and the building to help in writing the system of equations.

## Additional Answers

**31c.** There are 48 teams that are not from the U.S. and 18 teams that are from the U.S.

**31d.**

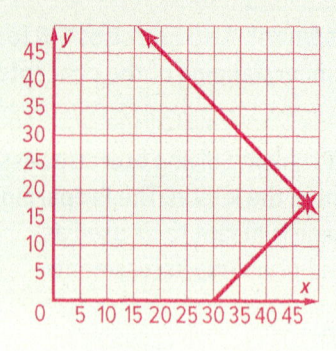

## Practice and Problem Solving

Extra Practice is on page R6.

**Examples 1, 3** Use elimination to solve each system of equations.
A.REI.6

7. $-v + w = 7$
   $v + w = 1$ $(-3, 4)$

8. $y + z = 4$
   $y - z = 8$ $(6, -2)$

9. $-4x + 5y = 17$
   $4x + 6y = -6$ $(-3, 1)$

10. $5m - 2p = 24$
    $3m + 2p = 24$ $(6, 3)$

11. $a + 4b = -4$
    $a + 10b = -16$ $(4, -2)$

12. $6r - 6t = 6$
    $3r - 6t = 15$ $(-3, -4)$

13. $6c - 9d = 111$
    $5c - 9d = 103$ $(8, -7)$

14. $11f + 14g = 13$
    $11f + 10g = 25$ $(5, -3)$

15. $9x + 6y = 78$
    $3x - 6y = -30$ $(4, 7)$

16. $3j + 4k = 23.5$
    $8j - 4k = 4$ $(2.5, 4)$

17. $-3x - 8y = -24$
    $3x - 5y = 4.5$ $(4, 1.5)$

18. $6x - 2y = 1$
    $10x - 2y = 5$ $(1, 2.5)$

**Example 2**
A.CED.2, A.REI.6

19. The sum of two numbers is 22, and their difference is 12. What are the numbers? **5, 17**

20. Find the two numbers with a sum of 41 and a difference of 9. **25, 16**

**(21)** Three times a number minus another number is −3. The sum of the numbers is 11. Find the numbers. **2, 9**

22. A number minus twice another number is 4. Three times the first number plus two times the second number is 12. What are the numbers? **4, 0**

**Example 4**
A.CED.2, A.REI.6

23. **MOVIES** The Blackwells and Joneses are going to a 3D movie. Find the adult price and the children's price of admission.

adult, $17.95; children, $13.95

| Family | Number of Adults | Number of Children | Total Cost |
|---|---|---|---|
| Blackwell | 2 | 5 | $105.65 |
| Jones | 2 | 3 | $77.75 |

**B** Use elimination to solve each system of equations.

24. $4(x + 2y) = 8$
    $4x + 4y = 12$ $(4, -1)$

25. $3x - 5y = 11$
    $5(x + y) = 5$ $(2, -1)$

26. $4x + 3y = 6$
    $3x + 3y = 7$ $(-1, 3\frac{1}{3})$

27. $6x - 7y = -26$
    $6x + 5y = 10$ $(-\frac{5}{6}, 3)$

28. $\frac{1}{2}x + \frac{2}{3}y = 2\frac{3}{4}$
    $\frac{1}{4}x - \frac{2}{3}y = 6\frac{1}{4}$ $(12, -4\frac{7}{8})$

29. $\frac{3}{5}x + \frac{1}{2}y = 8\frac{1}{3}$
    $-\frac{3}{5}x + \frac{3}{4}y = 8\frac{1}{3}$ $(2\frac{7}{9}, 13\frac{1}{3})$

30. **MP SENSE-MAKING** The total height of an office building $b$ and the granite statue that stands on top of it $g$ is 326.6 feet. The difference in heights between the building and the statue is 295.4 feet.

    a. How tall is the statue? **15.6 ft**
    b. How tall is the building? **311 ft**

31. **BIKE RACING** Professional Mountain Bike Racing currently has 66 teams. The number of non-U.S. teams is 30 more than the number of U.S. teams.

    a. Let $x$ represent the number of non-U.S. teams and $y$ represent the number of U.S. teams. Write a system of equations that represents the number of U.S. teams and non-U.S. teams. $x + y = 66; x = 30 + y$
    b. Use elimination to find the solution of the system of equations. **(48, 18)**
    c. Interpret the solution in the context of the situation. **See margin.**
    d. Graph the system of equations to check your solution. **See margin.**

## Differentiated Instruction OL BL

**Visual Learners** Have students find the missing value. The values represent the sums of each row and column. Hint: Let each symbol or set of symbols represent one variable.

| ※※ | ◆◆ | ◁▷ | ○○ | ? 15 |
|---|---|---|---|---|
| ※※ | ※※ | ◆◆ | ◆◆ | 10 |
| ◁▷ | ○○ | ◁▷ | ○○ | 20 |
| ◆◆ | ◆◆ | ◁▷ | ◁▷ | 16 |
| 14 | 11 | 20 | 16 | |

※※ = 3, ◆◆ = 2, ◁▷ = 6, ○○ = 4

## Go Online!

**eSolutions Manual**
Create worksheets, answer keys, and solutions handouts for your assignments.

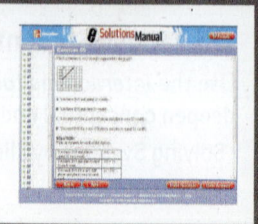

## Lesson 6-3 | Elimination Using Addition and Subtraction

**32. BUSINESS** David is analyzing the costs associated with each factory that he uses to ship his product. Assume that units shipped and total cost represent a linear relationship.

| Month | Factory A Units Shipped | Factory A Total Cost | Factory B Units Shipped | Factory B Total Cost |
|---|---|---|---|---|
| January | 386 | $39,160 | 415 | $45,200 |
| February | 421 | $41,260 | 502 | $52,160 |

a. Write a system of equations to represent the situation. **A: $y = 60x + 16,000$; B: $y = 80x + 12,000$**

b. Use elimination to find the solution to the system. **(200, 28,000)**

c. What does the solution represent? **Sample answer: When 200 units are shipped, the cost of shipping from Factory A is equal to the cost of shipping from Factory B.**

**33. MULTIPLE REPRESENTATIONS** Collect 9 pennies and 9 paper clips. For this game, you use objects to score points. Each paper clip is worth 1 point and each penny is worth 3 points. Let $p$ represent the number of pennies and $c$ represent the number of paper clips.

9 points = [2 pennies] + [3 paper clips] = $3p + c$ = $3(2) + 3$

a. **Concrete** Choose a combination of 9 objects and find your score. **Sample answer: If you choose 4 pennies and 5 paper clips, the score will be $4(3) + 5$ or 17**

b. **Analytical** Write and solve a system of equations to find the number of paper clips and pennies used for 15 points, if 9 total objects are used. **Sample answer: $p + c = 9$, $3p + c = 15$; $p = 3$, $c = 6$**

c. **Tabular** If 9 total objects are used, make a table showing the number of paper clips used and the total number of points when the number of pennies is 0, 1, 2, 3, 4, or 5. **See margin.**

d. **Verbal** Does the result in the table match the results in part **b**? Explain. **See margin.**

**37.** Sample answer: $-x + y = 5$; I used the solution to create another equation with the coefficient of the $x$-term being the opposite of its corresponding coefficient. A.CED.2, A.REI.6

### H.O.T. Problems   Use Higher-Order Thinking Skills

**34. REASONING** Describe the solution of a system of equations if after you added two equations the result was $0 = 0$. **If the result is a true statement such as $0 = 0$, then there would be an infinite number of solutions.**

**35. REASONING** What is the solution of a system of equations if the sum of the equations is $0 = 2$? **See margin.**

**36.** Sample answer: $2a + b = 5$, $a - b = 4$; a system that can be solved by using addition to eliminate one variable must have one variable with coefficients that are additive inverses.

**36. OPEN-ENDED** Create a system of equations that can be solved by using addition to eliminate one variable. Formulate a general rule for creating such systems.

**37. STRUCTURE** The solution of a system of equations is $(-3, 2)$. One equation in the system is $x + 4y = 5$. Find a second equation for the system. Explain how you derived this equation.

**38. CHALLENGE** The sum of the digits of a two-digit number is 8. The result of subtracting the units digit from the tens digit is $-4$. Define the variables and write the system of equations that you would use to find the number. Then solve the system and find the number. **See margin.**

**39. WRITING IN MATH** Describe when it would be most beneficial to use elimination to solve a system of equations. **See margin.**

### Standards for Mathematical Practice

| Emphasis On | Exercises |
|---|---|
| 1 Make sense of problems and persevere in solving them. | 30, 33 |
| 2 Reason abstractly and quantitatively. | 5, 34, 35, 38, 44 |
| 3 Construct viable arguments and critique the reasoning of others. | 39 |
| 4 Model with mathematics. | 6, 23, 31, 32 |
| 7 Look for and make use of structure. | 36, 37, 40–43, 45, 46 |

---

### MP Teaching the Mathematical Practices

**Structure** Mathematically proficient students look closely to discern a structure. In Exercise 37, encourage students to think about the lines that fit the criteria of the problem and how they can be represented mathematically. They should see that using the point-slope form with $(-3, 2)$ and any slope different than that of the given line will produce a correct system.

## Assess

**Ticket Out the Door** Have students write a system of equations that can be solved by using elimination with subtraction.

### Additional Answers

**33c.** Sample answer:

| Pennies ($p$) | 0 | 1 | 2 | 3 | 4 | 5 |
|---|---|---|---|---|---|---|
| Paper clips ($9 - p$) | 9 | 8 | 7 | 6 | 5 | 4 |
| Points ($3p + c$) | 9 | 11 | 13 | 15 | 17 | 19 |

**33d.** Yes; since the pennies are 3 points each, 3 of them makes 9 points. Add the 6 points from 6 paper clips and you get 15 points.

**35.** The result of the statement is false, so there is no solution.

**38.** Let $a =$ the tens digit of the number, and let $b =$ the ones digit of the number; $a - b = -4$; $a + b = 8$; 26 is the number.

**39.** Sample answer: It would be most beneficial when one variable has either the same coefficient or opposite coefficients in the equations.

# Lesson 6-3 | Elimination Using Addition and Subtraction

## Preparing for Assessment

Exercises 40–46 require students to use the skills they will need on standardized assessments. Each exercise is dual-coded with content standards and mathematical practice standards.

### Dual Coding

| Items | Content Standards | MP Mathematical Practices |
|---|---|---|
| 40 | A.CED.3, A.REI.6 | 4, 7 |
| 41 | A.CED.3, A.REI.6 | 7 |
| 42 | A.CED.3, A.REI.6 | 2, 7 |
| 43 | A.CED.3, A.REI.6 | 7 |
| 44 | A.CED.3, A.REI.6 | 2, 7 |
| 45 | A.CED.3, A.REI.6 | 4, 7 |
| 46 | A.CED.3, A.REI.6 | 2, 7 |

## Diagnose Student Errors

Survey student responses for each item. Class trends may indicate common errors and misconceptions.

**40.**

| A | Found allowance money earned in May |
|---|---|
| B | CORRECT |
| C | Found twice the allowance earned in May |
| D | Did not divide by 2 |

**42.**

| A | Found the sum and difference of the variables |
|---|---|
| B | Subtracted 16 from 52 instead of adding |
| C | CORRECT |
| D | Added 16 to 68 instead of subtracting |

**44.**

| A | Solved for $x$ and $y$ |
|---|---|
| B | Applied Distributive Property incorrectly |
| C | CORRECT |
| D | Used the wrong formula for perimeter |

## Go Online!

**Quizzes**
Students can use *Self-Check Quizzes* to check their understanding of this lesson. You can also give *Quiz 2*, which covers the content in Lessons 6-3 and 6-4.

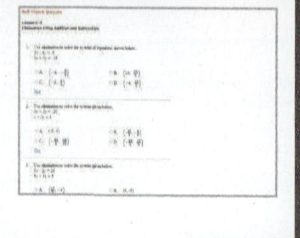

## Preparing for Assessment

**40.** In June, Kayla spent all of the $80 she had earned. The difference between her allowance in June and the money she had left over from May was $20. How much allowance money did she earn in June? MP 4, 7  A.CED.3, A.REI.6  **B**

- A  $30
- B  $50
- C  $60
- D  $100

**41.** What is the solution of the system of equations? MP 7  A.CED.3, A.REI.6  **D**

$$-2x + y = 4$$
$$2x + y = 10$$

- A  (0, 4)
- B  (5, −10)
- C  (1, 6)
- D  (1.5, 7)
- E  (6, −2)

**42.** The sum of two numbers is 120. Three times their difference is 48. What are the numbers? MP 2, 7  A.CED.3, A.REI.6  **C**

- A  16, 120
- B  36, 52
- C  52, 68
- D  68, 84

**43.** Find the values of $x$ and $y$ that make both equations true. MP 7  A.CED.3, A.REI.6  **E**

$$3x + y = 4$$
$$4x + y = 10$$

- A  $x = 2$ and $y = -2$
- B  $x = -1$ and $y = 7$
- C  $x = 1$ and $y = 6$
- D  $x = 12$ and $y = -32$
- E  $x = 6$ and $y = -14$

**44.** The perimeter of the parallelogram shown is 244 inches. The difference between the lengths of the given sides is 20 inches.

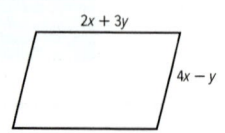

(parallelogram with sides labeled $2x + 3y$ and $4x - y$)

Assuming $2x + 3y$ is the longest side, find the length of each side. MP 2  A.CED.3, A.REI.6  **C**

- A  16 inches, 13 inches
- B  66 inches, 56 inches
- C  71 inches, 51 inches
- D  74 inches, 54 inches

**45.** Swati won a computer game that she played with Matt. The sum of their scores was 754. The difference of their scores was 176. How many points did each person score? MP 4, 7  A.CED.3, A.REI.6  **B**

- A  Matt 176; Swati 578
- B  Matt 289; Swati 465
- C  Matt 465; Swati 289
- D  Matt 754; Swati 578

46a. $x$: greater fraction
     $y$: lesser fraction

46b. $x + y = \frac{11}{12}$
     $x - y = \frac{5}{12}$

**46. MULTI-STEP** There are two fractions with one fraction that is greater than the other. The sum of the two fractions is $\frac{11}{12}$. If you subtract the lesser fraction from the greater fraction, the difference is $\frac{5}{12}$. Find both fractions. MP 2, 7  A.CED.3, A.REI.6

a. Define the variables.

b. Write a system of equations to represent the situation.

c. Use elimination to find the lesser and greater fractions. $x = \frac{2}{3}$   $y = \frac{1}{4}$

d. Explain how you used elimination to solve the system. **Answers may vary.**

## Differentiated Instruction  OL  BL

**Extension** Write two equations on the board with the constants missing. Have students find the missing constants that ensure the given solution. For example, write the system $3x + 2y = ?$, $5x - 2y = ?$ and tell students that the solution is (1.5, 0.25). $3x + 2y = 5, 5x - 2y = 7$

**45.**

| A | The first value is the difference and the second is the difference between 754 and 176 |
|---|---|
| B | CORRECT |
| C | Interchanged the answers for Swati and Matt |
| D | The first value is the sum and the second is the difference between 754 and 176 |

# LESSON 6-4
# Elimination Using Multiplication

**SUGGESTED PACING (DAYS)**

90 min. **0.5**
45 min. **1**
Instruction

## Track Your Progress

### Objectives

1. Solve systems of equations by using elimination with multiplication.
2. Solve real-world problems involving systems of equations.

### Mathematical Background
When the coefficients of one of the variables are neither the same nor additive inverses in a system of equations elimination using multiplication can be used to solve the system. Multiply one or both equations by a number to get the equations to have the same coefficient. Then solve the system using elimination using addition or subtraction.

### THEN

**A.CED.2** Create equations in two or more variables to represent relationships between quantities; graph equations on coordinate axes with labels and scales.

**A.REI.6** Solve systems of linear equations exactly, focusing on pairs of linear equations in two variables.

### NOW

**A.REI.5** Prove that, given a system of two equations in two variables, replacing one equation by the sum of that equation and a multiple of the other produces a system with the same solutions.

**A.REI.6** Solve systems of linear equations exactly, focusing on pairs of linear equations in two variables.

### NEXT

**A.REI.12** Graph the solutions to a linear inequality in two variables as a half-plane (excluding the boundary in the case of a strict inequality), and graph the solution set to a system of linear inequalities in two variables as the intersection of the corresponding half-planes.

---

**Go Online!** All of these resources and more are available at connectED.mcgraw-hill.com

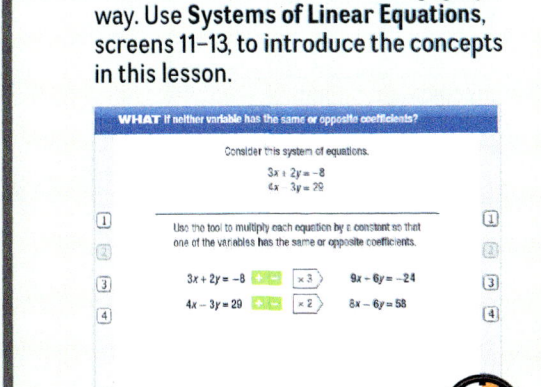

**eLessons** utilize the power of your interactive whiteboard in an engaging way. Use **Systems of Linear Equations**, screens 11–13, to introduce the concepts in this lesson.

*Use at Beginning of Lesson*

**Personal Tutors** (for every example) let students hear real teachers solve problems. Students can pause and repeat as many times as necessary.

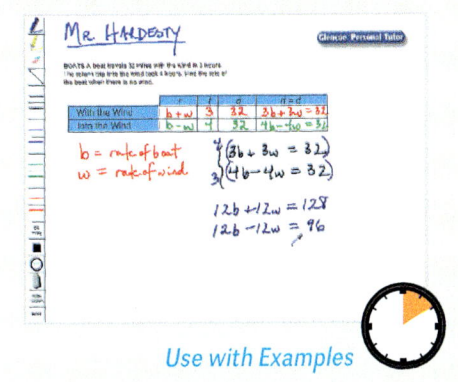

*Use with Examples*

**Chapter Projects** allow students to create and customize a project as a non-traditional method of assessment.

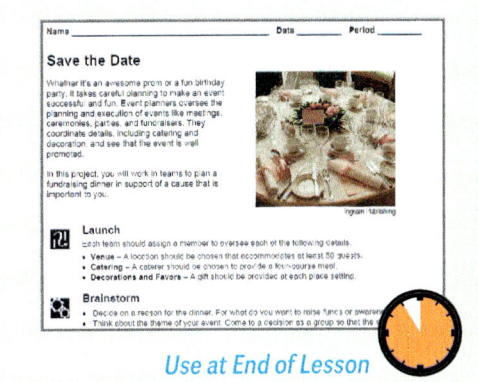

*Use at End of Lesson*

---

## Using Open Educational Resources

**Games** Have students play the Algebra Puzzle game on **mathplayground.com** to practice solving systems of equations. Students will solve systems of equations using elimination by multiplication, addition, and subtraction in order to complete a puzzle. *Use as homework or classwork*

connectED.mcgraw-hill.com  361A

# Differentiate Your Resources

**Extra Practice** Additional practice or homework; Skills Practice is best for approaching-level students and Practice is best for on-level and beyond-level students

## Skills Practice
AL  OL  ELL

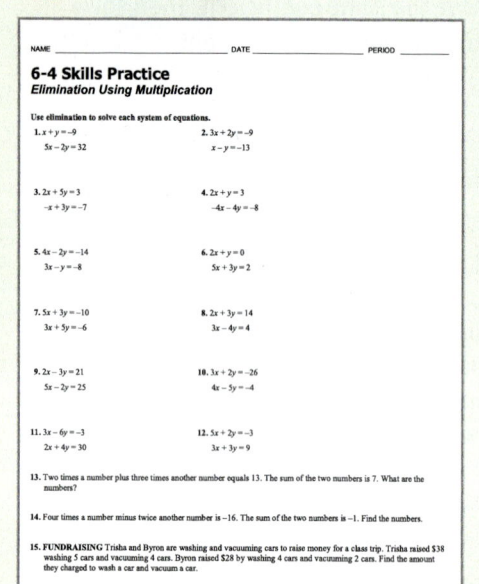

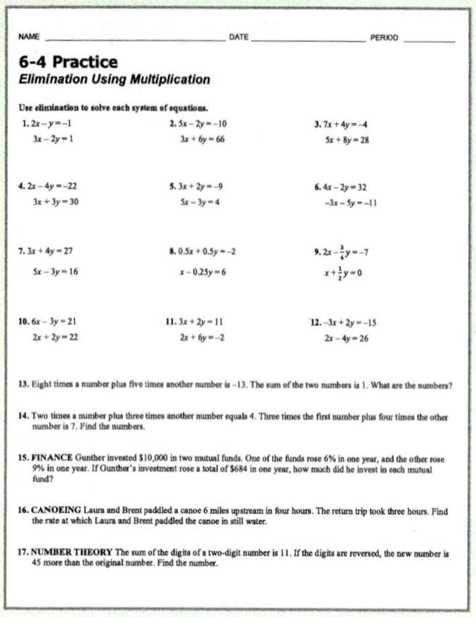

## Practice
AL  OL  BL  ELL

## Word Problem Practice
AL  OL  BL  ELL

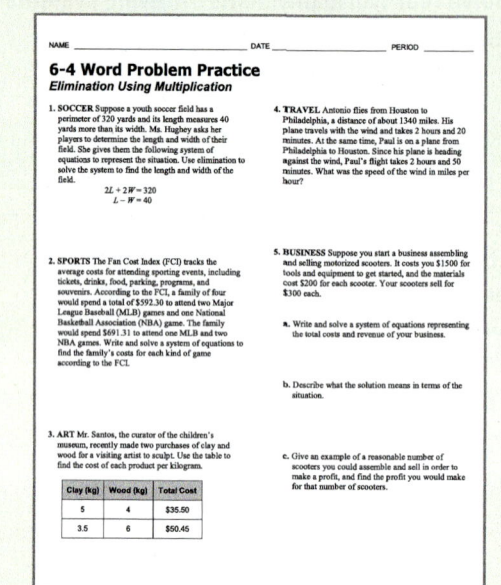

**Intervention** Reteaching and vocabulary activities that can be used with struggling or absent students and as ELL support

**Extension** Activities that can be used to extend lesson concepts

## Study Guide and Intervention
AL  OL  ELL

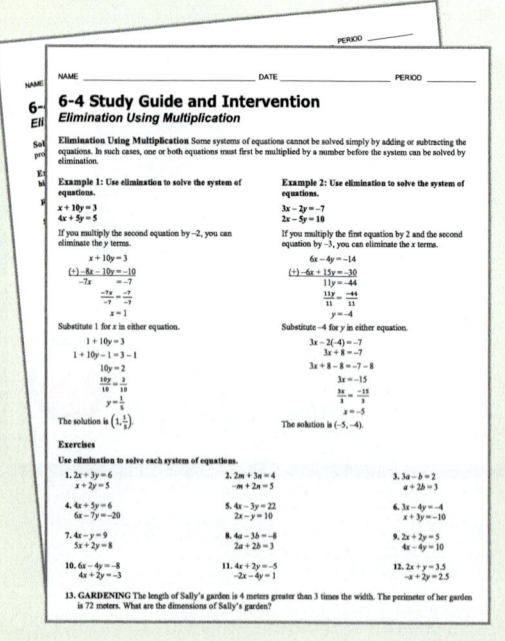

## Study Notebook
AL  OL  BL  ELL

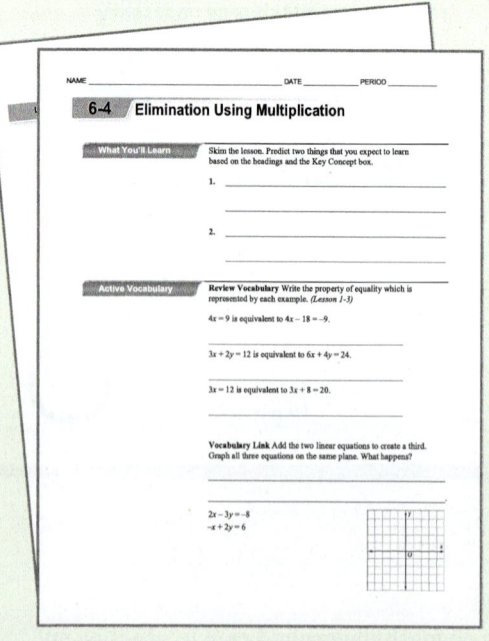

## Enrichment
OL  BL  ELL

361B | Chapter 6 | Elimination Using Multiplication

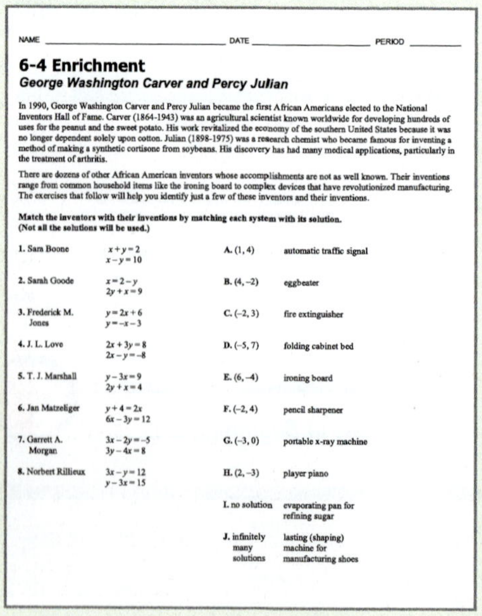

# LESSON 4
# Elimination Using Multiplication

**Then**
- You used elimination with addition and subtraction to solve systems of equations.

**Now**
1. Solve systems of equations by using elimination with multiplication.
2. Solve real-world problems involving systems of equations.

**Why?**
The table shows the number of cars at Scott's Auto Repair Shop for each type of service.

| Item | Repairs | Maintenance |
|---|---|---|
| body | 3 | 4 |
| engine | 2 | 2 |

The manager has allotted 1110 minutes for body work and 570 minutes for engine work. The system $3r + 4m = 1110$ and $2r + 2m = 570$ can be used to find the average time for each service.

## Mathematical Practices
1 Make sense of problems and persevere in solving them.

## Content Standards
**A.REI.5** Prove that, given a system of two equations in two variables, replacing one equation by the sum of that equation and a multiple of the other produces a system with the same solutions.

**A.REI.6** Solve systems of linear equations exactly and approximately (e.g., with graphs), focusing on pairs of linear equations in two variables.

## 1 Elimination Using Multiplication
In the system above, neither variable can be eliminated by adding or subtracting. You can use multiplication to solve.

### Key Concept — Solving by Elimination Using Multiplication
**Step 1** Multiply at least one equation by a constant to get two equations that contain opposite terms.
**Step 2** Add the equations, eliminating one variable. Then solve the equation.
**Step 3** Substitute the value from Step 2 into one of the equations and solve for the other variable. Write the solution as an ordered pair.

A.REI.5, A.REI.6

### Example 1  Multiply One Equation to Eliminate a Variable
Use elimination to solve the system of equations.
$5x + 6y = -8$
$2x + 3y = -5$

**Steps 1 and 2**

$5x + 6y = -8$ 
$2x + 3y = -5$ → Multiply each term by $-2$ → $5x + 6y = -8$
$(+) -4x - 6y = 10$  Add
$x = 2$  $y$ is eliminated.

**Step 3**
$2x + 3y = -5$  Second equation
$2(2) + 3y = -5$  Substitution, $x = 2$
$4 + 3y = -5$  Simplify.
$3y = -9$  Subtract 4 from each side and simplify.
$y = -3$  Divide each side by 3 and simplify.

The solution is $(2, -3)$.

### Guided Practice
**1A.** $6x - 2y = 10$
$3x - 7y = -19$  $(3, 4)$

**1B.** $9r + q = 13$
$3r + 2q = -4$  $(2, -5)$

## MP Mathematical Practices Strategies

**Make sense of problems and persevere in solving them.**
Help students understand how to multiply the equations in a system by a factor to eliminate a variable. For example, ask:

- What is a least common multiple (LCM)? What is the LCM of 6 and 7? The least common multiple of two numbers is the smallest number divisible by each of the numbers; the LCM of 6 and 7 is 42 since $42 \div 6 = 7$ and $42 \div 7 = 6$.

- Solve $\begin{cases} 2x + 6y = 5 \\ x - 7y = 7.5 \end{cases}$ by elimination.

- What numbers would you use to multiply each equation by to eliminate $y$? multiply the first equation by 7 and the second equation by 6

- What is the new system of equations? $\begin{cases} 14x + 42y = 35 \\ 6x - 42y = 45 \end{cases}$

- What is the solution? Solution: $(4, -0.5)$

- Substitute the solution into each equation. Show that both equations are true. $x - 7y = 4 - 7(-0.5) = 7.5$; $2x + 6y = 2(4) + 6(-0.5) = 5$

---

Lesson 6-4 | Elimination Using Multiplication

# Launch

Have students read the Why? section of the lesson. Ask:
- What does the first equation in the system of equations represent? the amount of time to perform body repair for 3 cars and body maintenance for 4 cars

- What does the second equation in the system represent? the amount of time to perform engine repair for 2 cars and engine maintenance for 2 cars

- What will the solution to the system represent? the average number of minutes allotted per repair and the average number of minutes allotted per maintenance

- If you multiply the second equation by $-2$, what variable could you eliminate by adding the equations? the variable $m$

# Teach

Ask the scaffolded questions for each example to build conceptual understanding for students at all levels.

## 1 Elimination Using Multiplication

### Example 1  Multiply One Equation to Eliminate a Variable

**AL** Which variable terms have coefficients with common multiples? $6y$ and $3y$

**OL** By what number can you multiply the second equation to eliminate a variable? 2 or $-2$ How will you eliminate the variable after multiplying? elimination by subtraction if multiplying by 2, elimination by addition if multiplying by $-2$

**BL** How can you check your solution? Sample answer: plug 2 in for $x$ and $-3$ in for $y$ in each equation and simplify

*(continued on the next page)*

## Go Online!

**Interactive Whiteboard**
Use the eLesson or Lesson Presentation to present this lesson.

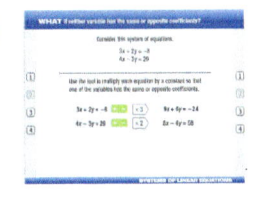

# Lesson 6-4 | Elimination Using Multiplication

**Need Another Example?**
Use elimination to solve the system of equations.
$2x + y = 23$
$3x + 2y = 37$ (9, 5)

## Example 2 Multiply Both Equations to Eliminate a Variable

**AL** Does it matter which variable is eliminated? no

**OL** What is the quickest way to find a common multiple for the constants? multiply the constants together

**BL** What do you notice about the solution to the system after solving using both methods in the example? Sample answer: the solution is the same using both methods.

**Need Another Example?**
Use elimination to solve the system of equations.
$4x + 3y = 8$
$3x - 5y = -23$ (−1, 4)

## 2 Solve Real-World Problems

### Example 3 Solve a System of Equations

**AL** What must be solved for? a What variable needs to be eliminated? w

**OL** What is the best method to eliminate w? multiply both equations to eliminate the variable

**BL** Once a is solved for, is the problem complete? Explain. yes; Sample answer: The problem only asks to determine the rate of the airplane, or a; therefore, it is not necessary to plug the solution back in and solve for w.

**Need Another Example?**
**Transportation** A fishing boat travels 10 miles downstream in 30 minutes. The return trip takes the boat 40 minutes. Find the rate in miles per hour of the boat in still water. 17.5 mi/h

---

Sometimes you have to multiply each equation by a different number in order to solve the system.

A.REI.5, A.REI.6

### Example 2 Multiply Both Equations to Eliminate a Variable

Use elimination to solve the system of equations.
$4x + 2y = 8$
$3x + 3y = 9$

**Study Tip**
**Perseverance** Unless the problem is asking for the value of a specific variable, you may use multiplication to eliminate either variable.

**Method 1** Eliminate $x$.

$4x + 2y = 8$ Multiply by 3.  $\quad 12x + 6y = 24$
$3x + 3y = 9$ Multiply by −4. $(+) -12x - 12y = -36$   Add equations.
$\qquad\qquad\qquad\qquad\qquad\qquad\quad -6y = -12$   $x$ is eliminated.
$\qquad\qquad\qquad\qquad\qquad\qquad\quad \frac{-6y}{-6} = \frac{-12}{-6}$   Divide each side by −6.
$\qquad\qquad\qquad\qquad\qquad\qquad\quad y = 2$   Simplify.

Now substitute 2 for $y$ in either equation to find the value of $x$.

$3x + 3y = 9$   Second equation
$3x + 3(2) = 9$   Substitute 2 for y.
$3x + 6 = 9$   Simplify.
$3x = 3$   Subtract 6 from each side and simplify.
$\frac{3x}{3} = \frac{3}{3}$   Divide each side by 3.
$x = 1$   The solution is (1, 2).

**Method 2** Eliminate $y$.

$4x + 2y = 8$ Multiply by 3. $\quad 12x + 6y = 24$
$3x + 3y = 9$ Multiply by −2. $(+) -6x - 6y = -18$   Add equations.
$\qquad\qquad\qquad\qquad\qquad\qquad\quad 6x \quad\;\;\; = \;\;\; 6$   $y$ is eliminated.
$\qquad\qquad\qquad\qquad\qquad\qquad\quad \frac{6x}{6} = \frac{6}{6}$   Divide each side by 6.
$\qquad\qquad\qquad\qquad\qquad\qquad\quad x = 1$   Simplify.

Now substitute 1 for $x$ in either equation to find the value of $y$.

$3x + 3y = 9$   Second equation
$3(1) + 3y = 9$   Substitute 1 for x.
$3 + 3y = 9$   Simplify.
$3y = 6$   Subtract 3 from each side and simplify.
$\frac{3y}{3} = \frac{6}{3}$   Divide each side by 3.
$y = 2$   Simplify.

The solution is (1, 2), which matches the result obtained with Method 1.

**CHECK** Substitute 1 for $x$ and 2 for $y$ in the first equation.

$4x + 2y = 8$   Original equation
$4(1) + 2(2) \stackrel{?}{=} 8$   Substitute (1, 2) for (x, y).
$4 + 4 \stackrel{?}{=} 8$   Multiply.
$8 = 8$ ✓   Add.

▶ **Guided Practice**

**2A.** $5x - 3y = 6$        **2B.** $6a + 2b = 2$
$\quad\;\; 2x + 5y = -10$ (0, −2)      $4a + 3b = 8$ (−1, 4)

---

**Watch Out!**
**Common Errors** When using elimination with multiplication, many students forget to multiply each term on both sides of the equation by the number. Suggest that they include an extra step in the solutions that shows the multiplication:

$3x + 2y = 7$
$2x - 7y = -12$
⇓
$2(3x + 2y) = 2(7)$
$-3(2x - 7y) = -3(-12)$

Lesson 6-4 | Elimination Using Multiplication

### Go Online!

Elimination using multiplication can be a challenging skill to master. Look for **worksheets** in the Resources if you need extra practice.

## 2 Solve Real-World Problems
Sometimes it is necessary to use multiplication before elimination in real-world problem solving too.

A.REI.5, A.REI.6

**Real-World Example 3** Solve a System of Equations

**FLIGHT** A personal aircraft traveling with the wind flies 520 miles in 4 hours. On the return trip, the airplane takes 5 hours to travel the same distance. Find the speed of the airplane if the air is still.

You are asked to find the speed of the airplane in still air.

Let $a$ = the rate of the airplane if the air is still.
Let $w$ = the rate of the wind.

|  | r | t | d | r • t = d |
|---|---|---|---|---|
| With the Wind | $a + w$ | 4 | 520 | $(a + w)4 = 520$ |
| Against the Wind | $a - w$ | 5 | 520 | $(a - w)5 = 520$ |

So, our two equations are $4a + 4w = 520$ and $5a - 5w = 520$.

$4a + 4w = 520$ Multiply by 5. $\quad 20a + 20w = 2600$
$5a - 5w = 520$ Multiply by 4. $\quad (+)\ 20a - 20w = 2080$
$\qquad\qquad\qquad\qquad\qquad\qquad\overline{\ 40a \qquad\quad = 4680}\quad$ *w is eliminated.*

$\dfrac{40a}{40} = \dfrac{4680}{40}\quad$ *Divide each side by 40.*

$a = 117\quad$ *Simplify.*

The rate of the airplane in still air is 117 miles per hour.

▶ **Guided Practice**

3. **CANOEING** A canoeist travels 4 miles downstream in 1 hour. The return trip takes the canoeist 1.5 hours. Find the rate of the boat in still water. $3\frac{1}{3}$ mi/h

**Check Your Understanding** ⬤ = Step-by-Step Solutions begin on page R11.

✓ Go Online! for a Self-Check Quiz

**Examples 1–2** Use elimination to solve each system of equations.

A.REI.5,
A.REI.6

1. $2x - y = 4$
$\phantom{1.\ }7x + 3y = 27$ (3, 2)

2. $2x + 7y = 1$
$\phantom{2.\ }x + 5y = 2$ (−3, 1)

**3.** $4x + 2y = -14$
$\phantom{3.\ }5x + 3y = -17$ (−4, 1)

4. $9a - 2b = -8$
$\phantom{4.\ }-7a + 3b = 12$ (0, 4)

**Example 3**
A.REI.5,
A.REI.6

5. **SENSE-MAKING** A kayaking group with a guide travels 16 miles downstream, stops for a meal, and then travels 16 miles upstream. The speed of the current remains constant throughout the trip. Find the speed of the kayak in still water. **6 mph**

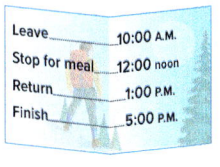

Leave _____ 10:00 A.M.
Stop for meal 12:00 noon
Return _____ 1:00 P.M.
Finish _____ 5:00 P.M.

6. 8 Hobbies and Recreation, 2 Soliloquies

6. **PODCASTS** Steve subscribed to 10 podcasts for a total of 340 minutes. He used his two favorite tags, Hobbies and Recreation, and Soliloquies. Each of the Hobbies and Recreation episodes lasted about 32 minutes. Each Soliloquies episode lasted 42 minutes. To how many of each tag did Steve subscribe?

---

## Differentiated Instruction AL ELL

**IF** students have trouble solving Exercises 1–4,

**THEN** suggest that they form groups of two or three to discuss the best strategy for solving each problem and then work through the solution together. Encourage all students to participate and remind students to check their solutions.

---

# Practice

**Formative Assessment** Use Exercises 1–6 to assess students' understanding of the concepts in this lesson.

The Practice and Problem Solving exercises assess the content taught in the lesson. The Preparing for Assessment page is meant to be used as preparation for end-of-course assessments.

### MP Teaching the Mathematical Practice

**Sense-Making** Mathematically proficient students consider analogous problems. In Exercise 5, encourage students to reference Example 3 as they write and solve the system of equations.

## Extra Practice

See page R6 for extra exercises for students who are approaching level or for on-level students who need additional reinforcement.

---

### Go Online! eBook

**Interactive Student Guide**
Use the *Interactive Student Guide* to deepen the conceptual understanding.

· Solving Systems by Elimination Using Multiplication

ALGEBRA 1
INTERACTIVE STUDENT GUIDE

connectED.mcgraw-hill.com 363

# Lesson 6-4 | Elimination Using Multiplication

## Teaching the Mathematical Practices

**Modeling** Mathematically proficient students apply the mathematics they know to solve problems arising in the workplace. In Exercise 25, encourage students to think about why this would be an important issue in a workplace setting.

**Critique Arguments** Mathematically proficient students critique the reasoning of others. In Exercise 30, tell students to look at the original system of equations before they evaluate Jason's and Daniela's work. Did they both find a way to eliminate one variable completely?

## Additional Answers

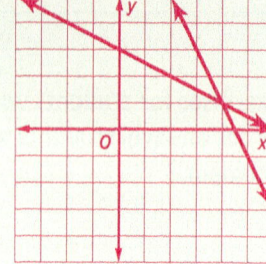

26b.

26c.

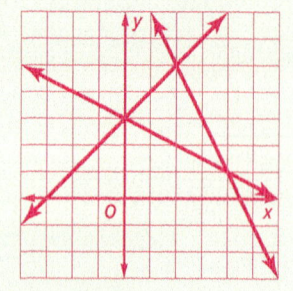

## Practice and Problem Solving

Extra Practice is on page R6.

**Examples 1–2** Use elimination to solve each system of equations.
A.REI.5, A.REI.6

7. $x + y = 2$
   $-3x + 4y = 15$  (−1, 3)

8. $x - y = -8$
   $7x + 5y = 16$  (−2, 6)

9. $x + 5y = 17$
   $-4x + 3y = 24$  (−3, 4)

10. $6x + y = -39$
    $3x + 2y = -15$  (−7, 3)

11. $2x + 5y = 11$
    $4x + 3y = 1$  (−2, 3)

12. $3x - 3y = -6$
    $-5x + 6y = 12$  (0, 2)

13. $3x + 4y = 29$
    $6x + 5y = 43$  (3, 5)

14. $8x + 3y = 4$
    $-7x + 5y = -34$  (2, −4)

15. $8x + 3y = -7$
    $7x + 2y = -3$  (1, −5)

16. $4x + 7y = -80$
    $3x + 5y = -58$  (−6, −8)

17. $12x - 3y = -3$
    $6x + y = 1$  (0, 1)

18. $-4x + 2y = 0$
    $10x + 3y = 8$  $\left(\frac{1}{2}, 1\right)$

**Example 3**
A.REI.5, A.REI.6

19. **NUMBER THEORY** Four times a number minus 5 times another number is equal to 21. Three times the sum of the two numbers is 36. What are the two numbers?  9, 3

20. **FOOTBALL** A field goal is 3 points and the extra point after a touchdown is 1 point. In the 2016 season, Adam Vinatieri of the Indianapolis Colts made a total of 71 field goals and extra points for 125 points. Find the number of field goals and extra points that he made.  27 field goals, 44 extra points.

**B** Use elimination to solve each system of equations.

21. $2.2x + 3y = 15.25$
    $4.6x + 2.1y = 18.325$  (2.5, 3.25)

22. $-0.4x + 0.25y = -2.175$
    $2x + y = 7.5$  (4.5, −1.5)

23. $\frac{1}{4}x + 4y = 2\frac{3}{4}$
    $3x + \frac{1}{2}y = 9\frac{1}{4}$  $\left(3, \frac{1}{2}\right)$

24. $\frac{2}{5}x + 6y = 24\frac{1}{5}$
    $3x + \frac{1}{2}y = 3\frac{1}{2}$  $\left(\frac{1}{2}, 4\right)$

25. **MULTI-STEP** Michelle and Julie work at a catering company. They need to bake 264 cookies for a birthday party that starts in a little over an hour and a half. Each tube of cookie dough claims to make 36 cookies, but Michelle eats about $\frac{1}{5}$ of every tube and Julie makes cookies that are 1.5 times as large as the recommended cookie size. It takes about 8 minutes to bake a container of cookies, but since Julie's cookies are larger, they take 12 minutes to bake.  See Ch. 6 Answer Appendix.

   a. How many tubes should each girl plan to bake? How long does each girl use the oven?
   b. Explain your solution process.
   c. What assumptions did you make?

26. **GEOMETRY** The graphs of $x + 2y = 6$ and $2x + y = 9$ contain two of the sides of a triangle. A vertex of the triangle is at the intersection of the graphs.

   a. What are the coordinates of the vertex? (4, 1)
   b. Draw the graph of the two lines. Identify the vertex of the triangle. See margin.
   c. The line that forms the third side of the triangle is the line $x - y = -3$. Draw this line on the previous graph. See margin.
   d. Name the other two vertices of the triangle. (0, 3); (2, 5)

## Differentiated Homework Options

| Levels | AL Basic | OL Core | BL Advanced |
|---|---|---|---|
| Exercises | 7–20, 29–31, 33–38 | 7–25 odd, 26–31, 33–38 | 28–33, 34–38 (optional) |
| 2-Day Option | 7–19 odd, 34–38 | 7–20 | |
| | 8–20 even, 29–31, 33 | 21–31, 33–38 | |

You can use ALEKS to provide additional remediation support with personalized instruction and practice.

**27. ENTERTAINMENT** At an entertainment center, two groups of people bought batting tokens and miniature golf games, as shown in the table.

**27a.** Let $x =$ the cost of each trip to the batting cage and let $y =$ the cost of a miniature golf game; $16x + 3y = 44.81$ and $22x + 5y = 67.73$.

| Group | Number of Trips to the Batting Cage | Number of Miniature Golf Games | Total Cost |
|---|---|---|---|
| A | 16 | 3 | $44.81 |
| B | 22 | 5 | $67.73 |

**27b.** (1.49, 6.99); A trip to the batting cage costs $1.49 and a game of miniature golf costs $6.99.

a. Define the variables, and write a system of linear equations for this situation.
b. Solve the system of equations, and explain what the solution represents.

**28. TESTS** Mrs. Henderson discovered that she had accidentally reversed the digits of a test score and did not give a student 36 points. Mrs. Henderson told the student that the sum of the digits was 14 and agreed to give the student his correct score plus extra credit if he could determine his actual score. What was his correct score? **95**

A.REI.5, A.REI.6

### H.O.T. Problems — Use Higher-Order Thinking Skills

**29. REASONING** Explain how you could recognize a system of linear equations with infinitely many solutions. **One of the equations will be a multiple of the other.**

**30. CRITIQUE ARGUMENTS** Jason and Daniela are solving a system of equations. Is either of them correct? Explain your reasoning.

**30.** Jason; in order to eliminate the $r$-terms, you can multiply the second equation by 2 and then subtract, or multiply the equation by $-2$ and then add. Daniela did not subtract the equations correctly.

**31.** Sample answer: $2x + 3y = 6, 4x + 9y = 5$

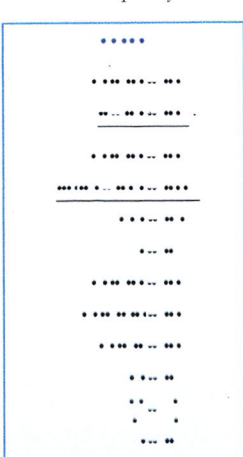

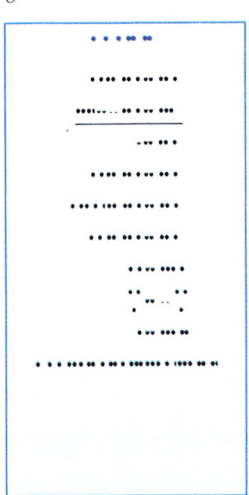

**31. OPEN-ENDED** Write a system of equations that can be solved by multiplying one equation by $-3$ and then adding the two equations together.

**32. CHALLENGE** The solution of the system $4x + 5y = 2$ and $6x - 2y = b$ is $(3, a)$. Find the values of $a$ and $b$. Discuss the steps that you used. **See margin.**

**33. WRITING IN MATH** Why is substitution sometimes more helpful than elimination, and vice versa? **See margin.**

### Standards for Mathematical Practice

| Emphasis On | Exercises |
|---|---|
| 1 Make sense of problems and persevere in solving them. | 5, 31, 38 |
| 2 Reason abstractly and quantitatively. | 19, 26, 29, 34, 35, 37 |
| 3 Construct viable arguments and critique the reasoning of others. | 30, 32, 33 |
| 4 Model with mathematics. | 6, 20, 25, 27, 28, 36 |

---

## Lesson 6-4 | Elimination Using Multiplication

### Levels of Complexity Chart

The levels of the exercises progress from 1 to 3 with Level 1 indicating the lowest level of complexity.

| Exercises | 7–20 | 21–26, 34–38 | 28–33 |
|---|---|---|---|
| Level 3 | | | ● |
| Level 2 | | ● | |
| Level 1 | ● | | |

### Follow-Up

Students have explored solving systems of linear equations by graphing and using algebra.
**Ask:**
What are the benefits of having different strategies for solving systems of equations? **Sample answer: You can use the strategy that is most efficient. For example, if an estimate of the solution is sufficient, graphing can be used. You can use algebraic methods to find exact solutions.**

### Additional Answers

**32.** $a = -2, b = 22$; Substitute 3 for $x$ and $a$ for $y$ in the first equation and then solve for $a$ to get $a = -2$. Then substitute 3 for $x$ and $-2$ for $y$ in the second equation and simplify to get $b = 22$.

**33.** Sample answer: It is more helpful to use substitution when one of the variables has a coefficient of 1 or if a coefficient can be reduced to 1 without turning other coefficients into fractions. Otherwise, elimination is more helpful because it will avoid the use of fractions when solving the system.

### Go Online!

**eSolutions Manual**
Create worksheets, answer keys, and solutions handouts for your assignments.

connectED.mcgraw-hill.com  365

## Lesson 6-4 | Elimination Using Multiplication

## Assess

**Crystal Ball** Have students write how they think what they learned today about using multiplication before elimination to solve a system of equations will help them in the next lesson on applying systems of linear equations.

## Preparing for Assessment

Exercises 34-38 require students to use the skills they will need on standardized assessments. Each exercise is dual-coded with content standards and mathematical practice standards.

### Dual Coding

| Items | Content Standards | MP Mathematical Practices |
|---|---|---|
| 34 | A.REI.5, A.REI.6 | 2, 4 |
| 35 | A.REI.5, A.REI.6 | 2, 7 |
| 36 | A.REI.5, A.REI.6 | 4, 7 |
| 37 | A.REI.5, A.REI.6 | 2, 4, 7 |
| 38 | A.REI.5, A.REI.6 | 1, 4 |

## Diagnose Student Errors

Survey student responses for each item. Class trends may indicate common errors and misconceptions.

**34.**
| A | CORRECT |
|---|---|
| B | Solved for the cost of an adult ticket |
| C | Substituted incorrectly |
| D | Did not multiply all terms of equations |

**35.**
| A | CORRECT |
|---|---|
| B | Substituted incorrectly |
| C | Reversed values for $x$ and $y$ |
| D | Applied the Distributive Property incorrectly |

## Go Online!

**Quizzes**
Students can use *Self-Check Quizzes* to check their understanding of this lesson.

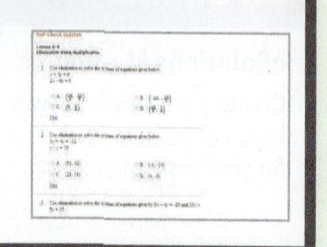

**36.**
| A | Number of Cal's double plays |
|---|---|
| B | Points for the single play $s$, instead of the double play $d$ |
| C | Sum of Cal's plays |
| D | CORRECT |

366 | Lesson 6-4 | Elimination Using Multiplication

---

## Preparing for Assessment

**34.** Seana and Mikayla sold tickets to the school play. Seana sold 3 adult tickets and 2 student tickets for $84. Mikayla sold 2 adult tickets and 5 student tickets for $100. How much did a student ticket cost? **MP** 2, 4  A.REI.5, A.REI.6  **A**

- ○ A  $12
- ○ B  $20
- ○ C  $36
- ○ D  $44

**35.** The perimeter of the first rectangle below is 120 inches. The perimeter of the second rectangle is 180 inches.

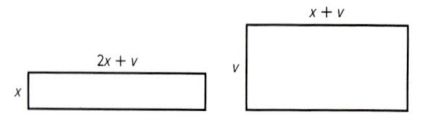

[Rectangle 1: $2x + v$ by $x$]  [Rectangle 2: $x + v$ by $v$]

What are the values of $x$ and $v$?  **MP** 2, 7  A.REI.5, A.REI.6  **A**

- ○ A  $x = 6; v = 42$
- ○ B  $x = 13; v = 42$
- ○ C  $x = 42; v = 6$
- ○ D  $x = 48; v = 28$

**36.** Nick and Cal are playing an electronic game. The table shows the number of each kind of play made by each player and each player's score at the end of the game. How many points does a player earn for a double play?  **MP** 4, 7  A.REI.5, A.REI.6  **D**

| Player | Single Plays | Double Plays | Total Points |
|---|---|---|---|
| Nick | 16 | 5 | 197 |
| Cal | 9 | 6 | 165 |

- ○ A  6 points
- ○ B  7 points
- ○ C  15 points
- ○ D  17 points

**37. FALL FUNDRAISER** A fundraiser by the football and track teams was held during the fall. Both teams sold boxes of cookies and candy bars, which were delivered in cartons. The total number of cartons of cookies and candy bars sold by each team is listed in the table below along with the total amount of money that was collected.  **MP** 2, 4, 7  A.REI.5, A.REI.6

| Team | Cartons of Cookies | Cartons of Candy Bars | Money Collected |
|---|---|---|---|
| Football | 9 | 12 | $1,845 |
| Track | 8 | 15 | $2,127.50 |

37a  $x$ = number of cartons of cookies; $y$ = number of cartons of candy bars
$9x + 12y = 1,845$
$8x + 15y = 2,127.50$

a. Define the variables and write a system of linear equations for this situation.

b. Solve the system of equations to determine the amount of money collected for each carton of cookies and candy bars sold.
cookies: $55/carton, candy: $112.50/carton

c. If a carton of cookies contained 10 boxes and a carton of candy bars contains 50 candy bars, what was the selling price for one box of cookies and for one candy bar?  $5.50, $2.25

**38. MULTI-STEP** Two cousins, Jay and Carolyn, planned to take flights and meet at a connecting city A and then fly together to a destination city B where they would meet their other cousin Diane for a vacation.  **MP** 1, 4  A.REI.5, A.REI.6

a. Jay's flight path followed the linear equation $y = 2x + 3$ and Carolyn's flight path followed the linear equation $y = -3x - 2$. Solve the system of equations to find the coordinates of city A.
$(-1, 1)$

b. Suppose Jay and Carolyn fly from city A to city B on a flight path of slope $-1$. What is the equation of Jay and Carolyn's flight path to city B?
$y = -x$

c. Suppose Diane is traveling from the north on a flight path following the linear equation $y = 2x + 24$. Find the coordinates of the destination city B where all three cousins would enjoy a vacation together.
$(-8, 8)$

---

## Differentiated Instruction  OL  BL

**Extension** Have students solve this system or a similar system using elimination.

$\frac{1}{2}x - \frac{2}{3}y = \frac{7}{3}$

$\frac{3}{2}x + 2y = -25$

$(-6, -8)$

# CHAPTER 6
## Mid-Chapter Quiz
Lessons 6-1 through 6-4

Use the graph to determine whether each system is *consistent* or *inconsistent* and whether it is *independent* or *dependent*. (Lesson 6-1)

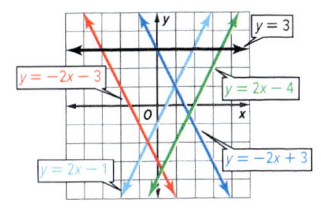

1. $y = 2x - 1$
   $y = -2x + 3$
   **consistent, independent**

2. $y = -2x + 3$
   $y = -2x - 3$
   **inconsistent**

Graph each system and determine the number of solutions that it has. If it has one solution, name it. (Lesson 6-1)

3. $y = 2x - 3$ **(7, 11)**
   $y = x + 4$

4. $x + y = 6$ **(5, 1)**
   $x - y = 4$

5. $x + y = 8$ **infinitely many solutions**
   $3x + 3y = 24$

6. $x - 4y = -6$ **(−10, −1)**
   $y = -1$

7. $3x + 2y = 12$ **no**
   $3x + 2y = 6$ **solutions**

8. $2x + y = -4$ **(−6, 8)**
   $5x + 3y = -6$

**3-8. See Ch. 6 Answer Appendix for graphs.**

Use substitution to solve each system of equations. (Lesson 6-2)

9. $y = x + 4$ **(4, 8)**
   $2x + y = 16$

10. $y = -2x - 3$ **(−12, 21)**
    $x + y = 9$

11. $x + y = 6$ **(7, −1)**
    $x - y = 8$

12. $y = -4x$ **(3, −12)**
    $6x - y = 30$

13. **PERSEVERANCE** The cost of two orders at a restaurant is shown in the table. (Lesson 6-2)

| Order | Total Cost |
|---|---|
| 3 tacos, 2 burritos | $23.35 |
| 4 tacos, 1 burrito | $19.80 |

a. Define variables to represent the cost of a taco and the cost of a burrito. **a–c. See margin.**

b. Write a system of equations to find the cost of a single taco and a single burrito.

c. Solve the systems of equations, and explain what the solution means.

d. How much would a customer pay for 2 tacos and 2 burritos? **$20.10**

14. **AMUSEMENT PARKS** The cost of two groups who recently visited an aquarium is shown in the table. (Lesson 6-3) **a–c. See Ch. 6 Answer Appendix.**

| Group | Total Cost |
|---|---|
| 4 adults, 2 children | $110 |
| 4 adults, 3 children | $123 |

a. Define variables to represent the cost of an adult ticket and the cost of a child ticket.

b. Write a system of equations to find the cost of an adult ticket and a child ticket.

c. Solve the system of equations, and explain what the solution means.

d. Which mathematical practice did you use to solve this problem? **See students' work.**

e. How much will a group of 3 adults and 5 children be charged for admission? **$128**

15. **MULTIPLE CHOICE** Angelina spent $16 for 12 pieces of candy to take to a meeting. Each chocolate bar costs $2, and each lollipop costs $1. Determine how many of each she bought. (Lesson 6-3) **B**

A  6 chocolate bars, 6 lollipops
B  4 chocolate bars, 8 lollipops
C  7 chocolate bars, 5 lollipops
D  3 chocolate bars, 9 lollipops

Use elimination to solve each system of equations. (Lessons 6-3 and 6-4)

16. $x + y = 9$ **(3, 6)**
    $x - y = -3$

17. $x + 3y = 11$ **(5, 2)**
    $x + 7y = 19$

18. $9x - 24y = -6$ **(2, 1)**
    $3x + 4y = 10$

19. $-5x + 2y = -11$ **(3, 2)**
    $5x - 7y = 1$

20. **MULTIPLE CHOICE** The Blue Mountain High School Drama Club is selling tickets to their spring musical. Adult tickets are $4 and student tickets are $1. A total of 285 tickets are sold for $765. How many of each type of ticket are sold? (Lesson 6-4) **D**

A  145 adult, 140 student
B  120 adult, 165 student
C  180 adult, 105 student
D  160 adult, 125 student

## Additional Answers

13a. Let $t =$ the cost of a taco and $b =$ the cost of a burrito.

13b. $3t + 2b = 23.35$
     $4t + b = 19.80$

13c. (3.25, 6.80); The cost of a single taco is $3.25 and the cost of a single burrito is $6.80.

# LESSON 6-5
# Applying Systems of Linear Equations

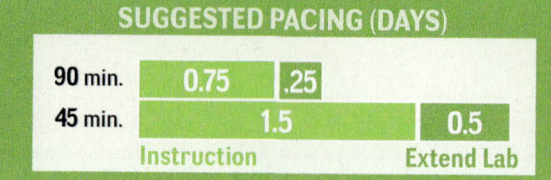

**SUGGESTED PACING (DAYS)**

| | | |
|---|---|---|
| 90 min. | 0.75 | .25 |
| 45 min. | 1.5 | 0.5 |
| Instruction | | Extend Lab |

## Track Your Progress

### Objectives

1. Determine the best method for solving systems of equations.
2. Apply systems of equations.

### Mathematical Background
Different methods work better for solving certain systems of equations. Graphing can be used when an estimate will do. Substitution is used when one of the variables in either equation has a coefficient of 1 or −1. Elimination using addition or subtraction can be used if the coefficients of one of the variables are the same or additive inverses. Elimination using multiplication can be used when none of the other methods will work.

### THEN
**F.IF.4** For a function that models a relationship between two quantities, interpret key features of graphs and tables in terms of the quantities, and sketch graphs showing key features given a verbal description of the relationship. *Key features include: intercepts; intervals where the function is increasing, decreasing, positive, or negative; relative maximums and minimums; symmetries; end behavior; and periodicity.*

**F.IF.7a** Graph linear and quadratic functions and show intercepts, maxima, and minima.

### NOW
**A.REI.6** Solve systems of linear equations exactly and approximately (e.g., with graphs), focusing on pairs of linear equations in two variables.

### NEXT
**A.REI.12** Graph the solutions to a linear inequality in two variables as a half-plane (excluding the boundary in the case of a strict inequality), and graph the solution set to a system of linear inequalities in two variables as the intersection of the corresponding half-planes.

## Go Online! All of these resources and more are available at connectED.mcgraw-hill.com

**eLessons** utilize the power of your interactive whiteboard in an engaging way. Use **Systems of Linear Equations**, screens 14–16, to introduce the concepts in this lesson.

*Use at Beginning of Lesson*

**Personal Tutors** (for every example) let students hear real teachers solve problems. Students can pause and repeat as many times as necessary.

*Use with Examples*

**Chapter Project** allows students to create and customize a project as a non-traditional method of assessment.

*Use at End of Lesson*

## Using Open Educational Resources

**Mnemonic Devices** Have students work individually to write a song or a poem using **Rhymes.net** to help them remember the ways of solving linear equations. Have students volunteer to perform their mnemonic device in front of the class. *Use as homework*

# Go Online!
connectED.mcgraw-hill.com

Worksheets

# Differentiate Your Resources

**Extra Practice** Additional practice or homework; Skills Practice is best for approaching-level students and Practice is best for on-level and beyond-level students

## Skills Practice
AL  OL  ELL

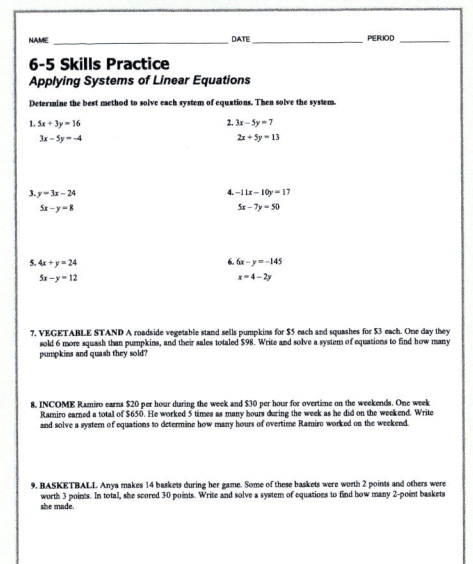

## Practice
AL  OL  BL  ELL

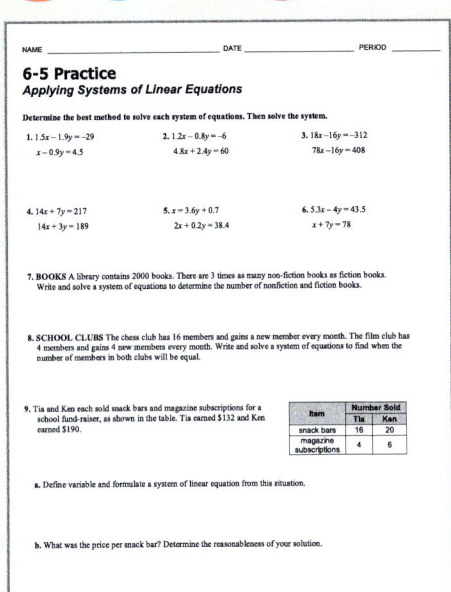

## Word Problem Practice
AL  OL  BL  ELL

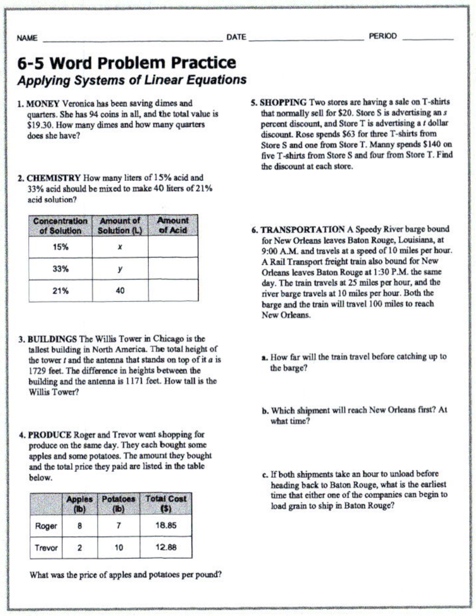

**Intervention** Reteaching and vocabulary activities that can be used with struggling or absent students and as ELL support

**Extension** Activities that can be used to extend lesson concepts

## Study Guide and Intervention
AL  OL  ELL

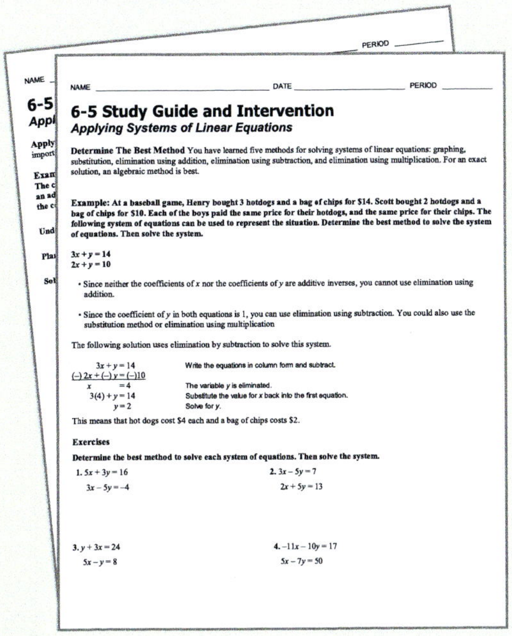

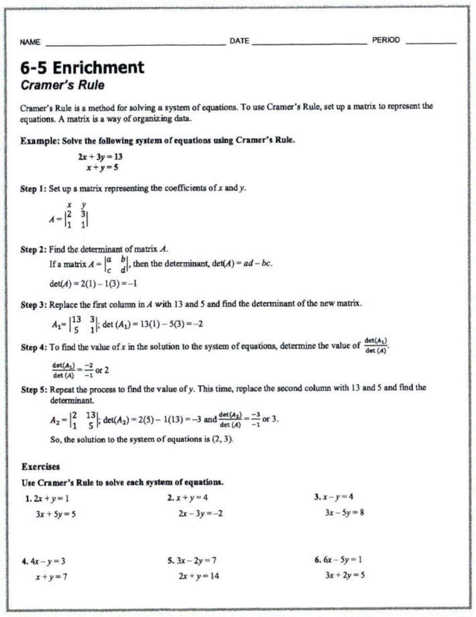

## Study Notebook
AL  OL  BL  ELL

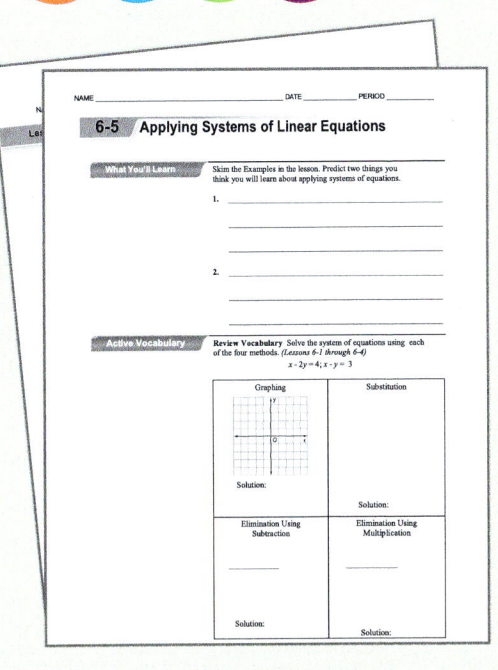

## Enrichment
OL  BL  ELL

connectED.mcgraw-hill.com  368B

Lesson 6-5 | Applying Systems of Linear Equations

# Launch

Have students read the Why? section of the lesson. Ask:

- What do the variables *x* and *y* represent in the problem? **The length of the official track is represented by *x* and the length of the short track by *y* in meters.**

- What method could you use to solve the system? **substitution or elimination**

**Go Online!**

**Interactive Whiteboard**
Use the *eLesson* or *Lesson Presentation* to present this lesson.

---

## LESSON 5
# Applying Systems of Linear Equations

**Then**
- You solved systems of equations by using substitution and elimination.

**Now**
1. Determine the best method for solving systems of equations.
2. Apply systems of equations.

**Why?**
- In speed skating, competitors race two at a time on a double track. Indoor speed skating rinks have two track sizes for race events: an official track and a short track.

| Speed Skating Tracks | |
|---|---|
| official track | x |
| short track | y |

The total length of the two tracks is 511 meters. The official track is 44 meters less than four times the short track. The total length is represented by $x + y = 511$. The length of the official track is represented by $x = 4y - 44$.

You can solve the system of equations to find the length of each track.

**Mathematical Practices**
2 Reason abstractly and quantitatively.
4 Model with mathematics.

**Content Standards**
**A.REI.6** Solve systems of linear equations exactly and approximately (e.g., with graphs), focusing on pairs of linear equations in two variables.

**1 Determine the Best Method** You have learned five methods for solving systems of linear equations. The table summarizes the methods and the types of systems for which each method works best.

**Concept Summary** Solving Systems of Equations

| Method | The Best Time to Use |
|---|---|
| Graphing | To estimate solutions, since graphing usually does not give an exact solution. |
| Substitution | If one of the variables in either equation has a coefficient of 1 or −1. |
| Elimination Using Addition | If one of the variables has opposite coefficients in the two equations. |
| Elimination Using Subtraction | If one of the variables has the same coefficient in the two equations. |
| Elimination Using Multiplication | If none of the coefficients are 1 or −1 and neither of the variables can be eliminated by simply adding or subtracting the equations. |

Substitution and elimination are algebraic methods for solving systems of equations. An algebraic method is best for an exact solution. Graphing, with or without technology, is a good way to estimate a solution.

A system of equations can be solved using each method. To determine the best approach, analyze the coefficients of each term in each equation.

---

**Mathematical Practices Strategies**

**Look for and make use of structure.**
Help students identify the structure within a system of equations that could help them choose a solution method. For example, ask:

- If one of the variables in either equation has a coefficient of 1 or −1, what method would you use? **substitution**

- If one of the variables has opposite coefficients in the two equations, would you use elimination using subtraction? Explain. **No, you would use elimination using addition.**

- If you only want to estimate the solution to a system of equations, what method would you use? **You would use graphing to estimate the solution to a system of equations.**

368 | Lesson 6-5 | Applying Systems of Linear Equations

# Lesson 6-5 | Applying Systems of Linear Equations

**A.REI.6**

### Example 1  Choose the Best Method

Determine the best method to solve the system of equations. Then solve the system.

$4x - 4y = 8$
$-8x + y = 19$

**Understand** To determine the best method to solve the system of equations, look closely at the coefficients of each term.

**Plan** Neither the coefficients of $x$ nor $y$ are the same or additive inverses, so you cannot add or subtract to eliminate a variable. Since the coefficient of $y$ in the second equation is 1, you can use substitution.

**Solve** First, solve the second equation for $y$.

| | |
|---|---|
| $-8x + y = 19$ | Second equation |
| $-8x + y + 8x = 19 + 8x$ | Add $8x$ to each side. |
| $y = 19 + 8x$ | Simplify. |

Next, substitute $19 + 8x$ for $y$ in the first equation.

| | |
|---|---|
| $4x - 4y = 8$ | First equation |
| $4x - 4(19 + 8x) = 8$ | Substitution |
| $4x - 76 - 32x = 8$ | Distributive Property |
| $-28x - 76 = 8$ | Simplify. |
| $-28x - 76 + 76 = 8 + 76$ | Add 76 to each side. |
| $-28x = 84$ | Simplify. |
| $\frac{-28x}{-28} = \frac{84}{-28}$ | Divide each side by $-28$. |
| $x = -3$ | Simplify. |

Last, substitute $-3$ for $x$ in the second equation.

| | |
|---|---|
| $-8x + y = 19$ | Second equation |
| $-8(-3) + y = 19$ | $x = -3$ |
| $y = -5$ | Simplify. |

The solution of the system of equations is $(-3, -5)$.

**Check** Use a graphing calculator to check your solution. If your algebraic solution is correct, then the graphs will intersect at $(-3, -5)$.

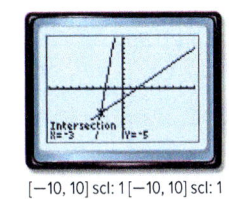

[−10, 10] scl: 1 [−10, 10] scl: 1

The system of equations can also be solved by using elimination with multiplication. You can multiply the first equation by 2 and then add to eliminate the $x$-term.

### Study Tip

**MP Reasoning** The system of equations in Example 1 can also be solved by using elimination with multiplication. You can multiply the first equation by 2 and then add to eliminate the $x$-term.

### Math History Link

**Leonardo Pisano** (1170–1250) Leonardo Pisano is better known by his nickname *Fibonacci*. His book introduced the Hindu-Arabic place-value decimal system. Systems of linear equations are studied in this work.

### Guided Practice

**1A.** $5x + 7y = 2$
$-2x + 7y = 9$

**1B.** $3x - 4y = -10$
$5x + 8y = -2$

**1C.** $5x - y = 17$
$3x + 2y = 5$

**1A.** elimination (−); (−1, 1)
**1B.** elimination (x); (−2, 1)
**1C.** substitution; (3, −2)

### MP Teaching the Mathematical Practice

**Reasoning** Mathematically proficient students know and flexibly use different properties of operations and objects. Encourage students to think of various ways to solve a system of equations before choosing the one that they think is best.

---

## Teach

Ask the scaffolded questions for each example to build conceptual understanding for students at all levels.

### 1 Determine the Best Method

**Example 1  Choose the Best Method**

**AL** What methods can be used to solve this system of equations? graphing, substitution, elimination by addition, elimination by subtraction, elimination by multiplication

**OL** Looking closely at the coefficients of each term, what do you think are the best two methods? Sample answer: substitution or elimination by multiplication Why did you choose those two methods? See students' work.

**BL** Solve the system of equations using a different method and explain which method you thought worked better for this example. See students' work.

**Need Another Example?**
Determine the best method to solve the system of equations. Then solve the system.
$2x + 3y = 23$
$4x + 2y = 34$
The best method is elimination using multiplication. The solution is (7, 3).

### Teaching Tip

**Problem Solving** Encourage students to take time to plan a problem-solving strategy before starting calculations. Advise students that taking this time may provide them with insight into the best strategy.

connectED.mcgraw-hill.com   369

**Lesson 6-5 | Applying Systems of Linear Equations**

## 2 Apply Systems of Linear Equations

**Example 2 Apply Systems of Linear Equations**

**AL** What is the *y*-intercept of each line?
blue (0, 20); red (0, 0)

**OL** How can Carter's equation begin at 20 hours?
Sample answer: The graph begins on Avery's first day. By the time Avery started practicing, Carter had already practiced for 20 hours.

**BL** What is another way to solve the system of equations other than determining the solution from looking at the graph? Sample answer: The system can be solved using substitution because both equations are already solved for *y*.

**Need Another Example?**

**Car Rental** The blue line represents the cost of renting a car from Ace Car Rental. The red line represents the cost of renting a car from Star Car Rental.

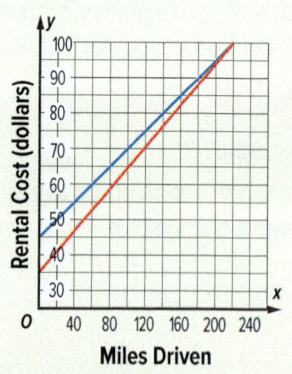

a. Write a system of linear equations based on the information in the graph. $y = 45 + 0.25x$ and $y = 35 + 0.30x$

b. Interpret the meaning of each equation. Ace has an initial charge of $45 and then charges $0.25 for each mile driven while Star Car has an initial charge of $35 and charges $0.30 for each mile driven.

c. Solve the system and describe its meaning in the context of the situation. The solution is (200, 95). This means that when the car has been driven 200 miles, the cost of renting a car will be the same ($95) at both rental companies.

---

**Study Tip**
**Substitution** You can check your solution by substituting 10 for *x* and 35 for *y* into the original equations.
$35 = 1.5(10) + 20$ ✓
$35 = 3.5(10)$ ✓

**Go Online!**
Highlighting information about each equation can be helpful as you solve problems involving systems. Log into your eStudent Edition to highlight important parts of your text. **ELL**

---

**2 Apply Systems of Linear Equations** When applying systems of linear equations to problems, it is important to analyze each solution in the context of the situation.

A.REI.6

**Example 2    Apply Systems of Linear Equations**

**PRACTICE** Carter has been practicing for a band concert. The blue line represents the total time he will spend practicing. Avery just joined the band today. The red line represents the time she will spend practicing for the concert.

a. Write a system of linear equations based on the information in the graph.
b. Interpret the meaning of each equation.
c. Solve the system and describe its meaning in the context of the situation.

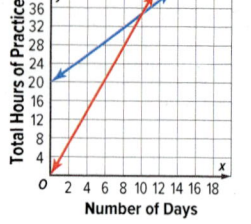

a. First, write an equation for the blue line.
Choose two points on the graph to find the slope of the line.

$m = \frac{y_2 - y_1}{x_2 - x_1}$   Slope-intercept formula

$m = \frac{32 - 20}{8 - 0}$   $(x_1, y_1) = (0, 20), (x_2, y_2) = (8, 32)$

$m = \frac{12}{8}$ or 1.5   Simplify.

The *y*-intercept is (0, 20), so $b = 20$.

The equation for the blue line is $y = 1.5x + 20$.

Now, write an equation for the red line.

$m = \frac{y_2 - y_1}{x_2 - x_1}$   Slope-intercept formula

$m = \frac{28 - 0}{8 - 0}$   $(x_1, y_1) = (0, 0), (x_2, y_2) = (8, 28)$

$m = \frac{28}{8}$ or 3.5   Simplify.

The *y*-intercept is (0, 0), so $b = 0$.

The equation for the red line is $y = 3.5x$.

b. Carter has already spent 20 hours practicing and he will spend an additional 1.5 hours each day practicing. Avery hasn't spent any time practicing yet, but she will spend 3.5 hours each day practicing.

c. The solution of this system of equations is the point where the two lines intersect, (10, 35). This means that after 10 days, Carter and Avery will have spent an equal amount of time practicing, 35 hours.

**Guided Practice**

2. **EMPLOYMENT** Jared has worked 70 hours this summer. He plans to work 2.5 hours in each of the coming days. Clementine just started working and plans to work 5 hours a day in the coming days. Write and solve a system of equations to find how long it will be before they will have worked the same number of hours. $y = 70 + 2.5x, y = 5x$, 28 days

---

**Differentiated Instruction** **AL** **OL** **ELL**

**IF** students have trouble writing the necessary equations for a system in a real-world situation,

**THEN** give them these steps to help them explore, plan, solve, and check.
- Determine the question.
- Describe the variables used for the unknowns.
- Translate the conditions in the problem into two equations.
- Solve the system by the best method.
- Analyze the solution in the context of the situation.

## Lesson 6-5 | Applying Systems of Linear Equations

### Check Your Understanding

= Step-by-Step Solutions begin on page R11.

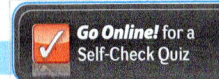

**Example 1**
A.REI.6

Determine the best method to solve each system of equations. Then solve the system.

1. $2x + 3y = -11$
   $-8x - 5y = 9$
   elim (×); (2, −5)

2. $3x + 4y = 11$
   $2x + y = -1$
   subst.; (−3, 5)

3. $3x - 4y = -5$
   $-3x + 2y = 3$
   elim (+); $(-\frac{1}{3}, 1)$

4. $3x + 7y = 4$
   $5x - 7y = -12$
   elim (+); (−1, 1)

**Example 2**
A.REI.6

5. **FINANCIAL LITERACY** The debate team is selling pizzas and subs. The blue line represents the total profit the debate team can earn and the red line represents the total number of items the debate team can sell.

   a. Write a system of linear equations based on the information in the graph. $y = -0.6x + 46.6, y = x - 11$
   
   b. Interpret the meaning of each equation. See margin.
   
   c. Solve the system and describe its meaning in the context of the situation.
   (36, 25); The debate team sold 36 subs and 25 pizzas.

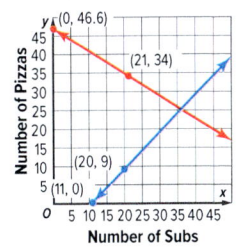

### Practice and Problem Solving

Extra Practice is on page R6.

**Example 1**
A.REI.6

Determine the best method to solve each system of equations. Then solve the system. 6–11. See margin.

6. $-3x + y = -3$
   $4x + 2y = 14$

7. $2x + 6y = -8$
   $x - 3y = 8$

8. $3x - 4y = -5$
   $-3x - 6y = -5$

9. $5x + 8y = 1$
   $-2x + 8y = -6$

10. $y + 4x = 3$
    $y = -4x - 1$

11. $-5x + 4y = 7$
    $-5x - 3y = -14$

**Example 2**
A.REI.6

12. Sample answer: $3x + 5y = 730$ and $x = y + 38$; the lacrosse team sold 115 individual tickets and 77 couples' tickets.

12. **FINANCIAL LITERACY** The lacrosse team is hosting a dance as a fundraiser. They sold 38 more individual tickets than couples' tickets and earned a total of $730. Write and solve a system of equations to represent this situation.

| Ticket Type | Selling Price |
|---|---|
| individual | $3 |
| couple | $5 |

13. **BAND** There are 40 students in band third period. The number of girls is 4 less than 3 times the number of boys. Write and solve a system of equations to find the number of girls and boys in third period band. $g + b = 40$ and $g = 3b - 4$; 29 girls, 11 boys

14. **CAVES** The Caverns of Sonora have two different tours: the Crystal Palace tour and the Horseshoe Lake tour. The total length of both tours is 3.25 miles. The Crystal Palace tour is a half-mile less than twice the length of the Horseshoe Lake tour. Determine the length of each tour. See margin.

15. **MODELING** The *break-even point* is the point at which income equals expenses. Ridgemont High School is paying $13,200 for the writing and research of their yearbook plus a printing fee of $25 per book. If they sell the books for $40 each, how many will they have to sell to break even? Explain. See margin.

Write a system of linear equations given each table. Then name the solution.

16.
| x | y₁ | y₂ |
|---|---|---|
| −6 | −10 | −12 |
| 0 | −6 | 4 |
| 6 | −2 | 20 |

$y = \frac{2}{3}x - 6$, $y = \frac{8}{3}x + 4$; $(-5, -\frac{28}{3})$

17.
| x | y₁ | y₂ |
|---|---|---|
| −4 | 11 | −7 |
| 4 | −5 | 1 |
| 8 | −13 | 5 |

$y = -2x + 3$, $y = x - 3$; (2, −1)

### Differentiated Homework Options

| Levels | AL Basic | OL Core | BL Advanced |
|---|---|---|---|
| Exercises | 6–17, 23, 24, 26–35 | 7–11 odd, 12–15, 17–24, 26–35 | 21–28, 29–35 (optional) |
| 2-Day Option | 7–17 odd, 29–35 | 6–17 | |
| | 6–16 even, 23, 24, 26–28 | 18–24, 26–35 | |

You can use ALEKS to provide additional remediation support with personalized instruction and practice.

### Additional Answers

5b. The profit from pizza sales is equal to $46.60 minus the profit from sub sales. The number of pizzas sold is equal to the number of subs sold minus 11.

6. subst.; (2, 3)

7. subst.; (2, −2)

8. elim (+); $(-\frac{1}{3}, 1)$

9. elim (−); $(1, -\frac{1}{2})$

10. subst.; no solution

11. elim (−); (1, 3)

14. Horseshoe Lake = 1.25 mi, Crystal Palace = 2 mi

15. 880 books; If they sell this number, then their income and expenses both equal $35,200.

---

## Practice

**Formative Assessment** Use Exercises 1–5 to assess students' understanding of the concepts in this lesson.

The Practice and Problem Solving exercises assess the content taught in the lesson. The Preparing for Assessment page is meant to be used as preparation for end-of-course assessments.

### Watch Out!

**Preventing Errors** Students frequently do not understand what their answers mean in the context of the problem. Stress that they should go back and read the problem again to make sure they have answered the question. It is also helpful to have them give their answers to a word problem in sentence form.

### MP Teaching the Mathematical Practices

**Modeling** Mathematically proficient students apply the mathematics they know to solve problems arising in everyday life. In Exercise 18, point out to students that because 400 paintballs is between 200 and 500, the cost of lunch and 400 paintballs should be between the cost for lunch and 200 paintballs and the cost for lunch and 500 paintballs.

### Extra Practice

See page R6 for extra exercises for students who are approaching level or for on-level students who need additional reinforcement.

### Levels of Complexity Chart

The levels of the exercises progress from 1 to 3, with Level 1 indicating the lowest level of complexity.

| Exercises | 6–17 | 18–20, 29–35 | 21–28 |
|---|---|---|---|
| Level 3 | | | ● |
| Level 2 | | ● | |
| Level 1 | ● | | |

connectED.mcgraw-hill.com  371

# Lesson 6-5 | Applying Systems of Linear Equations

## Teaching the Mathematical Practices

**Reasoning** Mathematically proficient students make sense of quantities and their relationships in problem situations. In Exercise 24, point out to students that a time of −1 makes sense if $x$ is hours since 12:00, but a negative number of hours riding a bike does not make sense.

## Additional Answers

**23.** Sample answer: $x + y = 12$ and $3x + 2y = 29$, where $x$ represents the cost of a student ticket for the basketball game and $y$ represents the cost of an adult ticket; substitution could be used to solve the system; (5, 7) means the cost of a student ticket is $5 and the cost of an adult ticket is $7.

**25.** Graphing: (2, 5);

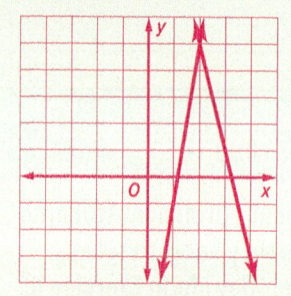

elimination by addition:
$$4x + y = 13$$
$$\underline{6x - y = 7}$$
$$10x = 20$$
$$x = 2$$
$$4(2) + y = 13$$
$$y = 5$$

substitution:
$$y = -4x + 13$$
$$6x - (-4x + 13) = 7$$
$$6x + 4x - 13 = 7$$
$$10x = 20$$
$$x = 2$$
$$4(2) + y = 13$$
$$y = 5$$

**26.** Sample answer: Would another method work better if one of the equations is in the form $y = mx + b$?

## Go Online!

**eSolutions Manual**
Create worksheets, answer keys, and solutions handouts for your assignments.

---

**18. PAINTBALL** Clara and her friends are planning a trip to a paintball park. Find the cost of lunch and the cost of each paintball. What would be the cost for 400 paintballs and lunch? **The cost of lunch is $10 and the cost of each paintball is $0.03. The cost of 400 paintballs and lunch would be $22.**

**Paintball in the Park**
- $25 for 500 paintballs
- $16 for 200 paintballs
- Lunch is included

**19a.** Let $x =$ the cost per pound of aluminum cans, and let $y =$ the cost per pound of newspaper; $9x + 26y = 3.77$ and $9x + 114y = 4.65$.

**19. RECYCLING** Mara and Ling each recycled aluminum cans and newspaper, as shown in the table. Mara earned $3.77, and Ling earned $4.65.

a. Define variables and write a system of linear equations from this situation.

b. What was the price per pound of aluminum? Determine the reasonableness of your solution. **$0.39; This solution is reasonable.**

| Materials | Pounds Recycled |  |
|---|---|---|
|  | Mara | Ling |
| aluminum cans | 9 | 9 |
| newspaper | 26 | 114 |

**20. BAKE SALE** The tennis team is having a bake sale. Cookies sell for $1 each, and cupcakes are $3 each. If Connie spends $15 for 7 items, how many cupcakes did she buy? **4**

**21. MUSIC** An online music club offers individual songs for one price or entire albums for another. Kendrick pays $14.90 to download 5 individual songs and 1 album. Geoffrey pays $21.75 to download 3 individual songs and 2 albums.

a. How much does the music club charge to download a song? **$1.15**

b. How much does the music club charge to download an entire album? **$9.15**

**22. CANOEING** Malik canoed against the current for 2 hours and then with the current for 1 hour before resting. Julio traveled against the current for 2.5 hours and then with the current for 1.5 hours before resting. If they traveled a total of 9.5 miles against the current, 20.5 miles with the current, and the current is 3 miles per hour, how fast do Malik and Julio travel in still water? **Malik: 4 mph; Julio: 6 mph**

A.REI.6

### H.O.T. Problems — Use Higher-Order Thinking Skills

**23. OPEN-ENDED** Formulate a system of equations that represents a situation in your school. Describe the method that you would use to solve the system. Then solve the system and explain what the solution means. **See margin.**

**24.** Sample answer: You should always check that the answer makes sense in the context of the original problem. If it does not, you may have made an incorrect calculation. If (−1, 7) was the solution, then it is probably incorrect, since time in this case cannot be a negative number. The solution should be recalculated.

**24. REASONING** In a system of equations, $x$ represents the time spent riding a bike, and $y$ represents the distance traveled. You determine the solution to be (−1, 7). Use this problem to discuss the importance of analyzing solutions in the context of real-world problems.

**25. CHALLENGE** Solve $4x + y = 13$ and $6x - y = 7$ by using three different methods. Show your work. **See margin.**

**26. WRITE A QUESTION** A classmate says that elimination is the best way to solve a system of equations. Write a question to challenge his conjecture. **See margin.**

**27. WHICH ONE DOESN'T BELONG?** Which system is different? Explain.

| $x - y = 3$ | $-x + y = 0$ | $y = x - 4$ | $y = x + 1$ |
|---|---|---|---|
| $x + \frac{1}{2}y = 1$ | $5x = 2y$ | $y = \frac{2}{x}$ | $y = 3x$ |

**27.** The third system; this system is the only one that is not a system of linear equations.

**28. WRITING IN MATH** How do you know what method to use when solving a system of equations? **See Ch. 6 Answer Appendix.**

## Standards for Mathematical Practice

| Emphasis On | Exercises |
|---|---|
| 1 Make sense of problems and persevere in solving them. | 5 |
| 2 Reason abstractly and quantitatively. | 24, 27, 29–32, 35 |
| 3 Construct viable arguments and critique the reasoning of others. | 26, 28 |
| 4 Model with mathematics. | 15, 18–22, 29–32, 34 |
| 6 Attend to precision. | 12, 14 |
| 7 Look for and make use of structure. | 13, 33 |

## Preparing for Assessment

**29.** Erin has dimes and nickels in her pocket. She has a total of 30 coins. The total value of the change in her pocket is $2.10. How many dimes does Erin have in her pocket? **MP** 2, 4  A.REI.6, A.CED.2

> 12

**30.** In right triangle $ABC$, angle $B$ measures $90°$. The measure of angle $A$ is 6 less than twice the measure of angle $C$. What is the measure of angle $A$? **MP** 2, 4
A.REI.6, A.CED.2  **B**

- A  $32°$
- B  $58°$
- C  $90°$
- D  $180°$

**31.** Terence and his father went on a driving trip. Terence drove 32 miles more than his father. Together, they drove 260 miles. How many miles did Terence drive? **MP** 2, 4  A.REI.6, A.CED.2  **C**

- A  82
- B  114
- C  146
- D  151

**32.** Kate bought 3 pounds of rice and 1 pound of beans for $4.50. Elise bought 4 pounds of rice and 2 pounds of beans for $7.00. Select all of the TRUE statements. **MP** 2, 4  A.REI.6, A.CED.2  **A, D, E, F**

- [ ] A  Beans cost more than rice.
- [ ] B  The cost of 1 pound of rice is $1.50.
- [ ] C  The cost of 1 pound of beans is $1.00.
- [ ] D  1 pound of rice and 1 pound beans would cost $2.50.
- [ ] E  The cost of 3 pounds of rice is $3.00.
- [ ] F  The equations used to solve the problem are $3x + y = 4.5$ and $4x + 2y = 7$.

**33.** For a graph of $x + y = 10$ and $y = 5x$, select all the statements that are TRUE for this system of equations. **MP** 7  A.REI.6  **C, D, G**

- [ ] A  Both linear equations have $y$-intercepts through the origin.
- [ ] B  $x$ and $y$ equal to 5 is one of the solutions.
- [ ] C  The solution is at $x$ equal to $\frac{5}{3}$.
- [ ] D  The solution is $(\frac{5}{3}, \frac{25}{3})$.
- [ ] E  The solution is $(5, 25)$.
- [ ] F  The solution is $(\frac{5}{3}, 25)$.
- [ ] G  The solution is the intersection point of the lines $x + y = 10$ and $y = 5x$.

**34. MULTI-STEP** There were 114 people that attended a particular seminar. There were twice as many women as men at the seminar. Use the variables $m$ for men and $w$ for women. **MP** 2, 4  A.REI.6, A.CED.2

  a. Write a system of equations to represent this situation.

  > $m + w = 114; 2m = w$

  b. Which equation shows the substitution method being used to solve the system of equations? **B**

  - A  $m + 2w = 114$
  - B  $m + 2m = 114$
  - C  $2m + w = 114$
  - D  $2m + 2w = 228$

  c. Solve the system of equations.  $m = 38; w = 76$

**35.** The sum of two numbers is 29. The difference of the same two numbers is 5. What are the numbers? **MP** 2  A.REI.6, A.CED.2

> 12, 17

---

**Lesson 6-5** | Applying Systems of Linear Equations

## Assess

**Yesterday's News**  Have students write how yesterday's concept of using elimination with multiplication to solve systems of equations helped them with today's concept of determining the best method for solving systems of equations.

## Preparing for Assessment

Exercises 29–35 require students to use the skills they will need on standardized assessments. Each exercise is dual-coded with content standards and mathematical practice standards.

| Dual Coding |||
|---|---|---|
| Items | Content Standards | **MP** Mathematical Practices |
| 29 | A.REI.6, A.CED.2 | 2, 4 |
| 30 | A.REI.6, A.CED.2 | 2, 4 |
| 31 | A.REI.6, A.CED.2 | 2, 4 |
| 32 | A.REI.6, A.CED.2 | 2, 4 |
| 33 | A.REI.6, A.CED.2 | 7 |
| 34 | A.REI.6, A.CED.2 | 2, 4 |
| 35 | A.REI.6, A.CED.2 | 2 |

## Diagnose Student Errors

Survey student responses for each item. Class trends may indicate common errors and misconceptions.

**30.**

| A | Made a sign error in the $y$-intercept |
|---|---|
| B | CORRECT |
| C | Interchanged the $x$- and $y$-intercepts |
| D | Identified the $y$-intercept as the $x$-intercept |

---

### Differentiated Instruction  OL  BL

**Extension**  Have students make up their own real-world problem that can be solved using a system of linear equations. This will help all students understand the concept of solving systems of linear equations.

**31.**

| A | Subtracted 32 from 114 instead of adding |
|---|---|
| B | Found the number of miles the father drove |
| C | CORRECT |
| D | Found the incorrect result of 238 when subtracting 32 from 260 |

## Go Online!

### Quizzes

Students can use *Self-Check Quizzes* to check their understanding of this lesson. You can also give *Quiz 3*, which covers the content in Lesson 6-5.

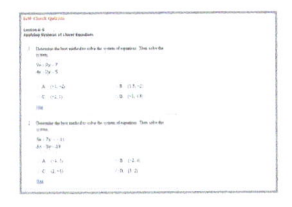

connectED.mcgraw-hill.com  373

Extend 6-5

# Launch

**Objective** Use matrices to solve systems of equations.

## Teaching Tips
- Remind students that multiplying a row does not involve multiplying every row in the matrix.
- Student may also perform row operations and check their solutions using a graphing calculator. Type [2nd] [MATRIX] to access the Matrix menu.

# Teach

**Working in Cooperative Groups** Divide the class into pairs. Work through Activity 1 as a class. Then ask students to work with their partners to complete Activity 2. **ELL**

**Practice** Have students complete Exercises 1–6.

---

# EXTEND 6-5
## Algebra Lab
# Using Matrices to Solve Systems of Equations

A **matrix** is a rectangular arrangement of numbers, called **elements**, in rows and columns enclosed in brackets. Usually named using an uppercase letter, a matrix can be described by its **dimensions** or by the number of rows and columns in the matrix. A matrix with $m$ rows and $n$ columns is an $m \times n$ matrix (read "$m$ by $n$").

$$A = \begin{bmatrix} 7 & -9 & 5 & 3 \\ -1 & 3 & -3 & 6 \\ 0 & -4 & 8 & 2 \end{bmatrix}$$ 3 rows

$A$ is a $3 \times 4$ matrix. 4 columns. The element 2 is in Row 3, Column 4.

**Mathematical Practices**
MP.7 Look for and make use of structure.

**Content Standards**
A.REI.6 Solve systems of linear equations exactly and approximately (e.g., with graphs), focusing on pairs of linear equations in two variables.

You can use an augmented matrix to solve a system of equations. An **augmented matrix** consists of the coefficients and the constant terms of a system of equations. Make sure that the coefficients of the $x$-terms are listed in one column, the coefficients of the $y$-terms are in another column, and the constant terms are in a third column. The coefficients and constant terms are usually separated by a dashed line.

| Linear System | Augmented Matrix |
|---|---|
| $x - 3y = 8$ <br> $-9x + 2y = -4$ | $\begin{bmatrix} 1 & -3 & \vdots & 8 \\ -9 & 2 & \vdots & -4 \end{bmatrix}$ |

### Activity 1  Write an Augmented Matrix

Work cooperatively. Write an augmented matrix for each system of equations.

**a.** $-2x + 7y = 11$
   $6x - 4y = 2$

Place the coefficients of the equations and the constant terms into a matrix.

$-2x + 7y = 11$
$6x - 4y = 2$ $\longrightarrow$ $\begin{bmatrix} -2 & 7 & \vdots & 11 \\ 6 & -4 & \vdots & 2 \end{bmatrix}$

**b.** $x - 2y = 5$
   $y = -4$

$x - 2y = 5$
$y = -4$ $\longrightarrow$ $\begin{bmatrix} 1 & -2 & \vdots & 5 \\ 0 & 1 & \vdots & -4 \end{bmatrix}$

You can solve a system of equations by using an augmented matrix. By performing row operations, you can change the form of the matrix. The operations are the same as the ones used when working with equations.

### Key Concept  Elementary Row Operations

The following operations can be performed on an augmented matrix.
- Interchange any two rows.
- Multiply all entries in a row by a nonzero constant.
- Replace one row with the sum of that row and a multiple of another row.

*(continued on the next page)*

# Extend 6-5

Row operations produce a matrix equivalent to the original system. **Row reduction** is the process of performing elementary row operations on an augmented matrix to solve a system. The goal is to get the coefficients portion of the matrix to have the form $\begin{bmatrix} 1 & 0 \\ 0 & 1 \end{bmatrix}$, which is called the **identity matrix**. The first row will give you the solution for $x$, because the coefficient of $y$ is 0. The second row will give you the solution for $y$, because the coefficient of $x$ is 0.

### Activity 2  Use Row Operations to Solve a System

Work cooperatively. Use an augmented matrix to solve the system of equations.
$-5x + 3y = 6$
$x - y = 4$

**Step 1** Write the augmented matrix: $\begin{bmatrix} -5 & 3 & | & 6 \\ 1 & -1 & | & 4 \end{bmatrix}$

**Step 2** Notice that the first element in the second row is 1. Interchange the rows so 1 can be in the upper left-hand corner.

$\begin{bmatrix} -5 & 3 & | & 6 \\ 1 & -1 & | & 4 \end{bmatrix}$ →Interchange $R_1$ and $R_2$→ $\begin{bmatrix} 1 & -1 & | & 4 \\ -5 & 3 & | & 6 \end{bmatrix}$

**Step 3** To make the first element in the second row a 0, multiply the first row by 5 and add the result to row 2.

$\begin{bmatrix} 1 & -1 & | & 4 \\ -5 & 3 & | & 6 \end{bmatrix}$ →$5R_1 + R_2$→ $\begin{bmatrix} 1 & -1 & | & 4 \\ 0 & -2 & | & 26 \end{bmatrix}$  $1(5) + (-5) = 0; -1(5) + 3 = -2;$ $4(5) + 6 = 26$

**Step 4** To make the second element in the second row a 1, multiply the second row by $-\frac{1}{2}$.

$\begin{bmatrix} 1 & -1 & | & 4 \\ 0 & -2 & | & 26 \end{bmatrix}$ →$-\frac{1}{2}R_2$→ $\begin{bmatrix} 1 & -1 & | & 4 \\ 0 & 1 & | & -13 \end{bmatrix}$  $0\left(-\frac{1}{2}\right) = 0; -2\left(-\frac{1}{2}\right) = 1;$ $26\left(-\frac{1}{2}\right) = -13$

**Step 5** To make the second element in the second row a 0, add the rows together.

$\begin{bmatrix} 1 & -1 & | & 4 \\ 0 & 1 & | & -13 \end{bmatrix}$ →$R_2 + R_1$→ $\begin{bmatrix} 1 & 0 & | & -9 \\ 0 & 1 & | & -13 \end{bmatrix}$  $1 + 0 = 1; -1 + 1 = 0;$ $4 + (-13) = -9$

The solution is $(-9, -13)$.

### Model and Analyze

Work cooperatively. Write an augmented matrix for each system of equations. Then solve the system. 1–6. See margin.

1. $x + y = -3$
   $x - y = 1$

2. $x - y = -2$
   $2x + 2y = 12$

3. $3x - 4y = -27$
   $x + 2y = 11$

4. $x + 4y = -6$
   $2x - 5y = 1$

5. $x - 3y = -2$
   $4x + y = 31$

6. $x + 2y = 3$
   $-3x + 3y = 27$

## Assess

**Formative Assessment** Use Exercises 1–6 to assess your students' knowledge of using matrices to solve systems of equations.

**From Concrete to Abstract** Ask students to summarize their favorite methods for using row operations on matrices to solve systems of equations. Have them explain when and why they prefer to use certain row operations.

## Teaching Tip

**Alternate Method** Row operations can be performed in different orders to arrive at the same result. In Activity 2, you could have started by multiplying the first row, $R_1$, by $-\frac{1}{5}$ instead of interchanging the rows.

## Follow-up

Students have explored matrices and matrix operations.

Ask:
- What are the advantages of using matrices to solve problems? **Sample answer: They provide a convenient way to organize data; they can be used to shorten notation.**

## Additional Answers

1. $\begin{bmatrix} 1 & 1 & | & -3 \\ 1 & -1 & | & 1 \end{bmatrix}$; $(-1, -2)$

2. $\begin{bmatrix} 1 & -1 & | & -2 \\ 2 & 2 & | & 12 \end{bmatrix}$; $(2, 4)$

3. $\begin{bmatrix} 3 & -4 & | & -27 \\ 1 & 2 & | & 11 \end{bmatrix}$; $(-1, 6)$

4. $\begin{bmatrix} 1 & 4 & | & -6 \\ 2 & -5 & | & 1 \end{bmatrix}$; $(-2, -1)$

5. $\begin{bmatrix} 1 & -3 & | & -2 \\ 4 & 1 & | & 31 \end{bmatrix}$; $(7, 3)$

6. $\begin{bmatrix} 1 & 2 & | & 3 \\ -3 & 3 & | & 27 \end{bmatrix}$; $(-5, 4)$

connectED.mcgraw-hill.com   375

# LESSON 6-6
# Systems of Inequalities

**SUGGESTED PACING (DAYS)**

| | | |
|---|---|---|
| 90 min. | 0.5 | 0.5 |
| 45 min. | 1 | 1 |
| | Instruction | Extend Lab |

## Track Your Progress

### Objectives
1. Solve systems of linear inequalities by graphing.
2. Apply systems of linear inequalities.

### Mathematical Background
A solution of a system of inequalities is the set of all points that satisfy both inequalities. To solve the system, graph each inequality and shade the region where the graphs overlap, or intersect. If the shaded regions have no points in common, then the lines are parallel and there is no solution. If the shaded regions overlap and extend on indefinitely then there are infinitely many solutions.

| THEN | NOW | NEXT |
|---|---|---|
| **A.REI.6** Solve systems of linear equations exactly and approximately (e.g., with graphs), focusing on pairs of linear equations in two variables. | **A.REI.12** Graph the solutions to a linear inequality in two variables as a half-plane (excluding the boundary in the case of a strict inequality), and graph the solution set to a system of linear inequalities in two variables as the intersection of the corresponding half-planes.<br><br>**A.CED.3** Represent constraints by equations of inequalities and by systems of equations and/or inequalities, and interpret solutions as viable or nonviable options in a modeling context. | **F.IF.7a** Graph linear and quadratic functions and show intercepts, maxima, and minima. |

## Go Online! All of these resources and more are available at connectED.mcgraw-hill.com

Use **The Geometer's Sketchpad** to solve systems of linear inequalities by graphing.

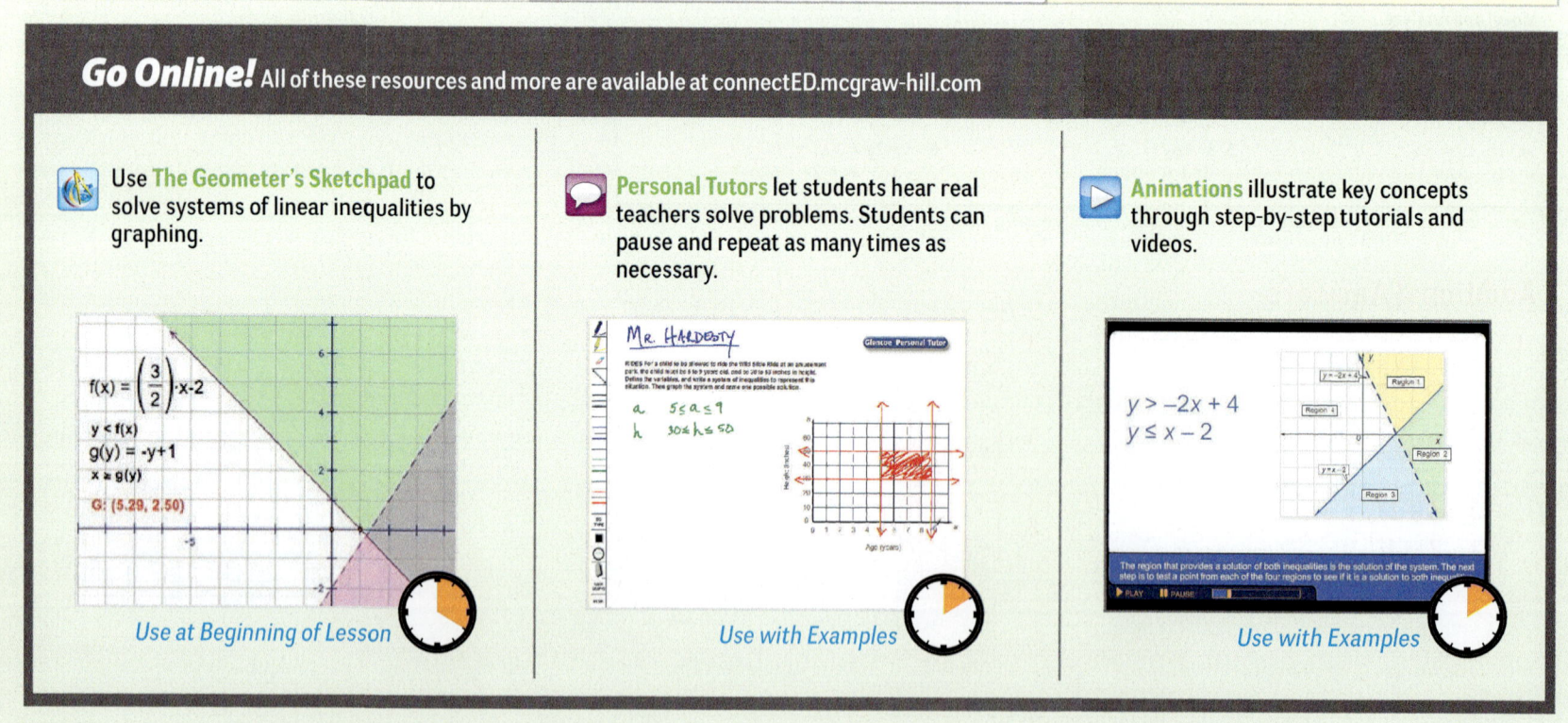

*Use at Beginning of Lesson*

**Personal Tutors** let students hear real teachers solve problems. Students can pause and repeat as many times as necessary.

*Use with Examples*

**Animations** illustrate key concepts through step-by-step tutorials and videos.

*Use with Examples*

## OER Using Open Educational Resources

**Extra Help** If students are still having trouble solving systems of linear inequalities, **Math Planet** is a site that allows them to see worked-out examples and check their understanding of the process.
*Use as homework or remediation*

376A | Lesson 6-6 | Systems of Inequalities

# Differentiate Your Resources

**Extra Practice** Additional practice or homework; Skills Practice is best for approaching-level students and Practice is best for on-level and beyond-level students

## Skills Practice
AL OL ELL

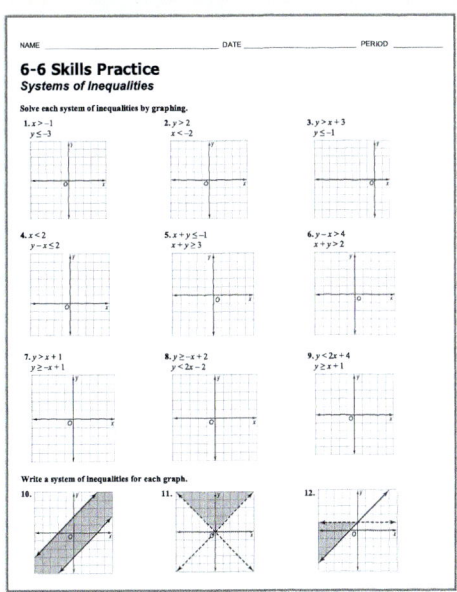

## Practice
AL OL BL ELL

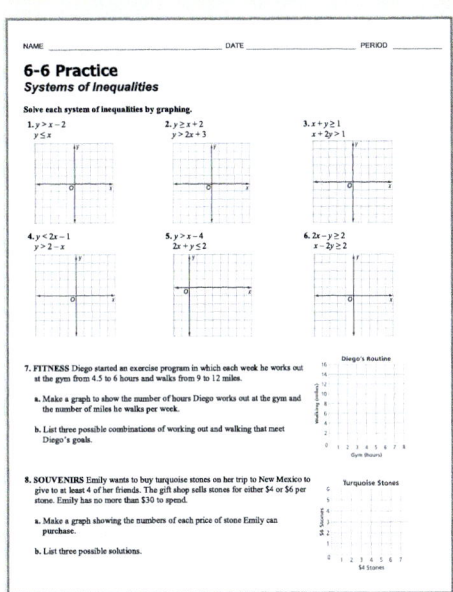

## Word Problem Practice
AL OL BL ELL

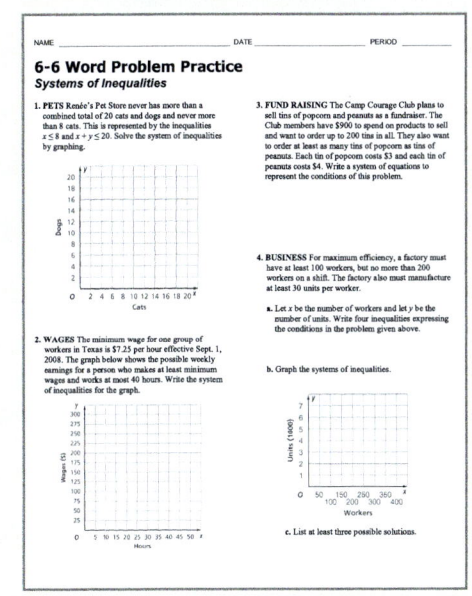

---

**Intervention** Reteaching and vocabulary activities that can be used with struggling or absent students and as ELL support

**Extension** Activities that can be used to extend lesson concepts

## Study Guide and Intervention
AL OL ELL

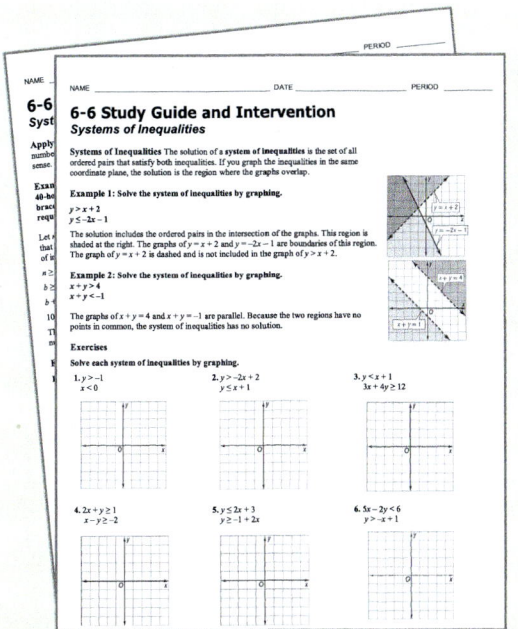

## Study Notebook
AL OL BL ELL

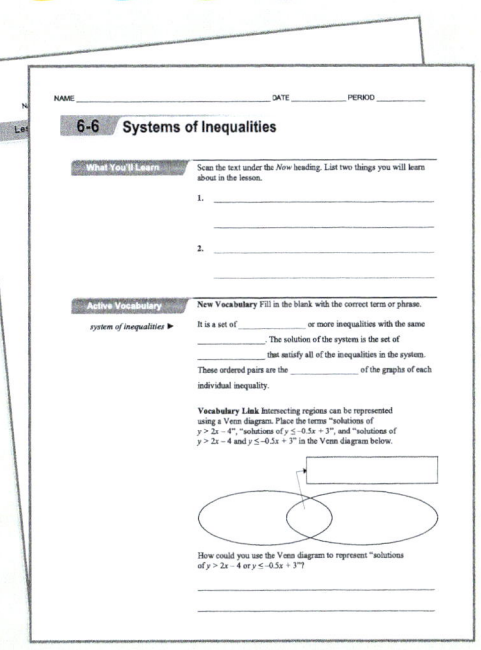

## Enrichment
OL BL ELL

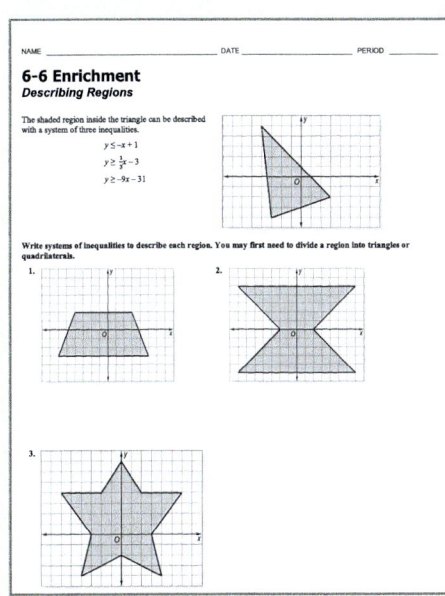

Lesson 6-6 | Systems of Inequalities

# Launch

Have students read the Why? section of the lesson. Ask:

- According to the graph, what are three possible heart rates in the preferred range for Jacui? **Sample answers: 110, 146, and 168**
- What are three heart rates that fall outside the range? **Sample answers: 100, 180, 195**
- What happens to the range of preferred heart rates as Jacui becomes older? **The lower and upper limits for the range decrease, and the overall range decreases.**

# Teach

Ask the scaffolded questions for each example to build conceptual understanding for students at all levels.

## 1 Systems of Inequalities

**Example 1 Solve by Graphing**

**AL** What kind of line represents the > symbol? **a dashed line** What kind of line represents the ≤ symbol? **a solid line**

**OL** Describe how to determine which side of the line to shade. **Choose a coordinate on the graph and plug it into the inequality. If the coordinate is a solution to the inequality, then the region where that coordinate lies should be shaded.**

**BL** How can you check to make sure you shaded the correct region of the graph for the solution to a system? **Sample answer: Choose a coordinate on the graph that is within the overlapping shaded region. Plug the coordinate into each of the inequalities. If the coordinate is a solution to both inequalities, then you know you shaded correctly.**

## Go Online!

**Interactive Whiteboard**
Use the *eLesson* or *Lesson Presentation* to present this lesson.

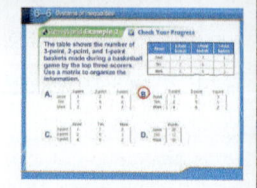

---

# LESSON 6
# Systems of Inequalities

**Then**
- You graphed and solved linear inequalities.

**Now**
1. Solve systems of linear inequalities by graphing.
2. Apply systems of linear inequalities.

**Why?**
Jacui is beginning an exercise program that involves an intense cardiovascular workout. Her trainer recommends that for a person her age, her heart rate should stay within the following range as she exercises.
- It should be higher than 102 beats per minute.
- It should not exceed 174 beats per minute.

The graph shows the maximum and minimum target heart rate for people ages 0 to 30 as they exercise. If the preferred range is in light green, how old do you think Jacui is?

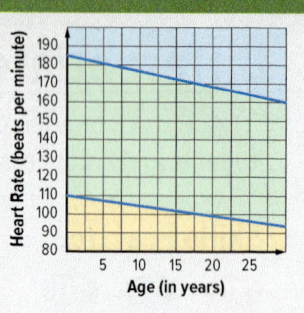

**New Vocabulary**
system of inequalities

**Mathematical Practices**
6 Attend to precision.
7 Look for and make use of structure.

**Content Standards**
A.REI.12 Graph the solutions to a linear inequality in two variables as a half-plane (excluding the boundary in the case of a strict inequality), and graph the solution set to a system of linear inequalities in two variables as the intersection of the corresponding half-planes.
A.CED.3 Represent constraints by equations or inequalities, and by systems of equations and/or inequalities, and interpret solutions as viable or nonviable options in a modeling context.

**1 Systems of Inequalities** The graph above is a graph of two inequalities. A set of two or more inequalities with the same variables is called a **system of inequalities**.

The solution of a system of inequalities with two variables is the set of ordered pairs that satisfy all of the inequalities in the system. The solution set is represented by the overlap, or intersection, of the graphs of the inequalities.

**Example 1**    Solve by Graphing

Solve the system of inequalities by graphing.

$y > -2x + 1$
$y \leq x + 3$

The graph of $y = -2x + 1$ is dashed and is not included in the graph of the solution. The graph of $y = x + 3$ is solid and is included in the graph of the solution.

The solution of the system is the set of ordered pairs in the intersection of the graphs of $y > -2x + 1$ and $y \leq x + 3$. This region is shaded in green.

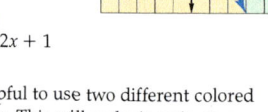

When graphing more than one region, it is helpful to use two different colored pencils or two different patterns for each region. This will make it easier to see where the regions intersect and find possible solutions.

▶ **Guided Practice**    1A–1D. See Ch. 6 Answer Appendix.

1A. $y \leq 3$
     $x + y \geq 1$

1B. $2x + y \geq 2$
     $2x + y < 4$

1C. $y \geq -4$
     $3x + y \leq 2$

1D. $x + y > 2$
     $-4x + 2y < 8$

Sometimes the regions never intersect. When this happens, there is no solution because there are no points in common.

---

**MP** **Mathematical Practices Strategies**

**Look for and express regularity in repeated reasoning.**
Help students identify the reasoning used to solve systems of inequalities. For example, ask:

- Is $y = 5$ a horizontal line or a vertical line? **horizontal line**
- Is the boundary line for the graph of $y \leq 5$ a dashed or solid line? **solid line**
- How can you rewrite $x + y = 3$ so that it is in slope intercept form? **$y = 3 - x$**
- Is the origin in the solution of $y \geq 3 - x$? **no**

376 | Lesson 6-6 | Systems of Inequalities

**Lesson 6-6 | Systems of Inequalities**

**Study Tip**

**Reasoning** A system of equations represented by parallel lines does not have a solution. However, a system of inequalities with parallel boundaries can have a solution. For example:

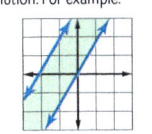

**Example 2** No Solution

A.REI.12

Solve the system of inequalities by graphing.
$3x - y \geq 2$
$3x - y < -5$

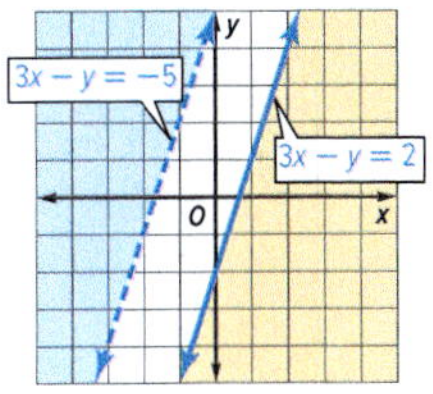

The graphs of $3x - y = 2$ and $3x - y = -5$ are parallel lines. The two regions do not intersect at any point, so the system has no solution.

**Guided Practice** 2A–2B. See Ch. 6 Answer Appendix.

2A. $y > 3$
$\phantom{2A.\ }y < 1$

2B. $x + 6y \leq 2$
$\phantom{2B.\ }y \geq -\frac{1}{6}x + 7$

**2 Apply Systems of Inequalities** When using a system of inequalities to describe constraints on the possible combinations in a real-world problem, sometimes only whole-number solutions will make sense.

A.CED.3, A.REI.12

**Real-World Example 3** Whole-Number Solutions

**ELECTIONS** Monifa is running for student council. The election rules say that for the election to be valid, at least 80% of the 900 students must vote. Monifa knows that she needs more than 330 votes to win.

**a.** Define the variables, and write a system of inequalities to represent this situation. Then graph the system.

Let $r$ = the number of votes required by the election rules; 80% of 900 students is 720 students. So $r \geq 720$.

Let $v$ = the number of votes that Monifa needs to win. So $v > 330$.

There are 900 students, so $r \leq 900$ and $v \leq 900$. The system of inequalities is $720 \leq r \leq 900$ and $330 < v \leq 900$.

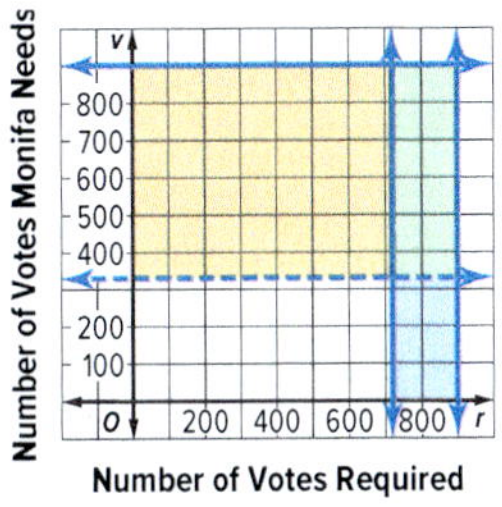

**Real-World Link**
Student government might be a good activity for you if you like to bring about change, plan events, and work with others.

**b.** Name one viable option.

Only whole-number solutions make sense in this problem. One possible solution is (800, 400); 800 students voted and Monifa received 400 votes.

**Guided Practice**

3. **FUNDRAISING** The Theater Club is selling shirts. They have only enough supplies to print 120 shirts. They will sell sweatshirts for $22 and T-shirts for $15, with a goal of at least $2000 in sales.

   **A.** Define the variables, and write a system of inequalities to represent this situation. 3A–3B. See Ch. 6 Answer Appendix.

   **B.** Then graph the system.

   **C.** Name one possible solution. Sample answer: 95 sweatshirts and 10 T-shirts

   **D.** Is (45, 30) a solution? Explain. No, the point does not fall in the overlapping region.

---

**Need Another Example?**

Solve the system of inequalities by graphing.
$y < 2x + 2$
$y \geq -x - 3$

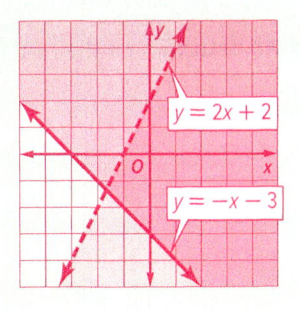

**Example 2** No Solution

**AL** What kind of line does the $\geq$ symbol represent in the first equation? a solid line What kind of line does the $<$ symbol represent in the second equation? a dashed line

**OL** What is the first step before graphing each inequality? solve each inequality for $y$

**BL** Without graphing the inequalities, how can you tell that there is no solution? Sample answer: Both inequalities contain $3x - y$, with one being greater than or equal to 2 and the other being less than $-5$, which is also less than 2. The lines are parallel and not shaded in between; therefore, there is no solution.

**Need Another Example?**

Solve the system of inequalities by graphing.
$y \geq -3x + 1$
$y \leq -3x - 2$  ∅

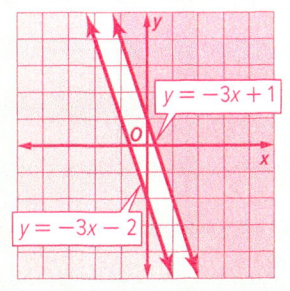

---

**Differentiated Instruction** AL OL

**IF** Students have difficulty graphing systems of inequalities,

**THEN** suggest that they graph each inequality on a separate coordinate graph and then put the two graphs together on the same coordinate graph by copying them over or tracing them.

---

**Go Online!**

The most up-to-date resources available for your program can be found at **connectED.mcgraw-hill.com**.

connectED.mcgraw-hill.com 377

# Lesson 6-6 | Systems of Inequalities

## 2 Apply Systems of Inequalities

### Example 3 Whole-Number Solutions

**AL** How can we determine if the function is proportional? Determine if the quantities are in a constant ratio.

**OL** Why are only whole number solutions included? Sample answer: Because the problem is talking about the number of people voting, the solution set cannot include a partial person.

**BL** Why does the graph have a solid vertical line at $r = 900$ and solid horizontal line at $v = 900$? Sample answer: The maximum number of votes is 900 because there are 900 students. Therefore, the shaded region cannot be above 900. A solid line is used to show that 900 is included in the possible solutions.

### Need Another Example?

**Service** A college service organization requires that its members maintain at least a 3.0 grade point average and volunteer at least 10 hours a week.

a. Define the variables and write a system of inequalities to represent this situation. Then graph the system. Let $g$ = grade point average; let $v$ = the number of volunteer hours; $g \geq 3.0$, $v \geq 10$.

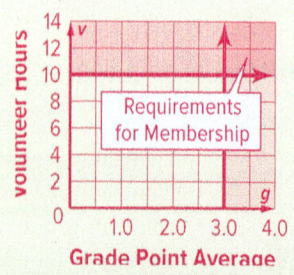

[Graph: Requirements for Membership, Volunteer hours vs. Grade Point Average]

b. Name one possible solution. (3.5, 12); a grade point average of 3.5 and 12 hours of volunteering

## Practice

**Formative Assessment** Use Exercises 1–9 to assess students' understanding of the concepts in this lesson.

The Practice and Problem Solving exercises assess the content taught in the lesson. The Preparing for Assessment page is meant to be used as preparation for end-of-course assessments.

## Extra Practice

See page R6 for extra exercises for students who are approaching level or for on-level students who need additional reinforcement.

378 | Lesson 6-6 | Systems of Inequalities

---

**Check Your Understanding** ○ = Step-by-Step Solutions begin on page R11.

**Go Online!** for a Self-Check Quiz

**Examples 1–2** Solve each system of inequalities by graphing. 1–8. See Ch. 6 Answer Appendix.
A.REI.12

1. $x \geq 4$
   $y \leq x - 3$

2. $y > -2$
   $y \leq x + 9$

3. $y < 3x + 8$
   $y \geq 4x$

4. $3x - y \geq -1$
   $2x + y \geq 5$

5. $y \leq 2x - 7$
   $y \geq 2x + 7$

6. $y > -2x + 5$
   $y \geq -2x + 10$

7. $2x + y \leq 5$
   $2x + y \leq 7$

8. $5x - y < -2$
   $5x - y > 6$

**Example 3** 9. **AUTO RACING** At a racecar driving school there are safety requirements.
A.CED.3,
A.REI.12

a. Define the variables, and write a system of inequalities to represent the height and weight requirements in this situation. Then graph the system. See Ch. 6 Answer Appendix.

b. Name one possible solution. **9b.** Sample answer: 72 in. and 220 lb

c. Is (50, 180) a solution? Explain. Yes, the point falls in the overlapping region.

[FAST DRIVING SCHOOL — RULES TO QUALIFY: 18 years of age or older, Good physical condition, Under 6 ft 7 in. tall, Under 295 lb]

### Practice and Problem Solving

Extra Practice is on page R6.

**Examples 1–2** Solve each system of inequalities by graphing. 10–24. See Ch. 6 Answer Appendix.
A.REI.12

10. $y < 6$
    $y > x + 3$

11. $y \geq 0$
    $y \leq x - 5$

12. $y \leq x + 10$
    $y > 6x + 2$

13. $y < 5x - 2$
    $y > -6x + 2$

14. $2x - y \leq 6$
    $x - y \geq -1$

15. $3x - y > -5$
    $5x - y < 9$

16. $y \geq x + 10$
    $y \leq x - 3$

17. $y < 5x - 5$
    $y > 5x + 9$

18. $y \geq 3x - 5$
    $3x - y > -4$

19. $4x + y > -1$
    $y < -4x + 1$

20. $3x - y \geq -2$
    $y < 3x + 4$

21. $y > 2x - 3$
    $2x - y \geq 1$

22. $5x - y < -6$
    $3x - y \geq 4$

23. $x - y \leq 8$
    $y < 3x$

24. $4x + y < -2$
    $y > -4x$

**Example 3** 25. **ICE RINKS** Ice resurfacers are used for rinks of at least 1000 square feet and up to 17,000 square feet. The price ranges from as little as $10,000 to as much as $150,000.
A.CED.3,
A.REI.12

a. Define the variables, and write a system of inequalities to represent this situation. Then graph the system. See Ch. 6 Answer Appendix.

b. Name one possible solution. **25b.** Sample answer: an ice resurfacer for a rink of 5000 ft² and a price of $20,000

c. Is (15,000, 30,000) a solution? Explain. Yes; the point satisfies each inequality.

26. **MP MODELING** Josefina works between 10 and 30 hours per week at a pizzeria. She earns $8.50 an hour, but can earn tips when she delivers pizzas.

a. Write a system of inequalities to represent the dollars $d$ she could earn for working $h$ hours in a week. $d \geq 8.50h$, $10 \leq h \leq 30$

b. Graph this system. See Ch. 6 Answer Appendix.

c. Josefina earned $195.50 last week. What range of hours could she have worked? $10 \leq h \leq 23$

### Differentiated Homework Options

| Levels | AL Basic | OL Core | BL Advanced |
|---|---|---|---|
| Exercises | 10–26, 39–48 | 11–25 odd, 26, 27–35 odd, 36, 37, 39–48 | 37–43, 44–48 (optional) |
| 2-Day Option | 11–25 odd, 44–48 | 10–26 | |
| | 10–26 even, 39–43 | 27–37, 39–48 | |

You can use ALEKS to provide additional remediation support with personalized instruction and practice.

Lesson 6-6 | Systems of Inequalities

**B** Solve each system of inequalities by graphing. **27–35. See Ch. 6 Answer Appendix.**

27. $x + y \geq 1$
    $x + y \leq 2$

28. $3x - y < -2$
    $3x - y < 1$

29. $2x - y \leq -11$
    $3x - y \geq 12$

30. $y < 4x + 13$
    $4x - y \geq 1$

31. $4x - y < -3$
    $y \geq 4x - 6$

32. $y \leq 2x + 7$
    $y < 2x - 3$

33. $y > -12x + 1$
    $y \leq 9x + 2$

34. $2y \geq x$
    $x - 3y > -6$

35. $x - 5y > -15$
    $5y \geq x - 5$

36. **CLASS PROJECT** An economics class formed a company to sell school supplies. They would like to sell at least 20 notebooks and 50 pens per week, with a goal of earning at least $150 per week. **a–b. See Ch. 6 Answer Appendix.**

    *School Supplies*
    Notebooks..........$2.50
    Pens..................$1.25

    a. Define the variables, and write a system of inequalities to represent this situation.
    b. Graph the system.
    c. Name one possible solution.
       **Sample answer: 25 notebooks and 100 pens**

**C** 37. **FINANCIAL LITERACY** Opal makes $15 per hour working for a photographer. She also coaches a competitive soccer team for $10 per hour. Opal needs to earn at least $90 per week, but she does not want to work more than 20 hours per week.

    a. Define the variables, and write a system of inequalities to represent this situation.
       **37a. Let $x$ = the hours worked for the photographer, let $y$ = the hours coaching, $x + y \leq 20$, $15x + 10y \geq 90$.**
    b. Graph this system.
       **See Ch. 6 Answer Appendix.**
    c. Give two possible solutions to describe how Opal can meet her goals.
       **37c. Sample answer: 6 hours at the photographer, 10 hours of coaching; 8 hours at the photographer, 10 hours of coaching**
    d. Is (2, 2) a solution? Explain.
       **37d. No; the point does not fall in the shaded region. She would not earn enough money.**

    A.CED.3, A.REI.12

**H.O.T. Problems** Use Higher-Order Thinking Skills

38. **CHALLENGE** Create a system of inequalities equivalent to $|x| \leq 4$. **$x \leq 4, x \geq -4$**

39. **REASONING** State whether the following statement is *sometimes*, *always*, or *never* true. Explain your answer with an example or counterexample. **See margin.**
    *Systems of inequalities with parallel boundaries have no solutions.*

40. **REASONING** Describe the graph of the solution of this system without graphing.
    $6x - 3y \leq -5$
    $6x - 3y \geq -5$  **It is the line $6x - 3y = -5$.**

41. **OPEN-ENDED** One inequality in a system is $3x - y > 4$. Write a second inequality so that the system will have no solution. **Sample answer: $3x - y < -4$**

42. **PRECISION** Graph the system of inequalities. Estimate the area of the solution.
    $y \geq 1$     **9 units²; See margin for graph.**
    $y \leq x + 4$
    $y \leq -x + 4$

43. **WRITING IN MATH** Refer to the beginning of the lesson. Explain what each colored region of the graph represents. Explain how shading in various colors can help to clearly show the solution set of a system of inequalities. **See margin.**

## Standards for Mathematical Practice

| Emphasis On | Exercises |
| --- | --- |
| 1 Make sense of problems and persevere in solving them. | 38 |
| 2 Reason abstractly and quantitatively. | 37, 39, 40 |
| 3 Construct viable arguments and critique the reasoning of others. | 43 |
| 4 Model with mathematics. | 9, 25, 26, 36, 44, 48 |
| 6 Attend to precision. | 42 |
| 7 Look for and make use of structure. | 41, 45–47 |

## Levels of Complexity Chart

The levels of the exercises progress from 1 to 3 with Level 1 indicating the lowest level of complexity.

| Exercises | 10–26 | 27–36, 44–48 | 37–43 |
| --- | --- | --- | --- |
| Level 3 | | | ● |
| Level 2 | | ● | |
| Level 1 | ● | | |

## Teaching the Mathematical Practices

**Modeling** Mathematically proficient students apply mathematics to problems arising in the workplace. In Exercise 26, point out to students that they can first consider just the wages and then add the tips.

**Precision** Mathematically proficient students calculate accurately and efficiently. In Exercise 42, stress that a precise graph will allow for a more accurate estimate.

## Additional Answers

39. Sometimes; sample answer: $y > 3$, $y < -3$ will have no solution, but $y > -3$, $y < 3$ will have solutions.

42.

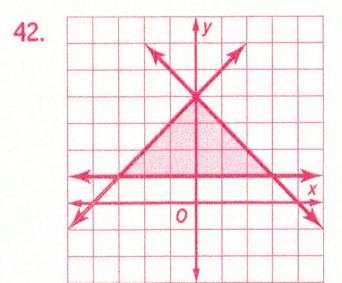

43. Sample answer: The yellow region represents the beats per minute below the target heart rate. The blue region represents the beats per minute above the target heart rate. The green region represents the beats per minute within the target heart rate. Shading in different colors clearly shows the overlapping solution set of the system of inequalities.

## Go Online!   eBook

**Interactive Student Guide**

Use the *Interactive Student Guide* to deepen conceptual understanding.
· Solving Systems of Inequalities

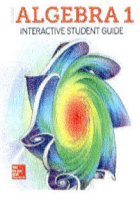

Lesson 6-6 | Systems of Inequalities

# Assess

**Name the Math** Have each student write the method for determining whether to shade above or below the line when graphing an inequality and how to determine the common solutions when graphing a system of inequalities.

# Preparing for Assessment

Exercises 44–48 require students to use the skills they will need on standardized assessments. Each exercise is dual-coded with content standards and mathematical practice standards.

| Dual Coding ||| 
|---|---|---|
| Items | Content Standards | MP Mathematical Practices |
| 44 | A.REI.12, A.CED.3 | 4 |
| 45 | A.REI.12 | 7 |
| 46 | A.REI.12 | 7 |
| 47 | A.REI.12 | 7 |
| 48 | A.REI.12, A.CED.3 | 2, 4 |

## Diagnose Student Errors

Survey student responses for each item. Class trends may indicate common errors and misconceptions.

**44.**

| | |
|---|---|
| A | Student may not realize from the graph that the triangular region is the solution area |
| B | Student may not realize from the graph that the triangular region is the solution area |
| C | CORRECT |
| D | Students should plug $x = 18$ and $y = 9$ into all the inequalities to see if the inequalities are all true, rather than looking at the graph since it is not obvious from the graph |

**45.**

| | |
|---|---|
| A | CORRECT |
| B | Used incorrect sign in first equation |
| C | Used incorrect sign in second equation |
| D | Used $-x$ in second equation |

## Go Online!

**Quizzes**
Students can use *Self-Check Quizzes* to check their understanding of this lesson. You can also give *Quiz 4*, which covers the content in Lesson 6-6.

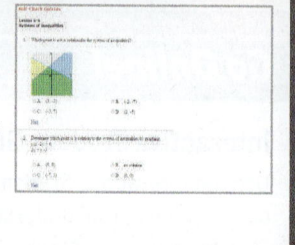

# Preparing for Assessment

**44.** Fernando wants to start an exercise regimen where each workout will include swimming and walking on a treadmill. He is going to spend at least 15 minutes swimming and at least 5 minutes walking on the treadmill. Each workout will be less than 25 minutes. He graphs a system of inequalities as shown to represent this situation, where $x$ represents the time spent swimming and $y$ represents the time spent walking. Which of the solutions could describe a workout, where $x$ is the number of minutes swimming and $y$ is the number of minutes walking? MP 4 A.REI.12, A.CED.3  **C**

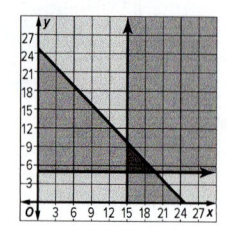

- A (11, 7)
- B (13, 10)
- C (15, 8)
- D (18, 9)

**45.** Which system of inequalities is shown on the graph? MP 7 A.REI.12  **A**

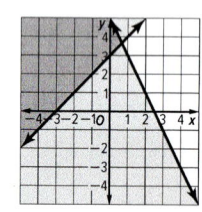

- A $2x + y \le 5$
  $-x + y \ge 3$
- B $2x + y \ge 5$
  $-x + y \ge 3$
- C $2x + y \ge 5$
  $-x + y \le 3$
- D $2x + y \le 5$
  $-x + y \le 3$

**46.** A system of inequalities is shown below. MP 7 A.REI.12

$y \le x + 5$
$y \le -3x + 3$
$y \ge 0$

What is the area, in square units, of the triangular region described by the system?  **13.5**

**47.** A system of inequalities is shown below. MP 7 A.REI.12

$-6x + 2y \ge 5$
$y < 3x + 2$

Which of the following is not a correct description of the graph of the system?  **C**

- A The graphs are parallel.
- B The point $(-3, 6)$ is included in the graph $-6x + 2y > 5$.
- C The solution to the system is the area between the two lines.
- D The point $(2, 5)$ is included in the graph $y < 3x + 2$.

**48. MULTI-STEP** A contractor was given constraints before creating a rectangular playground area at a city park. She was told that the width of one side of the fencing had to be at least 100 yards and the total fencing around the playground had to be no more than 350 yards. MP 2, 4 A.REI.12, A.CED.3

a. Define variables, and write a system of inequalities to represent this situation.
b. Select all possible solutions.  **A, D, E, F, G**
  - ☐ A Fence with length of 70 yards and width of 100 yards
  - ☐ B Fence with length of 75 yards and width of 105 yards
  - ☐ C Fence with length of 80 yards and width of 100 yards
  - ☐ D Fence with length of 85 yards and width of 85 yards
  - ☐ E (25, 150)
  - ☐ F (50, 125)
  - ☐ G (75, 100)
c. Is $(-50, 225)$ a solution? Explain.

48a. $x$ is length of rectangular playground
$y$ is width of rectangular playground
$y \ge 100$
$2x + 2y \le 350$
$x \ge 0$

48c. No. Even though it satisfies $y \ge 100$ and $2x + 2y \le 350$, $x$ must be greater than zero, since it is a length.

## Differentiated Instruction  OL  BL

**Extension** Have students graph $2 \le x \le 5$ and $1 \le y \le 4$ on the same coordinate plane. Have them describe what polygon is formed by their intersection. **a square with vertices at (2, 4), (5, 4), (2, 1) and (5, 1)**

**46.** Students should first graph the system and then determine the base and height to find the triangular area.

**47.**

| | |
|---|---|
| A | Found incorrect slopes |
| B | Shaded incorrectly |
| C | CORRECT |
| D | Shaded incorrectly |

**48.**
a. Sometimes it's hard for students to identify variables. Have students reread the problem and start with writing an expression for "the width of one side of the fencing had to be at least 100 yards."
b. Students must use the first number for $x$ and the second number for $y$ in each of the three inequalities. Students must understand that all three inequalities must be true for the point to be a solution.
c. Students must use the first number for $x$ and the second number for $y$ in each of the three inequalities. Students must understand that all three inequalities must be true for the point to be a solution. Also, negative numbers cannot be used for length.

**Extend 6-6**

# EXTEND 6-6
## Graphing Technology Lab
## Systems of Inequalities

You can use TI-Nspire technology to explore systems of inequalities. To prepare your calculator, add a new Graphs page from the Home screen.

**Mathematical Practices**
MP 5 Use appropriate tools strategically.

**Content Standards**
A.REI.12 Graph the solutions to a linear inequality in two variables as a half-plane (excluding the boundary in the case of a strict inequality), and graph the solution set to a system of linear inequalities in two variables as the intersection of the corresponding half-planes.

### Activity 1  Graph Systems of Inequalities

Mr. Jackson owns a car washing and detailing business. It takes 20 minutes to wash a car and 60 minutes to detail a car. He works at most 8 hours per day and does at most 4 details per day. Work cooperatively to write and graph a system of linear inequalities to represent this situation.

First, write a linear inequality that represents the time it takes for car washing and car detailing. Let $x$ represent the number of car washes, and let $y$ represent the number of car details. Then $20x + 60y \leq 480$.

To graph this using a graphing calculator, solve for $y$.

$20x + 60y \leq 480$   Original inequality
$60y \leq -20x + 480$   Subtract 20x from each side and simplify.
$y \leq -\frac{1}{3}x + 8$   Divide each side by 60 and simplify.

Mr. Jackson does at most 4 details per day. This means that $y \leq 4$.

**Step 1** Adjust the viewing window and then graph $y \leq 4$. Use the **Window Settings** option from the **Window/Zoom** menu to adjust the window to $-4$ to $30$ for $x$ and $-2$ to $10$ for $y$. Keep the scales as **Auto**. Then enter **del** $\leq 4$ **enter**.

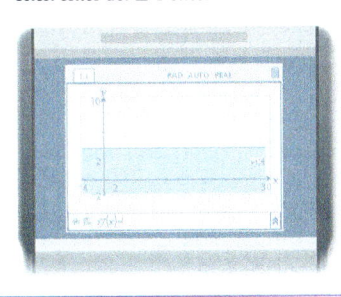

**Step 2** Graph $y \leq -\frac{1}{3}x + 8$. Press **tab del** $\leq$ and then enter $-\frac{1}{3}x + 8$.

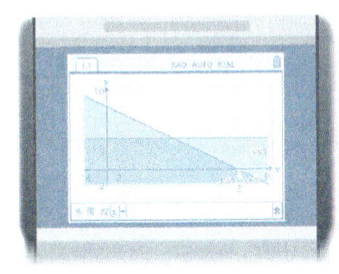

The darkest shaded region of the graph represents the solutions.

**Analyze the Results**
Work cooperatively.

1. If Mr. Jackson charges $75 for each car he details and $25 for each car wash, what is the maximum amount of money he could earn in one day? **$600**

2. What is the greatest number of washes that Mr. Jackson could do in a day? Explain. **24; It is the x-intercept of $y = -\frac{1}{3}x + 8$.**

---

# Launch

**Objective** Use a graphing calculator to explore systems of inequalities.

**Materials for Each Student**
- TI-Nspire Technology

**Teaching Tip**
To start a new document, students can press **CTRL N** and then select **ADD GRAPHS**.

# Teach

**Working in Cooperative Groups** Have students work in groups of two or three, mixing abilities, to complete Steps 1 and 2 of the Activity. **ELL**

- In Step 1, when changing the window settings, students should press tab or the down navigating arrow to move through the settings.
- If more or less contrast is needed to view the overlap of the graphs, press CTRL + for greater contrast and CTRL – for less contrast.

**Practice** Have students complete Exercises 1 and 2.

# Assess

**Formative Assessment** Use Exercise 1 to assess whether students can use a system of inequalities to solve a problem.

**From Concrete to Abstract** Exercise 2 asks students to interpret the graph of a system of inequalities to determine a maximum value.

# Go Online!

### Graphing Calculators
Students can use the Graphing Calculator Personal Tutors to review the use of the graphing calculator to represent functions. They can also use the Other Calculator Keystrokes, which cover lab content for students with calculators other than the TI-84 Plus.

# Chapter 6 Study Guide and Review

**FOLDABLES Study Organizer**

A completed Foldable for this chapter should include the Key Concepts related to systems of equations and inequalities.

## Key Vocabulary ELL

The page reference after each word denotes where that term was first introduced. If students have difficulty answering questions 1–10, remind them that they can use these page references to refresh their memories about the vocabulary terms.

Have students work together to review. Encourage them to reference and compare their notes from Chapter 6.

Then have students work together to answer each question in the Concept Check.

You can use the detailed reports in ALEKS to automatically monitor students' progress and pinpoint remediation needs prior to the chapter test.

---

## CHAPTER 6
## Study Guide and Review

*Go Online!* for Vocabulary Review Games and key vocabulary in 13 languages

### Study Guide

#### Key Concepts

**Systems of Equations** (Lessons 6-1 through 6-5)

- A system with a graph of two intersecting lines has one solution and is *consistent* and *independent*.
- Graphing a system of equations can only provide approximate solutions. For exact solutions, you must use algebraic methods.
- In the substitution method, one equation is solved for a variable and the expression substituted into the second equation to find the value of another variable.
- In the elimination method, one variable is eliminated by adding or subtracting the equations.
- Sometimes multiplying one or both equations by a constant makes it easier to use the elimination method.
- The best method for solving a system of equations depends on the coefficients of the variables.

**Systems of Inequalities** (Lesson 6-6)

- A system of inequalities is a set of two or more inequalities with the same variables.
- The solution of a system of inequalities is the intersection of the graphs.

**FOLDABLES Study Organizer**

Use your Foldable to review the chapter. Working with a partner can be helpful. Ask for clarification of concepts as needed.

#### Key Vocabulary

augmented matrix (p. 374)
consistent (p. 339)
dependent (p. 339)
dimension (p. 374)
element (p. 374)
elimination (p. 354)
inconsistent (p. 339)
independent (p. 339)
matrix (p. 374)
substitution (p. 348)
system of equations (p. 339)
system of inequalities (p. 376)

#### Vocabulary Check

State whether each sentence is *true* or *false*. If *false*, replace the underlined term to make a true sentence.

1. If a system has at least one solution, it is said to be <u>consistent</u>.  **true**
2. If a consistent system has exactly <u>two</u> solution(s), it is said to be independent.  **false; one**
3. If a consistent system has an infinite number of solutions, it is said to be <u>inconsistent</u>.  **false; dependent**
4. If a system has no solution, it is said to be <u>inconsistent</u>.  **true**
5. <u>Substitution</u> involves substituting an expression from one equation for a variable in the other.  **true**
6. In some cases, <u>dividing</u> two equations in a system together will eliminate one of the variables. This process is called elimination.  **false; adding or subtracting**
7. A set of two or more inequalities with the same variables is called a <u>system of equations</u>.  **false; system of inequalities**
8. When the graphs of the inequalities in a system of inequalities <u>do not intersect</u>, there are no solutions to the system.  **true**

#### Concept Check

9. Is only addition or subtraction used in the elimination method?  **No, multiplication and division may also be used depending on the coefficients in the system of equations.**
10. Does a consistent and independent system of equations have an infinite number of solutions?  **No, it has a unique solution.**

---

### Answering the Essential Question

Before answering the Essential Question, have students review their answers to the *Building on the Essential Question* exercises found throughout the chapter.

- How can you find the solution to a math problem? (p. 336)
- What are the advantages of using technology to solve systems of equations? (p. 346)
- What are the benefits of having different strategies for solving systems of equations? (p. 365)
- What are the advantages of using matrices to solve problems? (p. 375)

---

## Go Online!

**Vocabulary Review**

Students can use the *Vocabulary Review Games* to check their understanding of the vocabulary terms in this chapter. Students should refer to the *Student-Built Glossary* they have created as they went through the chapter to review important terms. You can also give a *Vocabulary Test* over the content of this chapter.

Chapter 6 Study Guide and Review

## Lesson-by-Lesson Review

### 6-1 Graphing Systems of Equations
A.CED.3, A.REI.6

Graph each system and determine the number of solutions that it has. If it has one solution, name it.
11–16. See margin for graphs.

11. $x - y = 1$
    $x + y = 5$  one; (3, 2)

12. $y = 2x - 4$
    $4x + y = 2$  one; (1, −2)

13. $2x - 3y = -6$
    $y = -3x + 2$  one; (0, 2)

14. $-3x + y = -3$
    $y = x - 3$  one; (0, −3)

15. $x + 2y = 6$
    $3x + 6y = 8$  no solution

16. $3x + y = 5$
    $6x = 10 - 2y$  infinitely many solutions

17. **MAGIC NUMBERS** Sean is trying to find two numbers with a sum of 14 and a difference of 4. Define two variables, write a system of equations, and solve by graphing. **See margin.**

**Example 1**

Graph the system and determine the number of solutions it has. If it has one solution, name it.

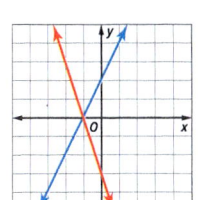

$y = 2x + 2$
$y = -3x - 3$

The lines appear to intersect at the point (−1, 0). You can check this by substituting −1 for x and 0 for y.

CHECK  $y = 2x + 2$         Original equation
       $0 \stackrel{?}{=} 2(-1) + 2$   Substitution
       $0 \stackrel{?}{=} -2 + 2$      Multiply.
       $0 = 0$ ✓

       $y = -3x - 3$         Original equation
       $0 \stackrel{?}{=} -3(-1) - 3$  Substitution
       $0 \stackrel{?}{=} 3 - 3$       Multiply.
       $0 = 0$ ✓

The solution is (−1, 0).

### 6-2 Substitution
A.CED.3, A.REI.6

Use substitution to solve each system of equations.

18. $x + y = 3$
    $x = 2y$  (2, 1)

19. $x + 3y = -28$
    $y = -5x$  (2, −10)

20. $3x + 2y = 16$
    $x = 3y - 2$  (4, 2)

21. $x - y = 8$
    $y = -3x$  (2, −6)

22. $y = 5x - 3$
    $x + 2y = 27$  (3, 12)

23. $x + 3y = 9$
    $x + y = 1$  (−3, 4)

24. **GEOMETRY** The perimeter of a rectangle is 48 inches. The length is 6 inches greater than the width. Define the variables, and write equations to represent this situation. Solve the system by using substitution. **Sample answer: Let w be the width and let ℓ be the length; $2w + 2\ell = 48$, $\ell = w + 6$; 9 is the width and 15 is the length.**

**Example 2**

Use substitution to solve the system.

$3x - y = 18$
$y = x - 4$

$3x - y = 18$            First equation
$3x - (x - 4) = 18$      Substitute $x - 4$ for y.
$2x + 4 = 18$            Simplify.
$2x = 14$                Subtract 4 from each side.
$x = 7$                  Divide each side by 2.

Use the value of x and either equation to find the value for y.

$y = x - 4$              Second equation
$= 7 - 4$ or 3           Substitute and simplify.

The solution is (7, 3).

---

## Lesson-by-Lesson Review

**Intervention** If the given examples are not sufficient to review the topics covered by the questions, remind students that the lesson references tell them where to review that topic in their textbook.

**Two-Day Option** Have students complete the Lesson-by-Lesson Review. Then you can use McGraw-Hill eAssessment to customize another review worksheet that practices all the objectives of this chapter or only the objectives on which your students need more help.

## Additional Answers

11.

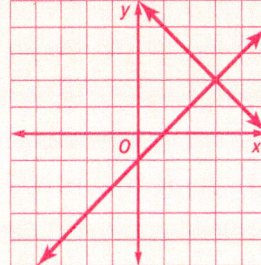

12.

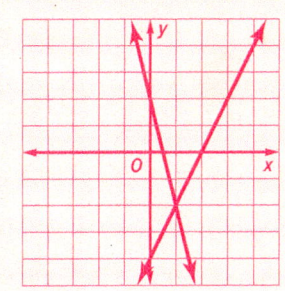

13.

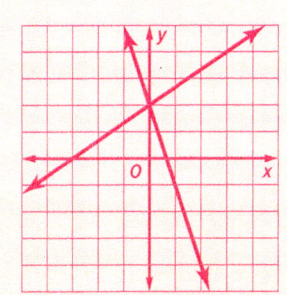

14.

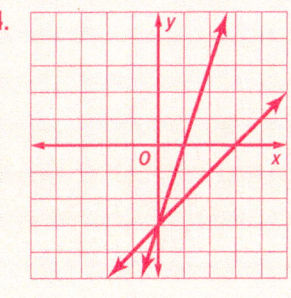

15.

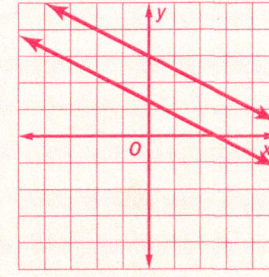

16.

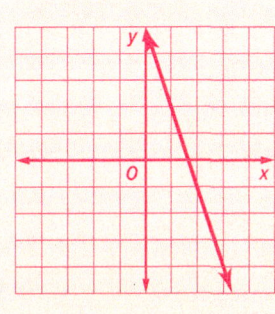

17. Sample answer: Let x be one number and y the other number; $x + y = 14$; $x - y = 4$; 9 and 5

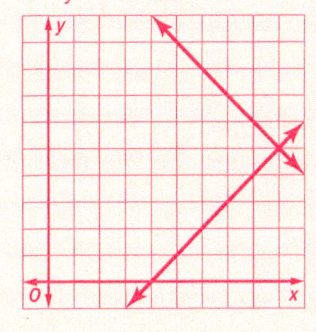

connectED.mcgraw-hill.com  383

# Chapter 6 Study Guide and Review

## Additional Answers

**33.** Sample answer: Let $f$ be the number of the first type of card, and let $c$ be the number of the second type of card; $f + c = 24$, $f + 3c = 50$; 11 $1 cards and 13 $3 cards

**42.** Sample answer: Let $c$ represent the number of cakes, and let $p$ represent the number of pies; $8c + 10p = 356$, $p + c = 40$; 22 cakes, 18 pies

---

# CHAPTER 6
# Study Guide and Review Continued

## 6-3 Elimination Using Addition and Subtraction
A.CED.2, A.REI.6

Use elimination to solve each system of equations.

**25.** $x + y = 13$
$x - y = 5$ **(9, 4)**

**26.** $-3x + 4y = 21$
$3x + 3y = 14$ $\left(-\dfrac{1}{3}, 5\right)$

**27.** $x + 4y = -4$
$x + 10y = -16$ **(4, −2)**

**28.** $2x + y = -5$
$x - y = 2$ **(−1, −3)**

**29.** $6x + y = 9$
$-6x + 3y = 15$ $\left(\dfrac{1}{2}, 6\right)$

**30.** $x - 4y = 2$
$3x + 4y = 38$ **(10, 2)**

**31.** $2x + 2y = 4$
$2x - 8y = -46$ **(−3, 5)**

**32.** $3x + 2y = 8$
$x + 2y = 2$ $\left(3, -\dfrac{1}{2}\right)$

**33. BASEBALL CARDS** Cristiano bought 24 baseball cards for $50. One type cost $1 per card, and the other cost $3 per card. Define the variables, and write equations to find the number of each type of card he bought. Solve by using elimination. **See margin.**

### Example 3

Use elimination to solve the system of equations.

$3x - 5y = 11$
$x + 5y = -3$

$\quad 3x - 5y = 11$
$(+)\; x + 5y = -3$
$\quad\quad 4x\quad = 8$    The variable $y$ is eliminated.
$\quad\quad\; x = 2$    Divide each side by 4.

Now, substitute 2 for $x$ in either equation to find the value of $y$.

$3x - 5y = 11$    First equation
$3(2) - 5y = 11$    Substitute.
$6 - 5y = 11$    Multiply.
$-5y = 5$    Subtract 6 from each side.
$y = -1$    Divide each side by −5.

The solution is (2, −1).

---

## 6-4 Elimination Using Multiplication
A.REI.5, A.REI.6

Use elimination to solve each system of equations.

**34.** $x + y = 4$
$-2x + 3y = 7$ **(1, 3)**

**35.** $x - y = -2$
$2x + 4y = 38$ **(5, 7)**

**36.** $3x + 4y = 1$
$5x + 2y = 11$ **(3, −2)**

**37.** $-9x + 3y = -3$
$3x - 2y = -4$ **(2, 5)**

**38.** $8x - 3y = -35$
$3x + 4y = 33$ **(−1, 9)**

**39.** $2x + 9y = 3$
$5x + 4y = 26$ **(6, −1)**

**40.** $-7x + 3y = 12$
$2x - 8y = -32$ **(0, 4)**

**41.** $8x - 5y = 18$
$6x + 6y = -6$ **(1, −2)**

**42. BAKE SALE** On the first day, a total of 40 items were sold for $356. Define the variables, and write a system of equations to find the number of cakes and pies sold. Solve by using elimination. **See margin.**

MONARCH MIDDLE SCHOOL
Bake Sale
Pies  $10
Cakes  $8

### Example 4

Use elimination to solve the system of equations.

$3x + 6y = 6$
$2x + 3y = 5$

Notice that if you multiply the second equation by −2, the coefficients of the $y$-terms are additive inverses.

$3x + 6y = 6$          $3x + 6y = 6$
$2x + 3y = 5$ Multiply by −2. $(+)\; -4x - 6y = -10$
                          $-x\quad\quad = -4$
                           $x = 4$

Now, substitute 4 for $x$ in either equation to find the value of $y$.

$2x + 3y = 5$    Second equation
$2(4) + 3y = 5$    Substitution
$8 + 3y = 5$    Multiply.
$3y = -3$    Subtract 8 from both sides.
$y = -1$    Divide each side by 3.

The solution is (4, −1).

## 6-5 Applying Systems of Linear Equations

A.REI.6

Determine the best method to solve each system of equations. Then solve the system.

43. $y = x - 8$    Subs.;
    $y = -3x$    (2, −6)

44. $y = -x$    Subs.;
    $y = 2x$    (0, 0)

45. $x + 3y = 12$    Subs.;
    $x = -6y$    (24, −4)

46. $x + y = 10$    Elim (+);
    $x - y = 18$    (14, −4)

47. $3x + 2y = -4$
    $5x + 2y = -8$

48. $6x + 5y = 9$    Elim (×);
    $-2x + 4y = 14$    (−1, 3)

49. $3x + 4y = 26$
    $2x + 3y = 19$

50. $11x - 6y = 3$    Elim (×);
    $5x - 8y = -25$    (3, 5)

47. Elim (−); (−2, 1)    49. Elim (×); (2, 5)

51. **COINS** Tionna has saved dimes and quarters in her piggy bank. Define the variables, and write a system of equations to determine the number of dimes and quarters. Then solve the system using the best method for the situation.

    $4.00
    25 coins

52. **FAIR** At a county fair, the cost for 4 slices of pizza and 2 orders of French fries is $21.00. The cost of 2 slices of pizza and 3 orders of French fries is $16.50. To find out how much a single slice of pizza and an order of French fries costs, define the variables and write a system of equations to represent the situation. Determine the best method to solve the system of equations. Then solve the system. (Lesson 6-5)

### Example 5

Determine the best method to solve the system of equations. Then solve the system.

$3x + 5y = 4$
$4x + y = -6$

The coefficient of $y$ is 1 in the second equation. So solving by substitution is a good method. Solve the second equation for $y$.

$4x + y = -6$      Second equation
$y = -6 - 4x$      Subtract 4x from each side.

Substitute $-6 - 4x$ for $y$ in the first equation.

$3x + 5(-6 - 4x) = 4$      Substitute.
$3x - 30 - 20x = 4$        Distributive Property
$-17x - 30 = 4$            Simplify.
$-17x = 34$                Add 30 to each side.
$x = -2$                   Divide by −17.

Last, substitute $-2$ for $x$ in either equation to find $y$.

$4x + y = -6$      Second equation
$4(-2) + y = -6$   Substitute.
$-8 + y = -6$      Multiply.
$y = 2$            Add 8 to each side.

The solution is $(-2, 2)$.

---

51. Sample answer: Let $d$ represent the dimes and $q$ represent the quarters; $d + q = 25$; $0.10d + 0.25q = 4$; 15 dimes, 10 quarters.

52. Let $p$ represent the cost of a slice of pizza and $f$ represent the cost of an order of French fries; $4p + 2f = 21$, $2p + 3f = 16.5$; Sample answer: elimination; pizza $3.75; French fries $3.

---

### Go Online!

**Anticipation Guide**

Students should complete the *Chapter 6 Anticipation Guide*, and discuss how their responses have changed now that they have completed Chapter 6.

# Chapter 6 Study Guide and Review

## Before the Test
Have students complete the Study Notebook Tie it Together activity to review topics and skills presented in the chapter.

## Additional Answers

53.

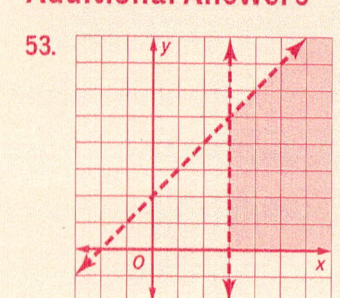

54.

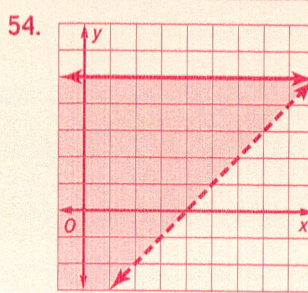

55.

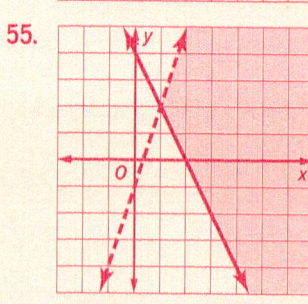

56.

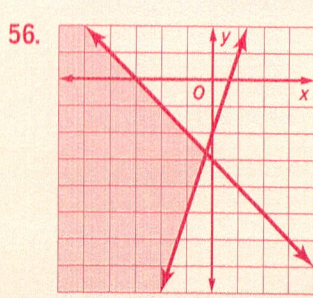

---

# CHAPTER 6
## Study Guide and Review Continued

### 6-6 Systems of Inequalities
A.CED.3, A.REI.12

Solve each system of inequalities by graphing.

53. $x > 3$
    $y < x + 2$

54. $y \leq 5$
    $y > x - 4$

55. $y < 3x - 1$
    $y \geq -2x + 4$

56. $y \leq -x - 3$
    $y \geq 3x - 2$

51–54. See margin.

57. **JOBS** Kishi makes $9 an hour working at the grocery store and $12 an hour delivering newspapers. She cannot work more than 20 hours per week. Graph two inequalities that Kishi can use to determine how many hours she needs to work at each job if she wants to earn at least $150 per week. See margin.

**Example 6**
Solve the system of inequalities by graphing.

$y < 3x + 1$
$y \geq -2x + 3$

The solution set of the system is the set of ordered pairs in the intersection of the two graphs. This portion is shaded in the graph below.

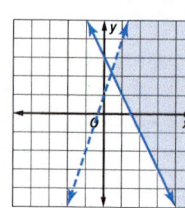

---

57. Jobs

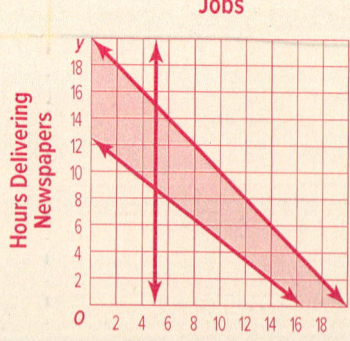

(graph with axes: Hours Delivering Newspapers vs Hours at the Grocery Store)

---

## Go Online!

### eAssessment
Customize and create multiple versions of chapter tests and answer keys that align to your standards. Tests can be delivered on paper or online.

# CHAPTER 6
## Practice Test

**Go Online!** for another Chapter Test

Graph each system and determine the number of solutions that it has. If it has one solution, name it.

1. $y = 2x$
   $y = 6 - x$  **one; (2, 4)**

2. $y = x - 3$
   $y = -2x + 9$  **one; (4, 1)**

3. $x - y = 4$
   $x + y = 10$  **one; (7, 3)**

4. $2x + 3y = 4$
   $2x + 3y = -1$  **no solution**

**1–4. See Ch. 6 Answer Appendix for graphs.**

Use substitution to solve each system of equations.

5. $y = x + 8$
   $2x + y = -10$  **(−6, 2)**

6. $x = -4y - 3$
   $3x - 2y = 5$  **(1, −1)**

7. **GARDENING** Corey has 42 feet of fencing around his garden. The garden is rectangular in shape, and its length is equal to twice the width minus 3 feet. Define the variables, and write a system of equations to find the length and width of the garden. Solve the system by using substitution. **See margin.**

[Diagram: rectangular garden with P = 42, labeled ℓ and w]

8. **MULTIPLE CHOICE** Use elimination to solve the system. **B**
   $6x - 4y = 6$
   $-6x + 3y = 0$
   
   A (5, 6)
   B (−3, −6)
   C (1, 0)
   D (4, −8)

9. Shelly has $300 to shop for jeans and sweaters. Each pair of jeans cost $65, each sweater costs $34, and she buys 7 items. Determine the number of pairs of jeans and sweaters Shelly bought. **9. 2 jeans, 5 sweaters**

Use elimination to solve each system of equations.

10. $x + y = 13$
    $x - y = 5$  **(9, 4)**

11. $3x + 7y = 2$
    $3x - 4y = 13$  **(3, −1)**

12. $x + y = 8$
    $x - 3y = -4$  **(5, 3)**

13. $2x + 6y = 18$
    $3x + 2y = 13$  **(3, 2)**

14. **MAGAZINES** Julie subscribes to a sports magazine and a fashion magazine. She received 24 issues this year. The number of fashion issues is 6 less than twice the number of sports issues. Define the variables, and write a system of equations to find the number of issues of each magazine. **See margin.**

Determine the best method to solve each system of equations. Then solve the system.

15. $y = 3x$
    $x + 2y = 21$  **Subst.; (3, 9)**

16. $x + y = 12$
    $y = x - 4$  **Subst.; (8, 4)**

17. $x + y = 15$
    $x - y = 9$  **Elim. (+); (12, 3)**

18. $3x + 5y = 7$
    $2x - 3y = 11$  **Elim. (×); (4, −1)**

**19.** $24p + 4c = 320$, $2p + c = 50$; Sample answer: subsitution; paper $7.50 per ream, inkjet cartridge, $35 each.

19. **OFFICE SUPPLIES** At a sale, Ricardo bought 24 reams of paper and 4 inkjet cartridges for $320. Britney bought 2 reams of paper and 1 inkjet cartridge for $50. The reams of paper were all the same price and the inkjet cartridges were all the same price. Write a system of equations to represent this situation. Determine the best method to solve the system of equations. Then solve the system.

**20–23. See Ch. 6 Answer Appendix.**
Solve each system of inequalities by graphing.

20. $x > 2$
    $y < 4$

21. $x + y \leq 5$
    $y \geq x + 2$

22. $3x - y > 9$
    $y > -2x$

23. $y \geq 2x + 3$
    $-4x - 3y > 12$

---

## Go Online!

### Chapter Tests
You can use premade leveled *Chapter Tests* to differentiate assessment for your students. Students can also take self-checking *Chapter Tests* to plan and prepare for chapter assessments.

MC = multiple-choice questions
FR = free-response questions

| Form | Type | Level |
|------|------|-------|
| 1 | MC | AL |
| 2A | MC | OL |
| 2B | MC | OL |
| 2C | FR | OL |
| 2D | FR | OL |
| 3 | FR | BL |
| Vocabulary Test | | |
| Extended-Response Test | | |

---

## Chapter 6 Practice Test

**RtI Response to Intervention**
Use the Intervention Planner to help you determine your Response to Intervention.

### Intervention Planner

**TIER 1 On Level OL**

**IF** students miss 25% of the exercises or less,
**THEN** choose a resource:
- SE Lessons 6-1 through 6-6
- **Go Online!**
  - Skills Practice
  - Chapter Project
  - Self-Check Quizzes

**TIER 2 Strategic Intervention AL**
Approaching grade level

**IF** students miss 50% of the exercises,
**THEN** choose a resource:
- *Quick Review Math Handbook*
- **Go Online!**
  - Study Guide and Intervention
  - Extra Examples
  - Personal Tutors
  - Homework Help

**TIER 3 Intensive Intervention**
2 or more grades below level

**IF** students miss 75% of the exercises,
**THEN** choose a resource:
- Use *Math Triumphs, Alg. 1*
- **Go Online!**
  - Extra Examples
  - Personal Tutors
  - Homework Help
  - Review Vocabulary

## Additional Answers

7. Sample answer: Let $w$ be the width and let $\ell$ be the length; $2w + 2\ell = 42$, $\ell = 2w - 3$; $w = 8$ ft, $\ell = 13$ ft

14. Let $f =$ the number of fashion magazines and $s =$ the number of sports magazines. Then $f + s = 24$, and $f = 2s - 6$. So $s = 10$ and $f = 14$.

# Chapter 6 Preparing for Assessment

## Launch

**Objective** Apply concepts and skills from this chapter in a real-world setting.

## Teach

**Ask:**
- How do you know that the sum of the two variables is equal to 0.385? **Sample answer: The total weight of the two metals is given as 0.385.**

- How do you know that the second equation is $y = 10x$ instead of $x = 10y$? **Sample answer: The phrase "10 times as much silver as gold" means that there is one-tenth the amount of gold as there is silver. If the whole amount were 11 parts, there would be 1 part of gold and 10 parts of silver.**

- How do you know that the best method to solve the system of equations is by using substitution? **Sample answer: When looking at the coefficients of the variables in the two equations, the second equation has a 1 for a coefficient meaning that 10x can easily be substituted for y in the first equation.**

The Performance Task focuses on the following content standards and standards for mathematical practice.

| Dual Coding |||
|---|---|---|
| Parts | Content Standards | Mathematical Practices |
| A | A.CED.3, A.REI.6 | 1, 4 |
| B | A.CED.3, A.REI.6 | 1, 4 |
| C | A.REI.1 | 2, 4 |
| D | A.REI.1, A.REI.6 | 3, 4 |
| E | A.REI.6 | 4, 7 |
| F | A.CED.3, A.REI.1 | 1, 7 |

## Go Online!  eBook

**Interactive Student Guide**
Refer to *Interactive Student Guide* for an additional Performance Task.

388 | Chapter 6 | Preparing for Assessment

---

# CHAPTER 6
# Preparing for Assessment

## Performance Task

Provide a clear solution to each part of the task. Be sure to show all of your work, include all relevant drawings, and justify your answers.

**SMARTPHONES** Smartphones contain gold and silver. A smartphone has 10 times as much silver as it has gold for a total weight of 0.385 gram. How much gold and silver is in that smartphone?

### Part A
1. **Sense Making** Define the variables, and write a system of linear equations to represent the situation.
   **Let $x$ represent the amount of gold and $y$ represent the amount of silver.**
   **System: $x + y = 0.385$; $y = 10x$**

### Part B
2. Interpret the meaning of each equation.
   **The equation $x + y = 0.385$ represents the fact that the total weight of the two metals in a smartphone is 0.385 gram. The equation $y = 10x$ represents the fact that there is 10 times as much silver in a smart phone as there is gold.**

### Part C
3. Determine the best method to solve the system of equations by looking closely at the coefficients of each term. Explain your reasoning.
   **Sample answer: Substitution is the best method to use to solve this system because the equation $y = 10x$ can easily be substituted into the equation $x + y = 0.385$ to find a solution.**

### Part D
4. **Tools** Solve the system and describe its meaning in the context of the situation. Show your work.

### Part E
5. Determine the number of solutions that the system has. If it has one solution, name it.
   **There is one solution to the system, (0.035, 0.35).**

### Part F
6. Use substitution to check the answer. Is the solution reasonable? Explain your reasoning.

---

4. Solve the first equation by substitution of the second equation.
$x + y = 0.385$
$x + 10x = 0.385$
$11x = 0.385$
$x = 0.035$
There is 0.035 gram of gold in a smartphone.
Solve for $y$ by substituting 0.35 for $x$.
$y = 10x$
$y = 10(0.035)$
$y = 0.35$
There is 0.35 gram of silver in a smartphone, which is 10 times as much as the gold.

6. Substitute the $x$- and $y$-values into the first equation in the system: $x + y = 0.385$; $0.035 + 0.35 = 0.385$; the solution is reasonable because when substituting the $x$- and $y$-values, (0.035, 0.35), the equation is true.

---

## Levels of Complexity Chart

The levels of the exercises progress from 1 to 3, with Level 1 indicating the lowest level of complexity.

| Parts | Level 1 | Level 2 | Level 3 |
|---|---|---|---|
| A | ● | | |
| B | | ● | |
| C | | | ● |
| D | | ● | |
| E | | | ● |
| F | | | ● |

# Chapter 6 Preparing for Assessment

## Test-Taking Strategy

### Example

Read the problem. Identify what you need to know. Then use the information in the problem to solve.

Solve $\begin{cases} 4x - 8y = 20 \\ -3x + 5y = -14 \end{cases}$

A  (5, 0)  
B  (4, −2)  
C  (3, −1)  
D  (−6, −5)

**Step 1** What do you need to find? The solution to the system of equations.
**Step 2** Is there enough information given to solve the problem? Yes.
**Step 3** What information, if any, is not needed to solve the problem? None that one can discern.
**Step 4** Are there any obvious wrong answers? No.
**Step 5** What is the correct answer? Find the answer choice that satisfies both equations of the system.

| Guess: (5, 0) | First Equation | Second Equation |
|---|---|---|
| | $4x - 8y = 20$ <br> $4(5) - 8(0) = 20$ ✓ | $-3x + 5y = -14$ <br> $-3(5) + 5(0) \neq -14$ ✗ |

| Guess: (4, −2) | First Equation | Second Equation |
|---|---|---|
| | $4x - 8y = 20$ <br> $4(4) - 8(-2) \neq 20$ ✗ | $-3x + 5y = -14$ <br> $-3(4) + 5(-2) \neq -14$ ✗ |

| Guess: (3, −1) | First Equation | Second Equation |
|---|---|---|
| | $4x - 8y = 20$ <br> $4(3) - 8(-1) = 20$ ✓ | $-3x + 5y = -14$ <br> $-3(3) + 5(-1) = -14$ ✓ |

The correct answer is C.

### Test-Taking Tip

**Guess and Check** It is very important to pace yourself and keep track of how much time you have left. If time is running short, or if you are unsure how to solve a problem, the guess and check strategy may help you determine the correct answer quickly.

### Apply the Strategy

Gina bought 5 hot dogs and 3 soft drinks at the ball game for $32.25. Renaldo bought 4 hot dogs and 2 soft drinks for $24.00. How much do a single hot dog and a single drink cost?  **B**

A  hot dog: $4.50 soft drink: $3.75  
B  hot dog: $3.75 soft drink: $4.50  
C  hot dog: $4.50 soft drink: $4.25  
D  hot dog: $3.75 soft drink: $4.25

Answer the questions below.

a. What do you need to find? The solution to the system of equations.
b. Is there enough information given to solve the problem? Yes
c. What information, if any, is not needed to solve the problem? No.
d. Are there any obvious wrong answers? No
e. What is the correct answer? B

## Test-Taking Strategy
## Guess and Check

**Step 1** Carefully look over each possible answer choice, and evaluate for reasonableness. Eliminate unreasonable answers.

**Step 2** For the remaining answer choices, use the guess and check method. Find the answer choice that satisfies both equations of the system.

**Step 3** Choose an answer choice and see if it satisfies the constraints of the problem statement. Identify the correct answer.

## Need Another Example?

Solve $2x + 5y = -18$
$-4x + 3y = 10$  **C**

A  (6, −6)  
B  (1, −4)  
C  (−4, −2)  
D  (−9, 0)

a. What do you need to find? The solution to the system of equations. No; all four answer choices are possible correct answers.
b. Is there enough information given to solve the problem? yes
c. What information, if any, is not needed to solve the problem? All the information is relevant.
d. Are there any obvious wrong answers? If so, which (one)s? Explain. none
e. What is the correct answer? C

## Differentiated Instruction

**Extension** In the Performance Task, what additional method could be used to solve the system of equations? Explain.

Sample answer: You could also use elimination using subtraction. You can rewrite $y = 10x$ as $-10x + y = 0$ and subtract it from $x + y = 0.385$:

$$\begin{array}{r} x + y = 0.385 \\ -(-10x + y) = 0 \\ \hline 11x\phantom{ + y} = 0.385 \end{array}$$

$x = 0.035$ and $y = 0.35$

connectED.mcgraw-hill.com   389

# Chapter 6 Preparing for Assessment

## Diagnose Student Errors

Survey student responses for each item. Class trends may indicate common errors and misconceptions.

| 2. | A | Wrote $y = 3x + 4$ for the second equation |
|---|---|---|
| | B | CORRECT |
| | C | Thought same slope implies infinite solutions |
| | D | Wrote incorrect $y$-intercept equations |

| 3a. | A | Found perimeter by adding the given sides |
|---|---|---|
| | B | Incorrectly subtracted 6 from 15 |
| | C | Solved for $x$ by subtracting 12 from both sides |
| | D | CORRECT |

| 4. | A | Added terms in numerator and denominator |
|---|---|---|
| | B | Wrote $2h - 3h$ in the denominator |
| | C | Wrote $6h - 3h$ in the numerator |
| | D | CORRECT |

| 6. | A | Incorrectly solved for $y$ |
|---|---|---|
| | B | CORRECT |
| | C | Eliminated $y$ by subtracting $x$ from $3x$ |
| | D | Subtracted $x$ from $3x$ and added 3 to 27 |

| 7. | A | CORRECT |
|---|---|---|
| | B | CORRECT |
| | C | CORRECT |
| | D | The graph does not pass through (2, 1) |

| 9. | A | Found the number of adult tickets sold |
|---|---|---|
| | B | Incorrectly calculated 300 divided by 2 |
| | C | CORRECT |
| | D | Found the total cost of 250 student tickets |

| 11. | A | CORRECT |
|---|---|---|
| | B | Forgot to subtract $2x$ from $9x$ |
| | C | CORRECT |
| | D | Incorrectly multiplied $-3$ by $-x$ in $-3(-x + 5)$ |

| 13. | A | Chose inequalities for quadrant IV |
|---|---|---|
| | B | Chose inequalities for quadrant II |
| | C | CORRECT |
| | D | Thought that $x > y$ described negative $x$-values |

## CHAPTER 6
## Preparing for Assessment
### Cumulative Review

Read each question. Then fill in the correct answer on the answer document provided by your teacher or on a sheet of paper.

1. What is the slope of a line that is perpendicular to $x + 2y = -3$? **2**

2. Which of the following statements is true of the system below? **B**

$$3x + y = -5$$
$$y + 3x = 4$$

 ○ A  The system has one solution at $\left(\frac{7}{6}, -\frac{1}{2}\right)$.
 ○ B  The system has no solution.
 ○ C  The system has infinitely many solutions.
 ○ D  The system has one solution at $\left(-\frac{5}{3}, 9\right)$.

3. Nadia's vegetable garden has a perimeter of 48 feet. Expressions for the dimensions of her garden are shown in the diagram below.

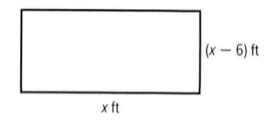

 [Rectangle with dimensions $(x-6)$ ft and $x$ ft]

 a. What are the dimensions of Nadia's vegetable garden? **D**
  ○ A  length = 27 feet, width = 21 feet
  ○ B  length = 15 feet, width = 8 feet
  ○ C  length = 9 feet, width = 3 feet
  ○ D  length = 15 feet, width = 9 feet

 b. What is the area of her garden? **135 ft²**

**Test-Taking Tip**
Question 5 To find 30% more than Ryan weighs, first let $r$ represent Ryan's weight. Then $0.3r$ is 30% of $r$, and $r + 0.3r$ or $1.3r$ is 30% more than what Ryan weighs.

4. If $(-3h, 2h)$ and $(2h, 6h)$ are two points on the graph of a line and $h$ is not equal to 0, what is the slope of the line? **D**
 ○ A  $-8$
 ○ B  $-4$
 ○ C  $\frac{3}{5}$
 ○ D  $\frac{4}{5}$

5. Ryan's brother, Joel, weighs 30% more than Ryan. Together, the boys weigh 276 pounds. In pounds, how much does Joel weigh? **156**

6. The sum of two numbers is $-3$. Three times the first number minus the second number is 27. What are the two numbers? **B**
 ○ A  6 and $-3$
 ○ B  6 and $-9$
 ○ C  12 and $-15$
 ○ D  $\frac{7}{15}$ and $-18$

7. Which of the following are accurate descriptions of the graph of $3x + y = -5$? **A–C**
 ☐ A  The graph contains the points $(-3, 4)$, $(-1, -2)$, and $(3, -14)$.
 ☐ B  The graph is a line with a $y$-intercept of $(0, -5)$ and a slope of $-3$.
 ☐ C  The graph has an $x$-intercept of $\left(-\frac{5}{3}, 0\right)$.
 ☐ D  The graph passes through the points $(1, -8)$ and $(2, 1)$.

8. Two friends own a total of 25 books. One owns four times as many books as the other. Find the number of books each friend owns. **5, 20**

## Go Online!

### Standardized Test Practice

Students can take self-checking tests in standardized format to plan and prepare for standardized assessments.

## Chapter 6 Preparing for Assessment

**Formative Assessment**
You can use these pages to benchmark student progress.

📄 Standardized Test Practice

**Answer Sheet Practice**
Have students simulate taking a standardized test by recording their answers on a practice recording sheet.

**Homework Option**

**Get Ready for Chapter 7** Assign students the exercises on p. 394 as homework to assess whether they possess the prerequisite skills needed for the next chapter.

---

Go Online! for Standardized Test Practice

9. The band at Jason's high school had a concert for charity. They raised $1500. Adult tickets were $5 each and student tickets were $3 each. The members of the band sold a total of 400 tickets. How many student tickets did they sell? **C**
   - A  150
   - B  175
   - C  250
   - D  750

10. The graph of a linear inequality is shown below.

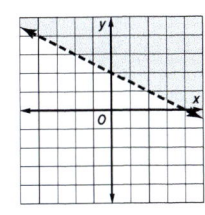

    What is the inequality that describes the graph?
    $y > -\frac{1}{2}x + 2$

11. Which expressions are equivalent to $-3(-x + 5) + 6(x - 4) - 2x$? **A, C**
    - A  $3(x - 5) + 4(x - 6)$
    - B  $9x - 39$
    - C  $7x - 39$
    - D  $x - 39$

12. At a garage sale, Julia sold a total of 32 T-shirts for $114. The prices she sold the shirts for are shown in the table.

    | T-shirt Prices | |
    |---|---|
    | Sports T-shirts | $5 |
    | Concert T-shirts | $3 |

    How many concert T-shirts did Julia sell? **23**

13. Which system of inequalities describes all points in Quadrant III? **C**
    - A  $x > 0, y < 0$
    - B  $x < 0, y > 0$
    - C  $x < 0, y < 0$
    - D  $x < y, y < 0$

14. There are 28 students in Lisa's Algebra 1 class. There are 12 more girls than boys. How many girls are in the class? **A**
    - A  20
    - B  18
    - C  16
    - D  8

15. The sum of the perimeters of two different equilateral triangles is 48 inches. The difference between the perimeters of the two triangles is 6 inches.
    a. Write the system of equations to find the side lengths for each triangle. $\begin{cases} 3x + 3y = 48 \\ 3x - 3y = 6 \end{cases}$
    b. What are the perimeters of the two triangles? **21 inches, 27 inches**

---

**Need Extra Help?**

| If you missed Question... | 1 | 2 | 3 | 4 | 5 | 6 | 7 | 8 | 9 | 10 | 11 | 12 | 13 | 14 | 15 |
|---|---|---|---|---|---|---|---|---|---|---|---|---|---|---|---|
| Go to Lesson... | 4-4 | 6-1 | 2-3 | 3-3 | 6-2 | 6-3 | 4-1 | 6-2 | 6-5 | 5-6 | 1-4 | 6-5 | 6-6 | 6-2 | 6-3 |

---

## LEARNSMART

Use LearnSmart as part of your test-preparation plan to measure student topic retention. You can create a student assignment in LearnSmart for additional practice on these **topics**.

· Solve Systems of Equations

---

## Go Online!

**eAssessment**

Customize and create multiple versions of chapter tests and answer keys that align to your standards. Tests can be delivered on paper or online.

## Lesson 6-1

16. 1 solution, (0, 3)
17. 1 solution, (−1, −2)
18. no solution
19. infinitely many
20. infinitely many
21. 1 solution, (5, −1)
22. 1 solution, (3, 2)
23. no solution
24. infinitely many
27.
28.
29.
30.
31.
32.
33.
34.
35.
36.
37.
38.

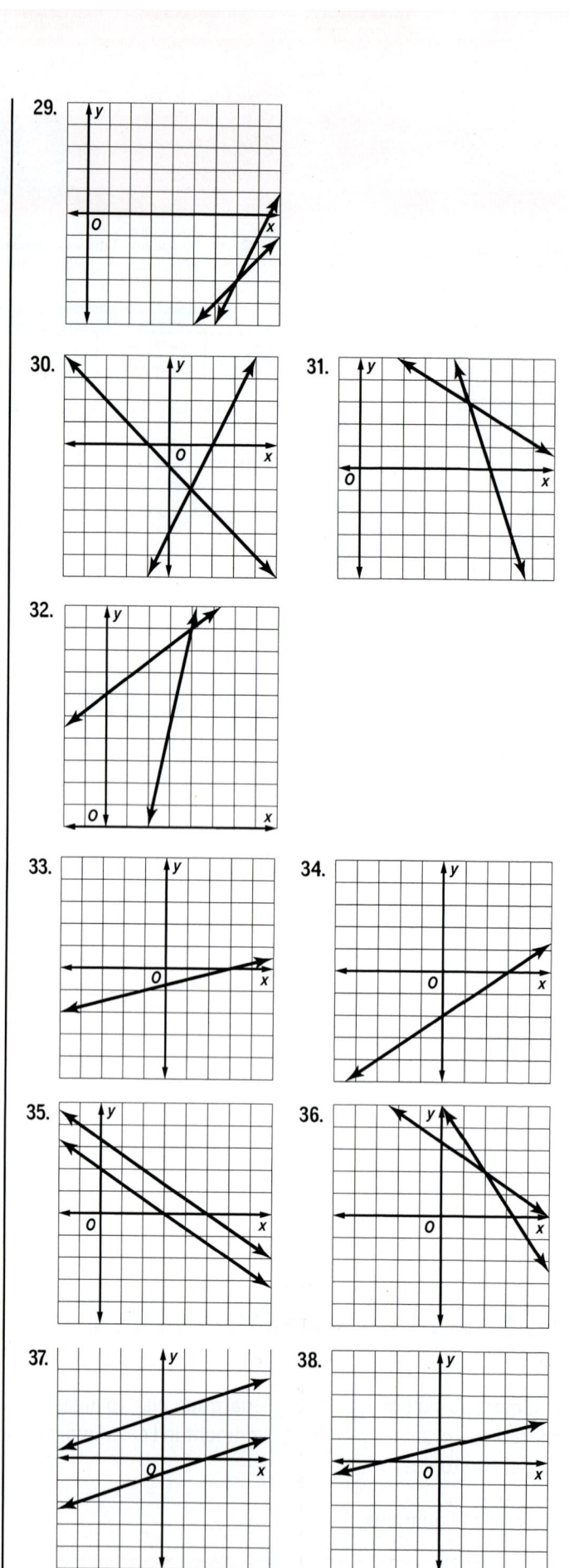

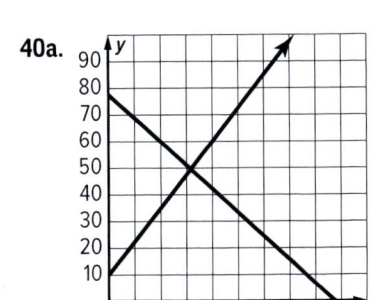

**40a.**

[Graph showing two lines intersecting around (4, 40) on axes with x from 0 to 9 and y from 0 to 90]

## Lesson 6-2

**26c.**

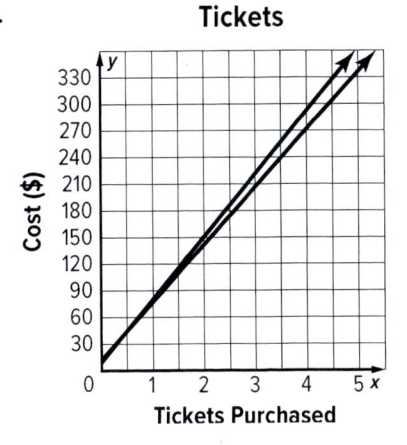

**Tickets**

[Graph with Cost ($) on y-axis from 0 to 330 and Tickets Purchased on x-axis from 0 to 5, showing two nearly overlapping lines]

## Lesson 6-4

**25a.** Michelle should bake 7 tubes of cookies for 56 minutes and Julie should bake 3 tubes of cookies in 36 minutes.

**25b.** Sample answer: Let $m$ = the number of tubes Michelle bakes. Let $j$ = the number of tubes Julie bakes. The number of cookies Michelle bakes plus the number of cookies Julie bakes should equal 264. Each batch Michelle makes produces 29 cookies. Each batch Julie makes produces 24 cookies. Therefore, $29m + 24j = 264$ relates the number of cookies baked to the number of tubes each girl uses. We also know that it takes Michelle's tubes 8 minutes to bake, and Julie's tubes take 12 minutes to bake. The girls have a little over 90 minutes to bake their cookies. So, $8m + 12j = 90$. Solve the system.

| | |
|---|---|
| $29m + 24j = 264$ | First equation |
| $(-)16m + 24j = 180$ | Multiply second equation by 2 and subtract. |
| $13m \phantom{+ 24j} = 84$ | $j$ is eliminated. |
| $m \approx 6.46$ | Divide each side by 13. |

It does not make sense for Michelle to bake part of a tube of cookie dough, so $m = 7$. Substituting 7 into the first equation, we determine that Julie bakes ≈2.54 or 3 tubes of cookies.

**25c.** Sample answer: I assumed that Michelle and Julie cannot bake their cookies simultaneously. I assumed that one of the cookies that Michelle makes per tube is slightly smaller than the others. (You could have assumed that she ate the leftover cookie dough.) I assumed that the girls were able to bake an entire tube of cookie dough in one batch. I assumed that the amount of time needed to transfer baked cookies from the oven and unbaked cookies to the oven is negligible.

## Mid-Chapter Quiz

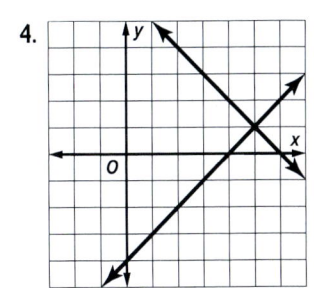

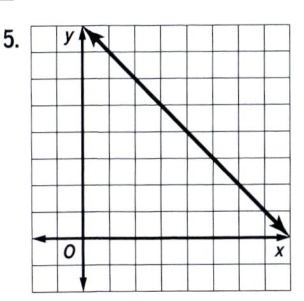

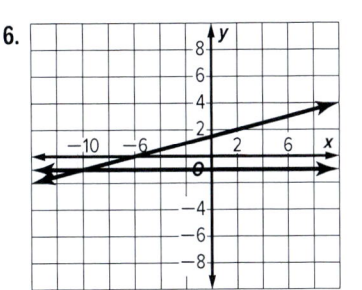

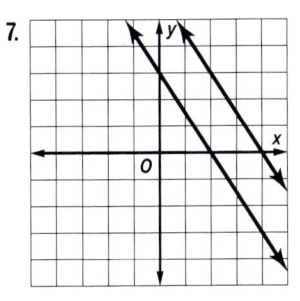

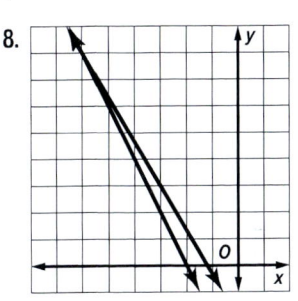

**3.** [Graph showing two lines intersecting, axes from −10 to 10 on x and up to 18 on y]

**4.** [Graph showing two intersecting lines through origin area]

**5.** [Graph showing a line with negative slope]

**6.** [Graph showing two nearly horizontal lines, x from −10 to 6]

**7.** [Graph showing two parallel lines with negative slope]

**8.** [Graph showing two lines with steep negative slope]

**14a.** Let $a$ = the cost of an adult ticket and $c$ = the cost of a child ticket.

**14b.** $4a + 2c = 110$
$4a + 3c = 123$

**14c.** (21, 13); The cost of an adult ticket is $21 and the cost of a child ticket is $13.

## Lesson 6-5

**28.** Sample answer: You can analyze the coefficients of the terms in each equation to determine which method to use. If one of the variables in either equation has a coefficient of 1 or −1, then substitution could be used. If a variable in both equations has coefficients that are opposites, then elimination using addition may be the most convenient method. If a variable in both equations has coefficients that are the same, elimination using subtraction may be the most convenient method. If none of these conditions are met, elimination using multiplication could be used. If both equations are written in slope-intercept form, graphing might be another solution option.

## Lesson 6-6 (Guided Practice)

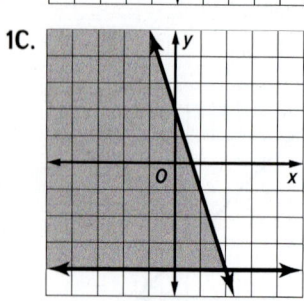

1A. [graph]  1B. [graph]

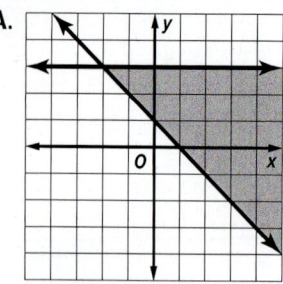

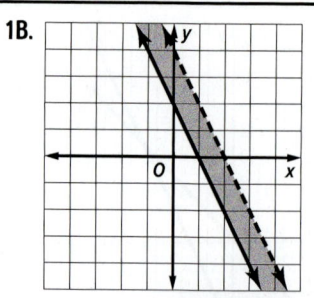

1C. [graph]  1D. [graph]

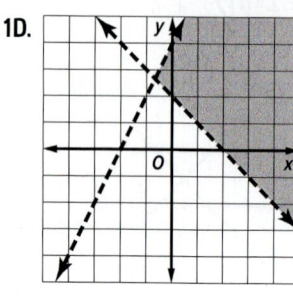

2A. no solution

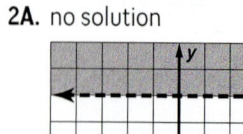

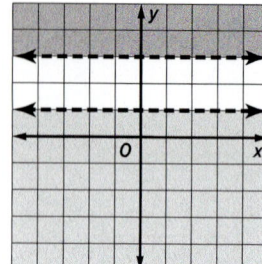

2B. no solution

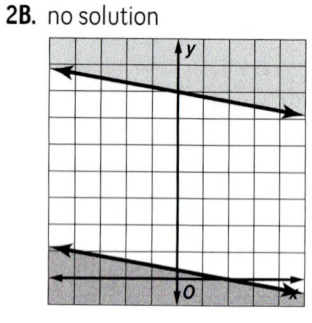

3A. Let $w$ = the number of sweatshirts and $t$ = the number of T-shirts; $w + t \leq 120$ and $22w + 15t \geq 2000$.

3B. [graph: Sweatshirts Sold vs T-Shirts Sold]

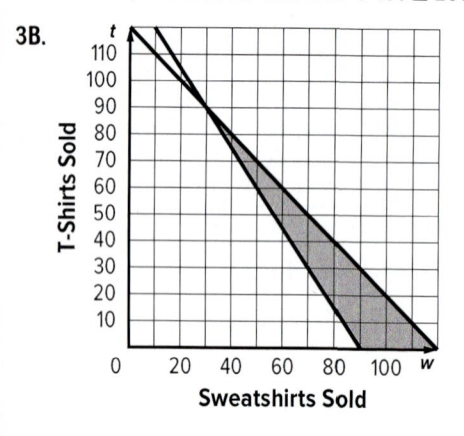

## Lesson 6-6

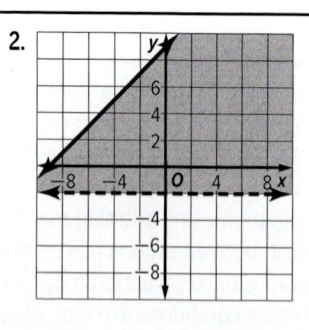

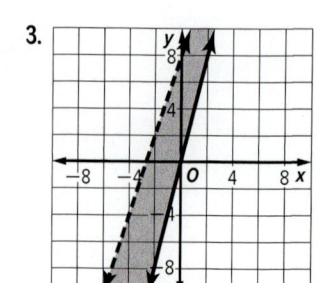

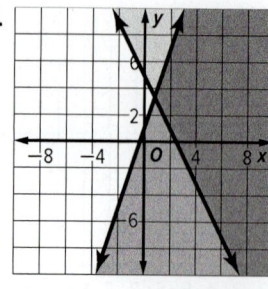

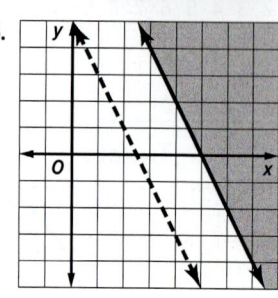

1. [graph]
2. [graph]
3. [graph]
4. [graph]
5. no solution 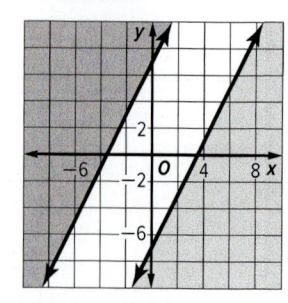
6. [graph]
7. [graph]
8. no solution

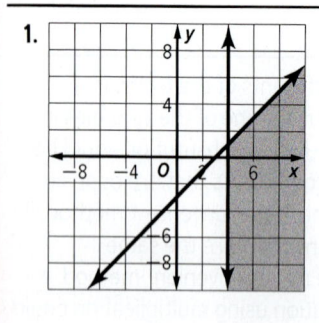

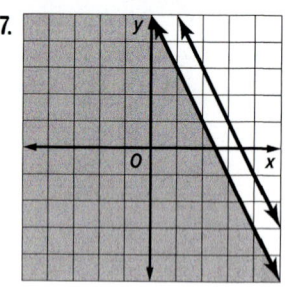

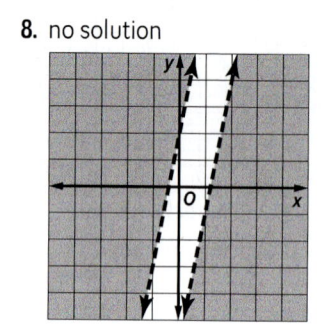

9a. Let $h$ = the height of the driver in inches and $w$ = the weight of the driver in pounds; $h < 79$ and $w < 295$.

**Driving Requirements** [graph: Height (in.) vs Weight (lb)]

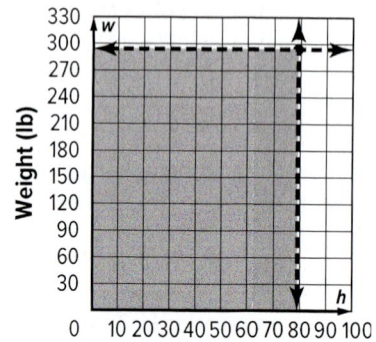

10. [graph]
11. [graph]

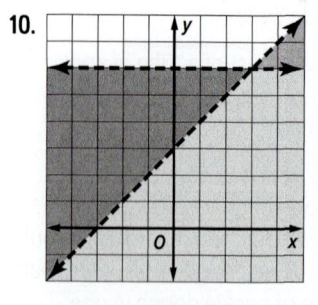

391C | Chapter 6 | Answer Appendix

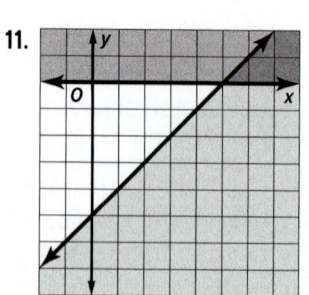

12.

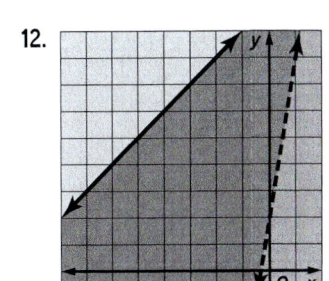

13.

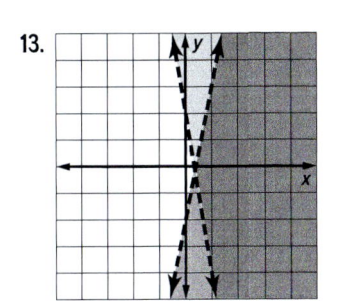

14.

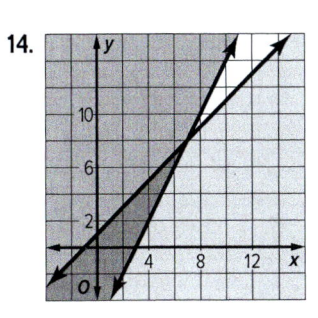

15.

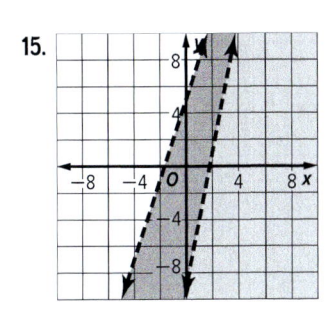

16. no solution

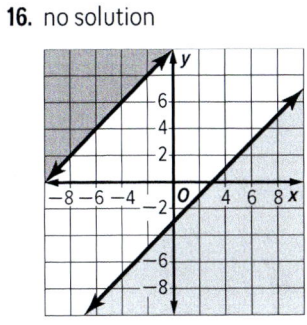

17. no solution

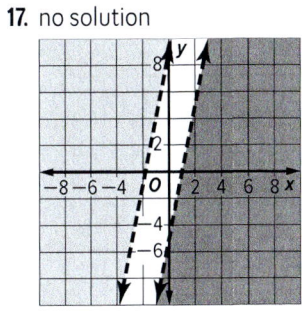

18.

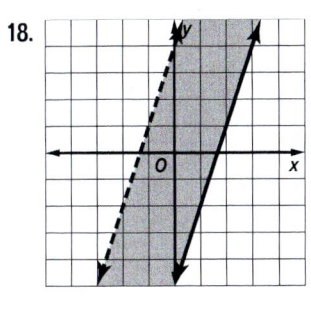

19.

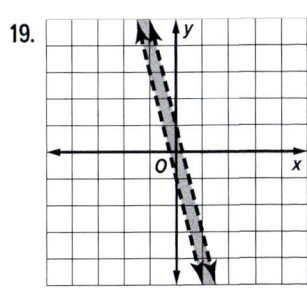

20.

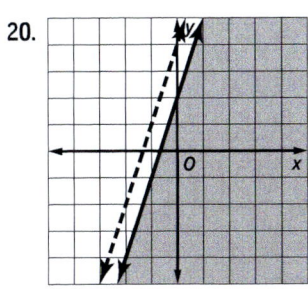

21.

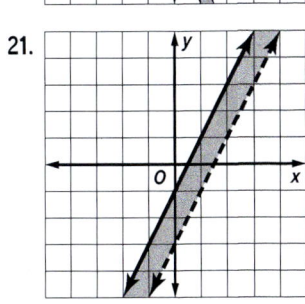

22.

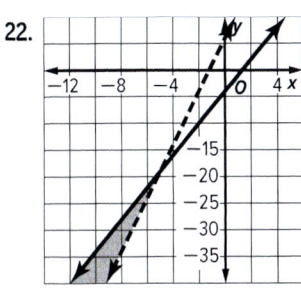

23.

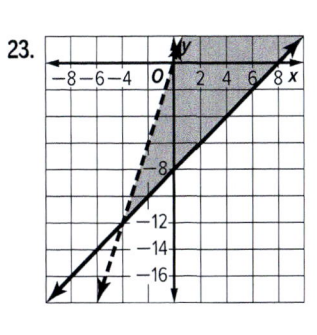

24. no solution

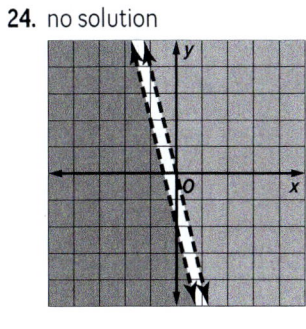

25a. Let $f$ = square footage and $p$ = price; $1000 \le f \le 17{,}000$ and $10{,}000 \le p \le 150{,}000$.

**Ice Rink Resurfacers**

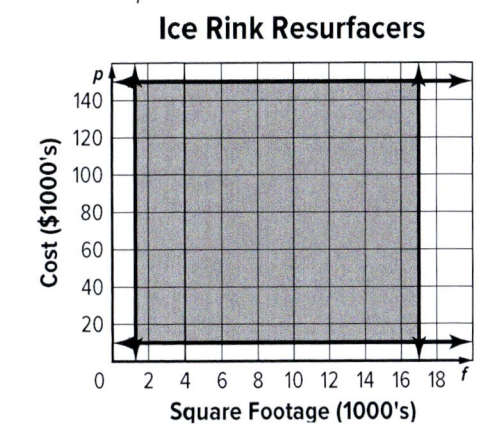

26b. **Earnings**

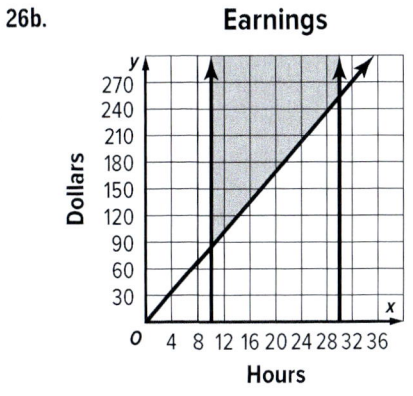

27.

28.

29.

30.

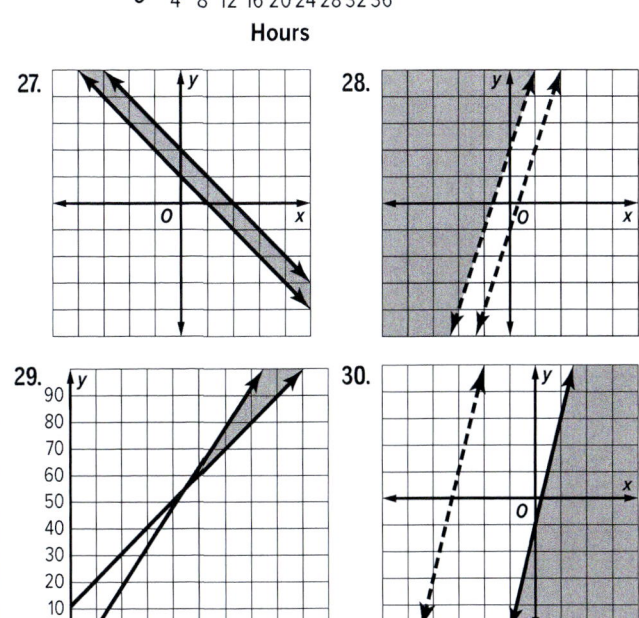

31.

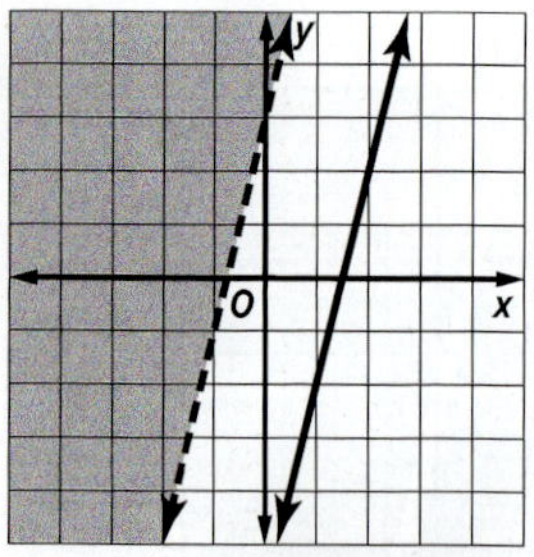

32.

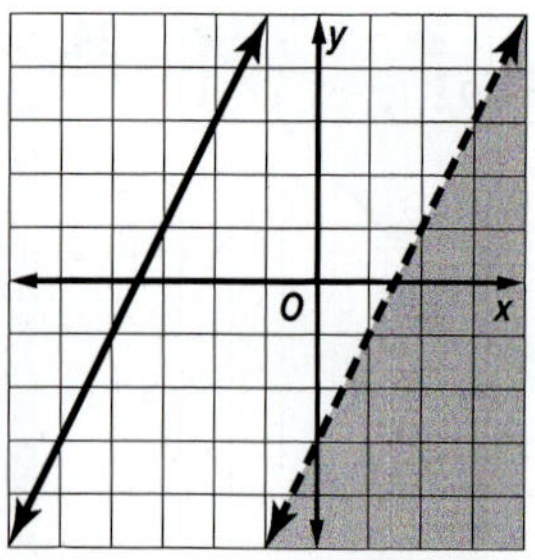

33.

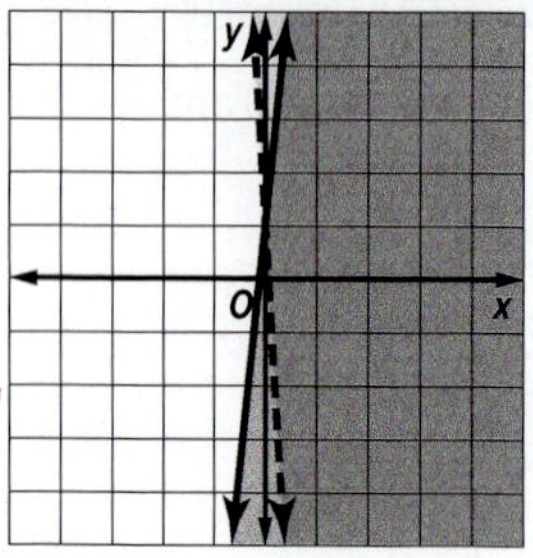

34.

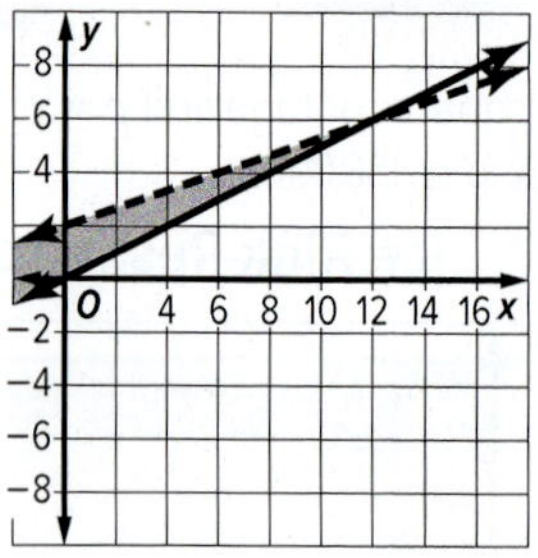


35.

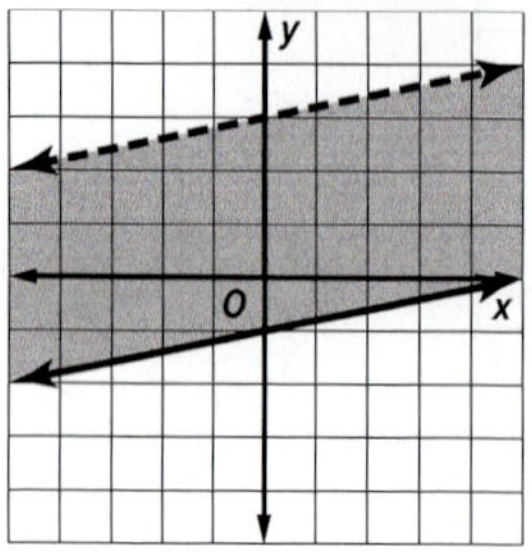

36a. Let $n$ = the number of notebooks and $p$ = the number of pens; $n \geq 20$, $p \geq 50$, $\$2.50n + \$1.25p \geq 150$

36b.

**Class Project**

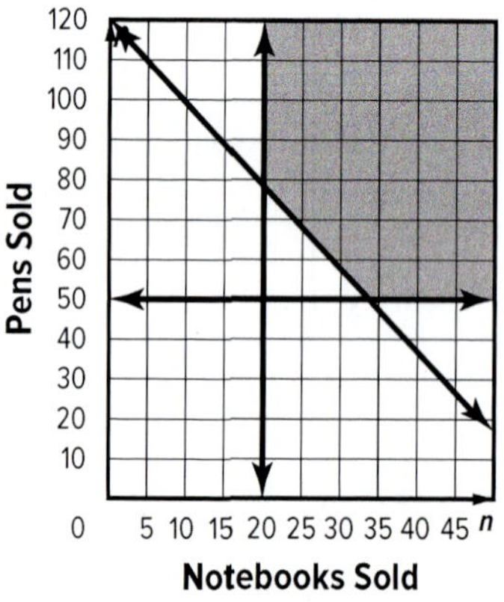


37b.

**Earnings**

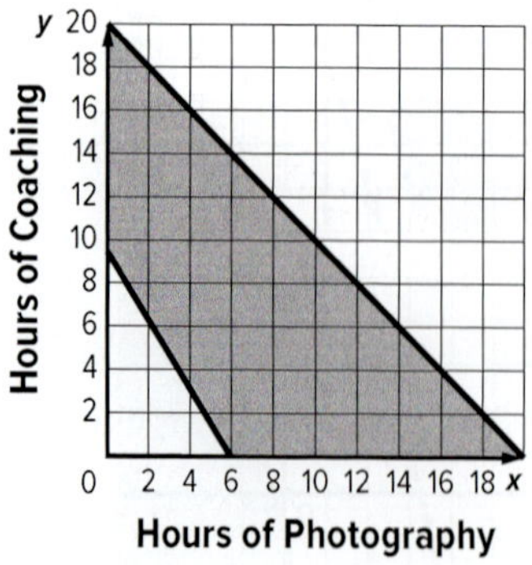


## Practice Test

1.

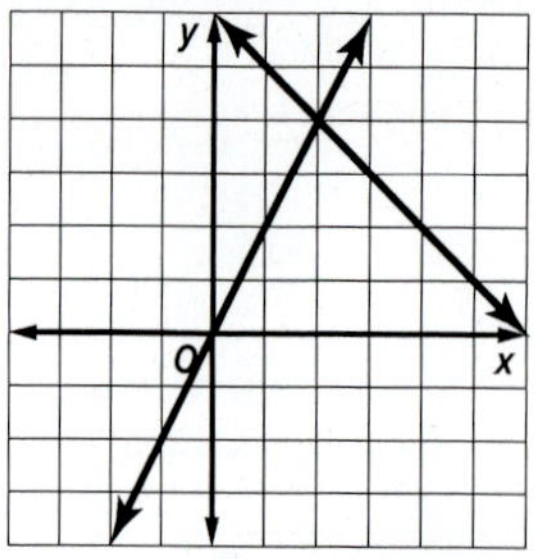

2.

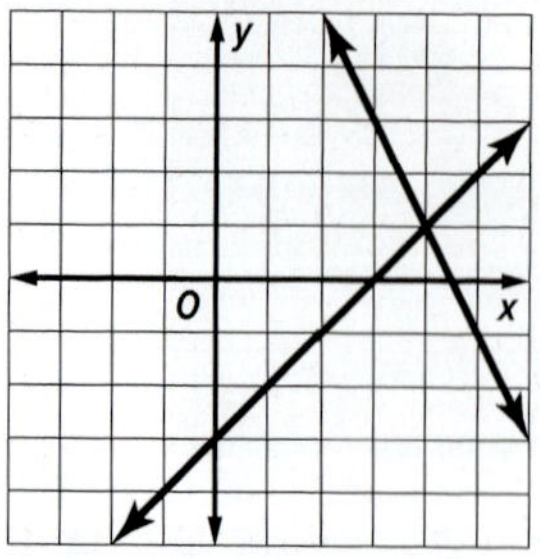

3.

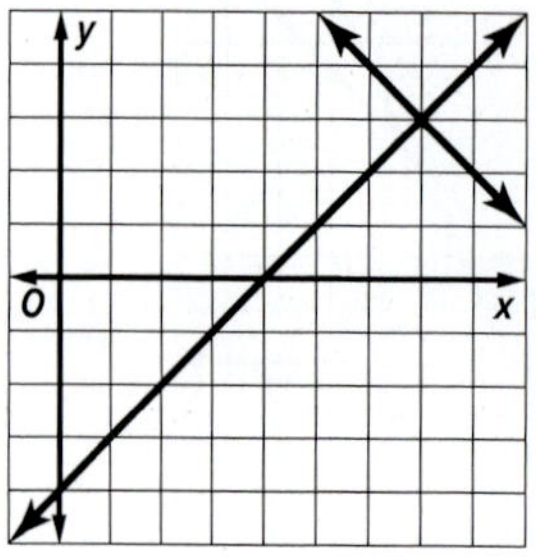

4.

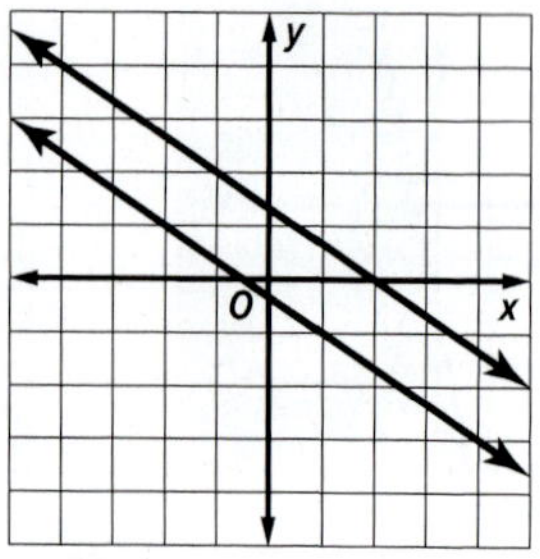

20.

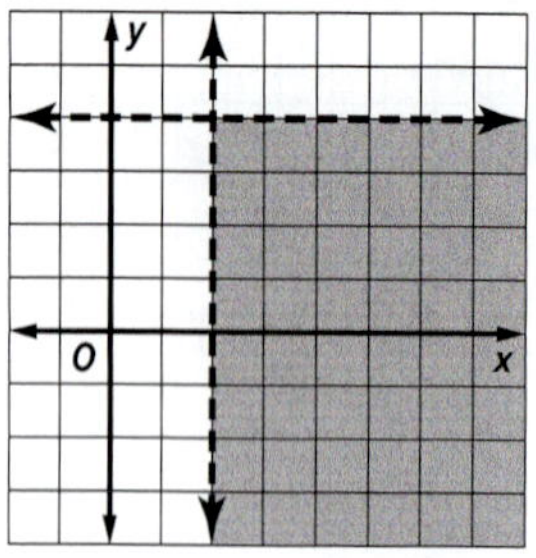

21.

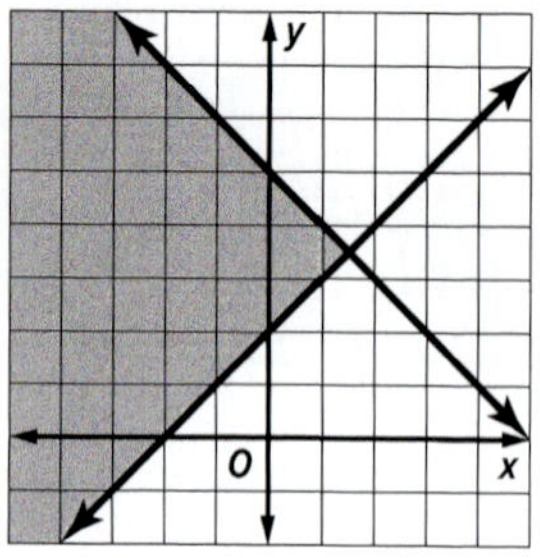

22.

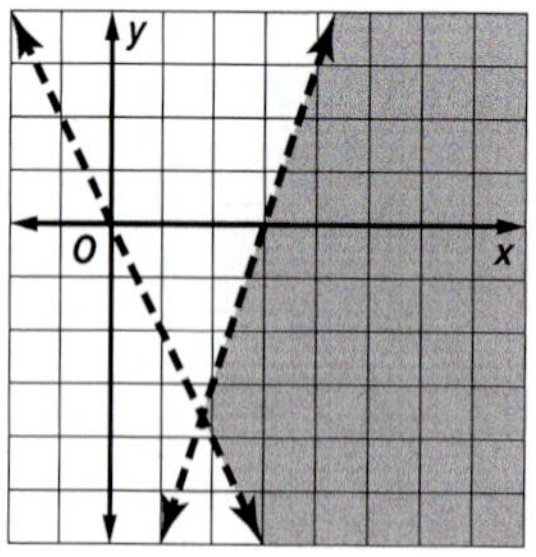

23.

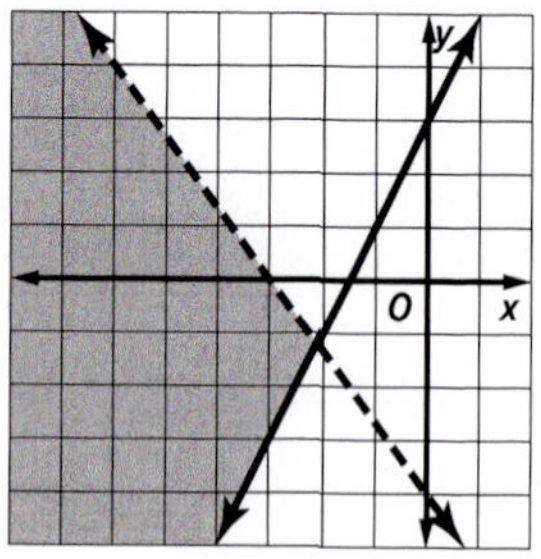

**NOTES**

# CHAPTER 7
# Exponents and Exponential Functions

**SUGGESTED PACING (DAYS)**

90 min.: 9 | 1
45 min.: 13 | 2
Instruction | Review & Assess

## Track Your Progress

This chapter focuses on content from the Interpreting Functions, The Real Number System, and Linear, Quadratic, and Exponential Models domains.

### THEN

**A.SSE.2** Use the structure of an expression to identify ways to rewrite it.

**F.IF.2** Use function notation, evaluate functions for inputs in their domains, and interpret statements that use function notation in terms of a context.

### NOW

**F.IF.7e** Graph exponential and logarithmic functions, showing intercepts and end behavior, and trigonometric functions, showing period, midline, and amplitude.

**F.LE.2** Construct linear and exponential functions, including arithmetic and geometric sequences, given a graph, a description of a relationship, or two input-output pairs (include reading these from a table).

**F.LE.5** Interpret the parameters in a linear or exponential function in terms of a context.

**N.RN.2** Rewrite expressions involving radicals and rational exponents using the properties of exponents.

### NEXT

**A.SSE.3b** Complete the square in a quadratic expression to reveal the maximum or minimum value of the function it defines.

**A.REI.4** Solve quadratic equations in one variable. (a) Use the method of completing the square to transform any quadratic equation in $x$ into an equation of the form $(x - p)^2 = q$ that has the same solutions. Derive the quadratic formula from this form. (b) Solve quadratic equations by inspection (e.g., for $x^2 = 49$), taking square roots, completing the square, the quadratic formula and factoring, as appropriate to the initial form of the equation. Recognize when the quadratic formula gives complex solutions and write them as $a \pm bi$ for real numbers $a$ and $b$.

**A.APR.1** Understand that polynomials form a system analogous to the integers, namely, they are closed under the operations of addition, subtraction, and multiplication; add, subtract, and multiply polynomials.

## Standards for Mathematical Practices

All of the Standards for Mathematical Practice will be covered in this chapter. The MP icon notes specific areas of coverage.

**MP** **Teaching the Mathematical Practices**
Help students develop the mathematical practices by asking questions like these.

**Questioning Strategies** As students approach problems in this chapter, help them develop mathematical practices by asking:

**Sense-Making**
- What is a geometric sequence and how does it relate to exponential functions?
- How can you tell if data display exponential behavior?

**Reasoning**
- How can you simplify and perform operations on expressions with exponents?
- How do you extend the properties of integer exponents to rational exponents?

**Tools**
- How can a graph be used to represent exponential functions?

**Structure**
- How do you find a pattern to write a recursive formula for a sequence?
- Can you simplify an expression using the multiplication properties of exponents?

### Go Online!

**StudySync: SMP Modeling Videos**

These demonstrate how to apply the Standards for Mathematical Practice to collaborate, discuss, and solve real-world math problems.

# Go Online!
connectED.mcgraw-hill.com

| LearnSmart | The Geometer's Sketchpad | Vocabulary | Personal Tutor | Tools | Calculator Resources | Self-Check Practice | Animations |

## Customize Your Chapter
Use the *Plan & Present*, *Assignment Tracker*, and *Assessment* tools in ConnectED to introduce lesson concepts, assign personalized practice, and diagnose areas of student need.

### Differentiated Instruction
Throughout the program, look for the icons to find specialized content designed for your students.

- **AL** Approaching Level
- **OL** On Level
- **BL** Beyond Level
- **ELL** English Language Learners

### Personalize

| Differentiated Resources | | | | |
|---|---|---|---|---|
| **FOR EVERY CHAPTER** | AL | OL | BL | ELL |
| ✓ Chapter Readiness Quizzes | ● | ● | ◐ | ● |
| ✓ Chapter Tests | ● | ● | ● | ● |
| ✓ Standardized Test Practice | ● | ● | ● | ● |
| abc Vocabulary Review Games | ● | ● | ● | ● |
| 📄 Anticipation Guide (English/Spanish) | ● | ● | ● | ● |
| 📄 Student-Built Glossary | ● | ● | ● | ● |
| 📄 Chapter Project | ● | ● | ● | ● |
| **FOR EVERY LESSON** | AL | OL | BL | ELL |
| 💬 Personal Tutors (English/Spanish) | ● | ● | ● | ● |
| 💬 Graphing Calculator Personal Tutors | ● | ● | ● | ● |
| ▷ Step-by-Step Solutions | ● | ● | ◐ | ● |
| ✓ Self-Check Quizzes | ● | ● | ● | ● |
| 📄 5-Minute Check | ● | ● | ● | ● |
| 📄 Study Notebook | ● | ● | ● | ● |
| 📄 Study Guide and Intervention | ● | ● | ◐ | ● |
| 📄 Skills Practice | ● | ◐ | ◐ | ● |
| 📄 Practice | ◐ | ● | ● | ● |
| 📄 Word Problem Practice | ● | ● | ● | ◐ |
| 📄 Enrichment | ◐ | ● | ● | ● |
| ✚ Extra Examples | ● | ◐ | ◐ | ◐ |
| 🖼 Lesson Presentations | ● | ● | ● | ● |

◐ Aligned to this group   ● Designed for this group

### Engage

**Featured IWB Resources**

**Geometer's Sketchpad** provides students with a tangible, visual way to learn. *Use with Lessons 7-1 and 7-2.*

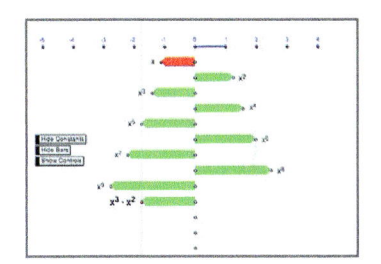

**eLessons** engage students and help build conceptual understanding of big ideas. *Use with Lessons 7-5, 7-6, and 7-7.*

**Brain POPs®** are animated, curricular content that engage students. *Use with Lesson 7-1 and 7-2.*

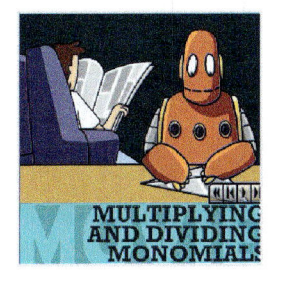

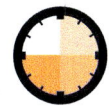

**Time Management** How long will it take to use these resources? Look for the clock in each lesson interleaf.

connectED.mcgraw-hill.com   392B

Chapter 7 | Exponents and Exponential Functions

# Introduce the Chapter

## Mathematical Background
Exponents and exponential functions have laws like all real numbers. Exponential growth and decay can be represented algebraically or by tables and graphs. Geometric sequences can be written as exponential functions and recursive formulas.

## Essential Question
At the end of this chapter, students should be able to answer the Essential Questions.

How can you make good decisions? What factors can affect good decision making? **Sample answers: Determine the available options, compare the pros and cons of each, analyze the outcomes, and choose the best. Some of the factors are time, process, environment, and people.**

## Apply Math to the Real World
**SPACE** In this activity, students will use what they already know to write equations to model the distance and orbit of objects in space. Have students work individually or in small groups to complete this project. **MP** 1, 4, 5

---

# CHAPTER 7
# Exponents and Exponential Functions

**THEN**
You evaluated expressions involving exponents.

**NOW**
In this chapter, you will:
- Simplify and perform operations on expressions involving exponents.
- Extend the properties of integer exponents to rational exponents.
- Write and transform exponential functions.
- Graph and use exponential functions.

**WHY**
**SPACE** NASA specializes in space exploration. NASA employees use math to, among other things, calculate travel time and to determine the solar power available to spacecraft as they travel away from the sun.

**Use the Mathematical Practices to complete the activity.**
1. **Using Tools** Use the Internet to find out how NASA calculated the solar energy available for the Juno spacecraft or another spacecraft and to learn the amount of solar power available on Earth (in w/m$^2$).
2. **Apply Math** Create a table to record distances from the sun (in AU) and then calculate the energy available to solar panels on the spacecraft at different points.
3. **Model with Mathematics** Use the Line Graph tool to graph the results of your research and calculations.

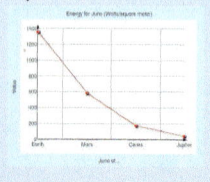

---

## Go Online!

**Chapter Project**
**INTERE$T-ing Thing About Credit Cards** Students can use what they have learned about exponential functions to complete a project. This project addresses financial literacy, as well as several specific skills identified as being essential to success by the Framework for 21st Century Learning. **MP** 1, 3, 7

---

## ALEKS®

**Your Student Success Tool** ALEKS is an adaptive, personalized learning environment that identifies precisely what each student knows and is ready to learn—ensuring student success at all levels.

- Formative Assessment: Dynamic, detailed reports monitor students' progress toward standards mastery.
- Automatic Differentiation: Strengthen prerequisite skills and target individual learning gaps.
- Personalized Instruction: Supplement in-class instruction with personalized assessment and learning opportunities.

Chapter 7 | Exponents and Exponential Functions

## Go Online to Guide Your Learning

### Explore & Explain

**The Geometer's Sketchpad**
With The Geometer's Sketchpad, you can multiply monomials using properties of exponents. It can also be used to simplify expressions containing negative and zero exponents, to simplify expressions containing negative and zero exponents, and to graph exponential functions.

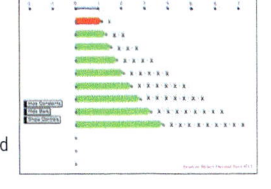

**Graphing Tools**
Explore **Exponential Functions** using the graphing tools to enhance your understanding in Lessons 7-5 and 7-7.

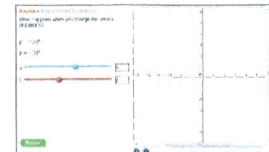

**eBook**
**Interactive Student Guide**
Before starting the chapter, answer the **Chapter Focus** preview questions. Check your answers as you complete each lesson. At the end of the chapter, try the **Performance Task**.

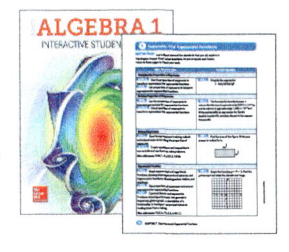

### Organize

**Foldables**
**Exponents and Exponential Functions** Make this Foldable to help you organize your notes about exponents and exponential functions. Begin with eleven sheets of notebook paper arranged in a stack and stapled. Starting with the second sheet of paper, cut along the right side to form tabs and label each tab with a lesson number.

### Collaborate

**Chapter Project**
In the **Interesting Thing About Credit Cards** project, you will use what you have learned about exponential functions to complete a project.

### Focus

**LEARNSMART**
Need help studying? Complete the **Linear and Exponential Relationships** domain in LearnSmart to review for the chapter test.

**ALEKS**
You can use the **Functions and Lines** and **Exponents** topics in ALEKS to find out what you know about linear functions and what you are ready to learn.*

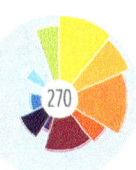

*Ask your teacher if this is part of your program.

## Dinah Zike's FOLDABLES

**Focus** Students create a tabbed book on which they organize information about polynomials.

**Teach** Have students make and label their Foldables as illustrated. Before beginning each lesson, ask students to think of one question that comes to mind as they skim through the lesson. Have them write the questions on the tabbed page of the corresponding lesson. As they read and work through the lesson, ask them to record the answers to their questions under the tabs.

**When to Use It** Encourage students to add to their Foldables as they work through the chapter and to use them to review for the chapter test.

### Go Online!

**Choosing Foldables**
How do you know which Foldables strategy to use? In this video, you will learn best practices to use when choosing Foldables. MP 5

connectED.mcgraw-hill.com

Chapter 7 | Exponents and Exponential Functions

# Get Ready for the Chapter

**RtI Response to Intervention**
Use the Concept Check results and the Intervention Planner chart to help you determine your Response to Intervention.

## Intervention Planner

### TIER 1 On Level OL

**IF** students miss 25% of the exercises or less,

**THEN** choose a resource:

**Go Online!**
- Skills Practice, Chapter 1
- Self-Check Quizzes

### TIER 2 Approaching Level AL

**IF** students miss 50% of the exercises,

**THEN** choose a resource:

**Go Online!**
- Study Guide and Intervention, Chapter 1
- Extra Examples
- Personal Tutors
- Homework Help

*Quick Review Math Handbook*

### TIER 3 Intensive Intervention

**IF** students miss 75% of the exercises,

**THEN** use *Math Triumphs, Alg. 1*

**Go Online!**
- Extra Examples
- Personal Tutors
- Homework Help
- Review Vocabulary

## Get Ready for the Chapter

*Go Online! for Vocabulary Review Games and key vocabulary in 13 languages.*

### Connecting Concepts

**Concept Check**
Review the concepts used in this chapter by answering the questions below.

1. What is a benefit of writing $4 \cdot 4 \cdot 4 \cdot 4 \cdot 4$ using exponents? **Using exponents simplifies expressions.**
2. If you know a photo is 4 inches on one side and 6 inches on the other, how would you determine the area of the photo? **You would multiply 4 by 6 to determine the area.**
3. What would the units be for the area of the picture in question 2? **The area would be in square inches.**
4. A cube is 5 feet on each side. How would you determine the volume of the cube? **Multiply the length by the width by the height.**
5. What would the units be for question 4? **The units would be feet cubed.**
6. Is the expression $-5^2$ different from $(-5)^2$? Why? **Yes, the first results in $-25$, the second in 25.**
7. Is the expression $2 + 5^2$ different from $(2 + 5)^2$? **Yes, these expressions are different.**
8. How would you evaluate the expression $\frac{5}{7} \times \frac{5}{7}$? How else could you write the expression? **Multiply 5 by 5 for numerator and 7 by 7 for denominator; or $\left(\frac{5}{7}\right)^2$.**

**Performance Task Preview**
You can use the concepts and skills in the chapter to solve problems in a real-world setting. Understanding exponents and exponential functions will help you finish the Performance Task at the end of the chapter.

**MP In this Performance Task you will:**
- make sense of problems
- look for and express regularity in reasoning

### New Vocabulary

| English | | Español |
|---|---|---|
| monomial | p. 395 | monomio |
| constant | p. 395 | constante |
| zero exponent | p. 403 | cero exponente |
| order of magnitude | p. 405 | orden de magnitud |
| rational exponent | p. 410 | exponent racional |
| cube root | p. 411 | raíz cúbica |
| $n$th root | p. 411 | raíz enésima |
| exponential equation | p. 413 | ecuación exponencial |
| radical expression | p. 419 | expresión radical |
| rationalizing the denominator | p. 421 | racionalizar el denominador |
| conjugate | p. 421 | conjugado |
| exponential function | p. 430 | función exponencial |
| exponential growth | p. 430 | crecimiento exponencial |
| exponential decay | p. 430 | desintegración exponencial |
| compound interest | p. 451 | interés es compuesta |
| geometric sequence | p. 462 | secuencia geométrica |
| common ratio | p. 462 | proporción común |
| recursive formula | p. 469 | fórmula recursiva |

### Review Vocabulary

**base** *base* In an expression of the form $x^n$, the base is $x$.

**Distributive Property** *Propiedad distributiva*
For any numbers $a$, $b$, and $c$, $a(b + c) = ab + ac$ and $a(b - c) = ab - ac$.

**exponent** *exponente*
In an expression of the form $x^n$, the exponent is $n$. It indicates the number of times $x$ is used as a factor.

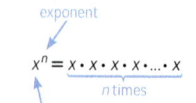

$$x^n = \underbrace{x \cdot x \cdot x \cdot x \cdot \ldots \cdot x}_{n \text{ times}}$$

## Key Vocabulary ELL

Introduce the key vocabulary in the chapter using the routine below.

**Define** A monomial is a number, a variable, or a product of a number and one or more variables.

**Example** $\frac{1}{20} n^2$

**Ask** Can you name another monomial? **Sample answer: $32ab$, 175, $x$, $y$**

# LESSON 7-1
# Multiplication Properties of Exponents

**SUGGESTED PACING (DAYS)**
90 min. 0.5
45 min. 1
Instruction

## Track Your Progress

### Objectives

1. Multiply monomials using the properties of exponents.

2. Simplify expressions using the multiplication properties of exponents.

### Mathematical Background
To multiply two powers that have the same base, add their exponents. To find the power of a power, multiply exponents. To find the power of a product, find the power of each factor and multiply. Monomial expressions are simplified when each base appears exactly once, there are no powers of powers, and all fractions are in simplest form.

| THEN | NOW | NEXT |
|---|---|---|
| **A.SSE.1a** Interpret parts of an expression, such as terms, factors, and coefficients. | **A.SSE.2** Use the structure of an expression to identify ways to rewrite it.<br><br>**F.IF.8b** Use the properties of exponents to interpret expressions for exponential functions. | **N.RN.1** Explain how the definition of the meaning of rational exponents follows from extending the properties of integer exponents to those values, allowing for a notation for radicals in terms of rational exponents. |

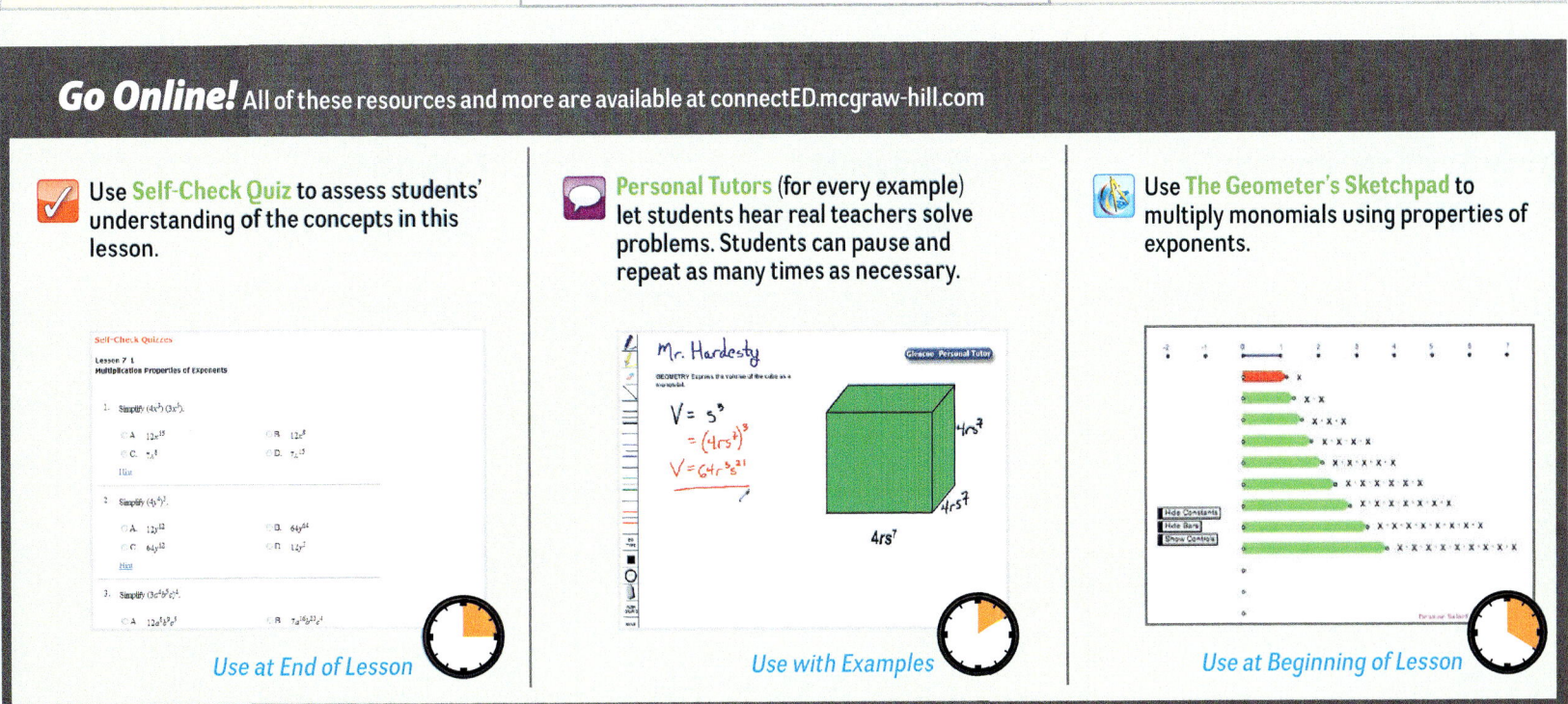

## Go Online! All of these resources and more are available at connectED.mcgraw-hill.com

Use **Self-Check Quiz** to assess students' understanding of the concepts in this lesson.

*Use at End of Lesson*

**Personal Tutors** (for every example) let students hear real teachers solve problems. Students can pause and repeat as many times as necessary.

*Use with Examples*

Use **The Geometer's Sketchpad** to multiply monomials using properties of exponents.

*Use at Beginning of Lesson*

## OER Using Open Educational Resources
**Games** Have students play the Otter Rush game on **Academic Skill Builders** to review properties of exponents before beginning this lesson. Up to 12 students can race each other at one time.
*Use before beginning the lesson*

connectED.mcgraw-hill.com 395A

# Go Online!
connectED.mcgraw-hill.com — Worksheets

# Differentiate Your Resources

**Extra Practice** Additional practice or homework; Skills Practice is best for approaching-level students and Practice is best for on-level and beyond-level students

## Skills Practice
AL  OL  ELL

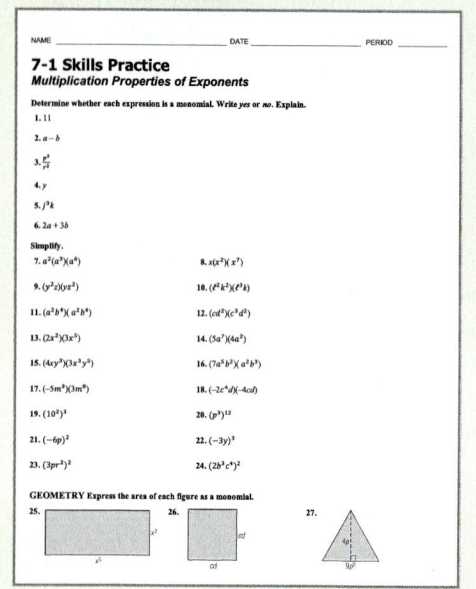

## Practice
AL  OL  BL  ELL

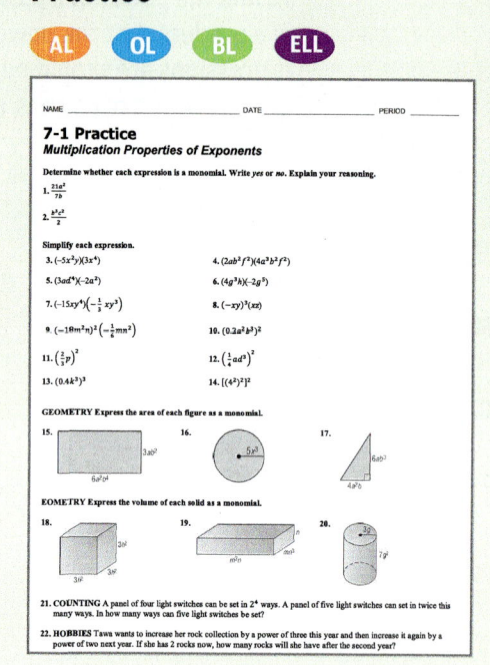

## Word Problem Practice
AL  OL  BL  ELL

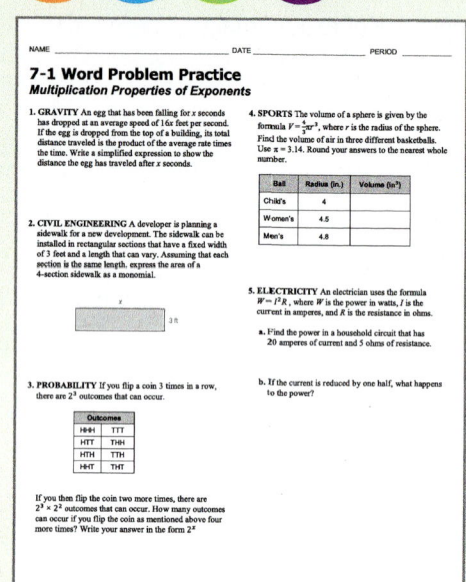

**Intervention** Reteaching and vocabulary activities that can be used with struggling or absent students and as ELL support

## Study Guide and Intervention
AL  OL  ELL

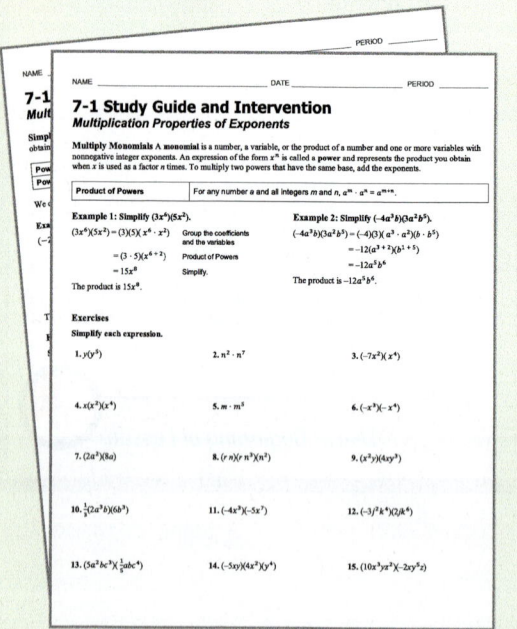

## Study Notebook
AL  OL  BL  ELL

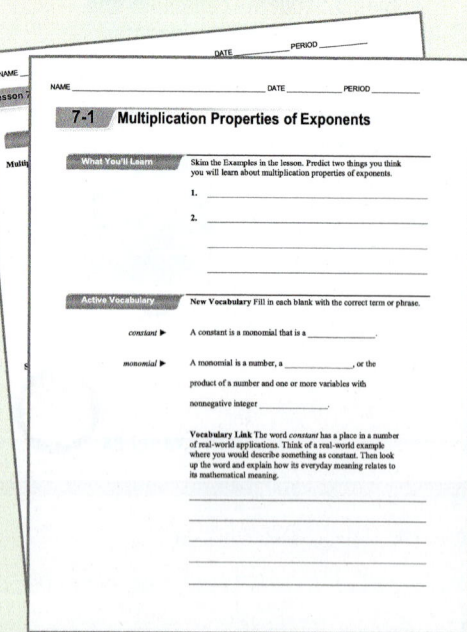

**Extension** Activities that can be used to extend lesson concepts

## Enrichment
OL  BL  ELL

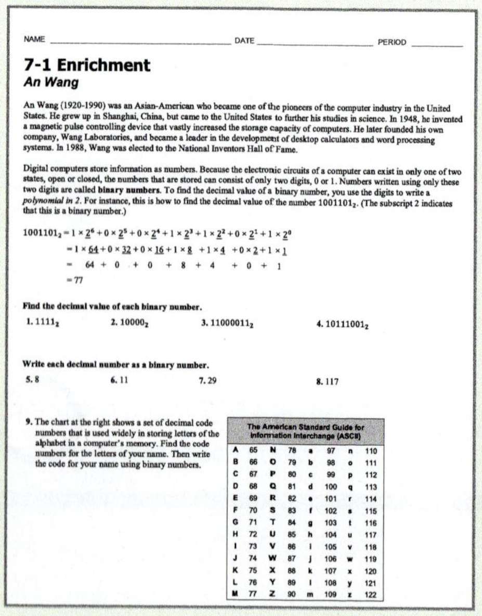

395B | Lesson 7-1 | Multiplication Properties of Exponents

# LESSON 1
# Multiplication Properties of Exponents

| Then | Now | Why? |
|---|---|---|
| • You evaluated expressions with exponents. | 1. Multiply monomials using the properties of exponents.<br>2. Simplify expressions using the multiplication properties of exponents. | Many formulas contain *monomials*. For example, the formula for the horsepower of a car is $H = w\left(\frac{v}{234}\right)^3$. $H$ represents the horsepower produced by the engine. $w$ equals the weight of the car with passengers, and $v$ is the velocity of the car at the end of a quarter of a mile. As the velocity increases, the horsepower increases. |

**New Vocabulary**
monomial
constant

**Mathematical Practices**
8 Look for and express regularity in repeated reasoning.

**Content Standards**
A.SSE.2 Use the structure of an expression to identify ways to rewrite it.
F.IF.8b Use the properties of exponents to interpret expressions for exponential functions.

**1 Multiply Monomials** A **monomial** is a number, a variable, or the product of a number and one or more variables with nonnegative integer exponents. It has only one term. In the formula to calculate the horsepower of a car, the term $w\left(\frac{v}{234}\right)^3$ is a monomial.

An expression that involves division by a variable, like $\frac{ab}{c}$, is not a monomial.

A **constant** is a monomial that is a real number. The monomial $3x$ is an example of a *linear expression* since the exponent of $x$ is 1. The monomial $2x^2$ is a *nonlinear expression* since the exponent is a positive number other than 1.

F.IF.8b

### Example 1  Identify Monomials

Determine whether each expression is a monomial. Write *yes* or *no*. Explain your reasoning.

a. 10  Yes; this is a constant, so it is a monomial.
b. $f + 24$  No; this expression has addition, so it has more than one term.
c. $h^2$  Yes; this expression is a product of variables.
d. $j$  Yes; single variables are monomials.

▸ **Guided Practice**

1A. $-x + 5$  **1A–1D. See margin.**    1B. $23abcd^2$
1C. $\frac{xyz^2}{2}$    1D. $\frac{mp}{n}$

Recall that an expression of the form $x^n$ is called a *power* and represents the result of multiplying $x$ by itself $n$ times. $x$ is the *base*, and $n$ is the *exponent*. The word *power* is also used sometimes to refer to the exponent.

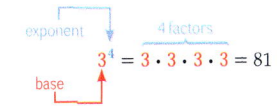

$$3^4 = 3 \cdot 3 \cdot 3 \cdot 3 = 81$$
(exponent, 4 factors, base)

---

## Mathematical Practices Strategies

**Look for and express regularity in repeated reasoning.**
Help students attend to the details of what makes an expression a simplified monomial while gaining understanding of the simplification process. For example, ask:

• What operations will you not see in a monomial? **addition and subtraction**

• Why is an expression that uses subtraction or addition not considered a monomial? **If an expression uses subtraction or addition, it has more than one term.**

• What is the greatest number of times the same variable variable will appear in a simplified monomial? **once**

---

## Lesson 7-1 | Multiplication Properties of Exponents

## Launch
Have students read the Why? section of the lesson. Ask:

• What two values do you need to know to be able to use the formula to find the horsepower of a car? **weight of the car, $w$, with passengers; velocity of the car, $v$, at the end of a quarter of a mile**

• Which values in the formula are raised to the third power? **$v$, 234**

• When would the horsepower of a car be equal to or greater than the weight of the passengers? **when $v \geq 234$**

## Teach
Ask the scaffolded questions for each example to build conceptual understanding for students at all levels.

### 1 Multiply Monomials

**Example 1  Identify Monomials**

**AL** In part **a**, what kind of term is given? **a constant**

**OL** In part **b**, what separates the terms? **an addition sign**

**BL** In part **b**, if it is not a monomial, what is it? **binomial**

*(continued on the next page)*

### Additional Answers

1A. No; the expression has addition and more than one term.
1B. Yes; this is a product of a number and variables.
1C. Yes; this is a product of variables, with a constant in the denominator.
1D. No; this expression has a variable in the denominator.

## Go Online!

**Interactive Whiteboard**
Use the *eLesson* or *Lesson Presentation* to present this lesson.

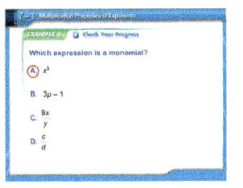

connectED.mcgraw-hill.com   395

# Lesson 7-1 | Multiplication Properties of Exponents

## Need Another Example?
Determine whether each expression is a monomial. Write *yes* or *no*. Explain your reasoning.
a. $17 - c$ **No; this expression involves subtraction, so it involves more than one term.**
b. $8f^2g$ **Yes; this expression is the product of a number and two variables.**
c. $\frac{3}{4}$ **Yes; the expression is a constant.**
d. $\frac{5}{t}$ **No; this expression has a variable in the denominator.**

## Example 2 Product of Powers

**AL** In part **a**, what are the coefficients of each factor? **6 and 2**

**OL** In part **b**, what do you do with the exponents if the bases are the same? **add them**

**BL** What properties allow you to regroup the factors so that the variables are together? **Commutative and Associative Properties of Multiplication**

## Need Another Example?
Simplify each expression.
a. $(r^4)(-12r^7)$ **$-12r^{11}$**
b. $(6cd^5)(5c^5d^2)$ **$30c^6d^7$**

## Example 3 Power of a Power

**AL** What are the exponents in the term? **3, 2, and 4**

**OL** When taking a power of a power, what do you do with the exponents? **multiply them**

**BL** Show and explain why multiplying the exponents makes sense. **Sample answer: $2^3$ is $2 \cdot 2 \cdot 2$. That entire quantity is to the second power, so write that as $(2 \cdot 2 \cdot 2)(2 \cdot 2 \cdot 2)$. That entire quantity is to the fourth power, so put it in brackets and multiply it four times. Count the number of twos being multiplied; there are 24, so the shorter way to write this is $2^{24}$. The same answer is obtained by multiplying 3 times 2 times 4.**

---

By applying the definition of a power, you can find the product of powers. Look for a pattern in the exponents.

$$2^2 \cdot 2^4 = \underbrace{2 \cdot 2}_{\text{2 factors}} \cdot \underbrace{2 \cdot 2 \cdot 2 \cdot 2}_{\text{4 factors}}$$
$$2 + 4 = 6 \text{ factors}$$

$$4^3 \cdot 4^2 = \underbrace{4 \cdot 4 \cdot 4}_{\text{3 factors}} \cdot \underbrace{4 \cdot 4}_{\text{2 factors}}$$
$$3 + 2 = 5 \text{ factors}$$

These examples demonstrate the property for the product of powers.

### Key Concept Product of Powers
**Words** To multiply two powers that have the same base, add their exponents.
**Symbols** For any real number $a$ and any integers $m$ and $p$, $a^m \cdot a^p = a^{m+p}$.
**Examples** $b^3 \cdot b^5 = b^{3+5}$ or $b^8$ $\qquad g^4 \cdot g^6 = g^{4+6}$ or $g^{10}$

A.SSE.2, F.IF.8b

### Example 2 Product of Powers
Simplify each expression.
a. $(6n^3)(2n^7)$
$(6n^3)(2n^7) = (6 \cdot 2)(n^3 \cdot n^7)$ Group the coefficients and the variables.
$= (6 \cdot 2)(n^{3+7})$ Product of Powers
$= 12n^{10}$ Simplify.

b. $(3pt^3)(p^3t^4)$
$(3pt^3)(p^3t^4) = (3 \cdot 1)(p \cdot p^3)(t^3 \cdot t^4)$ Group the coefficients and the variables.
$= (3 \cdot 1)(p^{1+3})(t^{3+4})$ Product of Powers
$= 3p^4t^7$ Simplify.

**Study Tip**
**Coefficients and Powers of 1** A variable with no exponent or coefficient shown can be assumed to have an exponent and coefficient of 1. For example, $x = 1x^1$.

▶ **Guided Practice**
2A. $(3y^4)(7y^5)$ **$21y^9$**
2B. $(-4rx^2t^3)(-6r^5x^2t)$ **$24r^6x^4t^4$**

We can use the Product of Powers Property to find the power of a power. In the following examples, look for a pattern in the exponents.

$$(3^2)^4 = \underbrace{(3^2)(3^2)(3^2)(3^2)}_{\text{4 factors}}$$
$$= 3^{2+2+2+2}$$
$$= 3^8$$

$$(r^4)^3 = \underbrace{(r^4)(r^4)(r^4)}_{\text{3 factors}}$$
$$= r^{4+4+4}$$
$$= r^{12}$$

These examples demonstrate the property for the power of a power.

### Key Concept Product of Powers
**Words** To find the power of a power, multiply the exponents.
**Symbols** For any real number $a$ and any integers $m$ and $p$, $(a^m)^p = a^{m \cdot p}$.
**Examples** $(b^3)^5 = b^{3 \cdot 5}$ or $b^{15}$ $\qquad (g^6)^7 = g^{6 \cdot 7}$ or $g^{42}$

---

### Differentiated Instruction OL BL

**Logical Learners** Give students an expression such as $240x^{12}y^8$ and challenge them to write ten different monomial expressions that, when simplified, equal the given expression.

Lesson 7-1 | Multiplication Properties of Exponents

**Study Tip**

**MP Regularity** The power rules are general methods. If you are unsure about when to multiply the exponents and when to add the exponents, write the expression in expanded form.

A.SSE.2, F.IF.8b

**Example 3** Power of a Power

Simplify $[(2^3)^2]^4$.

A $2^{24}$    B $2^{12}$    C $2^{10}$    D $2^9$

**Read the Item**

You need to apply the power of a power rule.

**Solve the Item**

$[(2^3)^2]^4 = (2^{3 \cdot 2})^4$    Power of a Power
$= (2^6)^4$    Simplify
$= 2^{6 \cdot 4}$ or $2^{24}$    Power of a Power

The correct choice is A.

▶ **Guided Practice**

3. Simplify $[(2^2)^2]^4$. **C**

   A $2^8$    B $2^{10}$    C $2^{16}$    D $2^{24}$

We can use the Product of Powers Property and the Power of a Power Property to find the power of a product. Look for a pattern in the exponents below.

3 factors

$(tw)^3 = (tw)(tw)(tw)$
$= (t \cdot t \cdot t)(w \cdot w \cdot w)$
$= t^3w^3$

3 factors

$(2yz^2)^3 = (2yz^2)(2yz^2)(2yz^2)$
$= (2 \cdot 2 \cdot 2)(y \cdot y \cdot y)(z^2 \cdot z^2 \cdot z^2)$
$= 2^3y^3z^6$ or $8y^3z^6$

These examples demonstrate the property for the power of a product.

**Key Concept** Power of a Product

| Words | To find the power of a product, find the power of each factor and multiply. |
| --- | --- |
| Symbols | For any real numbers $a$ and $b$ and any integer $m$, $(ab)^m = a^m b^m$. |
| Example | $(-2xy^3)^5 = (-2)^5 x^5 y^{15}$ or $-32x^5 y^{15}$ |

A.SSE.2, F.IF.8b

**Example 4** Power of a Product

**GEOMETRY** Express the area of the circle as a monomial.

Area $= \pi r^2$    Formula for the area of a circle
$= \pi(2xy^2)^2$    Replace $r$ with $2xy^2$.
$= \pi(2^2 x^2 y^4)$    Power of a Product
$= 4x^2 y^4 \pi$    Simplify.

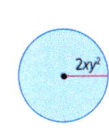

The area of the circle is $4x^2y^4\pi$ square units.

▶ **Guided Practice**

4A. Express the area of a square with sides of length $3xy^2$ as a monomial. **$9x^2y^4$**

4B. Express the area of a triangle with height $4a$ and base $5ab^2$ as a monomial. **$10a^2b^2$**

**Math-History Link**

**Albert Einstein (1879–1955)** Albert Einstein is perhaps the most well-known scientist of the 20th century. His formula, $E = mc^2$, where $E$ represents the energy, $m$ is the mass of the material, and $c$ is the speed of light, shows that if mass is accelerated enough, it could be converted into usable energy.

**Need Another Example?**

Simplify $[(2^3)^3]^2$. **D**

A $8^2$
B $8^4$
C $2^{11}$
D $2^{18}$

**Example 4** Power of a Product

**AL** What is the formula for the area of a circle?
$\pi r^2$

**OL** What is being raised to the second power?
2 and $x$ and $y^2$

**BL** In your own words, explain the Power of a Product Law. See students' work.

**Need Another Example?**

**Geometry** Express the volume of a cube with side length $5xyz$ as a monomial. $(5xyz)^3 = 125x^3y^3z^3$

**Watch Out!**

**Student Misconceptions** Students may simplify an expression such as $\frac{4}{6}(x^2y^5)^3[2(xy)^7]$ into $\frac{8}{6}x^{13}y^{22}$, not realizing that the simplification is incomplete because the fraction is not in the simplest form.

**MP Teaching the Mathematical Practices**

**Regularity** Mathematically proficient students notice if calculations are repeated, and look both for general methods and for shortcuts. Work through examples to demonstrate each property in this lesson to help students see the repeated reasoning in the calculations.

---

**Differentiated Instruction ELL**

**Beginning** Say key terms such as *monomial*, *constant*, *exponent*, *base*, and *power*, aloud, one at a time. Have students raise their hands if they have heard the term. Have them use a word or phrase to tell something about the term. Use yes/no questions and prompts to help students complete a KWL Chart like the one below.
**Intermediate** Have students work with a partner to complete the K section of the chart. Have partners seek clarification and discuss each term.
**Advanced** After students complete the L section of the chart, have them work in groups to compare answers and add each other's information to their own charts.

| Multiplication Properties of Exponents | |
| --- | --- |
| K—What I Already Know | |
| W—What I Want to Learn | |
| L—What I Learned | |

connectED.mcgraw-hill.com    397

**Lesson 7-1 | Multiplication Properties of Exponents**

## 2 Simplify Expressions

### Example 5 Simplify Expressions

**AL** How many terms are in the expression? **one**

**OL** To what factors does the exponent 2 need to be applied? **3, x, and y in first term and −2 and y in the second term**

**BL** When there is a power of a power of a product, does it matter if you do the power of a product first or if you do the power of a power first? Explain using the Example. **no; Sample answer: multiplication is commutative. You could do $(-2)^2$ first, which is 4, then take $4^3$, which is 64, or you could multiply the powers first to get $(-2)^6$, then simplify to get 64.**

### Need Another Example?
Simplify $[(8g^3h^4)^2]^2(2gh^5)^4$. **$65,536g^{16}h^{36}$**

## Teaching Tip

**Reasoning** Remind students that there is often more than one strategy that can be used to simplify an expression. For example, in Example 5, the first step could be to simplify $(-2y)^2$ first and then raise the product to the third power.

## Practice

**Formative Assessment** Use Exercises 1–20 to assess students' understanding of the concepts in this lesson.

The Practice and Problem Solving exercises assess the content taught in the lesson. The Preparing for Assessment page is meant to be used as preparation for end-of-course assessments.

## Go Online!   eBook

**Interactive Student Guide**
Use the *Interactive Student Guide* to deepen conceptual understanding.
· Multiplication Properties of Exponents

**398** | Lesson 7-1 | Multiplication Properties of Exponents

---

**Go Online!**

Remember that the order of operations applies when simplifying expressions involving monomials. **Search order of operations** in ConnectED to find resources to review.

## 2 Simplify Expressions
We can combine and use these properties to simplify expressions involving monomials.

### Key Concept Simplifying Monomial Expressions

To simplify a monomial expression, write an equivalent expression in which:
- each variable base appears exactly once,
- there are no powers of powers, and
- all fractions are in simplest form.

A.SSE.2, F.IF.8b

### Example 5 Simplify Expressions

Simplify $(3xy^4)^2[(-2y)^2]^3$.

$(3xy^4)^2[(-2y)^2]^3 = (3xy^4)^2(-2y)^6$   Power of a Power
$= (3)^2x^2(y^4)^2(-2)^6y^6$   Power of a Product
$= 9x^2y^8(64)y^6$   Power of a Power
$= 9(64)x^2 \cdot y^8 \cdot y^6$   Commutative
$= 576x^2y^{14}$   Product of Powers

### Guided Practice

5. Simplify $\left(\frac{1}{2}a^2b^2\right)^3[(-4b)^2]^2$. **$32a^6b^{10}$**

**Go Online!** for a Self-Check Quiz

### Check Your Understanding    ○ = Step-by-Step Solutions begin on page R11.

**Example 1** Determine whether each expression is a monomial. Write *yes* or *no*. Explain your reasoning. **1–6. See margin.**
F.IF.8b

1. 15
2. $2 - 3a$
3. $\frac{5c}{d}$
4. $-15g^2$
5. $\frac{r}{2}$
6. $7b + 9$

**Examples 2–3** Simplify each expression.
A.SSE.2, F.IF.8b

7. $k(k^3)$  **$k^4$**
8. $m^4(m^2)$  **$m^6$**
9. $2q^2(9q^4)$  **$18q^6$**
10. $(5u^4v)(7u^4v^3)$  **$35u^8v^4$**
11. $[(3^2)^2]^2$  **$3^8$ or 6561**
12. $(xy^4)^6$  **$x^6y^{24}$**
13. $(4a^4b^9c)^2$  **$16a^8b^{18}c^2$**
14. $(-2f^2g^3h^2)^3$  **$-8f^6g^9h^6$**
15. $(-3p^5t^6)^4$  **$81p^{20}t^{24}$**

**Example 4** 16. **GEOMETRY** The formula for the surface area of a cube is $SA = 6s^2$, where $SA$ is the surface area and $s$ is the length of any side.
A.SSE.2, F.IF.8b

a. Express the surface area of the cube as a monomial. **$6a^6b^2$**
b. What is the surface area of the cube if $a = 3$ and $b = 4$? **69,984 units$^2$**

($a^3b$)

**Example 5** Simplify each expression.
A.SSE.2, F.IF.8b

17. $(5x^2y)^2(2xy^3z)^3(4xyz)$  **$800x^8y^{12}z^4$**
18. $(-3d^2f^3g)^2[(-3d^2f)^3]^2$  **$6561d^{16}f^{12}g^2$**
19. $(-2g^3h)(-3gj^4)^2(-ghj)^2$  **$-18g^7h^3j^{10}$**
20. $(-7ab^4c)^3[(2a^2c)^2]^3$  **$-21,952a^{15}b^{12}c^9$**

### Additional Answers

1. Yes; constants are monomials.
2. No; there is subtraction and more than one term.
3. No; there is a variable in the denominator.
4. Yes; this is a product of a number and variables.
5. Yes; this is a product of a number and variables.
6. No; there is addition and more than one term.

## Practice and Problem Solving

Extra Practice is on page R7.

**Example 1** Determine whether each expression is a monomial. Write *yes* or *no*. Explain your reasoning. **21–26. See margin.**

21. 122
22. $3a^4$
23. $2c + 2$
24. $\dfrac{-2g}{4h}$
25. $\dfrac{5k}{10}$
26. $6m + 3n$

**Examples 2–3** Simplify each expression.

27. $(q^2)(2q^4)$  $2q^6$
28. $(-2u^2)(6u^6)$  $-12u^8$
29. $(9w^2x^8)(w^6x^4)$  $9w^8x^{12}$
30. $(y^6z^9)(6y^4z^2)$  $6y^{10}z^{11}$
31. $(b^8c^6d^5)(7b^6c^2d)$  $7b^{14}c^8d^6$
32. $(14fg^2h^2)(-3f^4g^2h^2)$  $-42f^5g^4h^4$
33. $(j^5k^7)^4$  $j^{20}k^{28}$
34. $(n^3p)^4$  $n^{12}p^4$
35. $[(2^2)^2]^2$  $2^8$ or 256
36. $[(3^2)^2]^4$  $3^{16}$ or 43,046,721
37. $[(4r^2t)^3]^2$  $4096r^{12}t^6$
38. $[(-2xy^2)^3]^2$  $64x^6y^{12}$

**Example 4** GEOMETRY Express the area of each triangle as a monomial.

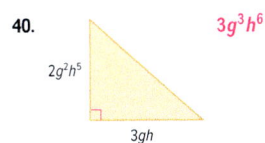

39. $20c^5d^5$ (triangle with sides $8c^2d^4$ and $5c^3d$)
40. $3g^3h^6$ (triangle with sides $2g^2h^5$ and $3gh$)

**Example 5** Simplify each expression.

41. $(2a^3)^4(a^3)^3$  $16a^{21}$
42. $(c^3)^2(-3c^5)^2$  $9c^{16}$
43. $(2gh^4)^3[(-2g^4h)^3]^2$  $512g^{27}h^{18}$
44. $(5k^2m)^3[(4km^2)^2]^2$  $32{,}000k^{10}m^{19}$
45. $(p^5r^2)^4(-7p^3r^4)^2(6pr^3)$  $294p^{27}r^{19}$
46. $(5x^2y)^2(2xy^3z)^3(4xyz)$  $800x^8y^{12}z^4$
47. $(5a^2b^3c^4)(6a^3b^4c^2)$  $30a^5b^7c^6$
48. $(10xy^5z^3)(3x^4y^6z^3)$  $30x^5y^{11}z^6$
49. $(0.5x^3)^2$  $0.25x^6$
50. $(0.4h^5)^3$  $0.064h^{15}$
51. $\left(-\dfrac{3}{4}c\right)^3$  $-\dfrac{27}{64}c^3$
52. $\left(\dfrac{4}{5}a^2\right)^2$  $\dfrac{16}{25}a^4$
53. $(8y^3)(-3x^2y^2)\left(\dfrac{3}{8}xy^4\right)$  $-9x^3y^9$
54. $\left(\dfrac{4}{7}m\right)^2(49m)(17p)\left(\dfrac{1}{34}p^5\right)$  $8m^3p^6$
55. $(-3r^3w^4)^3(2rw)^2(-3r^2)^3(4rw^2)^3(2r^2w^3)^4$  $2{,}985{,}984r^{28}w^{32}$
56. $(3ab^2c)^2(-2a^2b^4)^2(a^4c^2)^3(a^2b^4c^5)^2(2a^3b^2c^4)^3$  $288a^{31}b^{26}c^{30}$

57. **FINANCIAL LITERACY** Cleavon has money in an account that earns 3% simple interest. The formula for computing simple interest is $I = Prt$, where $I$ is the interest earned, $P$ represents the principal that he put into the account, $r$ is the interest rate (in decimal form), and $t$ represents time in years.

   a. Cleavon makes a deposit of $2c$ and leaves it for 2 years. Write a monomial that represents the interest earned. **0.12c**
   b. If $c$ represents a birthday gift of $250, how much will Cleavon have in this account after 2 years? **$280**

**TOOLS** Express the volume of each solid as a monomial.

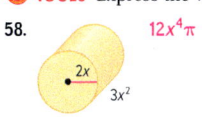

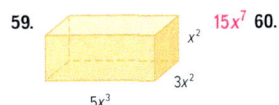

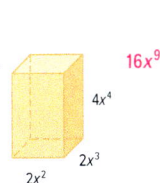

58. $12x^4\pi$ (cylinder: radius $2x$, height $3x^2$)
59. $15x^7$ (rectangular prism: $x^2$, $5x^3$, $3x^2$)
60. $16x^9$ (rectangular prism: $4x^4$, $2x^3$, $2x^2$)

---

## Differentiated Homework Options

| Levels | AL Basic | OL Core | BL Advanced |
|---|---|---|---|
| Exercises | 21–57, 65–74 | 21–57 odd, 59, 61–63, 65–74 | 55–67<br>68–74 (optional) |
| 2-Day Option | 21–57 odd, 59, 68–74 | 21–57, 68–74 | |
| | 22–56 even, 65–67 | 58–63, 65–67 | |

You can use ALEKS to provide additional remediation support with personalized instruction and practice.

---

**Lesson 7-1 | Multiplication Properties of Exponents**

### Exercise Alert
**Formulas** For Exercises 39, 40, and 58–60, students will need to know the formulas for area of a triangle and for volume of solids.

### Teaching the Mathematical Practices
**Tools** Mathematically proficient students are able to identify relevant external mathematical resources and use them to solve problems. In Exercises 58–60, point out that the volume formulas are listed inside the back cover of their book and can be found online.

### Extra Practice
See page R7 for extra exercises for students who are approaching level or for on-level students who need additional reinforcement.

### Levels of Complexity Chart

The levels of the exercises progress from 1 to 3, with Level 1 indicating the lowest level of complexity.

| Exercises | 21–54 | 55–63, 68–74 | 64–67 |
|---|---|---|---|
| Level 3 | | | ● |
| Level 2 | | ● | |
| Level 1 | ● | | |

### Additional Answers

21. Yes; constants are monomials.
22. Yes; this is a product of a number and variables.
23. No; there is addition and more than one term.
24. No; there is a variable in the denominator.
25. Yes; this can be written as the product of a number and a variable.
26. No; there is addition and more than one term.

### Go Online!
The most up-to-date resources available for your program can be found at connectED.mcgraw-hill.com.

## Lesson 7-1 | Multiplication Properties of Exponents

### Teaching the Mathematical Practices

**Perseverance** Mathematically proficient students try simpler forms of the original problem in order to gain insight into its solution. In Exercise 64, advise students to substitute integers for $a$ and $b$ and simplify the expressions before generalizing.

## Assess

**Ticket Out the Door** Make several copies each of five monomial expressions that need to be simplified. Give one expression to each student. As students leave the room, ask them to tell you the simplified versions of the expressions they possess.

### Additional Answers

**65b.**

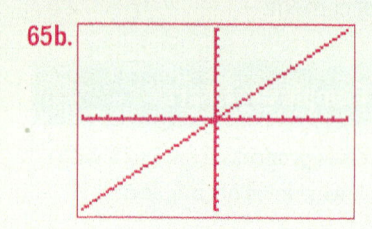

[−10, 10] scl: 1 by [−10, 10] scl: 1

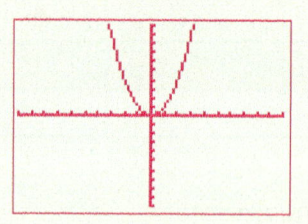

[−10, 10] scl: 1 by [−10, 10] scl: 1

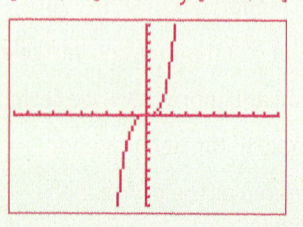

[−10, 10] scl: 1 by [−10, 10] scl: 1

**65d.** If the power of $x$ is 1, the equation or its related expression is linear. Otherwise, it is nonlinear.

### Go Online!

**eSolutions Manual**
Create worksheets, answer keys, and solutions handouts for your assignments.

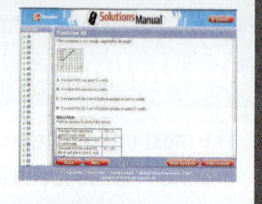

---

**61. PACKAGING** For a commercial art class, Aiko must design a new container for individually wrapped pieces of candy. The shape that she chose is a cylinder. The formula for the volume of a cylinder is $V = \pi r^2 h$.

a. The radius that Aiko would like to use is $2p^3$, and the height is $4p^3$. Write a monomial that represents the volume of her container. **$16\pi p^9$**

b. Make a table for five possible measures for the radius and height of a cylinder having the same volume.

c. What is the volume of Aiko's container if the height is doubled? **$32\pi p^9$**

**62. ENERGY** Albert Einstein's formula $E = mc^2$ shows that if mass is accelerated enough, it could be converted into usable energy. Energy $E$ is measured in joules, mass $m$ in kilograms, and the speed $c$ of light is about 300 million meters per second.

a. Complete the calculations to convert 3 kilograms of gasoline completely into energy. **270,000,000,000,000,000 joules**

b. What happens to the energy if the amount of gasoline is doubled? **The energy is also doubled.**

**63. MULTIPLE REPRESENTATIONS** In this problem, you will explore exponents.

a. **Tabular** Copy and use a calculator to complete the table.

| Power | $3^4$ | $3^3$ | $3^2$ | $3^1$ | $3^0$ | $3^{-1}$ | $3^{-2}$ | $3^{-3}$ | $3^{-4}$ |
|---|---|---|---|---|---|---|---|---|---|
| Value | 81 | 27 | 9 | 3 | 1 | $\frac{1}{3}$ | $\frac{1}{9}$ | $\frac{1}{27}$ | $\frac{1}{81}$ |

**61b. Sample answer:**

| Radius | Height |
|---|---|
| $4p$ | $p^7$ |
| $4p^2$ | $p^5$ |
| $2p^3$ | $4p^3$ |
| $2p^4$ | $4p$ |
| $2p$ | $4p^7$ |

b. **Analytical** What do you think the values of $5^0$ and $5^{-1}$ are? Verify your conjecture using a calculator. **1 and $\frac{1}{5}$**

c. **Analytical** Complete: For any nonzero number $a$ and any integer $n$, $a^{-n} = \underline{\frac{1}{a^n}}$.

d. **Verbal** Describe the value of a nonzero number raised to the zero power. **Any nonzero number raised to the zero power is 1.**

A.SSE.2, F.IF.8B

### H.O.T. Problems  Use Higher-Order Thinking Skills

**64. PERSEVERANCE** For any nonzero real numbers $a$ and $b$ and any integers $m$ and $t$, simplify the expression $\left(-\frac{a^m}{b^t}\right)^{2t}$ and describe each step. **64. Sample answer:** First use the power of a power rule to simplify the expression to $\frac{a^{2tm}}{b^{2t^2}}$.

**65. REGULARITY** Copy the table below.

| Equation | Related Expression | Power of x | Linear or Nonlinear |
|---|---|---|---|
| $y = x$ | $x$ | 1 | linear |
| $y = x^2$ | $x^2$ | 2 | nonlinear |
| $y = x^3$ | $x^3$ | 3 | nonlinear |

a. For each equation, write the related expression and record the power of $x$. See chart above.
b. Graph each equation using a graphing calculator. See margin.
c. Classify each graph as linear or nonlinear. See chart above.
d. Explain how to determine whether an equation, or its related expression, is linear or nonlinear without graphing. See margin.

**66. OPEN-ENDED** Write three different expressions that can be simplified to $x^6$. **Sample answer:** $x^4 \cdot x^2$; $x^5 \cdot x$; $(x^3)^2$

**67. WRITING IN MATH** Write two formulas that have monomial expressions in them. Explain how each is used in a real-world situation.

**67. Sample answer:** The area of a circle or $A = \pi r^2$, where the radius $r$ can be used to find the area of any circle. The area of a rectangle or $A = w \cdot \ell$, where $w$ is the width and $\ell$ is the length, can be used to find the area of any rectangle.

### Standards for Mathematical Practice

| Emphasis On | Exercises |
|---|---|
| 1 Make sense of problems and persevere in solving them. | 64 |
| 4 Model with mathematics. | 52, 61–62 |
| 5 Use appropriate tools strategically. | 58–60 |
| 7 Look for and make use of structure. | 68, 70, 73 |
| 8 Look for and express regularity in repeated reasoning. | 65, 69, 70, 72–74 |

## Preparing for Assessment

**68.** Which expression represents the area of the trapezoid? MP 7 A.SSE.2, F.IF.8b  **B**

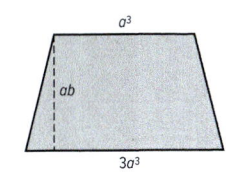

(trapezoid with top side $a^3$, bottom side $3a^3$, height $ab$)

- A  $1.5a^4b$
- B  $2a^4b$
- C  $3a^7b$
- D  $5a^3b$

**69.** Which expression is equivalent to $(3m^2np^4)(-5m^3n^2p)^2$? MP 8 A.SSE.2, F.IF.8b  **D**

- A  $-15m^5n^3p^6$
- B  $-15m^8n^5p^6$
- C  $75m^5n^3p^5$
- D  $75m^8n^5p^6$

**70.** WXYZ is a parallelogram.

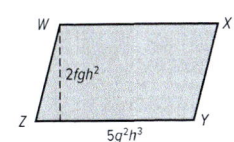

(parallelogram WXYZ with height $2fgh^2$ and base $5g^2h^3$)

What is the area of WXYZ, expressed as a monomial? MP 7 A.SSE.2, F.IF.8b  **C**

- A  $7fg^3h^5$
- B  $7fg^2h^6$
- C  $10fg^3h^5$
- D  $10fg^2h^6$

**71.** Which of the following expressions simplifies to the monomial $12h^6j^8$? MP 8 A.SSE.2, F.IF.8b  **C**

- A  $(4h^2j^4)(3h^3j^2)$
- B  $(2h^3j^2)(6h^3j^4)$
- C  $(3h^3j^6)(4h^3j^2)$
- D  $\dfrac{(8h^2j^4)(3h^3j^2)}{2}$

**72.** Identify which expressions are monomials. Select all that apply. MP 8 A.SSE.2, F.IF.8b  **A, C, D**

- ☐ A  $12$
- ☐ B  $\dfrac{3x+6}{9}$
- ☐ C  $\dfrac{5x^2y}{22}$
- ☐ D  $23a^3b^2c$
- ☐ E  $\dfrac{mn^4}{v}$

**73. MULTI-STEP** Consider the rectangular prism shown. MP 7, 8 A.SSE.2, F.IF.8b

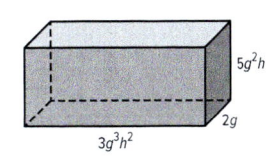

(rectangular prism with dimensions $3g^3h^2$, $2g$, $5g^2h$)

**a.** Which expression represents the area of the face with a length of $3g^3h^2$ and a width of $5g^2h$?  **B**

- A  $15g^5h^2$
- B  $15g^5h^3$
- C  $15g^6h^2$
- D  $15g^9h^2$

**b.** What is the volume of the prism?

$30g^6h^3$

**74.** Paul rewrote the expression $[(13^2)^5]^8$ as a power with a single exponent. He kept the base 13. What was the new exponent? MP 4, 8 A.SSE.2, F.IF.8b  **80**

---

### Differentiated Instruction  OL  BL

**Extension** Tell students that a sports car on a drag-racing strip can reach 100 miles per hour in a quarter mile. If s represents the speed in miles per hour, then the approximate number of feet that the driver must apply the brakes before stopping is $\dfrac{1}{20}s^2$. Calculate how far the car would travel on the drag strip, from start to stop, if the driver started braking when the car reached 100 miles per hour. A mile is 5280 feet. **1820 ft; the initial quarter mile (1320 ft) plus braking distance (500 ft)**

---

Lesson 7-1 | Multiplication Properties of Exponents

## Preparing for Assessment

Exercises 68–74 require students to use the skills they will need on standardized assessments. Each exercise is dual-coded with content standards and mathematical practice standards.

### Dual Coding

| Exercises | Content Standards | MP Mathematical Practices |
|---|---|---|
| 68 | A.SSE.2, F.IF.8 | 7 |
| 69 | A.SSE.2, F.IF.8 | 8 |
| 70 | A.SSE.2, F.IF.8 | 7 |
| 71 | A.SSE.2, F.IF.8 | 8 |
| 72 | A.SSE.2, F.IF.8 | 8 |
| 73 | A.SSE.2, F.IF.8 | 7, 8 |
| 74 | A.SSE.2, F.IF.8 | 4, 8 |

## Diagnose Student Errors

Survey student responses for each item. Class trends may indicate common errors and misconceptions.

**68.**

| A | Multiplied base and height then divided by 2 |
|---|---|
| B | CORRECT |
| C | Multiplied the bases and the height |
| D | Added and multiplied incorrectly |

**69.**

| A | Only squared $p$ in the second monomial |
|---|---|
| B | Did not square coefficient |
| C | Only squared the coefficient |
| D | CORRECT |

**70.**

| A | Added the coefficients |
|---|---|
| B | Added the coefficients and multiplied the exponents |
| C | CORRECT |
| D | Multiplied the exponents |

### Go Online!

**Quizzes**

Students can use *Self-Check Quizzes* to check their understanding of this lesson.

# LESSON 7-2
# Division Properties of Exponents

**SUGGESTED PACING (DAYS)**
90 min. 0.5
45 min. 1
Instruction

## Track Your Progress

### Objectives
1. Divide monomials using the properties of exponents.
2. Simplify expressions containing negative and zero exponents.

### Mathematical Background
To divide two powers that have the same base, subtract exponents. To find the power of a quotient, find the power of the numerator and the power of the denominator. A nonzero number raised to a negative integer power is the reciprocal of the same number with the opposite power. A fraction that has a negative exponent can be rewritten as its reciprocal with a positive power.

### THEN
**A.SSE.1a** Interpret parts of an expression, such as terms, factors, and coefficients.

**A.SSE.1b** Interpret complicated expressions by viewing one or more of their parts as a single entity.

### NOW
**A.SSE.2** Use the structure of an expression to identify ways to rewrite it.

**F.IF.8b** Use the properties of exponents to interpret expressions for exponential functions.

### NEXT
**N.RN.1** Explain how the definition of the meaning of rational exponents follows from extending the properties of integer exponents to those values, allowing for a notation for radicals in terms of rational exponents.

**N.RN.2** Rewrite expressions involving radicals and rational exponents using the properties of exponents.

---

**Go Online!** All of these resources and more are available at connectED.mcgraw-hill.com

**Brain POPs®** are animated, curricular content that engage students.

*Use with Examples*

**Personal Tutors** (for every example) let students hear real teachers solve problems. Students can pause and repeat as many times as necessary.

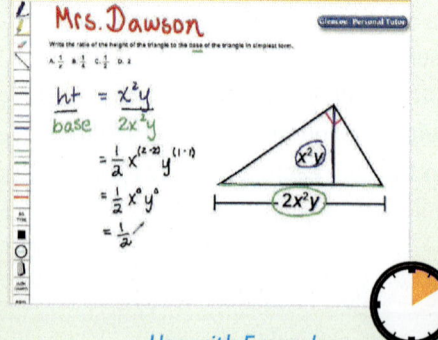

*Use with Examples*

Use **The Geometer's Sketchpad** to simplify expressions containing negative and zero exponents.

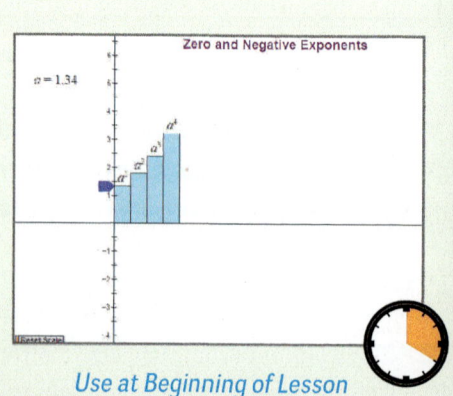

*Use at Beginning of Lesson*

---

### Using Open Educational Resources
**Tutorials** Have students watch the tutorial on **brightstorm** about multiplying and dividing exponents for quick help while doing homework. If you are unable to access brightstorm, try **MathATube**, **SchoolTube**, or **TeacherTube**. *Use as flipped learning or review*

# Go Online!
connectED.mcgraw-hill.com
**Worksheets**

# Differentiate Your Resources

**Extra Practice** Additional practice or homework; Skills Practice is best for approaching-level students and Practice is best for on-level and beyond-level students

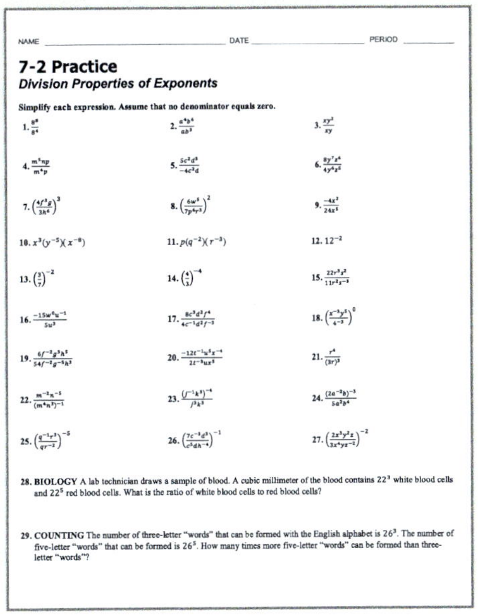

## Skills Practice
AL OL ELL

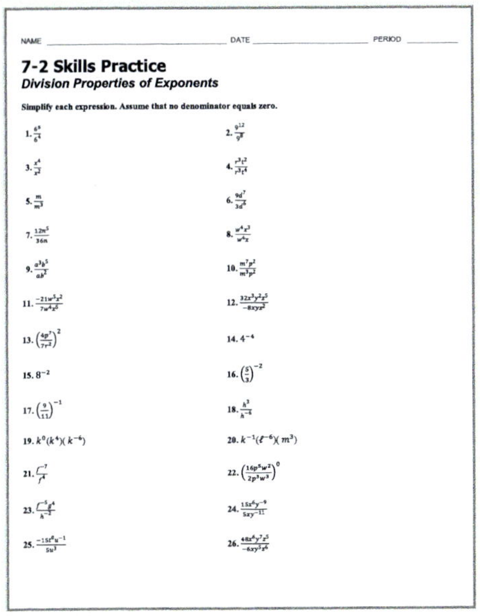

## Practice
AL OL BL ELL

## Word Problem Practice
AL OL BL ELL

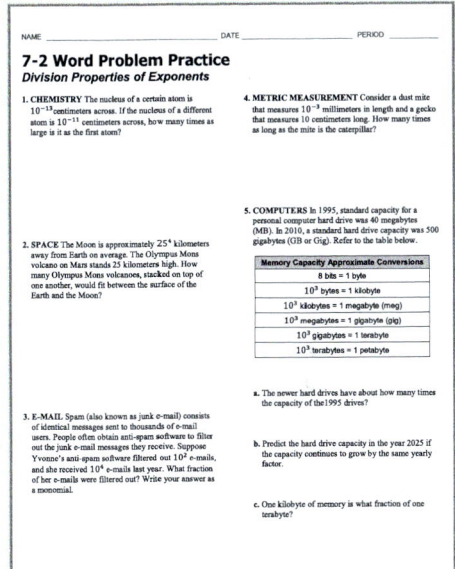

**Intervention** Reteaching and vocabulary activities that can be used with struggling or absent students and as ELL support

**Extension** Activities that can be used to extend lesson concepts

## Study Guide and Intervention
AL OL ELL

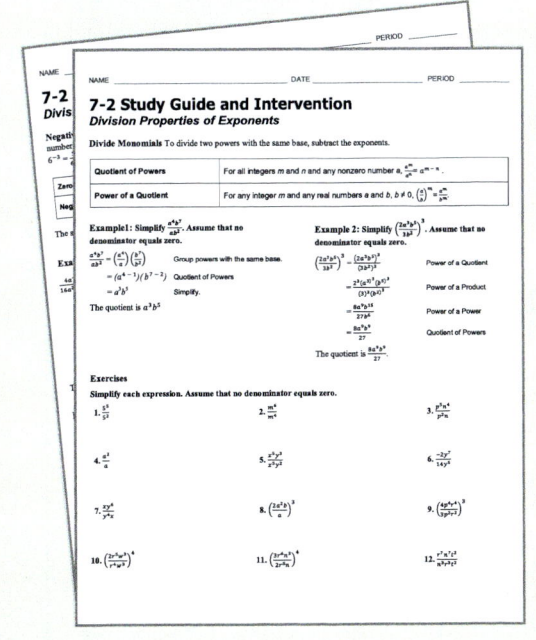

## Study Notebook
AL OL BL ELL

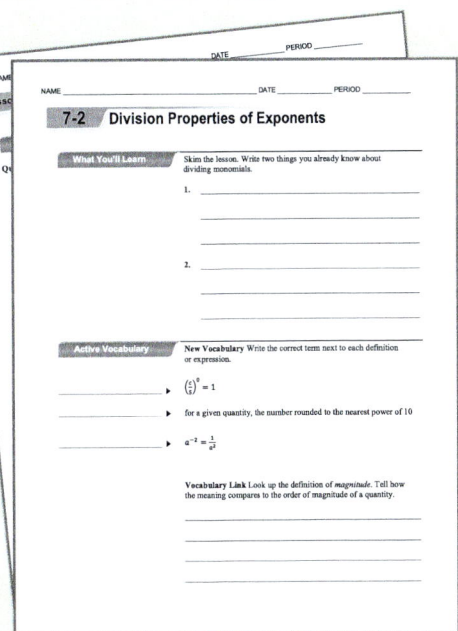

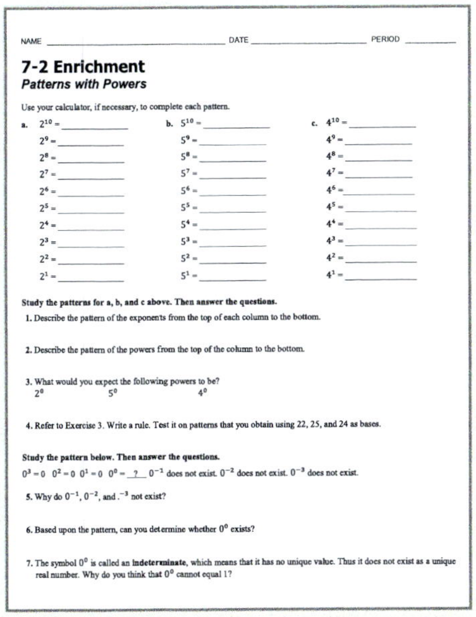

## Enrichment
OL BL ELL

connectED.mcgraw-hill.com  402B

Lesson 7-2 | Division Properties of Exponents

# Launch

Have students read the Why? section of the lesson. Ask:

- What is $10^2$ when simplified? **100**
- What is $10^1$ when simplified? **10**
- How could you use exponents to write the ratio of the height of the Statue of Liberty, which is about 93 meters tall, to the height of the tallest redwood tree? $\frac{10^2}{10^2}$ or $10^0$ or 1

# Teach

Ask the scaffolded questions for each example to build conceptual understanding for students at all levels.

## 1 Divide Monomials

**Example 1 Quotient of Powers**

**AL** What terms have the same base? $g^3$ and $g$, $h^5$ and $h^2$

**OL** How can the expression be rewritten as the multiplication of separate fractions? $\frac{g^3}{g}, \frac{h^5}{h^2}$

**BL** Show and explain why subtracting the exponents makes sense. Sample answer: $\frac{g^3 h^5}{gh^2} = \frac{g \cdot g \cdot g}{g} \cdot \frac{h \cdot h \cdot h \cdot h \cdot h}{h \cdot h}$; one $g$ cancels out and two $h$s cancel out when you divide, leaving $g \cdot g \cdot h \cdot h \cdot h$ or $g^2 h^3$. You can get this same result by subtracting exponents.

**Need Another Example?**
Simplify $\frac{x^7 y^{12}}{x^6 y^3}$. Assume that no denominator equals zero. $xy^9$

## Go Online!

**Interactive Whiteboard**
Use the eLesson or Lesson Presentation to present this lesson.

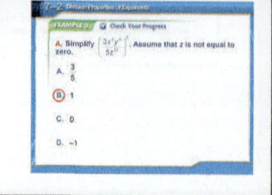

---

**LESSON 2**

# Division Properties of Exponents

**Then** · You multiplied monomials using the properties of exponents.

**Now** · **1** Divide monomials using the properties of exponents.
**2** Simplify expressions containing negative and zero exponents.

**Why?** · The tallest redwood tree is 112 meters or about $10^2$ meters tall. The average height of a redwood tree is 15 meters. The closest power of ten to 15 is $10^1$, so an average redwood is about $10^1$ meters tall. The ratio of the tallest tree's height to the average tree's height is $\frac{10^2}{10^1}$ or $10^1$. This means the tallest redwood tree is approximately 10 times as tall as the average redwood tree.

**New Vocabulary**
zero exponent
negative exponent
order of magnitude

**Mathematical Practices**
2 Reason abstractly and quantitatively.

**Content Standards**
A.SSE.2 Use the structure of an expression to identify ways to rewrite it.
F.IF.8b Use the properties of exponents to interpret expressions for exponential functions.

**1 Divide Monomials** We can use the principles for reducing fractions to find quotients of monomials like $\frac{10^2}{10^1}$. In the following examples, look for a pattern in the exponents.

$$\frac{2^7}{2^4} = \frac{\overset{7\ \text{factors}}{\overbrace{2 \cdot 2 \cdot 2 \cdot 2 \cdot 2 \cdot 2 \cdot 2}}}{\underset{4\ \text{factors}}{\underbrace{2 \cdot 2 \cdot 2 \cdot 2}}} = 2 \cdot 2 \cdot 2 \text{ or } 2^3 \qquad \frac{t^4}{t^3} = \frac{\overset{4\ \text{factors}}{\overbrace{t \cdot t \cdot t \cdot t}}}{\underset{3\ \text{factors}}{\underbrace{t \cdot t \cdot t}}} = t$$

These examples demonstrate the Quotient of Powers Rule.

**Key Concept** Quotient of Powers

Words: To divide two powers with the same base, subtract the exponents.
Symbols: For any nonzero number $a$, and any integers $m$ and $p$, $\frac{a^m}{a^p} = a^{m-p}$.
Examples: $\frac{c^{11}}{c^8} = c^{11-8}$ or $c^3$ $\qquad \frac{r^5}{r^2} = r^{5-2} = r^3$

**Example 1** Quotient of Powers

Simplify $\frac{g^3 h^5}{gh^2}$. Assume that no denominator equals zero.

$\frac{g^3 h^5}{gh^2} = \left(\frac{g^3}{g}\right)\left(\frac{h^5}{h^2}\right)$ Group powers with the same base.
$= (g^{3-1})(h^{5-2})$ Quotient of Powers
$= g^2 h^3$ Simplify.

**Guided Practice**
Simplify each expression. Assume that no denominator equals zero.

1A. $\frac{x^3 y^4}{x^2 y}$  $xy^3$
1B. $\frac{k^7 m^{10} p}{k^5 m^3 p}$  $k^2 m^7$

---

**MP Mathematical Practices Strategies**

**Look for and make use of structure.**
Breaking expressions into parts helps students analyze the structure of quotients that involve exponents. For example, ask:

- How can you rewrite an expression like $\frac{q^5 r^7}{q^4 r^2}$ as the product of fractions? group like variables; $\left(\frac{q^5}{q^4}\right)\left(\frac{r^7}{r^2}\right)$

- For the expression $\left(\frac{9mn^7}{7mn^2}\right)^0$, is it necessary to apply the exponent 0 to each individual part? No, you can treat the expression as one quantity raised to the power 0.

- In the expression $\frac{3x^a y}{x^b}$ with $a < b$, what is the exponent of the denominator? $b - a$

402 | Lesson 7-2 | Division Properties of Exponents

We can use the Product of Powers Property to find the powers of quotients for monomials. In the following example, look for a pattern in the exponents.

$$\left(\frac{3}{4}\right)^3 = \overbrace{\left(\frac{3}{4}\right)\left(\frac{3}{4}\right)\left(\frac{3}{4}\right)}^{3 \text{ factors}} = \frac{\overbrace{3 \cdot 3 \cdot 3}^{3 \text{ factors}}}{\underbrace{4 \cdot 4 \cdot 4}_{3 \text{ factors}}} = \frac{3^3}{4^3}$$

$$\left(\frac{c}{d}\right)^2 = \overbrace{\left(\frac{c}{d}\right)\left(\frac{c}{d}\right)}^{2 \text{ factors}} = \frac{\overbrace{c \cdot c}^{2 \text{ factors}}}{\underbrace{d \cdot d}_{2 \text{ factors}}} = \frac{c^2}{d^2}$$

### Study Tip

**MP Reasoning** The power rules apply to variables as well as numbers. For example, $\left(\frac{3a}{4b}\right)^3 = \frac{(3a)^3}{(4b)^3}$ or $\frac{27a^3}{64b^3}$.

### Key Concept  Power of a Quotient

**Words**  To find the power of a quotient, find the power of the numerator and the power of the denominator.

**Symbols**  For any real numbers $a$ and $b \neq 0$, and any integer $m$, $\left(\frac{a}{b}\right)^m = \frac{a^m}{b^m}$.

**Examples**  $\left(\frac{3}{5}\right)^4 = \frac{3^4}{5^4}$   $\left(\frac{r}{t}\right)^5 = \frac{r^5}{t^5}$

A.SSE.2

### Real-World Career

**Astronomer** An astronomer studies the universe and analyzes space travel and satellite communications. To be a technician or research assistant, a bachelor's degree is required.

### Example 2  Power of a Quotient

Simplify $\left(\frac{3p^3}{7}\right)^2$.

$\left(\frac{3p^3}{7}\right)^2 = \frac{(3p^3)^2}{7^2}$   Power of a Quotient

$= \frac{3^2(p^3)^2}{7^2}$   Power of a Product

$= \frac{9p^6}{49}$   Power of a Power

### Guided Practice

Simplify each expression.

**2A.** $\left(\frac{3x^4}{4}\right)^3$  $\frac{27x^{12}}{64}$   **2B.** $\left(\frac{5x^5y}{6}\right)^2$  $\frac{25x^{10}y^2}{36}$   **2C.** $\left(\frac{2y^2}{3z^3}\right)^2$  $\frac{4y^4}{9z^6}$   **2D.** $\left(\frac{4x^3}{5y^4}\right)^3$  $\frac{64x^9}{125y^{12}}$

A calculator can be used to explore expressions with 0 as the exponent. There are two methods to explain why a calculator gives a value of 1 for $3^0$.

**Method 1**

$\frac{3^5}{3^5} = 3^{5-5}$   Quotient of Powers

$= 3^0$   Simplify.

**Method 2**

$\frac{3^5}{3^5} = \frac{\cancel{3} \cdot \cancel{3} \cdot \cancel{3} \cdot \cancel{3} \cdot \cancel{3}}{\cancel{3} \cdot \cancel{3} \cdot \cancel{3} \cdot \cancel{3} \cdot \cancel{3}}$   Definition of Powers

$= 1$   Simplify.

Since $\frac{3^5}{3^5}$ can only have one value, we can conclude that $3^0 = 1$, which leads to the Zero Exponent Property. A **zero exponent** is any nonzero number raised to the zero power.

---

## Lesson 7-2 | Division Properties of Exponents

### Example 2  Power of a Quotient

**AL** To what factors does the exponent 2 need to be applied? 3, $p^3$, and 7

**OL** Will you do any dividing to simplify the expression? no

**BL** Write and solve your own Power of a Quotient expression. See students' work.

**Need Another Example?**

Simplify $\left(\frac{4c^3d^3}{5}\right)^3$.  $\frac{64c^9d^9}{125}$

### Watch Out!

**Preventing Errors** Remind students to also find the powers of the constant terms of the monomials.

### Teaching Tip

**Properties** Point out to students that the definition of the Quotient of Powers restricts $a$ to being nonzero. Ask why must $a$ be nonzero. If $a = 0$, we would be dividing by 0, which is undefined.

# Lesson 7-2 | Division Properties of Exponents

**Watch Out!**
**Preventing Errors** Ask students why the negative sign in Example 3a does not affect the outcome. Students should explain than any nonzero number raised to the zero power is 1, and a negative number is a nonzero number.

## Example 3 Zero Exponent

**AL** How many terms does the expression in part **a** include? 1

**OL** In part **a**, do you need to distribute the exponent to every factor before solving? Explain. no; Sample answer: everything inside the parentheses is a single quantity. Since any quantity to the zero power equals one, we know the solution is 1.

**BL** In part **b**, explain what is different about this expression from the expression in part **a**. Sample answer: part **b** does not have parentheses, so the zero power only applies to one factor.

**Need Another Example?**
Simplify each expression. Assume that no denominator equals zero.

a. $\left(\dfrac{12m^8 n^7}{8m^5 n^{10}}\right)^0$  1

b. $\dfrac{m^0 n^3}{n^2}$  $n$

## Teaching Tip

**Zero to the Zero Power** Ask students to write an expression involving division that will be equivalent to $0^0$, for example, $\dfrac{0^6}{0^6}$. Show students that $\dfrac{0^6}{0^6} = \dfrac{0}{0}$ and remind them that division by zero is undefined.

---

**Go Online!**
You have worked with exponents that are positive integers. Explore the concept of zero and negative exponents using the Geometer's Sketchpad® activity in ConnectED.

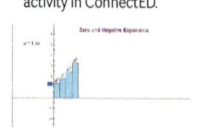

**Key Concept** Zero Exponent Property

| | |
|---|---|
| Words | Any nonzero number raised to the zero power is equal to 1. |
| Symbols | For any nonzero number $a$, $a^0 = 1$. |
| Examples | $15^0 = 1$    $\left(\dfrac{b}{c}\right)^0 = 1$    $\left(\dfrac{2}{7}\right)^0 = 1$ |

A.SSE.2

### Example 3  Zero Exponent

Simplify each expression. Assume that no denominator equals zero.

a. $\left(\dfrac{4n^2 q^5 r^2}{9n^3 q^2 r}\right)^0$

$\left(\dfrac{4n^2 q^5 r^2}{9n^3 q^2 r}\right)^0 = 1$     $a^0 = 1$

b. $\dfrac{x^5 y^0}{x^3}$

$\dfrac{x^5 y^0}{x^3} = \dfrac{x^5 (1)}{x^3}$     $a^0 = 1$

$= x^2$     Quotient of Powers

**Study Tip**
**Zero Exponent** Be careful of parentheses. The expression $(5x)^0$ is 1, but $5x^0 = 5$.

▶ **Guided Practice**

3A. $\dfrac{b^4 c^2 d^0}{b^2 c}$  $b^2 c$

3B. $\left(\dfrac{2f^4 g^7 h^3}{15f^3 g^9 h^6}\right)^0$  1

**2 Negative Exponents** Any nonzero real number raised to a negative power is a **negative exponent**. To investigate the meaning of a negative exponent, we can simplify expressions like $\dfrac{c^2}{c^5}$ using two methods.

**Method 1**

$\dfrac{c^2}{c^5} = c^{2-5}$     Quotient of Powers

$= c^{-3}$     Simplify.

**Method 2**

$\dfrac{c^2}{c^5} = \dfrac{\cancel{c} \cdot \cancel{c}}{\cancel{c} \cdot \cancel{c} \cdot c \cdot c \cdot c}$     Definition of Powers

$= \dfrac{1}{c^3}$     Simplify.

Since $\dfrac{c^2}{c^5}$ can only have one value, we can conclude that $c^{-3} = \dfrac{1}{c^3}$.

**Key Concept** Negative Exponent Property

| | |
|---|---|
| Words | For any nonzero number $a$ and any integer $n$, $a^{-n}$ is the reciprocal of $a^n$. Also, the reciprocal of $a^{-n}$ is $a^n$. |
| Symbols | For any nonzero number $a$ and any integer $n$, $a^{-n} = \dfrac{1}{a^n}$. |
| Examples | $2^{-4} = \dfrac{1}{2^4} = \dfrac{1}{16}$      $\dfrac{1}{j^{-4}} = j^4$ |

---

### Differentiated Instruction  AL  OL  ELL

**IF** students have difficulty relating the Key Concepts in this lesson to expressions,

**THEN** have students make flash cards to illustrate each Key Concept. Write an expression that is an example of a Key Concept on the board. Tell students to show their card that correlates to the example. Then ask a student to describe the process.

## Lesson 7-2 | Division Properties of Exponents

An expression is considered simplified when it contains only positive exponents, each base appears exactly once, there are no powers of powers, and all fractions are in simplest form.

A.SSE.2

**Example 4**    Negative Exponents

Simplify each expression. Assume that no denominator equals zero.

a. $\dfrac{n^{-5}p^4}{r^{-2}}$

$\dfrac{n^{-5}p^4}{r^{-2}} = \left(\dfrac{n^{-5}}{1}\right)\left(\dfrac{p^4}{1}\right)\left(\dfrac{1}{r^{-2}}\right)$    Write as a product of fractions.

$= \left(\dfrac{1}{n^5}\right)\left(\dfrac{p^4}{1}\right)\left(\dfrac{r^2}{1}\right)$    $a^{-n} = \dfrac{1}{a^n}$ and $\dfrac{1}{a^{-n}} = a^n$

$= \dfrac{p^4 r^2}{n^5}$    Multiply.

> **Study Tip**
> **Negative Signs** Be aware of where a negative sign is placed.
> $5^{-1} = \dfrac{1}{5}$, while $-5^1 \neq \dfrac{1}{5}$.

b. $\dfrac{5r^{-3}t^4}{-20r^2 t^7 u^{-5}}$

$\dfrac{5r^{-3}t^4}{-20r^2 t^7 u^{-5}} = \left(\dfrac{5}{-20}\right)\left(\dfrac{r^{-3}}{r^2}\right)\left(\dfrac{t^4}{t^7}\right)\left(\dfrac{1}{u^{-5}}\right)$    Group powers with the same base.

$= \left(-\dfrac{1}{4}\right)(r^{-3-2})(t^{4-7})(u^5)$    Quotient of Powers and Negative Exponents Property

$= -\dfrac{1}{4} r^{-5} t^{-3} u^5$    Simplify.

$= -\dfrac{1}{4}\left(\dfrac{1}{r^5}\right)\left(\dfrac{1}{t^3}\right)(u^5)$    Negative Exponent Property

$= -\dfrac{u^5}{4r^5 t^3}$    Multiply.

c. $\dfrac{2a^2 b^3 c^{-5}}{10a^{-3} b^{-1} c^{-4}}$

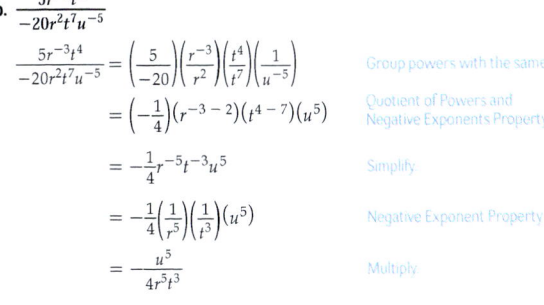

$\dfrac{2a^2 b^3 c^{-5}}{10a^{-3} b^{-1} c^{-4}} = \left(\dfrac{2}{10}\right)\left(\dfrac{a^2}{a^{-3}}\right)\left(\dfrac{b^3}{b^{-1}}\right)\left(\dfrac{c^{-5}}{c^{-4}}\right)$    Group powers with the same base.

$= \left(\dfrac{1}{5}\right)(a^{2-(-3)})(b^{3-(-1)})(c^{-5-(-4)})$    Quotient of Powers and Negative Exponents Property

$= \dfrac{1}{5} a^5 b^4 c^{-1}$    Simplify.

$= \dfrac{1}{5}(a^5)(b^4)\left(\dfrac{1}{c}\right)$    Negative Exponent Property

$= \dfrac{a^5 b^4}{5c}$    Multiply.

**Guided Practice**

Simplify each expression. Assume that no denominator equals zero.

4A. $\dfrac{v^{-3} w x^2}{w y^{-6}}$   $\dfrac{x^2 y^6}{v^3}$    4B. $\dfrac{32a^{-8} b^3 c^{-4}}{4a^3 b^5 c^{-2}}$   $\dfrac{8}{a^{11} b^2 c^2}$    4C. $\dfrac{5j^{-3} k^2 m^{-6}}{25k^{-4} m^{-2}}$   $\dfrac{k^6}{5j^3 m^4}$

> **Real-World Link**
> An adult human weighs about 70 kilograms and an adult dairy cow weighs about 700 kilograms. Their weights differ by 1 order of magnitude.

Order of magnitude is used to compare measures and to estimate and perform rough calculations. The **order of magnitude** of a quantity is the number rounded to the nearest power of 10. For example, the power of 10 closest to 95,000,000,000 is $10^{11}$, or 100,000,000,000. So the order of magnitude of 95,000,000,000 is $10^{11}$.

## 2 Negative Exponents

### Example 4 Negative Exponents

**AL** In part b, to divide powers with the same base, do you add, subtract, multiply, or divide the exponents? **subtract**

**OL** In part b, how would you rewrite $\dfrac{r^{-3}}{r^2}$ using positive exponents? $\dfrac{1}{r^5}$

**BL** In part c, why do we not leave the answer as $\dfrac{1}{5} a^5 b^4 c^{-1}$? **Sample answer: a simplified answer does not contain any negative exponents.**

**Need Another Example?**

Simplify each expression. Assume that no denominator equals zero.

a. $\dfrac{x^{-4} y^9}{z^{-6}}$    $\dfrac{y^9 z^6}{x^4}$

b. $\dfrac{75 p^3 m^{-5}}{15 p^5 m^{-4} r^{-8}}$    $\dfrac{5 r^8}{p^2 m}$

> **Watch Out!**
> **Preventing Errors** Have students look at the first step of each solution in Example 4. Point out that rewriting an expression as a product of fractions makes applying the Negative Exponent Property easier. Fractions that have negative exponents can be rewritten as their reciprocals.

### Teaching Tip

**Monomials** Point out to students that a variable with a negative exponent such as $x^{-2}$ is not a monomial. A monomial does not involve division by variables and $x^{-2} = \dfrac{1}{x^2}$.

connectED.mcgraw-hill.com    405

Lesson 7-2 | Division Properties of Exponents

## Example 5 Apply Properties of Exponents

**AL** What are we asked to determine? how many orders of magnitude taller is a man than an ant

**OL** Does the height of an ant need to be rounded? Explain. yes; Sample answer: the height of the ant should be a power of 10 so 0.0008 is rounded to 0.001

**BL** Why do we round the height of a man down to 1? Sample answer: to find order of magnitude, we have to round to the nearest power of 10. This means we have to round 1.7 down to 1 rather than up to 2 because 2 is not a power of 10.

### Need Another Example?

**Savings** Darin has $123,456 in his savings account. Tabo has $156 in his savings account. Determine the order of magnitude of Darin's account and Tabo's account. How many orders of magnitude as great is Darin's account as Tabo's account?
Darin: $10^5$, Tabo: $10^2$; Darin's account is 3 orders of magnitude as great as Tabo's account.

## Practice

**Formative Assessment** Use Exercises 1–18 to assess students' understanding of the concepts in this lesson.

The Practice and Problem Solving exercises assess the content taught in the lesson. The Preparing for Assessment page is meant to be used as preparation for end-of-course assessments.

---

**Real-World Link**
There are over 14,000 species of ants living all over the world. Some ants can carry objects that are 50 times their own weight.
Source: Maine Animal Coalition

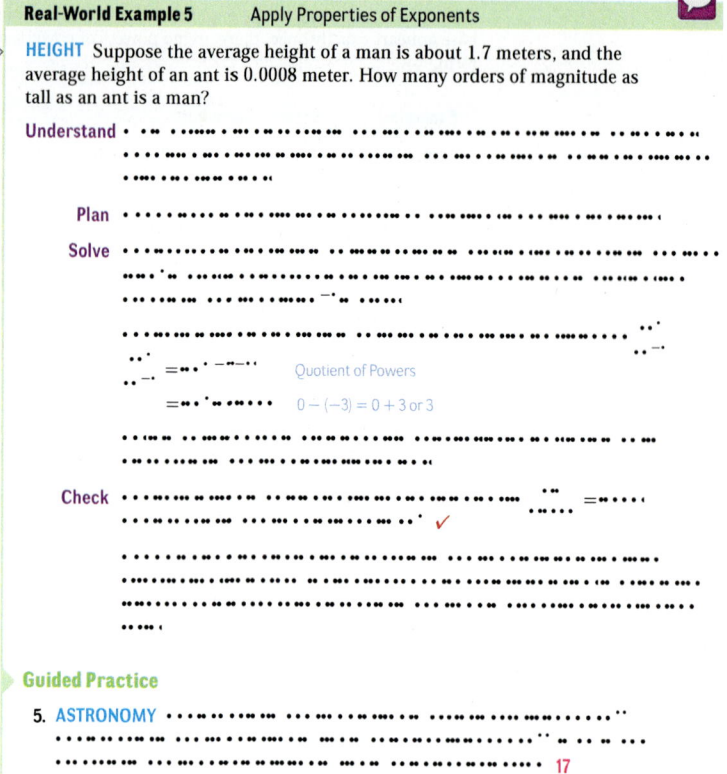

**Real-World Example 5** Apply Properties of Exponents

**HEIGHT** Suppose the average height of a man is about 1.7 meters, and the average height of an ant is 0.0008 meter. How many orders of magnitude as tall as an ant is a man?

Understand ...

Plan ...

Solve ...

Quotient of Powers

$0 - (-3) = 0 + 3$ or $3$

Check ...

### Guided Practice

5. **ASTRONOMY** ... 17

**Check Your Understanding** = Step-by-Step Solutions begin on page R11.

Examples 1–4 Simplify each expression. Assume that no denominator equals zero.
A.SSE.2

1. $\dfrac{t^4 u^*}{t^* u}$  $t^3 u^3$
2. $\dfrac{a^* b^* c^{**}}{a^* b^* c}$  $a^3 b^2 c^9$
3. $\dfrac{m^* r^* p^*}{m^* r^* p^*}$  $mr^3$
4. $\dfrac{b^* c^* f}{b^* c^* f}$  $c^3 f^3$
5. $\dfrac{g^* h^* m}{hg^*}$  $ghm$
6. $\dfrac{r^* t^* v^*}{t^* v^*}$  $r^4$
7. $\dfrac{x^* y^* z^*}{z^* x^* y}$  $xyz$
8. $\dfrac{n^* q^* w^*}{q^* n^* w}$  $nq^2 w^5$
9. $\left(\dfrac{\cdot a^* b^*}{\cdot}\right)^*$  $\dfrac{4 a^6 b^{10}}{9}$
10. $\dfrac{r^* v^{-*}}{t^{-*}}$  $\dfrac{r^3 t^7}{v^2}$
11. $\left(\dfrac{\cdot c^* d^*}{\cdot g}\right)^*$  $\dfrac{32 c^{15} d^{25}}{3125 g^{10}}$
12. $\left(-\dfrac{\cdot xy^* z^*}{x^* yz^*}\right)^*$  $1$
13. $\left(\dfrac{\cdot f g h^*}{\cdot \cdot f g^* h}\right)^*$  $1$
14. $\dfrac{\cdot r^* v^* t^*}{\cdot rt^*}$  $2 r t^2$
15. $\dfrac{f^{-*} g^*}{h^{-*}}$  $\dfrac{g^2 h^4}{f^3}$
16. $\dfrac{-\cdot x^* y^* z^{-*}}{\cdot \cdot x^* y^{-*} z^*}$  $\dfrac{-2 y^{15}}{3 x^2 z^{12}}$
17. $\dfrac{\cdot a^* b^{-*} c^{**}}{\cdot a^{-*} b^* c^{-*}}$  $\dfrac{a^5 c^{13}}{3 b^9}$

Example 5
A.SSE.2,
F.IF.8b

18. **POPULATION** ... $10^6$ or about 1,000,000

---

## Go Online!

**eBook**

**Interactive Student Guide**
Use the *Interactive Student Guide* to deepen conceptual understanding.
· Division Properties of Exponents

**Lesson 7-2 | Division Properties of Exponents**

**Practice and Problem Solving**  Extra Practice is on page R7.

Examples 1–4  Simplify each expression. Assume that no denominator equals zero.
A.SSE.2

19. $\dfrac{m^4 p^2}{m^2 p}$   $m^2 p$

20. $\dfrac{p^{12} t^3 r}{p^2 t r}$   $p^{10} t^2$

21. $\dfrac{3m^{-3} r^4 p^2}{12 t^4}$   $\dfrac{r^4 p^2}{4 m^3 t^4}$

22. $\dfrac{c^4 d^4 f^3}{c^2 d^4 f^3}$   $c^2$

23. $\left(\dfrac{3xy^4}{5z^2}\right)^2$   $\dfrac{9x^2 y^8}{25 z^4}$

24. $\left(\dfrac{3t^6 u^2 v^5}{9 tuv^{21}}\right)^0$   $1$

25. $\left(\dfrac{p^2 t^7}{10}\right)^3$   $\dfrac{p^6 t^{21}}{1000}$

26. $\dfrac{x^{-4} y^9}{z^{-2}} \cdot \dfrac{y^9 z^2}{x^4}$

27. $\dfrac{a^7 b^8 c^8}{a^5 b c^7}$   $a^2 b^7 c$

28. $\left(\dfrac{3np^3}{7q^2}\right)^2$   $\dfrac{9n^2 p^6}{49 q^4}$

29. $\left(\dfrac{2r^3 t^6}{5u^9}\right)^4$   $\dfrac{16 r^{12} t^{24}}{625 u^{36}}$

30. $\left(\dfrac{3m^5 r^3}{4p^8}\right)^4$   $\dfrac{81 m^{20} r^{12}}{256 p^{32}}$

31. $\left(-\dfrac{5 f^9 g^4 h^2}{f g^2 h^3}\right)^0$   $1$

32. $\dfrac{p^{12} t^7 r^2}{p^2 t^7 r}$   $p^{10} r$

33. $\dfrac{p^4 t^{-3}}{r^{-2}} \cdot \dfrac{p^4 r^2}{t^3}$

34. $-\dfrac{5c^2 d^5}{8cd^5 f^0}$   $-\dfrac{5c}{8}$

35. $\dfrac{-2 f^3 g^2 h^0}{8 f^2 g^2}$   $\dfrac{-f}{4}$

36. $\dfrac{12 m^{-4} p^2}{-15 m^3 p^{-9}}$   $\dfrac{4 p^{11}}{-5 m^7}$

37. $\dfrac{k^4 m^3 p^2}{k^2 m^2}$   $k^2 m p^2$

38. $\dfrac{14 f^{-3} g^2 h^{-7}}{21 k^3} \cdot \dfrac{2g^2}{3 f^3 h^7 k^3}$

39. $\dfrac{39 t^4 u v^{-2}}{13 t^{-3} u^7} \cdot \dfrac{3 t^7}{u^6 v^2}$

40. $\left(\dfrac{a^{-2} b^4 c^5}{a^{-4} b^{-4} c^3}\right)^2$   $a^4 b^{16} c^4$

41. $\dfrac{r^3 t^{-1} x^{-5}}{tx^5} \cdot \dfrac{r^3}{t^2 x^{10}}$

42. $\dfrac{g^0 h^7 j^{-2}}{g^{-5} h^0 j^{-2}}$   $g^5 h^7$

Example 5  43. **SOCIAL NETWORKING** In a recent year, a social networking site had about 750,000 servers. Suppose there were 1.15 billion active users on the site. Determine the order of magnitude for the servers and active users. Using the orders of magnitude, how many active users were there compared to servers?

43. $10^6$; $10^9$; about $10^3$ or 1000 times as many users as servers

**B**  44. **PROBABILITY** The probability of rolling a die and getting an even number is $\tfrac{1}{2}$. If you roll the die twice, the probability of getting an even number both times is $\left(\tfrac{1}{2}\right)\left(\tfrac{1}{2}\right)$ or $\left(\tfrac{1}{2}\right)^2$.
  a. What does $\left(\tfrac{1}{2}\right)^4$ represent? probability of all evens on 4 rolls
  b. Write an expression to represent the probability of rolling a die $d$ times and getting an even number every time. Write the expression as a power of 2. $\left(\tfrac{1}{2}\right)^d$; $2^{-d}$

Simplify each expression. Assume that no denominator equals zero.

45. $\dfrac{-4w^{12}}{12 w^3}$   $-\dfrac{w^9}{3}$

46. $\dfrac{13 r^7}{39 r^4}$   $\dfrac{r^3}{3}$

47. $\dfrac{(4 k^3 m^2)^3}{(5 k^2 m^{-3})^{-2}}$   $1600 k^{13}$

48. $\dfrac{3wy^{-2}}{(w^{-1} y)^3}$   $\dfrac{3w^4}{y^5}$

49. $\dfrac{20 q r^{-2} t^{-5}}{4 q^0 r^4 t^{-2}}$   $\dfrac{5q}{r^6 t^3}$

50. $\dfrac{-12 c^3 d^0 f^{-2}}{6 c^5 d^{-3} f^4}$   $\dfrac{-2 d^3}{c^2 f^6}$

51. $\dfrac{(2 g^3 h^{-2})^2}{(g^2 h^0)^{-3}}$   $\dfrac{4 g^{12}}{h^4}$

52. $\dfrac{(5 p r^{-2})^{-2}}{(3 p^{-1} r)^3}$   $\dfrac{pr}{675}$

53. $\left(\dfrac{-3 x^{-6} y^{-1} z^{-2}}{6 x^{-2} y z^{-5}}\right)^{-2}$   $\dfrac{4 x^8 y^4}{z^6}$

54. $\left(\dfrac{2 a^{-2} b^4 c^2}{-4 a^{-2} b^{-5} c^{-7}}\right)^{-1}$   $\dfrac{-2}{b^9 c^9}$

55. $\dfrac{(16 x^2 y^{-1})^0}{(4 x^0 y^{-4})^{-2}}$   $\dfrac{16 z^2}{y^8}$

56. $\left(\dfrac{4^0 c^2 d^3 f}{2 c^{-4} d^{-5}}\right)^{-3}$   $\dfrac{8}{c^{18} d^{24} f^3}$

57. **SENSE-MAKING** The processing speed of an older desktop computer is about $10^8$ instructions per second. A new computer can process about $10^{11}$ instructions per second. The newer computer is how many times as fast as the older one? **1000**

**Differentiated Homework Options**

| Levels | AL Basic | OL Core | BL Advanced |
|---|---|---|---|
| Exercises | 19–43, 61, 62, 64–72 | 19–43 odd, 44–55 odd, 57–62, 64–72 | 44–65  66–72 (optional) |
| 2-Day Option | 19–43 odd, 66–72  20–42 even, 61, 62, 64, 65 | 19–43, 66–72  44–62, 64, 65 | |

You can use **ALEKS** to provide additional remediation support with personalized instruction and practice.

**Teaching Tip**

**Checking Answers** When simplifying expressions, students may wish to check their answers by choosing nonzero values for the variables and evaluating the original and simplified expressions. If the results are the same, the expressions are likely to be equivalent.

**MP Teaching the Mathematical Practices**

**Sense-Making** Mathematically proficient students try simpler forms of the original problem in order to gain insight into its solution. In Exercise 57, encourage students to start by finding the number of instructions per second in a computer ten times as fast.

**Extra Practice**

See page R7 for extra exercises for students who are approaching level or for on-level students who need additional reinforcement.

**Levels of Complexity Chart**

The levels of the exercises progress from 1 to 3, with Level 1 indicating the lowest level of complexity.

| Exercises | 19–43 | 44–59, 66–72 | 60–65 |
|---|---|---|---|
| Level 3 | | | ● |
| Level 2 | | ● | |
| Level 1 | ● | | |

**Go Online!**

**eSolutions Manual**
Create worksheets, answer keys, and solutions handouts for your assignments.

connectED.mcgraw-hill.com  407

## Lesson 7-2 | Division Properties of Exponents

# Assess

**Yesterday's News** Ask students to write two ways in which the concepts of multiplying monomials helped them to understand dividing monomials.

### MP Teaching the Mathematical Practices

**Regularity** Mathematically proficient students notice if calculations are repeated. In Exercise 64, encourage students to examine the sequences $3^5, 3^4, 3^3, 3^2, \ldots$ and $243, 81, 27, 9, \ldots$

## Additional Answers

**64.** We can use the Quotient of Powers Property to show that $3^0 = 1$.

$$3^5 = 243$$
$$\frac{3^5}{3^5} = \frac{243}{3^5}$$
$$3^{5-5} = \frac{243}{3^5}$$
$$3^0 = \frac{243}{243}$$
$$3^0 = 1$$

**65.** The Quotient of Powers Property is used when dividing two powers with the same base. The exponents are subtracted. The Power of a Quotient Property is used to find the power of a quotient. You find the power of the numerator and the power of the denominator.

---

**58. ASTRONOMY** The brightness of a star is measured in magnitudes. The lower the magnitude, the brighter the star. A magnitude 9 star is 2.51 times as bright as a magnitude 10 star. A magnitude 8 star is 2.51 · 2.51 or $2.51^2$ times as bright as a magnitude 10 star. **$2.51^7$ or 627.647857**

a. How many times as bright is a magnitude 3 star as a magnitude 10 star?

b. Write an expression to compare a magnitude $m$ star to a magnitude 10 star. **$2.51^{10-m}$**

c. A full moon is considered to be magnitude −13, approximately. Does your expression make sense for this magnitude? Explain.

**59. PROBABILITY** The probability of rolling a die and getting a 3 is $\frac{1}{6}$. If you roll the die twice, the probability of getting a 3 both times is $\frac{1}{6} \cdot \frac{1}{6}$ or $\left(\frac{1}{6}\right)^2$.

a. Write an expression to represent the probability of rolling a die $d$ times and getting a 3 each time. $\left(\frac{1}{6}\right)^d$

b. Write the expression as a power of 6. **$6^{-d}$**

**60. MULTIPLE REPRESENTATIONS** To find the area of a circle, use $A = \pi r^2$. The formula for the area of a square is $A = s^2$.

a. **Algebraic** Find the ratio of the area of the circle to the area of the square. $\frac{\pi}{4}$

b. **Algebraic** If the radius of the circle and the length of each side of the square are doubled, find the ratio of the area of the circle to the area of the square. $\frac{\pi}{4}$

c. **Tabular** Copy and complete the table.

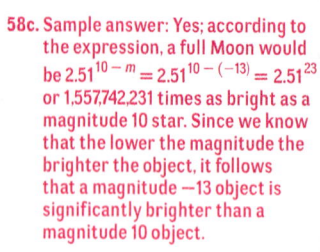

**58c.** Sample answer: Yes; according to the expression, a full Moon would be $2.51^{10-m} = 2.51^{10-(-13)} = 2.51^{23}$ or 1,557,742,231 times as bright as a magnitude 10 star. Since we know that the lower the magnitude the brighter the object, it follows that a magnitude −13 object is significantly brighter than a magnitude 10 object.

| Radius | Area of Circle | Area of Square | Ratio |
|---|---|---|---|
| $r$ | $\pi r^2$ | $4r^2$ | $\frac{\pi}{4}$ |
| $2r$ | $\pi 4r^2$ | $16r^2$ | $\frac{\pi}{4}$ |
| $3r$ | $\pi 9r^2$ | $36r^2$ | $\frac{\pi}{4}$ |
| $4r$ | $\pi 16r^2$ | $64r^2$ | $\frac{\pi}{4}$ |
| $5r$ | $\pi 25r^2$ | $100r^2$ | $\frac{\pi}{4}$ |
| $6r$ | $\pi 36r^2$ | $144r^2$ | $\frac{\pi}{4}$ |

**61.** Sometimes; Sample answer: The equation is true when $x = 1$, $y = 2$, and $z = 3$, but it is false when $x = 2$, $y = 3$, and $z = 4$.

d. **Analytical** What conclusion can be drawn from this? The ratio of the area of the circle to the area of the square will always be $\frac{\pi}{4}$.

A.SSE.2, F.IF.8b

### H.O.T. Problems  Use Higher-Order Thinking Skills

**61.** MP **REASONING** Is $x^y \cdot x^z = x^{yz}$ sometimes, always, or never true? Explain.

**62. OPEN-ENDED** Name two monomials with a quotient of $24a^2b^3$. Sample answer: $24a^4b^6$ and $a^2b^3$

**63. CHALLENGE** Use the Quotient of Powers Property to explain why $x^{-n} = \frac{1}{x^n}$. $\frac{1}{x^n} = \frac{x^0}{x^n} = x^{0-n} = x^{-n}$

**64.** MP **REGULARITY** Write a convincing argument to show why $3^0 = 1$. See margin.

**65. WRITING IN MATH** Explain how to use the Quotient of Powers Property and the Power of a Quotient Property. See margin.

---

### MP Standards for Mathematical Practice

| Emphasis On | Exercises |
|---|---|
| 1 Make sense of problems and persevere in solving them. | 57 |
| 2 Reason abstractly and quantitatively. | 61–62 |
| 3 Construct viable arguments and critique the reasoning of others. | 63, 65 |
| 4 Model with mathematics. | 18, 43–44, 58–59 |
| 7 Look for and make use of structure. | 66–72 |
| 8 Look for and express regularity in repeated reasoning. | 64, 69–72 |

## Preparing for Assessment

**66.** Which of the following is a simplified form of the given expression? Assume the denominator is not 0.  7, 8  A.SSE.2, F.IF.8b  **B**

$$\left(\frac{15n^5m^3}{5m^4n^2}\right)^2$$

- A  $\frac{3n^6}{m^2}$
- B  $\frac{9n^6}{m^2}$
- C  $\frac{9n^5}{m^3}$
- D  $9m^2n^2$

**67.** The value of the expression is 1. What is the value of $m$?  7  A.SSE.2, F.IF.8b  **C**

$$\left(\frac{x^2y^3}{y^2z}\right)^m$$

- A  $-2$
- B  $-1$
- C  $0$
- D  $1$

**68.** Simplify $(4^{-2} \cdot 5^0 \cdot 64)^3$.  7  A.SSE.2, F.IF.8b  **B**

- A  $\frac{1}{64}$
- B  $64$
- C  $1024$
- D  $8000$

**69.** Simplify the expression. Assume the denominator does not equal 0.  7, 8  A.SSE.2, F.IF.8b  **D**

$$\frac{3n^2v^{-4}w}{18n^{-3}v^5w^{-2}}$$

- A  $\frac{1}{6nvw}$
- B  $\frac{n^5w^2}{6v^9}$
- C  $\frac{v}{6nw}$
- D  $\frac{n^5w^3}{6v^9}$

**70.** In the simplified form of the expression $\left(\frac{8b^3c^4d^9}{2b^7c^2d^5}\right)^3$, what is the exponent of the variable $d$?  7, 8  A.SSE.2, F.IF.8b  **B**

- A  $4$
- B  $12$
- C  $14$
- D  $64$

**71.** Which expression simplifies to $\frac{q^{13}r^{10}}{3p^{13}}$?  7, 8  A.SSE.2, F.IF.8b  **C**

- A  $\frac{4p^5q^{-4}r^{-7}}{12p^{-8}q^9r^3}$
- B  $\frac{2p^{-5}q^4r^{-7}}{6p^8q^{-9}r^3}$
- C  $\frac{3p^{-5}q^4r^7}{9p^8q^{-9}r^{-3}}$
- D  $\frac{5p^5q^{-4}r^7}{15p^{-8}q^9r^{-3}}$

**72.** MULTI-STEP Find the product $(v^{-2}w^3x^2y^{-6})(vw^{-2}x^{-3}y^4)$ by following the steps below.  7, 8  A.SSE.2, F.IF.8b

a. Express the product as a fraction with all the current factors having positive exponents.

$$\frac{vw^3x^2y^4}{v^2w^2x^3y^6}$$

b. Which is the simplified form of the expression in part a? **D**

- A  $\frac{xy^2}{vw}$
- B  $\frac{wy^2}{vx}$
- C  $\frac{x}{vwy^2}$
- D  $\frac{w}{vxy^2}$

c. Examine the expression in part b. Which variables cannot equal 0? Select all that apply. **A, C, D**

- ☐  A  $v$
- ☐  B  $w$
- ☐  C  $x$
- ☐  D  $y$

---

**Differentiated Instruction** OL  BL

**Extension** Tell students that if you roll a color cube, the probability of getting red is $\frac{1}{2}$. If you roll the cube $n$ times, the probability of getting red each time is $\left(\frac{1}{2}\right)^n$. Ask them to determine how many times the cube is rolled if the probability of getting red each time is $\frac{1}{512}$. $\frac{1}{512} = \left(\frac{1}{2}\right)^9$, $n = 9$.

---

Lesson 7-2 | Division Properties of Exponents

## Preparing for Assessment

Exercises 66–72 require students to use the skills they will need on standardized assessments. Each exercise is dual-coded with content standards and mathematical practice standards.

### Dual Coding

| Items | Content Standards | Mathematical Practices |
|---|---|---|
| 66 | A.SSE.2, F.IF.8b | 7, 8 |
| 67 | A.SSE.2, F.IF.8b | 7 |
| 68 | A.SSE.2, F.IF.8b | 7 |
| 69 | A.SSE.2, F.IF.8b | 7, 8 |
| 70 | A.SSE.2, F.IF.8b | 7, 8 |
| 71 | A.SSE.2, F.IF.8b | 7, 8 |
| 72 | A.SSE.2, F.IF.8b | 7, 8 |

## Diagnose Student Errors

Survey student responses for each item. Class trends may indicate common errors and misconceptions.

**66.**

| A | Did not square the coefficient |
|---|---|
| B | CORRECT |
| C | Added the exponents |
| D | Divided powers with unlike bases |

**67.**

| A | Used exponent rules with unlike bases and added the exponents for power of powers |
|---|---|
| B | Used incorrect definition of negative exponent |
| C | CORRECT |
| D | Reversed the power and the value for a power of 0 |

**68.**

| A | Raised expression to the $-3$ power |
|---|---|
| B | CORRECT |
| C | Did not see negative exponent or power of 3 |
| D | Thought $5^0 = 5$ |

## Go Online!

### Quizzes

Students can use Self-Check Quizzes to check their understanding of this lesson. You can also give the Chapter Quiz, which covers the content in Lessons 7-1 and 7-2.

connectED.mcgraw-hill.com  **409**

# LESSON 7-3
# Rational Exponents

**SUGGESTED PACING (DAYS)**

| | |
|---|---|
| 90 min. | 0.5 |
| 45 min. | 1 |

Instruction

## Track Your Progress

### Objectives

1. Evaluate and rewrite expressions involving rational exponents.
2. Solve equations involving expressions with rational exponents.

### Mathematical Background
Exponents can be fractions. For any real numbers $a$ and $b$, and any positive integer $n$, if $a^n = b$, then $a$ is an $n$th root of $b$. If $a^n = b$, then $\sqrt[n]{b} = a$. For any positive real number $b$, and any integers $m$ and $n$ with $n > 1$, $b^{\frac{m}{n}} = \sqrt[n]{b^m}$.

### THEN

**A.SSE.2** Use the structure of an expression to identify ways to rewrite it.

**F.IF.8b** Use the properties of exponents to interpret expressions for exponential functions.

### NOW

**N.RN.1** Explain how the definition of the meaning of rational exponents follows from extending the properties of integer exponents to those values, allowing for a notation for radicals in terms of rational exponents.

**N.RN.2** Rewrite expressions involving radicals and rational exponents using the properties of exponents.

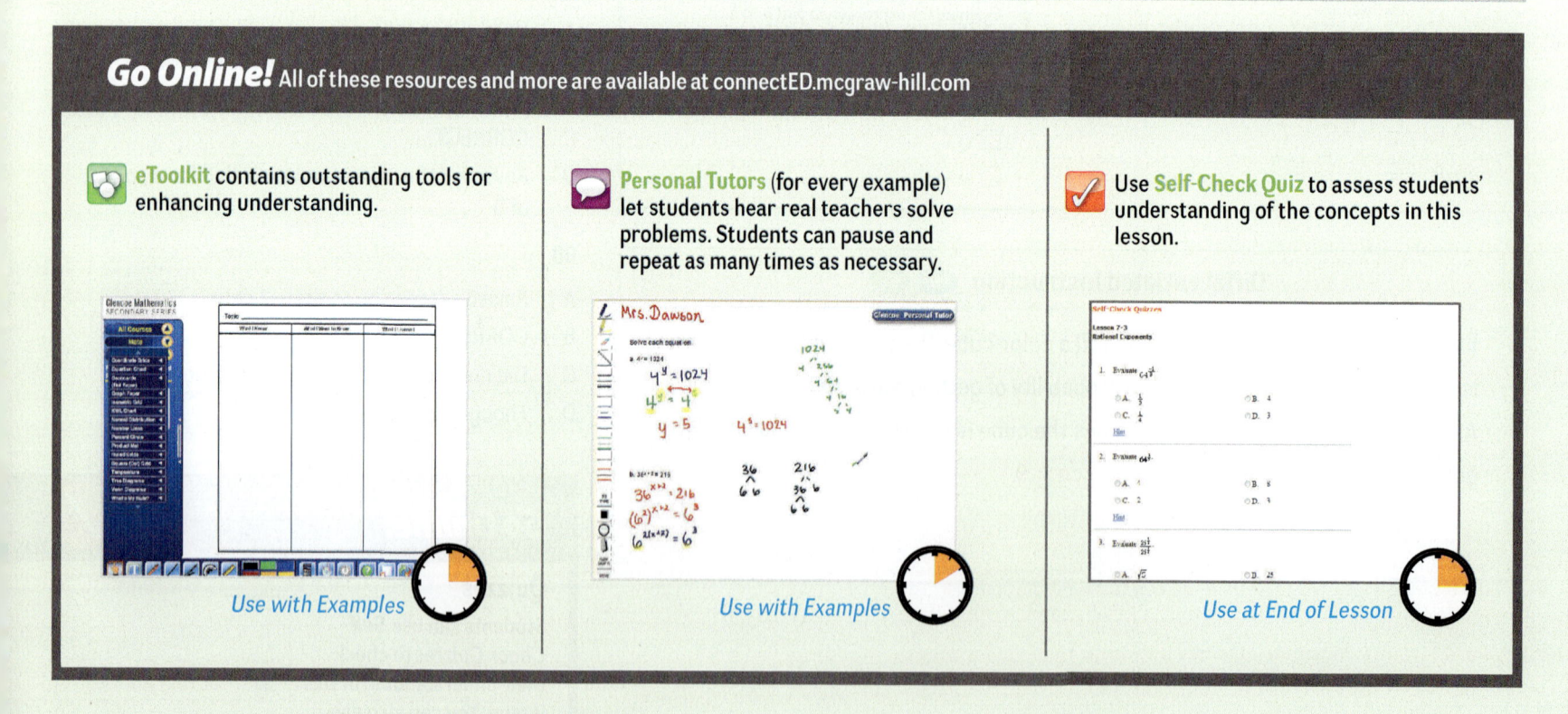

### NEXT

**F.IF.7e** Graph exponential and logarithmic functions, showing intercepts and end behavior, and trigonometric functions, showing period, midline, and amplitude.

## Go Online! All of these resources and more are available at connectED.mcgraw-hill.com

**eToolkit** contains outstanding tools for enhancing understanding.

*Use with Examples*

**Personal Tutors** (for every example) let students hear real teachers solve problems. Students can pause and repeat as many times as necessary.

*Use with Examples*

Use **Self-Check Quiz** to assess students' understanding of the concepts in this lesson.

*Use at End of Lesson*

## OER Using Open Educational Resources
**Apps** Have students use the **Buzz Math** app to practice solving problems with rational numbers and exponents before beginning the lesson objective on rational exponents. *Use as review before beginning the lesson*

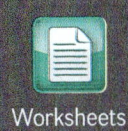

# Go Online!
connectED.mcgraw-hill.com — Worksheets

## Differentiate Your Resources

**Extra Practice** Additional practice or homework; Skills Practice is best for approaching-level students and Practice is best for on-level and beyond-level students

### Skills Practice
AL  OL  ELL

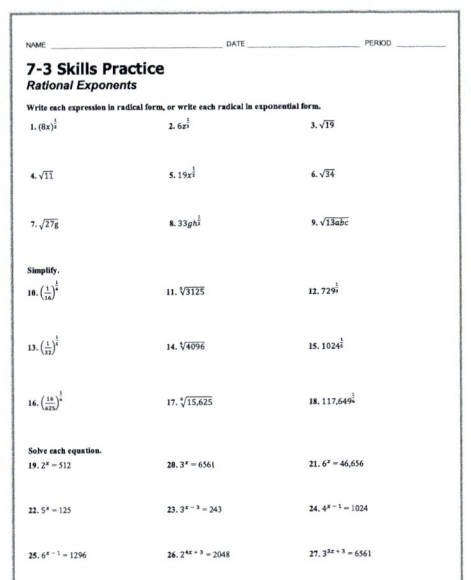

### Practice
AL  OL  BL  ELL

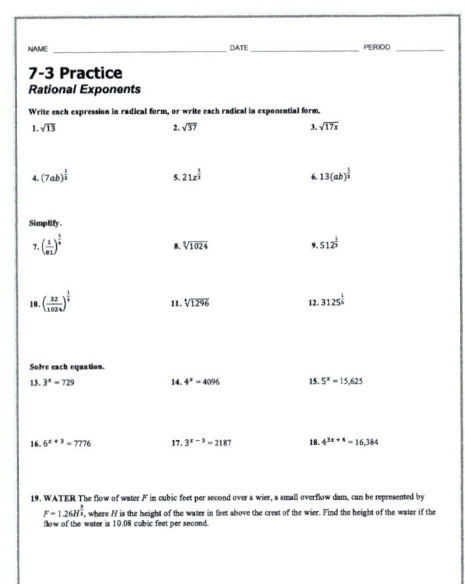

### Word Problem Practice
AL  OL  BL  ELL

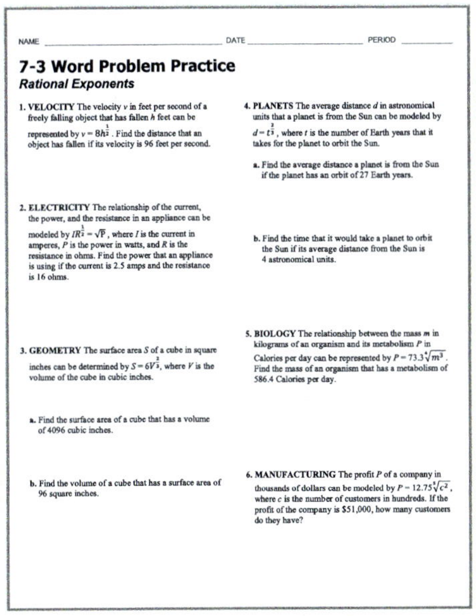

**Intervention** Reteaching and vocabulary activities that can be used with struggling or absent students and as ELL support

**Extension** Activities that can be used to extend lesson concepts

### Study Guide and Intervention
AL  OL  ELL

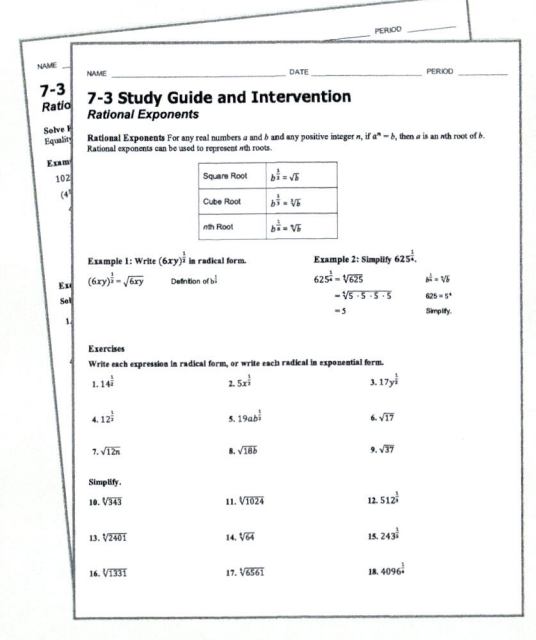

### Study Notebook
AL  OL  BL  ELL

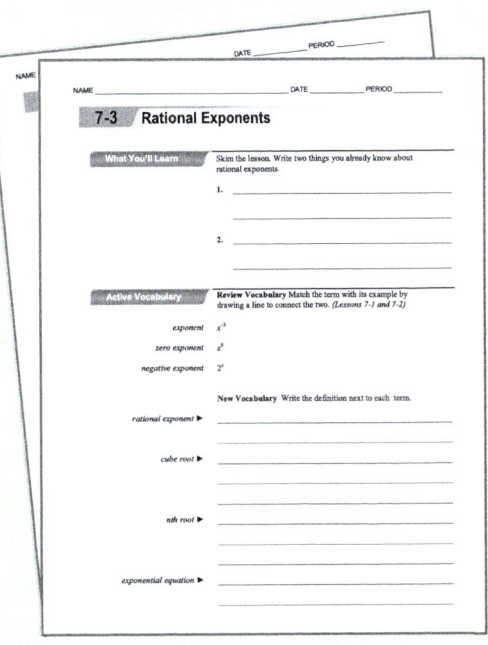

### Enrichment
OL  BL  ELL

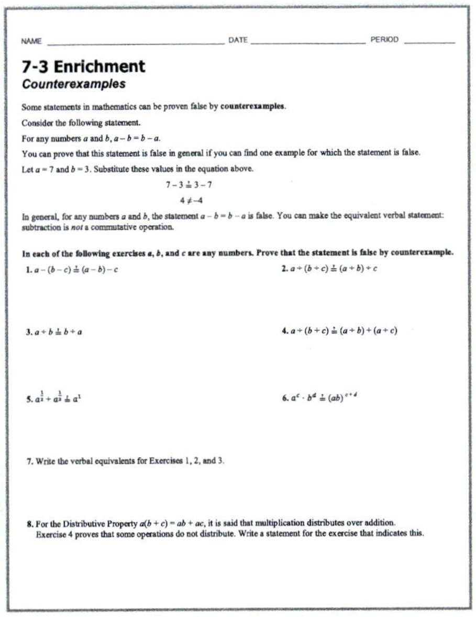

connectED.mcgraw-hill.com   410B

Lesson 7-3 | Rational Exponents

# Launch

Have students read the Why? section of the lesson. Ask:

- How does $50f^{0.2}$ differ from other exponential expressions students have seen before? **Sample answer: The exponent is not an integer.**

- Does a higher SPF offer more or less protection from sun damage? **more**

- What SPF numbers have you seen on sunscreens in stores? **Sample answer: 2, 4, 15, 30**

# Teach

Ask the scaffolded questions for each example to build conceptual understanding for students at all levels.

## 1 Rational Exponents

### Example 1 Radical and Exponential Forms

**AL** In part **b**, what fractional exponent can we use instead of the radical? $\frac{1}{2}$

**OL** In part **c**, does the exponent belong with the entire term? Explain. **no; Sample answer: only the x goes with the exponent, which means only the x goes under the radical.**

**BL** In part **d**, why does the $8p$ have to be in parentheses? **Sample answer: the entire term was under the radical. If I don't use parentheses when converting to exponential form, the exponent will only go with the p.**

### Go Online!

**Interactive Whiteboard**
Use the eLesson or Lesson Presentation to present this lesson.

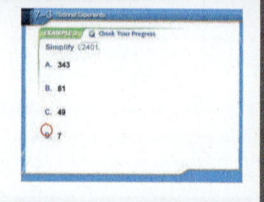

---

# LESSON 3
# Rational Exponents

**Then**
- You used the laws of exponents to find products and quotients of monomials.

**Now**
1. Evaluate and rewrite expressions involving rational exponents.
2. Solve equations involving expressions with rational exponents.

**Why?**
- It's important to protect your skin with sunscreen to prevent damage. The sun protection factor (SPF) of a sunscreen indicates how well it protects you. Sunscreen with an SPF of $f$ absorbs about $p$ percent of the UV-B rays, where $p = 50f^{0.2}$.

**New Vocabulary**
rational exponent
cube root
nth root
exponential equation

**Mathematical Practices**
5 Use appropriate tools strategically.

**Content Standards**
N.RN.1 Explain how the definition of the meaning of rational exponents follows from extending the properties of integer exponents to those values, allowing for a notation for radicals in terms of rational exponents.
N.RN.2 Rewrite expressions involving radicals and rational exponents using the properties of exponents.

## 1 Rational Exponents
You know that an exponent represents the number of times that the base is used as a factor. But how do you evaluate an expression with an exponent that is not an integer like the one above? Let's investigate **rational exponents** by assuming that they behave like integer exponents.

$\left(b^{\frac{1}{2}}\right)^2 = b^{\frac{1}{2}} \cdot b^{\frac{1}{2}}$   Write as a multiplication expression.

$= b^{\frac{1}{2} + \frac{1}{2}}$   Product of Powers

$= b^1$ or $b$   Simplify.

Thus, $b^{\frac{1}{2}}$ is a number with a square equal to $b$. So $b^{\frac{1}{2}} = \sqrt{b}$.

### Key Concept  $b^{\frac{1}{2}}$

Words    For any nonnegative real number $b$, $b^{\frac{1}{2}} = \sqrt{b}$.

Examples  $16^{\frac{1}{2}} = \sqrt{16}$ or 4        $38^{\frac{1}{2}} = \sqrt{38}$

N.RN.1, N.RN.2

### Example 1    Radical and Exponential Forms

Write each expression in radical form, or write each radical in exponential form.

a. $25^{\frac{1}{2}}$
$25^{\frac{1}{2}} = \sqrt{25}$   Definition of $b^{\frac{1}{2}}$
$= 5$   Simplify.

b. $\sqrt{18}$
$\sqrt{18} = 18^{\frac{1}{2}}$   Definition of $b^{\frac{1}{2}}$

c. $5x^{\frac{1}{2}}$
$5x^{\frac{1}{2}} = 5\sqrt{x}$   Definition of $b^{\frac{1}{2}}$

d. $\sqrt{8p}$
$\sqrt{8p} = (8p)^{\frac{1}{2}}$   Definition of $b^{\frac{1}{2}}$

**Guided Practice**
1A. $a^{\frac{1}{2}}$  $\sqrt{a}$    1B. $\sqrt{22}$  $22^{\frac{1}{2}}$    1C. $(7w)^{\frac{1}{2}}$  $\sqrt{7w}$    1D. $2\sqrt{x}$  $2x^{\frac{1}{2}}$

---

### Mathematical Practices Strategies

**Look for and make use of structure.**
Analyzing the parts that make up a rational expression helps students acquaint themselves with using fractions for exponents. For example, ask:

- Which part of a rational exponent indicates what the index of the radical symbol will be? **the denominator**

- How do the results compare when a base is raised to an improper fraction versus when a base is raised to a proper fraction? **With an improper fraction for an exponent, the result is greater than the base. With a proper fraction for an exponent, the result is less than the base.**

- How do you know whether an exponent applies to the coefficient or just to the variable? **If the coefficient is in parentheses with the variable, then the exponent is applied to the coefficient as well.**

410 | Lesson 7-3 | Rational Exponents

You know that to find the square root of a number $a$ you find a number with a square of $a$. In the same way, you can find other roots of numbers. If $a^3 = b$, then $a$ is the **cube root** of $b$, and if $a^n = b$ for a positive integer $n$, then $a$ is an **$n$th root** of $b$.

> **Key Concept** $n$th Root
>
> **Words** For any real numbers $a$ and $b$ and any positive integer $n$, if $a^n = b$, then $a$ is an $n$th root of $b$.
>
> **Symbols** If $a^n = b$, then $\sqrt[n]{b} = a$.
>
> **Example** Because $2^4 = 16$, 2 is a fourth root of 16; $\sqrt[4]{16} = 2$.

**Study Tip**

**MP Tools** You can use a graphing calculator to find $n$th roots. Enter $n$, then press MATH and choose $\sqrt[x]{\phantom{x}}$.

Since $3^2 = 9$ and $(-3)^2 = 9$, both 3 and $-3$ are square roots of 9. Similarly, since $2^4 = 16$ and $(-2)^4 = 16$, both 2 and $-2$ are fourth roots of 16. The positive roots are called *principal roots*. Radical symbols indicate principal roots, so $\sqrt{16} = 2$.

N.RN.2

**Example 2** $n$th Roots

Simplify.

a. $\sqrt[3]{27}$         b. $\sqrt[5]{32}$

$\sqrt[3]{27} = \sqrt[3]{3 \cdot 3 \cdot 3}$         $\sqrt[5]{32} = \sqrt[5]{2 \cdot 2 \cdot 2 \cdot 2 \cdot 2}$

$= 3$         $= 2$

**Guided Practice**

2A. $\sqrt[3]{64}$  4         2B. $\sqrt[4]{10{,}000}$  10

Like square roots, $n$th roots can be represented by rational exponents.

$\left(b^{\frac{1}{n}}\right)^n = \underbrace{b^{\frac{1}{n}} \cdot b^{\frac{1}{n}} \cdot \ldots \cdot b^{\frac{1}{n}}}_{n \text{ factors}}$   Write as a multiplication expression.

$= b^{\frac{1}{n} + \frac{1}{n} + \cdots + \frac{1}{n}}$   Product of Powers

$= b^1$ or $b$   Simplify

Thus, $b^{\frac{1}{n}}$ is a number with an $n$th power equal to $b$. So $b^{\frac{1}{n}} = \sqrt[n]{b}$.

> **Key Concept** $b^{\frac{1}{n}}$
>
> **Words** For any positive real number $b$ and any integer $n > 1$, $b^{\frac{1}{n}} = \sqrt[n]{b}$.
>
> **Example** $8^{\frac{1}{3}} = \sqrt[3]{8} = \sqrt[3]{2 \cdot 2 \cdot 2}$ or 2

---

**Lesson 7-3 | Rational Exponents**

**Need Another Example?**

Write each expression in radical form, or write each radical in exponential form.

a. $81^{\frac{1}{2}}$  9
b. $\sqrt{38}$  $38^{\frac{1}{2}}$
c. $12m^{\frac{1}{2}}$  $12\sqrt{m}$
d. $\sqrt{32w}$  $(32w)^{\frac{1}{2}}$

**Example 2** $n$th Roots

**AL** What does $\sqrt[3]{\phantom{x}}$ mean? cube root

**OL** In part **a**, how can you rewrite 27 as a product of factors where there are a total of 3 factors? $3 \cdot 3 \cdot 3$

**BL** In part **b**, explain how the solution is a prime factor of 32. Sample answer: when 32 is broken down into prime factorization, there are five 2s. Since we are looking for the fifth root, the solution is one of the prime factors.

**Need Another Example?**
Simplify.
a. $\sqrt[4]{256}$  4
b. $\sqrt[6]{15{,}625}$  5

**Teaching Tip**

**Order of Operations** Remind students to observe the order of operations with radical and exponential expressions. Emphasize that expressions like $5x^{\frac{1}{2}}$ and $(5x)^{\frac{1}{2}}$ are different.

**Notation** Emphasize that if there is no index on a radical symbol, it denotes a square root.

> **MP Teaching the Mathematical Practices**
>
> **Tools** Mathematically proficient students are familiar with the tools appropriate for their course and make sound decisions about when each might be helpful. Encourage students to use mental math to verify answers they find with a calculator.

# Lesson 7-3 | Rational Exponents

## Example 3 Evaluate $b^{\frac{1}{n}}$ Expressions

**AL** In part **a**, what does the exponent $\frac{1}{3}$ mean? We need to find the cube root.

**OL** For part **a**, rewrite 125 as a product of its factors where there are three total factors. $5 \cdot 5 \cdot 5$

**BL** In part **b**, why is 1296 not broken down into prime factors? Sample answer: we need to break the number down into a total of four factors because we need to find the fourth root. If I break down the number further, I will have more than four factors.

**Need Another Example?**
Simplify.
a. $1331^{\frac{1}{3}}$  11
b. $2401^{\frac{1}{4}}$  7

## Example 4 Evaluate $b^{\frac{m}{n}}$ Expressions

**AL** What is the exponent in part **a**? $\frac{2}{3}$

**OL** In part **a**, explain what the numerator and denominator of the exponent tell us to do. Sample answer: the denominator of 3 tells us we need to take the cube root, and then the numerator of 2 tells us to square the result.

**BL** In part **b**, explain how we got $6^3$. Sample answer: 36 broken down into two factors is $6 \cdot 6$. The square root of $6 \cdot 6$ is 6.

**Need Another Example?**
Simplify.
a. $32^{\frac{2}{5}}$  4
b. $81^{\frac{5}{2}}$  59,049

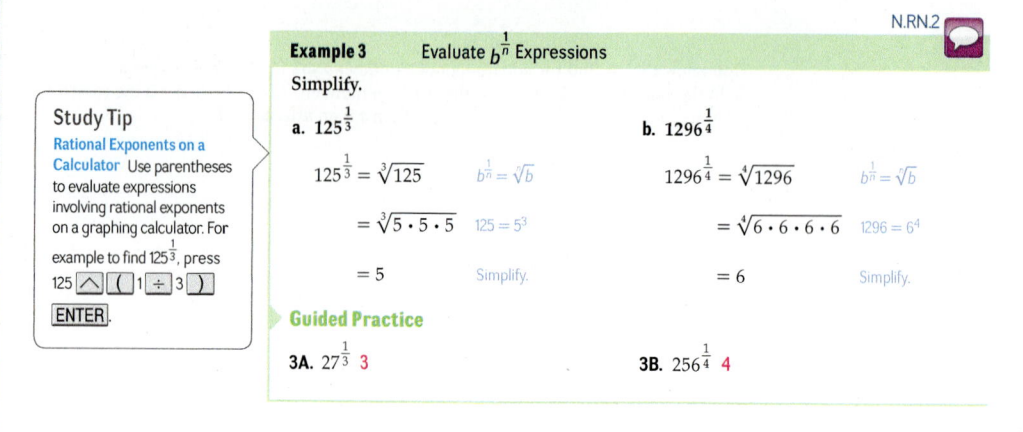

---

### Example 3 Evaluate $b^{\frac{1}{n}}$ Expressions

**Study Tip**
Rational Exponents on a Calculator Use parentheses to evaluate expressions involving rational exponents on a graphing calculator. For example to find $125^{\frac{1}{3}}$, press 125 ^ ( 1 ÷ 3 ) ENTER.

Simplify.

a. $125^{\frac{1}{3}}$
$125^{\frac{1}{3}} = \sqrt[3]{125}$   $b^{\frac{1}{n}} = \sqrt[n]{b}$
$= \sqrt[3]{5 \cdot 5 \cdot 5}$   $125 = 5^3$
$= 5$   Simplify.

b. $1296^{\frac{1}{4}}$
$1296^{\frac{1}{4}} = \sqrt[4]{1296}$   $b^{\frac{1}{n}} = \sqrt[n]{b}$
$= \sqrt[4]{6 \cdot 6 \cdot 6 \cdot 6}$   $1296 = 6^4$
$= 6$   Simplify.

**Guided Practice**
3A. $27^{\frac{1}{3}}$  3
3B. $256^{\frac{1}{4}}$  4

The Power of a Power Property allows us to extend the definition of $b^{\frac{1}{n}}$ to $b^{\frac{m}{n}}$.

$b^{\frac{m}{n}} = \left(b^{\frac{1}{n}}\right)^m$   Power of a Power
$= \left(\sqrt[n]{b}\right)^m$ or $\sqrt[n]{b^m}$   $b^{\frac{1}{n}} = \sqrt[n]{b}$

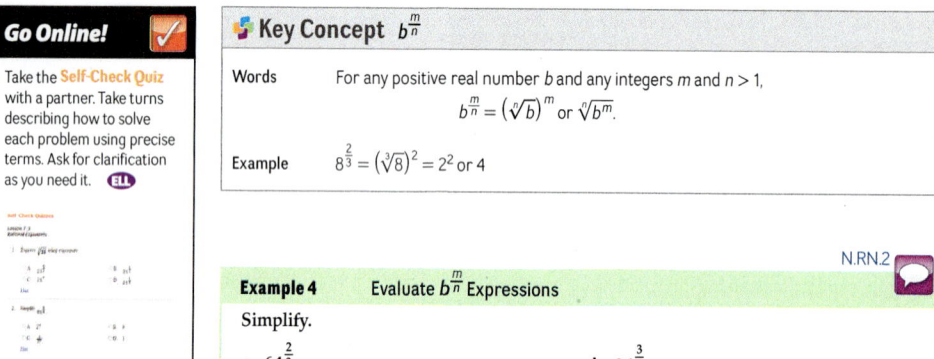

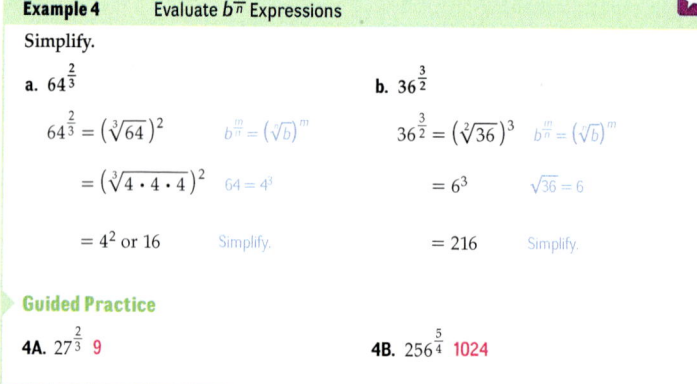

**Go Online!**
Take the Self-Check Quiz with a partner. Take turns describing how to solve each problem using precise terms. Ask for clarification as you need it. **ELL**

### Key Concept $b^{\frac{m}{n}}$

**Words** For any positive real number $b$ and any integers $m$ and $n > 1$,
$b^{\frac{m}{n}} = \left(\sqrt[n]{b}\right)^m$ or $\sqrt[n]{b^m}$.

**Example** $8^{\frac{2}{3}} = \left(\sqrt[3]{8}\right)^2 = 2^2$ or 4

### Example 4 Evaluate $b^{\frac{m}{n}}$ Expressions

Simplify.

a. $64^{\frac{2}{3}}$
$64^{\frac{2}{3}} = \left(\sqrt[3]{64}\right)^2$   $b^{\frac{m}{n}} = \left(\sqrt[n]{b}\right)^m$
$= \left(\sqrt[3]{4 \cdot 4 \cdot 4}\right)^2$   $64 = 4^3$
$= 4^2$ or 16   Simplify.

b. $36^{\frac{3}{2}}$
$36^{\frac{3}{2}} = \left(\sqrt{36}\right)^3$   $b^{\frac{m}{n}} = \left(\sqrt[n]{b}\right)^m$
$= 6^3$   $\sqrt{36} = 6$
$= 216$   Simplify.

**Guided Practice**
4A. $27^{\frac{2}{3}}$  9
4B. $256^{\frac{5}{4}}$  1024

---

### Differentiated Instruction AL OL BL ELL

**Interpersonal Learners** Divide the class into groups of two or three students. Have students discuss what they knew about exponents before starting the lesson and how it relates to rational exponents. For example, if an exponent tells you how many times to use the base as a factor, what does an exponent of $\frac{3}{2}$ represent? Encourage students to explore the concepts and ask each other for information.

## Lesson 7-3 | Rational Exponents

**2 Solve Exponential Equations** In an **exponential equation**, variables occur as exponents. The Power Property of Equality and the other properties of exponents can be used to solve exponential equations.

### Key Concept  Power Property of Equality

**Words**  For any real number $b > 0$ and $b \neq 1$, $b^x = b^y$ if and only if $x = y$.

**Examples**  If $5^x = 5^3$, then $x = 3$. If $n = \frac{1}{2}$, then $4^n = 4^{\frac{1}{2}}$.

N.RN.2

### Example 5  Solve Exponential Equations

Solve each equation.

**a.** $6^x = 216$

| | |
|---|---|
| $6^x = 216$ | Original equation |
| $6^x = 6^3$ | Rewrite 216 as $6^3$ |
| $x = 3$ | Property of Equality |

**CHECK**  $6^x = 216$
$6^3 \stackrel{?}{=} 216$
$216 = 216$ ✓

**b.** $25^{x-1} = 5$

| | |
|---|---|
| $25^{x-1} = 5$ | Original equation |
| $(5^2)^{x-1} = 5$ | Rewrite 25 as $5^2$ |
| $5^{2x-2} = 5^1$ | Power of a Power; Distributive Property |
| $2x - 2 = 1$ | Power Property of Equality |
| $2x = 3$ | Add 2 to each side |
| $x = \frac{3}{2}$ | Divide each side by 2 |

**CHECK**  $25^{x-1} = 5$
$25^{\frac{3}{2}-1} \stackrel{?}{=} 5$
$25^{\frac{1}{2}} = 5$ ✓

### Guided Practice

**5A.** $5^x = 125$  **3**

**5B.** $12^{2x+3} = 144$  $-\frac{1}{2}$

N.RN.2

### Real-World Example 6  Solve Exponential Equations

**SUNSCREEN** Refer to the beginning of the lesson. Find the SPF that absorbs 100% of UV-B rays.

| | |
|---|---|
| $p = 50f^{0.2}$ | Original equation |
| $100 = 50f^{0.2}$ | $p = 100$ |
| $2 = f^{0.2}$ | Divide each side by 50 |
| $2 = f^{\frac{1}{5}}$ | $0.2 = \frac{1}{5}$ |
| $(2^5)^{\frac{1}{5}} = f^{\frac{1}{5}}$ | $2 = 2^1 = (2^5)^{\frac{1}{5}}$ |
| $2^5 = f$ | Power Property of Equality |
| $32 = f$ | Simplify |

**Real-World Link**
Use extra caution near snow, water, and sand because they reflect the damaging rays of the Sun, which can increase your chance of sunburn.

**Source:** American Academy of Dermatology

### Guided Practice

**6. CHEMISTRY** The radius $r$ of the nucleus of an atom of mass number $A$ is $r = 1.2A^{\frac{1}{3}}$ femtometers. Find $A$ if $r = 3.6$ femtometers.  **27**

---

## 2 Solve Exponential Equations

### Example 5  Solve Exponential Equations

**AL** What do we need to determine? the exponent

**OL** For part **a**, into what can you break 216 down so that it is the product of three equal factors? $6 \cdot 6 \cdot 6$ or $6^3$.

**BL** For part **b**, how can you look at the problem and determine that the total exponent quantity must be $\frac{1}{2}$? Sample answer: I know that 5 is the square root of 25, which means that the exponent must equal $\frac{1}{2}$.

**Need Another Example?**
Solve each equation.
**a.** $9^x = 729$  **3**
**b.** $16^{2x-1} = 8$  $\frac{7}{8}$

### Example 6  Solve Exponential Equations

**AL** What value do we substitute for $p$? 100

**OL** Why do we change 0.2 to $\frac{1}{5}$? Sample answer: we need to work with the exponent as a fraction so that the numerator and denominator can tell us what to do.

**BL** Why can we change 2 to $(2^5)^{\frac{1}{5}}$? Sample answer: Because 5 and $\frac{1}{5}$ are inverses, I have not changed the value of 2 by using a power and a power of a power.

**Need Another Example?**
**Biology** The population $p$ of a culture that begins with 40 bacteria and doubles every 8 hours can be modeled by $p = 40(2)^{\frac{t}{8}}$, where $t$ is time in hours. Find $t$ if $p = 20,480$. 72 hours

connectED.mcgraw-hill.com  413

Lesson 7-3 | Rational Exponents

# Practice

**Formative Assessment** Use Exercises 1–16 to assess students' understanding of the concepts in this lesson.

The Practice and Problem Solving exercises assess the content taught in the lesson. The Preparing for Assessment page is meant to be used as preparation for end-of-course assessments.

## Teaching the Mathematical Practices

**Tools** Mathematically proficient students consider the available tools when solving a mathematical problem. In Exercise 16, discuss how a calculator could be used to solve and check the solution.

## Extra Practice

See page R7 for extra exercises for students who are approaching level or for on-level students who need additional reinforcement.

### Levels of Complexity Chart

The levels of the exercises progress from 1 to 3, with Level 1 indicating the lowest level of complexity.

| Exercises | 17–58 | 59–85, 94–102 | 86–93 |
|---|---|---|---|
| Level 3 | | | ● |
| Level 2 | | ● | |
| Level 1 | ● | | |

## Check Your Understanding

○ = Step-by-Step Solutions begin on page R11.

**Go Online!** for a Self-Check Quiz

**Example 1**
N.RN.1, N.RN.2

Write each expression in radical form, or write each radical in exponential form.

1. $12^{\frac{1}{2}}$  $\sqrt{12}$
2. $3x^{\frac{1}{2}}$  $3\sqrt{x}$
3. $\sqrt{33}$  $33^{\frac{1}{2}}$
4. $\sqrt{8n}$  $(8n)^{\frac{1}{2}}$

**Examples 2–4** Simplify.
N.RN.2

5. $\sqrt[3]{512}$  8
6. $\sqrt[5]{243}$  3
7. $343^{\frac{1}{3}}$  7
8. $\left(\frac{1}{16}\right)^{\frac{1}{4}}$  $\frac{1}{2}$

9. $343^{\frac{2}{3}}$  49
10. $81^{\frac{3}{4}}$  27
11. $216^{\frac{4}{3}}$  1296
12. $\left(\frac{1}{49}\right)^{\frac{3}{2}}$  $\frac{1}{343}$

**Example 5** Solve each equation.
N.RN.2

13. $8^x = 4096$  4
14. $3^{3x+1} = 81$  1
15. $4^{x-3} = 32$  5.5

**Example 6**
N.RN.2

16. **TOOLS** A weir is used to measure water flow in a channel. For a rectangular broad-crested weir, the flow $Q$ in cubic feet per second is related to the weir length $L$ in feet and height $H$ of the water by $Q = 1.6LH^{\frac{3}{2}}$. Find the water height for a weir that is 3 feet long and has flow of 38.4 cubic feet per second.  4 ft

## Practice and Problem Solving

Extra Practice is on page R7.

**Example 1**
N.RN.1, N.RN.2

Write each expression in radical form, or write each radical in exponential form.

17. $15^{\frac{1}{2}}$  $\sqrt{15}$
18. $24^{\frac{1}{2}}$  $\sqrt{24}$
19. $4k^{\frac{1}{2}}$  $4\sqrt{k}$
20. $(12y)^{\frac{1}{2}}$  $\sqrt{12y}$

21. $\sqrt{26}$  $26^{\frac{1}{2}}$
22. $\sqrt{44}$  $44^{\frac{1}{2}}$
23. $2\sqrt{ab}$  $2(ab)^{\frac{1}{2}}$
24. $\sqrt{3xyz}$  $(3xyz)^{\frac{1}{2}}$

**Examples 2–4** Simplify.
N.RN.2

25. $\sqrt[3]{8}$  2
26. $\sqrt[5]{1024}$  4
27. $\sqrt[3]{216}$  6
28. $\sqrt[4]{10{,}000}$  10

29. $\sqrt[3]{0.001}$  0.1
30. $\sqrt[4]{\frac{16}{81}}$  $\frac{2}{3}$
31. $1331^{\frac{1}{3}}$  11
32. $64^{\frac{1}{6}}$  2

33. $3375^{\frac{1}{3}}$  15
34. $512^{\frac{1}{9}}$  2
35. $\left(\frac{1}{81}\right)^{\frac{1}{4}}$  $\frac{1}{3}$
36. $\left(\frac{3125}{32}\right)^{\frac{1}{5}}$  $\frac{5}{2}$

37. $8^{\frac{2}{3}}$  4
38. $625^{\frac{3}{4}}$  125
39. $729^{\frac{5}{6}}$  243
40. $256^{\frac{3}{8}}$  8

41. $125^{\frac{4}{3}}$  625
42. $49^{\frac{5}{2}}$  16,807
43. $\left(\frac{9}{100}\right)^{\frac{3}{2}}$  $\frac{27}{1000}$
44. $\left(\frac{8}{125}\right)^{\frac{4}{3}}$  $\frac{16}{625}$

### Differentiated Homework Options

| Levels | **AL** Basic | **OL** Core | **BL** Advanced |
|---|---|---|---|
| Exercises | 17–58, 89, 90, 92–102 | 17–83 odd, 85–90, 92–102 | 59–93<br>94–102 (optional) |
| 2-Day Option | 18–58 even, 94–102 | 17–58, 94–102 | |
| | 17–57 odd, 89, 90, 92, 93 | 59–90, 92, 93 | |

You can use ALEKS to provide additional remediation support with personalized instruction and practice.

## Go Online!

**eBook**

**Interactive Student Guide**
Use the *Interactive Student Guide* to deepen conceptual understanding.
· Rational Exponents

**Example 5**
N.RN.2

Solve each equation.

45. $3^x = 243$   **5**
46. $12^x = 144$   **2**
47. $16^x = 4$   $\frac{1}{2}$
48. $27^x = 3$   $\frac{1}{3}$
49. $9^x = 27$   $\frac{3}{2}$
50. $32^x = 4$   $\frac{2}{5}$
51. $2^{x-1} = 128$   **8**
52. $4^{2x+1} = 1024$   **2**
53. $6^{x-4} = 1296$   **8**
54. $9^{2x+3} = 2187$   $\frac{1}{4}$
55. $4^{3x} = 512$   $\frac{3}{2}$
56. $128^{3x} = 8$   $\frac{1}{7}$

**Example 6**
N.RN.2

57. **CONSERVATION** Water collected in a rain barrel can be used to water plants and reduce city water use. Water flowing from an open rain barrel has velocity $v = 8h^{\frac{1}{2}}$, where $v$ is in feet per second and $h$ is the height of the water in feet. Find the height of the water if it is flowing at 16 feet per second.   **4 ft**

58. **ELECTRICITY** The radius $r$ in millimeters of a platinum wire $L$ centimeters long with resistance 0.1 ohm is $r = 0.059L^{\frac{1}{2}}$. How long is a wire with radius 0.236 millimeter?   **16 cm**

**B** Write each expression in radical form, or write each radical in exponential form.

59. $17^{\frac{1}{3}}$   $\sqrt[3]{17}$
60. $q^{\frac{1}{4}}$   $\sqrt[4]{q}$
61. $7b^{\frac{1}{3}}$   $7\sqrt[3]{b}$
62. $m^{\frac{2}{3}}$   $\sqrt[3]{m^2}$
63. $\sqrt[3]{29}$   $29^{\frac{1}{3}}$
64. $\sqrt[5]{h}$   $h^{\frac{1}{5}}$
65. $2\sqrt[3]{a}$   $2a^{\frac{1}{3}}$
66. $\sqrt[3]{xy^2}$   $x^{\frac{1}{3}}y^{\frac{2}{3}}$

Simplify.

67. $\sqrt[3]{0.027}$   **0.3**
68. $\sqrt[4]{\frac{n^4}{16}}$   $\frac{n}{2}$
69. $a^{\frac{1}{3}} \cdot a^{\frac{2}{3}}$   $a$
70. $c^{\frac{1}{2}} \cdot c^{\frac{3}{2}}$   $c^2$
71. $(8^2)^{\frac{2}{3}}$   **16**
72. $\left(y^{\frac{3}{4}}\right)^{\frac{1}{2}}$   $y^{\frac{3}{8}}$
73. $9^{-\frac{1}{2}}$   $\frac{1}{3}$
74. $16^{-\frac{3}{2}}$   $\frac{1}{64}$
75. $(3^2)^{-\frac{3}{2}}$   $\frac{1}{27}$
76. $\left(81^{\frac{1}{4}}\right)^{-2}$   $\frac{1}{9}$
77. $k^{-\frac{1}{2}}$   $\frac{1}{\sqrt{k}}$
78. $\left(d^{\frac{4}{3}}\right)^0$   **1**

Solve each equation.

79. $2^{5x} = 8^{2x-4}$   **12**
80. $81^{2x-3} = 9^{x+3}$   **3**
81. $2^{4x} = 32^{x+1}$   **−5**
82. $16^x = \frac{1}{2}$   $-\frac{1}{4}$
83. $25^x = \frac{1}{125}$   $-\frac{3}{2}$
84. $6^{8-x} = \frac{1}{216}$   **11**

85. **MODELING** The frequency $f$ in hertz of the $n$th key on a piano is $f = 440\left(2^{\frac{1}{12}}\right)^{n-49}$.

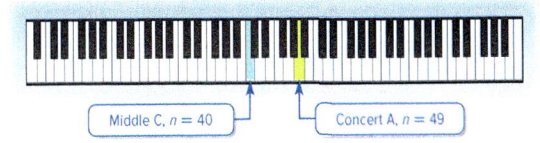

Middle C, $n = 40$    Concert A, $n = 49$

a. What is the frequency of Concert A?   **440 Hz**
b. Which note has a frequency of 220 Hz?   **the A below middle C, the 37th note**

## Teaching the Mathematical Practices

**Construct Arguments** Mathematically proficient students understand and use stated assumptions, definitions, and previously established results in constructing arguments. In Exercise 90, point out that $x$ is a nonnegative real number. Tell students to consider values of $x$ that are whole numbers and fractions that are greater than and less than 1. Point out that 1 is often a special case.

### Watch Out!

**Error Analysis** In Exercise 92, students should recognize that the bases of the expressions must have equal bases to apply the Power Property of Equality.

**86.** **RANDOM WALKS** Suppose you go on a walk where you choose the direction of each step at random. The path of a molecule in a liquid or a gas, the path of a foraging animal, and a fluctuating stock price are all modeled as random walks. The number of possible random walks $w$ of $n$ steps where you choose one of $d$ directions at each step is $w = d^n$.

a. How many steps have been taken in a 2-direction random walk if there are 4096 possible walks? 12

b. How many steps have been taken in a 4-direction random walk if there are 65,536 possible walks? 8

c. If a walk of 7 steps has 2187 possible walks, how many directions could be taken at each step? 3

**87.** **SOCCER** The radius $r$ of a ball that holds $V$ cubic units of air is modeled by $r = 0.62V^{\frac{1}{3}}$. What are the possible volumes of each size soccer ball? Size 3, 204.0 to 230.2 in$^3$; Size 4, 268.5 to 299.9 in$^3$; Size 5, 333.6 to 382.4 in$^3$

**Soccer Ball Dimensions**

| Size | Diameter (in.) |
|---|---|
| 3 | 7.3–7.6 |
| 4 | 8.0–8.3 |
| 5 | 8.6–9.0 |

**88.** **MULTIPLE REPRESENTATIONS** In this problem, you will explore the graph of an exponential function.

a. **TABULAR** Copy and complete the table below.

| $x$ | $-2$ | $-\frac{3}{2}$ | $-1$ | $-\frac{1}{2}$ | 0 | $\frac{1}{2}$ | 1 | $\frac{3}{2}$ | 2 |
|---|---|---|---|---|---|---|---|---|---|
| $f(x) = 4^x$ | $\frac{1}{16}$ | $\frac{1}{8}$ | $\frac{1}{4}$ | $\frac{1}{2}$ | 1 | 2 | 4 | 8 | 16 |

b. **GRAPHICAL** Graph $f(x)$ by plotting the points and connecting them with a smooth curve. b–c. See margin.

c. **VERBAL** Describe the shape of the graph of $f(x)$. What are its key features? Is it linear?

N.RN.1, N.RN.2

### H.O.T. Problems Use Higher-Order Thinking Skills

**89.** **OPEN-ENDED** Write two different expressions with rational exponents equal to $\sqrt{2}$. See margin.

**90.** **CONSTRUCT ARGUMENTS** Determine whether each statement is *always*, *sometimes*, or *never* true. Assume that $x$ is a nonnegative real number. Explain your reasoning. a–f. See Ch. 7 Answer Appendix.

a. $x^2 = x^{\frac{1}{2}}$ b. $x^{-2} = x^{\frac{1}{2}}$ c. $x^{\frac{1}{3}} = x^{\frac{1}{2}}$

d. $\sqrt{x} = x^{\frac{1}{2}}$ e. $\left(x^{\frac{1}{2}}\right)^2 = x$ f. $x^{\frac{1}{2}} \cdot x^2 = x$

**91.** **CHALLENGE** For what values of $x$ is $x = x^{\frac{1}{3}}$? −1, 0, 1

**92.** **ERROR ANALYSIS** Anna and Jamal are solving $128^x = 4$. Is either of them correct? Explain your reasoning. Anna; Jamal did not write the expressions with equal bases before applying the Power Property of Equality.

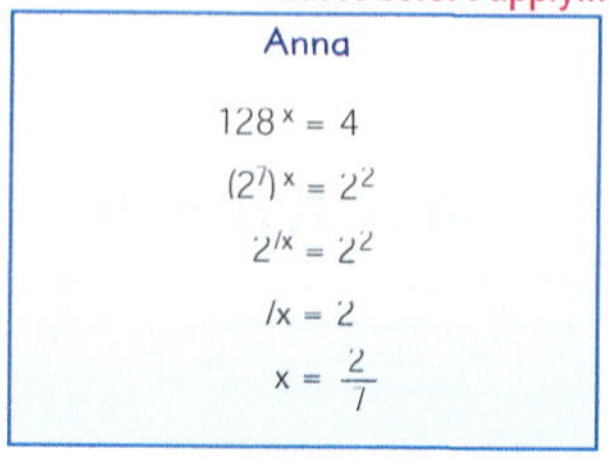
Anna
128ˣ = 4
(2⁷)ˣ = 2²
2⁷ˣ = 2²
7x = 2
x = 2/7

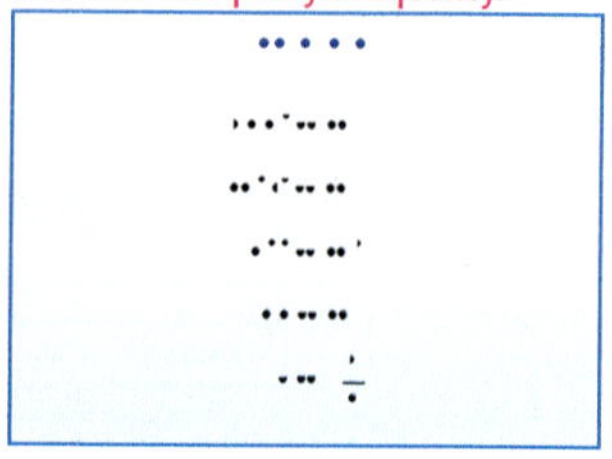

93. Sample answer: 2 is the principal fourth root of 16 because 2 is positive and $2^4 = 16$.

**93.** **WRITING IN MATH** Explain why 2 is the principal fourth root of 16.

## Assess

**Ticket Out the Door** Have each student tell you an exponential expression and an equivalent radical expression.

## Additional Answers

88b.

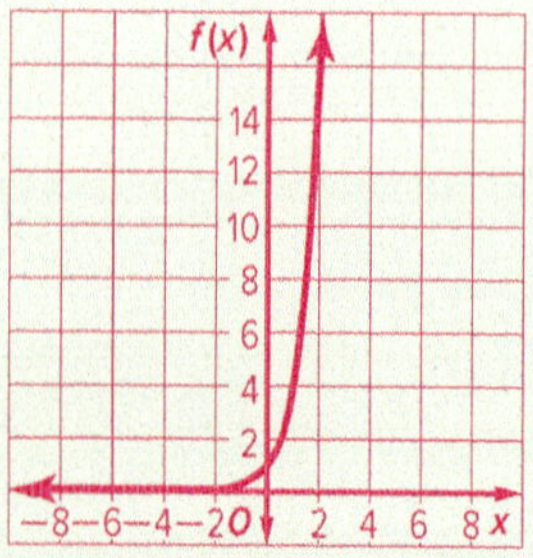
f(x)
14
12
10
8
6
4
2
−8 −6 −4 −2 O 2 4 6 8 x

88c. The graph of $f(x) = 4^x$ is a curve. It has no $x$-intercept, a $y$-intercept of 1, the domain is all reals, the range is all positive reals, it is increasing over the entire domain, as $x$ approaches infinity $f(x)$ approaches infinity, as $x$ approaches negative infinity $f(x)$ approaches 0.

89. Sample answer: $2^{\frac{1}{2}}$ and $4^{\frac{1}{4}}$

## Standards for Mathematical Practice

| Emphasis On | Exercises |
|---|---|
| 3 Construct viable arguments and critique the reasoning of others. | 90, 92–93 |
| 4 Model with mathematics. | 57–58, 85–87, 97 |
| 5 Use appropriate tools strategically. | 16 |
| 7 Look for and make use of structure. | 95–96, 99–100, 102 |
| 8 Look for and express regularity in repeated reasoning. | 94, 96, 98, 101 |

### Go Online!

The most up-to-date resources available for your program can be found at connectED.mcgraw-hill.com.

Lesson 7-3 | Rational Exponents

## Preparing for Assessment

94. ⬝⬝⬝⬝⬝⬝⬝⬝⬝⬝⬝⬝⬝⬝⬝⬝⬝⬝⬝ $x+$ ⬝ $=$ ⬝⬝⬝
    **MP** 8 N.RN.2  **A**
    ○ A ⬝
    ○ B ⬝
    ○ C ⬝
    ○ D ⬝

95. ⬝⬝⬝⬝⬝⬝ $c$ ⬝⬝ ⬝⬝⬝⬝⬝ ⬝⬝⬝ ⬝⬝⬝⬝ ⬝
    **MP** 7 N.RN.1, N.RN.2  **C**
    ○ A ⬝ $c$
    ○ B ⬝ $c$
    ○ C ⬝⬝ $c$
    ○ D ⬝⬝ $c$

96. ⬝⬝ $j =$ ⬝⬝⬝ × 32 × 32 × 32 × 32 ⬝⬝⬝ ⬝⬝⬝⬝⬝ ⬝ ⬝⬝⬝
    ⬝⬝ $j$ ⬝ ⬝⬝⬝⬝⬝ ⬝⬝⬝ **MP** 7, 8 N.RN.2  **B, C**
    ☐ A ⬝
    ☐ B ⬝⬝
    ☐ C ⬝⬝
    ☐ D ⬝⬝

97. **MULTI-STEP** ⬝⬝⬝⬝ ⬝ ⬝⬝⬝⬝⬝⬝⬝⬝⬝⬝⬝⬝⬝⬝
    ⬝⬝⬝⬝⬝⬝⬝ ⬝⬝ $t = \dfrac{\pi r}{g}$ ⬝⬝⬝ ⬝⬝⬝ ⬝ ⬝⬝⬝ ⬝⬝⬝⬝
    ⬝⬝⬝⬝ ⬝⬝⬝⬝⬝⬝ ⬝⬝⬝ $g$ ⬝⬝⬝⬝⬝ ⬝⬝⬝⬝⬝⬝
    ⬝⬝ ⬝ $r$ ⬝⬝⬝⬝⬝ **MP** 2, 4 N.RN.2
    a. ⬝⬝⬝ ⬝⬝⬝ ⬝⬝⬝⬝⬝⬝⬝⬝ ⬝⬝⬝⬝⬝ ⬝⬝⬝ ⬝⬝⬝⬝
       ⬝⬝ ⬝⬝⬝⬝⬝ ⬝ ⬝⬝⬝⬝ ⬝ ⬝⬝⬝⬝⬝⬝⬝⬝ ⬝⬝⬝ ⬝⬝⬝
       
       **17.95 s**

    b. ⬝⬝ ⬝⬝⬝⬝⬝⬝⬝⬝⬝⬝ ⬝⬝⬝⬝⬝⬝⬝⬝⬝ ⬝⬝⬝
       ⬝⬝⬝⬝ ⬝⬝ ⬝⬝⬝⬝ ⬝ ⬝⬝⬝⬝⬝

       **8.88 m/s²**

98. ⬝⬝⬝⬝⬝⬝⬝ ⬝⬝⬝⬝⬝⬝ ⬝⬝⬝⬝⬝ ⬝⬝⬝
    ⬝ $=$ ⬝⬝ $x+$ ⬝ **MP** 8 N.RN.2  **C**
    ○ A ⬝
    ○ B ⬝
    ○ C ⬝
    ○ D ⬝⬝

99. ⬝ ⬝⬝⬝⬝⬝⬝ ⬝⬝⬝⬝⬝⬝⬝⬝⬝ ⬝⬝⬝ ⬝⬝⬝⬝ ⬝ $x$ ⬝⬝
    **MP** 7 N.RN.1, N.RN.2  **B**
    ○ A $\sqrt{\cdot x}$
    ○ B ⬝$\sqrt{x}$
    ○ C ⬝ $x$
    ○ D ⬝⬝ $x$

100. ⬝ ⬝⬝⬝⬝ ⬝ ⬝⬝⬝⬝⬝⬝ ⬝⬝⬝⬝⬝ ⬝⬝⬝⬝⬝ ⬝ ⬝⬝
     **MP** 7 N.RN.2  **D**
     ○ A ⬝
     ○ B ⬝
     ○ C ⬝⬝
     ○ D ⬝⬝⬝

101. ⬝ ⬝⬝⬝⬝⬝⬝ ⬝⬝ ⬝⬝⬝ ⬝⬝ ⬝⬝⬝⬝⬝ ⬝ $x+$ ⬝ $=$ ⬝⬝ ⬝⬝
     **MP** 8 N.RN.2  **A**
     ○ A ⬝
     ○ B ⬝
     ○ C ⬝
     ○ D ⬝⬝

102. ⬝⬝ ⬝⬝⬝⬝⬝⬝ ⬝ ⬝⬝⬝⬝⬝ ⬝⬝⬝⬝⬝⬝ ⬝⬝⬝⬝ $\sqrt{\cdot \cdot n}$ ⬝⬝
     **MP** 7 N.RN.1, N.RN.2  **D**
     ○ A ⬝⬝⬝ $n$
     ○ B ⬝⬝ $n$
     ○ C ⬝⬝ $n$
     ○ D ⬝⬝ $n$

### Differentiated Instruction (BL)

**Extension** For Exercise 86, students found the numbers of random walks of a given length of steps. Have students investigate the applications of random walks.

**101.**

| | |
|---|---|
| A | CORRECT |
| B | Did not subtract 8 from each side when solving the equation |
| C | Used the fact that $2^4 = 16$ |
| D | Solved without first finding like bases |

**102.**

| | |
|---|---|
| A | Squared instead of taking the square root |
| B | Squared the variable and didn't include the coefficient |
| C | Applied exponent to variable only |
| D | CORRECT |

## Preparing for Assessment

Exercises 94–102 require students to use the skills they will need on standardized assessments. Each exercise is dual-coded with content standards and mathematical practice standards.

### Dual Coding

| Items | Content Standards | **MP** Mathematical Practices |
|---|---|---|
| 94 | N.RN.2 | 8 |
| 95 | N.RN.1, N.RN.2 | 7 |
| 96 | N.RN.2 | 7, 8 |
| 97 | N.RN.1, N.RN.2 | 1, 4 |
| 98 | N.RN.2 | 8 |
| 99 | N.RN.1, N.RN.2 | 7 |
| 100 | N.RN.2 | 7 |
| 101 | N.RN.2 | 8 |
| 102 | N.RN.1, N.RN.2 | 7 |

## Diagnose Student Errors

Survey student responses for each item. Class trends may indicate common errors and misconceptions.

**99.**

| | |
|---|---|
| A | Included coefficient under radical |
| B | CORRECT |
| C | Used the definition of a negative exponent |
| D | Squared instead of taking the square root |

**100.**

| | |
|---|---|
| A | Found the cube root |
| B | Only found the square root |
| C | Found the cube root and then squared it |
| D | CORRECT |

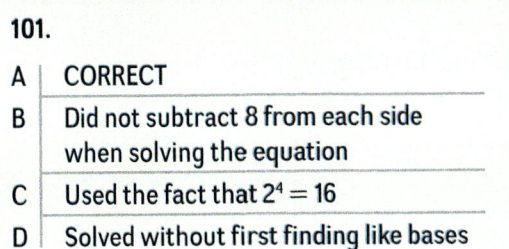

### Go Online!

**Quizzes**

Students can use *Self-Check Quizzes* to check their understanding of this lesson.

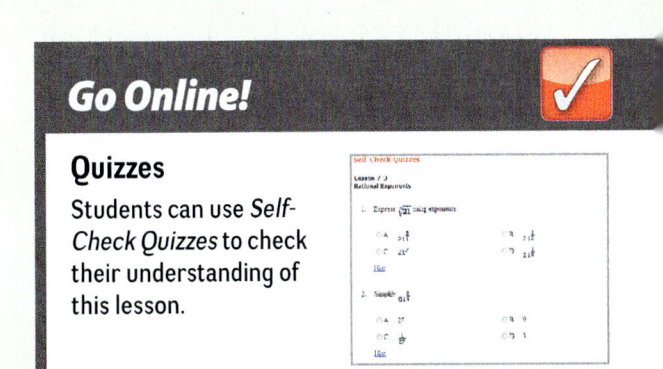

connectED.mcgraw-hill.com  **417**

# Chapter 7 Mid-Chapter Quiz

## Response to Intervention
Use the Intervention Planner to help you determine your Response to Intervention.

### Intervention Planner

**TIER 1 — On Level (OL)**

**IF** students miss 25% of the exercises or less,

**THEN** choose a resource:
- SE Lessons 7–1, 7–2 and 7–3

**Go Online!**
- Skills Practice
- Chapter Project
- Self-Check Quizzes

**TIER 2 — Strategic Intervention (AL)**
Approaching grade level

**IF** students miss 50% of the exercises,

**THEN** choose a resource:
- Quick Review Math Handbook

**Go Online!**
- Study Guide and Intervention
- Extra Examples
- Personal Tutors
- Homework Help

**TIER 3 — Intensive Intervention**
2 or more grades below level

**IF** students miss 75% of the exercises,

**THEN** choose a resource:
- Use *Math Triumphs, Alg. 1*

**Go Online!**
- Extra Examples
- Personal Tutors
- Homework Help
- Review Vocabulary

## Go Online!

### eAssessment
You can use the premade Mid-Chapter Test to assess students' progress in the first half of the chapter. Customize and create multiple versions of your Mid-Chapter Quiz and answer keys that align to the your standards. Tests can be delivered on paper or online.

---

# CHAPTER 7
## Mid-Chapter Quiz
### Lessons 7-1 through 7-3

**Simplify each expression.** (Lesson 7-1)

1. $(x^3)(4x^5)$  $4x^8$

2. $(m^2 p^5)^3$  $m^6 p^{15}$

3. $[(2xy^3)^2]^3$  $64x^6 y^{18}$

4. $(6ab^3 c^4)(-3a^2 b^3 c)$  $-18a^3 b^6 c^5$

5. **MULTIPLE CHOICE** Express the volume of the solid as a monomial. (Lesson 7-1)  **B**

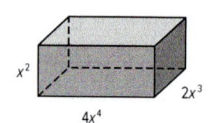

   A  $6x^9$
   B  $8x^9$
   C  $8x^{24}$
   D  $7x^{24}$

**Simplify each expression. Assume that no denominator equals 0.** (Lesson 7-2)

6. $\left(\dfrac{2a^4 b^3}{c^6}\right)^3$  $\dfrac{8a^{12} b^9}{c^{18}}$

7. $\dfrac{2xy^0}{6x}$  $\dfrac{1}{3}$

8. $\dfrac{m^7 n^4 p}{m^3 n^3 p}$  $m^4 n$

9. $\dfrac{p^4 t^{-2}}{r^{-5}}$  $\dfrac{p^4 r^5}{t^2}$

10. **ASTRONOMY** Physicists estimate that the number of stars in the universe has an order of magnitude of $10^{21}$. The number of stars in the Milky Way galaxy is around 100 billion. (Lesson 7-2)

    a. Using orders of magnitude, how many times as many stars are there in the universe as the Milky Way?  $10^{10}$

    b. Which mathematical practice did you use to solve this problem?  **See students' work.**

**Write each expression in radical form, or write each radical in exponential form.** (Lesson 7-3)

11. $42^{\frac{1}{2}}$  $\sqrt{42}$

12. $11x^{\frac{1}{2}}$  $11\sqrt{x}$

13. $(11g)^{\frac{1}{2}}$  $\sqrt{11g}$

14. $\sqrt{55}$  $55^{\frac{1}{2}}$

15. $\sqrt{5k}$  $(5k)^{\frac{1}{2}}$

16. $4\sqrt{p}$  $4p^{\frac{1}{2}}$

**Simplify.** (Lesson 7-3)

17. $\sqrt[3]{729}$  9

18. $\sqrt[4]{625}$  5

19. $1331^{\frac{1}{3}}$  11

20. $\left(\dfrac{16}{81}\right)^{\frac{1}{4}}$  $\dfrac{2}{3}$

21. $8^{\frac{2}{3}}$  4

22. $625^{\frac{3}{4}}$  125

23. $216^{\frac{5}{3}}$  7776

24. $\left(\dfrac{1}{4}\right)^{\frac{3}{2}}$  $\dfrac{1}{8}$

**Solve each equation.** (Lesson 7-3)

25. $4^x = 4096$  6

26. $5^{2x+1} = 125$  1

27. $4^{x-3} = 128$  6.5

## Foldables Study Organizer

### Dinah Zike's FOLDABLES

Before students complete the Mid-Chapter Quiz, encourage them to review the information for Lessons 7-1 through 7-3 in their Foldables. Give students time to ask any questions they have about the properties of exponents.

ALEKS can be used as a formative assessment tool to target learning gaps for those who are struggling, while providing enhanced learning for those who have mastered the concepts.

# LESSON 7-4
# Radical Expressions

**SUGGESTED PACING (DAYS)**

| | | |
|---|---|---|
| 90 min. | 0.5 | 0.5 |
| 45 min. | 1 | 0.5 |
| | Instruction | Explore Lab |

## Track Your Progress

### Objectives

1. Simplify square roots by using the Product and Quotient Properties of Square Roots.
2. Add, subtract, and multiply radical expressions.

### Mathematical Background

The Product Property and Quotient Property can be used to simplify radical expressions. For all nonnegative numbers, the product of each square root equals the square root of the products. For all nonnegative dividends and positive divisors, the square root of a quotient is equal to the quotient of the principal square roots of the numbers. Radical expressions can be added or subtracted only if the radicands are the same. Radical expressions can be multiplied when the radicands are the same or different.

### THEN

**F.IF.8b** Use the properties of exponents to interpret expressions for exponential functions.

**A.SSE.2** Use the structure of an expression to identify ways to rewrite it.

### NOW

**N.RN.2** Rewrite expressions involving radicals and rational exponents using the properties of exponents.

### NEXT

**F.IF.7e** Graph exponential and logarithmic functions, showing intercepts and end behavior, and trigonometric functions, showing period, midline, and amplitude.

**F.LE.5** Interpret the parameters in a linear or exponential function in terms of a context.

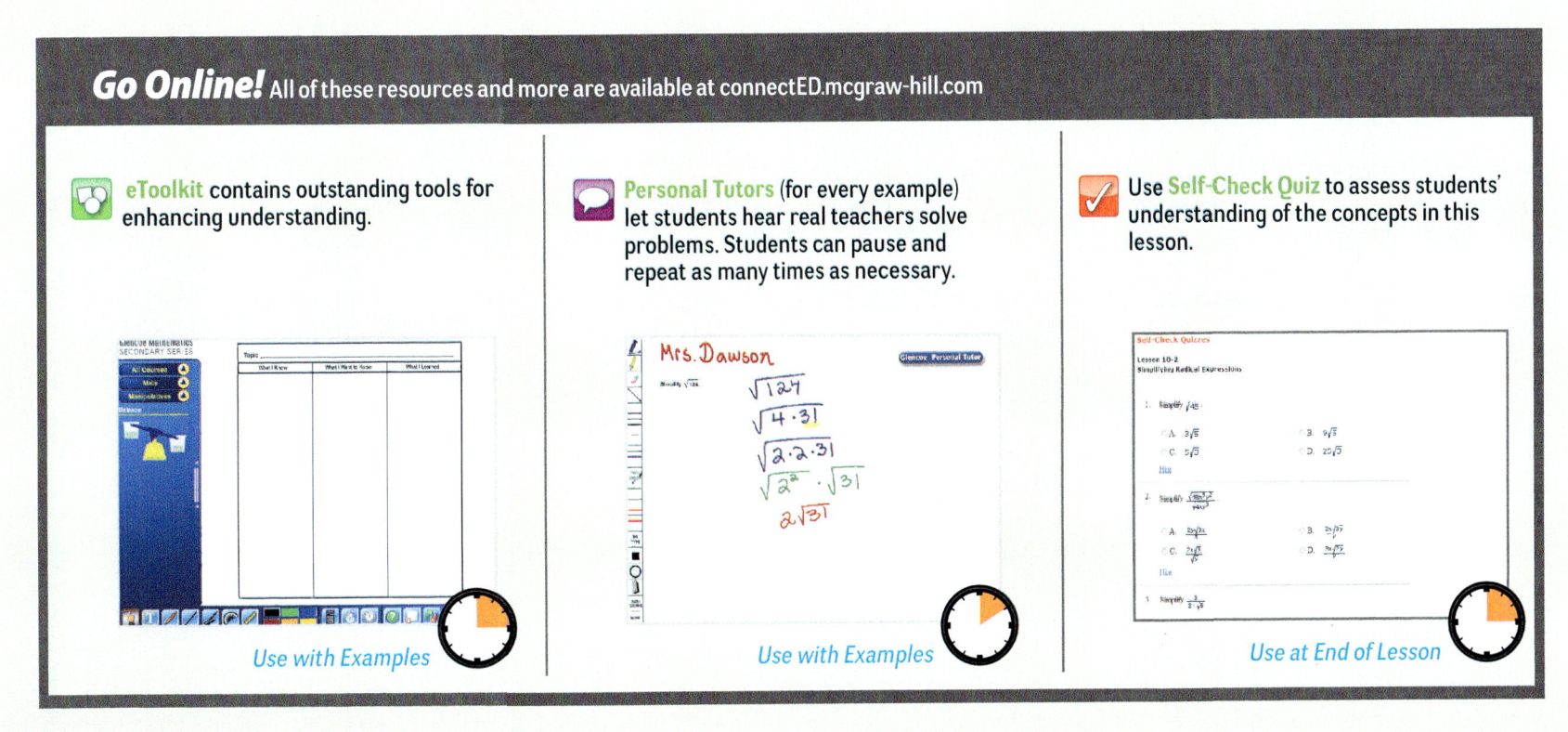

## Go Online! All of these resources and more are available at connectED.mcgraw-hill.com

**eToolkit** contains outstanding tools for enhancing understanding.

*Use with Examples*

**Personal Tutors** (for every example) let students hear real teachers solve problems. Students can pause and repeat as many times as necessary.

*Use with Examples*

Use **Self-Check Quiz** to assess students' understanding of the concepts in this lesson.

*Use at End of Lesson*

## OER Using Open Educational Resources

**Tutorials** Have students watch the tutorial on **NeoK12** about multiplying square roots to reinforce their knowledge of multiplying radical expressions. This tutorial can also be used to introduce students to the concept of multiplying square roots. *Use as flipped learning or an instructional aide*

connectED.mcgraw-hill.com    419A

# Go Online!
connectED.mcgraw-hill.com

Worksheets

# Differentiate Your Resources

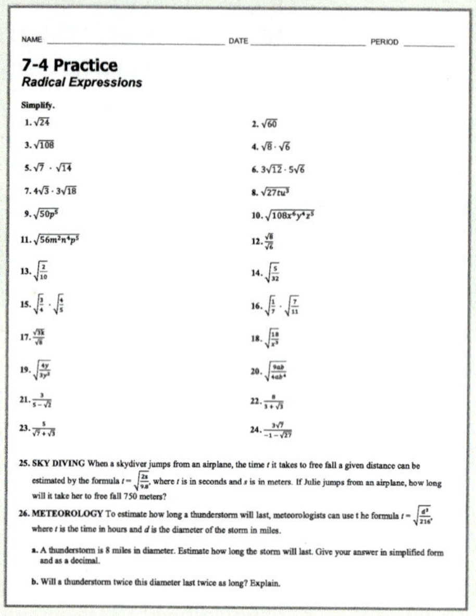

**Extra Practice** Additional practice or homework; Skills Practice is best for approaching-level students and Practice is best for on-level and beyond-level students

## Skills Practice
AL  OL  ELL

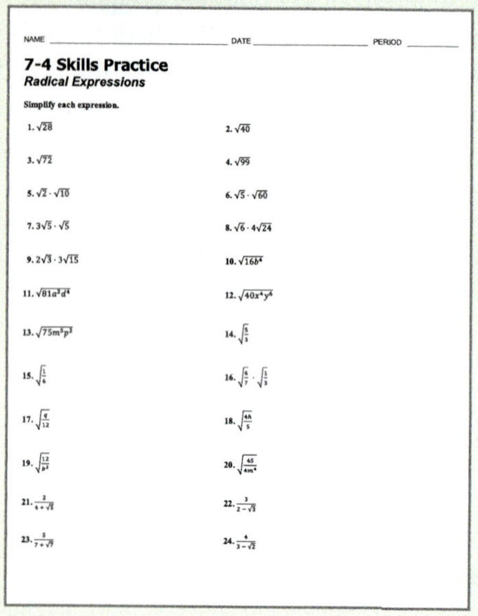

## Practice
AL  OL  BL  ELL

## Word Problem Practice
AL  OL  BL  ELL

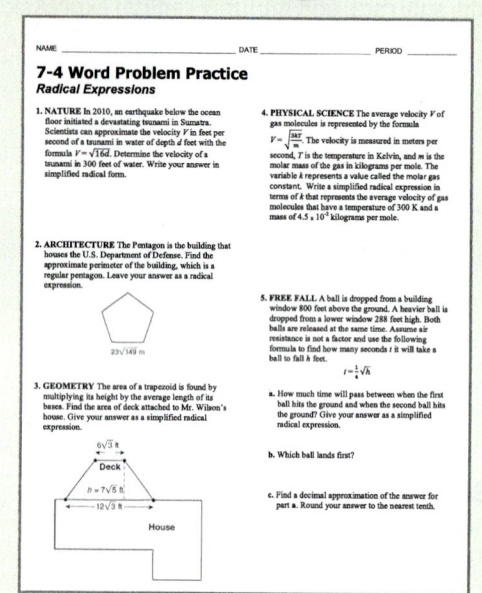

**Intervention** Reteaching and vocabulary activities that can be used with struggling or absent students and as ELL support

**Extension** Activities that can be used to extend lesson concepts

## Study Guide and Intervention
AL  OL  ELL

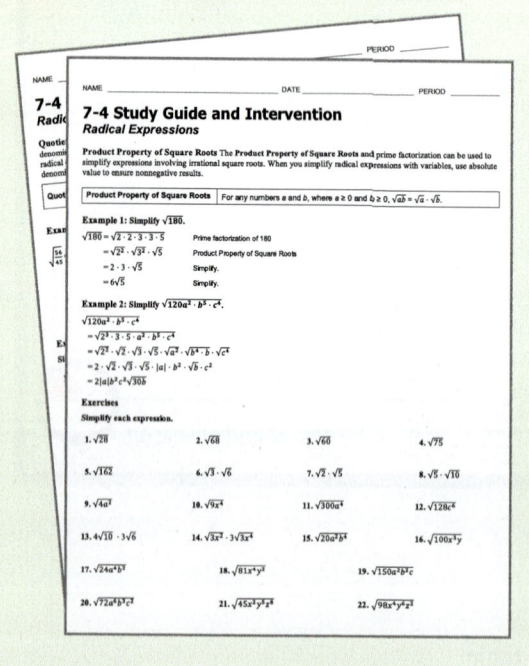

## Study Notebook
AL  OL  BL  ELL

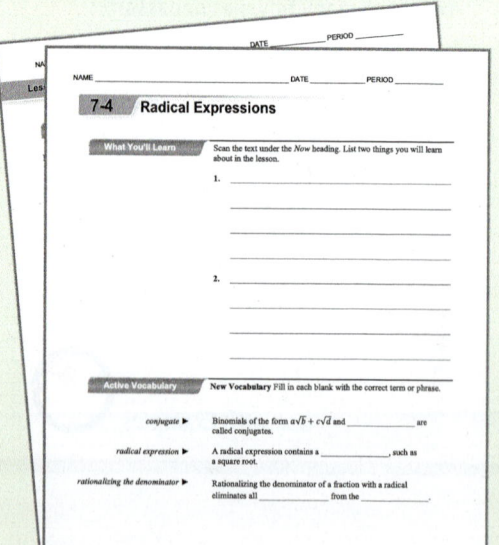

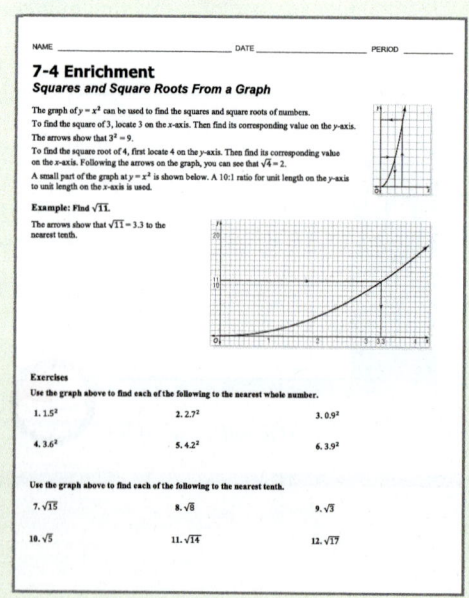

## Enrichment
OL  BL  ELL

419B | Lesson 7-4 | Radical Expressions

## LESSON 4
# Radical Expressions

**Then**
- You evaluated expressions involving rational exponents.

**Now**
1. Simplify square roots by applying the Product and Quotient Properties of Square Roots.
2. Add, subtract, and multiply radical expressions.

**Why?**
Ximena is going to run in her neighborhood to get ready for the soccer season. She plans to run the course that she has laid out three times each day.

How far does Ximena have to run to complete the course that she laid out?

How far does she run every day?

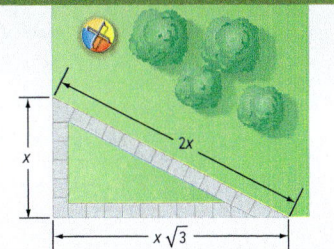

**New Vocabulary**
radical expression
rationalizing the denominator
conjugate

**Mathematical Practices**
6 Attend to precision.
7 Look for and make use of structure.
8 Look for and express regularity in repeated reasoning.

**Content Standards**
N.RN.2 Rewrite expressions involving radicals and rational exponents using the properties of exponents.

### 1 Properties of Square Roots
A **radical expression** contains a radical, such as a square root. The expression under the radical sign is called the radicand. A radicand under a square root symbol is in simplest form if the following three conditions are true.

- No radicands have perfect square factors other than 1.
- No radicands contain fractions.
- No radicals appear in the denominator of a fraction.

The following property can be used to simplify square roots.

**Key Concept** Product Property of Square Roots

| | |
|---|---|
| Words | For any nonnegative real numbers $a$ and $b$, the square root of $ab$ is equal to the square root of $a$ times the square root of $b$. |
| Symbols | $\sqrt{ab} = \sqrt{a} \cdot \sqrt{b}$, if $a \geq 0$ and $b \geq 0$ |
| Examples | $\sqrt{4 \cdot 9} = \sqrt{36}$ or $6$ $\qquad$ $\sqrt{4 \cdot 9} = \sqrt{4} \cdot \sqrt{9} = 2 \cdot 3$ or $6$ |

N.RN.2

**Example 1** Simplify Square Roots

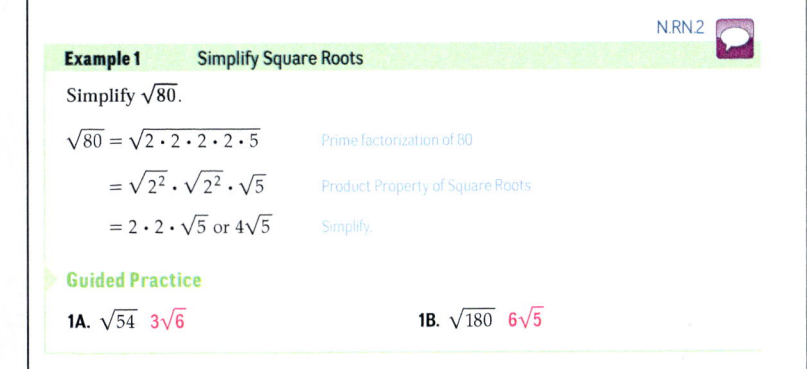

Simplify $\sqrt{80}$.

$\sqrt{80} = \sqrt{2 \cdot 2 \cdot 2 \cdot 2 \cdot 5}$ $\qquad$ Prime factorization of 80
$\quad\ \ = \sqrt{2^2} \cdot \sqrt{2^2} \cdot \sqrt{5}$ $\qquad$ Product Property of Square Roots
$\quad\ \ = 2 \cdot 2 \cdot \sqrt{5}$ or $4\sqrt{5}$ $\qquad$ Simplify.

▶ **Guided Practice**

1A. $\sqrt{54}$ $\quad 3\sqrt{6}$ $\qquad\qquad$ 1B. $\sqrt{180}$ $\quad 6\sqrt{5}$

---

### Lesson 7-4 | Radical Expressions

## Launch

Have students read the Why? section of the lesson. Ask:

- What expression shows the length of the course? $x + 2x + x\sqrt{3}$

- Which terms in the expression can be combined? Explain. $x$ and $2x$ because they are like terms

- Which term cannot be combined? Explain. $x\sqrt{3}$ because it contains a radical

- What expression shows how far Ximena will run each day? $3x(3 + \sqrt{3})$

## Teach

Ask the scaffolded questions for each example to build conceptual understanding for students at all levels.

### 1 Properties of Square Roots

**Example 1 Simplify Square Roots**

**AL** Why is $\sqrt{80}$ not in simplest form? It contains a perfect square factor.

**OL** What perfect square factor can you factor out of 80? 16 What is the square root of 16? 4

**BL** When might a simplified radical expression be more useful than the approximate value of the expression? Sample answer: when comparing two radical expressions

**Need Another Example?**
Simplify $\sqrt{52}$. $2\sqrt{13}$

---

## MP Mathematical Practices Strategies

**Construct viable arguments and critique the reasoning of others.**
Help students analyze and explain their answers. For example, ask:

- Why are finding factors of the radicand necessary for simplifying radicals?
When simplifying radicals, a factor that is a perfect square can be simplified.

- Why can't $3\sqrt{5} + 2\sqrt{3}$ be added to create one term?
The radicands must be alike to add radical expressions.

- Explain why the Distributive Property is not needed to simplify $\sqrt{3}(6\sqrt{5} + 2\sqrt{5})$.
The expression inside the parentheses can be added since the terms have like radicals. The sum inside the parentheses is then multiplied by the term outside the parentheses without using the Distributive Property.

---

## Go Online!

**Interactive Whiteboard**
Use the eLesson or Lesson Presentation to present this lesson.

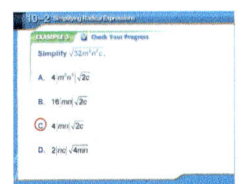

connectED.mcgraw-hill.com  419

# Lesson 7-4 | Radical Expressions

## Example 2 Multiply Square Roots

**AL** When simplifying a radical expression, which types of factors should you look for in the radicand? **perfect squares**

**OL** What perfect square factor can you factor out of 28? **4** What is the square root of 4? **2**

**BL** Tell about a real-life situation where you might use radicals. **Sample answer: electrical engineering, carpentry**

### Need Another Example?
Simplify $\sqrt{2} \cdot \sqrt{24}$. **$4\sqrt{3}$**

## Example 3 Simplify a Square Root with Variables

**AL** What is the square root of $y^4$? **$y^2$**

**OL** What perfect square factor can you factor out of 90? **9** What is the square root of 9? **3**

**BL** Why is there still an $x$ and $z$ in the simplified radicand? Explain. **Sample answer: we can only simplify factors with even exponents, so we were able to factor out $x^2$ and $z^4$, but not the remaining $x^1$ and $z^1$.**

### Need Another Example?
Simplify $\sqrt{45a^4b^5c^6}$. **$3a^2b^2|c^3|\sqrt{5b}$**

## Teaching Tip

**Intervention** In order to simplify square roots with the Product Property of Square Roots, students need to be able to find the prime factorization of the radicand. Take a few minutes to review finding prime factorizations so that students can focus on learning the new concept rather than trying to recall earlier material.

## e Follow-Up

Students have explored simplifying radical expressions.
Ask:
- Why do you think it is important to simplify radical expressions? **Sample answer: it is easier to evaluate, solve and enter in the calculator if the radical expressions are simplified first.**

---

**Example 2** Multiply Square Roots

Simplify $\sqrt{2} \cdot \sqrt{14}$.

$\sqrt{2} \cdot \sqrt{14} = \sqrt{2} \cdot \sqrt{2} \cdot \sqrt{7}$  Product Property of Square Roots

$\phantom{\sqrt{2} \cdot \sqrt{14}} = \sqrt{2^2} \cdot \sqrt{7}$ or $2\sqrt{7}$  Product Property of Square Roots

▶ **Guided Practice**

2A. $\sqrt{5} \cdot \sqrt{10}$  **$5\sqrt{2}$**  2B. $\sqrt{6} \cdot \sqrt{8}$  **$4\sqrt{3}$**

Consider the expression $\sqrt{x^2}$. It may seem that $x = \sqrt{x^2}$, but when finding the principal square root of an expression containing variables, you have to be sure that the result is not negative. Consider $x = -3$.

$\sqrt{x^2} \stackrel{?}{=} x$

$\sqrt{(-3)^2} \stackrel{?}{=} -3$  Replace $x$ with $-3$.

$\sqrt{9} \stackrel{?}{=} -3$  $(-3)^2 = 9$

$3 \neq -3$  $\sqrt{9} = 3$

Notice in this case, if the right hand side of the equation were $|x|$, the equation would be true. For expressions where the exponent of the variable inside a radical is even and the simplified exponent is odd, you must use absolute value.

$\sqrt{x^2} = |x|$  $\sqrt{x^3} = x\sqrt{x}$  $\sqrt{x^4} = x^2$  $\sqrt{x^6} = |x^3|$

**Example 3** Simplify a Square Root with Variables

Simplify $\sqrt{90x^3y^4z^5}$.

$\sqrt{90x^3y^4z^5} = \sqrt{2 \cdot 3^2 \cdot 5 \cdot x^3 \cdot y^4 \cdot z^5}$  Prime factorization

$= \sqrt{2} \cdot \sqrt{3^2} \cdot \sqrt{5} \cdot \sqrt{x^2} \cdot \sqrt{x} \cdot \sqrt{y^4} \cdot \sqrt{z^4} \cdot \sqrt{z}$  Product Property

$= \sqrt{2} \cdot 3 \cdot \sqrt{5} \cdot x \cdot \sqrt{x} \cdot y^2 \cdot z^2 \cdot \sqrt{z}$  Simplify.

$= 3y^2z^2x\sqrt{10xz}$  Simplify.

▶ **Guided Practice**

3A. $\sqrt{32r^2k^4t^5}$  **$4|r|k^2t^2\sqrt{2t}$**  3B. $\sqrt{56xy^{10}z^5}$  **$2|y^5|z^2\sqrt{14xz}$**

To divide square roots and simplify radical expressions, you can use the Quotient Property of Square Roots.

**Reading Math**

**Fractions in the Radicand** The expression $\sqrt{\frac{a}{b}}$ is read the square root of a over b, or the square root of the quantity of a over b.

**Key Concept** Quotient Property of Square Roots

**Words** For any real numbers $a$ and $b$, where $a \geq 0$ and $b > 0$, the square root of $\frac{a}{b}$ is equal to the square root of $a$ divided by the square root of $b$.

**Symbols** $\sqrt{\frac{a}{b}} = \frac{\sqrt{a}}{\sqrt{b}}$

You can use the properties of square roots to **rationalize the denominator** of a fraction with a radical. This involves multiplying the numerator and denominator by a factor that eliminates radicals in the denominator.

N.RN.2

### Example 4  Rationalize a Denominator

Which expression is equivalent to $\sqrt{\dfrac{35}{15}}$?

A $\dfrac{5\sqrt{21}}{15}$  B $\dfrac{\sqrt{21}}{3}$  C $\dfrac{\sqrt{525}}{15}$  D $\dfrac{\sqrt{35}}{15}$

**Read the Item** The radical expression needs to be simplified.

**Solve the Item**

$\sqrt{\dfrac{35}{15}} = \sqrt{\dfrac{7}{3}}$     Reduce $\dfrac{35}{15}$ to $\dfrac{7}{3}$

$= \dfrac{\sqrt{7}}{\sqrt{3}}$     Quotient Property of Square Roots

$= \dfrac{\sqrt{7}}{\sqrt{3}} \cdot \dfrac{\sqrt{3}}{\sqrt{3}}$     Multiply by $\dfrac{\sqrt{3}}{\sqrt{3}}$

$= \dfrac{\sqrt{21}}{3}$     Product Property of Square Roots

The correct choice is B.

**Study Tip**

**MP Structure** Look at the radicand to see if it can be simplified first. This may make your computations simpler.

**Guided Practice**

4. Simplify $\dfrac{\sqrt{6y}}{\sqrt{12}}$. C

A $\dfrac{\sqrt{y}}{2}$  B $\dfrac{\sqrt{y}}{4}$  C $\dfrac{\sqrt{2y}}{2}$  D $\dfrac{\sqrt{2y}}{4}$

Binomials of the form $a\sqrt{b} + c\sqrt{d}$ and $a\sqrt{b} - c\sqrt{d}$, where $a$, $b$, $c$, and $d$ are rational numbers, are called **conjugates**. For example, $2 + \sqrt{7}$ and $2 - \sqrt{7}$ are conjugates. The product of two conjugates is a rational number and can be found using the pattern for the difference of squares.

N.RN.2

### Example 5  Use Conjugates to Rationalize a Denominator

Simplify $\dfrac{3}{5 + \sqrt{2}}$.

$\dfrac{3}{5 + \sqrt{2}} = \dfrac{3}{5 + \sqrt{2}} \cdot \dfrac{5 - \sqrt{2}}{5 - \sqrt{2}}$     The conjugate of $5 + \sqrt{2}$ is $5 - \sqrt{2}$.

$= \dfrac{3(5 - \sqrt{2})}{5^2 - (\sqrt{2})^2}$     $(a - b)(a + b) = a^2 - b^2$

$= \dfrac{15 - 3\sqrt{2}}{25 - 2}$ or $\dfrac{15 - 3\sqrt{2}}{23}$     $(\sqrt{2})^2 = 2$

**Go Online!**

Working with a partner to complete the Self-Check Quiz can help your understanding. Take turns describing how to solve each problem using the graphs. Ask for clarification as you need it. **ELL**

**Guided Practice** Simplify each expression.

5A. $\dfrac{3}{2 + \sqrt{2}}$  $\dfrac{6 - 3\sqrt{2}}{2}$     5B. $\dfrac{7}{3 - \sqrt{7}}$  $\dfrac{21 + 7\sqrt{7}}{2}$

---

### Differentiated Instruction AL OL

**IF** students need further practice with conjugates,

**THEN** have students use their calculators to show that using conjugates produces equivalent expressions in Example 5. Have students show that $\dfrac{3}{5 + \sqrt{2}}$, $\dfrac{3(5 - \sqrt{2})}{(5 + \sqrt{2})(5 - \sqrt{2})}$, and $\dfrac{15 - 3\sqrt{2}}{23}$ are equivalent. They should find that each expression is about 0.4677.

---

**Lesson 7-4 | Radical Expressions**

### Example 4  Rationalize a Denominator

**AL** When there is a fraction in the radicand, what is the first thing we should try to do? Why? Reduce the fraction so that you can work with smaller numbers.

**OL** How can you get rid of the radical in the denominator? Multiply the numerator and the denominator by $\sqrt{3}$.

**BL** Why are we allowed to multiply by $\dfrac{\sqrt{3}}{\sqrt{3}}$? Explain. Sample answer: $\dfrac{\sqrt{3}}{\sqrt{3}}$ is equivalent to 1, so when we multiply, we are not changing the value of the fraction.

**Need Another Example?**

Which expression is equivalent to $\dfrac{\sqrt{3n}}{\sqrt{8}}$? D

A $\dfrac{\sqrt{3n}}{8}$  C $\dfrac{\sqrt{6n}}{2}$

B $\dfrac{\sqrt{3n}}{4}$  D $\dfrac{\sqrt{6n}}{4}$

### Example 5  Use Conjugates to Rationalize a Denominator

**AL** Can you recall the pattern for the difference of squares? What is it? $a^2 - b^2$

**OL** What is the conjugate of $5 + \sqrt{2}$? $5 - \sqrt{2}$

**BL** In your own words, explain how the pattern for the difference of squares helps eliminate the radical in the denominator of a fraction. See students' work.

**Need Another Example?**

Simplify $\dfrac{2}{4 - \sqrt{5}}$.  $\dfrac{8 + 2\sqrt{5}}{11}$

### MP Teaching the Mathematical Practices

**Structure** Mathematically proficient students can see complicated things as single objects or as being composed of several objects. Point out that simplifying the radicand first doesn't change the final answer, but it does simplify the calculations.

connectED.mcgraw-hill.com  421

# Lesson 7-4 | Radical Expressions

## 2 Operations with Radical Expressions

### Example 6 Add and Subtract Expressions with Like Radicands

**AL** In part b, what are the like terms? $10\sqrt{7}$ and $4\sqrt{7}$; $5\sqrt{11}$ and $-6\sqrt{11}$

**OL** In part a, what is being added or subtracted? $5 + 7 - 6$

**BL** Describe how adding and subtracting radical expressions is similar to adding and subtracting monomials. Sample answer: When adding or subtracting monomials, you group like terms and the variable does not change. When adding or subtracting radical expressions, you group like terms and the radicand does not change.

**Need Another Example?**
Simplify each expression.
a. $6\sqrt{5} + 2\sqrt{5} - 5\sqrt{5}$  $3\sqrt{5}$
b. $7\sqrt{2} + 8\sqrt{11} - 4\sqrt{11} - 6\sqrt{2}$  $\sqrt{2} + 4\sqrt{11}$

### Example 7 Add and Subtract Expressions with Unlike Radicands

**AL** Are all of the radicands in the expression in simplest form? no

**OL** What perfect square factor can be factored from 18? 9 from 32? 16 from 72? 36 What radicand remains in each term after factoring out the perfect squares? $\sqrt{2}$

**BL** Write your own real-world problem involving adding radicands. See students' work.

**Need Another Example?**
Simplify $6\sqrt{27} + 8\sqrt{12} + 2\sqrt{75}$.  $44\sqrt{3}$

---

**2 Operations with Radical Expressions** To add or subtract radical expressions, the radicands must be alike in the same way that monomial terms must be alike to add or subtract.

| Monomials | Radical Expressions |
|---|---|
| $4a + 2a = (4 + 2)a$ | $4\sqrt{5} + 2\sqrt{5} = (4 + 2)\sqrt{5}$ |
| $= 6a$ | $= 6\sqrt{5}$ |
| $9b - 2b = (9 - 2)b$ | $9\sqrt{3} - 2\sqrt{3} = (9 - 2)\sqrt{3}$ |
| $= 7b$ | $= 7\sqrt{3}$ |

Notice that when adding and subtracting radical expressions, the radicand does not change. This is the same as when adding or subtracting monomials.

N.RN.2

**Example 6** Add and Subtract Expressions with Like Radicands

Simplify each expression.

a. $5\sqrt{2} + 7\sqrt{2} - 6\sqrt{2}$
$5\sqrt{2} + 7\sqrt{2} - 6\sqrt{2} = (5 + 7 - 6)\sqrt{2}$  Distributive Property
$= 6\sqrt{2}$  Simplify.

b. $10\sqrt{7} + 5\sqrt{11} + 4\sqrt{7} - 6\sqrt{11}$
$10\sqrt{7} + 5\sqrt{11} + 4\sqrt{7} - 6\sqrt{11} = (10 + 4)\sqrt{7} + (5 - 6)\sqrt{11}$  Distributive Property
$= 14\sqrt{7} - \sqrt{11}$  Simplify.

**Guided Practice**
6A. $3\sqrt{2} - 5\sqrt{2} + 4\sqrt{2}$  $2\sqrt{2}$
6B. $6\sqrt{11} + 2\sqrt{11} - 9\sqrt{11}$  $-\sqrt{11}$
6C. $15\sqrt{3} - 14\sqrt{5} + 6\sqrt{5} - 11\sqrt{3}$  $4\sqrt{3} - 8\sqrt{5}$
6D. $4\sqrt{3} + 3\sqrt{7} - 6\sqrt{3} + 3\sqrt{7}$  $-2\sqrt{3} + 6\sqrt{7}$

Not all radical expressions have like radicands. Simplifying the expressions may make it possible to have like radicands so that they can be added or subtracted.

**Study Tip**
**MP Precision** Simplify each radical term first. Then perform the operations needed.

N.RN.2

**Example 7** Add and Subtract Expressions with Unlike Radicands

Simplify $2\sqrt{18} + 2\sqrt{32} + \sqrt{72}$.

$2\sqrt{18} + 2\sqrt{32} + \sqrt{72} = 2(\sqrt{3^2} \cdot \sqrt{2}) + 2(\sqrt{4^2} \cdot \sqrt{2}) + (\sqrt{6^2} \cdot \sqrt{2})$  Product Property
$= 2(3\sqrt{2}) + 2(4\sqrt{2}) + (6\sqrt{2})$  Simplify.
$= 6\sqrt{2} + 8\sqrt{2} + 6\sqrt{2}$  Multiply.
$= 20\sqrt{2}$  Simplify.

**Guided Practice**
7A. $4\sqrt{54} + 2\sqrt{24}$  $16\sqrt{6}$
7B. $4\sqrt{12} - 6\sqrt{48}$  $-16\sqrt{3}$
7C. $3\sqrt{45} + \sqrt{20} - \sqrt{245}$  $4\sqrt{5}$
7D. $\sqrt{24} - \sqrt{54} + \sqrt{96}$  $3\sqrt{6}$

Multiplying radical expressions is similar to multiplying monomial algebraic expressions. Let $x \geq 0$.

Monomials
$(2x)(3x) = 2 \cdot 3 \cdot x \cdot x$
$= 6x^2$

Radical Expressions
$(2\sqrt{x})(3\sqrt{x}) = 2 \cdot 3 \cdot \sqrt{x} \cdot \sqrt{x}$
$= 6x$

You can also apply the Distributive Property to radical expressions.

N.RN.2

### Example 8  Multiply Radical Expressions

Simplify each expression.

**a.** $3\sqrt{2} \cdot 2\sqrt{6}$

$3\sqrt{2} \cdot 2\sqrt{6} = (3 \cdot 2)(\sqrt{2} \cdot \sqrt{6})$  Associative Property
$= 6(\sqrt{12})$  Multiply.
$= 6(2\sqrt{3})$  Simplify.
$= 12\sqrt{3}$  Multiply.

**b.** $3\sqrt{5}(2\sqrt{5} + 5\sqrt{3})$

$3\sqrt{5}(2\sqrt{5} + 5\sqrt{3}) = (3\sqrt{5} \cdot 2\sqrt{5}) + (3\sqrt{5} \cdot 5\sqrt{3})$  Distributive Property
$= [(3 \cdot 2)(\sqrt{5} \cdot \sqrt{5})] + [(3 \cdot 5)(\sqrt{5} \cdot \sqrt{3})]$  Associative Property
$= [6(\sqrt{25})] + [15(\sqrt{15})]$  Multiply.
$= [6(5)] + [15(\sqrt{15})]$  Simplify.
$= 30 + 15\sqrt{15}$  Multiply.

**Watch Out!**
Multiplying Radicands Make sure that you multiply the radicands when multiplying radical expressions. A common mistake is to add the radicands rather than multiply.

▶ **Guided Practice**

**8A.** $2\sqrt{6} \cdot 7\sqrt{3}$  $42\sqrt{2}$
**8B.** $9\sqrt{5} \cdot 11\sqrt{15}$  $495\sqrt{3}$
**8C.** $3\sqrt{2}(4\sqrt{3} + 6\sqrt{2})$  $12\sqrt{6} + 36$
**8D.** $5\sqrt{3}(3\sqrt{2} - \sqrt{3})$  $15\sqrt{6} - 15$

You can also multiply radical expressions with more than one term in each factor. This is similar to multiplying two algebraic binomials with variables.

### Real-World Example 9  Multiply Radical Expressions

**GEOMETRY** Find the area of the rectangle in simplest form.

$A = (5\sqrt{2} - \sqrt{3})(\sqrt{5} + 4\sqrt{3})$  $A = \ell \cdot w$

$\underbrace{= (5\sqrt{2})(\sqrt{5})}_{\text{First Terms}} + \underbrace{(5\sqrt{2})(4\sqrt{3})}_{\text{Outer Terms}} + \underbrace{(-\sqrt{3})(\sqrt{5})}_{\text{Inner Terms}} + \underbrace{(-\sqrt{3})(4\sqrt{3})}_{\text{Last Terms}}$

$= 5\sqrt{10} + 20\sqrt{6} - \sqrt{15} - 4\sqrt{9}$  Multiply.
$= 5\sqrt{10} + 20\sqrt{6} - \sqrt{15} - 12$  Simplify.

Rectangle dimensions: $\sqrt{5} + 4\sqrt{3}$ and $5\sqrt{2} - \sqrt{3}$

**Review Vocabulary**
FOIL Method Multiply two binomials by finding the sum of the products of the First terms, the Outer terms, the Inner terms, and the Last terms.

▶ **Guided Practice**

**9. GEOMETRY** The area $A$ of a rhombus can be found using the equation $A = \frac{1}{2}d_1d_2$, where $d_1$ and $d_2$ are the lengths of the diagonals. What is the area of the rhombus at the right?  $A = 120 - 8\sqrt{15} - 9\sqrt{30} + 9\sqrt{2}$

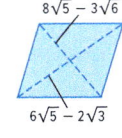

Rhombus diagonals: $8\sqrt{5} - 3\sqrt{6}$ and $6\sqrt{5} - 2\sqrt{3}$

---

Lesson 7-4 | Radical Expressions

### Example 8  Multiply Radical Expressions

**AL** What property allows us to regroup the parts of each factor in order to multiply? **Associative Property**

**OL** In part **a**, why do we not leave the solution as $6\sqrt{12}$? It is not in simplest form. We can factor out 4 from 12, which is a perfect square.

**BL** In part **b**, why do we not add the 30 and 15 to get a final solution of $45\sqrt{15}$? Sample answer: 30 is one term of the binomial and $15\sqrt{15}$ is the other term. They are not like terms because they do not have like radicands, therefore they cannot be added together.

**Need Another Example?**
Simplify each expression.
**a.** $2\sqrt{3} \cdot 4\sqrt{6}$  $24\sqrt{2}$
**b.** $4\sqrt{2}(3\sqrt{2} + 2\sqrt{6})$  $24 + 16\sqrt{3}$

### Example 9  Multiply Radical Expressions

**AL** What is the formula for the area of a rectangle? $A = \ell \times w$

**OL** What method do we need to use to multiply the length and the width of the rectangle? Sample answer: FOIL

**BL** Explain in your own words why we cannot simplify the expression any further than a solution with four terms.  See students' work.

**Need Another Example?**
Geometry Find the area of a rectangle in simplest form with a width of $4\sqrt{6} - 2\sqrt{10}$ and a length of $5\sqrt{3} + 7\sqrt{5}$.  $18\sqrt{30} - 10\sqrt{2}$

---

### Differentiated Instruction  AL  OL  ELL

**IF** students struggle with the FOIL method in Example 4,

**THEN** have students rewrite the example in five steps to facilitate an understanding of First terms, Outer terms, Inner terms, and Last terms. For example, in Step 1, have students write the original expression, underline the first terms, and then multiply to find the product of the first terms. Students continue this process until they have found the products of all of the terms. Then for the fifth step, have them combine and simplify the results of the previous four steps.

connectED.mcgraw-hill.com  **423**

# Lesson 7-4 | Radical Expressions

## Practice

**Formative Assessment** Use Exercises 1–29 to assess students' understanding of the concepts in this lesson.

The Practice and Problem Solving exercises assess the content taught in the lesson. The Preparing for Assessment page is meant to be used as preparation for end-of-course assessments.

## Extra Practice

See page R7 for extra exercises for students who are approaching level or for on-level students who need additional reinforcement.

### Levels of Complexity Chart

The levels of the exercises progress from 1 to 3, with Level 1 indicating the lowest level of complexity.

| Exercises | 30–61 | 62–74, 82–89 | 75–81 |
|---|---|---|---|
| Level 3 | | | ● |
| Level 2 | | ● | |
| Level 1 | ● | | |

### Concept Summary — Operations with Radical Expressions

| Operation | Symbols | Example |
|---|---|---|
| addition, $b \geq 0$ | $a\sqrt{b} + c\sqrt{b} = (a+c)\sqrt{b}$ like radicands | $4\sqrt{3} + 6\sqrt{3} = (4+6)\sqrt{3} = 10\sqrt{3}$ |
| subtraction, $b \geq 0$ | $a\sqrt{b} - c\sqrt{b} = (a-c)\sqrt{b}$ like radicands | $12\sqrt{5} - 8\sqrt{5} = (12-8)\sqrt{5} = 4\sqrt{5}$ |
| multiplication, $b \geq 0, g \geq 0$ | $a\sqrt{b}(f\sqrt{g}) = af\sqrt{bg}$ Radicands do not have to be like radicands. | $3\sqrt{2}(5\sqrt{7}) = (3 \cdot 5)(\sqrt{2 \cdot 7}) = 15\sqrt{14}$ |

**Check Your Understanding** ○ = Step-by-Step Solutions begin on page R11.

*Go Online! for a Self–Check Quiz*

**Examples 1–3** Simplify each expression.
N.RN.2

1. $\sqrt{24}$  $2\sqrt{6}$
2. $3\sqrt{16}$  $12$
3. $2\sqrt{25}$  $10$
4. $\sqrt{10} \cdot \sqrt{14}$  $2\sqrt{35}$
5. $\sqrt{3} \cdot \sqrt{18}$  $3\sqrt{6}$
6. $3\sqrt{10} \cdot 4\sqrt{10}$  $120$
7. $\sqrt{60x^4y^7}$  $2x^2y^3\sqrt{15y}$
8. $\sqrt{88m^3p^2r^5}$  $2m|p|r^2\sqrt{22mr}$
9. $\sqrt{99ab^5c^2}$  $3b^2|c|\sqrt{11ab}$

**Example 4**
N.RN.2

10. Which expression is equivalent to $\sqrt{\frac{45}{10}}$? **D**

    A $\frac{5\sqrt{2}}{10}$   B $\frac{\sqrt{45}}{10}$   C $\frac{\sqrt{50}}{10}$   D $\frac{3\sqrt{2}}{2}$

**Example 5** Simplify each expression.
N.RN.2

11. $\frac{3}{3+\sqrt{5}}$  $\frac{9-3\sqrt{5}}{4}$
12. $\frac{5}{2-\sqrt{6}}$  $\frac{10+5\sqrt{6}}{-2}$
13. $\frac{2}{1-\sqrt{10}}$  $\frac{2+2\sqrt{10}}{-9}$
14. $\frac{1}{4+\sqrt{12}}$  $\frac{2-\sqrt{3}}{2}$
15. $\frac{4}{6-\sqrt{7}}$  $\frac{24+4\sqrt{7}}{29}$
16. $\frac{6}{5+\sqrt{11}}$  $\frac{15-3\sqrt{11}}{7}$

**Examples 6–8** Simplify each expression.
A.RN.2

17. $3\sqrt{5} + 6\sqrt{5}$  $9\sqrt{5}$
18. $8\sqrt{3} + 5\sqrt{3}$  $13\sqrt{3}$
19. $\sqrt{7} - 6\sqrt{7}$  $-5\sqrt{7}$
20. $10\sqrt{2} - 6\sqrt{2}$  $4\sqrt{2}$
21. $4\sqrt{5} + 2\sqrt{20}$  $8\sqrt{5}$
22. $\sqrt{12} - \sqrt{3}$  $\sqrt{3}$
23. $\sqrt{8} + \sqrt{12} + \sqrt{18}$  $5\sqrt{2} + 2\sqrt{3}$
24. $\sqrt{27} + 2\sqrt{3} - \sqrt{12}$  $3\sqrt{3}$
25. $9\sqrt{2}(4\sqrt{6})$  $72\sqrt{3}$
26. $4\sqrt{3}(8\sqrt{3})$  $96$
27. $\sqrt{3}(\sqrt{7} + 3\sqrt{2})$  $\sqrt{21} + 3\sqrt{6}$
28. $\sqrt{5}(\sqrt{2} + 4\sqrt{2})$  $5\sqrt{10}$

**Example 9**
A.ARN.2

29. **GEOMETRY** The area $A$ of a triangle can be found by using the formula $A = \frac{1}{2}bh$, where $b$ represents the base and $h$ is the height. What is the area of the triangle at the right?  $14.5 + 3\sqrt{15}$ units$^2$

    (Triangle with legs $4\sqrt{3} + \sqrt{5}$ and $2\sqrt{3} + \sqrt{5}$)

### Differentiated Homework Options

| Levels | AL Basic | OL Core | BL Advanced |
|---|---|---|---|
| Exercises | 30–61, 82–89 | 30–76 odd, 64–72 | 62–81<br>82–89 (optional) |
| 2-Day Option | 31–61 odd, 82–89 | 30–52, 82–89 | |
| | 32–60 even, 64–72 | 53–77 | |

You can use ALEKS to provide additional remediation support with personalized instruction and practice.

424 | Lesson 7-4 | Radical Expressions

**Practice and Problem Solving**

Extra Practice is on page R7.

**Examples 1–3** Simplify each expression.
N.RN.2
30. $\sqrt{52}$  $2\sqrt{13}$
31. $\sqrt{56}$  $2\sqrt{14}$
32. $3\sqrt{18}$  $9\sqrt{2}$
33. $4\sqrt{2} \cdot 5\sqrt{8}$  $80$
34. $\sqrt{10} \cdot \sqrt{20}$  $10\sqrt{2}$
35. $5\sqrt{81q^5}$  $45q^2\sqrt{q}$
36. $\sqrt{28a^2b^3}$  $2|a|b\sqrt{7b}$
37. $\sqrt{75qr^3}$  $5r\sqrt{3qr}$
38. $7\sqrt{63m^3p}$  $21m\sqrt{7mp}$
39. $4\sqrt{66g^2h^4}$  $4|g|h^2\sqrt{66}$
40. $\sqrt{2ab^2} \cdot \sqrt{10a^5b}$  $2a^3b\sqrt{5b}$
41. $\sqrt{4c^3d^3} \cdot \sqrt{8c^3d}$  $4c^3d^2\sqrt{2}$

**Examples 4–5** Simplify each expression.
A.NR.2
42. $\sqrt{\dfrac{27}{m^5}}$  $\dfrac{3\sqrt{3m}}{m^3}$
43. $\sqrt{\dfrac{32}{t^4}}$  $\dfrac{4\sqrt{2}}{t^2}$
44. $\sqrt{\dfrac{3}{16}} \cdot \sqrt{\dfrac{9}{5}}$  $\dfrac{3\sqrt{15}}{20}$
45. $\dfrac{7}{5+\sqrt{3}}$  $\dfrac{35-7\sqrt{3}}{22}$
46. $\dfrac{9}{6-\sqrt{8}}$  $\dfrac{27+9\sqrt{2}}{14}$
47. $\dfrac{3\sqrt{3}}{-2+\sqrt{6}}$  $\dfrac{6\sqrt{3}+9\sqrt{2}}{2}$
48. $\dfrac{3}{\sqrt{7}-\sqrt{2}}$  $\dfrac{3\sqrt{7}+3\sqrt{2}}{5}$
49. $\dfrac{5}{\sqrt{6}+\sqrt{3}}$  $\dfrac{5\sqrt{6}-5\sqrt{3}}{3}$
50. $\dfrac{2\sqrt{5}}{2\sqrt{7}+3\sqrt{3}}$  $4\sqrt{35}-6\sqrt{15}$

51. **ROLLER COASTER** Starting from a stationary position, the velocity $v$ of a roller coaster in feet per second at the bottom of a hill can be approximated by $v = \sqrt{64h}$, where $h$ is the height of the hill in feet.
   a. Simplify the equation.  $v = 8\sqrt{h}$
   b. Determine the velocity of a roller coaster at the bottom of a 134-foot hill.  about 92.6 ft/s

52. Evaluate each expression for the given values. Write in simplest form. Assume all square roots are positive.
   a. $\sqrt{z}$ if $z = 36$  $6$
   b. $\sqrt{wx}$ if $w = 8$ and $x = 16$  $8\sqrt{2}$
   c. $2\sqrt[3]{w}$ if $w = 8$  $4$
   d. $\sqrt[3]{32x}$ if $x = 16$  $8$
   e. $\sqrt{w} \cdot \sqrt{y}$ if $w = 8$ and $y = 20$  $4\sqrt{10}$
   f. $\sqrt[3]{6z}$ if $z = 36$  $6$
   g. $a^2$ if $a = \sqrt{48}$  $48$
   h. $ab$ if $a = \sqrt{8}$ and $b = 2\sqrt{2}$  $8$
   i. $a + c$ if $a = \sqrt{49}$ and $c = \sqrt[3]{8}$  $9$
   j. $f - d$ if $f = \sqrt[3]{1}$ and $d = \sqrt[3]{27}$  $-2$
   k. $\sqrt[3]{ab}$ if $a = \sqrt{8}$ and $b = 2\sqrt{2}$  $2$
   l. $\sqrt[3]{b^2}$ if $b = 3\sqrt{3}$  $3$

**Examples 6–8** Simplify each expression. 56. $12\sqrt{3} + \sqrt{2}$
N.RN.2
53. $7\sqrt{5} + 4\sqrt{5}$  $11\sqrt{5}$
54. $3\sqrt{50} - 3\sqrt{32}$  $3\sqrt{2}$
55. $\sqrt{5}(\sqrt{2} + 4\sqrt{2})$  $5\sqrt{10}$
56. $7\sqrt{3} - 2\sqrt{2} + 3\sqrt{2} + 5\sqrt{3}$
57. $(\sqrt{3} - \sqrt{2})(\sqrt{15} + \sqrt{12})$  $3\sqrt{5} + 6 - \sqrt{30} - 2\sqrt{6}$
58. $5\sqrt{3}(6\sqrt{10} - 6\sqrt{3})$  $30\sqrt{30} - 90$
59. $(5\sqrt{2} + 3\sqrt{5})(2\sqrt{10} - 5)$  $5\sqrt{5} + 5\sqrt{2}$
60. $(3\sqrt{11} + 3\sqrt{15})(3\sqrt{3} - 2\sqrt{2})$  $9\sqrt{33} - 6\sqrt{22} + 27\sqrt{5} - 6\sqrt{30}$

**Example 9**
N.RN.2
61. **GEOMETRY** Find the perimeter and area of the rectangle.  $10\sqrt{7} + 2\sqrt{5}$ units; $12$ units$^2$

Rectangle with sides $3\sqrt{7} + 3\sqrt{5}$ and $2\sqrt{7} + 2\sqrt{5}$.

---

### Lesson 7-4 | Radical Expressions

**Watch Out!**

**Preventing Errors** Students may not think the simplified form of a fraction containing radicals looks any simpler than the original fraction. Explain that meeting the three conditions for simplifying a radical expression, not the look of the fraction, dictates whether a fraction containing radicals is simplified.

### Additional Answers

52b. No; Sample answer: The advertised pump will pump water only to a maximum height of about 76.6 feet.

52c. Yes; Sample answer: The advertised pump will pump water to a maximum height of about 92.6 feet.

# Lesson 7-4 | Radical Expressions

## Teaching the Mathematical Practices

**Construct Arguments** Mathematically proficient students make conjectures and build a logical progression of statements to explore the truth of their conjectures. In Exercise 88, suggest that students use a calculator to find several sums and products.

## Assess

**Crystal Ball** Ask students to write how they think what they learned today about adding, subtracting, and multiplying radical expressions will connect with the next lesson on radical equations.

## Additional Answers

**85.** Sample answer: Cross multiply then divide. Rationalize the denominator to find that $x = \frac{5\sqrt{3} + 9}{2}$.

**88.** True; $(x + y)^2 > \left(\sqrt{x^2 + y^2}\right)^2$
$x^2 + 2xy + y^2 > x^2 + y^2$
$2xy > 0$
Because $x > 0$ and $y > 0$, $2xy > 0$ is always true. So, $x + y > \sqrt{x^2 + y^2}$ is true for all $x > 0$ and $y > 0$.

**89.** Irrational; irrational; no rational number could be added to or multiplied by an irrational number so that the result is rational.

**90.** Sample answer: $\sqrt{12} + \sqrt{27} = 5\sqrt{3}$; When you simplify $\sqrt{12}$, you get $2\sqrt{3}$. When you simplify $\sqrt{27}$, you get $3\sqrt{3}$. Because $2\sqrt{3}$ and $3\sqrt{3}$ have the same radicand, you can add them.

**91.** Sample answer: You can use the FOIL method. You multiply the first terms within the parentheses. Then you multiply the outer terms within the parentheses. Then you multiply the inner terms within the parentheses. And, then you multiply the last terms within each parentheses. Combine any like terms and simplify any radicals. For example, $(\sqrt{2} + \sqrt{3})(\sqrt{5} + \sqrt{7}) = \sqrt{10} + \sqrt{14} + \sqrt{15} + \sqrt{21}$.

## Go Online!

**eSolutions Manual** Create worksheets, answer keys, and solutions handouts for your assignments.

---

**B** **62. ELECTRICITY** The amount of current in amperes $I$ that an appliance uses can be calculated using calculated using the formula $I = \sqrt{\frac{P}{R}}$, where $P$ is the power in watts and $R$ is the resistance in ohms.
  **a.** Simplify the formula. $I = \frac{\sqrt{PR}}{R}$
  **b.** How much current does an appliance use if the power used is 75 watts and the resistance is 5 ohms? **about 3.9 amps**

**63. KINETIC ENERGY** The speed $v$ of a ball can be determined by the equation $v = \sqrt{\frac{2k}{m}}$, where $k$ is the kinetic energy and $m$ is the mass of the ball.
  **a.** Simplify the formula if the mass of the ball is 3 kilograms. $v = \frac{\sqrt{6k}}{3}$
  **b.** If the ball is traveling 7 meters per second, what is the kinetic energy of the ball in Joules? **73.5 Joules**

Simplify each expression. **70.** $8\sqrt{5}$

**64.** $\sqrt{\frac{1}{9}} - \sqrt{2}$  $\frac{1}{3} - \sqrt{2}$   **65.** $\sqrt{2}(\sqrt{8} + \sqrt{6})$ $4 - 2\sqrt{3}$   **66.** $2\sqrt{3}(\sqrt{12} - 5\sqrt{3})$ $-18$

**67.** $\sqrt{\frac{1}{5}} - \sqrt{5}$  $\frac{-4\sqrt{5}}{5}$   **68.** $\sqrt{\frac{2}{3}} + \sqrt{6}$  $\frac{4\sqrt{6}}{3}$   **69.** $2\sqrt{\frac{1}{2}} + 2\sqrt{2} - \sqrt{8}$  $\sqrt{2}$

**70.** $8\sqrt{\frac{5}{4}} + 3\sqrt{20} - 10\sqrt{\frac{1}{5}}$   **71.** $(3 - \sqrt{5})^2$  $14 - 6\sqrt{5}$   **72.** $(\sqrt{2} + \sqrt{3})^2$  $5 + 2\sqrt{6}$

**73** **ROLLER COASTERS** The velocity $v$ in feet per second of a roller coaster at the bottom of a hill is related to the vertical drop $h$ in feet and the velocity $v_0$ of the coaster at the top of the hill by the formula $v_0 = \sqrt{v^2 - 64h}$.
  **a.** What velocity must a coaster have at the top of a 225-foot hill to achieve a velocity of 120 feet per second at the bottom? **0 ft/s**
  **b.** Explain why $v_0 = v - 8\sqrt{h}$ is not equivalent to the formula given.

**73b.** Sample answer: In the formula, we are taking the square root of the difference, not the square root of each term.

**74. FINANCIAL LITERACY** Tadi invests $225 in a savings account. In two years, Tadi has $232 in his account. You can use the formula $r = \sqrt{\frac{v_2}{v_0}} - 1$ to find the average annual interest rate $r$ that the account has earned. The initial investment is $v_0$, and $v_2$ is the amount in two years. What was the average annual interest rate that Tadi's account earned? **about 1.5%**

### H.O.T. Problems  Use Higher-Order Thinking Skills  N.RN.2

**C** **75.** **STRUCTURE** Explain how to solve $\frac{\sqrt{3} + 2}{x} = \frac{\sqrt{3} - 1}{\sqrt{3}}$. **See margin.**

**76.** **REASONING** Marge takes a number, subtracts 4, multiplies by 4, takes the square root, and takes the reciprocal to get $\frac{1}{2}$. What number did she start with? Write a formula to describe the process. **76. 5;** $\frac{1}{\sqrt{4(x-4)}} = \frac{1}{2}$

**77.** **OPEN-ENDED** Write two binomials of the form $a\sqrt{b} + c\sqrt{f}$ and $a\sqrt{b} - c\sqrt{f}$. Then find their product. **Sample answer:** $1 + \sqrt{2}$ and $1 - \sqrt{2}$; $(1 + \sqrt{2}) \cdot (1 - \sqrt{2}) = 1 - 2 = -1$

**78.** **CHALLENGE** Determine whether the following statement is *true* or *false*. Provide a proof or counterexample to support your answer. **See margin.**
$$x + y > \sqrt{x^2 + y^2} \text{ when } x > 0 \text{ and } y > 0$$

**79.** **CONSTRUCT ARGUMENTS** Make a conjecture about the sum of a rational number and an irrational number. Is the sum of a rational number and an irrational number *rational* or *irrational*? Is the product of a nonzero rational number and an irrational number *rational* or *irrational*? Explain your reasoning. **See margin.**

**80.** **OPEN-ENDED** Write an equation that shows a sum of two radicals with different radicands. Explain how you could combine these terms. **See margin.**

**81.** **WRITING IN MATH** Describe step by step how to multiply two radical expressions, each with two terms. Write an example to demonstrate your description. **See margin.**

## Standards for Mathematical Practice

| Emphasis On | Exercises |
| --- | --- |
| 1 Make sense of problems and persevere in solving them. | 83, 84, 86, 88, 89 |
| 2 Reason abstractly and quantitatively. | 51, 62, 63, 76, 80, 84, 87 |
| 3 Construct viable arguments and critique the reasoning of others. | 73, 78, 79 |
| 4 Model with mathematics. | 29, 51, 52, 61–63, 73, 74, 85, 89 |
| 6 Attend to precision. | 52, 61 |
| 7 Look for and make use of structure. | 75, 77 |
| 8 Look for and express regularity in repeated reasoning. | 81, 82 |

Lesson 7-4 | Radical Expressions

## Preparing for Assessment

82. Jason correctly simplified the product $\sqrt{2} \cdot \sqrt{22}$. Which of the following is most likely to be a step in the process he followed? **8 N.RN.2 A**
    - A  $\sqrt{2} \cdot \sqrt{22} = \sqrt{2} \cdot \sqrt{2} \cdot \sqrt{11}$
    - B  $\sqrt{2} \cdot \sqrt{22} = \sqrt{2} \cdot \sqrt{2} + \sqrt{11}$
    - C  $\sqrt{2} \cdot \sqrt{22} = (\sqrt{2 \cdot 2})^2 \cdot \sqrt{11}$
    - D  $\sqrt{2} \cdot \sqrt{22} = 4\sqrt{11}$

83. Which expression is equivalent to $\sqrt{\frac{40}{12}}$? **1 N.RN.2 B**
    - A  10
    - B  $\frac{\sqrt{30}}{3}$
    - C  $\frac{\sqrt{30}}{9}$
    - D  $\frac{\sqrt{30}}{\sqrt{3}}$

84. Which expression is equivalent to 4? **1,2 N.RN.2 D**
    - A  $\frac{\sqrt{8}}{2}$
    - B  $\frac{\sqrt{8}}{\sqrt{2}}$
    - C  $2\sqrt{8}$
    - D  $\sqrt{2} \cdot \sqrt{8}$

85. An object is dropped off a cliff into a deep canyon. The velocity $v$ of the object at the bottom of the canyon is given by the function $v = \sqrt{64h}$, where $h$ is the height of the cliff, in feet. Which equation is not equivalent to $v = \sqrt{64h}$? **4 N.RN.2 A**
    - A  $v = 2\sqrt{4h}$
    - B  $v = 2\sqrt{16h}$
    - C  $v = 2 \cdot 2\sqrt{4h}$
    - D  $v = 8\sqrt{h}$

86. Which expression is equivalent to $4\sqrt{27} + 2\sqrt{12} + \sqrt{75}$? **1 N.RN.2 C**
    - A  $12 + 9\sqrt{3}$
    - B  $10\sqrt{3}$
    - C  $21\sqrt{3}$
    - D  $25\sqrt{3}$

87. What is the area of the square in simplest form? **1,2 N.RN.2 C**

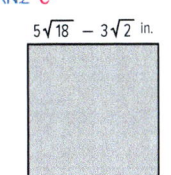

    $5\sqrt{18} - 3\sqrt{2}$ in.

    - A  252 in$^2$
    - B  270 in$^2$
    - C  288 in$^2$
    - D  648 in$^2$

88. For what value of $a$ does the expression $a\sqrt{5} \cdot 4\sqrt{20}$ equal 200? **1 N.RN.2**

    5

89. **MULTI-STEP** The annual interest rate $r$ earned by an investment can be found using the formula $r = \sqrt[t]{\frac{A}{P}} - 1$, where $P$ is the initial investment and $A$ is the value of the investment after $t$ years.

    a. What is the average interest rate for an initial investment of $5750 that has a value of $6080 after two years? Round to the nearest tenth of a percent.

    2.8%

    b. Solve the function for $P$.

    $P = \frac{A}{(1+r)^t}$

    c. Find the initial investment if its value is $4000 after 12 years at an interest rate of 3.4%.

    $2678.02

---

### Differentiated Instruction  OL  BL

**Extension** Write $\sqrt{6} + \sqrt{3} = \sqrt{9} = 3$ and $\sqrt{40} = \sqrt{36} + \sqrt{4} = 8$ on the board. Ask students to explain the errors. You cannot add radical expressions unless they can be written with the same radicand, and you cannot write the square root of a sum as the sum of the square roots.

---

## Preparing for Assessment

Exercises 92–99 require students to use the skills they will need on standardized assessments. Each exercise is dual-coded with content standards and mathematical practice standards.

| | Dual Coding | |
|---|---|---|
| Items | Content Standards | Mathematical Practices |
| 82 | N.RN.2 | 8 |
| 83 | N.RN.2 | 1 |
| 84 | N.RN.2 | 1, 2 |
| 85 | N.RN.2 | 4 |
| 86 | N.RN.2 | 1 |
| 87 | N.RN.2 | 1, 2 |
| 88 | N.RN.2 | 1 |
| 89 | N.RN.2 | 1, 4 |

### Diagnose Student Errors

Survey student responses for each item. Class trends may indicate common errors and misconceptions.

**86.**

| | |
|---|---|
| A | Simplified $\sqrt{27}$ as 3 |
| B | Forgot to multiply by constants after applying Product Property |
| C | CORRECT |
| D | Simplified $\sqrt{4}$ as 4 after applying Product Property to second term |

**87.**

| | |
|---|---|
| A | After using FOIL method, subtracted the last term (18) from 270 instead of adding it to 270 |
| B | After using FOIL method, forgot to add last term (18) to 270 |
| C | CORRECT |
| D | After using FOIL method, added middle terms to 450 instead of subtracting them from 450 |

### Go Online!

**Quizzes**

Students can use Self-Check Quizzes to check their understanding of this lesson. You can also give the Chapter Quiz which covers the content in Lesson 7-4.

connectED.mcgraw-hill.com  427

## Extend 7-4

### Launch

**Objective** Explain why the sums and products of rational and irrational numbers are irrational.

### Teach

**Working in Cooperative Groups** Have students work in groups of two or three, mixing abilities, to complete Activities 1–3 and Exercises 1 and 2.

**Teaching Tip** It is important for students to understand that by definition, an irrational number cannot be expressed as a fraction $\frac{a}{b}$, where $a$ and $b$ are integers and $b \neq 0$. An irrational number may be expressed as a decimal that does not repeat or is nonterminating. They can use this definition to prove that the product and sum of a rational and an irrational number is irrational.

### Assess

**Formative Assessment** Use Exercises 3–5 to assess whether students understand the relationships between the product and the sum of a rational and an irrational number.

### Go Online!

The most up-to-date resources available for your program can be found at **connectED.mcgraw-hill.com**.

---

## EXTEND 7-4

### Algebra Lab: Sums and Products of Rational and Irrational Numbers

Irrational numbers are the set of numbers that cannot be expressed as a terminating or repeating decimal. Irrational numbers cannot be written as a fraction $\frac{a}{b}$, where $a$ and $b$ are integers and $b \neq 0$.

**Real Numbers**

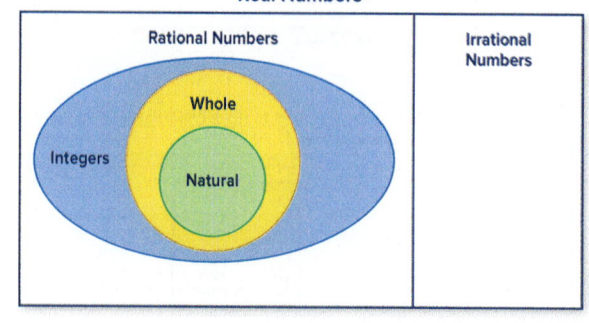

[Diagram: Rational Numbers containing Integers, Whole, and Natural (nested); Irrational Numbers separate]

**Mathematical Practices**
MP 2 Reason abstractly and quantitatively.

**Content Standards**
**N.RN.3** Explain why the sum or product of two rational numbers is rational; that the sum of a rational number and an irrational number is irrational; and that the product of a nonzero rational number and an irrational number is irrational.

### Activity 1  Sum and Product Rational Numbers and Irrational Numbers

Make a conjecture about whether the sum and product of a nonzero rational number and an irrational number by examining specific examples.

**Step 1** Make a table to examine the sum of $a$ and $b$, where $a$ is a nonzero rational number and $b$ is an irrational number.

| a | b | a + b | Rational or irrational? |
|---|---|-------|-------------------------|
| −2 | $\sqrt{2}$ | $-2 + \sqrt{2}$ | irrational |
| 0.4 | $2\sqrt{3}$ | $0.4 + 2\sqrt{3}$ | irrational |
| $\frac{1}{2}$ | $\pi$ | $\frac{1}{2} + \pi$ | irrational |
| −2 | $\sqrt{2}$ | $-2 + \sqrt{2}$ | irrational |

**Step 2** Make a table to examine the product of $a$ and $b$.

| a | b | ab | Rational or irrational? |
|---|---|----|-------------------------|
| −2 | $\sqrt{2}$ | $-2\sqrt{2}$ | irrational |
| 0.4 | $2\sqrt{3}$ | $0.8\sqrt{3}$ | irrational |
| $\frac{1}{2}$ | $\pi$ | $\frac{\pi}{2}$ | irrational |

Based on the examples, a conjecture can be made that the sum or product of a nonzero rational number and an irrational number is an irrational number.

It appears from the table that the sum and product of a rational and irrational number are always irrational. Listing examples is not sufficient to prove this conjecture. To prove this is true, it must be true for all rational and irrational numbers. That is, for any rational number $a$ and any irrational number $b$, $a + b$ and $ab$ are irrational.

---

428 | Extend 7-4 | Algebra Lab: Sums and Products of Rational and Irrational Numbers

## Extend 7-4

### Activity 2  Prove the Product of *a* and *b* Is Irrational

Prove that the product of a nonzero rational number and an irrational number is irrational.

Given: $a$ is a nonzero rational number and $b$ is an irrational number.

Prove: $ab$ is irrational.

Assume $ab$ is a rational number.

$ab = \frac{c}{d}$, where $c$ and $d$ are integers and $d \neq 0$     Definition of a rational number
$b = \frac{c}{ad}$     Divide each side by $a$.
$b \neq \frac{c}{ad}$     $b$ is an irrational number.

Because it is given that $b$ is irrational and an irrational number cannot be written as a fraction using integer values, the product $ab$ is not a rational number. Therefore, $ab$ is irrational.

### Activity 3  Prove the Sum of *a* and *b* Is Irrational

Prove that the sum of a nonzero rational number and an irrational number is irrational.

Given: $a$ is a nonzero rational number and $b$ is an irrational number.

Prove: $a + b$ is irrational.

Assume $a + b$ is a rational number.

$a + b = \frac{c}{d}$, where $c$ and $d$ are integers and $d \neq 0$     Definition of a rational number
$b = \frac{c}{d} - a$     Subtract each side by $a$.
$b = \frac{c - ad}{d}$     Multiply $a$ by $\frac{d}{d}$ and simplify.
$b \neq \frac{c - ad}{d}$     $b$ is an irrational number.

Because it is given that $b$ is irrational and an irrational number cannot be written as a fraction using integer values, the sum $a + b$ is not a rational number. Therefore, $a + b$ is irrational.

### Exercises

Give an example of each.

1. Two irrational numbers whose sum is rational  **Sample answer:** $\sqrt{2}, -\sqrt{2}$
2. Two irrational numbers whose sum is irrational  **Sample answer:** $\sqrt{3}, \pi$
3. Two irrational numbers whose product is rational  **Sample answer:** $\sqrt{5}, \sqrt{5}$
4. Two irrational numbers whose product is irrational  **Sample answer:** $\sqrt{2}, \sqrt{7}$
5. Recall that a set is closed under an operation if for any numbers in the set, the result of the operation is also in the set. Explain why the set of irrational numbers is not closed under addition or multiplication.
   **Sample answer:** Because the sum or product of two irrational numbers does not always result in an irrational number, the set of irrational numbers is not closed under these operations.

# LESSON 7-5
# Exponential Functions

**SUGGESTED PACING (DAYS)**

| | | |
|---|---|---|
| 90 min. | 0.5 | 0.75 |
| 45 min. | 1 | 0.75 |
| | Instruction | Extend Lab |

## Track Your Progress

### Objectives
1. Graph exponential functions.
2. Identify data that display exponential behavior.

### Mathematical Background
Exponential functions are nonlinear and nonquadratic. An exponential function has a variable as an exponent and can be described by an equation of the form $y = ab^x$, where $a \neq 0$, $b > 0$, and $b \neq 1$. When $b > 1$, the function is an exponential growth function, and when $0 < b < 1$, the function is an exponential decay function.

### THEN
**N.RN.2** Rewrite expressions involving radicals and rational exponents using the properties of exponents.

### NOW
**F.IF.7e** Graph exponential and logarithmic functions, showing intercepts and end behavior, and trigonometric functions, showing period, midline, and amplitude.

**F.LE.5** Interpret the parameters in a linear or exponential function in terms of a context.

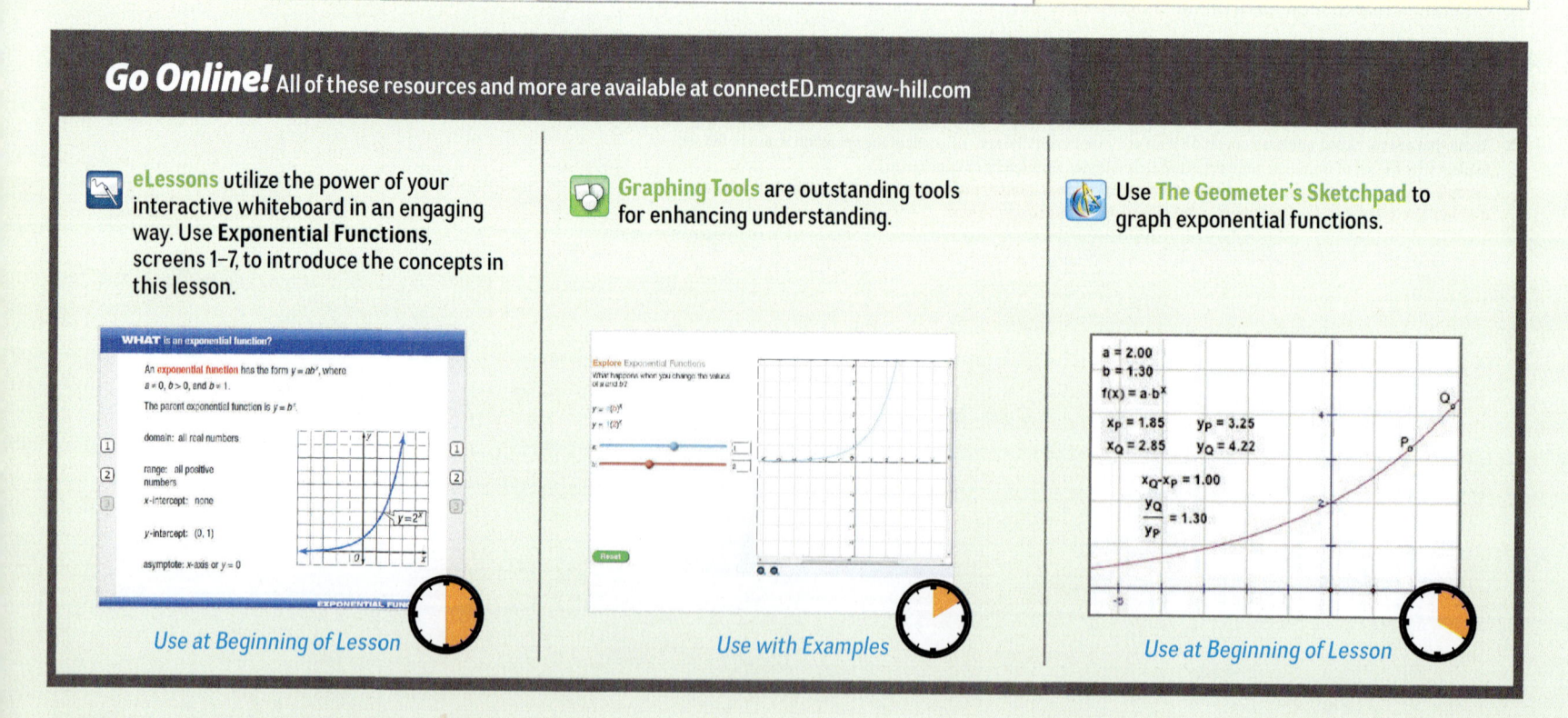

### NEXT
**A.SSE.3c** Use the properties of exponents to transform expressions for exponential functions.

## Go Online! All of these resources and more are available at connectED.mcgraw-hill.com

**eLessons** utilize the power of your interactive whiteboard in an engaging way. Use **Exponential Functions**, screens 1–7, to introduce the concepts in this lesson.

*Use at Beginning of Lesson*

**Graphing Tools** are outstanding tools for enhancing understanding.

*Use with Examples*

Use **The Geometer's Sketchpad** to graph exponential functions.

*Use at Beginning of Lesson*

## Using Open Educational Resources

**Homework Help** Have students review their work on **freemathhelp.com**. They can see the worked out step-by-step solution to any exponential function. If they make a mistake on a quiz or test they can use this site to see where they went wrong. *Use as homework*

# Go Online!
connectED.mcgraw-hill.com — Worksheets

# Differentiate Your Resources

**Extra Practice** Additional practice or homework; Skills Practice is best for approaching-level students and Practice is best for on-level and beyond-level students

## Skills Practice
AL  OL  ELL

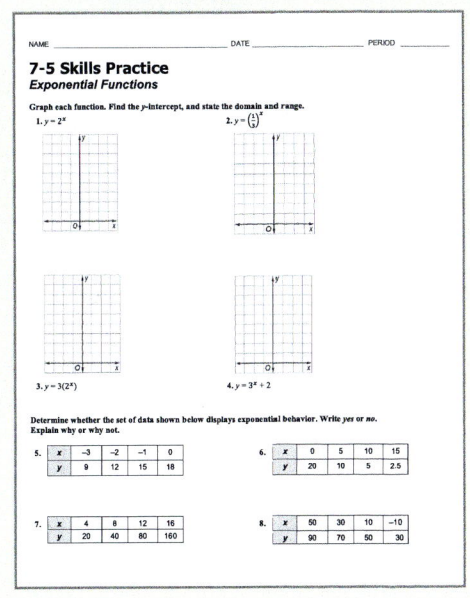

## Practice
AL  OL  BL  ELL

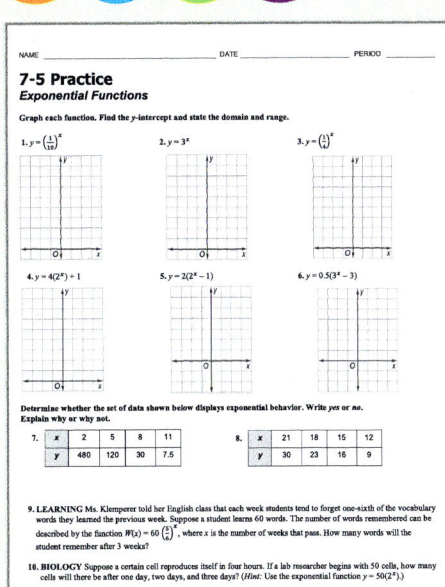

## Word Problem Practice
AL  OL  BL  ELL

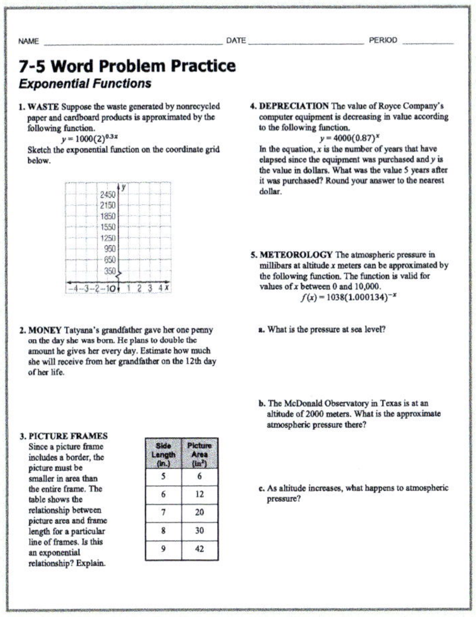

**Intervention** Reteaching and vocabulary activities that can be used with struggling or absent students and as ELL support

**Extension** Activities that can be used to extend lesson concepts

## Study Guide and Intervention
AL  OL  ELL

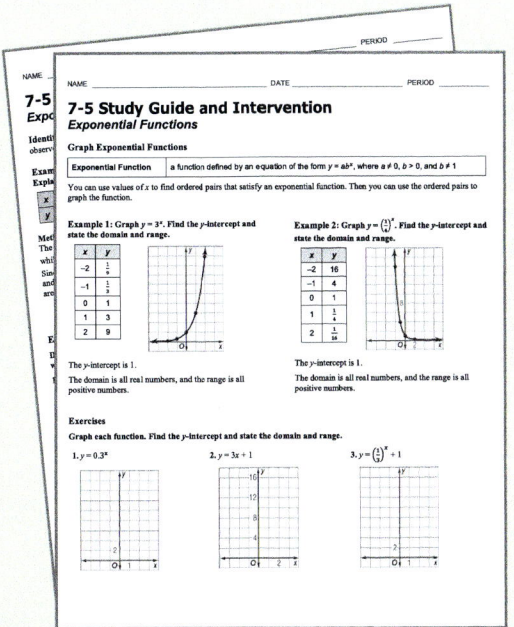

## Study Notebook
AL  OL  BL  ELL

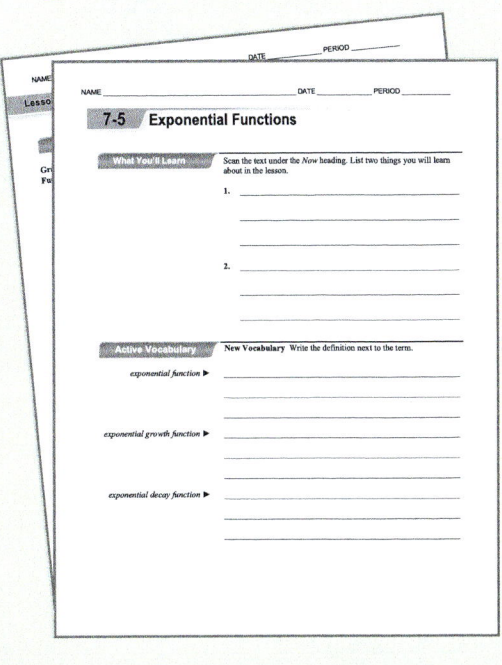

## Enrichment
OL  BL  ELL

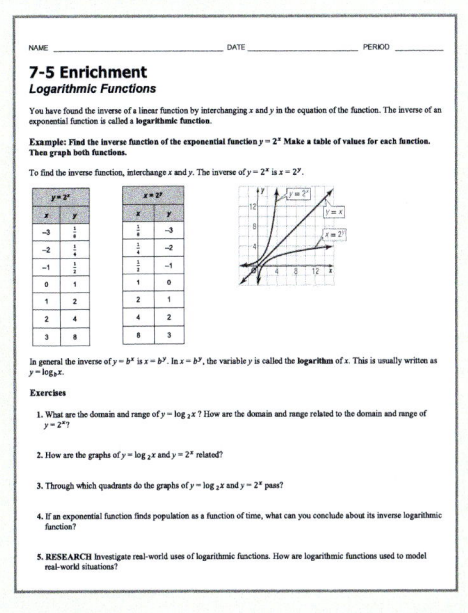

connectED.mcgraw-hill.com  430B

Lesson 7-5 | Exponential Functions

# Launch

Have students read the Why? section of the lesson. Ask:
- How is this equation different from a linear equation? **The independent variable $x$ is an exponent.**
- What is the value of $y$ when $x = 0$? **$y = 3$**
- Can the value of $y$ ever be 0? **no**

# Teach

Ask the scaffolded questions for each example to build conceptual understanding for students at all levels.

## 1 Graph Exponential Functions

**Example 1  Graph with $a > 0$ and $b > 1$**

**AL** How would you describe the shape of the graph of $y = 3^x$? **Sample answer: As you move from left to right, the graph rises slowly at first and then rises quickly.**

**OL** In the table, how does each whole number in the $y$-column compare to the previous whole number? **Each value is three times the previous value.** What does this tell you about the height of consecutive points on the graph? **Sample answer: As you move from left to right, the height of each point is three times the height of the previous point.**

**BL** In the definition of an exponential function, why is it required that $b \neq 1$? **If $b = 1$, then the function is a constant.**

## Go Online!

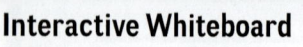

**Interactive Whiteboard**
Use the *eLesson* or *Lesson Presentation* to present this lesson.

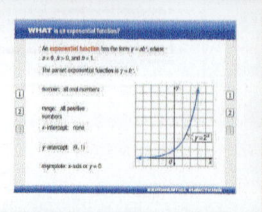

---

## LESSON 5
# Exponential Functions

| Then | Now | Why? |
|---|---|---|
| • Simplified radical expressions. | 1. Graph exponential functions.<br>2. Identify data that display exponential behavior. | Tarantulas can appear scary with their large hairy bodies and legs, but they are harmless to humans. The graph shows a tarantula spider population that increases over time. Notice that the graph is not linear.<br><br>The graph represents the function $y = 3(2)^x$. This is an example of an *exponential* function. |

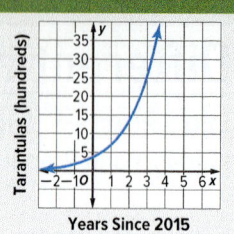

**New Vocabulary**
exponential function
asymptote
exponential growth function
exponential decay function

**Mathematical Processes**
1 Make sense of problems and persevere in solving them.

**Content Standards**
F.IF.7e Graph exponential and logarithmic functions, showing intercepts and end behavior, and trigonometric functions, showing periods, midline, and amplitude.
F.LE.5 Interpret the parameters in a linear or exponential function in terms of a context.

### 1 Graph Exponential Functions
An **exponential function** is a function of the form $y = ab^x$, where $a \neq 0$, $b > 0$, and $b \neq 1$. Notice that the base is a constant and the exponent is a variable. Exponential functions are nonlinear.

**Key Concept**  Exponential Function

| Words | An exponential function is a function that can be described by an equation of the form $y = ab^x$, where $a \neq 0$, $b > 0$, and $b \neq 1$. |
|---|---|
| Examples | $y = 2(3)^x$    $y = 4^x$    $y = \left(\frac{1}{2}\right)^x$ |

The graphs of exponential functions have a horizontal asymptote. An **asymptote** is a line that a graph approaches.

**Example 1**   Graph with $a > 0$ and $b > 1$

Graph $y = 3^x$. Find the $y$-intercept, and state the domain, range, and the equation of the horizontal asymptote.

The graph crosses the $y$-axis at 1, so the $y$-intercept is 1.
$D = \{-\infty < x < \infty\}$
$R = \{0 < y < \infty\}$

As $x$ decreases, the graph approaches the $x$-axis. So, the graph has a horizontal asymptote of $y = 0$.

Notice that the graph approaches the $x$-axis but there is no $x$-intercept. The graph is increasing on the entire domain.

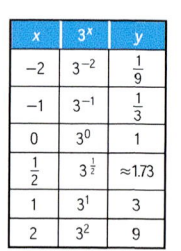

| $x$ | $3^x$ | $y$ |
|---|---|---|
| $-2$ | $3^{-2}$ | $\frac{1}{9}$ |
| $-1$ | $3^{-1}$ | $\frac{1}{3}$ |
| $0$ | $3^0$ | $1$ |
| $\frac{1}{2}$ | $3^{\frac{1}{2}}$ | $\approx 1.73$ |
| $1$ | $3^1$ | $3$ |
| $2$ | $3^2$ | $9$ |

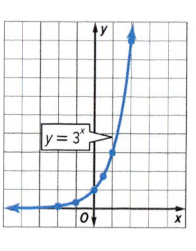

**Guided Practice**

1. Graph $y = 7^x$. Find the $y$-intercept, and state the domain, range, and the equation of the horizontal asymptote. **See Ch. 7 Answer Appendix.**

Functions of the form $y = ab^x$, where $a > 0$ and $b > 1$, are called **exponential growth functions** and all have the same shape as the graph in Example 1. Functions of the form $y = ab^x$, where $a > 0$ and $0 < b < 1$ are called **exponential decay functions** and also have the same general shape.

---

**MP  Teaching the Mathematical Practices**
Help students develop the mathematical practices by asking questions like these.

**Make sense of problems and persevere in solving them.**
Help students recognize general methods by asking them to make connections, graph exponential functions, and identify exponential behavior. For example, ask:

- In $y = b^x$, which variable represents the exponent in an exponential function?  **$x$**
- In $y = b^x$, which variable represents the base?  **$b$**
- When is a function in exponential growth?  **when the base is greater than 1**
- When is a function in exponential decay?  **when the base is between 0 and 1**
- What is the domain of an exponential function?  **The domain is all real numbers.**
- What is the range of an exponential function?  **The range is all values greater than the value of the asymptote.**

430 | Lesson 7-5 | Exponential Functions

Lesson 7-5 | Exponential Functions

### Example 2  Graph with $a > 0$ and $0 < b < 1$

Graph $y = \left(\frac{1}{3}\right)^x$. Find the $y$-intercept, and state the domain, range, and the equation of the asymptote.

The $y$-intercept is 1.
$D = \{-\infty < x < \infty\}$
$R = \{0 < y < \infty\}$
Notice that as $x$ increases, the $y$-values decrease less rapidly, but never touch the $x$-axis. The equation of the asymptote is $y = 0$.

| $x$ | $\left(\frac{1}{3}\right)^x$ | $y$ |
|---|---|---|
| $-2$ | $\left(\frac{1}{3}\right)^{-2}$ | 9 |
| 0 | $\left(\frac{1}{3}\right)^0$ | 1 |
| 2 | $\left(\frac{1}{3}\right)^2$ | $\frac{1}{9}$ |

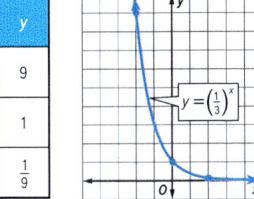

### Guided Practice

2. Graph $y = \left(\frac{1}{2}\right)^x - 1$. Find the $y$-intercept, and state the domain, range, and the equation of the asymptote.

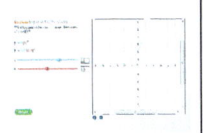

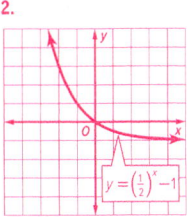

**2.** 0; $D = \{-\infty < x < \infty\}$; $R = \{-1 < y < \infty\}$; asymptote at $y = -1$

The key features of the graphs of exponential functions can be summarized as follows.

### Key Concept  Graphs of Exponential Functions

**Exponential Growth Functions**
Equation: $f(x) = ab^x, a > 0, b > 1$
Domain, Range: $\{-\infty < x < \infty\}; \{y > 0\}$
Intercepts: one $y$-intercept, no $x$-intercepts
Horizontal Asymptote: $y = 0$

**Exponential Decay Functions**
Equation: $f(x) = ab^x, a > 0, 0 < b < 1$
Domain, Range: $\{-\infty < x < \infty\}; \{y > 0\}$
Intercepts: one $y$-intercept, no $x$-intercepts
Horizontal Asymptote: $y = 0$

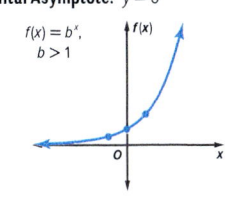

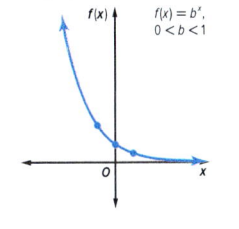

Exponential functions occur in many real world situations.

### Real-World Example 3  Use Exponential Functions to Solve Problems

**BOTTLED WATER** The function $W = 27.92(1.052)^t$ models the amount of bottled water consumed in the U.S. per capita, where $W$ is gallons of water and $t$ is the number of years since 2010.

a. Graph the function. What values of $W$ and $t$ are meaningful in the context of the problem?

Since $t$ represents time, $t > 0$. At $t = 0$, the consumption is 27.92 gallons. Therefore, in the context of this problem, $W > 27.92$ is meaningful.

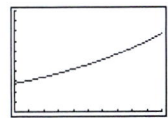

[0, 20] scl: 2 by [0, 100] scl: 10

**Real-World Link**
For several years, the consumption of soda in the U.S. has been decreasing. In 2016, the per capita consumption of bottled water was 39.3 gallons, surpassing 38.5 gallons of soda per capita.

Source: Fortune

### Watch Out!

**Student Misconceptions** Make sure students understand that the graphs of exponential functions never actually touch the $x$-axis. It is acceptable for hand-drawn graphs to show the graph just above and about parallel to the $x$-axis as long as students understand that the graph gets infinitely close to the axis without touching it.

---

### Go Online!

Investigate how changing the values in the equation of an exponential function affects the graph by using the Graphing Tools in ConnectED.

---

### Need Another Example?

Graph $y = 4^x$. Find the $y$-intercept and state the domain, range, and the equation of the asymptote. $y$-intercept: 1, $D = \{-\infty < x < \infty\}$; $R = \{0 < y < \infty\}$; $y = 0$

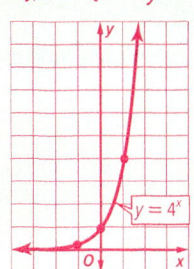

### Example 2  Graph with $a > 0$ and $0 < b < 1$

**AL** According to the table, what is the $y$-intercept? (0, 1)

**OL** Is there is any value of $x$ for which $\left(\frac{1}{3}\right)^x = 0$? Explain. No; $\left(\frac{1}{3}\right)^x > 0$ for all $x$.

**BL** Does the graph of $y = \left(\frac{1}{3}\right)^x$ ever intersect the $x$-axis? Why or why not? No; since $\left(\frac{1}{3}\right)^x > 0$ for all $x$, the graph never intersects the $x$-axis.

### Need Another Example?

Graph $y = \left(\frac{1}{4}\right)^x$. Find the $y$-intercept and state the domain, range, and the equation of the asymptote. $y$-intercept: 1; $D = \{-\infty < x < \infty\}$; $R = \{0 < y < \infty\}$; $y = 0$

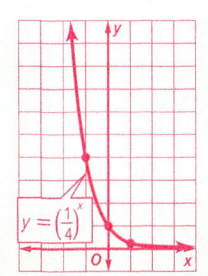

### Example 3  Use Exponential Functions to Solve Problems

**AL** What is the $y$-intercept of the function? What does it mean in the context of the situation? (0, 27.92), which represents 27.92 gallons of bottled water per person consumed in 2000.

**OL** What is the asymptote? What does it mean in the context of the situation? $y = 0$; Bottled water consumption can never be less than 0 liters.

**BL** Why might this graph be useful to a beverage company? See students' responses.

connectED.mcgraw-hill.com  431

## Lesson 7-5 | Exponential Functions

**Need Another Example?**

**Depreciation** The function $V = 25{,}000 \cdot 0.82^t$ models the depreciation in the value of a new car that originally cost $25,000. $V$ represents the value of the car and $t$ represents the time in years from the time of purchase.

a. Graph the function. What values of $V$ and $t$ are meaningful in the context of the problem?

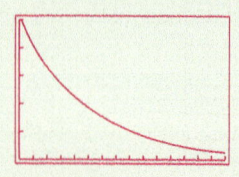

[0, 15] scl: 1 by [0, 25,000] scl: 500; Only values of $V \leq 25{,}000$ and $t \geq 0$ are meaningful.

b. What is the car's value after five years? about $9270

## 2 Identify Exponential Behavior

**Example 4 Identify Exponential Behavior**

**AL** What is the common factor between the y-values? $\frac{1}{2}$

**OL** How would you describe the shape of the graph? Sample answer: as $x$ increases, $y$ decreases.

**BL** Which method do you prefer to determine if the set of data displays exponential behavior? Explain. See students' work.

**Need Another Example?**

Determine whether the set of data shown below displays exponential behavior. Write *yes* or *no*. Explain why or why not.

| x | 0 | 10 | 20 | 30 |
|---|---|----|----|----|
| y | 10 | 25 | 62.5 | 156.25 |

The domain values are at regular intervals, and the range values have a common factor of 2.5, so the set is probably exponential.

---

b. Find and interpret the y-intercept and the equation of the asymptote.

The y-intercept, 27.92, represents that 27.92 gallons of bottled water was consumed in 2010, per capita. The asymptote $y = 0$ indicates that annual bottled water consumption can never be less than or equal to 0 gallons.

c. Predict the per capita consumption of bottled water in 2020.

$W = 27.92(1.052)^t$   Original equation
$= 27.92(1.052)^{10}$   $t = 10$
$\approx 46.35$   Use a calculator.

The predicted consumption of bottled water in 2020 is about 46.35 gallons in the U.S.

**Guided Practice**

3. **BIOLOGY** Beginning with 10 cells in a bacteria culture, the population can be represented by the function $B = 10(2)^t$, where $B$ is the number of cells and $t$ is time in 20 minute intervals. Graph the function. Then find and interpret the y-intercept and the equation of the asymptote. How many cells will there be after 2 hours? 640; See Ch. 7 Answer Appendix.

**2 Identify Exponential Behavior** Recall from Lesson 3-3 that linear functions have a constant rate of change. Exponential functions do not have constant rates of change, but they do have constant ratios.

F.LE.5

**Example 4   Identify Exponential Behavior**

Determine whether the set of data shown at right displays exponential behavior. Write *yes* or *no*. Explain why or why not.

| x | 0 | 5 | 10 | 15 | 20 | 25 |
|---|---|---|----|----|----|----|
| y | 64 | 32 | 16 | 8 | 4 | 2 |

**Method 1** Look for a pattern.

The domain values are at regular intervals of 5.
Look for a common factor among the range values.
The range values differ by the common factor of $\frac{1}{2}$.

Since the domain values are at regular intervals and the range values differ by a positive common factor, the data are probably exponential. Its equation may involve $\left(\frac{1}{2}\right)^x$.

**Method 2** Graph the data.

Plot the points and connect them with a smooth curve. The graph shows a rapidly decreasing value of $y$ as $x$ increases. This is a characteristic of exponential behavior in which the base is between 0 and 1.

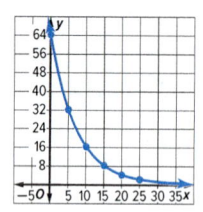

**Guided Practice**

4. No; the domain values are at regular intervals, but the range values have a common difference of 4.

4. Determine whether the set of data shown below displays exponential behavior. Write *yes* or *no*. Explain why or why not.

| x | 0 | 3 | 6 | 9 | 12 | 15 |
|---|---|---|---|---|----|----|
| y | 12 | 16 | 20 | 24 | 28 | 32 |

---

**Problem-Solving Tip**
**Make an Organized List**
Making an organized list of x-values and corresponding y-values is helpful in graphing the function. It can also help you identify patterns in the data.

---

## Differentiated Instruction OL BL

**Logical Learners** Ask students to write a comparison of an exponential function to a linear function.

## Lesson 7-5 | Exponential Functions

### Check Your Understanding

○ = Step-by-Step Solutions begin on page R11.

**Examples 1-2** Graph each function. Find the y-intercept and state the domain, range, and the
F.IF.7e equation of the asymptote.

1. $y = 2^x$
2. $y = -5^x$
3. $y = -\left(\frac{1}{5}\right)^x$
4. $y = 3\left(\frac{1}{4}\right)^x$
5. $f(x) = 6^x + 3$
6. $f(x) = 2 - 2^x$

1-6. See Ch. 7 Answer Appendix.

**Example 3** 7. **BIOLOGY** The function $f(t) = 100(1.05)^t$ models the growth of a fruit fly population,
F.LE.5 where $f(t)$ is the number of flies and $t$ is time in days.

a. What values for the domain and range are reasonable in the context of this
situation? Explain. D = {t| t ≥ 0}, the number of days is greater than or equal to 0;
R = {f(t) | f(t) ≥ 100}, the number of fruit flies is greater than or equal to 100.
b. After two weeks, approximately how many flies are in this population?
about 198 fruit flies

**Example 4** Determine whether the set of data shown below displays exponential behavior.
F.LE.5 Write *yes* or *no*. Explain why or why not. 8–9. See margin.

8.
| x | 1 | 2 | 3 | 4 | 5 | 6 |
|---|---|---|---|---|---|---|
| y | −4 | −2 | 0 | 2 | 4 | 6 |

9.
| x | 2 | 4 | 6 | 8 | 10 | 12 |
|---|---|---|---|---|---|---|
| y | 1 | 4 | 16 | 64 | 256 | 1024 |

### Practice and Problem Solving

Extra Practice is on page R7.

**Examples 1-2** Graph each function. Find the y-intercept and state the domain, range, and the
F.IF.7e equation of the asymptote. 10–19. See Ch. 7 Answer Appendix.

10. $y = 2 \cdot 8^x$
11. $y = 2 \cdot \left(\frac{1}{6}\right)^x$
12. $y = \left(\frac{1}{12}\right)^x$
13. $y = -3 \cdot 9^x$
14. $y = -4 \cdot 10^x$
15. $y = 3 \cdot 11^x$
16. $y = 4^x + 3$
17. $y = \frac{1}{2}(2^x - 8)$
18. $y = 5(3^x) + 1$
19. $y = -2(3^x) + 5$

**Example 3** 20. **MODELING** A population of rabbits increases according to the model
F.LE.5 $p = 200(2.5)^{0.02t}$, where $t$ is the number of months and $t = 0$ corresponds to January.

a. Use this model to estimate the number of rabbits in March. about 207
b. Graph the function. Find the equation of the asymptote, the y-intercept, and
describe what they represent in the context of the situation. State the domain and
range given the context of the situation. See Ch. 7 Answer Appendix.

**Example 4** Determine whether the set of data shown below displays exponential behavior.
F.LE.5 Write *yes* or *no*. Explain why or why not. 21–24. See margin.

21.
| x | −4 | 0 | 4 | 8 | 12 |
|---|---|---|---|---|---|
| y | 2 | −4 | 8 | −16 | 32 |

22.
| x | −6 | −3 | 0 | 3 |
|---|---|---|---|---|
| y | 5 | 10 | 15 | 20 |

23.
| x | −8 | −6 | −4 | −2 |
|---|---|---|---|---|
| y | 0.25 | 0.5 | 1 | 2 |

24.
| x | 20 | 30 | 40 | 50 | 60 |
|---|---|---|---|---|---|
| y | 1 | 0.4 | 0.16 | 0.064 | 0.0256 |

25. **PHOTOGRAPHY** Jameka is enlarging a photograph to make a poster for school. She
will enlarge the picture repeatedly at 150%. The function $P = 1.5^x$ models the new
size of the picture being enlarged, where $x$ is the number of enlargements. How
many times as big is the picture after 4 enlargements? about 506% bigger than the original

### Differentiated Homework Options

| Levels | AL Basic | OL Core | BL Advanced |
|---|---|---|---|
| Exercises | 10–24, 46–50 | 11–39 odd, 26, 40, 42–50 | 25–45<br>46–50 (optional) |
| 2-Day Option | 11–23 odd, 46–50 | 10–24, 46–50 | |
| | 10–24 even, 34–39 | 25–40, 34–45 | |

You can use ALEKS to provide additional remediation support with
personalized instruction and practice.

21. No; the domain values are at regular intervals, but the range values do not have a positive common factor.

22. No; the domain values are at regular intervals, but the range values have a common difference of 5.

23. Yes; the domain values are at regular intervals, and the range values have a common factor of 2.

24. Yes; the domain values are at regular intervals, and the range values have a common factor of 0.4.

---

## Practice

**Formative Assessment** Use Exercises 1–9 to assess students' understanding of the concepts in this lesson.

The Practice and Problem Solving exercises assess the content taught in the lesson. The Preparing for Assessment page is meant to be used as preparation for end-of-course assessments.

### Teaching the Mathematical Practices

**Modeling** Mathematically proficient students can use a function to describe how one quantity of interest depends on another. In Exercise 20, ask students how time will affect the population.

### Extra Practice

See page R7 for extra exercises for students who are approaching level or for on-level students who need additional reinforcement.

### Levels of Complexity Chart

The levels of the exercises progress from 1 to 3, with Level 1 indicating the lowest level of complexity.

| Exercises | 10–24 | 25–39, 46–50 | 40–45 |
|---|---|---|---|
| Level 3 | | | ● |
| Level 2 | | ● | |
| Level 1 | ● | | |

### Additional Answers

8. No; the domain values are at regular intervals, but the range values have a common difference of 2.

9. Yes; the domain values are at regular intervals, and the range values have a common factor of 4.

### Go Online!  eBook

**Interactive Student Guide**
Use the *Interactive Student Guide* to deepen conceptual understanding.
· Exponential Functions

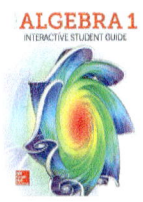

**Lesson 7-5 | Exponential Functions**

## e Follow-Up

Students have explored modeling using exponential functions.

Ask: How can mathematical models help you make good decisions? **Sample answer: Mathematical models can be used to compare different options that are available, as well as to predict the impact of an option if it is chosen.**

### MP Teaching the Mathematical Practices

**Perseverance** Mathematically proficient students analyze givens, constraints, relationships, and goals of a problem. In Exercise 41, students may struggle since they are given only two points. Suggest that they start with the general form of an exponential equation $y = ab^x$.

## Assess

**Crystal Ball** Ask students to write how they think exponential functions will connect with the next lesson, which involves real-world growth and decay problems.

## Additional Answers

43. Sample answer: The number of teams competing in a basketball tournament can be represented by $y = 2^x$, where the number of teams competing is $y$ and the number of rounds is $x$.

The $y$-intercept of the graph is 1. The graph increases rapidly for $x > 0$. With an exponential model, each team that joins the tournament will play all of the other teams. If the scenario were modeled with a linear function, each team that joined would play a fixed number of teams.

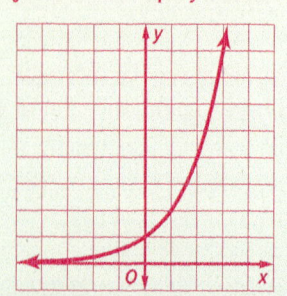

## Go Online!

**eSolutions Manual**
Create worksheets, answer keys, and solutions handouts for your assignments.

---

26. **MULTI-STEP** Daniel deposited $500 into a savings account that earns interest monthly. After 8 years, his investment is worth $807.07. Daniel wants to save a total of $1500 dollars in the next 5 years.
    a. How much should Daniel add to his savings account to ensure that he has $1500 in 5 years? **$304.99**
    b. Explain your solution process. **See Ch. 7 Answer Appendix.**
    c. What assumptions did you make? **I assumed that Daniel would not make any deposits or withdrawals from the account, and that the interest rate would stay the same.**

Identify each function as *linear*, *exponential*, or *neither*.

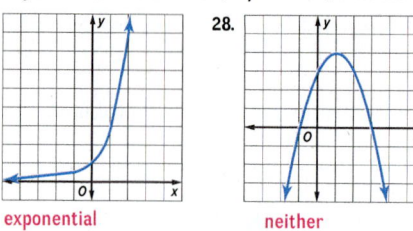

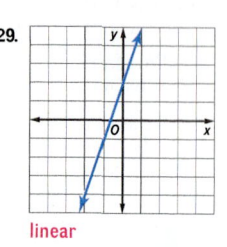

27. **exponential**   28. **neither**   29. **linear**

30. $y = 4^x$ **exponential**   31. $y = 2x(x - 1)$ **neither**   32. $5x + y = 8$ **linear**

33. **BUSINESS** The function $N = 380(0.85)^t$ models the number of desktop computers sold per year at a local technology store, where $t$ is the number of years since 2015 and $N$ is the number of desktop computers sold. How many desktop computers will be sold in 2025? **about 75**

Write the equation of the horizontal asymptote for each function.

34. $y = 2^x + 6$   35. $y = 5\left(\frac{2}{3}\right)^x$   36. $y = -\frac{1}{4}(2)^x$
    $y = 6$            $y = 0$

36. $y = 0$
39. $y = 1$

37. $y = -3 + 5^x$   38. $y = \left(\frac{2}{3}\right)^{x-1}$   39. $y = -5(3)^x + 1$
    $y = -3$             $y = 0$

40. **BACTERIA** The bacteria in a culture doubles every hour. At $t = 0$, there were 25 bacteria in the culture. The function $N = 25(2)^t$ models the number of bacteria $N$ in the culture $t$ hours after 7 A.M. Estimate the number of bacteria at 10 P.M. **819,200**

42. Never; the graph never crosses the $x$-axis because the powers of $b$ are always positive and $a \neq 0$. Thus, $ab^x$ is never 0.

F.IF.7e, F.LE.5

**H.O.T. Problems**   Use Higher-Order Thinking Skills

41. MP **PERSEVERANCE** Write an exponential function for which the graph passes through the points at (0, 3) and (1, 6). **$f(x) = 3(2)^x$**

42. MP **REASONING** Determine whether the graph of $y = ab^x$, where $a \neq 0$, $b > 0$, and $b \neq 1$, *sometimes*, *always*, or *never* has an $x$-intercept. Explain your reasoning. **See margin.**

43. **OPEN-ENDED** Find an exponential function that represents a real-world situation, and graph the function. Analyze the graph, and explain why the situation is modeled by an exponential function rather than a linear function. **See margin.**

44. MP **REASONING** Use tables and graphs to compare and contrast an exponential function $f(x) = ab^x + c$, where $a \neq 0$, $b > 0$, and $b \neq 1$, and a linear function $g(x) = ax + c$. Include intercepts, intervals where the functions are increasing, decreasing, positive, or negative, relative maxima and minima, symmetry, and end behavior. **See Ch. 7 Answer Appendix.**

45. **WRITING IN MATH** Explain how to determine whether a set of data displays exponential behavior. **See margin.**

---

### MP Standards for Mathematical Practice

| Emphasis On | Exercises |
|---|---|
| 1 Make sense of problems and persevere in solving them. | 1–6, 8–19, 21–24, 27–32, 34–39, 44, 46–50 |
| 3 Construct viable arguments and critique the reasoning of others. | 42, 45 |
| 4 Model with mathematics. | 7, 20, 25, 26, 33, 40, 43, 48, 50 |

---

42. Never; the graph never crosses the $x$-axis because the powers of $b$ are always positive and $a \neq 0$. Thus $ab^x$ is never 0.

45. Sample answer: First, look for a pattern by making sure that the domain values are at regular intervals and the range values differ by a common factor.

434 | Lesson 7-5 | Exponential Functions

## Lesson 7-5 | Exponential Functions

### Preparing for Assessment

**46.** Which of the following functions represents the data in the table? **MP 1 F.LE.5 C**

| x | 0 | 1 | 2 | 3 |
|---|---|---|---|---|
| y | 1 | 0.25 | 0.0625 | 0.015625 |

- A  $y = 0.25x + 1$
- B  $y = x - 0.75$
- C  $y = 0.25^x$
- D  $y = -0.25^x$

**47.** Which of following gives the equation and y-intercept of the graph shown? **MP 1 F.LE.5 D**

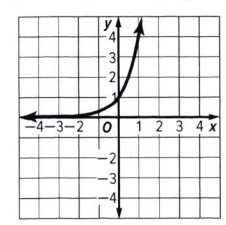

- A  $y = \left(\frac{1}{4}\right)^x$; 0
- B  $y = \left(\frac{1}{4}\right)^x$; 1
- C  $y = 4^x$; 0
- D  $y = 4^x$; 1

**48.** The function $f(t) = 2(2.25)^t$ models the growth of bacteria cells, where $f(t)$ is the number of bacteria cells and $t$ is time in days. After 10 days, approximately how many bacteria cells are there? **MP 4 F.LE.5 D**

- A  45
- B  1139
- C  $3325^x$
- D  6651

**49.** Given the function $y = \left(\frac{2}{3}\right)^x - 6$, **MP 1 F.LE.5**

a. What is the domain? **A**
- A  $-\infty < x < \infty$
- B  $-6 < x < 0$
- C  $-\infty < x < -6$
- D  $-6 < x < \infty$

b. What is the range? **D**
- A  $-\infty < y < \infty$
- B  $0 < y < \infty$
- C  $-\infty < y < -6$
- D  $-6 < y < \infty$

**50a.** 500; Sample answer: The patient initially received 500 mg of the medicine.

**50.** MULTI-STEP The concentration of a certain medicine in a patient's body is modeled by $f(x) = 500(0.85)^x$, where $f(x)$ is the medicine in milligrams in the patient's body after $x$ hours. **MP 1,4 F.LE.5, F.IF.7e**

a. Find the y-intercept and describe its meaning in the context of the situation.

b. In the context of the situation, what is the domain? **A**
- A  $x \geq 0$
- B  $x \geq 0.85$
- C  $0 < x \leq 500$
- D  all real numbers

c. In the context of the situation, what is the range? **C**
- A  $f(x) \geq 0$
- B  $f(x) \leq 500$
- C  $0 < f(x) \leq 500$
- D  all real numbers

d. Sketch the graph of the function with the appropriate labels and scales for the graph.

See Ch. 7 Answer Appendix.

---

### Preparing for Assessment

Exercises 46–50 require students to use the skills they will need on standardized assessments. Each exercise is dual-coded with content and process standards.

| Dual Coding |||
|---|---|---|
| Items | Content Standards | MP Mathematical Practices |
| 46 | F.LE.5 | 1 |
| 47 | F.LE.5, F.IF.7e | 1 |
| 48 | F.LE.5 | 1, 4 |
| 49 | F.LE.5 | 1 |
| 50 | F.LE.5, F.IF.7e | 1, 4 |

### Diagnose Student Errors

Survey student responses for each item. Class trends may indicate common errors and misconceptions.

**46.**

| A | Used a linear equation |
|---|---|
| B | Wrote a linear equation using (0, 1) and (1, 0.25) |
| C | CORRECT |
| D | Made 0.25 negative for the decreasing function |

**49a.**

| A | CORRECT |
|---|---|
| B | Used the x-coordinate of the y-intercept and the y-intercept |
| C | Used the reversed range values |
| D | Used the range values |

**49b.**

| A | Used the domain for the function |
|---|---|
| B | Used the range of the standard exponential function |
| C | Used the reversed range values |
| D | CORRECT |

---

### Differentiated Instruction OL BL

**Extension** Give students this scenario: a wise man asked his ruler to provide rice for his people. The wise man asked the ruler to give him 2 grains of rice for the first square on a chess board, 4 grains for the second, and so on, doubling the amount of rice with each square of the board.

Ask:
- How many grains of rice will the wise man receive for the 64th square on the chessboard? $2^{64}$ or about $1.84 \times 10^{19}$ grains
- If one pound of rice has approximately 24,000 grains, how many tons of rice will the wise man receive on the last day? (Hint: 1 ton = 2000 pounds) about $3.84 \times 10^{11}$ tons

### Go Online!

**Quizzes**
Students can use Self-Check Quizzes to check their understanding of this lesson.

connectED.mcgraw-hill.com  435

Extend 7-5

# Launch

**Objective** Prove that exponential functions grow by equal factors over equal intervals.

# Teach

**Working in Cooperative Groups** Have students work in groups of two or three, mixing abilities, to complete Activities 1–3 and Analyze the Results 1.

**Teaching Tip** It is important for students to understand that the equal factors over equal intervals can be found using three points on the function.

# Assess

**Formative Assessment** Use Analyze the Results 2–4 to assess whether students understand that exponential functions grow by equal factors over equal intervals.

---

**EXTEND 7-5**

Algebra Lab
## Exponential Growth Patterns

An exponential function grows in a predictable way. Given the equation of an exponential function, you can predict how the dependent variable will change over an interval of the independent variable.

**Mathematical Processes**
MP 8 Look for and express regularity in repeated reasoning.

**Content Standards**
F.LE.1a Prove that exponential functions grow by equal factors over equal intervals.

Let A, B, and C be any three points on the curve $y = ab^x$. You will explore the pattern for each of the intervals.

### Activity 1  $y = 2^x$

Work cooperatively. Explore the growth of $y = 2^x$ over $x$- and $y$-intervals.

**1A.** Complete the table below.

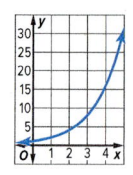

| Point | x | y |
|---|---|---|
| A | 2 | 4 |
| B | 3 | 8 |
| C | 4 | 16 |

**1B.** What is the relationship between the $x$- and $y$-values in the table?
The $x$-values have a common difference of 1 while the $y$-values increase by a factor of 2.

**1C.** Make a prediction about the $y$-values of points on the graph for $x$-values of 5, 6, and 7.
The $y$-values will continue to differ by a factor of 2.

**1D.** Check your prediction. Are you correct? Yes

**2C.** The $y$-values on each graph increase by a factor of 2 for each increase of 1 in the $x$-values. The $y$-values on the graph for Activity 2 are twice as great as those for the corresponding $x$-values in Activity 1.

### Activity 2  $y = 2(2)^x$

Work cooperatively. Explore the growth of $y = 2(2)^x$ over $x$- and $y$-intervals.

**2A.** Complete the table below.

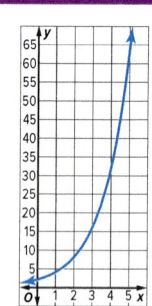

| Point | x | y |
|---|---|---|
| A | 2 | 8 |
| B | 3 | 16 |
| C | 4 | 32 |

**2B.** What is the relationship between the intervals for the $x$- and $y$-values?
The $x$-values differ by 1 while the $y$-values increase by a factor of 2.

**2C.** Compare and constrast the graphs from Activity 1 and Activity 2.

---

**Go Online!**

The most up-to-date resources available for your program can be found at **connectED.mcgraw-hill.com**.

436 | Extend 7-5 | Algebra Lab Exponential Growth Patterns

# Extend 7-5

### Activity 3  $y = 0.5(2)^x$

Work cooperatively. Explore the growth of $y = 0.5(2)^x$ over x- and y-intervals.

**3A.** Complete the table below.

| Point | x | y |
|---|---|---|
| A | 2 | 2 |
| B | 3 | 4 |
| C | 4 | 8 |

**3B.** What is the relationship between these intervals?
*The x-values differ by an interval of 1 while the y-values increase by an equal factor of 2.*

**3C.** Compare the graph from Activity 1 to the graph from Activity 3. What is a similarity? What is a difference?

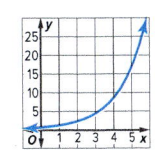

*3C. The similarity is that the y-values are increasing by an equal factor of 2. The difference is that the y-values are twice as great for the graph of Activity 1.*

### Analyze the Results
Work cooperatively.

1. Consider any points for the graph $y = b^x$.
   a. Suppose the first x-coordinate is n. If the length of the interval along the x-axis is 1, what is the x-coordinate of the second point? *n + 1*
   b. To find the first y-coordinate, substitute the x-coordinate of the first point into $y = b^x$. What is the y-coordinate? $y = b^n$
   c. To find the second y-coordinate, substitute the x-coordinate of the second point into $y = b^x$. What is the y-coordinate? $y = b^{n+1}$
   d. To find the ratio of the y-values, divide the two y-coordinates. What is the result? $\frac{b^{n+1}}{b^n} = b$
   e. Repeat this procedure for consecutive values of x such as $n + 2$, $n + 3$, and so on. What is your conclusion?
   
   *1e. Sample answer: The values of y would then be $b^{n+2}$, and $b^{n+3}$. The ratio between any consecutive y-values will then be b, so for equal intervals of x, the y-values grow by equal factors of b.*
   
   f. Repeat this procedure for a first x-coordinate of n and intervals of length m. How does the result prove that exponential functions grow by equal factors over equal intervals? *See margin.*

2. Examine the function $y = 3^x$.
   a. For an increase of 1 in the x-values, describe the change in the y-value. *increase by a factor of 3*
   b. If the x-value increases from 3 to 4, how do the corresponding y-values change? *It will increase by a factor of 3.*
   c. If the x-value increases from 3 to 6, how do the corresponding y-values change? *2c. See margin.*

3. Examine the function $y = \left(\frac{1}{3}\right)^x$.
   a. For an increase of 1 in the x-values, describe the change in the y-value. *3a. It will decrease by a factor of $\frac{1}{3}$.*
   b. If the x-value increases from 3 to 4, how do the corresponding y-values change? *3b. They decrease by a factor of $\frac{1}{3}$ from $\frac{1}{27}$ to $\frac{1}{81}$.*
   c. If the x-value increases from 3 to 6, how do the corresponding y-values change? *3 to 6 is three intervals of 1. That is why the y-values change by a factor $\left(\frac{1}{3}\right)^3$ or $\frac{1}{27}$ from $\frac{1}{27}$ to $\frac{1}{729}$.*

4. **CONSTRUCT ARGUMENTS** Is the function represented in the table an exponential function? Justify your answer by discussing intervals and differences. *No, for equal intervals of x, the y-values do not change by equal factors.*

| x | −3 | −2 | −1 | 0 | 1 | 2 | 3 |
|---|---|---|---|---|---|---|---|
| y | −8 | −5 | −2 | 1 | 4 | 7 | 10 |

## Additional Answers

**1f.** Sample answer: The x-values would, then, be $n$, $n + m$, $n + 2m$, $n + 3m$ and so on. The y-values would be $b^n$, $b^{n+m}$, $b^{n+2m}$, $b^{n+3m}$, and so on. It shows that for any two points with x-values that have a difference of m, the y-values would grow by factors of $b^m$.

**2c.** 3 to 6 is three intervals of 1. That is why the y-values change by a factor of $3^3$ or 27.

# LESSON 7-6
# Transformations of Exponential Functions

**SUGGESTED PACING (DAYS)**
| 90 min. | 1 |
| 45 min. | 1 |
Instruction

## Track Your Progress

### Objectives

1. Identify the effects on the graphs of exponential functions by replacing $f(x)$ with $f(x) + k$ and $f(x - h)$ for positive and negative values.

2. Identify the effect on the graphs of exponential functions by replacing $f(x)$ with $af(x)$, $f(ax)$, $-af(x)$ and $f(-ax)$.

### Mathematical Background
The form of an exponential function is $g(x) = ab^{(x-h)} + k$, where $a$, $h$, and $k$ are parameters. Changing the value of these parameters can cause the graph to be compressed, reflected across an axis, or translated left, right, up, or down. In this lesson, you will see how changing these parameters will affect the graph of the parent function.

### THEN

**F.IF.7e** Graph exponential and logarithmic functions, showing intercepts and end behavior, and trigonometric functions, showing period, midline, and amplitude.

**N.RN.2** Rewrite expressions involving radicals and rational exponents using the properties of exponents.

### NOW

**F.IF.7e** Graph exponential and logarithmic functions, showing intercepts and end behavior, and trigonometric functions, showing period, midline, and amplitude.

**F.BF.3** Identify the effect on the graph of replacing $f(x)$ by $f(x) + k$, $k f(x)$, $f(kx)$, and $f(x + k)$ for specific values of $k$ (both positive and negative); find the value of $k$ given the graphs. Experiment with cases and illustrate an explanation of the effects on the graph using technology.

**F.LE.5** Interpret the parameters in a linear or exponential function in terms of a context.

### NEXT

**F.LE.2** Construct linear and exponential functions, including arithmetic and geometric sequences, given a graph, a description of a relationship, or two input-output pairs (include reading these from a table).

**F.IF.8b** Use the properties of exponents to interpret expressions for exponential functions. For example, identify percent rate of change in functions such as $y = (1.02)^t$, $y = (0.97)^t$, $y = (1.01)12^t$, $y = (1.2)^{t}/10$, and classify them as representing exponential growth or decay.

## Go Online! All of these resources and more are available at connectED.mcgraw-hill.com

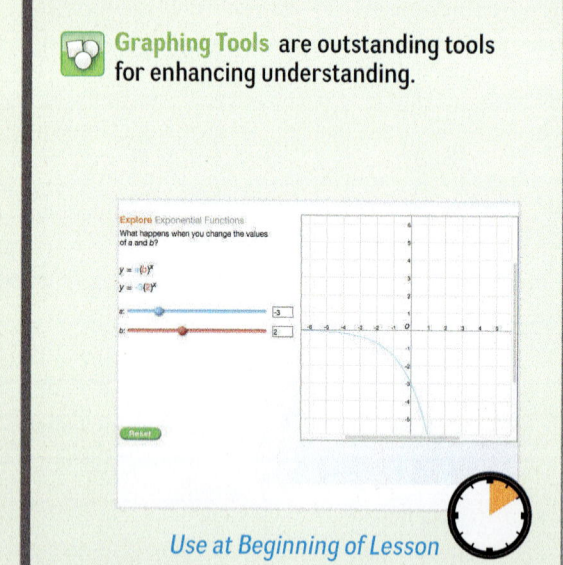

**Graphing Tools** are outstanding tools for enhancing understanding.
*Use at Beginning of Lesson*

**eLessons** utilize the power of your interactive whiteboard in an engaging way. Use **Exponential Functions**, screens 8–9, to introduce the concepts in this lesson. 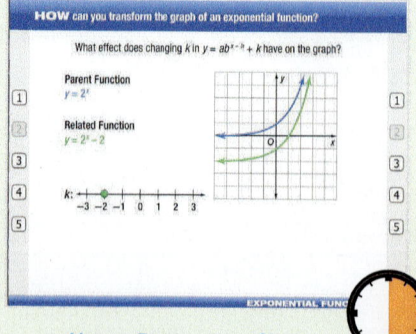
*Use at Beginning of Lesson*

Use **The Geometer's Sketchpad** to transform exponential functions.
*Use with Examples*

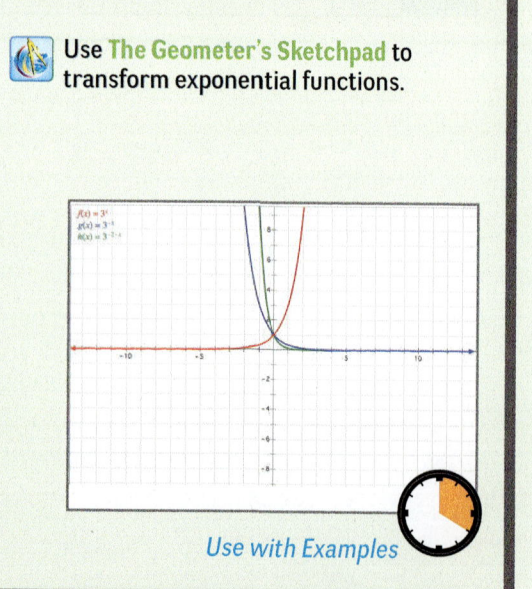

## OER Using Open Educational Resources
**Publishing** Have students work in groups to create a wiki on **Wikispaces** with everything they know about transformations of exponential functions. Then have groups review each other's wikis and post comments. *Use as homework or classwork*

438A | Lesson 7-6 | Transformations of Exponential Functions

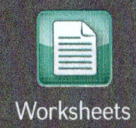

# Differentiate Your Resources

**Extra Practice** Additional practice or homework; Skills Practice is best for approaching-level students and Practice is best for on-level and beyond-level students

## Skills Practice
AL  OL  ELL

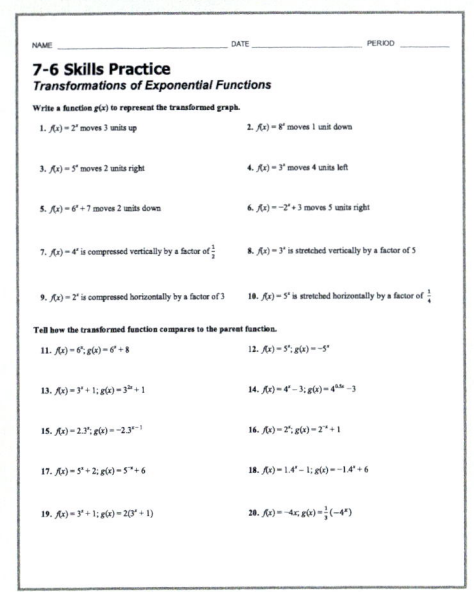

## Practice
AL  OL  BL  ELL

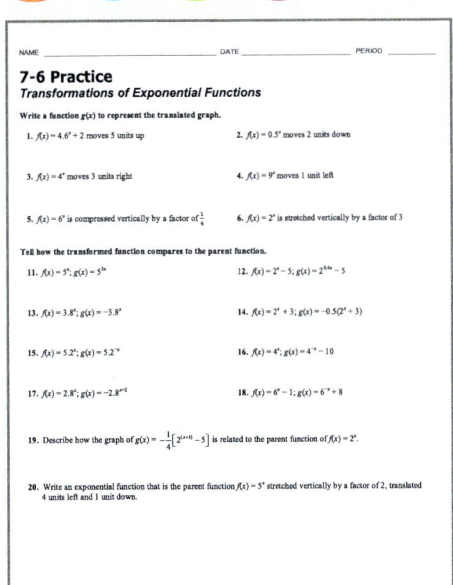

## Word Problem Practice
AL  OL  BL  ELL

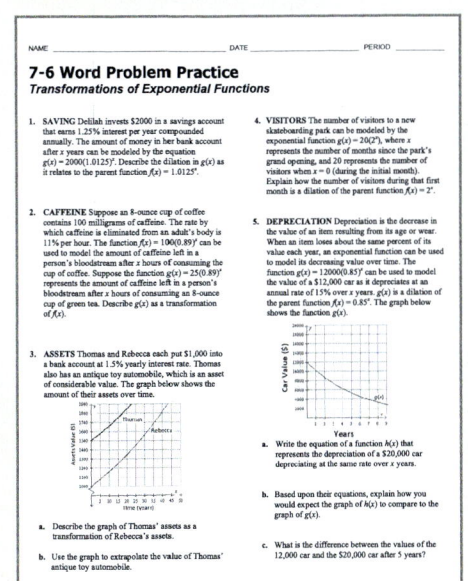

**Intervention** Reteaching and vocabulary activities that can be used with struggling or absent students and as ELL support

## Study Guide and Intervention
AL  OL  ELL

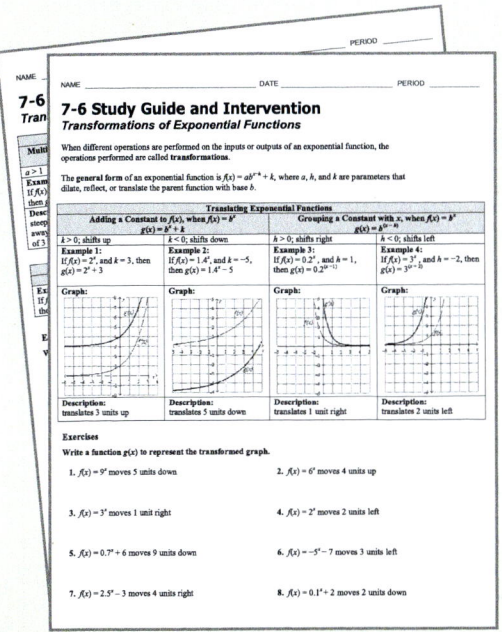

## Study Notebook
AL  OL  BL  ELL

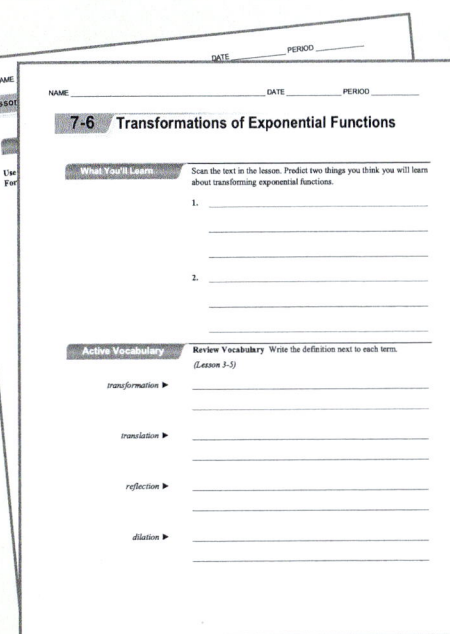

**Extension** Activities that can be used to extend lesson concepts

## Enrichment
OL  BL  ELL

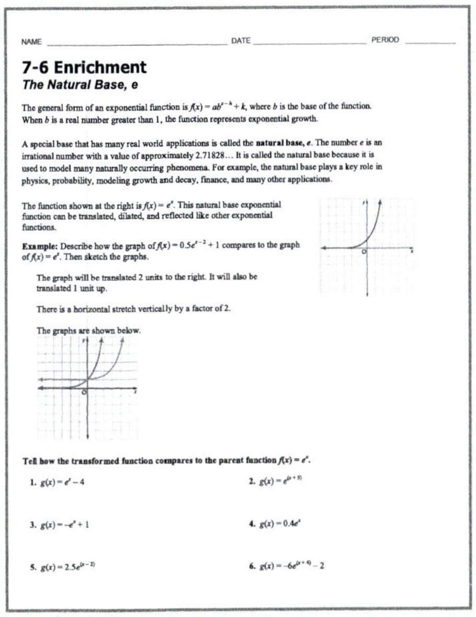

Lesson 7-6 | Transformations of Exponential Functions

# Launch

Have students read the Why? section of the lesson.

- Why do you think an exponential function is better than a linear function for the design of a half-pipe? **A line has the same slope at every point; an exponential function has a slope that changes, creating a curve. Each side can start steeply at the top and level out in the middle.**

- How could transformations of exponential functions be useful in designing a half-pipe? **You could use transformations to experiment with different values for the height, length, and curvature in order to find the best combination.**

# Teach

Ask the scaffolded questions for each example to build conceptual understanding for students at all levels.

## 1 Translations of Exponential Functions

**Example 1  Translations of Exponential Functions**

**AL** In which direction does adding or subtracting a value of $h$ translate a graph from the parent function? **left or right** $k$? **up or down**

**OL** If $k$ is a negative number, which way does the translated graph move from the graph of the parent function? **down**

**BL** Will the graph of a vertically translated function ever intersect the graph of the parent function? **No. Whatever the value of $f(x)$, the value of $f(x) + k$ is different.**

**Need Another Example?**
Describe the translation in each function as it relates to the graph of $f(x) = 5^x$.
a. $g(x) = 5^x + 1$ **translated up 1 unit**
b. $g(x) = 5^{x-7}$ **translated right 7 units**

## Go Online!

**Interactive Whiteboard**
Use the *eLesson* or *Lesson Presentation* to present this lesson.

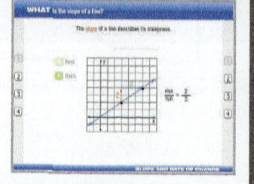

---

## LESSON 6
# Transformations of Exponential Functions

| Then | Now | Why? |
|---|---|---|
| • You defined, identified, and graphed exponential functions. | **1** Identify the effects on the graphs of exponential functions by replacing $f(x)$ with $f(x) + k$ and $f(x - h)$ for positive and negative values of $h$ and $k$. <br><br> **2** Identify the effects on the graph of exponential functions by replacing $f(x)$ with $af(x)$ and $f(ax)$ with positive and negative values of $a$. | In some extreme sports, such as skating, skateboarding, snowboarding, and skiing, athletes use a half-pipe to gain speed and perform tricks. A half-pipe is essentially two ramps, or quarter-pipes, facing each other with an extended flat bottom. Each quarter-pipe can be modeled by an exponential function. When placed together to form a half-pipe, one ramp can be described as a reflection of the other. |

**Mathematical Practices**
2 Reason abstractly and quantitatively.
4 Model with mathematics.

**Content Standards**
**F.IF.7e** Graph exponential and logarithmic functions, showing intercepts and end behavior, and trigonometric functions, showing period, midline, and amplitude.
**F.BF.3** Identify the effect on the graph of replacing $f(x)$ by $f(x) + k$, $kf(x)$, $f(kx)$, and $f(x + k)$ for specific values of $k$ (both positive and negative); find the value of $k$ given the graphs. Experiment with cases and illustrate an explanation of the efects of the graph using technology.
**F.LE.5** Interpret the parameters in a linear or exponential function in terms of a context.

## 1 Translations of Exponential Functions
The general form of an exponential function is $g(x) = ab^{x-h} + k$, where $a$, $h$, and $k$ are parameters that dilate, reflect, and translate a parent function $f(x) = b^x$. Recall that a translation moves a graph up, down, left, right, or in two directions.

The graph of the function $g(x) = b^x + k$ is the graph of $f(x)$ translated vertically. The graph of the function $g(x) = b^{x-h}$ is the graph of $f(x)$ translated horizontally.

### Key Concept  Translations of Exponential Functions

**Vertical Translations**

For $f(x) + k$:
- if $k > 0$, the graph of $f(x)$ is translated $k$ units up.
- if $k < 0$, the graph of $f(x)$ is translated $|k|$ units down.

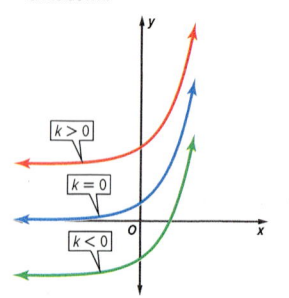

Every point on the graph of $f(x)$ moves vertically $k$ units.

**Horizontal Translations**

For $f(x - h)$:
- if $h > 0$, the graph of $f(x)$ is translated $h$ units right.
- if $h < 0$, the graph of $f(x)$ is translated $|h|$ units left.

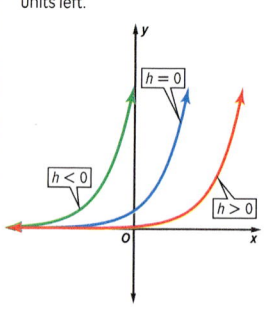

Every point on the graph of $f(x)$ moves horizontally $h$ units.

---

## Mathematical Practices Strategies

**Structure** Help students recognize patterns in exponential functions and their transformations. Ask:

- In which quadrants is the graph of $f(x) = b^x$ for any positive value of $b$? Explain. **Quadrants I and II; a positive number raised to a power must be positive; as $x$ gets smaller, $f(x)$ approaches but never reaches 0.**

- If $f(x) = b^x$ is multiplied by a negative number, in which quadrants will the graph be? Explain. **Quadrants III and IV; for any value of $x$, the $y$-value will be negative.**

- Why is the $y$-intercept a good reference point when determining a translation? **All graphs in the form $y = b^x$ have a $y$-intercept of 1.**

- Throughout the lesson, ask: How is the graph of this function related to the graph of the parent function?

438 | Lesson 7-6 | Transformations of Exponential Functions

## Lesson 7-6 | Transformations of Exponential Functions

**Study Tip**

**Structure** When an exponential function is translated, the function slides horizontally or vertically, but the shape of the curve stays the same.

### Example 1  Translations of Exponential Functions

Describe the translation in each function as it relates to the graph of $f(x) = 2^x$.

**a.** $g(x) = 2^x - 4$

The graph of $g(x) = 2^x - 4$ is a translation of the graph of $f(x) = 2^x$ down 4 units.

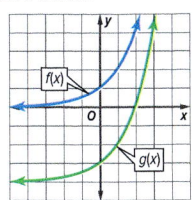

**b.** $q(x) = 2^{x+3}$

The graph of $q(x) = 2^{x+3}$ is a translation of the graph of $f(x) = 2^x$ left 3 units.

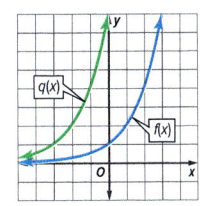

▶ **Guided Practice**

Describe the translation in each function as it relates to the graph of $f(x) = 4^x$.

**1A.** $g(x) = 4^x + 3$
translated up 3 units

**2B.** $q(x) = 4^{x-5}$
translated right 5 units

You can compare a graph to the parent function to write a function for the graph.

### Example 2  Analyze a Graph

The graph $g(x)$ is a translation of the parent function $f(x) = \left(\frac{1}{2}\right)^x$. Use the graph of the function to write an equation for $g(x)$.

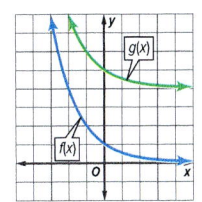

Notice that the horizontal asymptote of $g(x)$ is different than the horizontal asymptote of $f(x)$, implying a vertical translation of the form $g(x) = \left(\frac{1}{2}\right)^x + k$.

The parent graph has a $y$-intercept at $(0, 1)$ and the translated graph has a $y$-intercept at $(0, 5)$. The $y$-intercept is shifted 4 units up, so $k = 4$.

The graph is represented by $g(x) = \left(\frac{1}{2}\right)^x + 4$.

**Study Tip**

**Translations** Any exponential parent function has a $y$-intercept at $(0, 1)$ and an asymptote of $y = 0$. Examining the translation of these features will help you determine the values of $h$ and $k$.

▶ **Guided Practice**

2. The graph $g(x)$ is a translation of the parent function $f(x) = 1.5^x$. Use the graph of the function to write an equation for $g(x)$. $y = 1.5^{x-3}$

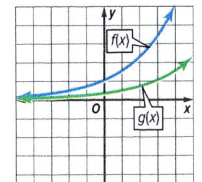

## Teaching Tip

**Horizontal Translations** It can be difficult for students to understand why adding a positive constant to $f(x)$ moves the function vertically in a positive direction, but adding a positive constant to $x$ moves the function horizontally in a *negative* direction. Help them to see that for a horizontal translation, the exponent for the general equation of an exponential function is $x - h$. So, an exponent of $x + h$ is really subtracting a negative $h$-value.

### Teaching the Mathematical Practices

**Sense-Making** Mathematically proficient students look for entry points to the solution of a problem and consider analogous problems to gain insight into the solution. In the case of transformations of exponential functions, students should apply what they know about transformations of linear functions, and look for similarities between the two.

### Example 2  Analyze a Graph

**AL** What is the asymptote of $g(x)$? $y = 4$

**OL** In a vertical translation, how can you use the values of the $y$-intercepts of $f(x)$ and $g(x)$ to find $k$? The difference between the $y$-intercepts is $k$.

**BL** What key features should you analyze in order to determine the equation of a translated function? the $y$-intercept and the asymptote

### Need Another Example?

The graph $g(x)$ is a translation of the parent function $f(x) = 3^x$. Use the graph of the function to write an equation for $g(x)$.
$g(x) = 3^x - 4$

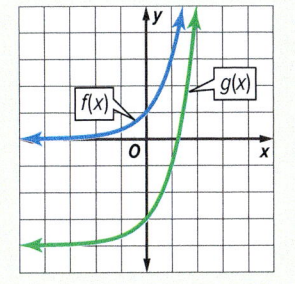

**Lesson 7-6** | Transformations of Exponential Functions

## 2 Dilations and Reflections

### Example 3 Analyze a Function

**AL** For what values of $a$ does the graph of $g(x) = ab^x$ rise more steeply than the graph of $f(x) = b^x$? $a > 1$

**OL** What values of $a$ cause the graph to be compressed vertically? $0 < a < 1$

**BL** What happens to the graph as the value of $a$ gets closer 0 for $g(x) = ab^x$? The graph becomes more vertically compressed, until finally it is almost a straight line. When $a = 0$, the equation is $g(x) = 0 \cdot b^x = 0$ and the graph is a straight line.

### Need Another Example?

**CARS** In 2015, Daeshawn purchased a new car for $27,500. The value of his car over time is modeled by $g(x) = 27,500(0.92)^x$, where $x$ is the number of years since he purchased the car. Describe the dilation in $g(x)$ as it relates to the parent function $f(x) = 0.92^x$. The graph of $f(x)$ is stretched vertically by a factor of 27,500.

### Teaching Tip

**Structure** To help students see more clearly the effects of dilations on exponential functions, encourage them to think of extreme cases. For example, if you multiply an exponential function by 100, the graph will increase very rapidly. Conversely, if you multiply the function by 0.001, the graph will increase much slower.

---

**2 Dilations and Reflections** Recall that a dilation stretches or compresses the graph of a function. A vertical dilation occurs when the parent function $f(x)$ is multiplied by a positive constant $a$ after the function is evaluated. The graph of $g(x) = a \cdot f(x)$ is the graph of $f(x)$ stretched or compressed vertically by a factor of $|a|$.

A horizontal dilation occurs when the parent function is multiplied by a constant $a$ before the function is evaluated. The graph of $g(x) = f(a \cdot x)$ is the graph of $f(x)$ stretched or compressed horizontally by a factor of $\frac{1}{|a|}$.

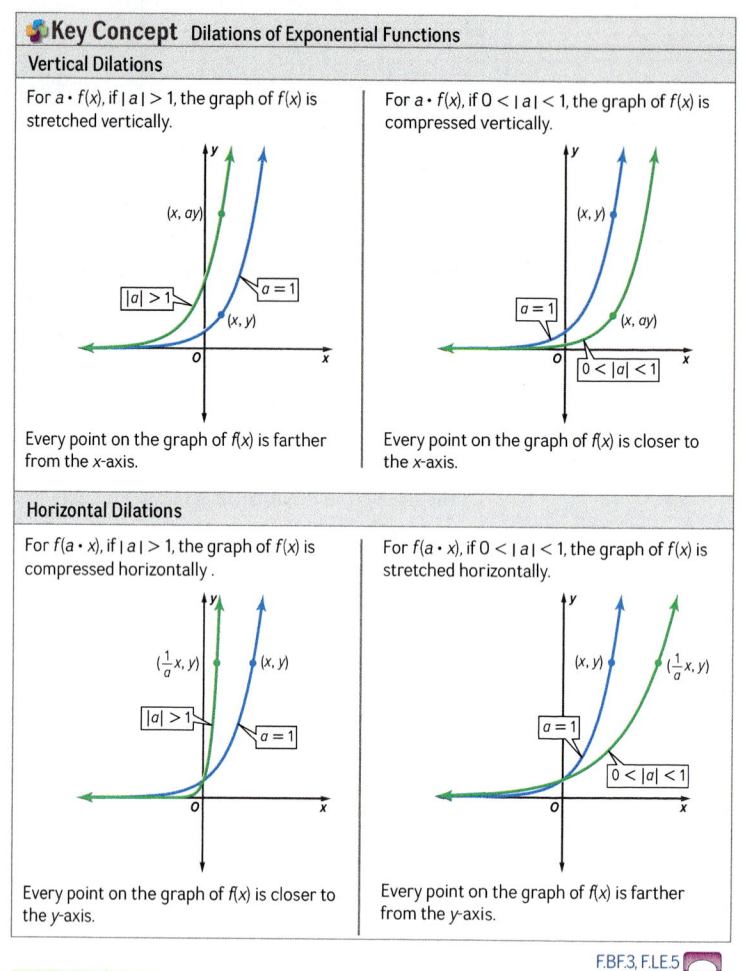

**Key Concept** Dilations of Exponential Functions

**Vertical Dilations**

For $a \cdot f(x)$, if $|a| > 1$, the graph of $f(x)$ is stretched vertically.

For $a \cdot f(x)$, if $0 < |a| < 1$, the graph of $f(x)$ is compressed vertically.

Every point on the graph of $f(x)$ is farther from the $x$-axis.

Every point on the graph of $f(x)$ is closer to the $x$-axis.

**Horizontal Dilations**

For $f(a \cdot x)$, if $|a| > 1$, the graph of $f(x)$ is compressed horizontally.

For $f(a \cdot x)$, if $0 < |a| < 1$, the graph of $f(x)$ is stretched horizontally.

Every point on the graph of $f(x)$ is closer to the $y$-axis.

Every point on the graph of $f(x)$ is farther from the $y$-axis.

**Study Tip**

**MP Reasoning** Stretch and compression dilations are related. The graph of a horizontal stretch may appear the same as a vertical compression.

**Real-World Example 3** Analyze a Function

**SOLAR ENERGY** Photovoltaic cells, or PV cells, convert solar energy into electricity. Since 2000, global solar PV capacity has been growing at a rate that can be approximated by the function $c(x) = 0.897(1.46)^x$, where $c(x)$ is the solar PV capacity in gigawatts, $x$ is the number of years since 2000, and 0.897 is the initial capacity. Describe the dilation in $c(x) = 0.897(1.46)^x$ as it is related to the parent function $f(x) = 1.46^x$.

F.BF.3, F.LE.5

Lesson 7-6 | Transformations of Exponential Functions

The function is multiplied by the positive constant $a$ after it has been evaluated and $|a|$ is between 0 and 1, so the graph of $f(x) = 1.46^x$ is compressed vertically by a factor of $|a|$, or 0.897.

▶ **Guided Practice**

3. Because $f(x) = 1.5^x$, $a = 50$. The graph of $f(x)$ is stretched vertically by a factor of 50.

3. **FISH** Suppose the population of a certain fish in a new pond can be modeled by $p(x) = 50(1.5)^x$, where $x$ is the number of generations after an initial stock of 50 fish were added to the pond. Describe the dilation in $p(x) = 50(1.5)^x$ as it relates to the parent function $p(x) = 1.5^x$.

When $a$ is negative, the graph of $a \cdot f(x)$ is reflected across the $x$-axis, while the graph of $f(a \cdot x)$ is reflected across the $y$-axis.

**Key Concept** Reflections of Linear Functions

| Reflection across $x$-axis | Reflection across $y$-axis |
|---|---|
| For $a \cdot f(x)$, if $a < 0$, the graph of $f(x)$ is reflected across the $x$-axis. | For $f(a \cdot x)$, if $a < 0$, the graph of $f(x)$ is reflected across the $y$-axis. |

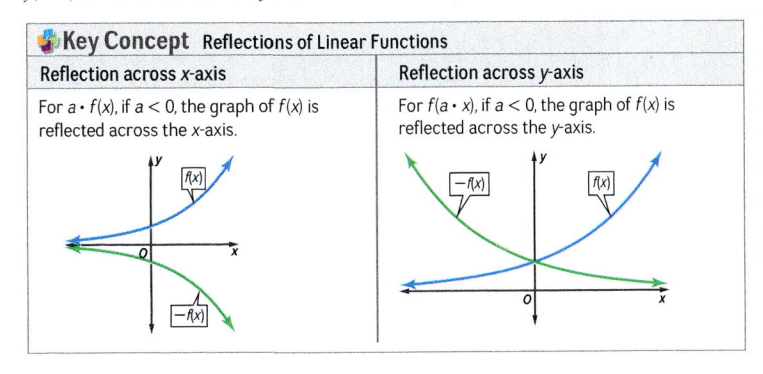

When $a$ is negative and $a \neq -1$, the graphs of $a \cdot f(x)$ and $f(a \cdot x)$ are both reflections and dilations.

F.BF.3

**Example 4** Reflections and Dilations of Exponential Functions

Describe the transformations in each function as it relates to the graph of $f(x) = 3^x$.

**a.** $g(x) = 3^{-4x}$

Because $a$ is negative in $g(x)$, the graph of $f(x)$ is reflected across the $y$-axis. Because $|a| > 1$, the graph is also compressed horizontally.

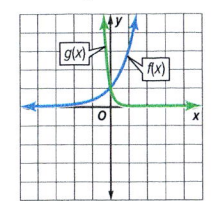

**b.** $q(x) = -\frac{1}{5}(3)^x$

Because $a$ is negative in $q(x)$, the graph of $f(x)$ is reflected across the $x$-axis. Because $0 < |a| < 1$, the graph is also compressed vertically.

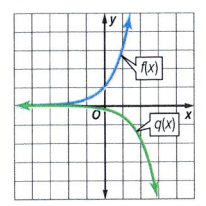

**Study Tip**
**Reflections** Determine whether multiplying by −1 occurs before or after the parent function is evaluated. Multiplying by −1 before the parent function is evaluated results in a reflection across the $y$-axis.

**Example 4 Reflections and Dilations of Exponential Functions**

**AL** Describe the graph when $a$ is negative. It is reflective across the $x$- or $y$-axis.

**OL** Why does multiplying $x$ by −1 result in a reflection across the $y$-axis? Every value $x$ becomes $-x$.

**BL** Suppose $0 < b < 1$, in which quadrants would $g(x) = -ab^x$ lie? Quadrants III and IV

**Need Another Example?**

Describe the transformation in each function as it related to the graph of $f(x) = \left(\frac{1}{3}\right)^x$.

**a.** $g(x) = \left(\frac{1}{3}\right)^{-0.5x}$ reflected across the $y$-axis and stretched horizontally

**b.** $g(x) = -10\left(\frac{1}{3}\right)^x$ reflected across the $x$-axis and stretched vertically

connectED.mcgraw-hill.com 441

# Lesson 7-6 | Transformations of Exponential Functions

## Example 5 Analyze Transformations

**AL** What happens to the graph when it is multiplied by a negative constant? **It is reflected across an axis.**

**OL** Describe how $g(x)$ would be different if $a = -\frac{1}{3}$. **$g(x)$ would be compressed vertically instead of stretched.**

**BL** Compare the key features of $f(x)$ and $g(x)$. **The $y$-intercept of $f(x)$ is 1 and the $y$-intercept of $g(x)$ is 3.5. $f(x)$ is increasing as $x \to \infty$, while $g(x)$ is decreasing as $x \to \infty$. $f(x)$ is positive for all values of $x$. $g(x)$ is positive when $x < 1.74$ and negative when $x > 1.74$.**

**Need Another Example?**
Describe the transformations in $g(x) = 0.7(2)^{x+4}$ as it relates to the graph of $f(x) = 2^x$. **compressed vertically and translated left 4 units**

## Example 6 Compare Graphs

**AL** How can you use a graphing calculator to check the transformations? **Graph each function and the parent function on the same coordinate plane. Make sure that the translations, dilations, and reflections match what you expected.**

**OL** How can you use the table function of a graphing calculator to compare the graphs? **You can use the table to find when $y = 0$ and $x = 0$ to determine the $x$- and $y$-intercepts. You can also use the table to determine whether the functions are increasing or decreasing as $x \to \infty$ and when the functions are positive and negative.**

**BL** Write $p(x)$ in terms of $f(x)$. **$p(x) = f(2x) - 3$**

**Need Another Example?**
Compare the graphs to the parent function $f(x) = 4^x$. Then, verify using technology.
$g(x) = (4)^{3x} - 2$, $p(x) = -3(4)^x + 2$, $q(x) = (4)^{x+3}$
**$g(x)$ is compressed horizontally and translated down 2 units.**

**$p(x)$ is reflected across the $x$-axis, stretched vertically, and translated up 2 units.**

**$q(x)$ is translated left 3 units.**

---

**Guided Practice** 4A reflected across $y$-axis and stretched horizontally  4B reflected across $x$-axis and stretched vertically

Describe the transformations in each function as it relates to the graph of $f(x) = 2^x$.

4A. $g(x) = 2^{-0.25x}$  4B. $q(x) = -1.75(2)^x$

Often more than one transformation is applied to a function. Examine the transformation to analyze how the graph is affected.

F.BF.3

### Example 5  Analyze Transformations

Describe the transformations in $g(x) = -3(2)^{x-1} + 5$ as it relates to the graph of $f(x) = 2^x$.

Because $f(x) = 2^x$, $g(x) = af(x - h) + k$, where $a = -3$, $h = 1$, and $k = 5$.

The graph of $f(x)$ is reflected across the $x$-axis because $a$ is negative. Because $|a| > 1$, the graph is stretched vertically. The graph is translated right 1 unit and up 5 units.

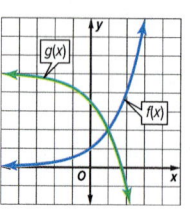

**Guided Practice**

5. Describe the transformations in $g(x) = -\left(\frac{3}{4}\right)^{x+5} - 4$ as it relates to the graph of $f(x) = \left(\frac{3}{4}\right)^x$. **reflected across $x$-axis, translated left 5 units and down 4 units**

You can examine a graph and compare it to the graph of the parent function to write a function for the graph.

F.IF.7e, F.BF.3

### Example 6  Compare Graphs

Compare the graphs to the parent function $f(x) = 2.5^x$. Then, verify using technology.

$g(x) = 0.5(2.5)^x + 1$, $p(x) = (2.5)^{2x} - 3$, $q(x) = -3(2.5)^{x+2} - 1$

$g(x)$ is compressed vertically by a factor of 0.5 and translated up 1 unit.

$p(x)$ is compressed horizontally by a factor of $\frac{1}{2}$ and translated down 3 units.

$q(x)$ is reflected across the $x$-axis, stretched vertically by a factor of 3, and translated left 2 units and down 1 unit.

To verify using a graphing calculator, enter each equation in the Y= list and graph. Use the features of the calculator to investigate the transformations. Note the change in asymptotes, $y$-intercepts, and relation to the axes.

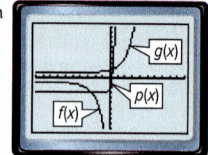

[−10, 10] scl: 1 by [−10, 10] scl: 1

**Watch Out!**
**Negative Values** Remember that the general form of an exponential function is $g(x) = a(b)^{x-h} + k$, so an exponent of $x + h$ indicates a negative value of $h$, which is a translation left.

**Guided Practice**

6. Compare the graphs to the parent function $f(x) = 3^x$. Then, verify using technology.

$g(x) = -4(3)^x$, $p(x) = 0.2(3)^{x+3} - 5$, $q(x) = (3)^{-x} + 2$  **See margin.**

## Additional Answers (Guided Practice)

6. $g(x)$ is reflected across $x$-axis and stretched vertically.

   $p(x)$ is compressed vertically and translated left 3 units and down 5 units.

   $q(x)$ is reflected across $y$-axis and translated up 2 units.

**Check Your Understanding**  = Step-by-Step Solutions begin on page R11.   Go Online! for a Self-Check Quiz

**Example 1**
F.BF.3
Describe the translation as it relates to the graph of $f(x) = 3^x$.

1. $g(x) = 3^x - 1$    translated down 1 unit
2. $g(x) = 3^{x-5}$    translated right 5 units
3. $g(x) = 3^{x+1}$    translated left 1 unit
4. $g(x) = 3^{x+2} - 4$    translated left 2 units and down 4 units

**Example 2**
F.BF.3
The graph $g(x)$ is a translation of the parent function $f(x) = 2^x$. Use the graph of the function to write an equation for $g(x)$.  5–6. See margin.

5.

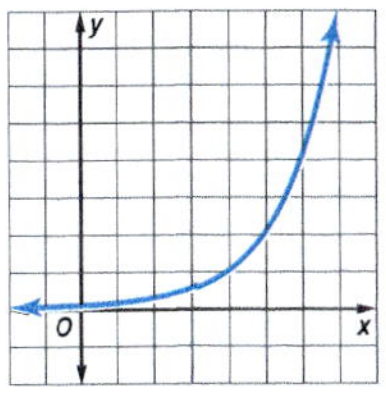

6.

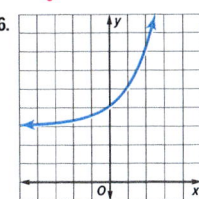

**Example 3**
F.BF.3, F.LE.5
7. **FINANCIAL LITERACY** Antonio invests $2200 in a savings account that earns 1.5% interest per year compounded annually. The amount of money in his bank account after $t$ years can be modeled by $g(t) = 2200(1.015)^t$. Describe the dilation in $g(t) = 2200(1.015)^t$ as it relates to the parent function $f(t) = 1.015^t$.    stretched vertically

**Example 4**
F.BF.3
Describe the transformations in each function as it relates to the graph of $f(x) = \left(\frac{1}{4}\right)^x$.

8. $g(x) = -\left(\frac{1}{4}\right)^x$    reflected across x-axis
9. $g(x) = \left(\frac{1}{4}\right)^{-3x}$    9–15. See margin.
10. $g(x) = -\frac{1}{8}\left(\frac{1}{4}\right)^x$
11. $g(x) = -9\left(\frac{1}{4}\right)^x$

**Examples 5, 6**
F.BF.3
Describe the transformations in each function as it relates to its parent function.

12. $g(x) = -2.7(3)^{x-8}$
13. $g(x) = \frac{2}{3}(5)^{-x} + 1$
14. $g(x) = -\left(\frac{4}{5}\right)^{x-2} - 6$
15. $g(x) = \left(\frac{5}{4}\right)^{4x} - 2$

**Practice and Problem Solving**   Extra Practice is found on page R7.

**Examples 1, 4**
F.BF.3
Describe the translation in each function as it relates to the parent function.

16. $g(x) = 4^x + 6$    translated up 6 units
17. $g(x) = \left(\frac{9}{5}\right)^{x-2}$    translated right 2 units
18. $g(x) = \left(\frac{2}{5}\right)^x - 1$    translated down 1 unit
19. $g(x) = 1.4^{0.3x}$    stretched horizontally
20. $g(x) = 1.5^{4x}$    compressed horizontally
21. $g(x) = 5(3)^x$    stretched vertically
22. $g(x) = 0.9^{3x}$    compressed horizontally
23. $g(x) = \frac{1}{3}\left(\frac{1}{5}\right)^x$    compressed vertically
24. $g(x) = 3^{-x}$    reflected across y-axis
25. $g(x) = -7^x$    reflected across x-axis

**Examples 5, 6**
F.BF.3
Describe the transformations in each function as it relates to the parent function.

26. $g(x) = 11^{x+2} + 8$
27. $g(x) = 0.5(2)^{x-1}$    26–35. See margin.
28. $g(x) = -3\left(\frac{1}{6}\right)^x + 2$
29. $g(x) = \left(\frac{2}{7}\right)^{-6x}$
30. $g(x) = -5^{x-8} + 4$
31. $g(x) = -\frac{1}{3}(2)^{x+3}$
32. $g(x) = 2(6.5)^{-x}$
33. $g(x) = 3^{5x} - 1$
34. $g(x) = -8\left(\frac{2}{3}\right)^{x+3} - 4$
35. $g(x) = -0.7(1.26)^{x+4} - 5$

**Differentiated Homework Options**

| Levels | AL Basic | OL Core | BL Advanced |
|---|---|---|---|
| Exercises | 16–44, 64–78 | 17–43 odd, 44, 45–57 odd, 62–78 | 62–70<br>71–78 (optional) |
| 2-Day Option | 17–43 odd, 71–78 | 16–44, 71–78 | |
|  | 16–44 even, 64–70 | 45–70 | |

---

**Lesson 7-6 | Transformations of Exponential Functions**

## Practice

**Formative Assessment** Use Exercises 1–15 to assess students' understanding of the concepts in the lesson.

The Practice and Problem Solving exercises assess the content taught in the lesson.

### Additional Answers

5. $g(x) = 2^{x-4}$
6. $g(x) = 2^x + 3$
9. reflected across y-axis and compressed horizontally
10. reflected across x-axis and compressed
11. reflected across x-axis and stretched vertically
12. reflected across x-axis, stretched vertically, and translated left 8 units
13. reflected across y-axis, compressed vertically, and translated up 1 unit
14. reflected across x-axis and translated right 2 units and down 6 units
15. compressed horizontally and translated down 2 units
26. translated left 2 units and down 8 units
27. compressed vertically and translated right 1 unit
28. reflected across x-axis, stretched vertically, and translated up 2 units
29. reflected across y-axis and compressed horizontally
30. reflected across x-axis and translated right 8 units and up 4 units
31. reflected across y-axis, compressed vertically, and translated left 3 units
32. reflected across x-axis and stretched vertically
33. compressed horizontally and translated down 1 unit
34. reflected across x-axis, stretched vertically, and translated left 3 units and down 4 units
35. reflected across x-axis, compressed vertically, and translated left 4 units and down 5 units

### Go Online!    eBook

**Interactive Student Guide**
Use the **Interactive Student Guide** to deepen conceptual understanding.
· Transforming Exponential Functions

ALGEBRA 1 INTERACTIVE STUDENT GUIDE

connectED.mcgraw-hill.com    443

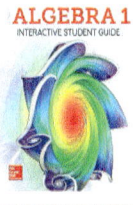

# Lesson 7-6 | Transformations of Exponential Functions

## Teaching Tip

**Analyze Relationships** Studying transformations of exponential equations can enhance students' ability to analyze graphs, discern relationships, and recognize how different operations performed on a function can change it. Challenge students to visualize about each transformation on the parent function as they describe it.

## Additional Answers

**44a.** Because $f(x) = 1.03^x$, $a = 130$. The graph of $f(x)$ is stretched vertically by a factor of 130.

**44b.** 95 pounds

**44c.** Both functions are stretched vertically. Since Tara started at a greater lifting weight, she adds more weight each week than Robert. The graph representing her weight is closer to the y-axis than Robert's graph.

**45.**

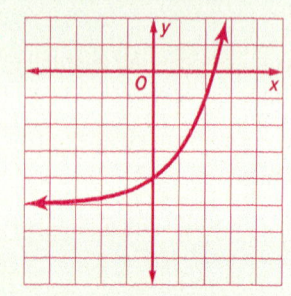

**46.**

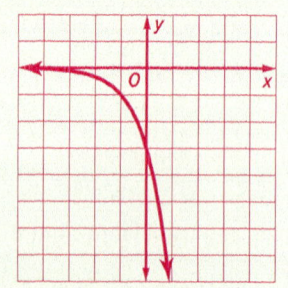

**47.**

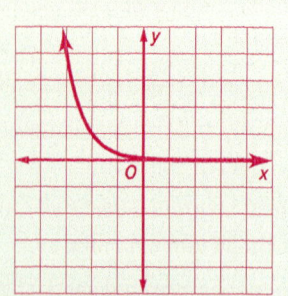

**48.**

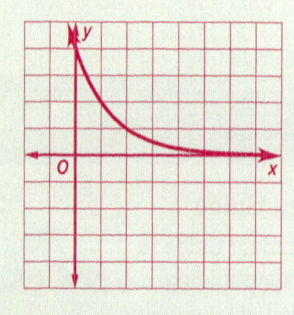

---

**Example 2** The graph of $g(x)$ is a transformation of the parent function $f(x) = \left(\frac{1}{2}\right)^x$. Find the value of $n$ to write an equation for $g(x)$.
F.BF.3

**36.** $g(x) = f(x) + n$  $-2$

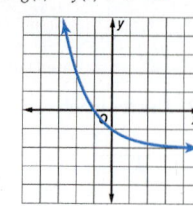

**37.** $g(x) = f(x + n)$  $-4$

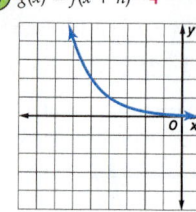

**38.** $g(x) = f(x) + n$  $1$

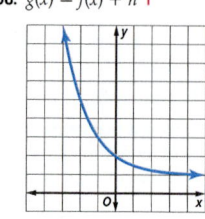

**39.** $g(x) = n \cdot f(x)$  $0.5$

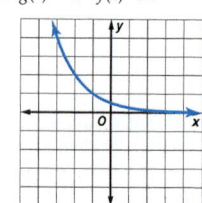

**40.** $g(x) = f(nx)$  $0.5$

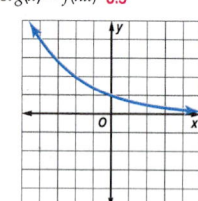

**41.** $g(x) = n \cdot f(x)$  $-3$

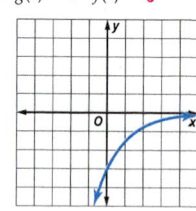

**42.** $g(x) = n \cdot f(x)$  $4$

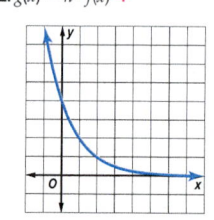

**43.** $g(x) = f(x) + n$  $3$

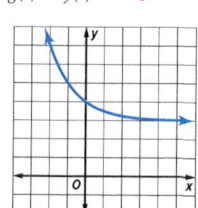

Examples 3, 6
F.BF.3, F.LE.5

**44. TRAINING** Tara can bench press 130 pounds. Each week of training, she wants to increase the weight by 3%. The function models the weight Tara can lift $w(x) = 130(1.03)^x$ where x represents the number of weeks of training. **See margin.**

a. Describe the transformation in $w(x)$ as it relates to the parent function $f(x) = 1.03x$.

b. Robert is also increasing the weight he lifts by 3% each week. The function $v(x) = 95(1.03)^x$ represents the weight he can lift after x weeks. What is his starting weight?

c. Compare $w(x)$ and $v(x)$.

### Levels of Complexity Chart

The levels of the exercises progress from 1 to 3, with Level 1 indicating the lowest level of complexity.

| Exercises | 16–44 | 45–64, 71–78 | 64–70 |
|---|---|---|---|
| ▶ Level 3 |  |  | ● |
| ▶ Level 2 |  | ● |  |
| Level 1 | ● |  |  |

444 | Lesson 7-6 | Transformations of Exponential Functions

**B** Use transformations to graph each function. 45–52. See margin.

45. $y = 2^x - 5$
46. $y = -3^{x+1}$
47. $y = \left(\frac{1}{4}\right)^{x+2}$
48. $y = 4(2)^{-x}$
49. $y = -0.5(4)^{x-3} + 1$
50. $y = 2^{0.5x} + 3$
51. $y = \left(\frac{2}{3}\right)^{4x}$
52. $y = -2(3)^x + 5$

Given the parent function $f(x) = \left(\frac{3}{2}\right)^x$ match each function to its graph.

$g(x) = -f(x) + 2$    $j(x) = -2f(x+2)$    $m(x) = f(x) - 2$
$n(x) = 5f(x)$    $p(x) = f(5x)$    $q(x) = \frac{1}{5}f(x)$

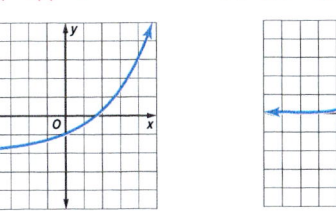

53. $m(x) = f(x) - 2$
54. $p(x) = f(5x)$ 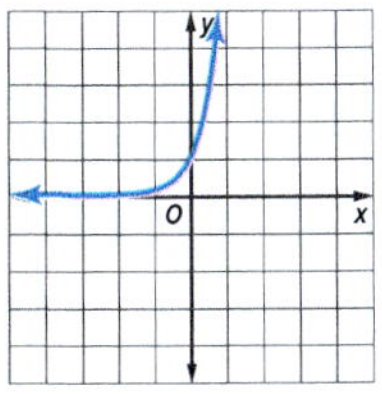
55. $j(x) = -2f(x+2)$

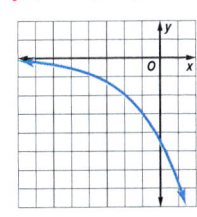

56. $q(x) = \frac{1}{5}f(x)$ 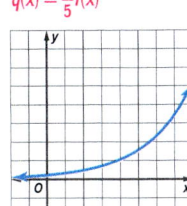
57. $n(x) = 5f(x)$ 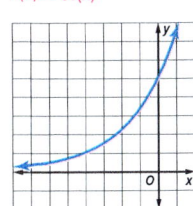
58. $g(x) = -f(x) + 2$

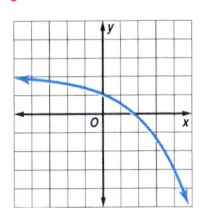

59. **MP REGULARITY** List the functions in order from the most vertically stretched to the least vertically stretched: $g(x) = 0.4(2)^x$, $j(x) = 4(2)^x$, $p(x) = \frac{1}{4}(2)^x$   $j(x), g(x), p(x)$

60. **MP TOOLS** Use a graphing calculator to compare the functions. 60. See margin.

$y = 3^x$      $y = 5(3)^x$      $y = \frac{1}{5}(3)^x$

a. Compare the graphs of $y = 3^x$ and $y = 5(3)^x$. Describe similarities and differences of the key features of the graphs.

b. Compare the graphs of $y = 3^x$ and $y = \frac{1}{5}(3)^x$. Describe similarities and differences of the key features of the graphs.

61. **MP TOOLS** Use a graphing calculator to compare the functions. 61. See margin.

$y = \left(\frac{2}{5}\right)^x$      $y = -\left(\frac{2}{5}\right)^x$      $y = \left(\frac{2}{5}\right)^{-x}$

a. Compare the graphs of $y = \left(\frac{2}{5}\right)^x$ and $y = -\left(\frac{2}{5}\right)^x$. Describe any similarities and differences you notice about the key features of the graphs.

b. Compare the graphs of $y = \left(\frac{2}{5}\right)^x$ and $y = \left(\frac{2}{5}\right)^{-x}$. Describe any similarities and differences you notice about the key features of the graphs.

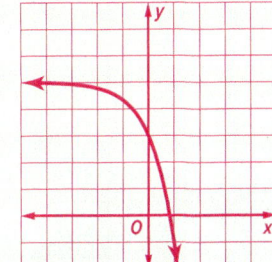

60a. They have the same asymptote and same end behavior. They are both increasing as $x \to \infty$. The y-intercept of $y = 3^x$ is 1. The function $y = 5(3)^x$ has a y-intercept of 5 and is stretched vertically.

b. They have the same asymptote and end behavior. They are both increasing as $x \to \infty$. The y-intercept of $y = 3^x$ is 1. The function $y = 5(3)^x$ has a y-intercept of 0.2 and is compressed vertically.

61a. They have the same asymptote and same end behavior as $x \to \infty$. For $y = \left(\frac{2}{5}\right)^x$, the function is decreasing as $x \to \infty$. The function $y = \left(\frac{2}{5}\right)^x$ is increasing as $x \to \infty$. The y-intercepts are 1 and −1, respectively.

b. They have the same asymptote and y-intercept. For $y = \left(\frac{2}{5}\right)^x$, the function is decreasing as $x \to \infty$. The function $y = \left(\frac{2}{5}\right)^{-x}$ is increasing as $x \to \infty$.

## Lesson 7-6 | Transformations of Exponential Functions

### Extra Practice
See page R7 for extra exercises for students who are approaching level or for on-level students who need additional reinforcement.

### MP Teaching the Mathematical Practices

**Tools** Mathematically proficient students can analyze graphs of functions generated using a graphing calculator. In Exercises 52 and 53, students can use a graphing calculator to verify the solutions they generated by first analyzing the values of a, h, and k of the exponential function.

### Additional Answers
49.
50.
51.
52.

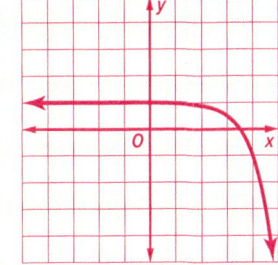

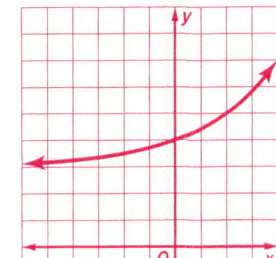

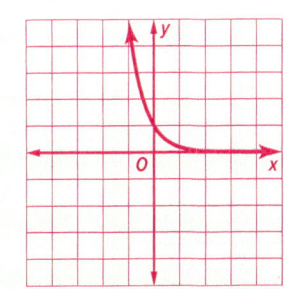

## Lesson 7-6 | Transformations of Exponential Functions

### Follow-Up

Students have explored the effects of performing operations on exponential functions. Ask students to summarize the transformations produced by performing various operations on the parent function. **Adding a constant to the function translates it up or down. Subtracting a constant to $x$ translates it left or right. Multiplying the function by a constant stretches or compresses it vertically. Multiplying $x$ by a constant stretches or compresses it horizontally. Multiplying the function by −1 reflects the graph across the $x$-axis. Multiplying $x$ by −1 reflects the graph across the $y$-axis.**

## Assess

**Ticket Out the Door** Write each of these transformations on a note card:

- Translate 3 units up.
- Translate 3 units down.
- Translate 3 units left.
- Translate 3 units right.
- Stretch vertically.
- Compress vertically.
- Stretch horizontally.
- Compress horizontally.
- Reflect over the $y$-axis.
- Reflect over the $x$-axis.

As students leave, give one card to each student and ask them to write a function for $g(x)$ that transforms the parent function $f(x) = b^x$ according to the instructions on the card.

### Additional Answers

**67.** Case 1: As $x$ decreases, $y$ still approaches 0, but as $x$ increases, $y$ goes to negative infinity instead of positive infinity. Case 2: As $x$ decreases, $y$ goes to negative infinity instead of positive infinity, but as $x$ increases, $y$ still approaches 0.

**68.** In each case, the end behavior switches. Case 1: As $x$ decreases, $y$ approaches infinity instead of 0, and as $x$ increases, $y$ approaches 0 instead of infinity. Case 2: As $x$ decreases, $y$ approaches 0 instead of infinity, and as $x$ increases, $y$ approaches infinity instead of 0.

### Go Online!

**eSolutions Manual**
Create worksheets, answer keys, and solutions handouts for your assignments.

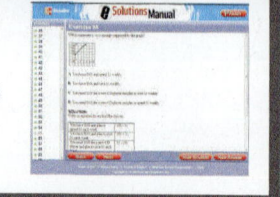

---

**62a.** While both amounts result in a vertically stretch, $P = 7500$ will cause the graph to be closer to the $y$-axis.

**63b.** The graph becomes more compressed and farther away from the $y$-axis.

**64.** (0, 1); Because $b^0 = 1$, any function in the form $f(x) = b^x$ will have a $y$-intercept of 1.

**65.** For any point $(x, y)$ of the function, the reflected function will have $(-x, y)$.

**62. FINANCIAL LITERACY** The balance of Laura's savings over time can be modeled by $g(x) = P(1.002)^t$, where $P$ is the initial amount in her bank account and $t$ is time in years.

a. Describe the graph of $g(x)$ when she invests $1000 and when she invests $7500 in her bank account.

b. Suppose Laura keeps $250 cash in a safe in her apartment, which she also considers part of her savings. How would this affect the graph of $g(x)$? **The graph is translated up 250 units.**

**63. MULTI-STEP** Consider transformations of the function $f(x) = b^x$.

a. What happens if the function is multiplied by a constant between 0 and 1? Describe the new graph compared to the parent function. **The graph is compressed vertically.**

b. Describe the graph as the constant approaches 0.

c. What happens is the function is multiplied by a constant greater than 1? Describe the new graph compared to the parent function. **The graph is stretched vertically.**

d. Describe the graph as the constant approaches ∞. **The graph becomes more stretched and closer to the $y$-axis.**

F.BF.3, F.LE.5

### H.O.T. Problems  Use Higher-Order Thinking Skills

**64. REASONING** What point is on the graph of every function of the form $f(x) = b^x$, where $b \neq 0$? Explain your answer.

**65. REASONING** Explain how a reflection across the $x$-axis affects each points on the graph of an exponential function.

**66. OPEN ENDED** Using $f(x) = 5^x$ as the parent function, write functions for the following transformations:

a. reflected across the $y$-axis and vertically compressed **Sample answer: $g(x) = 0.1(5)^{-x}$**

b. horizontally stretched and translated left and up **Sample answer: $g(x) = 5^{0.5x + 4} + 2$**

c. reflected across $x$-axis, vertically stretched, and translated right **Sample answer: $g(x) = -2(5)^{x-1}$**

d. horizontally compressed and translated left and down **Sample answer: $g(x) = 0.1(5)^{-x}$**

**67. SENSE-MAKING** Examine the following cases and describe the effect a reflection across the $x$-axis would have on the end behavior of the parent function $f(x) = ab^x$.

Case 1: $g(x) = -ab^x$ where $b > 1$  **See margin.**

Case 2: $g(x) = -ab^x$ where $0 < b < 1$

**68. SENSE-MAKING** Examine the following cases and describe the effect a reflection across the $y$-axis would have on the end behavior of the parent function $f(x) = ab^x$.

Case 1: $g(x) = ab^{-x}$ where $b > 1$  **See margin.**

Case 2: $g(x) = ab^{-x}$ where $0 < b < 1$

**69. OPEN ENDED** Write an exponential decay function that is reflected across the $x$-axis and has an asymptote of $y = -2$. **Sample answer: $y = \left(\frac{1}{4}\right)^x - 2$**

**70. REASONING** Are the following statements *always*, *sometimes*, or *never* true? Explain.

a. The graph of $g(x) = b^x + k$ has an asymptote of $y = k$. **70a-c. See margin.**

b. The graph of $j(x) = b^{x-h}$ passes through $(h, 1)$.

c. The graph of $p(x) = -b^x$ decreases as $x \to \infty$.

---

### Standards for Mathematical Practice

| Emphasis On | Exercises |
| --- | --- |
| 1 Make sense of problems and persevere in solving them. | 67, 68, 76, 77 |
| 2 Reason abstractly and quantitatively. | 63–65, 70, 74–76, 78 |
| 4 Model with mathematics. | 7, 44, 62 |
| 5 Use appropriate tools strategically. | 60, 61 |
| 7 Look for and make use of structure. | 66, 71–75, 77, 78 |

**70a.** Always; because $f(x) = b^x$ has an asymptote of $y = 0$, vertically translating $k$ units will result in an asymptote of $y = 0 + k$ or $y = k$.

**70b.** Always; because $f(x) = b^x$ has a $y$-intercept of (0, 1), horizontally translating the parent function $h$ units will result in a $y$-intercept of $(0 + h, 1)$ or $(h, 1)$.

**70c.** Sometimes; When $b > 1$, the graph decreases as $x \to \infty$. When $0 < b < 1$, the graph increases as $x \to \infty$.

Lesson 7-6 | Transformations of Exponential Functions

## Preparing for Assessment

71. The graph of $g(x) = 5^{x-2} + 8$ is a translation of the graph $f(x) = 5^x$. Which of the following describes the translation? **MP 7 F.BF.3 C**

    ○ A  translated left 2 units and down 8 units
    ○ B  translated left 2 units and up 8 units
    ○ C  translated right 2 units and up 8 units
    ○ D  translated right 8 units and down 2 units

72. Which function is a translation of the parent function $f(x) = 3^x$ right 4 units? **MP 7 F.BF.3 D**

    ○ A  $g(x) = 3^x + 4$
    ○ B  $g(x) = 3^x - 4$
    ○ C  $g(x) = 3^{x+4}$
    ○ D  $g(x) = 3^{x-4}$

73. Which of the following transformations represents a vertical compression of the parent function $f(x) = 2^x$? **MP 7 F.BF.3 B, C**

    ☐ A  $g(x) = -5(2)^x$
    ☐ B  $g(x) = -0.4(2)^x$
    ☐ C  $g(x) = \frac{1}{6}(2)^x$
    ☐ D  $g(x) = 2^{-7x}$
    ☐ E  $g(x) = 2^{0.1x}$
    ☐ F  $g(x) = 2^{6x}$

74. The graph of $f(x) = \frac{1}{2}^x$ is reflected across the $y$-axis and translated up 5 units. Write the transformed function $g(x)$. **MP 2, 7 F.BF.3**

    $g(x) = -\frac{1}{2}^{-x} + 5$

75. The graph of $f(x) = \frac{1}{2}^x$ is reflected across the $x$-axis, stretched vertically by a factor of 7, and translated left 1 unit. Write the transformed function $g(x)$. **MP 2, 7 F.BF.3**

    $g(x) = -7\left(\frac{1}{2}\right)^{x+1}$

76. The graph of $g(x)$ is a translation of the parent function $f(x) = 1.5^x$. Which is the equation for the graph? **MP 1, 2 F.BF.3 B**

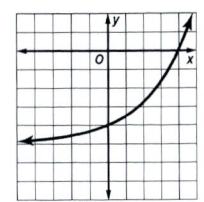

    A  $g(x) = 1.5^x + 5$
    B  $g(x) = 1.5^x - 5$
    C  $g(x) = 1.5^{x+5}$
    D  $g(x) = 1.5^{x-5}$

77. **MULTI-STEP** Consider the function $g(x) = -2(3)^{x+1} + 4$ as it relates to the parent function $f(x) = 3^x$. **MP 1, 7 F.BF.3, F.IF.7e**

    a. Identify the value of a and its affect on the parent function. **See margin.**
    b. Identify the value of h and its affect on the parent function. **See margin.**
    c. Identify the value of k and its affect on the parent function. **See margin.**
    d. Graph $g(x)$. **See margin.**

78. If $f(x) = b^x$, which function represents $g(x)$? **MP 2, 7 F.BF.3, F.IF.7e C**

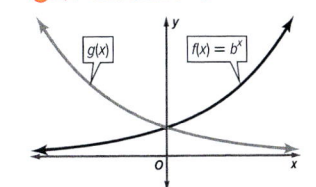

    ○ A  $g(x) = -b^x$
    ○ B  $g(x) = b^x - 1$
    ○ C  $g(x) = b^{-x}$
    ○ D  $g(x) = -b^{-x}$

## Preparing for Assessment

Exercises 71–78 require students to use the skills they will need on assessments. Each exercise is dual-coded with content.

### Dual Coding

| Items | Content Standards | MP Mathematical Practices |
|---|---|---|
| 71–73 | F.BF.3 | 7 |
| 74, 75 | F.BF.3 | 2, 7 |
| 76 | F.BF.3 | 1, 2 |
| 77 | F.BF.3, F.IF.7e | 1, 7 |
| 78 | F.BF.3, F.IF.7e | 2, 7 |

## Diagnose Student Errors

**71.**

| A | Confused the meaning of the signs for h and k |
|---|---|
| B | Confused the meaning of the sign for h |
| C | CORRECT |
| D | Transposed h and k |

**72.**

| A | Confused with a vertical translation |
|---|---|
| B | Confused with a vertical translation |
| C | Used the wrong operation |
| D | CORRECT |

**78.**

| A | Chose a reflection over the x-axis instead of the y-axis |
|---|---|
| B | Chose a translation instead of a reflection |
| C | CORRECT |
| D | Chose a reflection over the x-axis and the y-axis |

## Additional Answers

**77a.** −2; reflects across the x-axis and vertically stretches the graph

**77b.** −1; translates the graph left 1 unit

**77c.** 4; translates the graph up 4 units

**77d.**

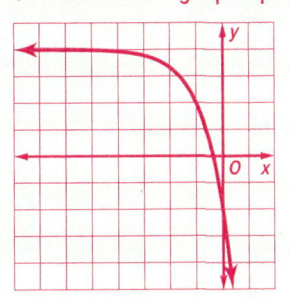

### Go Online!

**Quizzes**

Students can use Self-Check Quizzes to check their understanding of this lesson.

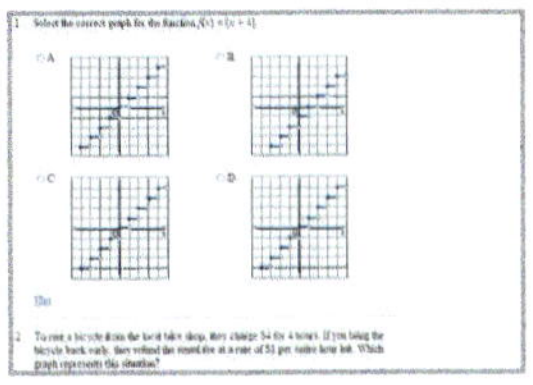

connectED.mcgraw-hill.com  447

# LESSON 7-7
# Writing Exponential Functions

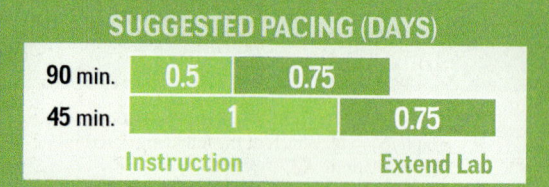

**SUGGESTED PACING (DAYS)**

| | | |
|---|---|---|
| 90 min. | 0.5 | 0.75 |
| 45 min. | 1 | 0.75 |
| | Instruction | Extend Lab |

## Track Your Progress

### Objectives

1. Write exponential functions by using a graph, a description, or two points.
2. Solve problems involving exponential growth and decay.

### Mathematical Background

Exponential growth can be modeled using the general equation $y = a(1 + r)^t$, and exponential decay by $y = a(1 - r)^t$, where $y$ is the final amount, $a$ is the initial amount, $r$ is the rate of change expressed as a decimal and $r > 0$, and $t$ is time. Compound interest is an example of exponential growth and depreciation is an example of exponential decay.

### THEN

**F.IF.7e** Graph exponential and logarithmic functions, showing intercepts and end behavior, and trigonometric functions, showing period, midline, and amplitude.

**F.BF.3** Identify the effect on the graph of replacing $f(x)$ by $f(x) + k$, $kf(x)$, $f(kx)$, and $f(x + k)$ for specific values of $k$ (both positive and negative); find the value of $k$ given the graphs. Experiment with cases and illustrate an explanation of the effects on the graph using technology. Include recognizing even and odd functions from their graphs and algebraic expressions for them.

### NOW

**A.CED.2** Create equations in two or more variables to represent relationships between quantities; graph equations on coordinate axes with labels and scales.

**F.LE.2** Construct linear and exponential functions, including arithmetic and geometric sequences, given a graph, a description of a relationship, or two input-output pairs (include reading these from a table).

**F.LE.5** Interpret the parameters in a linear or exponential function in terms of a context.

### NEXT

**F.BF.2** Write arithmetic and geometric sequences both recursively and with an explicit formula, use them to model situations, and translate between the two forms.

## Go Online! All of these resources and more are available at connectED.mcgraw-hill.com

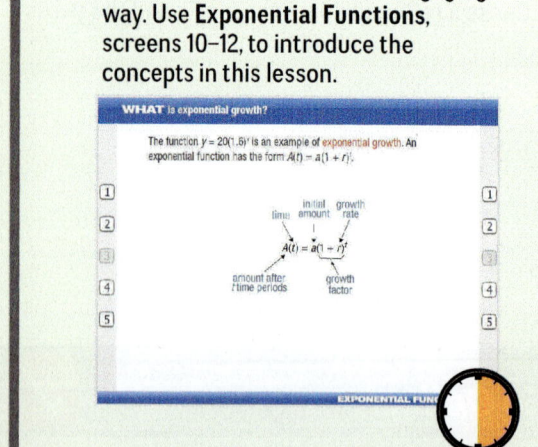

**eLessons** utilize the power of your interactive whiteboard in an engaging way. Use **Exponential Functions**, screens 10–12, to introduce the concepts in this lesson.

*Use at Beginning of Lesson*

**Personal Tutors** (for every example) let students hear real teachers solve problems. Students can pause and repeat as many times as necessary.

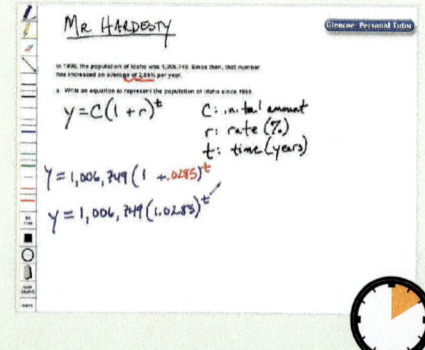

*Use with Examples*

The **Chapter Project** allows students to create and customize a project as a nontraditional method of assessment.

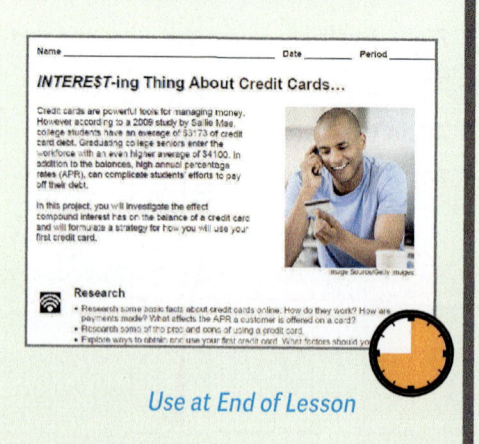

*Use at End of Lesson*

## OER Using Open Educational Resources

**Games** Look at the Growth and Decay Stations Game on scribd.com for cool new ways to introduce exponential growth and decay. Set this game up as five different stations around the room with your class divided into five small groups. *Use as classwork*

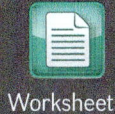

# Differentiate Your Resources

**Extra Practice** Additional practice or homework; Skills Practice is best for approaching-level students and Practice is best for on-level and beyond-level students

## Skills Practice
AL  OL  ELL

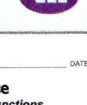

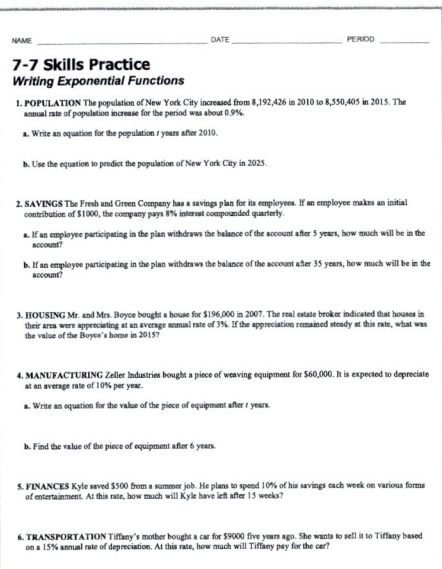

## Practice
AL  OL  BL  ELL

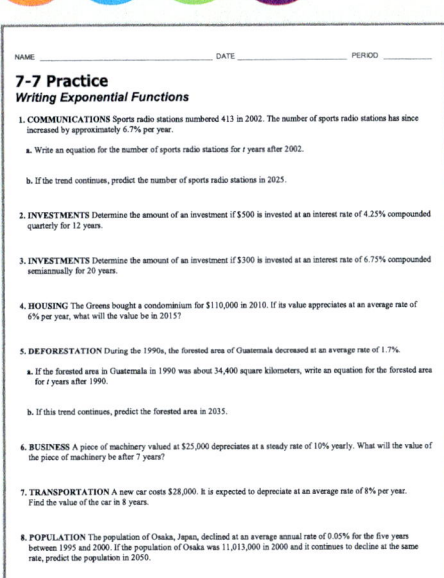

## Word Problem Practice
AL  OL  BL  ELL

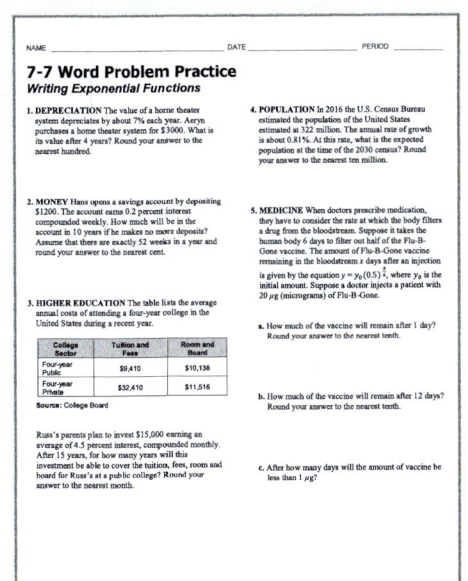

---

**Intervention** Reteaching and vocabulary activities that can be used with struggling or absent students and as ELL support

**Extension** Activities that can be used to extend lesson concepts

## Study Guide and Intervention
AL  OL  ELL

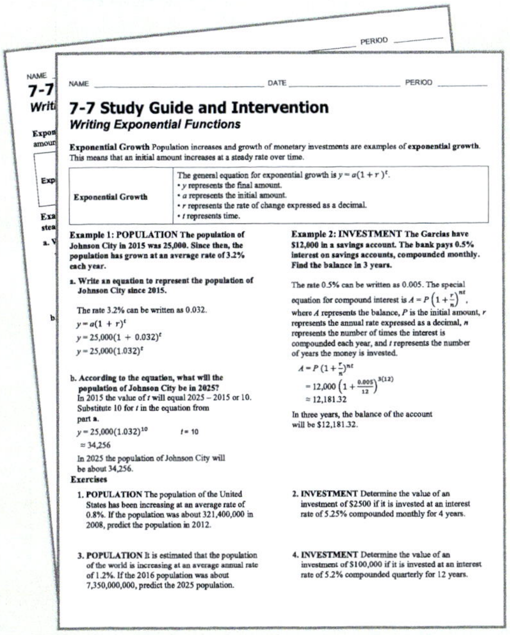

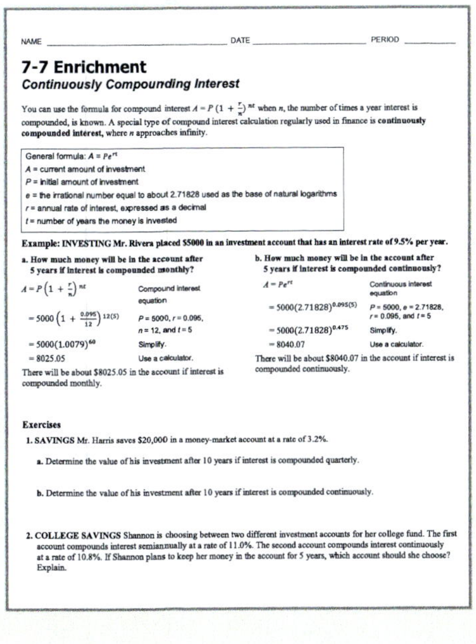

## Study Notebook
AL  OL  BL  ELL

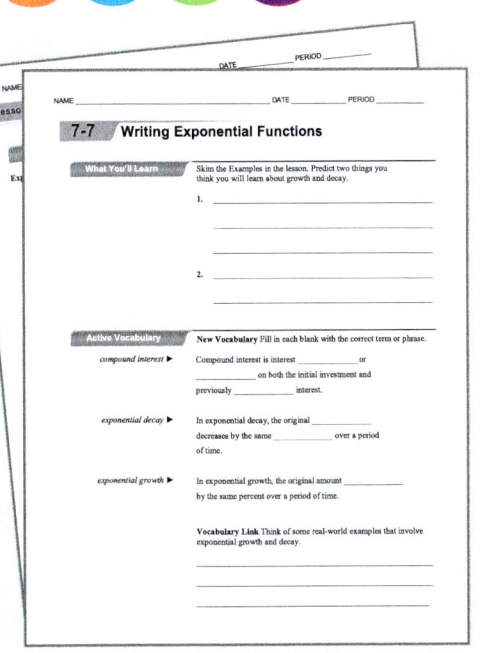

## Enrichment
OL  BL  ELL

connectED.mcgraw-hill.com

Lesson 7-7 | Writing Exponential Functions

# Launch

Have students read the Why? section of the lesson. Ask:

- Looking at the equation, how do you know the function is not linear? **Time, represented by $t$, is an exponent, so the function is not linear.**

- Use the equation to predict the number of podcasts in June 2013. **about 42,000**

# Teach

Ask the scaffolded questions for each example to build conceptual understanding for students at all levels.

## 1 Write Exponential Functions

**Example 1 Write an Exponential Function Given a Graph**

**AL** What happens to the value of $y$ as $x$ approaches $-\infty$? **It approaches, but does not reach 0.**

**OL** How can you check that your equation is correct? **Substitute the given $x$-values into the equation and check if the resulting $y$-values correspond.**

**BL** Based on the graph, is $b > 1$ or is $0 < b < 1$? **$b > 1$**

**Need Another Example?**
Write an exponential function for the graph.
$y = 6\left(\dfrac{1}{3}\right)^x$

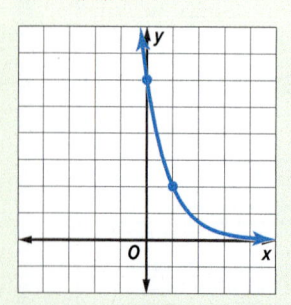

### Go Online!

**Interactive Whiteboard**
Use the *eLesson* or *Lesson Presentation* to present this lesson.

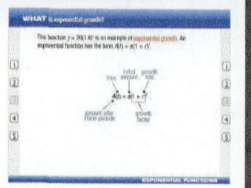

---

## LESSON 7
# Writing Exponential Functions

**Then** — You analyzed exponential functions.

**Now** —
1. Write exponential functions by using a graph, a description, or two points.
2. Create equations and solve problems involving exponential growth and decay.

**Why?** — The first podcast premiered in 2003 and the number of podcasts continues to grow. Since 2010, the number of active podcasts has been increasing by about 17.3% each year. The growth of podcasts can be modeled by $y = 26.5(1 + 0.173)^t$ or $y = 26.5(1.173)^t$, where $y$ represents the number of podcasts in thousands and $t$ is the number of years since 2010.

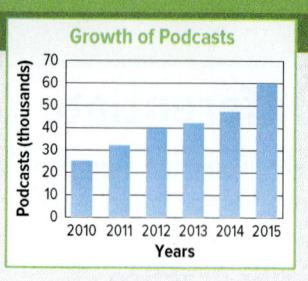

**New Vocabulary**
compound interest

**Mathematical Practices**
1 Make sense of problems and persevere in solving them.

**Content Standards**
A.CED.2 Create equations in two or more variables to represent relationships between quantities; graph equations on coordinate axes with labels and scales.
F.LE.2 Construct linear and exponential functions, including arithmetic and geometric sequences, given a graph, a description of a relationship, or two input-output pairs (including reading these from a table).
F.LE.5 Interpret the parameters in a linear or exponential function in terms of a context.

### 1 Write Exponential Functions
Given two points, a graph, or description, an exponential function in the form $y = ab^x$, where $a \neq 0$, $b > 0$, and $b \neq 1$, can be written.
F.LE.2

**Example 1** Write an Exponential Function Given a Graph

Write an exponential function for the graph.

**Step 1** Select two points on the graph.

Any two points on the graph can be used to write a system of two equations using $y = ab^x$. Selecting the $y$-intercept will make writing the equation of the function easier because $y = ab^0 = a$.

Use $(0, 0.5)$ and $(4, 8)$ to write the function.

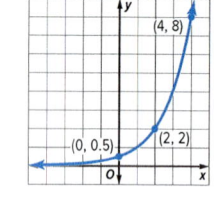

**Step 2** Solve for $a$.

| $y = ab^x$ | Exponential equation |
| $0.5 = ab^0$ | $(x, y) = (0, 0.5)$ |
| $0.5 = a \cdot 1$ | $b^0 = 1$ |
| $0.5 = a$ | Simplify. |

**Step 3** Solve for $b$.

Use the value of $a$ from Step 2 and the second point to solve for $b$.

| $y = ab^x$ | Exponential equation |
| $8 = (0.5)b^4$ | $(x, y) = (4, 8)$, $a = 0.5$ |
| $16 = b^4$ | Divide each side by 0.5. |
| $2 = b$ | Simplify. |

**Step 4** Write the function.

| $y = ab^x$ | Exponential equation |
| $y = 0.5(2)^x$ | $a = 0.5$, $b = 2$ |

The graph is represented by the function $y = 0.5(2)^x$.

---

### Mathematical Practices Strategies

**Make sense of problems and persevere in solving them.**
Help students make sense of growth trends, represent a situation symbolically, and understand the meaning of the symbols in a problem. For example, ask:

What does a graph of exponential growth look like? **From left to right, the graph rises slowly at first and then rises rapidly.**

How can you tell that an exponential equation represents growth? **The value of $b$ is greater than 1.**

How can you tell that an exponential equation represents decay? **The value of $b$ is less than 1.**

▶ **Guided Practice** $y = 2(4)^x$

1. Write an exponential function for the graph.

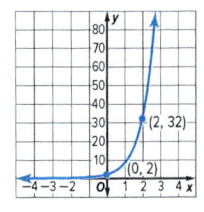

F.LE.2

**Example 2**    Write an Exponential Function Given Two Points

Write an exponential function for a graph that passes through (1, 6) and (3, 24).

Substitute the $x$- and $y$-coordinates into $y = ab^x$ to get a system of two equations.

    Equation 1:   $6 = ab^1$      $(x_1, y_1) = (1, 6)$
    Equation 2: $24 = ab^3$      $(x_2, y_2) = (3, 24)$

Solve the system for $a$ and $b$ using substitution.

**Step 1** Solve for a variable.

    $6 = ab^1$      Equation 1
    $\dfrac{6}{b} = a$      Divide each side by $b$.

**Step 2** Substitute.

Substitute $\dfrac{6}{b}$ for $a$ in Equation 2 and solve for $b$.

    $24 = ab^3$      Equation 2
    $24 = \left(\dfrac{6}{b}\right)b^3$      Substitute $\dfrac{6}{b}$ for $a$.
    $24 = \dfrac{6b^3}{b}$      Multiply.
    $24 = 6b^2$      Quotient of Powers
    $4 = b^2$      Divide each side by 6.
    $2 = b$      Simplify.

**Step 3** Solve for $a$.

Substitute 2 for $b$ in either equation to find $a$.

    $6 = ab^1$      Equation 1
    $6 = a(2)^1$      $b = 2$
    $3 = a$      Divide each side by 2.

$y = 3 \cdot 2^x$ is an exponential function that passes through (1, 6) and (3, 24).

**Study Tip**

**MP Sense-Making** You can check the equation by making sure that the two given points are solutions of the equation.

▶ **Guided Practice**    2. $y = 2 \cdot 4^x$

2. Write an exponential function for a graph that passes through (1, 8) and (4, 512).

**2 Exponential Growth and Decay** The equation for the number of podcasts over time at the beginning of this lesson is in the form $y = a(1 + r)^t$. This is the general equation for exponential growth.

---

**Lesson 7-7** | **Writing Exponential Functions**

**Example 2**    Write an Exponential Function Given Two Points

**AL** What is the general form of an exponential function? $y = ab^x$

**OL** Would it have been possible to write the function by first solving for $b$ in Step 1? Explain. Yes; Sample answer: Because there is only one exponential function $y = ab^x$ passing through the points, the result is the same.

**BL** Does it matter which equation you use to solve for $a$ in terms of $b$? Explain. No; you can use either equation, but the first equation results in a simpler expression.

**Need Another Example?**

Write an exponential function for a graph that passes through (1, 4) and (2, 1.6). $y = 10(0.4)^x$

connectED.mcgraw-hill.com    **449**

# Lesson 7-7 | Writing Exponential Functions

## 2 Exponential Growth and Decay

### Example 3 Interpret the Parameters of an Equation

**AL** How many users were there in 2013? Explain how you determined this. 1.5 million; Sample answer: the year 2013 is year 0 and in year 0 the initial amount is 1.5.

**OL** If $(1 + r) = 1.40$, what is $r$? 0.40

**BL** Is it reasonable to think that the number of users will continue to increase by 40% forever? Sample answer: no; technology has trends and while the site might be growing rapidly now, eventually its use will probably go back down as people start using a new site.

**Need Another Example?**

**Sales** The number of purchases in millions made on a website can be represented by the equation $y = 2.5(1.13)^t$, where $t$ is the number of years after 2014.
a. Interpret the parameters of the equation. initial: 2.5 million; rate of change: 13%
b. Make a prediction about the future number of purchases made on the website. If this growth rate continues, then the number of purchases made on the website will rapidly increase, exceeding 10 million by 2026.

### Example 4 Exponential Growth

**AL** What is the initial value of the prize? $100 For what value should this be substituted in the equation? $a$

**OL** What value do we substitute for $r$? Explain how you determined this number. 0.025; Sample answer: we are given the rate as a percent, 2.5%, but the value of $r$ must be expressed as a decimal, so I converted 2.5% to 0.025.

**BL** How much will the prize be worth on the 15th day if someone won on the 10th day? $110.38

**Need Another Example?**

**Population** In 2013, the town of Flat Creek had a population of about 280,000 and a growth rate of 0.85% per year.
a. Write an equation to represent the population of Flat Creek since 2013. $y = 280,000\,(1.0085)^t$
b. According to the equation, what will be the population of Flat Creek in the year 2023? about 304,731

---

**Key Concept** General Equation for Exponential Growth

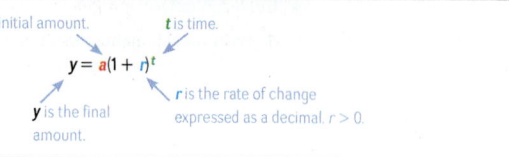

$a$ is the initial amount.  $t$ is time.

$$y = a(1 + r)^t$$

$y$ is the final amount.  $r$ is the rate of change expressed as a decimal. $r > 0$.

F.LE.5

### Example 3 Interpret the Parameters of an Equation

**SOCIAL NETWORKS** The number of users, in millions, that belong to a social networking site can be represented by $y = 1.5(1.40)^t$, where $t$ is the number of years since 2013. Interpret the parameters of the equation. Make a prediction about when the number of users will exceed 40 million.

The initial number of users $a$ is 1.5 million.

The rate of change $r$ is $1.40 - 1$, or 0.40. Therefore, the number of users is growing at a rate of 40% per year.

If this growth rate continues, then the number of users that belong to the networking site will rapidly increase. By 2023, more than 40 million users are expected because $1.5(1.40)^{10} \approx 43.4$.

**Guided Practice**

3. **SMARTPHONE APPS** The number of apps, in thousands, that are on the market for a smartphone can be represented by $y = 320(1.22)^t$, where $t$ is the number of years since 2012. Interpret the parameters of the equation. Make a prediction about when the future number of apps on the market will exceed 1 million.

3. Initial: 320,000; rate of change: 22% per year; If this growth rate continues, then the number of apps will continue to rapidly increase, exceeding 1 million in 6 years.

F.LE.2

### Real-World Example 4 Exponential Growth

**CONTEST** The prize for a radio station contest begins with a $100 cash prize. Once a day, a name is announced. The person has 15 minutes to call or the prize increases by 2.5% for the next day.

a. Write an equation to represent the amount of the cash prize in dollars after $t$ days with no winners.

$y = a(1 + r)^t$     Equation for exponential growth
$y = 100(1 + 0.025)^t$     $a = 100$ and $r = 2.5\%$ or 0.025
$y = 100(1.025)^t$     Simplify.

In the equation $y = 100(1.025)^t$, $y$ is the amount of the prize and $t$ is the number of days since the contest began.

b. How much will the prize be worth if no one wins after 10 days?

$y = 100(1.025)^t$     Equation for amount of gift card
$\phantom{y} = 100(1.025)^{10}$     $t = 10$
$\phantom{y} \approx 128.01$     Use a calculator.

In 10 days, the cash prize will be worth $128.01.

**Guided Practice**

4. **TUITION** A college's tuition has risen 5% each year since 2010. If the tuition in 2010 was $10,850, write an equation for the amount of the tuition $t$ years after 2010. Predict the cost of tuition for this college in 2025. $y = 10,850(1.05)^t$; about $22,556.37

---

**Watch Out!**

**Growth Factor** Be careful when you convert the percent to a decimal. The base of the exponential equation is 1.025, not 1.25.

**Watch Out!**

**Student Misconceptions** Remind students that in growth and decay equations, the amount inside the parentheses will be greater than 1 for growth and less than 1 for decay.

450 | Lesson 7-7 | Writing Exponential Functions

## Lesson 7-7 | Writing Exponential Functions

In exponential decay, the original amount decreases by the same percent over time. A variation of the growth equation can be used to represent exponential decay.

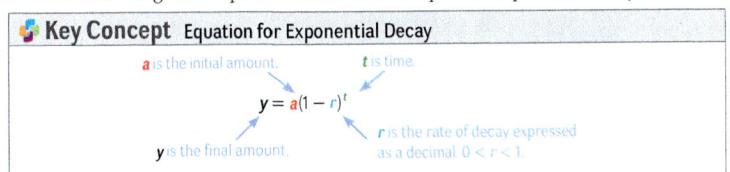

### Key Concept  Equation for Exponential Decay

$a$ is the initial amount.  
$t$ is time.  
$$y = a(1 - r)^t$$
$y$ is the final amount.  
$r$ is the rate of decay expressed as a decimal. $0 < r < 1$.

F.LE.2

### Real-World Example 5  Exponential Decay

**SWIMMING** A fully inflated child's raft for a pool is losing 6.6% of its air every day. The raft originally contained 4500 cubic inches of air.

a. Write an equation to represent the loss of air.

$y = a(1 - r)^t$    Equation for exponential decay
$= 4500(1 - 0.066)^t$    $a = 4500$ and $r = 6.6\%$ or $0.066$
$= 4500(0.934)^t$    Simplify.

b. Estimate the amount of air in the raft after 7 days.

$y = 4500(0.934)^t$    Equation for air loss
$= 4500(0.934)^7$    $t = 7$
$\approx 2790$    Use a calculator.

The amount of air in the raft after 7 days will be about 2790 cubic inches.

**Study Tip**

**MP Reasoning** Since $r$ is added to 1, the value inside the parentheses will be greater than 1 for exponential growth functions. For exponential decay functions, this value will be less than 1 since $r$ is subtracted from 1.

### Guided Practice

5. **POPULATION** The population of Campbell County, Kentucky, has been decreasing at an average rate of about 0.3% per year. In 2010, its population was 88,647. Write an equation to represent the population since 2010. If the trend continues, predict the population in 2020.    $y = 88,647(1 - 0.003)^t$; about 86,023

**Compound interest** is interest earned or paid on both the initial investment and previously earned interest. It is an application of exponential growth.

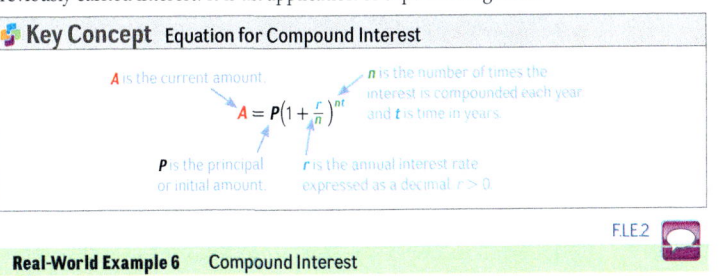

### Key Concept  Equation for Compound Interest

$A$ is the current amount.  
$n$ is the number of times the interest is compounded each year.  
$$A = P\left(1 + \frac{r}{n}\right)^{nt}$$
$P$ is the principal or initial amount.  
$r$ is the annual interest rate expressed as a decimal. $r > 0$.  
$t$ is time in years.

F.LE.2

**Real-World Career**

**Financial Advisor** Financial advisors help people plan their financial futures. A good financial advisor has mathematical, problem-solving, and communication skills. A bachelor's degree is strongly preferred but not required.

### Real-World Example 6  Compound Interest

**FINANCE** Maria's parents invested $14,000 at 6% per year compounded monthly. How much money will there be in the account after 10 years?

$A = P\left(1 + \frac{r}{n}\right)^{nt}$    Compound interest equation
$= 14,000\left(1 + \frac{0.06}{12}\right)^{12(10)}$    $P = 14000, r = 6\%$ or $0.06, n = 12$, and $t = 10$
$= 14,000(1.005)^{120}$    Simplify.
$\approx 25,471.55$    Use a calculator.

There will be $25,471.55 in 10 years.

### Example 5  Exponential Decay

**AL** What values will you substitute for each of the variables in the equation? $a = 4500$; $r = 0.066$; $t$ will still be $t$ because we aren't given a value.

**OL** In part **a**, can we solve the equation? Why or why not? Sample answer: no, there are two variables — $y$ and $t$ — so we cannot solve for one of them. We are only asked to write the equation.

**BL** Why is the solution to part **b** an estimate? Sample answer: we have to round the answer we get after multiplying, so the solution is not exact.

### Need Another Example?

**Charity** During an economic recession, a charitable organization found that its donations dropped by 1.1% per year. Before the recession, its donations were $390,000.

a. Write an equation to represent the charity's donations since the beginning of the recession. $A = 390,000(0.989)^t$

b. Estimate the amount of the donations 5 years after the start of the recession. about $369,017

### Example 6  Compound Interest

**AL** For what variable will you solve? $A$ What values will you substitute for each of the other variables in the equation? $P = 14,000$; $r = 0.06$; $n = 12$; $t = 10$

**OL** How did you determine the value of $n$? Sample answer: since the interest is compounded monthly and there are 12 months in a year, I determined that 12 is the value of $n$.

**BL** Determine the values of $n$ and $t$ if interest is compounded monthly for 6 months. $n = 12$; $t = \frac{1}{2}$

### Need Another Example?

**College** When Jing May was born, her grandparents invested $1000 in a fixed rate savings account at a rate of 7% compounded annually. Jing May will receive the money when she turns 18 to help with her college expenses. What amount of money will Jing May receive from the investment? She will receive about $3380.

connectED.mcgraw-hill.com    451

Lesson 7-7 | Writing Exponential Functions

# Practice

**Formative Assessment** Use Exercises 1–8 to assess students' understanding of the concepts in this lesson.

The Practice and Problem Solving exercises assess the content taught in the lesson. The Preparing for Assessment page is meant to be used as preparation for end-of-course assessments.

## Teaching the Mathematical Practices

**Precision** Mathematically proficient students express answers with a degree of precision appropriate for the problem context. In Exercise 23, point out that an estimate is appropriate to answer the question. In Exercise 31, students can approximate by graphing on a graphing calculator or by guessing and checking on a scientific calculator.

### Levels of Complexity Chart

The levels of the exercises progress from 1 to 3, with Level 1 indicating the lowest level of complexity.

| Exercises | 9–26 | 27, 35–42 | 28–34 |
|---|---|---|---|
| Level 3 |  |  | ● |
| Level 2 |  | ● |  |
| Level 1 | ● |  |  |

## Go Online!   eBook

**Interactive Student Guide**
Use the *Interactive Student Guide* to deepen conceptual understanding.
· Modeling: Exponential Functions

452 | Lesson 7-7 | Writing Exponential Functions

---

▶ **Guided Practice**

6. **FINANCE** Determine the amount of an investment if $300 is invested at an interest rate of 3.5% compounded monthly for 22 years. **about $647.20**

**Check Your Understanding**   ○ = Step-by-Step Solutions begin on page R11.

**Go Online!** for a Self-Check Quiz

**Example 1** Use the information in each graph to construct an exponential function to model the graph.
F.LE.2

1. 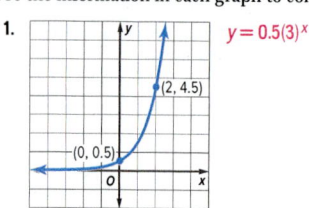 $y = 0.5(3)^x$ (2, 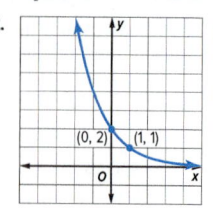 4.5), (0, 0.5)

2. $y = 2(0.5)^x$ (0, 2), (1, 1)

**Example 2** Write an exponential function for a graph that passes through each pair of points.
F.LE.2

3. (1, 12) and (3, 108)  $y = 4(3)^x$

4. (1, 1.5) and (3, 54)  $y = \frac{1}{4}(6)^x$

**Example 3**
F.LE.5

5. **ADVERTISING** The number of people that have "liked" Mindy's Candy Store website can be represented by $y = 270(1.65)^t$, where $t$ is the number of weeks after a review in a national magazine. Interpret the parameters of the equation. Make a prediction about the future number of likes on the website. **Initial: 270; rate of change: 65% per week; if this growth rate continues, then the number of likes will continue to increase, exceeding 100,000 in 12 weeks.**

**Example 4**
F.LE.2

6. **SALARY** Ms. Acosta received a job as a teacher with a starting salary of $34,000. According to her contract, she will receive a 1.5% increase in her salary every year. How much will Ms. Acosta earn in 7 years? **about $37,734.73**

**Example 5**
F.LE.2

7. **ENROLLMENT** In 2015, 2200 students attended Polaris High School. The enrollment has been declining 2% annually.

   a. Write an equation for the enrollment of Polaris High School $t$ years after 2015. $y = 2200(0.98)^t$

   b. If this trend continues, how many students will be enrolled in 2030? **about 1625**

**Example 6**
F.LE.2

8. **MONEY** Paul invested $400 into an account with a 5.5% interest rate compounded monthly. How much will Paul's investment be worth in 8 years? **about $620.46**

**Practice and Problem Solving**   Extra Practice is on page R7.

**Example 1** Use the information in each graph to construct an exponential function to model the graph.
F.LE.2

9. 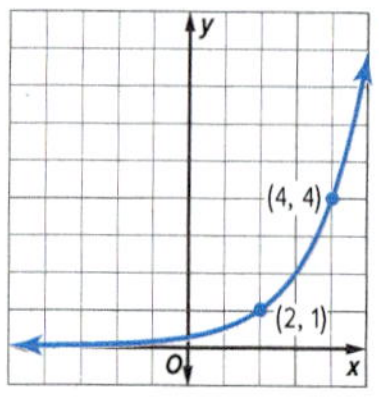 $y = 0.25(2)^x$ (4, 4), (2, 1)

10. 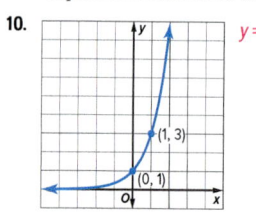 $y = 3^x$ (1, 3), (0, 1)

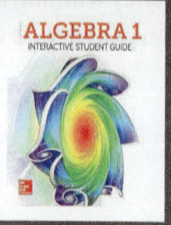

### Differentiated Homework Options

| Levels | AL Basic | OL Core | BL Advanced |
|---|---|---|---|
| Exercises | 9–24, 27, 30–42 | 9–25 odd, 27, 28, 30–42 | 27–34<br>35–42 (optional) |
| 2-Day Option | 9–23 odd, 35–42 | 9–27, 35–42 |  |
|  | 10–24 even, 27, 30–34 | 28–34 |  |

You can use ALEKS to provide additional remediation support with personalized instruction and practice.

Lesson 7-7 | Writing Exponential Functions

11. 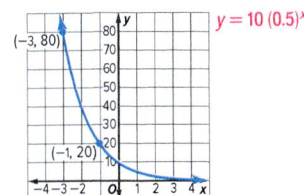 $y = 10(0.5)^x$  (graph through (−3, 80), (−1, 20))

12. 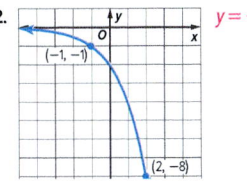 $y = -2(2)^x$  (graph through (−1, −1), (2, −8))

13. $y = -0.5(4)^x$  (graph through (0, −0.5), (1, −2))

14. $y = -0.1^x$  (graph through (−1, 10), (0, 1))

18. Initial: 6.7 million; rate of change: 8% per month; If this growth rate continues, then the number of downloads will continue to rapidly increase, exceeding 10 million in 6 months.

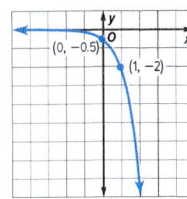

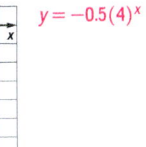

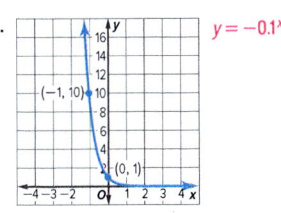

**Examples 2**
F.LE.2

Write an exponential function for a graph that passes through each pair of points.

15. (1, 10) and (5, 160)  $y = 5(2)^x$

16. (1, 1) and (4, 64)  $y = 0.25(4)^x$

**Example 3**
F.LE.5

17. **ANTS** A colony of ants are looting a food source. The amount of food at the source, in grams, can be represented by $y = 82(0.65)^t$, where $t$ is the number of minutes after the looting began. Interpret the parameters of the equation. Make a prediction about the future amount of food left at the source. **See margin.**

18. **MUSIC DOWNLOADS** The total number of songs, in millions, that have been downloaded from a music sharing site can be represented by $y = 6.7(1.08)^t$, where $t$ is the number of months after the site began. Interpret the parameters of the equation. Make a prediction about the future number of songs downloaded from the site.

**Example 4**
F.LE.2

19. **COINS** Camilo purchased a rare coin from a dealer for $300. The value of the coin increases 5% each year. Determine the value of the coin in 5 years.  **about $382.88**

20. **GAME SHOWS** The jackpot on a game show starts with a value of $1000. For every question that a contestant answers correctly, the jackpot's value increases by 25%. Determine the value of the jackpot if the contestant correctly answers 12 questions in a row.  **about $14,552**

**Example 5**
F.LE.2

21. **INVESTMENTS** Jin's investment of $4500 has been losing its value at a rate of 2.5% each year. What will his investment be worth in 5 years?  **about $3964.93**

22. **MUSEUMS** The director of a science museum begins a membership drive with the goal of signing up 850 new members. The number of new members that remain to be signed up decreases by 11.6% every month. Determine the number of new members that still need to be signed up at the end of one year.  **about 194**

**Example 6**
F.LE.2

23. **PRECISION** Brooke is saving money for a trip to the Bahamas that costs $295.99. She puts $150 into a savings account that pays 7.25% interest compounded quarterly. Will she have enough money in the account after 4 years? Explain.

23. Sample answer: No; she will have about $199.94 in the account in 4 years.

24. **FINANCE** Determine the value of an investment of $650 that is invested at an annual interest rate of 1.5% compounded monthly for 16 years.  **$826.19**

25. **CARS** Leonardo purchases a car for $18,995. The car depreciates at a rate of 18% annually. After 6 years, Manuel offers to buy the car for $4500. Should Leonardo sell the car? Explain.  **Sample answer: No; the car is worth about $5774.61.**

26. **POPULATION** In the years from 2010 to 2015, the population of the District of Columbia decreased an average of 0.9% annually. In 2010, the population was about 530,000. What is the population of the District of Columbia expected to be in 2030?  **about 442,333**

## Extra Practice

See page R7 for extra exercises for students who are approaching level or for on-level students who need additional reinforcement.

## Follow-Up

Students have explored growth and decay.
Ask: How can financial literacy help you make good decisions? Sample answer: If you are financially literate, you understand the vocabulary of financial terms and know how to analyze data and trends. Successfully applying these skills when considering your available options can help you to make good decisions in many real-world situations such as opening a bank account, applying for college loans, and buying a car.

## Assess

**Ticket Out the Door** Make several copies each of five equations for exponential growth or decay. Give one equation to each student. As students leave the room, ask them to tell you whether their equations are for growth or decay.

## Additional Answer

17. initial: 82 grams of food; rate of change: decreasing by 35% per minute; if this decay rate continues, there will be less than a gram of food after 11 minutes.

## Standards for Mathematical Practices

| Emphasis On | Exercises |
| --- | --- |
| 1 Make sense of problems and persevere in solving them. | 15, 16, 19–22, 19, 33, 34, 38, 39 |
| 2 Reason abstractly and quantitatively. | 23–25, 30, 32, 35, 36, 38, 40, 41 |
| 4 Model with mathematics. | 17–29, 35–37, 39, 40, 42 |
| 6 Attend to precision. | 23, 31 |
| 7 Look for and make use of structure. | 17, 18, 29 |
| 8 Look for and express regularity in repeated reasoning. | 29, 30 |

### Go Online!

**eSolutions Manual**
Create worksheets, answer keys, and solutions handouts for your assignments.

connectED.mcgraw-hill.com   453

# Lesson 7-7 | Writing Exponential Functions

## Additional Answers

**33.** Sample answer: Exponential models can grow without bound, which is usually not the case of the situation that is being modeled. For instance, a population cannot grow without bound due to space and food constraints. Therefore, when using a model, the situation that is being modeled should be carefully considered when used to make decisions.

**34.** The exponential growth formula is $y = a(1 + r)^t$, where $a$ is the initial amount, $t$ is time, $y$ is the final amount, and $r$ is the rate of change expressed as a decimal. The exponential decay formula is the same except $r$ represents the rate of decay and it is subtracted from 1.

---

**B** **27. HOUSING** The median house price in the United States decreased an average of 4.25% each year starting in 2010 and continuing through 2012. Assume that this pattern continues.

a. Write an equation for the median house price for $t$ years after 2010. $I = 194,375(1 - 0.0425)^t$

b. Predict the median house price in 2030. **about $81,549**

**C** **28. ELEMENTS** A radioactive element's half-life is the time it takes for one half of the element's quantity to decay. The half-life of Plutonium-241 is 14.4 years. The number of grams $A$ of Plutonium-241 left after $t$ years can be modeled by $A = p(0.5)^{\frac{t}{14.4}}$, where the variable $p$ is the original amount of the element.

**Median House Price**

| | |
|---|---|
| 2010 | $194,375 |
| 2011 | $180,046 |
| 2012 | $178,005 |

**Source:** Real Estate Journal

a. How much of a 0.2-gram sample remains after 72 years? **0.00625 g**

b. How much of a 5.4-gram sample remains after 1095 days? **≈4.7 g**

**29c.** $C(t) = 300t + 19,000(0.995)^t$; The function represents the number of gallons of water in the pool at any time after the hose is turned on.

**29. COMBINING FUNCTIONS** A swimming pool holds a maximum volume of 20,500 gallons of water. It evaporates at a rate of 0.5% per hour. The pool currently contains a volume of 19,000 gallons of water.

a. Write an exponential function $w(t)$ to express the amount of water remaining in the pool after time $t$. The variable $t$ represents the number of hours after the pool has reached the volume of 19,000 gallons. $w(t) = 19,000(0.995)^t$

b. At this same time, a hose is turned on to refill the pool at a net rate of 300 gallons per hour. Write a function $p(t)$ where $t$ is time in hours the hose is running to express the amount of water that is pumped into the pool. $p(t) = 300t$

c. Find $C(t) = p(t) + w(t)$. What does this new function represent?

d. Use the graph of $C(t)$ to determine how long the hose must run to fill the pool to its maximum capacity. **about 7.3 h**

A.CED.2, F.LE.2, F.LE.5

### H.O.T. Problems — Use Higher-Order Thinking Skills

**30. REASONING** Determine the growth rate (as a percent) of a population that quadruples every year. Explain. **300%; Solving $y = a(1 + r)^t$ for $y = 4$, $a = 1$, and $t = 1$ gives $r = 3$ or 300%.**

**31. PRECISION** Santos invested $1200 into an account with an interest rate of 8% compounded monthly. Use a calculator to approximate how long it will take for Santos's investment to reach $2500. **about 9.2 yr**

**32. REASONING** The amount of water in a container doubles every minute. After 8 minutes, the container is full. After how many minutes was the container half full? Explain. **7; Sample answer: Since the amount of water doubles every minute, the container would be half full a minute before it was full.**

**33. WRITING IN MATH** What should you consider when using exponential models to make decisions? **See margin.**

**34. WRITING IN MATH** Compare and contrast the exponential growth formula and the exponential decay formula. **See margin.**

---

## Go Online!

### Quizzes

Students can use *Self-Check Quizzes* to check their understanding of this lesson. You can also give the *Chapter Quiz*, which covers the content in Lesson 7-7.

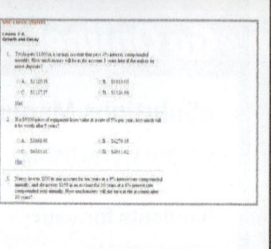

## Preparing for Assessment

**35.** A small town currently has a population of 15,000. Each year the population decreases by 2.25%. What will the population be in 8 years? ⓜ 2, 4  A.CED.2, F.LE.5
**A**
- ⓐ A  12,503
- ⓑ B  14,343
- ⓒ C  14,982
- ⓓ D  17,922

**36.** Robert invested $25,000 in an account that pays 4.5% interest compounded weekly. After 3 years, what is the balance of his account? ⓜ 2, 4  A.CED.2, F.LE.5  **B**
- ⓐ A  $3611.75
- ⓑ B  $28,611.75
- ⓒ C  $70,877.18
- ⓓ D  $95,877.18

**37.** The value $m$ of a baseball card is given by the equation $m = 240(1.25)^t$, where $t$ is the number of years after 2012. What is the value of the card $m$ in dollars in the year 2012? ⓜ 2, 4  F.LE.5  **240**

**38.** Which exponential function passes through (0, 2) and (2, 12.5)? ⓜ 1, 2  F.LE.2  **D**
- ⓐ A  $y = 2(2.5)^x$
- ⓑ B  $y = 2(12.5)^x$
- ⓒ C  $y = 2(3.125)^x$
- ⓓ D  $y = 2.5(2)^x$

**39.** The population of a small town, in thousands, can be represented by $y = 2.6(1.02)^t$, where $t$ is the number of years since 2010. Based on this model, which statements are true? ⓜ 1, 4  F.LE.5  **A, B, D**
- ☐ A  The population in 2010 was 2600.
- ☐ B  The population of the town is increasing.
- ☐ C  The population increases by 20% each year.
- ☐ D  The population in 2020 will be greater than 3100.
- ☐ E  The population in 2015 was about 3500.
- ☐ F  From 2010 to 2011, the population increased by 102%.

**40.** Gina just purchased an antique desk worth $8000. The equation $v = 8000(1.03)^t$ models the value of the desk after $t$ years. Which of the following best describes the meaning of 1.03 in the equation? ⓜ 2, 4  F.LE.5  **C**
- ⓐ A  The value of the desk will decrease by 1.03% each year.
- ⓑ B  The value of the desk will increase by 1.03% each year.
- ⓒ C  The value of the desk will increase by 3% each year.
- ⓓ D  The value of the desk will decrease by 3% each year.

**41.** Which exponential function represents the graph shown? ⓜ 1, 2  F.LE.2  **A**

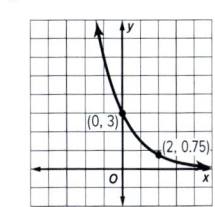

(0, 3)  (2, 0.75)

- ⓐ A  $y = 3(0.5)^x$
- ⓑ B  $y = 3(1.5)^x$
- ⓒ C  $y = 0.5(3)^x$
- ⓓ D  $y = 1.5(3)^x$

**42b.** The ball loses air every 2 months and the total time period is 9 months, the exponent will be $\frac{9}{2}$.

**42.** MULTI-STEP  Miguel filled a ball with 1.6 pounds of air. Every 2 months, the ball loses half its air. ⓜ 4, 8  F.LE.2, F.LE.5

**a.** Which equation represents the pounds of air $p$ that remain in the ball after 9 months?  **B**
- ⓐ A  $p = 0.8(0.5)^9$
- ⓑ B  $p = 1.6(0.5)^{\frac{9}{2}}$
- ⓒ C  $p = 1.6(0.5)^9$
- ⓓ D  $p = 1.6(1.5)^{\frac{9}{2}}$

**42c.** The amount of air is decreasing by half every 2 months, the base is (1 − 0.5), or 0.5.

**b.** Explain how you determined the exponent in the equation.

**c.** Explain how you determined the base of the exponential expression in the equation.

---

### Differentiated Instruction  AL  OL

**IF** you think students need a challenge in this lesson,

**THEN** ask students to write their own exponential growth or decay problems, using data from periodicals or the Internet. Have students share their problems with the class when they are complete.

---

Lesson 7-7 | Writing Exponential Functions

## Preparing for Assessment

Exercises 35–42 require students to use the skills they will need on standardized assessments. Each exercise is dual-coded with content standards and mathematical practice standards.

### Dual Coding

| Exercises | Content Standards | ⓜ Mathematical Practices |
|---|---|---|
| 35, 36 | A.CED.2, F.LE.5 | 2, 4 |
| 37 | F.LE.5 | 2, 4 |
| 38 | F.LE.2 | 1, 2 |
| 39 | F.LE.5 | 1, 4 |
| 40 | F.LE.5 | 2, 4 |
| 41 | F.LE.2 | 1, 2 |
| 42 | F.LE.2, F.LE.5 | 8, 4 |

## Diagnose Student Errors

Survey student responses for each item. Class trends may indicate common errors and misconceptions.

**35.**

| A | CORRECT |
|---|---|
| B | Used 2.25 instead of 0.0225 and subtracted instead of multiplying |
| C | Subtracted 2.25 for each year |
| D | Found an increase of 2.25% each year |

**36.**

| A | Found the interest earned |
|---|---|
| B | CORRECT |
| C | Found interest earned at 45% instead of 4.5% |
| D | Used 45% instead of 4.5% |

**40.**

| A | Didn't convert the ratio correctly or notice the ratio is greater than 1, which is an increase |
|---|---|
| B | Didn't convert the ratio correctly |
| C | CORRECT |
| D | Didn't notice the ratio is greater than 1, which is an increase |

# Extend 7-7

## Focus

**Objective** Use a graphing calculator to solve exponential equations and inequalities.

### Materials
- TI-Nspire Technology

### Teaching Tips
- For Activity 1, remind students that to enter $3^x + 4$, they will need to use the ^ key for the exponent and use the down arrow before entering the + 4.
- When changing the windows settings, use the **tab** key to move from one field to another.
- In Activity 2, students will need to use the **tab** key to move the curser to the entry line to type f2(x).

## Teach

**Working in Cooperative Groups** Divide the class into pairs. Work through Activity 1 and Activity 2 as a class. Then ask students to work with their partners to complete Exercises 1–9 and Activities 3 and 4.

**Practice** Have students complete Exercises 10–12.

---

### MP Teaching the Mathematical Practices

**Tools** Mathematically proficient students are sufficiently familiar with tools appropriate to make sound decisions about when each of these tools might be helpful, recognizing both the insight to be gained and their limitations. Point out that Activities 2, 3, and 4 offer various methods for solving equations and inequalities. Discuss when to use the methods and the technology tools available.

---

# EXTEND 7-7

## Graphing Technology Lab
## Solving Exponential Equations and Inequalities

You can use TI-Nspire Technology to solve exponential equations and inequalities by graphing and by using tables.

**Mathematical Practices**
MP 5 Use appropriate tools strategically.

**Content Standards**
A.REI.11 Explain why the x-coordinates of the points where the graphs of the equations $y = f(x)$ and $y = g(x)$ intersect are the solutions of the equation $f(x) = g(x)$; find the solutions approximately, e.g., using technology to graph the functions, make tables of values, or find successive approximations. Include cases where $f(x)$ and/or $g(x)$ are linear, polynomial, rational, absolute value, exponential, and logarithmic functions.

### Activity 1 Graph an Exponential Equation

Graph $y = 3^x + 4$ using a graphing calculator.

**Step 1** Add a new **Graphs** page.

**Step 2** Enter $3^x + 4$ as **f1(x)**.

**Step 3** Use the **Window Settings** option from the **Window/Zoom** menu to adjust the window so that $x$ is from $-10$ to 10 and $y$ is from $-100$ to 100. Keep the scales as **Auto**.

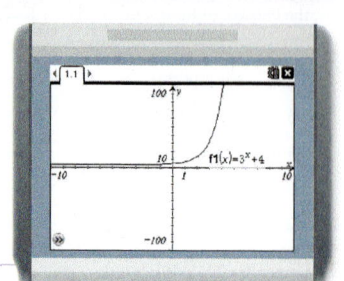

To solve an equation by graphing, graph both sides of the equation and locate the point(s) of intersection.

### Activity 2 Solve an Exponential Equation by Graphing

Solve $2^{x-2} = \frac{3}{4}$.

**Step 1** Add a new **Graphs** page.

**Step 2** Enter $2^{x-2}$ as **f1(x)** and $\frac{3}{4}$ as **f2(x)**.

**Step 3** Use the **Intersection Point(s)** tool from the **Points & Lines** menu to find the intersection of the two graphs. Select the graph of **f1(x) enter** and then the graph of **f2(x) enter**.

The graphs intersect at about (1.58, 0.75).

Therefore, the solution of $2^{x-2} = \frac{3}{4}$ is 1.58.

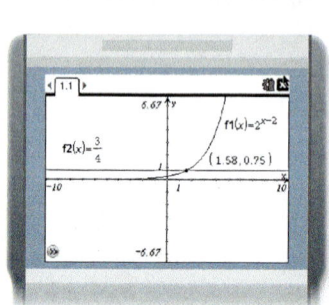

### Exercises

**TOOLS** Use a graphing calculator to solve each equation.

1. $\left(\frac{1}{3}\right)^{x-1} = \frac{3}{4}$  ≈1.26
2. $2^{2x-1} = 2x$  0.5, 1
3. $\left(\frac{1}{2}\right)^{2x} = 2^{2x}$  0
4. $5^{\frac{1}{3}x+2} = -x$  ≈−3.61
5. $\left(\frac{1}{8}\right)^{2x} = -2x + 1$  0, ≈0.409
6. $2^{\frac{1}{4}x-1} = 3^{x+1}$  ≈−1.94
7. $2^{3x-1} = 4^x$  1
8. $4^{2x-3} = 5^{-x+1}$  ≈1.32
9. $3^{2x-4} = 2^x + 1$  3

### Activity 3 Solve an Exponential Equation by Using a Table

Solve $2\left(\frac{1}{2}\right)^{x+2} = \frac{1}{4}$ using a table.

**Step 1** Add a new **Lists & Spreadsheet** page.

**Step 2** Label column A as $x$. Enter values from $-4$ to $4$ in cells A1 to A9.

**Step 3** In column B in the formula row, enter the left side of the rational equation. In column C of the formula row, enter $= \frac{1}{4}$. Specify **Variable Reference** when prompted.

Scroll until you see where the values in Columns B and C are equal. This occurs at $x = 1$. Therefore, the solution of $2\left(\frac{1}{2}\right)^{x+2} = \frac{1}{4}$ is 1.

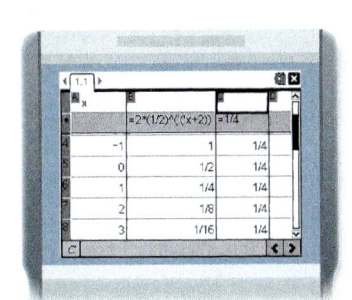

You can also use a graphing calculator to solve exponential inequalities.

### Activity 4 Solve an Exponential Inequality

Solve $4^{x-3} \leq \left(\frac{1}{4}\right)^{2x}$.

**Step 1** Add a new **Graphs** page.

**Step 2** Enter the left side of the inequality into **f1(x)**. Press **del**, select $\geq$, and enter $4^{x-3}$. Enter the right side of the inequality into **f2(x)**. Press **tab del** $\leq$, and enter $\left(\frac{1}{4}\right)^{2x}$.

The $x$-values of the points in the region where the shading overlap is the solution set of the original inequality.
Therefore, the solution of $4^{x-3} \leq \left(\frac{1}{4}\right)^{2x}$ is $x \leq 1$.

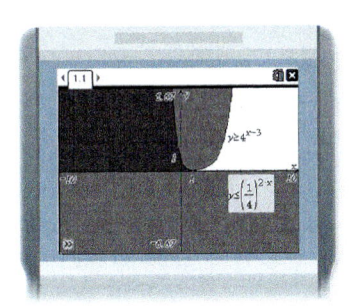

### Exercises

**TOOLS** Use a graphing calculator to solve each equation or inequality.

10. $\left(\frac{1}{3}\right)^{3x} = 3^x$  0

11. $\left(\frac{1}{6}\right)^{2x} = 4^x$  0

12. $3^{1-x} \leq 4^x$  $x \geq 0.442$

13. $4^{3x} \leq 2x + 1$  $-0.409 \leq x \leq 0$

14. $\left(\frac{1}{4}\right)^x > 2^{x+4}$  $x < -1.33$

15. $\left(\frac{1}{3}\right)^{x-1} \geq 2^x$  $x \leq 0.613$

---

## Assess

**Formative Assessment** Use Exercises 13–15 to assess each student's knowledge of solving exponential equations and inequalities.

**From Concrete to Abstract** Ask students to summarize the use of technology to find the solutions to exponential equations and inequalities.

# LESSON 7-8
# Transforming Exponential Expressions

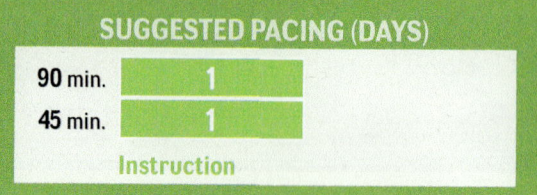

**SUGGESTED PACING (DAYS)**
90 min. — 1
45 min. — 1
Instruction

## Track Your Progress

### Objective
1. Transform and interpret expressions of exponential functions by applying the properties of exponents.

### Mathematical Background
One application for using exponential functions in the real world is applying it to the growth and depreciation of money. The formula $y = P(1 + r)^t$ can be used to find out how the amount of money $P$ increases over a period of time $t$, at a given percent rate of change $r$. Changing the formula to $y = P(1 - r)^t$ will show how the amount of money $P$ depreciates. When looking at a graph, in order to find the pecent rate of change one simply takes the (new value − original value)/(original value).

### THEN
**F.LE.2** Construct linear and exponential functions, including arithmetic and geometric sequences, given a graph, a description of a relationship, or two input-output pairs (include reading these from a table).

**F.LE.5** Interpret the parameters in a linear or exponential function in terms of a context.

### NOW
**A.SSE.3c** Use the properties of exponents to transform expressions for exponential functions.

**F.IF.8b** Use the properties of exponents to interpret expressions for exponential functions.

### NEXT
**F.BF.2** Write arithmetic and geometric sequences both recursively and with an explicit formula, use them to model situations, and translate between the two forms.

**F.IF.3** Recognize that sequences are functions, sometimes defined recursively, whose domain is a subset of the integers.

## Go Online! All of these resources and more are available at connectED.mcgraw-hill.com

**eLessons** utilize the power of your interactive whiteboard in an engaging way. Use **Exponential Functions**, screens 10–11, to review exponential growth and decay and to introduce the concepts in this lesson.

*Use at Beginning of Lesson*

Use **The Geometer's Sketchpad** to explore transformations of exponential expressions and compare their graphs.

*Use with Examples*

The **Chapter Project** allows students to create and customize a project as a nontraditional method of assessment.

*Use at End of Lesson*

## Using Open Educational Resources
**Professional Development** Exponential expressions can be a difficult subject to teach. For new ideas, visit Google Classroom Lessons and Resources. *Use during planning*

# Go Online!
connectED.mcgraw-hill.com — Worksheets

# Differentiate Your Resources

**Extra Practice** Additional practice or homework; Skills Practice is best for approaching-level students and Practice is best for on-level and beyond-level students

## Skills Practice
AL  OL  ELL

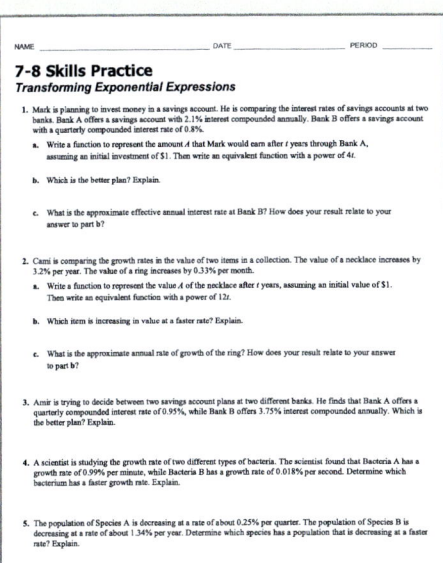

## Practice
AL  OL  BL  ELL

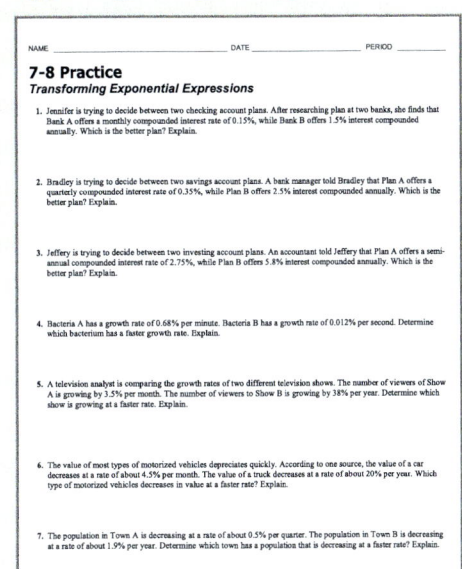

## Word Problem Practice
AL  OL  BL  ELL

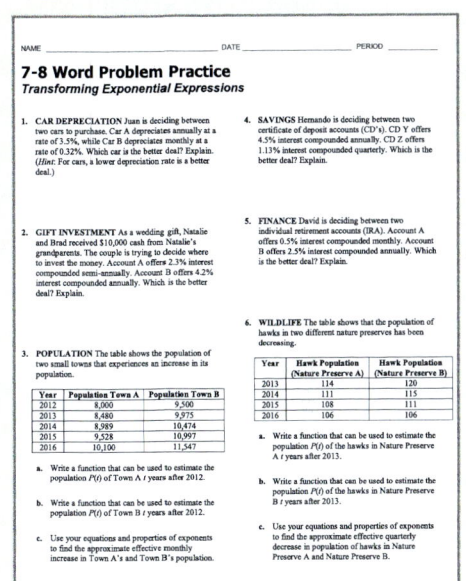

**Intervention** Reteaching and vocabulary activities that can be used with struggling or absent students and as ELL support

**Extension** Activities that can be used to extend lesson concepts

## Study Guide and Intervention
AL  OL  ELL

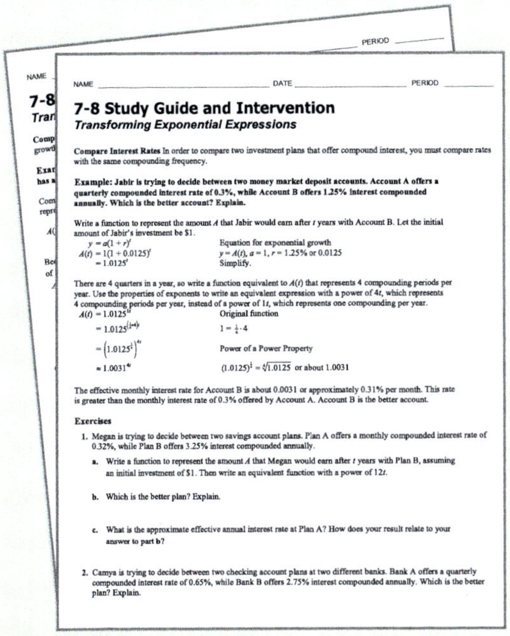

## Study Notebook
AL  OL  BL  ELL

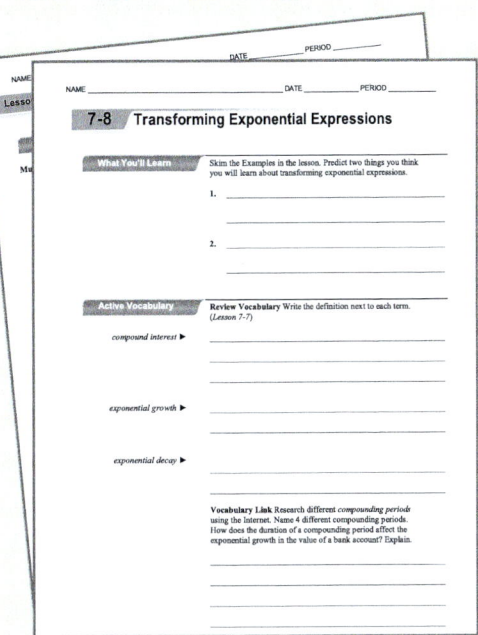

## Enrichment
OL  BL  ELL

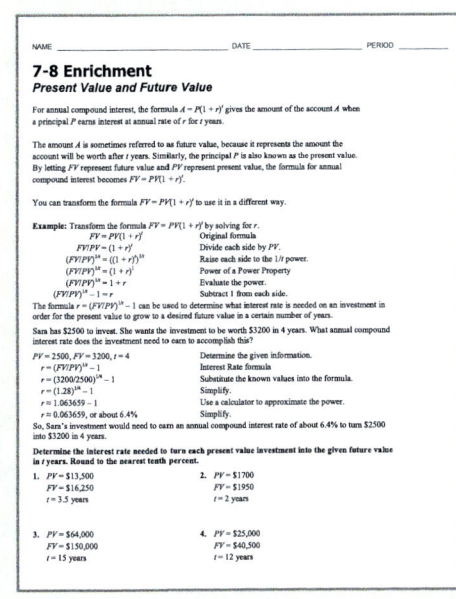

connectED.mcgraw-hill.com  458B

Lesson 7-8 | Transforming Exponential Expressions

# Launch

Have students read the Why? section of the lesson.

- **What is meant by the compounding frequency?** The compounding frequency describes how often the interest is applied to the existing balance in the account.

- **When all other parameters are the same, how does the compounding frequency affect the total balance in an account?** The greater the compounding frequency is, the faster the growth of the total account balance.

# Teach

Ask the scaffolded questions for each example to build conceptual understanding for students at all levels.

## 1 Transform and Interpret Exponential Expressions

### Example 1 Compare Rates

**AL** What process do you use to write the interest rate 2.5% as a decimal? Delete the percent symbol and move the decimal point two places to the left: 2.5% = 0.025.

**OL** Would the effective monthly interest rate be same if a different value for *a* was used? Explain. Yes; because the interest rate affects the base of the exponential expression, the constant multiplier *a* does not affect the interest rate.

**BL** How would you transform the function $A(t)$ if you wanted to compare Plan B to a plan that offers quarterly compounded interest?

$A(t) = 1.025^{(\frac{1}{4} \cdot 4)t} = \left(1.025^{\frac{1}{4}}\right)^{4t} \approx 1.0062^{4t}$

## Go Online!

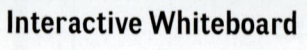

**Interactive Whiteboard**
Use the eLesson or Lesson Presentation to present this lesson.

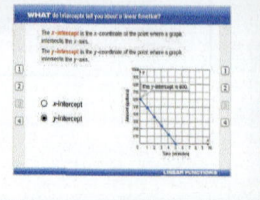

---

# LESSON 8
# Transforming Exponential Expressions

| :: Then | :: Now | :: Why? |
|---|---|---|
| • You wrote and graphed exponential functions. | 1 Transform and interpret expressions of exponential functions by applying the properties of exponents. | • When you invest money at the bank, it earns compound interest. To compare two different investment plans, you need to be able to compare interest rates and compounding frequencies to make the best investment with your money. |

**MP Mathematical Practices**
4 Model with mathematics.
7 Look for and make use of structure.

**Content Standards**
A.SSE.3c Use the properties of exponents to transform expressions for exponential functions.
F.IF.8b Use the properties of exponents to interpret expressions for exponential functions.

**1 Transform and Interpret Exponential Expressions** You can use the properties of exponents to transform exponential functions into other forms in order to solve real-world problems. Compound interest plans that have different numbers of compoundings per year can be compared by writing their exponential growth functions with the same exponent. For instance, to compare a plan compounded annually with another compounded monthly, write the function for the annual plan in terms of the exponent for the monthly plan. The *effective* monthly interest rate of the annual plan can then be determined from the resulting function.

F.IF.8b

### Example 1 Compare Rates

Monique is trying to decide between two savings account plans. Plan A offers a monthly compounded interest rate of 0.25%, while Plan B offers 2.5% interest compounded annually. Which is the better plan? Explain.

Write a function to represent the amount $A(t)$ that Monique would earn after $t$ years with Plan B. For convenience, let the initial amount of Monique's investment be $1.

$y = a(1 + r)^t$    Equation for exponential growth
$A(t) = 1(1 + 0.025)^t$    $y = A(t), a = 1, r = 2.5\%$ or $0.025$
$= 1.025^t$    Simplify.

Now write a function equivalent to $A(t)$ that represents 12 compoundings per year.

Use the properties of exponents to write an equivalent expression with a power of $12t$, which represents 12 compoundings per year, instead of a power of $1t$, which represents one compounding per year.

$A(t) = 1.025^{1t}$    Original function
$= 1.025^{(\frac{1}{12} \cdot 12)t}$    $1 = \frac{1}{12} \cdot 12$
$= \left(1.025^{\frac{1}{12}}\right)^{12t}$    Power of a Power Property
$\approx 1.0021^{12t}$    $(1.025)^{\frac{1}{12}} = \sqrt[12]{1.025}$ or about $1.0021$

From this equivalent function, you can determine that the effective monthly interest rate for Plan B is about 0.0021 or approximately 0.21% per month. This rate is less than the monthly interest rate of 0.25% offered by Plan A.

Plan A is the better plan.

**Guided Practice**   1. $y = (1.05)^t$; $y = (1.0123)^{4t}$; 1.23%; The second plan is better.

1. Let Me Grow Your Money offers two investment programs. The first plan has a 5% rate of interest compounded annually while the second plan has a quarterly compounded rate of 1.3%. Write a function to represent the amount earned by the first plan after $t$ years. Then, write an equivalent function that represents the first plan with 4 compoundings per year. What is the *effective* quarterly rate of interest of the first plan? Which is the better plan?

---

## MP Mathematical Practices Strategies

**Look for and make use of structure.** Help students use properties to see complicated things, such as some exponential expressions, as single objects or as composed of several objects. For example, ask:

- **What is the Power of a Power Property?** For any nonzero real number $a$, $(a^m)^n = a^{mn}$.

- **State the Power of a Power Property in your own words?** To find a power raised to another power, you can find the base raised to the product of the exponents.

- **What is the equation for exponential growth? What do the variables in the equation represent?** $y = a(1 + r)^t$, where $y$ is the final amount, $a$ is the initial amount, $r$ is the growth rate written as a decimal, and $t$ is time.

- **How do you modify the equation for exponential decay?** The equation becomes $y = a(1 - r)^t$, where $r$ is the decay rate written as a decimal.

You can use the properties of exponents and the compound interest equation to compare rates of exponential growth in a variety of real-world situations.

A.SSE.3c, F.IF.8b

**Real-World Example 2** — Compare Rates by Using Compound Interest Equation

**BACTERIA** *Lactobacillus acidophilus* is a bacteria used in the production of yogurt and some cheeses. It has a growth rate of 0.92% per minute. The bacteria *bacillus megaterium* is synthesized to create penicillin and has a growth rate of 0.046% per second. Determine the effective rate of growth per minute for the *bacillus megaterium*. Which bacteria grows at a faster rate? Explain.

The compound interest equation can be used to express a growth rate in a greater unit of time. Use the compound interest equation to write a function to represent the amount $A(t)$ of *bacillus megaterium* after $t$ minutes. Let the initial amount of bacteria equal $x$ and the number of compoundings equal 60.

$A(t) = P\left(1 + \frac{r}{n}\right)^{nt}$  Compound interest equation
$A(t) = x(1 + 0.00046)^{60t}$  $P = x, \frac{r}{n} = 0.046\%$ or $0.00046$, $n = 60$
$\quad = x(1.00046)^{60t}$  Simplify.

The amount of *bacillus megaterium* after t minutes is given by $A(t) = 1.00046^{60t}$.

To determine the per minute rate, write an equivalent equation representing only one compounding per minute by using the Product of Powers Property to change the exponent from $60t$ to $1t$.

$A(t) = x(1.00046)^{60t}$  Original function
$\quad = x(1.00046^{60})^{t}$  Product of Powers Property
$\quad \approx x(1.028)^{t}$  $1.00046^{60} \approx 1.028$

The equivalent equation representing only one compounding per minute is $A(t) = x(1.028)^{t}$. The effective rate of growth per minute for the *bacillus megaterium* is 0.028 or about 2.8%. *Bacillus megaterium* grows faster because its effective growth rate is 2.8% per minute and the growth rate of *lactobacillus acidophilus* is 0.92% per minute.

**Study Tip**

**MP Sense-Making** In one equation, $t$ is a variable representing minutes and in the other equation, $t$ is a variable representing seconds. Therefore, $t$ is not equivalent in the two equations and must be converted to represent the same unit.

**Guided Practice**

2. $A(t) = (1.015)^{12t}$; $A(t) = (1.196)^{t}$; 19.6%; App A is growing at a faster rate.

2. **SOCIAL MEDIA** A business analyst is comparing the growth rates of two social media apps. The number of subscribers to App A is growing by 1.5% per month. The number of subscribers to App B is growing by 16% per year. Write a function to represent the number of subscribers to App A after $t$ years. Then, write an equivalent function that represents the subscribers to App A with only 1 compounding per year. What is the *effective* yearly rate of growth of the subscribers to App A? Which app is growing at a faster rate?

**Check Your Understanding** ○ = Step-by-Step Solutions begin on page R11.   Go Online! for a Self-Check Quiz

Example 1
A.SSE.3c,
F.IF.8b

1. Tareq is planning to invest money in a savings account. Oak Hills Financial offers 3.1% interest compounded annually. First City Bank has a savings account with a quarterly compounded interest rate of 0.7%.

    a. Write a function to represent the amount $A(t)$ that Tareq would earn after $t$ years through Oak Hills Financial, assuming an initial investment of $1. Then write an equivalent function with an exponent of $4t$.  $A(t) = (1.031)^{t}$; $A^{(t)} = (1.0077)^{4t}$
    b. What is the effective quarterly interest rate at Oak Hills Financial?  0.77%
    c. Which is the better plan? Explain.  Oak Hill Financial; The effective quarterly rate of 0.77% is greater than the 0.7% quarterly rate at First City Bank.

### Differentiated Homework Options

| Levels | AL Basic | OL Core | BL Advanced |
|---|---|---|---|
| Exercises | 3, 4, 8–18 | 3–5, 7–18 | 5–11<br>12–18 (optional) |
| 2-Day Option | 3, 12–18 | 3–7 | |
| | 4, 8–11 | 8–18 | |

You can use ALEKS to provide additional remediation support with personalized instruction and practice.

---

**Lesson 7-8 | Transforming Exponential Expressions**

**Need Another Example?**
Tyrell is comparing two savings account plans. Plan X offers an annually compounded interest rate of 1.9%. Plan Y offers a monthly compounded interest rate of 0.13%. Which is the better plan? Explain.

Plan X; Plan X has an effective monthly interest rate of about 0.16%, which is greater than the monthly interest rate for Plan Y.

**Example 2 Compare Rates by Using the Compound Interest Equation**

**AL** Is it possible to solve the problem just by comparing the growth rates given in the problem, 0.92% and 0.046%? Why or why not? No; the compounding frequencies are different for the two bacteria.

**OL** How do you find the growth rate from the expression $(1.0280)^{\frac{t}{60}}$? Write the value in parentheses in the form $1 + r$, or $1 + 0.0280$. The growth rate $r$ is 0.0280 or 2.8%.

**BL** What is another way to compare the growth rates in this problem? Compare the growth rates per second. The rate for *bacillus megaterium* is already in this form. For *lactobacillus acidophilus*, the effective rate is about $1.0092^{\frac{1}{60}} \approx 0.00015$, or about 0.015% per second. This is less than the rate for *lactobacillus acidophilus*, so *bacillus megaterium* has a faster growth rate.

**Need Another Example?**
**POPULATION** In 2014, the population of New Zealand grew at a rate of about 0.83% per year. The population of Canada grew at a rate of about 0.063% per month. Determine which country grew at a faster rate. Explain. New Zealand grew at a faster rate because its effective monthly growth rate is about 0.069%, which is greater than the rate for Canada.

## Practice

**Formative Assessment** Use Exercises 1 and 2 to assess students' understanding of the concepts in the lesson.

The Practice and Problem Solving exercises assess the content taught in the lesson. The Preparing for Assessment page is meant to be used as preparation for end-of-course assessment.

connectED.mcgraw-hill.com 459

# Lesson 7-8 | Transforming Exponential Expressions

## Teaching the Mathematical Practices

**Precision** Mathematically proficient students calculate carefully and efficiently, and express numerical answers with a degree of precision appropriate for the problem context. Exercise 5 gives students a chance to consider how to best round their answer. Students should understand that expressing the effective monthly growth rate as 0.64% is more useful than rounding to 1% and understand that an answer like 0.643403% is more precise than is warranted by the given information in the problem.

## Extra Practice

See page R7 for extra exercises for students who are approaching level or for on-level students who need additional reinforcement.

## Assess

**Ticket Out the Door** Write several different pairs of savings account interest rates and their compounding frequencies on sheets of paper. Make copies of the interest rates and give one pair of rates to each student. As students leave the room, have them explain which rate is a better deal.

### Levels of Complexity Chart

The levels of the exercises progress from 1 to 3, with Level 1 indicating the lowest level of complexity.

| Exercises | 3–4 | 5–6, 12–18 | 7–11 |
|---|---|---|---|
| Level 3 |  |  | ● |
| Level 2 |  | ● |  |
| Level 1 | ● |  |  |

## Additional Answers

**6.** The graphs of the two functions are almost identical on a large interval of their domain. This shows that $A = P[(1+r)^{\frac{1}{n}}]^{nt}$ gives a good approximation for $A = P(1+\frac{r}{n})^{nt}$.

**8.** Sample answer: monthly compounded rate of 0.41% and annual compounded rate of 5%; for the 5% annual rate, $(1+0.05)^t = 1.05^t = (1.05^{\frac{1}{12}})^{12t} \approx (1.0041)^{12t}$, so the effective monthly rate is approximately 0.41%.

**9.** No; in the expression, $0.987^{12t}$, the base of the expression represents a decrease of 1.3%, so the effective monthly decrease is about 1.3%, not 98.7%.

---

**Example 2**
A.SSE.3c, F.IF.8b

**2. COLLECTIBLES** Jemma is comparing the growth rates in the value of two items in her baseball collection. The value of a rare baseball card increases by 8.1% per year. The value of a mitt increases by 0.57% each month. Write a function to represent the value of the mitt after $t$ years with an initial value of $650. Then, write an equivalent function that represents the value of the mitt with only one compounding per year. What is the *effective* yearly rate of growth of the mitt? Which item in Jemma's collection is increasing in value at a faster rate?
$A(t) = (1.0057)^{12t}$; $A(t) = (1.071)^t$; 7.1%; the baseball card

### Practice and Problem Solving
Extra Practice is on page R7.

**Example 1**
A.SSE.3c, F.IF.8b

**3. CREDIT** Adam is choosing between two credit cards. World Mutual offers a credit card with 15.3% interest compounded annually. Super City Card has a monthly compounded interest rate of 1.4%. Which card is the better offer? Explain.

**Example 2**
A.SSE.3c, F.IF.8b

**4. TECHNOLOGY** According to one source, the value of a desktop computer decreases at a rate of about 6% per month. The value of a laptop computer decreases at a rate of about 66% per year. Write a function to represent the value of the desktop computer after $t$ years with an initial value of $450. Then, write an equivalent function that represents the value of the desktop computer with only 1 compounding per year. What is the effective yearly rate of the decrease in value of the desktop computer? Which type of computer is decreasing in value at a faster rate?

**3.** World Mutual; World Mutual has an effective monthly interest rate of about 1.19%, which is lower than the monthly rate for Super City Card.

**5. POPULATION** The table shows the population of a small town that experiences a rapid increase in its population. The function $P(t) = 10,200(1.08)^t$ represents the growth of the population using an annual growth rate. Find the approximate effective monthly increase in the town's population. ≈ 0.64%

| Year | Population |
|---|---|
| 2012 | 10,200 |
| 2013 | 11,016 |
| 2014 | 11,897 |
| 2015 | 12,849 |
| 2016 | 13,877 |

**6. TOOLS** Using the compound interest equation $A = P(1+\frac{r}{n})^{nt}$ choose fixed and reasonable values for $P$, $n$, and $r$, and use a graphing calculator to graph $A(t)$. Using the same graphing window and the same values for $P$, $n$, and $r$, graph $A = P[(1+r)^{\frac{1}{n}}]^{nt}$. What do you notice? What conclusions can you make? **See margin.**

**7. HYPERINFLATION** Hyperinflation is extremely rapid inflation within the economy of a country. In Austria in 1922, the inflation rate for one year was 1426%.

**a.** What was the effective monthly inflation rate in Austria in 1922? ≈ 25.5%

**b.** During this period, Austria used the crown as their unit of currency. Suppose an item cost 10 crowns at the beginning of May, 1922. What would have been the expected cost of the item at the beginning of August, 1922? ≈ 19.77 crowns

A.SSE.3c, F.IF.8b

### H.O.T. Problems  Use Higher-Order Thinking Skills

**4.** $A(t) = (0.94)^{12t}$; $A(t) = (0.476)^t$; 52.4%; Since 66% is greater than 52.4%, the laptop computer is decreasing at a faster rate

**8. OPEN-ENDED** Give an example of two interest rates—one that is compounded monthly and one that is compounded annually—such that the effective monthly interest rates are approximately equal. Justify your answer. **See margin.**

**9. CRITIQUE ARGUMENTS** Celia said that an annual decrease of 10% is equivalent to an effective monthly decrease of about 98.7% because $(1-r)^t = (1-0.1)^t = 0.9^t = (0.9^{\frac{1}{12}})^{12t} \approx 0.987^{12t}$. Do you agree? Explain. **See margin.**

**10. ERROR ANALYSIS** A savings account offers an annual interest rate of 2.1%. Yoshio was asked to estimate the effective quarterly interest rate. His work is shown here. Based on his work, he concluded that the effective quarterly rate is about 8.7%. Is Yoshio correct? If not, explain his error and show how to solve the problem correctly. **See margin.**

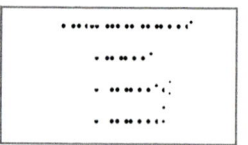

**11. WRITING IN MATH** In Example 1, why does it make sense to assume that the initial investment is $1? Would a different initial investment change the result of the problem? Explain. **See margin.**

---

### Standards for Mathematical Practice

| Emphasis On | Exercises |
|---|---|
| 3 Construct viable arguments and critique the reasoning of others | 9, 10 |
| 4 Model with mathematics. | 1–7, 10, 12–18 |
| 5 Use appropriate tools strategically. | 6 |
| 6 Attend to precision. | 5, 8, 11 |

**10.** No; Yoshio should have written the expression as $A(t) = (1.021^{\frac{1}{4}})^{4t} \approx 1.0052^{4t}$, which shows that the effective quarterly rate is about 0.52%.

**11.** Sample answer: Any amount can be used for the initial investment, but $1 is simply a convenient amount. In the expression $1(1+0.00046)^t$, the key quantity to analyze is $(1+0.00046)^t$; the initial amount, which appears outside the parentheses, does not affect the analysis, so choosing a different initial amount would not change the result.

Lesson 7-8 | Transforming Exponential Expressions

## Preparing for Assessment

**12.** The number of players of an online video game is growing according to the model $V = 2000(1.09)^t$, where $t$ is the time in years. Which equation can be used to model the monthly growth in the number of players? MP 4, 7  A.SSE.3c  **D**

- A  $V = 2000(2.8167)^{12t}$
- B  $V = 166.7(2.8167)^{12t}$
- C  $V = 166.7(1.0072)^{12t}$
- D  $V = 2000(1.0072)^{12t}$

**13.** The population of a certain bacteria is given by $A(t) = 200(1.4)^t$, where $t$ is time in hours.

| Time (h) | Population |
|---|---|
| 0 | 200 |
| 1 | 280 |
| 2 | 392 |
| 3 | 549 |

Which statements about the growth of the bacteria are true? MP 4, 7  A.SSE.3c, F.IF.8b  **B, C, E**

- A  After 20 minutes, the predicted population of bacteria is 167,337.
- B  The number of bacteria grows by 40% per hour.
- C  The number of bacteria is modeled by $A(t) = 200(1.0056)^{60t}$, where $t$ is the time in minutes.
- D  The number of bacteria grows by about 5.6% per second.
- E  The initial population of bacteria was 200.

**14.** The total number of followers of a musician on a social network is growing at a rate of 24% per year. Which is the best estimate of the effective weekly growth rate in the number of followers? MP 4, 7  A.SSE.3c, F.IF.8b  **A**

- A  0.045%
- B  0.41%
- C  11.1%
- D  46%

**15.** The population of a town is increasing at a rate of 0.14% per month. Which equation best models the growth of the population, where $P_0$ is the initial population and $t$ is the time in years? MP 4, 7  A.SSE.3c, F.IF.8b  **B**

- A  $P(t) = P_0(1.0001)^t$
- B  $P(t) = P_0(1.017)^t$
- C  $P(t) = P_0(1.68)^t$
- D  $P(t) = P_0(4.82)^t$

**16.** The population of an invasive species of mussels in a river is increasing at a rate of about 1.63% each month. Which is the effective annual growth rate of the mussel population in the river? MP 4, 7  **D**  A.SSE.3c, F.IF.8b

- A  0.13%
- B  6.12%
- C  19.56%
- D  21.41%

18d. Capital Street; the effective annual rate for Capital Street is about 2.2%, which is greater than the annual rates at the other banks.

**17.** The value of a tablet computer decreases by 14% per year. Determine the approximate effective monthly rate of decrease in the value. Write your answer rounded to the nearest tenth. MP 4, 7
**1.1%**
A.SSE.3c, F.IF.8b

**18.** MULTI-STEP Sharonda is comparing the saving accounts that are offered at the two banks. MP 4, 7  A.SSE.3c, F.IF.8b

| Bank | Interest Rate |
|---|---|
| XYZ Savings | 1.8% compounded annually |
| Statewide Financial | 0.12% compounded monthly |

a. Determine the effective monthly interest rate for the savings account at XYZ Savings. ≈ **0.15%**

b. Which of the two banks offers a better deal? Explain. **XYZ Savings; the effective monthly rate 0.15% is greater than the monthly rate at Statewide Financial.**

c. What is the effective annual interest rate for the savings account at Statewide Financial? ≈ **1.45%**

d. Sharonda learns that Capital Street Bank offers a savings account with 0.55% interest compounded quarterly. Which of the three banks should she choose? Explain. **See margin.**

## Preparing for Assessment

Exercises 12–18 require students to use the skills they will need on assessments. Each exercise is dual-coded with content standards and standards for mathematical practice.

### Dual Coding

| Items | Content Standards | MP Mathematical Practices |
|---|---|---|
| 12 | A.SSE.3c | 4, 7 |
| 13 | A.SSE.3c, F.IF.8b | 4, 7 |
| 14 | A.SSE.3c, F.IF.8b | 4, 7 |
| 15 | A.SSE.3c, F.IF.8b | 4, 7 |
| 16 | A.SSE.3c, F.IF.8b | 4, 7 |
| 17 | A.SSE.3c, F.IF.8b | 4, 7 |
| 18 | A.SSE.3c, F.IF.8b | 4, 7 |

### Diagnose Student Errors

Survey student responses for each item. Class trends may indicate common errors and misconceptions.

**12.**

| A | Evaluated $1.09^{12}$ instead of $1.09^{\frac{1}{12}}$ |
|---|---|
| B | Evaluated $1.09^{12}$ instead of $1.09^{\frac{1}{12}}$ and divided the initial amount by 12 |
| C | Divided the initial amount by 12 |
| D | CORRECT |

**14.**

| A | Incorrectly converted the growth rate to a percent |
|---|---|
| B | CORRECT |
| C | Divided 24% by 52 |
| D | Calculated $1.24^{52}$ |

**15.**

| A | Chose an equation that models the growth where $t$ is the time in years |
|---|---|
| B | CORRECT |
| C | Calculated $1.028^{12}$ |
| D | Multiplied 0.028 by 12 |

**16.**

| A | Calculated $1.22^{12}$ |
|---|---|
| B | Interchanged months and years |
| C | CORRECT |
| D | Incorrectly interpreted the value of $1.22^{\frac{1}{12}}$ |

**18d.** Capital Street Bank; the effective annual rate is 2.2%, which is greater than the other banks.

### Go Online!

**Quizzes**
Students can use Self-Check Quizzes to check their understanding of this lesson.

connectED.mcgraw-hill.com  461

# LESSON 7-9
# Geometric Sequences as Exponential Functions

**SUGGESTED PACING (DAYS)**

| | Instruction | Extend Lab |
|---|---|---|
| 90 min. | 0.5 | 0.5 |
| 45 min. | 1 | 0.5 |

## Track Your Progress

### Objectives
1. Identify and generate geometric sequences.
2. Relate geometric sequences to exponential functions.

### Mathematical Background
In a geometric sequence, each term after the first is found by multiplying the previous term by a common ratio, $r$. The common ratio can be found by dividing any term by its previous term. Successive terms can be found by multiplying the previous term by $r$. The $n$th term can be found using the formula $a_n = a_1 \cdot r^{n-1}$, where $n$ is any positive integer.

### THEN
**A.SSE.3c** Use the properties of exponents to transform expressions for exponential functions.

**F.IF.8b** Use the properties of exponents to interpret expressions for exponential functions.

### NOW
**F.BF.2** Write arithmetic and geometric sequences both recursively and with an explicit formula, use them to model situations, and translate between two forms.

**F.LE.2** Construct linear and exponential functions, including arithmetic and geometric sequences, given a graph, a description of the relationship, or two input-output pairs (including reading these from a table).

### NEXT
**F.IF.3** Recognize that sequences are functions, sometimes defined recursively, whose domain is a subset of the integers.

**F.BF.2** Write arithmetic and geometric sequences both recursively and with an explicit formula, use them to model situations, and translate between the two forms.

---

**Go Online!** All of these resources and more are available at connectED.mcgraw-hill.com

Use **Self-Check Quiz** to assess students' understanding of the concepts in this lesson. 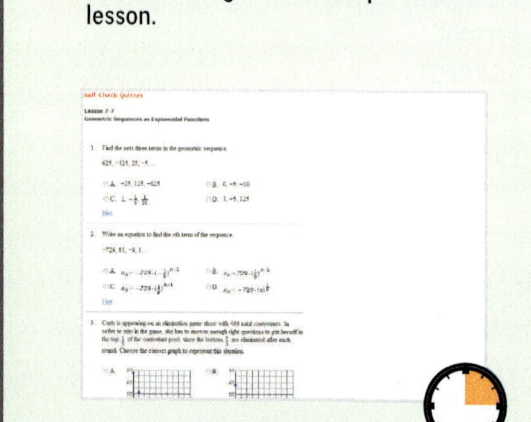
*Use at End of Lesson*

**Personal Tutors** (for every example) let students hear real teachers solve problems. Students can pause and repeat as many times as necessary 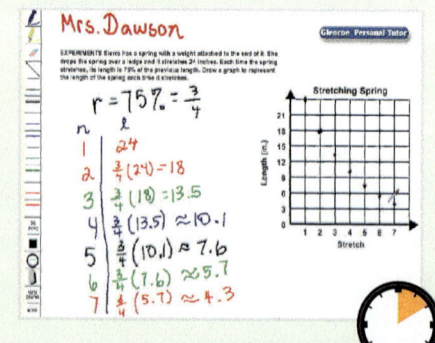
*Use with Examples*

The **Chapter Project** allows students to create and customize a project as a non-traditional method of assessment 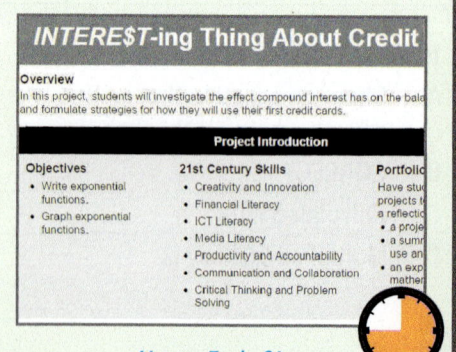
*Use at End of Lesson*

---

**OER Using Open Educational Resources**
**Lesson Planning** For new ways to introduce the concepts in this lesson, visit the geometric sequences and exponential functions page on **ck12.org**. This page contains videos, exploration activities, practice problems, and sample quizzes. *Use as planning*

# Go Online!
connectED.mcgraw-hill.com — Worksheets

## Differentiate Your Resources

**Extra Practice** Additional practice or homework, with Skills Practice best for approaching-level students and Practice best for on-level students

### Skills Practice
AL  OL  ELL

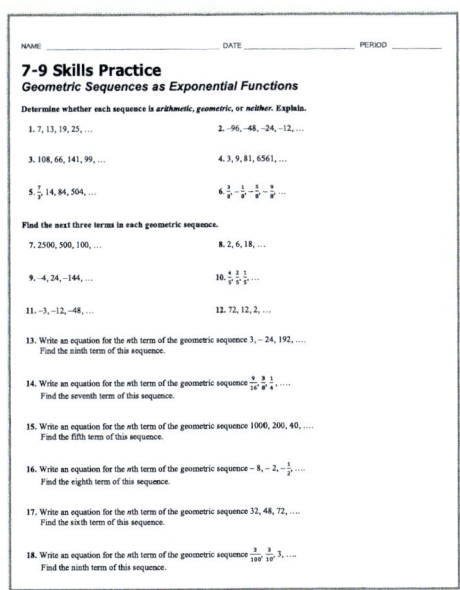

### Practice
AL  OL  BL  ELL

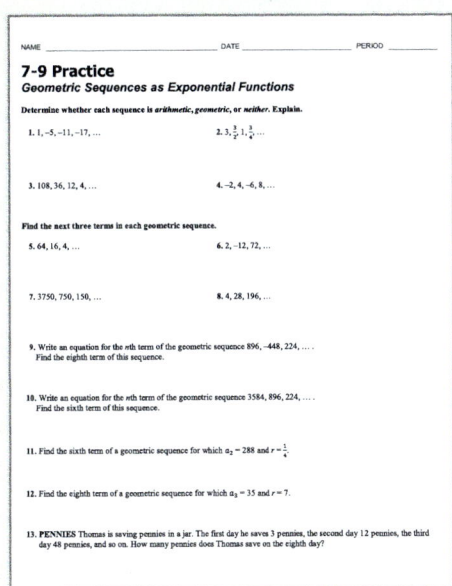

### Word Problem Practice
AL  OL  BL  ELL

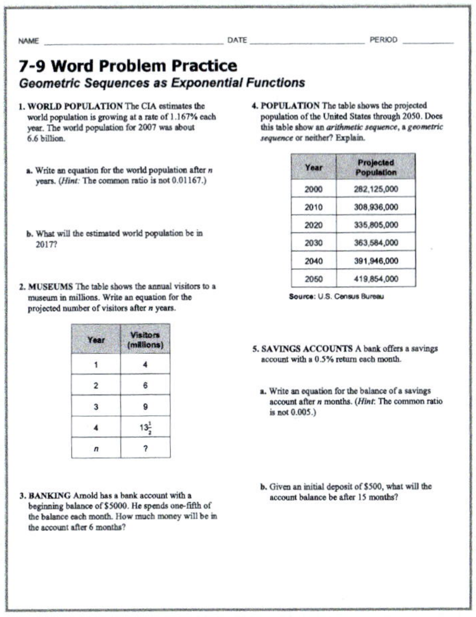

---

**Intervention** Reteaching and vocabulary activities that can be used with struggling or absent students and as ELL support

### Study Guide and Intervention
AL  OL  ELL

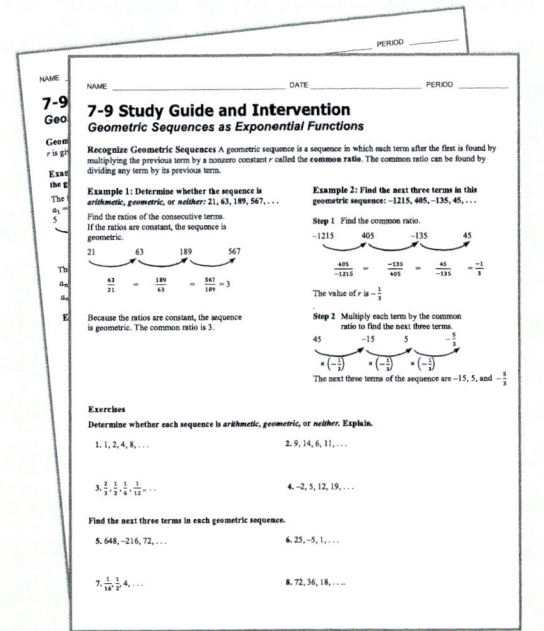

### Study Notebook
AL  OL  BL  ELL

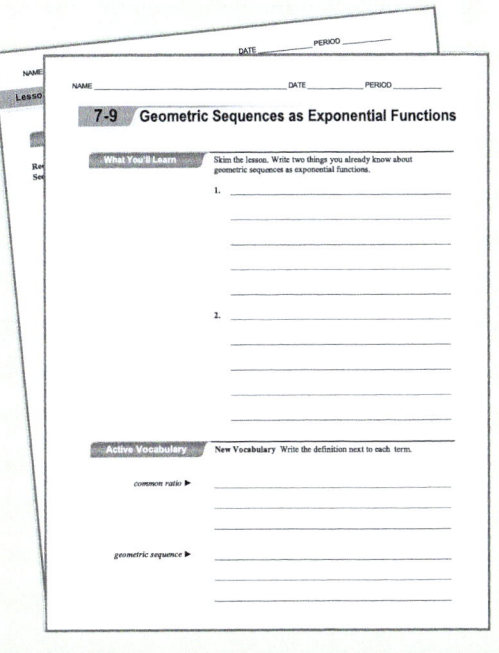

**Extension** Activities that can be used to extend lesson concepts.

### Enrichment
OL  BL  ELL

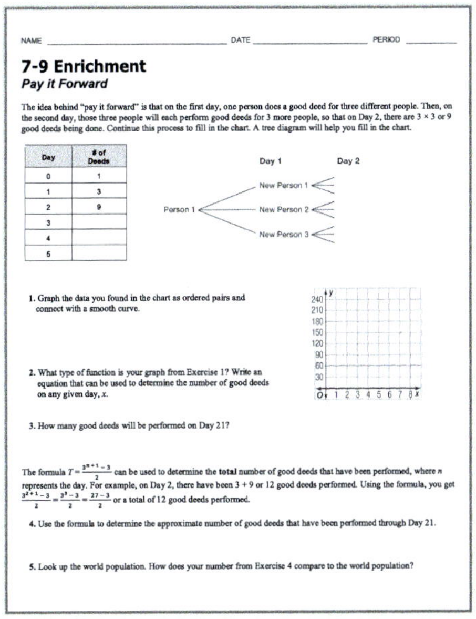

**Lesson 7-9 | Geometric Sequences as Exponential Functions**

## Launch

Have students read the Why? section of the lesson. Ask:

- If Genevieve posted one link on her page, how many people reposted links during the second round? the third round? the fourth round? **5, 25, 125**

- How do you determine the number of people who repost the link in each subsequent round? **Multiply the previous number by 5.**

- What is the equation to find the number of people who repost the link y after x rounds? **y = 5$^x$**

## Teach

Ask the scaffolded questions for each example to build conceptual understanding for students at all levels.

### 1 Recognize Geometric Sequences

**Example 1 Identify Geometric Sequences**

**AL** How do you determine each next term in an arithmetic sequence? **Add the same number to the previous term.** a geometric sequence? **Multiply the previous term by the same number.**

**OL** In part a, what is the relationship between the terms? **Each term is $\frac{1}{2}$ of the previous term. Since the ratios are constant, it is a geometric sequence.**

**BL** Will the sequence in part a ever reach zero? Explain. **no; Sample answer: because you are taking $\frac{1}{2}$ of the previous term, the sequence will continue to get smaller and smaller, but never reach zero.**

### Go Online!

**Interactive Whiteboard**
Use the *eLesson* or *Lesson Presentation* to present this lesson.

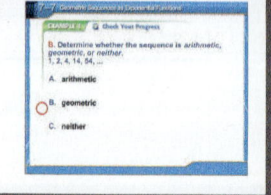

---

## LESSON 9
# Geometric Sequences as Exponential Functions

**Then** · You related arithmetic sequences to linear functions.

**Now** 
1. Identify and generate geometric sequences.
2. Relate geometric sequences to exponential functions.

**Why?** · Genevieve posts a fundraising link on her social media page. Five of her friends repost the link. Each of those five friends have five friends who repost the link. The number of reposts generated forms a geometric sequence.

**New Vocabulary**
geometric sequence
common ratio

**Mathematical Practices**
7 Look for and make use of structure.

**Content Standards**
F.BF.2 Write arithmetic and geometric sequences both recursively and with an explicit formula, use them to model situations, and translate between two forms.
F.LE.2 Construct linear and exponential functions, including arithmetic and geometric sequences given a graph, a description of the relationship, or two input-output pairs (including reading these from a table).

### 1 Recognize Geometric Sequences
The first person generates 5 reposts. If each of these people repost 5 times, 25 links are generated. If each of the 25 people repost 5 times, 125 links are generated. The sequence of reposts generated, 1, 5, 25, 125, ... is an example of a **geometric sequence**.

In a geometric sequence, the first term is nonzero and each term after the first is found by multiplying the previous term by a nonzero constant r called the **common ratio**. The common ratio can be found by dividing any term by its previous term.

F.BF.2

**Example 1    Identify Geometric Sequences**

Determine whether each sequence is *arithmetic*, *geometric*, or *neither*. Explain.

**a.** 256, 128, 64, 32, ...

Find the ratios of consecutive terms.

256   128   64   32

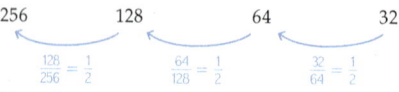

$\frac{128}{256} = \frac{1}{2}$    $\frac{64}{128} = \frac{1}{2}$    $\frac{32}{64} = \frac{1}{2}$

Since the ratios are constant, the sequence is geometric. The common ratio is $\frac{1}{2}$.

**b.** 4, 9, 12, 18, ...

Find the ratios of consecutive terms.

4    9    12    18

$\frac{9}{4} = 2\frac{1}{4}$    $\frac{12}{9} = 1\frac{1}{3}$    $\frac{18}{12} = 1\frac{1}{2}$

The ratios are not constant, so the sequence is not geometric.

Find the differences of consecutive terms.

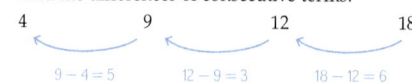

4    9    12    18

$9 - 4 = 5$    $12 - 9 = 3$    $18 - 12 = 6$

There is no common difference, so the sequence is not arithmetic. Thus, the sequence is neither geometric nor arithmetic.

▶ **Guided Practice**    1A–1C. See margin.

**1A.** 1, 3, 9, 27, ...    **1B.** −20, −15, −10, −5, ...    **1C.** 2, 8, 14, 22, ...

---

**MP** **Teaching the Mathematical Practices**
Help students develop the mathematical practices by asking questions like these.

**Look for and make use of structure.**
Help students recognize the structure of a geometric sequence. For example, ask:

- How can you tell whether a sequence is arithmetic?
  **There is the same common difference between all terms.**

- How can you tell whether a sequence is geometric?
  **There is the same common ratio between all terms.**

- How can you find the terms of a geometric sequence?
  **Multiply the previous term by the common ratio.**

- What is needed to find the *n*th term of a sequence?
  **the first term and the common ratio**

Once the common ratio is known, more terms of a sequence can be generated.

**Example 2  Find Terms of Geometric Sequences**

Find the next three terms in each geometric sequence.

a. 1, −4, 16, −64, …

**Step 1** Find the common ratio.

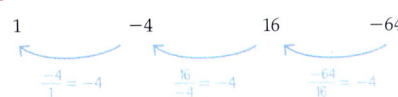

1    −4    16    −64

$\frac{-4}{1} = -4$    $\frac{16}{-4} = -4$    $\frac{-64}{16} = -4$

**Step 2** Multiply each term by the common ratio to find the next three terms.

−64    256    −1024    4096

×(−4)    ×(−4)    ×(−4)

The next three terms are 256, −1024, and 4096.

b. 9, 3, 1, $\frac{1}{3}$ …

**Step 1** Find the common ratio.

9    3    1    $\frac{1}{3}$

$\frac{3}{9} = \frac{1}{3}$    $\frac{1}{3} = \frac{1}{3}$    $\frac{\frac{1}{3}}{1} = \frac{1}{3}$

The value of $r$ is $\frac{1}{3}$.

**Step 2** Multiply each term by the common ratio to find the next three terms.

$\frac{1}{3}$    $\frac{1}{9}$    $\frac{1}{27}$    $\frac{1}{81}$

×$\frac{1}{3}$    ×$\frac{1}{3}$    ×$\frac{1}{3}$

The next three terms are $\frac{1}{9}, \frac{1}{27},$ and $\frac{1}{81}$.

▶ **Guided Practice**  2A. −1875, 9375, −46,875    2B. 121.5, 182.25, 273.375

2A. −3, 15, −75, 375, …    2B. 24, 36, 54, 81, …

**Study Tip**

**MP** **Structure** If the terms of a geometric sequence alternate between positive and negative terms or vice versa, the common ratio is negative.

**Go Online!**

You can quickly generate terms of geometric sequences using a graphing calculator or the **calculator** in the eToolkit. Enter the first term and press ENTER or =. Then multiply by the common ratio and press ENTER or = repeatedly.

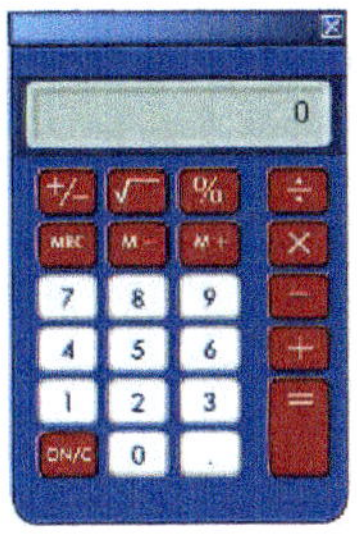

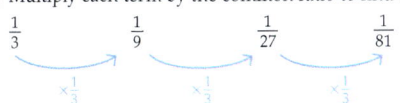

**2 Geometric Sequences and Functions** Finding the $n$th term of a geometric sequence would be tedious if we used the above method. The table below shows a rule for finding the $n$th term of a geometric sequence.

| Position, $n$ | 1 | 2 | 3 | 4 | … | $n$ |
|---|---|---|---|---|---|---|
| Term, $a_n$ | $a_1$ | $a_1 r$ | $a_1 r^2$ | $a_1 r^3$ | … | $a_1 r^{n-1}$ |

Notice that the common ratio between the terms is $r$. The table shows that to get the $n$th term, you multiply the first term by the common ratio $r$ raised to the power $n − 1$. A geometric sequence can be defined by an exponential function in which $n$ is the independent variable, $a_n$ is the dependent variable, and $r$ is the base. The domain is the counting numbers.

---

**Differentiated Instruction** ELL

**Interpersonal Learners** Put students in groups of mixed language and math abilities. Have groups discuss the differences between arithmetic and geometric sequences. Suggest they help each other organize clear, concise, and accurate notes about these and other concepts taught in this lesson. ELL

---

**Lesson 7-9 | Geometric Sequences as Exponential Functions**

**Need Another Example?**

Determine whether each sequence is *arithmetic*, *geometric*, or *neither*. Explain.

a. 0, 8, 16, 24, 32, … arithmetic; the common difference is 8.

b. 64, 48, 36, 27, … geometric; the common ratio is $\frac{3}{4}$.

**Example 2  Find Terms of Geometric Sequences**

**AL** In part **b**, what appears to be happening to each term? Sample answer: each term is being divided by 3.

**OL** In part **a**, will the term 16,384 ever appear? Explain. no; Sample answer: the next term after 4096 has to be negative, so it will be −16,384.

**BL** For part **a**, how do you know the common ratio has to be a negative number? Sample answer: the sequence is positive, negative, positive, negative, and so on, and the only way to achieve that is multiplying by a negative number.

**Need Another Example?**

Find the next three terms of each geometric sequence.

a. 1, −8, 64, −512, … 4096; −32,768; 262,144

b. 40, 20, 10, 5, … $\frac{5}{2}, \frac{5}{4}, \frac{5}{8}$

**MP  Teaching the Mathematical Practices**

**Structure** Mathematically proficient students look closely to discern a pattern or structure. Ask students to explain why the common ratio is negative when the signs of the terms alternate.

**Additional Answers (Guided Practice)**

**1A.** geometric; the common ratio is 3.

**1B.** arithmetic; the common difference is 5.

**1C.** neither; there is no common ratio or common difference.

**Lesson 7-9 | Geometric Sequences as Exponential Functions**

## 2 Geometric Sequences and Functions

### Example 3 Find the nth Term of a Geometric Sequence

**AL** For part **a**, what values do we substitute for $a_1$ in the equation? **−6** For part **b**, what value do we substitute for $n$ in the equation? **9**

**OL** What value do we substitute for $r$? Explain. **−2; Sample answer: each term is multiplied by −2 to determine the next term.**

**BL** Why is it important to enclose the −2 in parentheses? **Sample answer: the variable $r$ has an exponent with it. When $r$ is negative, like in this example, if we do not use parentheses, then the exponent only goes with the 2, not the negative sign.**

**Need Another Example?**
a. Write an equation for the $n$th term of the geometric sequence 1, −2, 4, −8, ...
$a_n = 1 \cdot (-2)^{n-1}$
b. Find the 12th term of this sequence. **−2048**

### Example 4 Graph a Geometric Sequence

**AL** What is the value for $r$ that is given to us in the problem? $\frac{1}{2}$

**OL** Should we connect the plotted points with a line? Explain. **no; Sample answer: there are no values in between each point, so they should not be connected.**

**BL** Would it make sense for there to be a seventh round in the tournament? Explain. **no; Sample answer: after the sixth round, there is one team left. It would not make sense for one team to play, and then be left with $\frac{1}{2}$ of a team.**

**Need Another Example?**
**Art** A 50-pound ice sculpture is melting at a rate in which 80% of its weight remains each hour. Draw a graph to represent how many pounds of the sculpture is left at each hour.

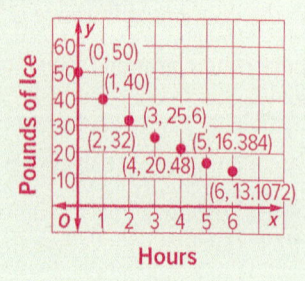

[Graph: Pounds of Ice vs. Hours, points: (0, 50), (1, 40), (2, 32), (3, 25.6), (4, 20.48), (5, 16.384), (6, 13.1072)]

---

**Watch Out!**
**Negative Common Ratio** If the common ratio is negative, as in Example 3, make sure to enclose the common ratio in parentheses. $(-2)^8 \neq -2^8$

**Real-World Link**
The first NCAA Division I women's basketball tournament was held in 1982. The Final Four ticket prices were $5 and $7 in 1982. They now cost as much as $800.
**Source:** NCAA Sports

**Teaching Tip**
**The Power of $r$** Make sure students raise $r$ to the power $(n − 1)$ in their equations for the $n$th term of a geometric sequence, instead of $n$.

**MP Teaching the Mathematical Practices**
**Reasoning** Mathematically proficient students attend to the meaning of quantities, not just how to compute them. In Exercise 32, ask students why the common ratio is not 0.2. How does the common ratio relate to the meaning of the sequence?

---

**Key Concept** $n$th term of a Geometric Sequence

The $n$th term $a_n$ of a geometric sequence with first term $a_1$ and common ratio $r$ is given by the following formula, where $n$ is any positive integer and $a_1, r \neq 0$.

$$a_n = a_1 r^{n-1}$$

F.BF.2, F.LE.2

**Example 3** Find the $n$th Term of a Geometric Sequence

a. Write an equation for the $n$th term of the sequence −6, 12, −24, 48, ... .

The first term of the sequence is −6. So, $a_1 = -6$. Now find the common ratio.

−6, 12, −24, 48

$\frac{12}{-6} = -2$, $\frac{-24}{12} = -2$, $\frac{48}{-24} = -2$

The common ratio is −2.

$a_n = a_1 r^{n-1}$    Formula for $n$th term
$a_n = -6(-2)^{n-1}$    $a_1 = -6$ and $r = 2$

b. Find the ninth term of this sequence.

$a_n = a_1 r^{n-1}$    Formula for $n$th term
$a_9 = -6(-2)^{9-1}$    For the $n$th term, $n = 9$.
$= -6(-2)^8$    Simplify.
$= -6(256)$    $(-2)^8 = 256$
$= -1536$

**Guided Practice** 3. $a_n = 96 \cdot \left(\frac{1}{2}\right)^{n-1}; \frac{3}{16}$

3. Write an equation for the $n$th term of the geometric sequence 96, 48, 24, 12, ... . Then find the tenth term of the sequence.

**Real-World Example 4** Graph a Geometric Sequence   F.BF.2, F.LE.2

**BASKETBALL** The NCAA women's basketball tournament begins with 64 teams. In each round, one half of the teams are left to compete, until only one team remains. Draw a graph to represent how many teams are left in each round.

Compared to the previous rounds, one half of the teams remain. So, $r = \frac{1}{2}$. Therefore, the geometric sequence that models this situation is 64, 32, 16, 8, 4, 2, 1. So in round two, 32 teams compete, in round three 16 teams compete, and so forth. Use this information to draw a graph.

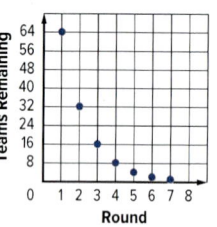

[Graph: Teams Remaining vs. Round]

**Guided Practice**
4. **TENNIS** A tennis ball is dropped from a height of 12 feet. Each time the ball bounces back to 80% of the height from which it fell. Draw a graph to represent the height of the ball after each bounce. **See margin.**

**Additional Answer (Guided Practice)**

4. **Tennis Ball**

[Graph: Height of Rebounds (ft) vs. Number of Rebounds]

464 | Lesson 7-9 | Geometric Sequences as Exponential Functions

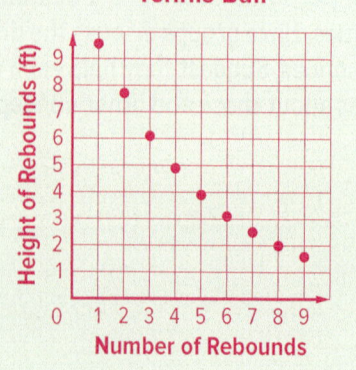

Lesson 7-9 | Geometric Sequences as Exponential Functions

## Check Your Understanding

= Step-by-Step Solutions begin on page R11.

**Example 1**
F.BF.2, F.LE.2

Determine whether each sequence is *arithmetic*, *geometric*, or *neither*. Explain.

1. 200, 40, 8, ...
2. 2, 4, 16, ...
3. −6, −3, 0, 3, ...
4. 1, −1, 1, −1, ...

**Example 2**
F.BF.2, F.LE.2

Find the next three terms in each geometric sequence. 5–8. See margin.

5. 10, 20, 40, 80, ...
6. 100, 50, 25, ...
7. 4, −1, $\frac{1}{4}$, ...
8. −7, 21, −63, ...

1. Geometric; the common ratio is $\frac{1}{5}$.

**Example 3**
F.BF.2, F.LE.2

Write an equation for the *n*th term of each geometric sequence, and find the indicated term.

9. the fifth term of −6, −24, −96, ... $a_n = -6 \cdot (4)^{n-1}$; −1536
10. the seventh term of −1, 5, −25, ... $a_n = -1 \cdot (-5)^{n-1}$; −15,625
11. the tenth term of 72, 48, 32, ... $a_n = 72 \cdot \left(\frac{2}{3}\right)^{n-1}$; $\frac{4096}{2187}$
12. the ninth term of 112, 84, 63, ... $a_n = 112 \cdot \left(\frac{3}{4}\right)^{n-1}$; $\frac{45,927}{4096}$

2. Neither; there is no common ratio or difference.
3. Arithmetic; the common difference is 3.
4. Geometric; the common ratio is −1.

**Example 4**
F.BF.2, F.LE.2

13. **EXPERIMENT** In a physics class experiment, Diana drops a ball from a height of 16 feet. Each bounce has 70% the height of the previous bounce. Draw a graph to represent the height of the ball after each bounce. **See margin.**

32b. 2.0736; The map will be magnified at approximately 207% of the original size after the fourth click.

## Practice and Problem Solving

Extra Practice is on page R7

**Example 1**
F.BF.2, F.LE.2

Determine whether each sequence is *arithmetic*, *geometric*, or *neither*. Explain.

14–19. See margin.

14. 4, 1, 2, ...
15. 10, 20, 30, 40, ...
16. 4, 20, 100, ...
17. 212, 106, 53, ...
18. −10, −8, −6, −4, ...
19. 5, −10, 20, 40, ...

**Example 2**
F.BF.2, F.LE.2

Find the next three terms in each geometric sequence. 20–25. See margin.

20. 2, −10, 50, ...
21. 36, 12, 4, ...
22. 4, 12, 36, ...
23. 400, 100, 25, ...
24. −6, −42, −294, ...
25. 1024, −128, 16, ...

**Example 3**
F.BF.2, F.LE.2

26. The first term of a geometric series is 1 and the common ratio is 9. What is the 8th term of the sequence? **4,782,969**
27. The first term of a geometric series is 2 and the common ratio is 4. What is the 14th term of the sequence? **134,217,728**
28. What is the 15th term of the geometric sequence −9, 27, −81, ...? **−43,046,721**
29. What is the 10th term of the geometric sequence 6, −24, 96, ...? **−1,572,864**

**Example 4**
F.BF.2, F.LE.2

30. **PENDULUM** The first swing of a pendulum is shown. On each swing after that, the arc length is 60% of the length of the previous swing. Draw a graph that represents the arc length after each swing. **See Ch. 7 Answer Appendix.** 24 ft

31. Find the eighth term of a geometric sequence for which $a_3 = 81$ and $r = 3$. **19,683**

32. **REASONING** At an online mapping site, Mr. Mosley notices that when he clicks a spot on the map, the map zooms in on that spot. The magnification increases by 20% each time. a. $a_n = 1.2^n$

   a. Write a formula for the *n*th term of the geometric sequence that represents the magnification of each zoom level for the map.
   (*Hint:* The common ratio is not just 0.2.)
   b. What is the fourth term of this sequence? What does it represent?

| Differentiated Homework Options |
|---|
| Levels | **AL** Basic | **OL** Core | **BL** Advanced |
| Exercises | 14–31, 39–49 | 15–31 odd, 32–37, 39–49 | 32–42<br>43–49 (optional) |
| 2-Day Option | 15–31 odd, 43–49 | 14–31, 43–49 | |
| | 14–30 even, 39–42 | 32–37, 39–42 | |

You can use ALEKS to provide additional remediation support with personalized instruction and practice.

20. −250, 1250, −6250
21. $\frac{4}{3}, \frac{4}{9}, \frac{4}{27}$
22. 108, 324, 972
23. $\frac{25}{4}, \frac{25}{16}, \frac{25}{64}$
24. −2058; −14,406; −100,842
25. −2, $\frac{1}{4}$, −$\frac{1}{32}$

## Practice

**Formative Assessment** Use Exercises 1–13 to assess students' understanding of the concepts in this lesson.

The Practice and Problem Solving exercises assess the content taught in the lesson. The Preparing for Assessment page is meant to be used as preparation for end-of-course assessments.

### Exercise Alert

**Grid Paper** For Exercises 13, 30, 37, and 48, students will need grid paper.

### Extra Practice

See page R7 for extra exercises for students who are approaching level or for on-level students who need additional reinforcement.

### Additional Answers

5. 160, 320, 640
6. 12.5, 6.25, 3.125
7. $\frac{1}{16}, \frac{1}{64}, -\frac{1}{256}$
8. 189, −567, 1701

13.

**Experiment**

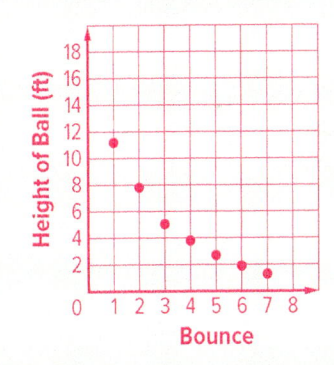

14. Neither; there is no common ratio or difference.
15. Arithmetic; the common difference is 10.
16. Geometric; the common ratio is 5.
17. Geometric; the common ratio is $\frac{1}{2}$.
18. Arithmetic; the common difference is 2.
19. Neither; there is no common ratio or difference.

### Go Online! eBook

**Interactive Student Guide**
Use the *Interactive Student Guide* to deepen conceptual understanding.
- Geometric Sequences as Exponential Functions

connectED.mcgraw-hill.com

**Lesson 7-9 | Geometric Sequences as Exponential Functions**

## Levels of Complexity Chart

The levels of the exercises progress from 1 to 3, with Level 1 indicating the lowest level of complexity.

| Exercises | 14–31 | 32, 43–49 | 32–42 |
|---|---|---|---|
| Level 3 | | | ● |
| Level 2 | | ● | |
| Level 1 | ● | | |

### Teaching the Mathematical Practices

**Critique Arguments** Mathematically proficient students can read the arguments of others and decide whether they make sense. In Exercise 39, advise students to start by comparing the arguments line by line to find the differences.

## Assess

**Name the Math** Give each student one of five different geometric sequences. Ask students to explain how to find the common ratio for their sequence.

## Additional Answers

**37c.** The graph appears to be exponential. The rate of change between any two points does not match any others.

**39.** Neither; Haro calculated the exponent incorrectly. Matthew did not calculate $(-2)^8$ correctly.

**41.** Sample answer: When graphed, the terms of a geometric sequence lie on a curve that can be represented by an exponential function. They are different in that the domain of a geometric sequence is the set of natural numbers, while the domain of an exponential function is all real numbers. Thus, geometric sequences are discrete, while exponential functions are continuous.

## Go Online!

**eSolutions Manual**
Create worksheets, answer keys, and solutions handouts for your assignments.

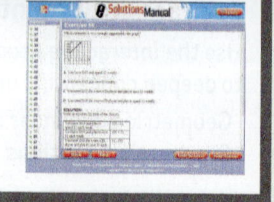

---

**33. MULTI-STEP** Sarina's parents have offered her two different allowance options for her 9-week summer vacation. She can get paid $30 a week, or she can get paid $1 the first week, $2 the second week, $4 the third week, and so on.

a. Which option should Sarina choose? **the second option**

b. Explain your solution process.

**33b.** She should choose whichever option would earn her the most money over the summer. For the first option, $30 a week for 9 weeks would yield $270. The second option is a geometric sequence with a common ratio of 2, totaling $511 over 9 weeks. So, although this option starts out slow, it ends up being the most profitable.

**34. SIERPINSKI'S TRIANGLE** Consider the inscribed equilateral triangles at the right. The perimeter of each triangle is one half of the perimeter of the next larger triangle. What is the perimeter of the smallest triangle? **7.5 cm**

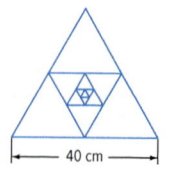

**35.** If the second term of a geometric sequence is 3 and the third term is 1, find the first and fourth terms of the sequence. $9; \frac{1}{3}$

**36.** If the third term of a geometric sequence is $-12$ and the fourth term is 24, find the first and fifth terms of the sequence. $-3; -48$

**37. EARTHQUAKES** The Richter scale is used to measure the force of an earthquake. The table shows the increase in magnitude for the values on the Richter scale.

| Richter Number (x) | Increase in Magnitude (y) | Rate of Change (slope) |
|---|---|---|
| 1 | 1 | — |
| 2 | 10 | 9 |
| 3 | 100 | 90 |
| 4 | 1000 | 900 |
| 5 | 10,000 | 9000 |

a. Copy and complete the table. Remember that the rate of change is the change in $y$ divided by the change in $x$.

b. Plot the ordered pairs (Richter number, increase in magnitude). **See Ch. 7 Answer Appendix.**

c. Describe the graph that you made of the Richter scale data. Is the rate of change between any two points the same? **See margin.**

d. Write an exponential equation that represents the Richter scale. $y = 1 \cdot (10)^{x-1}$

F.BF.2, F.LE.2

**H.O.T. Problems** Use Higher-Order Thinking Skills

**38. CHALLENGE** Write a sequence that is both geometric and arithmetic. Explain your answer.

**38.** 1, 1, 1, 1, ...; The common ratio is 1 making it a geometric sequence, but the common difference is 0 making it an arithmetic sequence as well.

**39.** **CRITIQUE ARGUMENTS** Haro and Matthew are finding the ninth term of the geometric sequence $-5, 10, -20, \ldots$. Is either of them correct? Explain your reasoning.

**Haro**
$r = \frac{10}{-5}$ or $-2$
$a_9 = -5(-2)^{9-1}$
$= -5(512)$
$= -2560$

**Matthew**
$r = \frac{10}{-5}$ or $-2$
$a_9 = -5 \cdot (-2)^{9-1}$
$= -5 \cdot -256$
$= 1280$

**40.** Sample answer: 1, 4, 9, 16, 25, 36, ...; This is the sequence of squares of counting numbers.

**40. STRUCTURE** Write a sequence of numbers that form a pattern but are neither arithmetic nor geometric. Explain the pattern.

**41. WRITING IN MATH** How are graphs of geometric sequences and exponential functions similar? different? **See margin.**

**42. WRITING IN MATH** Summarize how to find a specific term of a geometric sequence. **See margin.**

### Standards for Mathematical Practice

| Emphasis On | Exercises |
|---|---|
| 3 Construct viable arguments and critique the reasoning of others. | 39 |
| 4 Model with mathematics. | 13, 30, 33, 34, 37, 47–49 |
| 7 Look for and make use of structure. | 1–12, 14–29, 31, 35, 36, 38, 40–49 |

**42.** Sample answer: First find the common ratio. Then use the formula $a_n = a_1 \cdot r^{n-1}$. Substitute the first term for $a_1$ and the common ratio for $r$. Let $n$ be equal to the number of the term you are finding. Then solve the equation.

## Lesson 7-9 | Geometric Sequences as Exponential Functions

### Preparing for Assessment

43. The formula for the $n$th term of a sequence is $a_n = 5(3)^{n-1}$. If the sequence were graphed on a coordinate plane, which point would be on the curve? **MP 7** F.LE.2 **C**

- A (3, 225)
- B (4, 400)
- C (5, 405)
- D (6, 243)

44. Find the formula for the sequence 5125, 1025, 205, 41 . . . **MP 7** F.BF.2, F.LE.2 **B**

- A $a_n = 5125\left(\frac{1}{5}\right)^n$
- B $a_n = 5125\left(\frac{1}{5}\right)^{n-1}$
- C $a_n = 5125(5)^n$
- D $a_n = 41(5)^{n-1}$

45. Which formula represents the terms of this table? **MP 7** F.BF.2, F.LE.2 **B**

| n   | 1 | 2  | 3   | 4    |
|-----|---|----|-----|------|
| $a_n$ | 3 | 27 | 243 | 2187 |

- A $a_n = 3(9)^n$
- B $a_n = 3(9)^{n-1}$
- C $a_n = 9(3)^n$
- D $a_n = 9(3)^{n-1}$

46. What is the 8th term of the sequence 64,768, 16,192, 4048, 1012. . .? **MP 7** F.BF.2 **C**

- A 63.25
- B 15.8125
- C 3.953125
- D 0.25

47. A coffee shop increases the price of their coffee by 15% each year. If the initial price of a cup of coffee is $1.32, which points are on the graph that represents the price of coffee over time? **MP 4,7** F.BF.2, F.LE.2 **A, D, E**

- ☐ A (1, 1.32)
- ☐ B (2, 1.65)
- ☐ C (3, 1.83)
- ☐ D (4, 2.01)
- ☐ E (5, 2.31)
- ☐ F (6, 2.84)

48. **MULTI-STEP** In 2015, the average attendance for a college's basketball games was 6500. Since then, the attendance has dropped by an average of 7% each year. **MP 4,7** F.BF.2, F.LE.2

a. Which formula can be used to find the nth term of a geometric sequence that represents the attendance over time? **D**

- A $a_n = 6500(1.07)^n$
- B $a_n = 6500(1.07)^{n-1}$
- C $a_n = 6500(0.93)^n$
- D $a_n = 6500(0.93)^{n-1}$

b. What is the sixth term in the sequence?

**3911**

c. What does the eighth term of the sequence represent?

**In 2022, the expected average attendance is 3911.**

d. Sketch a graph of the sequence.
**See Answer Appendix for Graph.**

49. The population of a town was 10,250 in 2016 and is expected to decline by 1.5% each year. What is the predicted population in 2030? **MP 4,7** F.BF.2 **8,295**

### Preparing for Assessment

Exercises 43–49 require students to use the skills they will need on standardized assessments. Each exercise is dual-coded with content standards and mathematical practice standards.

#### Dual Coding

| Items | Content Standards | MP Mathematical Practices |
|-------|-------------------|---------------------------|
| 43    | F.LE.2            | 7                         |
| 44    | F.BF.2, F.LE.2    | 7                         |
| 45    | F.BF.2, F.LE.2    | 7                         |
| 46    | F.BF.2            | 7                         |
| 47    | F.BF.2, F.LE.2    | 4, 7                      |
| 48    | F.BF.2, F.LE.2    | 4, 7                      |
| 49    | F.BF.2            | 4, 7                      |

### Diagnose Student Errors

Survey student responses for each item. Class trends may indicate common errors and misconceptions.

**47.**

| A | CORRECT |
|---|---------|
| B | Calculated the point incorrectly |
| C | Calculated the point incorrectly |
| D | CORRECT |
| E | CORRECT |
| F | Calculated the point incorrectly |

**48a.**

| A | Used $n$ instead of $n-1$ as the power and added the percentage |
|---|---|
| B | Added the percentage |
| C | Used $n$ instead of $n-1$ as the power |
| D | CORRECT |

### Differentiated Instruction OL BL

**Extension** Often sequences of numbers do not appear, on first calculations, to have a pattern. Sometimes the differences between terms themselves create a sequence that can be used to determine the next term in the original sequence. Ask students to determine the sixth term in the sequence 4, 7, 14, 25, 40, .... Have them also explain how they found the term.

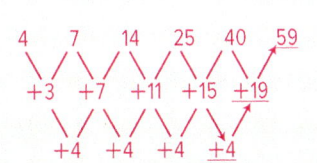

```
  4    7    14   25   40   59
   \  / \  / \  / \  / \  /
   +3   +7  +11  +15  +19
     \  / \  / \  / \  /
     +4   +4   +4   +4
```

### Go Online!

**Quizzes**

Students can use *Self-Check Quizzes* to check their understanding of this lesson. You can also give the *Chapter Quiz*, which covers the content in Lessons 7–8 and 7–9.

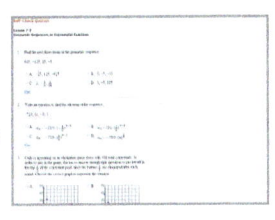

# Extend 7-9

## Focus

**Objective** Calculate and interpret the average rate of change of an exponential function.

**Materials for Each Student**
- grid paper

**Easy to Make Manipulatives** *Teaching Algebra with Manipulatives* Template for grid paper, p. 1

## Teach

**Working in Cooperative Groups** Put students in groups of two or three, mixing abilities. Have groups complete the activity.

- Discuss how the length of time of the investment affects the comparison of the plans.
- Have students describe the shape of the graph for each plan and discuss how the shape is related to the average rates of change.

**Practice** Have students complete Exercises 1–4.

## Assess

**Formative Assessment** Use Exercises 1–3 to assess whether students can calculate and interpret an average rate of change.

**From Concrete to Abstract** After students have completed Exercise 4, have them discuss what characteristics of a graph they can determine by examining the rate of change for a function.

---

# EXTEND 7-9
## Algebra Lab
# Average Rate of Change of Exponential Functions

You know that the rate of change of a linear function is the same for any two points on the graph. The rate of change of an exponential function is not constant.

**Mathematical Practices**
MP 4 Model with Mathematics
**Content Standards**
F.IF.6 Calculate and interpret the average rate of change of a function (presented symbolically or as a table) over a specified interval. Estimate the rate of change from a graph.

### Activity  Evaluating Investment Plans

John has $2000 to invest in one of two plans. Plan 1 offers to increase his principal by $75 each year, while Plan 2 offers to pay 3.6% interest compounded monthly. The dollar value of each investment after $t$ years is given by $A_1 = 2000 + 75t$ and $A_2 = 2000(1.003)^{12t}$, respectively. Use the function values, the average rate of change, and the graphs of the equations to interpret and compare the plans.

**Step 1** Copy and complete the table below by finding the missing values for $A_1$ and $A_2$.

| $t$ | 0 | 1 | 2 | 3 | 4 | 5 |
|---|---|---|---|---|---|---|
| $A_1$ | 2000 | 2075 | 2150 | 2225 | 2300 | 2375 |
| $A_2$ | 2000 | 2073.2 | 2149.08 | 2227.74 | 2309.27 | 2393.79 |

**Step 2** Find the average rate of change for each plan from $t = 0$ to 1, $t = 3$ to 4, and $t = 0$ to 5.

Plan 1: $\frac{2075 - 2000}{1 - 0}$ or 75     $\frac{2300 - 2225}{4 - 3}$ or 75     $\frac{2375 - 2000}{5 - 0}$ or 75

Plan 2: $\frac{2073.2 - 2000}{1 - 0}$ or 73.2     $\frac{2309.27 - 2227.74}{4 - 3}$ or about 82     $\frac{2393.79 - 2000}{5 - 0}$ or about 79

**Step 3** Graph the ordered pairs for each function. Connect each set of points with a smooth curve.

**Step 4** Use the graph and the rates of change to compare the plans. Both graphs have a rate of change for the first year of about $75 per year. From year 3 to 4, Plan 1 continues to increase at $75 per year, but Plan 2 grows at a rate of more than $81 per year. The average rate of change over the first five years for Plan 1 is $75 per year and for Plan 2 is over $78 per year. This indicates that as the number of years increases, the investment in Plan 2 grows at an increasingly faster pace. This is supported by the widening gap between their graphs.

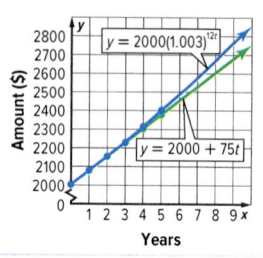

3. Sample answer: The value of the equipment decreases at a slower rate as the number of years increases.

### Exercises

The value of a company's piece of equipment decreases over time due to depreciation. The function $y = 16{,}000(0.985)^{2t}$ represents the value after $t$ years.

1. What is the average rate of change over the first five years?  **−$448.86 per year**
2. What is the average rate of change of the value from year 5 to year 10?  **−$386 per year**
3. What conclusion about the value can we make based on these average rates of change?
4. **MP REGULARITY** Copy and complete the table for $y = x^4$.

| $x$ | −3 | −2 | −1 | 0 | 1 | 2 | 3 |
|---|---|---|---|---|---|---|---|
| $y$ | 81 | 16 | 1 | 0 | 1 | 16 | 81 |

Compare and interpret the average rate of change for $x = -3$ to 0 and for $x = 0$ to 3.

4. Sample answer: The average rate of change for $x = -3$ to 0 is −27 while the average rate of change for $x = 0$ to 3 is 27. This would indicate that the graph of the function was going down and then changed to going up.

---

### MP Teaching the Mathematical Practices

**Regularity** Mathematically proficient students notice if calculations are repeated, and look both for general methods and for shortcuts. In Exercise 4, advise students to look for and use regularity in their calculations.

468 | Extend 7-9 | Algebra Lab: Average Rate of Change of Exponential Functions

# LESSON 7-10
# Recursive Formulas

**SUGGESTED PACING (DAYS)**
90 min. 0.5
45 min. 1
Instruction

## Track Your Progress

### Objectives
1. Use a recursive formula to list terms in a sequence.
2. Write recursive formulas for arithmetic and geometric sequences.

### Mathematical Background
In a recursive formula, each term is defined in terms of one or more previous terms. To write a recursive formula, give the first term and a formula for successive terms. For example, a recursive formula for 2, 5, 8, 11,... is $a_1 = 2$, $a_n = a_{n-1} + 3$. A recursive formula for the geometric sequence 3, 6, 12, 24,... is $a_1 = 3$, $a_n = 2 \cdot a_{n-1}$.

### THEN
**A.SSE.3c** Use the properties of exponents to transform expressions for exponential functions.

**F.IF.8b** Use the properties of exponents to interpret expressions for exponential functions.

### NOW
**F.IF.3** Recognize that sequences are functions, sometimes defined recursively, whose domain is a subset of the integers.

**F.BF.2** Write arithmetic and geometric sequences both recursively and with an explicit formula, use them to model situations, and translate between the two forms.

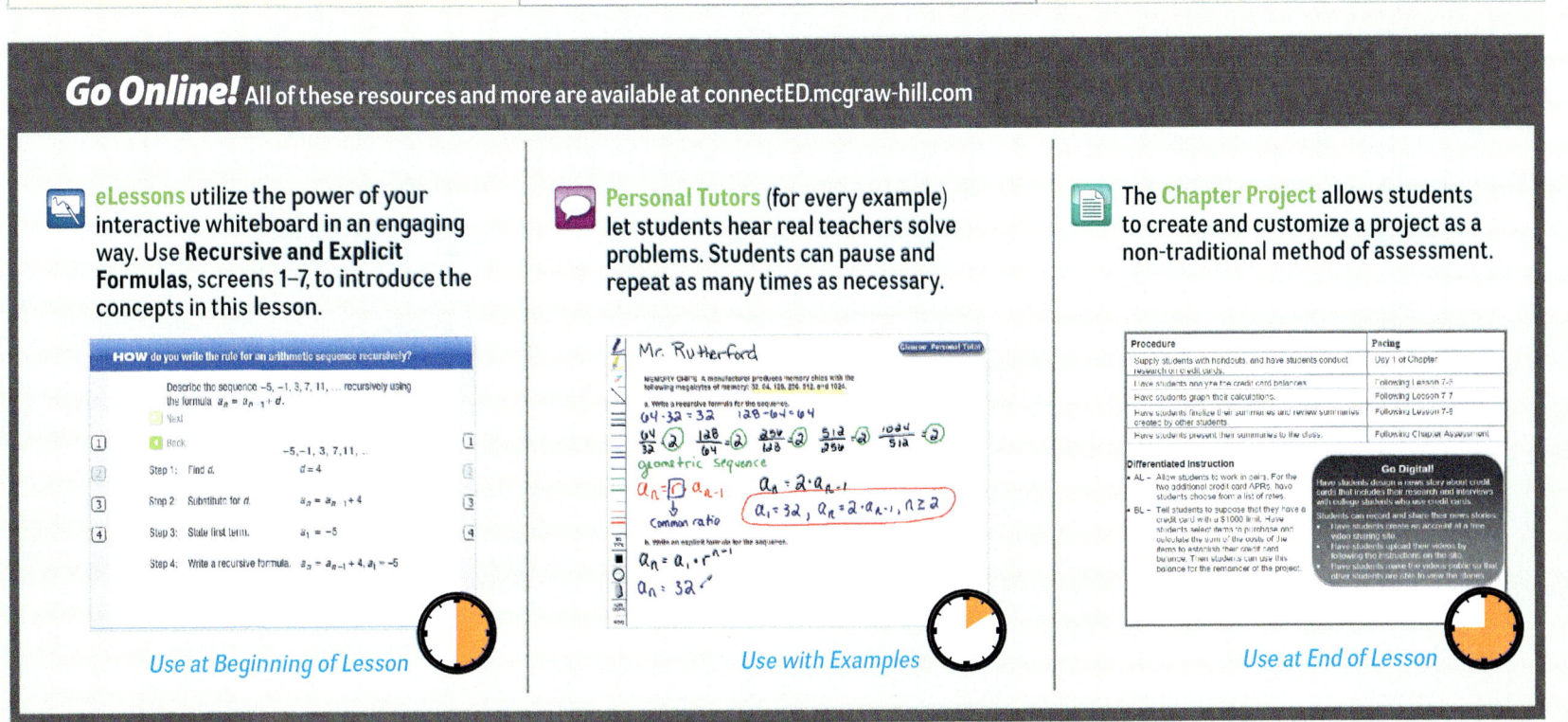

### NEXT
**A.APR.1** Understand that polynomials form a system analogous to the integers, namely, they are closed under the operations of addition, subtraction, and multiplication; add, subtract, and multiply polynomials.

---

**Go Online!** All of these resources and more are available at connectED.mcgraw-hill.com

**eLessons** utilize the power of your interactive whiteboard in an engaging way. Use **Recursive and Explicit Formulas**, screens 1–7, to introduce the concepts in this lesson.

*Use at Beginning of Lesson*

**Personal Tutors** (for every example) let students hear real teachers solve problems. Students can pause and repeat as many times as necessary.

*Use with Examples*

The **Chapter Project** allows students to create and customize a project as a non-traditional method of assessment.

*Use at End of Lesson*

---

## OER Using Open Educational Resources

**Communicating** Use **Remind 101** to send text message reminders to parents that Chapter 7 has been completed and there will be a test shortly. This is a free service where parents can opt in to an established group. *Use as reminder*

connectED.mcgraw-hill.com  469A

# Go Online!
connectED.mcgraw-hill.com — Worksheets

# Differentiate Your Resources

**Extra Practice** Additional practice or homework; Skills Practice is best for approaching-level students and Practice is best for on-level and beyond-level students

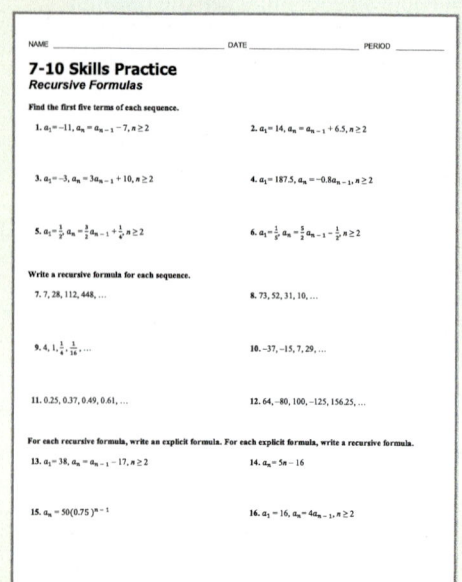

## Skills Practice
AL  OL  ELL

## Practice
AL  OL  BL  ELL

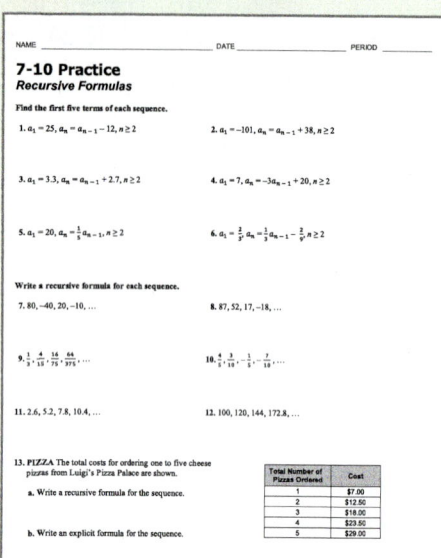

## Word Problem Practice
AL  OL  BL  ELL

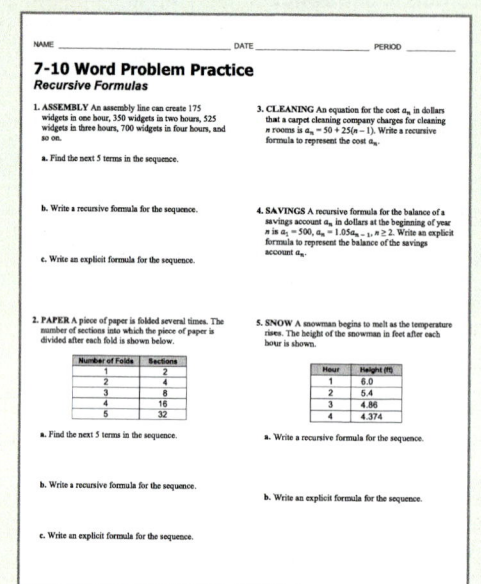

**Intervention** Reteaching and vocabulary activities that can be used with struggling or absent students and as ELL support

## Study Guide and Intervention
AL  OL  ELL

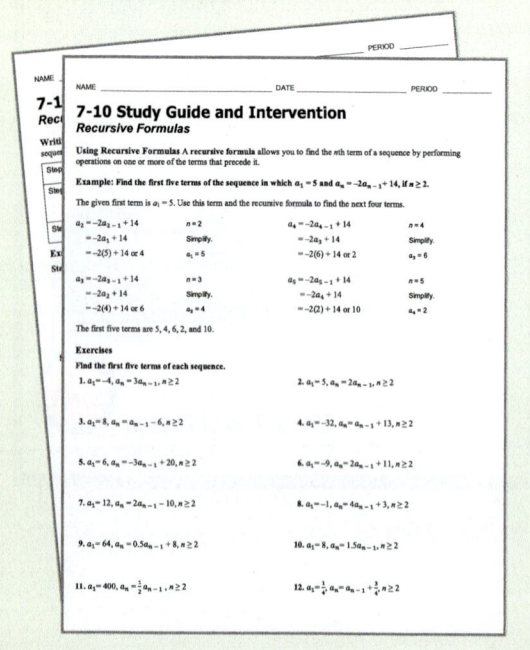

## Study Notebook
AL  OL  BL  ELL

**Extension** Activities that can be used to extend lesson concepts

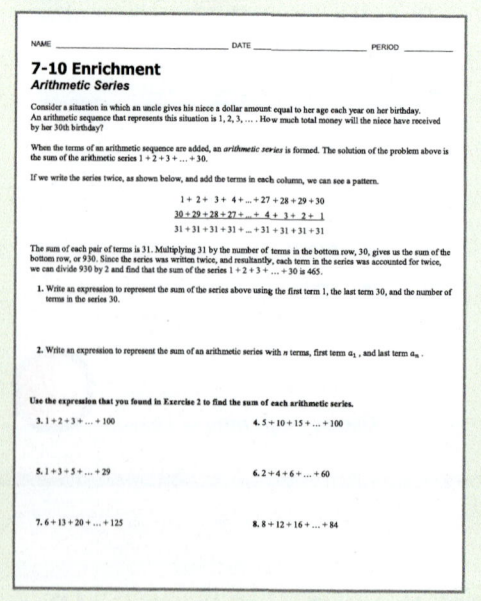

## Enrichment
OL  BL  ELL

469B | Lesson 7-10 | Recursive Formulas

# LESSON 10
# Recursive Formulas

**::Then**
- You wrote explicit formulas to represent arithmetic and geometric sequences.

**::Now**
1. Use a recursive formula to list terms in a sequence.
2. Write recursive formulas for arithmetic and geometric sequences.

**::Why?**
Clients of a shuttle service get picked up from their homes and driven to premium outlet stores for shopping. The total cost of the service depends on the total number of customers. The costs for the first six customers are shown.

| Number of Customers | Cost ($) |
|---|---|
| 1 | 25 |
| 2 | 35 |
| 3 | 45 |
| 4 | 55 |
| 5 | 65 |
| 6 | 75 |

**New Vocabulary**
recursive formula

**Mathematical Practices**
3 Construct viable arguments and critique the reasoning of others.

**Content Standards**
F.IF.3 Recognize that sequences are functions, sometimes defined recursively, whose domain is a subset of the integers.
F.BF.2 Write arithmetic and geometric sequences both recursively and with an explicit formula, use them to model situations, and translate between the two forms.

**1 Using Recursive Formulas** An explicit formula allows you to find any term $a_n$ of a sequence by using a formula written in terms of $n$. For example, $a_n = 2n$ can be used to find the fifth term of the sequence 2, 4, 6, 8, 10 or $a_5 = 2(5) = 10$.

A **recursive formula** allows you to find the $n$th term of a sequence by performing operations to one or more of the preceding terms. Since each term in the sequence above is 2 greater than the term that preceded it, we can add 2 to the fourth term to find that the fifth term is $8 + 2$ or 10. We can then write a recursive formula for $a_n$.

$$a_1 = \phantom{a_1 + 2 \text{ or } 2 + 2} = 2$$
$$a_2 = a_1 + 2 \text{ or } 2 + 2 = 4$$
$$a_3 = a_2 + 2 \text{ or } 4 + 2 = 6$$
$$a_4 = a_3 + 2 \text{ or } 6 + 2 = 8$$
$$\vdots$$
$$a_n = a_{n-1} + 2$$

A recursive formula for the sequence above is $a_1 = 2$, $a_n = a_{n-1} + 2$, for $n \geq 2$ where $n$ is an integer. The term denoted $a_{n-1}$ represents the term immediately before $a_n$. Notice that the first term $a_1$ is given, along with the domain for $n$.

F.BF.2

**Example 1** Use a Recursive Formula

Find the first five terms of each sequence.

**a.** $a_1 = 7$ and $a_n = a_{n-1} - 12$, if $n \geq 2$

Use $a_1 = 7$ and the recursive formula. A table can help organize the results.

| $n$ | $a_n = a_{n-1} - 12$ | $a_n$ |
|---|---|---|
| 1 | — | 7 |
| 2 | $a_2 = 7 - 12$ | −5 |
| 3 | $a_3 = -5 - 12$ | −17 |
| 4 | $a_4 = -17 - 12$ | −29 |
| 5 | $a_5 = -29 - 12$ | −41 |

The first five terms are 7, −5, −17, −29, and −41.

---

## MP Mathematical Practices Strategies

**Reason abstractly and quantitatively.**
Help students analyze the patterns and represent the sequence symbolically. For example, ask:

- Why is $a_{n+1} = a_n + 4$ not enough information to write the sequence? **The first term is not given. The sequence could be 0, 4, 8, 12… or 2, 6, 10, 14… Both sequences make the given formula true, but they have different first terms.**
- Which term is represented by $a_{n+2}$? **The term that is two terms after $a_n$.**
- If the terms in a sequence are increasing, will the recursive formula always use addition? Explain. **No, it is possible to multiply by a number greater than 1 and have a sequence that is increasing.**

---

**Lesson 7-10 | Recursive Formulas**

## Launch

Have students read the Why? section of the lesson. Ask:

- How does the total cost of the shuttle service change as a customer is added? **Sample answer: The total cost increases by $10.**
- Is this sequence *arithmetic*, *geometric*, or *neither*? **arithmetic**
- How much would it cost for 9 customers? **$105**

## Teach

Ask the scaffolded questions for each example to build conceptual understanding for students at all levels.

### 1 Using Recursive Formulas

**Example 1 Use a Recursive Formula**

**AL** What values do we substitute for $n$? **2, 3, 4, and 5**

**OL** Why are we given $a_1$? **Sample answer: we have to have a number to start from when generating the sequence.**

**BL** Can you now find the 20th term using the formula without finding terms 6 through 19? Explain. **No; Sample answer: this kind of formula requires that you know the previous term.**

**Need Another Example?**
Find the first five terms of the sequence in which $a_1 = -8$ and $a_n = -2a_{n-1} + 5$, if $n \geq 2$. **−8, 21, −37, 79, −153**

### Go Online!

**Interactive Whiteboard**
Use the *eLesson* or *Lesson Presentation* to present this lesson.

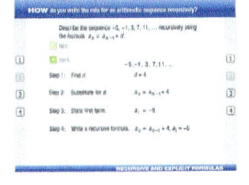

connectED.mcgraw-hill.com  469

## Lesson 7-10 | Recursive Formulas

# 2 Writing Recursive Formulas

### Example 2 Write Recursive Formulas

**AL** What in the domain of $n$? $n \geq 2$

**OL** Is the sequence arithmetic or geometric? geometric

**BL** How can you check to make sure your recursive formula is correct? Sample answer: use the formula to determine terms two, three, and four and make sure they are the same as the given sequence.

### Need Another Example?
Write a recursive formula for each sequence.
a. 23, 29, 35, 41, … $a_1 = 23$, $a_n = a_{n-1} + 6$, $n \geq 2$
b. 7, −21, 63, −189, … $a_1 = 7$, $a_n = -3a_{n-1}$, $n \geq 2$

---

b. $a_1 = 3$ and $a_n = 4a_{n-1}$, if $n \geq 2$

Use $a_1 = 3$ and the recursive formula to find the next four terms.

| $n$ | $a_n = 4a_{n-1}$ | $a_n$ |
|---|---|---|
| 1 | — | 3 |
| 2 | $a_2 = 4(3)$ | 12 |
| 3 | $a_3 = 4(12)$ | 48 |
| 4 | $a_4 = 4(48)$ | 192 |
| 5 | $a_5 = 4(192)$ | 768 |

The first five terms are 3, 12, 48, 192, and 768.

▶ **Guided Practice**

**1A.** $a_1 = -2$ and $a_n = a_{n-1} + 4$, if $n \geq 2$   −2, 2, 6, 10, 14

**1B.** $a_1 = -4$ and $a_n = \frac{3}{2}a_{n-1}$, if $n \geq 2$   $-4, -6, -9, -\frac{27}{2}, -\frac{81}{4}$

**2 Writing Recursive Formulas** To write a recursive formula for an arithmetic or geometric sequence, complete the following steps.

### 🧩 Key Concept  Writing Recursive Formulas

**Step 1** Determine whether the sequence is arithmetic or geometric by finding a common difference or a common ratio.

**Step 2** Write a recursive formula.

| Arithmetic Sequences | $a_n = a_{n-1} + d$, where $d$ is the common difference |
| Geometric Sequences | $a_n = r \cdot a_{n-1}$, where $r$ is the common ratio |

**Step 3** State the first term and domain for $n$.

F.BF.2

**Study Tip**
Defining $n$ For the $n$th term of a sequence, the value of $n$ must be a positive integer. Although we must still state the domain of $n$, from this point forward, we will assume that $n$ is an integer.

### Example 2  Write Recursive Formulas

Write a recursive formula for the sequence 6, 24, 96, 384… .

**Step 1** First subtract each term from the term that follows it.

$24 - 6 = 18$     $96 - 24 = 72$     $384 - 96 = 288$

There is no common difference. Check for a common ratio by dividing each term by the term that precedes it.

$\frac{24}{6} = 4$     $\frac{96}{24} = 4$     $\frac{384}{96} = 4$

There is a common ratio of 4. The sequence is geometric.

**Step 2** Use the formula for a geometric sequence.

$a_n = r \cdot a_{n-1}$   Recursive formula for geometric sequence
$a_n = 4a_{n-1}$   $r = 4$

**Study Tip**
**MP Perseverance** The domain for $n$ is decided by the given terms. Since the first term is already given, it makes sense that the first term to which the formula would apply is the 2nd term of the sequence, or when $n = 2$.

**Step 3** The first term $a_1$ is 6, and $n \geq 2$.

A recursive formula for the sequence is $a_1 = 6$, $a_n = 4a_{n-1}$, $n \geq 2$.

▶ **Guided Practice**

**2.** Write a recursive formula for the sequence 4, 10, 25, 62.5, …
$a_1 = 4$, $a_n = 2.5a_{n-1}$, $n \geq 2$

---

## Go Online!

The most up-to-date resources available for your program can be found at connectED.mcgraw-hill.com.

A sequence can be represented by both an explicit formula and a recursive formula.

F.BF.2

### Example 3 Write Recursive and Explicit Formulas

**COST** Refer to the beginning of the lesson. Let $n$ be the number of customers.

**a.** Write a recursive formula for the sequence.

**Steps 1 and 2** First subtract each term from the term that follows it.

$35 - 25 = 10$   $45 - 35 = 10$   $55 - 45 = 10$

There is a common difference of 10. The sequence is arithmetic.

**Step 3** Use the formula for an arithmetic sequence.

$a_n = a_{n-1} + d$    Recursive formula for arithmetic sequence
$a_n = a_{n-1} + 10$    $d = 10$

**Step 4** The first term $a_1$ is 25, and $n \geq 2$.

A recursive formula for the sequence is $a_1 = 25$, $a_n = a_{n-1} + 10$, $n \geq 2$.

**b.** Write an explicit formula for the sequence.

**Step 1** The common difference is 10.

**Step 2** Use the formula for the $n$th term of an arithmetic sequence.

$a_n = a_1 + (n-1)d$    Formula for the $n$th term
$= 25 + (n-1)10$    $a_1 = 25$ and $d = 10$
$= 25 + 10n - 10$    Distributive Property
$= 10n + 15$    Simplify.

An explicit formula for the sequence is $a_n = 10n + 15$.

### Guided Practice

**3. SAVINGS** The money that Ronald has in his savings account earns interest each year. He does not make any withdrawals or additional deposits. The account balance at the beginning of each year is $10,000, $10,300, $10,609, $10,927.27, and so on. Write a recursive formula and an explicit formula for the sequence.
$a_1 = 10,000$, $a_n = 1.03a_{n-1}$, $n \geq 2$; $a_n = 10,000(1.03)^{n-1}$

If several successive terms of a sequence are needed, a recursive formula may be useful, whereas if just the $n$th term of a sequence is needed, an explicit formula may be useful. Thus, it is sometimes beneficial to translate between the two forms.

F.BF.2

### Example 4 Translate between Recursive and Explicit Formulas

**a.** Write a recursive formula for $a_n = 6n + 3$.

$a_n = 6n + 3$ is an explicit formula for an arithmetic sequence with $d = 6$ and $a_1 = 6(1) + 3$ or 9. Therefore, a recursive formula for $a_n$ is $a_1 = 9$, $a_n = a_{n-1} + 6$, $n \geq 2$.

**b.** Write an explicit formula for $a_1 = 120$, $a_n = 0.8a_{n-1}$, $n \geq 2$.

$a_n = 0.8a_{n-1}$ is a recursive formula for a geometric sequence with $a_1 = 120$ and $r = 0.8$. Therefore, an explicit formula for $a_n$ is $a_n = 120(0.8)^{n-1}$.

### Guided Practice

**4A.** Write a recursive formula for $a_n = 4(3)^{n-1}$.  $a_1 = 4$, $a_n = 3a_{n-1}$, $n \geq 2$.
**4B.** Write an explicit formula for $a_1 = -16$, $a_n = a_{n-1} - 7$, $n \geq 2$.  $a_n = -7n - 9$.

---

**Real-World Career**

**Buses** Intercity busses are becoming more popular in the U.S. with 60.9 million passengers in 2015. That is a 22% increase from the previous year, according to the U.S. Department of Transportation. One bus line logged more than 5.4 billion passenger miles in 2015 alone.

Source: skift.com

**Study Tip**

**Geometric Sequences** Recall that the formula for the $n$th term of a geometric sequence is $a_n = a_1 r^{n-1}$.

---

## Differentiated Instruction  AL  OL

**Interpersonal Learners** Divide the class into groups of two or three students. Have each student write a sequence on one note card and the recursive formula for the sequence on another note card. Repeat the process for 10 sequences. Then, have the students lay the cards face down. Each student should take turns flipping over two cards, attempting to find a match between a sequence and its recursive formula.

---

## Lesson 7-10 | Recursive Formulas

### Example 3 Write Recursive and Explicit Formulas

**AL** Is the sequence arithmetic or geometric? **arithmetic**

**OL** What is the common difference? **10** In part **b**, where do we substitute 10 in to the formula? **for the variable $d$**

**BL** Why might an explicit formula be useful? **Sample answer: it can be used to find the $n$th term when you don't know previous terms.**

### Need Another Example?

**Cars** The price of a car depreciates at the end of each year.

| Year | Price ($) |
|---|---|
| 1 | 12,000 |
| 2 | 7200 |
| 3 | 4320 |
| 4 | 2592 |

**a.** Write a recursive formula for the sequence.
$a_1 = 12,000$, $a_n = 0.6a_{n-1}$, $n \geq 2$

**b.** Write an explicit formula for the sequence.
$a_n = 12,000(0.6)^{n-1}$

### Example 4 Translate between Recursive and Explicit Formulas

**AL** For part **a**, in what form is the formula currently written? **explicit**

**OL** In part **a**, what value is the variable $d$? **6**

**BL** Write your own arithmetic sequence. Then write the recursive and explicit formulas to find the $n$th term. **See students' work.**

### Need Another Example?

**a.** Write a recursive formula for $a_n = 2n - 4$.
$a_1 = -2$, $a_n = a_{n-1} + 2$, $n \geq 2$

**b.** Write an explicit formula for $a_1 = 84$, $a_n = 1.5a_{n-1}$, $n \geq 2$.  $a_n = 84(1.5)^{n-1}$

### Teaching Tip

**Terms** The first term of a sequence is occasionally denoted as $a_0$.

connectED.mcgraw-hill.com  471

# Lesson 7-10 | Recursive Formulas

## Practice

**Formative Assessment** Use Exercises 1–9 to assess students' understanding of the concepts in this lesson.

The Practice and Problem Solving exercises assess the content taught in the lesson. The Preparing for Assessment page is meant to be used as preparation for end-of-course assessments.

### Teaching the Mathematical Practices

**Modeling** Mathematically proficient students use tools such as diagrams to map the relationships of important quantities in a practical situation. In Exercise 22, tell students that they can sketch the patio to help them see the pattern.

## Extra Practice

See page R7 for extra exercises for students who are approaching level or for on-level students who need additional reinforcement.

### Levels of Complexity Chart

The levels of the exercises progress from 1 to 3, with Level 1 indicating the lowest level of complexity.

| Exercises | 10–26 | 27–29, 36–40 | 30–35 |
|---|---|---|---|
| Level 3 |  |  | ● |
| Level 2 |  | ● |  |
| Level 1 | ● |  |  |

**Go Online!**   **eBook**

**Interactive Student Guide**
Use the *Interactive Student Guide* to deepen conceptual understanding.
· Recursive Formulas

---

### Check Your Understanding

○ = Step-by-Step Solutions begin on page R11.

**Go Online!** for a Self-Check Quiz

**Example 1** Find the first five terms of each sequence.

1. $a_1 = 16, a_n = a_{n-1} - 3, n \geq 2$
   16, 13, 10, 7, 4
2. $a_1 = -5, a_n = 4a_{n-1} + 10, n \geq 2$
   −5, −10, −30, −110, −430

**Example 2** Write a recursive formula for each sequence.

3. 1, 6, 11, 16, ...
   $a_1 = 1, a_n = a_{n-1} + 5, n \geq 2$
4. 4, 12, 36, 108, ...
   $a_1 = 4, a_n = 3a_{n-1}, n \geq 2$

**Example 3**

5. **BALL** A ball is dropped from an initial height of 10 feet. The maximum heights the ball reaches on the first three bounces are shown. 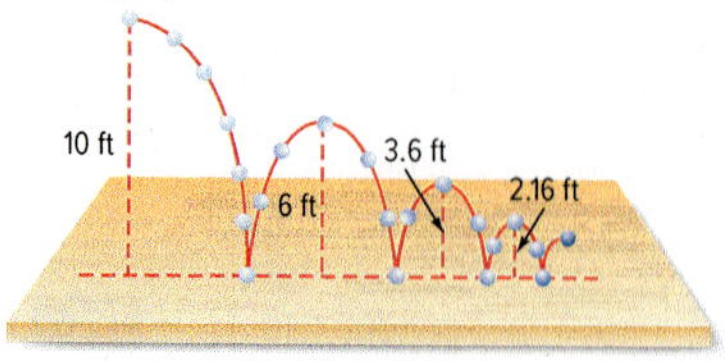
   a. Write a recursive formula for the sequence.
   b. Write an explicit formula for the sequence.
   5a. $a_1 = 10, a_n = 0.6a_{n-1}, n \geq 2$
   5b. $a_n = 10(0.6)^{n-1}$

**Example 4** For each recursive formula, write an explicit formula. For each explicit formula, write a recursive formula.

6. $a_1 = 4, a_n = a_{n-1} + 16, n \geq 2$
   $a_n = 16n - 12$
7. $a_n = 5n + 8$
   $a_1 = 13, a_n = a_{n-1} + 5, n \geq 2$
8. $a_n = 15(2)^{n-1}$
   $a_1 = 15, a_n = 2a_{n-1}, n \geq 2$
9. $a_1 = 22, a_n = 4a_{n-1}, n \geq 2$
   $a_n = 22(4)^{n-1}$

### Practice and Problem Solving

Extra Practice is on page R7.

**Example 1** Find the first five terms of each sequence.

10. $a_1 = 23, a_n = a_{n-1} + 7, n \geq 2$
    23, 30, 37, 44, 51
11. $a_1 = 48, a_n = -0.5a_{n-1} + 8, n \geq 2$
    48, −16, 16, 0, 8
12. $a_1 = 8, a_n = 2.5a_{n-1}, n \geq 2$
    8, 20, 50, 125, 312.5
13. $a_1 = 12, a_n = 3a_{n-1} - 21, n \geq 2$
    12, 15, 24, 51, 132
14. $a_1 = 13, a_n = -2a_{n-1} - 3, n \geq 2$
    13, −29, 55, −113, 223
15. $a_1 = \frac{1}{2}, a_n = a_{n-1} + \frac{3}{2}, n \geq 2$
    $\frac{1}{2}, 2, \frac{7}{2}, 5, \frac{13}{2}$

**Example 2** Write a recursive formula for each sequence.

16. 12, −1, −14, −27, ...
    $a_1 = 12, a_n = a_{n-1} - 13, n \geq 2$
17. 27, 41, 55, 69, ...
    $a_1 = 27, a_n = a_{n-1} + 14, n \geq 2$
18. 2, 11, 20, 29, ...
    $a_1 = 2, a_n = a_{n-1} + 9, n \geq 2$
19. 100, 80, 64, 51.2, ...
    $a_1 = 100, a_n = 0.8a_{n-1}, n \geq 2$
20. 40, −60, 90, −135, ...
    $a_1 = 40, a_n = -1.5a_{n-1}, n \geq 2$
21. 81, 27, 9, 3, ...
    $a_1 = 81, a_n = \frac{1}{3}a_{n-1}, n \geq 2$

**Example 3**

22. **MODELING** A landscaper is building a brick patio. Part of the patio includes a pattern constructed from triangles. The first four rows of the pattern are shown. 22a. $a_1 = 15, a_n = a_{n-1} - 2, n \geq 2$ 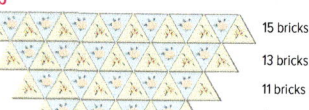
    - 15 bricks
    - 13 bricks
    - 11 bricks
    - 9 bricks
    a. Write a recursive formula for the sequence.
    b. Write an explicit formula for the sequence. $a_n = 17 - 2n$

**Example 4** For each recursive formula, write an explicit formula. For each explicit formula, write a recursive formula.

23. $a_n = 3(4)^{n-1}$   $a_1 = 3, a_n = 4a_{n-1}, n \geq 2$
24. $a_1 = -2, a_n = a_{n-1} - 12, n \geq 2$   $a_n = -12n + 10$
25. $a_n = 38\left(\frac{1}{2}\right)^{n-1}$   $a_1 = 38, a_n = \frac{1}{2}a_{n-1}, n \geq 2$
26. $a_n = -7n + 52$   $a_1 = 45, a_n = a_{n-1} - 7, n \geq 2$

### Differentiated Homework Options

| Levels | AL Basic | OL Core | BL Advanced |
|---|---|---|---|
| Exercises | 10–26, 31, 33–40 | 11–25 odd, 27–31, 33–40 | 27–35<br>36–40 (optional) |
| 2-Day Option | 11–25 odd, 36–40 | 10–26, 36–40 |  |
|  | 10–26 even, 31, 33–35 | 27–31, 33–35 |  |

You can use ALEKS to provide additional remediation support with personalized instruction and practice.

**Lesson 7-10 | Recursive Formulas**

**27. PHOTO SHARING** Barbara shares a photo with five of her friends. Each of her friends shared the photo with five more friends, and so on.

a. Find the first five terms of the sequence representing the number of people who receive the photo in the $n$th round. **1, 5, 25, 125, 625**

b. Write a recursive formula for the sequence. **$a_1 = 1, a_n = 5a_{n-1}, n \geq 2$**

c. If Barbara represents $a_1$, find $a_8$. **78,125**

**28. GEOMETRY** Consider the pattern below. The number of blue boxes increases according to a specific pattern.

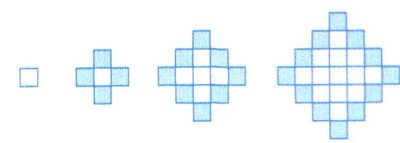

**28a.** $a_1 = 0, a_n = a_{n-1} + 4, n \geq 2$

a. Write a recursive formula for the sequence of the number of blue boxes in each figure.

b. If the first box represents $a_1$, find the number of blue boxes in $a_8$. **28**

**29. TREE** The heights of a certain type of tree over the past four years are shown.

10 ft    11 ft    12.1 ft    13.31 ft

**30a.** Sample answer: The first two terms are 1. Starting with the third term, the two previous terms are added together to get the next term; 13, 21, 34, 55, 89.

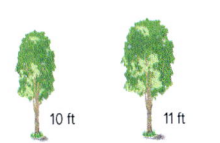

a. Write a recursive formula for the height of the tree. $a_1 = 10, a_n = 1.1a_{n-1}, n \geq 2$

b. If the pattern continues, how tall will the tree be in two more years? Round your answer to the nearest tenth of a foot. **16.1 ft**

**30. MULTIPLE REPRESENTATIONS** The Fibonacci sequence is neither arithmetic nor geometric and can be defined by a recursive formula. The first terms are 1, 1, 2, 3, 5, 8, …

a. **Logical** Determine the relationship between the terms of the sequence. What are the next five terms in the sequence? **30b.** $a_1 = 1, a_2 = 1, a_n = a_{n-2} + a_{n-1}, n \geq 3$

b. **Algebraic** Write a formula for the $n$th term if $a_1 = 1, a_2 = 1$, and $n \geq 3$.

c. **Algebraic** Find the 15th term. **610**

d. **Analytical** Explain why the Fibonacci sequence is not an arithmetic sequence. Sample answer: There is no common difference.

F.IF.3, F.BF.2

**H.O.T. Problems** Use Higher-Order Thinking Skills

**31. ERROR ANALYSIS** Patrick and Lynda are working on a math problem that involves the sequence 2, −2, 2, −2, 2, … . Patrick thinks that the sequence can be written as a recursive formula. Lynda believes that the sequence can be written as an explicit formula. Is either of them correct? Explain. **See margin.**

**32. CHALLENGE** Find $a_1$ for the sequence in which $a_4 = 1104$ and $a_n = 4a_{n-1} + 16$. **12**

**33. CONSTRUCT ARGUMENTS** Determine whether the following statement is *true* or *false*. Justify your reasoning. **33, 35. See margin.**

*There is only one recursive formula for every sequence.*

**34. CHALLENGE** Find a recursive formula for 4, 9, 19, 39, 79, … . $a_1 = 4, a_n = 2a_{n-1} + 1, n \geq 2$

**35. WRITING IN MATH** Explain the difference between an explicit formula and a recursive formula.

---

### Watch Out!

**Error Analysis** In Exercise 31, students should recognize that the sequence is geometric with a common ratio of −1. Therefore, the sequence can be represented as both an explicit formula and a recursive formula.

### Teaching the Mathematical Practices

**Construct Arguments** Mathematically proficient students can recognize and use counterexamples. In Exercise 33, tell students that they can start with a sequence for which they know a recursive formula and see if they can write another recursive formula that fits the sequence to see if they can find a counterexample to the statement.

## Assess

**Ticket Out the Door** Have each student create a sequence by writing the first five terms. Then have them write an explicit formula and a recursive formula for the sequence.

### Additional Answers

**31.** Both; Sample answer: The sequence can be written as the recursive formula $a_1 = 2, a_n = (-1)a_{n-1}, n \geq 2$. The sequence can also be written as the explicit formula $a_n = 2(-1)^{n-1}$.

**33.** False; Sample answer: A recursive formula for the sequence 1, 2, 3, … can be written as $a_1 = 1, a_{n-1} + 1, n \geq 2$ or as $a_1 = 1, a_2 = 2, a_n = a_{n-2} + 2, n \geq 3$.

**35.** Sample answer: In an explicit formula, the $n$th term $a_n$ is given as a function of $n$. In a recursive formula, the $n$th term of $a_n$ is found by performing operations to one or more of the terms that precede it.

### Go Online!

**eSolutions Manual**
Create worksheets, answer keys, and solutions handouts for your assignments.

---

## Standards for Mathematical Practices

| Emphasis On | Exercises |
| --- | --- |
| 1 Make sense of problems and persevere in solving them. | 36, 39, 40 |
| 2 Reason abstractly and quantitatively. | 27, 30, 32, 34, 36–40 |
| 3 Construct viable arguments and critique the reasoning of others. | 31, 33 |
| 4 Model with mathematics. | 22, 27–29 |
| 7 Look for and make use of structure. | 36 |

connectED.mcgraw-hill.com 473

# Lesson 7-10 | Recursive Formulas

## Preparing for Assessment

Exercises 36–40 require students to use the skills they will need on standardized assessments. Each exercise is dual-coded with content standards and mathematical practice standards.

| | Dual Coding | |
|---|---|---|
| Items | Content Standards | Mathematical Practices |
| 36 | F.IF.3, F.BF.2 | 1, 2, 7 |
| 37 | F.IF.3 | 2 |
| 38 | F.BF.2 | 2 |
| 39 | F.IF.3 | 2 |
| 40 | F.IF.3 | 1, 2 |

## Diagnose Student Errors

Survey student responses for each item. Class trends may indicate common errors and misconceptions.

**36c.**

| A | Did not use the previous term in the formula |
|---|---|
| B | CORRECT |
| C | Subtracted 1 instead of added 1 to the formula |
| D | Found a formula that is only valid for the first two terms |

**37.**

| A | Found the amount of increase after the first year |
|---|---|
| B | Found a decrease in population rather than an increase |
| C | CORRECT |
| D | Added next year's population to the original population |

**39a.**

| A | CORRECT |
|---|---|
| B | Multiplied by positive 2 instead of negative 2 |
| C | Used 2 as an exponent rather than a coefficient |
| D | Did not use a recursive formula |

## Go Online!

**Quizzes**

Students can use *Self-Check Quizzes* to check their understanding of this lesson. You can also give the *Chapter Quiz* which covers the content in Lessons 7-9 and 7-10.

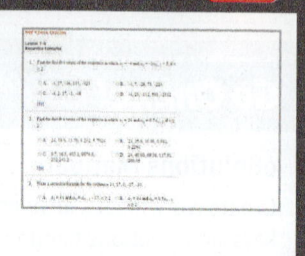

474 | Lesson 7-10 | Recursive Formulas

## Preparing for Assessment

**36. MULTI-STEP** The table shows several terms of a sequence. MP 1, 2, 7 F.IF.3

| n | 1 | 2 | 3 | 4 |
|---|---|---|---|---|
| $a_n$ | −12 | 25 | −49 | 99 |

a. Which type of sequence is shown? **A**
- A recursive
- B explicit
- C arithmetic
- D geometric
- E none of the above

b. What is the next term in the sequence?
**−197**

c. Which formula describes the sequence? **B**
- A $a_1 = -12$, $a_n = -12n - 1$, $n \geq 2$
- B $a_1 = -12$, $a_n = -2a_{n-1}$, $n \geq 2$
- C $a_1 = -12$, $a_n = -2a_{n-1}$, $n \geq 2$
- D $a_1 = -12$, $a_n = a_{n-1} + 37$, $n \geq 2$

d. Can an explicit formula be written as a linear equation for the sequence? Explain.
**No; because the sequence can only be described by using the previous term, an explicit formula cannot be written.**

**37.** The number of students at Superior Middle School over a three-year period is shown in the table.

| Year | 1 | 2 | 3 |
|---|---|---|---|
| Number of Students | 700 | 770 | 847 |

Use a recursive formula to find the number of students that will be at the school in the next year. MP 2 F.IF.3 **C**
- A 70
- B 630
- C 932
- D 1632

**38.** Which of the following are formulas for the sequence 8, 16, 24, 32, …? MP 2 F.BF.2 **B, C**
- ☐ A $a_1 = 8$, $a_n = 2a_{n-1}$, $n \geq 2$
- ☐ B $a_1 = 8$, $a_n = 2a_{n-1} + 8$, $n \geq 2$
- ☐ C $a_n = 8n$
- ☐ D $a_n = 8n + 8$
- ☐ E $a_n = 8^{3n}$

**39.** The table shows several terms of a sequence.

| n | 1 | 2 | 3 | 4 |
|---|---|---|---|---|
| $a_n$ | −2 | 4 | −8 | 16 |

a. Which of the following is a recursive formula for the sequence? MP 2 F.IF.3 **A**
- A $a_1 = -2$, $a_n = -2a_{n-1}$
- B $a_1 = -2$, $a_n = 2a_{n-1}$
- C $a_1 = -2$, $a_n = -1(a_{n-1})^2$
- D $a_1 = -2$, $a_n = -1(2^n)$

b. What is the tenth term of the given sequence from the table? MP 1, 2 F.IF.3
**1024**

**40.** Mrs. Rodriguez wrote the first four terms of a sequence in a table on the board.

| n | 1 | 2 | 3 | 4 |
|---|---|---|---|---|
| $a_n$ | −2 | −9 | −37 | −149 |

What is the sixth term of the given sequence from the table? MP 1, 2 F.IF.3 **A**
- A −2389
- B −2388
- C −597
- D −596

## Differentiated Instruction BL

**Extension** For Exercise 30, students wrote a recursive formula for the Fibonacci sequence, which is neither arithmetic nor geometric. Have students write a recursive formula for another sequence that is neither arithmetic nor geometric.

**40.**

| A | CORRECT |
|---|---|
| B | Didn't subtract 1 from $4a_{n-1}$ |
| C | Found the fifth term |
| D | Stopped at the fifth term and added 1 |

# CHAPTER 7

## Study Guide and Review

*Go Online!* for Vocabulary Review Games and key vocabulary in 13 languages

## Study Guide

### Key Concepts

**Multiplication and Division Properties of Exponents** (Lessons 7-1 and 7-2)

For any nonzero real numbers $a$ and $b$ and any integers $m$, $n$, and $p$, the following are true.

- Product of Powers: $a^m \cdot a^n = a^{m+n}$
- Power of a Power: $(a^m)^n = a^{m \cdot n}$
- Power of a Product: $(ab)^m = a^m b^m$
- Quotient of Powers: $\frac{a^m}{a^p} = a^{m-p}$
- Power of a Quotient: $\left(\frac{a}{b}\right)^m = \frac{a^m}{b^m}$
- Zero Exponent: $a^0 = 1$
- Negative Exponent: $a^{-n} = \frac{1}{a^n}$ and $\frac{1}{a^{-n}} = a^n$

**Rational Exponents** (Lesson 7-3)

For any positive real number $b$ and any integers $m$ and $n > 1$, the following are true:

$b^{\frac{1}{2}} = \sqrt{b}$  $\qquad b^{\frac{1}{n}} = \sqrt[n]{b}$  $\qquad b^{\frac{m}{n}} = \left(\sqrt[n]{b}\right)^m$ or $\sqrt[n]{b^m}$

**Radical Expressions** (Lesson 7-4)

For real numbers $a$, $b$, $c$, and $d$,

- $\sqrt{ab} = \sqrt{a} \cdot \sqrt{b}$, if $a \geq 0$ and $b \geq 0$.
- $\sqrt{\frac{a}{b}} = \frac{\sqrt{a}}{\sqrt{b}}$, if $a \geq 0$ and $b > 0$.
- $a\sqrt{b} \pm c\sqrt{b} = (a \pm c)\sqrt{b}$, if $b \geq 0$.
- $a\sqrt{b}(c\sqrt{d}) = (ac)\sqrt{bd}$, if $b \geq 0$ and $d \geq 0$.

**Transformations of Exponential Functions** (Lesson 7-6)

- An exponential function can be written as $f(x) = ab^{x-h} + k$. The parameters $a$, $h$, and $k$ are parameters that dilate, reflect, or translate a parent function with base $b$.

**Exponential Functions** (Lesson 7-7)

- Exponential growth: $y = a(1 + r)^t$, where $r > 0$
- Exponential decay: $y = a(1 - r)^t$, where $0 < r < 1$

**FOLDABLES Study Organizer**

Use your Foldable to review the chapter. Working with a partner can be helpful. Ask for clarification of concepts as needed.

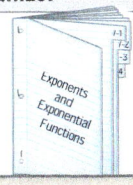

### Key Vocabulary

| | |
|---|---|
| asymptote (p. 430) | geometric sequence (p. 462) |
| common ratio (p. 462) | monomial (p. 395) |
| compound interest (p. 451) | negative exponent (p. 404) |
| conjugates (p. 421) | $n$th root (p. 411) |
| constant (p. 395) | order of magnitude (p. 405) |
| cube root (p. 411) | radical expression (p. 419) |
| exponential decay (p. 430) | rational exponent (p. 410) |
| exponential equation (p. 413) | rationalizing the denominator (p. 421) |
| exponential function (p. 430) | recursive formula (p. 469) |
| exponential growth (p. 430) | zero exponent (p. 403) |

### Vocabulary Check

Choose the word or term that best completes each sentence.

1. 2 is a(n) _____ of 8. **cube root**

2. The rules for operations with exponents can be extended to **rational exponent** apply to expressions with a(n) _____ such as $7^{\frac{2}{3}}$.

3. $f(x) = 3^x$ is an example of a(n) _____. **exponential function**

4. $a_1 = 4$ and $a_n = 3a_{n-1} - 12$, if $n \geq 2$, is a(n) _____ for the sequence 4, −8, −20, −32, … . **recursive formula**

5. $2^{3x-1} = 16$ is an example of a(n) _____. **exponential equation**

6. The equation for _____ is $y = C(1 - r)^t$. **exponential decay**

### Concept Check

7. How can you determine whether a set of data in a table display exponential behavior? **See margin.**

8. Explain how to find the common ratio of a geometric sequence. **See margin.**

---

## Chapter 7 Study Guide and Review

**FOLDABLES Study Organizer**

A completed Foldable for this chapter should include the Key Concepts related to exponents and exponential functions.

### Key Vocabulary  ELL

The page reference after each word denotes where that term was first introduced. If students have difficulty answering questions 1–8, remind them that they can use these page references to refresh their memories about the vocabulary terms.

Have students work with a partner to complete the Vocabulary Check using the vocabulary list. Encourage them to seek clarification of each vocabulary term as needed.

You can use the detailed reports in ALEKS to automatically monitor students' progress and pinpoint remediation needs prior to the chapter test.

### Additional Answers

7. If the domain values are at equal intervals, find the common factors among each consecutive pair of range values. If the common factors are the same, the data are exponential.

8. Divide a term by the previous term.

---

## Answering the Essential Question

Before answering the Essential Question, have students review their answers to the *Building on the Essential Question* exercises found throughout the chapter.

- How can you make good decisions? What factors can affect good decision making? (p. 392)

- Why do you think it is important to simplify radical expressions? (p. 420)

- How can mathematical models help you make good decisions? (p. 434)

- Summarize the transformations produced by performing various operations on the parent exponential function? (p. 446)

- How can being financially literate help you to make good decisions? (p. 453)

### Go Online!

**Vocabulary Review**

Students can use the *Vocabulary Review Games* to check their understanding of the vocabulary terms in this chapter. Students should refer to the *Student-Built Glossary* they have created as they went through the chapter to review important terms. You can also give a *Vocabulary Test* over the content of this chapter.

connectED.mcgraw-hill.com  475

# Chapter 7 Study Guide and Review

## Lesson-by-Lesson Review

**Intervention** If the given examples are not sufficient to review the topics covered by the questions, remind students that the lesson references tell them where to review that topic in their textbook.

**Two-Day Option** Have students complete the Lesson-by-Lesson Review. Then you can use McGraw-Hill eAssessment to customize another review worksheet that practices all the objectives of this chapter or only the objectives on which your students need more help.

---

# CHAPTER 7
# Study Guide and Review *Continued*

## Lesson-by-Lesson Review

### 7-1 Multiplication Properties of Exponents
A.SSE.2, F.IF.8b

Simplify each expression.

9. $x \cdot x^3 \cdot x^5$   $x^9$
10. $(2xy)(-3x^2y^5)$   $-6x^3y^6$
11. $(-4ab^4)(-5a^5b^2)$   $20a^6b^6$
12. $(6x^3y^2)^2$   $36x^6y^4$
13. $[(2r^3t)^3]^2$   $64r^{18}t^6$
14. $(-2u^3)(5u)$   $-10u^4$
15. $(2x^2)^3(x^3)^3$   $8x^{15}$
16. $\frac{1}{2}(2x^3)^3$   $4x^9$

17. **GEOMETRY** Use the formula $V = \pi r^2 h$ to find the volume of the cylinder.   $45\pi x^4$

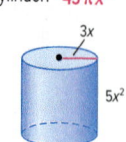

(cylinder: height $3x$, radius $5x^2$)

**Example 1**

Simplify $(5x^2y^3)(2x^4y)$.

$(5x^2y^3)(2x^4y)$
$= (5 \cdot 2)(x^2 \cdot x^4)(y^3 \cdot y)$   Commutative Property
$= 10x^6y^4$   Product of Powers

**Example 2**

Simplify $(3a^2b^4)^3$.

$(3a^2b^4)^3 = 3^3(a^2)^3(b^4)^3$   Power of a Product
$= 27a^6b^{12}$   Simplify.

---

### 7-2 Division Properties of Exponents
A.SSE.2, F.IF.8b

Simplify each expression. Assume that no denominator equals zero.

18. $\frac{(3x)^0}{2a}$   $\frac{1}{2a}$
19. $\left(\frac{3xy^3}{2z}\right)^3$   $\frac{27x^3y^9}{8z^3}$
20. $\frac{12y^{-4}}{3y^{-5}}$   $4y$
21. $a^{-3}b^0c^6$   $\frac{c^6}{a^3}$
22. $\frac{-15x^7y^8z^4}{-45x^3y^5z^3}$   $\frac{x^4y^3z}{3}$
23. $\frac{(3x^{-1})^{-2}}{(3x^2)^{-2}}$   $x^6$
24. $\left(\frac{6xy^{11}z^9}{48x^6yz^{-7}}\right)^0$   $1$
25. $\left(\frac{12}{2}\right)\left(\frac{x}{y^5}\right)\left(\frac{y^4}{x^4}\right)$   $\frac{6}{yx^3}$

26. **GEOMETRY** The area of a rectangle is $25x^2y^4$ square feet. The width of the rectangle is $5xy$ feet. What is the length of the rectangle?   $5xy^3$ ft

(rectangle labeled $5xy$)

**Example 3**

Simplify $\frac{2k^4m^3}{4k^2m}$. Assume that no denominator equals zero.

$\frac{2k^4m^3}{4k^2m} = \left(\frac{2}{4}\right)\left(\frac{k^4}{k^2}\right)\left(\frac{m^3}{m}\right)$   Group powers with the same base.
$= \left(\frac{1}{2}\right)k^{4-2}m^{3-1}$   Quotient of Powers
$= \frac{k^2m^2}{2}$   Simplify.

**Example 4**

Simplify $\frac{t^4uv^{-2}}{t^{-3}u^7}$. Assume that no denominator equals zero.

$\frac{t^4uv^{-2}}{t^{-3}u^7} = \left(\frac{t^4}{t^{-3}}\right)\left(\frac{u}{u^7}\right)(v^{-2})$   Group the powers with the same base.
$= (t^{4+3})(u^{1-7})(v^{-2})$   Quotient of Powers
$= t^7u^{-6}v^{-2}$   Simplify.
$= \frac{t^7}{u^6v^2}$   Simplify.

---

**476** | Chapter 7 | Study Guide and Review

## Chapter 7 Study Guide and Review

### 7-3 Rational Exponents
N.RN.1, N.RN.2

Simplify.

27. $\sqrt[3]{343}$  7
28. $\sqrt[6]{729}$  3
29. $625^{\frac{1}{4}}$  5
30. $\left(\frac{8}{27}\right)^{\frac{1}{3}}$  $\frac{2}{3}$
31. $256^{\frac{3}{4}}$  64
32. $32^{\frac{2}{5}}$  4
33. $343^{\frac{4}{3}}$  2401
34. $\left(\frac{4}{49}\right)^{\frac{3}{2}}$  $\frac{8}{343}$

Solve each equation.

35. $6^x = 7776$  5
36. $4^{4x-1} = 32$  $\frac{7}{8}$

**Example 5**
Simplify $125^{\frac{2}{3}}$.
$125^{\frac{2}{3}} = \left(\sqrt[3]{125}\right)^2$     $b^{\frac{m}{n}} = \left(\sqrt[n]{b}\right)^m$
$= \left(\sqrt[3]{5 \cdot 5 \cdot 5}\right)^2$     $125 = 5^3$
$= 5^2$ or $25$     Simplify.

**Example 6**
Solve $9^{x-1} = 729$.
$9^{x-1} = 729$     Original equation
$9^{x-1} = 9^3$     Rewrite 729 as $9^3$.
$x - 1 = 3$     Power Property of Equality
$x = 4$     Add 1 to each side.

### 7-4 Radical Expressions
N.RN.2

Simplify.

37. $\sqrt{36x^2y^7}$  $6|x|y^3\sqrt{y}$
38. $\sqrt{20ab^3}$  $2|b|\sqrt{5ab}$
39. $\sqrt{3} \cdot \sqrt{6}$  $3\sqrt{2}$
40. $2\sqrt{3} \cdot 3\sqrt{12}$  36
41. $(4 - \sqrt{5})^2$  $21 - 8\sqrt{5}$
42. $(1 + \sqrt{2})^2$  $2\sqrt{2} + 3$
43. $\sqrt{\frac{50}{a^2}}$  $\frac{5\sqrt{2}}{|a|}$
44. $\sqrt{\frac{2}{5}} \cdot \sqrt{\frac{3}{4}}$  $\frac{\sqrt{30}}{10}$
45. $\frac{3}{2-\sqrt{5}}$  $-6 - 3\sqrt{5}$
46. $\frac{5}{\sqrt{7}+6}$  $\frac{30 - 5\sqrt{7}}{29}$
47. $\sqrt{6} - \sqrt{54} + 3\sqrt{12} + 5\sqrt{3}$  $-2\sqrt{6} + 11\sqrt{3}$
48. $2\sqrt{6} - \sqrt{48}$  $2\sqrt{6} - 4\sqrt{3}$
49. $\sqrt{2}(5 + 3\sqrt{3})$  $5\sqrt{2} + 3\sqrt{6}$
50. $(2\sqrt{3} - \sqrt{5})(\sqrt{10} + 4\sqrt{6})$  $-2\sqrt{30} + 19\sqrt{2}$
51. $(6\sqrt{5} + 2)(4\sqrt{2} + \sqrt{3})$  $24\sqrt{10} + 8\sqrt{2} + 6\sqrt{15} + 2\sqrt{3}$

52. **MOTION** The velocity of a dropped object when it hits the ground can be found using $v = \sqrt{2gd}$, where $v$ is the velocity in feet per second, $g$ is the acceleration due to gravity, and $d$ is the distance in feet the object drops. Find the speed of a penny when it hits the ground, after being dropped from 984 feet. Use 32 feet per second squared for $g$.  about 250.95 ft/s

**Example 7**
Simplify $\frac{2}{4+\sqrt{3}}$.

$\frac{2}{4+\sqrt{3}}$     Original expression

$= \frac{2}{4+\sqrt{3}} \cdot \frac{4-\sqrt{3}}{4-\sqrt{3}}$     Rationalize the denominator.

$= \frac{2(4) - 2\sqrt{3}}{4^2 - (\sqrt{3})^2}$     $(a - b)(a + b) = a^2 - b^2$

$= \frac{8 - 2\sqrt{3}}{16 - 3}$     $(\sqrt{3})^2 = 3$

$= \frac{8 - 2\sqrt{3}}{13}$     Simplify.

**Example 8**
Simplify $2\sqrt{6} - \sqrt{24}$.
$2\sqrt{6} - \sqrt{24} = 2\sqrt{6} - \sqrt{4 \cdot 6}$     Product Property
$= 2\sqrt{6} - 2\sqrt{6}$     Simplify.
$= 0$     Simplify.

**Example 9**
Simplify $(\sqrt{3} - \sqrt{2})(\sqrt{3} + 2\sqrt{2})$.
$(\sqrt{3} - \sqrt{2})(\sqrt{3} + 2\sqrt{2})$
$= (\sqrt{3})(\sqrt{3}) + (\sqrt{3})(2\sqrt{2}) + (-\sqrt{2})(\sqrt{3}) + (\sqrt{2})(2\sqrt{2})$
$= 3 + 2\sqrt{6} - \sqrt{6} + 4$
$= 7 + \sqrt{6}$

## Go Online!

### Anticipation Guide

Students should complete the *Chapter 7 Anticipation Guide*, and discuss how their responses have changed now that they have completed Chapter 7.

connectED.mcgraw-hill.com  477

# Chapter 7 Study Guide and Review

## Additional Answers

**53.** *y*-intercept 1; D = {all real numbers}; R = {*y* | *y* > 0}; *y* = 0

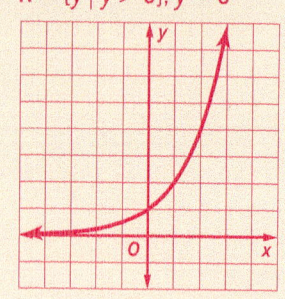

**54.** *y*-intercept 2; D = {all real numbers}; R = {*y* | *y* > 1}; *y* = 1

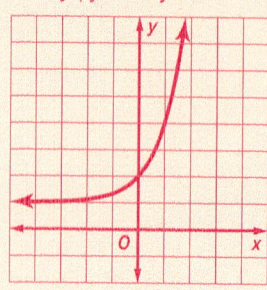

**55.** *y*-intercept 3; D = {all real numbers}; R = {*y* | *y* > 2}; *y* = 2

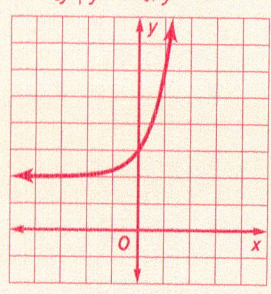

**56.** *y*-intercept −2; D = {all real numbers}; R = {*y* | *y* > −3}; *y* = −3

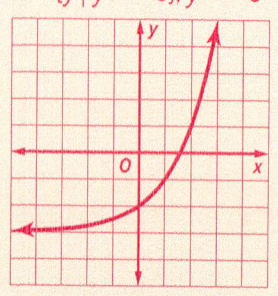

**58.** stretched vertically

**59.** reflected across the *x*-axis, compressed vertically, and translated down 2 units

---

# CHAPTER 7
## Study Guide and Review *Continued*

### 7-5 Exponential Functions
F.IF.7e, F.LE.5

Graph each function. Find the *y*-intercept, and state the domain, range, and the equation of the asymptote.

**53.** $y = 2^x$  53–56. See margin.
**54.** $y = 3^x + 1$
**55.** $y = 4^x + 2$
**56.** $y = 2^x - 3$

**57. BIOLOGY** The population of bacteria in a petri dish increases according to the model $p = 550(2.7)^{0.008t}$, where *t* is the number of hours and *t* = 0 corresponds to 1:00 P.M. Use this model to estimate the number of bacteria in the dish at 5:00 P.M. **about 568**

#### Example 10
Graph $y = 3^x + 6$. Find the *y*-intercept, and state the domain and range.

| x  | $3^x + 6$    | y    |
|----|--------------|------|
| −3 | $3^{-3} + 6$ | 6.04 |
| −2 | $3^{-2} + 6$ | 6.11 |
| −1 | $3^{-1} + 6$ | 6.33 |
| 0  | $3^0 + 6$    | 7    |
| 1  | $3^1 + 6$    | 9    |

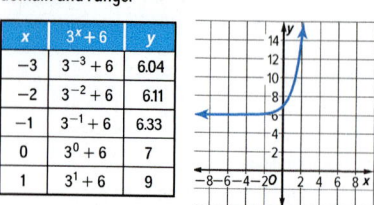

The *y*-intercept is (0, 7). The domain is all real numbers, and the range is all real numbers greater than 6.

### 7-6 Transformations of Exponential Functions
F.IF.7e, F.BF.3, F.LE.5

Describe the transformation *g*(*x*) as it relates to the parent function *f*(*x*). **58–61. See margin.**

**58.** $f(x) = 2^x$; $g(x) = 5(2)^x$
**59.** $f(x) = 3^x$; $g(x) = -0.5(3)^x - 2$
**60.** Compare the key features of $f(x) = 2^x$, $g(x) = 2^x - 3$, and $p(x) = 2^x + 4$.
**61.** Write the function *g*(*x*) given that *g*(*x*) is a reflection and vertical dilation of $f(x) = 4^x$.

#### Example 11
Compare the graphs of $f(x) = 2^x$ and $g(x) = \frac{1}{3}(2^x)$.

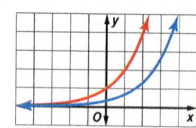

For both functions, the asymptote is *y* = 0. The function *f*(*x*) has *y*-intercept 1 and *g*(*x*) has *y*-intercept $\frac{1}{3}$. Both functions are increasing as $x \rightarrow \infty$. Compared to *f*(*x*), *g*(*x*) is vertically compressed.

---

**60.** The end behavior is the same for all three functions and they are all increasing as $x \rightarrow \infty$. *f*(*x*) has an asymptote of *y* = 0 and a *y*-intercept of 1. *g*(*x*) has an asymptote of *y* = −3 and a *y*-intercept of −2. *p*(*x*) has an asymptote of *y* = 4 and a *y*-intercept of 5.

**61.** $g(x) = -2(4)^x$

---

## Go Online!
### eAssessment
Customize and create multiple versions of chapter tests and answer keys that align to your standards. Tests can be delivered on paper or online.

# Chapter 7 Study Guide and Review

**Additional Answer**

67. For First Bank, the effective yearly rate is about 1.8%, which is greater than yearly rate of 1.5% at Main Street Bank.

## 7-7 Writing Exponential Functions
A.CED.2, F.LE.2, F.LE.5, F.IF.8b, F.IF.7e

62. Write an exponential function in the form $y = ab^x$ for the graph.

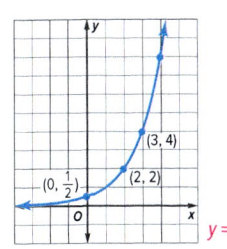

$y = \frac{1}{2}(2^x)$

Write an exponential function for a graph that passes through the given points.

63. (1, 12) and (2, 36)   $y = 4(3)^x$

64. (2, 20) and (4, 80)   $y = 5(2)^x$

65. Find the final value of $2500 invested at an annual interest rate of 2% compounded monthly for 10 years.  **$3053.00**

66. **COMPUTERS** Zita's computer is depreciating at a rate of 3% per year. She bought the computer for $1200.
    a. Write an equation to represent this situation.
    b. What will the computer's value be after 5 years?
    
    $1030.48     66a. $y = 1200(1 - 0.03)^t$

### Example 12
Write an exponential function for the graph.

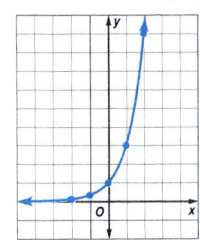

$1 = ab^0$ — Use the points (0, 1) and (2, 9) to write a
$9 = ab^2$ — system of equations.

$1 = a \cdot 1$ — Simplify the first equation and solve for $a$.
$a = 1$

$9 = 1 \cdot b^2$ — Substitute in the second equation and solve
$9 = b^2$ — for $b$.
$3 = b$

$y = ab^x$
$y = 1 \cdot 3^x$
$y = 3^x$ — Write the exponential function.

### Example 13
Find the final value of $2000 invested at an interest rate of 3% compounded quarterly for 8 years.

$A = P\left(1 + \frac{r}{n}\right)^{nt}$ — Compound interest equation

$= 2000\left(1 + \frac{0.03}{4}\right)^{4(8)}$ — $P = 2000, r = 0.03, n = 4, \text{ and } t = 8$

$\approx \$2540.22$ — Use a calculator.

## 7-8 Transforming Exponential Expressions
A.SSE.3c, F.IF.8b

67. First Bank's savings accounts have an interest rate of 0.15% compounded monthly. Main Street Bank has a 1.5% yearly rate. Compare the yearly rates for both banks. **See margin.**

68. What is the effective daily interest rate for a credit card that has a 24% yearly interest rate?  **about 0.06%**

69. Mark deposits $2000 in a savings account with an interest rate of 1.75% compounded annually. Find the effective quarterly interest rate.  **about 0.6%**

### Example 14
What is the effective annual interest rate for a credit card that has a monthly compounded interest rate of 1.8%?

$A(t) = (1.018)^{12t}$, where $t$ is the time in years

$= (1.018^{12})^t$ — Product of Powers

$\approx 1.239^t$ — Simplify.

The effective annual interest rate is about 23.9%.

# Chapter 7 Study Guide and Review

## Additional Answers

76. **Basketball Rebound**

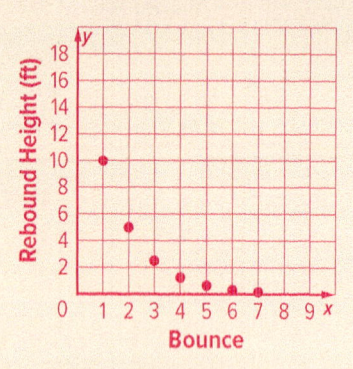

[Graph: Rebound Height (ft) vs. Bounce, showing decreasing points from about 10 ft at bounce 1 down toward 0]

## Before the Test

Have students complete the Study Notebook Tie it Together activity to review topics and skills presented in the chapter.

---

# CHAPTER 7
# Study Guide and Review Continued

### 7-9 Geometric Sequences as Exponential Functions
F.BF.2, F.LE.2

Find the next three terms in each geometric sequence.

70. $-1, 1, -1, 1, \ldots$  **$-1, 1, -1$**
71. $3, 9, 27, \ldots$  **$81, 243, 729$**
72. $256, 128, 64, \ldots$  **$32, 16, 8$**

Write the equation for the $n$th term of each geometric sequence.

73. $-1, 1, -1, 1, \ldots$  **$a_n = -1(-1)^{n-1}$**
74. $3, 9, 27, \ldots$  **$a_n = 3(3)^{n-1}$**
75. $256, 128, 64, \ldots$  **$a_n = 256\left(\frac{1}{2}\right)^{n-1}$**

76. **SPORTS** A basketball is dropped from a height of 20 feet. It bounces to $\frac{1}{2}$ its height after each bounce. Draw a graph to represent the situation. **See margin.**

**Example 15**
Find the next three terms in the geometric sequence $2, 6, 18, \ldots$ .

**Step 1** Find the common ratio. Each number is 3 times the previous number, so $r = 3$.

**Step 2** Multiply each term by the common ratio to find the next three terms.
$18 \times 3 = 54, 54 \times 3 = 162, 162 \times 3 = 486$
The next three terms are 54, 162, and 486.

**Example 16**
Write the equation for the $n$th term of the geometric sequence $-3, 12, -48, \ldots$ .

The common ratio is $-4$. So $r = -4$.

$a_n = a_1 r^{n-1}$   Formula for the $n$th term
$a_n = -3(-4)^{n-1}$   $a_1 = -3$ and $r = -4$

### 7-10 Recursive Formulas
F.IF.3, F.BF.2

Find the first five terms of each sequence.

77. $a_1 = 11, a_n = a_{n-1} - 4, n \geq 2$  **$11, 7, 3, -1, -5$**
78. $a_1 = 3, a_n = 2a_{n-1} + 6, n \geq 2$  **$3, 12, 30, 66, 138$**

Write a recursive formula for each sequence.

79. $2, 7, 12, 17, \ldots$  **$a_1 = 2, a_n = a_{n-1} + 5, n \geq 2$**
80. $32, 16, 8, 4, \ldots$  **$a_1 = 32, a_n = 0.5a_{n-1}, n \geq 2$**
81. $2, 5, 11, 23, \ldots$  **$a_1 = 2, a_n = 2a_{n-1} + 1, n \geq 2$**

**Example 17**
Write a recursive formula for $3, 1, -1, -3, \ldots$ .

**Step 1** First subtract each term from the term that follows it.
$1 - 3 = -2, -1 - 1 = -2, -3 - (-1) = -2$
There is a common difference of $-2$. The sequence is arithmetic.

**Step 2** Use the formula for an arithmetic sequence.
$a_n = a_{n-1} + d$   Recursive formula
$a_n = a_{n-1} + (-2)$   $d = -2$

**Step 3** The first term $a_1$ is 3, and $n \geq 2$.
A recursive formula is $a_1 = 3, a_n = a_{n-1} - 2, n \geq 2$.

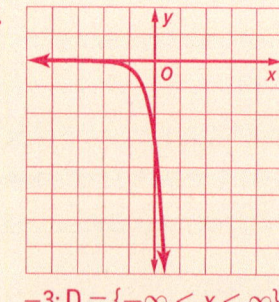

---

## Additional Answers (Practice Test)

26. [Graph of increasing exponential curve]
$2; D = \{-\infty < x < \infty\}$;
$R = \{0 < y < \infty\}; y = 0$

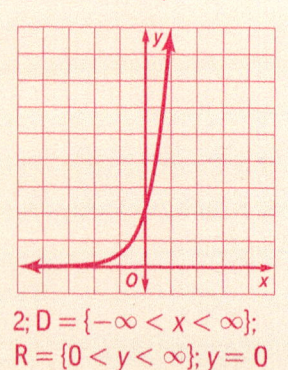

27. [Graph of decreasing curve going to $-\infty$]
$-3; D = \{-\infty < x < \infty\}$;
$R = \{-\infty < y < 0\}$;
$y = 0$

28. [Graph of increasing exponential curve above $y=2$]
$3; D = \{-\infty < x < \infty\}$;
$R = \{2 < y < \infty\}$;
$y = 2$

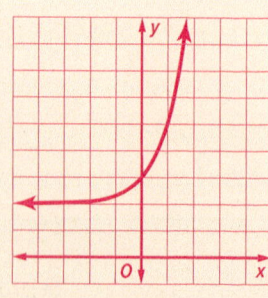

# CHAPTER 7
## Practice Test

**Go Online!** for another Chapter Test

Simplify each expression.

1. $(x^2)(7x^8)$   $7x^{10}$
2. $(5a^7bc^2)(-6a^2bc^5)$   $-30a^9b^2c^7$
3. **MULTIPLE CHOICE** Express the volume of the solid as a monomial. **A**

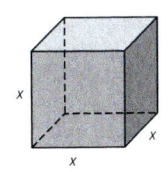

   A $x^3$
   B $6x$
   C $6x^3$
   D $x^6$

Simplify each expression. Assume that no denominator equals 0.

4. $\dfrac{x^6y^8}{x^2}$   $x^4y^8$
5. $\left(\dfrac{2a^4b^3}{c^6}\right)^0$   $1$
6. $\dfrac{2xy^{-7}}{8x}$   $\dfrac{1}{4y^7}$

Simplify.

7. $\sqrt[3]{1000}$   $10$
8. $\sqrt[5]{3125}$   $5$
9. $1728^{\frac{1}{3}}$   $12$
10. $\left(\dfrac{16}{81}\right)^{\frac{1}{2}}$   $\dfrac{4}{9}$
11. $27^{\frac{2}{3}}$   $9$
12. $10{,}000^{\frac{3}{4}}$   $1000$
13. $27^{\frac{5}{3}}$   $243$
14. $\left(\dfrac{1}{121}\right)^{\frac{3}{2}}$   $\dfrac{1}{1331}$

Solve each equation.

15. $12^x = 1728$   $3$
16. $7^{x-1} = 2401$   $5$
17. $9^{x-3} = 729$   $6$

Simplify.

18. $\sqrt{98}$   $7\sqrt{2}$
19. $\sqrt{6} \cdot \sqrt{18}$   $6\sqrt{3}$
20. $\sqrt{50x^3y^2}$   $5|xy|\sqrt{x}$
21. $\sqrt{\dfrac{10}{25}}$   $\dfrac{\sqrt{10}}{5}$

Simplify.

22. $5\sqrt{3} - 15\sqrt{3} + 8\sqrt{5}$   $-10\sqrt{3} + 8\sqrt{5}$
23. $7\sqrt{2} + 4\sqrt{18} - 6\sqrt{50}$   $-11\sqrt{2}$
24. $(2\sqrt{3} + 5\sqrt{2})(4\sqrt{3} - 6\sqrt{2})$   $-36 + 8\sqrt{6}$
25. $(3\sqrt{8} - 5\sqrt{12})^2$   $372 - 120\sqrt{6}$

Graph each function. Find the y-intercept, and state the domain, range, and the equation of the asymptote.

26. $y = 2(5)^x$   **26–28. See margin of previous page.**
27. $y = -3(11)^x$
28. $y = 3^x + 2$
29. Describe the transformations in $g(x) = -0.15(3)^{x+2}$ and $p(x) = (3)^{-x} - 1$ as each relates to the parent function $f(x) = 3^x$. **See margin.**

Find the next three terms in each geometric sequence.

30. $2, -6, 18, \ldots$   $-54, 162, -486$
31. $1000, 500, 250, \ldots$   $125, 62.5, 31.25$
32. $32, 8, 2, \ldots$   $\dfrac{1}{2}, \dfrac{1}{8}, \dfrac{1}{32}$

33. **MULTIPLE CHOICE** Lynne invested $500 into an account with a 6.5% interest rate compounded monthly. How much will Lynne's investment be worth in 10 years? **C**

   A $600.00
   B $938.57
   C $956.09
   D $957.02

34. **INVESTMENTS** Shelly's investment of $3000 has been losing value at a rate of 3% each year. What will her investment be worth in 6 years?  **$2498.92**

35. **RATES** Which offers a better rate for a savings account, 2% compounded yearly or 0.2% compounded monthly? Why?  **See margin.**

Find the first five terms of each sequence.

36. $a_1 = 18, a_n = a_{n-1} - 4, n \geq 2$   $18, 14, 10, 6, 2$
37. $a_1 = -2, a_n = 4a_{n-1} + 5, n \geq 2$   $-2, -3, -7, -23, -87$

## Go Online!

### Chapter Tests

You can use premade leveled *Chapter Tests* to differentiate assessment for your students. Students can also take self-checking *Chapter Tests* to plan and prepare for chapter assessments.

MC = multiple-choice questions
FR = free-response questions

| Form | Type | Level |
|------|------|-------|
| 1 | MC | AL |
| 2A | MC | OL |
| 2B | MC | OL |
| 2C | FR | OL |
| 2D | FR | OL |
| 3 | FR | BL |
| Vocabulary Test | | |
| Extended-Response Test | | |

## Response to Intervention

Use the Intervention Planner to help you determine your Response to Intervention.

### Intervention Planner

**TIER 1 — On Level OL**

IF students miss 25% of the exercises or less,
THEN choose a resource:
- SE  Lessons 7-1 through 7-10

**Go Online!**
- Skills Practice
- Chapter Project
- Self-Check Quizzes

**TIER 2 — Strategic Intervention AL**
Approaching grade level

IF students miss 50% of the exercises,
THEN choose a resource:
- Quick Review Math Handbook

**Go Online!**
- Study Guide and Intervention
- Extra Examples
- Personal Tutors
- Homework Help

**TIER 3 — Intensive Intervention**
2 or more grades below level

IF students miss 75% of the exercises,
THEN choose a resource:
- Use *Math Triumphs, Alg. 1*

**Go Online!**
- Extra Examples
- Personal Tutors
- Homework Help
- Review Vocabulary

### Additional Answers

29. y-intercept of $f(x)$ is 1, of $g(x)$ is $-1$, and of $h(x)$ is 2; $f(x)$ and $h(x)$ have the asymptote of $x = 0$; $g(x)$ has asymptote of $x = -2$; $f(x)$ and $g(x)$ have the same steepness; $h(x)$ is steeper than $f(x)$.

35. 0.2%; because the effective yearly is about 2.43%, which is greater than 2%

# Chapter 7 Preparing for Assessment

## Launch

**Objective** Apply concepts and skills from this chapter in a real-world setting.

## Teach

**Ask:**

- What kind of equation is appropriate here? Is the bacterium increasing by a common difference or common ratio? **Sample answer: The bacterium is increasing by a common ratio, 3, so an exponential equation is appropriate.**

- What is the best way to graph an exponential function? **Sample answer: Pick numbers for $x$ (or $t$) and solve for $y$, then plot and connect the points.**

- What form does an exponential growth or decay question take? **Sample answer: $y = (1 \pm r)^t$, where $+$ indicates growth and $-$ indicates decay.**

- What is a geometric sequence? How is it different from an arithmetic sequence? What does that difference mean in the context of a graph? **Sample answer: A geometric sequence is a series of numbers with a common ratio, while an arithmetic sequence has terms with a common difference. A geometric sequence forms an exponential graph, while an arithmetic sequence forms a linear graph.**

The Performance Task focuses on the following content standards and standards for mathematical practice.

### Dual Coding

| Parts | Content Standards | MP Mathematical Practices |
|---|---|---|
| A | N.Q.1, A.CED.2, F.IF.7, F.LE.2 | 1, 2, 4, 5 |
| B | A.CDE.2, F.BF.1, F.LE.1, F.LE.2 | 1, 2, 4 |
| C | F.LE.1, F.LE.2, F.LE.5 | 1, 2, 4, 6, 7 |

### Go Online! eBook

**Interactive Student Guide**
Refer to *Interactive Student Guide* for an additional Performance Task.

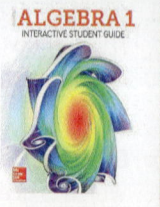

---

# CHAPTER 7
# Preparing for Assessment

## Performance Task

Provide a clear solution to each part of the task. Be sure to show all of your work, include all relevant drawings, and justify your answers.

**BACTERIAL GROWTH** A scientist is studying a newly discovered bacterium thought to be responsible for a recent decline in a certain kind of tree.

### Part A

The scientist creates a culture of the bacteria in a petri dish and monitors it over time. She notices that the number of bacterial cells triples every hour.

1. Determine what kind of expression would accurately model the bacteria's growth over time. **See margin.**
2. **Sense-Making** Write an equation that models the bacteria's growth over time.
3. **Regularity** Graph the equation. **See margin.**
4. Assuming the scientist started with one bacterial cell, determine the number of cells that will be in the petri dish after 6 hours. **See margin.**

### Part B

The scientist determines that the tree population is decreasing by 20% every year due to this bacterium.

5. Write an equation that models the tree population over time. **See margin.**
6. Determine after about how many years will the tree population be about half of its starting population. **See margin.**

### Part C

The scientist formulates an antibiotic to counteract the bacterium. She and her team apply it to a sample region of the affected trees and record the tree population over the course of four years. Results are gathered one year after her experiment begins and are recorded in the table below.

| Year | 1 | 2 | 3 | 4 |
|---|---|---|---|---|
| Number of Trees | 1,000 | 900 | 810 | 729 |

7. Assuming the pattern continues, write an equation that represents the geometric sequence in the table. **See margin.**
8. Determine whether or not the antibiotic is effective. Explain your answer. **See margin.**

---

### Levels of Complexity Chart

The levels of the exercises progress from 1 to 3, with Level 1 indicating the lowest level of complexity.

| Parts | Level 1 | Level 2 | Level 3 |
|---|---|---|---|
| A | | ● | |
| B | | ● | |
| C | | ● | |

### Differentiated Instruction

**Extension** Give students an example of exponential growth in finance. Explain the concept of compound interest. Write the equation $A = P\left(1 + \frac{r}{n}\right)^{nt}$ and label each part as follows: $A$ is the current amount, $P$ is the principal or initial amount, $r$ is the annual interest rate, $t$ is the time in years, and $n$ is the number of times per year that the interest is compounded. Ask students to give a sample principal, rate, and how often the interest is compounded. Then, showing your work step-by-step, see how much the current amount would be after 1, 3, 5, and 10 years.

482 | Chapter 7 | Preparing for Assessment

# Chapter 7 Preparing for Assessment

## Test-Taking Strategy

### Example

Read the problem. Identify what you need to know. Then use the information in the problem to solve.

According to one Website, the value of a new smart TV depreciates by 30% each year. David bought a 65-inch smart TV for $1149. What is the value of his TV after 6 years?

A  $135.18    C  $344.70
B  $804.30    D  $5546.00

**Step 1** What do you need to find?
The value of David's TV 6 years after he purchased it.

**Step 2** Is there enough information to solve the problem?
Yes, there is.

**Step 3** What information, if any, is not needed to solve the problem?
The size of the TV is not needed to solve the problem.

**Step 4** Are there any obvious wrong answers?
D is obviously wrong because the value is greater than the original cost.

**Step 5** What is the correct answer?
To solve the problem, I need to write an equation. Because the value is depreciating, I will use the equation for exponential decay, $y = a(1 - r)^t$. In this problem, the initial amount $a$ is $1149, the rate $r$ of depreciation is 30% or 0.30, the time $t$ is 6 years, and $y$ is the value of the TV. So, $y = 1149(1 - 0.30)^6$. I can enter the expression into a calculator, and the result is 135.178701. If I round to the nearest cent, the value of the TV after 6 years is $135.18.

The correct answer is A.

**Test-Taking Tip**
Using a Scientific or Graphing Calculator
Knowing *when* to use a calculator to solve a problem is just as important as knowing *how* to use one.

### Apply the Strategy

In 2017, the estimated population of the United States was 325,900,000, and the population was increasing at an annual rate of 0.73%. Estimate the population of the United States in 2030.  **C**

A  296,291,126
B  328,279,070
C  358,219,482
D  405,367,922

## Test-Taking Strategy

**Step 1** Determine whether or not to use your calculator.

**Step 2** Solve the problem using your calculator.

**Step 3** Evaluate the result for reasonableness.

### Need Another Example?

A 16-ounce glass of iced tea has 52 milligrams of caffeine. Each hour, the amount of caffeine in a person's body decreases by 14%. How much caffeine is in a person's body 4 hours after drinking a 16-ounce glass of iced tea?  **B**

A  7.3 mg
B  8.8 mg
C  28.4 mg
D  87.8 mg

a. What do you need to find?  the amount of caffeine in a person's body after 4 hours

b. Is there enough information given to solve the problem?  yes

c. What information, if any, is not needed to solve the problem?  the size of the drink

d. Are there any obvious wrong answers? If so, which (one)s? Explain.  D; because the caffeine is decreasing over time

e. What is the correct answer?  B

---

### Part A
1. exponential
2. $y = 3^x$
3. [graph showing exponential curve through points, x-axis 0 to 4, y-axis 0 to 27]
4. 729

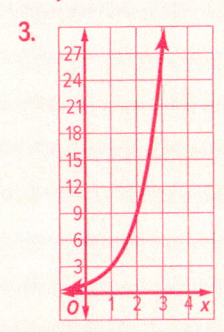

### Part B
5. $y = a(1 - 0.2)^t$ or $y = a(0.8)^t$
6. about 3 years

### Part C
7. $y = a(1 - 0.1)^t$ or $y = a(0.9)^t$
8. Yes, because the tree population is now decreasing at a slower rate (10%, which is less than 20%).

**Go Online!**
The most up-to-date resources available for your program can be found at connectED.mcgraw-hill.com.

## Chapter 7 Preparing for Assessment

### Diagnose Student Errors
Survey student responses for each item. Class trends may indicate common errors and misconceptions.

| 1. | A | Did not rewrite powers with like bases |
|---|---|---|
| | B | CORRECT |
| | C | Did not divide by the coefficient, 2 |
| | D | Multiplied by 2 instead of dividing when solving the equation |

| 2. | A | Confused the *greater than* sign with the *greater than or equal to* sign |
|---|---|---|
| | B | CORRECT |
| | C | Did not reverse the inequality sign after dividing by a negative number |
| | D | Did not understand the meaning of the inequality |

| 3. | A | CORRECT |
|---|---|---|
| | B | Divided the coefficient by 3 |
| | C | Added the coefficients and multiplied the exponents |
| | D | Found the volume |

| 5. | A | CORRECT |
|---|---|---|
| | B | Confused domain and range |
| | C | CORRECT |
| | D | CORRECT |
| | E | Confused exponential growth and decay |

| 6. | A | Canceled the *r* in the numerator and denominator |
|---|---|---|
| | B | Used 0 as exponent for the *r* in the denominator |
| | C | Did not raise the coefficient to fourth power |
| | D | CORRECT |

| 7. | A | Found a percent increase |
|---|---|---|
| | B | Did not add 0.065 to 1 |
| | C | CORRECT |
| | D | Showed a decrease of 6.5% each year |

| 8. | A | CORRECT |
|---|---|---|
| | B | Reversed the coefficient and the base |
| | C | Wrote equation with the *y*-intercept like a linear equation |
| | D | Used the first term as the base and did not include the ratio |

| 9. | A | Did not square the coefficient |
|---|---|---|
| | B | Doubled the coefficient |
| | C | Added 2 to each exponent |
| | D | CORRECT |

### Go Online!
**Standardized Test Practice**
Students can take self-checking tests in standardized format to plan and prepare for standardized assessments.

---

# CHAPTER 7
## Preparing for Assessment
### Cumulative Review

Read each question. Then fill in the correct answer on the answer document provided by your teacher or on a sheet of paper.

1. Solve $3^{2x} = 27^{\frac{4}{3}}$. N.RN.2 **B**
   - A  $\frac{2}{3}$
   - B  2
   - C  4
   - D  8

2. To find the number of games the team needs to win in order to get to the playoffs, Robert solved the following inequality. Let *x* be the number of games. Which statement describes the solution? A.CED.1 **B**
   $$-3x < -15$$
   - A  The team must win at least 5 games.
   - B  The team must win more than 5 games.
   - C  The team must win less than 5 games.
   - D  The team has already won 5 more games than they need to in order to make the playoffs.

3. A rectangular prism has length $3t^7$, width $t^2z^4$, and height $9z^8$. What is the length of a side of a cube with the same volume? A.SSE.2 **A**
   - A  $3t^3z^4$
   - B  $9t^3z^4$
   - C  $12t^{14}z^{32}$
   - D  $27t^9z^{12}$

4. Simplify $\sqrt{32x^4y}$. N.RN.2
   $$4x^2\sqrt{2y}$$

**Test-Taking Tip**
*Question 1* Rewrite each power with the same base before solving the equation. Substitute the answer choice into the original equation to check your answer.

5. Select all true statements about the graphs of $y = 3^x$ and $y = \left(\frac{1}{3}\right)^x$. F.IF.7e **A, C, D**
   - A  Both graphs have a domain of all real numbers.
   - B  Both graphs have a range of all real numbers.
   - C  Both graphs have a *y*-intercept of 1.
   - D  The graph of $y = 3^x$ increases as *x* increases.
   - E  The graph of $y = \left(\frac{1}{3}\right)^x$ shows exponential growth.

6. Simplify the expression. Assume the denominator is not 0. A.SSE.2 **D**
   $$\left(\frac{3b^6 r^8}{24b^{11}r}\right)^4$$
   - A  $\frac{1}{b^{20}}$
   - B  $\frac{r^{32}}{4096b^{20}}$
   - C  $\frac{r^{28}}{8b^{20}}$
   - D  $\frac{r^{28}}{4096b^{20}}$

7. Juaquim's starting salary at a company is $42,000 per year. On average, employees receive a 6.5% increase in salary every year. Which equation shows what his salary will be in *x* years? A.CED.2 **C**
   - A  $s = (1 + 0.065)^x$
   - B  $s = 42,000(0.065)^x$
   - C  $s = 42,000(1 + 0.065)^x$
   - D  $s = 42,000(1 - 0.065)^x$

8. Which function represents the data in the table? F.LE.2 **A**

   | x | 0 | 1 | 2 | 3 |
   |---|---|---|---|---|
   | y | 6 | 12 | 24 | 48 |

   - A  $y = 6(2^x)$
   - B  $y = 2(6^x)$
   - C  $y = 2^x + 6$
   - D  $y = 6^x$

**Chapter 7 Preparing for Assessment**

**Formative Assessment**

You can use these pages to benchmark student progress.

📄 Standardized Test Practice

**Test Item Formats**

In the Cumulative Review, students will encounter different formats for assessment questions to prepare them for standardized tests.

| Question Type | Exercises |
|---|---|
| Multiple-Choice | 1–3, 6–8, 9, 11, 13, 14 |
| Multiple Correct Answers | 5 |
| Type Entry: Short Response | 4, 10, 12, 15 |
| Type Entry: Extended Response | 15 |

**Answer Sheet Practice**

Have students simulate taking a standardized test by recording their answers on a practice recording sheet.

**Homework Option**

*Get Ready for Chapter 8* Assign students the exercises on p. 488 as homework to assess whether they possess the prerequisite skills needed for the next chapter.

---

*Go Online! for Standardized Test Practice*

9. For which expression is $3xy^3z^6$ the square root? N.RN.2  **D**
   - A  $3x^2y^6z^{12}$
   - B  $6xy^3z^6$
   - C  $9x^3y^5z^8$
   - D  $9x^2y^6z^{12}$

10. What is the 7th term of the sequence? F.BF.2
    13, 52, 208, 832, …
    
    **53,248**

11. On the coordinate plane, Roger draws a line that is parallel to the given line and passes through the point $(-4, 1)$.

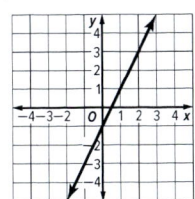

    What is the equation of Roger's line? F.LE.2  **A**
    - A  $y = 2x + 9$
    - B  $y = 2x - 1$
    - C  $y = \frac{1}{2}x + 3$
    - D  $y = -\frac{1}{2}x - 1$

12. Find the value of the exponent. A.SSE.2
    
    $\left(\dfrac{a^5bc^3}{b^4c^2}\right)^? = \dfrac{b^{27}}{a^{45}c^9}$
    
    **−9**

13. Which equation represents the $n$th term of the arithmetic sequence? F.BF.2  **D**
    
    $-6, -1, 4, 9, \ldots$
    - A  $a_n = -5n + 11$
    - B  $a_n = -5n - 1$
    - C  $a_n = 5n - 1$
    - D  $a_n = 5n - 11$

14. What is the recursive formula for the sequence represented in the table? F.IF.3, F.BF.2  **C**

    | $n$ | 1 | 2 | 3 | 4 |
    |---|---|---|---|---|
    | $a_n$ | 405 | 135 | 45 | 15 |
    
    - A  $a_1 = 405, a_n = 3a_{n-1}$
    - B  $a_1 = 405, a_n = \frac{1}{3}a_1$
    - C  $a_1 = 405, a_n = \frac{1}{3}a_{n-1}$
    - D  $a_1 = 405, a_n = 405 - \left(\frac{1}{3}\right)^{n-1}$

15. There were originally 460 spores in a petri dish. After each inspection, there were 10% fewer spores. F.IF.3
    - A  Write a function for the number of spores $S(t)$ after $t$ inspections.
      
      $S(t) = 460(0.9)^t$
    - B  Approximately how many spores were in the petri dish after 4 inspections?
      
      **302**
    - C  What mathematical practice did you use to solve this problem? See students' work.

**Need Extra Help?**

| If you missed Question… | 1 | 2 | 3 | 4 | 5 | 6 | 7 | 8 | 9 | 10 | 11 | 12 | 13 | 14 | 15 |
|---|---|---|---|---|---|---|---|---|---|---|---|---|---|---|---|
| Go to Lesson… | 7-3 | 5-2 | 7-1 | 7-4 | 7-5 | 7-2 | 7-7 | 7-7 | 7-1 | 7-9 | 4-3 | 7-2 | 3-6 | 7-10 | 7-9 |

---

# LS LEARNSMART®

Use LearnSmart as part of your test-preparation plan to measure student topic retention. You can create a student assignment in LearnSmart for additional practice on these topics.

- Extend the Properties of Exponents to Rational Exponents
- Build Linear and Exponential Functions Models

**Go Online!**

eAssessment

Customize and create multiple versions of chapter tests and answer keys that align to your standards. Tests can be delivered on paper or online.

connectED.mcgraw-hill.com   485

## Lesson 7-3

**90a.** sometimes; true only when $x = 1$
**b.** sometimes; true only when $x = 1$
**c.** sometimes; true only when $x = 1$
**d.** always; by definition of $x^{\frac{1}{2}}$
**e.** always; $\left(x^{\frac{1}{2}}\right)^2 = x^{\frac{1}{2} \cdot 2} = x^1$ or $x$
**f.** sometimes; true only when $x = 1$

## Lesson 7-5 (Guided Practice)

**1.** $y = 7^x$
$1; D = \{-\infty < x < \infty\};$
$R = \{0 < y < \infty\};$ asymptote at $y = 0$

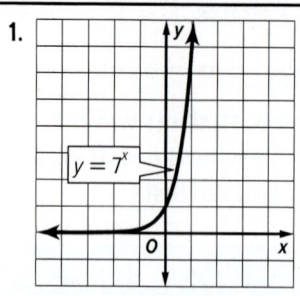

**3.** The $y$-intercept, 10, represents the original 10 cells in a bacteria culture. The asymptote $y = 0$ indicates that the number of bacteria will never be 0. There will be 640 cells after 2 hours.

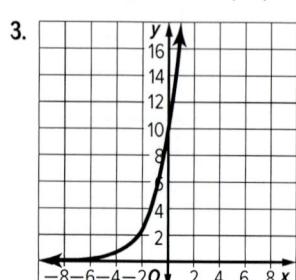

## Lesson 7-5

**1.** $1; D = \{-\infty < x < \infty\};$
$R = \{y \mid y > 0\}; y = 0$

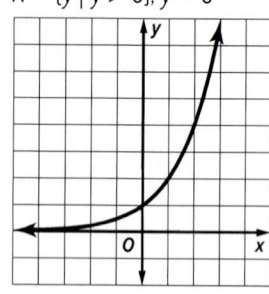

**2.** $-1; D = \{-\infty < x < \infty\};$
$R = \{y \mid y < 0\}; y = 0$

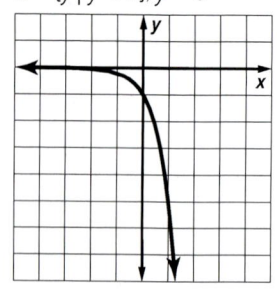

**3.** $-1; D = \{-\infty < x < \infty\};$
$R = \{y \mid y < 0\}; y = 0$

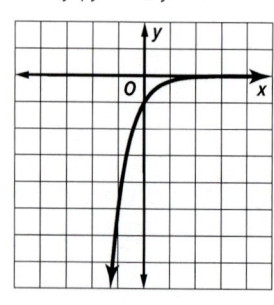

**4.** $3; D = \{-\infty < x < \infty\};$
$R = \{y \mid y > 0\}; y = 0$

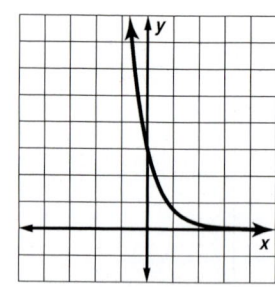

**5.** $4; D = \{-\infty < x < \infty\};$
$R = \{y \mid y > 3\}; y = 3$

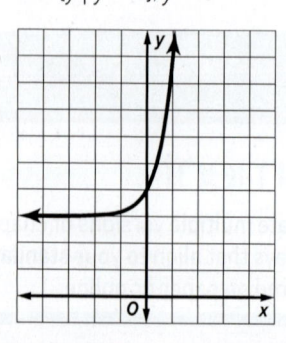

**6.** $1; D = \{-\infty < x < \infty\};$
$R = \{y \mid y < 2\}; y = 2$

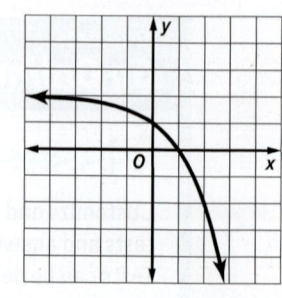

**10.** $2; D = \{-\infty < x < \infty\};$
$R = \{y \mid y > 0\}; y = 0$

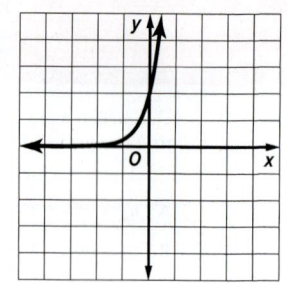

**11.** $2; D = \{-\infty < x < \infty\};$
$R = \{y \mid y > 0\}; y = 0$

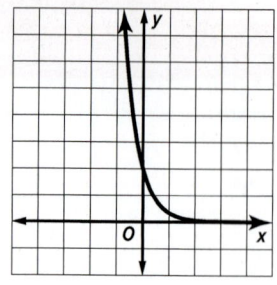

**12.** $1; D = \{-\infty < x < \infty\};$
$R = \{y \mid y > 0\}; y = 0$

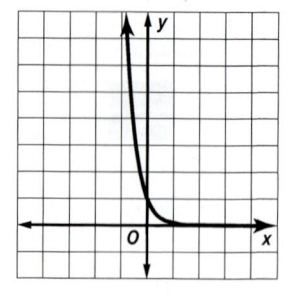

**13.** $-3; D = \{-\infty < x < \infty\};$
$R = \{y \mid y < 0\}; y = 0$

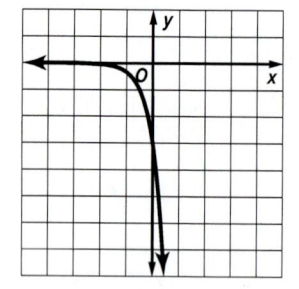

**14.** $-4; D = \{-\infty < x < \infty\};$
$R = \{y \mid y < 0\}; y = 0$

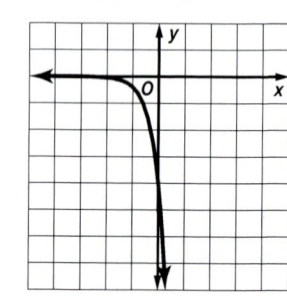

**15.** $3; D = \{-\infty < x < \infty\};$
$R = \{y \mid y > 0\}; y = 0$

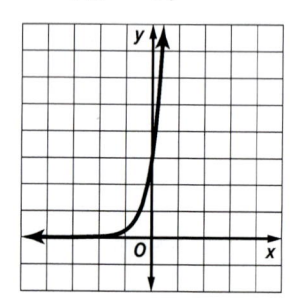

**16.** $4; D = \{-\infty < x < \infty\};$
$R = \{y \mid y > 3\}; y = 3$

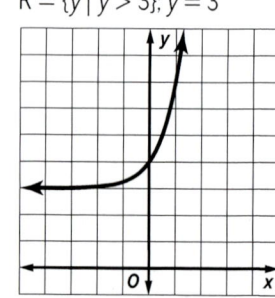

**17.** $-3.5; D = \{-\infty < x < \infty\};$
$R = \{y \mid y > -4\}; y = -4$

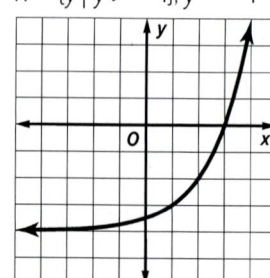

**18.** $6; D = \{-\infty < x < \infty\};$
$R = \{y \mid y > 1\}; y = 1$

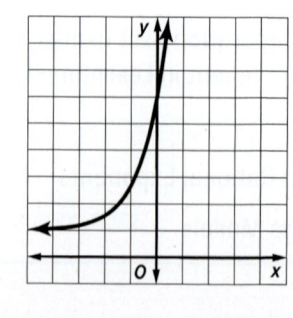

**19.** $3; D = \{-\infty < x < \infty\};$
$R = \{y \mid y < 5\}; y = 5$

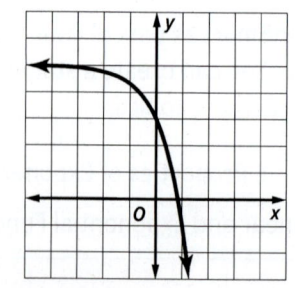

**20b.** 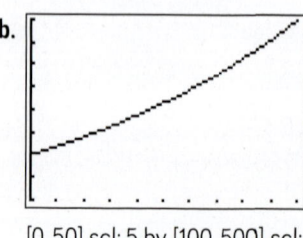 [0, 50] scl: 5 by [100, 500] scl: 50

The $p$-intercept is 200; there are 200 rabbits in January;
$D = \{t \mid t \geq 0\}$
$R = \{p \mid p \geq 200\}$

**26b.** I know that Daniel initially invested $500 into his savings account. I know that his savings account earned interest monthly for 8 years and that interest was added to his original investment every month. I need to find his interest rate.

$$807.07 = 500x^{12(8)}$$

$$\frac{807.07}{500} = x^{96}$$

$$1.61414 = x^{96}$$

$$(1.61414)^{\frac{1}{96}} = x^{96 \times \left(\frac{1}{96}\right)}$$

$$1.005 = x$$

How can Daniel get to $1500 in 5 years?

$$1500 = (x + 807.07)(1.005)^{12(5)}$$

$$1500 = (x + 807.07)(1.35)$$

$$1112.06 = x + 807.07$$

$$304.99 = x$$

Daniel needs to deposit $304.99 into his account to have $1500 in his account in 5 years.

**44.** Sample answer: Let $a = 3$, $b = 2$, and $c = 1$.

| x | $f(x) = 3(2)^x + 1$ | $g(x) = 3x + 1$ |
|---|---|---|
| −5 | 1.09375 | −14 |
| −4 | 1.1875 | −11 |
| −3 | 1.375 | −8 |
| −2 | 1.75 | −5 |
| −1 | 2.5 | −2 |
| 0 | 4 | 1 |
| 1 | 7 | 4 |
| 2 | 13 | 7 |
| 3 | 25 | 10 |
| 4 | 49 | 13 |
| 5 | 97 | 16 |

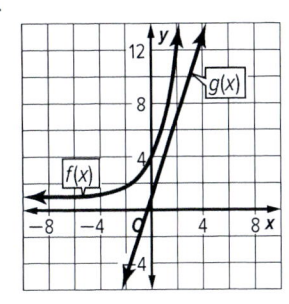

The y-intercept of $f(x)$ is 4 and the y-intercept of $g(x)$ is 1. Both $f(x)$ and $g(x)$ increase as x increases. All function values for $f(x)$ are positive, while $g(x)$ has both positive and negative values. Neither $f(x)$ nor $g(x)$ have maximum or minimum points, and neither has symmetry.

**50F.** Medicine Concentration

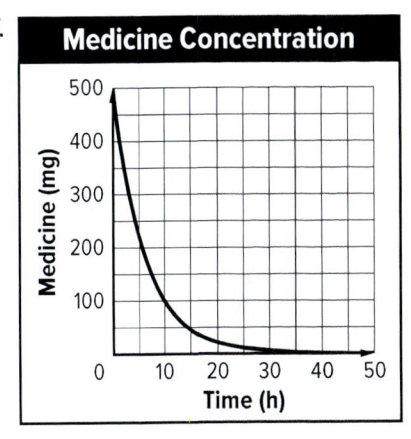

### Lesson 7-9

**30.** Pendulum

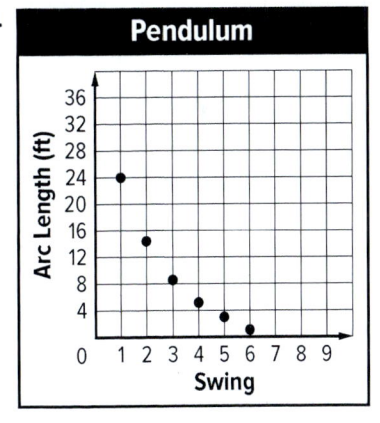

**37b.** Earthquakes

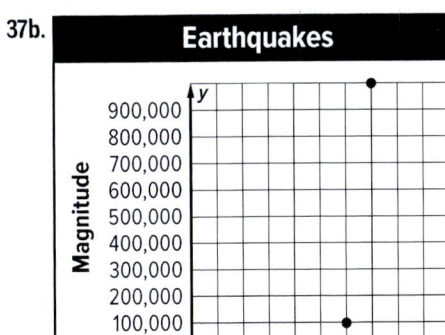

**48d.** Basketball Attendance

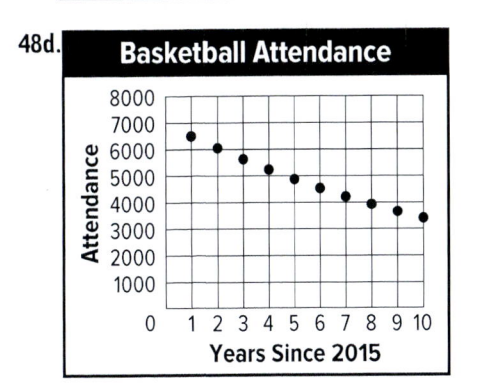

# CHAPTER 8
# Polynomials

**SUGGESTED PACING (DAYS)**

90 min. — Instruction 7 | Review & Assess 1
45 min. — Instruction 11 | Review & Assess 2

## Track Your Progress

### Content Standards
This chapter focuses on content from the Arithmetic with Polynomials and Rational Expressions and the Seeing Structure in Expressions domains.

| THEN | NOW | NEXT |
|---|---|---|
| **F.LE.2** Construct linear and exponential functions, including arithmetic and geometric sequences, given a graph, a description of a relationship, or two input-output pairs (include reading these from a table). **A.SSE.2** Use the structure of an expression to identify ways to rewrite it. | **A.SSE.2** Use the structure of an expression to identify ways to rewrite it. **A.APR.1** Understand that polynomials form a system analogous to the integers, namely, they are closed under the operations of addition, subtraction, and multiplication; add subtract, and multiply polynomials. | **F.IF.7a** Graph linear and quadratic functions and show intercepts, maxima, and minima. **A.REI.4b** Solve quadratic equations by inspection, taking square roots, completing the square, the quadratic formula and factoring. |

### Standards for Mathematical Practice
All of the Standards for Mathematical Practice will be covered in this chapter. The MP icon notes specific areas of coverage.

**MP — Teaching the Mathematical Practices**
Help students develop the mathematical practices by asking questions like these.

**Questioning Strategies** As students approach problems in this chapter, help them develop mathematical practices by asking:

**Reasoning**
- Why is it important to be able to identify a graph by its shape?

**Tools**
- How can your graphing calculator help you solve problems?
- What problems can arise when you use a calculator without writing anything on the page?

**Precision**
- Recall some previous mathematical errors that you have made. How can you avoid making these errors in the future?

**Structure**
- How does the structure of a polynomial help you decide what strategy you will use to factor it?
- How could you use color coding to help you to add and subtract polynomials?

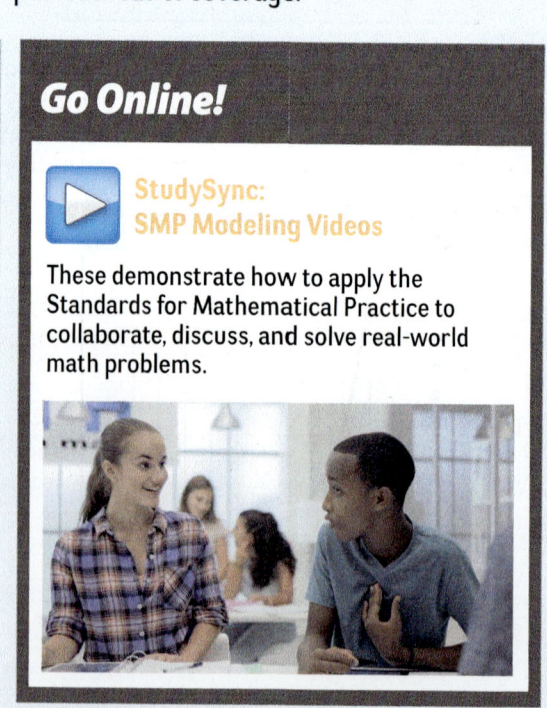

### Go Online!

**StudySync: SMP Modeling Videos**

These demonstrate how to apply the Standards for Mathematical Practice to collaborate, discuss, and solve real-world math problems.

# Go Online!
connectED.mcgraw-hill.com

LearnSmart | The Geometer's Sketchpad | Vocabulary | Personal Tutor | Tools | Calculator Resources | Self-Check Practice | Animations

## Customize Your Chapter

Use the *Plan & Present*, *Assignment Tracker*, and *Assessment* tools in ConnectED to introduce lesson concepts, assign personalized practice, and diagnose areas of student need.

**Differentiated Instruction** Throughout the program, look for the icons to find specialized content designed for your students.

- AL — Approaching Level
- OL — On Level
- BL — Beyond Level
- ELL — English Language Learners

### Personalize

| Differentiated Resources | AL | OL | BL | ELL |
|---|---|---|---|---|
| **FOR EVERY CHAPTER** | | | | |
| ✓ Chapter Readiness Quizzes | ● | ● | ◐ | ● |
| ✓ Chapter Tests | ● | ● | ● | ● |
| ✓ Standardized Test Practice | ● | ● | ● | ● |
| Vocabulary Review Games | ● | ● | ● | ● |
| Anticipation Guide (English/Spanish) | ● | ● | ● | ● |
| Student-Built Glossary | ● | ● | ● | ● |
| Chapter Project | ● | ● | ● | ● |
| **FOR EVERY LESSON** | AL | OL | BL | ELL |
| Personal Tutors (English/Spanish) | ● | ● | ● | ● |
| Graphing Calculator Personal Tutors | ● | ● | ● | ● |
| Step-by-Step Solutions | ● | ● | ◐ | ● |
| ✓ Self-Check Quizzes | ● | ● | ● | ● |
| 5-Minute Check | ● | ● | ● | ● |
| Study Notebook | ● | ● | ● | ● |
| Study Guide and Intervention | ● | ● | ◐ | ● |
| Skills Practice | ● | ◐ | ◐ | ● |
| Practice | ◐ | ● | ● | ● |
| Word Problem Practice | ● | ● | ● | ◐ |
| Enrichment | ◐ | ● | ● | ● |
| Extra Examples | ● | ◐ | ◐ | ◐ |
| Lesson Presentations | ● | ● | ● | ● |

● Aligned to this group    ◐ Designed for this group

### Engage

**Featured IWB Resources**

**Geometer's Sketchpad** provides students with a tangible, visual way to learn. *Use with Lesson 8-4.*

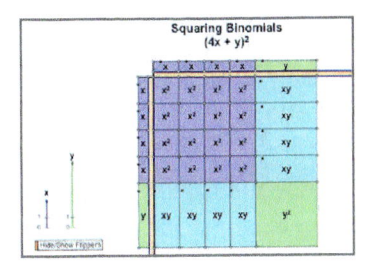

**eLessons** engage students and help build conceptual understanding of big ideas. *Use with Lessons 8-1 through 8-7.*

**Animations** help students make important connections through motion. *Use with Lesson 8-4.*

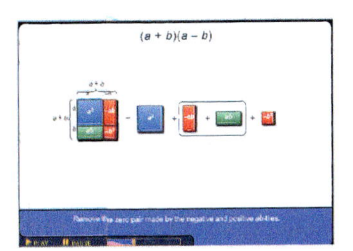

**Time Management** How long will it take to use these resources? Look for the clock in each lesson interleaf.

connectED.mcgraw-hill.com    486B

Chapter 8 | Polynomials

# Introduce the Chapter

## Mathematical Background
Polynomials can be added, subtracted, and multiplied. Polynomials can sometimes be factored in problem situations.

## Essential Question
At the end of this chapter, students should be able to answer the Essential Question.

When could a nonlinear function be used to model a real-world situation? Sample answer: When the relationship that is modeled has a rate of change that is not constant, and thus is nonlinear.

## Apply Math to the Real World
**ARCHITECTURE** In this activity, students will use what they already know about quadratic expressions to model the height of a firework that is fired off of a bridge. Have students complete this activity individually or in small groups. MP 3, 8

---

## Go Online!

### Chapter Project
**Greens Going Green** Students can use what they have learned about polynomial expressions, including quadratic expressions, to complete a project. This chapter project addresses environmental literacy, as well as several specific skills identified as being essential to success by the Framework for 21st Century Learning.
MP 1, 2, 3, 4, 5

---

# CHAPTER 8
# Polynomials

**THEN**
You applied the laws of exponents and explored exponential functions.

**NOW**
In this chapter, you will:
- Add, subtract, and multiply polynomials.
- Factor trinomials.
- Factor differences of squares.
- Factor perfect squares.

**WHY**
**PROJECTILES** Polynomial expressions, including quadratic expressions, can be used to model the heights of objects during a period of motion. These models are used to model many situations such as firing fireworks off of the ground or off of a structure like a bridge. The quadratic expression $-16t^2 + v_0t + h_0$ is used model the height in feet of an object where $v_0$ is the initial velocity of the object in feet per second, $h_0$ is the initial height in feet, and $t$ is time in seconds. These expressions can then be factored.

Use the mathematical practices to complete the activity.
1. **Use Tools** Use the Internet to find a bridge. Choose one to model and find its height.
2. **Model with Mathematics** Suppose a firework was fired off of the bridge you chose with an initial velocity of 30 feet per second. Write a quadratic expression that represents the situation.
3. **Discuss** Compare your expression with that of your classmates. What similarities and differences do you notice?
4. **Use Structure** Is there a common factor between the terms in your expression? If so, what is it?

---

## ALEKS

**Your Student Success Tool** ALEKS is an adaptive, personalized learning environment that identifies precisely what each student knows and is ready to learn—ensuring student success at all levels.

- Formative Assessment: Dynamic, detailed reports monitor students' progress toward standards mastery.
- Automatic Differentiation: Strengthen prerequisite skills and target individual learning gaps.
- Personalized Instruction: Supplement in-class instruction with personalized assessment and learning opportunities.

Chapter 8 | Polynomials

## Go Online to Guide Your Learning

### Explore & Explain

**Product Mat and Algebra Tiles**
Use the Product Mat and the Algebra Tiles tool to model multiplying polynomials. This will enhance your understanding of the math concepts discussed in Lesson 8-3.

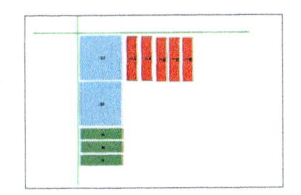

**The Geometer's Sketchpad**
Use The Geometer's Sketchpad to find squares and products of sums and differences.

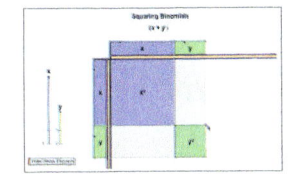

**eBook**
**Interactive Student Guide**
Before starting the chapter, answer the **Chapter Focus** preview questions. Check your answers as you complete each lesson. At the end of the chapter, try the **Performance Task**.

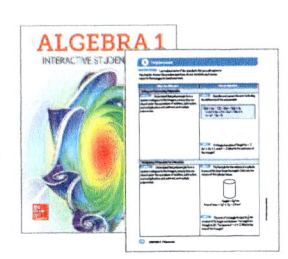

### Organize

**Foldables**
Make this Foldable to help you organize your notes about quadratic expressions. Fold 5 sheets of grid paper in half. Label the first page with the title of the chapter. Label the rest of the pages with lesson numbers.

### Collaborate

**Chapter Project**
In the **Greens Going Green** project, you will use what you have learned about factoring and multiplying polynomials to complete a project that addresses environmental literacy.

### Focus

**LEARNSMART**
Need help studying? Complete the **Expressions and Equations** domain in LearnSmart to review for the chapter test.

**ALEKS**
You can use the **Polynomials and Factoring** topic in ALEKS to find out what you know about linear functions and what you are ready to learn.*

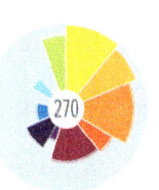

*Ask your teacher if this is part of your program.

## Dinah Zike's FOLDABLES

**Focus** Students write notes about factoring polynomials as they are presented in the lessons of this chapter.

**Teach** Have students make and label their Foldables as illustrated. Suggest that students use their Foldables to take notes, record concepts, and define terms. They can also use them to record the direction and progress of learning, to describe positive and negative experiences during learning, to write about personal associations and experiences, and to list examples of ways in which this new knowledge has been or will be used in their daily lives.

**When to Use It** Encourage students to add to their Foldables as they work through the chapter and to use them to review for the chapter test.

## Go Online!

**Notebooking with Foldables**
Save a tree! In this video, you will learn tips for decreasing the amount of paper used when creating notebooks with Foldables. MP 6, 8

connectED.mcgraw-hill.com

Chapter 8 | Polynomials

# Get Ready for the Chapter

## RtI Response to Intervention
Use the Concept Check results and the Intervention Planner chart to help you determine your Response to Intervention.

### Intervention Planner

**TIER 1 On Level OL**

**IF** students miss 25% of the exercises or less,

**THEN** choose a resource:

*Go Online!*
- Skills Practice, Chapter 1
- Skills Practice, Chapter 7

**TIER 2 Approaching Level AL**

**IF** students miss 50% of the exercises,

**THEN** choose a resource:

*Go Online!*
- Study Guide and Intervention, Chapter 1
- Study Guide and Intervention, Chapter 7
- Extra Examples
- Personal Tutors
- Homework Help
- Quick Review Math Handbook

**TIER 3 Intensive Intervention**

**IF** students miss 75% of the exercises,

**THEN** use *Math Triumphs, Alg. 1*

*Go Online!*
- Extra Examples
- Personal Tutors
- Homework Help
- Review Vocabulary

---

# Get Ready for the Chapter

## Connecting Concepts

**Concept Check**

Review the concepts used in this chapter by answering each question below. **1-8. See margin.**

1. State the Distributive Property.
2. How would you use the Distributive Property to simplify $6x(3+x)$?
3. How do you determine the area of a rectangle? Write that equation for the rectangle shown.

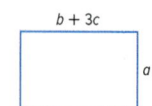

   [rectangle with sides $b + 3c$ and $a$]

4. Use the Distributive Property to simplify the expression for the area of the rectangle shown.
5. Two numbers have a product of $-12$ and a sum of 4. What are the numbers?
6. Can $5a - 2 + 6a$ be simplified? If so, find the simplified expression.
7. Can $(-2y^3)(9y^4)$ be simplified? If so, find the simplified expression.
8. Write the exponent law for multiplying powers, for example $m^3(m^4)$.

**Performance Task Preview**

You will use the concepts and skills in this chapter to solve problems about installing hot tubs and swimming pools. Understanding polynomials will help you finish the Performance Task at the end of the chapter.

**In this Performance Task you will:**
- make sense of problems
- look for and express regularity in repeated reasoning

## New Vocabulary

*Go Online! for Vocabulary Review Games and key vocabulary in 13 languages.*

| English | | Español |
|---|---|---|
| polynomial | p. 491 | polinomio |
| binomial | p. 491 | binomio |
| trinomial | p. 491 | trinomio |
| degree of a monomial | p. 491 | grado de un monomio |
| degree of a polynomial | p. 491 | grado de un polinomio |
| standard form of a polynomial | p. 492 | forma estándar de polinomio |
| leading coefficient | p. 492 | coeficiente lider |
| FOIL method | p. 507 | método foil |
| quadratic expression | p. 507 | expression cuadrática |
| factoring | p. 520 | factorización |
| factoring by grouping | p. 521 | factorización por agrupamiento |
| prime polynomial | p. 535 | polinomio primo |
| difference of two squares | p. 539 | diferencia de cuadrados |
| perfect square trinomial | p. 540 | trinomio cuadrado perfecto |

### Review Vocabulary

**absolute value** *valor absoluto* the absolute value of any number $n$ is the distance the number is from zero on a number line and is written $|n|$

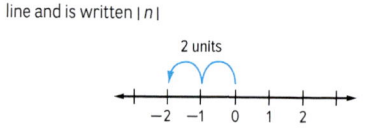

[number line showing 2 units from $-2$ to 0, marks at $-2, -1, 0, 1, 2$]

The absolute value of $-2$ is 2 because it is 2 units from 0.

**perfect square** *cuadrado perfecto* a number with a square root that is a rational number

## Key Vocabulary ELL

Introduce the key vocabulary in the chapter using the routine below.

**Define** Writing an integer or polynomial in factored form is writing it as a product of its prime factors.

**Example** The factored form of 12 is $2^2 \cdot 3$. The factored form of $x^2 + 5x + 6$ is $(x + 2)(x + 3)$.

**Ask** What is the factored form of 48? $2^4 \cdot 3$

### Additional Answers.

1. A factor in front of parentheses multiplies across all the terms inside the parentheses: $a(b + c) = ab + ac$
2. Distribute $6x$ by multiplying it by 3 and by the $x$ inside the parentheses.
3. Multiply length by width; area $= a(b + 3c)$
4. $ab + 3ac$
5. $-2, 6$
6. Yes; $11a - 2$
7. Yes; $-18y^7$
8. $a^m(a^n) = a^{m+n}$

488 | Chapter 8 | Polynomials

# LESSON 8-1
# Adding and Subtracting Polynomials

**SUGGESTED PACING (DAYS)**

| | | |
|---|---|---|
| 90 min. | 0.5 | 0.5 |
| 45 min. | 0.5 | 0.5 |
| | Explore Lab | Instruction |

## Track Your Progress

### Objectives
1. Write polynomials in standard form.
2. Add and subtract polynomials.

### Mathematical Background
A polynomial is a monomial or sum or difference of monomials. To add polynomials, add the coefficients of like terms using the rules of adding real numbers. To subtract polynomials, replace each term of the polynomial being subtracted with its additive inverse and combine like terms.

### THEN
**F.BF.1.A** Determine an explicit expression, a recursive process, or steps for calculation from a context.

### NOW
**A.APR.1** Understand that polynomials form a system analogous to the integers, namely, they are closed under the operations of addition, subtraction, and multiplication; add, subtract, and multiply polynomials.

**A.SSE.1a** Interpret parts of an expression, such as terms, factors, and coefficients.

### NEXT
**A.APR.4** Prove polynomial identities and use them to describe numerical relationships. For example, the polynomial identity $(x^2 + y^2)^2 = (x^2 - y^2)^2 + (2xy)^2$ can be used to generate Pythagorean triples.

---

**Go Online!** All of these resources and more are available at connectED.mcgraw-hill.com

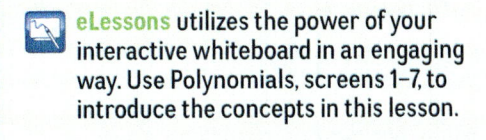

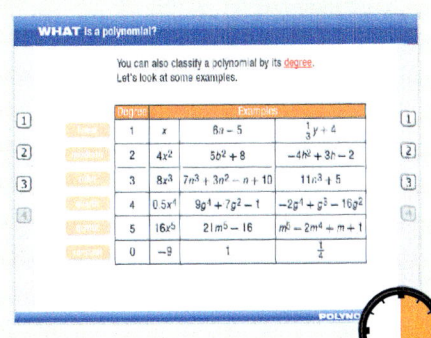

**eLessons** utilizes the power of your interactive whiteboard in an engaging way. Use Polynomials, screens 1–7, to introduce the concepts in this lesson.

*Use at Beginning of Lesson*

**eToolkit** allows students to explore and enhance their understanding of math concepts. Use the Algebra Tiles tool to model adding polynomials.

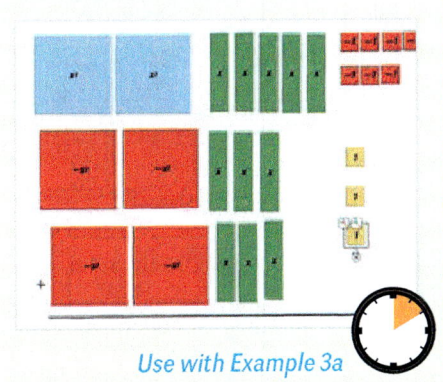

*Use with Example 3a*

**Brain POPs®** are animated, curricular content that engages students.

*Use with Examples*

---

## OER Using Open Educational Resources

**Practice** Have students work individually to read the information on adding and subtracting polynomials posted on **MathIsFun.com**. They can try practice problems after reading the examples and watching the animations. *Use as homework or flipped learning*

# Go Online!
connectED.mcgraw-hill.com — Worksheets

## Differentiate Your Resources

**Extra Practice** Additional practice or homework; Skills Practice is best for approaching-level students and Practice is best for on-level and beyond-level students

### Skills Practice
AL  OL  ELL

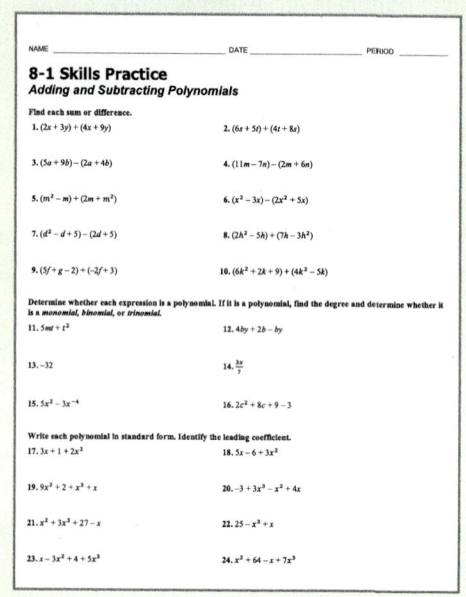

### Practice
AL  OL  BL  ELL

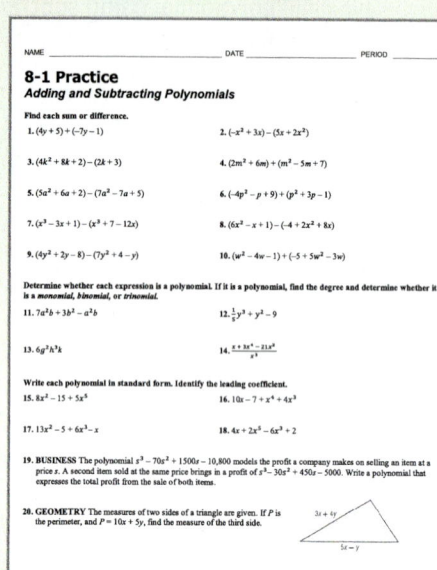

### Word Problem Practice
AL  OL  BL  ELL

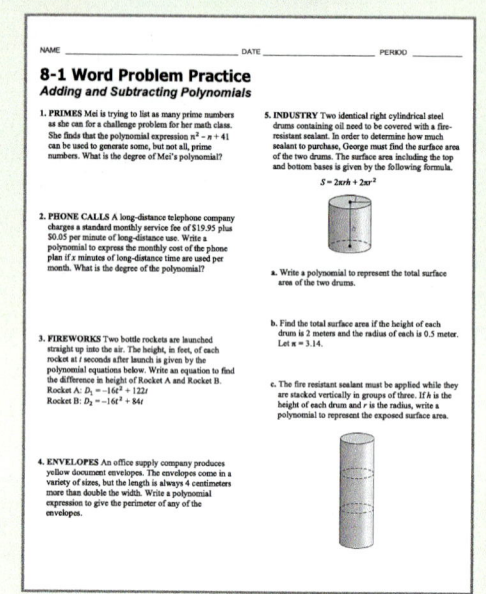

**Intervention** Reteaching and vocabulary activities that can be used with struggling or absent students and as ELL support

### Study Guide and Intervention
AL  OL  ELL

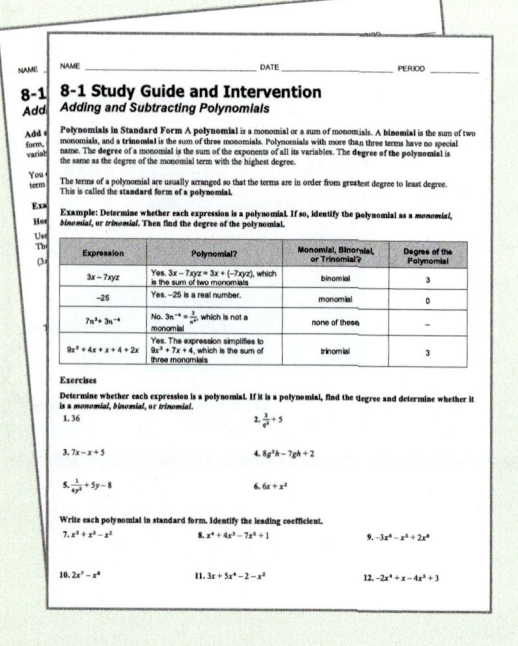

### Study Notebook
AL  OL  BL  ELL

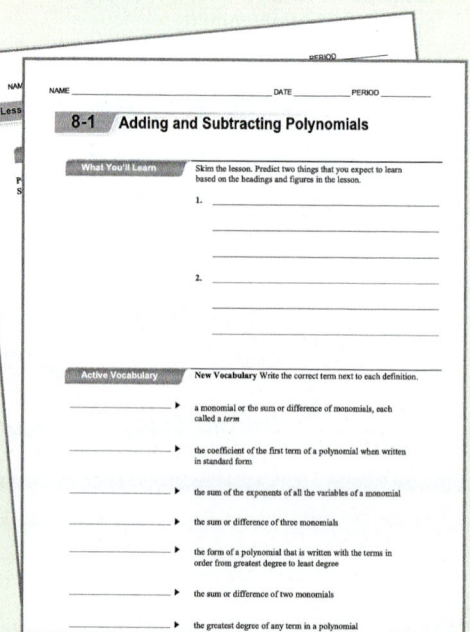

**Extension** Activities that can be used to extend lesson concepts

### Enrichment
OL  BL  ELL

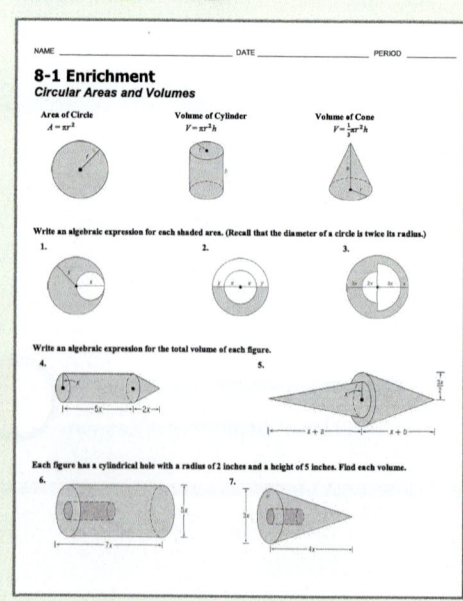

489B | Lesson 8-1 | Adding and Subtracting Polynomials

# EXPLORE 8-1

## Algebra Lab
## Adding and Subtracting Polynomials

Algebra tiles can be used to model polynomials. A polynomial is a monomial or the sum of monomials. The diagram below shows the models.

**Mathematical Practices**
5 Use appropriate tools strategically.

**Content Standards**
A.APR.1 Understand that polynomials form a system analogous to the integers, namely, they are closed under the operations of addition, subtraction, and multiplication; add, subtract, and multiply polynomials.

### Polynomial Models

- Polynomials are modeled using three types of tiles.
- Each tile has an opposite.

### Activity 1  Model Polynomials

Work cooperatively. Use algebra tiles to model each polynomial.

$5x$

To model this polynomial, you will need five green $x$-tiles.

$-2x^2 + x + 3$

To model this polynomial, you will need two red $-x^2$-tiles, one green $x$-tile, and three yellow 1-tiles.

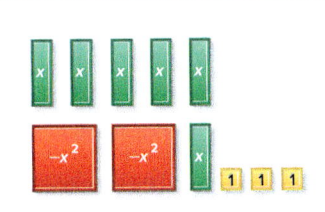

Monomials such as $3x$ and $-2x$ are called *like terms* because they have the same variable to the same power.

### Polynomial Models

- Like terms are represented by tiles that have the same shape and size.
- A *zero pair* may be formed by pairing one tile with its opposite. You can remove or add zero pairs without changing the polynomial.

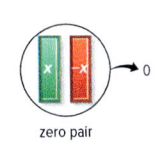

like terms          zero pair

### Activity 2  Add Polynomials

Work cooperatively. Use algebra tiles to find $(2x^2 - 3x + 5) + (x^2 + 6x - 4)$.

**Step 1**
Model each polynomial.

$2x^2 \quad + \quad -3x \quad + \quad 5$

$x^2 \quad + \quad 6x \quad + \quad -4$

---

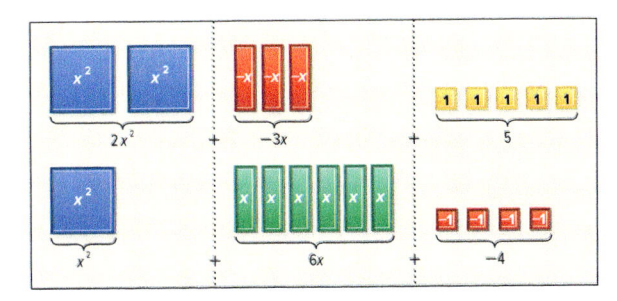

# Launch

**Objective** Use algebra tiles to add and subtract polynomials.

### Materials for Each Student
- algebra tiles

### Easy-to-Make Manipulatives
*Teaching Algebra with Manipulatives*
Template for algebra tiles, pp. 10–11

### Teaching Tip
**Zero Pairs** Prior to Activities 2 and 3, discuss the concept of a zero pair. Have students form zero pairs using 1-tiles, $x$-tiles, and $x^2$-tiles and their opposites.

# Teach

**Working in Cooperative Groups** Put students in groups of two or three, mixing abilities. Have groups complete Activity 1 and Exercises 1–3.

- Make sure students understand that the number of $x$-tiles and $x^2$-tiles represent the coefficients of $x$ and $x^2$, respectively. The number of 1-tiles represents the constant in the expression.

- Tell students to be careful to use the tiles with the correct colors. It is easy to incorrectly substitute an $x$-tile for a $-x$-tile.

- Talk about like terms in the context of the tiles. Tiles with the same shape and size represent like terms.

- For Activity 2, tell students that it is easier to model the polynomials if they arrange the tiles in the same order as the monomials within each polynomial. In this case, the monomials are arranged in descending order of degree. Therefore, students should arrange the tiles in descending order from left to right.

# Explore 8-1

- After groups have completed Activity 2, write the addition of the two polynomials vertically so students can see that the coefficients of like terms are added.

- For Activity 3, explain that adding a zero pair to the polynomial does not change its value because the zero pair is equal to zero.

- After groups complete Activity 3, write the difference vertically so students can see that coefficients of like terms are subtracted.

**Practice** Have students complete Exercises 6 and 7.

## Assess

### Formative Assessment
Use Exercise 8 to assess whether students can use models to compare polynomials.

### From Concrete to Abstract
Write a polynomial addition or subtraction problem on the board. Have students determine the sum or difference without using tiles. If they answer incorrectly, have them use their tiles to help them find their errors.

## Go Online!

### eToolkit
The eToolkit allows students to explore and enhance their understanding of math concepts. Use the Algebra Tiles tool to demonstrate adding and subtracting polynomials.

490 | Explore 8-1 | Adding and Subtracting Polynomials

---

## EXPLORE 8-1
### Algebra Lab
# Adding and Subtracting Polynomials *Continued*

**Step 2** Combine like terms and remove zero pairs.

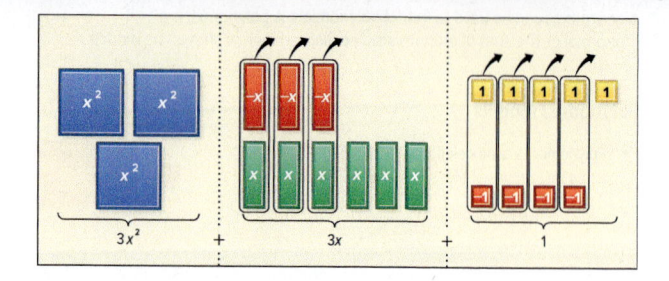

$3x^2 \quad + \quad 3x \quad + \quad 1$

**Step 3** Write the polynomial. $(2x^2 - 3x + 5) + (x^2 + 6x - 4) = 3x^2 + 3x + 1$

### Activity 3  Subtract Polynomials

Work cooperatively. Use algebra tiles to find $(4x + 5) - (-3x + 1)$.

**Step 1** Model the polynomial $4x + 5$.

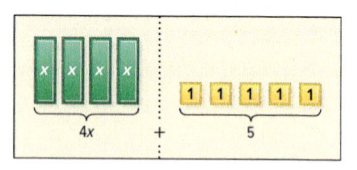

$4x \quad + \quad 5$

**Step 2** To subtract $-3x + 1$, remove three red $-x$-tiles and one yellow 1-tile. You can remove the 1-tile, but there are no $-x$-tiles. Add 3 zero pairs of $x$-tiles. Then remove the three red $-x$-tiles.

**Step 3** Write the polynomial. $(4x + 5) - (-3x + 1) = 7x + 4$

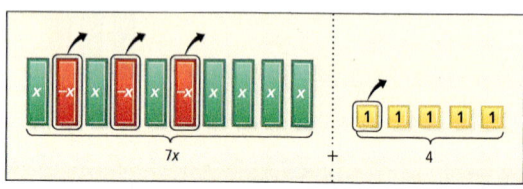

$7x \quad + \quad 4$

### Model and Analyze

Use algebra tiles to model each polynomial. Then draw a diagram of your model. 1–3. See margin.

1. $-2x^2$
2. $5x - 4$
3. $x^2 - 4x$

Write an algebraic expression for each model.

4. 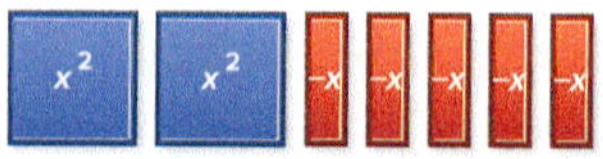 $2x^2 - 5x$
5. 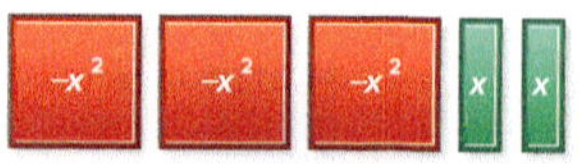 $-3x^2 + 2x + 1$

Use algebra tiles to find each sum or difference. 6. $4x^2 + 3x + 4$  7. $x^2 + 12x + 3$  8. $-5x^2 - 4x$

6. $(x^2 + 5x - 2) + (3x^2 - 2x + 6)$
7. $(2x^2 + 8x + 1) - (x^2 - 4x - 2)$
8. $(-4x^2 + x) - (x^2 + 5x)$

## Additional Answers

1. 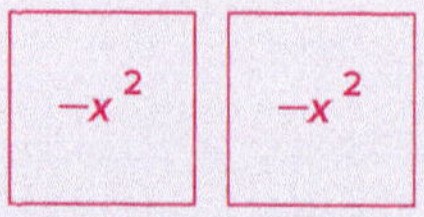 $-x^2 \quad -x^2$

2. $x \ x \ x \ x \ x \quad -1 \ -1 \ -1 \ -1$

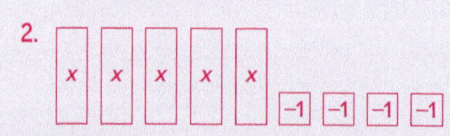

3. 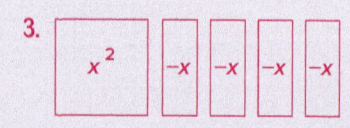 $x^2 \quad -x \ -x \ -x \ -x$

# LESSON 1
## Adding and Subtracting Polynomials

**∴ Then**
- You identified monomials and their characteristics.

**∴ Now**
1. Write polynomials in standard form.
2. Add and subtract polynomials.

**∴ Why?**
- The sales data of digital audio players can be modeled by the equation $U = -2.7t^2 + 49.4t + 128.7$, where $U$ is the number of units shipped in millions and $t$ is the number of years since 2005.

  The expression $-2.7t^2 + 49.4t + 128.7$ is an example of a polynomial. Polynomials can be used to model situations.

### New Vocabulary
polynomial
binomial
trinomial
degree of a monomial
degree of a polynomial
standard form of a polynomial
leading coefficient

### Mathematical Practices
3 Construct viable arguments and critique the reasoning of others.

### Content Standards
**A.SSE.1a** Interpret parts of an expression, such as terms, factors, and coefficients.

**A.APR.1** Understand that polynomials form a system analogous to the integers, namely, they are closed under the operations of addition, subtraction, and multiplication; add, subtract, and multiply polynomials.

## 1 Polynomials in Standard Form
A **polynomial** is a monomial or the sum of monomials, each called a *term* of the polynomial. Some polynomials have special names. A **binomial** is the sum of *two* monomials, and a **trinomial** is the sum of *three* monomials.

| Monomial | Binomial | Trinomial |
|---|---|---|
| $5x$ | $2x^2 + 7$ | $x^3 - 10x + 1$ |

The **degree of a monomial** is the sum of the exponents of all its variables. A nonzero constant term has degree 0, and zero has no degree.

The **degree of a polynomial** is the greatest degree of any term in the polynomial. You can find the degree of a polynomial by finding the degree of each term. Polynomials are named based on their degree.

| Degree | Name |
|---|---|
| 0 | constant |
| 1 | linear |
| 2 | quadratic |
| 3 | cubic |
| 4 | quartic |
| 5 | quintic |
| 6 or more | 6th degree, 7th degree, and so on |

A.SSE.1a

### Example 1 Identify Polynomials
Determine whether each expression is a polynomial. If it is a polynomial, find the degree and determine whether it is a *monomial*, *binomial*, or *trinomial*.

| Expression | Is it a polynomial? | Degree | Monomial, binomial, or trinomial? |
|---|---|---|---|
| a. $4y - 5$ | Yes; $4y - 5$ is the sum of $4y$ and $-5$. | 1 | binomial |
| b. $-6.5$ | Yes; $-6.5$ is a real number. | 0 | monomial |
| c. $7a^{-3} + 9b$ | No; $7a^{-3} = \frac{7}{a^3}$, which is not a monomial. | — | — |
| d. $6x^3 + 4x + x + 3$ | Yes; $6x^3 + 4x + x + 3 = 6x^3 + 5x + 3$, the sum of three monomials. | 3 | trinomial |

**Guided Practice**

1A. $x$ yes; 1; monomial
1B. $-3y^2 - 2y + 4y - 1$ yes; 2; trinomial
1C. $5rx + 7tuv$ yes; 3; binomial
1D. $10x^{-4} - 8x^a$

1D. No; $10x^{-4} = \frac{10}{x^4}$, which is not a monomial, and $8x^a$ has a variable exponent.

---

### MP Mathematical Practices Strategies

**Attend to precision.**
Help students understand the precise vocabulary to use when discussing polynomials. Make sure they understand each term well enough to both describe it in words and give an example. For example, ask:

- What is the difference between a binomial and a trinomial? **An expression with two terms is a binomial. An expression with three terms is a trinomial.**
- Name three possible parts of a monomial and give an example. **Sample answer: a coefficient, a variable, and an exponent; $-3y^4$**
- Write an example of a quadratic expression. **Sample answer: $x^2 - 8x + 12$**
- Write an example of a linear expression. **Sample answer: $8x + 12$**
- Describe the difference between a quadratic expression and a linear expression. **The greatest degree in a quadratic expression is 2. The greatest degree in a linear expression is 1.**

---

**Lesson 8-1 | Adding and Subtracting Polynomials**

## Launch

Have students read the Why? section of the lesson. Ask:

- What is the value of $t$ for the year 2007? **2**
- What would be the value of $t$ for the year 2010? **5**
- Using the equation, find the value of $U$ for the year 2007. **216.7**
- How many monomials make up the expressions that equals $U$? **3**
- What are they? **$-2.7t^2$; $49.4t$; $128.7$**

## Teach

Ask the scaffolded questions for each example to build conceptual understanding for students at all levels.

### 1 Polynomials in Standard Form

**Example 1 Identify Polynomials**

**AL** What symbols could be used to determine if the expression is a polynomial? **addition or subtraction sign**

**OL** Name the polynomial in each example. **A.** linear; **B.** constant; **C.** not a polynomial; **D.** cubic

**BL** What cannot be contained in the expression for it to be a polynomial? Explain. **division by a variable or a variable with a negative exponent; Sample answer: The definition of a monomial states that there can be a product of a number and one or more variables, but it does not say there can be a quotient. A negative exponent results in a quotient.**

### Go Online!

**Interactive Whiteboard**
Use the *eLesson* or *Lesson Presentation* to present this lesson.

Lesson 8-1 | Adding and Subtracting Polynomials

**Need Another Example?**
Determine whether each expression is a polynomial. If it is a polynomial, find the degree and determine whether it is a *monomial*, *binomial*, or *trinomial*.
a. $6x - 4$ yes; 1; binomial
b. $x^2 + 2xy - 7$ yes; 2; trinomial
c. $\dfrac{14d + 19e^3}{5d^4}$ no
d. $26b^2$ yes; 2; monomial

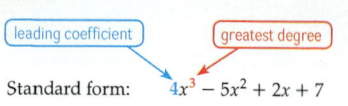

**Example 2 Standard Form of a Polynomial**

**AL** In part a, is the expression a polynomial? yes in part b? yes

**OL** In part a, what is the degree of the polynomial? 5 in part b? 4

**BL** Why does a number with no variable go at the end of the polynomial rather than the beginning? Sample answer: A term with no variable has a degree of 0, so it is the least degree.

**Need Another Example?**
Write each polynomial in standard form. Identify the leading coefficient.
a. $9x^2 + 3x^6 - 4x$  $3x^6 + 9x^2 - 4x$; 3
b. $12 + 5y + 6xy + 8xy^2$  $8xy^2 + 6xy + 5y + 12$; 8

## 2 Add and Subtract Polynomials

**Example 3 Add Polynomials**

**AL** In part a, what are the pairs of like terms? $2x^2$ and $-4x^2$; $5x$ and $6x$; $-7$ and $3$

**OL** How can rewriting the polynomials in standard form be helpful when adding? Sample answer: Writing the terms in order can help find like terms and group them together.

**BL** Which method do you prefer? Explain. See students' work.

**Need Another Example?**
Find each sum.
a. $(7y^2 + 2y - 3) + (2 - 4y + 5y^2)$  $12y^2 - 2y - 1$
b. $(4x^2 - 2x + 7) + (3x - 7x^2 - 9)$  $-3x^2 + x - 2$

---

**Go Online!**
Watch and listen to the animation to learn about polynomials. Summarize what you hear for a partner, and have them ask you questions to help your understanding. **ELL**

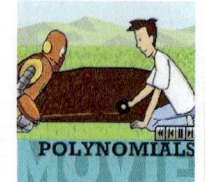

The terms of a polynomial can be written in any order. However, polynomials in one variable are usually written in standard form. The **standard form of a polynomial** has the terms in order from greatest to least degree. In this form, the coefficient of the first term is called the **leading coefficient**.

Standard form: $4x^3 - 5x^2 + 2x + 7$

(leading coefficient) — $4x^3$; (greatest degree) — $x^3$

A.SSE.1a

**Example 2  Standard Form of a Polynomial**

Write each polynomial in standard form. Identify the leading coefficient.

a. $3x^2 + 4x^5 - 7x$

Find the degree of each term.
Degree:  2  5  1

Polynomial: $3x^2 + 4x^5 - 7x$

The greatest degree is 5. Therefore, the polynomial can be rewritten as $4x^5 + 3x^2 - 7x$, with a leading coefficient of 4.

b. $5y - 9 - 2y^4 - 6y^3$

Find the degree of each term.
Degree:  1  0  4  3

Polynomial: $5y - 9 - 2y^4 - 6y^3$

The greatest degree is 4. Therefore, the polynomial can be rewritten as $-2y^4 - 6y^3 + 5y - 9$, with a leading coefficient of $-2$.

▶ **Guided Practice**
2A. $4x^4 - 2x^2 - 3x + 8$; 4     2B. $-7y^6 + 5y^3 - 2y^2 + y + 10$; $-7$
2A. $8 - 2x^2 + 4x^4 - 3x$     2B. $y + 5y^3 - 2y^2 - 7y^6 + 10$

**2 Add and Subtract Polynomials** Adding polynomials involves adding like terms. You can group like terms by using a horizontal or vertical format.
A.APR.1

**Example 3  Add Polynomials**

Find each sum.

a. $(2x^2 + 5x - 7) + (3 - 4x^2 + 6x)$

**Horizontal Method**

Group and combine like terms.

$(2x^2 + 5x - 7) + (3 - 4x^2 + 6x)$
$= [2x^2 + (-4x^2)] + [5x + 6x] + [-7 + 3]$  Group like terms.
$= -2x^2 + 11x - 4$  Combine like terms.

b. $(3y + y^3 - 5) + (4y^2 - 4y + 2y^3 + 8)$

**Vertical Method**

Align like terms in columns and combine.

$\phantom{(+)\ }y^3 + 0y^2 + 3y - 5$  Insert a placeholder to help align the terms.
$\underline{(+)\ 2y^3 + 4y^2 - 4y + 8}$  Align and combine like terms.
$\phantom{(+)\ }3y^3 + 4y^2 - y + 3$

▶ **Guided Practice**
3A. $(5x^2 - 3x + 4) + (6x - 3x^2 - 3)$   $2x^2 + 3x + 1$
3B. $(y^4 - 3y + 7) + (2y^3 + 2y - 2y^4 - 11)$   $-y^4 + 2y^3 - y - 4$

**Study Tip**

**MP Sense-Making** When finding the additive inverse of a polynomial, you are multiplying every term by −1.

You can subtract a polynomial by adding its additive inverse. To find the additive inverse of a polynomial, write the opposite of each term, as shown.

$$-(3x^2 + 2x - 6) = \underbrace{-3x^2 - 2x + 6}_{\text{Additive Inverse}}$$

A.APR.1

**Example 4** Subtract Polynomials

Find each difference.

a. $(3 - 2x + 2x^2) - (4x - 5 + 3x^2)$

**Horizontal Method**

Subtract $4x - 5 + 3x^2$ by adding its additive inverse.

$(3 - 2x + 2x^2) - (4x - 5 + 3x^2)$

$= (3 - 2x + 2x^2) + (-4x + 5 - 3x^2)$   The additive inverse of $4x - 5 + 3x^2$ is $-4x + 5 - 3x^2$

$= [2x^2 + (-3x^2)] + [(-2x) + (-4x)] + [3 + 5]$   Group like terms.

$= -x^2 - 6x + 8$   Combine like terms.

b. $(7p + 4p^3 - 8) - (3p^2 + 2 - 9p)$

**Vertical Method**

Align like terms in columns and subtract by adding the additive inverse.

$\quad 4p^3 + 0p^2 + 7p - 8 \qquad\qquad 4p^3 + 0p^2 + 7p - 8$
$(-)\qquad\quad 3p^2 - 9p + 2$   Add the opposite   $(+)\quad\; -3p^2 + 9p - 2$
$\qquad\qquad\qquad\qquad\qquad\qquad\qquad\qquad\qquad 4p^3 - 3p^2 + 16p - 10$

**Guided Practice**

4. $(4x^3 - 3x^2 + 6x - 4) - (-2x^3 + x^2 - 2)$   $6x^3 - 4x^2 + 6x - 2$

Adding or subtracting integers results in an integer, so the set of integers is closed under addition and subtraction. Similarly, adding or subtracting polynomials results in a polynomial, so the set of polynomials is closed under addition and subtraction.

A.APR.1

**Real-World Example 5** Add and Subtract Polynomials

**ELECTRONICS** The equations $P = 7m + 137$ and $C = 4m + 78$ represent the number of smartphones $P$ and gaming consoles $C$ sold in $m$ months at an electronics store. Write an equation for the total monthly sales $T$ of phones and gaming consoles. Then predict the number of phones and gaming consoles sold in 10 months.

To write an equation that represents the total sales $T$, add the equations that represent the number of smartphones $P$ and gaming consoles $C$.

$T = 7m + 137 + 4m + 78$

$= 11m + 215$

Substitute 10 for $m$ to predict the number of phones and gaming consoles sold in 10 months.

$T = 11(10) + 215$

$= 110 + 215$ or 325

Therefore, a total of 325 smartphones and gaming consoles will be sold in 10 months.

$D(m) = 3m + 59; 131$

**Guided Practice**

5. Use the information above to write an equation that represents the difference in the monthly sales of smartphones and the monthly sales of gaming consoles. Use the equation to predict the difference in monthly sales in 24 months.

**Real-World Link**

Sales of smartphones have been steadily increasing by a rate of 4% each year.

**Source:** Business Wire

---

**Lesson 8-1 | Adding and Subtracting Polynomials**

**Example 4** Subtract Polynomials

**AL** In part **b**, what is the degree of each polynomial? 3; 2

**OL** What is the first step for finding the difference of the polynomials? Sample answer: Find the additive inverse of each term in the polynomial after the subtraction sign.

**BL** In part **b**, why is it important to add a $0p^2$ to the first polynomial? Sample answer: Because the first polynomial does not have a term with a degree of 2, $0p^2$ is added as a placeholder.

**Need Another Example?**

Find each difference.

a. $(6y^2 + 8y^4 - 5y) - (9y^4 - 7y + 2y^2)$   $-y^4 + 4y^2 + 2y$
b. $(6n^2 + 11n^3 + 2n) - (4n - 3 + 5n^2)$   $11n^3 + n^2 - 2n + 3$

**Watch Out!**

**Preventing Errors** Some students may find it helpful to mark through like terms as they mentally combine them. This saves time spent on rewriting to group like terms.

**Example 5** Add and Subtract Polynomials

**AL** What information is needed to find the number of phones and game consoles sold in 10 months? the total sales of phones and game consoles in one month

**OL** What terms need to be combined? $7m$ and $4m$; 137 and 78

**BL** How many smartphones will be sold in month 10? 207 game consoles? 118

**Need Another Example?**

**Video Games** The total amount of toy sales $T$ (in billions of dollars) consists of two groups: sales of video games $V$ and sales of traditional toys $R$. In recent years, the sales of traditional toys and total sales could be represented by the following equations, where $n$ is the number of years since 2010.
$R = 0.46n^3 - 1.9n^2 + 3n + 19$
$T = 0.45n^3 - 1.85n^2 + 4.4n + 22.6$

a. Write an equation that represents the sales of video games $V$.   $V = -0.01n^3 + 0.05n^2 + 1.4n + 3.6$
b. Use the equation to predict the amount of video game sales in the year 2022.   10.32 billion dollars

connectED.mcgraw-hill.com   493

# Lesson 8-1 | Adding and Subtracting Polynomials

## Practice

**Formative Assessment** Use Exercises 1–19 to assess students' understanding of the concepts in this lesson.

The Practice and Problem Solving exercises assess the content taught in the lesson. The Preparing for Assessment page is meant to be used as preparation for end-of-course assessment.

### Teaching the Mathematical Practices

**Sense-Making** Mathematically proficient students start by explaining to themselves the meaning of a problem. In Exercise 19, ask students to explain how the given polynomials relate to the number of students who drove.

## Extra Practice

See page R8 for extra exercises for students who are approaching level or for on-level students who need additional reinforcement.

### Levels of Complexity Chart

The levels of the exercises progress from 1 to 3, with Level 1 indicating the lowest level of complexity.

| Exercises | 20–44 | 45–58, 66–72 | 59–65 |
|---|---|---|---|
| Level 3 | | | ● |
| Level 2 | | ● | |
| Level 1 | ● | | |

## Check Your Understanding

= Step-by-Step Solutions begin on page R11.

**Go Online!** for a Self-Check Quiz

**Example 1** — A.SSE.1a
Determine whether each expression is a polynomial. If it is a polynomial, find the degree and determine whether it is a *monomial, binomial,* or *trinomial*.

1. $7ab + 6b^2 - 2a^3$  yes; 3; trinomial
2. $2y - 5 + 3y^2$  yes; 2; trinomial
3. $3x^2$  yes; 2; monomial
4. $\frac{4m}{3p}$  No; a monomial cannot have a variable in the denominator.
5. $5m^2p^3 + 6$  yes; 5; binomial
6. $5q^{-4} + 6q$  No; $5q^{-4} = \frac{5}{q^4}$, and a monomial cannot have a variable in the denominator.

**Example 2** — A.SSE.1a
Write each polynomial in standard form. Identify the leading coefficient.

7. $2x^5 - 12 + 3x$  $2x^5 + 3x - 12; 2$
8. $-4d^4 + 1 - d^2$  $-4d^4 - d^2 + 1; -4$
9. $4z - 2z^2 - 5z^4$  $-5z^4 - 2z^2 + 4z; -5$
10. $2a + 4a^3 - 5a^2 - 1$  $4a^3 - 5a^2 + 2a - 1; 4$

**Examples 3–4** — A.APR.1
Find each sum or difference.  13. $-a^2 + 6a - 3$   15. $-8z^3 - 3z^2 - 2z + 13$   16. $-2d^2 + 6d - 20$

11. $(6x^3 - 4) + (-2x^3 + 9)$  $4x^3 + 5$
12. $(g^3 - 2g^2 + 5g + 6) - (g^2 + 2g)$  $g^3 - 3g^2 + 3g + 6$
13. $(4 + 2a^2 - 2a) - (3a^2 - 8a + 7)$
14. $(8y - 4y^2) + (3y - 9y^2)$  $-13y^2 + 11y$
15. $(-4z^3 - 2z + 8) - (4z^3 + 3z^2 - 5)$
16. $(-3d^2 - 8 + 2d) + (4d - 12 + d^2)$
17. $(y + 5) + (2y + 4y^2 - 2)$  $4y^2 + 3y + 3$
18. $(3n^3 - 5n + n^2) - (-8n^2 + 3n^3)$  $9n^2 - 5n$

**Example 5** — A.APR.1

19. **SENSE-MAKING** The total number of students $T$ who traveled for spring break consists of two groups: students who flew to their destinations $F$ and students who drove to their destination $D$. The number (in thousands) of students who flew and the total number of students who flew or drove can be modeled by the following equations, where $n$ is the number of years since 2010.

$$T = 14n + 21 \qquad F = 8n + 7$$

a. Write an equation that models the number of students who drove to their destination for this time period.  $D(n) = 6n + 14$
b. Predict the number of students who will drive to their destination in 2027.  116,000 students
c. How many students will drive or fly to their destination in 2030?  301,000 students

## Practice and Problem Solving

Extra Practice is on page R8.

**Example 1** — A.SSE.1a
Determine whether each expression is a polynomial. If it is a polynomial, find the degree and determine whether it is a *monomial, binomial,* or *trinomial*.

20. No; a monomial cannot have a variable in the denominator.

20. $\frac{5y^3}{x^2} + 4x$
21. $21$  yes; 0; monomial
22. $c^4 - 2c^2 + 1$  yes; 4; trinomial
23. $d + 3d^c$  No; the exponent is a variable.
24. $a - a^2$  yes; 2; binomial
25. $5n^3 + nq^3$  yes; 4; binomial

**Example 2** — A.SSE.1a
Write each polynomial in standard form. Identify the leading coefficient.

26. $5x^2 - 2 + 3x$  $5x^2 + 3x - 2; 5$
27. $8y + 7y^3$  $7y^3 + 8y; 7$
28. $4 - 3c - 5c^2$  $-5c^2 - 3c + 4; -5$
29. $-y^3 + 3y - 3y^2 + 2$  $-y^3 - 3y^2 + 3y + 2; -1$
30. $11t + 2t^2 - 3 + t^5$  $t^5 + 2t^2 + 11t - 3; 1$
31. $2 + r - r^3$  $-r^3 + r + 2; -1$
32. $\frac{1}{2}x - 3x^4 + 7$  $-3x^4 + \frac{1}{2}x + 7; -3$
33. $-9b^2 + 10b - b^6$  $-b^6 - 9b^2 + 10b; -1$

### Differentiated Homework Options

| Levels | **AL** Basic | **OL** Core | **BL** Advanced |
|---|---|---|---|
| Exercises | 20–44, 60–62, 65–72 | 21–49 odd, 58–72 | 45–65, (optional: 66–72) |
| 2-Day Option | 21–43 odd, 60, 66–72 | 20–44, 66–72 | |
| | 20–44 even, 61, 62, 65 | 45–51, 54–62, 65 | |

You can use ALEKS to provide additional remediation support with personalized instruction and practice.

## Go Online! — eBook

### Interactive Student Guide

Use the *Interactive Student Guide* to deepen conceptual understanding.
· Adding and Subtracting Polynomials

Lesson 8-1 | Adding and Subtracting Polynomials

**Examples 3–4** Find each sum or difference.
A.SSE.1a

34. $(2c^2 + 6c + 4) + (5c^2 - 7)$  $7c^2 + 6c - 3$
35. $(2x + 3x^2) - (7 - 8x^2)$  $11x^2 + 2x - 7$
36. $(3c^3 - c + 11) - (c^2 + 2c + 8)$  $3c^3 - c^2 - 3c + 3$
37. $(z^2 + z) + (z^2 - 11)$  $2z^2 + z - 11$
38. $(2x - 2y + 1) - (3y + 4x)$  $-2x - 5y + 1$
39. $(4a - 5b^2 + 3) + (6 - 2a + 3b^2)$  $-2b^2 + 2a + 9$
40. $(x^2y - 3x^2 + y) + (3y - 2x^2y)$  $-x^2y - 3x^2 + 4y$
41. $(-8xy + 3x^2 - 5y) + (4x^2 - 2y + 6xy)$  $7x^2 - 2xy - 7y$
42. $(5n - 2p^2 + 2np) - (4p^2 + 4n)$  $-6p^2 + 2np + n$
43. $(4rxt - 8r^2x + x^2) - (6rx^2 + 5rxt - 2x^2)$  $3x^2 - rxt - 8r^2x - 6rx^2$

**Example 5**
A.SSE.1a,
A.APR.1

44. **PETS** From 2006 through 2016, suppose the number of dogs $D$ and the number of cats $C$ (in hundreds) adopted from animal shelters in a region of the United States are modeled by the equations $D = 2n + 3$ and $C = n + 4$, where $n$ is the number of years since 2006.
    a. Write a function that models the total number $T$ of dogs and cats adopted in hundreds for this time period.  $T(n) = 3n + 7$
    b. If this trend continues, how many dogs and cats will be adopted in 2020?  **4900 dogs and cats**

B Classify each polynomial according to its degree and number of terms.

45. $4x - 3x^2 + 5$  **quadratic trinomial**
46. $11z^3$  **cubic monomial**
47. $9 + y^4$  **quartic binomial**
48. $3x^3 - 7$  **cubic binomial**
49. $-2x^5 - x^2 + 5x - 8$  **quintic polynomial**
50. $10t - 4t^2 + 6t^3$  **cubic trinomial**

51. **ENROLLMENT** In a rapidly growing school system, the number (in hundreds) of total students is represented by $N$ and the number of students in kindergarten through fifth grade is represented by $P$. The equations $N = 1.25t^2 - t + 7.5$ and $P = 0.7t^2 - 0.95t + 3.8$ model the number of students enrolled from 2006 to 2015, where $t$ is the number of years since 2006.
    a. Write an equation modeling the number of students $S$ in grades 6 through 12 enrolled for this time period.  $S = 0.55t^2 - 0.05t + 3.7$
    b. How many students were enrolled in grades 6 through 12 in the school system in 2013?  **3030 students**

52. **REASONING** The perimeter of the triangle can be represented by the expression $3x^2 - 7x + 2$. Write a polynomial that represents the measure of the third side.  $4x$

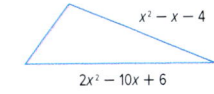

    (Triangle with sides: $x^2 - x - 4$, $2x^2 - 10x + 6$)

53. **GEOMETRY** Consider the rectangle.
    a. What does $(4x^2 + 2x - 1)(2x^2 - x + 3)$ represent?
    b. What does $2(4x^2 + 2x - 1) + 2(2x^2 - x + 3)$ represent?
    53a. the area of the rectangle  53b. the perimeter of the rectangle

    (Rectangle with sides: $4x^2 + 2x - 1$ and $2x^2 - x + 3$)

Find each sum or difference.

54. $(4x + 2y - 6z) + (5y - 2z + 7x) + (-9z - 2x - 3y)$  $9x + 4y - 17z$
55. $(5a^2 - 4) + (a^2 - 2a + 12) + (4a^2 - 6a + 8)$  $10a^2 - 8a + 16$
56. $(3c^2 - 7) + (4c + 7) - (c^2 + 5c - 8)$  $2c^2 - c + 8$
57. $(3n^3 + 3n - 10) - (4n^2 - 5n) + (4n^3 - 3n^2 - 9n + 4)$  $7n^3 - 7n^2 - n - 6$

58. **FOOTBALL** A school district has two high schools, North and South. From 2010 through 2015, the total attendance $T$ at games for both schools and at games for North High School $N$ can be modeled by the following equations, where $x$ is the number of years since 2010.
    $T = -0.69x^3 + 55.38x^2 + 643.31x + 10{,}538$   $N = -3.78x^3 + 58.96x^2 + 265.96x + 5257$
    Estimate how many people attended South High School football games in 2015.  **7465**

Lesson 8-1 | Adding and Subtracting Polynomials

## Teaching the Mathematical Practices

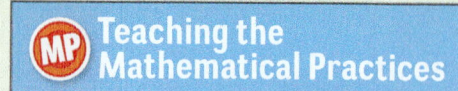

**Critique Arguments** Mathematically proficient students distinguish correct logic or reasoning from that which is flawed, and explain any flaws in an argument. In Exercise 61, have students check each step of each solution. Remind students that both need to add the additive inverse of each term in the polynomial being subtracted.

# Assess

**Ticket Out the Door** Make several copies each of five polynomial expressions. Give one expression to each student. As the students leave the room, ask them to tell you the degree of their expressions.

## Additional Answers

**60a.** Sample answer:

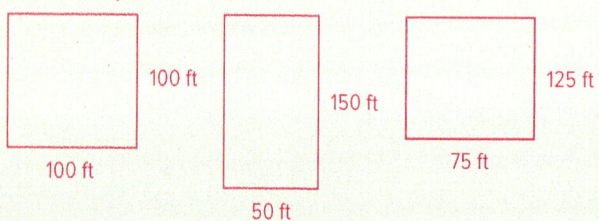

[Three rectangles: 100 ft × 100 ft; 150 ft × 50 ft; 125 ft × 75 ft]

**60c.**

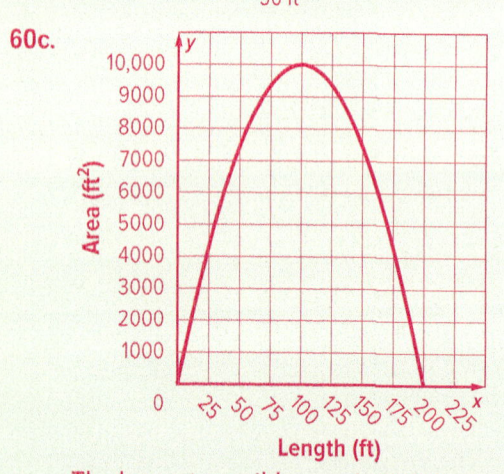

[Graph of parabola with x-axis labeled Length (ft) from 0 to 225 and y-axis labeled Area (ft²) from 0 to 10,000. Peak at (100, 10000).]

The largest possible area is 10,000 square feet.

**64.** Sample answer: When you add or subtract two or more polynomial equations, like terms are combined, which reduces the number of terms in the resulting equation. This could help minimize the number of operations performed when using the equations

---

**59. CAR RENTAL** The cost to rent a car for a day is $25 plus $0.35 for each mile driven.

a. Write a polynomial that represents the cost of renting a car for $m$ miles. **$25 + 0.35m$**

b. If a car is driven 145 miles for one day, how much would it cost to rent? **$75.75**

c. If a car is driven 105 miles each day for four days, how much would it cost to rent a car? **$247**

d. If a car is driven 220 miles each day for seven days, how much would it cost to rent a car? **$714**

**60. MULTIPLE REPRESENTATIONS** In this problem, you will explore perimeter and area.

a. **Geometric** Draw three rectangles that each have a perimeter of 400 feet. **See margin.**

b. **Tabular** Record the width and length of each rectangle in a table like the one shown below. Find the area of each rectangle.

| Rectangle | Length | Width | Area |
|---|---|---|---|
| 1 | 100 ft | 100 ft | 10,000 ft² |
| 2 | 50 ft | 150 ft | 7500 ft² |
| 3 | 75 ft | 125 ft | 9375 ft² |
| 4 | $x$ ft | $(200 - x)$ ft | $x(200 - x)$ ft² |

c. **Graphical** On a coordinate system, graph the area of rectangle 4 in terms of the length, $x$. Use the graph to determine the largest area possible. **See margin.**

d. **Analytical** Determine the length and width that produce the largest area. **The length and width of the rectangle must be 100 feet each to have the largest area.**

A.SSE.1a, A.APR.1

**H.O.T. Problems** Use Higher-Order Thinking Skills

**61. ERROR ANALYSIS** Cheyenne and Sebastian are finding $(2x^2 - x) - (3x + 3x^2 - 2)$. Is either of them correct? Explain your reasoning. **Neither; neither of them found the additive inverse correctly. All terms should have been multiplied by −1.**

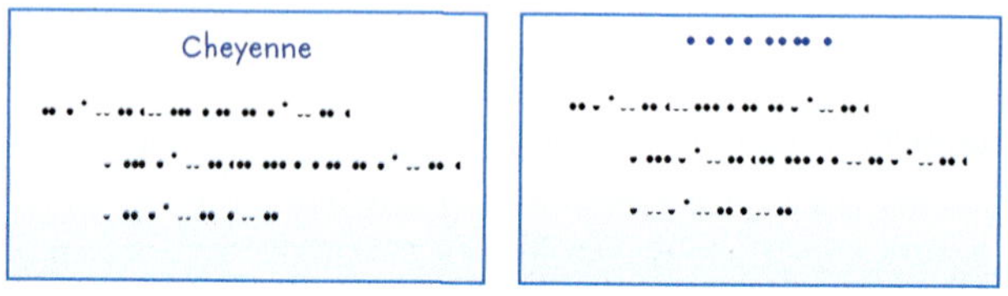

[Two work samples shown: Cheyenne and Sebastian]

**62. REASONING** Determine whether each of the following statements is *true* or *false*. Explain your reasoning.

a. A binomial can have a degree of zero. **62a. False; Sample answer: a binomial must have at least one monomial term with degree greater than zero.**

b. The order in which polynomials are subtracted does not matter. **False; sample answer: $(2x - 3) - (4x - 3) = -2x$, but $(4x - 3) - (2x - 3) = 2x$.**

**63. CHALLENGE** Write a polynomial that represents the sum of an odd integer $2n + 1$ and the next two consecutive odd integers. **$6n + 9$**

**64. WRITING IN MATH** Why would you add or subtract equations that represent real-world situations? Explain. **See margin.**

**65. WRITING IN MATH** Describe how to add and subtract polynomials using both the vertical and horizontal formats. **See margin.**

## Standards for Mathematical Practices

| Emphasis On | Exercises |
|---|---|
| 1 Make sense of problems and persevere in solving them. | 19, 44, 51–53, 58, 68, 70, 72 |
| 2 Reason abstractly and quantitatively. | 1–6, 52, 54–57, 62, 66, 71 |
| 6 Attend to precision. | 20–33, 45–50 |
| 7 Look for and make use of structure. | 7–18, 66, 67, 69 |

**65.** Sample answer: To add polynomials in a horizontal format, you combine like terms. For the vertical format, you write the polynomials in standard form, align like terms in columns, and combine like terms. To subtract polynomials in a horizontal format you find the additive inverse of the polynomial you are subtracting, and then combine like terms. For the vertical format you write the polynomials in standard form, align like terms in columns, and subtract by adding the additive inverse.

Lesson 8-1 | Adding and Subtracting Polynomials

## Preparing for Assessment

**66.** An equilateral triangle is shown below.

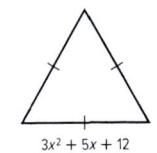

(triangle with side labels, base $3x^2 + 5x + 12$)

What is its perimeter? **MP** 2  A.APR.1  **C**

- A  $24x + 36$
- B  $6x^2 + 10x + 24$
- C  $9x^2 + 15x + 36$
- D  $9x^6 + 15x^3 + 36$

**70d.** $(4x^2 + 16x + 20) - (2x) = 4x^2 + 14x + 20$

**67.** Which expression is equivalent to
$3x^2 + 14 - (7x - 6) + 29 + (3x^2 + 5x) + 9x$?
**MP** 7  A.APR.1  **C**

- A  $3x^2 + 21x + 37$
- B  $6x^2 - 2x + 37$
- C  $6x^2 + 7x + 49$
- D  $6x^2 + 21x + 49$

**70e.** No; when the rectangles are combined in d, you must subtract 2 sides of width x, or 2x, from the total perimeter in c since they are not included in the perimeter of the combined rectangle.

**68.** The perimeter of a triangle is $3x^2 + 6x - 9$. If two of the side lengths add to $2x^2 + 4x - 6$, what is the length of the third side? **MP** 1  A.APR.1

$x^2 + 2x - 3$

**69.** The age of a child is $2x - 5$. The age of her mother is $x^2 + 4$. Find an expression for the sum of their ages 5 years from now. **MP** 7  A.APR.1

$x^2 + 2x + 9$

**70. MULTI-STEP** Two rectangles are shown. **MP** 1, 3  A.APR.1

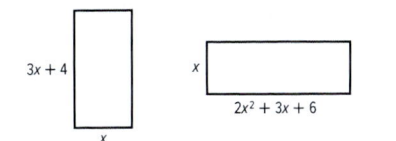

(rectangles: left one $3x + 4$ by $x$; right one $x$ by $2x^2 + 3x + 6$)

a. What is an expression for the perimeter of the rectangle on the left? **D**

- A  $3x + 5$
- B  $4x + 4$
- C  $6x + 8$
- D  $8x + 8$

b. Write an expression for the perimeter of the rectangle on the right. $4x^2 + 8x + 12$

c. Write an expression for the total perimeter of both rectangles.

$4x^2 + 16x + 20$

d. The two rectangles are combined to form a longer rectangle with width $x$. Find the perimeter of the new rectangle. Show your work.

e. Is the total perimeter of both rectangles you found in c the same as the perimeter you found after combining the two rectangles in d? Why or why not?

f. If $x = 1.5$ cm, what is the perimeter of the new rectangle you found in part d?

50 cm

**71.** A board game has square and equilateral triangle pieces. If the square and equilateral triangle pieces both have the same side length of $2x + 1$, find the sum of the perimeters of one square and one equilateral triangle. **MP** 2  A.APR.1

$14x + 7$

**72.** A library holds $x^2 - 2x + 10$ books. Alice checks out $3x - 5$ of them. How many books are left in the library? **MP** 1  A.APR.1

$x^2 - 5x + 15$

---

### Differentiated Instruction  OL  BL

**Extension**  Tell students the equations for the monthly unit sales of cars $C$ and trucks $D$ are $C = 7m + 87$ and $D = 9m + 152$, where $m$ represents time in months since a dealership opened. Suppose the total monthly sales of cars, trucks, and vans is represented by $T = 15m + 248$. Write an equation that can be used to calculate monthly van sales $V$. How many vans did the dealership sell in the sixth month when $m = 5$? $V = -m + 9$; 4

---

## Preparing for Assessment

Exercises 66–72 require students to use the skills they will need on standardized assessments. Each exercise is dual-coded with content standards and mathematical practice standards.

| Dual Coding |||
|---|---|---|
| Exercises | Content Standards | **MP** Mathematical Practices |
| 66 | A.APR.1 | 2 |
| 67 | A.APR.1 | 7 |
| 68 | A.APR.1 | 1 |
| 69 | A.APR.1 | 7 |
| 70 | A.APR.1 | 1, 3 |
| 71 | A.APR.1 | 2 |
| 72 | A.APR.1 | 1 |

## Diagnose Student Errors

Survey student responses for each item. Class trends may indicate common errors and misconceptions.

**66.**

| A | Added the coefficients of $x^2$ to the coefficients of $x$ |
|---|---|
| B | Added only two sides of the triangle |
| C | CORRECT |
| D | Found the sum of the degrees of each $x$ term |

**67.**

| A | Found only one $x^2$ term and did not add the additive inverse of $7x + 6$ |
|---|---|
| B | Subtracted rather than added 6 |
| C | CORRECT |
| D | Added rather than subtracting $7x$ |

**70a.**

| A | Incorrectly added $3x + 4$ and $x$ and only added each side length once |
|---|---|
| B | Forgot to add each side length twice |
| C | Added $3x + 4$ twice |
| D | CORRECT |

### Go Online!

**Quizzes**

Students can use *Self-Check Quizzes* to check their understanding of this lesson.

connectED.mcgraw-hill.com  497

# LESSON 8-2
## Multiplying a Polynomial by a Monomial

**SUGGESTED PACING (DAYS)**
- 90 min. — 0.5
- 45 min. — 1.0

Instruction

## Track Your Progress

### Objectives
1. Multiply a polynomial by a monomial.
2. Solve equations involving the products of monomials and polynomials.

### Mathematical Background
The Distributive Property can be used to find the product of a polynomial and a monomial. Multiply each term of the polynomial by the monomial and simplify the product by combining like terms. Equations often contain polynomials that must be added, subtracted, or multiplied before they can be solved. Simplify each side and then apply the rules of solving multi-step equations.

### THEN
**A.SSE.1** Interpret expressions that represent a quantity in terms of its context.

**A.SEE.1a.** Interpret parts of an expression, such as terms, factors, and coefficients.

### NOW
**A.APR.1** Understand that polynomials form a system analogous to the integers, namely, they are closed under the operations of addition, subtraction, and multiplication; add, subtract, and multiply polynomials.

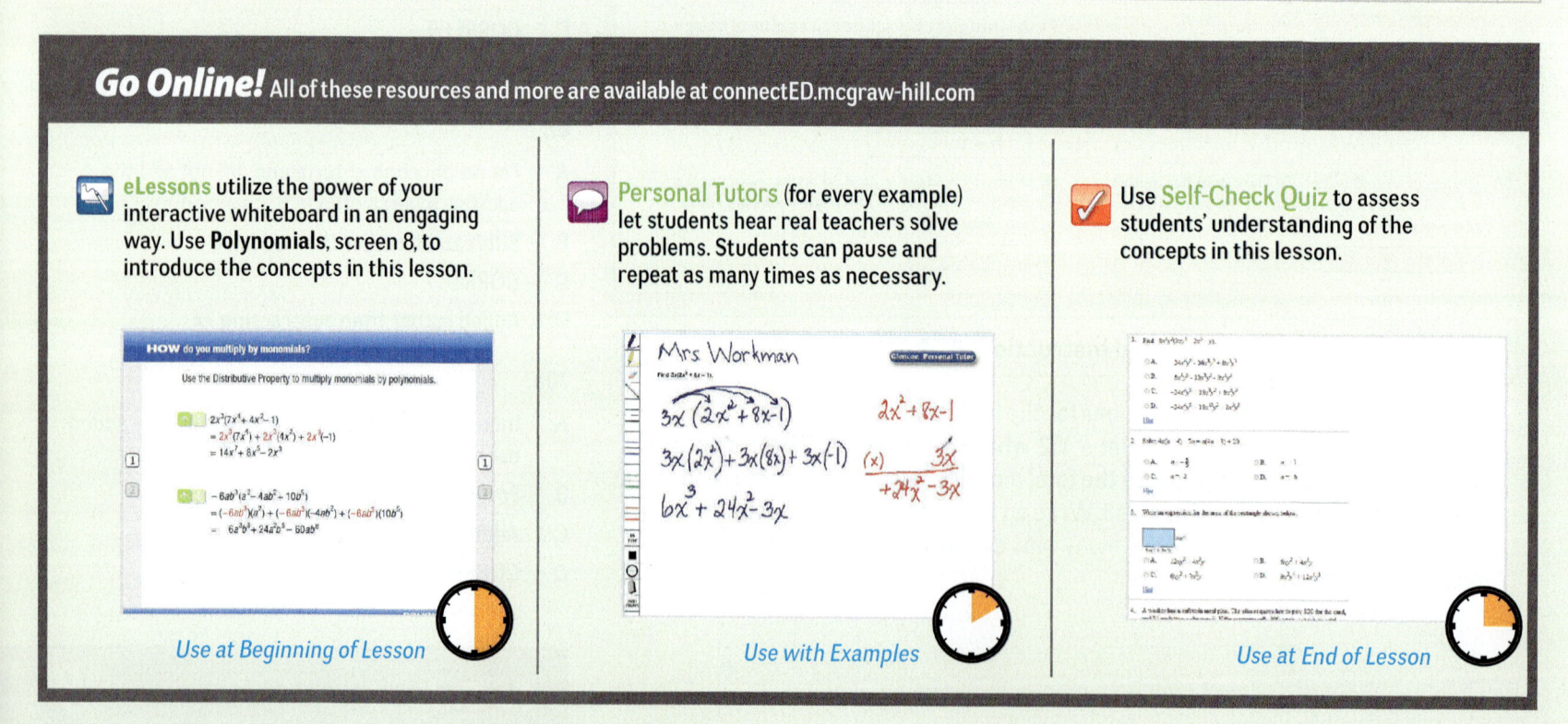

### NEXT
**A.SSE.2** Use the structure of an expression to identify ways to rewrite it.

---

## Go Online! All of these resources and more are available at connectED.mcgraw-hill.com

**eLessons** utilize the power of your interactive whiteboard in an engaging way. Use **Polynomials**, screen 8, to introduce the concepts in this lesson.

*Use at Beginning of Lesson*

**Personal Tutors** (for every example) let students hear real teachers solve problems. Students can pause and repeat as many times as necessary.

*Use with Examples*

Use **Self-Check Quiz** to assess students' understanding of the concepts in this lesson.

*Use at End of Lesson*

---

## OER Using Open Educational Resources
**Tutorials** Have students watch the step-by-step worked out examples on **mathwarehouse.com** about multiplying polynomials by monomials. The examples are broken out by steps so the students can watch the solutions at their own pace. If you are unable to access mathwarehouse, try **YouTube, Khan Academy,** or **cK12**.org. *Use as homework help or flipped learning*

# Go Online!
connectED.mcgraw-hill.com

**LearnSmart**

## Differentiate Your Resources

**Extra Practice** Additional practice or homework; Skills Practice is best for approaching-level students and Practice is best for on-level and beyond-level students

### Skills Practice
AL  OL  ELL

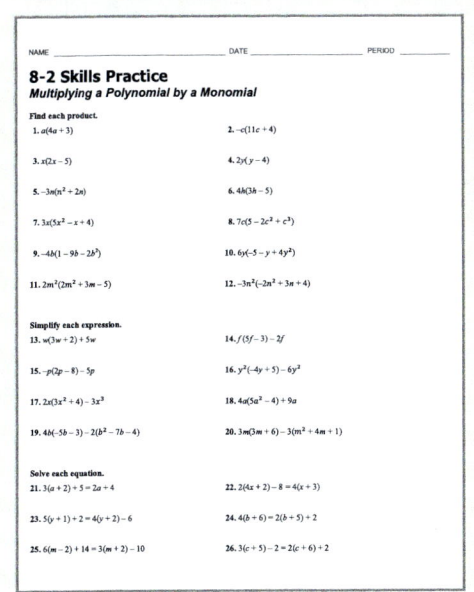

### Practice
AL  OL  BL  ELL

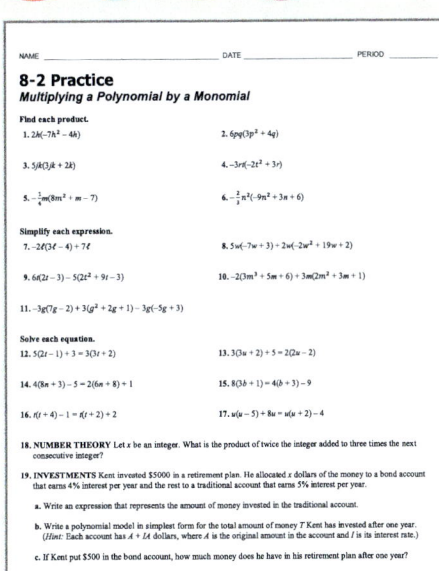

### Word Problem Practice
AL  OL  BL  ELL

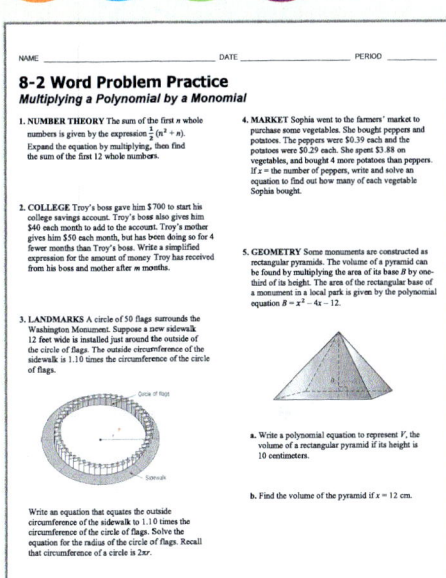

**Intervention** Reteaching and vocabulary activities that can be used with struggling or absent students and as ELL support

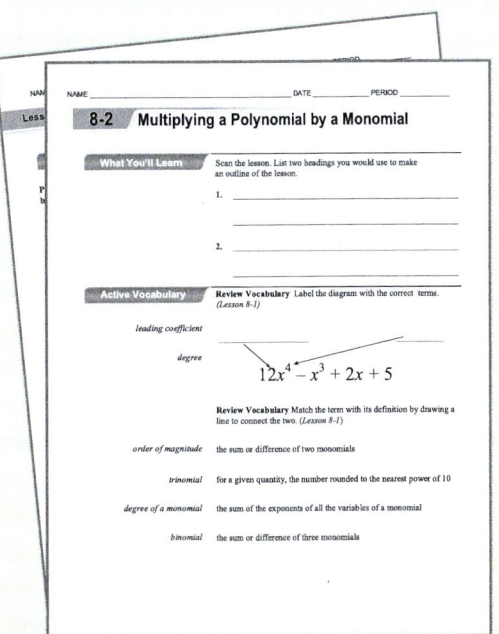

**Extension** Activities that can be used to extend lesson concepts

### Study Guide and Intervention
AL  OL  ELL

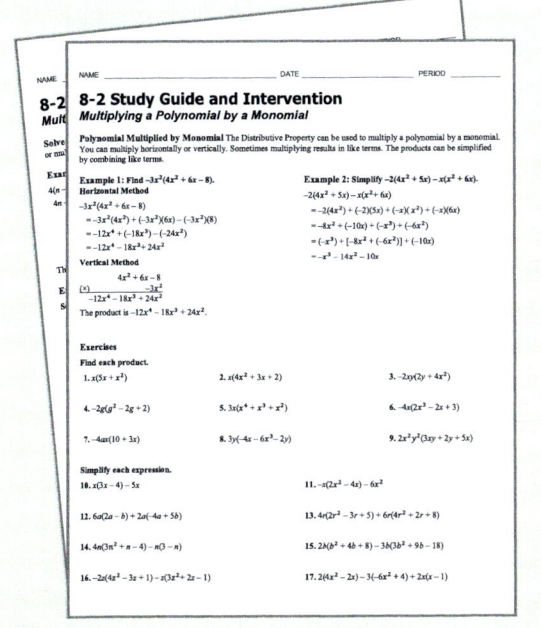

### Study Notebook
AL  OL  BL  ELL

### Enrichment
OL  BL  ELL

connectED.mcgraw-hill.com

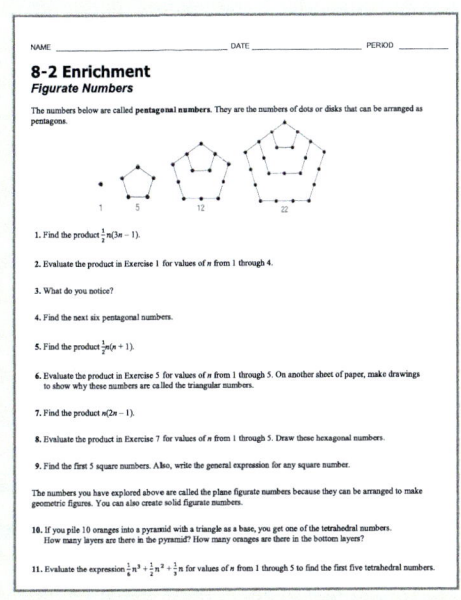

**Lesson 8-2** | Multiplying a Polynomial by a Monomial

## Launch

Have students read the Why? section of the lesson. Ask:

- What is the formula for finding the area of a rectangle? $A = \ell w$, where $\ell$ is the length and $w$ is the width.

- What are $\ell$ and $w$ for the expression shown? $\ell$ is $(3w + 8)$ and $w$ is $w$.

- Which of the dimensions is a monomial? $w$

- Describe how you would find the area of the room if the width is 20 feet. $20(3 \cdot 20 + 8) = 20(60 + 8) = 1200 + 160 = 1360$ ft²

## Teach

Ask the scaffolded questions for each example to build conceptual understanding for students at all levels.

### 1 Polynomial Multiplied by Monomial

**Example 1 Multiply a Polynomial by a Monomial**

**AL** By what is each term in the polynomial being multiplied? $-3x^2$

**OL** When multiplying $-3x^2$ and $7x^2$, what parts of the terms are being multiplied? $-3$ and $7$; $x^2$ and $x^2$ What property is used to multiply the variables? Product of a Power

**BL** Which method do you prefer? See students' work.

**Need Another Example?**
Find $6y(4y^2 - 9y - 7)$. $24y^3 - 54y^2 - 42y$

### Go Online!

**Interactive Whiteboard**
Use the *eLesson* or *Lesson Presentation* to present this lesson.

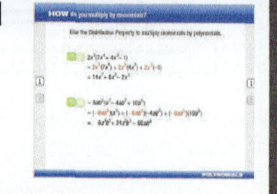

---

## LESSON 2
# Multiplying a Polynomial by a Monomial

| Then | Now | Why? |
|---|---|---|
| You multiplied monomials. | **1** Multiply a polynomial by a monomial. **2** Solve equations involving the products of monomials and polynomials. | Charmaine Brooks is opening a fitness club. She tells the contractor that the length of the fitness room should be three times the width plus 8 feet. To cover the floor with mats for exercise classes, Ms. Brooks needs to know the area of the floor. So she multiplies the width times the length, $w(3w + 8)$. |

**MP Mathematical Practices**
3 Construct viable arguments and critique the reasoning of others.
8 Look for and express regularity in repeated reasoning.

**Content Standards**
**A.APR.1** Understand that polynomials form a system analogous to the integers, namely, they are closed under the operations of addition, subtraction, and multiplication; add, subtract, and multiply polynomials.

**1 Polynomial Multiplied by Monomial** To find the product of a polynomial and a monomial, you can use the Distributive Property.

A.APR.1

**Example 1** Multiply a Polynomial by a Monomial

Find $-3x^2(7x^2 - x + 4)$.

**Horizontal Method**

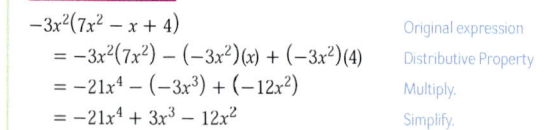

$-3x^2(7x^2 - x + 4)$     Original expression
$= -3x^2(7x^2) - (-3x^2)(x) + (-3x^2)(4)$     Distributive Property
$= -21x^4 - (-3x^3) + (-12x^2)$     Multiply.
$= -21x^4 + 3x^3 - 12x^2$     Simplify.

**Vertical Method**

$$\begin{array}{r} 7x^2 - x + 4 \\ (\times) \phantom{xxxx} -3x^2 \\ \hline -21x^4 + 3x^3 - 12x^2 \end{array}$$

Distributive Property
Multiply.

▶ **Guided Practice**

Find each product.    $-20a^4 + 10a^3 - 35a^2$      $-18a^7 + 12d^6 + 6d^4 - 54d^3$
**1A.** $5a^2(-4a^2 + 2a - 7)$      **1B.** $-6d^3(3d^4 - 2d^3 - d + 9)$

We can use this same method more than once to simplify large expressions.

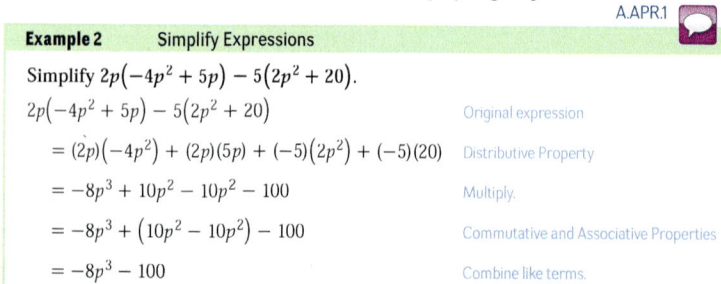

A.APR.1

**Example 2** Simplify Expressions

Simplify $2p(-4p^2 + 5p) - 5(2p^2 + 20)$.

$2p(-4p^2 + 5p) - 5(2p^2 + 20)$     Original expression
$= (2p)(-4p^2) + (2p)(5p) + (-5)(2p^2) + (-5)(20)$     Distributive Property
$= -8p^3 + 10p^2 - 10p^2 - 100$     Multiply.
$= -8p^3 + (10p^2 - 10p^2) - 100$     Commutative and Associative Properties
$= -8p^3 - 100$     Combine like terms.

---

**MP Mathematical Practices Strategies**

**Reason abstractly and quantitatively.**
Help students understand the meaning of multiplying polynomials and the higher degree variables. For example, ask:

- How would you illustrate $x$ and $x^2$? $x$ could be a bar or line, and $x^2$ could be a square.
- How would you illustrate $x^3$? $x^3$ would be a cube.
- How would $3x^3$ look different from $3x^2$? One would be three cubes, the other would be three squares.
- Is it possible for the product of two polynomials to have a lower degree than either of the factors? Explain. No; Sample answer: Since the terms of a polynomial must have positive, whole number exponents, the degree of the product of two polynomials will be the same as or greater than its factors.

498 | Lesson 8-2 | Multiplying a Polynomial by a Monomial

**Guided Practice**

2A. $-7x^3 + 13x^2 + 9x - 12$
2B. $150y^3t^6 + 73y^2t^2 - 8y^3$

2A. $3(5x^2 + 2x - 4) - x(7x^2 + 2x - 3)$  2B. $15t(10y^3t^5 + 5y^2t) - 2y(yt^2 + 4y^2)$

We can use the Distributive Property to multiply monomials by polynomials and solve real-world problems.

A.APR.1

**Real-World Example 3** Write and Evaluate a Polynomial Expression

DANCE ••••••••••••••••••••••••••••••••
••••••••••••••••••••••••••••••••••••
••••••••••••••••••••••••••••••••••••
••••••••••••••••••••••••••••••••••••
••••••••••••••••••••••••••••••••••••
••••••••••••••

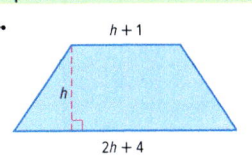

Trapezoid with top base $h+1$, height $h$, bottom base $2h+4$.

**Study Tip**

**MP Tools** Choosing the right formula is an important step in solving problems. If you don't already have a formula sheet, you might want to make one of your own to use while studying.

**Understand** You want to find the area of the poster board.

**Plan** Since the poster board is a trapezoid, Area $= \frac{1}{2} \cdot$ height $\cdot$ (base$_1$ + base$_2$).

**Solve** Let $h$ = the height of the poster, $b_1 = h + 1$.
Let $b_1 = h + 1$, let $b_2 = 2h + 4$ and let $h$ = height of the trapezoid.

$A = \frac{1}{2}h(b_1 + b_2)$   Area of a trapezoid

$= \frac{1}{2}h[(h + 1) + (2h + 4)]$   $b_1 = h + 1$ and $b_2 = 2h + 4$

$= \frac{1}{2}h(3h + 5)$   Add and simplify.

$= \frac{3}{2}h^2 + \frac{5}{2}h$   Distributive Property

$= \frac{3}{2}\cdots^2 + \frac{5}{2}\cdots$   $h = 18$

$= 531$   Simplify.

Kana will need 531 square inches of metallic paper.

**Check** $b_1 = h + 1 = \cdots + 1 = 19$ and $b_2 = 2h + 4 = 2\cdots + 4 = 40$
$A = \frac{1}{2}h(b_1 + b_2) = \frac{1}{2}(18)(19 + 40) = \frac{1}{2}(18)(59) = 531$ ✓

**Guided Practice**

3. **DANCE** Kachima is making triangular bandanas for the dogs and cats in her pet club. The base of the bandana is the length of the collar with 4 inches added to each end to tie it on. The height is $\frac{1}{2}$ of the collar length.

   A. If Kachima's dog has a collar length of 12 inches, how much fabric does she need in square inches? 60

   B. If Kachima makes a bandana for her friend's cat with a 6-inch collar, how much fabric does Kachima need in square inches? 21

**Real-World Link**
In a recent year, the pet supply business hit an estimated $7.05 billion in sales. This business ranges from gourmet food to rhinestone tiaras, pearl collars, and cashmere coats.

**Source:** Entrepreneur Magazine

---

**Differentiated Instruction** AL OL

**Visual/Spatial Learners** Have students group algebra tiles to form a rectangle with a width of $2x$ and a length of $x + 3$ using 2 blue $x^2$-tiles and 6 green $x$-tiles. Ask students to use their models to write an expression for the area of the rectangle. Then ask students to use the formula for area to calculate the area. $2x^2 + 6x$; $2x(x + 3) = 2x^2 + 6x$

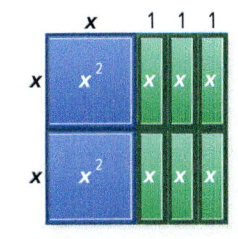

---

## Lesson 8-2 | Multiplying a Polynomial by a Monomial

**Example 2 Simplify Expressions**

**AL** What is the first step in simplifying the expressions? distribute

**OL** After distributing, what are the like terms? $10p^2$ and $-10p^2$

**BL** Knowing that you use the Distributive Property to multiply a monomial and a polynomial, how do you think you might use this to multiply a polynomial by a polynomial? See students' work.

**Need Another Example?**
Simplify $3(2t^2 - 4t - 15) + 6t(5t + 2)$. $36t^2 - 45$

**Example 3 Write and Evaluate a Polynomial Expression**

**AL** What is the formula for the area of a trapezoid? $A = \frac{1}{2}h(b_1 + b_2)$

**OL** What expressions are substituted for $b_1$ and $b_2$? $h + 1$ and $2h + 4$ What terms can be combined? $h$ and $2h$; 1 and 4 What is the next step? distribute

**BL** Are there any shortcuts to the problem? Explain. See students' work.

**Need Another Example?**
**RIDES** Admission to the Super Fun Amusement Park is $10. Once in the park, super rides are an additional $3 each, and regular rides are an additional $2. Wyome goes to the park and rides 15 rides, of which $s$ of those 15 are super rides. Find the cost in dollars if Wyome rode 9 super rides. $49

**MP Teaching the Mathematical Practices**

**Tools** Mathematically proficient students are able to use relevant external mathematical resources. Point out that there are different ways to represent the same formula, and formula sheets may show a different representation than one they learned. For example, the area of a trapezoid may also be shown as $A = \frac{h}{2}(b_1 + b_2)$ or $A = h\left(\frac{b_1 + b_2}{2}\right)$.

Lesson 8-2 | Multiplying a Polynomial by a Monomial

## 2 Solve Equations with Polynomial Expressions

**Example 4** Equations with Polynomials on Both Sides

**AL** What is the first step? distribute

**OL** At what point will you combine like terms from opposite sides of the equals sign? after combining like terms on each side of the equals sign

**BL** Besides the given Study Tip, what is another way you can keep the like terms organized? Sample answer: use colored pencils to highlight the terms.

**Need Another Example?**
Solve $b(12 + b) - 7 = 2b + b(-4 + b)$. $\frac{1}{2}$

## Practice

**Formative Assessment** Use Exercises 1–17 to assess students' understanding of the concepts in this lesson.

The Practice and Problem Solving exercises assess the content taught in the lesson. The Preparing for Assessment page is meant to be used as preparation for end-of-course-assessments.

## Teaching Tip

**Multiplying by a Negative Monomial** If students are having difficulty multiplying by a monomial with a negative sign, you may want to have them apply the negative first by multiplying all terms by −1 and then multiply by the rest of the monomial.

**Multiplication Strategies** In Exercises 18–29, some students may prefer using the horizontal method for multiplying a polynomial by a monomial. Others may prefer the vertical method. Since these two methods are equivalent, either may be used.

---

## 2 Solve Equations with Polynomial Expressions
We can use the Distributive Property to solve equations that involve the products of monomials and polynomials.

A.APR.1

### Example 4 Equations with Polynomials on Both Sides

Solve $2a(5a - 2) + 3a(2a + 6) + 8 = a(4a + 1) + 2a(6a - 4) + 50$.

| | |
|---|---|
| $2a(5a - 2) + 3a(2a + 6) + 8 = a(4a + 1) + 2a(6a - 4) + 50$ | Original equation |
| $10a^2 - 4a + 6a^2 + 18a + 8 = 4a^2 + a + 12a^2 - 8a + 50$ | Distributive Property |
| $16a^2 + 14a + 8 = 16a^2 - 7a + 50$ | Combine like terms. |
| $14a + 8 = -7a + 50$ | Subtract $16a^2$ from each side. |
| $21a + 8 = 50$ | Add $7a$ to each side. |
| $21a = 42$ | Subtract 8 from each side. |
| $a = 2$ | Divide each side by 21. |

**CHECK**

| | |
|---|---|
| $2a(5a - 2) + 3a(2a + 6) + 8 = a(4a + 1) + 2a(6a - 4) + 50$ | |
| $2(2)[5(2) - 2] + 3(2)[2(2) + 6] + 8 \stackrel{?}{=} 2[4(2) + 1] + 2(2)[6(2) - 4] + 50$ | |
| $4(8) + 6(10) + 8 \stackrel{?}{=} 2(9) + 4(8) + 50$ | Simplify. |
| $32 + 60 + 8 \stackrel{?}{=} 18 + 32 + 50$ | Multiply. |
| $100 = 100$ ✓ | Add and subtract. |

**Guided Practice**

Solve each equation.

**4A.** $2x(x + 4) + 7 = (x + 8) + 2x(x + 1) + 12$   $2\frac{3}{5}$

**4B.** $d(d + 3) - d(d - 4) = 9d - 16$   $8$

---

1. $-15w^3 + 10w^2 - 20w$   2. $18g^5 + 24g^4 + 60g^3 - 6g^2$   3. $32k^2m^4 + 8k^3m^3 + 20k^2m^2$

**Check Your Understanding** ◯ = Step-by-Step Solutions begin on page R11.

**Example 1**
A.APR.1

Find each product.   5. $14a^5b^3 + 2a^6b^2 - 4a^2b$   4. $-6p^6r^7 + 18p^{10}r^6 + 15p^4r^3$

1. $5w(-3w^2 + 2w - 4)$
2. $6g^2(3g^3 + 4g^2 + 10g - 1)$
3. $4km^2(8km^2 + 2k^2m + 5k)$
4. $-3p^4r^3(2p^2r^4 - 6p^6r^3 - 5)$
5. $2ab(7a^4b^2 + a^5b - 2a)$
6. $c^2d^3(5cd^7 - 3c^3d^2 - 4d^3)$   $5c^3d^{10} - 3c^5d^5 - 4c^2d^6$

**Example 2**
A.APR.1

Simplify each expression.   7. $4t^3 + 15t^2 - 8t + 4$

7. $t(4t^2 + 15t + 4) - 4(3t - 1)$
8. $x(3x^2 + 4) + 2(7x - 3)$   $3x^3 + 18x - 6$
9. $-2d(d^3c^2 - 4dc^2 + 2d^2c) + c^2(dc^2 - 3d^4)$   $-5d^4c^2 + 8d^2c^2 - 4d^3c + dc^4$
10. $-5w^2(8w^2x - 11wx^2) + 6x(9wx^4 - 4w - 3x^2)$   $-40w^4x + 55w^3x^2 + 54wx^5 - 24wx - 18x^3$

**Example 3**
A.APR.1

11. **TELEVISIONS** Marlene is buying a new LED television. The height of the screen of the television is one half the width plus 10 inches. The width is 40 inches. Find the height of the screen in inches.   30

**Example 4**
A.APR.1

Solve each equation.

12. $-6(11 - 2c) = 7(-2 - 2c)$   $2$
13. $t(2t + 3) + 20 = 2t(t - 3)$   $-\frac{20}{9}$
14. $-2(w + 1) + w = 7 - 4w$   $3$
15. $3(y - 2) + 2y = 4y + 14$   $20$
16. $a(a + 3) + a(a - 6) + 35 = a(a - 5) + a(a + 7)$   $7$
17. $n(n - 4) + n(n + 8) = n(n - 13) + n(n + 1) + 16$   $1$

---

## Go Online!

**Interactive Student Guide**

Use the Interactive Student Guide to deepen conceptual understanding.
- Multiplying a Polynomial by a Monomial

**Lesson 8-2** | Multiplying a Polynomial by a Monomial

## Practice and Problem Solving

Extra Practice is on page R8.

**Example 1**
A.APR.1

Find each product. 18–23. See margin.

18. $b(b^2 - 12b + 1)$
19. $f(f^2 + 2f + 25)$
20. $-3m^3(2m^3 - 12m^2 + 2m + 25)$
21. $2j^2(5j^3 - 15j^2 + 2j + 2)$
22. $2pr^2(2pr + 5p^2r - 15p)$
23. $4t^3u(2t^2u^2 - 10tu^4 + 2)$

**Example 2**
A.APR.1

Simplify each expression. 24. $-13x^2 - 9x - 27$  26. $-20d^3 + 55d + 35$  25. $-8a^3 + 20a^2 + 4a - 12$

24. $-3(5x^2 + 2x + 9) + x(2x - 3)$
25. $a(-8a^2 + 2a + 4) + 3(6a^2 - 4)$
26. $-4d(5d^2 - 12) + 7(d + 5)$
27. $-9g(-2g + g^2) + 3(g^2 + 4)$ $-9g^3 + 21g^2 + 12$
28. $2j(7j^2k^2 + jk^2 + 5k) - 9k(-2j^2k^2 + 2k^2 + 3j)$ $14j^3k^2 + 2j^2k^2 - 17jk + 18j^2k^3 - 18k^3$
29. $4n(2n^3p^2 - 3np^2 + 5n) + 4p(6n^2p - 2np^2 + 3p)$ $8n^4p^2 + 12n^2p^2 + 20n^2 - 8np^3 + 12p^2$

**Example 3**
A.APR.1

30. **DAMS** A new dam being built has the shape of a trapezoid. The length of the base at the bottom of the dam is 2 times the height. The length of the base at the top of the dam is $\frac{1}{5}$ times the height minus 30 feet.

    a. Write an expression to find the area of the trapezoidal cross section of the dam. $\frac{11}{10}h^2 - 15h$
    b. If the height of the dam is 180 feet, find the area of this cross section. $32,940 \text{ ft}^2$

**Example 4**
A.APR.1

Solve each equation.

31. $7(t^2 + 5t - 9) + t = t(7t - 2) + 13$   $2$
32. $w(4w + 6) + 2w = 2(2w^2 + 7w - 3)$   $1$
33. $5(4z + 6) - 2(z - 4) = 7z(z + 4) - z(7z - 2) - 48$   $\frac{43}{6}$
34. $9c(c - 11) + 10(5c - 3) = 3c(c + 5) + c(6c - 3) - 30$   $0$
35. $2f(5f - 2) - 10(f^2 - 3f + 6) = -8f(f + 4) + 4(2f^2 - 7f)$   $\frac{30}{43}$
36. $2k(-3k + 4) + 6(k^2 + 10) = k(4k + 8) - 2k(2k + 5)$   $-6$

Simplify each expression. 37. $20np^4 + 6n^3p^3 - 8np^2$   38. $6r^5t + 3r^3t^4 + 9r^2t^3$

37. $\frac{2}{3}np^2(30p^2 + 9n^2p - 12)$
38. $\frac{3}{5}r^2t(10r^3 + 5rt^3 + 15t^2)$
39. $-5q^2w^3(4q + 7w) + 4qw^2(7q^2w + 2q) - 3qw(3q^2w^2 + 9)$   39. $-q^3w^3 - 35q^2w^4 + 8q^2w^2 - 27qw$
40. $-x^2z(2z + 4xz^3) + xz^2(xz + 5x^3z) + x^2z^3(3x^2z + 4xz)$   $-x^2z^3 + 5x^4z^3 + 3x^4z^4$

41. **PARKING** A parking garage charges $30 per month plus $1.50 per daytime hour and $1.00 per hour during nights and weekends. Suppose Trent parks in the garage for 47 hours in January and $h$ of those are night and weekend hours.

    a. Find an expression for Trent's January bill. $100.5 - 0.5h$
    b. Find the cost if Trent had 12 hours of night and weekend hours. $94.50

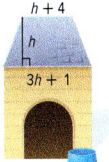

42. **MODELING** Che is building a dog house for his new puppy. The upper face of the dog house is a trapezoid. If the height of the trapezoid is 12 inches, find the area of the face of this piece of the dog house. $318 \text{ in}^2$

### Differentiated Homework Options

| Levels | AL Basic | OL Core | BL Advanced |
|---|---|---|---|
| Exercises | 18–36, 45, 48–54 | 19–39 odd, 41–45, 48–54 | 37–50, (optional: 51–54) |
| 2-Day Option | 19–35 odd, 51–54 | 18–36, 51–54 | |
| | 18–36 even, 45, 48–50 | 37–45, 48–50 | |

You can use ALEKS to provide additional remediation support with personalized instruction and practice.

## Teaching the Mathematical Practices

**Modeling** Mathematically proficient students routinely interpret their mathematical results in the context of the situation. In Exercise 42, ask students what their result means and why it might be useful to Che.

### Exercise Alert

**Formula** For Exercises 30 and 42, suggest students refer to Example 3 to help them with the formula for the area of a trapezoid.

### Extra Practice

See page R8 for extra exercises for students who are approaching level or for on-level students who need additional reinforcement.

### Levels of Complexity Chart

The levels of the exercises progress from 1 to 3, with Level 1 indicating the lowest level of complexity.

| Exercises | 18–36 | 37–43, 51–54 | 44–50 |
|---|---|---|---|
| Level 3 | | | ● |
| Level 2 | | ● | |
| Level 1 | ● | | |

### Additional Answers

18. $b^3 - 12b^2 + b$
19. $f^3 + 2f^2 + 25f$
20. $-6m^6 + 36m^5 - 6m^4 - 75m^3$
21. $10j^5 - 30j^4 + 4j^3 + 4j^2$
22. $4p^2r^3 + 10p^3r^3 - 30p^2r^2$
23. $8t^5u^3 - 40t^4u^5 + 8t^3u$

### Go Online!

**eSolutions Manual**
Create worksheets, answer keys, and solutions handouts for your assignments.

connectED.mcgraw-hill.com  501

Lesson 8-2 | Multiplying a Polynomial by a Monomial

**Watch Out!**

**Error Analysis** For Exercise 45, point out to students that Pearl's method should draw a critical eye because her final polynomial has only two terms. When a monomial is multiplied by a polynomial with three different degree terms, the result will be a polynomial with three terms.

## Assess

### Crystal Ball

Ask students to write a sentence predicting how learning to multiply a polynomial by a monomial will help them to learn to multiply polynomials by other polynomials in the next lesson.

### Teaching the Mathematical Practices

**Perseverance** Mathematically proficient students consider analogous problems. In Exercise 46, students may be intimidated by the variables in the exponents. Encourage them to think of how they treat numbers as exponents.

---

**43. TENNIS** The tennis club is building a new tennis court with a path around it.

a. Write an expression for the area of the tennis court. $1.5x^2 + 24x$

b. Write an expression for the area of the path. $x^2 - 9x$

c. If $x = 36$ feet, what is the perimeter of the outside of the path? 264 ft

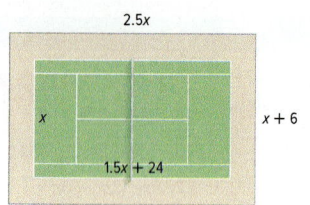

(Diagram labels: 2.5x, x, x + 6, 1.5x + 24)

**44. MULTIPLE REPRESENTATIONS** In this problem, you will investigate the degree of the product of a monomial and a polynomial.

a. **Tabular** Write three monomials of different degrees and three polynomials of different degrees. Determine the degree of each monomial and polynomial. Multiply the monomials by the polynomials. Determine the degree of each product. Record your results in a table like the one shown below. **Sample answer:**

| Monomial | Degree | Polynomial | Degree | Product of Monomial and Polynomial | Degree |
|---|---|---|---|---|---|
| $2x$ | 1 | $x^2 - 1$ | 2 | $2x^3 - 2x$ | 3 |
| $3x^2$ | 2 | $x^5 + 1$ | 5 | $3x^7 + 3x^2$ | 7 |
| $4x^3$ | 3 | $x^6 + 1$ | 6 | $4x^9 + 4x^3$ | 9 |

b. **Verbal** Make a conjecture about the degree of the product of a monomial and a polynomial. What is the degree of the product of a monomial of degree $a$ and a polynomial of degree $b$? The degree of the product is the sum of the degree of the monomial and the degree of the polynomial; $a + b$.

A.APR.1

**H.O.T. Problems** Use Higher-Order Thinking Skills

**45. ERROR ANALYSIS** Pearl and Ted both worked on this problem. Is either of them correct? Explain your reasoning. Ted; Pearl used the Distributive Property incorrectly.

50. Sample answer: To multiply a polynomial by a monomial, use the Distributive Property. Multiply each term of the polynomial by the monomial. Then simplify by multiplying the coefficients together and using the Product of Powers Property for the variables.

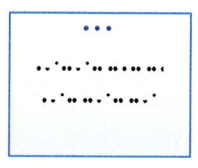

Pearl
$2x^2(3x^2 + 4x + 2)$
$6x^4 + 8x^2 + 4x^2$
$6x^4 + 12x^2$

**46. PERSEVERANCE** Find $p$ such that $3x^p(4x^{2p+3} + 2x^{3p-2}) = 12x^{12} + 6x^{10}$. 3

**47. CHALLENGE** Simplify $4x^{-3}y^2(2x^5y^{-4} + 6x^{-7}y^6 - 4x^0y^{-2})$.
$8x^2y^{-2} + 24x^{-10}y^8 - 16x^{-3}$

**48. REASONING** Is there a value for $x$ that makes the statement $(x + 2)^2 = x^2 + 2^2$ true? If so, find a value for $x$. Explain your reasoning. Yes; 0; when 0 is substituted in for $x$ in the equation, both sides are $2^2$ or 4, which makes the equation true.

**49. OPEN-ENDED** Write a monomial and a polynomial using $n$ as the variable. Find their product. Sample answer: $3n$, $4n + 1$; $12n^2 + 3n$

**50. WRITING IN MATH** Describe the steps to multiply a polynomial by a monomial.

---

### Standards for Mathematical Practice

| Emphasis On | Exercises |
|---|---|
| 1 Make sense of problems and persevere in solving them. | 11, 30, 41, 42, 43, 49–54 |
| 2 Reason abstractly and quantitatively. | 1–10, 50 |
| 4 Model with mathematics. | 51, 53 |
| 6 Attend to precision. | 12–17, 37–40, 44 |
| 7 Look for and make use of structure. | 24–29, 31–36 |

### Differentiated Instruction OL BL

**Extension** Give students this problem: Nate multiplied a polynomial by a monomial and got $6x^8 - 3x^4 + 9x^2$. If the polynomial factor was $2x^6 - x^2 + 3$, what was the monomial factor? $3x^2$

502 | Lesson 8-2 | Multiplying a Polynomial by a Monomial

## Lesson 8-2 | Multiplying a Polynomial by a Monomial

### Preparing for Assessment

**51.** Brian has 3 less than 5 times as many quarters as his sister has dimes. Brian used $n$ for his sister's number of dimes and wrote the expression below to represent the total amount of money they have.

$$0.25(5n - 3) + 0.10n$$

Which expression is an equivalent form for this total? **MP 2, 4  A.APR.1  B**

- A  $1.35n - 3$
- B  $1.35n - 0.75$
- C  $0.115n + 0.85$
- D  $0.135n^2 - 0.075n$

**52.** A student is simplifying an expression.

**Step 1:** $3x(x^2 + 5x + 12) - 4x(2 - x)$

**Step 2:** ?

Which expression could be an equivalent form written for Step 2? **MP 2  A.APR.1  D**

- A  $3x(18x^3) + 4x(x)$
- B  $3x(60x^3) + 4x(-2x)$
- C  $3x^3 + 15x^2 + 36 - 8x - 4x^2$
- D  $3x^3 + 15x^2 + 36x - 8x + 4x^2$

**53. MULTI-STEP** Katie is making different colored pennants for her 30 classmates. Each pennant is the same size with a base of length $x$ and a height of $2x^2 + 3x + 6$. **MP 2, 4  A.APR.1**

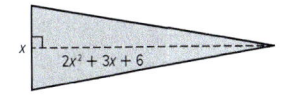

(triangle pennant diagram with base $x$ and height $2x^2 + 3x + 6$)

**a.** Which expression represents the number of square units of fabric Katie needs to make each pennant? **A**

- A  $x^3 + 1.5x^2 + 3x$
- B  $2x^3 + 3x^2 + 6x$
- C  $3x^3 + 4.5x^2 + 9x$
- D  $4x^3 + 6x^2 + 12x$

**b.** Which expression represents the number of square units of fabric Katie needs to make all the pennants? **C**

- A  $10x^3 + 15x^2 + 30x$
- B  $20x^3 + 30x^2 + 60x$
- C  $30x^3 + 45x^2 + 90x$
- D  $60x^3 + 90x^2 + 180x$

**c.** The city's sports team asked Katie to make 30 larger pennants to hang in the city's sports dome. If the pennants have a base of 2 meters what is the area of one of the larger pennants?

$$20\ m^2$$

**d.** What is the total area of the fabric that Katie will need to make the 30 pennants for the city? Show your work.  $30x^3 + 45x^2 + 90x = 240 + 180 + 180 = 600\ m^2$

**e.** If fabric costs $10/m^2$, what is the total cost of the fabric that Katie will need to make pennants for the city? Show your work.

$600\ m^2 \times \$10/m^2 = \$6000$

**54.** Which statements about the expressions are true? **MP 2  A.APR.1  B, D, E, F**

- I.  $2.5x^2(2x^3 - 10x^2 + x)$
- II.  $6yz^3(9y^2 - 4z)$
- III.  $v^2w(w^3 + 3v^2 - 180v^5)$

- ☐ A  The value of expression I when $x = 1$ is greater than 0.
- ☐ B  Two of the expressions have a degree greater than 5.
- ☐ C  When $v = -2$ and $w = 2$, the value of the expression III is negative.
- ☐ D  Expression III has the greatest degree.
- ☐ E  The simplified form of expression II is $54y^3z^3 - 24yz^4$.
- ☐ F  Expression I has a degree of 5.

### Preparing for Assessment

Exercises 51–54 require students to use the skills they will need on standardized assessments. Each exercise is dual-coded with content standards and mathematical practice standards.

| Dual Coding |||
|---|---|---|
| Items | Content Standards | MP Mathematical Practices |
| 51 | A.APR.1 | 2, 4 |
| 52 | A.APR.1 | 2 |
| 53 | A.APR.1 | 2, 4 |
| 54 | A.APR.1 | 2 |

### Diagnose Student Errors

Survey student responses for each item. Class trends may indicate common errors and misconceptions.

**51.**

| | |
|---|---|
| A | Did not distribute 0.25 to $-3$ |
| B | CORRECT |
| C | Incorrectly distributed and used the wrong sign on the 3 |
| D | Incorrectly distributed 0.025 |

**52.**

| | |
|---|---|
| A | Added the coefficients and constants inside the parentheses |
| B | Multiplied the terms inside the parentheses |
| C | Multiplied $-x$ by $4x$ instead of by $-4x$ and incorrectly multiplied 3 by 12 |
| D | CORRECT |

**53a.**

| | |
|---|---|
| A | CORRECT |
| B | Did not divide by 2 |
| C | Multiplied by $\frac{3}{2}$ instead of $\frac{1}{2}$ |
| D | Multiplied by 2 instead of dividing |

**53b.**

| | |
|---|---|
| A | Multiplied by 10 instead of 30 |
| B | Multiplied by 20 instead of 30 |
| C | CORRECT |
| D | Used incorrect expression for area of one pennant |

### Go Online!

**Quizzes**

Students can use *Self-Check Quizzes* to check their understanding of this lesson. You can also give the *Chapter Quiz*, which covers the content in Lessons 8-1 and 8-2.

# LESSON 8-3
# Multiplying Polynomials

**SUGGESTED PACING (DAYS)**

| | Explore Lab | Instruction |
|---|---|---|
| 90 min. | 0.25 | 0.75 |
| 45 min. | 0.5 | 1.5 |

## Track Your Progress

**Objectives**
1. Multiply binomials by using the FOIL method.
2. Multiply polynomials by using the Distributive Property.

**Mathematical Background**
The FOIL method only works for multiplying two binomials. Use the Distributive Property to multiply other polynomials. To use the FOIL method, find the sum of the products of the first terms (F), the outer terms (O), the inner terms (I), and the last terms (L).

### THEN
**A.SSE.1** Interpret expressions that represent a quantity in terms of its context.

### NOW
**A.APR.1** Understand that polynomials form a system analogous to the integers, namely, they are closed under the operations of addition, subtraction, and multiplication; add, subtract, and multiply polynomials.

### NEXT
**A.SSE.2** Use the structure of an expression to identify ways to rewrite it.

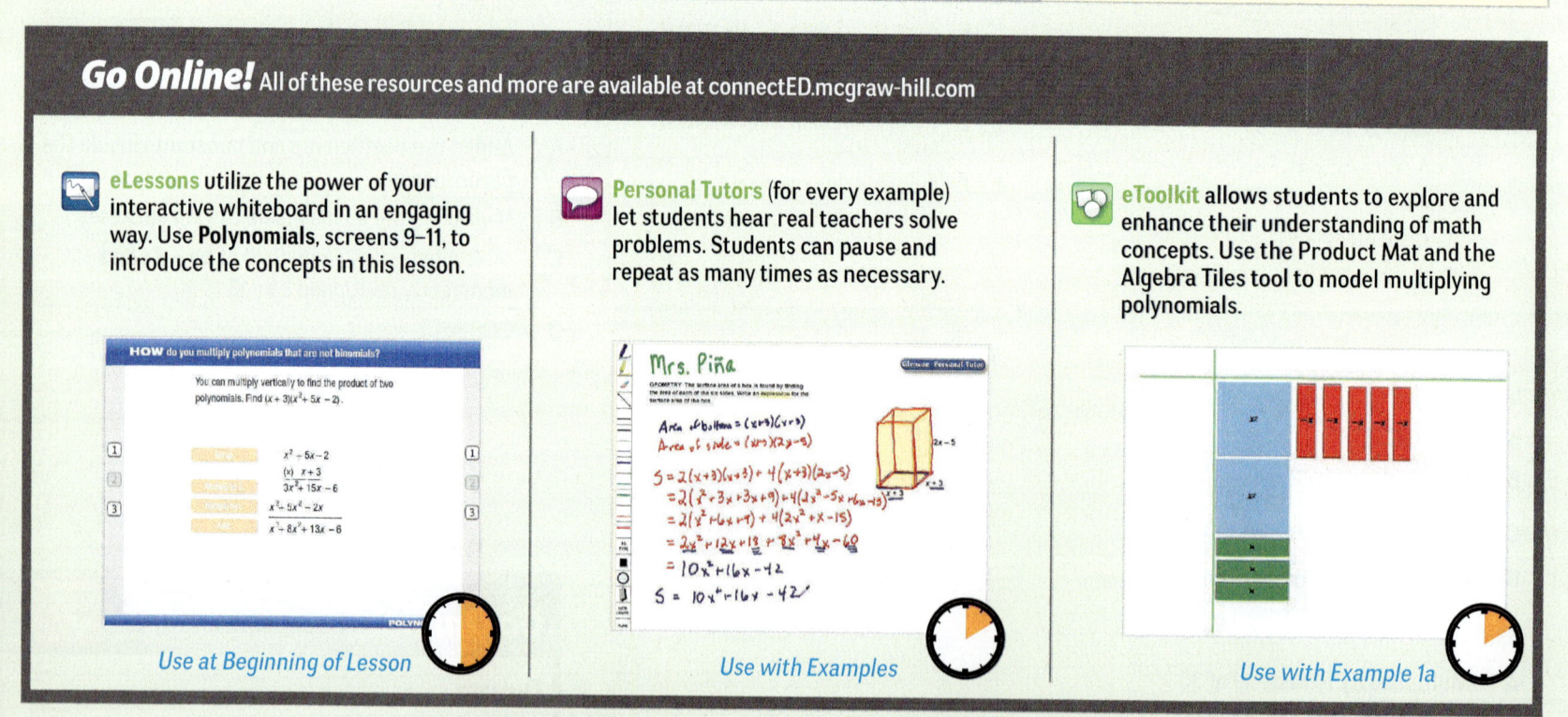

## Go Online! All of these resources and more are available at connectED.mcgraw-hill.com

**eLessons** utilize the power of your interactive whiteboard in an engaging way. Use **Polynomials**, screens 9–11, to introduce the concepts in this lesson.

*Use at Beginning of Lesson*

**Personal Tutors** (for every example) let students hear real teachers solve problems. Students can pause and repeat as many times as necessary.

*Use with Examples*

**eToolkit** allows students to explore and enhance their understanding of math concepts. Use the Product Mat and the Algebra Tiles tool to model multiplying polynomials.

*Use with Example 1a*

## OER Using Open Educational Resources

**Games** Download the Grid Game for Multiplying Polynomials from the **regentsprep.org** site. Have students work in pairs to create a grid as demonstrated on the site. They can then cut out the squares and switch with their partner who is responsible for putting the grid back together. The goal is for all touching sides to be equivalent. Be sure to have students make a copy of their original, so they have an answer key to check their partner's work. *Use as classwork*

# Go Online!
connectED.mcgraw-hill.com — Worksheets

# Differentiate Your Resources

**Extra Practice** Additional practice or homework; Skills Practice is best for approaching-level students and Practice is best for on-level and beyond-level students

## Skills Practice
AL  OL  ELL

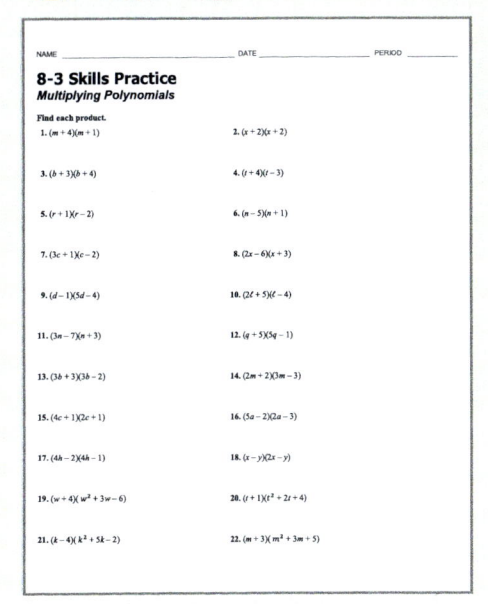

## Practice
AL  OL  BL  ELL

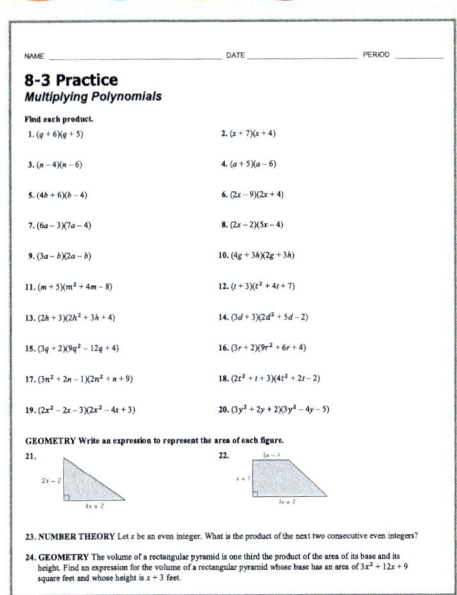

## Word Problem Practice
AL  OL  BL  ELL

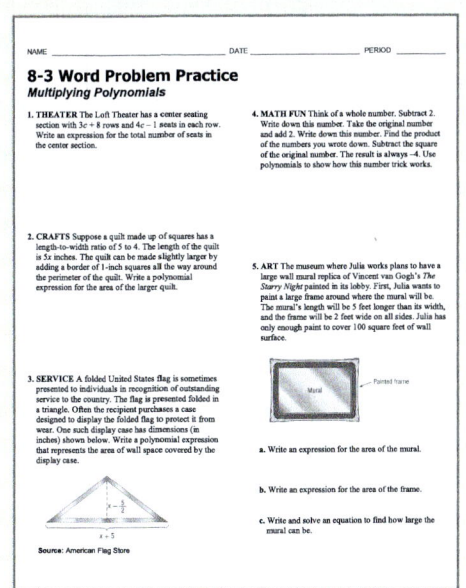

**Intervention** Reteaching and vocabulary activities that can be used with struggling or absent students and as ELL support

## Study Guide and Intervention
AL  OL  ELL

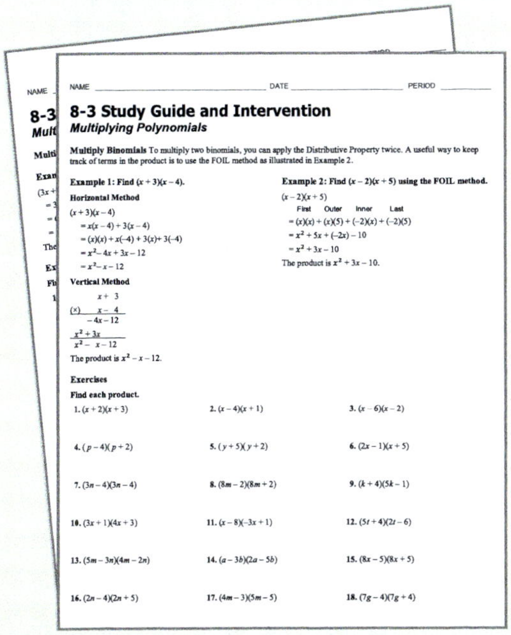

## Study Notebook
AL  OL  BL  ELL

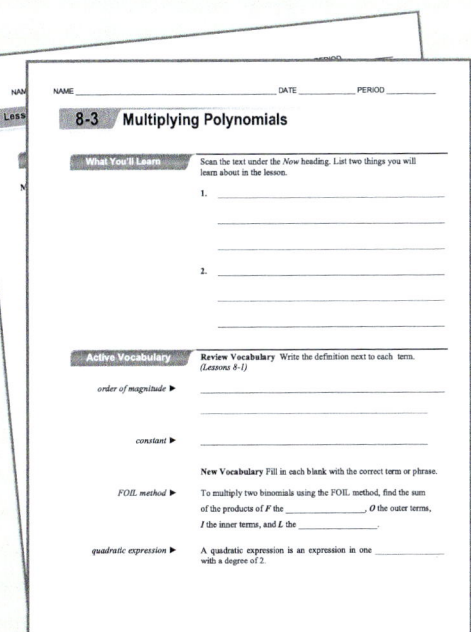

**Extension** Activities that can be used to extend lesson concepts

## Enrichment
OL  BL  ELL

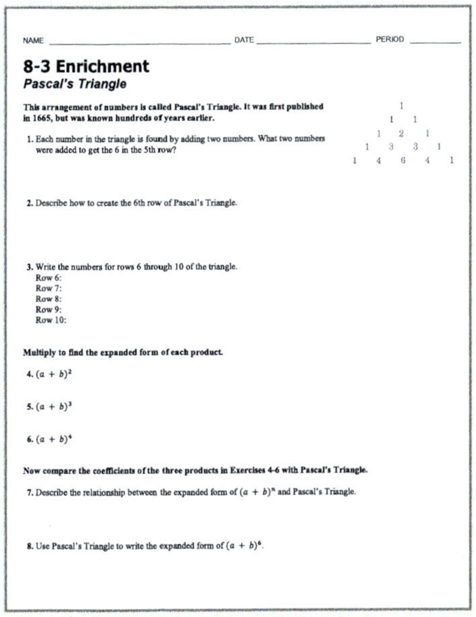

connectED.mcgraw-hill.com  504B

# Explore 8-3

## Launch

**Objective** Use algebra tiles to multiply polynomials.

### Materials for Each Student
- algebra tiles
- product mat

### Easy to Make Manipulatives
*Teaching Algebra with Manipulatives* Template for:
- algebra tiles, pp. 10–11
- product mat, p. 17

### Teaching Tip
Some students may benefit from laying tiles along the top and side of the product mat to model each expression. Have them remove the two factors before determining their final product.

## Teach

### Working in Cooperative Groups  ELL

Put students in groups of two or three, mixing abilities. Have groups complete Activities 1–3 and Exercise 1.
- For Activity 1, make sure groups mark the dimensions properly on the product mat. Since $x$-tiles are rectangular, remind students that the long side is the correct side to use to mark a value of $x$ on the mat.
- When students are filling in the mats with the tiles, remind them to look carefully at the horizontal and vertical dimensions of each tile on the product mat. If both dimensions have a value of $x$, then use an $x^2$-tile. If one dimension is $x$ and the other is 1, then use an $x$-tile. If both dimensions are 1, then use a 1-tile.

## Go Online!

### eToolkit
The eToolkit allows students to explore and enhance their understanding of math concepts. Use the Algebra Tiles tool to demonstrate multiplying polynomials.

---

# EXPLORE 8-3
## Algebra Lab
# Multiplying Polynomials

You can use algebra tiles to find the product of two binomials.

**Mathematical Practices**
- MP 3 Construct viable arguments and critique the reasoning of others.
- 8 Look for and express regularity in repeated reasoning.

**Content Standards**
**A.APR.1** Understand that polynomials form a system analogous to the integers, namely, they are closed under the operations of addition, subtraction, and multiplication; add, subtract, and multiply polynomials.

### Activity 1  Multiply Binomials

Work cooperatively. Use algebra tiles to find $(x + 3)(x + 4)$.

The rectangle will have a width of $x + 3$ and a length of $x + 4$. Use algebra tiles to mark off the dimensions on a product mat. Then complete the rectangle with algebra tiles.

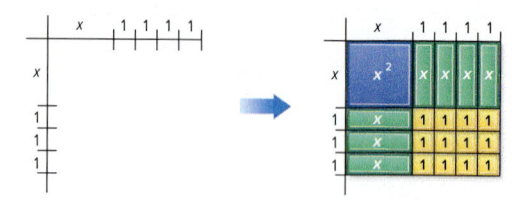

The rectangle consists of one blue $x^2$-tile, seven green $x$-tiles, and 12 yellow 1-tiles. The area of the rectangle is $x^2 + 7x + 12$. So, $(x + 3)(x + 4) = x^2 + 7x + 12$.

### Activity 2  Multiply Binomials

Work cooperatively. Use algebra tiles to find $(x - 2)(x - 5)$.

**Step 1** The rectangle will have a width of $x - 2$ and a length of $x - 5$. Use algebra tiles to mark off the dimensions on a product mat. Then begin to make the rectangle with algebra tiles.

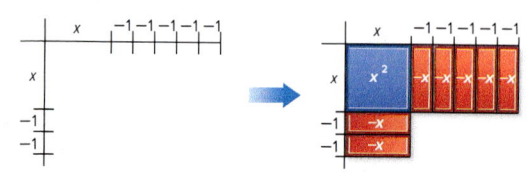

**Step 2** Determine whether to use 10 yellow 1-tiles or 10 red $-1$-tiles to complete the rectangle. The area of each yellow tile is the product of $-1$ and $-1$. Fill in the space with 10 yellow 1-tiles to complete the rectangle.

The rectangle consists of one blue $x^2$-tile, seven red $-x$-tiles, and 10 yellow 1-tiles. The area of the rectangle is $x^2 - 7x + 10$. So, $(x - 2)(x - 5) = x^2 - 7x + 10$.

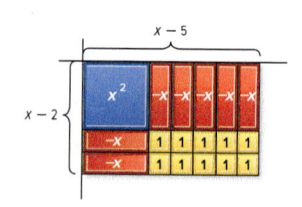

*(continued on the next page)*

Explore 8-3

### Activity 3  Multiply Binomials

Work cooperatively. Use algebra tiles to find $(x-4)(2x+3)$.

**Step 1** The rectangle will have a width of $x-4$ and a length of $2x+3$. Use algebra tiles to mark off the dimensions on a product mat. Then begin to make the rectangle with algebra tiles.

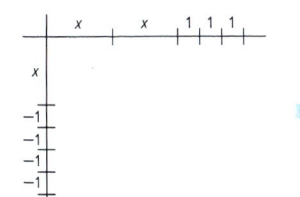

**Step 2** Determine what color $x$-tiles and what color 1-tiles to use to complete the rectangle. The area of each red $x$-tile is the product of $x$ and $-1$. The area of each red $-1$-tile is represented by $1(-1)$ or $-1$.

Complete the rectangle with four red $x$-tiles and 12 red $-1$-tiles.

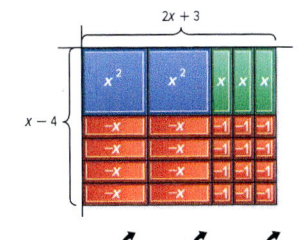

**Step 3** Rearrange the tiles to simplify the polynomial you have formed. Notice that 3 zero pairs are formed by three positive and three negative $x$-tiles.

There are two blue $x^2$-tiles, five red $-x$-tiles, and 12 red $-1$-tiles left. In simplest form, $(x-4)(2x+3) = 2x^2 - 5x - 12$.

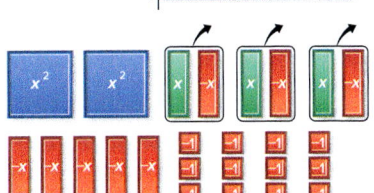

### Model and Analyze

Work cooperatively. Use algebra tiles to find each product.

1. $(x+1)(x+4)$   $x^2 + 5x + 4$
2. $(x-3)(x-2)$   $x^2 - 5x + 6$
3. $(x+5)(x-1)$   $x^2 + 4x - 5$
4. $(x+2)(2x+3)$   $2x^2 + 7x + 6$
5. $(x-1)(2x-1)$   $2x^2 - 3x + 1$
6. $(x+4)(2x-5)$   $2x^2 + 3x - 20$

7–8. See margin for drawings.

Is each statement *true* or *false*? Justify your answer with a drawing of algebra tiles.

7. $(x-4)(x-2) = x^2 - 6x + 8$   **true**
8. $(x+3)(x+5) = x^2 + 15$   **false**

9. **WRITING IN MATH** You can also use the Distributive Property to find the product of two binomials. The figure at the right shows the model for $(x+4)(x+5)$ separated into four parts. Write a sentence or two explaining how this model shows the use of the Distributive Property.
By the Distributive Property, $(x+4)(x+5) = x(x+5) + 4(x+5)$. The top row represents $x(x+5)$ or $x^2 + 5x$. The bottom row represents $4(x+5)$ or $4x + 20$.

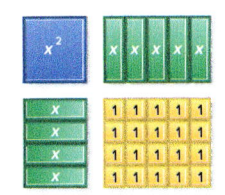

### Additional Answers

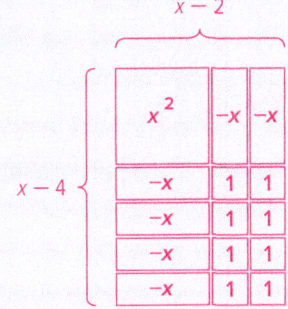

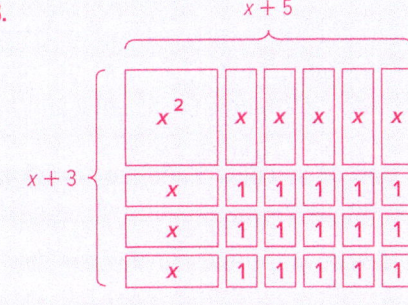

7.

| | $x-2$ | |
|---|---|---|
| $x^2$ | $-x$ | $-x$ |
| $-x$ | 1 | 1 |
| $-x$ | 1 | 1 |
| $-x$ | 1 | 1 |
| $-x$ | 1 | 1 |

$x - 4$

8.

$x+3$, $x+5$ rectangle with $x^2$, five $x$ tiles across top; then three rows of $x$ and five 1-tiles.

---

- For Activity 2, Step 2, have students pay close attention to whether the dimensions for each tile are positive or negative, as this affects which tile to use. If both dimensions are positive, then the tile is positive. If one is positive and the other is negative, the tile is negative. If both are negative, then the tile is positive.

- For Activity 3, as an alternative to removing zero pairs, have students write the expression based on the tiles without removing zero pairs. They can then simplify the expression by combining like terms.

**Practice** Have students complete Exercises 1–9.

## Assess

### Formative Assessment
Use Exercise 9 to assess whether students can model a product correctly.

### From Concrete to Abstract
After students have completed Exercise 9, help them to see that when using the Distributive Property to multiply polynomials, each term from the first polynomial is multiplied by each term from the second polynomial.

### Extending the Concept
Ask students to model $(x-3)(x+2)$ using algebra tiles. Then ask students to write the expression based on the tiles without removing zero pairs. $x^2 + 2x - 3x - 6$

Finally, ask students to find the sum of the product of the two first terms, outer terms, inner terms, and last terms of $(x-3)(x+2)$ and then compare the results to the expression they wrote. **The two expressions are the same.**

connectED.mcgraw-hill.com  505

# Lesson 8-3 | Multiplying Polynomials

## Launch

Have students read the Why? section of the lesson. Ask:

- What expression would you get if you multiplied the first term in $(h - 32)$ times $\left(\frac{1}{2}h + 11\right)$? $\frac{1}{2}h^2 + 11h$

- What expression would you get if you multiplied the second term in $(h - 32)$ by $\left(\frac{1}{2}h + 11\right)$? $-16h - 352$

- What expression do you get when you add these two answers together? $\frac{1}{2}h^2 - 5h - 352$

## Teach

Ask the scaffolded questions for each example to build conceptual understanding for students at all levels.

### 1 Multiply Binomials

**Example 1 The Distributive Property**

- **AL** In part **b**, what could be done to the expression to make multiplying easier? change $(x - 2)$ to $(x + (-2))$

- **OL** What needs to be multiplied to find the product of these two expressions? Explain. Both terms in the first binomial need to be multiplied by both terms in the second binomial.

- **BL** Which method do you prefer? Explain. See students' work.

**Need Another Example?**
Find each product.
a. $(y + 8)(y - 4)$  $y^2 + 4y - 32$
b. $(2x + 1)(x + 6)$  $2x^2 + 13x + 6$

### Go Online!
**Interactive Whiteboard**
Use the *eLesson* or *Lesson Presentation* to present this lesson.

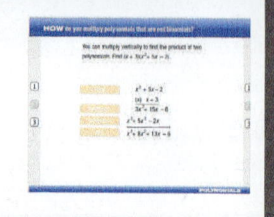

---

# LESSON 3
## Multiplying Polynomials

**Then**
- You multiplied polynomials by monomials.

**Now**
1. Multiply binomials by using the FOIL method.
2. Multiply polynomials by using the Distributive Property.

**Why?**
Bodyboards, which are used to ride waves, are made of foam and are more rectangular than surfboards. A bodyboard's dimensions are determined by the height and skill level of the user.

The length of Ann's bodyboard should be Ann's height $h$ minus 32 inches or $h - 32$. The board's width should be half of Ann's height plus 11 inches or $\frac{1}{2}h + 11$. To approximate the area of the bodyboard, you need to find $(h - 32)\left(\frac{1}{2}h + 11\right)$.

**New Vocabulary**
FOIL method
quadratic expression

**Mathematical Practices**
6 Attend to precision.
8 Look for and express regularity in repeated reasoning.

**Content Standards**
A.APR.1 Understand that polynomials form a system analogous to the integers, namely, they are closed under the operations of addition, subtraction, and multiplication; add, subtract, and multiply polynomials.

### 1 Multiply Binomials
To multiply two binomials such as $h - 32$ and $\frac{1}{2}h + 11$, the Distributive Property is used. Binomials can be multiplied horizontally or vertically.

A.APR.1

**Example 1** The Distributive Property
Find each product.

**a.** $(2x + 3)(x + 5)$

**Vertical Method**

Multiply by 5.
$2x + 3$
$(\times)\ x + 5$
$\overline{10x + 15}$

$5(2x + 3) = 10x + 15$

Multiply by $x$.
$2x + 3$
$(\times)\ x + 5$
$\overline{10x + 15}$
$2x^2 + 3x$

$x(2x + 3) = 2x^2 + 3x$

Combine like terms.
$2x + 3$
$(\times)\ x + 5$
$\overline{10x + 15}$
$2x^2 + 3x$
$\overline{2x^2 + 13x + 15}$

**Horizontal Method**
$(2x + 3)(x + 5) = 2x(x + 5) + 3(x + 5)$     Rewrite as the sum of two products.
$= 2x^2 + 10x + 3x + 15$     Distributive Property
$= 2x^2 + 13x + 15$     Combine like terms.

**b.** $(x - 2)(3x + 4)$

**Vertical Method**

Multiply by 4.
$x - 2$
$(\times)\ 3x + 4$
$\overline{4x - 8}$

$4(x - 2) = 4x - 8$

Multiply by $3x$.
$x - 2$
$(\times)\ 3x + 4$
$\overline{4x - 8}$
$3x^2 - 6x$

$3x(x - 2) = 3x^2 - 6x$

Combine like terms.
$x - 2$
$(\times)\ 3x + 4$
$\overline{4x - 8}$
$3x^2 - 6x$
$\overline{3x^2 - 2x - 8}$

**Horizontal Method**
$(x - 2)(3x + 4) = x(3x + 4) - 2(3x + 4)$     Rewrite as the difference of two products.
$= 3x^2 + 4x - 6x - 8$     Distributive Property
$= 3x^2 - 2x - 8$     Combine like terms.

---

## Mathematical Practices Strategies

**Look for and express regularity in repeated reasoning.**
Help students maintain oversight of the process of multiplying polynomials by attending to single terms and combining like terms. For example, ask:

- How do you determine the new term after multiplying two terms of a polynomial? Multiply the coefficients and add the exponents.

- How do you know you have finished the multiplication? When you have distributed the last term in one set of parentheses over the last term in the other set of parentheses

- What are like terms? Terms that contain the same variables, with corresponding variables having the same exponent.

- How do you combine like terms? By adding their coefficients

### Guided Practice

**1A.** $(3m + 4)(m + 5)$   $3m^2 + 19m + 20$    **1B.** $(5y - 2)(y + 8)$   $5y^2 + 38y - 16$

A shortcut version of the Distributive Property for multiplying binomials is called the **FOIL method**.

#### Key Concept  FOIL Method

**Words**   To multiply two binomials, find the sum of the products of **F** the First terms, **O** the Outer terms, **I** the Inner terms, and **L** the Last terms.

**Example**

$$(x + 4)(x - 2) = (x)(x) + (x)(-2) + (4)(x) + (4)(-2)$$
$$= x^2 - 2x + 4x - 8$$
$$= x^2 + 2x - 8$$

**Go Online!** ELL

The expression $(x + 4)(x - 2)$ is read *the quantity x plus four times the quantity x minus 2*. To hear more pronunciations of expressions, log into your *eStudent Edition*. Ask your teacher or a partner for clarification as you need it.

A.APR.1

### Example 2  FOIL Method

Find each product.

**a.** $(2y - 7)(3y + 5)$

$(2y - 7)(3y + 5) = (2y)(3y) + (2y)(5) + (-7)(3y) + (-7)(5)$   FOIL method
$\phantom{(2y - 7)(3y + 5)} = 6y^2 + 10y - 21y - 35$   Multiply.
$\phantom{(2y - 7)(3y + 5)} = 6y^2 - 11y - 35$   Combine like terms.

**b.** $(4a - 5)(2a - 9)$

$(4a - 5)(2a - 9)$
$= (4a)(2a) + (4a)(-9) + (-5)(2a) + (-5)(-9)$   FOIL method
$= 8a^2 - 36a - 10a + 45$   Multiply.
$= 8a^2 - 46a + 45$   Combine like terms.

### Guided Practice

**2A.** $(x + 3)(x - 4)$   $x^2 - x - 12$    **2B.** $(4b - 5)(3b + 2)$   $12b^2 - 7b - 10$
**2C.** $(2y - 5)(y - 6)$   $2y^2 - 17y + 30$    **2D.** $(5a + 2)(3a - 4)$   $15a^2 - 14a - 8$

Notice that when two linear expressions are multiplied, the result is a quadratic expression. A **quadratic expression** is an expression in one variable with a degree of 2. When three linear expressions are multiplied, the result has a degree of 3.

The FOIL method can be used to find an expression that represents the area of a rectangular object when the lengths of the sides are given as binomials.

---

## Lesson 8-3 | Multiplying Polynomials

### Example 2  FOIL Method

**AL** What does FOIL stand for? **First, Outer, Inner, Last**

**OL** For both part **a** and part **b**, what do you notice about the terms after you multiply? **Sample answer: There are like terms that need to be combined.**

**BL** How does using the FOIL method help keep you organized? **Sample answer: The FOIL method helps keep my work organized by grouping together the terms that I need to multiply.**

### Need Another Example?

Find each product.
**a.** $(z - 6)(z - 12)$   $z^2 - 18z + 72$
**b.** $(5x - 4)(2x + 8)$   $10x^2 + 32x - 32$

### Teaching Tip

**FOIL** Point out that FOIL is a memory tool. The order in which the terms are multiplied is not important, as long as all four products are found.

---

### Differentiated Instruction  OL  BL

**IF** students are less familiar with the Distributive Property,

**THEN** they may wish to use the vertical method for multiplying binomials because it is similar to multiplying two-digit numbers. Suggest that students use the method with which they are most comfortable.

connectED.mcgraw-hill.com    507

**Lesson 8-3 | Multiplying Polynomials**

**Example 3 FOIL Method**

**AL** What is the formula for the area of a rectangle? length times width

**OL** What is the total length of the pool and deck together? $x + 20 + x$ or $2x + 20$ What is the total width of the pool and deck together? $x + 15 + x$ or $2x + 15$

**BL** How did learning about multiplying monomials help you with multiplying polynomials? How might multiplying binomials help you understand multiplying trinomials? See students' work.

**Need Another Example?**

**Patio** A patio in the shape of the triangle shown below is being built in Lavali's backyard. The dimensions given are in feet. The area $A$ of the triangle is one half the height $h$ times the base $b$. Write an expression for the area of the patio.

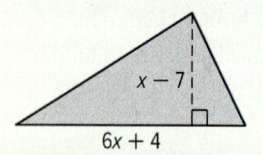

$x - 7$
$6x + 4$

$(3x^2 - 19x - 14)$ ft²

## 2 Multiply Polynomials

**Example 4 The Distributive Property**

**AL** What types of polynomials are being multiplied in part a? a binomial and a trinomial

**OL** Explain how to multiply the terms in part a. Sample answer: I will multiply $6x$ by each term in the trinomial, then I will multiply 5 by each term in the trinomial.

**BL** How might the study tip "Multiplying Polynomials" help you when multiplying large polynomials? Sample answer: I can count the number of terms I end up with after multiplying and make sure that I have the correct number of terms. If I don't have enough terms, I know I forgot to multiply something.

**Need Another Example?**

Find each product.
a. $(3a + 4)(a^2 - 12a + 1)$ $3a^3 - 32a^2 - 45a + 4$
b. $(2b^2 + 7b + 9)(b^2 + 3b - 1)$ $2b^4 + 13b^3 + 28b^2 + 20b - 9$

---

**Real-World Link**
The cost of a swimming pool depends on many factors, including the size of the pool, whether the pool is an above-ground or an in-ground pool, and the material used.
**Source:** American Dream Homes

**Real-World Example 3 FOIL Method**

**SWIMMING POOL** A contractor is building a deck around a rectangular swimming pool. The deck is $x$ feet from every side of the pool. Write an expression for the total area of the pool and deck.

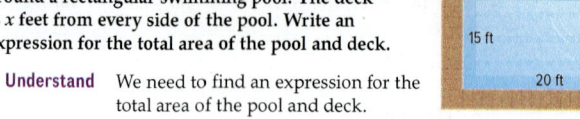

**Understand** We need to find an expression for the total area of the pool and deck.

**Plan** Find the product of the length and width of the pool with the deck.

**Solve** Since the deck is the same distance from every side of the pool, the length and width of the pool are $2x$ longer. So, the length can be represented by $2x + 20$ and the width can be represented by $2x + 15$.

Area = length · width          Area of a rectangle
= $(2x + 20)(2x + 15)$          Substitution
= $(2x)(2x) + (2x)(15) + (20)(2x) + (20)(15)$   FOIL Method
= $4x^2 + 30x + 40x + 300$      Multiply.
= $4x^2 + 70x + 300$            Combine like terms.

So, the total area of the deck and pool is $4x^2 + 70x + 300$.

**Check** Choose a value for $x$. Substitute this value into $(2x + 20)(2x + 15)$ and $4x^2 + 70x + 300$. The result should be the same for both expressions.

▶ **Guided Practice** $4x^2 + 90x + 500$

3. If the pool is 25 feet long and 20 feet wide, find the area of the pool and deck.

**2 Multiply Polynomials** The Distributive Property can also be used to multiply any two polynomials.

**Example 4 The Distributive Property**

Find each product.

a. $(6x + 5)(2x^2 - 3x - 5)$
$(6x + 5)(2x^2 - 3x - 5)$
$= 6x(2x^2 - 3x - 5) + 5(2x^2 - 3x - 5)$   Distributive Property
$= 12x^3 - 18x^2 - 30x + 10x^2 - 15x - 25$   Multiply.
$= 12x^3 - 8x^2 - 45x - 25$   Combine like terms.

b. $(2y^2 + 3y - 1)(3y^2 - 5y + 2)$
$(2y^2 + 3y - 1)(3y^2 - 5y + 2)$
$= 2y^2(3y^2 - 5y + 2) + 3y(3y^2 - 5y + 2) - 1(3y^2 - 5y + 2)$   Distributive Property
$= 6y^4 - 10y^3 + 4y^2 + 9y^3 - 15y^2 + 6y - 3y^2 + 5y - 2$   Multiply.
$= 6y^4 - y^3 - 14y^2 + 11y - 2$   Combine like terms.

**Study Tip**

**MP Structure** If a polynomial with $c$ terms and a polynomial with $d$ terms are multiplied together, there will be $c \cdot d$ terms before simplifying. In Example 4a, there are 2 · 3 or 6 terms before simplifying.

▶ **Guided Practice**    $6x^3 + 11x^2 - 59x + 40$    $4m^4 + m^3 - 21m^2 + 31m - 15$
4A. $(3x - 5)(2x^2 + 7x - 8)$       4B. $(m^2 + 2m - 3)(4m^2 - 7m + 5)$

---

**Additional Answers**

25. $2y^3 - 17y^2 + 37y - 22$
26. $36a^3 + 71a^2 - 14a - 49$
27. $m^4 + 2m^3 - 34m^2 + 43m - 12$
28. $5x^4 + 19x^3 - 34x^2 + 11x - 1$
29. $6b^5 - 3b^4 - 35b^3 - 10b^2 + 43b + 63$
30. $18z^5 - 15z^4 - 18z^3 - 14z^2 + 24z + 8$

## Lesson 8-3 | Multiplying Polynomials

### Check Your Understanding

= Step-by-Step Solutions begin on page R11.

**Examples 1–2** Find each product.
A.APR.1
1. $(x+5)(x+2)$  $x^2 + 7x + 10$
2. $(y-2)(y+4)$  $y^2 + 2y - 8$
3. $(b-7)(b+3)$  $b^2 - 4b - 21$
4. $(4n+3)(n+9)$  $4n^2 + 39n + 27$
5. $(8h-1)(2h-3)$  $16h^2 - 26h + 3$
6. $(2a+9)(5a-6)$  $10a^2 + 33a - 54$

**Example 3** 7. **FRAME** Hugo is designing a frame as shown at the right. The frame has a width of $x$ inches all the way around. Write an expression that represents the total area of the picture and frame. $4x^2 + 72x + 320$

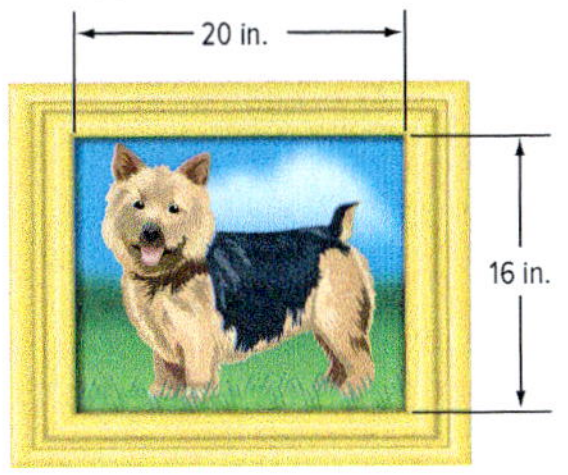

**Example 4** Find each product.
A.APR.1
8. $(2a-9)(3a^2+4a-4)$  $6a^3 - 19a^2 - 44a + 36$
9. $(4y^2-3)(4y^2+7y+2)$  $16y^4 + 28y^3 - 4y^2 - 21y - 6$
10. $(x^2-4x+5)(5x^2+3x-4)$  $5x^4 - 17x^3 + 9x^2 + 31x - 20$
11. $(2n^2+3n-6)(5n^2-2n-8)$  $10n^4 + 11n^3 - 52n^2 - 12n + 48$

16. $15y^2 - 17y + 4$
17. $24d^2 - 62d + 35$
18. $6m^2 + 19m + 15$
19. $49n^2 - 84n + 36$

### Practice and Problem Solving

Extra Practice is on page R8.

**Examples 1–2** Find each product.
A.APR.1
12. $(3c-5)(c+3)$  $3c^2 + 4c - 15$
13. $(g+10)(2g-5)$  $2g^2 + 15g - 50$
14. $(6a+5)(5a+3)$  $30a^2 + 43a + 15$
15. $(4x+1)(6x+3)$  $24x^2 + 18x + 3$
16. $(5y-4)(3y-1)$
17. $(6d-5)(4d-7)$
18. $(3m+5)(2m+3)$
19. $(7n-6)(7n-6)$
20. $(12t-5)(12t+5)$  $144t^2 - 25$
21. $(5r+7)(5r-7)$  $25r^2 - 49$
22. $(8w+4x)(5w-6x)$  $40w^2 - 28wx - 24x^2$
23. $(11z-5y)(3z+2y)$  $33z^2 + 7yz - 10y^2$

**Example 3** 24. **GARDEN** A walkway surrounds a rectangular garden. The width of the garden is 8 feet, and the length is 6 feet. The width $x$ of the walkway around the garden is the same on every side. Write an expression that represents the total area of the garden and walkway. $4x^2 + 28x + 48$

**Example 4** Find each product. 25–30. See margin.
A.APR.1
25. $(2y-11)(y^2-3y+2)$
26. $(4a+7)(9a^2+2a-7)$
27. $(m^2-5m+4)(m^2+7m-3)$
28. $(x^2+5x-1)(5x^2-6x+1)$
29. $(3b^3-4b-7)(2b^2-b-9)$
30. $(6z^2-5z-2)(3z^3-2z-4)$

**B** Simplify.
31. $(m+2)[(m^2+3m-6)+(m^2-2m+4)]$  $2m^3 + 5m^2 - 4$
32. $[(t^2+3t-8)-(t^2-2t+6)](t-4)$  $5t^2 - 34t + 56$

**MP STRUCTURE** Find an expression to represent the area of each shaded region.

33. $4\pi x^2 + 12\pi x + 9\pi - 3x^2 - 5x - 2$

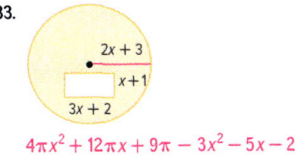

34. $24x^2 - \frac{3}{2}$

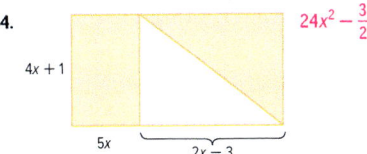

### Differentiated Homework Options

| Levels | **AL** Basic | **OL** Core | **BL** Advanced |
|---|---|---|---|
| Exercises | 12–30, 45, 48–59 | 13–33 odd, 35, 36, 37–41 odd, 43–45, 48–59 | 31–49, (optional: 50–59) |
| 2-Day Option | 13–29 odd, 50–59 | 12–30, 50–59 | |
| | 12–30 even, 45, 48, 49 | 31–45, 48, 49 | |

You can use ALEKS to provide additional remediation support with personalized instruction and practice.

## Practice

**Formative Assessment** Use Exercises 1–11 to assess students' understanding of the concepts in this lesson.

The Practice and Problem Solving exercises assess the content taught in the lesson. The Preparing for Assessment page is meant to be used as preparation for end-of-course-assessment.

### MP Teaching the Mathematical Practices

**Structure** Mathematically proficient students can see complicated things as single objects or as being composed of several objects. In Exercises 33 and 34, ask students what shapes they see in each diagram, and how the area of each shape can be represented.

### Extra Practice

See page R8 for extra exercises for students who are approaching level or for on-level students who need additional reinforcement.

### Levels of Complexity Chart

The levels of the exercises progress from 1 to 3, with Level 1 indicating the lowest level of complexity.

| Exercises | 12–30 | 31–43, 50–59 | 44–49 |
|---|---|---|---|
| Level 3 | | | ● |
| Level 2 | | ● | |
| Level 1 | ● | | |

### Watch Out!

**Common Errors** When students multiply polynomials horizontally, they often try to combine terms that are not like terms. For students who are having difficulty finding the products in Exercises 12–23, suggest that they try multiplying the polynomials in vertical form, aligning like terms.

## Go Online!    eBook

**Interactive Student Guide**
Use the *Interactive Student Guide* to deepen conceptual understanding.
· Multiplying Polynomials

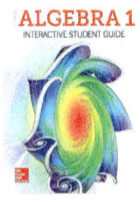

## Lesson 8-3 | Multiplying Polynomials

### Teaching the Mathematical Practices

**Regularity** Mathematically proficient students look both for general methods and for shortcuts. In Exercise 48, tell students to start by multiplying a three-digit number and a two-digit number to analyze the procedure.

## Assess

### Name the Math
Ask students to name the mathematical procedures they use when multiplying two binomials with the FOIL method.

### Additional Answers

37. $a^2 - 4ab + 4b^2$
38. $9c^2 + 24cd + 16d^2$
39. $x^2 - 10xy + 25y^2$
40. $8r^3 - 36r^2t + 54rt^2 - 27t^3$
41. $125g^3 + 150g^2h + 60gh^2 + 8h^3$
42. $64y^3 - 48y^2z - 36yz^2 + 27z^3$
43a. $x > 4$; If $x = 4$ the width of the rectangular sandbox would be zero and if $x < 4$ the width of the rectangular sandbox would be negative.
44b. The first term of the square of a sum is the first term of the sum squared. The middle term of the sum is two times the first term of the sum multiplied by the last term of the sum. The third term of the square of the sum is the last term of the sum squared.
48. The three monomials that make up the trinomial are similar to the three digits that make up the 3-digit number. The single monomial is similar to a 1-digit number. With each procedure you perform 3 multiplications. The difference is that polynomial multiplication involves variables and the resulting product is often the sum of two or more monomials, while numerical multiplication results in a single number.

### Go Online!

**eSolutions Manual**
Create worksheets, answer keys, and solutions handouts for your assignments.

510 | Lesson 8-3 | Multiplying Polynomials

---

35. **VOLLEYBALL** The dimensions of a sand volleyball court are represented by a width of $6y - 5$ feet and a length of $3y + 4$ feet.
    a. Write an expression that represents the area of the court. $18y^2 + 9y - 20$
    b. The length of a sand volleyball court is 31 feet. Find the area of the court. $1519 \text{ ft}^2$

36. **GEOMETRY** Write an expression for the area of a triangle with a base of $2x + 3$ and a height of $3x - 1$. $3x^2 + \frac{7}{2}x - \frac{3}{2}$

**Find each product.** 37–42. See margin.

37. $(a - 2b)^2$
38. $(3c + 4d)^2$
39. $(x - 5y)^2$
40. $(2r - 3t)^3$
41. $(5g + 2h)^3$
42. $(4y + 3z)(4y - 3z)^2$

43. **CONSTRUCTION** A sandbox kit allows you to build a square sandbox or a rectangular sandbox as shown.
    a. What are the possible values of $x$? Explain. See margin.
    b. Which shape has the greater area? square
    c. What is the difference in areas between the two? $4 \text{ ft}^2$

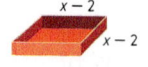

44. **MULTIPLE REPRESENTATIONS** In this problem, you will investigate the square of a sum.
    a. **Tabular** Copy and complete the table for each sum.

    | Expression | (Expression)² |
    |---|---|
    | $x + 5$ | $x^2 + 10x + 25$ |
    | $3y + 1$ | $9y^2 + 6y + 1$ |
    | $z + q$ | $z^2 + 2zq + q^2$ |

    b. **Verbal** Make a conjecture about the terms of the square of a sum. See margin.
    c. **Symbolic** For a sum of the form $a + b$, write an expression for the square of the sum. $a^2 + 2ab + b^2$

45. Always; by grouping two adjacent terms a trinomial can be written as a binomial a sum of two quantities, and apply the FOIL method. For example, $(2x + 3)(x^2 + 5x + 7) = (2x + 3)[x^2 + (5x + 7)] = 2x(x^2) + 2x(5x + 7) + 3(x^2) + 3(5x + 7)$. Then use the Distributive Property and simplify. A.APR.1

### H.O.T. Problems — Use Higher-Order Thinking Skills

45. **ARGUMENTS** Determine if the following statement is *sometimes*, *always*, or *never* true. Explain your reasoning.

    The FOIL method can be used to multiply a binomial and a trinomial.

46. **STRUCTURE** Find $(x^m + x^p)(x^{m-1} - x^{1-p} + x^p)$. $x^{2m-1} - x^{m-p+1} + x^{m+p} + x^{m+p-1} - x + x^{2p}$

47. **CHALLENGE** Write a binomial and a trinomial involving a single variable. Then find their product. Sample answer: $x - 1$, $x^2 - x - 1$; $(x - 1)(x^2 - x - 1) = x^3 - 2x^2 + 1$

48. **REGULARITY** Compare and contrast the procedure used to multiply a trinomial by a binomial using the vertical method with the procedure used to multiply a three-digit number by a two-digit number. See margin.

49. **WRITING IN MATH** Summarize the methods that can be used to multiply polynomials. See margin.

### Standards for Mathematical Practices

| Emphasis On | Exercises |
|---|---|
| 1 Make sense of problems and persevere in solving them. | 1–23 |
| 2 Reason abstractly and quantitatively. | 33–36 |
| 3 Construct viable arguments and critique the reasoning of others. | 44, 45, 49 |
| 4 Model with mathematics. | 24, 43, 50 |
| 5 Use appropriate tools strategically. | 25–30 |
| 6 Attend to precision. | 31, 32, 51 |
| 7 Look for and make use of structure. | 46, 47, 52–53 |
| 8 Look for and express regularity in repeated reasoning. | 1–30, 48, 54–58 |

49. The Distributive Property can be used with a vertical or horizontal format by distributing, multiplying, and combining like terms. The FOIL method is used with a horizontal format. You multiply the first, outer, inner, and last terms of the binomials and then combine like terms. A rectangular method can also be used by writing the terms of the polynomials along the top and left side of a rectangle and then multiplying the terms and combining like terms.

**Lesson 8-3 | Multiplying Polynomials**

## Preparing for Assessment

50. Maria is framing an 8 inch-by-10 inch photograph with matting as shown. **MP 4 A.APR.1 D**

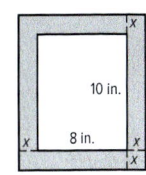

    Which is an expression for the area of the photograph with the mat?
    - A  $4x + 18$
    - B  $4x^2 + 80$
    - C  $4x^2 + 28x + 80$
    - D  $4x^2 + 36x + 80$

51. **MULTI-STEP** Consider a square with side length $x$ and another square with side length $x + 3$. **MP 6 A.APR.1**
    a. Write an expression for the area of each square. $x^2$, $(x+3)^2$ or $x^2 + 6x + 9$
    b. The area of the larger square is 15 square units more than the area of the smaller square. Find the value of $x$. **1**
    c. Use the value of $x$ you found in b to find the area of each square. **1 unit², 16 units²**

52. The area of the shaded region shown below is 214 units².

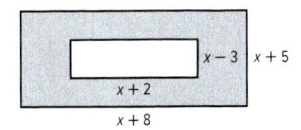

    What is the value of $x$? **MP 7 A.APR.1 12**

53. Which expression is equivalent to the square of the quantity $4x$ plus 8? **MP 7 A.APR.1 D**
    - A  $8x + 16$
    - B  $16x^2 + 64$
    - C  $16x^2 + 24x + 64$
    - D  $16x^2 + 64x + 64$

54. What is the product of $(2x + 5)$ and $(4x - 7)$? **MP 8 A.APR.1 D**
    - A  $6x - 2$
    - B  $8x^2 - 27x - 35$
    - C  $8x^2 - 35$
    - D  $8x^2 + 6x - 35$

55. The sum of two polynomials is $4x^2 + 3x - 1$. One of the polynomials is $2x + 1$. What is the product of the polynomials? **MP 8 A.APR.1**
    
    $8x^3 + 6x^2 - 3x - 2$

56. The side length of a square is $0.5x^2 - 2x$. **MP 8 A.APR.1**
    a. Find the perimeter of the square.
    
    $2x^2 - 8x$
    
    b. Find the area of the square.
    
    $0.25x^4 - 2x^3 + 4x^2$

57. Find each product. **MP 8 A.APR.1**
    a. $(2x - 3)(x^2 - 4x + 1)$
    
    $2x^3 - 11x^2 + 14x - 3$
    
    b. $(3x - 3)(2x^2 - 4x + 1)$
    
    $6x^3 + 16x^2 + 15x - 3$

58. A rectangle has a length of $x^3 - 2x^2 + 5x$ and a width of $2x^2 - 9x - 5$. **MP 8 A.APR.1**
    a. Find the perimeter of the rectangle.
    
    $2x^3 - 8x - 10$
    
    b. Find the area of the rectangle.
    
    $2x^5 - 13x^4 + 23x^3 - 35x^2 - 25x$

59. Bob is twice as old as his sister. The sum of their ages is $6x + 12$. Find the product of their ages. **MP 1 A.APR.1**
    
    $8x^2 + 32x + 32$

## Preparing for Assessment

Exercises 50–59 require students to use the skills they will need on standardized assessments. Each exercise is dual-coded with content standards and mathematical practice standards.

| Dual Coding |||
|---|---|---|
| Items | Content Standards | MP Mathematical Practices |
| 50 | A.APR.1 | 4 |
| 51 | A.APR.1 | 6 |
| 52–53 | A.APR.1 | 7 |
| 54–58 | A.APR.1 | 8 |
| 59 | A.APR.1 | 1 |

### Diagnose Student Errors

Survey student responses for each item. Class trends may indicate common errors and misconceptions.

**50.**
| A | Added length and width |
|---|---|
| B | Multiplied first terms and last terms |
| C | Used $x + 10$ and $x + 8$ as length and width |
| D | CORRECT |

**53.**
| A | Doubled $4x + 8$ |
|---|---|
| B | Squared each term individually |
| C | Added like terms incorrectly |
| D | CORRECT |

**54.**
| A | Found the sum instead of the product |
|---|---|
| B | Added like terms incorrectly |
| C | Multiplied only the like terms |
| D | CORRECT |

### Go Online!

**Quizzes**

Students can use Self-Check Quizzes to check their understanding of this lesson. You can also give the Chapter Quiz, which covers the content in Lessons 8–1 through 8–3.

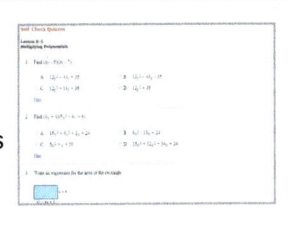

---

**Differentiated Instruction** OL BL

**Extension** Tell students that one way to multiply 25 and 18 mentally is to multiply $(20 + 5)$ by $(20 - 2)$. Have them show how the FOIL method can be used to find each product.

a. **35(19)** $35(19) = (30 + 5)(10 + 9) = (30)(10) + (30)(9) + 5(10) + 5(9) = 300 + 270 + 50 + 45 = 665$

b. **67(102)** $67(102) = (60 + 7)(100 + 2) = (60)(100) + (60)(2) + 7(100) + 7(2) = 6000 + 120 + 700 + 14 = 6834$

# LESSON 8-4
## Special Products

**SUGGESTED PACING (DAYS)**
- 90 min. — 0.5
- 45 min. — 1

Instruction

## Track Your Progress

### Objectives
1. Find squares of sums and differences.
2. Find the product of a sum and a difference.

### Mathematical Background
When multiplying binomials, there are a few special instances where finding the product does not need to involve multiplying each term in the first binomial by each term in the second binomial. Instead, a pattern can be followed, reducing the chance for making a mathematical error.

| THEN | NOW | NEXT |
|---|---|---|
| **A.SSE.2** Use the structure of an expression to identify ways to rewrite it. | **A.APR.1** Understand that polynomials form a system analogous to the integers, namely, they are closed under the operations of addition, subtraction, and multiplication; add, subtract, and multiply polynomials. | **A.SSE.3a** Factor a quadratic expression to reveal the zeros of the function it defines. |

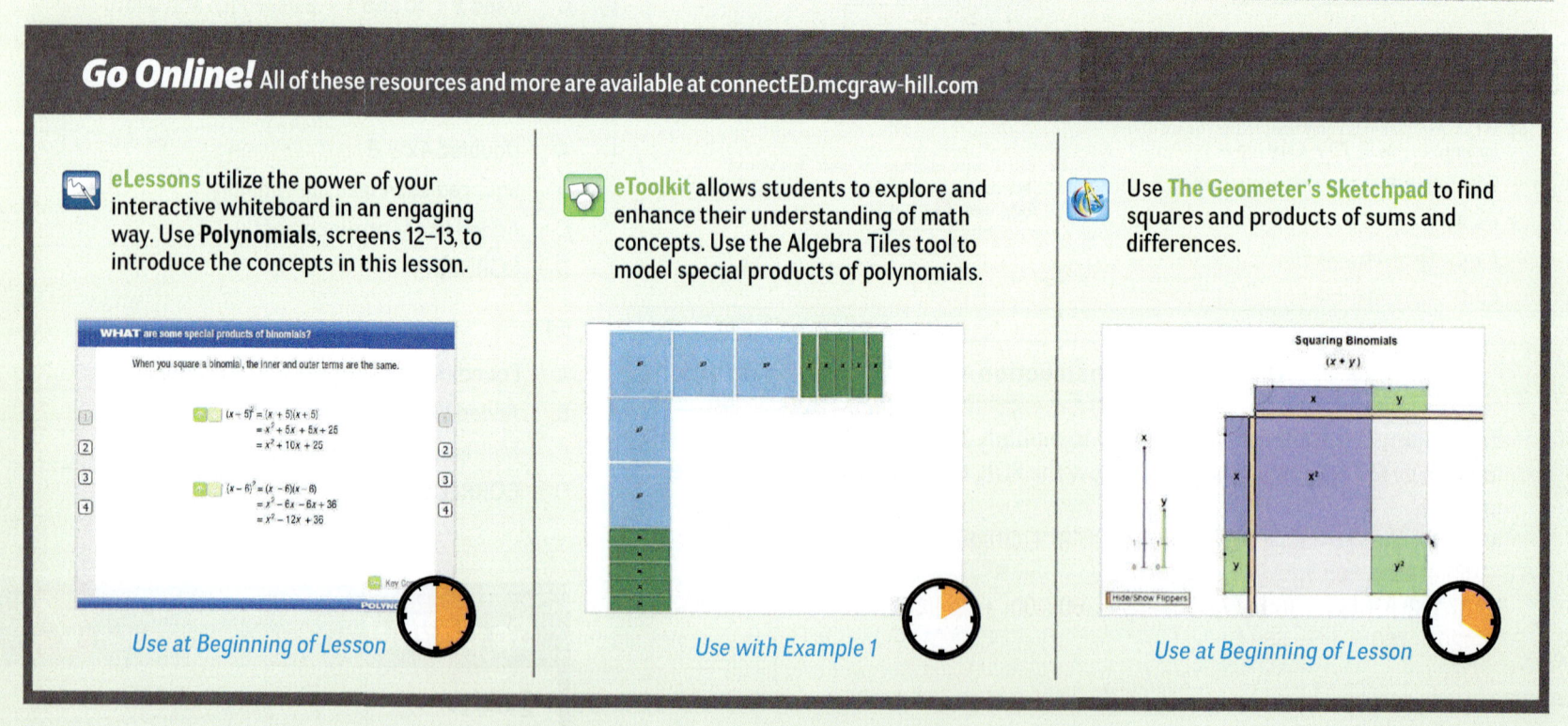

### Go Online! All of these resources and more are available at connectED.mcgraw-hill.com

**eLessons** utilize the power of your interactive whiteboard in an engaging way. Use **Polynomials**, screens 12–13, to introduce the concepts in this lesson.

*Use at Beginning of Lesson*

**eToolkit** allows students to explore and enhance their understanding of math concepts. Use the Algebra Tiles tool to model special products of polynomials.

*Use with Example 1*

Use **The Geometer's Sketchpad** to find squares and products of sums and differences.

*Use at Beginning of Lesson*

### OER Using Open Educational Resources

**Lesson Sharing** Teachers can use **Twitter** to explore new ways of teaching, learn about new apps, or swap materials with other educators on special products. Educators can set up dedicated classroom Twitter accounts and use hashtag chats. Students in flipped classrooms are using Twitter to build out their peer network and work in a more collaborative way. *Use as professional development or flipped learning*

# Go Online!
connectED.mcgraw-hill.com — Worksheets

# Differentiate Your Resources

**Extra Practice** Additional practice or homework; Skills Practice is best for approaching-level students and Practice is best for on-level and beyond-level students

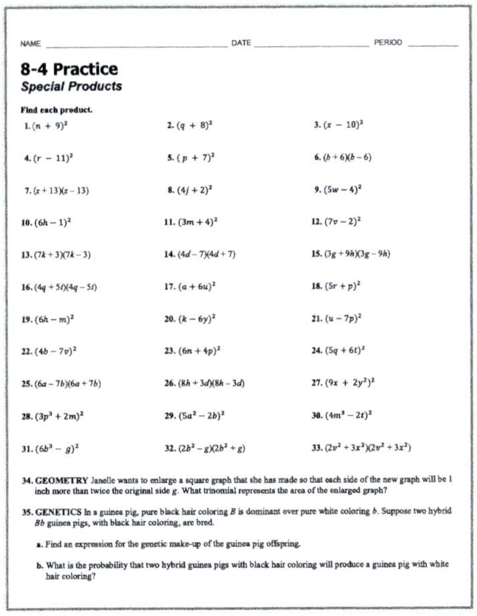

## Skills Practice
AL  OL  ELL

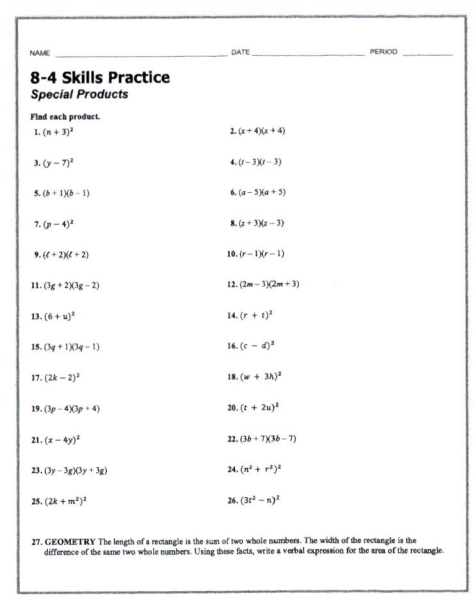

## Practice
AL  OL  BL  ELL

## Word Problem Practice
AL  OL  BL  ELL

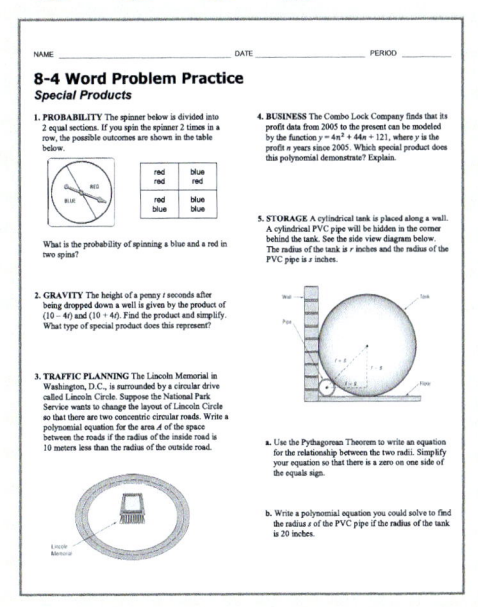

**Intervention** Reteaching and vocabulary activities that can be used with struggling or absent students and as ELL support

## Study Guide and Intervention
AL  OL  ELL

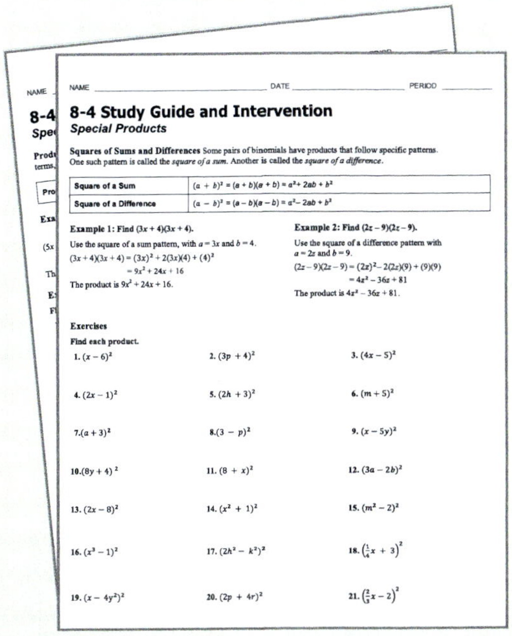

## Study Notebook
AL  OL  BL  ELL

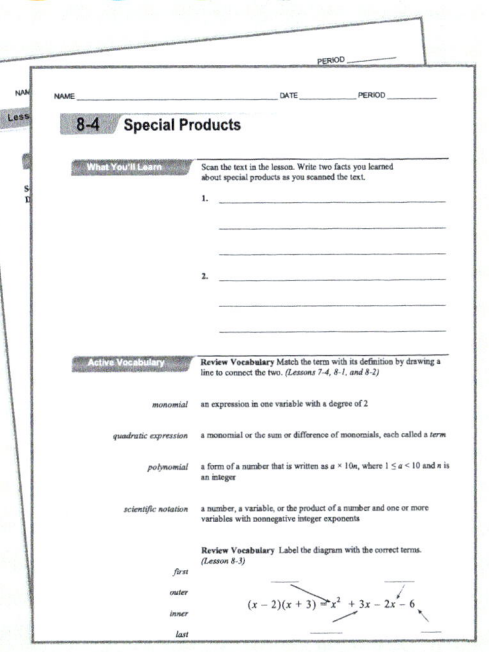

**Extension** Activities that can be used to extend lesson concepts

## Enrichment
OL  BL  ELL

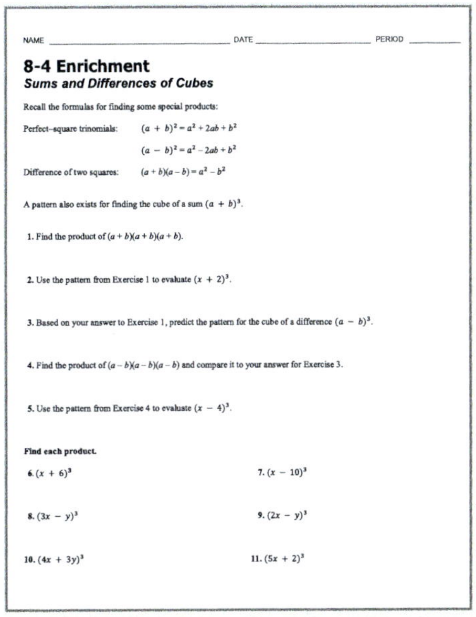

connectED.mcgraw-hill.com   512B

# Lesson 8-4 | Special Products

## Launch

Have students read the Why? section of the lesson. Ask:

- What two factors equal $(2r + 24)^2$ when multiplied? $(2r + 24)(2r + 24)$
- Use the FOIL method to find $(2r + 24)^2$. $4r^2 + 96r + 576$
- If $2r = a$ and $24 = b$, does $(2r + 24)^2$ equal $a^2 + 2ab + b^2$? Explain. Yes; $(2r + 24)^2 = (2r)^2 + 2(2r)(24) + 24^2 = 4r^2 + 96r + 576$.

## Teach

Ask the scaffolded questions for each example to build conceptual understanding for students at all levels.

### 1 Squares of Sums and Differences

**Example 1 Square of a Sum**

**AL** Rewrite $(3x + 5)^2$ without the exponent. $(3x + 5)(3x + 5)$

**OL** What values are substituted for $a$ and $b$? $a: 3x$; $b: 5$

**BL** What could you do if you don't remember the Square of a Sum rule? Sample answer: use FOIL

**Need Another Example?**
Find $(7z + 2)^2$. $49z^2 + 28z + 4$

---

### Go Online!

**Interactive Whiteboard**
Use the *eLesson* or *Lesson Presentation* to present this lesson.

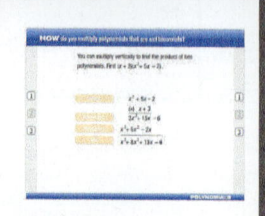

---

## LESSON 4
# Special Products

**Then** • You multiplied binomials by using the FOIL method.

**Now**
1. Find squares of sums and differences.
2. Find the product of a sum and a difference.

**Why?** Colby wants to attach a dartboard to a square piece of corkboard. If the radius of the dartboard is $r + 12$, how large does the square corkboard need to be?
Colby knows that the diameter of the dartboard is $2(r + 12)$ or $2r + 24$. Each side of the square also measures $2r + 24$. To find how much corkboard is needed, Colby must find the area of the square: $A = (2r + 24)^2$.

**Mathematical Practices**
8 Look for and express regularity in repeated reasoning.

**Content Standards**
A.APR.1 Understand that polynomials form a system analogous to the integers, namely, they are closed under the operations of addition, subtraction, and multiplication; add, subtract, and multiply polynomials.

### 1 Squares of Sums and Differences
Some pairs of binomials, such as squares like $(2r + 24)^2$, have products that follow a specific pattern. Using the pattern can make multiplying easier. The square of a sum, $(a + b)^2$ or $(a + b)(a + b)$, is one of those products.

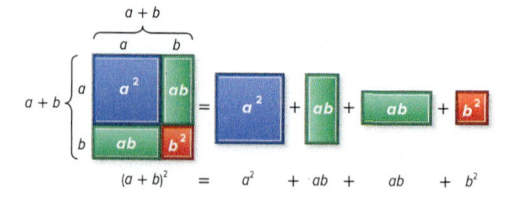

$(a + b)^2 = a^2 + ab + ab + b^2$

**Key Concept** Square of a Sum

**Words** The square of $a + b$ is the square of $a$ plus twice the product of $a$ and $b$ plus the square of $b$.

**Symbols** $(a + b)^2 = (a + b)(a + b)$
$= a^2 + 2ab + b^2$

**Example** $(x + 4)^2 = (x + 4)(x + 4)$
$= x^2 + 8x + 16$

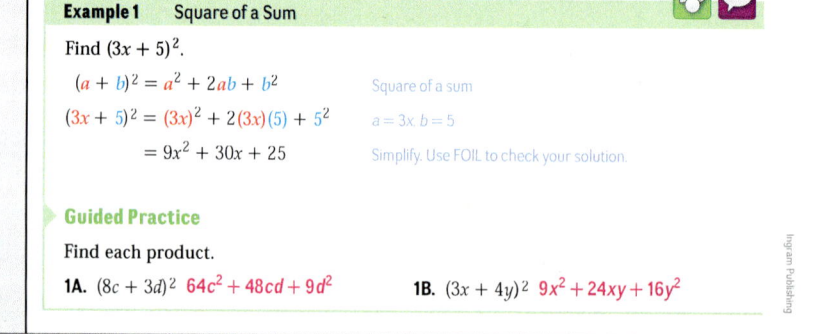

**Example 1** Square of a Sum

Find $(3x + 5)^2$.

$(a + b)^2 = a^2 + 2ab + b^2$    Square of a sum
$(3x + 5)^2 = (3x)^2 + 2(3x)(5) + 5^2$    $a = 3x, b = 5$
$= 9x^2 + 30x + 25$    Simplify. Use FOIL to check your solution.

**Guided Practice**

Find each product.

**1A.** $(8c + 3d)^2$   $64c^2 + 48cd + 9d^2$    **1B.** $(3x + 4y)^2$   $9x^2 + 24xy + 16y^2$

---

### MP Mathematical Practices Strategies

**Look for and express regularity in repeated reasoning.**
Help students maintain oversight of the process of using special products. For example, ask:

- How do you determine the pattern for a special product? The two end terms are perfect squares and the middle term is twice the product of the first and last term in the binomial.

- Why is the difference of squares not a perfect square? Because you have a sum and a difference of two terms in it, not the same term multiplied by itself.

- Can you use FOIL with these special products? Yes, however these special products and their rules can help reduce the amount of time spent solving the problems.

Lesson 8-4 | Special Products

There is also a pattern for the *square of a difference*. Write $a - b$ as $a + (-b)$ and square it using the square of a sum pattern.

$$(a - b)^2 = [a + (-b)]^2$$
$$= a^2 + 2(a)(-b) + (-b)^2 \quad \text{Square of a sum}$$
$$= a^2 - 2ab + b^2 \quad \text{Simplify}$$

**Key Concept** Square of a Difference

Words: The square of $a - b$ is the square of $a$ minus twice the product of $a$ and $b$ plus the square of $b$.

Symbols: $(a - b)^2 = (a - b)(a - b)$
$= a^2 - 2ab + b^2$

Example: $(x - 3)^2 = (x - 3)(x - 3)$
$= x^2 - 6x + 9$

A.APR.1

**Watch Out!**

**MP Regularity** Remember that $(x - 7)^2$ does not equal $x^2 - 7^2$, or $x^2 - 49$.

$(x - 7)^2$
$= (x - 7)(x - 7)$
$= x^2 - 14x + 49$

**Example 2** Square of a Difference

Find $(2x - 5y)^2$.

$(a - b)^2 = a^2 - 2ab + b^2$ Square of a difference
$(2x - 5y)^2 = (2x)^2 - 2(2x)(5y) + (5y)^2$ $a = 2x$ and $b = 5y$
$= 4x^2 - 20xy + 25y^2$ Simplify

**Guided Practice**

Find each product.

2A. $(6p - 1)^2$  $36p^2 - 12p + 1$     2B. $(a - 2b)^2$  $a^2 - 4ab + 4b^2$

The product of the square of a sum or the square of a difference is called a *perfect square trinomial*. We can use these to find patterns to solve real-world problems.

A.APR.1

**Real-World Example 3** Square of a Difference

**PHYSICAL SCIENCE** Each edge of a cube of aluminum is 4 centimeters less than each edge of a cube of copper. Write an equation to model the surface area of the aluminum cube.

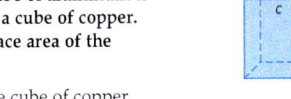

Let $c$ = the length of each edge of the cube of copper. So, each edge of the cube of aluminum is $c - 4$.

$SA = 6s^2$ Formula for surface area of a cube
$SA = 6(c - 4)^2$ Replace $s$ with $c - 4$.
$SA = 6[c^2 - 2(4)(c) + 4^2]$ Square of a difference
$SA = 6(c^2 - 8c + 16)$ Simplify

**Guided Practice**

3. **GARDENING** Alano has a garden that is $g$ feet long and $g$ feet wide. He wants to add 3 feet to the length and the width.
   A. Show how the new area of the garden can be modeled by the square of a binomial. $(g + 3)^2$
   B. Find the square of this binomial. $g^2 + 6g + 9$

**Differentiated Instruction** AL ELL

**IF** students have trouble remembering the pattern for special products studied in this lesson,

**THEN** have them write the symbols for and examples of each Key Concept in this lesson on separate index cards. They can use their note cards for a quick reminder on how to proceed when they are finding products of squares of sums or differences or the product of a sum and a difference.

**Example 2 Square of a Difference**

**AL** Rewrite $(2x - 5y)^2$ without the exponent. $(2x - 5y)(2x - 5y)$

**OL** What values are substituted for $a$ and $b$? $a: 2x; b: 5y$

**BL** Why do we not need to change the subtraction to addition and add the opposite of $5y$? Sample answer: Because the pattern that we follow for the square of a difference includes the subtraction sign, we only need to substitute in the values, so we don't need to worry about forgetting to multiply the subtraction through.

**Need Another Example?**
Find $(3c - 4)^2$. $9c^2 - 24c + 16$

**Teaching Tip**

**Alternative Method** Even though it is important to learn the special products, point out to students that they can always find these products using methods from previous lessons in the chapter.

**Example 3 Square of a Difference**

**AL** What is the formula for the surface area of a cube? $6s^2$

**OL** What shortcut can be used to solve Example 3? square of a difference

**BL** Why do you not need to distribute the 6? Sample answer: factored form is in simplest form

**Need Another Example?**
**Geometry** Write an expression that represents the area of a square that has a side length of $3x + 12$ units. $(9x^2 + 72x + 144)$ units$^2$

**MP Teaching the Mathematical Practices**

**Regularity** Mathematically proficient students notice if calculations are repeated, and look both for general methods and for shortcuts. Encourage students to look for patterns in the Examples and step through finding each pattern before you present the formulas in the Key Concepts.

connectED.mcgraw-hill.com  513

# Lesson 8-4 | Special Products

## 2 Product of a Sum and a Difference

### Example 4 Product of a Sum and a Difference

**AL** Why are the binomials in this example not written as a square? *One has addition and one has subtraction so it is not a binomial being multiplied by itself.*

**OL** How can you check to make sure you used the pattern correctly? *Sample answer: use FOIL to verify the pattern.*

**BL** How might these patterns help us later? *Sample answer: if we have to work backwards we can use the patterns to factor the polynomials.*

**Need Another Example?**
Find $(9d + 4)(9d - 4)$. $81d^2 - 16$

## Practice

**Formative Assessment** Use Exercises 1–11 to assess students' understanding of the concepts in this lesson.

The Practice and Problem Solving exercises assess the content taught in the lesson. The Preparing for Assessment page is meant to be used as preparation for end-of-course-assessment.

### MP Teaching the Mathematical Practices

**Sense-Making** Mathematically proficient students can explain the correspondences between equations, verbal descriptions, and diagrams. In Exercise 45, ask students to explain how to write and simplify the polynomial based on the diagram.

### Exercise Alert

**Formula** For Exercises 46 and 55, students will need to know the formula for the area of a circle, $A = \pi r^2$.

### Go Online! | eBook

**Interactive Student Guide**
Use the *Interactive Student Guide* to deepen conceptual understanding.
- Special Products

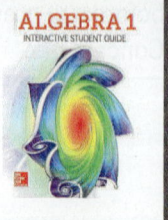

---

**Go Online!** Watch and listen to the animation to see how to model the product of a sum and a difference with algebra tiles. **ELL**

## 2 Product of a Sum and a Difference
Now we will see what the result is when we multiply a sum and a difference, or $(a + b)(a - b)$. Recall that $a - b$ can be written as $a + (-b)$.

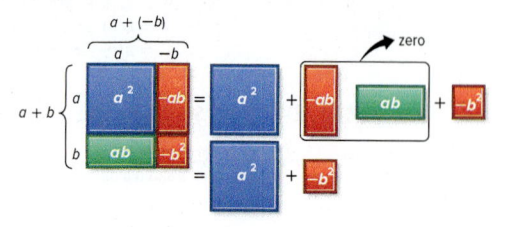

Notice that the middle terms are opposites and add to a zero pair.
So $(a + b)(a - b) = a^2 - ab + ab - b^2 = a^2 - b^2$.

**Study Tip**
**Patterns** When using any of these patterns, $a$ and $b$ can be numbers, variables, or expressions with numbers and variables.

### Key Concept — Product of a Sum and a Difference

| Words | The product of $a + b$ and $a - b$ is the square of $a$ minus the square of $b$. |
|---|---|
| Symbols | $(a + b)(a - b) = (a - b)(a + b)$ $= a^2 - b^2$ |

A.APR.1

### Example 4 Product of a Sum and a Difference

Find $(2x^2 + 3)(2x^2 - 3)$.

$(a + b)(a - b) = a^2 - b^2$   Product of a sum and a difference
$(2x^2 + 3)(2x^2 - 3) = (2x^2)^2 - (3)^2$   $a = 2x^2$ and $b = 3$
$= 4x^4 - 9$   Simplify.

### Guided Practice
Find each product.
**4A.** $(3n + 2)(3n - 2)$  $9n^2 - 4$
**4B.** $(4c - 7d)(4c + 7d)$  $16c^2 - 49d^2$

---

### Check Your Understanding
● = Step-by-Step Solutions begin on page R11.

**Go Online!** for a Self-Check Quiz

**Examples 1–2** Find each product.
A.APR.1
1. $(x + 5)^2$  $x^2 + 10x + 25$
2. $(11 - a)^2$  $121 - 22a + a^2$
3. $(2x + 7y)^2$  3. $4x^2 + 28xy + 49y^2$
4. $(3m - 4)(3m - 4)$
5. $(g - 4h)(g - 4h)$
6. $(3c + 6d)^2$
4. $9m^2 - 24m + 16$  5. $g^2 - 8gh + 16h^2$  6. $9c^2 + 36cd + 36d^2$

**Example 3** 7. **GENETICS** The color of a Labrador retriever's fur is genetic. Dark genes $D$ are dominant over yellow genes $y$. A dog with genes $DD$ or $Dy$ will have dark fur. A dog with genes $yy$ will have yellow fur. Pepper's genes for fur color are $Dy$, and Ramiro's are $Dy$.

|   | D | y |
|---|---|---|
| D | DD | Dy |
| y | Dy | yy |

**a.** Use the square of a sum to write an expression that models the possible fur colors of the dogs' puppies.
**7a.** $D^2 + 2Dy + y^2$
**b.** What is the probability that a puppy will have dark fur?  75%

---

### Follow-Up
Students have explored operations on polynomials.
Ask: Why would you add, subtract, or multiply equations that represent real-world situations? *Sample answer: When you add, subtract, or multiply two or more polynomial equations, like terms are combined, which reduces the number of terms in the resulting equation. This could help minimize the number of operations performed when using the equations.*

### Additional Answers
22. $u^2 - 9$
23. $b^2 - 49$
24. $4 - x^2$
25. $16 - x^2$
26. $4q^2 - 25r^2$
27. $9a^4 - 49b^2$
28. $25y^2 + 70y + 49$
29. $64 - 160a + 100a^2$
30. $100x^2 - 4$
31. $9t^2 - 144$
32. $a^2 + 8ab + 16b^2$
33. $9q^2 - 30qr + 25r^2$

Lesson 8-4 | Special Products

**Example 4**
A.APR.1

Find each product.

8. $(a-3)(a+3)$  $a^2 - 9$
9. $(x+5)(x-5)$  $x^2 - 25$
10. $(6y-7)(6y+7)$  $36y^2 - 49$
11. $(9t+6)(9t-6)$  $81t^2 - 36$

## Practice and Problem Solving

Extra Practice is on page R8.

**Examples 1–2**
A.APR.1

Find each product.

12. $(a+10)(a+10)$  $a^2 + 20a + 100$
13. $(b-6)(b-6)$  $b^2 - 12b + 36$
14. $(h+7)^2$  $h^2 + 14h + 49$
15. $(x+6)^2$  $x^2 + 12x + 36$
16. $(8-m)^2$  $64 - 16m + m^2$
17. $(9-2y)^2$  $81 - 36y + 4y^2$
18. $(2b+3)^2$  $4b^2 + 12b + 9$
19. $(5t-2)^2$  $25t^2 - 20t + 4$
20. $(8h-4n)^2$  $64h^2 - 64hn + 16n^2$

**Example 3**
A.APR.1

21. **GENETICS** The ability to roll your tongue is inherited genetically if either parent has the dominant trait T.

    a. Show how the combinations can be modeled by the square of a sum. $(T+t)^2 = T^2 + 2Tt + t^2$

    |   | T  | t  |
    |---|----|----|
    | T | TT | Tt |
    | t | Tt | tt |

    b. Predict the percent of children that will have both dominant genes, one dominant gene, and both recessive genes. TT: 25%; Tt: 50%; tt: 25%

**Example 4**
A.APR.1

Find each product. 22–44. See margin.

22. $(u+3)(u-3)$
23. $(b+7)(b-7)$
24. $(2+x)(2-x)$
25. $(4-x)(4+x)$
26. $(2q+5r)(2q-5r)$
27. $(3a^2+7b)(3a^2-7b)$
28. $(5y+7)^2$
29. $(8-10a)^2$
30. $(10x-2)(10x+2)$
31. $(3t+12)(3t-12)$
32. $(a+4b)^2$
33. $(3q-5r)^2$
34. $(2c-9d)^2$
35. $(g+5h)^2$
36. $(6y-13)(6y+13)$
37. $(3a^4-b)(3a^4+b)$
38. $(5x^2-y^2)^2$
39. $(8a^2-9b^3)(8a^2+9b^3)$
40. $\left(\frac{3}{4}k+8\right)^2$
41. $\left(\frac{2}{5}y-4\right)^2$
42. $(7z^2+5y^2)(7z^2-5y^2)$
43. $(2m+3)(2m-3)(m+4)$
44. $(r+2)(r-5)(r-2)(r+5)$

45. **SENSE-MAKING** Write a polynomial that represents the area of the figure at the right. $2x^2 + 2x + 5$

46. **MULTI-STEP** A flying disk has a radius of 4.9 inches. Jolanda discovers that adding a hole with a radius of 3.75 inches to the center of the disk reduces the weight of the disk, so it travels farther. Jolanda wants to experiment with the size of the disk to increase flying distance. In the new disks she creates, the sizes of the disks and holes are increased or decreased by the same amount.

    a. Write an expression for the area of the top of the new flying disks. $\pi(2.3x + 9.9475)$
    b. Explain your solution process. See margin.

**GEOMETRY** Find the area of each shaded region.

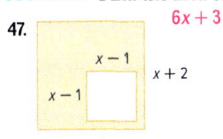

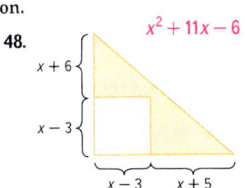

47. (figure: outer square side $x+2$, inner square side $x-1$, with labels $6x+3$ and $x-1$, $x+2$)

48. (figure: triangle with segments $x+6$, $x-3$, and base divided as $x-3$, $x+5$; area $x^2 + 11x - 6$)

## Differentiated Homework Options

| Levels | AL Basic | OL Core | BL Advanced |
|---|---|---|---|
| Exercises | 12–28, 56, 57, 60–70 | 13–53 odd, 55–57, 60–70 | 46–61, (optional: 62–70) |
| 2-Day Option | 13–27 odd, 62–70 | 12–45, 62–70 | |
|  | 12–28 even, 56, 57, 60, 61 | 46–57, 60, 61 | |

You can use ALEKS to provide additional remediation support with personalized instruction and practice.

## Extra Practice

See page R8 for extra exercises for students who are approaching level or for on-level students who need additional reinforcement.

### Levels of Complexity Chart

The levels of the exercises progress from 1 to 3, with Level 1 indicating the lowest level of complexity.

| Exercises | 12–27 | 28–55, 62–70 | 56–61 |
|---|---|---|---|
| Level 3 |  |  | ● |
| Level 2 |  | ● |  |
| Level 1 | ● |  |  |

## Additional Answers

34. $4c^2 - 36cd + 81d^2$
35. $g^2 + 10gh + 25h^2$
36. $36y^2 - 169$
37. $9a^8 - b^2$
38. $25x^4 - 10x^2y^2 + y^4$
39. $64a^4 - 81b^6$
40. $\frac{9}{16}k^2 + 12k + 64$
41. $\frac{4}{25}y^2 - \frac{16}{5}y + 16$
42. $49z^4 - 25y^4$
43. $4m^3 + 16m^2 - 9m - 36$
44. $r^4 - 29r^2 + 100$

46b. I am given the radius of the original disk and I know that Jolanda is going to increase the size by some increment x. So the radius of each new disk is given by the equation $r = x + 4.9$. The size of the hole cut in the original disk has a radius of 3.75 inches and increases as the size of the disk increases. The equation for the radius of the hole is $r = x + 3.75$. We then find the area for the disk without a hole and subtract the area of the hole to find the final answer.

## Go Online!

### eSolutions Manual

Create worksheets, answer keys, and solutions handouts for your assignments.

**Lesson 8-4 | Special Products**

---

### Teaching the Mathematical Practices

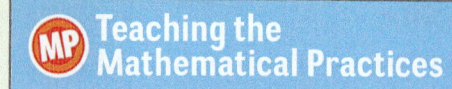

**Structure** Mathematically proficient students look closely to discern a pattern or structure. In Exercise 58a, point out that $(a + b)(a + b)(a + b)$ can be rewritten as $(a + b)(a^2 + 2ab + b^2)$.

### Exercise Alerts

**Construction Paper** For Exercise 56, students will need a square piece of construction paper.

**Isometric Dot Paper** For Exercise 58, students may want to use isometric dot paper to draw their models for the cube of a sum.

## Assess

**Ticket Out the Door** Make several copies each of five squares of sums that need to be multiplied. Give one expression to each student. As the students leave the room, ask them to tell you the products of the expressions.

### Additional Answers

**56a.** The area of the larger square is $a^2$. The area of the smaller square is $b^2$.

**57.** Sample answer: $(2c + d) \cdot (2c - d)$; The product of these binomials is a difference of two squares and does not have a middle term. The other three do.

**61.** Sample answer: To find the square of a sum, apply the FOIL method or apply the pattern. The square of the sum of two quantities is the first quantity squared plus two times the product of the two quantities plus the second quantity squared. The square of the difference of two quantities is the first quantity squared minus two times the product of the two quantities plus the second quantity squared. The product of the sum and difference of two quantities is the square of the first quantity minus the square of the second quantity.

---

Find each product. **49.** $c^3 + 3c^2d + 3cd^2 + d^3$ **50.** $8a^3 - 12a^2b + 6ab^2 - b^3$ **51.** $f^3 + f^2g - fg^2 - g^3$

**49.** $(c + d)(c + d)(c + d)$ **50.** $(2a - b)^3$ **51.** $(f + g)(f - g)(f + g)$

**52.** $(k - m)(k + m)(k - m)$ **53.** $(n - p)^2(n + p)$ **54.** $(q + r)^2(q - r)$ $q^3 + q^2r - qr^2 - r^3$
$k^3 - k^2m - km^2 + m^3$ $n^3 - n^2p - np^2 + p^3$

**55. WRESTLING** A high school wrestling mat must be a square with 38-foot sides and contain two circles as shown. Suppose the inner circle has a radius of $r$ feet, and the radius of the outer circle is 9 feet longer than the inner circle. **55a.** about $(3.14r^2 + 56.52r + 254.34)$ ft$^2$

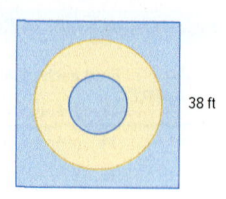

a. Write an expression for the area of the larger circle.

b. Write an expression for the area of the portion of the square outside the larger circle.
about $(1189.66 - 3.14r^2 - 56.52r)$ ft$^2$

**56. MULTIPLE REPRESENTATIONS** In this problem, you will investigate a pattern. Begin with a square piece of construction paper. Label each edge of the paper $a$. In any of the corners, draw a smaller square and label the edges $b$.

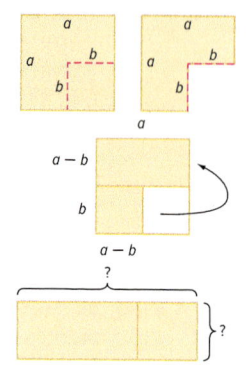

a. **Numerical** Find the area of each of the squares. See margin.

b. **Concrete** Cut the smaller square out of the corner. What is the area of the shape? $a^2 - b^2$

c. **Analytical** Remove the smaller rectangle on the bottom. Turn it and slide it next to the top rectangle. What is the length of the new arrangement? What is the width? What is the area? $a + b, a - b, (a + b)(a - b)$

d. **Analytical** What pattern does this verify?
$(a + b)(a - b) = a^2 - b^2$

A.APR.1

**H.O.T. Problems** Use Higher-Order Thinking Skills

**57. WHICH ONE DOESN'T BELONG?** Which expression does not belong? Explain. See margin.

| $(2c - d)(2c - d)$ | $(2c + d)(2c - d)$ | $(2c + d)(2c + d)$ | $(c + d)(c + d)$ |

**58. CHALLENGE** Does a pattern exist for $(a + b)^3$?

a. Investigate this question by finding the product $(a + b)(a + b)(a + b)$. $a^3 + 3a^2b + 3ab^2 + b^3$

b. Use the pattern you discovered in part **a** to find $(x + 2)^3$. $x^3 + 6x^2 + 12x + 8$

c. Draw a diagram of a geometric model for $(a + b)^3$.

d. What is the pattern for the cube of a difference, $(a - b)^3$? $a^3 - 3a^2b + 3ab^2 - b^3$

**59. CHALLENGE** Find $c$ such that $25x^2 - 90x + c$ is a perfect square trinomial. **81**

**58c.** Sample answer:

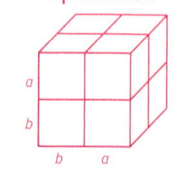

**60. OPEN-ENDED** Write two binomials with a product that is a binomial. Then write two binomials with a product that is not a binomial.
Sample answer: $(x - 2)(x + 2) = x^2 - 4$ and $(x - 2)(x - 2) = x^2 - 4x + 4$

**61. WRITING IN MATH** Describe how to square the sum of two quantities, square the difference of two quantities, and how to find the product of a sum of two quantities and a difference of two quantities.
See margin.

---

### Standards for Mathematical Practice

| Emphasis On | Exercises |
| --- | --- |
| 1 Make sense of problems and persevere in solving them. | 1–20, 63, 68 |
| 2 Reason abstractly and quantitatively. | 47–54, 70 |
| 3 Construct viable arguments and critique the reasoning of others. | 55, 56, 64 |
| 4 Model with mathematics. | 21, 46, 62 |
| 5 Use appropriate tools strategically. | 22–45 |
| 6 Attend to precision. | 37–44, 64 |
| 7 Look for and make use of structure. | 57–59, 66, 69 |
| 8 Look for and express regularity in repeated reasoning. | 60, 61, 65, 67–68 |

Lesson 8-4 | Special Products

## Preparing for Assessment

62. Danielle is making a quilt with triangles and squares. There are 4 colored triangles inside each square. The base and height of each triangle is $x + 4$ inches as shown. Which polynomial represents the area of the colored triangles in square inches? MP 4 A.APR.1  **C**

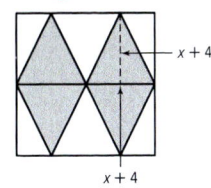

- A  $\frac{1}{2}x + 2$
- B  $\frac{1}{2}x^2 + 4x + 8$
- C  $2x^2 + 16x + 32$
- D  $4x^2 + 32x + 64$

63. **MULTI-STEP** Write an expression the for the area of each figure. MP 1 A.APR.1

   a. a square with a side length of $2x + 4$ units

   $x^2 + 16x + 16$

   b. a rectangle with a length of $5x + 3$ units and a width of $5x - 3$ units

   $(25x^2 - 9)$

   c. What is the difference in area between the square and rectangle?

   $-21x^2 + 16x + 7$

64. How do the patterns used to find the square of a sum and the square of a difference differ? MP 3, 6 A.APR.1

   The middle term, $2ab$, is added when finding the square of a sum and subtracted when finding the square of a difference.

65. For a right triangle, the Pythagorean theorem says the sum of the squares of the lengths of the legs equals the square of the hypotenuse, or $a^2 + b^2 = c^2$.

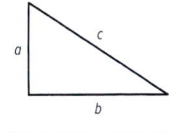

   The legs of a right triangle have lengths of $2x + 3$ and $4x - 1$. Which polynomial is equivalent to $c^2$? MP 8 A.APR.1  **C**

   - A  $6x + 2$
   - B  $4x^2 + 12x + 9$
   - C  $20x^2 + 4x + 10$
   - D  $36x^2 + 24x + 4$

66. The top of a square coffee table has an area of $64x^2 - 48x + 9$. Which is the perimeter of the tabletop? MP 7 A.APR.1  **C**

   - A  $8x - 3$
   - B  $16x - 9$
   - C  $32x - 12$
   - D  $20x$

67. What is the expanded form of $\left(\frac{1}{5} + 4x\right)^2$? MP 8 A.APR.1  **C**

   - A  $8x + \frac{2}{5}$
   - B  $16x^2 + \frac{8}{10}x + \frac{1}{10}$
   - C  $16x^2 + \frac{8}{5}x + \frac{1}{25}$
   - D  $16x^2 + \frac{8}{5}x + \frac{1}{10}$

68. Find each product. MP 1, 8 A.APR.1

   a. $(2x - 3)^2$  $4x^2 - 12x + 9$
   b. $(2x + 3)^2$  $4x^2 + 12x + 9$
   c. $(2x + 3)(2x - 3)$  $4x^2 - 9$

69. A square field has perimeter $4x^2 + 8$ yards. Find the area of this field. MP 7 A.APR.1  $x^4 + 4x^2 + 4$ yd$^2$

70. The area of an isosceles right triangle is $8x^2 + 4x + 0.5$. Find the square of the length of the hypotenuse. MP 2 A.APR.1  $32x^2 + 16x + 2$

### Differentiated Instruction  OL  BL

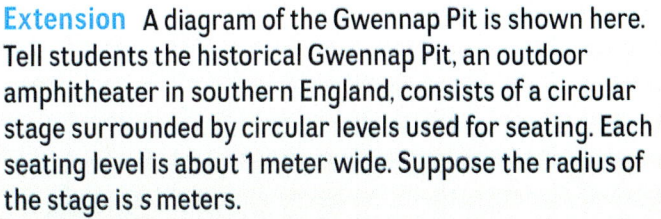

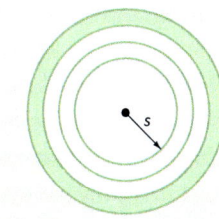

**Extension** A diagram of the Gwennap Pit is shown here. Tell students the historical Gwennap Pit, an outdoor amphitheater in southern England, consists of a circular stage surrounded by circular levels used for seating. Each seating level is about 1 meter wide. Suppose the radius of the stage is $s$ meters.

a. Find binomial representations for the radii of the second and third seating levels. $s + 2, s + 3$

b. Find the area of the shaded region representing the third seating level.
about $(6.3s + 15.7)$ m$^2$

67.

| | |
|---|---|
| A | Multiplied each term by 2 instead of finding the square |
| B | Added the denominators of the fractions |
| C | CORRECT |
| D | Multiplied the denominator of $\frac{1}{5}$ by 2 instead of squaring |

## Preparing for Assessment

Exercises 62–70 require students to use the skills they will need on standardized assessments. Each exercise is dual-coded with content standards and mathematical practice standards.

| Dual Coding |||
|---|---|---|
| Items | Content Standards | MP Mathematical Practices |
| 62 | A.APR.1 | 4 |
| 63 | A.APR.1 | 1 |
| 64 | A.APR.1 | 3, 6 |
| 65, 67, 68 | A.APR.1 | 8 |
| 66, 69 | A.APR.1 | 7 |
| 70 | A.APR.1 | 2 |

### Diagnose Student Errors

Survey student responses for each item. Class trends may indicate common errors and misconceptions.

62.

| | |
|---|---|
| A | Found $\frac{1}{2}b$ for one triangle |
| B | Found area of one triangle |
| C | CORRECT |
| D | Did not divide by 2 when finding the area of one triangle |

65.

| | |
|---|---|
| A | Added the side lengths |
| B | Squared only the length $2x + 3$ |
| C | CORRECT |
| D | Squared the sum of the lengths instead of finding the sum of the squares of the lengths |

66.

| | |
|---|---|
| A | Found the length of each side of the square |
| B | Found the incorrect side length |
| C | CORRECT |
| D | Subtracted 3 from $8x$ and got $5x$, then multiplied $5x$ by 4 |

### Go Online!

**Quizzes**
Students can use *Self-Check Quizzes* to check their understanding of this lesson.

## Chapter 8 Mid-Chapter Quiz

### RtI Response to Intervention
Use the Intervention Planner to help you determine your Response to Intervention.

**Intervention Planner**

**TIER 1 — On Level (OL)**

IF students miss 25% of the exercises or less,
THEN choose a resource:
- SE Lessons 8-1, 8-2, 8-3, and 8-4

**Go Online!**
- Skills Practice
- Chapter Project
- Self-Check Quizzes

**TIER 2 — Strategic Intervention (AL)**
Approaching grade level

IF students miss 50% of the exercises,
THEN choose a resource:
- Quick Review Math Handbook

**Go Online!**
- Study Guide and Intervention
- Extra Examples
- Personal Tutors
- Homework Help

**TIER 3 — Intensive Intervention**
2 or more grades below level

IF students miss 75% of the exercises,
THEN choose a resource:
- Use *Math Triumphs, Alg. 1*

**Go Online!**
- Extra Examples
- Personal Tutors
- Homework Help
- Review Vocabulary

### Go Online!

**eAssessment**

You can use the premade Mid-Chapter Test to assess students' progress in the first half of the chapter. Customize and create multiple versions of your Mid-Chapter Quiz and answer keys that align to your standards can be delivered on paper or online.

---

## CHAPTER 8 Mid-Chapter Quiz
### Lessons 8-1 through 8-4

Determine whether each expression is a polynomial. If it is a polynomial, find the degree and determine whether it is a *monomial*, *binomial*, or *trinomial*. (Lesson 8-1)

1. $3y^2 - 2$  **yes; 2; binomial**
2. $4t^5 + 3t^2 + t$  **yes; 5; trinomial**
3. $\frac{3x}{5y}$  **not a polynomial**
4. $ax^{-3}$  **not a polynomial**
5. $3b^2$  **yes; 2; monomial**
6. $2x^{-3} - 4x + 1$  **not a polynomial**

7. **POPULATION** The table shows the population density for Nevada for various years. (Lesson 8-1)

| Year | Years Since 1930 | People/Square Mile |
|------|------------------|--------------------|
| 1930 | 0  | 0.8  |
| 1960 | 30 | 2.6  |
| 1980 | 50 | 7.3  |
| 1990 | 60 | 10.9 |
| 2010 | 80 | 24.6 |

a. The population density $d$ of Nevada from 1930 to 2010 can be modeled by $d = 0.005n^2 - 0.142n + 1$, where $n$ represents the number of years since 1930. Classify the polynomial. **quadratic trinomial**
b. What is the degree of the polynomial? **2**
c. Predict the population density of Nevada for 2020 and for 2030. Explain your method. **See margin.**
d. MP What mathematical practice did you use to solve this problem? **See students' work.**

Find each sum or difference. (Lesson 8-1)

8. $(y^2 + 2y + 3) + (y^2 + 3y - 1)$  **$2y^2 + 5y + 2$**
9. $(3n^3 - 2n + 7) - (n^2 - 2n + 8)$  **$3n^3 - n^2 - 1$**
10. $(5d + d^2) - (4 - 4d^2)$  **$5d^2 + 5d - 4$**
11. $(x + 4) + (3x + 2x^2 - 7)$  **$2x^2 + 4x - 3$**
12. $(3a - 3b + 2) - (4a + 5b)$  **$-a - 8b + 2$**
13. $(8x - y^2 + 3) + (9 - 3x + 2y^2)$  **$5x + y^2 + 12$**

Find each product. (Lesson 8-2)

14. $6y(y^2 + 3y + 1)$  **$6y^3 + 18y^2 + 6y$**
15. $3n(n^2 - 5n + 2)$  **$3n^3 - 15n^2 + 6n$**
16. $d^2(-4 - 3d + 2d^2)$  **$-4d^2 - 3d^3 + 2d^4$**
17. $-2xy(3x^2 + 2xy - 4y^2)$  **$-6x^3y - 4x^2y^2 + 8xy^3$**
18. $ab^2(12a + 5b - ab)$  **$12a^2b^2 + 5ab^3 - a^2b^3$**
19. $x^2y^4(3xy^2 - x + 2y^2)$  **$3x^3y^6 - x^3y^4 + 2x^2y^6$**

20. **MULTIPLE CHOICE** Simplify $x(4x + 5) + 3(2x^2 - 4x + 1)$. (Lesson 8-2) **B**
  A $10x^2 + 17x + 3$
  B $10x^2 - 7x + 3$
  C $2x^2 - 7x + 3$
  D $2x^2 + 17x + 3$

Find each product. (Lesson 8-3)

21. $(x + 2)(x + 5)$  **$x^2 + 7x + 10$**
22. $(3b - 2)(b - 4)$  **$3b^2 - 14b + 8$**
23. $(n - 5)(n + 3)$  **$n^2 - 2n - 15$**
24. $(4c - 2)(c + 2)$  **$4c^2 + 6c - 4$**
25. $(k - 1)(k - 3k^2)$  **$-3k^3 + 4k^2 - k$**
26. $(8d - 3)(2d^2 + d + 1)$  **$16d^3 + 2d^2 + 5d - 3$**

27. **MANUFACTURING** A company is designing a box for dry pasta in the shape of a rectangular prism. The length is 2 inches more than twice the width, and the height is 3 inches more than the length. Write an expression, in terms of the width, for the volume of the box. (Lesson 8-3) **$4w^3 + 14w^2 + 10w$**

Find each product. (Lesson 8-4)

28. $(x + 2)^2$  **$x^2 + 4x + 4$**
29. $(n - 11)^2$  **$n^2 - 22n + 121$**
30. $(4b - 2)^2$  **$16b^2 - 16b + 4$**
31. $(6c + 3)^2$  **$36c^2 + 36c + 9$**
32. $(5d - 3)(5d + 3)$  **$25d^2 - 9$**
33. $(9k + 1)(9k - 1)$  **$81k^2 - 1$**

34. **DISC GOLF** The discs approved for use in disc golf vary in size. (Lesson 8-4)

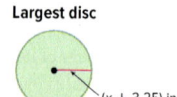

Smallest disc: $x$ in.
Largest disc: $(x + 3.25)$ in.

a. Write two different expressions for the area of the largest disc. **34a. $\pi(x + 3.25)^2$ in$^2$, $(\pi x^2 + 6.5\pi x + 10.5625\pi)$ in$^2$**
b. If $x$ is 10.5, what are the areas of the smallest and largest discs? **346.4 in$^2$; 594.0 in$^2$**

---

## Foldables Study Organizer

**Dinah Zike's FOLDABLES**

Before students complete the Mid-Chapter Quiz, encourage them to review the information for Lessons 8-1 through 8-4 in their Foldables. Ask students to share the items they have added to their Foldables that have been helpful as they study Chapter 8.

ALEKS can be used as a formative assessment tool to target learning gaps for those who are struggling, while providing enhanced learning for those who have mastered the concepts.

### Additional Answer

7c. 28.72 people/mi$^2$; Since 2020 is 90 years after 1930, substitute 90 for $n$ in the equation; 36.8 people/mi$^2$; Since 2030 is 100 years after 1930, substitute 100 for $n$ in the equation.

# LESSON 8-5
# Using the Distributive Property

**SUGGESTED PACING (DAYS)**

| | | |
|---|---|---|
| 90 min. | .25 | 0.75 |
| 45 min. | 0.5 | 1.5 |
| | Explore Lab | Instruction |

## Track Your Progress

### Objectives
1. Use the Distributive Property to factor polynomials.
2. Factor polynomials by grouping.

### Mathematical Background
The Distributive Property is used to find the product of a monomial and a polynomial. Reverse the process to factor a polynomial whose terms have a GCF. If the polynomial has four or more terms, factor by grouping.

---

**THEN**

**A.SSE.2** Use the structure of an expression to identify ways to rewrite it.

**NOW**

**A.SSE.2** Use the structure of an expression to identify ways to rewrite it.

**Prep. for A.SSE.3a** Factor a quadratic expression to reveal the zeros of the function it defines.

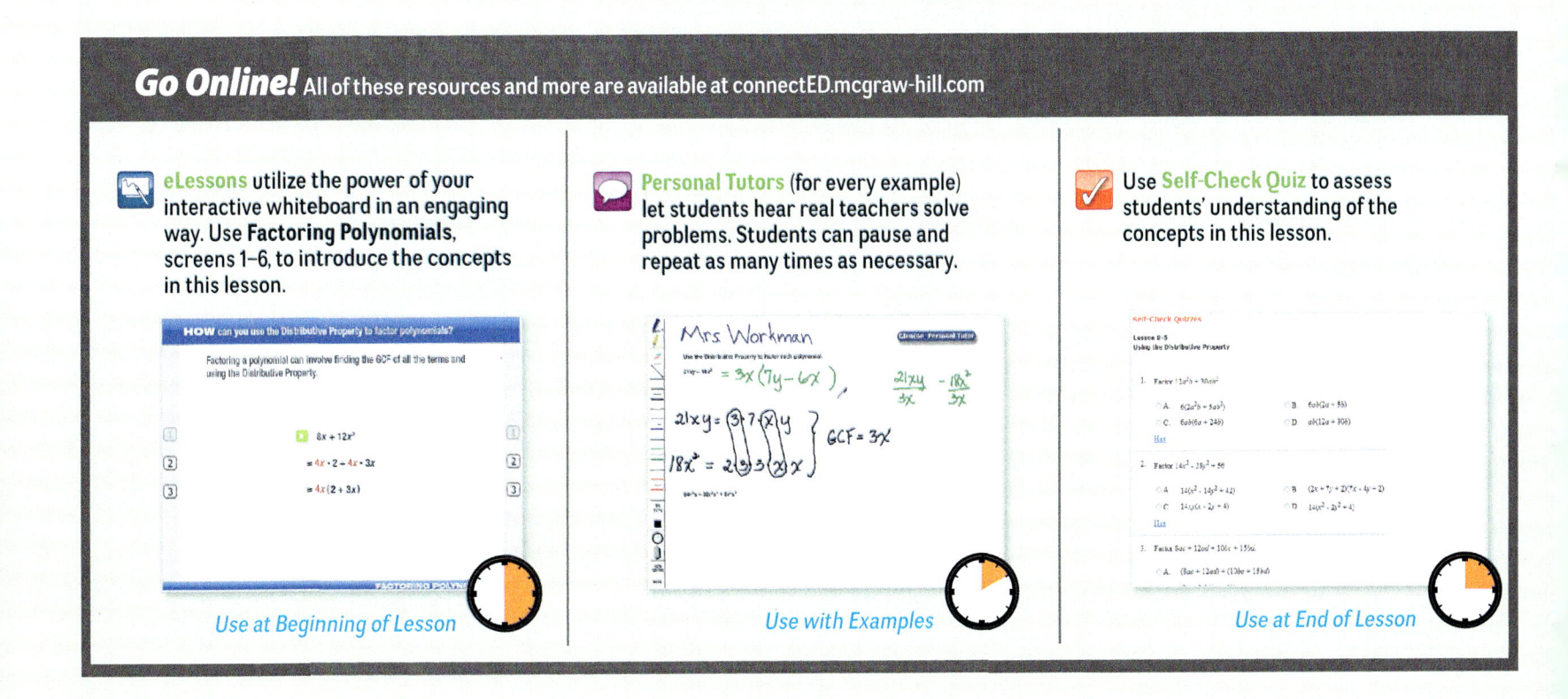

**NEXT**

**A.REI.4** Solve quadratic equations in one variable.

---

## Go Online! All of these resources and more are available at connectED.mcgraw-hill.com

**eLessons** utilize the power of your interactive whiteboard in an engaging way. Use **Factoring Polynomials**, screens 1–6, to introduce the concepts in this lesson.

*Use at Beginning of Lesson*

**Personal Tutors** (for every example) let students hear real teachers solve problems. Students can pause and repeat as many times as necessary.

*Use with Examples*

Use **Self-Check Quiz** to assess students' understanding of the concepts in this lesson.

*Use at End of Lesson*

---

## OER Using Open Educational Resources
**Apps** Have students play **DragonBox Algebra 12+** to practice distribution and factorization. Students can also review algebra skills they learned previously before moving on to factoring. *Use as homework or classwork*

# Go Online!
connectED.mcgraw-hill.com — Worksheets

## Differentiate Your Resources

**Extra Practice** Additional practice or homework; Skills Practice is best for approaching-level students and Practice is best for on-level and beyond-level students

### Skills Practice
AL  OL  ELL

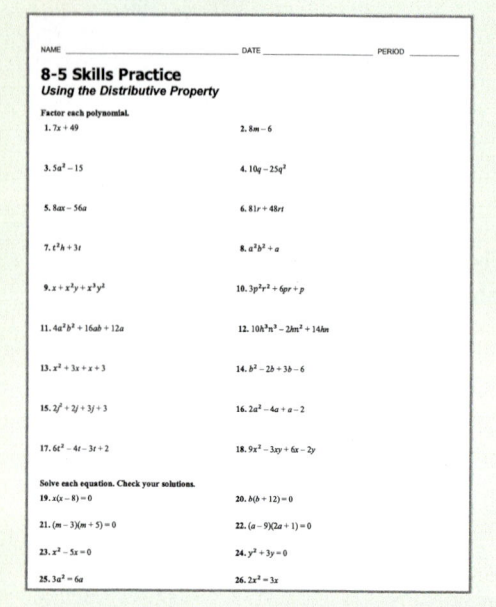

### Practice
AL  OL  BL  ELL

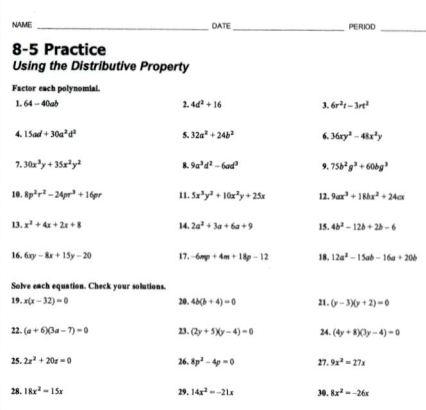

### Word Problem Practice
AL  OL  BL  ELL

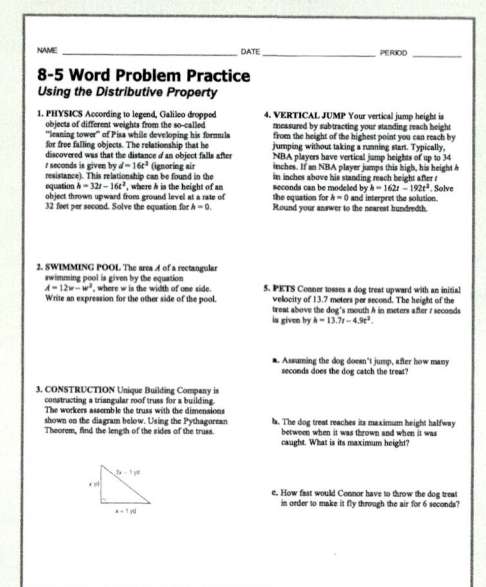

**Intervention** Reteaching and vocabulary activities that can be used with struggling or absent students and as ELL support

**Extension** Activities that can be used to extend lesson concepts

### Study Guide and Intervention
AL  OL  ELL

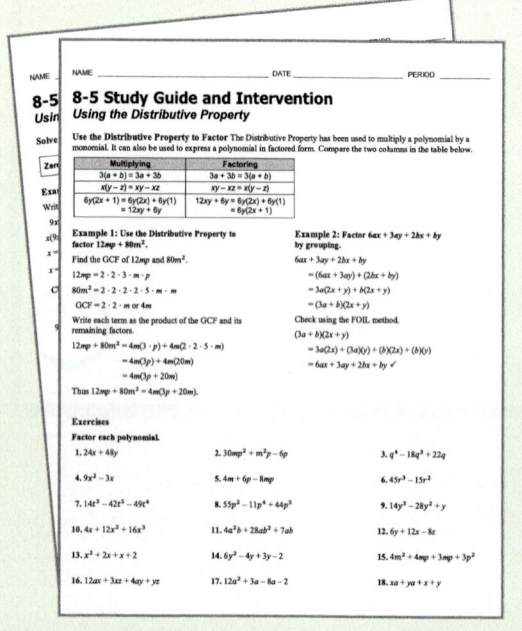

### Study Notebook
AL  OL  BL  ELL

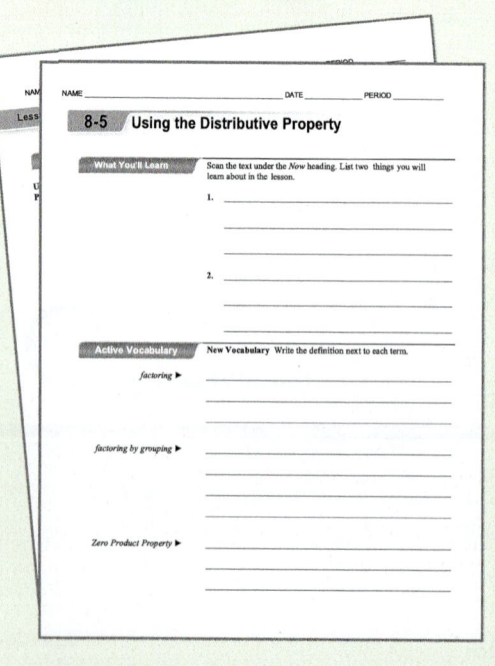

### Enrichment
OL  BL  ELL

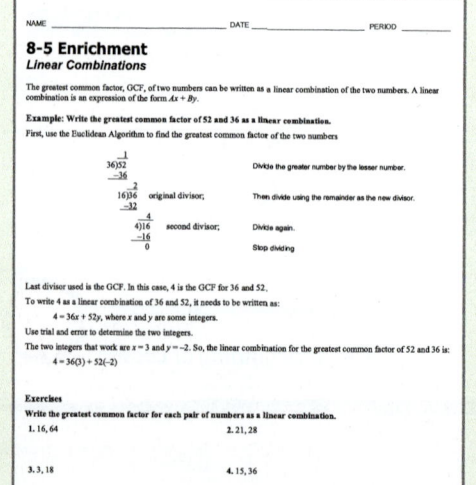

519B | Lesson 8-5 | Using the Distributive Property

# EXPLORE 8-5
## Algebra Lab
# Factoring Using the Distributive Property

When two or more numbers are multiplied, these numbers are *factors* of the product. Sometimes you know the product of binomials and are asked to find the factors. This is called factoring. You can use algebra tiles and a product mat to factor binomials.

**Mathematical Practices**
MP 3 Construct viable arguments and critique the reasoning of others.

**Content Standards**
A.SSE.2 Use the structure of an expression to identify ways to rewrite it.

Work cooperatively to complete Activities 1 and 2.

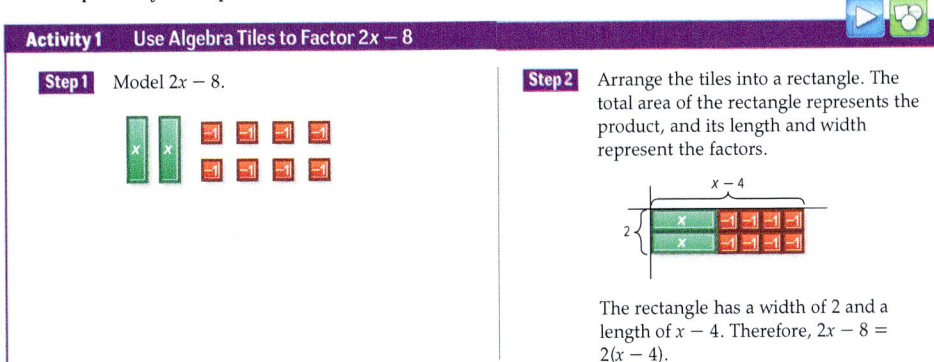

### Activity 1  Use Algebra Tiles to Factor $2x - 8$

**Step 1** Model $2x - 8$.

**Step 2** Arrange the tiles into a rectangle. The total area of the rectangle represents the product, and its length and width represent the factors.

The rectangle has a width of 2 and a length of $x - 4$. Therefore, $2x - 8 = 2(x - 4)$.

### Activity 2  Use Algebra Tiles to Factor $x^2 + 3x$

**Step 1** Model $x^2 + 3x$.

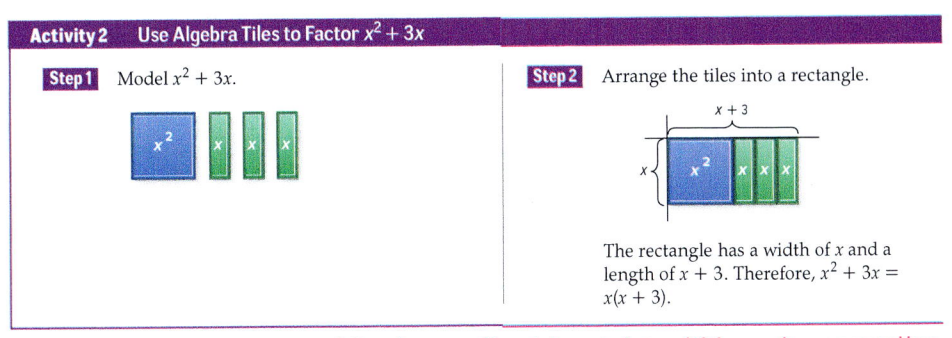

**Step 2** Arrange the tiles into a rectangle.

The rectangle has a width of $x$ and a length of $x + 3$. Therefore, $x^2 + 3x = x(x + 3)$.

**Model and Analyze**
Work cooperatively. Use algebra tiles to factor each binomial.

1. $4x + 12$  $4(x + 3)$
2. $4x - 6$  $2(2x - 3)$
3. $3x^2 + 4x$  $x(3x + 4)$
4. $10 - 2x$  $2(5 - x)$

Determine whether each binomial can be factored. Justify your answer with a drawing.

5. $6x - 9$  yes
6. $5x - 4$  no
7. $4x^2 + 7$  no
8. $x^2 + 3x$  yes

5–8. See Ch. 8 Answer Appendix for drawings.

9. **WRITING IN MATH** Write a paragraph that explains how you can use algebra tiles to determine whether a binomial can be factored. Include an example of one binomial that can be factored and one that cannot.

9. Sample answer: Binomials can be factored if they can be represented by a rectangle. Examples: $3x + 3$ can be factored and $3x + 2$ cannot be factored.

## From Concrete to Abstract

Write $x^2 + 5x$ on the board. Have students factor the binomial without using tiles. If they answer incorrectly, have them use their tiles to help them find their errors.

---

Explore 8-5

# Launch

**Objective** Use algebra tiles to model using the Distributive Property to factor binomials.

**Materials for Each Student**
- algebra tiles and product mats

**Easy to Make Manipulatives**
*Teaching Algebra with Manipulatives* Template for:
- algebra tiles, pp. 10–11
- product mat, p. 17

# Teach

**Working in Cooperative Groups** Put students in groups of two or three, mixing abilities. Have groups complete Activities 1 and 2 and Exercises 1–4.

- For Exercises 1–4, students should recognize that they must arrange the tiles into a rectangle with a width greater than 1 in order to find the factors of the polynomial.
- Point out that the area of the rectangle represents the polynomial, and the length and width represent the factors of the polynomial.

**Practice** Have students complete Exercises 5–9.

- For Exercises 5–9, emphasize that if a binomial can only be modeled with a width of 1, it cannot be factored.

# Assess

**Formative Assessment**

Use Exercise 3 to assess whether students can use algebra tiles to factor a binomial.

## Go Online!

**eToolkit**

The eToolkit allows students to explore and enhance their understanding of math concepts. Use the Algebra Tiles tool to demonstrate factoring using the Distributive Property.

connectED.mcgraw-hill.com  519

Lesson 8-5 | Using the Distributive Property

# Launch

Have students read the Why? section of the lesson. Ask:

- What is the formula for the area of rectangle? $A = \ell \times w$

- What do you multiply $w$ by to get $1.6w^2 + 6w$? $1.6w + 6$

- What is the area for Mr. Cole's store expressed as a monomial times a polynomial? $w(1.6w + 6)$

- What is the area when $w = 50$? $4300$ ft²

# Teach

Ask the scaffolded questions for each example to build conceptual understanding for students at all levels.

## 1 Use the Distributive Property to Factor

**Example 1 Use the Distributive Property**

**AL** How is this example different from previous lessons? Sample answer: We are working backward now, pulling out what the terms have in common.

**OL** In part **a**, what factors does each term have in common? 9 and $y$ In part **b**? 2, $a$, and $b$

**BL** What properties are used to factor the polynomial? Sample answer: Distributive Property and Quotient of a Power

**Need Another Example?**
Use the Distributive Property to factor each polynomial.
a. $15x + 25x^2$  $5x(3 + 5x)$
b. $12xy + 24xy^2 - 30x^2y^4$  $6xy(2 + 4y - 5xy^3)$

## Go Online!

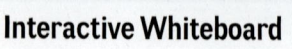

**Interactive Whiteboard**
Use the eLesson or Lesson Presentation to present this lesson.

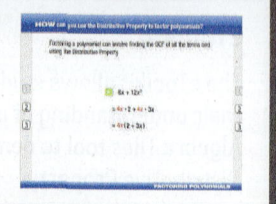

---

# LESSON 5
# Using the Distributive Property

**Then** · Used the Distributive Property to evaluate expressions.

**Now** · 1 Use the Distributive Property to factor polynomials.
2 Factor polynomials by grouping.

**Why?** · The cost of rent for Mr. Cole's store is determined by the square footage of the space. The area of the store can be modeled by the expression $1.6w^2 + 6w$, where $w$ is the width of the store in feet. We can use factoring to simplify the expression.

**New Vocabulary**
factoring
factoring by grouping
Zero Product Property

**Mathematical Practices**
2 Reason abstractly and quantitatively.

**Content Standards**
A.SSE.2 Use the structure of an expression to identify ways to rewrite it.
Prep. for A.SSE.3a Factor a quadratic expression to reveal the zeros of the function it defines.

**1 Use the Distributive Property to Factor** You have used the Distributive Property to multiply a monomial by a polynomial. You can work backward to express a polynomial as the product of a monomial factor and a polynomial factor.

$$1.6w^2 + 6w = 1.6w(w) + 6(w)$$
$$= w(1.6w + 6)$$

So, $w(1.6w + 6)$ is the *factored form* of $1.6w^2 + 6w$. **Factoring** a polynomial involves finding the *completely* factored form.

**Example 1** Use the Distributive Property

Use the Distributive Property to factor each polynomial.

a. $27y^2 + 18y$

Find the GCF of each term.

$27y^2 = ③ \cdot ③ \cdot 3 \cdot ⓨ \cdot y$   Factor each term.
$18y = 2 \cdot ③ \cdot ③ \cdot ⓨ$   Circle common factors.
GCF = $3 \cdot 3 \cdot y$ or $9y$

Write each term as the product of the GCF and the remaining factors. Use the Distributive Property to *factor out* the GCF.

$27y^2 + 18y = 9y(3y) + 9y(2)$   Rewrite each term using the GCF.
$= 9y(3y + 2)$   Distributive Property

b. $-4a^2b - 8ab^2 + 2ab$

$-4a^2b = -1 \cdot ② \cdot 2 \cdot ⓐ \cdot a \cdot ⓑ$   Factor each term.
$-8ab^2 = -1 \cdot ② \cdot 2 \cdot 2 \cdot ⓐ \cdot ⓑ \cdot b$   Circle common factors.
$2ab = ② \cdot ⓐ \cdot ⓑ$
GCF = $2 \cdot a \cdot b$ or $2ab$

$-4a^2b - 8ab^2 + 2ab = 2ab(-2a) - 2ab(4b) + 2ab(1)$   Rewrite each term using the GCF.
$= 2ab(-2a - 4b + 1)$   Distributive Property

**Guided Practice**

1A. $15w - 3v$  $3(5w - v)$     1B. $7u^2t^2 + 21ut^2 - ut$  $ut(7ut + 21t - 1)$

---

**MP Mathematical Practices Strategies**

**Make sense of problems and persevere in solving them.**
Help students analyze given polynomials in order to transform the expressions. For example, ask:

- What is always the first step when factoring a polynomial?
look for and factor out the GCF

- When factoring by grouping, what must be true about the common factor?
They must be the same or additive inverses.

- How can you check that you have correctly factored a polynomial?
multiply the factored form to see if the product is the same as the original expression

520 | Lesson 8-5 | Using the Distributive Property

## Real-World Example 2 Use Factoring

**AGILITY** ...

Find the GCF of each term.

$20t = \textcircled{2} \cdot \textcircled{2} \cdot 5 \cdot \textcircled{t}$     Factor each term.
$-16t^2 = -1 \cdot \textcircled{2} \cdot \textcircled{2} \cdot 2 \cdot 2 \cdot \textcircled{t} \cdot t$     Circle common factors.
GCF $= 2 \cdot 2 \cdot t$ or $4t$

$20t - 16t^2 = 4t(5) + 4t(-4t)$     Rewrite each term using the GCF.
$\phantom{20t - 16t^2} = 4t(5 - 4t)$     Distributive Property

The height of Penny's leap after $t$ seconds can be modeled by $4t(5 - 4t)$.

Substitute 0.5 for $t$ to find the height after 0.5 second.
$4t(5 - 4t)$
$= 4(0.5)[5 - 4(0.5)]$     Substitution, $t = 0.5$
$= 2(5 - 2)$     Simplify.
$= 2(3)$     Subtract.
$= 6$     Multiply.

After 0.5 second, Penny will be 6 inches above the ground.

### Guided Practice

2. **KANGAROOS** The height of a kangaroo's hop in inches after $t$ seconds can be modeled by $24t - 16t^2$. Factor the expression. Then find the height of the kangaroo after 1 second. $8t(3 - 2t)$; 8 in.

## 2 Factoring by Grouping

Using the Distributive Property to factor polynomials with four or more terms is called ⬚⬚⬚⬚⬚⬚ because terms are put into groups and then factored. The Distributive Property is then applied to a common binomial factor.

### Key Concept  Factoring by Grouping

**Words**    A polynomial can be factored by grouping only if all of the following conditions exist.
- There are four of more terms.
- Terms have common factors that can be grouped together.
- There are two common factors that are identical or additive inverses of each other.

**Symbols**    $ax + bx + ay + by = (ax + bx) + (ay + by)$
$\phantom{ax + bx + ay + by} = x(a + b) + y(a + b)$
$\phantom{ax + bx + ay + by} = (x + y)(a + b)$

---

**Real-World Link**
Dog agility tests a person's skills as a trainer and handler. Competitors race through an obstacle course that includes hurdles, tunnels, a see-saw, and line poles.
**Source:** United States Dog Agility Association

---

Lesson 8-5 | Using the Distributive Property

### Example 2  Use Factoring

**AL** What value is substituted for $t$? 0.5

**OL** What is the GCF of the binomial? $4t$

**BL** Describe another method that you could use to find the height after 0.5 seconds. Sample answer: Substitute 0.5 for $t$ in the original expression. Because the original expression and factored expression are equivalent, they will result in the same height.

### Need Another Example?
**Football** A football is kicked into the air. The height of the football, in feet, after $x$ seconds can be modeled by the expression $-16x^2 + 16x$. Factor the expression. Then find the height of the football after 0.3 seconds. $-16x(x - 1)$; 3.36 feet

Lesson 8-5 | Using the Distributive Property

## 2 Factor by Grouping

### Example 3 Factor by Grouping

**AL** Do all four terms have a common factor in part **a** or part **b**? no

**OL** What is the common factor after the terms are grouped in part **a**? $(q+2)$

**BL** Is there another way to group the terms in part **a**? If so, explain it and how it affects the solution. Yes; Sample answer: I could group $4qr$ and $3q$ because they have a common factor of $q$, and group $8r$ and $6$ because they have a common factor of 2. I still get the same solution.

**Need Another Example?**
Factor $2xy + 7x - 2y - 7$. $(x-1)(2y+7)$

### Teaching Tip

**Reasoning** Sometimes students find the monomial that is the GCF of the terms of the polynomial but do not know how to get the other polynomial factor. One way to find the remaining factor is to divide each term of the polynomial by the GCF. Tell students to check their answers by multiplying their factors using the Distributive Property.

### Example 4 Factor by Grouping with Additive Inverses

**AL** What terms can be grouped so that they have common factors in part **a**? Sample answer: $2mk$ and $-12m$; $42$ and $-7k$.

**OL** What binomials are inverses of each other?
a. $k-6$ and $6-k$; b. $7-a$ and $a-7$

**BL** How can you check your solutions? Sample answer: I can use the FOIL method to multiply the binomials together and make sure I get the same polynomial I started with.

**Need Another Example?**
Factor $15a - 3ab + 4b - 20$. $(-3a+4)(b-5)$ or $(3a-4)(5-b)$

### Go Online!

The most up-to-date resources available for your program can be found at **connectED.mcgraw-hill.com**.

---

**Example 3** Factor by Grouping

Factor each polynomial.
a. $4qr + 8r + 3q + 6$

$4qr + 8r + 3q + 6$   Original expression
$= (4qr + 8r) + (3q + 6)$   Group terms with common factors.
$= 4r(q + 2) + 3(q + 2)$   Factor the GCF from each group.

Notice that $(q + 2)$ is common in both groups, so it becomes the GCF.
$= (4r + 3)(q + 2)$

**Study Tip**
**MP Reasoning** To check your factored answers, multiply your factors out. You should get your original expression as a result.

**Check** You can check this result by multiplying the two factors. The product should be equal to the original expression.

$(4r + 3)(q + 2) = 4qr + 8r + 3q + 6$ ✓ FOIL Method

b. $2u^2v - 15 - 6u^2 + 5v$

$2u^2v - 15 - 6u^2 + 5v$   Original expression
$= 2u^2v - 6u^2 + 5v - 15$   Commutative Property
$= (2u^2v - 6u^2) + (5v - 15)$   Group terms with common factors.
$= 2u^2(v - 3) + 5(v + 3)$   Factor the GCF from each group.
$= (2u^2 + 5)(v - 3)$   Distributive Property

▶ **Guided Practice**
Factor each polynomial.
**3A.** $rn + 5n - r - 5$   $(r+5)(n-1)$    **3B.** $3np + 15p - 4n - 20$   $(n+5)(3p-4)$
**3C.** $tw^3 - 2w^3 + 10t - 20$   $(w^3+10)(t-2)$    **3D.** $4ab^2 + 21 + 12b^2 + 7a$   $(4b^3+7)(a+3)$

It can be helpful to recognize when binomials are additive inverses of each other. For example, $6 - a = -1(a - 6)$.

**Example 4** Factor by Grouping with Additive Inverses

Factor each polynomial.
a. $2mk - 12m + 42 - 7k$

$2mk - 12m + 42 - 7k$
$= (2mk - 12m) + (42 - 7k)$   Group terms with common factors.
$= 2m(k - 6) + 7(6 - k)$   Factor the GCF from each group.
$= 2m(k - 6) + 7[(-1)(k - 6)]$   $6 - k = -1(k-6)$
$= 2m(k - 6) - 7(k - 6)$   Associative Property
$= (2m - 7)(k - 6)$   Distributive Property

---

### Differentiated Instruction AL OL

**IF** students have trouble factoring quadratic expressions like Example 1a or 2,

**THEN** you may wish to allow students to use algebra tiles. They can use the methods from Explore 8-5 to factor the quadratic expression.

**Lesson 8-5 | Using the Distributive Property**

b. $21b^4 - 3ab^4 + 4a - 28$

$21b^4 - 3ab^4 + 4a - 28$
$= (21b^4 - 3ab^4) + (4a - 28)$ — Group terms with common factors.
$= 3b^4(7 - a) + 4(a - 7)$ — Factor the GCF from each group.
$= 3b^4[(-1)(a - 7)] + 4(a - 7)$ — $7 - k = -1(k - 7)$
$= -3b^4(a - 7) + 4(a - 7)$ — Associative Property
$= (-3b^4 + 4)(a - 7)$ — Distributive Property

▶ **Guided Practice**
Factor each polynomial.
**4A.** $c - 2cd + 8d - 4$
**4B.** $3p - 2p^2 - 18p + 27$

**4A.** $(-c + 4)(2d - 1)$ or $(c - 4)(1 - 2d)$
**4B.** $(p + 9)(3 - 2p)$ or $(-p - 9)(2p - 3)$

## Practice

**Formative Assessment** Use Exercises 1–14 to assess students' understanding of the concepts in this lesson.

The Practice and Problem Solving exercises assess the content taught in the lesson. The Preparing for Assessment page is meant to be used as preparation for end-of-course assessment.

### MP Teaching the Mathematical Practices

**Reasoning** Mathematically proficient students attend to the meaning of quantities, not just how to compute them. In Exercise 14, tell students that the path of a projectile or other body moving through space is called a trajectory. Quadratic expressions are always used to represent motion of a trajectory.

## Extra Practice

See page R8 for extra exercises for students who are approaching level or for on-level students who need additional reinforcement.

---

**Check Your Understanding** ● = Step-by-Step Solutions begin on page R11. ✓ Go Online! for a Self-Check Quiz

**Example 1**
A.SSE.2

Use the Distributive Property to factor each polynomial.
1. $21b - 15a$   $3(7b - 5a)$
2. $14c^2 + 2c$   $2c(7c + 1)$
3. $10g^2h^2 + 9gh^2 - g^2h$   $gh(10gh + 9h - g)$
4. $12jk^2 + 6j^2k + 2j^2k^2$   $2jk(6k + 3j + jk)$

**Examples 3–4**
A.SSE.2

Factor each polynomial.
5. $np + 2n + 8p + 16$   $(n + 8)(p + 2)$
6. $xy - 7x + 7y - 49$   $(x + 7)(y - 7)$
7. $3bc - 2b - 10 + 15c$   $(b + 5)(3c - 2)$
8. $9fg - 45f - 7g + 35$   $(9f - 7)(g - 5)$
9. $3km - 21k + 2m - 14$   $(3k + 2)(m - 7)$
10. $4cd^2 + 3cd - 20d - 15$   $(cd - 5)(4d + 3)$
11. $10p^3q - 5p^4 + 3pq - 6q^2$   $(5p^3 - 3q)(2q - p)$
12. $5ab - b - 6 + 30a$   $(b + 6)(5a - 1)$

**Example 2**
A.SSE.3a

13. **SPIDERS** Jumping spiders can commonly be found in homes and barns throughout the United States. The height of a jumping spider's jump in feet after $t$ seconds can be modeled by the expression $12t - 16t^2$.

   a. Write the factored form of the expression for the height of the spider after $t$ seconds. **$4t(3 - 4t)$**
   b. What is the spider's height after 0.1 second? **1.04 ft**
   c. What is the spider's height after 0.5 second? **2 ft**

14. **MP REASONING** At a Fourth of July celebration, a rocket is launched straight up with an initial velocity of 128 feet per second. The height of the rocket in feet above sea level is modeled by the expression $128t - 16t^2$, where $t$ is the time in seconds after the rocket is launched.

   a. What is the GCF of each term of the expression? **16 ft**
   b. Write the factored form of the expression for the height of the rocket after $t$ seconds. **$16t(8 - t)$**
   c. Will the height of the rocket be greater after 4 seconds or after 6 seconds? **after 4 seconds**

### Levels of Complexity Chart

The levels of the exercises progress from 1 to 3, with Level 1 indicating the lowest level of complexity.

| Exercises | 15–45 | 46–48, 57–65 | 49–56 |
|---|---|---|---|
| C Level 3 | | | ● |
| B Level 2 | | ● | |
| A Level 1 | ● | | |

---

### Differentiated Homework Options

| Levels | AL Basic | OL Core | BL Advanced |
|---|---|---|---|
| Exercises | 15–45, 52, 54–65 | 15–45 odd, 46–52, 54–65 | 46–56, (optional: 57–65) |
| 2-Day Option | 15–45 odd, 57–65 | 15–45, 57–65 | |
| | 16–44 even, 52, 54–56 | 46–52, 54–56 | |

You can use ALEKS to provide additional remediation support with personalized instruction and practice.

### Go Online! / eBook

**Interactive Student Guide**
Use the *Interactive Student Guide* to deepen conceptual understanding.
· Solving by Factoring: The Distributive Property

ALGEBRA 1 INTERACTIVE STUDENT GUIDE

connectED.mcgraw-hill.com  523

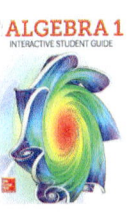

# Lesson 8-5 | Using the Distributive Property

## Teaching the Mathematical Practices

**Sense-Making** Mathematically proficient students check their answers to problems using a different method. In Exercise 45, have students check their answers by substituting values for *a* and *b* and finding the areas directly and by using their expressions.

**Critique** Mathematically proficient students compare the effectiveness of two plausible arguments. In Exercise 52, remind students that they can check the factorization by multiplying the factors.

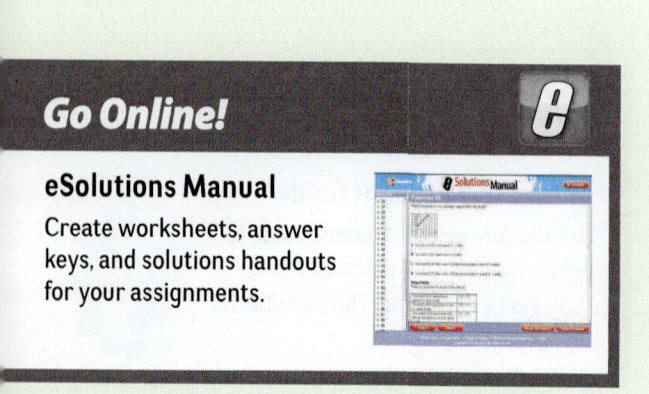

## Practice and Problem Solving

Extra Practice is on page R8.

**Example 1** Use the Distributive Property to factor each polynomial.
A.SSE.2

15. ··t−···y   8(2t−5y)
16. ··v+···x   10(3v+5x)
17. ·k·+··k   2k(k+2)
18. ·z·+···z   5z(z+2)
19. ·a·b·+··a·b−··ab·   2ab(2ab+a−5b)
20. ·c·v−···c·v·+··c·v·   5c²v(1−3v+v²)

**Examples 3–4** Factor each polynomial.   25. (9q−10)(5p−3)   35. 3cd(9d−6cd+1)   37. 2(8u−15)(3t+2)
A.SSE.2

21. fg−··g+··f−··   (g+4)(f−5)
22. a·−··a−··+··a   (a−4)(a+6)
23. hj−··h+··j−··   (h+5)(j−2)
24. xy−··x−··+y   (x+1)(y−2)
25. ··pq−···q−··p+··
26. ··ty−···t+··y−··   (6t+1)(4y−3)
27. ·dt−···d+··−··t   (3d−5)(t−7)
28. ·r·+···r   4r(2r+3)
29. ··th−··t−··h+··   (3t−5)(7h−1)
30. vp+···v+··p+···   (v+8)(p+12)
31. ·br−···b+··r−···   (r−5)(5b+2)
32. ·nu−···u+··n−···   (2u+3)(n−4)
33. ·gf·+g·f+··gf   gf(5f+g+15)
34. rp−··r+··p−···   (r+9)(p−9)
35. ··cd·−···c·d·+··cd
36. ··r·t·+···r·t·−··r·t   6r²t(3rt+2t−1)
37. ··tu−···t+··u−···
38. ··gh+···g−··h−··   (8g−1)(2h+3)
39. ··p·+···p+··pr+··r   (5p+2r)(4p+3)
40. ·bc−···b+··c−··c·   (2b−c)(2c−5)
41. ·mk·−···k·+···−··m   (3k²−2)(3m−5)
42. x·y·−···x·y+··xy−···   (x³y+3)(xy−7)
43. ··fg+··g−··f−··   (6f+1)(8g−3)
44. 2a³b−8a²+12b−3ab²   (2a²−3b)(ab−4)

**Example 2** 45. SENSE-MAKING ···
A.SSE.3a
  a. ··· ab
  b. ··· (a+6)(b+6)
  c. ··· 6(a+b+6)

46. FIREWORKS ···
  a. ··· 8t(33−2t)
  b. ··· 0 ft
  c. ···
  d. ··· 1089 ft

46c. The shell was on the ground at the time of launch, t = 0, and then returned to the ground after 16.5 seconds.

47. VOLCANOS ···
  a. ··· 8t(55−2t)
  b. ··· 2800 ft, 2400 ft
  c. ··· 3025 ft

## Go Online!

**eSolutions Manual**
Create worksheets, answer keys, and solutions handouts for your assignments.

**48. RIDES** Suppose the height in feet of a rider $t$ seconds after being dropped on a drop tower thrill ride can be modeled by the expression $-16t^2 - 96t + 160$.
  a. Write an expression to represent the height in factored form. $16(-t^2 - 6t + 10)$
  b. From what height is the rider initially dropped? **160 ft**
  c. At what height will the rider be after 3 seconds of falling? Is this possible? Explain. **−272 ft; No, the rider cannot be a negative number of feet in the air.**

**49. ARCHERY** The height in feet of an arrow can be modeled by the expression $72t - 16t^2$, where $t$ is time in seconds after the arrow is released. Write the expression in factored form and find the height of the arrow after 3 seconds. $16t(5 - t)$; **96 ft, 64 ft**

**50. TENNIS** A tennis player hits a tennis ball upward with an initial velocity of 80 feet per second. The height in feet of the tennis ball can be modeled by the expression $80t - 16t^2$, where $t$ is time in seconds. Write the expression in factored form and find the height of the tennis ball 2 seconds and 4 seconds after being hit. **96 ft, 64 ft**

**51. MULTIPLE REPRESENTATIONS** In this problem, you will explore the *box method* of factoring. To factor $x^2 + x - 6$, write the first term in the top left-hand corner of the box, and then write the last term in the lower right-hand corner.
  a. **Analytical** Determine which two factors have a product of −6 and a sum of 1. **3 and −2**
  b. **Symbolic** Write each factor in an empty square in the box. Include the positive or negative sign and variable. **See margin.**
  c. **Analytical** Find the factor for each row and column of the box. What are the factors of $x^2 + x - 6$? **See margin.**
  d. **Verbal** Describe how you would use the box method to factor $x^2 - 3x - 40$.

**51d.** Sample answer: Place $x^2$ in the top left-hand corner and place −40 in the lower right-hand corner. Then determine which two factors have a product of −40 and a sum of −3. Then place these factors in the box. Then find the factor of each row and column. The factors will be listed on the very top and far left of the box. A.SSE.2, A.SSE.3a

### H.O.T. Problems  Use Higher-Order Thinking Skills

**52. ERROR ANALYSIS** Hernando and Rachel are factoring $2mp - 6p + 27 - 9m$. Is either of them correct? Explain your reasoning. **Rachel; since $3 - m$ and $m - 3$ are additive inverses, you must multiply one of the binomials by −1 to find the common factor.**

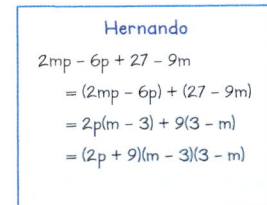

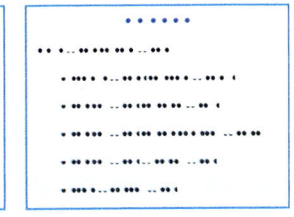

Hernando
$2mp - 6p + 27 - 9m$
$= (2mp - 6p) + (27 - 9m)$
$= 2p(m - 3) + 9(3 - m)$
$= (2p + 9)(m - 3)(3 - m)$

**53. CHALLENGE** Given the expression $4yz^2 + 24z + 5yz + 30$, show two different ways to group the terms and factor the polynomial. **See margin.**

**54. OPEN-ENDED** Write a four-term polynomial that can be factored by grouping. Then factor the polynomial. **Sample answers:** $x^2 + 2xy + 3x + 6y, (x + 3)(x + 2y)$

**55. REASONING** Given the expression $ab^2c^3 + a^7bc^2 + a^3c^6$, identify the GCF of the terms. Then factor the expression. $ac^2$; $ac^2(b^2c + a^6b + a^2c^4)$

**56. WRITING IN MATH** Explain why you cannot use factoring by grouping on a polynomial with three terms. **See margin.**

---

**Lesson 8-5 | Using the Distributive Property**

## Assess

**Yesterday's News** Have students write how yesterday's concept of finding the GCF of a set of monomials helped them with today's new material.

### Follow-Up
Students have explored modeling using quadratic expressions.

**Ask:** What are the advantages of using quadratic expressions for modeling? **Sample answer:** They have well known and understood properties; they can be used to model situations that have both increasing and decreasing behavior; the computation that is done to make predictions is relatively easy to perform.

### Additional Answers

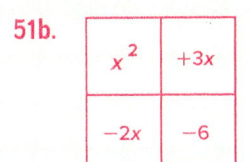

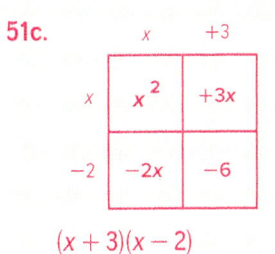

**51b.**

|       |      |
|-------|------|
| $x^2$ | $+3x$ |
| $-2x$ | $-6$ |

**51c.**

|    | $x$   | $+3$  |
|----|-------|-------|
| $x$  | $x^2$ | $+3x$ |
| $-2$ | $-2x$ | $-6$  |

$(x + 3)(x - 2)$

**53.** Sample answer:
$4yz^2 + 24z + 5yz + 30$
$= (4yz^2 + 24z) + (5yz + 30)$
$= 4z(yz + 6) + 5(yz + 6)$
$= (4z + 5)(yz + 6)$

$4yz^2 + 24z + 5yz + 30$
$= 4yz^2 + 5yz + 24z + 30$
$= (4yz^2 + 5yz) + (24z + 30)$
$= yz(4z + 5) + 6(4z + 5)$
$= (yz + 6)(4z + 5)$

**56.** Since you cannot create two groups with common binomial factors, the grouping method will not work.

---

### Standards for Mathematical Practice

| Emphasis On | Exercises |
|---|---|
| 1 Make sense of problems and persevere in solving them. | 15-45, 54, 58, 63 |
| 2 Reason abstractly and quantitatively. | 46-48, 55, 58, 64 |
| 3 Construct viable arguments and critique the reasoning of others. | 51, 52, 56 |
| 4 Model with mathematics. | 45-50, 59, 61 |
| 7 Look for and make use of structure. | 53-55, 57, 60, 65, 63, 64 |

# Lesson 8-5 | Using the Distributive Property

## Preparing for Assessment

Exercises 57–65 require students to use the skills they will need on standardized assessments. Each exercise is dual-coded with content and process standards and mathematical practice standards.

### Dual Coding

| Exercises | Content Standards | MP Mathematical Practices |
|---|---|---|
| 57, 60, 65 | A.SSE.2 | 7 |
| 58 | A.SSE.2 | 1, 2 |
| 59, 61 | A.SSE.2 | 4 |
| 62 | A.SSE.2 | 6 |
| 63 | A.SSE.2 | 1, 7 |
| 64 | A.SSE.2 | 2, 7 |

## Diagnose Student Errors

Survey student responses for each item. Class trends may indicate common errors and misconceptions.

**59.**

| | |
|---|---|
| A | CORRECT |
| B | Did not factor 2 out of the second term |
| C | Used the length |
| D | Did not include $x$ in the GCF |

**60.**

| | |
|---|---|
| A | Used the common factor twice |
| B | CORRECT |
| C | Found the common factor |
| D | Switched signs in the factors |

**61.**

| | |
|---|---|
| A | Did not completely factor each term |
| B | Did not completely factor each term |
| C | Did not include $t$ in the GCF |
| D | CORRECT |

## Go Online!

### Quizzes

Students can use *Self-Check Quizzes* to check their understanding of this lesson. You can also give the *Chapter Quiz*, which covers the content in Lessons 8-1 through 8-5.

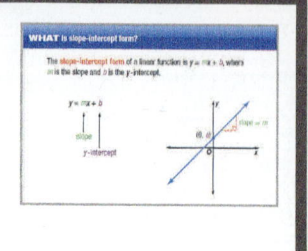

## Preparing for Assessment

**57.** The area of a rectangle is represented by $10x^3 + 15x^2 + 4x + 6$. Its dimensions are binomials in $x$ with prime coefficients. What are the dimensions of the rectangle? MP 7 A.SSE.3a

$(5x^2 + 2)$ and $(2x + 3)$

**58. MULTI-STEP** A rectangle has a length of $4x$ units and a width of $(6x + 7)$ units. A right triangle has legs of length $x$ and $(8x - 4)$ units. MP 1, 2 A.SSE.3a

a. Express the areas of the rectangle and the right triangle, respectively.

$4x^2 + 28x, 4x^2 + 2x$

b. Find the factored form of the difference of the area of the rectangle and triangle.

$10x(2x + 3)$

**59.** Jill is designing her name using geometric shapes. She is using a vertical rectangle with length $2x$ for the letter "i," as shown.

If the area of the rectangle is represented by $6x^2 - 14x$, which expression represents the width? MP 4 A.SSE.3a  **A**

- A  $3x - 7$
- B  $3x - 14$
- C  $2x$
- D  $3x^2 - 7x$

**60.** Factor $6b^2c - 9bc + 22bd - 33d$. MP 7 A.SSE.2  **B**

- A  $(3bc + 11d)(2b - 3)^2$
- B  $(3bc + 11d)(2b - 3)$
- C  $(2b - 3)$
- D  $(3bc - 11d)(2b + 3)$

**61.** The height of a golf ball in feet after $t$ seconds can be modeled by $104t - 16t^2$. Which is the GCF of the terms in the polynomial? MP 4 A.SSE.2  **D**

- A  $2t$
- B  $4t$
- C  $8$
- D  $8t$

**62.** Which expression is equivalent to the following polynomial? MP 6 A.SSE.2  **C**

$$2x^3 - 10x^2 + 15 - 3x$$

- A  $(2x^2 + 3)(x - 5)$
- B  $(2x^2 + 3)(x - 5)(5 - x)$
- C  $(2x^2 - 3)(x - 5)$
- D  $(2x^2 - 3)(x - 5)^2$

**63.** Factor the following expressions completely: MP 1, 7 A.SSE.3a

a. $6y^3 + 21y^2 + 4y + 14$

$(3y^2 + 2)(2y + 7)$

b. $12a^2 - 20ab + 9ay - 15by$

$(4a + 3y)(3a - 5b)$

c. $4w^3 + 3wz - 8w^2 - 6z$

$(w - 2)(4w^2 + 3z)$

d. $15m^2n + 27mn^2$

$3mn(5m - 9n)$

e. $8wy + 12xy + 10wz + 15xz$

$(4y + 5z)(2w + 3x)$

**64.** The area of two rectangles can be represented by $x^3 + 4x^2 + 2x + 8$ and $x^3 - 3x^2 + 2x - 6$. If the widths of the rectangles are equal, find the width of the rectangles. MP 2, 7 A.SSE.3a

$x^2 + 2$

**65.** The product of two binomials $2x^3 - 3x^2 - 10x + 15$. Find the sum of these two binomials. MP 7 A.SSE.3a

$x^2 + 2x - 8$

## Differentiated Instruction OL BL

**Extension** Write the following polynomial on the board: $c^2xy - c^3 - x^2y + cx$. Ask students to factor it by grouping. $(c^2 - x)(xy - c)$, or $(x - c^2)(c - xy)$

**62.**

| | |
|---|---|
| A | Used incorrect sign in first factor |
| B | Did not recognize additive inverses |
| C | CORRECT |
| D | Included the common factor twice |

526 | Lesson 8-5 | Using the Distributive Property

# EXTEND 8-5
## Algebra Lab
# Proving the Elimination Method

The elimination method for solving systems of equations requires adding one equation in a system with a multiple of the other equation. You can prove that this process does not affect the solution and is therefore is a valid method by using the steps outlined below.

**Claim:** Given a system of linear equations in two variables, replacing one equation by the sum of that equation and a multiple of the other produces a system with the same solutions.

**Mathematical Practices**
MP 3 Look for viable arguments and critique the reasoning of others.
7 Look for and make use of structure.

**Content Standards**
A.REI.5 Prove that, given a system of two equations in two variables, replacing one equation by the sum of that equation and a multiple of the other produces a system with the same solutions.

### Step 1 Solve a General System of Equations

Solve the general system of linear equations by using substitution.

$ax + by = c$

$dx + ey = f$

Solve each equation for $x$ to get $x = \frac{c - by}{a}$ and $x = \frac{f - ey}{d}$. Use substitution to set the right sides of the equations equal and find an expression for $y$. Solving $\frac{c - by}{a} = \frac{f + ey}{d}$ for $y$ gives $y = \frac{af - dc}{-bd + ae}$.

Repeat this process to find $x$. Solving each equation in the system for $y$ gives $y = \frac{c - ax}{b}$ and $y = \frac{f - dx}{e}$.

Substitute to find an expression for $x$. Solving $\frac{c - ax}{b} = \frac{f - dx}{e}$ for $x$ gives $x = \frac{bf - ec}{-ae + bd}$.

The solution of the general system is $\left(\frac{bf - ec}{-ae + bd}, \frac{af - dc}{-bd + ae}\right)$.

Now that you know the solution of the general system, replace one of the equations with the sum of that equation and a multiple of the other to create a new system of equations. If the systems have the same solution, then the method of elimination is valid.

### Step 2 Create a New System of Equations

Replace an equation in the general system of equations with the sum of that equation and a multiple of the other.

Replace $dx + ey = f$ with the sum of $dx + ey = f$ and a multiple of $ax + by = c$. To ensure that this will be true for any case, use a general real number $g$ as the factor.

| | |
|---|---|
| $dx + ey = f$ | First equation |
| $(+)\ agx + bgy = gc$ | Multiple of second equation |
| $dx + agx + ey + bgy = f + gc$ | Add. |
| $(d + ag)x + (e + bg)y = f + gc$ | Factor. |

So, the new system is $\begin{cases} ax + by = c \\ (d + ag)x + (e + bg)y = f + gc. \end{cases}$

---

# Launch

**Objective** Prove the elimination method for solving a system of equations simultaneously.

**Teaching Tip** Because there are so many variables, students may struggle when solving for $x$ and $y$ in Steps 1 and 3. As students work through the steps, direct them to show all of the work required to find $x$ and $y$. Remind students that the properties of equality can be used to solve equations with more than one variable.

# Teach

**Working in Cooperative Groups** Have students work in groups of two or four, mixing abilities, to complete Steps 1–3 and Exercises 1–7.

Extend 8-5

## Assess

### From Concrete to Abstract

Students may have trouble understanding why they can add two equations. Point out that the two sides of any equation are equal and that by the Addition Property of Equality, adding the same value to each side of an equation maintains equality.

## Additional Answers

1. (5, 2)
   Sample answer:

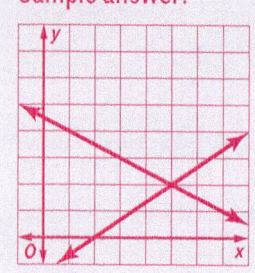

2. (7, 8)
   Sample answer:

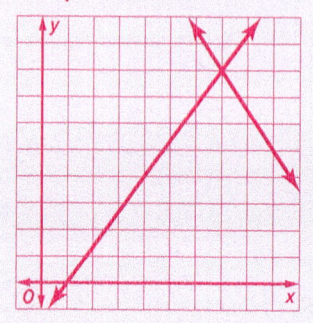

3. (3, 4)
   Sample answer:

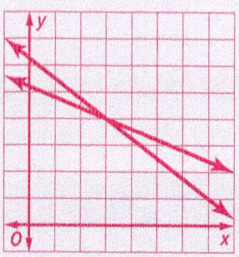

4. (−1, 5)
   Sample answer:

## Go Online!

The most up-to-date resources available for your program can be found at connected.mcgraw-hill.com.

528 | Extend 8-5 | Algebra Lab: Proving the Elimination Method

---

# EXTEND 8-5
## Algebra Lab
## Proving the Elimination Method Continued

### Step 3 Solve the New System of Equations

Solve the new system of linear equations by using substitution.

$ax + by = c$

$(d + ag)x + (e + bg)y = f + gc$

Solve each equation for $x$ to get $x = \dfrac{c - by}{a}$ and $x = \dfrac{f + gc - ey - bgy}{d + ag}$. Substitute to find $y$.

$\dfrac{c - by}{a} = \dfrac{f + gc - ey - bgy}{d + ag}$   Substitution

$dc - bdy + acg - abgy = af + acg - aey - abgy$   Cross multiply and distribute.

$dc - bdy = af - aey$   Subtract $acg$ and add $abgy$ to each side.

$-bdy + aey = af - dc$   Isolate the $y$-terms.

$y(-bd + ae) = af - dc$   Factor out $y$.

$y = \dfrac{af - dc}{-bd + ae}$   Divide each side by $-bd + ae$.

Repeat this process to find $x$. Solving each equation in the system for $y$ gives $y = \dfrac{c - ax}{b}$ and $y = \dfrac{f + gc - dx - agx}{e + bg}$. Substitute to find an expression for $x$. Solving $\dfrac{c - ax}{b} = \dfrac{f + gc - dx - agx}{e + bg}$ for $x$ gives $x = \dfrac{bf - ec}{-ae + bd}$.

The solution of the new system is $\left(\dfrac{bf - ec}{-ae + bd}, \dfrac{af - dc}{-bd + ae}\right)$.

The solutions of the original system in Step 1 and the new system are the same. Therefore, the claim is true.

### Exercises

Solve the system of equations by elimination. Verify the solution using a different method.

1. $2x - 3y = 4$
   $x + 2y = 9$

2. $4x - 3y = 4$
   $3x + 2y = 37$

3. $2x + 5y = 26$
   $-3x - 4y = -25$

   1–6 See margin.

4. $5x + y = 0$
   $5x + 2y = 30$

5. $2x - 3y = -7$
   $3x + y = -5$

6. $4x - 3y = -1$
   $3x - y = -2$

7. Consider the system $\begin{array}{l} ax - 3y = 4 \\ 3x + 2y = 3a \end{array}$.

   a. Solve the system of equations by the process of elimination. Leave your solution in terms of $a$.

   b. Is your solution in part a the same as the solution you get using the substitution method? $\left(\dfrac{9a + 8}{2a + 9}, \dfrac{3a^2 - 12}{2a + 9}\right)$
      yes

---

5. (−2, 1)
   Sample answer:

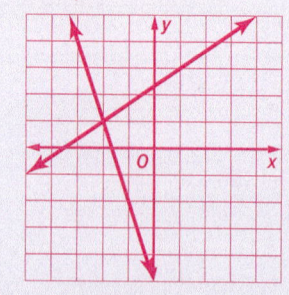

6. (−1, −1)
   Sample answer:

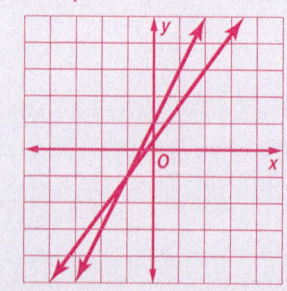

# LESSON 8-6
# Factoring Quadratic Trinomials

**SUGGESTED PACING (DAYS)**

| | | |
|---|---|---|
| 90 min. | .25 | 0.75 |
| 45 min. | 0.5 | 1.5 |
| | Explore Lab | Instruction |

## Track Your Progress

### Objectives

1. Factor trinomials of the form $x^2 + bx + c$.
2. Factor trinomials of the form $ax^2 + bx + c$.

### Mathematical Background

A trinomial of the form $ax^2 + bx + c$ may or may not be factorable into binomial factors. If the trinomial is factorable, then the factors of $ac$ must be two integers $m$ and $p$ such that $m + p = b$ and $mp = ac$.

When $a = 1$, the trinomial is in the form $x^2 + bx + c$ and the factors of $c$ must be two integers, $m$ and $p$, such that $m + p = b$, and $mp = c$.

### THEN

**A.APR.1** Understand that polynomials form a system analogous to the integers, namely, they are closed under the operations of addition, subtraction, and multiplication; add, subtract, and multiply polynomials.

### NOW

**A.SSE.2** Use the structure of an expression to identify ways to rewrite it.

### NEXT

**A.SSE.B.3.a** Factor a quadratic expression to reveal zeros of the function it defines.

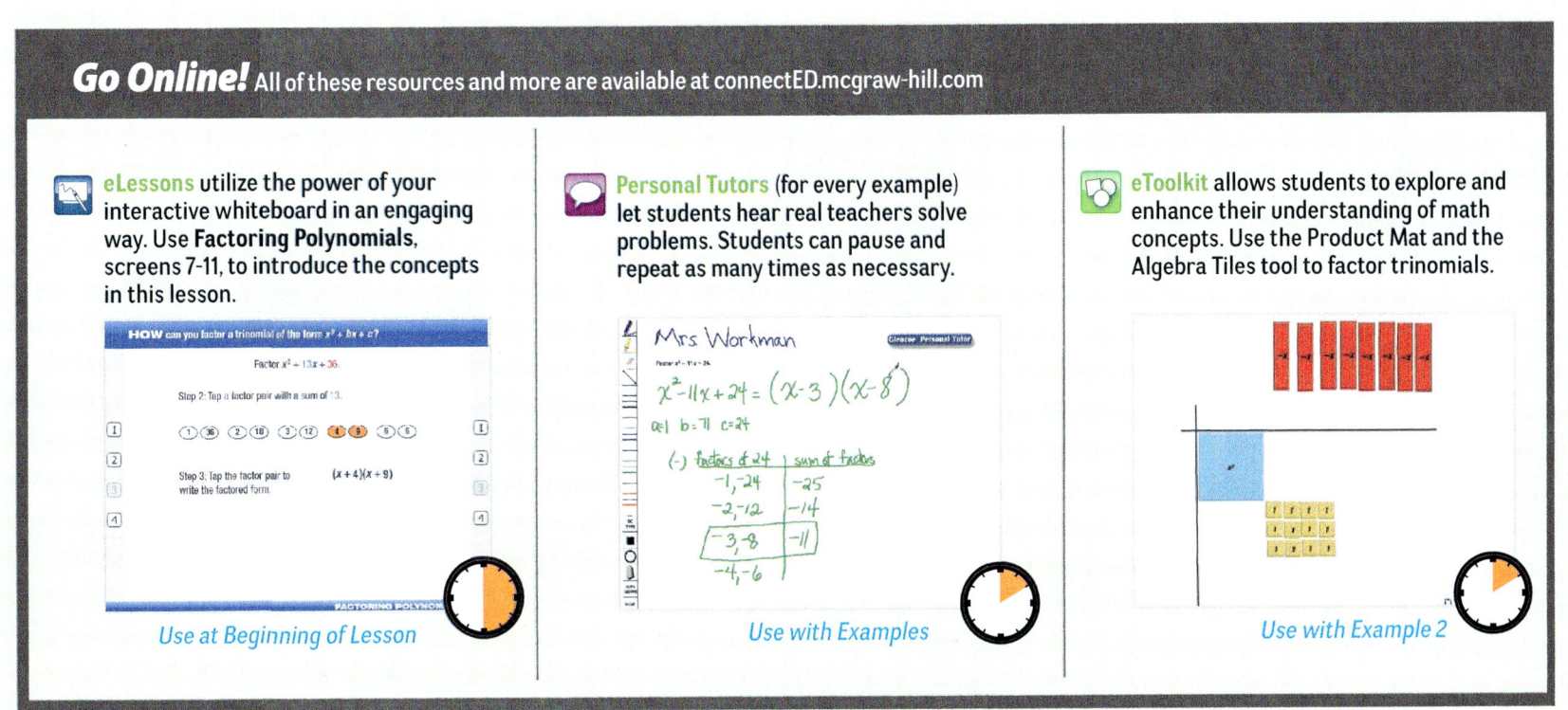

---

**Go Online!** All of these resources and more are available at connectED.mcgraw-hill.com

**eLessons** utilize the power of your interactive whiteboard in an engaging way. Use **Factoring Polynomials**, screens 7-11, to introduce the concepts in this lesson.

*Use at Beginning of Lesson*

**Personal Tutors** (for every example) let students hear real teachers solve problems. Students can pause and repeat as many times as necessary.

*Use with Examples*

**eToolkit** allows students to explore and enhance their understanding of math concepts. Use the Product Mat and the Algebra Tiles tool to factor trinomials.

*Use with Example 2*

---

### OER Using Open Educational Resources

**Games** Have students play the factoring trinomials game on **XP Math**. This is an arcade like game where the goal is to find two numbers whose sum is $b$ and product is $c$. Students receive visuals for each problem to reinforce the relationship between the polynomial and its factor. *Use as homework or classwork*

# Go Online!
connectED.mcgraw-hill.com

## Differentiate Your Resources

**Extra Practice** Additional practice or homework; Skills Practice is best for approaching-level students and Practice is best for on-level and beyond-level students

### Skills Practice
AL  OL  ELL

### Practice
AL  OL  BL  ELL

### Word Problem Practice
AL  OL  BL  ELL

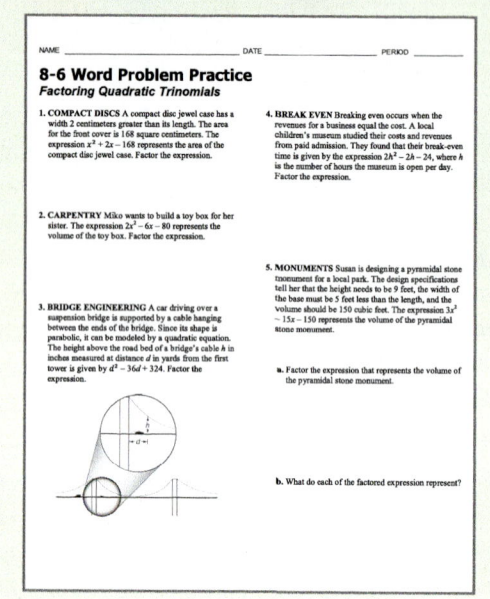

**Intervention** Reteaching and vocabulary activities that can be used with struggling or absent students and as ELL support

### Study Guide and Intervention
AL  OL  ELL

### Study Notebook
AL  OL  BL  ELL

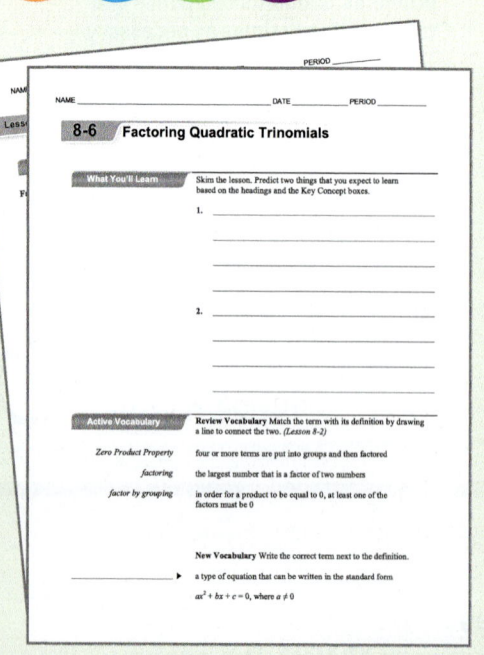

**Extension** Activities that can be used to extend lesson concepts

### Enrichment
OL  BL  ELL

529B | Lesson 8-6 | Factoring Quadratic Trinomials

# EXPLORE 8-6

## Algebra Lab
# Factoring Trinomials

You can use algebra tiles to factor trinomials. If a polynomial represents the area of a rectangle formed by algebra tiles, then the rectangle's length and width are *factors* of the area. If a rectangle cannot be formed to represent the trinomial, then the trinomial is not factorable.

**Mathematical Practices**
MP 3 Construct viable arguments and critique the reasoning of others.

**Content Standards**
A.SSE.2 Use the structure of an expression to identify ways to rewrite it.

### Activity 1   Factor $x^2 + bx + c$

Work cooperatively. Use algebra tiles to factor $x^2 + 4x + 3$.

**Step 1**  Model $x^2 + 4x + 3$.

**Step 2**  Place the $x^2$-tile at the corner of the product mat. Arrange the 1-tiles into a rectangular array. Because 3 is prime, the three tiles can be arranged in a rectangle in one way, a 1-by-3 rectangle.

**Step 3**  Complete the rectangle with the $x$-tiles.

The rectangle has a width of $x + 1$ and a length of $x + 3$.

Therefore, $x^2 + 4x + 3 = (x + 1)(x + 3)$.

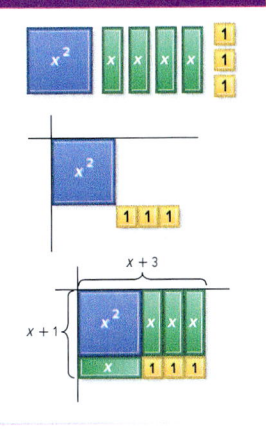

### Activity 2   Factor $x^2 + bx + c$

Use algebra tiles to factor $x^2 + 8x + 12$.

**Step 1**  Model $x^2 + 8x + 12$.

**Step 2**  Place the $x^2$-tile at the corner of the product mat. Arrange the 1-tiles into a rectangular array. Since $12 = 3 \times 4$, try a 3-by-4 rectangle. Try to complete the rectangle. Notice that there is an extra $x$-tile.

**Step 3**  Arrange the 1-tiles into a 2-by-6 rectangular array. This time you can complete the rectangle with the $x$-tiles.

The rectangle has a width of $x + 2$ and a length of $x + 6$.

Therefore, $x^2 + 8x + 12 = (x + 2)(x + 6)$.

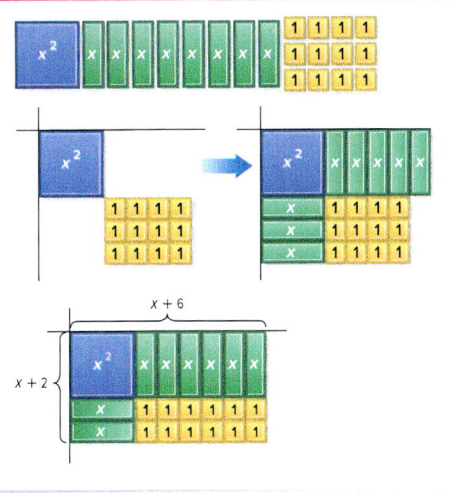

*(continued on the next page)*

## Launch

**Objective** Use algebra tiles to model factoring trinomials.

### Materials for Each Group
- algebra tiles
- product mat

### Easy to Make Manipulatives
*Teaching Algebra with Manipulatives* Template for:
- algebra tiles, pp. 10–11
- product mat, p. 17

### Teaching Tip
You may want to remind students that the area of the rectangle represents the polynomial, and the length and width of the rectangle represent the factors of the polynomial.

## Teach

**Working in Cooperative Groups** Place students in groups of two or three, mixing abilities. Have groups complete Activities 1–4 and Exercises 1–4. **ELL**

- Ask students to name the shape they must form with the tiles in order to factor a polynomial.  *rectangle*

- In Activity 1, remind students to read the width of the tiles along the edge of the rectangle. The $x^2$-tiles have a width of $x$ and the $x$-tiles have a width of one.

- For Activity 2, encourage students to try several different arrangements until they can form a rectangle. While the $x^2$-tile should be in the corner, there is more than one correct way to arrange the tiles into a rectangle.

## Go Online!

### eLesson
The eToolkit allows students to explore and enhance their understanding of math concepts. Use the Algebra Tiles tool to demonstrate factoring trinomials.

# Explore 8-6

- As students work through Activity 3, remind them to pay close attention to the sign of each tile.

- As students work through Activity 4, remind them to be careful to add one x-tile and one −x-tile when they add a zero pair.

**Practice** Have students complete Exercises 5–13.

## Assess

### Formative Assessment
Use Exercises 7 and 8 to assess whether students can use algebra tiles to factor trinomials.

### From Concrete to Abstract
After students complete Exercises 1–8, ask them whether they noticed a correlation between the need to use zero pairs to factor the trinomial and the appearance of the resulting factors. Sample answer: When zero pairs are used, the signs of the constant terms of the factors are opposite. When zero pairs are not used, the signs of the constant terms of factors are the same.

### Extending the Concept
Ask students what they notice about the sum of the constant terms in the factors of the trinomials in Exercises 1–8. Sample answer: Their sum equals the coefficient of the middle term of the trinomial.

### Additional Answers
1. $(x+1)(x+2)$
2. $(x+2)(x+4)$
3. $(x-1)(x+4)$
4. $(x-3)(x-4)$
5. $(x+2)(x+5)$
6. $(x-1)(x-1)$
7. $(x+4)(x-3)$
8. $(x-3)(x-5)$

---

# EXPLORE 8-6
## Algebra Lab
# Factoring Trinomials Continued

### Activity 3  Factor $x^2 - bx + c$

Work cooperatively. Use algebra tiles to factor $x^2 - 5x + 6$.

**Step 1** Model $x^2 - 5x + 6$.

**Step 2** Place the $x^2$-tile at the corner of the product mat. Arrange the 1-tiles into a 2-by-3 rectangular array as shown.

**Step 3** Complete the rectangle with the x-tiles. The rectangle has a width of $x - 2$ and a length of $x - 3$.
Therefore, $x^2 - 5x + 6 = (x-2)(x-3)$.

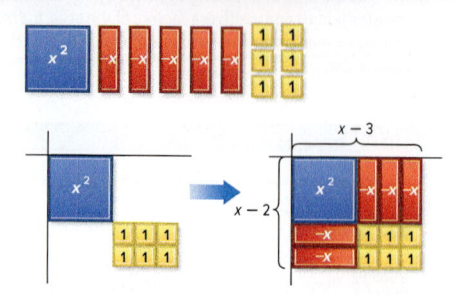

### Activity 4  Factor $x^2 - bx - c$

Use algebra tiles to factor $x^2 - 4x - 5$.

**Step 1** Model $x^2 - 4x - 5$.

**Step 2** Place the $x^2$-tile at the corner of the product mat. Arrange the 1-tiles into a 1-by-5 rectangular array as shown.

**Step 3** Place the x-tiles as shown. Recall that you can add zero pairs without changing the value of the polynomial. In this case, add a zero pair of x-tiles.
The rectangle has a width of $x + 1$ and a length of $x - 5$.
Therefore, $x^2 - 4x - 5 = (x+1)(x-5)$.

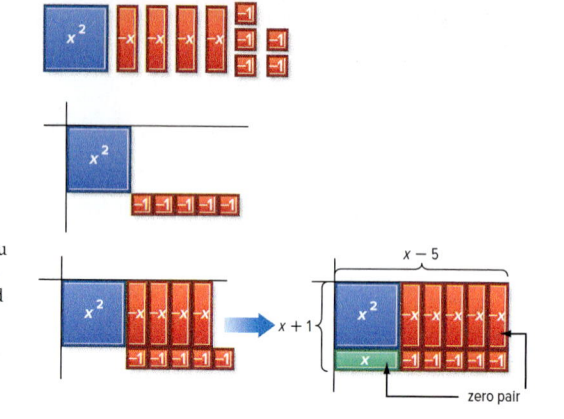

### Model and Analyze

Work cooperatively. Use algebra tiles to factor each trinomial.  1–8. See margin.

1. $x^2 + 3x + 2$
2. $x^2 + 6x + 8$
3. $x^2 + 3x - 4$
4. $x^2 - 7x + 12$
5. $x^2 + 7x + 10$
6. $x^2 - 2x + 1$
7. $x^2 + x - 12$
8. $x^2 - 8x + 15$

Tell whether each trinomial can be factored. Justify your answer with a drawing.

9. $x^2 + 3x + 6$  no
10. $x^2 - 5x - 6$  yes
11. $x^2 - x - 4$  no
12. $x^2 - 4$  yes

13. **WRITING IN MATH** How can you use algebra tiles to determine whether a trinomial can be factored?

13. Trinomials can be factored if they can be represented by a rectangle. Sample answers: $x^2 + 4x + 4$ can be factored, and $x^2 + 6x + 4$ cannot be factored.

9–12. See Ch. 8 Answer Appendix for drawings.

---

## MP Teaching the Mathematical Practices

**Tools** Mathematically proficient students consider the available tools when solving a mathematical problem. Encourage students to use algebra tiles as they progress in the chapter. Provide tiles or paper tiles that they can keep in their notebooks so they are always available.

# LESSON 6
## Factoring Quadratic Trinomials

**Then**
- You multiplied binomials by using the FOIL method.

**Now**
1. Factor trinomials of the form $x^2 + bx + c$.
2. Factor trinomials of the form $ax^2 + bx + c$.

**Why?**
Diana is having a rectangular hot tub installed near her pool. A 24-foot fence is to surround the hot tub. If the hot tub will cover an area of 36 square feet, what will be the dimensions of the hot tub?

To solve this problem, the landscape architect needs to find two numbers that have a product of 36 and a sum of 12, half the perimeter of the hot tub.

**New Vocabulary**
prime polynomial

**Mathematical Practices**
8 Look for and express regularity in repeated reasoning.

**Content Standards**
A.SSE.2 Use the structure of an expression to identify ways to rewrite it.

**1 Factor $x^2 + bx + c$** You have learned how to multiply two binomials using the FOIL method. Each of the binomials was a factor of the product. The pattern for multiplying two binomials can be used to factor certain types of trinomials.

$(x + 3)(x + 4) = x^2 + 4x + 3x + 3 \cdot 4$     Use the FOIL method.
$= x^2 + (4 + 3)x + 3 \cdot 4$     Distributive Property
$= x^2 + 7x + 12$     Simplify.

Notice that the coefficient of the middle term, $7x$, is the sum of 3 and 4, and the last term, 12, is the product of 3 and 4.

Observe the following pattern in this multiplication.

$(x + 3)(x + 4) = x^2 + (4 + 3)x + (3 \cdot 4)$
$(x + m)(x + p) = x^2 + (p + m)x + mp$     Let $3 = m$ and $4 = p$.
$\phantom{(x + m)(x + p)} = x^2 + (m + p)x + mp$     Commutative (+)
$\phantom{(x + m)(x + p) =} x^2 + \phantom{xx} bx \phantom{xx} + \phantom{xx} c$     $b = m + p$ and $c = mp$

Notice that the coefficient of the middle term is the sum of $m$ and $p$, and the last term is the product of $m$ and $p$. This pattern can be used to factor trinomials of the form $x^2 + bx + c$.

### Key Concept   Factoring $x^2 + bx + c$

**Words**    To factor trinomials in the form $x^2 + bx + c$, find two integers, $m$ and $p$, with a sum of $b$ and a product of $c$. Then write $x^2 + bx + c$ as $(x + m)(x + p)$.

**Symbols**    $x^2 + bx + c = (x + m)(x + p)$ when $m + p = b$ and $mp = c$.

**Example**    $x^2 + 6x + 8 = (x + 2)(x + 4)$, because $2 + 4 = 6$ and $2 \cdot 4 = 8$.

When $c$ is positive, its factors have the same signs. Both of the factors are positive or negative based upon the sign of $b$. If $b$ is positive, the factors are positive. If $b$ is negative, the factors are negative.

---

## Lesson 8-6 | Factoring Quadratic Trinomials

## Launch

Have students read the Why? section of the lesson. Ask:

- Why do you need to find two numbers that have a product of 36 to find the dimensions of the hot tub? The hot tub is a rectangle, so the area is equal to the length times the width. Since the hot tub's area is 36 ft², the length and width must be two numbers with a product of 36.

- What pairs of whole numbers have a product of 36? 1 and 36; 2 and 18; 3 and 12; 4 and 9; 6 and 6

- Which pair has a sum of 12? 6 and 6

- What are the dimensions of the hot tub? 6 ft by 6 ft

---

## MP Mathematical Practices Strategies

### Look for and express regularity in repeated reasoning.

Help students maintain oversight of the process of factoring quadratic expressions. For example, ask:

- When $a$ is 1, how can you check to see if you chose the correct pair of factors? Add them to see if they equal the coefficient of the middle term.
- When the last term is positive but the coefficient of the middle term is negative, what are the signs of the two factors? negative
- When the last term is negative, how can you check to see if you chose the right factors? Make sure one is positive and the other is negative and that the factor with the greater absolute value has the same sign as $b$.
- When $a$ is not 1, how do we start to look for factor pairs? Look for factors of the product $ac$.
- What does it mean for a trinomial to be prime? there are no integral factors of the constant term that add to the coefficient of the middle term
- How can you check to see if your factorization is correct? Multiply the two binomials or use a graphing calculator.

### Go Online!

**Interactive Whiteboard**
Use the eLesson or Lesson Presentation to present this lesson.

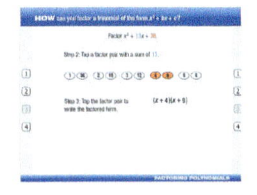

connectED.mcgraw-hill.com    531

# Lesson 8-6 | Factoring Quadratic Trinomials

## Teach

Ask the scaffolded questions for each example to build conceptual understanding for students at all levels.

### 1 Factor $x^2 + bx + c$

**Example 1** *b* and *c* are Positive

**AL** Can the factors be negative? **No; because 20 and 9 are both positive, both factors must be positive.**

**OL** What is needed to factor the trinomial? **factors of 20 that when added together equal 9**

**BL** How can I check my work? **Sample answer: Multiply the binomials and make sure the solution is equal to the original trinomial expression.**

**Need Another Example?**
Factor $x^2 + 7x + 12$. $(x + 3)(x + 4)$

**Example 2** *b* is Negative and *c* is Positive

**AL** Can the factors be positive? **No; because −8 is negative and 12 is positive, both factors must be negative.**

**OL** What is needed to factor the trinomial? **negative factors of 12 that when added together equal −8**

**BL** What is another way you can check the solution? **Sample answer: Graph the trinomial as well as your solution on a graphing calculator. If there is only one graph shown, that means the two graphs are equivalent.**

**Need Another Example?**
Factor $x^2 - 12x + 27$. $(x - 3)(x - 9)$

---

**Watch Out!**

**Preventing Errors** If students use a graphing calculator to check their factoring, make sure they clear all other functions from the Y= list and clear all other drawings from the draw menu. Caution students that while two graphs may appear to coincide in the standard viewing window, they may not. Suggest that students use the TABLE feature to verify identical *y*-values.

---

**Problem-Solving Tip**

**Guess and Check** When factoring a trinomial, make an educated guess, check for reasonableness, and then adjust the guess until the correct answer is found.

**Example 1** *b* and *c* are Positive  A.SSE.2

Factor $x^2 + 9x + 20$.

In this trinomial, $b = 9$ and $c = 20$. Since *c* is positive and *b* is positive, you need to find two positive factors with a sum of 9 and a product of 20. Make an organized list of the factors of 20, and look for the pair of factors with a sum of 9.

| Factors of 20 | Sum of Factors |
|---|---|
| 1, 20 | 21 |
| 2, 10 | 12 |
| 4, 5 | 9 |

The correct factors are 4 and 5.

$x^2 + 9x + 20 = (x + m)(x + p)$  Write the pattern.
$= (x + 4)(x + 5)$  $m = 4$ and $p = 5$

**CHECK** You can check this result by multiplying the two factors. The product should be equal to the original expression.

$(x + 4)(x + 5) = x^2 + 5x + 4x + 20$  FOIL Method
$= x^2 + 9x + 20$ ✓  Simplify.

**Guided Practice**

Factor each polynomial.

**1A.** $d^2 + 11d + 24$ $(d + 3)(d + 8)$     **1B.** $9 + 10t + t^2$ $(t + 9)(t + 1)$

When factoring a trinomial in which *b* is negative and *c* is positive, use what you know about the product of binomials to narrow the list of possible factors.

**Example 2** *b* is Negative and *c* is Positive  A.SSE.2

Factor $x^2 - 8x + 12$. Confirm your answer using a graphing calculator.

In this trinomial, $b = -8$ and $c = 12$. Since *c* is positive and *b* is negative, you need to find two negative factors with a sum of −8 and a product of 12.

**Study Tip**

**MP Regularity** Once the correct factors are found, it is not necessary to test any other factors. In Example 2, −2 and −6 are the correct factors, so −3 and −4 do not need to be tested.

| Factors of 12 | Sum of Factors |
|---|---|
| −1, −12 | −13 |
| −2, −6 | −8 |
| −3, −4 | −7 |

The correct factors are −2 and −6.

$x^2 - 8x + 12 = (x + m)(x + p)$  Write the pattern.
$= (x - 2)(x - 6)$  $m = -2$ and $p = -6$

**CHECK** Graph $y = x^2 - 8x + 12$ and $y = (x - 2)(x - 6)$ on the same screen. Since only one graph appears, the two graphs must coincide. Therefore, the trinomial has been factored correctly. ✓

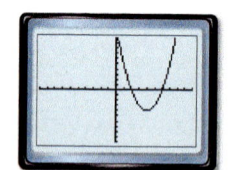

[−10, 10] scl: 1 by [−10, 10] scl: 1

**Guided Practice**

Factor each polynomial.

**2A.** $21 - 22m + m^2$ $(m - 1)(m - 21)$     **2B.** $w^2 - 11w + 28$ $(w - 7)(w - 4)$

**Teaching Tip**

**Calculators** Have students change the line type on the graph of the second equation so they can see it as it is graphed on the first graph.

**Review Vocabulary**
**Absolute Value** the distance a number is from zero on a number line, written $|n|$

When $c$ is negative, its factors have opposite signs. To determine which factor is positive and which is negative, look at the sign of $b$. The factor with the greater absolute value has the same sign as $b$.

A.SSE.2

### Example 3   $c$ is Negative

Factor each polynomial. Confirm your answers using a graphing calculator.

**a.** $x^2 + 2x - 15$

In this trinomial, $b = 2$ and $c = -15$. Since $c$ is negative, the factors $m$ and $p$ have opposite signs. So either $m$ or $p$ is negative, but not both. Since $b$ is positive, the factor with the greater absolute value is also positive.

List the factors of $-15$, where one factor of each pair is negative. Look for the pair of factors with a sum of 2.

| Factors of $-15$ | Sum of Factors |
|---|---|
| $-1, 15$ | 14 |
| $-3, 5$ | 2 |

The correct factors are $-3$ and 5.

$x^2 + 2x - 15 = (x + m)(x + p)$   Write the pattern.
$\phantom{x^2 + 2x - 15} = (x - 3)(x + 5)$   $m = -3$ and $p = 5$

**CHECK** $(x - 3)(x + 5) = x^2 + 5x - 3x - 15$   FOIL Method
$\phantom{(x - 3)(x + 5)} = x^2 + 2x - 15$ ✓   Simplify

**b.** $x^2 - 7x - 18$

In this trinomial, $b = -7$ and $c = -18$. Either $m$ or $p$ is negative, but not both. Since $b$ is negative, the factor with the greater absolute value is also negative.

List the factors of $-18$, where one factor of each pair is negative. Look for the pair of factors with a sum of $-7$.

| Factors of $-18$ | Sum of Factors |
|---|---|
| $1, -18$ | $-17$ |
| $2, -9$ | $-7$ |
| $3, -6$ | $-3$ |

The correct factors are 2 and $-9$.

$x^2 - 7x - 18 = (x + m)(x + p)$   Write the pattern.
$\phantom{x^2 - 7x - 18} = (x + 2)(x - 9)$   $m = 2$ and $p = -9$

**CHECK** Graph $y = x^2 - 7x - 18$ and $y = (x + 2)(x - 9)$ on the same screen.

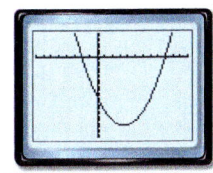

$[-10, 15]$ scl: 1 by $[-40, 20]$ scl: 1

The graphs coincide. Therefore, the trinomial has been factored correctly. ✓

**Go Online!**
Many graphing calculators allow you to choose the way that each graph is displayed. Using different thicknesses or colors can allow you to see whether two graphs coincide.

▶ **Guided Practice**

**3A.** $y^2 + 13y - 48$   $(y - 3)(y + 16)$
**3B.** $r^2 - 2r - 24$   $(r + 4)(r - 6)$

---

Lesson 8-6 | Factoring Quadratic Trinomials

### Example 3   $c$ is Negative

**AL** For part **a**, are the factors both positive, both negative, or one positive and one negative?
Because $-15$ is negative and 2 is positive, one factor is positive and the other is negative.

**OL** In part **a**, what are the possible factors of $-15$?
$-3$ and 5

**BL** Write your own trinomial and then factor it.
See students' work.

### Need Another Example?
Factor each polynomial.
**a.** $x^2 + 3x - 18$   $(x + 6)(x - 3)$
**b.** $x^2 - x - 20$   $(x - 5)(x + 4)$

### Teaching the Mathematical Practices

**Regularity** Mathematically proficient students maintain oversight of the process, while attending to the details of solving a problem. Stress that exactly one pair of factors correctly factors a trinomial of the form $x^2 + bx + c$ that can be factored.

---

### Differentiated Instruction  AL OL

**IF** the concept of factoring trinomials seems somewhat abstract to some students,

**THEN** whenever you introduce abstract concepts, it is a good idea to reinforce them with concrete examples. After introducing factoring trinomials, refer students to the lesson-opener problem. Ask students to describe any similarities they notice between finding the dimensions of the pool and factoring a trinomial. Some students may benefit from modeling some problems using algebra tiles. They can then use this method to solve quadratic equations.

## Lesson 8-6 | Factoring Quadratic Trinomials

## 2 Factor $ax^2 + bx + c$

**Example 4** Factor $ax^2 + bx + c$

**AL** In part c, what should you look to factor first? **the GCF**

**OL** In part a, what are the factors you need to find? **two positive factors of 28 that have a sum of 29**

**BL** Why does factoring out the GCF first help? **Sample answer: If you can factor out the GCF first, you might be able to get rid of the $x^2$ constant, which makes it much easier to factor the trinomial**

**Need Another Example?**

Factor each trinomial.
a. $5x^2 + 27x + 10$ $(5x + 2)(x + 5)$
b. $4x^2 + 24x + 32$ $4(x + 2)(x + 4)$

---

**2 Factor $ax^2 + bx + c$** You can apply the factoring methods to quadratic expressions where $a$ is not 1. The factoring by grouping method is necessary when factoring these quadratic expressions.

### Key Concept  Factoring $ax^2 + bx + c$

**Words**    To factor trinomials of the form $ax^2 + bx + c$, find two integers, $m$ and $p$, with a sum of $b$ and a product of $ac$. Then write $ax^2 + bx + c$ as $ax^2 + mx + px + c$, and factor by grouping.

**Example**
$$5x^2 - 13x + 6 = 5x^2 - 10x - 3x + 6 \quad m = -10 \text{ and } p = -3$$
$$= 5x(x - 2) + (-3)(x - 2)$$
$$= (5x - 3)(x - 2)$$

A.SSE.2

**Example 4**    Factor $ax^2 + bx + c$

Factor each trinomial.

**a.** $7x^2 + 29x + 4$

In this trinomial, $a = 7$, $b = 29$, and $c = 4$. You need to find two numbers with a sum of 29 and a product of $7 \cdot 4$ or 28. Make a list of the factors of 28 and look for the pair of factors with the sum of 29.

| Factors of 28 | Sum of Factors |
|---|---|
| 1, 28 | 29 |

The correct factors are 1 and 28.

$7x^2 + 29x + 4 = 7x^2 + mx + px + 4$    Write the pattern.
$\quad\quad\quad\quad\quad\quad = 7x^2 + 1x + 28x + 4$    $m = 1$ and $p = 28$
$\quad\quad\quad\quad\quad\quad = (7x^2 + 1x) + (28x + 4)$    Group terms with common factors.
$\quad\quad\quad\quad\quad\quad = x(7x + 1) + 4(7x + 1)$    Factor the GCF.
$\quad\quad\quad\quad\quad\quad = (x + 4)(7x + 1)$    $7x + 1$ is the common factor.

**Study Tip**

**MP Regularity** Always look for a GCF of the terms of a polynomial before you factor.

**b.** $5x^2 + 12x - 9$

In this trinomial, $a = 5$, $b = 12$, and $c = -9$. You need to find two numbers with a sum of 12 and a product of $5(-9)$ or $-45$. Make a list of factors of $-45$ and look for the pair of factors with the sum of 12.

| Factors of $-45$ | Sum of Factors |
|---|---|
| $-1$, 45 | 44 |
| $-3$, 15 | 12 |

The correct factors are $-3$ and 15.

$5x^2 + 12x - 9 = 5x^2 + 15x - 3x - 9$    $m = 15$ and $p = -3$
$\quad\quad\quad\quad\quad\quad = (5x^2 + 15x) + (-3x - 9)$    Group terms with common factors.
$\quad\quad\quad\quad\quad\quad = 5x(x + 3) + (-3)(x + 3)$    Factor the GCF.
$\quad\quad\quad\quad\quad\quad = (5x - 3)(x + 3)$    Distributive Property

**c.** $3x^2 + 15x + 18$

The GCF of the terms $3x^2$, $15x$, and 18 is 3. Factor this first.

$3x^2 + 15x + 18 = 3(x^2 + 5x + 6)$    Distributive Property
$\quad\quad\quad\quad\quad\quad = 3(x + 3)(x + 2)$    Find two factors of 6 with a sum of 5.

▶ **Guided Practice**

**4A.** $5x^2 + 13x + 6$ $(5x + 3)(x + 2)$    **4B.** $6x^2 + 22x - 8$ $2(3x - 1)(x + 4)$

**Lesson 8-6** | **Factoring Quadratic Trinomials**

Sometimes the coefficient of the x-term is negative.

A.SSE.2

**Example 5** Factor $ax^2 - bx + c$

Factor $3x^2 - 17x + 20$.

In this trinomial, $a = 3$, $b = -17$, and $c = 20$. Since $b$ is negative, $m + p$ will be negative. Since $c$ is positive, $mp$ will be positive.

To determine $m$ and $p$, list the negative factors of $ac$ or 60. The sum of $m$ and $p$ should be $-17$.

| Factors of 60 | Sum of Factors |
|---|---|
| $-2, -30$ | $-32$ |
| $-3, -20$ | $-23$ |
| $-4, -15$ | $-19$ |
| $-5, -12$ | $-17$ |

The correct factors are $-5$ and $-12$.

$3x^2 - 17x + 20 = 3x^2 - 12x - 5x + 20$    $m = -12$ and $p = -5$
$= (3x^2 - 12x) + (-5x + 20)$    Group terms with common factors.
$= 3x(x - 4) + (-5)(x - 4)$    Factor the GCF.
$= (3x - 5)(x - 4)$    Distributive Property

▶ **Guided Practice**

**5A.** $2n^2 - n - 1$ $(n - 1)(2n + 1)$    **5B.** $10y^2 - 35y + 30$ $5(2y - 3)(y - 2)$

A polynomial that cannot be written as a product of two polynomials with integral coefficients is called a **prime polynomial**.

A.SSE.2

**Example 6** Determine Whether a Polynomial Is Prime

Factor $4x^2 - 3x + 5$, if possible. If the polynomial cannot be factored using integers, write *prime*.

| Factors of 20 | Sum of Factors |
|---|---|
| $-20, -1$ | $-21$ |
| $-4, -5$ | $-9$ |
| $-2, -10$ | $-12$ |

In this trinomial, $a = 4$, $b = -3$, and $c = 5$. Since $b$ is negative, $m + p$ is negative. Since $c$ is positive, $mp$ is positive. So, $m$ and $p$ are both negative. Next, list the factors of 20. Look for the pair with a sum of $-3$.

There are no factors with a sum of $-3$. So the quadratic expression cannot be factored using integers. Therefore, $4x^2 - 3x + 5$ is prime.

▶ **Guided Practice**

Factor each polynomial, if possible. If the polynomial cannot be factored using integers, write *prime*.

**6A.** $4t^2 - t + 7$ prime    **6B.** $2x^2 + 3x - 5$ $(2x + 5)(x - 1)$

**Real-World Career**

**Urban Planner** Urban planners design the layout of an area. They take into consideration the available land and geographical and environmental factors to design an area that benefits the community the most. City planners have a bachelor's degree in planning and almost half have a master's degree.

---

**Example 5** Factor $ax^2 - bx + c$

**AL** Is there a GCF of the terms that is greater than one? no

**OL** What are the factors you need to find? two negative factors of 60 that have a sum of $-17$

**BL** How can you keep yourself organized when guessing and checking the factors? Sample answer: you can make a table of the factors of 60 and then find the sum of each pair of factors.

**Need Another Example?**
Factor $24x^2 - 22x + 3$. $(4x - 3)(6x - 1)$

**Watch Out!**
**Preventing Errors** Many students forget to included the GCF that they factored from the trinomial. Remind students to put the GCF in front of the other two factors.

**Watch Out!**
**Preventing Errors** Make sure students list all possible factors of $mp$, including both positive and negative factors, before they decide the polynomials is prime.

**Example 6** Determine Whether a Polynomial Is Prime

**AL** Is there a GCF of the terms that is greater than one? no

**OL** What are the factors you need to find? two negative factors of 20 that have a sum of $-3$

**BL** In your own words, write what it means for a trinomial to be prime. See students' work.

**Need Another Example?**
Factor $3x^2 + 7x - 5$. If the polynomial cannot be factored using integers, write *prime*. prime

---

**Differentiated Instruction** **AL**

**IF** some students have trouble factoring polynomials,

**THEN** place students in groups to factor polynomials such as those in Examples 4 and 5. Depending on the number of factors and number of students in each group, have each student find one or two factors for $mp$. By dividing the labor, students should be able to find the factors for $mp$ that sum to $m + p$ quickly. Once they have found the factors, have students complete the factoring as a group.

connectED.mcgraw-hill.com   535

# Lesson 8-6 | Factoring Quadratic Trinomials

## Practice

**Formative Assessment** Use Exercises 1–10 to assess students' understanding of the concepts in this lesson.

The Practice and Problem Solving exercises assess the content taught in the lesson. The Preparing for Assessment page is meant to be used as preparation for end-of-course assessments.

### Teaching the Mathematical Practices

**Structure** Mathematically proficient students look closely to discern a pattern or structure. In Exercises 31–34, point out that the same patterns that apply to polynomials that contain numbers apply to variables.

## Extra Practice

See page R8 for extra exercises for students who are approaching level or for on-level students who need additional reinforcement.

### Watch Out!

**Student Misconceptions** For Exercises 1–6, students may need to be reminded that the order in which they record the factors does not matter. So, $(x + m)(x + p)$ and $(x + p)(x + m)$ are both correct.

### Go Online! / eBook

**Interactive Student Guide**
Use the *Interactive Student Guide* to deepen conceptual understanding.
- Solving by Factoring $x^2 + bx + c = 0$
- Solving by Factoring $ax^2 + bx + c = 0$

---

### Check Your Understanding
= Step-by-Step Solutions begin on page R11.
**Go Online!** for a Self-Check Quiz

**Examples 1–3** Factor each polynomial. Confirm your answers using a graphing calculator.
A.SSE.2
1. $x^2 + 14x + 24$  $(x+2)(x+12)$
2. $y^2 - 7y - 30$  $(y-10)(y+3)$
3. $n^2 + 4n - 21$  $(n+7)(n-3)$
4. $m^2 - 15m + 50$  $(m-5)(m-10)$

**Examples 4–6** Factor each polynomial, if possible. If the polynomial cannot be factored using integers, write *prime*.
A.SSE.2
5. $4x^2 - 30x + 36$  $2(2x-3)(x-6)$
6. $6x^2 - x - 14$  prime
7. $3x^2 + 17x + 10$  $(3x+2)(x+5)$
8. $2x^2 + 22x + 56$  $2(x+4)(x+7)$
9. $5x^2 - 3x + 4$  prime
10. $3x^2 - 11x - 20$  $(3x+4)(x-5)$

### Practice and Problem Solving
Extra Practice is on page R8.

**Examples 1–3** Factor each polynomial. Confirm your answers using a graphing calculator.
A.SSE.2
11. $x^2 + 17x + 42$  $(x+3)(x+14)$
12. $y^2 - 17y + 72$  $(y-9)(y-8)$
13. $a^2 + 8a - 48$  $(a-4)(a+12)$
14. $n^2 - 2n - 35$  $(n-7)(n+5)$
15. $44 + 15h + h^2$  $(h+4)(h+11)$
16. $40 - 22x + x^2$  $(x-2)(x-20)$
17. $-24 - 10x + x^2$  $(x+2)(x-12)$
18. $-42 - m + m^2$  $(m+6)(m-7)$

**Examples 4–6** Factor each polynomial, if possible. If the polynomial cannot be factored using integers, write *prime*.  19. $(2x+3)(x+8)$  20. $(5x+4)(x+6)$  22. $2(2x+5)(x+7)$  23. $(2x+3)(x-3)$
A.SSE.2
19. $2x^2 + 19x + 24$
20. $5x^2 + 34x + 24$
21. $4x^2 + 22x + 10$  $2(2x+1)(x+5)$
22. $4x^2 + 38x + 70$
23. $2x^2 - 3x - 9$
24. $4x^2 - 13x + 10$  $(4x-5)(x-2)$
25. $2x^2 + 3x + 6$  prime
26. $5x^2 + 3x + 4$  prime
27. $12x^2 + 69x + 45$  $3(4x+3)(x+5)$
28. $4x^2 - 5x + 7$  prime
29. $5x^2 + 23x + 24$  $(5x+8)(x+3)$
30. $3x^2 - 8x + 15$  prime

**STRUCTURE** Factor each polynomial.
31. $q^2 + 11qr + 18r^2$  $(q+2r)(q+9r)$
32. $x^2 - 14xy - 51y^2$  $(x+3y)(x-17y)$
33. $x^2 - 6xy + 5y^2$  $(x-y)(x-5y)$
34. $a^2 + 10ab - 39b^2$  $(a+13b)(a-3b)$

**GEOMETRY** Find an expression for the perimeter of a rectangle with the given area.
35. $A = x^2 + 24x - 81$  $4x + 48$
36. $A = x^2 + 13x - 90$  $4x + 26$

Factor each polynomial, if possible. If the polynomial cannot be factored using integers, write *prime*.
37. $-6x^2 - 23x - 20$  $-(2x+5)(3x+4)$
38. $-4x^2 - 15x - 14$  $-(x+2)(4x+7)$
39. $-5x^2 + 18x + 8$  $-(x-4)(5x+2)$
40. $-6x^2 + 31x - 35$  $-(2x-7)(3x-5)$
41. $-4x^2 + 5x - 12$  prime
42. $-12x^2 + x + 20$  $-(3x-4)(4x+5)$

43. **MULTIPLE REPRESENTATIONS** In this problem, you will explore factoring a special type of polynomial.

   a. **Geometric** Draw a square and label the sides $a$. Within this square, draw a smaller square that shares a vertex with the first square. Label the sides $b$. What are the areas of the two squares?  $a^2$ and $b^2$

---

### Differentiated Homework Options

| Levels | AL Basic | OL Core | BL Advanced |
|---|---|---|---|
| Exercises | 11–30, 44, 49, 50, 52–62 | 11–41 odd, 44, 49, 50, 52–62 | 31–53, (optional: 54–62) |
| 2-Day Option | 11–29 odd, 54–62 | 11–30, 54–62 | |
| | 12–30 even, 44, 49, 50, 52, 53 | 31–44, 49, 50, 52, 53 | |

You can use ALEKS to provide additional remediation support with personalized instruction and practice.

**Lesson 8-6** | Factoring Quadratic Trinomials

43e. $(a-b)(a+b)$; the figure with area $a^2 - b^2$ and the rectangle with area $(a-b)(a+b)$ have the same area, so $a^2 - b^2 = (a-b)(a+b)$.

b. **Geometric** Cut and remove the small square. What is the area of the remaining region? $a^2 - b^2$

c. **Analytical** Draw a diagonal line between the inside corner and outside corner of the figure, and cut along this line to make two congruent pieces. Then rearrange the two pieces to form a rectangle. What are the dimensions? width: $a-b$, length: $a+b$

d. **Analytical** Write the area of the rectangle as the product of two binomials. $(a-b)(a+b)$

e. **Verbal** Complete this statement: $a^2 - b^2 = \ldots$ Why is this statement true?

A.SSE.2, A.SSE.3a

**H.O.T. Problems** Use Higher-Order Thinking Skills

44. **ERROR ANALYSIS** Jerome and Charles have factored $x^2 + 6x - 16$. Is either of them correct? Explain your reasoning. Charles; Jerome's answer once multiplied is $x^2 - 6x - 16$. The middle term should be positive.

| Jerome | Charles |
|---|---|
| $x^2 + 6x - 16 = (x + 2)(x - 8)$ | $x^2 + 6x - 16 = (x - 2)(x + 8)$ |

**CONSTRUCT ARGUMENTS** Find all values of $k$ so that each polynomial can be factored using integers.

45. $x^2 + kx + 14$   $-15, -9, 9, 15$

46. $2x^2 + kx + 12$   $\pm 25, \pm 14, \pm 11, \pm 10$

47. $x^2 - 8x + k, k > 0$   $7, 12, 15, 16$

48. $2x^2 - 5x + k, k > 0$   $2, 3$

49. **REASONING** For any factorable trinomial, $x^2 + bx + c$, will the absolute value of $b$ sometimes, always, or never be less than the absolute value of $c$? Explain. See margin.

50. **REASONING** A square has an area of $9x^2 + 30xy + 25y^2$ square inches. The dimensions are binomials with positive integer coefficients. What is the perimeter of the square? Explain. See margin.

51. **CHALLENGE** Factor $(4y - 5)^2 + 3(4y - 5) - 70$.
51. $(4y-5)^2 + 3(4y-5) - 70 = [(4y-5) + 10] \cdot [(4y-5) - 7]$
$= (4y+5)(4y-12) = 4(4y+5)(y-3)$

52. **WRITING IN MATH** Compare the method for factoring trinomials of the form $ax^2 + bx + c$ to the method for factoring trinomials of the form $x^2 + bx + c$. Discuss how to find the signs of the factors of $c$. See margin.

53. **FLAGS** The official flag of most countries is rectangular in shape, but the flag of Switzerland is a square. However, Swiss naval vessels fly a rectangular flag. The area of the square flag is $x^2 - 6x + 9$ square feet and the area of the rectangular naval flag is $x^2 - 2x - 3$. How many feet longer is the naval flag than the square flag? 4 feet

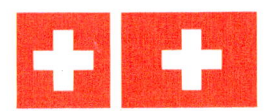

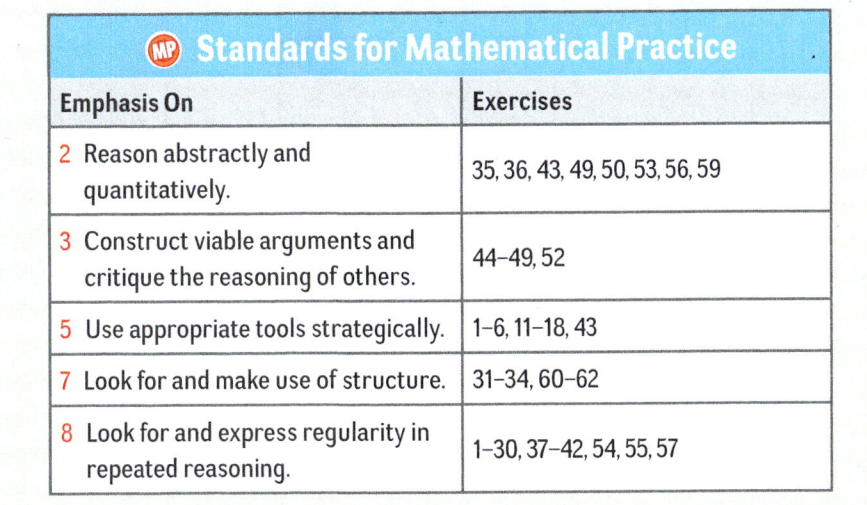

## Standards for Mathematical Practice

| Emphasis On | Exercises |
|---|---|
| 2 Reason abstractly and quantitatively. | 35, 36, 43, 49, 50, 53, 56, 59 |
| 3 Construct viable arguments and critique the reasoning of others. | 44–49, 52 |
| 5 Use appropriate tools strategically. | 1–6, 11–18, 43 |
| 7 Look for and make use of structure. | 31–34, 60–62 |
| 8 Look for and express regularity in repeated reasoning. | 1–30, 37–42, 54, 55, 57 |

## Levels of Complexity Chart

The levels of the exercises progress from 1 to 3, with Level 1 indicating the lowest level of complexity.

| Exercises | 11–30 | 31–42, 54–62 | 43–53 |
|---|---|---|---|
| Level 3 | | | ● |
| Level 2 | | ● | |
| Level 1 | ● | | |

**Watch Out!**

**Error Analysis** For Exercise 44, remind students that they can check the correctness of the factors by multiplying them to see if the result is the original polynomial. They could also use their graphing calculators for the procedure shown in Example 2.

## Teaching the Mathematical Practices

**Construct Arguments** Mathematically proficient students make conjectures and build a logical progression of statements to explore the truth of their conjectures. In Exercises 45–48, tell students to start by thinking about how they would factor each polynomial if $k$ is given. How is $k$ related to the other constants or coefficients in the polynomial?

## Additional Answers

49. sometimes; Sample answer: The trinomial $x^2 + 10x + 9 = (x + 1)(x + 9)$ and $10 > 9$. The trinomial $x^2 + 7x + 10 = (x + 2)(x + 5)$ and $7 < 10$.

50. $(12x + 20y)$ in.; The area of the square equals $(3x + 5y) \cdot (3x + 5y)$ in$^2$, so the length of one side is $(3x + 5y)$ in. The perimeter is $4(3x + 5y)$ or $(12x + 20y)$ in.

52. Sample answer: To factor $ax^2 + bx + c$, find integers $m$ and $n$ such that $m + n = b$ and $mn = ac$. If $b$ and $c$ are positive, then $m$ and $n$ are positive. The process for factoring $x^2 + bx + c$ is the same, but because $a = 1$, $mn = c$. If $b$ is negative and $c$ is positive, then $m$ and $n$ are negative. When $ac$ is negative, $m$ and $n$ have different signs and the factor with the greatest absolute value has the same sign as $b$.

connectED.mcgraw-hill.com  537

**Lesson 8-6 | Factoring Quadratic Trinomials**

# Preparing for Assessment

Exercises 54–62 require students to use the skills they will need on standardized assessments. Each exercise is dual-coded with content standards and mathematical practice standards.

## Dual Coding

| Items | Content Standards | MP Mathematical Practices |
|---|---|---|
| 54, 55, 57 | A.SSE.2 | 8 |
| 56 | A.SSE.2 | 2 |
| 58 | A.SSE.2 | 1 |
| 59 | A.SSE.2 | 1, 4 |
| 60–62 | A.SSE.2 | 7 |

## Diagnose Student Errors

Survey student responses for each item. Class trends may indicate common errors and misconceptions.

**54.**

| A | Factors of −56 are correct, but the sum is not 10 |
|---|---|
| B | Correct factors, but signs are reversed |
| C | CORRECT |
| D | Factors of −56 are correct, but the sum is not 10 |

**55.**

| A | Factored the polynomial as $(x - 9)(x + 12)$ |
|---|---|
| B | CORRECT |
| C | Factored the polynomial as $(x + 4)(x - 27)$ |
| D | Factored the polynomial as $(x - 27)(x + 4)$ |

**56.**

| A | Placed signs incorrectly |
|---|---|
| B | CORRECT |
| C | Used incorrect sign in $3x + 6$ |
| D | Used a negative sign in both binomials |

## Go Online!

**Quizzes**
Students can use *Self-Check Quizzes* to check their understanding of this lesson. You can also give the *Chapter Quiz*, which covers the content in Lessons 8-1 through 8-6.

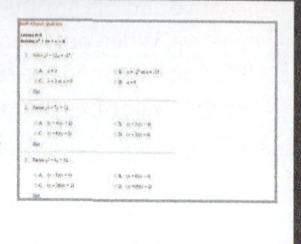

**57.**

| A | CORRECT |
|---|---|
| B | Used incorrect sign in $5x + 2y$ |
| C | Used incorrect sign in $x - 3y$ |
| D | Omitted the $y$ in $5x - 2$ |

---

# Preparing for Assessment

**54.** Which expression shows the factored form of $x^2 + 10x - 56$? MP 8 A.SSE.2 **C**

- A $(x - 7)(x + 8)$
- B $(x - 14)(x + 4)$
- C $(x + 14)(x - 4)$
- D $(x + 7)(x - 8)$

**55.** Which expression is a factor of $x^2 - 3x - 108$? MP 8 A.SSE.2 **B**

- A $(x - 9)$
- B $(x - 12)$
- C $(x + 4)$
- D $(x - 27)$

**56.** The area of a rectangle is represented by the polynomial $6x^2 + 3x - 30$. Which expressions could represent the sides of the rectangle? MP 2 A.SSE.2 **B**

- A $2x - 5$ and $3x + 6$
- B $2x + 5$ and $3x - 6$
- C $2x + 5$ and $3x + 6$
- D $2x - 5$ and $3x - 6$

**57.** Which is a factor of the polynomial $5x^2 + 13xy - 6y^2$? MP 8 A.SSE.2 **A**

- A $(5x - 2y)$
- B $(5x + 2y)$
- C $(x - 3y)$
- D $(5x - 2)$

**58.** Factor each polynomial. MP 2 A.SSE.2

a. $x^2 - 7x + 10$    $(x - 2)(x - 5)$

b. $3x^2 - 6x + 3$    $3(x - 1)^2$

**59. MULTI-STEP** A city has commissioned the building of a rectangular park. The area of the park will be $660x^2 + 524x + 85$ square feet. MP 1, 4 A.SSE.2

a. To factor the expression for the area of the park, what must be the sum of $m$ and $p$? the product?

   524; 56,100

b. Which expressions could represent the dimensions of the park? **D**

- A $22x(30x + 17)$ and $5(30x + 17)$
- B $22x - 5$ and $30x - 17$
- C $30x + 17$ and $30x + 17$
- D $22x + 5$ and $30x + 17$

c. The city plans to build a pavilion in the park that will require an area of $27x^2 - 30x - 13$. Factor the expression to find the dimensions of the pavilion. $(9x - 13)$ ft and $(3x + 1)$ ft

d. If $x = 5$, what will be the dimensions of the park and the pavilion? 115 ft by 167 ft, 16 ft by 32 ft

**60.** The perimeter of a rectangle is the same as the perimeter of a square. If the area of this rectangle is $x^2 + 4x - 21$ and the side lengths are integers, find: MP 7 A.SSE.2

a. the perimeter of the rectangle    $4x + 8$

b. the area of the square    $x^2 + 4x + 4$

**61.** Find a common factor between the polynomials $4x^2 + 5x + 9$ and $x^2 - 4x + 3$. MP 7 A.SSE.2

   $x - 1$

**62.** Factor each polynomial. MP 7 A.SSE.2

a. $z^2 + 2yz - 3y^2$    $(z + 3y)(z - y)$

b. $3z^2 + 2yz - y^2$    $(3z - y)(z + y)$

c. $27y^2 - 12z^2$    $3(3y - 2z)(3y + 2z)$

---

## Differentiated Instruction  OL  BL

**Extension** Write the trinomials $x^2 + x - 6$ and $x^2 - x - 6$ on the board. Ask students to compare the two trinomials. How are the trinomials related? How are their factors related? The trinomials are the same except for the sign of the middle term. When factored, $x^2 + x - 6$ is $(x + 3)(x - 2)$, while $x^2 - x - 6$ is $(x - 3)(x + 2)$. The factors have opposite signs in the constant term.

# LESSON 8-7
# Factoring Special Products

**SUGGESTED PACING (DAYS)**
90 min. 0.5
45 min. 1
Instruction

## Track Your Progress

### Objectives
1. Factor binomials that are differences of squares.
2. Factor trinomials that are perfect squares.

### Mathematical Background
Factoring a polynomial can be done more quickly if the polynomial fits one of these special product patterns: difference of two squares: $a^2 - b^2 = (a + b)(a - b)$; perfect square trinomials: $a^2 + 2ab + b^2 = (a + b)(a + b)$ or $a^2 - 2ab + b^2 = (a - b)(a - b)$. If the terms of the original polynomial have a greatest common factor, factor it out before applying any other factoring technique.

### THEN
**A.APR.1** Understand that polynomials form a system analogous to the integers, namely, they are closed under the operations of addition, subtraction, and multiplication; add subtract, and multiply polynomials.

**A.SSE.1a** Interpret parts of an expression, such as terms, factors, and coefficients.

### NOW
**A.SSE.2** Use the structure of an expression to identify ways to rewrite it.

### NEXT
**A.REI.4b** Solve quadratic equations by inspection (e.g., for $x^2 = 49$), taking square roots, completing the square, the quadratic formula and factoring, as appropriate to the initial form of the equation. Recognize when the quadratic formula gives complex solutions and write them as $a \pm bi$ for real numbers $a$ and $b$.

**F.IF.8a** Use the process of factoring and completing the square in a quadratic function to show zeros, extreme values, and symmetry of the graph, and interpret these in terms of a context.

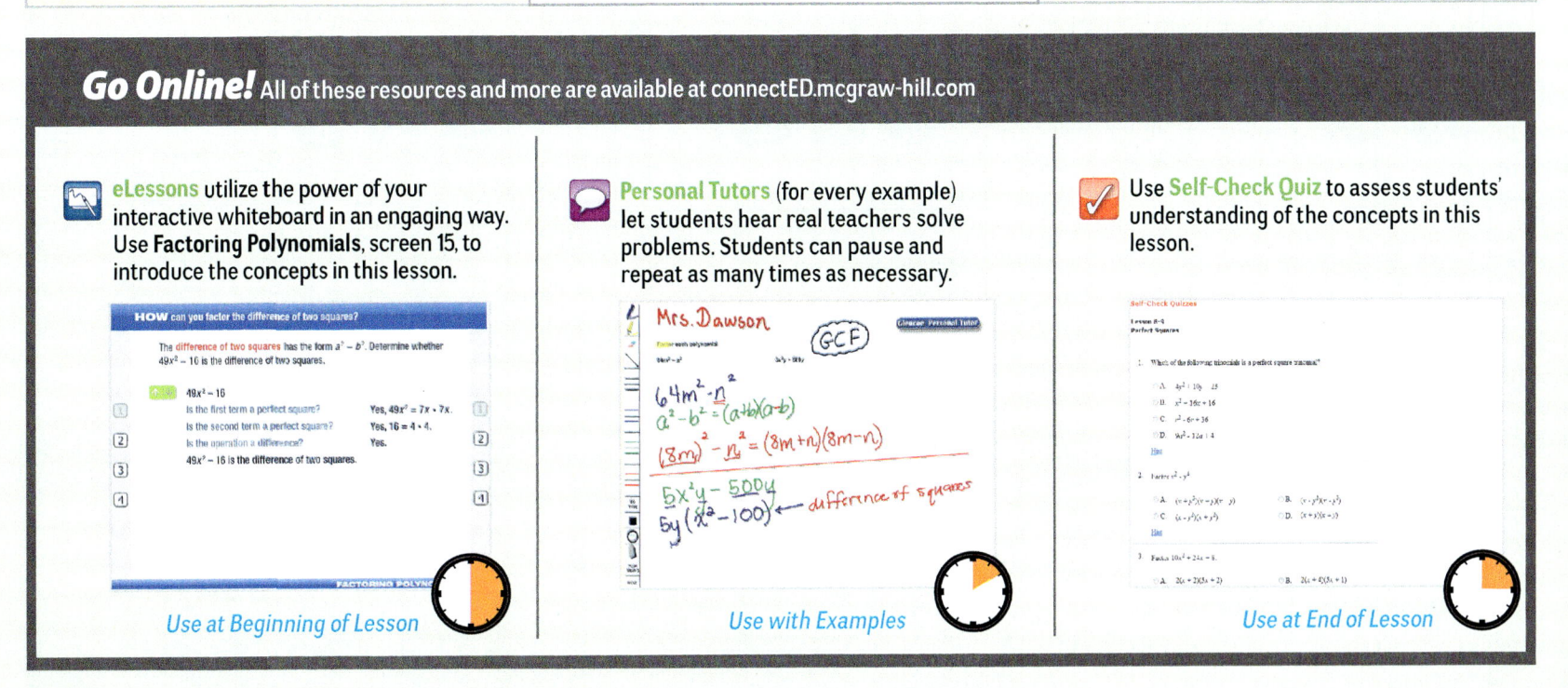

---

## Go Online! All of these resources and more are available at connectED.mcgraw-hill.com

**eLessons** utilize the power of your interactive whiteboard in an engaging way. Use **Factoring Polynomials**, screen 15, to introduce the concepts in this lesson.

*Use at Beginning of Lesson*

**Personal Tutors** (for every example) let students hear real teachers solve problems. Students can pause and repeat as many times as necessary.

*Use with Examples*

Use **Self-Check Quiz** to assess students' understanding of the concepts in this lesson.

*Use at End of Lesson*

### OER Using Open Educational Resources
**Games** Have students play the difference of two squares game on **Manga High** individually. This is an interactive game that has different levels of difficulty. Students answer a variety of question types to reinforce the relationship between the polynomial and its factors. *Use as homework or classwork*

connectED.mcgraw-hill.com 539A

# Differentiate Your Resources

**Extra Practice** Additional practice or homework; Skills Practice is best for approaching-level students and Practice is best for on-level and beyond-level students

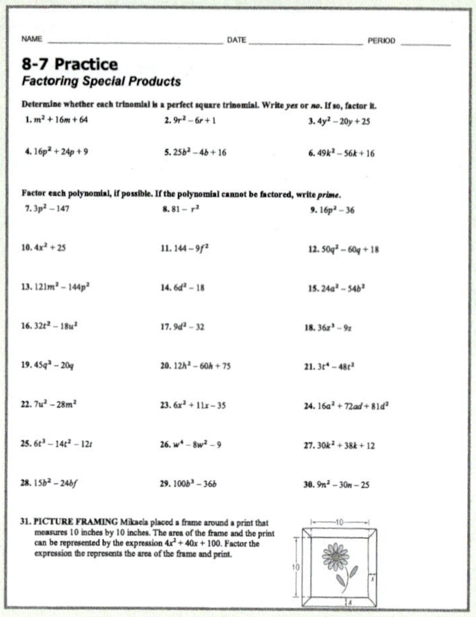

## Skills Practice
AL  OL  ELL

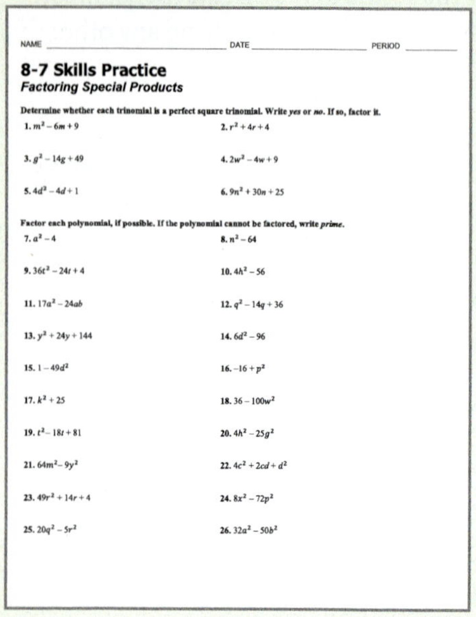

## Practice
AL  OL  BL  ELL

## Word Problem Practice
AL  OL  BL  ELL

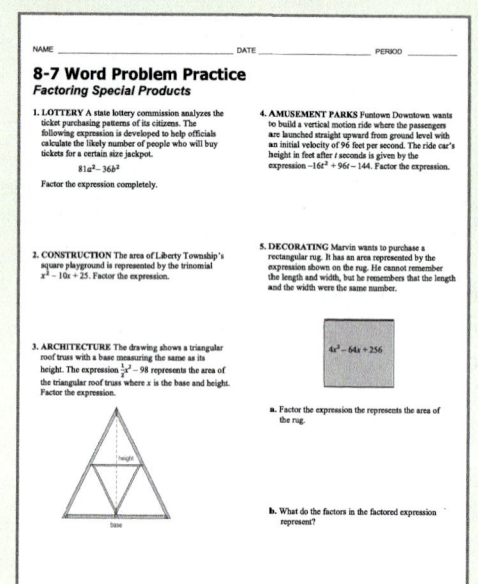

**Intervention** Reteaching and vocabulary activities that can be used with struggling or absent students and as ELL support

**Extension** Activities that can be used to extend lesson concepts

## Study Guide and Intervention
AL  OL  ELL

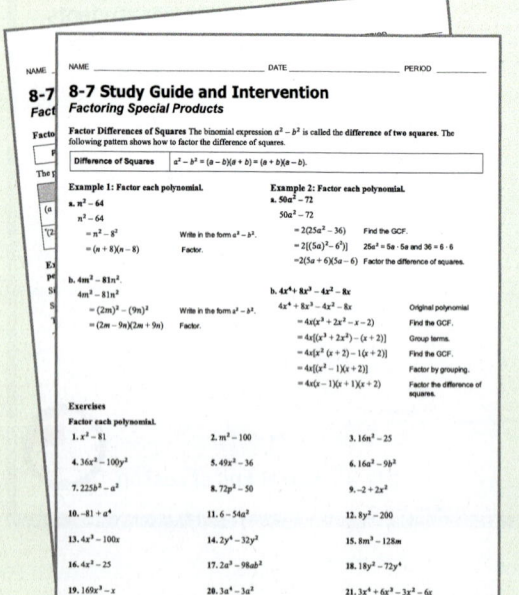

## Study Notebook
AL  OL  BL  ELL

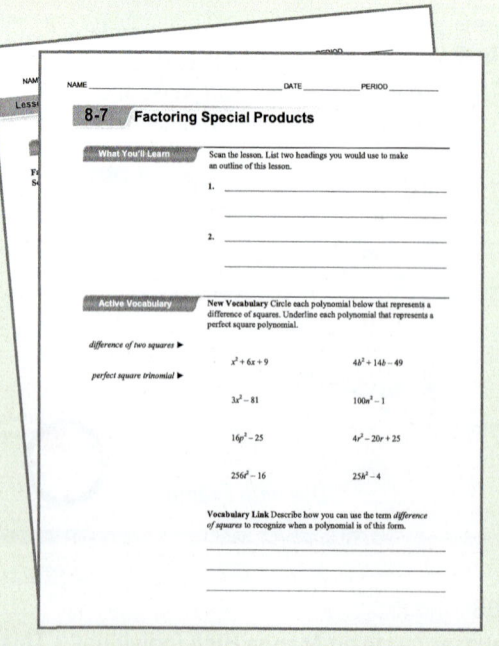

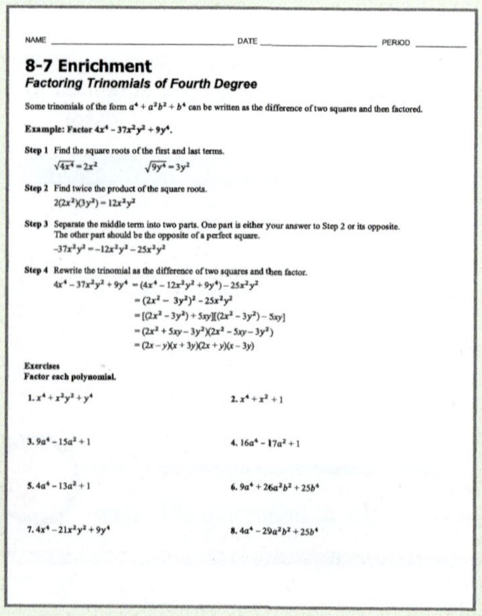

## Enrichment
OL  BL  ELL

539B | Lesson 8-7 | Factoring Special Products

# LESSON 7
## Factoring Special Products

**∴Then**
- You factored trinomials into two binomials.

**∴Now**
1. Factor binomials that are the difference of squares.
2. Factor trinomials that are perfect squares.

**∴Why?**
Computer graphics designers use a combination of art and mathematics skills to design images and videos. Factoring can help to determine the dimensions and shapes of the figures they design.

**New Vocabulary**
difference of two squares
perfect square trinomial

**Mathematical Practices**
1 Make sense of problems.
4 Model with mathematics.
7 Make use of structure.
8 Look for repeated reasoning.

**Content Standards**
A.SSE.2 Use the structure of an expression to identify ways to rewrite it.

### 1 Factor Differences of Squares
You have previously learned about the product of the sum and difference of two quantities. This resulting product is referred to as the **difference of two squares**. So, the factored form of the difference of squares is called the product of the sum and difference of the two quantities.

**Key Concept** Factoring Differences of Squares

Symbols  $a^2 - b^2 = (a + b)(a - b)$ or $(a - b)(a + b)$

Examples  $x^2 - 25 = (x + 5)(x - 5)$ or $(x - 5)(x + 5)$
$t^2 - 64 = (t + 8)(t - 8)$ or $(t - 8)(t + 8)$

A.SSE.2

**Example 1** Factor Differences of Squares

Determine whether each polynomial is a difference of squares. Write *yes* or *no*. If so, factor it.

a. $16h^2 - 9a^2$

Yes; Since $16h^2 = (4h)^2$ and $9a^2 = (3a)^2$, this is the difference of squares.

$16h^2 - 9a^2 = (4h)^2 - (3a)^2$     Write in the form of $a^2 - b^2$.
$= (4h + 3a)(4h - 3a)$     Factor the difference of squares.

b. $121c^4 - 25d^3$

No; While $121c^4 = (11c^2)^2$, $25d^3$ does not equal the square of any monomial.

**Guided Practice**

1A. $81 - c^2$   yes; $(9 + c)(9 - c)$     1B. $36n^2 - 27m^4$   no
1C. $25y^2 + 1$   no     1D. $64g^2 - h^2$   yes; $(8g + h)(8g - h)$

To factor a polynomial completely, a technique may need to be applied more than once. This also applies to the difference of squares pattern.

---

**MP Mathematical Practices Strategies**

**Look for and make use of structure.**
Help students look closely at polynomials to discern whether they can be factored by using a pattern. For example, ask:

- How can you factor any polynomial of the form $a^2 - b^2$? It will follow the pattern $a^2 - b^2 = (a + b)(a - b)$.

- Describe how the middle term of a perfect square trinomial of the form $a^2 - 2ab + b^2$ relates to the first and last terms. The middle term is twice the product of the square roots of the first and last terms.

- In Guided Practice 1B, why does the polynomial not follow the difference of two squares pattern? The coefficient 27 is not a perfect square, so the term $27m^4$ is not a perfect square.

---

**Lesson 8-7** | Factoring Special Products

## Launch

Have students read the Why? section of the lesson. Ask:

- What is an example of an algebraic expression that can model the dimensions of a geometric figure? Sample answer: The dimensions of a rectangle can be $(x + 8)$ by $(x + 12)$ units.

- How can you use factoring to determine the dimensions a rectangle with area $x^2 + 7x + 12$ square units? $x^2 + 7x + 12$ factors into $(x + 4)(x + 3)$, so the rectangle may have dimensions $(x + 4)$ units and $(x + 3)$ units.

## Teach

Ask the scaffolded questions for each example to build conceptual understanding for students at all levels.

### 1 Factor Differences of Squares

**Example 1 Factor Differences of Squares**

**AL** What pattern can you use to factor the binomial if it is a difference of squares? the difference of squares pattern

**OL** How can we determine if the binomial is a difference of squares? Sample answer: if both the variable and the constant of each term can be written as a square, then each term is a square. In part a, 16 is the square of 4, and 9 is the square of 3. The variables in each term are squared so, we know they are squares. It also must be a subtraction. Therefore part **a** is a difference of squares.

**BL** In part **b**, why is $c^4$ a perfect square? Explain. Sample answer: a variable with an exponent of 4 is the square of a square.

**Need Another Example?**
Determine whether each polynomial is a difference of squares. Write *yes* or *no*. If so, factor it.
a. $m^2 - 64$   yes; $(m + 8)(m - 8)$
b. $16y^2 - 81z^2$   yes; $(4y + 9z)(4y - 9z)$
c. $9b^4 - 25$   yes; $(3b^2 + 5)(3b^2 - 5)$

### Go Online!

**Interactive Whiteboard**
Use the *eLesson* or *Lesson Presentation* to present this lesson.

connectED.mcgraw-hill.com   **539**

## Lesson 8-7 | Factoring Special Products

## 2 Factor Perfect Square Trinomials

**Example 2 Recognize and Factor Perfect Square Trinomials**

**AL** In part **a**, are the first and last terms perfect squares? **yes**

**OL** What are the three conditions that must be satisfied for the trinomial to be a perfect square trinomial? **The first and last terms are perfect squares and the middle term is equal to $2(2y)(3)$.**

**BL** How can you quickly determine that the trinomial is not a perfect square? **Sample answer: If the last term–the constant term– is negative, the trinomial is not a perfect square because that would mean the constant is the product of a negative number and a positive number.**

### Need Another Example?
Determine whether each trinomial is a perfect square trinomial. Write *yes* or *no*. If it is a perfect square, factor it.
**a.** $25x^2 - 30x + 9$ **yes; $(5x - 3)^2$**
**b.** $49y^2 + 42y + 36$ **no**

### Watch Out!
**Preventing Errors** Students should be reminded to look closely at the coefficient of the second term of a perfect square trinomial. Its sign determines whether the factors are in the form $(a + b)$ or $(a - b)$.

### Teaching Tip
Remind students that they should always check their factors by multiplying them using the FOIL method. Point out that the sum of squares $a^2 + b^2$ does not factor into $(a + b)(a + b)$. The sum of squares is a prime polynomial and cannot be factored.

---

**2 Factor Perfect Square Trinomials** You have learned the patterns for the products of the binomials $(a + b)^2$ and $(a - b)^2$. Recall that these are special products that follow specific patterns.

$$(a + b)^2 = (a + b)(a + b) \qquad (a - b)^2 = (a - b)(a - b)$$
$$= a^2 + ab + ab + b^2 \qquad = a^2 - ab - ab + b^2$$
$$= a^2 + 2ab + b^2 \qquad = a^2 - 2ab + b^2$$

These products are called **perfect square trinomials**, because they are the squares of binomials. The above patterns can help you factor perfect square trinomials.

For a trinomial to be factorable as a perfect square, the first and last terms must be perfect squares and the middle term must be two times the square roots of the first and last terms.

> **Key Concept** Factoring Perfect Square Trinomials
>
> **Symbols** $a^2 + 2ab + b^2 = (a + b)(a + b) = (a + b)^2$
> $a^2 - 2ab + b^2 = (a - b)(a - b) = (a - b)^2$
>
> **Examples** $x^2 + 8x + 16 = (x + 4)(x + 4)$ or $(x + 4)^2$
> $x^2 - 6x + 9 = (x - 3)(x - 3)$ or $(x - 3)^2$

**Study Tip**
**Reasoning** If the constant term of the trinomial is negative, the trinomial is not a perfect square trinomial, so it is not necessary to check the other conditions.

**Example 2** Recognize and Factor Perfect Square Trinomials

Determine whether each trinomial is a perfect square trinomial. Write *yes* or *no*. If so, factor it.

**a.** $4y^2 + 12y + 9$

1. Is the first term a perfect square? Yes, $4y^2 = (2y)^2$
2. Is the last term a perfect square? Yes, $9 = 3^2$
3. Is the middle term equal to $2(2y)(3)$? Yes, $12y = 2(2y)(3)$

Since all three conditions are satisfied, $4y^2 + 12y + 9$ is a perfect square trinomial.

$$4y^2 + 12y + 9 = (2y)^2 + 2(2y)(3) + 3^2 \quad \text{Write as } a^2 + 2ab + b^2.$$
$$= (2y + 3)^2 \quad \text{Factor using the pattern.}$$

**b.** $9x^2 - 6x + 4$

1. Is the first term a perfect square? Yes, $9x^2 = (3x)^2$
2. Is the last term a perfect square? Yes, $4 = 2^2$
3. Is the middle term equal to $-2(3x)(2)$? No, $-6x \neq -2(3x)(2)$

Since the middle term does not satisfy the required condition, $9x^2 - 6x + 4$ is not a perfect square trinomial.

▶ **Guided Practice**

**2A.** $9y^2 + 24y + 16$ **yes; $(3y + 4)(3y + 4)$**  **2B.** $2a^2 + 10a + 25$ **no**

---

**MP Teaching the Mathematical Practices**

**Structure** In Examples 1 and 2, factoring polynomials is simplified if the polynomials have the difference of two square pattern or the perfect square trinomial pattern. Encourage students to learn these patterns well.

Lesson 8-7 | Factoring Special Products

A polynomial is completely factored when it is written as a product of prime polynomials. More than one method might be needed to factor a polynomial completely. When completely factoring a polynomial, the Concept Summary can help you decide where to start.

Remember, if the polynomial does not fit any pattern or cannot be factored, the polynomial is prime.

### Watch Out!
**Preventing Errors** Students should notice that after the GCF has been factored out and the difference of squares factoring technique has been applied once, one of the factors should be prime.

**Go Online!**
You will use the factoring methods in this Concept Summary frequently. Log into your eStudent Edition to bookmark this page.

### Concept Summary  Factoring Methods

| Steps | Number of Terms | Examples |
|---|---|---|
| **Step 1** Factor out the GCF. | any | $4x^3 + 2x^2 - 6x = 2x(2x^2 + x - 3)$ |
| **Step 2** Check for a difference of squares or a perfect square trinomial. | 2 or 3 | $9x^2 - 16 = (3x + 4)(3x - 4)$ <br> $16x^2 + 24x + 9 = (4x + 3)^2$ <br> $25x^2 - 10x + 1 = (5x - 1)^2$ |
| **Step 3** Apply the factoring patterns for $x^2 + bx + c$ or $ax^2 + bx + c$ (general trinomials), or factor by grouping. | 3 or 4 | $x^2 - 8x + 12 = (x - 2)(x - 6)$ <br> $2x^2 + 13x + 6 = (2x + 1)(x + 6)$ <br> $12y^2 + 17y + 6$ <br> $= 12y^2 + 9y + 8y + 6$ <br> $= (12y^2 + 9y) + (8y + 6)$ <br> $= 3y(4y + 3) + 2(4y + 3)$ <br> $= (4y + 3)(3y + 2)$ |

### Example 3  Find Dimensions

**AL** In part a, why must the area of the clear squares be subtracted from $8a^2$? $8a^2$ represents the total area of the window. The colored area does not include the clear squares, so their area must be subtracted from the total area.

**OL** In part b, why is the GCF factored out first? It will make the remaining polynomial a difference of two squares, which then can be factored using the pattern.

**BL** If the pieces of clear glass were rectangular, could you use the difference of squares pattern to factor the polynomial? Explain. only if the total area of the clear glass is still a perfect square

### Real-World Example 3  Find Dimensions

**DESIGN** An artist is designing a stained glass window. It will be made up of pieces of colored glass and 2 clear squares as shown. Each clear square has a side length of 8 inches.

**a.** Write a polynomial that represents the area of the colored glass.

The total area of the window is $4a \cdot 2a$, or $8a^2$ square inches.

The area of the clear squares is $2 \cdot 8^2$, or 128 square inches.

The area of the colored glass is the total area of the window minus the area of the clear glass. So, the area of the colored glass is $8a^2 - 128$ square inches.

**b.** Factor the polynomial completely.

$8a^2 - 128 = 8(a^2 - 16)$     Factor out the GCF.
$= 8(a - 4)(a + 4)$     Factor the difference of squares.

### Guided Practice

3. Write the factored form of a polynomial that represents the area of the colored glass if the clear squares have a side length of 6 inches. $8(a - 3)(a + 3)$

### Need Another Example?
**Design** A graphic designer drew the logo below with dimensions as shown.

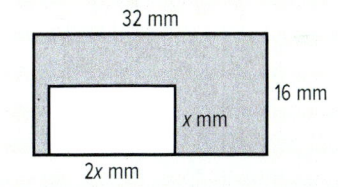

32 mm
16 mm
$x$ mm
$2x$ mm

a. Write a polynomial that represents the shaded area. $512 - 2x^2$

b. Factor the polynomial completely. $2(16 - x)(16 + x)$

### Teaching Tip
**Sense-Making** Students may not be used to thinking of fractions as perfect squares. Remind them that if both the numerator and denominator are perfect squares, then the fraction is a perfect square.

---

### Differentiated Instruction  AL  OL  BL  ELL

**Visual/Spatial Learners** Draw the geometric model shown below on the board. Ask students to use their own paper square and scissors to make the model for $a^2$ when the $b^2$ square is removed. Then ask students to explain how their models show that $(a - b)(a + b) = a^2 - b^2$.

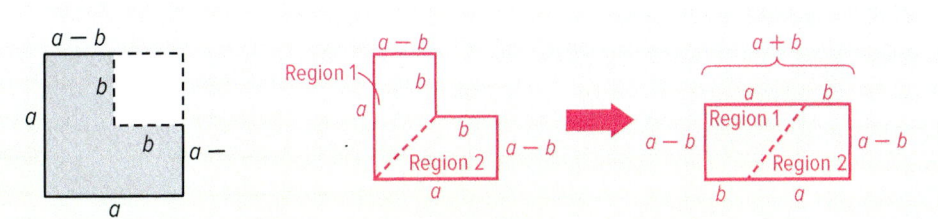

connectED.mcgraw-hill.com   541

# Lesson 8-7 | Factoring Special Products

## Example 4 Factor Completely

**AL** Is there a GCF greater than 1 in each of the examples? If so, what is the GCF? **a. yes, 5; b. no; c. yes**

**OL** In part **a**, after factoring out the GCF, what should you check for next? **if it is a difference of squares**

**BL** How can you check your work? **Sample answer: FOIL and graph the original expression and the factored expression on the same coordinate plane to see if the graphs coincide.**

### Need Another Example?
Factor each polynomial if possible. If the polynomial cannot be factored, write *prime*.
a. $6x^2 - 96$  $6(x+4)(x-4)$
b. $16y^2 + 8y - 15$  $(4y+5)(4y-3)$

## Example 5 Apply a Technique More Than Once

**AL** Is each term of the binomial a perfect square? **yes**

**OL** Why do we need to factor the difference of square twice? **Sample answer: when we factor the polynomial, we get a binomial factor that is also a difference of squares, so we can factor this part again.**

**BL** Explain in your own words why the sum of squares cannot be factored. **See students' work.**

### Need Another Example?
Factor each polynomial.
a. $y^4 - 625$  $(y^2+25)(y+5)(y-5)$
b. $256 - n^4$  $(16+n^2)(4-n)(4+n)$

---

**MP Teaching the Mathematical Practices**

**Structure** Mathematically proficient students can see complicated things as being composed of several objects. In example 3, students should first find the areas of the entire window and of the clear glass before writing the polynomial expression.

---

### Example 4 Factor Completely

Factor each polynomial, if possible. If the polynomial cannot be factored, write *prime*.

a. $5x^2 - 80$

**Step 1** The GCF of $5x^2$ and $-80$ is 5, so factor it out.

**Step 2** Since there are two terms, check for a difference of squares.

$5x^2 - 80 = 5(x^2 - 16)$    5 is the GCF of the terms.
$\quad\quad\quad\quad = 5(x^2 - 4^2)$    $x^2 = x \cdot x$ and $16 = 4 \cdot 4$
$\quad\quad\quad\quad = 5(x - 4)(x + 4)$    Factor the difference of squares.

b. $9x^2 - 6x - 35$

**Step 1** The GCF of $9x^2$, $-6x$, and $-35$ is 1.

**Step 2** Since 35 is not a perfect square, this is not a perfect square trinomial.

**Step 3** Factor using the pattern $ax^2 + bx + c$. Are there two numbers with a product of $9(-35)$ or $-315$ and a sum of $-6$? Yes, the product of 15 and $-21$ is $-315$, and the sum is $-6$.

$9x^2 - 6x - 35 = 9x^2 + mx + px - 35$    Write the pattern.
$\quad\quad\quad\quad\quad\quad = 9x^2 + 15x - 21x - 35$    $m = 15$ and $p = -21$
$\quad\quad\quad\quad\quad\quad = (9x^2 + 15x) + (-21x - 35)$    Group terms with common factors.
$\quad\quad\quad\quad\quad\quad = 3x(3x + 5) - 7(3x + 5)$    Factor out the GCF from each grouping.
$\quad\quad\quad\quad\quad\quad = (3x + 5)(3x - 7)$    $3x + 5$ is the common factor.

c. $18x^4 + 24x^2 + 8$

**Step 1** The GCF of $18x^4$, $24x^2$, and 8 is 2, so factor it out.

**Step 2** Since there are three terms, check for a perfect square trinomial.

$18x^4 + 24x^2 + 8 = 2(9x^4 + 12x^2 + 4)$    2 is the GCF of the terms.
$\quad\quad\quad\quad\quad\quad = 2[(3x^2)^2 + 2(3x^2)(2) + 2^2]$    Write as $a^2 + 2ab + b^2$.
$\quad\quad\quad\quad\quad\quad = 2(3x^2 + 2)^2$    Factor the perfect square trinomial.

**Guided Practice**

4A. $2x^2 - 32$  $2(x-4)(x+4)$      4B. $12x^2 + 5x - 25$  $(4x-5)(3x+5)$

---

**Study Tip**

**MP Perseverance** You can check your answer by:
- Using the FOIL method.
- Using the Distributive Property.
- Graphing the original expression and factored expression and comparing the graphs.

If the product of the factors does not match the original expression exactly, the answer is incorrect.

---

### Example 5 Apply a Technique More Than Once

Factor $625 - x^4$.

$625 - x^4 = (25)^2 - (x^2)^2$    Write $625 - x^4$ in $a^2 - b^2$ form.
$\quad\quad\quad = (25 + x^2)(25 - x^2)$    Factor the difference of squares.
$\quad\quad\quad = (25 + x^2)(5^2 - x^2)$    Write $25 - x^2$ in $a^2 - b^2$ form.
$\quad\quad\quad = (25 + x^2)(5 - x)(5 + x)$    Factor the difference of squares.

**Guided Practice**

5A. $2y^4 - 50$  $2(y^2-5)(y^2+5)$      5B. $6x^4 - 96$  $6(x-2)(x+2)(x^2+4)$

5C. $2m^3 + m^2 - 50m - 25$  $(2m+1)(m+5)(m-5)$      5D. $16y^4 - 1$  $(4y^2+1)(2y-1)(2y+1)$

---

**Watch Out!**

**Sum of Squares** The sum of squares, $a^2 + b^2$, does not factor into $(a + b)(a + b)$. The sum of squares is a prime polynomial and cannot be factored.

---

**Watch Out!**

**Factoring** Students often fail to factor polynomials completely. Point out to students that $4x^2 - 36$ is a difference of squares and can be factored as $(2x - 6)(2x + 6)$. Remind students that a polynomial is not considered completely factored if the terms of any of its factors have a GCF greater than 1.

---

542 | Lesson 8-7 | Factoring Special Products

# Lesson 8-7 | Factoring Special Products

## Check Your Understanding

= Step-by-Step Solutions begin on page R11.

**Go Online!** for a Self-Check Quiz

**Example 1**
A.SSE.2

Determine whether each polynomial is a difference of squares. Write *yes* or *no*. If so, factor it.

1. $q^2 - 121$ yes; $(q + 11)(q - 11)$
2. $4a^2 - 25$ yes; $(2a + 5)(2a - 5)$
3. $9n^2 + 1$ no
4. $x^2 - 16y^3$ no
5. $16m^2 - k^4$ yes; $(4m + k^2)(4m - k^2)$
6. $r^2 - 9t^2$ yes; $(r + 3t)(r - 3t)$

**Example 2**
A.SSE.2

Determine whether each polynomial is a perfect square trinomial. Write *yes* or *no*. If so, factor it. 15. $(c + 1)(c - 1)(2c + 3)$ 17. $(t + 4)(t - 4)(3t + 2)$

7. $25x^2 + 60x + 36$ yes; $(5x + 6)^2$
8. $6x^2 + 30x + 36$ no
9. $y^4 + 2y^2 + 1$ yes; $(y^2 + 1)^2$
10. $25x^2y^2 - 20xy + 4y^2$ yes; $(5x - 2y)^2$

**Examples 4-5**
A.SSE.2

Factor each polynomial, if possible. If the polynomial cannot be factored, write *prime*.

11. $u^4 - 81$ $(u + 3)(u - 3)(u^2 + 9)$
12. $2d^4 - 32f^4$ $2(d^2 + 4f^2)(d + 2f)(d - 2f)$
13. $20r^4 - 45n^4$ $5(2r^2 - 3n^2)(2r^2 + 3n^2)$
14. $256n^4 - c^4$ $(16n^2 + c^2)(4n + c)(4n - c)$
15. $2c^3 + 3c^2 - 2c - 3$
16. $3f^2 - 24f + 48$ $3(f - 4)^2$
17. $3t^3 + 2t^2 - 48t - 32$
18. $w^3 - 3w^2 - 9w + 27$ $(w - 3)(w + 3)(w - 3)$

**Example 3**
A.SSE.2

19. **MODELING** The drawing at the right is a square with a square cut out of it.
    a. Write an expression that represents the area of the shaded region. $(4n + 1)^2 - 5^2$
    b. Find the dimensions of a rectangle with the same area as the shaded region in the drawing. Assume that the dimensions of the rectangle must be represented by binomials with integral coefficients. $(4n + 6)$ by $(4n - 4)$

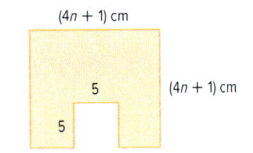

$(4n + 1)$ cm
5
5
$(4n + 1)$ cm

## Practice and Problem Solving

Extra Practice begins on page R8.

**Examples 1, 2, 4, 5**
A.SSE.2

Factor each polynomial, if possible. If the polynomial cannot be factored, write *prime*.

20. $16a^2 - 121b^2$ $(4a - 11b)(4a + 11b)$
21. $12m^3 - 22m^2 - 70m$ $2m(2m - 7)(3m + 5)$
22. $8c^2 - 88c + 242$ $2(2c - 11)^2$
23. $12x^2 - 84x + 147$ $3(2x - 7)^2$
24. $w^4 - 625$ $(w^2 + 25)(w^2 - 25)$
25. $12p^3 - 3p$ $3p(2p + 1)(2p - 1)$
26. $8x^2 + 10x - 21$ prime
27. $a^2 - 49$ $(a + 7)(a - 7)$
28. $4m^3 + 9m^2 - 36m - 81$ $(m + 3)(m - 3)(4m + 9)$
29. $3m^4 + 243$ $3(m^4 + 81)$
30. $3x^3 + x^2 - 75x - 25$ $(x + 5)(x - 5)(3x + 1)$
31. $12a^3 + 2a^2 - 192a - 32$ $2(a + 4)(a - 4)(6a + 1)$

**Example 3**
A.SSE.2

32. **REASONING** The area of a square is represented by $9x^2 - 42x + 49$. Find the length of each side. $3x - 7$

33. **MODELING** Zelda is building a deck in her backyard. The area of the deck in square feet can be represented by the expression $576 - x^2$. Find expressions for the dimensions of the deck. Then, find the perimeter of the deck when $x = 8$ feet. $(24 + x)$ by $(24 - x)$; 96 ft

**B** 34. **REASONING** The volume of a rectangular prism is represented by the expression $8y^3 + 40y^2 + 50y$. Find the possible dimensions of the prism if the dimensions are represented by polynomials with integer coefficients. **Sample answer:** $2y, 2y + 5, 2y + 5$

## Differentiated Homework Options

| Levels | AL Basic | OL Core | BL Advanced |
|---|---|---|---|
| Exercises | 20–33, 47–49, 51–62 | 21–45 odd, 47–49, 51–62 | 45–53, (optional: 54–62) |
| 2-Day Option | 21–33 odd, 54–62 | 20–33, 54–62 | |
| | 20–32 even, 47–49, 51–53 | 35–49, 51–53 | |

You can use ALEKS to provide additional remediation support with personalized instruction and practice.

## Practice

**Formative Assessment** Use Exercises 1–19 to assess students' understanding of the concepts in this lesson.

The Practice and Problem Solving exercises assess the content taught in the lesson. The Preparing for Assessment page is meant to be used as preparation for end-of-course assessments.

### Watch Out!
**Factoring** Remind students that any of the factoring methods they have studied thus far can be used in the exercises.

### MP Teaching the Mathematical Practices

**Modeling** Encourage students to work with a partner to think of several ways to interpret the drawing in Exercise 19. Have them use the difference of two squares pattern to justify their factorization.

## Extra Practice

See page R8 for extra exercises for students who are approaching level or for on-level students who need additional reinforcement.

### Levels of Complexity Chart

The levels of the exercises progress from 1 to 3, with Level 1 indicating the lowest level of complexity.

| Exercises | 20–33 | 34–44, 54–62 | 45–53 |
|---|---|---|---|
| Level 3 | | | ● |
| Level 2 | | ● | |
| Level 1 | ● | | |

## Go Online! eBook

### Interactive Student Guide
Use the *Interactive Student Guide* to deepen conceptual understanding.
· Solving by Factoring: Differences of Squares
· Solving by Factoring: Perfect Squares

connectED.mcgraw-hill.com 543

# Lesson 8-7 | Factoring Special Products

## Watch Out!

**Error Analysis** For Exercise 47, make sure students can explain what Elizabeth did wrong. Stress that her error is a common one. Ask students what they can do to avoid making the same mistake themselves.

## MP Teaching the Mathematical Practices

**Perseverance** Mathematically proficient students analyze givens, constraints, relationships, and goals. In Exercise 50, tell students to make sure their final polynomial is completely simplified.

## Assess

**Ticket Out the Door** Prepare two bags with slips of paper. One bag should contain polynomials that can be factored using the difference of two squares pattern, and the other bag should contain polynomials that can be factored because they are perfect square trinomials. Have students select and factor an expression from each bag.

## Additional Answers

**51.** When the difference of squares pattern is multiplied together using the FOIL method, the outer and inner terms are opposites of each other. When these terms are added together, the sum is zero.

**52.** Determine if the first and last terms are perfect squares. Then determine if the middle term is equal to $\pm 2$ times the product of the principal square roots of the first and last terms. If these three criteria are met, the trinomial is a perfect square trinomial.

**53.** $4x^2 + 10x + 4$ because it is the only expression that is not a perfect square trinomial.

---

Factor each polynomial, if possible. If the polynomial cannot be factored, write *prime*. **35.** $(x+2y)(x-2)(x+2)$ **37.** $(r-6)(r+6)(2r-1)$ **39.** $2cd(c^2+d^2)(2c-5)$

**35.** $x^3 + 2x^2y - 4x - 8y$

**36.** $2a^2b^2 - 2a^2 - 2ab^2 + 2ab$  $2a(a-b)(b+1)(b-1)$

**37.** $2r^3 - r^2 - 72r + 36$

**38.** $3k^3 - 24k^2 + 48k$  $3k(k-4)(k-4)$

**39.** $4c^4d - 10c^3d + 4c^2d^3 - 10cd^3$

**40.** $g^2 + 2g - 3h^2 + 4h$  prime

**41.** $x(x+6)^2(x-6)$  **41.** $x^4 + 6x^3 - 36x^2 - 216x$

**42.** $15m^3 + 12m^2 - 375m - 300$  $3(m-5)(m+5)(5m+4)$

**43.** $y^8 - 256$  $(y-2)(y+2)(y^2+4)(y^2+16)$

**44.** $k^3 - 5k^2 - 100k + 500$  $(k-5)(k-10)(k+10)$

**45b.** Yes; since length cannot be negative, the value of $h$ has to be greater than 6 so that $(h-6)$ is greater than 0. So, the sides of the box are each greater than 6 in.

**45. MODELING** For the student council elections, Franco is building a voting box shown. The volume of the box in cubic inches is given by the polynomial $h^3 - 12h^2 + 36h$.

a. Factor the polynomial. What are the possible dimensions of the box in terms of $h$? $h(h-6)^2$; $h$, $(h-6)$, $(h-6)$

b. Franco says that your value of $h$ must be greater than 6 inches. Is he correct? Justify your reasoning.

**46. SENSE-MAKING** Determine whether each of the following statements are *true* or *false*. Give an example or counterexample to justify your answer.

a. All binomials that have a perfect square in each of the two terms can be factored.  false; $a^2 + b^2$

b. All trinomials that have a perfect square in the first and last terms can be factored.  false; $a^2 + ab + b^2$

A.SSE.2

### H.O.T. Problems  Use Higher-Order Thinking Skills

**47. ERROR ANALYSIS** Elizabeth and Lorenzo are factoring an expression. Is either of them correct? Explain your reasoning.

**47.** Lorenzo; Sample answer: Checking Elizabeth's answer gives us $16x^2 - 25y^2$. The exponent on $x$ in the final product should be 4.

| Elizabeth | Lorenzo |
|---|---|
| $16x^4 - 25y^2 =$ | $16x^4 - 25y^2 =$ |
| $(4x - 5y)(4x + 5y)$ | $(4x^2 - 5y)(4x^2 + 5y)$ |

**50.** $[3 + (k+3)][3 - (k+3)] = (k+6)(-k) = -k^2 - 6k$

**48. MP PERSEVERENCE** Factor $x^n + 6 + x^n + 2 + x^n$ completely.  $x^n(x^6 + x^2 + 1)$

**49. MP STRUCTURE** Write a perfect square trinomial equation in which the coefficient of the middle term is negative and the last term is a fraction. Solve the equation.
Sample answer: $x^2 - 3x + \frac{9}{4} = 0$; $\left\{\frac{3}{2}\right\}$

**50. MP STRUCTURE** Factor and simplify $9 - (k+3)^2$, a difference of squares.

**51. WRITING IN MATH** Describe why the difference of squares pattern has no middle term with a variable.  See margin.

**52. WRITING IN MATH** Explain how to determine whether a trinomial is a perfect square trinomial.  See margin.

**53. WHICH ONE DOESN'T BELONG** Identify the trinomial that does not belong. Explain.  See margin.

| $4x^2 - 36x + 81$ | $25x^2 + 10x + 1$ | $4x^2 + 10x + 4$ | $9x^2 - 24x + 16$ |

### MP Standards for Mathematical Practice

| Emphasis On | Exercises |
|---|---|
| 1 Make sense of problems and persevere in solving them. | 46, 48, 54 |
| 2 Reason abstractly and quantitatively. | 32, 34, 54, 56–58, 61 |
| 3 Construct viable arguments and critique the reasoning of others. | 46, 47, 51–53, 55, 60 |
| 4 Model with mathematics. | 33, 45, 54, 61 |
| 7 Look for and make use of structure. | 20–31, 35–44, 48–50, 55, 59, 62 |

**Lesson 8-7 | Factoring Special Products**

## Preparing for Assessment

**54. MULTI-STEP** A water tank has a volume represented by $2x^2y^4 - 32x^2$.

a. Is the polynomial a difference of two squares, yes or no? **MP 2, 7** A.SSE.2

   **no**

b. Which expression shows how to factor out the GCF? **B**

   - A  $2(x^2y^4 - 16x^2)$
   - B  $2x^2(y^4 - 16)$
   - C  $2x^2(y^4 - 32)$
   - D  $2x(xy^4 - 32x)$

c. Which, if any, pattern can be used to factor the resulting expression in part **b**?

   **difference of squares**

d. Write the complete factorization of the polynomial.

   $2x^2(y^2 + 4)(y - 2)(y + 2)$

**55.** Carli is factoring the expression $6x^5 - 54x$. Which of the following statements is true? **MP 7** A.SSE.2 **A**

- A  $x^2 - 3$ is not a difference of squares.
- B  $x^4 - 9$ is not a difference of squares.
- C  The complete factored form of the expression is $6x(x^4 - 9)$.
- D  The expression cannot be factored.

**56.** The volume of the prism shown below is represented by the polynomial $\ell^3 - 16\ell^2 + 60\ell$ where $\ell$ is the length of the prism. **MP 2** A.SSE.2

a. Write the polynomial in factored form.

   $\ell(\ell - 10)(\ell - 6)$

b. Use the factors to write an expression for the height and width in terms of the length.

   **height** $= \ell - 6$; **width** $= \ell - 10$

**57.** The area of a square is represented by $16x^2 + 40x + 25$. Find the length of each side. **MP 2** A.SSE.2

   $|4x + 5|$

**58.** The area of a circle is $A = \pi(25x^2 - 40x + 16)$. Which represents the radius of the circle? **MP 2** A.SSE.2 **C**

- A  $5x + 4$
- B  $(5x + 4)^2$
- C  $5x - 4$
- D  $(5x - 4)^2$

**59.** Which expression is equivalent to $x^{16} - y^{16}$? **MP 1, 2, 7** A.SSE.2 **B**

- A  $(x^4 + y^4)(x^4 - y^4)$
- B  $(x^8 + y^8)(x^4 + y^4)(x^2 + y^2)(x + y)(x - y)$
- C  $(x^8 - y^8)(x^8 - y^8)$
- D  $(x^8 + y^8)(x^4 + y^4)(x^2 + y^2)(x + y)(x + y)$

**60.** Determine whether the following statement is *true* or *false*. Give an example or counterexample to justify your answer. **MP 2, 3** A.SSE.2

   *For a trinomial to be a perfect square trinomial, the middle term must be positive.*

   **False; Sample answer:** $a^2 - 2ab + b^2$

**61.** The volume of a swimming pool is represented by the expression $12x^3 - 84x^2 + 147x$. Find the possible dimensions of the pool if the dimensions are represented by polynomials with integer coefficients. **MP 2, 4**

   **Sample answer:** $3x$, $(2x - 7)$, $(2x - 7)$

**62.** Which expression is equivalent to $3r^3 - 7r^2 - 3r + 7$? **MP 7** A.SSE.2 **C**

- A  $(r - 1)(r - 1)(3r + 7)$
- B  $(r - 1)(r + 1)(3r + 7)$
- C  $(r - 1)(r + 1)(3r - 7)$
- D  $r(3r - 7)(r - 1)$

## Preparing for Assessment

Exercises 54–62 require students to use the skills they will need on standardized assessments. Each exercise is dual-coded with content standards and mathematical practice standards.

### Dual Coding

| Items | Content Standards | Mathematical Practices |
|---|---|---|
| 54 | A.SSE.2 | 2, 7 |
| 55, 62 | A.SSE.2 | 7 |
| 56–58 | A.SSE.2 | 2 |
| 59 | A.SSE.2 | 1, 7 |
| 60 | A.SSE.2 | 3 |
| 61 | A.SSE.2 | 2, 4 |

## Diagnose Student Errors

Survey student responses for each item. Class trends may indicate common errors and misconceptions.

**54b.**

| A | Did not include $x^2$ in the GCF |
|---|---|
| B | CORRECT |
| C | Did remove 2 from the second term |
| D | Used $2x$ instead of $2x^2$ as the GCF |

**55.**

| A | CORRECT |
|---|---|
| B | Did not think of the fourth power of $x$ as a square |
| C | Forgot to factor the expression in the parentheses |
| D | Needs to review how to factor |

**58.**

| A | Used incorrect sign when factoring |
|---|---|
| B | Used incorrect sign when factoring and used the radius squared |
| C | CORRECT |
| D | Used the radius squared |

---

## Differentiated Instruction  OL  BL

**Extension** Factor $(x^4 + 8x^2 + 16) - x^2$ completely. $(x^2 + x + 4)(x^2 - x + 4)$

**59.**

| A | Found the square root of 16 instead of half |
|---|---|
| B | CORRECT |
| C | Did not use the difference of two squares pattern |
| D | Did not factor $x^2 - y^2$ correctly in the last step of factoring |

**60.** If the answer is incorrect, remind students that they need to factor out the GCF and then use the perfect square trinomial pattern to factor the polynomial in parentheses.

**61.**

| A | Did not use the difference of two squares pattern correctly when factoring |
|---|---|
| B | Did not factor correctly by using the grouping technique |
| C | CORRECT |
| D | Did not use the difference of two squares pattern correctly when factoring |

## Go Online!

**Quizzes**

Students can use *Self-Check Quizzes* to check their understanding of this lesson. You can also give the *Chapter Quiz*, which covers the content in Lessons 8-1 through 8-7.

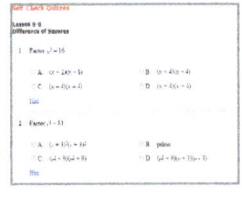

# Chapter 8 Study Guide and Review

## FOLDABLES Study Organizer

A completed Foldable for this chapter should include the Key Concepts related to quadratic expressions and equations.

## Key Vocabulary ELL

The page reference after each word denotes where that term was first introduced. If students have difficulty answering questions 1–10, remind them that they can use these page references to refresh their memories about the vocabulary terms.

Have students work together to determine whether each sentence in the Vocabulary Check is true or false. Have students take turns saying each sentence aloud while the other student listens carefully.

You can use the detailed reports in ALEKS to automatically monitor students' progress and pinpoint remediation needs prior to the chapter test.

---

# CHAPTER 8
# Study Guide and Review

**Go Online!** for Vocabulary Review Games and key vocabulary in 13 languages

## Study Guide

### Key Concepts

**Operations with Polynomials** (Lessons 8-1 through 8-4)
- To add or subtract polynomials, add or subtract like terms.
- To multiply polynomials, use the Distributive Property.
- Special products:
  $(a + b)^2 = a^2 + 2ab + b^2$
  $(a - b)^2 = a^2 - 2ab + b^2$
  $(a + b)(a - b) = a^2 - b^2$

**Factoring Using the Distributive Property** (Lesson 8-5)
- Using the Distributive Property to factor polynomials with four or more terms is called factoring by grouping.
  $ax + bx + ay + by = (ax + bx) + (ay + by)$
  $= x(a + b) + y(a + b)$
  $= (a + b)(x + y)$

**Factoring Quadratic Trinomials** (Lesson 8-6)
- To factor $x^2 + bx + c$, find two integers, $m$ and $p$, with a sum of $b$ and a product of $c$. Then write $x^2 + bx + c$ as $(x + m)(x + p)$.
- To factor $ax^2 + bx + c$, find two integers, $m$ and $p$, with a sum of $b$ and a product of $ac$. Then write as $ax^2 + mx + px + c$ and factor by grouping.

**Factoring Special Products** (Lesson 8-7)
- $a^2 - b^2 = (a - b)(a + b)$
- For a trinomial to be a perfect square, the first and last terms must be perfect squares, and the middle term must be twice the product of the square roots of the first and last terms.
- When factoring, begin by looking for a GCF that can be factored out.

### FOLDABLES Study Organizer

Use your Foldable to review the chapter. Working with a partner can be helpful. Ask for clarification of concepts as needed.

### Key Vocabulary

| | |
|---|---|
| binomial (p. 491) | perfect square trinomial (p. 540) |
| degree of a monomial (p. 491) | polynomial (p. 491) |
| degree of a polynomial (p. 491) | prime polynomial (p. 535) |
| difference of two squares (p. 539) | quadratic expression (p. 507) |
| factoring (p. 520) | standard form of a polynomial (p. 492) |
| factoring by grouping (p. 521) | trinomial (p. 491) |
| FOIL method (p. 507) | |
| leading coefficient (p. 492) | |

### Vocabulary Check

State whether each sentence is *true* or *false*. If *false*, replace the underlined phrase or expression to make a true sentence.

1. $x^2 + 5x + 6$ is an example of a prime polynomial. **false; sample answer: $x^2 + 5x + 7$**
2. $(x + 5)(x - 5)$ is the factorization of a difference of squares. **true**
3. $4x^2 - 2x + 7$ is a polynomial of degree 2. **true**
4. $(x + 5)(x - 2)$ is the factored form of $x^2 - 3x - 10$. **false; $(x - 5)(x + 2)$**
5. Expressions with four or more unlike terms can sometimes be factored by grouping. **true**
6. A polynomial is in standard form when the terms are in order from least to greatest. **false; greatest to least**
7. $x^2 - 12x + 36$ is an example of a perfect square trinomial. **true**
8. The leading coefficient of $1 + 6a + 9a^2$ is 1. **false; 9**
9. $x^2 - 16$ is an example of a(n) perfect square trinomial. **false; difference of squares**
10. The FOIL method is used to multiply two trinomials. **10. false; binomials**

### Concept Check

Classify each polynomial as *prime polynomial*, *difference of squares*, or *perfect square trinomial*.

11. $x^2 - 25$      12. $(16x^2 + 40x + 25)$
    **difference of squares**    **perfect square trinomial**
13. $x^2 + 25$   **prime polynomial**

---

## Answering the Essential Question

Before answering the Essential Question, have students review their answers to the *Building on the Essential Question* exercises found throughout the chapter.

- When could a nonlinear function be used to model a real-world situation? (p. 486)
- Why would you add, subtract, or multiply equations that represent real-world situations? (p. 514)
- What are the advantages of using quadratic expressions for modeling? (p. 525)

---

## Go Online!

### Vocabulary Review

Students can use the *Vocabulary Review Games* to check their understanding of the vocabulary terms in this chapter. Students should refer to the *Student-Built Glossary* they have created as they went through the chapter to review important terms. You can also give a *Vocabulary Test* over the content of this chapter.

Chapter 8 Study Guide and Review

## Lesson-by-Lesson Review

**Intervention** If the given examples are not sufficient to review the topics covered by the questions, remind students that the lesson references tell them where to review that topic in their textbook.

**Two-Day Option** Have students complete the Lesson-by-Lesson Review. Then you can use McGraw-Hill eAssessment to customize another review worksheet that practices all the objectives of this chapter or only the objectives on which your students need more help.

## Lesson-by-Lesson Review

### 8-1 Adding and Subtracting Polynomials
A.SSE.1a, A.APR.1

Write each polynomial in standard form. **14.** $3x^2 + x + 2$

14. $x + 2 + 3x^2$
15. $1 - x^4$   $-x^4 + 1$
16. $2 + 3x + x^2$   $x^2 + 3x + 2$
17. $3x^5 - 2 + 6x - 2x^2 + x^3$   $3x^5 + x^3 - 2x^2 + 6x - 2$

Find each sum or difference.
18. $(x^3 + 2) + (-3x^3 - 5)$   $-2x^3 - 3$
19. $a^2 + 5a - 3 - (2a^2 - 4a + 3)$   $-a^2 + 9a - 6$
20. $(4x - 3x^2 + 5) + (2x^2 - 5x + 1)$   $-x^2 - x + 6$

21. **PICTURE FRAMES** Jean is framing a painting that is a rectangle. What is the perimeter of the frame? (5x + 3 by 2x² − 3x + 1)
$4x^2 + 4x + 8$

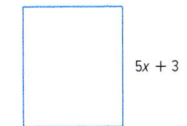

**Example 1**

Write $3 - x^2 + 4x$ in standard form.

**Step 1** Find the degree of each term.
3:   degree 0
$-x^2$:   degree 2
$4x$:   degree 1

**Step 2** Write the terms in descending order of degree.
$3 - x^2 + 4x = -x^2 + 4x + 3$

**Example 2**

Find $(8r^2 + 3r) - (10r^2 - 5)$.
$(8r^2 + 3r) - (10r^2 - 5)$
$= (8r^2 + 3r) + (-10r^2 + 5)$   Use the additive inverse.
$= (8r^2 - 10r^2) + 3r + 5$   Group like terms.
$= -2r^2 + 3r + 5$   Add like terms.

### 8-2 Multiplying a Polynomial by a Monomial
A.APR.1

Solve each equation.
22. $x^2(x + 2) = x(x^2 + 2x + 1)$   0
23. $2x(x + 3) = 2(x^2 + 3)$   1
24. $2(4w + w^2) - 6 = 2w(w - 4) + 10$   1

25. **GEOMETRY** Find the area of the rectangle. (3x by x² + x − 7)
$3x^3 + 3x^2 - 21x$

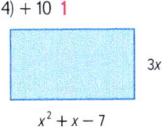

**Example 3**

Solve $m(2m - 5) + m = 2m(m - 6) + 16$.
$m(2m - 5) + m = 2m(m - 6) + 16$
$2m^2 - 5m + m = 2m^2 - 12m + 16$
$2m^2 - 4m = 2m^2 - 12m + 16$
$-4m = -12m + 16$
$8m = 16$
$m = 2$

### 8-3 Multiplying Polynomials
A.APR.1

Find each product. **26.** $x^2 + 4x - 21$   **27.** $18a^2 + 3a - 10$
26. $(x - 3)(x + 7)$
27. $(3a - 2)(6a + 5)$
28. $(3r - 7t)(2r + 5t)$   $6r^2 + rt - 35t^2$
29. $(2x + 5)(5x + 2)$   $10x^2 + 29x + 10$

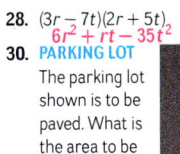

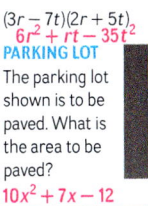

30. **PARKING LOT** The parking lot shown is to be paved. What is the area to be paved? (2x + 3 by 5x − 4)
$10x^2 + 7x - 12$

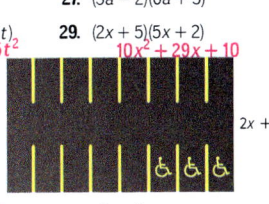

**Example 4**

Find $(6x - 5)(x + 4)$.
$(6x - 5)(x + 4)$
    F     O     I     L
$= (6x)(x) + (6x)(4) + (-5)(x) + (-5)(4)$
$= 6x^2 + 24x - 5x - 20$   Multiply.
$= 6x^2 + 19x - 20$   Combine like terms.

connectED.mcgraw-hill.com   547

# Chapter 8 Study Guide and Review

## CHAPTER 8
## Study Guide and Review Continued

### 8-4 Special Products
A.APR.1

**Find each product.**

31. $(x + 5)(x - 5)$  $x^2 - 25$
32. $(3x - 2)^2$  $9x^2 - 12x + 4$
33. $(5x + 4)^2$  $25x^2 + 40x + 16$
34. $(2x - 3)(2x + 3)$  $4x^2 - 9$
35. $(2r + 5t)^2$  $4r^2 + 20rt + 25t^2$
36. $(3m - 2)(3m + 2)$
37. **GEOMETRY** Write an expression to represent the area of the shaded region. $3x^2 - 21$

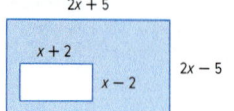

$2x + 5$
$x + 2$
$x - 2$
$2x - 5$

**Example 5**
Find $(x - 7)^2$.

$(a - b)^2 = a^2 - 2ab + b^2$  Square of a Difference
$(x - 7)^2 = x^2 - 2(x)(7) + (-7)^2$  $a = x$ and $b = 7$
$= x^2 - 14x + 49$  Simplify.

**Example 6**
Find $(5a - 4)(5a + 4)$.

$(a + b)(a - b) = a^2 - b^2$  Product of a Sum and Difference
$(5a - 4)(5a + 4) = (5a)^2 - (4)^2$  $a = 5a$ and $b = 4$
$= 25a^2 - 16$  Simplify.

### 8-5 Using the Distributive Property
A.SSE.2, A.SSE.3a

**Use the Distributive Property to factor each polynomial.**

38. $12x + 24y$  $12(x + 2y)$
39. $14x^2y - 21xy + 35xy^2$  $7xy(2x - 3 + 5y)$
40. $8xy - 16x^3y + 10y$  $2y(4x - 8x^3 + 5)$
41. $a^2 - 4ac + ab - 4bc$  $(a + b)(a - 4c)$
42. $2x^2 - 3xz - 2xy + 3yz$  $(2x - 3z)(x - y)$
43. $24am - 9an + 40bm - 15bn$  $(3a + 5b)(8m - 3n)$

**Factor each expression.**

44. $24ab + 54a - 20b - 45$  $(6a - 5)(4b + 9)$
45. $3r^3 - 12r^2 + 4p - pr$  $(3r^2 + p)(r - 4)$
46. $6c^2d + 30cd$  $6cd(c + 5)$
47. $18f^4g^5 - 3f^6g^2 + 9f^4g^3$  $3f^4g^2(6g^3 - f^2 + 3g)$

48. **GEOMETRY** The area of the rectangle shown is $x^3 - 2x^2 + 5x$ square units. What is the length? $x^2 - 2x + 5$

$x$

**Example 7**
Factor $12y^2 + 9y + 8y + 6$.

$12y^2 + 9y + 8y + 6$
$= (12y^2 + 9y) + (8y + 6)$  Group terms with common factors.
$= 3y(4y + 3) + 2(4y + 3)$  Factor the GCF from each group.
$= (4y + 3)(3y + 2)$  Distributive Property

**Example 8**
Factor $5a^2 - 10ab + 6b - 3a$.

$5a^2 - 10ab + 6b - 3a$
$= (5a^2 - 10ab) + (6b - 3a)$  Group terms with common factors.
$= 5a(a - 2b) + 3(2b - a)$  Factor the GCF from each group.
$= 5a(a - 2b) + 3[(-1)(a - 2b)]$  $2b - a = -1(a - 2b)$
$= 5a(a - 2b) - 3(a - 2b)$  Associative Property
$= (5a - 3)(a - 2b)$  Distributive Property

548 | Chapter 8 | Study Guide and Review

## 8-6 Factoring Quadratic Trinomials

A.SSE.2

Factor each polynomial.

49. $x^2 - 8x + 15$  $(x - 5)(x - 3)$
50. $x^2 + 9x + 20$  $(x + 5)(x + 4)$
51. $x^2 - 5x - 6$  $(x - 6)(x + 1)$
52. $x^2 + 3x - 18$  $(x + 6)(x - 3)$
53. $x^2 + 5x - 50$  $(x + 10)(x - 5)$
54. $x^2 - 6x + 8$  $(x - 2)(x - 4)$
55. $x^2 + 12x + 32$  $(x + 4)(x + 8)$
56. $x^2 - 2x - 48$  $(x + 6)(x - 8)$
57. $x^2 + 11x + 10$  $(x + 10)(x + 1)$

58. **ART** An artist is working on a painting. If the area of the canvas is represented by $x^2 + 2x - 24$, what are the dimensions of the canvas? $(x + 4)$ and $(x - 6)$

Factor each trinomial, if possible. If the trinomial cannot be factored, write *prime*.

59. $12x^2 + 22x - 14$  $2(2x - 1)(3x + 7)$
60. $2y^2 - 9y + 3$  prime
61. $3x^2 - 6x - 45$  $3(x - 5)(x + 3)$
62. $2a^2 + 13a - 24$  $(2a - 3)(a + 8)$
63. $20x^2 + x - 12$  $(4x - 3)(5x + 4)$
64. $2x^2 - 3x - 20$  $(x - 4)(2x + 5)$
65. $3x^2 - 13x - 10$  $(3x + 2)(x - 5)$
66. $6x^2 - 7x - 5$  $(3x + 5)(2x + 1)$

67. **GEOMETRY** The area of the rectangle shown is $6x^2 + 11x - 7$ square units. What is the width of the rectangle? $3x + 7$

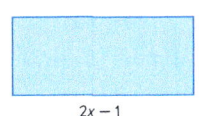

$2x - 1$

### Example 9

Factor $x^2 + 10x + 21$.

$b = 10$ and $c = 21$, so $m + p$ is positive and $mp$ is positive. Therefore, $m$ and $p$ must both be positive. List the positive factors of 21, and look for the pair of factors with a sum of 10.

| Factors of 21 | Sum of 10 |
|---|---|
| 1, 21 | 22 |
| 3, 7 | 10 |

The correct factors are 3 and 7.

$x^2 + 10x + 21 = (x + m)(x + p)$   Write the pattern.
$\quad\quad\quad\quad\quad\; = (x + 3)(x + 7)$   $m = 3$ and $p = 7$

### Example 10

Factor $12a^2 + 17a + 6$.

$a = 12$, $b = 17$, and $c = 6$. Since $b$ is positive, $m + p$ is positive. Since $c$ is positive, $mp$ is positive. So, $m$ and $p$ are both positive. List the factors of 12(6) or 72, where both factors are positive.

| Factors of 72 | Sum of 17 |
|---|---|
| 1, 72 | 73 |
| 2, 36 | 38 |
| 3, 24 | 27 |
| 4, 18 | 22 |
| 6, 12 | 18 |
| 8, 9 | 17 |

The correct factors are 8 and 9.

$12a^2 + 17a + 6 = 12a^2 + ma + pa + 6$
$\quad\quad\quad\quad\quad\;\; = 12a^2 + 8a + 9a + 6$
$\quad\quad\quad\quad\quad\;\; = (12a^2 + 8a) + (9a + 6)$
$\quad\quad\quad\quad\quad\;\; = 4a(3a + 2) + 3(3a + 2)$
$\quad\quad\quad\quad\quad\;\; = (3a + 2)(4a + 3)$

So, $12a^2 + 17a + 6 = (3a + 2)(4a + 3)$.

# Go Online!

## Anticipation Guide

Students should complete the Chapter 8 *Anticipation Guide*, and discuss how their responses have changed now that they have completed Chapter 8.

## Chapter 8 Study Guide and Review

### Before the Test
Have students complete the Study Notebook Tie it Together activity to review topics and skills presented in the chapter.

---

**CHAPTER 8**

## Study Guide and Review Continued

### 8-7 Factoring Special Products
A.SSE.2

Factor each polynomial.
68. $y^2 - 81$  $(y + 9)(y - 9)$
69. $64 - 25x^2$  $(8 + 5x)(8 - 5x)$
70. $16a^2 - 21b^2$  prime
71. $3x^2 - 3$  $3(x + 1)(x - 1)$
72. $a^2 - 25$  $(a + 5)(a - 5)$
73. $9x^2 - 25$  $(3x - 5)(3x + 5)$

Factor each polynomial, if possible. If the polynomial cannot be factored, write *prime*.
74. $x^2 + 12x + 36$  $(x + 6)^2$
75. $x^2 + 5x + 25$  prime
76. $9y^2 - 12y + 4$  $(3y - 2)^2$
77. $4 - 28a + 49a^2$  $(2 - 7a)^2$
78. $x^4 - 1$  $(x^2 + 1)(x + 1)(x - 1)$
79. $x^4 - 16x^2$  $x^2(x + 4)(x - 4)$
80. $9x^2 + 25$  prime
81. $-3x^2 - 12x - 12$  $-3(x + 2)^2$

**Example 11**
Factor $x^4 - 16$.

$x^4 - 16$    Original expression
$= (x^2)^2 - 4^2$    Difference of squares
$= (x^2 - 4)(x^2 + 4)$    Factor the difference of squares.
$= (x + 2)(x - 2)(x^2 + 4)$    Factor the second difference of squares.

**Example 12**
Factor $4x^2 + 16x + 16$.

$4x^2 + 16x + 16 = (2x)^2 + 2(2x)(4) + 4^2$    Write as $a^2 + 2ab + b^2$.
$= (2x + 4)^2$    Factor using the pattern.

---

### Go Online!

**eAssessment**

Customize and create multiple versions of chapter tests and answer keys. Tests can be delivered on paper or online.

# CHAPTER 8
## Practice Test

*Go Online! for another Chapter Test*

**Find each sum or difference.**

1. $(x + 5) + (x^2 - 3x + 7)$   $x^2 - 2x + 12$
2. $(7m - 8n^2 + 3n) - (-2n^2 + 4m - 3n)$   $3m - 6n^2 + 6n$

3. **MULTIPLE CHOICE** Antonia is carpeting two of the rooms in her house. The dimensions are shown. Which expression represents the total area to be carpeted? **B**

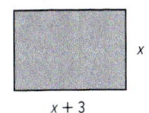

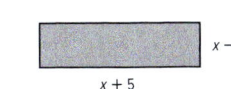

   Rectangle 1: $x$ by $x + 3$
   Rectangle 2: $x - 2$ by $x + 5$

   A $x^2 + 3x$
   B $2x^2 + 6x - 10$
   C $x^2 + 3x - 5$
   D $8x + 12$

**Find each product.**

4. $a(a^2 + 2a - 10)$   $a^3 + 2a^2 - 10a$
5. $(2a - 5)(3a + 5)$   $6a^2 - 5a - 25$
6. $(x - 3)(x^2 + 5x - 6)$   $x^3 + 2x^2 - 21x + 18$
7. $(x + 3)^2$   $x^2 + 6x + 9$
8. $(2b - 5)(2b + 5)$   $4b^2 - 25$

9. **FINANCIAL LITERACY** Suppose you invest $4000 in a 2-year certificate of deposit (CD).

   a. If the interest rate is 5% per year, the expression $4000(1 + 0.05)^2$ can be evaluated to find the total amount of money after two years. Explain the numbers in this expression. **See margin.**
   
   b. Find the amount at the end of two years. **$4410**
   
   c. Suppose you invest $10,000 in a CD for 4 years at an annual rate of 6.25%. What is the total amount of money you will have after 4 years? **about $12,744**

10. **MULTIPLE CHOICE** The area of the rectangle shown below is $2x^2 - x - 15$ square units. What is the width of the rectangle? **C**

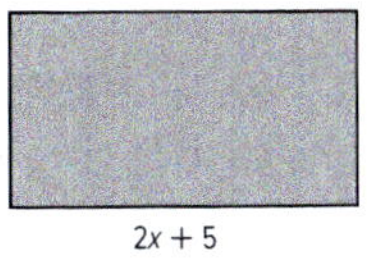

    (Rectangle with length $2x + 5$)

    A $x - 5$
    B $x + 3$
    C $x - 3$
    D $2x - 3$

**Solve each equation.**

11. $5(t^2 - 3t + 2) = t(5t - 2)$   $\frac{10}{13}$
12. $3x(x + 2) = 3(x^2 - 2)$   $-1$

**Factor each polynomial.**

13. $5xy - 10x$   $5x(y - 2)$
14. $7ab + 14ab^2 + 21a^2b$   $7ab(1 + 2b + 3a)$
15. $4x^2 + 8x + x + 2$   $(4x + 1)(x + 2)$
16. $10a^2 - 50a - a + 5$   $(10a - 1)(a - 5)$

**Factor each polynomial.**

17. $x^3 + x^2 - x - 1$   $(x - 1)(x + 1)^2$
18. $x^2 - x - 2$   $(x - 2)(x + 1)$
19. $a^2 - 2a$   $a(a - 2)$

20. **MULTIPLE CHOICE** Chantel is carpeting a room that has an area of $x^2 - 100$ square feet. If the width of the room is $x - 10$ feet, what is the length of the room? **B**

    A $x - 10$ ft
    B $x + 10$ ft
    C $x - 100$ ft
    D $10$ ft

**Factor each trinomial.**

21. $x^2 + 7x + 6$
22. $x^2 - 3x - 28$
23. $10x^2 - x - 3$
24. $15x^2 + 7x - 2$
25. $x^2 - 25$   $(x + 5)(x - 5)$
26. $4x^2 - 81$   $(2x + 9)(2x - 9)$
27. $9x^2 - 12x + 4$   $(3x - 2)(3x - 2)$
28. $16x^2 + 40x + 25$   $(4x + 5)(4x + 5)$

21. $(x + 6)(x + 1)$   22. $(x - 7)(x + 4)$
23. $(5x - 3)(2x + 1)$   24. $(3x + 2)(5x - 1)$

**Factor each polynomial.**

29. $x^3 + 4x^2 - x - 4$   $(x - 1)(x + 1)(x + 4)$
30. $x^4 - x^3 - 2x^2$   $x^2(x - 2)(x + 1)$
31. $a^2 - 2a + 1$   $(a - 1)^2$
32. $2x^2 - 13x + 20$   $(2x - 5)(x - 4)$

33. **MULTIPLE CHOICE** Which choice is a factor of $x^4 - 1$ when it is factored completely? **B**

    A $x^2 - 1$
    B $x - 1$
    C $x$
    D $1$

## Go Online!

### Chapter Tests

You can use premade leveled *Chapter Tests* to differentiate assessment for your students. Students can also take self-checking *Chapter Tests* to plan and prepare for chapter assessments.

| Form | Type | Level |
|---|---|---|
| 1 | MC | AL |
| 2A | MC | OL |
| 2B | MC | OL |
| 2C | FR | OL |
| 2D | FR | OL |
| 3 | FR | BL |
| Vocabulary Test | | |
| Extended-Response Test | | |

MC = multiple-choice questions
FR = free-response questions

---

## Chapter 8 Practice Test

**RtI Response to Intervention**

Use the Intervention Planner to help you determine your Response to Intervention.

### Intervention Planner

**TIER 1 — On Level OL**

IF students miss 25% of the exercises or less,
THEN choose a resource:
- SE Lessons 8-1 through 8-9
- **Go Online!**
  - Skills Practice
  - Chapter Project
  - Self-Check Quizzes ✓

**TIER 2 — Strategic Intervention AL**
Approaching grade level

IF students miss 50% of the exercises,
THEN choose a resource:
- *Quick Review Math Handbook*
- **Go Online!**
  - Study Guide and Intervention
  - Extra Examples
  - Personal Tutors
  - Homework Help

**TIER 3 — Intensive Intervention**
2 or more grades below level

IF students miss 75% of the exercises,
THEN choose a resource:
- Use *Math Triumphs, Alg. 1*
- **Go Online!**
  - Extra Examples
  - Personal Tutors
  - Homework Help
  - Review Vocabulary

### Additional Answer

9a. $4000 is the amount of the investment; 1 will add the amount of the investment to the interest; 0.05 is the interest rate as a decimal; 2 is the number of the years of the investment

connectED.mcgraw-hill.com    551

# Chapter 8 Preparing for Assessment

## Launch

**Objective** Apply concepts and skills from this chapter in a real-world setting.

## Teach

Ask:
- What is the first step you'll need to take to solve Part A? **Sample answer: Write two expressions, one for the length and one for the width.**
- What steps do you need to take in order to answer the third question in Part B? **Sample answer: First, I need to find the side length of the hot tub. Then, I can use this information to find the width of the pool. I can use that information to find the length, and finally multiply to get the area.**
- What kind of expressions do you expect to get when you factor the expression in Part C? **Sample answer: I should get two identical expressions because the area of a square is side length squared.**
- What would the area of the client in Part E's pool be if he reduced the length and removed the hot tub? **243 square feet**

The Performance Task focuses on the following content standards and standards for mathematical practice.

### Dual Coding

| Parts | Content Standards | MP Mathematical Practices |
|---|---|---|
| A | A.APR.1 | 1, 2, 8 |
| B | A.APR.1 | 1, 2, 8 |
| C | A.SSE.2 | 1, 7 |
| D | A.APR.1, A.SSE.2 | 1, 2, 7 |
| E | A.APR.1 | 3, 4 |

### Go Online!

**Interactive Student Guide**
Refer to *Interactive Student Guide* for an additional Performance Task.

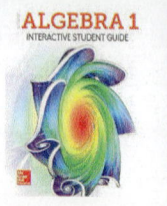

**eBook**

---

# CHAPTER 8
# Preparing for Assessment

## Performance Task

Provide a clear solution to each part of the task. Be sure to show all of your work, include all relevant drawings, and justify your answers.

**DESIGN** Henri owns a company that builds swimming pools. He is currently working on several projects.

### Part A

One of Henri's customers is thinking about having a pool installed in their backyard. Henri marks off the maximum area the pool could occupy, which is in the shape of a square, with side length s, measured in feet. The client decides they want to maintain some of their existing yard space, so she asks Henri to subtract 2 feet from the width and 8 feet from the length.

1. **Model** Write expressions that represent the perimeter and area of the pool the client has requested. **1–7. See margin.**

### Part B

Another of Henri's clients wants his pool to have a length that is 6 less than 4 times its width. He wants a hot tub installed directly beside the pool, such that it overflows and heats the pool. The hot tub will be a square with a side length that is 3 feet less than the width of the pool.

2. Write an expression that represents the area of the pool.
3. Write an expression that represents the area of the pool and hot tub together.
4. **Reasoning** If the area of the hot tub is 36 square feet, find the area of the pool.

### Part C

Another client only wants a hot tub installed. The expression that represents the total area of the square hot tub is $x^2 + 8x + 16$.

5. Write an expression that represents the side length of the hot tub.

### Part D

Another client is a professional swimmer and wants a lap pool installed. The lap pool will occupy an area of $9x^2 - 400$ square feet.

6. Find the dimensions of the lap pool if $x = 9$.

### Part E

**Construct an Argument** The cost of Henri's services is directly proportional to the area of the installation. The client in Part B wants to save some money so he decides to make the size of his pool smaller. He decides to either reduce the length of the pool so that it is twice the width plus 8 feet or not build the hot tub.

7. If the width of the pool remains the same as in Part B, which option should the client choose? Justify your answer.

### Levels of Complexity Chart

The levels of the exercises progress from 1 to 3, with Level 1 indicating the lowest level of complexity.

| Parts | Level 1 | Level 2 | Level 3 |
|---|---|---|---|
| A |  | ● |  |
| B |  |  | ● |
| C | ● |  |  |
| D |  | ● |  |
| E |  |  | ● |

552 | Chapter 8 | Preparing for Assessment

# Chapter 8 Preparing for Assessment

## Test-Taking Strategy

### Example

Read the problem. Identify what you need to know. Then use the information in the problem to solve.

The city wants to pave a walking path around the rectangular pond shown. If $x = 21$, what will be the length of the path in yards?

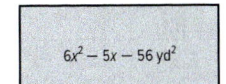

$6x^2 - 5x - 56$ yd$^2$

A  106 yards
B  208 yards
C  212 yards
D  2485 yards

**Step 1** What do you need to find?
The length of the walking path, which is the perimeter of the pond.

**Step 2** What are the steps needed to solve the problem?
I need to factor the given expression to find the dimensions of the pond. Then, I need to write an expression for the perimeter and evaluate the expression for $x = 21$.

**Step 3** What is the correct answer?
Factoring the expression for area gives $6x^2 - 5x - 56 = (2x - 7)(3x + 8)$. So, the pond is $2x - 7$ yards by $3x + 8$ yards. The perimeter is twice the width plus twice the length. So, $P = 2(2x - 7) + 2(3x + 8)$, or $10x + 2$ yards. Substituting 21 for $x$ in the expression gives a perimeter of 212 yards.

**Step 4** How can you check that your answer is correct?
I can substitute 21 for $x$ in the expressions for the length and width to find that the pond is 35 yards by 71 yards. Using these dimensions, the perimeter should be $2(35) + 2(71)$, or 212 yards. The correct answer is C.

### Test-Taking Tip
**Solve Multi-Step Problems** Some problems that you will encounter on standardized tests require you to solve multiple parts in order to come up with the final solution. To solve these problems, you'll need to read the problem carefully and organize your approach.

## Apply the Strategy

Read the problem. Identify what you need to know. Then use the information in the problem to solve.

What is the area of the square?

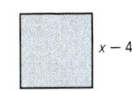

A  $x^2 + 16$
B  $4x - 16$
C  $x^2 - 8x - 16$
D  $x^2 - 8x + 16$

$x - 4$

Answer the questions below.

a. What do you need to find? the area of the square
b. What are the steps needed to solve the problem? Square the expression for the side length.
c. What is the correct answer? $(x - 4)^2 = x^2 - 8x + 16$
d. How can you check that your answer is correct? Factor the expression from part c to make sure I get $x - 4$ as the side length of the square.

## Test-Taking Strategy

**Solve Multi-Step Problems**

**Step 1** Identify what is being asked for.

**Step 2** Determine the steps you need to solve the problem.

**Step 3** Use the steps to solve the problem.

**Step 4** Check that your answer is correct.

### Need Another Example?

A square has an area of $64g^2 + 96gh + 36h^2$. Which one of the following expressions represents the length of each side of the square?

A  $8g + 6h$
B  $8g - 6h$
C  $8g^2 + 6h^2$
D  $8g^2 - 6h^2$

a. What do you need to find? the length of each side of the square
b. What are the steps needed to solve the problem? Factor the expression for the area using the pattern for a perfect square trinomial.
c. What is the correct answer? $64g^2 + 96gh + 36h^2 = (8g + 6h)^2$. So, the side length of the square is $8g + 6h$. The correct answer is A.
d. How can you check that your answers is correct? Multiply the square of a sum from part c to make sure it is the same as the original expression.

## Additional Answers (Performance Task)

**Part A**
1. $4s - 20$, $s^2 - 10s + 16$

**Part B**
2. $4w^2 - 6w$
3. $5w^2 - 12w + 9$
4. 270 square feet

**Part C**
5. $x + 4$

**Part D**
6. width = 7 feet and length = 47 feet

**Part E**
7. Either; both options result in a total area of 270 square feet.

### Go Online!

The most up-to-date resources available for your program can be found at connectED.mcgraw-hill.com.

connectED.mcgraw-hill.com  553

# Chapter 8 Preparing for Assessment

## Diagnose Student Errors

Survey student responses for each item. Class trends may indicate common errors and misconceptions.

| 1. | A | Did not write $x + 10$ as a quantity |
|---|---|---|
| | B | CORRECT |
| | C | Added $x$ and $\frac{2}{3}$ |
| | D | Did not add $x$ to 10 |

| 2. | A | Did not add $x^2$ and $2x^2$ |
|---|---|---|
| | B | Did not multiply by 2 |
| | C | CORRECT |
| | D | Only multiplied the length by 2 |

| 4. | A | Used wrong signs in the factors |
|---|---|---|
| | B | Used the common factor squared. |
| | C | Used wrong signs in the factors |
| | D | CORRECT |

| 5. | A | Added 4 instead of subtracting |
|---|---|---|
| | B | CORRECT |
| | C | Added $6x$ instead of subtracting |
| | D | Dropped the negative sign |

| 6. | A | Made a sign error when multiplying |
|---|---|---|
| | B | Made a sign error when multiplying |
| | C | CORRECT |
| | D | Did not multiply the constant terms |

| 7. | A | Named the constant term |
|---|---|---|
| | B | Found the $y$-intercept |
| | C | CORRECT |
| | D | Forgot to solve for $y$ first |

| 8. | A | CORRECT |
|---|---|---|
| | B | Forgot to multiply area of base by 3 |
| | C | Did not square $(x + 3)$ correctly |
| | D | Did not square $(x + 3)$ correctly |

| 9. | A | Found the area |
|---|---|---|
| | B | Forgot to double the sum |
| | C | CORRECT |
| | D | Made a sign error when adding |

| 10. | A | Found wrong $y$-intercepts and slopes |
|---|---|---|
| | B | Found a wrong $y$-intercept and slope |
| | C | CORRECT |
| | D | Found a wrong slope |

## Go Online!

### Standardized Test Practice

Students can take self-checking tests in standardized format to plan and prepare for your state assessments.

---

# CHAPTER 8
# Preparing for Assessment
## Cumulative Review

Read each question. Then fill in the correct answer on the answer document provided by your teacher or on a sheet of paper.

1. Two thirds of $x$ plus 16 is as much as 24 times $x$ minus the quantity $x$ plus 10. Which equation models this relationship? A.CED.1  **B**

   ○ A  $\frac{2}{3}x + 16 = 24x - x + 10$
   ○ B  $\frac{2}{3}x + 16 = 24x - (x + 10)$
   ○ C  $\frac{2}{3} + x + 16 = 24x - (x + 10)$
   ○ D  $\frac{2}{3}x + 16 = 24x - 10$

2. A rectangle has a length of $2x^2 + 3x + 7$ and a width of $x^2 + 5$. What is the perimeter of the rectangle? A.APR.1  **C**

   ○ A  $2(2x^2 + 3x + 12)$
   ○ B  $3(x^2 + x + 4)$
   ○ C  $6(x^2 + x + 4)$
   ○ D  $5x^2 + 6x + 19$

3. A rectangular garden has a perimeter of 108 feet. The length of the garden is 8 feet more than the width. A.SSE.2

   a. Write expressions that represent the perimeter and area of the garden in terms of the width, $w$.
   
   Perimeter: $4w + 16$; Area: $w(w + 8)$ or $w^2 + 8w$

   b. What is the width of the garden?
   
   **23** feet

   c. (MP) What mathematical practice did you use to solve this problem? See students' work.

4. Which expression is a factor of $8x^2 - 26x - 45$? A.SSE.2  **D**

   ○ A  $4x - 5$      ○ C  $2x + 9$
   ○ B  $(2x + 9)^2$  ○ D  $2x - 9$

5. Solve: $3x(x - 6) + 4 = x(3x + 6) + 18$  A.APR.1  **B**

   ○ A  $-\frac{11}{12}$
   ○ B  $-\frac{7}{12}$
   ○ C  $-\frac{7}{6}$
   ○ D  $\frac{7}{12}$

6. Jennifer plans a triangular garden to grow fruits and vegetables. She needs to know the area. The measurements of the garden are shown below.

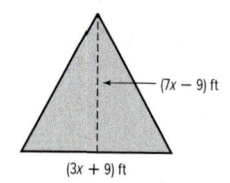

   $(7x - 9)$ ft
   $(3x + 9)$ ft

   Which expression should Jennifer use to find the area of the garden? A.APR.1  **C**

   ○ A  $\frac{1}{2}(21x^2 + 90x - 81)$
   ○ B  $\frac{1}{2}(21x^2 - 90x - 81)$
   ○ C  $\frac{1}{2}(21x^2 + 36x - 81)$
   ○ D  $\frac{1}{2}(21x^2 + 36x)$

7. What is the slope of the line $3y = 12x - 6$? F.LE.2  **C**

   ○ A  $-6$
   ○ B  $-2$
   ○ C  $4$
   ○ D  $12$

### Test-Taking Tip

**Question 7** Rewrite a linear equation in slope-intercept form to find the slope of the line. In this form, the coefficient of $y$ is 1.

---

554 | Chapter 8 | Preparing for Assessment

Chapter 8 Preparing for Assessment

**Formative Assessment**

You can use these pages to benchmark student progress.

**Standardized Test Practice**

**Test Item Formats**

In the Cumulative Review, students will encounter different formats for assessment questions to prepare them for standardized tests.

| Question Type | Exercises |
| --- | --- |
| Multiple-Choice | 1–2, 4–10, 12–13 |
| Multiple Correct Answers | 11 |
| Short Response | 3 |
| Extended Response | 3 |

**Answer Sheet Practice**

Have students simulate taking a standardized test by recording their answers on a practice recording sheet.

**Homework Option**

**Get Ready for Chapter 9** Assign students the exercises on p. 558 as homework to assess whether they possess the prerequisite skills needed for the next chapter.

---

**Go Online!** for Standardized Test Practice

8. A prism has a volume of 243 cubic centimeters. If it has a height of 3 centimeters and its base is a square with sides of $(x + 3)$ centimeters, which equation represents the area of the base of the prism? A.APR.1  **A**

   ○ A   $81 = x^2 + 6x + 9$
   ○ B   $243 = x^2 + 6x + 9$
   ○ C   $81 = x^2 + 6x + 6$
   ○ D   $81 = x^2 + 3x + 9$

9. The area of a rectangular room is $x^2 + 2x - 224$. The length of the room is $x + 16$. Which expression represents the perimeter of the room? A.SSE.2  **C**

   ○ A   $(x + 16)(x - 14)$
   ○ B   $2(x + 1)$
   ○ C   $4(x + 1)$
   ○ D   $4(x - 1)$

10. Some values for two linear equations are shown in the tables below.

| x | y |
| --- | --- |
| −2 | 13 |
| −1 | 14 |
| 0 | 15 |
| 3 | 18 |

| x | y |
| --- | --- |
| −6 | 8 |
| 0 | 5 |
| 4 | 3 |
| 6 | 2 |

What are the two equations that make up the system represented by these tables? F.IF.2  **C**

   ○ A   $x - y = 15, x + y = 5$
   ○ B   $2x + y = 13, x + 2y = 10$
   ○ C   $x - y = -15, x + 2y = 10$
   ○ D   $x - y = -15, x + y = 5$

11. Consider the polynomial expression below.

$$3x^2 + 18x + 27$$

Select all of the factoring methods needed to completely factor the polynomial. **B, E**

   ☐ A   Factor a difference of squares.
   ☐ B   Factor a perfect square trinomial.
   ☐ C   Factor out the GCF.
   ☐ D   Factor by grouping.
   ☐ E   The polynomial cannot be factored.

12. During the coin toss, the referee at a football game dropped the coin to the ground from a height of 4 feet. The expression $-16t^2 + h_0$ can be used to represents the height of the coin in feet $t$ seconds after it is tossed, where $h_0$ is the initial height. Which is the factored form of the expression that represents the height of the coin tossed by the referee after $t$ seconds? A.REI.4b  **B**

   ○ A   $-16t^2 + 4$
   ○ B   $-4(4t^2 - 1)$
   ○ C   $-4t(4t - 1)$
   ○ D   $-2(8t^2 - 2)$

13. Which expression is equivalent to $8x^4 - 128$? A.SSE.2  **A**

   ○ A   $8(x + 2)(x - 2)(x^2 + 4)$
   ○ B   $8(x + 2)^3(x - 2)$
   ○ C   $(x + 2)(x - 2)(x^2 + 4)$
   ○ D   $(x - 2)^2(x^2 + 4)$

**Need Extra Help?**

| If you missed Question... | 1 | 2 | 3 | 4 | 5 | 6 | 7 | 8 | 9 | 10 | 11 | 12 | 13 |
| --- | --- | --- | --- | --- | --- | --- | --- | --- | --- | --- | --- | --- | --- |
| Go to Lesson... | 2-1 | 8-1 | 8-2 | 8-6 | 8-2 | 8-3 | 4-1 | 8-4 | 8-6 | 4-2 | 8-7 | 8-5 | 8-7 |

---

**LS LEARNSMART**

Use LearnSmart as part of your test-preparation plan to measure student topic retention. You can create a student assignment in LearnSmart with additional practice on these topics.

· Perform Arithmetic Operations on Polynomials

---

**Go Online!**

**eAssessment**

Customize and create multiple versions of chapter tests and answer keys that align to your standards. Tests can be delivered on paper or online.

## Explore 8-5

**5.**

$3\Big\{$ | $2x-3$ |||||
|---|---|---|---|---|
| $x$ | $x$ | $-1$ | $-1$ | $-1$ |
| $x$ | $x$ | $-1$ | $-1$ | $-1$ |
| $x$ | $x$ | $-1$ | $-1$ | $-1$ |

**6.**

| $x$ | $-1$ |
|---|---|
| $x$ | $-1$ |
| $x$ | $-1$ |
| $x$ | $-1$ |
| $x$ | |

**7.**

| $x^2$ | $x^2$ | $x^2$ | $x^2$ | 1 | 1 |
|---|---|---|---|---|---|
| | | | | 1 | 1 |
| | | | | 1 | 1 |
| | | | | 1 | 1 | 1 |

**8.**

$x\Big\{$ | $x+3$ ||||
|---|---|---|---|
| $x^2$ | $x$ | $x$ | $x$ |

## Explore 8-6

**9.**

| $x^2$ | $x$ | $x$ | $x$ |
|---|---|---|---|
| | 1 | 1 | 1 |
| | 1 | 1 | 1 |

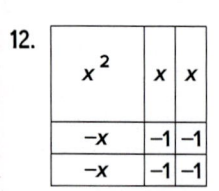

**10.**

| $x^2$ | $-x$ | $-x$ | $-x$ | $-x$ | $-x$ | $-x$ |
|---|---|---|---|---|---|---|
| $x$ | $-1$ | $-1$ | $-1$ | $-1$ | $-1$ | $-1$ |

**11.**

| $x^2$ | $-x$ |
|---|---|
| | $-1$ | $-1$ |
| | $-1$ | $-1$ |

**12.**

| $x^2$ | $x$ | $x$ |
|---|---|---|
| $-x$ | $-1$ | $-1$ |
| $-x$ | $-1$ | $-1$ |

**NOTES**

# CHAPTER 9
# Quadratic Functions and Equations

**SUGGESTED PACING (DAYS)**

90 min. — 9 Instruction, 1 Review & Assess
45 min. — 14 Instruction, 2 Review & Assess

## Track Your Progress

### Content Standards
This chapter focuses on content from Interpreting Functions and the Seeing Structure in Expressions domains.

| THEN | NOW | NEXT |
|---|---|---|
| **A.SSE.3a** Factor a quadratic expression to reveal the zeros of the function it defines.<br><br>**A.APR.1** Understand that polynomials form a system analogous to the integers, namely, they are closed under the operations of addition, subtraction, and multiplication; add, subtract, and multiply polynomials. | **F.IF.7a** Graph linear and quadratic functions and show intercepts, maxima, and minima.<br><br>**A.REI.4b** Solve quadratic equations by inspection, taking square roots, completing the square, the quadratic formula, and factoring.<br><br>**F.IF.6** Calculate and interpret the average rate of change of a function over a specified interval. Estimate the rate of change from a given graph. | **S.ID.1** Represent data with plots on the real number line.<br><br>**S.ID.2** Use statistics appropriate to the shape of the data distribution to compare center and spread of two or more different data sets.<br><br>**S.ID.3** Interpret differences in shape, center, and spread in the context of the data sets, accounting for possible effects of extreme data points (outliers). |

### Standards for Mathematical Practice
All of the Standards for Mathematical Practice will be covered in this chapter. The MP icon notes specific areas of coverage.

**MP Teaching the Mathematical Practices**
Help students develop the mathematical practices by asking questions like these.

**Questioning Strategies** As students approach problems in this chapter, help them develop mathematical practices by asking:

**Sense-Making**
- When seeing an unfamiliar function, how can you get a sense of what the curve will look like?
- How does memorizing the shapes of certain functions help you to solve new problems?

**Construct Arguments**
- Which method for solving quadratic equations do you prefer to use? Why?
- What is the benefit of solving by more than one method?

**Modeling**
- How are quadratic equations used by engineers, designers, financial forecasters, and others in their jobs?

**Tools**
- How can you use your graphing calculator to solve quadratic equations?
- What are the limitations of solving a quadratic equation by graphing?

**Precision**
- What are some techniques you can use to eliminate common errors when solving quadratic equations?

**Structure**
- How does the graph of a quadratic equation differ from that of a linear equation?

### Go Online!

**StudySync: SMP Modeling Videos**

These demonstrate how to apply the Standards for Mathematical Practice to collaborate, discuss, and solve real-world math problems.

# Go Online!
connectED.mcgraw-hill.com

LearnSmart | The Geometer's Sketchpad | Vocabulary | Personal Tutor | Tools | Calculator Resources | Self-Check Practice | Animations

## Customize Your Chapter
Use the *Plan & Present*, *Assignment Tracker*, and *Assessment* tools in ConnectED to introduce lesson concepts, assign personalized practice, and diagnose areas of student need.

### Differentiated Instruction
Throughout the program, look for the icons to find specialized content designed for your students.

- **AL** Approaching Level
- **OL** On Level
- **BL** Beyond Level
- **ELL** English Language Learners

## Personalize

| Differentiated Resources | | | | |
|---|---|---|---|---|
| **FOR EVERY CHAPTER** | AL | OL | BL | ELL |
| ✓ Chapter Readiness Quizzes | ● | ● | ◐ | ● |
| ✓ Chapter Tests | ● | ● | ● | ● |
| ✓ Standardized Test Practice | ● | ● | ● | ● |
| abc Vocabulary Review Games | ● | ● | ● | ● |
| Anticipation Guide (English/Spanish) | ● | ● | ● | ● |
| Student-Built Glossary | ● | ● | ● | ● |
| Chapter Project | ● | ● | ● | ● |
| **FOR EVERY LESSON** | AL | OL | BL | ELL |
| 💬 Personal Tutors (English/Spanish) | ● | ● | ● | ● |
| 💬 Graphing Calculator Personal Tutors | ● | ● | ● | ● |
| ▷ Step-by-Step Solutions | ● | ● | ◐ | ● |
| ✓ Self-Check Quizzes | ● | ● | ● | ● |
| 5-Minute Check | ● | ● | ● | ● |
| Study Notebook | ● | ● | ● | ● |
| Study Guide and Intervention | ● | ● | ◐ | ● |
| Skills Practice | ● | ◐ | ◐ | ● |
| Practice | ◐ | ● | ● | ● |
| Word Problem Practice | ● | ● | ● | ◐ |
| Enrichment | ◐ | ● | ● | ● |
| Extra Examples | ● | ◐ | ◐ | ◐ |
| Lesson Presentations | ● | ● | ● | ● |

◐ Aligned to this group    ○ Designed for this group

## Engage

### Featured IWB Resources

**Geometer's Sketchpad** provides students with a tangible, visual way to learn. *Use with Lessons 9-1 through 9-3.*

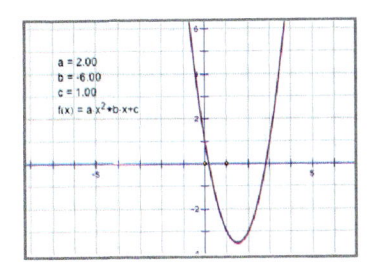

**eLessons** engage students and help build conceptual understanding of big ideas. *Use with Lessons 9-1 through 9-6.*

**Animations** help students make important connections through motion. *Use with Lessons 9-1 and 9-3.*

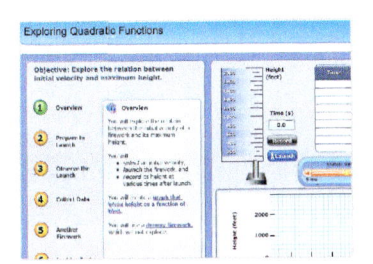

**Time Management** How long will it take to use these resources? Look for the clock in each lesson interleaf.

connectED.mcgraw-hill.com     556B

# Introduce the Chapter

## Mathematical Background
Quadratic functions are nonlinear functions and can be written in the form $f(x) = ax^2 + bx + c$. Quadratic equations can be solved by graphing or by using algebraic methods. The quadratic parent function can be used to sketch related graphs.

## Essential Question
At the end of this chapter, students should be able to answer the Essential Question.

Why do we use different methods to solve math problems? **Sample answer: Depending on the information given, one method may be easier to use than another. It also depends on whether you need an approximate or exact answer.**

## Apply Math to the Real World
**FINANCE** In this activity, students will use what they already know about graphing functions to explore the financial market and how stocks can be represented by quadratic functions. Have students complete this activity individually or in small groups. MP 1, 4, 5

## Go Online!

### Chapter Project
**3…2…1…Blast Off!** Students can use what they have learned about quadratic equations to complete a project. This chapter project addresses business literacy, as well as several specific skills identified as being essential to success by the Framework for 21st Century Learning. MP 1, 2, 3, 4

---

# CHAPTER 9
# Quadratic Functions and Equations

**THEN**
You factored polynomial expressions, including quadratic expressions.

**NOW**
In this chapter, you will:
- Solve quadratic equations by factoring, graphing, completing the square, and using the Quadratic Formula.
- Analyze functions with successive differences and ratios.
- Identify and graph special functions.
- Solve systems of linear and quadratic equations.

**WHY**
**FINANCE** The value of a company's stock is extremely important to the company as well as to shareholders.
Use the mathematical practices to complete the activity.

1. **Use Tools** Use the Internet to find information on financial markets and functions that have been used to model them. Then, select a stock that increased and then decreased over a period of time and record its value each day during this interval.

2. **Model with Mathematics** Use graphing technology to plot the data you recorded. Let $x$ be the time, in days, since the beginning of the interval, and $y$ be the value, in dollars, of the stock. Describe the shape of the graph.

3. **Discuss** Compare your graphs to those of your classmates. What do you think the graph would look like if the value of the stock decreased and then increased over a period of time?

---

## ALEKS

**Your Student Success Tool** ALEKS is an adaptive, personalized learning environment that identifies precisely what each student knows and is ready to learn—ensuring student success at all levels.

- **Formative Assessment:** Dynamic, detailed reports monitor students' progress toward standards mastery.
- **Automatic Differentiation:** Strengthen prerequisite skills and target individual learning gaps.
- **Personalized Instruction:** Supplement in-class instruction with personalized assessment and learning opportunities.

Chapter 9 | Quadratic Functions and Equations

## Go Online to Guide Your Learning

### Explore & Explain

**Equation Chart Mat**
Use the Equation Chart Mat and the Algebra Tiles tool to explore and enhance your understanding of solving quadratic equations by completing the square, as discussed in Lesson 9-5.

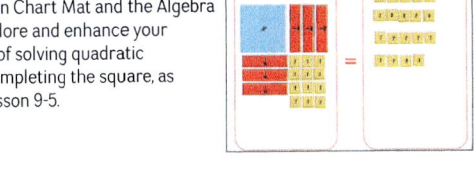

**The Geometer's Sketchpad**
Use The Geometer's Sketchpad to apply dilations and reflections to quadratic functions (Lesson 9-2) and to solve quadratic equations by graphing (Lesson 9-3).

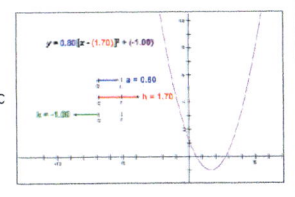

**eBook**
**Interactive Student Guide**
Before starting the chapter, answer the **Chapter Focus** preview questions. Check your answers as you complete each lesson. At the end of the chapter, try the **Performance Task**.

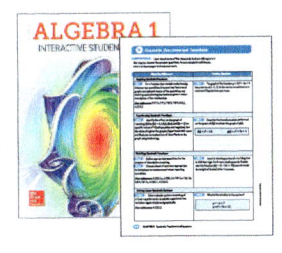

### Organize

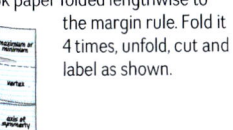

**Foldables**
Make this Foldable to help you organize your notes about quadratic functions. Begin with a sheet of notebook paper folded lengthwise to the margin rule. Fold it 4 times, unfold, cut and label as shown.

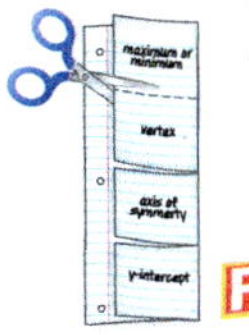

### Collaborate

**Chapter Project**
In the **3... 2... 1... Blast Off!** project, you can use what you learn about quadratic equations to complete a project that addresses business literacy.

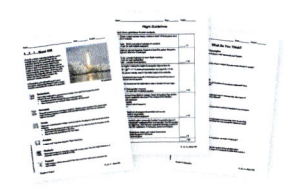

### Focus

**LEARNSMART**
Need help studying? Complete the **Quadratic Functions and Modeling** and **Expressions and Equations** domains in LearnSmart to review for the chapter test.

**ALEKS**
You can use the **Functions and Lines** and **Quadratic Functions and Equations** topics in ALEKS to find out what you know about linear functions and what you are ready to learn.*

*Ask your teacher if this is part of your program.

### Dinah Zike's FOLDABLES

**Focus** Students write notes about the characteristics of quadratic functions as they are presented in the lessons of this chapter.

**Teach** Have students make and label their Foldables as illustrated. Students should use the appropriate section to fill in examples of each characteristic listed. Under each example have students write a brief explanation as to how that characteristic relates to the other three concepts in the Foldable.

**When to Use It** Encourage students to add to their Foldables as they work through the chapter and to use them to review for the chapter test.

### Go Online!

**Vocabulary Learning with VKVs and Foldables**
Looking for ways to immerse your students in math vocabulary? Find out how to use flashcards and Foldables to introduce new terms, model math terms and symbols, and organize related vocabulary terms. MP 6, 8

connectED.mcgraw-hill.com 557

## Chapter 9 | Quadratic Functions and Equations

# Get Ready for the Chapter

**RtI Response to Intervention**
Use the Concept Check results and the Intervention Planner chart to help you determine your Response to Intervention.

### Intervention Planner

**TIER 1 On Level OL**

**IF** students miss 25% of the exercises or less,

**THEN** choose a resource:

**Go Online!**
- Skills Practice, Chapter 2
- Skills Practice, Chapter 3
- Skills Practice, Chapter 8
- Chapter Project
- Self-Check Quiz

**TIER 2 Approaching Level AL**

**IF** students miss 50% of the exercises,

**THEN** choose a resource:

**Go Online!**
- Study Guide and Intervention, Chapter 2
- Study Guide and Intervention, Chapter 3
- Study Guide and Intervention, Chapter 8
- Extra Examples
- Personal Tutors
- Homework Help

*Quick Review Math Handbook*

**TIER 3 Intensive Intervention**

**IF** students miss 75% of the exercises,

**THEN** Use *Math Triumphs, Alg. 1*

**Go Online!**
- Extra Examples
- Personal Tutors
- Homework Help
- Review Vocabulary

### Additional Answers

1. A table can organize data so that you can determine points to use for creating a graph.
2. Choose $x$-values and then use the equation to calculate each corresponding $y$-value.

---

# Get Ready for the Chapter

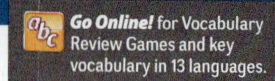

## Connecting Concepts

**Concept Check**
Review the concepts used in this chapter by answering each question below. **1-8. See margin.**

1. Explain how a table of values could be used to graph a function.
2. How do you create a table of values from a new equation?
3. Explain how to draw the graph of a line from a table of values.
4. Explain how to draw the graph of a line from an equation without using a table of values.
5. Refer to the graph at the right. Create a table of values for this graph.
6. Write an equation of the line graphed at the right. 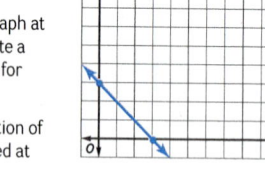
7. Explain how to factor a trinomial with a coefficient of 1.
8. Describe the difference between solving an absolute value equation and a linear equation.

**Performance Task Preview**
You will use the concepts and skills in this chapter to solve problems about the mechanics of baseball. Understanding quadratic equations will help you finish the Performance Task at the end of the chapter.

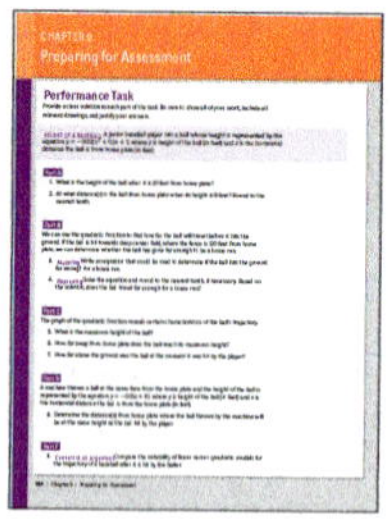

**MP In this Performance Task you will:**
- model with mathematics
- make sense of problems and persevere in solving them
- reason abstractly and quantitatively
- look for and make use of structure

## New Vocabulary

| English | | Español |
|---|---|---|
| quadratic function | p. 559 | función cuadrática |
| parabola | p. 559 | parábola |
| axis of symmetry | p. 559 | eje de simetría |
| vertex | p. 559 | vértice |
| minimum | p. 559 | mínimo |
| maximum | p. 559 | máximo |
| vertex form | p. 575 | forma de vértice |
| double root | p. 581 | doble raíz |
| completing the square | p. 596 | completar el cuadrado |
| Quadratic Formula | p. 606 | Fórmula cuadrática |
| discriminant | p. 610 | discriminante |

### Review Vocabulary

**domain** *dominio* all the possible values of the independent variable, $x$

**factoring** *factorización* to express a polynomial as the product of monomials and polynomials

**leading coefficient** *coeficiente delantero* the coefficient of the first term of a polynomial written in standard form

**range** *rango* all the possible values of the dependent variable, $y$

In the function represented by the table, the domain is {0, 2, 4, 6}, and the range is {3, 5, 7, 9}.

| x | y |
|---|---|
| 0 | 3 |
| 2 | 5 |
| 4 | 7 |
| 6 | 9 |

**transformation** *transformación* the movement of a graph on the coordinate plane

---

## Key Vocabulary ELL

Introduce the key vocabulary in the chapter using the routine below.

**Define** The axis of symmetry is the vertical line containing the vertex of a parabola.

**Example**

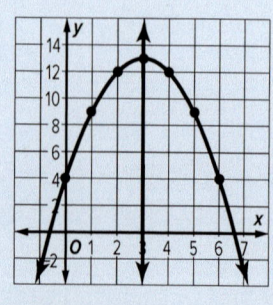

**Ask** How many equal parts of a parabola does an axis of symmetry create? **two**

3. Plot each point given by the $(x, y)$ pairs, then draw a straight line through all the points.

4. Plot the $y$-intercept and use the slope of the equation to plot a second point. Check that your points satisfy the equation. Draw a straight line through the points.

5. 
| x | y |
|---|---|
| 0 | 3 |
| 1 | 2 |
| 2 | 1 |

6. $y = -x + 3$

7. Sample answer: find two integers that have a sum of $b$ and a product of $c$.

8. For linear equations, just isolate the variable. For absolute value equations, first write the case where the expression inside the absolute value symbol is positive and the case where it is negative and then isolate the variable in each case.

# LESSON 9-1
# Graphing Quadratic Functions

**SUGGESTED PACING (DAYS)**

90 min. — 1 | .25
45 min. — 1.5 | 0.5
Instruction | Extend

## Track Your Progress

### Objectives
1. Analyze the characteristics of graphs of quadratic functions.
2. Graph quadratic functions.

### Mathematical Background
The maximum or minimum point of a parabola is called the vertex. When a quadratic function is written in standard form $y = ax^2 + bx + c$, and $a$ is positive, the parabola opens upward, and the vertex is a minimum. When $a$ is negative, the parabola opens downward, and the vertex is a maximum.

### THEN
**A.CED.2** Create equations in two or more variables to represent relationships between quantities; graph equations on coordinate axes with labels and scales.

**A.SSE.3a** Factor a quadratic expression to reveal the zeros of the function it defines.

### NOW
**F.IF.4** For a function that models a relationship between two quantities, interpret key features of graphs and tables in terms of quantities, and sketch graphs showing key features given a verbal description of the relationship.

**F.IF.7a** Graph a linear and quadratic function and show intercepts, maxima, and minima.

### NEXT
**S.ID.6a** Fit a function to the data; use functions fitted to data to solve problems in the context of the data. Use given functions or choose a function suggested by the context. Emphasize linear, quadratic, and exponential models.

---

**Go Online!** All of these resources and more are available at connectED.mcgraw-hill.com

**eLessons** utilize the power of your interactive whiteboard in an engaging way. Use **Quadratic Functions**, screens 1–6, to introduce the concepts in this lesson.

*Use at Beginning of Lesson*

**Personal Tutors** (for every example) let students hear real teachers solve problems. Students can pause and repeat as many times as necessary.

*Use with Examples*

**Graphing Tools** are outstanding tools for enhancing understanding.

*Use with Examples*

---

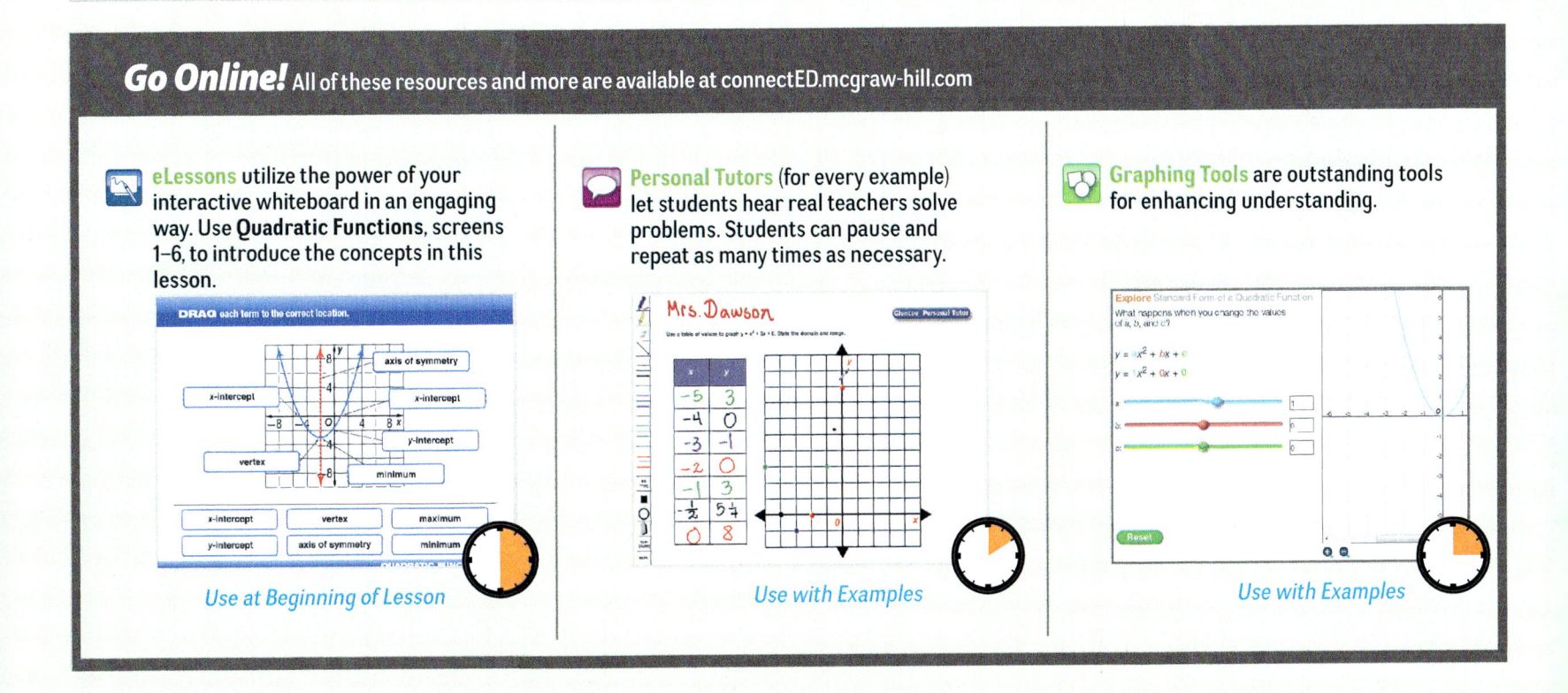

## OER Using Open Educational Resources

**Vocabulary Resources** Direct students struggling with the vocabulary in this lesson to use **WordHippo** to find synonyms and antonyms of the words. Students can also look up definitions and pronouncations. WordHippo can also be used as a translator. *Use as homework*

connectED.mcgraw-hill.com    559A

# Go Online!
connectED.mcgraw-hill.com — Worksheets

# Differentiate Your Resources

**Extra Practice** Additional practice or homework; Skills Practice is best for approaching-level students and Practice is best for on-level and beyond-level students

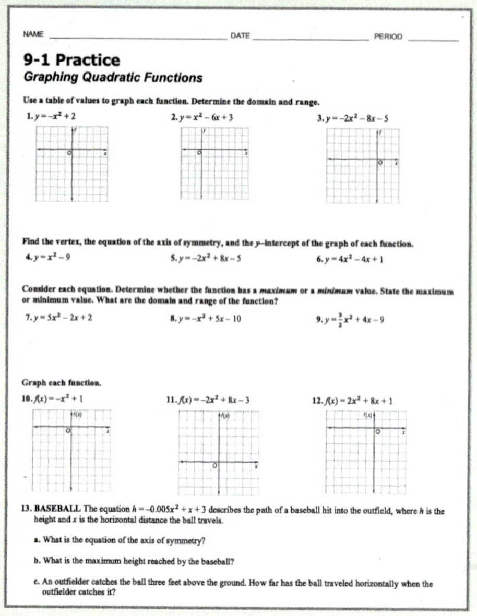

## Skills Practice
AL  OL  ELL

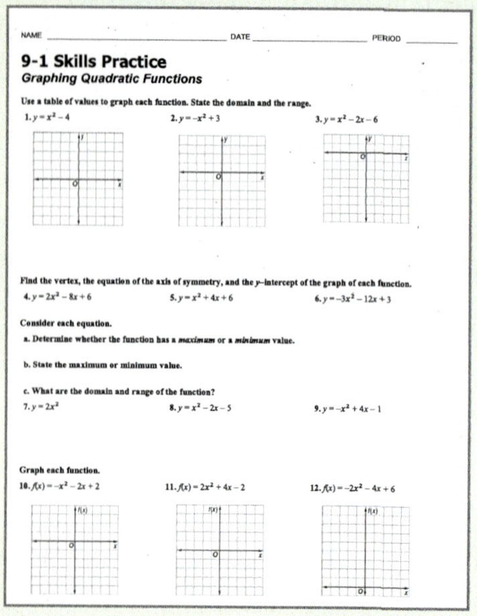

## Practice
AL  OL  BL  ELL

## Word Problem Practice
AL  OL  BL  ELL

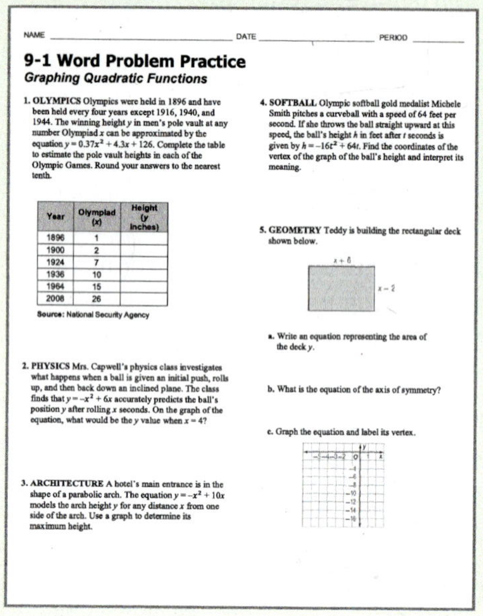

**Intervention** Reteaching and vocabulary activities that can be used with struggling or absent students and as ELL support

**Extension** Activities that can be used to extend lesson concepts

## Study Guide and Intervention
AL  OL  ELL

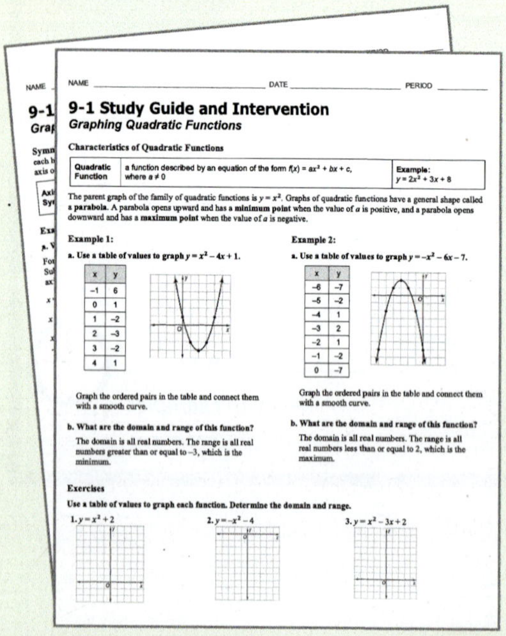

## Study Notebook
AL  OL  BL  ELL

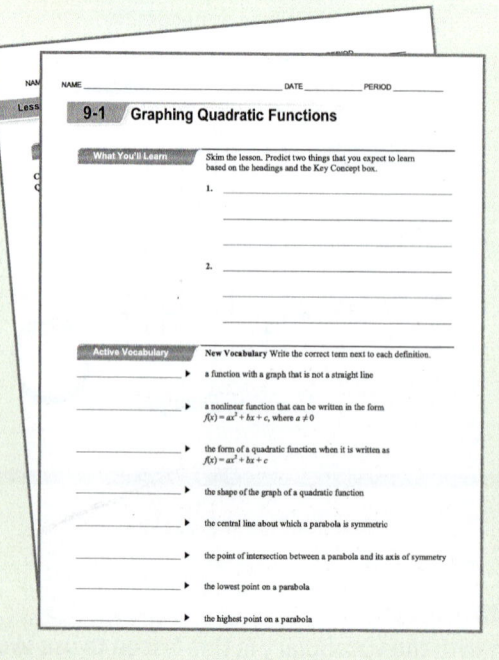

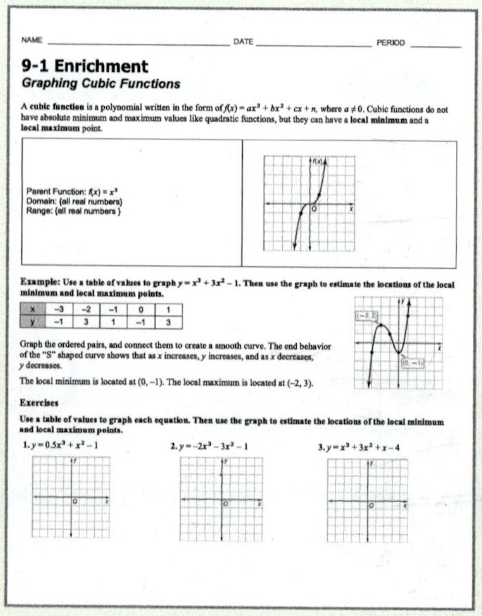

## Enrichment
OL  BL  ELL

559B | Lesson 9-1 | Graphing Quadratic Functions

Lesson 9-1 | Graphing Quadratic Functions

# LESSON 1
# Graphing Quadratic Functions

:·Then·:  :·Now·:  :·Why?·:

- You graphed linear and exponential functions.

1. Analyze the characteristics of graphs of quadratic functions.
2. Graph quadratic functions.

The Innoventions Fountain in Epcot's Futureworld in Orlando, Florida, is an elaborate display of water, light, and music. The sprayers shoot water in shapes that can be modeled by quadratic equations. You can use graphs of these equations to show the path of the water.

### New Vocabulary
quadratic function
standard form
parabola
axis of symmetry
vertex
minimum
maximum

### Mathematical Practices
2 Reason abstractly and quantitatively.

### Content Standards
**F.IF.4** For a function that models a relationship between two quantities, interpret key features of graphs and tables in terms of the quantities, and sketch graphs showing key features given a verbal description of the relationship.

**F.IF.7a** Graph linear and quadratic functions and show intercepts, maxima, and minima.

## 1 Characteristics of Quadratic Functions
**Quadratic functions** are nonlinear and can be written in the form $f(x) = ax^2 + bx + c$, where $a \neq 0$. This form is called the **standard form** of a quadratic function.

The shape of the graph of a quadratic function is called a **parabola**. Parabolas are symmetric about a central line called the **axis of symmetry**. The axis of symmetry intersects a parabola at only one point, called the **vertex**.

### Key Concept  Quadratic Functions

| Parent Function: | $f(x) = x^2$ |
| Standard Form: | $f(x) = ax^2 + bx + c$ |
| Type of Graph: | parabola |
| Axis of Symmetry: | $x = -\dfrac{b}{2a}$ |
| y-intercept: | $c$ |

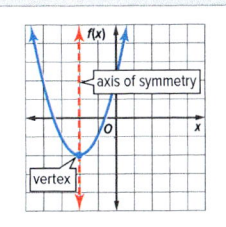

When $a > 0$, the graph of $y = ax^2 + bx + c$ opens upward. The lowest point on the graph is the **minimum**. When $a < 0$, the graph opens downward. The highest point is the **maximum**. The maximum or minimum is the vertex.

F.IF.4, F.IF.7a

### Example 1  Graph a Parabola
Use a table of values to graph $y = 3x^2 + 6x - 4$. State the domain and range.

| x | y |
|---|---|
| 1 | 5 |
| 0 | -4 |
| -1 | -7 |
| -2 | -4 |
| -3 | 5 |

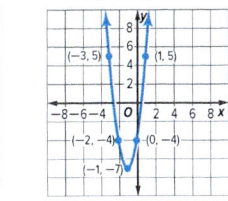

Graph the ordered pairs, and connect them to create a smooth curve. The parabola extends to infinity. The domain is all real numbers or $\{-\infty < x < \infty\}$. The range is $\{y \mid y \geq -7\}$, because $-7$ is the minimum.

### Guided Practice
1. Use a table of values to graph $y = x^2 + 3$. State the domain and range.

See Ch. 9 Answer Appendix.

## Launch
Have students read the Why? section of the lesson. Ask:

- The shape of the water's path can be modeled by $y = -8x^2 - 49x - 75$. Is this a linear equation? Explain. **No; it has an $x^2$-term.**

- Is the path of the water a line? **no**

- How would you describe the shape the water makes when it is shot from the sprayer? **Sample answer: a symmetrical curve called a parabola**

## Teach
Ask the scaffolded questions for each example to build conceptual understanding for students at all levels.

### 1 Characteristics of Quadratic Functions

#### Example 1  Graph a Parabola

**AL** When have you previously seen quadratic expressions? What did you learn about them? **Sample answer: Chapter 8; I factored them.**

**OL** What are the coordinates of the vertex of the parabola? **(−1, −7)**

**BL** What do you notice about the y-values in the table as the x-values decrease? Explain. **Sample answer: The y-values start out decreasing, but then increase after (−1, −7).**

*(continued on the next page)*

---

### MP Mathematical Practices Strategies

**Make sense of problems and persevere in solving them.**
Help students analyze the characteristics of the graphs of quadratic functions and their equations. For example, ask:

- What two characteristics of a parabola are found by using the expression $-\dfrac{b}{2a}$? **the x-coordinate of the vertex and the axis of symmetry**

- Will the domain of every quadratic function be all real numbers? Explain. **Yes, the graph of every quadratic function extends forever in both the positive and negative horizontal directions.**

- Will a parabola that opens up have a maximum? Explain. **No, as x increases or decreases, y goes to infinity, so there is a minimum.**

### Go Online!

**Interactive Whiteboard**
Use the *eLesson* or *Lesson Presentation* to present this lesson.

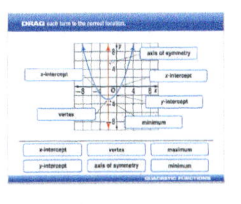

connectED.mcgraw-hill.com   559

# Lesson 9-1 | Graphing Quadratic Functions

**Need Another Example?**

Use a table of values to graph $y = x^2 - x - 2$. State the domain and range.

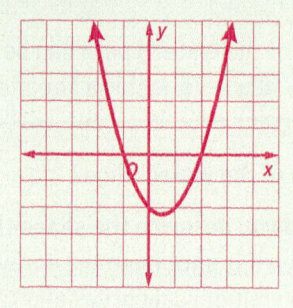

domain: all real numbers; range: $\left\{y \mid y \geq -2\frac{1}{4}\right\}$

## Example 2 Identify Characteristics from Graphs

**AL** What is a vertex? the maximum or minimum point on the parabola

**OL** In part **a**, does the parabola have a max or a min? min part **b**? max

**BL** What do you notice about the equation for the axis of symmetry and the x-coordinate of the vertex? Sample answer: The x-value is the same, which means the line of symmetry goes through the vertex.

**Need Another Example?**

Find the vertex, the equation of the axis of symmetry, and y-intercept.

a.

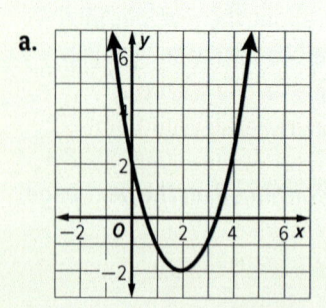

vertex: (2, −2); axis of symmetry: $x = 2$; y-intercept: 2

b.

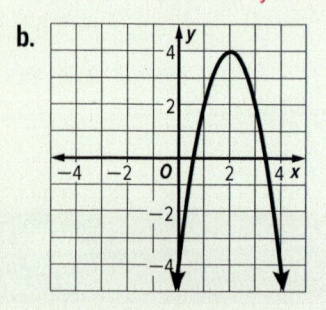

vertex: (2, 4); axis of symmetry: $x = 2$; y-intercept: −4

---

Recall that figures with symmetry are those in which each half of the figure matches exactly. A parabola is symmetric about the axis of symmetry. Every point on the parabola to the left of the axis of symmetry has a corresponding point on the other half. The function is increasing on one side of the axis of symmetry and decreasing on the other side.

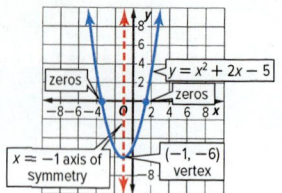

When identifying characteristics from a graph, it is often easiest to locate the vertex first. It is either the maximum or minimum point of the graph. Then locate the x-intercepts. They are the zeros of a quadratic function.

F.IF.4

### Example 2  Identify Characteristics from Graphs

Find the vertex, the equation of the axis of symmetry, the y-intercept, and the zeros of each graph.

a.

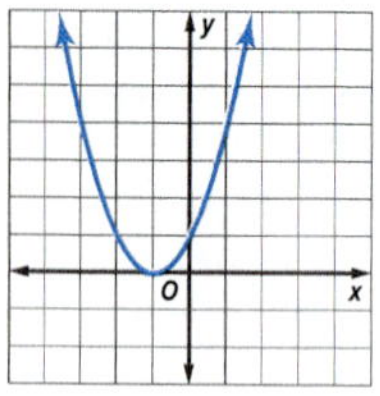

**Step 1** Find the vertex. The parabola opens upward, so the vertex is located at the minimum point, (−1, 0).

**Step 2** Find the axis of symmetry. The axis of symmetry is the line that goes through the vertex and divides the parabola into congruent halves. It is located at $x = -1$.

**Step 3** Find the y-intercept. The parabola crosses the y-axis at (0, 1), so the y-intercept is 1.

**Step 4** Find the zeros. The zeros are the points where the graph intersects the x-axis. This graph only intersects the x-axis once, at (−1, 0).

b.

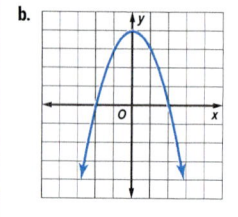

**Step 1** Find the vertex. The parabola opens downward, so the vertex is located at its maximum point, (0, 4).

**Step 2** Find the axis of symmetry. The axis of symmetry is located at $x = 0$.

**Step 3** Find the y-intercept. The parabola crosses the y-axis at (0, 4), so the y-intercept is 4.

**Step 4** Find the zeros. The zeros are the points where the graph intersects the x-axis, (−2, 0) and (2, 0).

**Go Online!**

Investigate how changing the values of a, b, and c in the equation of a quadratic function affects the graph by using the Graphing Tools in ConnectED.

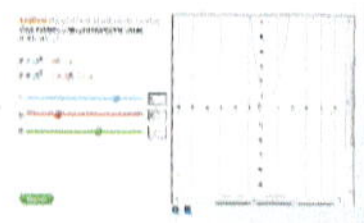

▶ **Guided Practice**

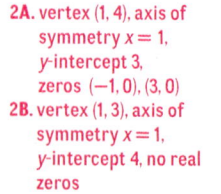

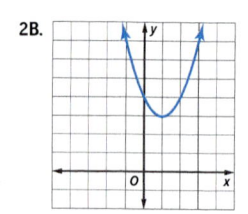

2A. vertex (1, 4), axis of symmetry $x = 1$, y-intercept 3, zeros (−1, 0), (3, 0)

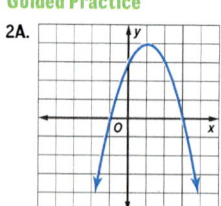

2B. vertex (1, 3), axis of symmetry $x = 1$, y-intercept 4, no real zeros

2A.

2B.

---

### Differentiated Instruction ELL

**Beginning** Define the vocabulary words in the lesson in English and provide examples and explanations. Say the terms out loud and have students repeat the words. Then have students write the word in their notes.

**Advanced/Advanced High** Allow students to use a search engine to find images for each vocabulary term. Have pairs of students choose a representative image for each term to share with the class. Ask them to explain why their image represents the term. ELL

## Lesson 9-1 | Graphing Quadratic Functions

**Study Tip**

**Perseverance** When identifying characteristics of a function, it is often easiest to locate the axis of symmetry first.

### Example 3  Identify Characteristics from Functions

Find the vertex, the equation of the axis of symmetry, and the $y$-intercept of each function.

**a.** $y = 2x^2 + 4x - 3$

$x = -\dfrac{b}{2a}$     Formula for the equation of the axis of symmetry

$x = -\dfrac{4}{2 \cdot 2}$     $a = 2$ and $b = 4$

$x = -1$     Simplify.

The equation for the axis of symmetry is $x = -1$.

To find the vertex, use the value you found for the axis of symmetry as the $x$-coordinate of the vertex. Find the $y$-coordinate using the original equation.

$y = 2x^2 + 4x - 3$     Original equation

$\phantom{y} = 2(-1)^2 + 4(-1) - 3$     $x = -1$

$\phantom{y} = -5$     Simplify.

The vertex is at $(-1, -5)$.

The $y$-intercept always occurs at $(0, c)$. So, the $y$-intercept is $-3$.

**b.** $y = -x^2 + 6x + 4$

$x = -\dfrac{b}{2a}$     Formula for the equation of the axis of symmetry

$x = -\dfrac{6}{2(-1)}$     $a = -1$ and $b = 6$

$x = 3$     Simplify.

The equation of the axis of symmetry is $x = 3$.

$y = -x^2 + 6x + 4$     Original equation

$\phantom{y} = -(3)^2 + 6(3) + 4$     $x = 3$

$\phantom{y} = 13$     Simplify.

The vertex is at $(3, 13)$.

The $y$-intercept is 4.

**Study Tip**

$y$-intercept The $y$-intercept is also the constant term ($c$) of the quadratic function in standard form.

### Guided Practice

**3A.** $y = -3x^2 + 6x - 5$     **3B.** $y = 2x^2 + 2x + 2$

**3A.** vertex $(1, -2)$, axis of symmetry $x = 1$, $y$-intercept $-5$

**3B.** vertex $\left(-\dfrac{1}{2}, \dfrac{3}{2}\right)$, axis of symmetry $x = -\dfrac{1}{2}$, $y$-intercept 2

Next you will learn how to identify whether the vertex is a maximum or a minimum.

### Key Concept  Maximum and Minimum Values

**Words**

The graph of $f(x) = ax^2 + bx + c$, where $a \neq 0$:
- opens upward and has a minimum value when $a > 0$, and
- opens downward and has a maximum value when $a < 0$.

The range of a quadratic function is all real numbers greater than or equal to the minimum, or all real numbers less than or equal to the maximum.

**Examples**    $a$ is positive.      $a$ is negative.

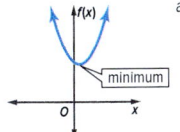

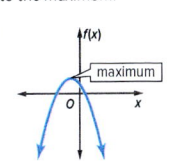

### Example 3  Identify Characteristics from Functions

**AL** What would happen if a parabola was folded along its axis of symmetry? Sample answer: The two halves of the parabola would match up perfectly.

**OL** In general, how is the equation of a vertical line written? $x = c$, where $c$ is a constant

**BL** Can the graph of a quadratic function have more than one $x$-intercept? Why or why not? Yes; the shape of the graph allows it to cross the $x$-axis up to twice. Can the graph of a quadratic function have more than one $y$-intercept? Why or why not? No; a function cannot cross a vertical line more than once.

### Need Another Example?

Find the vertex, the equation of the axis of symmetry, and $y$-intercept.

**a.** $y = -2x^2 - 8x - 2$
$(-2, 6), x = -2, -2$

**b.** $y = 3x^2 + 6x - 2$
$(-1, -5), x = -1, -2$

### Go Online!

The most up-to-date resources available for your program can be found at **connectED.mcgraw-hill.com**.

connectED.mcgraw-hill.com    561

# Lesson 9-1 | Graphing Quadratic Functions

## Example 4 Maximum and Minimum Values

**AL** How can the vertex of a parabola be located? **Sample answers: Draw the axis of symmetry and find the point at which it intersects the parabola. Then find the turning point of the parabola.**

**OL** How can you tell, without graphing, whether the graph opens up or down? **Sample answer: Look at the $a$ value—the coefficient of $x^2$. If it is positive the graph opens up; if it is negative, the graph opens down.**

**BL** Do you think it is possible for the range of a quadratic function to be all real numbers? Why or why not? **Sample answer: No; the graph of a quadratic function changes direction, so it will always have a maximum or minimum value.**

### Need Another Example?

Consider $f(x) = -x^2 - 2x - 2$.

a. Determine whether the function has a maximum or a minimum value. **maximum**
b. State the maximum or minimum value of the function. **−1**
c. State the domain and range of the function. **domain: all real numbers; range: $\{y \mid y \leq -1\}$**

---

### Example 4 Maximum and Minimum Values

F.IF.4

Consider $f(x) = -2x^2 - 4x + 6$.

a. Determine whether the function has a *maximum* or *minimum* value.

 For $f(x) = -2x^2 - 4x + 6$, $a = -2$, $b = -4$, and $c = 6$.

 Because $a$ is negative the graph opens down, so the function has a maximum value.

b. State the maximum or minimum value of the function.

 The maximum value is the $y$-coordinate of the vertex.

 The $x$-coordinate of the vertex is $\frac{-b}{2a}$ or $\frac{4}{2(-2)}$ or $-1$.

 $f(x) = -2x^2 - 4x + 6$     Original function
 $f(-1) = -2(-1)^2 - 4(-1) + 6$     $x = -1$
 $f(-1) = 8$     Simplify.

 The maximum value is 8.

c. State the domain and range of the function.

 The domain is all real numbers. The range is all real numbers less than or equal to the maximum value, or $\{y \mid y \leq 8\}$.

**Watch Out!**
**Minimum and Maximum Values** Don't forget to find both coordinates of the vertex $(x, y)$. The minimum or maximum value is the $y$-coordinate.

**Review Vocabulary**
**Domain and Range** The domain is the set of all of the possible values of the independent variable $x$. The range is the set of all of the possible values of the dependent variable $y$.

### Guided Practice

Consider $g(x) = 2x^2 - 4x - 1$.

**4A.** Determine whether the function has a *maximum* or *minimum* value. **minimum**

**4B.** State the maximum or minimum value. **−3**

**4C.** State the domain and range of the function. **D = {all real numbers}, R = {$y \mid y \geq -3$}**

---

## 2 Graph Quadratic Functions
You have learned how to find several important characteristics of quadratic functions.

### Key Concept   Graph Quadratic Functions

**Step 1** Find the equation of the axis of symmetry.
**Step 2** Find the vertex, and determine whether it is a maximum or minimum.
**Step 3** Find the $y$-intercept.
**Step 4** Use symmetry to find additional points on the graph, if necessary.
**Step 5** Connect the points with a smooth curve.

**Lesson 9-1 | Graphing Quadratic Functions**

## Study Tip

**MP Sense-Making** When locating points that are on opposite sides of the axis of symmetry, not only are the points equidistant from the axis of symmetry, they are also equidistant from the vertex.

**Example 5** Graph Quadratic Functions

Graph $f(x) = x^2 + 4x + 3$.

**Step 1** Find the equation of the axis of symmetry.

$x = \dfrac{-b}{2a}$  Formula for the equation of the axis of symmetry

$x = \dfrac{-4}{2 \cdot 1}$ or $-2$    $a = 1$ and $b = 4$

**Step 2** Find the vertex, and determine whether it is a maximum or minimum.

$f(x) = x^2 + 4x + 3$    Original equation

$= (-2)^2 + 4(-2) + 3$    $x = -2$

$= -1$    Simplify.

The vertex lies at $(-2, -1)$. Because $a$ is positive the graph opens up, and the vertex is a minimum.

**Step 3** Find the $y$-intercept.

$f(x) = x^2 + 4x + 3$    Original equation

$= (0)^2 + 4(0) + 3$    $x = 0$

$= 3$    The $y$-intercept is 3.

**Step 4** The axis of symmetry divides the parabola into two equal parts. So if there is a point on one side, there is a corresponding point on the other side that is the same distance from the axis of symmetry and has the same $y$-value.

**Step 5** Connect the points with a smooth curve.

**5A.**

**5B.**

**Guided Practice** Graph each function.

**5A.** $f(x) = -2x^2 + 2x - 1$    **5B.** $f(x) = 3x^2 - 6x + 2$

There are general differences between linear, exponential, and quadratic functions.

|  | Linear Functions | Exponential Functions | Quadratic Functions |
|---|---|---|---|
| Equation | $y = mx + b$ | $y = ab^x, a \neq 0, b > 0, b \neq 1$ | $y = ax^2 + bx + c, a \neq 0$ |
| Degree | 1 | $x$ | 2 |
| Graph | line | curve | parabola |
| Increasing/ Decreasing | $m > 0$: $y$ is increasing on the entire domain. $m < 0$: $y$ is decreasing on the entire domain. | $a > 0, b > 1$ or $a < 0$, $0 < b < 1$: $y$ is increasing on the entire domain. $a > 0, 0 < b < 1$ or $a < 0$, $b > 1$: $y$ is decreasing on the entire domain. | $a > 0$: $y$ is decreasing to the left of the axis of symmetry and increasing on the right. $a < 0$: $y$ is increasing to the left of the axis of symmetry and decreasing on the right. |
| End Behavior | $m > 0$: as $x$ increases, $y$ increases; as $x$ decreases, $y$ decreases $m < 0$: as $x$ increases, $y$ decreases; as $x$ decreases, $y$ increases | $b > 1$: as $x$ decreases, $y$ approaches 0; $a > 0$, as $x$ increases, $y$ increases; $a < 0$, as $x$ increases, $y$ decreases. $0 < b < 1$: as $x$ increases, $y$ approaches 0; $a > 0$, as $x$ decreases, $y$ increases; $a < 0$, as $x$ decreases, $y$ decreases. | $a > 0$: as $x$ increases, $y$ increases; as $x$ decreases, $y$ increases. $a < 0$: as $x$ increases, $y$ decreases; as $x$ decreases, $y$ decreases |

---

**Differentiated Instruction** AL OL

**IF** students need a visual to understand the concept of a vertex and an axis of symmetry,

**THEN** ask students to make a table of values and graph $y = x^2 + 6x + 8$ on grid paper. Have students hold their paper up to the light and fold their parabola in half so that the two sides match exactly. Have students use the crease on their unfolded paper to locate the vertex, axis of symmetry, and minimum value. $(-3, -1); x = -3; -1$

---

## 2 Graph Quadratic Functions

**Watch Out!**

**Preventing Errors** Tell students that their sketches of parabolas do not have to be perfect. However, students should not "connect the dots" with straight lines. The important thing is for the curve to pass through the graphed ordered pairs.

**Teaching Tip**

**Symmetry and Graphing** When students use symmetry to graph parabolas, they only need to find a few points and then reflect those points across the line of symmetry. You may want to suggest that students occasionally check their reflected points by substituting them into the original equation.

**Example 5 Graph Quadratic Functions**

**AL** How is the graph of a quadratic function different from the graph of a linear function? Sample answer: The graph of a quadratic function is curved and symmetrical, while the graph of a linear function is a straight line.

**OL** After determining the axis of symmetry, how can the coordinates of the vertex be determined? Substitute the value of the axis of symmetry in for $x$ in the function and solve for $y$.

**BL** To graph a linear function, two points on the line are needed. How many points are needed to graph a quadratic function? Explain. 3; Sample answer: Because a quadratic function is curved, we need the vertex and one point on each side of the vertex. It is easiest to find the $y$-intercept, and then find the corresponding point to the $y$-intercept on the other side of the axis of symmetry.

**Need Another Example?**

Graph $f(x) = -x^2 + 5x - 2$.

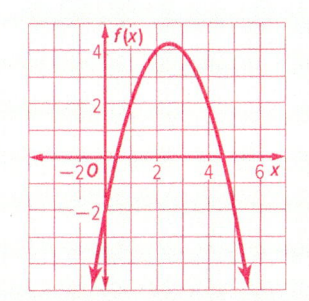

# Lesson 9-1 | Graphing Quadratic Functions

## Example 6 Use a Graph of a Quadratic Function

**AL** What is the axis of symmetry? $x = \frac{3}{2}$ What is the vertex? $\left(\frac{3}{2}, 42\right)$

**OL** What does the vertex tell you about the height of the T-shirt? The maximum height the T-shirt reached before starting to fall back down was 42 feet.

**BL** Why is the y-intercept at 6 and not 0? Sample answer: The T-shirt was not launched from the ground. It was launched by a person, so the height from which the person launched it is the starting height.

## Need Another Example?

**Archery** Ben shoots an arrow. The height of the arrow can be modeled by $y = -16x^2 + 100x + 4$, where y represents the height in feet of the arrow x seconds after it is shot into the air.
a. Graph the height of the arrow.

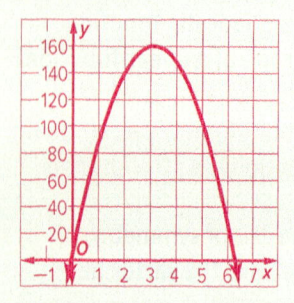

b. At what height was the arrow shot? **4 feet**
c. What is the maximum height of the arrow? **$160\frac{1}{4}$ feet**

## Additional Answers

13a. maximum  13b. 1
13c. $D = \{-\infty < x < \infty\}; R = \{y \mid y \leq 1\}$
14a. maximum  14b. 3
14c. $D = \{-\infty < x < \infty\}; R = \{y \mid y \leq 3\}$
15a. maximum  15b. 6
15c. $D = \{-\infty < x < \infty\}; R = \{y \mid y \leq 6\}$
16a. maximum  16b. 2
16c. $D = \{-\infty < x < \infty\}; R = \{y \mid y \leq 2\}$
17.

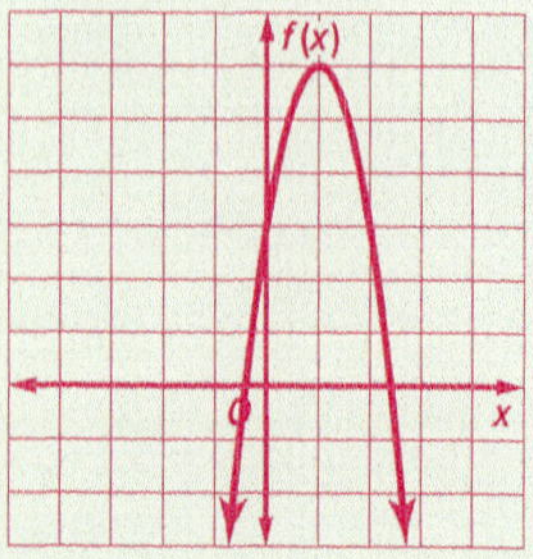

---

You have used what you know about quadratic functions, parabolas, and symmetry to create graphs. You can analyze these graphs to solve real-world problems.

F.IF.4, F.IF.7a

**Real-World Example 6** Use a Graph of a Quadratic Function

**SCHOOL SPIRIT** The cheerleaders at Lake High School launch T-shirts into the crowd every time the Lakers score a touchdown. The height of the T-shirt can be modeled by the function $h(x) = -16x^2 + 48x + 6$, where $h(x)$ represents the height in feet of the T-shirt after x seconds.

a. Graph the function.

$x = -\frac{b}{2a}$   Equation of the axis of symmetry

$x = -\frac{48}{2(-16)}$ or $\frac{3}{2}$   $a = -16$ and $b = 48$

The equation of the axis of symmetry is $x = \frac{3}{2}$. Thus, the x-coordinate for the vertex is $\frac{3}{2}$.

$y = -16x^2 + 48x + 6$   Original equation

$= -16\left(\frac{3}{2}\right)^2 + 48\left(\frac{3}{2}\right) + 6$   $x = \frac{3}{2}$

$= -16\left(\frac{9}{4}\right) + 48\left(\frac{3}{2}\right) + 6$   $\left(\frac{3}{2}\right)^2 = \frac{9}{4}$

$= -36 + 72 + 6$ or 42   Simplify.

The vertex is at $\left(\frac{3}{2}, 42\right)$.

Let's find another point. Choose an x-value of 0 and substitute. Our new point is at (0, 6). The point paired with it on the other side of the axis of symmetry is (3, 6).

Repeat this and choose an x-value of 1 to get (1, 38) and its corresponding point (2, 38). Connect these points and create a smooth curve.

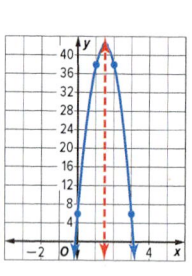

b. At what height was the T-shirt launched?

The T-shirt is launched when time equals 0, or at the y-intercept.

So, the T-shirt was launched 6 feet from the ground.

c. What is the maximum height of the T-shirt? When was the maximum height reached?

The maximum height of the T-shirt occurs at the vertex.

So the T-shirt reaches a maximum height of 42 feet. The time was $\frac{3}{2}$ or 1.5 seconds after launch.

**Guided Practice**

6. **TRACK** Emilio is competing in the javelin throw. The height of the javelin can be modeled by the equation $y = -16x^2 + 64x + 6$, where y represents the height in feet of the javelin after x seconds.

A. Graph the path of the javelin.
B. At what height is the javelin thrown? **6 ft**
C. What is the maximum height of the javelin? **70 ft**

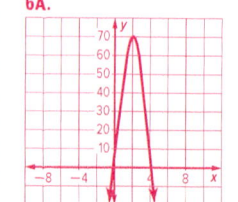

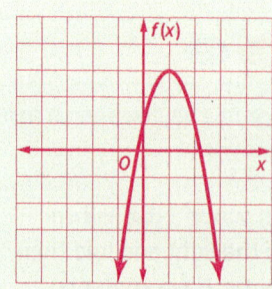

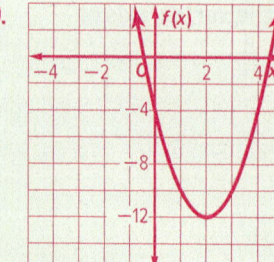

**Real-World Link**
Only 3–4% of high school football players get the opportunity to play college football. Only 1.7% of college players play professionally, and only 0.08% of high school players will move on to the NFL.
Source: Business Insider

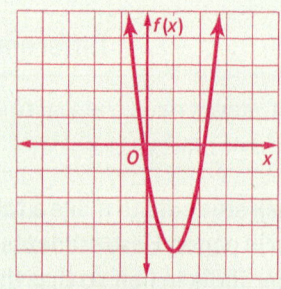

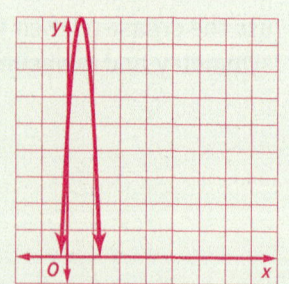

Lesson 9-1 | Graphing Quadratic Functions

## Check Your Understanding

= Step-by-Step Solutions begin on page R11.

Go Online! for a Self-Check Quiz

**Example 1**
F.IF.4, F.IF.7a

Use a table of values to graph each equation. State the domain and range. 1–4. See Ch. 9 Answer Appendix.

1. $y = 2x^2 + 4x - 6$
2. $y = x^2 + 2x - 1$
3. $y = x^2 - 6x - 3$
4. $y = 3x^2 - 6x - 5$

**Example 2**
F.IF.4

Find the vertex, the equation of the axis of symmetry, the $y$-intercept, and the zeros of each graph.

5. 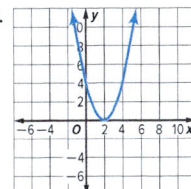 
6. 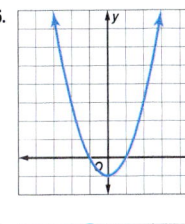 
7. 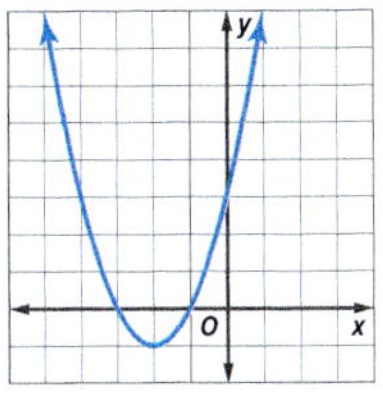 
8. 

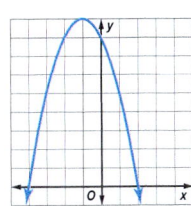

5. vertex (2, 0), axis of symmetry $x = 2$, $y$-intercept 4, zero (2, 0)
6. vertex (0, −1), axis of symmetry $x = 0$, $y$-intercept −1, zeros (−1, 0), (1, 0)
7. vertex (−2, −1), axis of symmetry $x = −2$, $y$-intercept 3, zeros (−1, 0), (−3, 0)
8. vertex (−1, 9), axis of symmetry $x = −1$, $y$-intercept 8, zeros (−4, 0), (2, 0)
9. vertex (1, 2), axis of symmetry $x = 1$, $y$-intercept −1
10. vertex (1, 2), axis of symmetry $x = 1$, $y$-intercept 1
11. vertex (2, 1), axis of symmetry $x = 2$, $y$-intercept 5
12. vertex (1, 5), axis of symmetry $x = 1$, $y$-intercept 9

**Example 3**
F.IF.4

Find the vertex, the equation of the axis of symmetry, and the $y$-intercept of the graph of each function.

9. $y = -3x^2 + 6x - 1$
10. $y = -x^2 + 2x + 1$
11. $y = x^2 - 4x + 5$
12. $y = 4x^2 - 8x + 9$

**Example 4**
F.IF.4

Consider each function. 13–16. See margin.
a. Determine whether the function has *maximum* or *minimum* value.
b. State the maximum or minimum value.
c. What are the domain and range of the function?

13. $y = -x^2 + 4x - 3$
14. $y = -x^2 - 2x + 2$
15. $y = -3x^2 + 6x + 3$
16. $y = -2x^2 + 8x - 6$

**Example 5**
F.IF.7a

Graph each function. 17–20. See margin.

17. $f(x) = -3x^2 + 6x + 3$
18. $f(x) = -2x^2 + 4x + 1$
19. $f(x) = 2x^2 - 8x - 4$
20. $f(x) = 3x^2 - 6x - 1$

**Example 6**
F.IF.4, F.IF.7a

21. **REASONING** A juggler is tossing a ball into the air. The height of the ball in feet can be modeled by the equation $y = -16x^2 + 16x + 5$, where $y$ represents the height of the ball at $x$ seconds.
    a. Graph this equation. See margin.
    b. At what height is the ball thrown? 5 ft
    c. What is the maximum height of the ball? 9 ft

## Practice

**Formative Assessment** Use Exercises 1–21 to assess students' understanding of the concepts in this lesson.

The Practice and Problem Solving exercises assess the content taught in the lesson. The Preparing for Assessment page is meant to be used as preparation for end-of-course assessments.

### Exercise Alert

**Grid Paper** For Exercises 1–4, 17–27, 52–58, 63, and 67 students will need grid paper.

### Teaching the Mathematical Practices

**Reasoning** Mathematically proficient students make sense of quantities and their relationships in problem situations. In Exercise 21, discuss with students why the coefficient of $x^2$ is negative.

### Extra Practice

See page R9 for extra exercises for students who are approaching level or for on-level students who need additional reinforcement.

### Levels of Complexity Chart

The levels of the exercises progress from 1 to 3 with Level 1 indicating the lowest level of complexity.

| Exercises | 22–58 | 59–65, 75–80 | 66–74 |
|---|---|---|---|
| Level 3 | | | ● |
| Level 2 | | ● | |
| Level 1 | ● | | |

### Differentiated Homework Options

| Levels | AL Basic | OL Core | BL Advanced |
|---|---|---|---|
| Exercises | 22–58, 68, 69, 71–80 | 23–63 odd, 64–69, 71–80 | 59–74, (optional: 75–80) |
| 2-Day Option | 23–57 odd, 75–80 | 22–58, 75–80 | |
| | 22–58 even, 68, 69, 71–74 | 59–69, 71–74 | |

You can use ALEKS to provide additional remediation support with personalized instruction and practice.

## Go Online!    eBook

**Interactive Student Guide**
Use the *Interactive Student Guide* to deepen conceptual understanding.
· Graphic Quadratic Functions

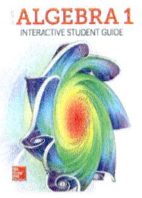

connectED.mcgraw-hill.com    565

Lesson 9-1 | Graphing Quadratic Functions

## Teaching the Mathematical Practices

**Structure** Encourage students to make a habit of stepping back for an overview and shifting perspective. In Exercise 66, ask students how the quadratic and square root functions are related.

**Sense-Making** In Exercise 71, guide students to think about the characteristics of a parabola that can help them solve the problem.

## Additional Answers

34. vertex $(-4, -6)$, axis of symmetry $x = -4$, y-intercept 10

35. vertex $(-3, -8)$, axis of symmetry $x = -3$, y-intercept 10

36. vertex $(-1, 10)$, axis of symmetry $x = -1$, y-intercept 7

37. vertex $(-3, 4)$, axis of symmetry $x = -3$, y-intercept $-5$

38. vertex $(-2, -10)$, axis of symmetry $x = -2$, y-intercept 10

39. vertex $(2, -14)$, axis of symmetry $x = 2$, y-intercept 14

40. vertex $(3, -12)$, axis of symmetry $x = 3$, y-intercept 6

41. vertex $(1, -15)$, axis of symmetry $x = 1$, y-intercept $-18$

42. vertex $(5, 12)$, axis of symmetry $x = 5$, y-intercept $-13$

43a. maximum  43b. 9
43c. $D = \{-\infty < x < \infty\}$, $R = \{y \mid y \leq 9\}$

44a. minimum  44b. $-9$
44c. $D = \{-\infty < x < \infty\}$, $R = \{y \mid y \geq -9\}$

45a. minimum  45b. $-48$
45c. $D = \{-\infty < x < \infty\}$, $R = \{y \mid y \geq -48\}$

46a. maximum  46b. 50
46c. $D = \{-\infty < x < \infty\}$, $R = \{y \mid y \leq 50\}$

47a. maximum  47b. 33
47c. $D = \{-\infty < x < \infty\}$, $R = \{y \mid y \leq 33\}$

## Go Online!

**eSolutions Manual**
Create worksheets, answer keys, and solutions handouts for your assignments.

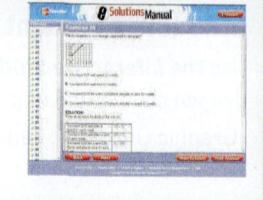

---

**Practice and Problem Solving**  Extra Practice is on page R9.
22–27. See Ch. 9 Answer Appendix.

**Example 1** Use a table of values to graph each equation. State the domain and range.
F.IF.4, F.IF.7a
22. $y = x^2 + 4x + 6$
23. $y = 2x^2 + 4x + 7$
24. $y = 2x^2 - 8x - 5$
25. $y = 3x^2 + 12x + 5$
26. $y = 3x^2 - 6x - 2$
27. $y = x^2 - 2x - 1$

**Example 2** Find the vertex, the equation of the axis of symmetry, the y-intercept, and the zeros
F.IF.4 of each graph.

28. 29. 30. 31. 32. 33.

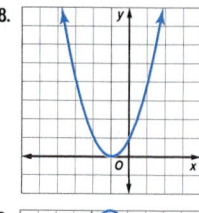

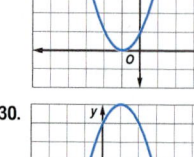

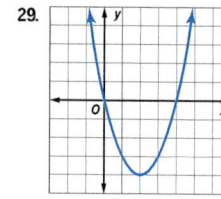

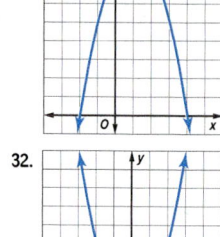

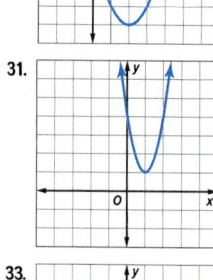

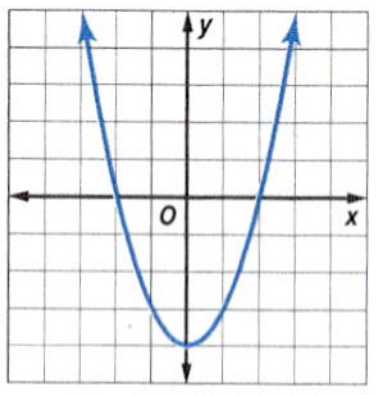

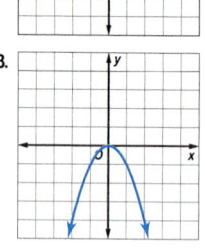

28. vertex $(-1, 0)$, axis of symmetry $x = -1$, y-intercept 1, zero $(-1, 0)$
29. vertex $(2, -4)$, axis of symmetry $x = 2$, y-intercept 0, zeros $(0, 0)$, $(4, 0)$
30. vertex $(1, 9)$, axis of symmetry $x = 1$, y-intercept 8, zeros $(-2, 0)$, $(4, 0)$
31. vertex $(1, 1)$, axis of symmetry $x = 1$, y-intercept 4, no real zeros
32. vertex $(0, -4)$, axis of symmetry $x = 0$, y-intercept $-4$, zeros $(-2, 0)$, $(2, 0)$
33. vertex $(0, 0)$, axis of symmetry $x = 0$, y-intercept 0, zero $(0, 0)$

**Example 3** Find the vertex, the equation of the axis of symmetry, and the y-intercept
F.IF.4 of each function. 34–42. See margin.
34. $y = x^2 + 8x + 10$
35. $y = 2x^2 + 12x + 10$
36. $y = -3x^2 - 6x + 7$
37. $y = -x^2 - 6x - 5$
38. $y = 5x^2 + 20x + 10$
39. $y = 7x^2 - 28x + 14$
40. $y = 2x^2 - 12x + 6$
41. $y = -3x^2 + 6x - 18$
42. $y = -x^2 + 10x - 13$

**Example 4** Consider each function. 43–51. See margin.
F.IF.4
a. Determine whether the function has a *maximum* or *minimum* value.
b. State the maximum or minimum value.
c. What are the domain and range of the function?

43. $y = -2x^2 - 8x + 1$
44. $y = x^2 + 4x - 5$
45. $y = 3x^2 + 18x - 21$
46. $y = -2x^2 - 16x + 18$
47. $y = -x^2 - 14x - 16$
48. $y = 4x^2 + 40x + 44$
49. $y = -x^2 - 6x - 5$
50. $y = 2x^2 + 4x + 6$
51. $y = -3x^2 - 12x - 9$

**Example 5** Graph each function. 52–57. See Ch. 9 Answer Appendix.
F.IF.7a
52. $y = -3x^2 + 6x - 4$
53. $y = -2x^2 - 4x - 3$
54. $y = -2x^2 - 8x + 2$
55. $y = x^2 + 6x - 6$
56. $y = x^2 - 2x + 2$
57. $y = 3x^2 - 12x + 5$

48a. minimum  48b. $-56$
48c. $D = \{-\infty < x < \infty\}$, $R = \{y \mid y \geq -56\}$

49a. maximum  49b. 4
49c. $D = \{-\infty < x < \infty\}$, $R = \{y \mid y \leq 4\}$

50a. minimum  50b. 4
50c. $D = \{-\infty < x < \infty\}$, $R = \{y \mid y \geq 4\}$

51a. maximum  51b. 3
51c. $D = \{-\infty < x < \infty\}$, $R = \{y \mid y \leq 3\}$

58a. 63a.

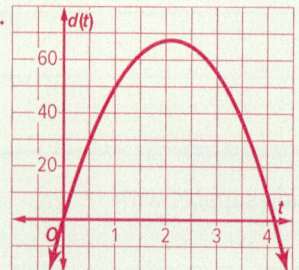

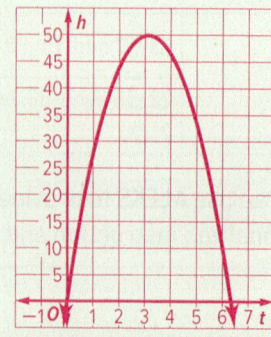

Where $h > 0$, the ball is above the ground. The height of the ball decreases as more time passes.

Lesson 9-1 | Graphing Quadratic Functions

Example 6
F.IF.4,
F.IF.7a

**58. BOATING** Miranda has her boat docked on the west side of Casper Point. She is boating over to Casper Marina, which is located directly east of where her boat is docked. The equation $d = -16t^2 + 66t$ models the distance she travels north of her starting point, where $d$ is the number of feet and $t$ is the time traveled in minutes.

a. Graph this equation. **See margin.**

b. What is the maximum number of feet north that she traveled? **about 68 ft**

c. How long did it take her to reach Casper Marina? **about 4 min**

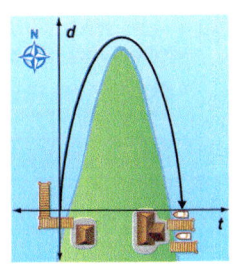

**B GRAPHING CALCULATOR** Graph each equation. Use the **TRACE** feature to find the vertex on the graph. Round to the nearest thousandth if necessary. **59–62. See Ch. 9 Answer Appendix for graphs.**

59. $y = 4x^2 + 10x + 6$ (−1.25, −0.25)
60. $y = 8x^2 − 8x + 8$ (0.5, 6)
61. $y = −5x^2 − 3x − 8$ (−0.3, −7.55)
62. $y = −7x^2 + 12x − 10$ (0.857, −4.857)

**63. GOLF** The average amateur golfer can hit a ball with an initial upward velocity of 31.3 meters per second. The height can be modeled by the equation $h = −4.9t^2 + 31.3t$, where $h$ is the height of the ball, in meters, after $t$ seconds.

a. Graph this equation. What do the portions of the graph where $h > 0$ represent in the context of the situation? What does the end behavior of the graph represent? **See margin.**

b. At what height is the ball hit? **0 m**

c. What is the maximum height of the ball? **about 50.0 m**

d. How long did it take for the ball to hit the ground? **about 6.4 s**

e. State a reasonable range and domain for this situation. $D = \{t \mid 0 \leq t \leq 6.4\}$; $R = \{h \mid 0 \leq h \leq 50.0\}$

**64. MULTI-STEP** The marching band is selling poinsettias to raise money for new uniforms. In last year's sale, the band charged $5 each, and they sold 150 poinsettias. They want to increase the price this year, and they expect to sell 10 fewer poinsettias for each $1 that they increase the price. The nursery will donate 50 poinsettias and charge them $1.50 for each additional poinsettia.

a. If new band uniforms cost $925, will the band be able to raise enough money to buy them? If so, what is the lowest price they could charge for each poinsettia and still have enough money to buy the new uniforms? **yes; $10 per poinsettia**

b. Describe your solution process. **See margin.**

**65. FOOTBALL** A football is kicked up from ground level at an initial upward velocity of 90 feet per second. The equation $h = −16t^2 + 90t$ gives the height $h$ of the football after $t$ seconds.

a. What is the height of the ball after one second? **74 ft**

b. When is the ball 126 feet high? **2.625 seconds and 3 seconds**

c. When is the height of the ball 0 feet? What do these points represent in the context of the situation? $t = 0$, $t = 5.625$; before the ball is kicked, and when the ball hits the ground after the kick

**C 66. STRUCTURE** Let $f(x) = x^2 − 9$.

a. What is the domain of $f(x)$? **{all real numbers}**

b. What is the range of $f(x)$? $\{f(x) \mid f(x) \geq −9\}$

c. For what values of $x$ is $f(x)$ negative? $\{x \mid −3 < x < 3\}$

d. When $x$ is a real number, what are the domain and range of $f(x) = \sqrt{x^2 − 9}$?

66d. $D = \{x \mid x \leq −3 \text{ or } x \geq 3\}$.
$R = \{y \mid y \geq 0\}$

**Watch Out!**

**Error Analysis** For Exercise 69, suggest students identify the values for $a$, $b$, and $c$ for the quadratic function in the graph, being careful to include the signs of the values. Then have them substitute those values into $x = -\frac{b}{2a}$, using parentheses as necessary.

**Additional Answer**

**64b.** Sample answer: We can determine their expected profit by multiplying the number of poinsettias they expect to sell by the price they charge per poinsettia and then subtracting the cost of each poinsettia they sell. They plan to increase their price from $5 by $1 increments, so the price they will charge is $5 + x$. They expect to sell $150 − 10x$ poinsettias, or 10 fewer for every $1 price increase. They are charged $1.50 for every poinsettia they have to buy, which is 1.50 minus the loss in sales caused by the price increase, $10x$, minus the 50 poinsettias donated by the nursery, or $1.50(100 − 10x)$.

So, their expected profit is given by the equation $y = (5 + x)(150 − 10x) − 1.50(100 − 10x)$, where $x$ is the number of $1 price increases. First, I wrote the function in standard form $y = −10x^2 + 115x + 600$. The maximum amount of money the band can raise is the $y$-coordinate of the vertex, so I found the $x$-coordinate of the vertex first, $-\frac{115}{2(-10)}$ or 5.75. Then I used this value to find the $y$-coordinate, $−10(5.75^2) + 115(5.75) + 600$ or 930.63. So, the maximum amount of revenue the band can make is $930.63. Then, I graphed the quadratic equation and the line $y = 925$. The lowest amount they could charge is $5 + the $x$-value of the point of intersection between the two equations, or $x = 5$. So the lowest amount they could charge and still purchase their uniforms is $10 per poinsettia. 1/2

connectED.mcgraw-hill.com 567

**Lesson 9-1 | Graphing Quadratic Functions**

# Assess

**Ticket Out the Door** Make several copies each of the five graphs of quadratic functions. Give one graph to each student. As the students leave the room, ask them to tell you the coordinates of the parabolas' vertices and to identify them as maximums or minimums.

## Additional Answers

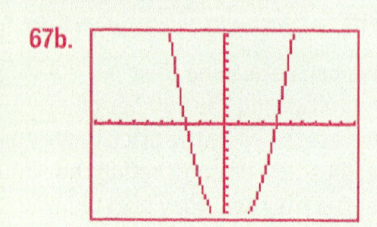

**67b.**

[−10, 10] scl: 1 by [−10, 10] scl: 1
$y = x^2 - x - 12$

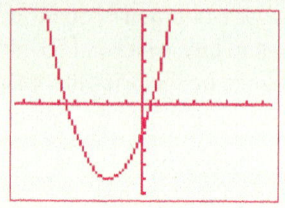

[−15, 15] scl: 2 by [−30, 30] scl: 5
$y = x^2 + 8x - 9$

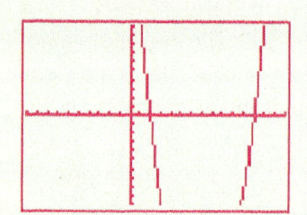

[−10, 15] scl: 1 by [−10, 10] scl: 1
$y = x^2 - 14x + 24$

[−15, 10] scl: 1 by [−10, 10] scl: 1
$y = x^2 + 16x + 28$

**68.** Sample answer: $y = 4x^2 + 3x + 5$; Write the equation of the axis of symmetry, $x = \frac{-b}{2a}$. From the equation, $b = 3$ and $2a = 8$, so $a = 4$. Substitute these values for $a$ and $b$ into the equation $y = ax^2 + bx + c$.

**71.** (−1, 9); Sample answer: I graphed the points given, and sketched the parabola that goes through them. I counted the spaces over and up from the vertex and did the same on the opposite side of the line $x = 2$.

---

**67 MULTIPLE REPRESENTATIONS** In this problem, you will investigate solving quadratic equations using tables.

**a. Algebraic** Determine the related function for each equation. Copy and complete the first two columns of the table below.

| Equation | Related Function | Zeros | y-Values |
|---|---|---|---|
| $x^2 - x = 12$ | $y = x^2 - x - 12$ | −3, 4 | −3; 8, −6; 4; −6, 8 |
| $x^2 + 8x = 9$ | $y = x^2 + 8x - 9$ | −9, 1 | −9; 11, −9; 1; −9, 11 |
| $x^2 = 14x - 24$ | $y = x^2 - 14x + 24$ | 2, 12 | 2; 11, −9; 12; −9, 11 |
| $x^2 + 16x = -28$ | $y = x^2 + 16x + 28$ | −14, −2 | −14; 13, −11; −2; −11, 13 |

**b. Graphical** Graph each related function with a graphing calculator. **See margin.**

**c. Analytical** The number of zeros is equal to the degree of the related function. Use the table feature on your calculator to determine the zeros of each related function. Record the zeros in the table above. Also record the values of the function one unit less than and one unit more than each zero. **See table.**

**d. Verbal** Examine the function values for x-values just before and just after a zero. What happens to the sign of the function value before and after a zero? **The function values have opposite signs just before and just after a zero.**

F.IF.4, F.IF.7a

**H.O.T. Problems** Use Higher-Order Thinking Skills

**68. SENSE-MAKING** Write a quadratic function for which the graph has an axis of symmetry of $x = -\frac{3}{8}$. Summarize your steps. **See margin.**

**69. ERROR ANALYSIS** Jade thinks that the parabolas represented by the graph and the description have the same axis of symmetry. Chase disagrees. Who is correct? Explain your reasoning. **Chase; the lines of symmetry are $x = 2$ and $x = 1.5$.**

a parabola that opens downward, passing through (0, 6) and having a vertex at (2, 2)

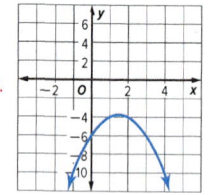

**70. CHALLENGE** Using the axis of symmetry, the y-intercept, and one x-intercept, write an equation for the graph shown. $y = -x^2 + 6x + 16$

**71. SENSE-MAKING** The graph of a quadratic function has a vertex at (2, 0). One point on the graph is (5, 9). Find another point on the graph. Explain how you found it. **See margin.**

**72. REASONING** Describe a real-world situation that involves a quadratic equation. Explain what the vertex represents. **Sample answer: A football is kicked during a game. The vertex gives the maximum height of the ball.**

**73. CONSTRUCT ARGUMENTS** Provide a counterexample that is a specific case to show that the following statement is false. *The vertex of a parabola is always the minimum of the graph.* **Sample answer: The function $y = -x^2 - 4$ has a vertex at (0, −4), but it is a maximum.**

**74. WRITING IN MATH** Use tables and graphs to compare and contrast an exponential function $f(x) = ab^x + c$, where $a \neq 0$, $b > 0$, and $b \neq 1$, a quadratic function $g(x) = ax^2 + c$, and a linear function $h(x) = ax + c$. Include intercepts, portions of the graph where the functions are increasing, decreasing, positive, or negative, relative maxima and minima, symmetries, and end behavior. Which function eventually increases at a faster rate than the others? **See Ch. 9 Answer Appendix.**

## Standards for Mathematical Practice

| Emphasis On | Exercises |
|---|---|
| 1 Make sense of problems and persevere in solving them. | 22–57, 67, 68, 70, 71, 75–80 |
| 2 Reason abstractly and quantitatively. | 72 |
| 3 Construct viable arguments and critique the reasoning of others. | 69, 73, 74 |
| 4 Model with mathematics. | 21, 58, 63–65 |
| 5 Use appropriate tools strategically. | 59–62 |
| 7 Look for and make use of structure. | 66, 77, 79, 80 |

Lesson 9-1 | Graphing Quadratic Functions

## Preparing for Assessment

**75.** Which function has a range of $\{y \mid y \leq 4\}$?  **MP** 1  F.IF.4  **B**
- A  $y = x^2 - 8x + 20$
- B  $y = -x^2 + 8x - 12$
- C  $y = x^2 - 8x + 12$
- D  $y = -x^2 + 8x - 20$

**76.** Which statement best describes the function graphed below?  **MP** 1  F.IF.4, F.IF.7a  **D**

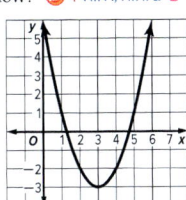

- A  The equation of the axis of symmetry is $x = -3$.
- B  The $y$-intercept is 1.
- C  The maximum value is 6.
- D  The range is $\{y \mid y \geq -3\}$.

**77.** What is the $y$-intercept of $f(x) = x^2 - 2x + 3$?
**MP** 1, 7  F.IF.7a

3

**78.** Rachel correctly identified the vertex and the equation of the axis of symmetry for the function graphed below. Which of the following were most likely Rachel's answers?  **MP** 1  F.IF.4, F.IF.7a  **D**

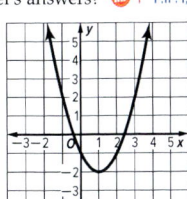

- A  vertex: $(-2, 1)$; axis of symmetry: $x = 1$
- B  vertex: $(-2, 1)$; axis of symmetry: $y = -2$
- C  vertex: $(1, -2)$; axis of symmetry: $y = -2$
- D  vertex: $(1, -2)$; axis of symmetry: $x = 1$

**79.** MULTI-STEP Use the function $y = x^2 + 4x + 3$ to answer the questions.  **MP** 1, 7  F.IF.4, F.IF.7a

a. What is the domain of the function?  **D**
- A  $\{x \mid x \geq 1\}$
- B  $\{x \mid x \leq -3\}$
- C  $\{x \mid x \geq -1\}$
- D  all real numbers

b. What is the range of the function?  **A**
- A  $\{y \mid y \geq -1\}$
- B  $\{y \mid y \leq -1\}$
- C  $\{y \mid y \geq 1\}$
- D  all real numbers

c. What is the vertex of the function?  **C**
- A  $(-4, 3)$
- B  $(-2, 1)$
- C  $(-2, -1)$
- D  $(0, 3)$

d. Which points lie on the parabola? Select all that apply.  **B, C, D, E, F**
- A  $(-4, 0)$
- B  $(-3, 0)$
- C  $(-2, 1)$
- D  $(-1, 0)$
- E  $(0, 3)$
- F  $(1, 8)$
- G  $(2, 12)$

**80.** Ed is graphing the function $y = x^2 - 2x + c$. For what value of $c$ will the range of the function be $\{y \mid y \geq -7\}$?  **MP** 1, 7  F.IF.4

−6

## Preparing for Assessment

Exercises 75–80 require students to use the skills they will need on standardized assessments. Each exercise is dual-coded with content standards and mathematical practice standards.

| Dual Coding | | |
|---|---|---|
| Exercises | Content Standards | **MP** Mathematical Practices |
| 75 | F.IF.4 | 1 |
| 76, 78 | F.IF.4, F.IF.7a | 1 |
| 77 | F.IF.7a | 1, 7 |
| 79 | F.IF.4, F.IF.7a | 1, 7 |
| 80 | F.IF.4 | 1, 7 |

## Diagnose Student Errors

Survey student responses for each item. Class trends may indicate common errors and misconceptions.

**75.**

| A | Confused minimum value and maximum value when determining range |
|---|---|
| B | CORRECT |
| C | Identified function with range $\{y \mid y \geq -4\}$ |
| D | Identified function with range $\{y \mid y \leq -4\}$ |

**76.**

| A | Did not understand concept of axis of symmetry |
|---|---|
| B | Did not understand concept of $y$-intercept |
| C | Did not understand how to find the maximum |
| D | CORRECT |

**78.**

| A | Confused the $x$- and $y$-coordinates of the vertex |
|---|---|
| B | Confused the $x$- and $y$-coordinates of the vertex and found the wrong axis of symmetry |
| C | Found the wrong axis of symmetry |
| D | CORRECT |

## Differentiated Instruction  OL  BL

**Extension**  Tell students that a fireworks rocket is designed to explode at its highest point. The height is given by the equation $h = -4.9t^2 + 34.2t + 1.6$, where $h$ is the rocket's height in meters after $t$ seconds. At what time and height will the rocket explode?  The rocket will explode after about 3.5 seconds at a height of approximately 61 meters.

## Go Online!

**Quizzes**

Students can use Self-Check Quizzes to check their understanding of this lesson. You can also give Quiz 1, which covers the content in Lessons 9-1 through 9-3.

connectED.mcgraw-hill.com  569

# Extend 9-1

## Focus

**Objective** Use a given quadratic function to investigate the rate of change of a quadratic function.

### Materials for Each Student
- grid paper

### Teaching Tip
Students will need to make a table with 19 columns for Step 1 of the Activity.

## Teach  ELL

**Working in Cooperative Groups** Put students in groups of two or three, mixing abilities. Have groups complete the Activity.

**Ask:**
- What maximum and minimum *x*- and *y*-axis values do you need when drawing the coordinate system for the function in the Activity? *x*: 0 to 9, *y*: 0 to 324
- How does the rate of change of a quadratic function differ from the rate of change of a linear function? A linear function has a constant rate of change, so it never changes sign. The rate of change of a quadratic function is positive on some intervals and negative on others.

**Practice** Have students complete Exercises 1–4.

## Assess

### Formative Assessment
Use Exercises 1–3 to assess whether students understand the rate of change in a quadratic function.

### From Concrete to Abstract
Have students examine the situation in the Activity. Ask students how the equation and its rate of change will be affected if the model rocket is launched from the ground with an upward velocity of 44 meters per second. The equation (with rounded coefficients) will be $h = -5t^2 + 44t$. Since a meter is about 3.3 times as long as a foot, the coefficients in the equation for feet per second are divided by 3.3. The variables are not affected. The rates of change will change accordingly.

---

# EXTEND 9-1
## Algebra Lab
# Rate of Change of a Quadratic Function

**Mathematical Practices**
MP 4 Model with mathematics

**Content Standards**
F.IF.6 Calculate and interpret the average rate of change of a function (presented symbolically or as a table) over a specified interval. Estimate the rate of change from a graph.

A model rocket is launched from the ground with an upward velocity of 144 feet per second. The function $y = -16x^2 + 144x$ models the height $y$ of the rocket in feet after $x$ seconds. Using this function, we can investigate the rate of change of a quadratic function.

### Activity

**Step 1** Copy the table below.

| x | 0 | 0.5 | 1.0 | 1.5 | ... | 9.0 |
|---|---|-----|-----|-----|-----|-----|
| y | 0 |     |     |     |     |     |
| Rate of Change | — |  |  |  |  |  |

**Step 2** Find the value of $y$ for each value of $x$ from 0 through 9.

**Step 3** Graph the ordered pairs $(x, y)$ on grid paper. Connect the points with a smooth curve. Notice that the function *increases* when $0 < x < 4.5$ and *decreases* when $4.5 < x < 9$.

**Step 4** Recall that the *rate of change* is the change in $y$ divided by the change in $x$. Find the rate of change for each half second interval of $x$ and $y$.
136, 120, 104, 88, 72, 56, 40, 24, 8, −8, −24, −40, −56, −72, −88, −104, −120, −136

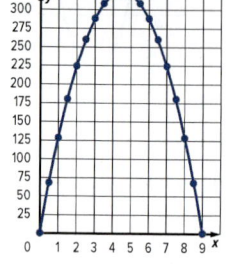

(0.5, 68), (1, 128), (1.5, 180), (2, 224), (2.5, 260), (3, 288), (3.5, 308), (4, 320), (4.5, 324), (5, 320), (5.5, 308), (6, 288), (6.5, 260), (7, 224), (7.5, 180), (8, 128), (8.5, 68), (9, 0)

### Exercises

Use the quadratic function $y = x^2$. 1–4. See Ch. 9 Answer Appendix.

1. Make a table, similar to the one in the Activity, for the function using $x = -4, -3, -2, -1, 0, 1, 2, 3,$ and 4. Find the values of $y$ for each $x$-value.

2. Graph the ordered pairs on grid paper. Connect the points with a smooth curve. Describe where the function is increasing and where it is decreasing.

3. Find the rate of change for each column starting with $x = -3$. Compare the rates of change when the function is increasing and when it is decreasing.

4. **CHALLENGE** If an object is dropped from 100 feet in the air and air resistance is ignored, the object will fall at a rate that can be modeled by the function $f(x) = -16x^2 + 100$, where $f(x)$ represents the object's height in feet after $x$ seconds. Make a table like that in Exercise 1, selecting appropriate values for $x$. Fill in the $x$-values, the $y$-values, and rates of change. Compare the rates of change. Describe any patterns that you see.

# LESSON 9-2
# Transformations of Quadratic Functions

**SUGGESTED PACING (DAYS)**
90 min.  0.75
45 min.  1.5
*Instruction*

## Track Your Progress

### Objectives
1. Apply translations to quadratic functions.
2. Apply dilations and reflections to quadratic functions.

### Mathematical Background
A family of functions is a group of functions whose graphs share the same basic characteristics. The parent graph is the simplest graph in a family. All other graphs in the family are transformations of the parent function. Translations, dilations, and reflections can be performed on the parent function of quadratics.

### THEN
**F.IF.6** Calculate and interpret the average rate of change of a function (presented symbolically or as a table) over a specified interval. Estimate the rate of change from a graph.

### NOW
**F.IF.7a** Graph linear and quadratic functions and show intercepts, maxima, and minima.

**F.BF.3** Identify the effect on the graph of replacing $f(x)$ by $f(x) + k$, $kf(x)$, $f(kx)$, and $f(x + k)$ for specific values of $k$ (both positive and negative); find the value of $k$ given the graphs.

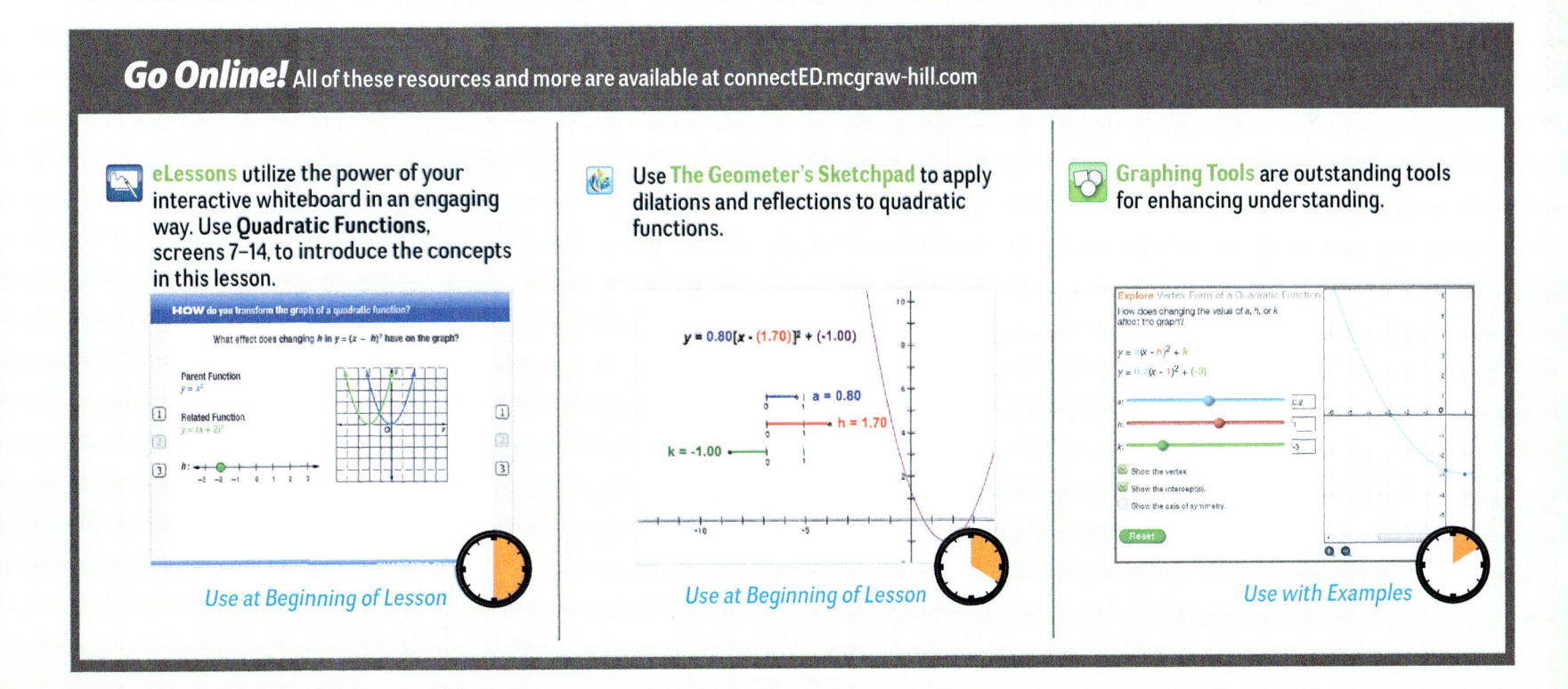

### NEXT
**A.REI.4b** Solve quadratic equations by inspection (e.g., for $x^2 = 49$), taking square roots, completing the square, the quadratic formula and factoring, as appropriate to the initial form of the equation.

---

## Go Online! All of these resources and more are available at connectED.mcgraw-hill.com

**eLessons** utilize the power of your interactive whiteboard in an engaging way. Use **Quadratic Functions**, screens 7–14, to introduce the concepts in this lesson.

*Use at Beginning of Lesson*

Use **The Geometer's Sketchpad** to apply dilations and reflections to quadratic functions.

*Use at Beginning of Lesson*

**Graphing Tools** are outstanding tools for enhancing understanding.

*Use with Examples*

---

### OER Using Open Educational Resources
**Interactives** Before beginning this lesson, have students use the Quadratic Equation Graph on **softschools.com** to see what happens to the graph of a quadratic function when you change the values of $a$, $b$, and $c$. Have students begin with the function $f(x) = x^2 + x + 1$. Then, have them look at the impact on the graph when the values of $a$, $b$, and $c$ are increased and decreased. *Use as introductory activity*

# Go Online!
connectED.mcgraw-hill.com
Worksheets

# Differentiate Your Resources

**Extra Practice** Additional practice or homework; Skills Practice is best for approaching-level students and Practice is best for on-level and beyond-level students

## Skills Practice
AL  OL  ELL

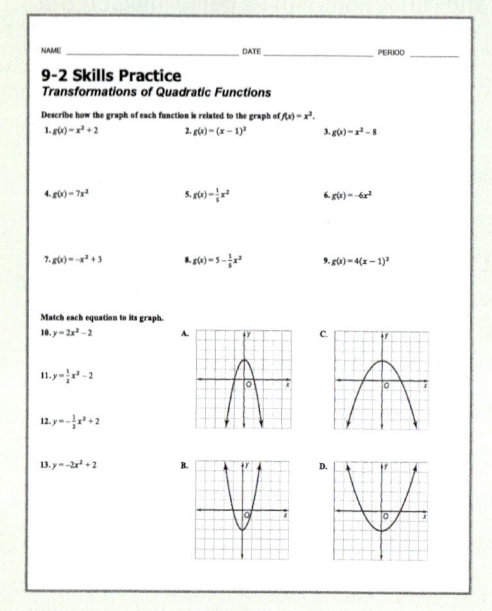

## Practice
AL  OL  BL  ELL

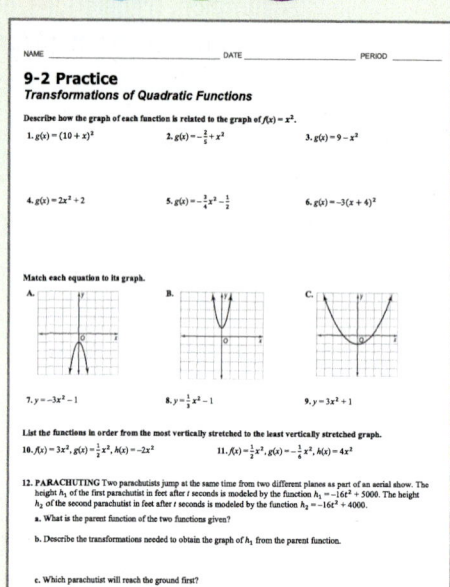

## Word Problem Practice
AL  OL  BL  ELL

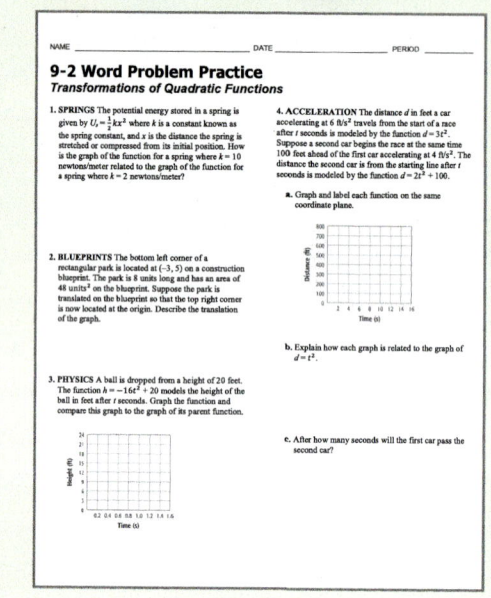

**Intervention** Reteaching and vocabulary activities that can be used with struggling or absent students and as ELL support

**Extension** Activities that can be used to extend lesson concepts

## Study Guide and Intervention
AL  OL  ELL

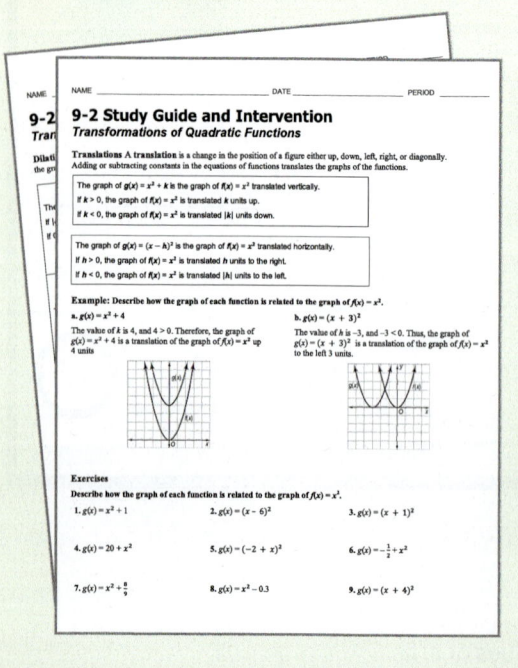

## Study Notebook
AL  OL  BL  ELL

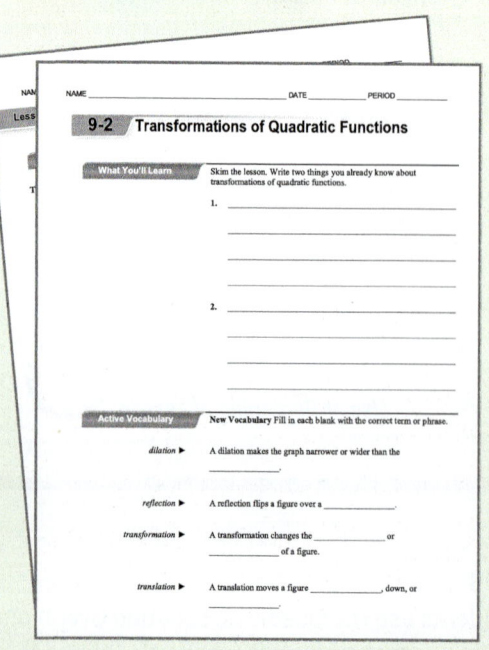

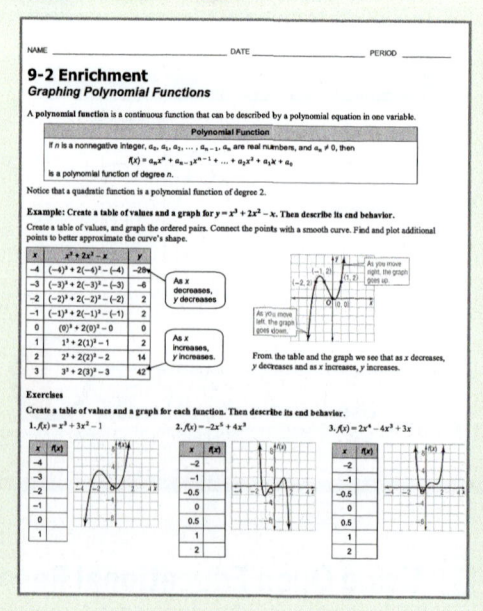

## Enrichment
OL  BL  ELL

571B | Lesson 9-2 | Transformations of Quadratic Functions

# LESSON 2
## Transformations of Quadratic Functions

**::Then** · You graphed quadratic functions by using the vertex and axis of symmetry.

**::Now**
1. Apply translations to quadratic functions.
2. Apply dilations and reflections to quadratic functions.

**::Why?** · The graphs of the parabolas shown at the right are the same size and shape, but notice that the vertex of the red parabola is higher on the y-axis than the vertex of the blue parabola. Shifting a parabola up and down is an example of a transformation.

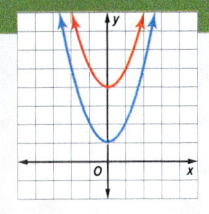

**New Vocabulary**
vertex form

**Mathematical Practices**
1 Make sense of problems and persevere in solving them.
8 Look for and express regularity in repeated reasoning.

**Content Standards**
F.IF.7a Graph linear and quadratic functions and show intercepts, maxima, and minima.
F.BF.3 Identify the effect on the graph of replacing f(x) by f(x) + k, kf(x), f(kx), and f(x + k) for specific values of k (both positive and negative); find the value of k given the graphs. Experiment with cases and illustrate an explanation of the effects on the graph using technology.

### 1 Translations
The graph of $f(x) = x^2$ represents the parent graph of the quadratic functions. The parent graph can be translated up, down, left, right, or in two directions.

When a constant $k$ is added to the quadratic function $f(x)$, the result is a vertical translation.

**Key Concept** Vertical Translations

The graph of $g(x) = x^2 + k$ is the graph of $f(x) = x^2$ translated vertically.

If $k > 0$, the graph of $f(x) = x^2$ is translated $|k|$ units **up**.

If $k < 0$, the graph of $f(x) = x^2$ is translated $|k|$ units **down**.

F.BF.3

**Example 1** Vertical Translations of Quadratic Functions

Describe the translation in each function as it relates to the graph of $f(x) = x^2$.

a. $h(x) = x^2 + 3$
$k = 3$ and $3 > 0$
$h(x)$ is a translation of the graph of $f(x) = x^2$ up 3 units.

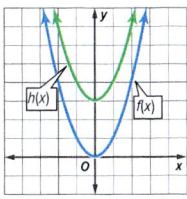

b. $g(x) = x^2 - 4$
$k = -4$ and $-4 < 0$
$g(x)$ is a translation of the graph of $f(x) = x^2$ down 4 units.

**Guided Practice**
1A. $g(x) = x^2 - 7$ translated down 7
1B. $g(x) = 5 + x^2$ translated up 5
1C. $g(x) = -5 + x^2$ translated down 5
1D. $g(x) = x^2 + 1$ translated up 1

---

**MP Mathematical Practices Strategies**

**Make sense of problems and persevere in solving them.**
Help students understand the effect of performing operations on the function $f(x) = x^2$ and explain the correspondences between the equations and graphs of transformed functions. For example, ask:

- What happens when you add 5 to a quantity? *The value increases by 5.*

- What happens to the vertex of a parabola when you add 5 to the y-coordinate? *The vertex moves up 5 units.* when you subtract 5 from the y-coordinate? *The vertex moves down 5 units.*

- Compare the graphs of $y = ax^2$ when $a > 0$ and when $a < 0$. *When $a > 0$, the graph opens up and when $a < 0$, it opens down.*

---

**Lesson 9-2** | Transformations of Quadratic Functions

## Launch

Have students read the Why? section of the lesson. Ask:

- What is the vertex of each parabola? *blue: (0, 1); red: (0, 4)*

- What do we know about the coefficient of the $x^2$ term for the equations of these graphs? *They are both positive.*

- What is the axis of symmetry? *the y-axis*

- What do we know about the $b$ value in the equations for these graphs? *Because $-\frac{b}{2a} = 0$, we know that $b$ must equal 0 for both.*

## Teach

Ask the scaffolded questions for each example to build conceptual understanding for students at all levels.

### 1 Translations

**Example 1 Vertical Translations of Quadratic Functions**

**AL** What is the parent function of the family of quadratics? $f(x) = x^2$

**OL** What is the vertex of $y = x^2$? *(0, 0)* $y = x^2 + 3$? *(0, 3)* $y = x^2 - 4$? *(0, -4)*

**BL** How is adding a constant to the parent function of the family of quadratics similar to adding a constant to a linear function? *Sample answer: In both functions, adding a constant moves the y-intercept up or down the y-axis.*

**Need Another Example?**
Describe the translation in each function as it relates to the graph of $f(x) = x^2$.
a. $g(x) = 10 + x^2$ *translated up 10 units*
b. $h(x) = x^2 - 8$ *translated down 8 units*

**Go Online!**

**Interactive Whiteboard**
Use the *eLesson* or *Lesson Presentation* to present this lesson.

connectED.mcgraw-hill.com 571

# Lesson 9-2 | Transformations of Quadratic Functions

## Example 2 Horizontal Translations of Quadratic Functions

**AL** In part a, what do you notice about the movement of the graph from $f(x)$ to $g(x)$? **The graph moved to the right.** In part b? **The graph moved to the left.**

**OL** In part a, how does the translation affect the coordinates of the vertex? **The x-coordinate increased by 2 units, the y-coordinate stayed at 0.** In part b? **The x-coordinate decreased by 1 unit, the y-coordinate stayed at 0.**

**BL** How can you remember which direction the graph moves, left or right, just by looking at the equation of the function? **Sample answer: If there is a subtraction sign inside the parentheses, it means h is positive and the graph moves right. If there is an addition sign inside the parentheses, it means h is negative (subtracting a negative number is the same as adding a positive) and the graph moves left.**

### Need Another Example?
Describe the translation in each function as it relates to the graph of $f(x) = x^2$.
a. $g(x) = (x + 1)^2$ **translated left 1 unit**
b. $h(x) = (x - 4)^2$ **translated right 4 units**

## Example 3 Multiple Translations of Quadratic Functions

**AL** In part a, what do you notice about the movement of the graph from $f(x)$ to $g(x)$? **The graph moved to the right and up.** in part b? **The graph moved to the left and down.**

**OL** Without looking at the graph, can you give the coordinates of the vertex in part a? Explain. **(3, 2); Sample answer: The graph starts at (0, 0), then moves to the right 3 units. Because h is 3, add 3 to the x-coordinate. It also moves up 2, because k is 2. So add 2 to the y-coordinate.**

**BL** Write a function that moves the graph to the left and up. **Sample answer:** $f(x) = (x + 3)^2 + 4$
Write a function that moves the graph to the right and down. **Sample answer:** $f(x) = (x - 3)^2 - 4$

### Need Another Example?
Describe the translations in each function is it relates to the graph of $f(x) = x^2$.
a. $g(x) = (x + 1)^2 + 1$ **translated left 1 unit and up 1 unit**
b. $h(x) = (x - 2)^2 + 6$ **translated right 2 units and up 6 units**

---

When a constant $h$ is subtracted from the x-value before the function $f(x)$ is performed, the result is a horizontal translation.

**Key Concept** Horizontal Translations

The graph of $g(x) = (x - h)^2$ is the graph of $f(x) = x^2$ translated horizontally.

If $h > 0$, the graph of $f(x) = x^2$ is translated $h$ units to the **right**.

If $h < 0$, the graph of $f(x) = x^2$ is translated $|h|$ units to the **left**.

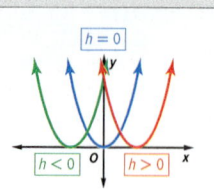

F.BF.3

**Example 2** Horizontal Translations of Quadratic Functions

Describe the translation in each function as it relates to the graph of $f(x) = x^2$.

a. $g(x) = (x - 2)^2$
$k = 0, h = 2$ and $2 > 0$
$g(x)$ is a translation of the graph of $f(x) = x^2$ to the right 2 units.

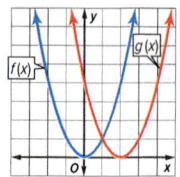

b. $g(x) = (x + 1)^2$
$k = 0, h = -1$ and $-1 < 0$
$g(x)$ is a translation of the graph of $f(x) = x^2$ to the left 1 unit.

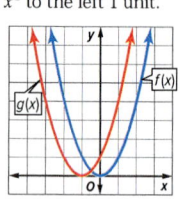

**Guided Practice**

2A. $g(x) = (x - 3)^2$ **translated right 3**
2B. $g(x) = (x + 2)^2$ **translated left 2**

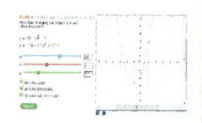

When constants $h$ and $k$ are present in the function, the parent function is translated in both the horizontal and vertical directions.

F.BF.3

**Example 3** Multiple Translations of Quadratic Functions

Describe the translations in each function as it relates to the graph of $f(x) = x^2$.

a. $g(x) = (x - 3)^2 + 2$
$k = 2, h = 3$ and $3 > 0$
$g(x)$ is a translation of the graph of $f(x) = x^2$ to the right 3 units and up 2 units.

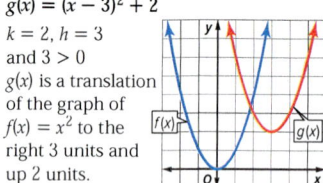

b. $g(x) = (x + 3)^2 - 1$
$k = -1, h = -3$ and $-3 < 0$
$g(x)$ is a translation of the graph of $f(x) = x^2$ to the left 3 units and down 1 unit.

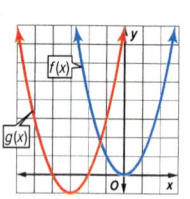

**Guided Practice**

3A. $g(x) = (x + 2)^2 + 3$
3B. $g(x) = (x - 4)^2 - 4$

**Go Online!**
Investigate transformations of the graphs of quadratic functions by using the Graphing Tools in ConnectED.

3A. **translated left 2 units and up 3 units**
3B. **translated right 4 units and down 4 units**

**Lesson 9-2 | Transformations of Quadratic Functions**

## 2 Dilations and Reflections

When a quadratic function $f(x)$ is multiplied by a positive constant $a$, the result $a \cdot f(x)$ is a vertical dilation. The function is stretched or compressed vertically by a factor of $|a|$. When $x$ is multiplied by a positive constant $a$ before the quadratic function $f(x)$ is evaluated, the result $f(ax)$ is a horizontal dilation. The function is stretched or compressed vertically by a factor of $\frac{1}{|a|}$.

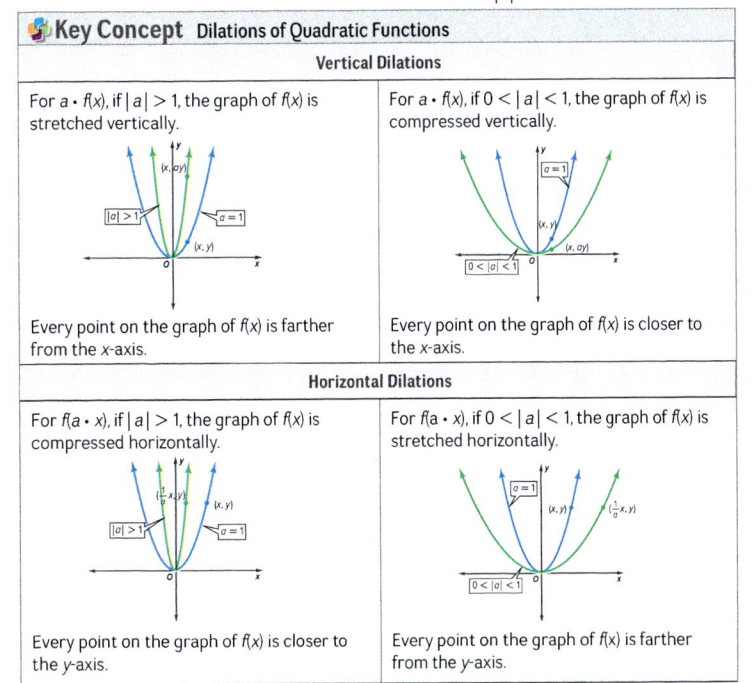

### Key Concept — Dilations of Quadratic Functions

**Vertical Dilations**

| For $a \cdot f(x)$, if $|a| > 1$, the graph of $f(x)$ is stretched vertically. | For $a \cdot f(x)$, if $0 < |a| < 1$, the graph of $f(x)$ is compressed vertically. |
|---|---|
| Every point on the graph of $f(x)$ is farther from the x-axis. | Every point on the graph of $f(x)$ is closer to the x-axis. |

**Horizontal Dilations**

| For $f(a \cdot x)$, if $|a| > 1$, the graph of $f(x)$ is compressed horizontally. | For $f(a \cdot x)$, if $0 < |a| < 1$, the graph of $f(x)$ is stretched horizontally. |
|---|---|
| Every point on the graph of $f(x)$ is closer to the y-axis. | Every point on the graph of $f(x)$ is farther from the y-axis. |

**Study Tip**

*Sense-Making* When the graph of a quadratic function is stretched vertically, the shape of the graph is narrower than that of the parent function. When it is compressed vertically, the graph is wider than the parent function.

### Example 4 — Dilations of Quadratic Functions

Describe the dilation in each function as it relates to the graph of $f(x) = x^2$.

a. $p(x) = \frac{1}{2}x^2$

$a = \frac{1}{2}$ and $0 < \frac{1}{2} < 1$
$p(x)$ is a dilation of the graph of $f(x) = x^2$ that is compressed vertically.

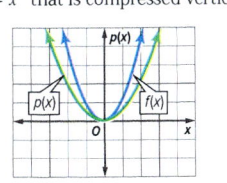

b. $g(x) = (3x)^2$

$a = 3$ and $3 > 1$
$g(x)$ is a dilation of the graph of $f(x) = x^2$ that is compressed horizontally.

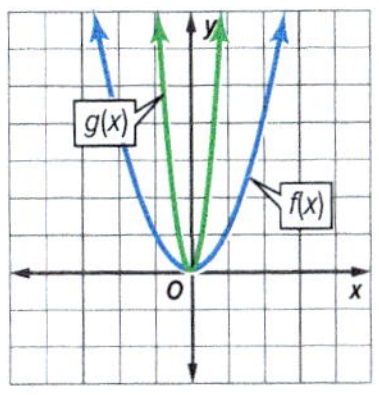

**Guided Practice**

4A. $j(x) = 2x^2$    4B. $g(x) = \left(\frac{2}{5}x\right)^2$

4A. stretched vertically
4B. stretched horizontally

---

## 2 Dilations and Reflections

### Example 4 Dilations of Quadratic Functions

**AL** In part **a**, what are the similarities and differences of $f(x)$ and $p(x)$? Sample answer: their vertices are in the same place, but $p(x)$ is wider than $f(x)$.

**OL** How does the equation of the function in part **a** compare to the equation of the function in part **b**? Sample answer: Both equations have a coefficient for $x^2$ but part **a** has a coefficient less than 1, while part **b** has a coefficient greater than 1. Part **b** is also multiplied by the constant before the function is evaluated.

**BL** What is the difference between what the value of $a$ does to the graph of the function, compared to what the values of $h$ and $k$ can do to the graph of the function? Sample answer: The values of $h$ and $k$ make the vertex of the graph change location on the coordinate plane, but the value of $a$ does not move the vertex; it changes the shape of the graph.

**Need Another Example?**

Describe the dilation in each function as it relates to the graph of $f(x) = x^2$.

a. $g(x) = \frac{1}{3}x^2$ compressed vertically

b. $h(x) = (2x)^2$ compressed horizontally

### MP Teaching the Mathematical Practices

**Sense-Making** Point out that students may have studied transformations of figures in previous math courses. Stress that the dilations of graphs in this lesson are with respect to the x- or y-axis, not with respect to a point as in dilations with respect to the origin.

---

**Differentiated Instruction** AL OL

**IF** students need help comparing and contrasting a function's graph and its parent function's graph,

**THEN** have students write and graph three different functions on the same coordinate plane. Using a red pencil, on this same coordinate plane, have students graph the parent function $f(x) = x^2$. Finally, have students create observation notebooks in which to record their thoughts on how each function's graph is similar to or different from the parent function's graph.

# Lesson 9-2 | Transformations of Quadratic Functions

## Example 5 Reflections and Dilations of Quadratic Functions

**AL** What happens to the graph when $a$ is negative and multiplied by the function after it has been evaluated? Sample answer: The graph is flipped so it opens down.

**OL** Why is the graph reflected across the $x$-axis in part **a** and the $y$-axis in part **b**? In part **a**, the function is multiplied by $-1$ after it is evaluated. In part **b**, function is multiplied by $-1$ before it is evaluated.

**BL** How does a reflection across the $x$-axis affect the end behavior of the graph? Sample answer: As $x$ increases or decreases, $y$ approaches negative infinity instead of positive infinity.

**Need Another Example?**
Describe the transformations in each function as it relates to the graph of $f(x) = x^2$.
a. $g(x) = -3x^2$ The graph is reflected across the $x$-axis and stretched vertically.
b. $h(x) = \left(-\frac{1}{2}x\right)^2$ The graph is reflected across the $y$-axis and stretched horizontally.

## Example 6 Analyze a Function

**AL** What are the values of $a$, $h$, and $k$ in $g(x)$?
$-0.0018$, $251.5$, $118$

**OL** What is the vertex of the graph of $g(x)$?
$(251.5, 118)$

**BL** What is the horizontal distance between each end of the bridge? Explain. 503 meters; Sample answer: Since the center of the bridge is 251.5 meters from one end, the total distance between each end of the bridge is $251.5 + 251.5 = 503$ meters.

**Need Another Example?**
**DANCE** The path of the center of gravity of a dancer during a leap can be modeled by $g(x) = -0.17(x - 3)^2 + 1.53$, where $x$ is the horizontal distance in feet of the center of gravity from the starting point and $h(x)$ is the vertical distance in feet. Describe the transformations in $g(x)$ as it relates to the graph of the parent function. The graph is reflected across the $x$-axis, compressed vertically, and translated right 3 units and up 1.53 units.

---

When a quadratic function $f(x)$ is multiplied by $-1$ after the function has been evaluated, the result is a reflection across the $x$-axis. When $f(x)$ is multiplied by $-1$ before the function has been evaluated, the result is a reflection across the $y$-axis.

### Key Concept Reflections

The graph of $-f(x)$ is the reflection of the graph of $f(x) = x^2$ across the $x$-axis.

The graph of $f(-x)$ is the reflection of the graph of $f(x) = x^2$ across the $y$-axis. Because $f(x)$ is symmetric about the $y$-axis, $f(-x)$ appears the same as $f(x)$.

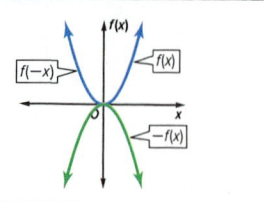

**Study Tip**
Reflection A reflection of $f(x) = x^2$ across the $y$-axis results in the same function, because $f(-x) = (-x)^2 = x^2$.

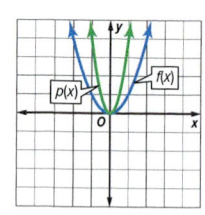

**Watch Out!**
Transformations The graph of $f(x) = -ax^2$ can result in two transformations of the graph of $f(x) = x^2$: a reflection across the $x$-axis if $a > 0$ and either a compression or expansion depending on the absolute value of $a$.

**5A.** stretched vertically and reflected across the $y$-axis
**5B.** compressed vertically and reflected across the $x$-axis
**5C.** reflected across $y$-axis and compressed horizontally

### Example 5 Reflections and Dilations of Quadratic Functions

Describe the transformation in each function as it relates to the graph of $f(x) = x^2$.

**a.** $g(x) = -2x^2$

$a = -2$, $-2 < 0$, and $|-2| > 1$, so the graph is reflected across the $x$-axis and the graph is vertically stretched.

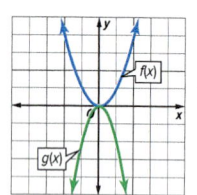

**b.** $p(x) = (-2x)^2$

$a = -2$, $-2 < 0$, so the graph is reflected across the $y$-axis and horizontally compressed.

**Guided Practice**
**5A.** $r(x) = 2(-x)^2$
**5B.** $g(x) = -\frac{1}{5}x^2$
**5C.** $j(x) = (-2x)^2$

### Real-World Example 6 Analyze a Function

**BRIDGES** The lower arch of the Sydney Harbor Bridge can be modeled by $g(x) = -0.0018(x - 251.5)^2 + 118$, where $x$ is horizontal distance in meters and $g(x)$ is height in meters. Describe the transformations in $g(x)$ as it relates to the graph of the parent function.

The values of $a$, $h$, and $k$ are $-0.0018$, $251.5$, and $118$, respectively. Because $a$ is negative and $0 < |a| < 1$, $g(x)$ is reflected across the $x$-axis and compressed vertically. The graph is translated right 251.5 units and up 118 units.

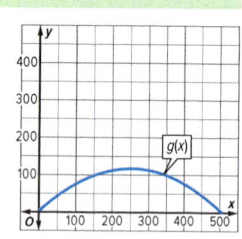

### Lesson 9-2 | Transformations of Quadratic Functions

**Guided Practice**

6. **FOOTBALL** In order to allow water to drain, football fields rise from each sideline to the center of the field. The cross section of a football field can be modeled by $g(x) = -0.000234(x - 80)^2 + 1.5$ where $g(x)$ is the height of the field and $x$ is the distance from the sideline in feet. Describe how $g(x)$ is related to the graph of $f(x) = x^2$. **See margin.**

A quadratic function written in the form $f(x) = a(x - h)^2 + k$ is in **vertex form**.

**Concept Summary** Transformations of Quadratic Functions

$$f(x) = a(x - h)^2 + k$$

**h, Horizontal Translation**
- $h$ units to the right if $h$ is positive
- $|h|$ units to the left if $h$ is negative

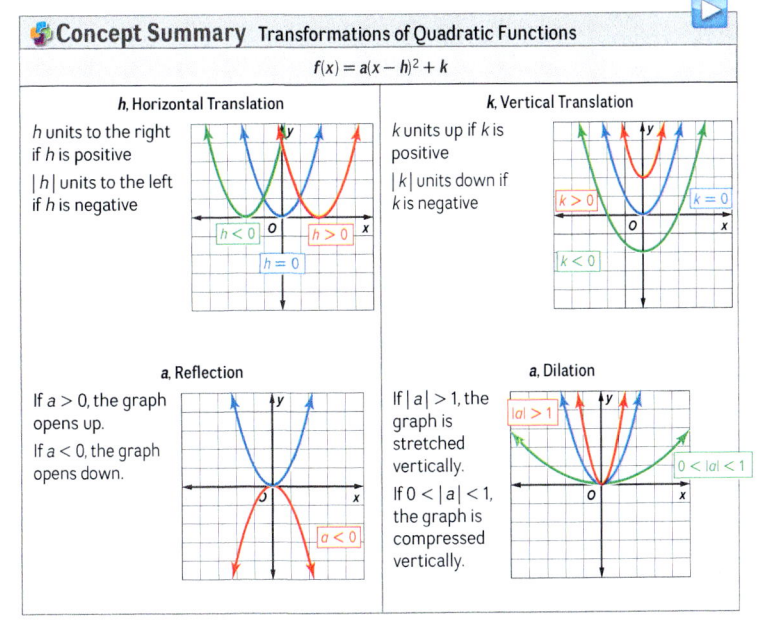

**k, Vertical Translation**
- $k$ units up if $k$ is positive
- $|k|$ units down if $k$ is negative

**a, Reflection**
- If $a > 0$, the graph opens up.
- If $a < 0$, the graph opens down.

**a, Dilation**
- If $|a| > 1$, the graph is stretched vertically.
- If $0 < |a| < 1$, the graph is compressed vertically.

You can use a graphing calculator to verify descriptions of transformations.

**Study Tip**
A graph that opens upward has a positive leading coefficient. A graph that opens downward has a negative leading coefficient.

**Example 7 Compare Graphs**

Compare the graphs to the parent function $f(x) = x^2$. Then, verify using technology.

$$g(x) = -2.1(x + 3)^2 \qquad j(x) = (-0.4x)^2 + 6$$

$g(x)$ is reflected across the $x$-axis, stretched vertically, and translated left 3 units. $j(x)$ is reflected across the $y$-axis, stretched horizontally and translated up 6 units.

To verify using a graphing calculator, enter each equation in the Y= list and graph. Use the features of the calculator to investigate the transformations. Note the change in the vertex, intercepts, and relation to the axes.

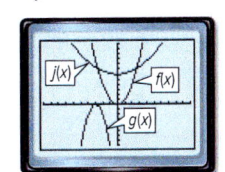

[−10, 10] scl: 1 by [−8, 12] scl: 1

---

**Example 7 Compare Graphs**

**AL** What types of transformations occur in $g(x)$ and $j(x)$? $g(x)$: reflection, dilation, horizontal translation $j(x)$: dilation and vertical translation

**OL** Compare the reflection in $g(x)$ and the reflection in $j(x)$. $g(x)$ is reflected across the $x$-axis, so it opens down. $j(x)$ is reflected across the $y$-axis, so it opens up.

**BL** How do you expect the positions and shapes of $g(x)$ and $j(x)$ to compare? Sample answer: The graph of $g(x)$ will open down and be narrower than $j(x)$. Its vertex will be farther to the left and have a lesser $y$-coordinate than $j(x)$.

**Need Another Example?**
Compare the graphs to the parent function $f(x) = x^2$. Then, verify using technology.
$$g(x) = 4(x + 1)^2 + 3 \qquad h(x) = -\frac{2}{3}(x - 6)^2 + 2$$
$g(x)$ is stretched vertically and translated left 1 unit and up 3 units. $h(x)$ is reflected across the $x$-axis, compressed vertically, and translated right 6 units and up 2 units.

**Teaching Tip**

**Sense-Making** Students often mistake a greater value for $|a|$ to imply a wider parabola. Suggest that students think of the value of $a$ in $y = ax^2 + k$ the same way they thought of the value of the slope $m$ in $y = mx + b$. A greater value for $m$ means a steeper line. A greater value for $|a|$ means that the sides of the parabola are steeper.

---

### Additional Answer (Guided Practice)

6. reflected across the $x$-axis, compressed vertically, and translated right 80 units and up 1.5 units

**Lesson 9-2** | Transformations of Quadratic Functions

**Example 8  Identify Equations Given the Vertex and a Point**

**AL** What is the vertex form of any quadratic function? $f(x) = a(x - h)^2 + k$

**OL** What does a vertex at (2, 4) tell you about the graph? It was translated right 2 units and up 4 units. What variables do we plug in the vertex coordinates? $2 = h; 4 = k$.

**BL** How can you solve for $a$? Substitute the other given point, (3, 6), into the equation and solve for $a$.

**Need Another Example?**
Write a quadratic function, in vertex and standard form, which contains (−4, 1) and vertex (−2, 5).
$f(x) = (x + 4)^2 + 1; f(x) = x^2 + 8x + 17$

## Practice

**Formative Assessment** Use Exercises 1–9 to assess students' understanding of the concepts in this lesson.

The Practice and Problem Solving exercises assess the content taught in the lesson. The Preparing for Assessment page is meant to be used as preparation for end-of-course assessments.

## Extra Practice

See page R9 for extra exercises for students who are approaching level or for on-level students who need additional reinforcement.

### Levels of Complexity Chart

The levels of the exercises progress from 1 to 3 with Level 1 indicating the lowest level of complexity

| Exercises | 10–23 | 24–30, 47–51 | 31–46 |
|---|---|---|---|
| Level 3 | | | ● |
| Level 2 | | ● | |
| Level 1 | ● | | |

**Go Online!**      **eBook**

**Interactive Student Guide**
Use the *Interactive Student Guide* to deepen conceptual understanding.
· Transforming Quadratic Equations

576 | Lesson 9-2 | Transformations of Quadratic Functions

---

**Study Tip**
The vertex form of any quadratic function is $f(x) = a(x - h)^2 + k$ where the vertex is $(h, k)$. The vertex being at (2, 4) means that the graph was translated 2 units right and 4 units up.

8A. $f(x) = -(x - 5)^2 - 1$
$f(x) = -x^2 + 10x - 26$
8B. $f(x) = \frac{5}{16}(x - 0)^2 + 3$
$f(x) = \frac{5}{16}x^2 + 3$
8C. $f(x) = \frac{2}{3}(x - 1)^2 + 2$
$f(x) = \frac{2}{3}x^2 - \frac{4}{3}x + \frac{8}{3}$

▶ **Guided Practice**

7. ... See margin.

**Example 8  Identify Equations Given the Vertex and a Point**   F.IF.7a

Write a quadratic function in vertex and standard form that contains (3, 6) and vertex (2, 4).

$f(x) = a(x - h)^2 + k$   Vertex form of a quadratic function
$\phantom{f(x)} = a \cdot \_\_ + \_\_$   $[x, f(x)] = [(3, 6), (h, k) = (2, 4)]$
$\phantom{f(x)} = a + \_\_$   $(3 - 2)^2 = 1$
$\phantom{f(x)} = a$   Subtract 4 from each side.

$f(x) = \_\_ x - \_\_ + \_\_$   Vertex form
$f(x) = \_\_ x^2 - \_\_ x + \_\_ + \_\_$   Expand $(x - 2)^2$
$f(x) = \_\_ x^2 - \_\_ x + \_\_$   Simplify.

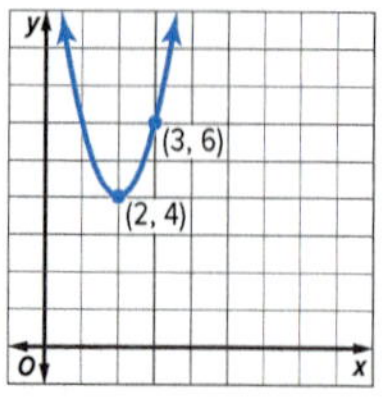

▶ **Guided Practice**

Write a quadratic function in vertex and standard form that contains vertex $V$ and point $P$.

8A. $V$...     8B. $V$...     8C. $V$...

**Check Your Understanding**   ● = Step-by-Step Solutions begin on page R11.

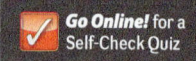

**Examples 1–5**   Describe the transformations in each function as it relates to the graph of $f(x) = x^2$. 1–6. See margin.
F.BF.3
1. $g(x) = x^2 - \_\_$     2. $h(x) = \_\_ x - \_\_$     3. $h(x) = -x^2 + \_\_$
4. $g(x) = \_\_ x^2$     5. $g(x) = -\_\_ x + \_\_$     6. $h(x) = -x^2 - \_\_$

**Example 6–7**   7. **BASKETBALL** ...
F.BF.3, F.IF.7a
See margin.

**Example 8**   Write a quadratic function in vertex and standard form that contains vertex $V$ and point $P$.
F.BF.3
8. $V$...     9. $V$...
8. $f(x) = \frac{2}{3}(x - 3)^2 - 2$; $f(x) = \frac{2}{3}x^2 - 4x + 4$     9. $f(x) = 4(x - 1)^2 - 1$; $f(x) = 4x^2 - 8x + 3$

**Practice and Problem Solving**   Extra Practice is on page R9.

**Examples 1–5**   Describe the transformations in each function as it relates to the graph of $f(x) = x^2$. 10–17. See margin.
F.BF.3
10. $g(x) = -\_\_ + x^2$     11. $h(x) = -\_\_ - x^2$     12. $g(x) = \_\_ x - \_\_ + \_\_$
13. $h(x) = \_\_ + \_\_ x^2$     14. $g(x) = -\_\_ \cdot \_\_ x^2$     15. $h(x) = \_\_ + \_\_ x^2$
16. $g(x) = \_\_ x^2 - \_\_$     17. $h(x) = \_\_ x + \_\_ + \_\_$

### Differentiated Homework Options

| Levels | **AL** Basic | **OL** Core | **BL** Advanced |
|---|---|---|---|
| Exercises | 10–23, 42, 44–51 | 11–29 odd, 42, 44–51 | 31–46, (optional: 47–51) |
| 2-Day Option | 11–23 odd, 47–51 | 10–23, 47–51 | |
| | 10–22 even, 42, 44–46 | 24–30, 34–42, 44–46 | |

You can use ALEKS to provide additional remediation support with personalized instruction and practice.

**Additional Answer (Guided Practice)**

7. $p(x)$: reflected across $x$-axis, compressed vertically, and translated left 2 units

$q(x)$: reflected across the $y$-axis, compressed horizontally and translated down 5 units

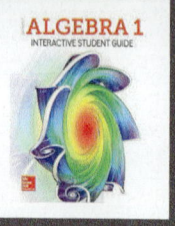

**Example 6–7**
**F.BF.3, F.IF.7a**

**18. DIVING** The height of a diver in meters after $x$ seconds is modeled by the function $g(x) = -7.2(x - 0.75)^2 + 10$.

a. Find the values $a$, $h$, and $k$ in $g(x)$. **−7.2, 0.75, 10**
b. Describe the transformations in $g(x)$ as it relates to the graph of the parent function. **See margin.**
c. Use technology to estimate the height of the diver after 1 second. **9.55 m**

**19. ROCKETS** Candice and Paulo both launch a rocket during science club. The path of Candice's rocket is modeled by $c(x) = -16(x - 2.5)^2 + 105$ and Paulo's rocket is modeled by $p(x) = -16(x - 2.8)^2 + 126.5$, where $c(x)$ and $p(x)$ represent the height of each rocket after $x$ seconds.

a. Use technology to compare the graphs of $c(x)$ and $p(x)$ to the parent function. **See margin.**
b. Whose rocket went higher? **Paulo's**
c. Whose rocket was in the air for a longer time? **Paulo's**

**Example 8**
**F.IF.7a**

Write a quadratic function in vertex and standard forms that contains vertex $V$ and point $P$.

**20.** $V(-2, -2)$, $P(-4, -10)$
 $f(x) = -2(x + 2)^2 - 2$
 $f(x) = -2x^2 - 8x - 10$

**21.** $V(-1, -4)$, $P(2, 0)$
 $f(x) = \frac{4}{9}(x + 1)^2 - 4$
 $f(x) = \frac{4}{9}x^2 + \frac{8}{9}x - \frac{32}{9}$

**22.** $V(7, -2)$, $P(4, 4)$
 $f(x) = \frac{2}{3}(x - 7)^2 - 2$
 $f(x) = \frac{2}{3}x^2 - \frac{28}{3}x + \frac{92}{3}$

**23.** $V(-5, -9)$, $P(-2, -6)$
 $f(x) = \frac{1}{3}(x + 5)^2 - 9$
 $f(x) = \frac{1}{3}x^2 + \frac{10}{3}x - \frac{2}{3}$

**B  24. SQUIRRELS** A squirrel 12 feet above the ground drops an acorn from a tree. The function $h = -16t^2 + 12$ models the height of the acorn above the ground in feet after $t$ seconds. Graph the function, and compare it to its parent graph. **The graph is stretched vertically, reflected across the x-axis, and translated up. See margin for graph.**

**25 ROCKS** A rock falls from a cliff 300 feet above the ground. At the same time, another rock falls from a cliff 700 feet above the ground.

a. Write functions that model the height $h$ of each rock after $t$ seconds. **a.** $h = -16t^2 + 300$ and $h = -16t^2 + 700$
b. If the rocks fall at the same time, how much sooner will the first rock reach the ground? **about 2.3 seconds**

**26. SPRINKLERS** The path of water from a sprinkler can be modeled by quadratic functions. The following functions model paths for three different sprinklers.

Sprinkler A: $y = -0.35x^2 + 3.5$
Sprinkler B: $y = -0.21x^2 + 1.7$
Sprinkler C: $y = -0.08x^2 + 2.4$

a. Which sprinkler will send water the farthest? Explain. **Sprinkler C because the graph is compressed vertically the most.**
b. Which sprinkler will send water the highest? Explain. **Sprinkler A because it is translated up the most.**
c. Which sprinkler will produce the narrowest path? Explain. **Sprinkler A because it is expanded the least.**

**27a.** The graph of $g(x)$ is the graph of $f(x)$ translated 200 yards right and 20 yards up, compressed vertically, and reflected across the x-axis.

**27. GOLF** The path of a drive can be modeled by a quadratic function where $g(x)$ is the vertical distance in yards of the ball from the ground and $x$ is the horizontal distance in yards.

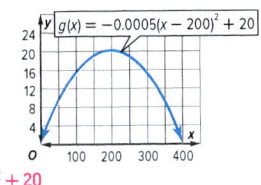

a. How can you obtain $g(x)$ from the graph of $f(x) = x^2$.
b. A second golfer hits a ball from the red tee, which is 30 yards closer to the hole. What function $h(x)$ can be used to describe the second golfer's shot? $h(x) = 0.0005(x - 230)^2 + 20$

Graph shown: $g(x) = -0.0005(x - 200)^2 + 20$

**18b.** The parent function is stretched vertically by a factor of 7.2 and translated right 0.75 units and up 10 units.

**19b.** $c(x)$ is reflected across the x-axis, stretched vertically, and translated right 2.5 units and up 105 units. $p(x)$ is reflected across the x-axis, stretched vertically, and translated right 2.8 units and up 126.5 units.

**24.** [Graph showing function $h$ with vertex at (0, 12), narrow parabola opening downward]

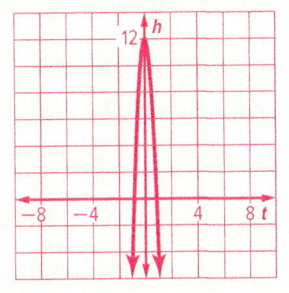

---

**Lesson 9-2 | Transformations of Quadratic Functions**

**MP Teaching the Mathematical Practices**

**Construct Arguments** Mathematically proficient students are able to analyze situations by breaking them into cases. In Exercise 44, suggest that students look back at the graphs that appear throughout the lesson to see various types of graphs.

**Exercise Alert**

**Grid Paper** For Exercise 24 students will need grid paper.

**Additional Answers**

1. translated down 11 units
2. translated right 2 units and compressed vertically
3. reflected across the x-axis, translated up 8 units
4. compressed horizontally or stretched vertically
5. reflected across the x-axis, translated left 3 units and stretched vertically
6. reflected across the x-axis, translated down 2 units
7. Because $a = -0.204$, the parent function is reflected across the x-axis and compressed vertically. $h = 6.2$ and $k = 13.8$, so the parent function is translated right 6.2 units and up 13.8 units.
10. translated down 10 units
11. reflected across the x-axis, translated down 7 units
12. translated right 3 units and up 8 units and stretched vertically
13. compressed vertically, translated up 6 units
14. reflected across the x-axis, stretched vertically, translated down 5 units
15. stretched vertically, translated up 3 units
16. compressed vertically, translated down 1.1 unit
17. translated left 1 unit and up 2.6 units and stretched vertically

**Go Online!**

**eSolutions Manual**
Create worksheets, answer keys, and solutions handouts for your assignments.

connectED.mcgraw-hill.com

**Lesson 9-2 | Transformations of Quadratic Functions**

## Assess

**Ticket Out the Door** Write five different quadratic functions of the form $f(x) = ax^2 + k$ on pieces of paper. Give each student one. Have students tell you how the graph of each function is related to the graph of $y = x^2$ as they walk out the door.

## Additional Answers

**42a.** Sometimes; this only occurs if $k = 0$. For any other value, the graph will be translated up or down.

**42b.** Always; the reflection does not affect the width. Both graphs are dilated by a factor of $a$.

**42c.** Never; if the graph with a vertex at $(0, -3)$ opens upward, it will have a lower minimum. If it opens downward, it will have a maximum. The first graph has a minimum.

**44.** Sample answer: Not all reflections over the y-axis produce the same graph. If the vertex of the original graph is not on the y-axis, the graph will not have the y-axis as its axis of symmetry and its reflection across the y-axis will be a different parabola.

**46.** Sample answer: For $y = ax^2$, the parent graph is stretched vertically if $a > 1$ or compressed vertically if $0 < a < 1$. The y-values in the table will all be multiplied by a factor of $a$. For $y = x^2 + k$, the parent graph is translated up if $k$ is positive and moved down if $k$ is negative. The y-values in the table will all have the constant $k$ added to them or subtracted from them. For $y = ax^2 + k$, the graph will either be stretched vertically or compressed vertically based upon the value of $a$ and then will be translated up or down depending on the value of $k$. The y-values in the table will be multiplied by a factor of $a$ and the constant $k$ added to them.

---

Describe the transformations to obtain the graph of $g(x)$ from the graph of $f(x)$.

**28.** $f(x) = x^2 + 3$
$g(x) = x^2 - 2$
Translate the graph of $f(x)$ down 5 units.

**29.** $f(x) = x^2 - 4$
$g(x) = (x - 2)^2 + 7$
Translate the graph of $f(x)$ up 11 units and to the right 2 units.

**30.** $f(x) = -6x^2$
$g(x) = -3x^2$
Compress vertically the graph of $f(x)$.

**C 31. MULTIPLE REPRESENTATIONS** In this problem you will analyze $f(x)$ if $f(x) = x^2$.
  **a.** Graphical Graph $f(bx)$ for $b = -3, -1, 0.25, 0.5, 1, 2,$ and $3$. a-e. See Ch. 9 Answer Appendix.
  **b.** Analytical Describe the transformation from $f(x)$ to $f(bx)$ when $b > 1$.
  **c.** Analytical Describe the transformation from $f(x)$ to $f(bx)$ when $0 < b < 1$.
  **d.** Analytical Describe the transformation from $f(x)$ to $f(bx)$ when $b < 0$.
  **e.** Graphical Graph $af(x)$ for $a = -2, -0.5, 0.5,$ and $3$.

Tell which equation matches Graph 1 and Graph 2.

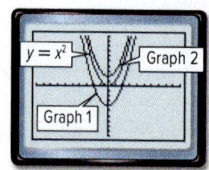

**32.** $[-10, 10]$ scl: 1 by $[-10, 10]$ scl: 1
$y = x^2 + 2$ Graph 1: $y = x^2 - 4$
$y = x^2 - 4$ Graph 2: $y = x^2 + 2$

**33.** $[-10, 10]$ scl: 1 by $[-10, 10]$ scl: 1
$y = -\frac{1}{3}x^2$ Graph 1: $y = -\frac{1}{3}x^2$
$y = -2x^2$ Graph 2: $y = -2x^2$

Compare and contrast each pair of functions. Use a graphing calculator to confirm. 34–41. See Ch. 9 Answer Appendix.

**34.** $y = x^2, y = x^2 + 3$
**35.** $y = \frac{1}{2}x^2, y = 3x^2$
**36.** $y = x^2, y = (x - 5)^2$
**37.** $y = 3x^2, y = -3x^2$
**38.** $y = x^2, y = -4x^2$
**39.** $y = x^2 - 1, y = x^2 + 2$
**40.** $y = \frac{1}{2}x^2 + 3, y = -2x^2$
**41.** $y = x^2 - 4, y = (x - 4)^2$

F.IF.7a, F.BF.3

**H.O.T. Problems** Use Higher-Order Thinking Skills

**42. CONSTRUCT ARGUMENTS** Are the following statements *sometimes*, *always*, or *never* true? Explain. See margin.
  **a.** The graph of $y = x^2 + k$ has its vertex at the origin.
  **b.** The graphs of $y = ax^2$ and its reflection over the x-axis are the same width.
  **c.** The graph of $y = x^2 + k$, where $k \geq 0$, and the graph of a quadratic with vertex at $(0, -3)$ have the same maximum or minimum point.

**43. CHALLENGE** Write a function of the form $y = ax^2 + k$ with a graph that passes through the points $(-2, 3)$ and $(4, 15)$. $y = x^2 - 1$

**44. CONSTRUCT ARGUMENTS** Determine whether all quadratic functions that are reflected across the y-axis produce the same graph. Explain your answer. See margin.

**45. OPEN-ENDED** Write a quadratic function that opens downward and is wider than the parent graph. Sample answer: $f(x) = -\frac{1}{2}x^2$

**46. WRITING IN MATH** Describe how the values of $a$ and $k$ affect the graphical and tabular representations for the functions $y = ax^2$, $y = x^2 + k$, and $y = ax^2 + k$. See margin.

### Standards for Mathematical Practice

| Emphasis On | Exercises |
|---|---|
| 3 Construct viable arguments and critique the reasoning of others. | 26–31, 32, 33, 42, 44, 50 |
| 4 Model with mathematics. | 18, 19, 24–27, 51 |

# Lesson 9-2 | Transformations of Quadratic Functions

## Preparing for Assessment

**47.** Luis graphed the parent quadratic function as shown. Then he graphed a second function that is a translation of the parent graph 2 units down and 3 units to the left. Which is an equation for the second graph? **MP** 1, 8  F.IF.7a, F.BF.3  **C**

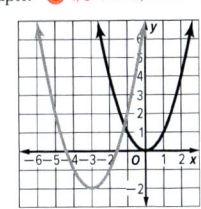

- ○ A  $f(x) = x^2 - 6x + 7$
- ○ B  $f(x) = x^2 - 6x + 11$
- ○ C  $f(x) = x^2 + 6x + 7$
- ○ D  $f(x) = x^2 + 6x + 11$

**48.** The graph of the function $f(x) = x^2$ is reflected across the x-axis and compressed vertically. Which of the following could be the equation for the graph? Select all that apply.  **MP** 1, 8  F.BF.3  **D, E**

- ☐ A  $f(x) = \frac{3}{2}x^2$
- ☐ B  $f(x) = -\frac{3}{2}x^2$
- ☐ C  $f(x) = \frac{1}{3}x^2$
- ☐ D  $f(x) = -\frac{1}{3}x^2$
- ☐ E  $f(x) = -\frac{2}{3}x^2$
- ☐ F  $f(x) = \frac{2}{3}x^2$

**49.** The graph of $f(x) = x^2$ is reflected across the x-axis and translated to the left 4 units. What is the value of h when the equation of the transformed graph is written in vertex form?  **MP** 1, 8  F.BF.3  **−4**

**50.** The graph of a quadratic function $g(x)$ is shown below.

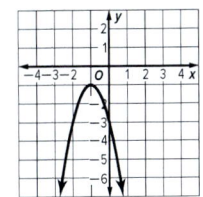

Which statement about the relationship between the graph of $g(x)$ and the graph of the parent function $f(x) = x^2$ is not true?  **MP** 1, 8  F.IF.7a, F.BF.3  **C**

- ○ A  $g(x)$ is a vertical stretch of the graph of $f(x) = x^2$.
- ○ B  $g(x)$ is a reflection across the x-axis of the graph of $f(x) = x^2$.
- ○ C  In vertex form, the equation of the function is $g(x) = -2(x-1)^2 - 1$.
- ○ D  In standard form, the equation of the function is $g(x) = -2x^2 - 4x - 3$.

**51. MULTI-STEP** The ideal weight of a kitten in pounds is modeled by the function $g(x) = 0.0009(x + 7.05)^2 - 0.071$, where x is the age of the kitten in weeks.  **MP** 1, 7  F.BF.3

a. Determine the value of $a$ in $g(x)$. **0.009**

b. Determine the value of $h$ in $g(x)$. **−7.05**

c. Determine the value of $k$ in $g(x)$. **−0.071**

d. Select all transformations in $g(x)$ as it relates to the parent function. **B, E, F**
- ☐ A  reflected across the x-axis
- ☐ B  compressed vertically
- ☐ C  stretched vertically
- ☐ D  translated right
- ☐ E  translated left
- ☐ F  translated up

## Preparing for Assessment

Exercises 47–51 require students to use the skills they will need on standardized assessments. Each exercise is dual-coded with content standards and mathematical practice standards.

### Dual Coding

| Items | Content Standards | MP Mathematical Practices |
|---|---|---|
| 47 | F.IF.7a, F.BF.3 | 1, 8 |
| 48 | F.BF.3 | 1, 8 |
| 49 | F.BF.3 | 1, 8 |
| 50 | F.IF.7a, F.BF.3 | 1, 8 |
| 51 | F.BF.3 | 1, 7 |

## Diagnose Student Errors

Survey student responses for each item. Class trends may indicate common errors and misconceptions.

**47.**

| A | Used $h = 3$ instead of $h = -3$ in vertex form |
|---|---|
| B | Used $h = 3$ instead of $h = -3$ and $k = 2$ instead of $k = -2$ in vertex form |
| C | CORRECT |
| D | Used $k = 2$ instead of $k = -2$ in vertex form |

**49.**

| A | Confused vertical stretches and compressions |
|---|---|
| B | Did not understand how a reflection affects the graph of $f(x)$ |
| C | CORRECT |
| D | Made an error going from vertex form to standard form |

---

## Differentiated Instruction  OL  BL

**Extension** Now that students have described transformations as they relate to the parent graph, $f(x) = x^2$, have students to describe the transformations in functions as they relate to other quadratic functions. Ask students to describe the transformations in $h(x) = 4(x + 1)^2 - 6$ as it relates to the graph of $g(x) = (x - 3)^2 + 2$. $h(x)$ is the graph of $g(x)$ stretched vertically and translated left 4 units and down 8 units.

### Go Online!

**Quizzes**

Students can use *Self-Check Quizzes* to check their understanding of this lesson. You can also give the *Chapter Quiz*, which covers the content in Lessons 9–1 and 9–2.

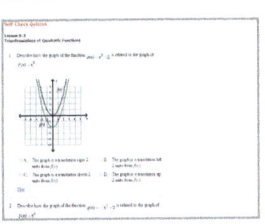

connectED.mcgraw-hill.com  579

# LESSON 9-3
# Solving Quadratic Equations by Graphing

**SUGGESTED PACING (DAYS)**

| | | |
|---|---|---|
| 90 min. | 0.75 | 0.5 |
| 45 min. | 1 | 0.5 |
| | Instruction | Extend Lab |

## Track Your Progress

### Objectives
1. Solve quadratic equations by graphing.
2. Estimate solutions of quadratic equations by graphing.

### Mathematical Background
The solutions of a quadratic equation are called roots. All quadratic equations have two roots. When the parabola crosses the $x$-axis at two distinct points, the roots are real. When the vertex of the parabola is on the $x$-axis, the roots are a double real root. When the parabola does not intersect the $x$-axis, the roots are imaginary.

### THEN
**F.IF.4** For a function that models a relationship between two quantities, interpret key features of graphs and tables in terms of quantities, and sketch graphs showing key features given a verbal description of the relationship.

**A.REI.11** Explain why the $x$-coordinates of the points where the graphs of the equations $y = f(x)$ and $y = g(x)$ intersect are the solutions of the equation $f(x) = g(x)$; find the solutions approximately.

### NOW
**A.REI.4** Solve quadratic equations in one variable.

**F.IF.7a** Graph a linear and quadratic function and show intercepts, maxima, and minima.

### NEXT
**A.SSE.3a** Factor a quadratic expression to reveal the zeros of the function it defines.

**A.REI.4** Solve quadratic equations in one variable.

**F.IF.8a** Use the process of factoring and completing the square in a quadratic function to show zeros, extreme values, and symmetry of the graph, and interpret these in terms of a context.

## Go Online! All of these resources and more are available at connectED.mcgraw-hill.com

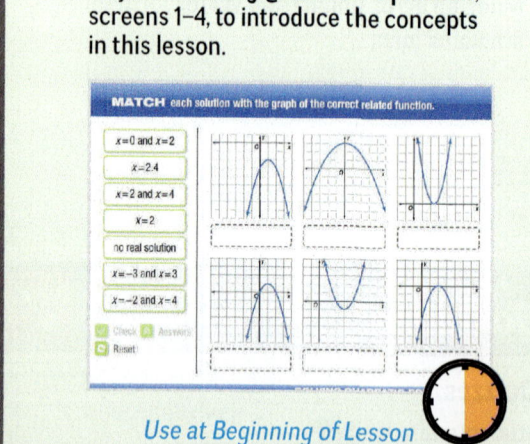

**eLessons** utilize the power of your interactive whiteboard in an engaging way. Use **Solving Quadratic Equations**, screens 1–4, to introduce the concepts in this lesson.

*Use at Beginning of Lesson*

**Animations** illustrate key concepts through step-by-step tutorials and videos.

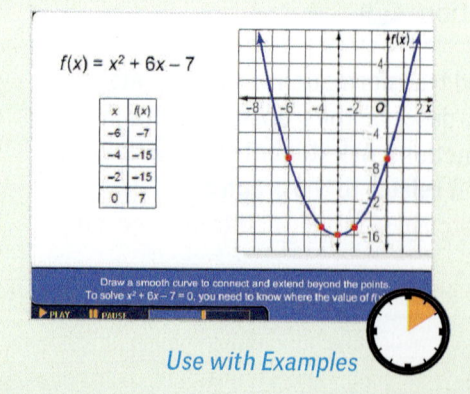

*Use with Examples*

Use **The Geometer's Sketchpad** to solve quadratic equations by graphing.

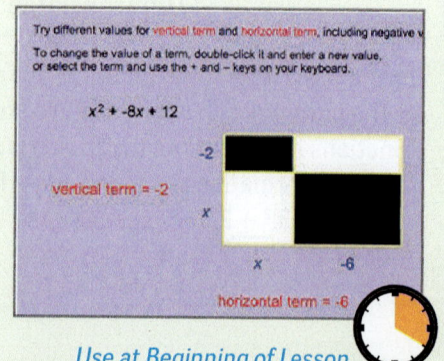

*Use at Beginning of Lesson*

### OER Using Open Educational Resources
**Activities** Before beginning the lesson, have students complete the Matching Graphs to Quadratic Equations activity on **teacherspayteachers.com** to reinforce their understanding of graphing quadratic functions. Have students begin by graphing the equation on their own. Then, have them look for the matching graph. *Use as classwork or review*

# Go Online!
connectED.mcgraw-hill.com
Worksheets

## Differentiate Your Resources

**Extra Practice** Additional practice or homework; Skills Practice is best for approaching-level students and Practice is best for on-level and beyond-level students

### Skills Practice
AL  OL  ELL

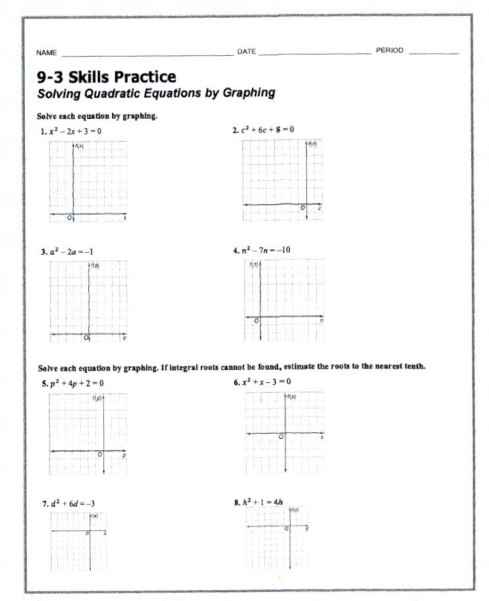

### Practice
AL  OL  BL  ELL

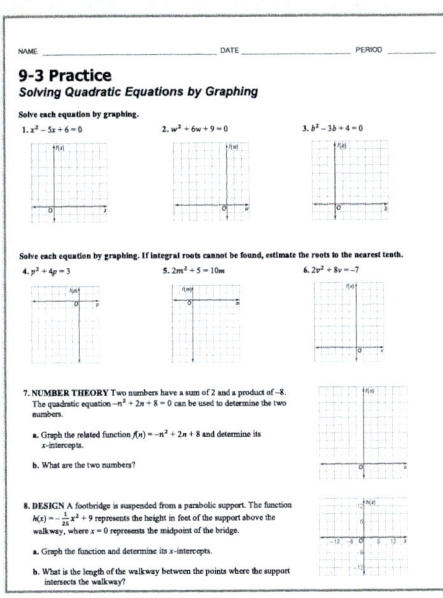

### Word Problem Practice
AL  OL  BL  ELL

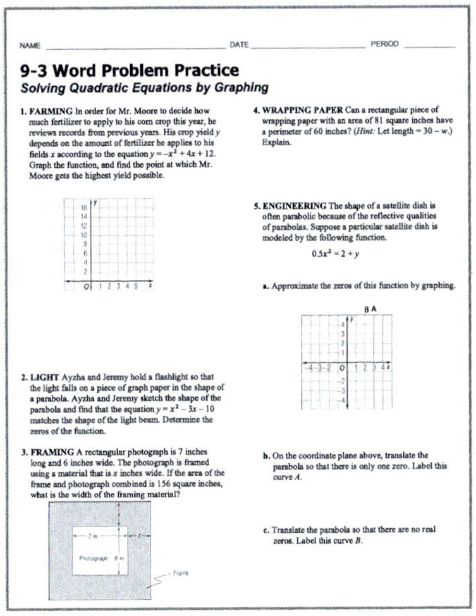

**Intervention** Reteaching and vocabulary activities that can be used with struggling or absent students and as ELL support

### Study Guide and Intervention
AL  OL  ELL

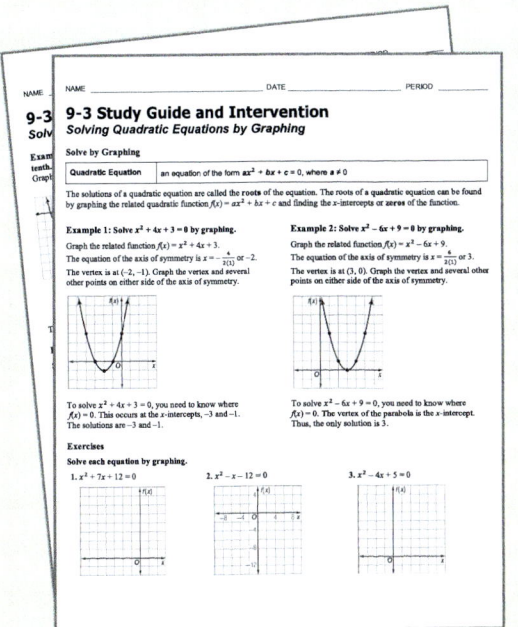

### Study Notebook
AL  OL  BL  ELL

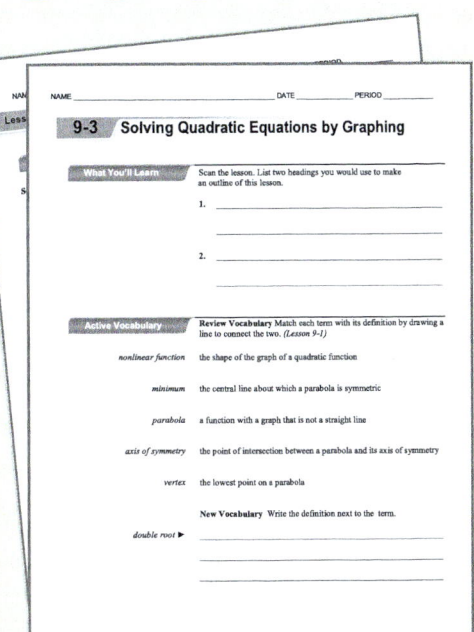

**Extension** Activities that can be used to extend lesson concepts

### Enrichment
OL  BL  ELL

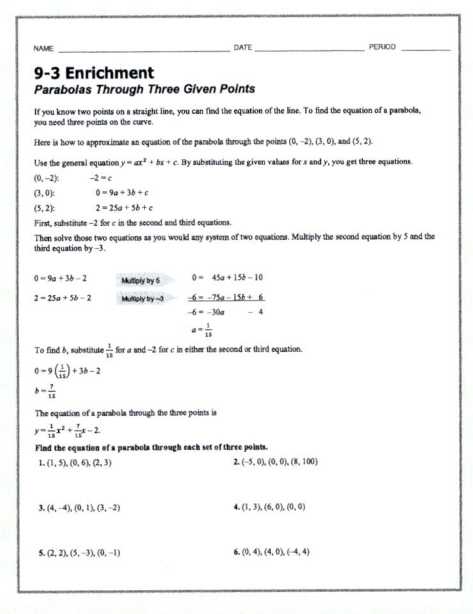

connectED.mcgraw-hill.com  580B

Lesson 9-3 | Solving Quadratic Equations by Graphing

# Launch

Have students read the Why? section of the lesson. Ask:

- If the *x*-intercepts represent where the parabola meets the ground, what represents the ground? **the *x*-axis**

- What are the *x*-intercepts of the graph of the equation? **0, 127**

- What is the equation of the axis of symmetry? **$x = 63.5$**

- What is the distance between the points where the parabola meets the ground? **127 feet**

# Teach

Ask the scaffolded questions for each example to build conceptual understanding for students at all levels.

## 1 Solve by Graphing

**Example 1 Two Roots**

**AL** How can you determine points of the function to graph? **Sample answer: create a table of values**

**OL** What are the solutions of the equation graphed? **the *x*-intercepts**

**BL** How can you check that the graph was drawn correctly and the roots are accurate? **Sample answer: Substitute the roots into the original equation to check that it results in a true statement.**

## Go Online!

**Interactive Whiteboard**
Use the *eLesson* or *Lesson Presentation* to present this lesson.

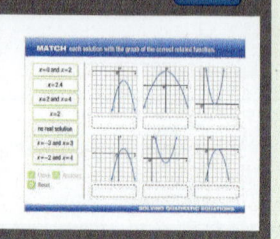

---

## LESSON 3
# Solving Quadratic Equations by Graphing

**Then**
- You solved quadratic equations by factoring.

**Now**
1. Solve quadratic equations by graphing.
2. Estimate solutions of quadratic equations by graphing.

**Why?**
- Dorton Arena at the state fairgrounds in Raleigh, North Carolina, has a shape created by two intersecting parabolas. The shape of one of the parabolas can be modeled by $y = -x^2 + 127x$, where *x* is the width of the parabola in feet, and *y* is the length of the parabola in feet. The *x*-intercepts of the graph of this function can be used to find the distance between the points where the parabola meets the ground.

**New Vocabulary**
double root

**Mathematical Practices**
3 Construct viable arguments and critique the reasoning of others.
5 Use appropriate tools strategically.

**Content Standards**
A.REI.4 Solve quadratic equations in one variable.
F.IF.7a Graph linear and quadratic functions and show intercepts, maxima, and minima.

**1 Solve by Graphing** A quadratic equation can be written in the standard form $ax^2 + bx + c = 0$, where $a \neq 0$. To write a quadratic function as an equation, replace *y* or *f(x)* with 0. Quadratic equations may have two, one, or no real solutions.

### Key Concept  Solutions of Quadratic Equations

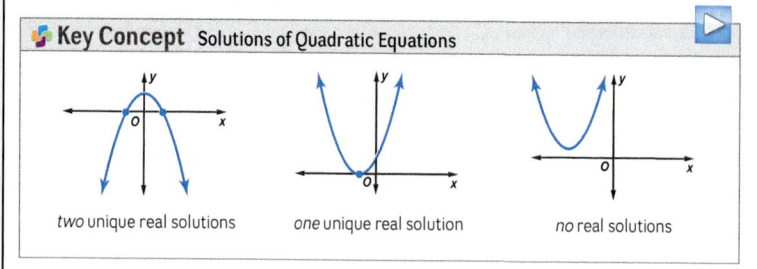

two unique real solutions | one unique real solution | no real solutions

Recall that the solutions or roots of an equation can be identified by finding the *x*-intercepts of the related graph.

**Quadratic Function**
$f(x) = x^2 - x - 6$

$f(-2) = (-2)^2 - (-2) - 6$ or 0
$f(3) = 3^2 - 3 - 6$ or 0

−2 and 3 are zeros of the function.

**Quadratic Equation**
$x^2 - x - 6 = 0$

$(-2)^2 - (-2) - 6$ or 0
$3^2 - 3 - 6$ or 0

−2 and 3 are roots of the equation.

**Graph of Function**

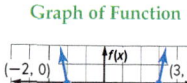

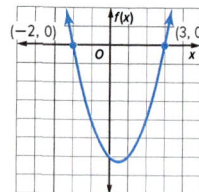

The *x*-intercepts are −2 and 3.

---

### MP Mathematical Practices Strategies

**Attend to precision.**
Help students explain the connection between the *x*-intercepts of the graphs of quadratic functions and the solutions of the related equations. For example, ask:

- Why do the *x*-intercepts give the solutions of a quadratic equation? **Since the related function is found by replacing 0 with *y* in the equation, we need to find the points where $y = 0$. These are the *x*-intercepts of the graph.**

- Can the exact solutions of a quadratic equation be found by sketching the graph? Explain. **Possibly. If the graph was sketched using the exact *x*-intercepts, then yes. But, if the graphs were sketched using other points, and the *x*-intercepts estimated, then no. However, you can estimate and then check using the equation.**

- When will a quadratic equation have exactly one real solution? no solution? **when the vertex lies on the *x*-axis; when the parabola does not intersect the *x*-axis**

## Lesson 9-3 | Solving Quadratic Equations by Graphing

F.IF.7a, A.REI.4

**Example 1** Two Roots

Solve each equation by graphing.
a. $x^2 - 2x - 8 = 0$

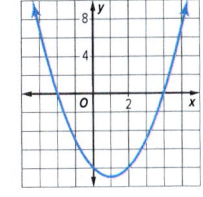

**Step 1** Rewrite the equation in standard form. This equation is written in standard form.

**Step 2** Graph the related function $f(x) = x^2 - 2x - 8$.

**Step 3** Locate the $x$-intercepts of the graph. The $x$-intercepts of the graph appear to be at $-2$ and $4$, so the solutions are $-2$ and $4$.

**Watch Out!**

**MP Precision** Solutions found from the graph of an equation may appear to be exact. Check them in the original equation to be sure.

**CHECK** Check each solution in the original equation.

$x^2 - 2x - 8 = 0$    Original equation    $x^2 - 2x - 8 = 0$
$(-2)^2 - 2(-2) - 8 \stackrel{?}{=} 0$    $x = -2$ or $x = 4$    $(4)^2 - 2(4) - 8 \stackrel{?}{=} 0$
$0 = 0$ ✓    Simplify    $0 = 0$ ✓

b. $-2x^2 + 8 = -6x$

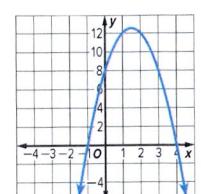

**Step 1** Rewrite the equation in standard form.

$-2x^2 + 8 = -6x$    Original equation
$-2x^2 + 6x + 8 = 0$    Add $6x$ to each side.

**Step 2** Graph the related function $f(x) = -2x^2 + 6x + 8$.

**Step 3** Locate the $x$-intercepts of the graph. The $x$-intercepts of the graph appear to be at $-1$ and $4$, so the solutions are $-1$ and $4$.

**Guided Practice**

Solve each equation by graphing. 1A-1B. See Ch. 9 Answer Appendix for graphs.

**1A.** $-x^2 - 3x + 18 = 0$   3, -6     **1B.** $x^2 - 4x + 3 = 0$   1, 3

The solutions in Example 1 were two distinct numbers. Sometimes the two roots are the same number, called a **double root**.

F.IF.7a, A.REI.4

**Example 2** Double Root

Solve each equation by graphing.
a. $-x^2 + 4x - 4 = 0$

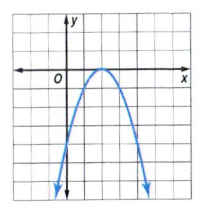

**Step 1** Rewrite the equation in standard form. This equation is written in standard form.

**Step 2** Graph the related function $f(x) = -x^2 + 4x - 4$.

**Step 3** Locate the $x$-intercepts of the graph. Notice that the vertex of the parabola is the only $x$-intercept. Therefore, there is only one solution, 2.

---

**Need Another Example?**

Solve $x^2 - 3x - 10 = 0$ by graphing. $-2, 5$

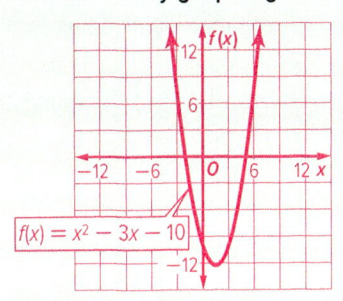

**Example 2 Double Root**

**AL** At what point does the graph touch the $x$-axis? the vertex

**OL** What does a double root tell you about the related equation? There is only one value of $x$ that satisfies the equation.

**BL** What type of trinomial is each function? perfect square trinomial

**Need Another Example?**

Solve $x^2 + 6x = -9$ by graphing. $-3$

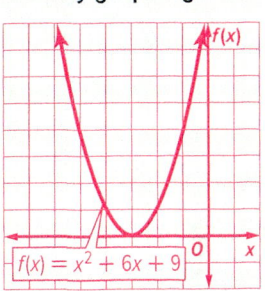

---

**Differentiated Instruction** AL OL

**IF** students assume that the vertex of a parabola always has coordinates that are integers,

**THEN** point out that in Example 3, the $y$-value for the vertex of the graph is greater than 3 and somewhat less than 4.

connectED.mcgraw-hill.com    **581**

Lesson 9-3 | Solving Quadratic Equations by Graphing

## Example 3 No Real Roots

**AL** Does the graph open up or down? up

**OL** What are the solutions of the equation graphed? The graph does not cross the x-axis, so there are no solutions.

**BL** How many solutions would the equation have if $a = -2$? Explain. 2; Sample answer: The graph of the related function would open down and cross the x-axis twice.

**Need Another Example?**
Solve $x^2 + 2x + 3 = 0$ by graphing. ∅

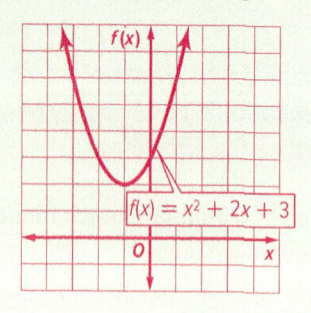

## Additional Answers (Guided Practice)

2A.

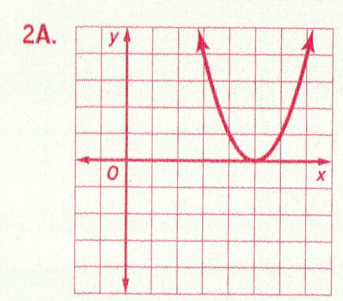

2B.

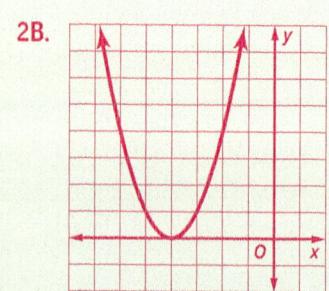

---

**Go Online!**

In *Personal Tutor* videos for this lesson, teachers describe how to solve quadratic equations. Watch with a partner, then try describing how to solve a problem for them. Have them ask questions to help your understanding. **ELL**

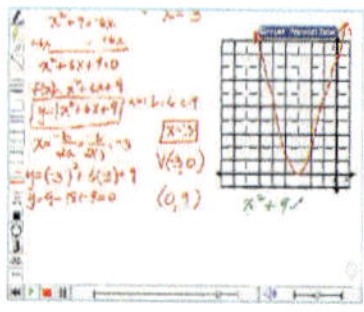

**CHECK** Check the solution in the original equation.
$$-x^2 + 4x - 4 = 0 \quad \text{Original equation}$$
$$-(2)^2 + 4(2) - 4 = 0 \quad x = 2$$
$$0 = 0 \checkmark \quad \text{Simplify.}$$

b. $x^2 - 6x = -9$

**Step 1** Rewrite the equation in standard form.
$$x^2 - 6x = -9 \quad \text{Original equation}$$
$$x^2 - 6x + 9 = 0 \quad \text{Add 9 to each side.}$$

**Step 2** Graph the related function $f(x) = x^2 - 6x + 9$.

**Step 3** Locate the x-intercepts of the graph. Notice that the vertex of the parabola is the only x-intercept. Therefore, there is only one solution, 3.

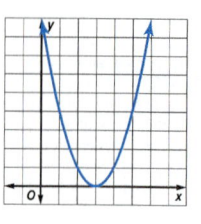

▶ **Guided Practice**

Solve each equation by graphing. 2A–2B. See margin for graphs.

**2A.** $x^2 + 25 = 10x$   5

**2B.** $x^2 = -8x - 16$   −4

Sometimes the roots are not real numbers. Quadratic equations with solutions that are not real numbers lead us to extend the number system to allow for solutions of these equations. These numbers are called *complex numbers*. You will study complex numbers in Algebra 2.

F.IF.7a, A.REI.4

### Example 3  No Real Roots

Solve $2x^2 - 3x + 5 = 0$ by graphing.

**Step 1** Rewrite the equation in standard form. This equation is written in standard form.

**Step 2** Graph the related function $f(x) = 2x^2 - 3x + 5$.

**Step 3** Locate the x-intercepts of the graph. This graph has no x-intercepts. Therefore, this equation has no real number solutions. The solution set is ∅.

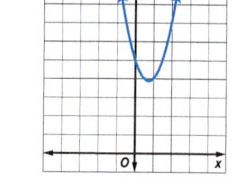

▶ **Guided Practice**

Solve each equation by graphing. 3A–3B. See margin for graphs.

**3A.** $-x^2 - 3x = 5$   ∅

**3B.** $-2x^2 - 8 = 6x$   ∅

3A.

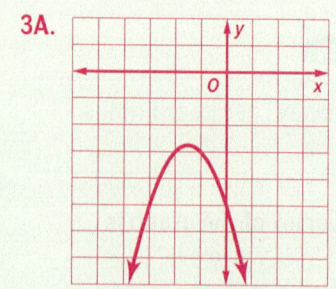

3B.

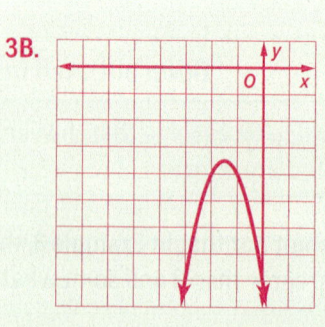

582 | Lesson 9-3 | Solving Quadratic Equations by Graphing

Lesson 9-3 | Solving Quadratic Equations by Graphing

## 2 Estimate Solutions

**Estimate Solutions** The real roots found thus far have been integers. However, the roots of quadratic equations are usually not integers. In these cases, use estimation to approximate the roots of the equation.

F.IF.7a, A.REI.4b

**Example 4** Approximate Roots with a Table

Solve $x^2 + 6x + 6 = 0$ by graphing. If integral roots cannot be found, estimate the roots to the nearest tenth.

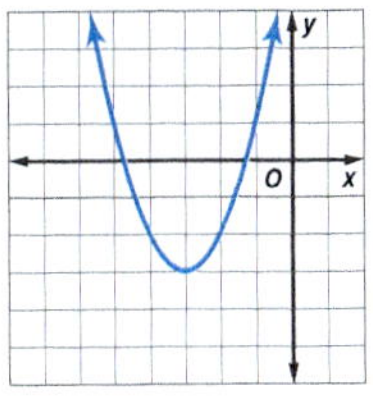

Graph the related function $f(x) = x^2 + 6x + 6$.

The $x$-intercepts are located between $-5$ and $-4$ and between $-2$ and $-1$.

Make a table using an increment of 0.1 for the $x$-values located between $-5$ and $-4$ and between $-2$ and $-1$.

Look for a change in the signs of the function values. The function value that is closest to zero is the best approximation for a zero of the function.

| x | −4.9 | −4.8 | −4.7 | −4.6 | −4.5 | −4.4 | −4.3 | −4.2 | −4.1 |
|---|---|---|---|---|---|---|---|---|---|
| y | 0.61 | 0.24 | −0.11 | −0.44 | −0.75 | −1.04 | −1.31 | −1.56 | −1.79 |

| x | −1.9 | −1.8 | −1.7 | −1.6 | −1.5 | −1.4 | −1.3 | −1.2 | −1.1 |
|---|---|---|---|---|---|---|---|---|---|
| y | −1.79 | −1.56 | −1.31 | −1.04 | −0.75 | −0.44 | −0.11 | 0.24 | 0.61 |

For each table, the function value that is closest to zero when the sign changes is −0.11. Thus, the roots are approximately −4.7 and −1.3.

**Study Tip**

*Location of Zeros* Since quadratic functions are continuous, there must be a zero between two $x$-values for which the corresponding $y$-values have opposite signs.

> **Guided Practice**
>
> **4.** Solve $2x^2 + 6x - 3 = 0$ by graphing. If integral roots cannot be found, estimate the roots to the nearest tenth. **0.4, −3.4**

Approximating the $x$-intercepts of graphs is helpful for real-world applications.

F.IF.7a

**Real-World Example 5** Approximate Roots with a Calculator

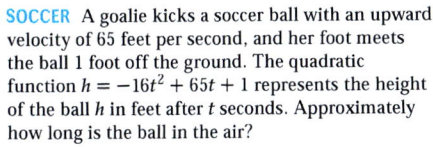

**SOCCER** A goalie kicks a soccer ball with an upward velocity of 65 feet per second, and her foot meets the ball 1 foot off the ground. The quadratic function $h = -16t^2 + 65t + 1$ represents the height of the ball $h$ in feet after $t$ seconds. Approximately how long is the ball in the air?

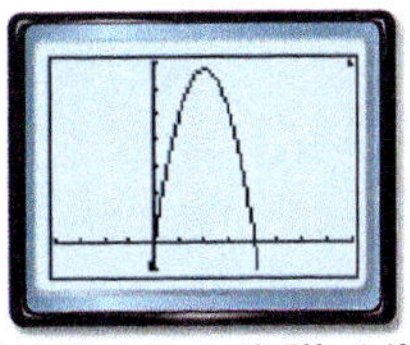

You need to find the roots of the equation $-16t^2 + 65t + 1 = 0$. Use a graphing calculator to graph the related function $f(t) = -16t^2 + 65t + 1$.

[−4, 7] scl: 1 by [−10, 70] scl: 10

The positive $x$-intercept of the graph is approximately 4. Therefore, the ball is in the air for approximately 4 seconds.

**Real-World Link**

The game of soccer, called "football" outside of North America, began in 1863 in Britain when the Football Association was founded. Soccer is played on every continent of the world.

**Source:** Sports Know How

> **Guided Practice**
>
> **5.** If the goalie kicks the soccer ball with an upward velocity of 55 feet per second and his foot meets the ball 2 feet off the ground, approximately how long is the ball in the air? **3.5 seconds**

---

**(MP) Teaching the Mathematical Practices**

**Precision** Mathematically proficient students calculate accurately and efficiently. Discuss with students why a graph may not give an exact answer.

---

## 2 Estimate Solutions

### Example 4 Approximate Roots with a Table

**AL** Can you determine the exact roots from the graph? **no**

**OL** How precisely can you find the roots by using this graph? **Sample answer: nearest unit**

**BL** How could you approximate the roots to the nearest hundredth? **Find the function values for {−4.79, −4.78, ..., −4.71} and {−1.29, −1.28, ..., −1.21} and look for the sign change.**

**Need Another Example?**

Solve $x^2 - 4x + 2 = 0$ by graphing. If integral roots cannot be found, estimate the roots to the nearest tenth. **0.6, 3.4**

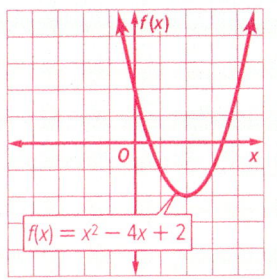

$f(x) = x^2 - 4x + 2$

### Example 5 Approximate Roots with a Calculator

**AL** Does the second root have meaning in the situation? **No, negative time doesn't make sense in the context.**

**OL** Can you tell how far the ball has traveled in the 4 seconds from the graph? **No, the graph represents time and height but not horizontal distance.**

**BL** What did you assume in concluding that the ball lands in 4 seconds? **Sample answer: The ball isn't kicked again before it lands.**

**Need Another Example?**

**Model Rockets** Consuela built a model rocket for her science project. The equation $h = -16t^2 + 250t$ models the flight of the rocket launched from ground level at a velocity of 250 feet per second, where $h$ is the height of the rocket in feet after $t$ seconds. Approximately how long was Consuela's rocket in the air? **15.6 seconds**

connectED.mcgraw-hill.com 583

# Lesson 9-3 | Solving Quadratic Equations by Graphing

## Practice

**Formative Assessment** Use Exercises 1–9 to assess students' understanding of the concepts in this lesson.

The Practice and Problem Solving exercises assess the content taught in the lesson. The Preparing for Assessment page is meant to be used as preparation for end-of-course assessments.

### Exercise Alert

**Grid Paper** For Exercises 1–4, 7, 8, 10–21, 25–27, and 38, students will need grid paper.

### Extra Practice

See page R9 for extra exercises for students who are approaching level or for on-level students who need additional reinforcement.

### Levels of Complexity Chart

The levels of the exercises progress from 1 to 3, with Level 1 indicating the lowest level of complexity.

| Exercises | 10–28 | 29–36, 45–50 | 37–44 |
|---|---|---|---|
| Level 3 |   |   | ● |
| Level 2 |   | ● |   |
| Level 1 | ● |   |   |

### Additional Answers

38a.

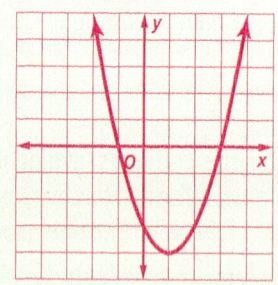

---

### Check Your Understanding
= Step-by-Step Solutions begin on page R11.

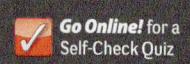

**Go Online!** for a Self-Check Quiz

**Examples 1–3** Solve each equation by graphing. 1–4. See Ch. 9 Answer Appendix for graphs.
F.IF.7a, A.REI.4

1. $x^2 + 3x - 10 = 0$   2, −5
2. $2x^2 - 8x = 0$   0, 4
3. $x^2 + 4x = -4$   −2
4. $x^2 + 12 = -8x$   −6, −2

Solve each equation by examining the given graph of the related function.

5. $2x^2 + 14x - 16 = 0$
6. $\frac{1}{4}x^2 - 2x + 3 = 0$

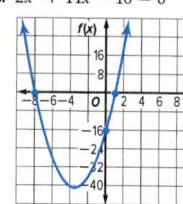

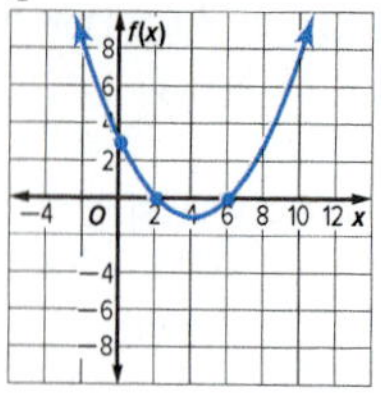

−8, 1      2, 6

**Example 4** Solve each equation by graphing. If integral roots cannot be found, estimate the roots
F.IF.7a, A.REI.4b to the nearest tenth. 7–8. See Ch. 9 Answer Appendix for graphs.

7. $x^2 = 25$   5, −5
8. $x^2 - 8x = -9$   6.6, 1.4

**Example 5** 9. **SCIENCE FAIR** Ricky built a model rocket. Its flight can be modeled by
F.IF.7a the equation shown, where $h$ is the height of the rocket in feet after $t$ seconds. About how long was Ricky's rocket in the air? **about 8.4 seconds**

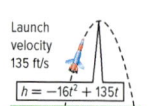

Launch velocity 135 ft/s
$h = -16t^2 + 135t$

### Practice and Problem Solving
Extra Practice is on page R9.

**Examples 1–3** Solve each equation by graphing. 10–24. See Ch. 9 Answer Appendix for graphs.
F.IF.7a, A.REI.4

10. $x^2 + 7x + 14 = 0$   ∅
11. $x^2 + 2x - 24 = 0$   4, −6
12. $x^2 - 16x + 64 = 0$   8
13. $x^2 - 5x + 12 = 0$   ∅
14. $x^2 + 14x = -49$   −7
15. $x^2 = 2x - 1$   1
16. $x^2 - 10x = -16$   2, 8
17. $-2x^2 - 8x = 13$   ∅
18. $2x^2 - 16x = -30$   3, 5
19. $2x^2 = -24x - 72$   −6
20. $-3x^2 + 2x = 15$   ∅
21. $x^2 = -2x + 80$   8, −10
22. $3x^2 - 6 = 3x$   −1, 2
23. $4x^2 + 24x = -36$   −3
24. $-2x^2 - 9 = 8x$   ∅

**Example 4** Solve each equation by graphing. If integral roots cannot be found, estimate the roots to
F.IF.7a, A.REI.4 the nearest tenth. 25–27. See Ch. 9 Answer Appendix for graphs.

25. $x^2 + 2x - 9 = 0$   2.2, −4.2
26. $x^2 - 4x = 20$   6.9, −2.9
27. $x^2 + 3x = 18$   3, −6

**Example 5** 28. **SOFTBALL** The equation $h = -16t^2 + 47t + 3$ models the height $h$, in feet, of a ball that
F.IF.7a Sofia hits after $t$ seconds.
a. How long is the ball in the air?   **3 seconds**
b. Find the $y$-intercept and describe its meaning in the context of the situation.   **3; The ball was hit from a height of 3 feet.**

29. **RIDES** The Terror Tower launches riders straight up and returns straight down. The equation $h = -16t^2 + 122t$ models the height $h$, in feet, of the riders from their starting position after $t$ seconds. How long is it until the riders return to the bottom?   **about 7.6 seconds**

---

### Differentiated Homework Options

| Levels | **AL** Basic | **OL** Core | **BL** Advanced |
|---|---|---|---|
| Exercises | 10–28, 42, 44–50 | 11–33 odd, 34–40, 42, 44–50 | 29–44, (optional: 45–50) |
| 2-Day Option | 11–27 odd, 45–50 | 10–28, 45–50 |   |
|   | 10–28 even, 42, 44 | 29–40, 42, 44 |   |

You can use ALEKS to provide additional remediation support with personalized instruction and practice.

---

**Go Online!**   eBook

**Interactive Student Guide**
Use the Interactive Student Guide to deepen conceptual understanding.
· Solving by Graphing

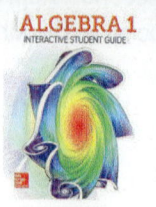

584 | Lesson 9-3 | Solving Quadratic Equations by Graphing

Use factoring to determine how many times the graph of each function intersects the $x$-axis. Identify each zero.

30. $y = x^2 - 8x + 16$  1; 4
31. $y = x^2 + 4x + 4$  1; −2
32. $y = x^2 + 2x - 24$  2; −6, 4
33. $y = x^2 + 12x + 32$  2; −4, −8

34. **NUMBER THEORY** Use a quadratic equation to find two numbers that have a sum of 9 and a product of 20.  4, 5

35. **NUMBER THEORY** Use a quadratic equation to find two numbers that have a sum of 1 and a product of −12.  −3, 4

36. **MODELING** The height of a golf ball in the air can be modeled by the equation $h = -16t^2 + 76t$, where $h$ is the height in feet of the ball after $t$ seconds.
    a. How long was the ball in the air?  4.75 seconds
    b. What is the ball's maximum height?  about 90 ft
    c. When will the ball reach its maximum height?  about 2.4 seconds

37. **SKIING** Stefanie is in a freestyle aerial competition. The equation $h = -16t^2 + 30t + 10$ models Stefanie's height $h$, in feet, $t$ seconds after leaving the ramp.
    a. How long is Stefanie in the air?  about 2.2 seconds
    b. When will Stefanie reach a height of 15 feet?  about 1.7 seconds and 0.2 seconds
    c. To earn bonus points in the competition, you must reach a height of 20 feet. Will Stefanie earn bonus points?  Yes; Stefanie's maximum height is about 24 ft.

38. **MULTIPLE REPRESENTATIONS** In this problem, you will explore the relationship between the factors of a quadratic equation and the zeros of the related graph.
    a. **Graphical** Graph $y = x^2 - 2x - 3$.  See margin.
    b. **Analytical** Name the zeros of the function.  −1, 3
    c. **Algebraic** Factor the related equation $x^2 - 2x - 3 = 0$.  $(x-3)(x+1) = 0$
    d. **Analytical** Set each factor equal to zero and solve. What are the values of $x$?  −1, 3
    e. **Analytical** What conclusions can you draw about the factors of quadratic equations?

**GRAPHING CALCULATOR** Solve each equation by graphing.

39. $x^3 - 3x^2 - 6x + 8 = 0$  −2, 1, 4
40. $x^3 - 8x^2 + 15x = 0$  0, 3, 5

38e. Setting each factor of a quadratic equation equal to zero and solving the resulting equations gives the zeros of the related quadratic function.

A.REI.4, F.IF.7a

**H.O.T. Problems** Use Higher-Order Thinking Skills

41. **CHALLENGE** Describe a real-world situation in which a thrown object travels in the air. Write an equation that models the height of the object with respect to time, and determine how long the object travels in the air.  See margin.

42. **STRUCTURE** For the quadratic function $y = ax^2 + bx + c$, determine the number of $x$-intercepts of the function if:
    a. $a < 0$ and the vertex lies below the $x$-axis  0
    b. $a > 0$ and the vertex lies below the $x$-axis  2
    c. $a < 0$ and the vertex lies above the $x$-axis  2
    d. $a > 0$ and the vertex lies above the $x$-axis  0

43. **CHALLENGE** Find the roots of $x^2 = 2.25$ without using a calculator. Explain your strategy.  See margin.

44. **WRITING IN MATH** Explain how the roots of a quadratic equation are related to the graph of a quadratic function.  See margin.

## Standards for Mathematical Practice

| Emphasis On | Exercises |
| --- | --- |
| 1 Make sense of problems and persevere in solving them. | 38, 45–47, 49, 50 |
| 4 Model with mathematics. | 9, 28, 29, 36, 37, 41 |
| 5 Use appropriate tools strategically. | 39, 40 |
| 6 Attend to precision. | 25–27, 34, 35, 44–46, 48 |
| 7 Look for and make use of structure. | 42, 47 |

---

**Lesson 9-3 | Solving Quadratic Equations by Graphing**

## Teaching the Mathematical Practices

**Modeling** Mathematically proficient students are able to identify important quantities. In Exercise 36, point out that in the functions that model projectiles the coefficient of the $x^2$-term is always −16 when the speed is measured in feet per second.

## Follow-Up

Students have explored graphs of quadratic functions and solved quadratic equations by graphing.
Ask:
• How can the graph of a quadratic function help you solve the corresponding quadratic equation? Sample answer: From the placement of the graph, you can see whether there are 0, 1, or 2 real solutions. The symmetry of the graph helps you identify a second solution when one solution is found.

## Assess

**Yesterday's News** Ask students to write two ways in which learning to graph quadratic functions in the previous lesson helped them to solve quadratic equations in this lesson.

## Additional Answers

41. Sample answer: A tennis ball being hit in the air; an equation is $h = -16t^2 + 25t + 2$. The ball is in the air for about 1.6 seconds.

43. 1.5 and −1.5; Sample answer: Make a table of values for $x$ from −2.0 to 2.0. Use increments of 0.1.

44. The $x$-intercepts of the graph of the related function $f(x)$ are the zeros of the function, and are the solutions of the equation $f(x) = 0$.

## Go Online!

**eSolutions Manual**
Create worksheets, answer keys, and solutions handouts for your assignments.

connectED.mcgraw-hill.com  585

Lesson 9-3 | Solving Quadratic Equations by Graphing

# Preparing for Assessment

Exercises 45–50 require students to use the skills they will need on standardized assessments. Each exercise is dual-coded with content standards and mathematical practice standards.

| Dual Coding | | |
|---|---|---|
| Items | Content Standards | MP Mathematical Practices |
| 45 | A.REI.4, F.IF.7a | 1, 6 |
| 46 | F.IF.7a | 1 |
| 47 | A.REI.4, F.IF.7a | 1, 7 |
| 48 | A.REI.4b, F.IF.7a | 6, 7 |
| 49 | A.REI.4, F.IF.7a | 1 |
| 50 | F.IF.7a | 1, 6 |

## Diagnose Student Errors

Survey student responses for each item. Class trends may indicate common errors and misconceptions.

**45b.**

| A | Identified the graph of $y = x^2 - x$ |
|---|---|
| B | CORRECT |
| C | Identified the graph of $y = x^2 - x + 6$ |
| D | Identified the graph of $y = x^2 + x - 6$ |

**46.**

| A | Found the x-coordinate of the y-intercept |
|---|---|
| B | Found the x-coordinate of the vertex |
| C | CORRECT |
| D | Does not understand the relationship between zeros and solutions |

## Go Online!

**Quizzes** Students can use *Self-Check Quizzes* to check their understanding of this lesson. You can also give the *Chapter Quiz*, which covers the content in Lessons 9-1 through 9-3.

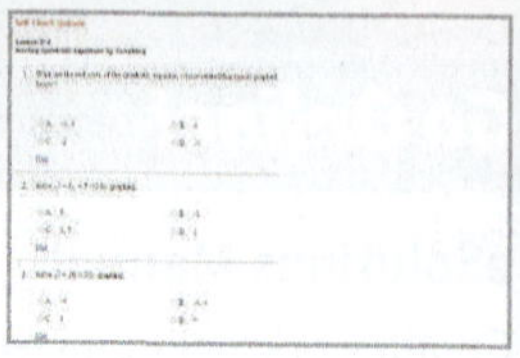

586 | Lesson 9-3 | Solving Quadratic Equations by Graphing

# Preparing for Assessment

45. ••••••• $x^2 - x = ••$  MP 1, 6  F.IF.7a, A.REI.4

a. ••••••••••••••••••••••••••••••••• B
- A [graph]
- B [graph]
- C [graph]
- D [graph]

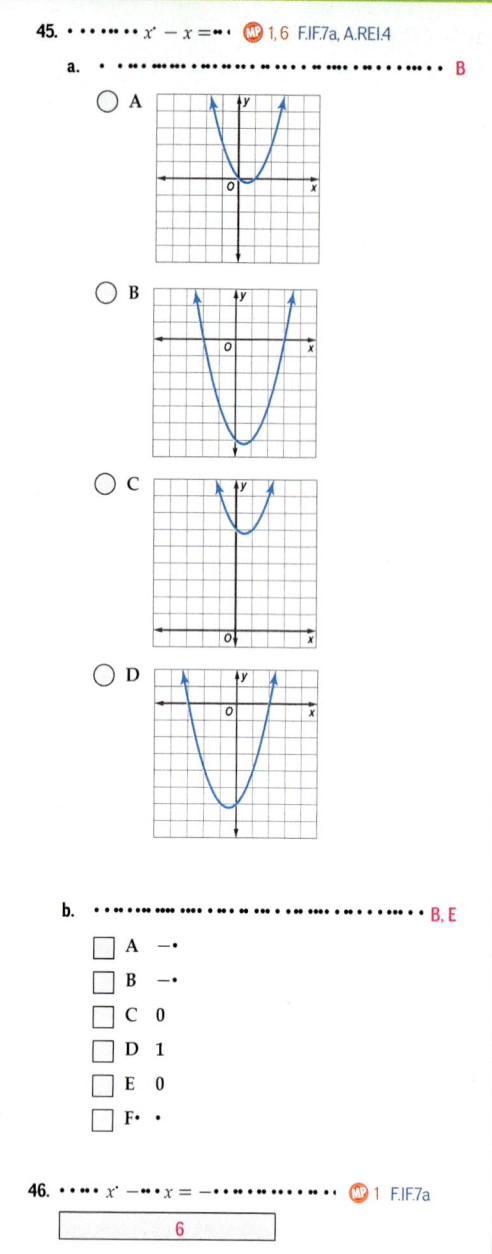

b. ••••••••••••••••••••••••••• B, E
- A —•
- B —•
- C 0
- D 1
- E 0
- F• •

46. •••• $x^2 - •• x = -••$••••••••••• MP 1  F.IF.7a

[ 6 ]

47. **MULTI-STEP** ••••••••••••••• $y = x^2 - ••x + c$••• MP 1, 7  A.REI.4, F.IF.7a

a. •• $c = -••$ • $y = ••$•• ••••••••• •••••• •• 2

b. •• $c = -••$ • $y = -••$ •• •••• •••• •• 1

c. •••• ••••• •• •••• $c$•• •••• •••••• •• •• •••• •• $y = $••• ••• •••• •••• •• C, E
- A $c = -1$
- B $c = 0$
- C $c = ••$
- D $c = ••$
- E $c = ••$

48. •••••• •• •••• •• •••••• •• ••••••• •••• •• ••••• •••• • ••• ••• •••  MP 6, 7  A.REI.4b, F.IF.7a  D
- A ••• •••• •• •••• • $x$••• •••••• •••
- B ••• •••• •• •••• • $x$••• •••••• ••
- C ••• •••• •• •••• • $x$••• •••••• ••• •••
- D ••• •••• •• •••• • $x$••• •••••• •••

49. • ••• ••• •••••• •• •••• •• ••• •• ••• •••• -•. •••••• •• •••••• •••••• ••• ••• •• MP 1  A.REI.4, F.IF.7a  A, C, E
- A $x^2 - ••x - •• = ••$
- B $x^2 + ••x = ••$•
- C• • $x^2 - ••x = ••$•
- D $-x^2 + ••• = •• x$
- E $-x^2 + ••x + •• = ••$

50. • •••••• ••• ••••• ••• •••••• •• •• ••• $x^2 - ••x + •• = ••$•• ••••••••••• •••• •• ••• •••• ••••••• •••• •• MP 1, 6  F.IF.7a  C

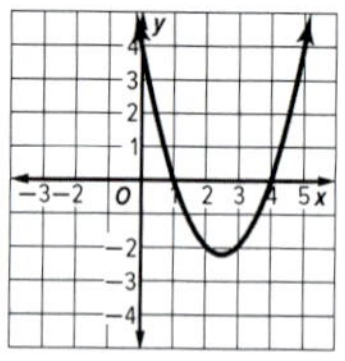

[graph]
- A •
- B •••
- C •••
- D •

## Differentiated Instruction BL

**Extension** Tell students that in a computer golf game, the function $y = -0.002x^2 + 0.22x$ models the path of a golf ball, where y is the height of the ball and x is the horizontal distance in yards the ball has traveled. The green lies uphill from the tee, 90 yards away horizontally, atop a hill that has a steady incline of 1 yard per 10 yards of distance. Ask, "Will the ball reach the green without hitting the ground first? Explain your answer." No; the ball will land on the slope when it is 60 yards horizontally from the tee.

# EXTEND 9-3
## Graphing Technology Lab
# Quadratic Inequalities

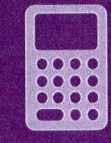

Recall that the graph of a linear inequality consists of the boundary and the shaded half-plane. The solution set of the inequality lies in the shaded region of the graph. Graphing quadratic inequalities is similar to graphing linear inequalities.

**Mathematical Practices**
MP 5 Use appropriate tools strategically.

### Activity 1  Shade Inside a Parabola

Work cooperatively. Graph $y \geq x^2 - 5x + 4$ in a standard viewing window.

First, clear all functions from the **Y=** list.

To graph $y \geq x^2 - 5x + 4$, enter the equation in the **Y=** list. Then use the left arrow to select =. Press **ENTER** until shading above the line is selected.

**KEYSTROKES:** ◀ ◀ ENTER ENTER ▶ ▶ X,T,θ,n $x^2$ − 5 X,T,θ,n + 4 ZOOM 6

[−10, 10] scl: 1 by [−10, 10] scl: 1

All ordered pairs for which $y$ is *greater than or equal to* $x^2 - 5x + 4$ lie *above or on* the line and are solutions.

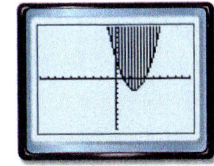

A similar procedure will be used to graph an inequality in which the shading is outside of the parabola.

### Activity 2  Shade Outside a Parabola

Work cooperatively. Graph $y - 4 \leq x^2 - 5x$ in a standard viewing window.

First, clear the graph that is displayed.

**KEYSTROKES:** Y CLEAR

Then rewrite $y - 4 \leq x^2 - 5x$ as $y \leq x^2 - 5x + 4$, and graph it.

**KEYSTROKES:** ◀ ◀ ENTER ENTER ENTER ▶ ▶ X,T,θ,n $x^2$ − 5 X,T,θ,n + 4 GRAPH

[−10, 10] scl: 1 by [−10, 10] scl: 1

All ordered pairs for which $y$ is *less than or equal to* $x^2 - 5x + 4$ lie *below or on* the line and are solutions.

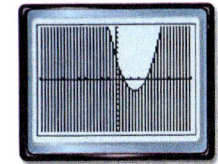

### Exercises

1. Compare and contrast the two graphs shown above.
   **1.** The graph of $y \geq x^2 - 5x + 4$ is a parabola with the inside shaded. The graph of $y \leq x^2 - 5x + 4$ is a parabola with the outside shaded.

2. Graph $y - 2x + 6 \geq 5x^2$ in the standard viewing window. Name three solutions of the inequality. **Sample answer:** (0, −6), (1, 1), (2, 18)

3. Graph $y - 6x \leq -x^2 - 3$ in the standard viewing window. Name three solutions of the inequality. **Sample answer:** (0, −3), (1, 2), (2, 5)

---

# Launch

**Objective** Use a graphing calculator to investigate quadratic inequalities.

### Materials for Each Group
- TI-83/84 Plus or other graphing calculator

### Teaching Tip
Remind students that the $x^2$ key squares the quantity but does not enter $x^2$ into an equation. To enter $5x^2$, press 5 X,T,θ,n $x^2$.

# Teach

**Working in Cooperative Groups** Put students in groups of two or three, mixing their abilities. Have groups complete the Activities and Exercise 1. **ELL**

**Ask:**
- Where are the solutions for the inequalities? *all ordered pairs in the shaded area of the graph, including the graph of the related function itself*
- How many solutions does each inequality have? *infinite number*
- Are any of the solutions to the first graph the same as for the second graph? *yes, the solutions that lie on the graph of the function itself*

**Practice** Have students complete Exercises 2 and 3.

# Assess

### Formative Assessment
Use Exercise 3 to assess whether students understand how to use a graphing calculator to solve an inequality

### From Concrete to Abstract
Have students examine the graph they made for the inequality in Activity 1. Then have them explain how the solution set to this inequality is the same as, or different from, the solution set for $y > x^2 - 5x + 4$. *The solution set for $y > x^2 - 5x + 4$ does not include the values on the graph of the related function, while the solution set for $y \geq x^2 - 5x + 4$ does.*

## Go Online!

### Graphing Calculators
Students can use the Graphing Calculator Personal Tutors to review the use of the graphing calculator to represent functions. They can also use the Other Calculator Keystrokes, which cover lab content for students with calculators other than the TI-84 Plus.

# LESSON 9-4
# Solving Quadratic Equations by Factoring

**SUGGESTED PACING (DAYS)**
- 90 min. 0.5
- 45 min. 1

Instruction

## Track Your Progress

### Objectives
1. Solve quadratic equations by using the Square Root Property.
2. Solve quadratic equations by factoring.

### Mathematical Background
Quadratic equations of the form $ax^2 + bx + c = 0$ can be factored by finding two integers with a product of $c$ and a sum of $b$. Setting the two factors of the quadratic expression equal to zero gives you the solutions of a quadratic equation.

### THEN

**A.REI.4b** Solve quadratic equations by inspection (e.g, for $x^2 = 49$), taking square roots, completing the square, the quadratic formula, and factoring, as appropriate to the initial form of the equation. Recognize when the quadratic formula gives complex solutions and write them as $a \pm bi$ for real numbers $a$ and $b$.

**F.IF.7a** Graph linear and quadratic functions and show intercepts, maxima, and minima.

### NOW

**A.SSE.3a** Factor a quadratic expression to reveal the zeros of the function it defines.

**A.REI.4b** Solve quadratic equations by inspection (e.g., for $x^2 = 49$), taking square roots, competing the square, the quadratic formula and factoring, as appropriate to the initial form of the equation. Recognize when the quadratic formula gives complex solutions and write them as $a \pm bi$ for real numbers $a$ and $b$.

### NEXT

**A.SSE.3b** Complete the square in a quadratic expression to reveal the maximum or minimum value of the function it defines.

**F.IF.8a** Use the process of factoring and completing the square in a quadratic function to show zeros, extreme values, and symmetry of the graph, and interpret these in terms of a context.

---

**Go Online!** All of these resources and more are available at connectED.mcgraw-hill.com

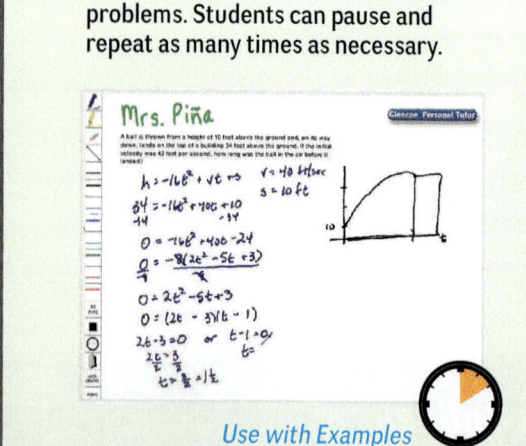

**Personal Tutors** (for every example) let students hear real teachers solve problems. Students can pause and repeat as many times as necessary.

*Use with Examples*

**eToolkit** allows students to explore and enhance their understanding of math concepts. Use the Product Mat and the Algebra Tiles tool to factor trinomials.

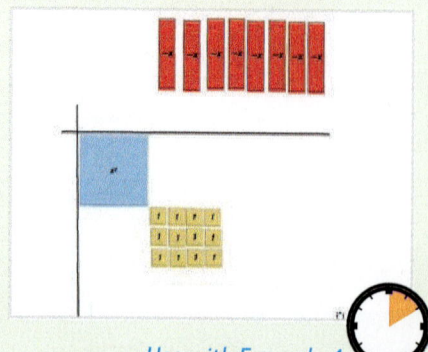

*Use with Example 4*

Use **Self-Check Quiz** to assess students' understanding of the concepts in this lesson.

*Use at End of Lesson*

---

### OER Using Open Educational Resources
**Apps** Have students access Purple Math to get a visual representation of how to solve quadratic equations by factoring. As an educator, this tool will offer interactive ways to explain the material. *Use as planning tool*

# Go Online!
connectED.mcgraw-hill.com — Worksheets

# Differentiate Your Resources

**Extra Practice** Additional practice or homework; Skills Practice is best for approaching-level students and Practice is best for on-level and beyond-level students

## Skills Practice
AL  OL  ELL

## Practice
AL  OL  BL  ELL

## Word Problem Practice
AL  OL  BL  ELL

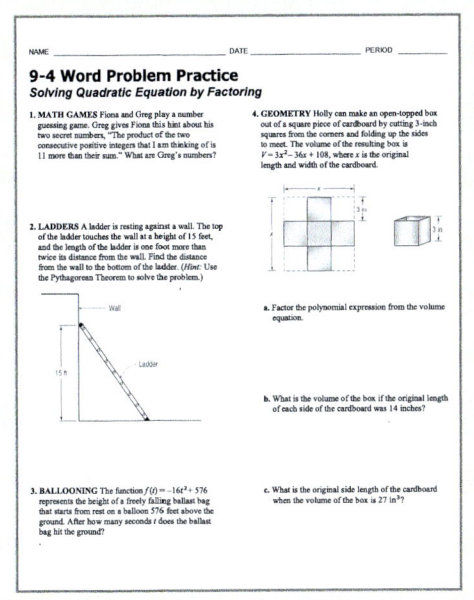

**Intervention** Reteaching and vocabulary activities that can be used with struggling or absent students and as ELL support

**Extension** Activities that can be used to extend lesson concepts

## Study Guide and Intervention
AL  OL  ELL

## Study Notebook
AL  OL  BL  ELL

## Enrichment
OL  BL  ELL

connectED.mcgraw-hill.com   588B

Lesson 9-4 | Solving Quadratic Equations by Factoring

# Launch

Have students read the Why? section of the lesson. Ask:
- How would you factor the expression? **Find the greatest common factor.**
- What could a height of 0 feet represent? **the froghopper being on the ground when it begins and completes it jump**

# Teach

Ask the scaffolded questions for each example to build conceptual understanding for students at all levels.

## 1 The Square Root Property

### Example 1 Use the Square Root Property

- **AL** In part **a**, is each side of the equation a perfect square? **yes**
- **OL** In part **b**, why are the solutions estimated? **Sample answer: 21 is not a perfect square, so an exact answer cannot be written. It is rounded.**
- **BL** How could you have solved part **a** mentally? **Sample answer: Since $8^2 = 64$ and $(-8)^2 = 64$, we need to find values of $x$ such that $x - 3 = 8$ or $x - 3 = -8$. So, $x = 11$ or $x = -5$.**

### Need Another Example?
Solve each equation. Check the solutions.
A. $(b - 7)^2 = 36$ **1, 13**
B. $(x + 9)^2 = 8$ **$-9 \pm 2\sqrt{2}$**

### Teaching Tip
**Structure** Explain to students the positive and negative solutions are true for the equation when solving by the square root method.

## Go Online!

**Interactive Whiteboard**
Use the eLesson or Lesson Presentation to present this lesson.

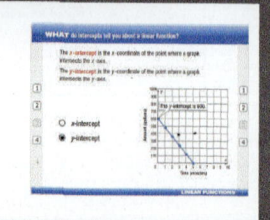

---

# LESSON 4
# Solving Quadratic Equations by Factoring

**Then** | **Now** | **Why?**
--- | --- | ---
You solved quadratic equations by graphing. | 1 Solve quadratic equations by using the Square Root Property. 2 Solve quadratic equations by factoring. | Froghoppers are insects commonly found in Africa, Europe, and North America. They are only about 6 millimeters long, but they can jump up to 70 times their body height. A froghopper's jump can be modeled by the equation $h = 12t - 16t^2$, where $t$ is the time in seconds and $h$ is the height in feet. You can use factoring and the Zero Product Property to determine when the froghopper will complete its jump.

**New Vocabulary**
Square Root Property
Zero Product Property

**Mathematical Practices**
1 Make sense of problems and persevere in solving them.
6 Attend to precision.

**Content Standards**
A.SSE.3a Choose and produce an equivalent form of an expression to reveal and explain properties of the quantity represented by the expression. Factor a quadratic expression to reveal the zeros of the function it defines.
A.REI.4b Solve quadratic equations in one variable. Solve quadratic equations by inspection (e.g., for $x^2 = 49$), taking square roots, completing the square, the quadratic formula and factoring, as appropriate to the initial form of the equation.

Quadratic equations can be solved using a variety of methods. In addition to solving quadratic equations by graphing, you can solve quadratic equations algebraically.

**1 Square Root Property** A quadratic equation in the form $x^2 = n$ can be solved by using the **square root property**. The equation can be solved by applying the square root to each sides of the equation.

### Key Concept — Square Root Property

| Words | To solve a quadratic equation in the form $x^2 = n$, take the square root of each side. |
|---|---|
| Symbols | For any number $n \geq 0$, if $x^2 = n$, then $x = \pm\sqrt{n}$. |
| Example | $x^2 = 25$ <br> $x = \pm\sqrt{25}$ or $\pm 5$ |

In the equation $x^2 = n$, if $n$ is not a perfect square, you need to approximate the square root. Use a calculator to find an approximation. If $n$ is a perfect square, you will have an exact answer.

A.REI.4b

### Example 1 — Use the Square Root Property

Solve each equation. Check your solutions.

a. $(x - 3)^2 = 64$

$(x - 3)^2 = 64$    Original Equation
$x - 3 = \pm\sqrt{64}$    Square Root Property
$x - 3 = \pm 8$    $64 = 8(8)$ or $-8(-8)$
$x = 3 \pm 8$    Add 3 to each side.
$x = 3 + 8$ or $x = 3 - 8$    Separate into two equations.
$x = 11$    $x = -5$    Simplify.

The roots are 11 and $-5$.

## MP Mathematical Practices Strategies

### Attend to precision.
Help students communicate precisely by allowing them to discuss their reasoning with others. For example, in Guided Practice 4C, ask:

- What technique is used to factor the expression on the left side of the equation? **factoring quadratic trinomials**
- Why are there two solutions? **When substituted back into the original equation, both solutions result in a true statement.**
- When will a quadratic equation have one solution? **when the factors are exactly the same**

---

588 | Lesson 9-4 | Solving Quadratic Equations by Factoring

**CHECK** Substitute −11 and 5 for x in the original equation.

$(x - 3)^2 = 64$     $(x - 3)^2 = 64$
$(11 - 3)^2 \stackrel{?}{=} 64$     $[(-5) - 3]^2 \stackrel{?}{=} 64$
$(8)^2 \stackrel{?}{=} 64$     $(-8)^2 \stackrel{?}{=} 64$
$64 = 64$ ✓     $64 = 64$ ✓

**b.** $(y + 5)^2 + 7 = 28$

| | |
|---|---|
| $(y + 5)^2 + 7 = 28$ | Original Equation |
| $(y + 5)^2 = 21$ | Subtract 7 from each side. |
| $y + 5 = \pm\sqrt{21}$ | Square Root Property |
| $y = -5 \pm \sqrt{21}$ | Subtract −5 from each side. |
| $y = -5 + \sqrt{21}$ or $y = -5 - \sqrt{21}$ | Separate into two equations. |

The roots are $-5 + \sqrt{21}$ and $-5 - \sqrt{21}$.
Using a calculator, $-5 + \sqrt{21} \approx -0.42$ and $-5 - \sqrt{21} \approx -9.58$.

> **Study Tip**
>
> **MP Persevere** Equations involving square roots can often be solved mentally. For $x^2 = n$, think *The square of what number is n?* When n is a perfect square, x is rational. Otherwise, x is irrational.

▶ **Guided Practice**

Solve each equation. Round to the nearest hundredth if necessary.

**1A.** $(x + 6)^2 = 81$  **3, −15**

**1B.** $(a - 8)^2 - 3 = 10$   **8 ± √13 or about 11.61 and 4.39**

**1C.** $4(m + 1)^2 = 36$  **2, −4**

When solving real-world problems using the Square Root Property, it is important to determine whether both solutions make sense in the context of the situation.

A.REI.4b

**Real-World Example 2** Solve an Equation by Using the Square Root Property

**PHYSICAL SCIENCE** During an experiment, a ball is dropped from a height of 205 feet. The formula $h = -16t^2 + h_0$ can be used to approximate the number of seconds t it takes for the ball to reach height h from an initial height of $h_0$ in feet. Find the time it takes the ball to reach the ground.

At ground level, $h = 0$ and the initial height is 205, so $h_0 = 205$.

| | |
|---|---|
| $h = -16t^2 + h_0$ | Original formula |
| $0 = -16t^2 + 205$ | Replace h with 0 and $h_0$ with 205. |
| $-205 = -16t^2$ | Subtract 205 from each side. |
| $12.8125 = t^2$ | Divide each side by −16. |
| $\pm 3.6 \approx t$ | Use the Square Root Property. |

Since a negative number does not make sense in this situation, the solution is 3.6. It takes about 3.6 seconds for the ball to reach the ground.

▶ **Guided Practice**

**2.** Find the time it takes a ball to reach the ground if it is dropped from a height that is half as high as the one described above. **about 2.5 seconds**

---

**Lesson 9-4 | Solving Quadratic Equations by Factoring**

**Example 2 Solve an Equation by Using the Square Root Property**

🔴 **AL** What is the given value of $h_0$? **205** What is the value of h? **0**

🔵 **OL** What would be the easiest way to solve for t? **Sample answer: Because the binomial is not a difference of squares and cannot easily be factored, it would be easiest to subtract 205 from both sides and take the square root of each side.**

🟢 **BL** What do the roots of the equation mean in the context of the situation? **Sample answer: The roots tell us how long it took for the ball to hit the ground. Since a negative time does not make sense, only use the positive root. The ball took about 3.6 seconds to hit the ground.**

**Need Another Example?**
**PHYSICAL SCIENCE** A book falls from a shelf that is 5 feet above the floor. A model for the height h in feet if an object dropped from an initial height of $h_0$ feet is $h = -16t^2 + h_0$, where t is the time in seconds after the object is dropped. Use this model to determine approximately how long it took for the book to reach the ground. **0.56 s**

---

**MP Teaching the Mathematical Practices**

**Structure** Mathematically proficient students can see complicated expressions as being composed of several objects. Make sure students set each factor of a quadratic equation equal to zero to find all of the solutions.

**Lesson 9-4** | Solving Quadratic Equations by Factoring

## 2 Solve Quadratic Equations by Factoring

**Example 3 Solve Equations by Factoring**

**AL** In part **a**, what must be set equal to 0? $2d + 6$ and $3d - 15$

**OL** Is it possible for both roots to be solutions at the same time? Explain. Yes; Sample answer: If the roots are substituted into the original equation, we get $0 \times 0$, which equals 0.

**BL** How can you check the roots in part **a** to make sure they are solutions? Sample answer: Substitute in the first root for $d$ and solve, then substitute the second root for $d$ and solve. They should both result in a true statement.

**Need Another Example?**
Solve each equation. Check the solutions.
A. $(b - 7)(3b + 4) = 0$  $-\frac{4}{3}, 7$
B. $c^2 = 10c$  $10, 0$

---

**2 Solve Quadratic Equations by Factoring** Some equations can be solved by factoring. A quadratic equation will have roots if its factored form has real numbers. Consider the following:

$7(0) = 0$    $0(4 - 3 - 1) = 0$    $-71(0) = 0$    $(3.59)0 = 0$

Notice that in each case, at least one of the factors is 0. These examples demonstrate the **Zero Product Property**.

> **Key Concept** Zero Product Property
>
> **Words**   If the product of two factors is 0, then at least one of the factors must be 0.
>
> **Symbols**   For any numbers $a$ and $b$, if $ab = 0$, then $a = 0$, $b = 0$, or both $a$ and $b$ equal zero.

A.SSE.3a, A.REI.4b

**Watch Out!**
**Unknown Value** It may be tempting to solve an equation by dividing each side by the variable. However, the variable has an unknown value, so you may be dividing by 0, which is undefined.

**Example 3**   Solve Equations by Factoring

Solve each equation. Check your solutions.

**a.** $(2d + 6)(3d - 15) = 0$

$(2d + 6)(3d - 15) = 0$     Original equation
$2d + 6 = 0$  or  $3d - 15 = 0$     Zero Product Property
$2d = -6$         $3d = 15$     Solve each equation.
$d = -3$          $d = 5$     Divide.

The roots are $-3$ and 5.

**CHECK** Substitute $-3$ and 5 for $d$ in the original equation.

$(2d + 6)(3d - 15) = 0$                    $(2d + 6)(3d - 15) = 0$
$[2(-3) + 6][3(-3) - 15] \stackrel{?}{=} 0$    $[2(5) + 6][3(5) - 15] \stackrel{?}{=} 0$
$(-6 + 6)(-9 - 15) \stackrel{?}{=} 0$          $(10 + 6)(15 - 15) \stackrel{?}{=} 0$
$(0)(-24) \stackrel{?}{=} 0$                   $16(0) \stackrel{?}{=} 0$
$0 = 0$ ✓                                     $0 = 0$ ✓

**b.** $c^2 = 3c$

$c^2 = 3c$     Original equation
$c^2 - 3c = 0$     Subtract $3c$ from each side to get 0 on one side of the equation.
$c(c - 3) = 0$     Factor by using the GCF to get the form $ab = 0$.
$c = 0$  or  $c - 3 = 0$     Zero Product Property
$c = 3$     Solve each equation.

The roots are 0 and 3.     Check by substituting 0 and 3 for $c$.

**Guided Practice**

**4A.** $3n(n + 2) = 0$  $0, -2$     **4B.** $8b^2 - 40b = 0$  $0, 5$     **4C.** $x^2 = -10x$  $0, -10$

Recall that there are different ways to factor an equation. After an equation is factored, it can be solved. First factor the equation. Then use the properties of equality to isolate the variable for which you are trying to solve the equation. Check by substituting in the original equation.

**Key Concept** Methods of Factoring

| Method | Symbols |
|---|---|
| Factor Using the Distributive Property | $ax + bx + ay + by = x(a + b) + y(a + b) = (a + b)(x + y)$ |
| Factor Quadratic Trinomials | $x^2 + bx + c = (x + m)(x + p)$ |
| Factor Differences of Squares | $a^2 - b^2 = (a + b)(a - b)$ |
| Factor Perfect Squares | $a^2 + 2ab + b^2 = (a + b)^2$ <br> $a^2 - 2ab + b^2 = (a - b)^2$ |

A.SSE.3a, A.REI.4b

**Example 4** Solve Quadratic Equations by Factoring

Solve each equation. Check your solutions.

a. $y^2 + 5y - 24 = 0$

$y^2 + 5y - 24 = 0$     Original equation
$(y + 8)(y - 3) = 0$     Factor.
$y + 8 = 0$ or $y - 3 = 0$     Zero Product Property
$y = -8$ or $y = 3$     Solve each equation.

The roots are −8 and 3.

**CHECK** Substitute −8 and 3 for $y$ in the original equation.

$y^2 + 5y - 24 = 0$          $y^2 + 5y - 24 = 0$
$(-8)^2 + 5(-8) - 24 \stackrel{?}{=} 0$     $(3)^2 + 5(3) - 24 \stackrel{?}{=} 0$
$64 - 40 - 24 \stackrel{?}{=} 0$         $9 + 15 - 24 \stackrel{?}{=} 0$
$0 = 0$ ✓                $0 = 0$ ✓

b. $3x^2 - 12 = 9x$

$3x^2 - 12 = 9x$     Original equation
$3x^2 - 9x - 12 = 0$     Subtract 9x from each side.
$(3x - 12)(x + 1) = 0$     Factor.
$3x - 12 = 0$ or $x + 1 = 0$     Zero Product Property
$x = 4$    $x = -1$     Solve each equation.

The roots are 4 and −1.

**Guided Practice**

Solve each equation. Check your solutions.

4A. $49 - x^2 = 0$    −7, 7
4B. $25y^2 - 60y + 81 = 45$    $\frac{6}{5}$
4C. $x^2 - 13x + 15 = -21$    4, 9

**Example 4** Solve Quadratic Equations by Factoring

**AL** In part **b**, why must you first subtract 9x from each side? *so the Zero Product Property can be applied*

**OL** In part **a**, what are the two factors that have a sum of 5 and product of −24? *−3 and 8*

**BL** When would a quadratic equation have only one solution when solving by factoring? *If it is a perfect square trinomial, then the factors are the same. So, when each factor is set equal to 0, they have the same solution.*

**Need Another Example?**
Solve each equation. Check your solutions.
A. $y^2 + 10 + 24 = 0$   −6, −4

A. $4x^2 + 36x = -81$   $-\frac{9}{2}$

**Lesson 9-4** | Solving Quadratic Equations by Factoring

### Example 5 Write Quadratic Functions Given Their Graphs

**AL** How many points on the graph must be identified in order to write the equation of the graph? **3**

**OL** What two points should be identified first? **the roots**

**BL** Can you choose another point on the graph in order to write the function? Explain. **Sample answer: Yes; any point on the graph will give us the same equation.**

**Need Another Example?**
Write a quadratic function for the given graph.
$y = -2(x-1)(x+1)$ or $y = -2x^2 + 2$

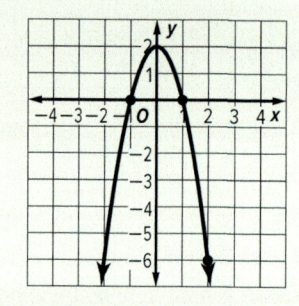

## Practice

**Formative Assessment** Use Exercises 1–16 to assess students' understanding of the concepts in the lesson.

The Practice and Problem Solving exercises assess the content taught in the lesson.

---

A.SSE.3a

**Example 5** Write Quadratic Functions Given Their Graphs

Write a quadratic function for the given graph.

**Step 1** Find the factors of the related expression.

The two zeros of the graph are −4 and 3, so $(x + 4)$ and $(x − 3)$ are factors of the related expression.

There are many quadratic functions of different shapes that have these factors, so we need to identify this specific shape. The function $f(x) = a(x + 4)(x − 3)$ represents the graph.

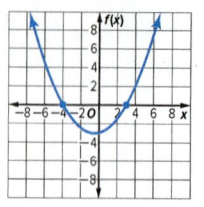

**Step 2** Determine whether $a$ is positive or negative.

The graph opens upward, so $a$ must be positive.

**Step 3** Determine the value of $a$.

Use another point on the graph to determine the value of $a$. The point (4, 2) is on the graph.

$f(x) = a(x + 4)(x − 3)$  Quadratic function with roots of −4 and 3
$2 = a(4 + 4)(4 − 3)$  $[x, f(x)] = (4, 2)$
$2 = 8a$  Simplify.
$\frac{1}{4} = a$  Divide each side by 8.

The function is $f(x) = \frac{1}{4}(x + 4)(x − 3)$ or $f(x) = \frac{1}{4}x^2 + \frac{1}{4}x − 3$.

**Guided Practice**

Write a quadratic function that has a graph that contains the given points.

5A. (−2, 0), (4, −12), (5, 0)   $f(x) = 2x^2 − 6x − 20$
5B. (−5, 0), (−1, 0), (1, 6)   $f(x) = 0.5x^2 + 3x + 2.5$
5C. (−4, 0), (0, 4), (4, 0)   $f(x) = −0.25x^2 + 4$

---

**Check Your Understanding**   ◯ = Step-by-Step Solutions begin on page R11.

**Go Online!** for a Self-Check Quiz

**Example 1**   Solve each equation. Check your solutions.
A.REI.4b
1. $x^2 = 88$   $\pm 2\sqrt{22}$
2. $4x^2 = 36$   $\pm 3$
3. $(x + 1)^2 = 16$   3, −5
4. $(x − 3)^2 = 10$   $3 \pm \sqrt{10}$

**Example 2**
A.REI.4b
5. **MP REASONING** While painting his bedroom, Nick drops his paintbrush off his ladder from a height of 6 feet. Use the formula $h = −16t^2 + h_0$ to approximate the number of seconds to the nearest tenth it takes for the paintbrush to hit the floor. **0.6 second**

6. **SEWING** Stefani is making a square quilt with a side length of $x + 2$ feet. If the quilt will have an area of 36 feet, what is the value of $x$? **4**

---

### Differentiated Homework Options

| Levels | AL Basic | OL Core | BL Advanced |
|---|---|---|---|
| Exercises | 17–43, 50, 52–55, 59, 61–68 | 17–47 odd, 48–50, 52–55, 59, 61–68 | 48–61, (optional: 62–68) |
| 2-Day Option | 17–43 odd, 62–68 | 17–43, 62–68 | |
| | 18–42 even, 44, 48, 50, 52–55, 59, 61 | 44–50, 52–55, 59, 61 | |

You can use ALEKS to provide additional remediation support with personalized instruction and practice.

**Lesson 9-4 | Solving Quadratic Equations by Factoring**

**Examples 3-4**
A.SSE.3a,
A.REI.4b

Solve each equation. Check your solutions.

7. $3k(k + 10) = 0$   $k = -10, k = 0$
8. $(4m + 2)(3m + 9) = 0$   $m = -3, m = -\frac{1}{2}$
9. $20p^2 - 15p = 0$   $p = \frac{3}{4}, p = 0$
10. $r^2 = 14r$   $r = 0, r = 14$
11. $a^2 - 10a + 9 = 0$   $a = 1, a = 9$
12. $b^2 + 7b - 30 = 0$   $b = -10, b = 3$
13. $5x^2 - 12x - 7 = 14$   $x = 1, x = \frac{7}{5}$
14. $2y^2 = y + 1$   $y = -\frac{1}{2}, y = 1$

**Example 5**
A.REI.4b

**SENSE-MAKING** Write a quadratic function that has a graph that contains the given points.

15. $(-3, 0), (4, 0), (8, -11)$   $f(x) = -\frac{1}{4}x^2 + \frac{1}{4}x + 3$
16. $(-6, 0), (6, 24), (9, 0)$   $f(x) = -\frac{2}{3}x^2 + 2x + 36$

**Practice and Problem Solving**     Extra Practice is found on page R9.

**Example 1**
A.REI.4b

Solve each equation. Check your solutions.

17. $9a^2 = 81$   $\pm 3$
18. $2b^2 = 66$   $\pm\sqrt{33}$
19. $c^2 - 10 = 90$   $\pm 10$
20. $d^2 + 16 = 60$   $\pm 2\sqrt{11}$
21. $m^2 = 72$   $\pm 6\sqrt{2}$
22. $n^2 = 169$   $\pm 13$
23. $(j + 5)^2 = 20$   $-5 \pm 2\sqrt{5}$
24. $(p - 15)^2 = 121$   $4, 26$
25. $(x - 2)^2 = 16$   $-2$
26. $(y + 18)^2 = 77$   $-18 \pm \sqrt{77}$

**Examples 3-4**
A.SSE.3a,
A.REI.4b

Solve each equation. Check your solutions.

27. $3b(9b - 27) = 0$   $0, 3$
28. $2n(3n + 3) = 0$   $0, -1$
29. $(8z + 4)(5z + 10) = 0$   $-2, -\frac{1}{2}$
30. $(7x + 3)(2x - 6) = 0$   $-\frac{3}{7}, 3$
31. $b^2 = -3b$   $-3, 0$
32. $a^2 = 4a$   $0, 4$
33. $x^2 - 18x + 80 = 0$   $8, 10$
34. $2y^2 - 26y + 80 = 0$   $5, 8$
35. $z^2 - 5z - 66 = 0$   $-6, 11$
36. $3a^2 + 18a = 81$   $-3, 9$
37. $16b^2 + 24b + 20 = 15$   $-\frac{5}{4}, -\frac{1}{4}$
38. $8c^2 + 7c = 1$   $-1, \frac{1}{8}$
39. $48x^2 + 68x + 24 = 0$   $-\frac{3}{4}, -\frac{2}{3}$
40. $14y^2 = -2y + 16$   $-\frac{8}{7}, 1$

**Example 5**
A.REI.4b

Write a quadratic function that has a graph that contains the given points.

41. $(-2, 0), (6, 0), (8, 15)$   $f(x) = \frac{3}{4}x^2 - 3x - 9$
42. $(-8, 0), (0, -16), (1, 0)$   $f(x) = 2x^2 + 14x - 16$

**Example 2**
A.REI.4b

43. **SCREENS** The area $A$ in square feet of a projected picture on a movie screen can be modeled by the equation $A = 0.25 \, d^2$, where $d$ represents the distance from a projector to a movie screen. At what distance will the projected picture have an area of 100 square feet?   **20 feet**

44. **CHECKERBOARD** A standard checkerboard is square and made up of 64 equally sized smaller squares. The equation $576 = (x + 4)^2$ can be used to model the total area of a checkerboard. Find the length of the side of the checkerboard.   **24 inches**

Solve each equation. Check your solutions.

45. $a^2 + 8a + 16 = 25$   $-9, 1$
46. $4b^2 = 80b - 400$   $10$

47. Write a quadratic function for the given graph.   $f(x) = \frac{1}{4}x^2 - 2x + 3$

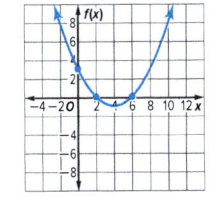

**Teaching Tip**

**Structure** Discuss with students when it is best to use the square root and factoring methods to solve a quadratic equation.

**Extra Practice**

See page R9 for extra exercises for students who are approaching level or for on-level students who need additional reinforcement.

**Levels of Complexity Chart**

The levels of the exercises progress from 1 to 3, with Level 1 indicating the lowest level of complexity.

| Exercises | 17–43 | 44–47, 62–68 | 50–61 |
|---|---|---|---|
| Level 3 | | | ● |
| Level 2 | | ● | |
| Level 1 | ● | | |

connectED.mcgraw-hill.com    593

**Lesson 9-4** | Solving Quadratic Equations by Factoring

# Assess

**Ticket Out the Door** Ask students to write a quadratic equation that can be solved by using by the Square Root Property and another than can be solved by factoring.

## Additional Answers

**48b.** I found the dimensions of Model A, 42 in. × 20 ft × 25 ft. For every 10 feet, the cost of heating the pool increases by 25%. Ichiro can afford to spend $50 more on heating than his neighbor. I know increasing the length and width by a combined 10 ft results in a 25% cost increase. Because he can afford 33%, I set up a proportion to determine the possible increase in length. This is a 33% increase in cost. That means that the pool can have dimensions that are $\frac{10 \times 0.33}{0.25} = 13.3$ ft larger than his neighbor's pool. I divided 13.3 by 2 and added the result to the length and width to determine the dimensions of Ichiro's pool.

**48c.** I assumed that the cost of heating the pool rises proportionally with its size. I also assumed that we wanted Ichiro's pool to have the same length-to-width relationship.

**51.** The quadratic is a perfect square, so I can factor it as $(x + 4)^2 = 0$. Then, I can use the Square Root Property to solve. So, $x = -4$.

---

**48. MULTI-STEP** Ichiro wants to build an indoor swimming pool. Model A is 42 inches deep and holds 1750 cubic feet of water. The length of the pool is 5 feet more than the width. Ichiro has budgeted $200 dollars per month to heat the pool. His neighbor owns Model A, and she spends about $150 per month to heat her pool. Ichiro wants a larger pool with a depth of 42 inches. Changing the length and width by a combined 10 feet increases the heating cost by 25%. The depth does not affect the cost of heating.

a. What size pool should Ichiro have built to fit his budget? (*Hint:* Find the percent of increase in cost.) **Sample answer: 42 in. by 26.65 ft by 31.65 ft**
b. Explain your solution process. **See margin.**
c. What assumptions did you make? **See margin.**

**49. PROM** For prom court voting, Jen is building a ballot box that is *h* inches tall. The width of the box is 2 inches shorter than the height and the length is 8 inches longer than the height. If the volume is 96 cubic inches, what are the dimensions of the box?
**4 in. by 12 in. by 2 in.**  A.SSE.3a, A.REI.4b

### H.O.T. Problems  Use Higher-Order Thinking Skills

**50. ERROR ANALYSIS** Kerry says that the given graphs can be represented by the same quadratic function because they have the same zeros. Is she correct? Explain.

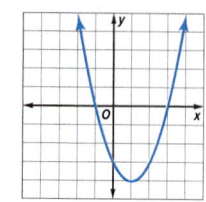

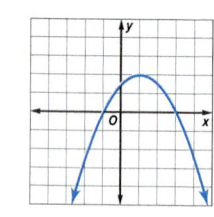

**No; because the graphs open in different directions, $a > 0$ for the first function and $a < 0$ for the second function.**

**51.** **SENSE-MAKING** Explain how you could you solve $x^2 + 8x + 16 = 0$ by using the Square Root Property. **See margin.**

**52.** **STRUCTURE** Given the equation $c = a^2 - ab$, for what values of *a* and *b* does $c = 0$?
**Sample answer: $a = 0$ or $a = b$ for any real values of *a* and *b***

**53.** **SENSE-MAKING** Given the equation $(ax + b)(ax - b) = 0$, solve for *x*. What do we know about the values of *a* and *b*? **Since the solutions are $-\frac{b}{a}$ and $\frac{b}{a}$, $a \neq 0$ and *b* is any real number.**

**54. WRITING IN MATH** Explain how to solve a quadratic equation by using the Zero Product Property.

54. Rewrite the equation to have zero on one side of the equals sign. Then factor the other side. Set each factor equal to zero, and then solve each equation.

**55.** **SENSE-MAKING** The polynomial $2x^2 - 5x - 3$ has $(2x + 1)$ and $(x - 3)$ as its factors. What are the solutions to the equation $2x^2 - 5x - 3 = 0$? $x = -\frac{1}{2}$ and $x = 3$

Write an equation that has the given roots.

**56.** $-6, 4$  $x^2 + 2x - 24 = 0$   **57.** $0, 7$  $x^2 - 7x = 0$   **58.** $-3, 1, 6$  $x^3 - 4x^2 - 15x + 18 = 0$

**59. SHOTPUT** An athlete throws a shot put with an initial upward velocity of 29 feet per second and from an initial height of 6 feet.
a. Write an equation that models the height of the shot put in feet with respect to time in seconds. $h = -16t^2 + 29t + 6$
b. After how many seconds will the shot put hit the ground? **2 seconds**

**60.** For what value of *c* does $4x^2 - 32x + c = 0$ have only one real solution? **64**

**61.** Solve $x^4 - 18x^2 + 81 = 0$. $-3, 3$

---

### Standards for Mathematical Practice

| Emphasis On | Exercises |
|---|---|
| 1 Make sense of problems and persevere in solving them. | 17–42, 48, 51, 53, 55, 62, 63, 67 |
| 6 Attend to precision. | 17–26, 47, 48, 64–67 |
| 7 Look for and make use of structure. | 27–42, 45, 46, 52, 56–58, 60, 64, 68 |

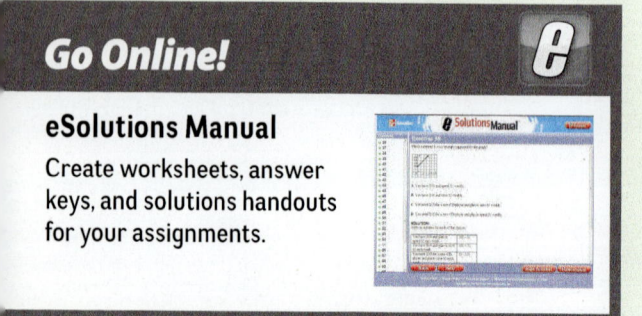

---

## Go Online!

**eSolutions Manual**
Create worksheets, answer keys, and solutions handouts for your assignments.

Lesson 9-4 | Solving Quadratic Equations by Factoring

## Preparing for Assessment

62. The rectangle shown has an area of 60 square centimeters. **MP** 5  A.SSE.3a, A.REI.4b  **B**

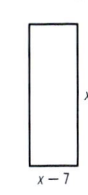

    [rectangle with width $x$ and height $x - 7$]

    a. What is the length $x$ of the rectangle?
       - A  5 cm
       - B  12 cm
       - C  33.5 cm
       - D  46 cm

    b. Find the perimeter of the rectangle.

       34 cm

63. A soccer ball is kicked into the air. The height of the soccer ball can be modeled by the equation $h = -16t^2 + 24t$, where $h$ is the height of the ball at $t$ seconds. What is the value of $t$ when $h = 0$? Select all that apply. **MP** 1, 4  A.SSE.3a, A.REI.4b  **A, D**
    - A  0 s
    - B  0.5 s
    - C  1 s
    - D  1.5 s
    - E  2 s

64. Write a quadratic function that has a graph that contains the points $(-7, 0)$, $(4, 0)$, and $(5, 18)$. **MP** 6, 7  A.SSE.3a

    $f(x) = \frac{3}{2}x^2 + \frac{9}{2}x - 42$

65. Solve $2(x^2 + 8) = 16$. **MP** 6, 7  A.SSE.3a, A.REI.4b  **0**

66. Suppose a maple tree has a leaf that is 60 feet from the ground. The equation $h = -16t^2 + 60$ describes the height $h$, in feet, of the leaf $t$ seconds after it falls from the tree. How many seconds will it take the leaf to fall to the ground? **MP** 4, 6  A.REI.4b  **D**
    - A  $-\frac{\sqrt{15}}{2}$ s
    - B  $\frac{15}{4}$ s
    - C  $\frac{\sqrt{15}}{4}$ s
    - D  $\frac{\sqrt{15}}{2}$ s

67. **MULTI-STEP** The area of the triangle shown, in square inches, is equivalent to the area of another triangle represented by the expression $2x^2 - 16x - 31$. **MP** 1, 6  A.SSE.3a, A.REI.4b

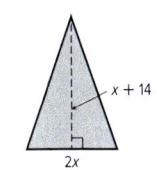

    [triangle with base $2x$ and height $x + 14$]

    a. Write an equation relating the areas of the triangles.

       $2x^2 - 16x - 31 = \frac{1}{2}(2x)(x + 14)$

    b. Find the height of the triangle shown.

       45 inches

    c. Are both solutions valid? Explain.
       No; the negative value cannot be used for the solution since length cannot be negative and $2(-1) = -2$.

68. Which are solutions of the equation $(x - 2)^2 = 16$? Select all that apply. **MP** 7  A.SSE.3a, A.REI.4b  **B, E**
    - A  $-4$
    - B  $-2$
    - C  2
    - D  4
    - E  6

## Preparing for Assessment

Exercises 62–68 require students to use the skills they will need on assessments. Each exercise is dual-coded with content.

### Dual Coding

| Items | Content Standards | **MP** Mathematical Practices |
|-------|-------------------|-------------------------------|
| 62 | A.SSE.3a, A.REI.4b | 5 |
| 63 | A.SSE.3a, A.REI.4b | 1, 4 |
| 64 | A.SSE.3a | 6, 7 |
| 65 | A.SSE.3a, A.REI.4b | 6, 7 |
| 66 | A.REI.4b | 4, 6 |
| 67 | A.SSE.3a, A.REI.4b | 1, 6 |
| 68 | A.SSE.3a, A.REI.4b | 7 |

## Diagnose Student Errors

Survey student responses for each item. Class trends may indicate common errors and misconceptions.

**62.**

| A | Found the width |
| B | CORRECT |
| C | Set the sum of the sides equal to the area |
| D | Used the Zero Product Property incorrectly |

**66.**

| A | Selected the negative solution |
| B | Did not take the square root |
| C | Took the square root of the numerator, but not the denominator |
| D | CORRECT |

## Go Online!

### Quizzes

Students can use *Self-Check Quizzes* to check their understanding of this lesson. You can also give *Quiz 2*, which covers the content in Lessons 9-4 and 9-5.

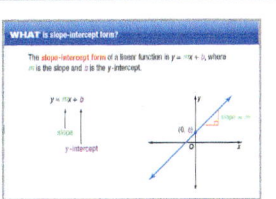

connectED.mcgraw-hill.com    595

# LESSON 9-5
# Solving Quadratic Equations by Completing the Square

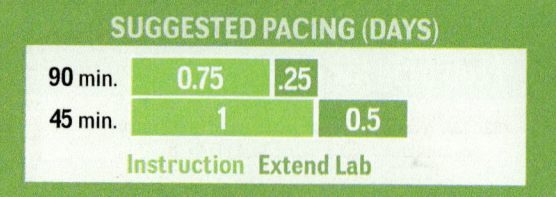

**SUGGESTED PACING (DAYS)**

| | | |
|---|---|---|
| 90 min. | 0.75 | .25 |
| 45 min. | 1 | 0.5 |

Instruction  Extend Lab

## Track Your Progress

### Objectives

1. Solve quadratic equations by completing the square.
2. Identify key features of quadratic functions by writing quadratic equations in vertex form.

### Mathematical Background
Completing the square to solve a quadratic equation does not mean that the solutions will be integers. If the equation already has a nonzero constant term, it is likely that after completing the square, the constant will not be a perfect square, and the solutions will be irrational.

### THEN

**F.IF.7a** Graph linear and quadratic functions and show intercepts, maxima, and minima.

**A.REI.4b** Solve quadratic equations by inspection (e.g. for $x^2 = 49$), taking square roots, completing the square, the quadratic formula and factoring, as appropriate to the initial form of the equation. Recognize when the quadratic formula gives complex solutions and write them as $a \pm bi$ form for real numbers $a$ and $b$.

### NOW

**A.REI.4** Solve quadratic equations in one variable.

**F.IF.8a** Use the process of factoring and completing the square in a quadratic function to show zeros, extreme values, the symmetry of the graph, and interpret these in terms of a context.

### NEXT

**A.CED.2** Create equations in two or more variables to represent relationships between quantities; graph equations on coordinate axes with labels and scales.

**F.IF.6** Calculate and interpret the average rate of change of a function (represented symbolically or as a table) over a specified interval. Estimate the rate of change from a graph.

## Go Online! All of these resources and more are available at connectED.mcgraw-hill.com

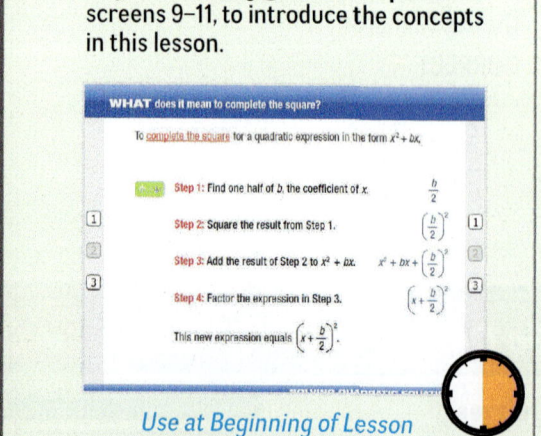

**eLessons** utilize the power of your interactive whiteboard in an engaging way. Use **Solving Quadratic Equations**, screens 9–11, to introduce the concepts in this lesson.

*Use at Beginning of Lesson*

**Personal Tutors** (for every example) let students hear real teachers solve problems. Students can pause and repeat as many times as necessary.

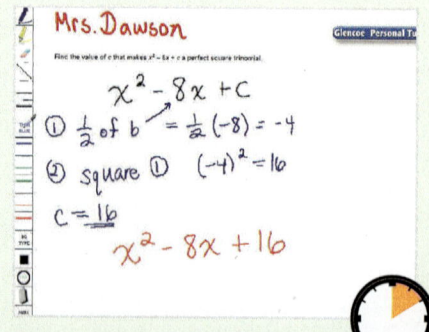

*Use with Examples*

**eToolkit** allows students to explore and enhance their understanding of math concepts. Use the Equation Chart Mat and the Algebra Tiles tool to complete the square.

*Use with Example 2*

## OER Using Open Educational Resources

**Tutorials** Have students watch the tutorials on **LEARNZILLION** about solving quadratic equations by completing the square to reinforce their understanding of the process. If you are unable to access LEARNZILLION, try **KidsTube**, **MathATube**, **SchoolTube**, or **TeacherTube**. *Use as homework or flipped learning*

# Go Online!
connectED.mcgraw-hill.com

Worksheets

# Differentiate Your Resources

**Extra Practice** Additional practice or homework; Skills Practice is best for approaching-level students and Practice is best for on-level and beyond-level students

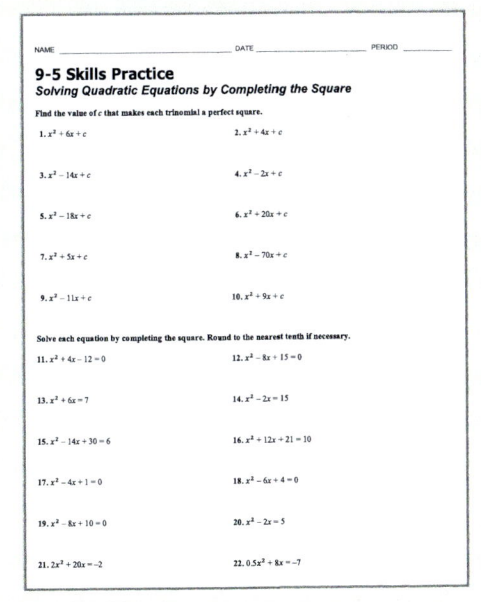

## Skills Practice
AL OL ELL

## Practice
AL OL BL ELL

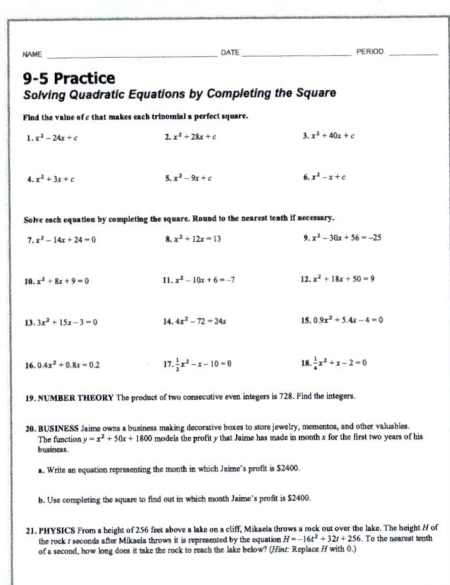

## Word Problem Practice
AL OL BL ELL

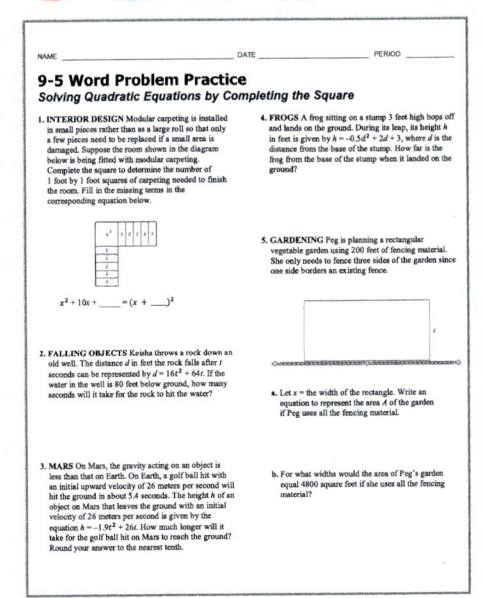

**Intervention** Reteaching and vocabulary activities that can be used with struggling or absent students and as ELL support

## Study Guide and Intervention
AL OL ELL

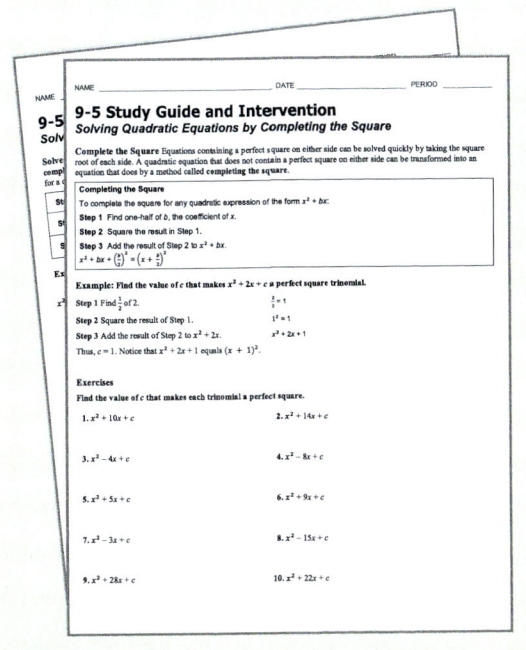

## Study Notebook
AL OL BL ELL

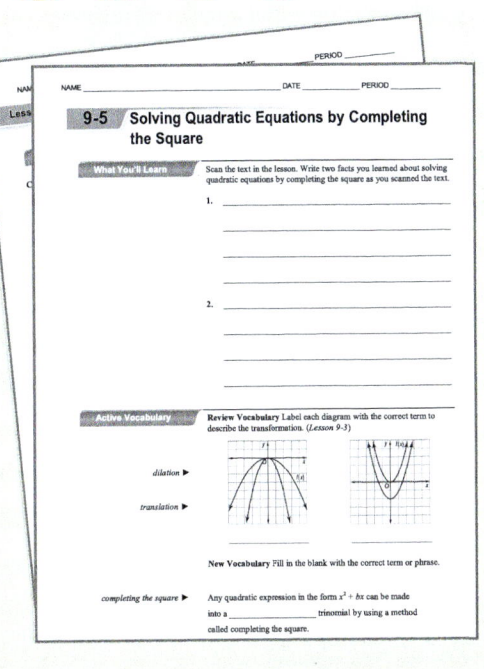

**Extension** Activities that can be used to extend lesson concepts

## Enrichment
OL BL ELL

connectED.mcgraw-hill.com   596B

**Lesson 9-5** | Solving Quadratic Equations by Completing the Square

# Launch

Have students read the Why? section of the lesson. Ask:

- Look at the equation. Is 25 a perfect square? **yes**
- Is $-16t^2 + 20t + 12$ a perfect square? **no**
- Can you solve the equation by taking the square root of each side of the equation? **no**

# Teach

Ask the scaffolded questions for each example to build conceptual understanding for students at all levels.

## 1 Complete the Square

**Example 1 Complete the Square**

**AL** Is the quadratic expression in the form $x^2 + bx$? **yes**

**OL** Describe how to find the value of $c$. **Take half of 4 (the value of $b$), then square it.**

**BL** Could this same formula be used if the first term was $3x^2$? Explain. **No; Sample answer: The pattern only works when the leading coefficient is 1.**

**Need Another Example?**
Find the value of $c$ that makes $x^2 - 12x + c$ a perfect square trinomial. **36**

### Go Online!

**Interactive Whiteboard**
Use the *eLesson* or *Lesson Presentation* to present this lesson.

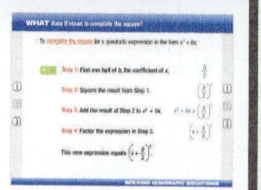

---

## LESSON 5
# Solving Quadratic Equations by Completing the Square

**:Then**
- You solved quadratic equations by using the Square Root Property and by factoring.

**:Now**
1. Solve quadratic equations by completing the square.
2. Identify key features of quadratic functions by writing quadratic equations in vertex form.

**:Why?**
- In competitions, skateboarders may launch themselves from a half pipe into the air to perform tricks. The equation $h = -16t^2 + 20t + 12$ can be used to model their height, in feet, after $t$ seconds.

To find how long a skateboarder is in the air if he is 25 feet above the half pipe, you can solve $25 = -16t^2 + 20t + 12$ by using a method called completing the square.

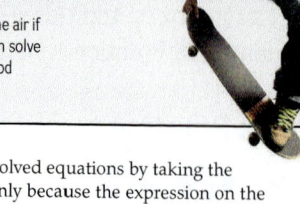

**New Vocabulary**
completing the square

**MP Mathematical Practices**
1 Make sense of problems and persevere in solving them.
2 Reason abstractly and quantitatively.
4 Model with mathematics.
7 Look for and make use of structure.

**Content Standards**
**A.SSE.3b** Complete the square in a quadratic expression to reveal the maximum or minimum value of the function it defines.
**A.REI.4** Solve quadratic equations in one variable.
**F.IF.8a** Use the process of factoring and completing square in a quadratic function to show zeros, extreme values, and symmetry of the graph, and interpret these in terms of a context.

### 1 Complete the Square

You have previously solved equations by taking the square root of each side. This method worked only because the expression on the left-hand side was a perfect square. In perfect square trinomials in which the leading coefficient is 1, there is a relationship between the **coefficient of the $x$-term** and the **constant term**.

$$(x + 5)^2 = x^2 + 2(5)(x) + 5^2$$
$$= x^2 + 10x + 25$$

Notice that $\left(\frac{10}{2}\right)^2 = 25$. To get the constant term, divide the coefficient of the $x$-term by 2 and square the result. Any quadratic expression in the form $x^2 + bx$ can be made into a perfect square by using a method called **completing the square**.

**Key Concept** Completing the Square

**Words** To complete the square for any quadratic expression of the form $x^2 + bx$, follow the steps below.

**Step 1** Find one half of $b$, the coefficient of $x$.
**Step 2** Square the result in Step 1.
**Step 3** Add the result of Step 2 to $x^2 + bx$.

**Symbols** $x^2 + bx + \left(\frac{b}{2}\right)^2 = \left(x + \frac{b}{2}\right)^2$

**Example 1** Complete the Square

Find the value of $c$ that makes $x^2 + 4x + c$ a perfect square trinomial.

**Method 1** Use algebra tiles.

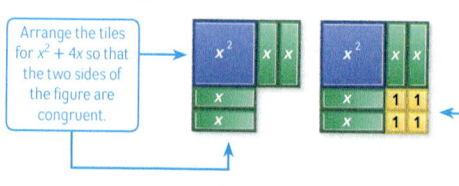

Arrange the tiles for $x^2 + 4x$ so that the two sides of the figure are congruent. → To make the figure a square, add 4 positive 1-tiles.

---

**MP Mathematical Practices Strategies**

**Look for and express regularity in repeated reasoning.**
Help students recognize general methods by asking them to analyze the steps for completing the square for any quadratic expression of the form $x^2 + bx$. For example, ask:

- How do you find the value of $c$ to make $x^2 + bx + c$ a perfect square trinomial? **Find one half of $b$, the coefficient of $x$, and square it.**

- If the $x^2$-term does not have a coefficient of 1, is the completing the square method the same? **No; First you have to divide each term of the quadratic equation by the coefficient of the $x^2$-term, and then isolate the $x^2$-term and the $bx$-term before completing the square.**

- Why is completing the square a useful method for solving quadratic equations? **Unlike using the Square Root Property or factoring, it can be used to solve any quadratic equation.**

596 | Lesson 9-5 | Solving Quadratic Equations by Completing the Square

**Study Tip**

**MP Structure** An algorithm is a series of steps for carrying out a procedure or solving a problem.

**Method 2** Use the complete the square algorithm.

**Step 1** Find $\frac{1}{2}$ of 4. $\quad\quad \frac{4}{2} = 2$
**Step 2** Square the result in Step 1. $\quad 2^2 = 4$
**Step 3** Add the result of Step 2 to $x^2 + 4x$. $\quad x^2 + 4x + 4$

Thus, $c = 4$. Notice that $x^2 + 4x + 4 = (x + 2)^2$.

▶ **Guided Practice**

1. Find the value of $c$ that makes $r^2 - 8r + c$ a perfect square trinomial. **16**

You can complete the square to solve quadratic equations. First, you must isolate the $x^2$- and $bx$-terms.

A.REI.4

**Example 2**    Solve an Equation by Completing the Square

Solve $x^2 - 6x + 12 = 19$ by completing the square.

| | |
|---|---|
| $x^2 - 6x + 12 = 19$ | Original equation |
| $x^2 - 6x = 7$ | Subtract 12 from each side. |
| $x^2 - 6x + 9 = 7 + 9$ | Since $\left(\frac{-6}{2}\right)^2 = 9$, add 9 to each side. |
| $(x - 3)^2 = 16$ | Factor $x^2 - 6x + 9$. |
| $x - 3 = \pm 4$ | Take the square root of each side. |
| $x = 3 \pm 4$ | Add 3 to each side. |
| $x = 3 + 4$ or $x = 3 - 4$ | Separate the solutions. |
| $= 7 \quad\quad = -1$ | The solutions are 7 and −1. |

▶ **Guided Practice**

2. Solve $x^2 - 12x + 3 = 8$ by completing the square. **about −0.4, 12.4**

To solve a quadratic equation in which the leading coefficient is not 1, divide each term by the coefficient. Then isolate the $x^2$- and $x$-terms and complete the square.

A.REI.4

**Watch Out!**

**MP Structure** Remember that the leading coefficient has to be 1 before you can complete the square.

**Example 3**    Equation with $a \neq 1$

Solve $-2x^2 + 8x - 18 = 0$ by completing the square.

| | |
|---|---|
| $-2x^2 + 8x - 18 = 0$ | Original equation |
| $\dfrac{-2x^2 + 8x - 18}{-2} = \dfrac{0}{-2}$ | Divide each side by −2. |
| $x^2 - 4x + 9 = 0$ | Simplify. |
| $x^2 - 4x = -9$ | Subtract 9 from each side. |
| $x^2 - 4x + 4 = -9 + 4$ | Since $\left(\frac{-4}{2}\right)^2 = 4$, add 4 to each side. |
| $(x - 2)^2 = -5$ | Factor $x^2 - 4x + 4$. |

No real number has a negative square. So, this equation has no real solutions.

▶ **Guided Practice**

3. Solve $3x^2 - 9x - 3 = 21$ by completing the square. **about −1.7, 4.7**

---

**Lesson 9-5 | Solving Quadratic Equations by Completing the Square**

**Example 2 Solve an Equation by Completing the Square**

**AL** Describe how to get the $x^2$- and $bx$-terms by themselves. **subtract 12 from each side**

**OL** What value should be added to each side? Explain. **9; Half of −6 is −3 and −3 squared is 9.**

**BL** What does the solution tell you about the graph of the function? **The graph crosses the $x$-axis at the coordinates (−1, 0) and (7, 0).**

**Need Another Example?**
Solve $x^2 + 6x + 5 = 12$ by completing the square. **−7, 1**

**Teaching Tip**

**Reasoning** Students should always check their solutions by graphing the related function or by substituting the solutions into the original equation. For instance, tell students that in Example 2, substituting 7 and −1 into $x^2 - 6x + 12$ should produce 19.

**Example 3 Equation with $a \neq 1$**

**AL** Describe how to get rid of the leading coefficient. **divide each side by the coefficient, −2**

**OL** Is $x^2 - 4x + 9$ a perfect square trinomial? **no** How can you make the equation become a perfect square trinomial? **subtract 9 from each side, then add 4 to each side**

**BL** Without separating the factors, how do you know that there are no real solutions? **Sample answer: Take the square root of each side. There are no real solutions to the square root of a negative number, so there are no real solutions.**

**Need Another Example?**
Solve $-2x^2 + 36x - 10 = 24$ by completing the square. **17, 1**

connectED.mcgraw-hill.com    597

**Lesson 9-5** | Solving Quadratic Equations by Completing the Square

**Example 4 Solve a Problem by Completing the Square**

**AL** What are we asked to determine? how many jerseys can be purchased for $430

**OL** Why is it necessary to approximate each value of x? because 4194 is not a perfect square

**BL** Why is the solution 52 jerseys and not about 52.8 jerseys or 53 jerseys? Sample answer: It is not possible to buy a partial jersey, so the solution must be a whole number. Rounding up to 53 will not work because they will not have quite enough money.

**Need Another Example?**
**Canoeing** Suppose the rate of flow of an 80-foot-wide river is given by the equation $r = -0.01x^2 + 0.8x$, where r is the rate in miles per hour and x is the distance from the shore in feet. Joacquim does not want to paddle his canoe against a current that is faster than 5 miles per hour. At what distance from the river bank must he paddle in order to avoid a current of 5 miles per hour? up to 7 ft from either bank **Note:** The solutions of the equation are about 7 feet and about 73 feet. Since the river is 80 feet wide, 80 − 73 = 7. Both ranges are within 7 feet of one bank or the other.

**Teaching Tip**
**Student Misconceptions** Tell students not to throw out negative solutions to real-world problems. Remind them that they must first examine the problem to see if the solution fits the situation.

**MP Teaching the Mathematical Practices**

**Tools** Mathematically proficient students can use technology to visualize assumptions and analyze graphs of functions. Encourage students to use their graphing calculators to determine or verify key features found from a function in vertex form.

**Go Online!** **eBook**

**Interactive Student Guide**
Use the *Interactive Student Guide* to deepen conceptual understanding.
· Solving by Completing the Square

---

**Real-World Link**
The annual "Battle for Paul Bunyan's Axe" is one of the oldest Division 1-A college football rivalry games. It takes place between the University of Minnesota and the University of Wisconsin. The rivalry began in 1890 and the trophy was introduced in 1948.
**Source:** Bleacher Report

---

A.REI.4

**Real-World Example 4** Solve a Problem by Completing the Square

**JERSEYS** The senior class at Bay High School buys jerseys to wear to the football games. The cost of the jerseys can be modeled by the equation $C = 0.1x^2 + 2.4x + 25$, where C is the amount it costs to buy x jerseys. How many jerseys can they purchase for $430?

The seniors have $430, so set the equation equal to 430 and complete the square.

| | |
|---|---|
| $0.1x^2 + 2.4x + 25 = 430$ | Original equation |
| $\dfrac{0.1x^2 + 2.4x + 25}{0.1} = \dfrac{430}{0.1}$ | Divide each side by 0.1. |
| $x^2 + 24x + 250 = 4300$ | Simplify. |
| $x^2 + 24x + 250 - 250 = 4300 - 250$ | Subtract 250 from each side. |
| $x^2 + 24x = 4050$ | Simplify. |
| $x^2 + 24x + 144 = 4050 + 144$ | Since $\left(\dfrac{24}{2}\right)^2 = 144$, add 144 to each side. |
| $x^2 + 24x + 144 = 4194$ | Simplify. |
| $(x + 12)^2 = 4194$ | Factor $x^2 + 24x + 144$. |
| $x + 12 = \pm\sqrt{4194}$ | Take the square root of each side. |
| $x = -12 \pm \sqrt{4194}$ | Subtract 12 from each side. |

Use a calculator to approximate each value of x.

$x = -12 + \sqrt{4194}$ or $x = -12 - \sqrt{4194}$  Separate the solutions.
$\approx 52.8$ $\approx -76.8$  Evaluate.

Since you cannot buy a negative number of jerseys, the negative solution is not reasonable. The seniors can afford to buy 52 jerseys.

▶ **Guided Practice**

66 jerseys
4. If the senior class were able to raise $620, how many jerseys could they buy?

**2 Vertex Form** In Lesson 9-1, you graphed quadratic functions written in standard form. A quadratic function can also be written in vertex form, $y = a(x − h)^2 + k$, where $a \neq 0$. In this form, you can identify key features of the graph of the function.

**Key Concept** Vertex Form

For a quadratic function in vertex form, $y = a(x − h)^2 + k$, the following are true.

- The vertex of the graph is the point (h, k).
- The graph opens up and has a minimum value of k when a > 0.
- The graph opens down and has a maximum value of k when a < 0.
- The axis of symmetry is the line $x = h$.
- The zeros are the x-intercepts of the graph.

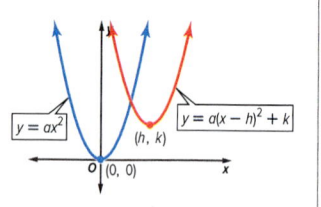

---

**Differentiated Instruction** AL OL ELL

**Kinesthetic Learners** Some students may benefit from using algebra tiles to complete the square when solving quadratic equations like those in Examples 2 and 3. Have students use an equation mat. Remind them to add or remove the same number of tiles to or from each side of the mat.

**Example 5**  Write Functions in Vertex Form

Write $y = -2x^2 + 20x - 42$ in vertex form.

| | |
|---|---|
| $y = -\cdot x^2 + \cdot\cdot \cdot x - \cdot\cdot$ | Original Equation |
| $y + \cdot\cdot = -\cdot x^2 + \cdot\cdot \cdot x$ | Add 42 to each side. |
| $y + \cdot\cdot = -\cdot\cdot x^2 - \cdot\cdot x\cdot$ | Factor out $-2$. |
| $y + \cdot\cdot - \cdot\cdot = -\cdot\cdot x^2 - \cdot\cdot x + \cdot\cdot$ | Since $-2\left(\frac{10}{2}\right)^2 = -50$, add $-50$ to each side. |
| $y - \cdot\cdot = -\cdot\cdot x - \cdot\cdot$ | Factor $x^2 - 10x + 25$. |
| $y = -\cdot\cdot x - \cdot\cdot + \cdot$ | Add 8 to each side. |
| $\cdots\cdots\cdots\cdots\cdot\cdot\cdots\cdots\cdots\cdots\cdots y = -\cdot\cdot x - \cdot\cdot + \cdot\cdot$ | |

▶ **Guided Practice**

5. $y = x^2 + \cdot\cdot x + \cdot\cdot$  $y = (x + 1)^2 + 3$

---

**Check Your Understanding**   ⬤ = Step-by-Step Solutions begin on page R11.

*Go Online! for a Self-Check Quiz*

**Example 1**  Find the value of $c$ that makes each trinomial a perfect square.
A.SSE.3b

   **1.** $x^2 - \cdot\cdot x + c$  81     **2.** $x^2 + \cdot\cdot x + c$  121

   **3.** $x^2 + \cdot\cdot x + c$  $\frac{81}{4}$     **4.** $x^2 - \cdot\cdot x + c$  $\frac{49}{4}$

**Examples 2–3**  Solve each equation by completing the square. Round to the nearest tenth
A.REI.4 if necessary.

   **5.** $x^2 + \cdot\cdot x = \cdot\cdot$  $-5.2, 1.2$     **6.** $x^2 - \cdot\cdot x = -\cdot$  $1.4, 6.6$

   **7.** $\cdot x^2 + \cdot\cdot x - \cdot\cdot = \cdot\cdot$  $-2.4, 0.1$     **8.** $-\cdot x^2 + \cdot\cdot x + \cdot\cdot = \cdot\cdot$  $-1.4, 6.4$

**Example 4**  **9.** 🅜🅟 **MODELING** · · · · · · · · · · · · · · · · · · · · · · · · · · · · ·
A.REI.4 · · · · · · · · · · · · · · · · · · · · · · · · · · · · · · · · · · · · · · · · · · · · · · ·
· · · · · · · · · · · · · · · 8 ft by 18 ft

**Example 5**  Write each quadratic function in vertex form.
F.IF.8a

   **10.** $y = x^2 - \cdot\cdot x + \cdot\cdot$  $y = (x - 6)^2 - 20$

   **11.** $y = x^2 + \cdot\cdot x + \cdot\cdot$  $y = (x + 9)^2 - 45$

   **12.** $y = \cdot\cdot x^2 + \cdot\cdot x - \cdot\cdot$  $y = 3(x + 2)^2 - 51$

---

**Practice and Problem Solving**   *Extra Practice is on page R9.*

**Example 1**  Find the value of $c$ that makes each trinomial a perfect square.
A.SSE.3b

   **13.** $x^2 + \cdot\cdot x + c$  169     **14.** $x^2 - \cdot\cdot x + c$  144     **15.** $x^2 - \cdot\cdot x + c$  $\frac{361}{4}$

   **16.** $x^2 + \cdot\cdot x + c$  $\frac{289}{4}$     **17.** $x^2 + \cdot\cdot x + c$  $\frac{25}{4}$     **18.** $x^2 - \cdot\cdot x + c$  $\frac{169}{4}$

---

Lesson 9-5 | Solving Quadratic Equations by Completing the Square

## 2 Vertex Form

**Example 5  Write Functions in Vertex Form**

**AL**  What is the factored form of the trinomial $x^2 - 10x + 25$? $(x - 5)^2$

**OL**  Why is $-2\left[\left(\frac{10}{2}\right)^2\right]$ added to each side? Since the trinomial $x^2 - 10x + 25$ is multiplied by $-2$, we are actually adding $-2 \times \left(\frac{10}{2}\right)^2$ in order to complete the square.

**BL**  Based on your solution, what is the vertex of the graph of the function? Explain. $(5, 8)$; in vertex form, $(h, k)$ represents the vertex of the graph.

**Need Another Example?**
Write $y = x^2 + 2x - 5$ in vertex form.
$y = (x + 1)^2 - 6$

## Practice

**Formative Assessment**  Use Exercises 1–12 to assess students' understanding of the concepts in this lesson.

The Practice and Problem Solving exercises assess the content taught in the lesson. The Preparing for Assessment page is meant to be used as preparation for end-of-course assessments.

## Extra Practice

See page R9 for extra exercises for students who are approaching level or for on-level students who need additional reinforcement.

connectED.mcgraw-hill.com  **599**

# Lesson 9-5 | Solving Quadratic Equations by Completing the Square

## Levels of Complexity Chart

The levels of the exercises progress from 1 to 3, with Level 1 indicating the lowest level of complexity.

| Exercises | 13–32 | 33–40, 46–53 | 42–45 |
|---|---|---|---|
| Level 3 |  |  | ● |
| Level 2 |  | ● |  |
| Level 1 | ● |  |  |

## Watch Out!

**Error Analysis** For Exercises 19-26, remind students that the amount that they add to one side of the equation to complete the square must also be added to the other side of the equation.

**Formula** For Exercises 28 and 30, students will need to know the formula for the area of a triangle, $A = \frac{1}{2}bh$.

## MP Teaching the Mathematical Practices

**Modeling** Mathematically proficient students can analyze relationships mathematically. In Exercise 9, suggest that students draw a diagram to represent the situation.

## e Follow-Up

Students have explored transformations of quadratic functions and solving quadratic equations by completing the square.

Ask:
- How is the symmetry of the graph of a quadratic function reflected in the solutions found by completing the square? **Sample answer: The solutions of the equation are expressed as conjugates, which shows how they are the same distance from the axis of symmetry.**

## Go Online!

### eSolutions Manual

Create worksheets, answer keys, and solutions handouts for your assignments.

---

**Examples 2-3** Solve each equation by completing the square. Round to the nearest tenth if necessary.
A.REI.4

19. $x^2 + 6x - 16 = 0$  **−8, 2**
20. $x^2 - 2x - 14 = 0$  **−2.9, 4.9**
21. $x^2 - 8x - 1 = 8$  **−1, 9**
22. $x^2 + 3x + 21 = 22$  **−3.3, 0.3**
23. $x^2 - 11x + 3 = 5$  **−0.2, 11.2**
24. $5x^2 - 10x = 23$  **−1.4, 3.4**
25. $2x^2 - 2x + 7 = 5$  **∅**
26. $3x^2 + 12x + 81 = 15$  **∅**

**Example 4**
A.REI.4

27. **MODELING** The price $p$ in dollars for a particular stock can be modeled by the quadratic equation $p = 3.5t - 0.05t^2$, where $t$ represents the number of days after the stock is purchased. When is the stock worth $60? **on the 30th and 40th days after purchase**

**MODELING** Find the value of $x$ for each figure. Round to the nearest tenth if necessary.

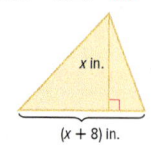

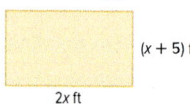

28. $A = 45$ in² **6.3**

    ($x$ in., $(x + 8)$ in.)

29. $A = 110$ ft² **5.3**

    ($(x + 5)$ ft, $2x$ ft)

30. **MODELING** Find the area of the triangle below. **216 m²**

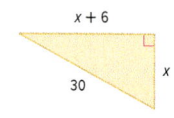

    ($x + 6$, $x$, 30)

**Example 5**
F.IF.8a

Write each quadratic function in vertex form.

31. $y = x^2 - 8x - 10$  $y = (x - 4)^2 - 26$
32. $y = -x^2 + 2x - 3$  $y = -(x - 1)^2 - 2$

Solve each equation by completing the square. Round to the nearest tenth if necessary.

33. $0.2x^2 - 0.2x - 0.4 = 0$  **−1, 2**
34. $0.5x^2 = 2x - 0.3$  **0.2, 3.8**
35. $2x^2 - \frac{11}{5}x = -\frac{3}{10}$  **0.2, 0.9**
36. $\frac{2}{3}x^2 - \frac{4}{3}x = \frac{5}{6}$  **−0.5, 2.5**

37. **MODELING** The height of an object $t$ seconds after it is dropped is given by the equation $h = -\frac{1}{2}gt^2 + h_0$, where $h_0$ is the initial height and $g$ is the acceleration due to gravity. The acceleration due to gravity near the surface of Mars is 3.73 m/s², while on Earth it is 9.8 m/s². Suppose an object is dropped from an initial height of 120 meters above the surface of each planet.
   a. On which planet would the object reach the ground first? **Earth**
   b. How long would it take the object to reach the ground on each planet? Round each answer to the nearest tenth. **Earth: 4.9 seconds, Mars: 8.0 seconds**
   c. Do the times that it takes the object to reach the ground seem reasonable? Explain your reasoning.

37c. Sample answer: Yes; the acceleration due to gravity is much greater on Earth than on Mars, so the time to reach the ground should be much less.

Write each quadratic function in vertex form.

38. $y = -2x^2 + 16x - 29$
    $y = -2(x - 4)^2 + 3$
39. $y = 3x^2 + 24x + 45$
    $y = 3(x + 4)^2 - 3$

## Differentiated Homework Options

| Levels | AL Basic | OL Core | BL Advanced |
|---|---|---|---|
| Exercises | 13–32, 42, 44–53 | 13–39 odd, 40, 42, 44–53 | 41–45, (optional: 46–53) |
| 2-Day Option | 13–31 odd, 46–53 | 13–32, 46–53 | |
|  | 14–32 even, 42, 44, 45 | 33–40, 42, 44, 45 | |

You can use ALEKS to provide additional remediation support with personalized instruction and practice.

600 | Lesson 9-5 | Solving Quadratic Equations by Completing the Square

**40. MODELING** Before she begins painting a picture, Donna stretches her canvas over a wood frame. The frame has a length of 60 inches and a width of 4 inches. She has enough canvas to cover 480 square inches. Donna decides to increase the dimensions of the frame. If the increase in the length is 10 times the increase in the width, what will the dimensions of the frame be? **6 in. by 80 in.**

**41. MULTIPLE REPRESENTATIONS** In this problem, you will investigate a property of quadratic equations.

a. **Tabular** Copy the table shown and complete the second column.

b. **Algebraic** Set each trinomial equal to zero, and solve the equation by completing the square. Complete the last column of the table with the number of real roots of each equation.

| Trinomial | $b^2 - 4ac$ | Number of Roots |
|---|---|---|
| $x^2 - 8x + 16$ | 0 | 1 |
| $2x^2 - 11x + 3$ | 97 | 2 |
| $3x^2 + 6x + 9$ | −72 | 0 |
| $x^2 - 2x + 7$ | −24 | 0 |
| $x^2 + 10x + 25$ | 0 | 1 |
| $x^2 + 3x - 12$ | 57 | 2 |

**41c.** If $b^2 - 4ac$ is negative, the equation has no real solutions. If $b^2 - 4ac$ is zero, the equation has one real solution. If $b^2 - 4ac$ is positive, the equation has 2 real solutions.

c. **Verbal** Compare the number of real roots of each equation to the result in the $b^2 - 4ac$ column. Is there a relationship between these values? If so, describe it.

d. **Analytical** Predict how many real solutions $2x^2 - 9x + 15 = 0$ will have. Verify your prediction by solving the equation. **0 because $b^2 - 4ac$ is negative. The equation has no real solutions because taking the square root of a negative number does not produce a real number.**

A.REI.4, F.IF.8a

**H.O.T. Problems** Use Higher-Order Thinking Skills

**42. ARGUMENTS** Given $y = ax^2 + bx + c$ with $a \neq 0$, derive the equation for the axis of symmetry by completing the square and rewriting the equation in the form $y = a(x - h)^2 + k$. **42–45. See margin.**

**43. CHALLENGE** Determine the number of solutions $x^2 + bx = c$ has if $c < -\left(\frac{b}{2}\right)^2$. Explain.

**44. WHICH ONE DOESN'T BELONG?** Identify the expression that does not belong with the other three. Explain your reasoning.

| $n^2 - n + \frac{1}{4}$ | $n^2 + n + \frac{1}{4}$ | $n^2 - \frac{2}{3}n + \frac{1}{9}$ | $n^2 + \frac{1}{3}n + \frac{1}{9}$ |

**45. REASONING** Write an equation in vertex form for the parabola shown below. Justify your reasoning.

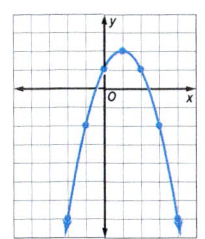

---

## Standards for Mathematical Practice

| Emphasis On | Exercises |
|---|---|
| 2 Reason abstractly and quantitatively. | 31, 32, 38, 39, 41, 45, 50, 51 |
| 3 Construct viable arguments and critique the reasoning of others. | 42, 44, 47, 52 |
| 4 Model with mathematics. | 27–30, 37, 40 |
| 7 Look for and make use of structure. | 13–16, 33–36, 41, 43, 46–49 |

---

**Lesson 9-5 | Solving Quadratic Equations by Completing the Square**

## Assess

**Ticket Out the Door** Make several copies of each of the exercises on solving quadratic equations by completing the square. Include some equations that have no solution. As the students leave the room, ask them to describe how to solve the equation using the completing the square method.

## Additional Answers

**42.** $y = ax^2 + bx + c$

$y = a\left(x^2 + \frac{b}{a}x\right) + c$

$y = a\left[x^2 + \frac{b}{a}x + \left(\frac{b}{2a}\right)^2\right] + c - a\left(\frac{b}{2a}\right)^2$

$y = a\left[x - \left(-\frac{b}{2a}\right)\right]^2 + \frac{4ac - b^2}{4a}$

If $h = -\frac{b}{2a}$ and $k = \frac{4ac - b^2}{4a}$, then $y = a(x - h)^2 + k$. The axis of symmetry is $x = -\frac{b}{2a}$ or $x = h$.

**43.** None; sample answer: If you add $\left(\frac{b}{2}\right)^2$ to each side of the equation and each side of the inequality, you get $x^2 + bx + \left(\frac{b}{2}\right)^2 = c + \left(\frac{b}{2}\right)^2$ and $c + \left(\frac{b}{2}\right)^2 < 0$. Since the left side of the last equation is a perfect square, it cannot equal the negative number $c + \left(\frac{b}{2}\right)^2$. So, there are no real solutions.

**44.** $n^2 + \frac{1}{3}n + \frac{1}{9}$; It is the trinomial that is not a perfect square.

**45.** $y = -(x - 1)^2 + 2$; The function is of the form $y = a(x - h)^2 + k$. Using the coordinates given in the graph, the vertex is (1, 2). So, the function $y = a(x - 1)^2 + 2$. Substitute the coordinates of one of the points to get $-7 = a(-2 - 1)^2 + 2$, or $-7 = 9a + 2$. Solve for $a$ to get $a = -1$. So the function is $y = -(x - 1)^2 + 2$.

## Go Online!

**Quizzes**

Students can use *Self-Check Quizzes* to check their understanding of this lesson. You can also give the *Chapter Quiz*, which covers the content in Lessons 9-1 through 9-5.

**Lesson 9-5 | Solving Quadratic Equations by Completing the Square**

# Preparing for Assessment

Exercises 46–53 require students to use the skills they will need on standardized assessments. Each exercise is dual-coded with content standards and mathematical practice standards.

### Dual Coding

| Items | Content Standards | MP Mathematical Practices |
|---|---|---|
| 46, 48, 49 | A.SSE.3b, A.REI.4 | 1, 7 |
| 47 | A.SSE.3b, A.REI.4 | 3, 7 |
| 50 | F.IF.8a | 2 |
| 52 | F.IF.8a | 2 |
| 53 | A.REI.4 | 6 |

## Diagnose Student Errors

Survey student responses for each item. Class trends may indicate common errors and misconceptions.

**46b.**

| A | Thought $\left(-\frac{8}{2}\right)^2 = -16$ |
|---|---|
| B | Did not square $-\frac{8}{2}$ |
| C | CORRECT |
| D | Did not divide the equation by 3 first |

**47.**

| A | Forgot to isolate $x^2$- and $bx$-terms |
|---|---|
| B | CORRECT |
| C | Forgot to add $\left(\frac{b}{2}\right)^2$ to both sides |
| D | Forgot to factor the trinomial |

**48.**

| A | Incorrectly had a negative number on the right side of the equation after completing the square |
|---|---|
| B | Incorrectly had a negative number on the right side of the equation after completing the square |
| C | Incorrectly had a negative number on the right side of the equation after completing the square |
| D | CORRECT |

**49.**

| A | Did not divide $b$ in $bx$-term by 2 |
|---|---|
| B | Added 7 to both sides instead of subtracting |
| C | CORRECT |
| D | Forgot to divide right side of equation by 2 |

**53.**

| A | Did not add 3 to $\pm\sqrt{17}$ |
|---|---|
| B | Did not add $\pm\sqrt{17}$ to 3 |
| C | CORRECT |
| D | Incorrectly had a negative number on the right side of the equation after completing the square |

# Preparing for Assessment

**46. MULTI-STEP** Consider $3x^2 - 24x = 51$. **MP 1, 7** A.SSE.3b, A.REI.4

a. What is the first step to solve the equation by completing the square? **divide the equation by 3**

b. What value should be added to each side to complete the square? **C**
- ○ A −16
- ○ B −4
- ○ C 16
- ○ D 144

c. Which of the following are solutions to the equation? Select all solutions. **C, E**
- ☐ A 3
- ☐ B $\frac{17}{3}$
- ☐ C $4 + \sqrt{33}$
- ☐ D $-4 + \sqrt{33}$
- ☐ E $4 - \sqrt{33}$
- ☐ F $-4 - \sqrt{33}$

d. Which is a key feature of the graph of the equation? **A**
- ○ A The axis of symmetry is $x = 4$.
- ○ B The vertex is $(-4, -33)$.
- ○ C The graph opens down.
- ○ D The minimum value is 33.

**47.** Which of the following is least likely to be a step in solving the equation $x^2 + 10x + 10 = 66$ by completing the square? **MP 3, 7** A.SSE.3b, A.REI.4 **B**
- ○ A Subtract 10 from both sides of the equation.
- ○ B Subtract 25 from both sides of the equation.
- ○ C Add $\left(\frac{10}{2}\right)^2$ to both sides of the equation.
- ○ D Factor $x^2 + 10x + 25$.

**48.** Which equation has no real solutions? **MP 1, 7** A.SSE.3b, A.REI.4 **D**
- ○ A $x^2 - 2x = 2$
- ○ B $x^2 - 2x = 3$
- ○ C $x^2 - 3x = -2$
- ○ D $x^2 + 3x = -3$

**49.** Which equation has solutions of −1 and 3? **MP 1, 7** A.SSE.3b, A.REI.4 **C**
- ○ A $x^2 - x - 7 = -4$
- ○ B $x^2 - 2x + 7 = -4$
- ○ C $x^2 - 2x - 7 = -4$
- ○ D $2x^2 - 4x - 14 = -4$

**50.** What value should be added to $x^2 - 20$ in order to complete the square? **MP 2** F.IF.8a

**100**

**51.** Solve $-3x + 30x - 72 = 0$ by completing the square. **MP 2** F.IF.8a

**4, 6**

**52.** Determine whether the following statement is *true* or *false*. Justify your answer. **MP 3** F.IF.8a **See margin.**

*To complete the square for $x^2 + bx$, add $\frac{b}{2}$ to the expression.*

**53.** Chris solved the equation $x^2 - 6x - 2 = 6$ by completing the square. Which best describes the solution or solutions? **MP 6** A.REI.4 **C**
- ○ A Rounded to the nearest tenth, the solutions are −4.1 and 4.1.
- ○ B The solution is 3.
- ○ C Rounded to the nearest tenth, the solutions are −1.1 and 7.1.
- ○ D This equation has no real solutions.

### Differentiated Instruction BL

**Extension** Have students solve $\frac{1}{3}x^2 - \frac{7}{6}x + \frac{1}{2} = 0$ by completing the square. Ask them how this strategy compares to factoring and graphing. $\frac{1}{2}$, 3; the equation can be solved more easily by factoring. Graphing may not produce an exact answer.

### Additional Answer

**52.** False. To complete the square, you need to square half of $b$ and add it to $x^2 + bx$. So, the $b^2$ term on the left is incorrect, and the equation should be $x^2 + bx + \left(\frac{b}{2}\right)^2 = \left(x + \frac{b}{2}\right)^2$.

# EXTEND 9-5

## Algebra Lab
# Finding the Maximum or Minimum Value

In Lesson 9-5, we learned about the vertex form of the equation of a quadratic function. You will now learn how to write equations in vertex form and use them to identify key characteristics of the graphs of quadratic functions.

**Mathematical Practices**
MP 8 Look for and express regularity in repeated reasoning

**Content Standards**
**A.SSE.3b** Complete the square in a quadratic expression to reveal the maximum or minimum value of the function it defines.
**F.IF.8a** Use the process of factoring and completing the square in a quadratic function to show zeros, extreme values, and symmetry of the graph, and interpret these in terms of a context.

### Activity 1  Find a Minimum

Work cooperatively. Write $y = x^2 + 4x - 10$ in vertex form. Identify the axis of symmetry, extrema, and zeros. Then graph the function.

**Step 1** Complete the square to write the function in vertex form.

| | |
|---|---|
| $y = x^2 + 4x - 10$ | Original function |
| $y + 10 = x^2 + 4x$ | Add 10 to each side. |
| $y + 10 + 4 = x^2 + 4x + 4$ | Since $\left(\frac{4}{2}\right)^2 = 4$, add 4 to each side. |
| $y + 14 = (x + 2)^2$ | Factor $x^2 + 4x + 4$. |
| $y = (x + 2)^2 - 14$ | Subtract 14 from each side to write in vertex form. |

**Step 2** Identify the axis of symmetry and extrema based on the equation in vertex form. The vertex is at $(h, k)$ or $(-2, -14)$. Since there is no negative sign before the $x^2$-term, the parabola opens up and has a minimum at $(-2, -14)$. The equation of the axis of symmetry is $x = -2$.

**Step 3** Solve for $x$ to find the zeros.

| | |
|---|---|
| $(x + 2)^2 - 14 = 0$ | Vertex form, $y = 0$ |
| $(x + 2)^2 = 14$ | Add 14 to each side. |
| $x + 2 = \pm\sqrt{14}$ | Take square root of each side. |
| $x \approx -5.74$ or $1.74$ | Subtract 2 from each side. |

The zeros are approximately $-5.74$ and $1.74$.

**Step 4** Use the key features to graph the function.

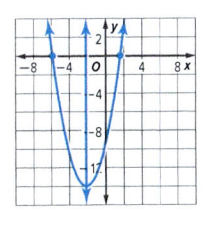

There may be a negative coefficient before the quadratic term. When this is the case, the parabola will open down and have a maximum.

### Activity 2  Find a Maximum

Work cooperatively. Write $y = -x^2 + 6x - 5$ in vertex form. Identify the axis of symmetry, extrema, and zeros. Then graph the function.

**Step 1** Complete the square to write the equation of the function in vertex form.

| | |
|---|---|
| $y = -x^2 + 6x - 5$ | Original function |
| $y + 5 = -x^2 + 6x$ | Add 5 to each side. |
| $y + 5 = -(x^2 - 6x)$ | Factor out $-1$. |
| $y + 5 - 9 = -(x^2 - 6x + 9)$ | Since $\left(\frac{6}{2}\right)^2 = 9$, add $-9$ to each side. |
| $y - 4 = -(x - 3)^2$ | Factor $x^2 - 6x + 9$. |
| $y = -(x - 3)^2 + 4$ | Add 4 to each side to write in vertex form. |

## Launch

**Objective** Complete the square in a quadratic expression to find the maximum or minimum value of the related function.

## Teach

**Working in Cooperative Groups** Divide the class into pairs. Work through Activity 1 as a class. Then ask students to work with their partners to complete Activities 2 and 3. **ELL**

**Practice** Have students complete Exercises 1–8 and 11.

## Go Online!

The most up-to-date resources available for your program can be found at connectED.mcgraw-hill.com.

# Extend 9-5

## Assess

**Formative Assessment**

Use Exercises 9 and 10 to assess each student's knowledge of vertex form and finding zeros, the line of symmetry, and extrema.

**From Concrete to Abstract**

Ask students to summarize how to write an equation in vertex form and find the zeros, the line of symmetry, and extrema.

## Additional Answers

1. Sample answer: In vertex form, the *x* only appears once. You must use completing the square to create a perfect square trinomial, so that it can be factored and reduce the *x*-terms to one.

11.

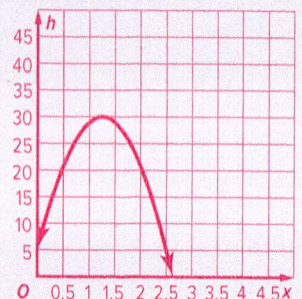

30 ft; $t \approx 1.25$; 2.62 seconds

---

# EXTEND 9-5
## Algebra Lab
### Finding the Maximum or Minimum Value *Continued*

**Step 2** Identify the axis of symmetry and extrema based on the equation in vertex form. The vertex is at $(h, k)$ or $(3, 4)$. Since there is a negative sign before the $x^2$-term, the parabola opens down and has a maximum at $(3, 4)$. The equation of the axis of symmetry is $x = 3$.

**Step 3** Solve for $x$ to find the zeros.

$0 = -(x - 3)^2 + 4$    Vertex form, $y = 0$
$(x - 3)^2 = 4$         Add $(x - 3)^2$ to each side.
$x - 3 = \pm 2$         Take the square root of each side.
$x = 5$ or $1$          Add 3 to each.

**Step 4** Use the key features to graph the function.

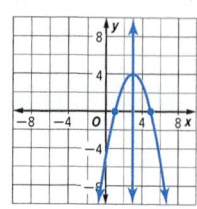

### Analyze the Results

Work cooperatively.

1. Why do you need to complete the square to write the equation of a quadratic function in vertex form? **See margin.**

Write each function in vertex form. Identify the axis of symmetry, extrema, and zeros. Then graph the function. **2–10. See Ch. 9 Answer Appendix.**

2. $y = x^2 + 6x$
3. $y = x^2 - 8x + 6$
4. $y = x^2 + 2x - 12$
5. $y = x^2 + 6x + 8$
6. $y = x^2 - 4x + 3$
7. $y = x^2 - 2.4x - 2.2$
8. $y = -4x^2 + 16x - 11$
9. $y = 3x^2 - 12x + 5$
10. $y = -x^2 + 6x - 5$

### Activity 3  Use Extrema in the Real World

**DIVING** Alexis jumps from a diving platform upward and outward before diving into the pool. The function $h = -9.8t^2 + 4.9t + 10$, where $h$ is the height of the diver in meters above the pool after $t$ seconds approximates Alexis's dive. Graph the function, then find the maximum height that she reaches and the equation of the axis of symmetry.

**Step 1** Graph the function.

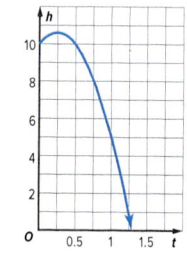

**Step 2** Complete the square to write the eqution of the function in vertex form.
$h = -9.8t^2 + 4.9t + 10$
$h = -9.8(t - 0.25)^2 + 10.6125$

**Step 3** The vertex is at $(0.25, 10.6125)$, so the maximum height is $10.6125$ meters. The equation of the axis of symmetry is $x = 0.25$.

### Exercises

11. **SOFTBALL** Jenna throws a ball in the air. The function $h = -16t^2 + 40t + 5$, where $h$ is the height in feet and $t$ represents the time in seconds, approximates Jenna's throw. Graph the function, then find the maximum height of the ball and the equation of the axis of symmetry. When does the ball hit the ground? **See margin.**

604 | Extend 9-5 | Algebra Lab: Finding the Maximum or Minimum Value

# CHAPTER 9
## Mid-Chapter Quiz
Lessons 9-1 through 9-5

Use a table of values to graph each equation. State the domain and range. (Lesson 9-1)
**1–4. See Ch. 9 Answer Appendix.**
1. $y = x^2 + 3x + 1$
2. $y = 2x^2 - 4x + 3$
3. $y = -x^2 - 3x - 3$
4. $y = -3x^2 - x + 1$

Consider $y = x^2 - 5x + 4$. (Lesson 9-1)

5. Write the equation of the axis of symmetry. **$x = 2.5$**
6. Find the coordinates of the vertex. Is it a maximum or minimum point? **(2.5, −2.25); minimum**
7. Graph the function. **See Ch. 9 Answer Appendix.**
8. **SOCCER** A soccer ball is kicked from ground level with an initial upward velocity of 90 feet per second. The equation $h = -16t^2 + 90t$ gives the height $h$ of the ball after $t$ seconds. (Lesson 9-1)
   a. What is the height of the ball after one second? **74 ft**
   b. How many seconds will it take for the ball to reach its maximum height? **2.8125 s**
   c. When is the height of the ball 0 feet? What do these points represent in this situation?
   d. **MP** What mathematical practice did you use to solve this problem? **See students' work.**

Describe how the graph of each function is related to the graph of $f(x) = x^2$. (Lesson 9-2)
**8c. $t = 0$, $t = 5.625$; before the ball is kicked, and when the ball hits the ground after the kick**
9. $g(x) = x^2 + 3$ **translated up 3 units**
10. $h(x) = 2x^2$ **stretched vertically**
11. $g(x) = x^2 - 6$ **translated down 6 units**
12. $h(x) = \frac{1}{5}x^2$ **compressed vertically**
13. $g(x) = -x^2 + 1$ **reflected over x-axis and translated up 1 unit**
14. $h(x) = -\frac{5}{8}x^2$ **reflected over x-axis and compressed vertically**

15. **CONSTRUCTION** Christopher is repairing the roof on a shed. He accidentally dropped a box of nails from a height of 14 feet. This is represented by the equation $h = -16t^2 + 14$, where $h$ is the height in feet and $t$ is the time in seconds. Describe how the graph is related to $h = t^2$. (Lesson 9-2)
**15. compressed vertically and shifted up 14 units**

16. **MULTIPLE CHOICE** Which is an equation for the function shown in the graph? (Lesson 9-2) **D**

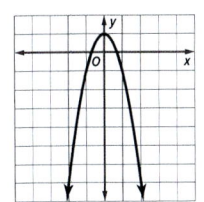

   A $y = -2x^2$
   B $y = 2x^2 + 1$
   C $y = x^2 - 1$
   D $y = -2x^2 + 1$

Solve each equation by graphing. If integral roots cannot be found, estimate the roots to the nearest tenth. (Lesson 9-3)
17. $x^2 + 5x + 6 = 0$ **−3, −2**
18. $x^2 + 8 = -6x$ **−4, −2**
19. $-x^2 + 3x - 1 = 0$ **0.4, 2.6**
20. $x^2 = 12$ **−3.5, 3.5**

21. **PARTIES** Della's parents are throwing a Sweet 16 party for her. At 10:00, a ball will slide 25 feet down a pole and light up. A function that models the drop is $h = -t^2 + 5t + 25$, where $h$ is height in feet of the ball after $t$ seconds. How many seconds will it take for the ball to reach the bottom of the pole? (Lesson 9-3) **≈8.1 s**

Solve each equation by factoring. (Lesson 9-4)
22. $x^2 + 14x + 49 = 0$ **−7**
23. $x^2 - 2x = 48$ **−6, 8**
24. $2x^2 + 5x - 12 = 0$ **−4**

Solve each equation by completing the square. Round to the nearest tenth. (Lesson 9-5)
25. $x^2 + 4x + 2 = 0$ **−3.4, −0.6**
26. $x^2 - 2x - 10 = 0$ **−2.3, 4.3**
27. $2x^2 + 4x - 5 = 7$ **−3.6, 1.6**

## Foldables Study Organizer
### Dinah Zike's FOLDABLES

Before students complete the Mid-Chapter Quiz, encourage them to review the information for Lessons 9-1 through 9-5 in their Foldables. Students may benefit from sharing their Foldable with a partner and taking turns summarizing what they have learned about quadratic functions and equations, while the other partner listens carefully. They should seek clarification of any concepts, as needed.

**ALEKS** can be used as a formative assessment tool to target learning gaps for those who are struggling, while providing enhanced learning for those who have mastered the concepts.

## RtI Response to Intervention
Use the Intervention Planner to help you determine your Response to Intervention.

### Intervention Planner

**TIER 1 On Level OL**

**IF** students miss 25% of the exercises or less,
**THEN** choose a resource:
- SE Lessons 9-1 through 9-5

**Go Online!**
- Skills Practice
- Chapter Project
- Self-Check Quizzes

**TIER 2 Strategic Intervention AL**
Approaching grade level

**IF** students miss 50% of the exercises,
**THEN** choose a resource:

**Go Online!**
- Study Guide and Intervention
- Extra Examples
- Personal Tutors
- Homework Help

**TIER 3 Intensive Intervention**
2 or more grades below level

**IF** students miss 75% of the exercises,
**THEN** choose a resource:
- Use *Math Triumphs, Alg. 1*

**Go Online!**
- Extra Examples
- Personal Tutors
- Homework Help
- Review Vocabulary

## Go Online!
### eAssessment

You can use the premade Mid-Chapter Test to assess students' progress in the first half of the chapter. Customize and create multiple versions of your Mid-Chapter Quiz and answer keys that align to your standards can be delivered on paper or online.

**LESSON 9-6**
# Solving Quadratic Equations by Using the Quadratic Formula

**SUGGESTED PACING (DAYS)**
- 90 min. | 1
- 45 min. | 1.5

Instruction

## Track Your Progress

### Objectives
1. Solve quadratic equations by using the Quadratic Formula.
2. Use the discriminant to determine the number of solutions of a quadratic equation.

### Mathematical Background
Even though the Quadratic Formula may not be the easiest way to solve some quadratic equations, it always works. The expression under the radical sign is called the discriminant. The discriminant can be used to determine the number of real solutions of a quadratic equation.

### THEN
**F.IF.8a** Use the process of factoring and completing the square in a quadratic function to show zeros, extreme values, and symmetry of the graph, and interpret these in terms of a context.

**A.SSE.3b** Complete the square in a quadratic expression to reveal the maximum or minimum value of the function it defines.

### NOW
**A.CED.1** Create equations and inequalities in one variable and use them to solve problems. Include equations arising from linear and quadratic functions, and simple rational and exponential functions.

**A.REI.4a** Use the method of completing the square to transform any quadratic equation in $x$ into an equation of the form $(x - p)^2 = q$ that has the same solutions. Derive the quadratic formula from this form.

### NEXT
**A.CED.2** Create equations in two or more variables to represent relationships between quantities; graph equations on coordinate axes with labels and scales.

**A.REI.7** Solve a simple system consisting of a linear equation and a quadratic equation in two variables algebraically and graphically.

---

**Go Online!** All of these resources and more are available at connectED.mcgraw-hill.com

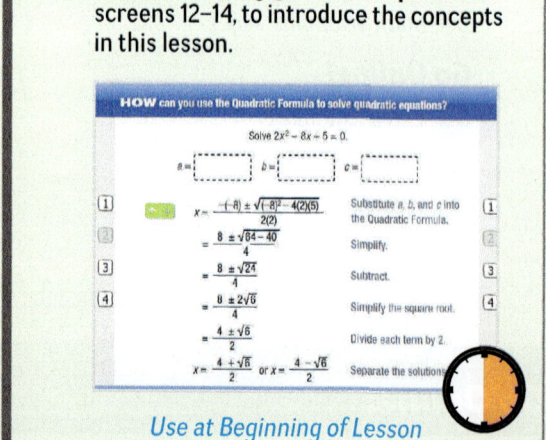

**eLessons** utilize the power of your interactive whiteboard in an engaging way. Use **Solving Quadratic Equations**, screens 12–14, to introduce the concepts in this lesson.

*Use at Beginning of Lesson*

**Personal Tutors** (for every example) let students hear real teachers solve problems. Students can pause and repeat as many times as necessary.

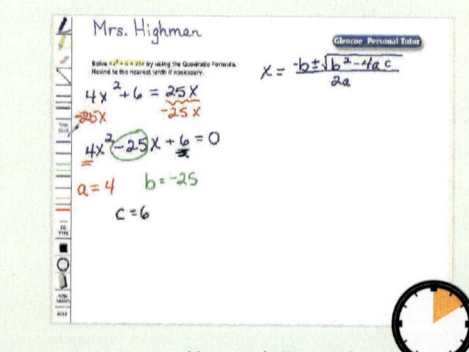

*Use with Examples*

**eToolkit** contains outstanding tools for enhancing understanding.

*Use with Examples*

---

**OER Using Open Educational Resources**
**Video Sharing** Have students work in groups to record and upload songs on **SchoolTube** about solving quadratic equations using the quadratic formula to check their understanding of the solution process. Have them watch videos of others for ideas. If you are unable to access **SchoolTube**, try **KidsTube**, **MathATube**, **YouTube**, or **TeacherTube**. *Use as homework*

# Differentiate Your Resources

**Extra Practice** Additional practice or homework; Skills Practice is best for approaching-level students and Practice is best for on-level and beyond-level students

## Skills Practice

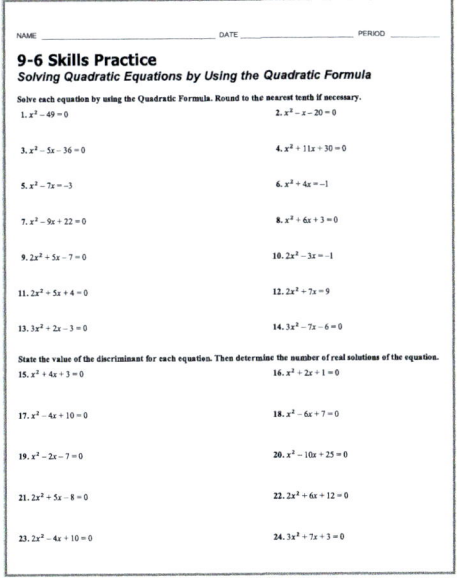

## Practice

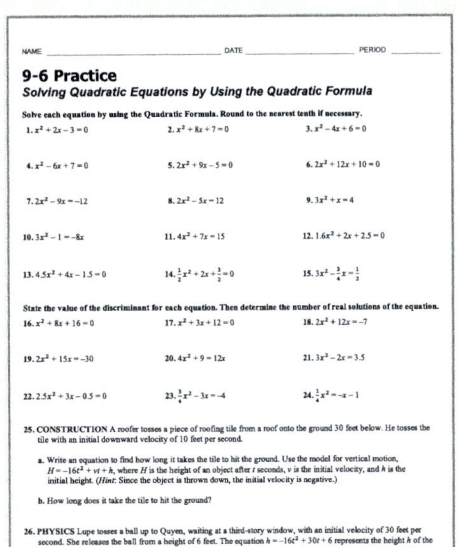

## Word Problem Practice

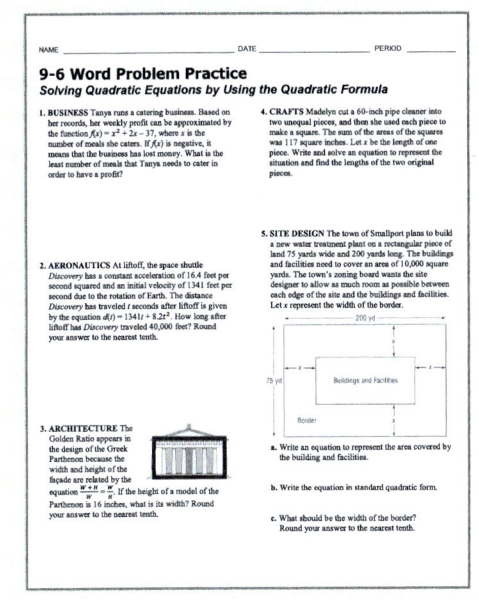

**Intervention** Reteaching and vocabulary activities that can be used with struggling or absent students and as ELL support

## Study Guide and Intervention

## Study Notebook

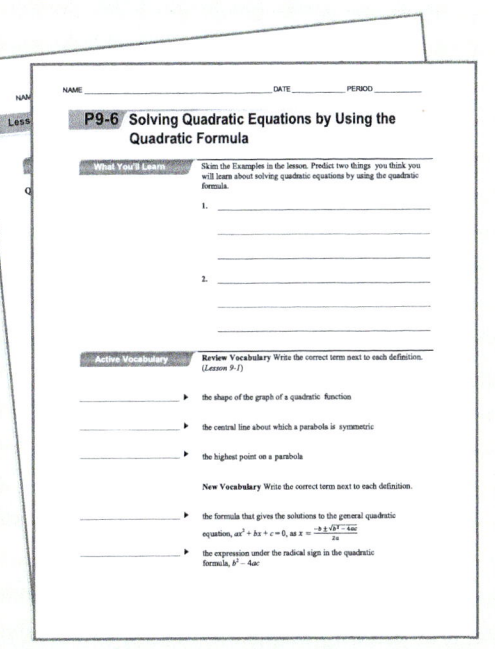

**Extension** Activities that can be used to extend lesson concepts

## Enrichment

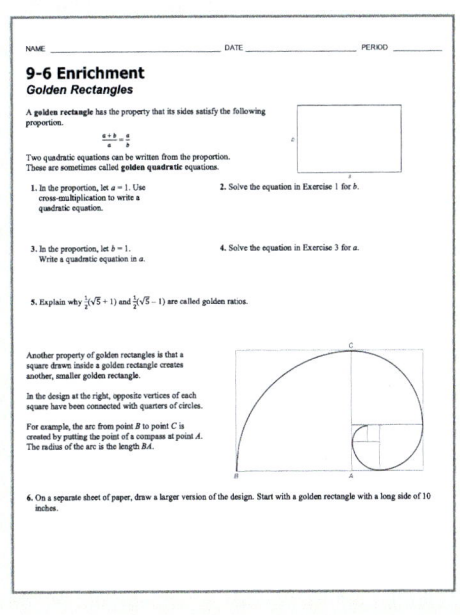

Lesson 9-6 | Solving Quadratic Equations by Using the Quadratic Formula

# Launch

Have students read the Why? section of the lesson. Ask:
- What equation would you need to solve to find the age of a woman whose systolic blood pressure is 120? $120 = 0.01a^2 + 0.05a + 107$
- Write this equation in standard form. $0 = 0.01a^2 + 0.05a - 13$
- What are the values for $a$, $b$, and $c$? $a = 0.01$, $b = 0.05$, $c = -13$
- Why would the equation be difficult to solve using factoring or completing the square? The decimal values of the coefficients may make using the methods difficult, if not impossible.

# Teach

Ask the scaffolded questions for each example to build conceptual understanding for students at all levels.

## 1 Quadratic Formula

**Example 1 Use the Quadratic Formula**

**AL** What does the ± symbol tell us to do? both add and subtract

**OL** In part a, what are the values of $a$, $b$, and $c$ in the equation? $a = 1$; $b = -12$; $c = 20$

**BL** Of the methods you have learned to solve quadratic equations, which do you prefer? Explain. See students' work.

**Need Another Example?**
Solve $x^2 - 2x = 35$ by using the Quadratic Formula. $-5, 7$

## Go Online!

**Interactive Whiteboard**
Use the eLesson or Lesson Presentation to present this lesson.

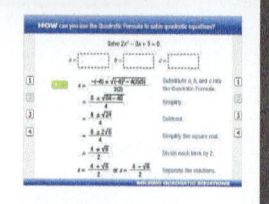

---

# LESSON 6
# Solving Quadratic Equations by Using the Quadratic Formula

**Then** You solved quadratic equations by completing the square.

**Now**
1. Solve quadratic equations by using the Quadratic Formula.
2. Use the discriminant to determine the number of solutions of a quadratic equation.

**Why?** For adult women, the normal systolic blood pressure $P$ in millimeters of mercury (mm Hg) can be modeled by $P = 0.01a^2 + 0.05a + 107$, where $a$ is age in years. This equation can be used to approximate the age of a woman with a certain systolic blood pressure. However, it would be difficult to solve by factoring, graphing, or completing the square.

**New Vocabulary**
Quadratic Formula

**Mathematical Practices**
6 Attend to precision.

**Content Standards**
A.REI.4 Solve quadratic equations in one variable.
a. Use the method of completing the square to transform any quadratic equation in $x$ into an equation of the form $(x - p)^2 = q$ that has the same solutions. Derive the quadratic formula from this form.
b. Solve quadratic equations by inspection (e.g., for $x^2 = 49$), taking square roots, completing the square, the quadratic formula and factoring, as appropriate to the initial form of the equation. Recognize when the quadratic formula gives complex solutions and write them as $a \pm bi$ for real numbers $a$ and $b$.
A.CED.1 Create equations and inequalities in one variable and use them to solve problems.

**1 Quadratic Formula** By completing the square for the quadratic equation $ax^2 + bx + c = 0$, $a \neq 0$, you can derive the Quadratic Formula. The **Quadratic Formula** can be used to find the solutions of any quadratic equation.

$ax^2 + bx + c = 0$    Standard quadratic equation

$x^2 + \frac{b}{a}x + \frac{c}{a} = 0$    Divide each side by $a$.

$x^2 + \frac{b}{a}x = -\frac{c}{a}$    Subtract $\frac{c}{a}$ from each side.

$x^2 + \frac{b}{a}x + \left(\frac{b}{2a}\right)^2 = -\frac{c}{a} + \left(\frac{b}{2a}\right)^2$    Complete the square by adding $\left(\frac{b}{2a}\right)^2$ to each side.

$x^2 + \frac{b}{a}x + \left(\frac{b}{2a}\right)^2 = -\frac{c}{a} + \frac{b^2}{4a^2}$    Simplify the right side.

$\left(x + \frac{b}{2a}\right)^2 = -\frac{c}{a} + \frac{b^2}{4a^2}$    Factor the left side.

$\left(x + \frac{b}{2a}\right)^2 = \frac{b^2 - 4ac}{4a^2}$    Find a common denominator and simplify the right side.

$x + \frac{b}{2a} = \pm \frac{\sqrt{b^2 - 4ac}}{2a}$    Square Root Property

$x = -\frac{b}{2a} \pm \frac{\sqrt{b^2 - 4ac}}{2a}$    Subtract $\frac{b}{2a}$ from each side.

$x = \frac{-b \pm \sqrt{b^2 - 4ac}}{2a}$    Simplify.

The equation $x = \frac{-b \pm \sqrt{b^2 - 4ac}}{2a}$ is the Quadratic Formula.

**Key Concept** The Quadratic Formula

The solutions of a quadratic equation $ax^2 + bx + c = 0$, where $a \neq 0$, are given by the Quadratic Formula.    $x = \frac{-b \pm \sqrt{b^2 - 4ac}}{2a}$

---

**MP Mathematical Practices Strategies**

**Look for and express regularity in repeated reasoning.**
Help students understand how to use the Quadratic Formula to solve quadratic equations while maintaining oversight of the process.

- Why is the symbol ± necessary in the Quadratic Formula? The formula is derived by taking the square root of each side of an equation. The ± symbol represents the postive and negative roots.

- How does finding the discriminant help when solving a quadratic equation? The discriminant tells you whether there are 0, 1, or 2 real solutions. Knowing the number of solutions can help you maintain oversight of the process and check your answer(s).

- What is an advantage of using the Quadratic Formula? It works for every quadratic equation and gives exact solutions.

**Example 1** Use the Quadratic Formula

Solve each equation by using the Quadratic Formula.

**a.** $x^2 - 12x = -20$

**Step 1** Rewrite the equation in standard form.

$x^2 - 12x = -20$    Original equation

$x^2 - 12x + 20 = 0$    Add 20 to each side.

**Step 2** Apply the Quadratic Formula.

$x = \dfrac{-b \pm \sqrt{b^2 - 4ac}}{2a}$    Quadratic Formula

$= \dfrac{-(-12) \pm \sqrt{(-12)^2 - 4(1)(20)}}{2(1)}$    $a = 1, b = -12,$ and $c = 20$

$= \dfrac{12 \pm \sqrt{144 - 80}}{2}$    Multiply.

$= \dfrac{12 \pm \sqrt{64}}{2}$ or $\dfrac{12 \pm 8}{2}$    Subtract and take the square root.

$x = \dfrac{12 - 8}{2}$ or $x = \dfrac{12 + 8}{2}$    Separate the solutions.

$= 2$    $= 10$    The solutions are 2 and 10.

**b.** $x^2 - 6x = 7$

**Step 1** Rewrite the equation in standard form.

$x^2 - 6x = 7$    Original equation

$x^2 - 6x - 7 = 0$    Subtract 7 from each side.

**Step 2** Apply the Quadratic Formula.

$x = \dfrac{-b \pm \sqrt{b^2 - 4ac}}{2a}$    Quadratic Formula

$= \dfrac{-(-6) \pm \sqrt{(-6)^2 - 4(1)(7)}}{2(1)}$    $a = 1, b = -6,$ and $c = -7$

$= \dfrac{6 \pm \sqrt{36 - 28}}{2}$    Multiply.

$= \dfrac{6 \pm \sqrt{64}}{2}$ or $\dfrac{6 \pm 8}{2}$    Subtract and take the square root.

$x = \dfrac{6 + 8}{2}$ or $x = \dfrac{6 - 8}{2}$    Separate the solutions.

$x = 7$ or $x = -1$    The solutions are −1 and 7.

**Guided Practice**

**1A.** $2x^2 + 9x = 18$   $-6, \dfrac{3}{2}$

**1B.** $-3x^2 + 5x = -2$   $2, -\dfrac{1}{3}$

**1C.** $x^2 + 14x + 40 = 0$   $-4, -10$

**1D.** $3x^2 + 2x - 1 = 0$   $-1, \dfrac{1}{3}$

---

**Lesson 9-6** | Solving Quadratic Equations by Using the Quadratic Formula

### Teaching the Mathematical Practices

**Precision** Mathematically proficient students pay attention to details. Common errors when evaluating the quadratic formula by using a calculator include using incorrect signs, omitting a grouping symbol, or substituting incorrect values for $a$, $b$, or $c$. Encourage students to start by writing down the values of $a$, $b$, and $c$ with the appropriate positive or negative sign. Then use those values to find $b^2 - 4ac$. Doing this step with pencil and paper can help students determine if solutions found using a calculator are reasonable.

**Lesson 9-6 | Solving Quadratic Equations by Using the Quadratic Formula**

## Example 2 Use the Quadratic Formula

**AL** What are the values of $a$, $b$, and $c$ in the equation for part a? $a = 3$; $b = 5$; $c = -12$

**OL** In part a, are the solutions exact or approximations? exact in part b? approximations

**BL** What are the exact solutions of the equation in part b? $\dfrac{5 + \sqrt{1025}}{20}$ and $\dfrac{5 - \sqrt{1025}}{20}$

**Need Another Example?**
Solve each equation by using the Quadratic Formula. Round to the nearest tenth if necessary.
a. $2x^2 - 2x - 5 = 0$  2.2, −1.2
b. $5x^2 - 8x = 4$  2, −0.4

---

**Watch Out!**

**Preventing Errors** Some students may notice that the trinomial in Example 1 can be factored. Explain that an equation that could be factored was used to demonstrate that the Quadratic Formula produces the correct solutions. Tell students that if they see an easier way (such as factoring) to solve a quadratic equation, they should use the easier method.

---

**Go Online!**
You will want to reference the Quadratic Formula often. Log into your eStudent Edition to bookmark this lesson.

**Study Tip**

**MP Precision** In Example 2, the number $\sqrt{1025}$ is irrational, so the calculator can only give you an approximation of its value. So, the exact answer in Example 2 is $\dfrac{5 \pm \sqrt{1025}}{20}$. The numbers −1.4 and 1.9 are approximations.

---

The solutions of quadratic equations are not always integers.

A.REI.4b

**Example 2** Use the Quadratic Formula

Solve each equation by using the Quadratic Formula. Round to the nearest tenth if necessary.

a. $3x^2 + 5x - 12 = 0$

For this equation, $a = 3$, $b = 5$, and $c = -12$.

$x = \dfrac{-b \pm \sqrt{b^2 - 4ac}}{2a}$  Quadratic Formula

$= \dfrac{-(5) \pm \sqrt{(5)^2 - 4(3)(-12)}}{2(3)}$  $a = 3, b = 5,$ and $c = -12$

$= \dfrac{-5 \pm \sqrt{25 + 144}}{6}$  Multiply.

$= \dfrac{-5 \pm \sqrt{169}}{6}$ or $\dfrac{-5 \pm 13}{6}$  Add and simplify.

$x = \dfrac{-5 - 13}{6}$ or $x = \dfrac{-5 + 13}{6}$  Separate the solutions.

$= -3 \qquad \qquad = \dfrac{4}{3}$  Simplify.

The solutions are $-3$ and $\dfrac{4}{3}$.

b. $10x^2 - 5x = 25$

**Step 1** Rewrite the equation in standard form.

$10x^2 - 5x = 25$  Original equation

$10x^2 - 5x - 25 = 0$  Subtract 25 from each side.

**Step 2** Apply the Quadratic Formula.

$x = \dfrac{-b \pm \sqrt{b^2 - 4ac}}{2a}$  Quadratic Formula

$= \dfrac{-(-5) \pm \sqrt{(-5)^2 - 4(10)(-25)}}{2(10)}$  $a = 10, b = -5,$ and $c = -25$

$= \dfrac{5 \pm \sqrt{25 + 1000}}{20}$  Multiply.

$= \dfrac{5 \pm \sqrt{1025}}{20}$  Add.

$= \dfrac{5 - \sqrt{1025}}{20}$ or $\dfrac{5 + \sqrt{1025}}{20}$  Separate the solutions.

$\approx -1.4 \qquad \approx 1.9$  Simplify.

The solutions are about $-1.4$ and $1.9$.

▶ **Guided Practice**

2A. $4x^2 - 24x + 35 = 0$  $\dfrac{7}{2}, \dfrac{5}{2}$

2B. $3x^2 - 2x - 9 = 0$  2.1, −1.4

You can solve quadratic equations by using one of many equivalent methods. No one way is always best.

## Lesson 9-6 | Solving Quadratic Equations by Using the Quadratic Formula

**Watch Out!**

**Solutions** Remember that the methods of factoring, completing the square, or using the quadratic formula to solve quadratic equations are equivalent. Each method should produce the same solution.

### Example 3  Solve Quadratic Equations Using Different Methods

A.REI.4a, b

Solve $x^2 - 4x = 12$.

**Method 1** Graphing

Rewrite the equation in standard form.

$x^2 - 4x = 12$    Original equation

$x^2 - 4x - 12 = 0$    Subtract 12 from each side.

Graph the related function $f(x) = x^2 - 4x - 12$. Locate the $x$-intercepts of the graph.

The solutions are $-2$ and $6$.

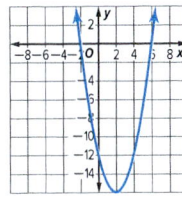

**Method 2** Factoring

$x^2 - 4x = 12$    Original equation

$x^2 - 4x - 12 = 0$    Subtract 12 from each side.

$(x - 6)(x + 2) = 0$    Factor.

$x - 6 = 0$ or $x + 2 = 0$    Zero Product Property

$x = 6 \qquad x = -2$    Solve for $x$.

**Method 3** Completing the Square

The equation is in the correct form to complete the square, since the leading coefficient is 1 and the $x^2$ and $x$ terms are isolated.

$x^2 - 4x = 12$    Original equation

$x^2 - 4x + 4 = 12 + 4$    Since $\left(\frac{-4}{2}\right)^2 = 4$, add 4 to each side.

$(x - 2)^2 = 16$    Factor $x^2 - 4x + 4$.

$x - 2 = \pm 4$    Take the square root of each side.

$x = 2 \pm 4$    Add 2 to each side.

$x = 2 + 4$ or $x = 2 - 4$    Separate the solutions.

$= 6 \qquad\qquad = -2$    Simplify.

**Method 4** Quadratic Formula

From Method 1, the standard form of the equation is $x^2 - 4x - 12 = 0$.

$x = \dfrac{-b \pm \sqrt{b^2 - 4ac}}{2a}$    Quadratic Formula

$= \dfrac{-(-4) \pm \sqrt{(-4)^2 - 4(1)(-12)}}{2(1)}$    $a = 1, b = -4,$ and $c = -12$.

$= \dfrac{4 \pm \sqrt{16 + 48}}{2}$    Multiply.

$= \dfrac{4 \pm \sqrt{64}}{2}$ or $\dfrac{4 \pm 8}{2}$    Add and simplify.

$x = \dfrac{4 - 8}{2}$ or $x = \dfrac{4 + 8}{2}$    Separate the solutions.

$= -2 \qquad\qquad = 6$    Simplify.

### Guided Practice

Solve each equation.

**3A.** $2x^2 - 17x + 8 = 0$    $8, \dfrac{1}{2}$

**3B.** $4x^2 - 4x - 11 = 0$    about $2.2, -1.2$

---

### Teaching the Mathematical Practices

**Precision** Mathematically proficient students express numerical answers with a degree of precision appropriate for the problem context. Discuss with students how to know when an exact answer is required.

### Teaching Tip

**Using the Quadratic Formula** Tell students that when they use the Quadratic Formula, it is best to simplify one step at a time. For example, students should first simplify under the radical sign, then find the square root, then simplify the numerator and denominator, and finally perform the division. Skipping steps may introduce errors.

### Example 3  Solve Quadratic Equations Using Different Methods

**AL** What are the four methods we have learned for solving quadratic equations? graphing, factoring, completing the square, and the Quadratic Formula

**OL** Does each method produce the same solution? yes

**BL** Give examples of when you would use each method and explain why you would choose each method. See students' work.

### Need Another Example?

Solve $3x^2 - 5x = 12$.   $3, -\dfrac{4}{3}$

connectED.mcgraw-hill.com    **609**

**Lesson 9-6** | Solving Quadratic Equations by Using the Quadratic Formula

## Teaching Tip

**Sense-Making** While the chart on this page offers suggestions for when to use each method, do not hold students to using all methods. Some students do not yet have the mathematical maturity to analyze each equation and determine the best method to use to save time.

## 2 The Discriminant

**Example 4 Use the Discriminant**

**AL** What is the formula for the discriminant?
$b^2 - 4ac$

**OL** What can the discriminant tell us about the solutions of the quadratic equation? The discriminant can tell us the number of real solutions.

**BL** Write an equation with a positive discriminant? See students' work.

**Need Another Example?**
State the value of the discriminant of $3x^2 + 10x = 12$. Then determine the number of real solutions of the equation. 244; 2 real solutions

---

**Watch Out!**

**Preventing Errors** Remind students to be careful to include any negative signs when finding the discriminant. One missed negative sign can turn the discriminant from positive to negative and yield an incorrect result.

---

**Concept Summary** Solving Quadratic Equations

| Method | When to Use |
|---|---|
| Factoring | Use when the constant term is 0 or if the factors are easily determined. Not all equations are factorable. |
| Graphing | Use when an approximate solution is sufficient. |
| Using Square Roots | Use when an equation can be written in the form $x^2 = n$. Can only be used if the equation has no $x$-term. |
| Completing the Square | Can be used for any equation $ax^2 + bx + c = 0$, but is simplest to apply when $b$ is even and $a = 1$. |
| Quadratic Formula | Can be used for any equation $ax^2 + bx + c = 0$. |

**2 The Discriminant** In the Quadratic Formula, the expression under the radical sign, $b^2 - 4ac$, is called the **discriminant**. The discriminant can be used to determine the number of real solutions of a quadratic equation.

**Study Tip**

**Sense-Making** Recall that when the left side of the standard form of an equation is a perfect square trinomial, there is only one solution. Therefore, the discriminant of a perfect square trinomial will always be zero.

**Key Concept** Using the Discriminant

| Equation | $x^2 + 2x + 5 = 0$ | $x^2 + 10x + 25 = 0$ | $2x^2 - 7x + 2 = 0$ |
|---|---|---|---|
| Discriminant | $b^2 - 4ac = -16$ negative | $b^2 - 4ac = 0$ zero | $b^2 - 4ac = 33$ positive |
| Graph of Related Function | 0 $x$-intercepts | 1 $x$-intercept | 2 $x$-intercepts |
| Real Solutions | 0 | 1 | 2 |

A.REI.4b

**Example 4** Use the Discriminant

State the value of the discriminant of $4x^2 + 5x = -3$. Then determine the number of real solutions of the equation.

**Step 1** Rewrite in standard form. $4x^2 - 5x = -3 \longrightarrow 4x^2 - 5x + 3 = 0$

**Step 2** Find the discriminant.

$b^2 - 4ac = (-5)^2 - 4(4)(3)$    $a = 4, b = -5,$ and $c = 3$
$= -23$    Simplify.

Since the discriminant is negative, the equation has no real solutions.

**Guided Practice**    **4A.** 1; two real solutions    **4B.** 0; one real solution
**4A.** $2x^2 + 11x + 15 = 0$    **4B.** $9x^2 - 30x + 25 = 0$

---

**Differentiated Instruction** AL OL BL ELL

**IF** you think students would benefit from being creative in this lesson,

**THEN** have students come up with a riddle, poem, or some other mnemonic device to help them remember the Quadratic Formula and how to use the discriminant to determine the number of real solutions. Ask students to share with the class.

## Lesson 9-6 | Solving Quadratic Equations by Using the Quadratic Formula

### Check Your Understanding
= Step-by-Step Solutions begin on page R11.

**Examples 1–2** Solve each equation by using the Quadratic Formula. Round to the nearest tenth.
A.REI.4b
1. $x^2 - 2x - 15 = 0$  −3, 5
2. $x^2 - 10x + 16 = 0$  2, 8
3. $x^2 - 8x = -10$  6.4, 1.6
4. $x^2 + 3x = 12$  2.3, −5.3
5. $10x^2 - 31x + 15 = 0$  0.6, 2.5
6. $5x^2 + 5 = -13x$  −0.5, −2.1

**Example 3** Solve each equation. State which method you used. See students' work for method.
A.REI.4a, b
7. $2x^2 + 11x - 6 = 0$  $-6, \frac{1}{2}$
8. $2x^2 - 3x - 6 = 0$  2.6, −1.1
9. $9x^2 = 25$  $\pm\frac{5}{3}$
10. $x^2 - 9x = -19$  5.6, 3.4

**Example 4** State the value of the discriminant for each equation. Then determine the number of real solutions of the equation.
A.REI.4b
11. $x^2 - 9x + 21 = 0$  −3; no real solutions
12. $2x^2 - 11x + 10 = 0$  41; two real solutions
13. $9x^2 + 24x = -16$  0; one real solution
14. $3x^2 - x = 8$  97; two real solutions

15. **JAGUARUNDI** A jaguarundi springs from a fence post to swat at a low flying bird. Her height $h$ in feet can be modeled by the equation $h = -16t^2 + 22.3t + 2$, where $t$ is time in seconds. Use the discriminant to determine if the jaguarundi will reach the bird if the bird is flying at a height of 10 feet. Explain.

15. The discriminant is −14.71, so the equation has no real solution. Thus, the jaguarundi will not reach a height of 10 feet.

### Practice and Problem Solving
Extra Practice is on page R9.

**Examples 1–2** Solve each equation by using the Quadratic Formula. Round to the nearest tenth if necessary.
A.REI.4b
16. $4x^2 + 5x - 6 = 0$  $-2, \frac{3}{4}$
17. $x^2 + 16 = 0$  ∅
18. $6x^2 - 12x + 1 = 0$  1.9, 0.1
19. $5x^2 - 8x = 6$  2.2, −0.6
20. $2x^2 - 5x = -7$  ∅
21. $5x^2 + 21x = -18$  $-3, -\frac{6}{5}$
22. $81x^2 = 9$  $\pm\frac{1}{3}$
23. $8x^2 + 12x = 8$  0.5, −2
24. $4x^2 = -16x - 16$  −2
25. $10x^2 = -7x + 6$  0.5, −1.2
26. $-3x^2 = 8x - 12$  1.1, −3.7
27. $2x^2 = 12x - 18$  3

28. **SQUIRRELS** A flying squirrel drops 60 feet from a tree before leveling off. A function that approximates this drop is $h = -16t^2 + 60$, where $h$ is the distance it drops in feet and $t$ is the time in seconds. About how many seconds does it take for the squirrel to drop 60 feet?  about 1.9 seconds

**Example 3** Solve each equation. State which method you used. See students' work for method.
A.REI.4a, b
29. $2x^2 - 8x = 12$  −1.2, 5.2
30. $3x^2 - 24x = -36$  2, 6
31. $x^2 - 3x = 10$  −2, 5
32. $4x^2 + 100 = 0$  ∅
33. $x^2 = -7x - 5$  −6.2, −0.8
34. $12 - 12x = -3x^2$  2

**Example 4** State the value of the discriminant for each equation. Then determine the number of real solutions of the equation. 35–40. See margin.
A.REI.4b
35. $0.2x^2 - 1.5x + 2.9 = 0$
36. $2x^2 - 5x + 20 = 0$
37. $x^2 - \frac{4}{5}x = 3$
38. $0.5x^2 - 2x = -2$
39. $2.25x^2 - 3x = -1$
40. $2x^2 = \frac{5}{2}x + \frac{3}{2}$

41. **MODELING** The percent of customers who have a store's loyalty card $h$ can be estimated by $h = -0.2n^2 + 7.2n + 1.5$, where $n$ is the number of years since 2000.
   a. Use the Quadratic Formula to determine when 20% of the store's customers will have a loyalty card.  2036
   b. Is a quadratic equation a good model for this information? Explain. See margin.

### Differentiated Homework Options

| Levels | AL Basic | OL Core | BL Advanced |
|---|---|---|---|
| Exercises | 16–41, 52, 57–64 | 17–41 odd, 42, 43–49 odd, 52, 57–64 | 42–58, (optional: 59–64) |
| 2-Day Option | 17–41 odd, 59–64 | 16–41, 59–64 | |
| | 16–40 even, 52, 57, 58 | 42–50, 52, 57, 58 | |

You can use ALEKS to provide additional remediation support with personalized instruction and practice.

---

## Practice

**Formative Assessment** Use Exercises 1–15 to assess students' understanding of the concepts in this lesson.

The Practice and Problem Solving exercises assess the content taught in the lesson. The Preparing for Assessment page is meant to be used as preparation for end-of-course assessments.

## Extra Practice

See page R9 for extra exercises for students who are approaching level or for on-level students who need additional reinforcement.

## Additional Answers

35. −0.07; no real solution
36. −135; no real solutions
37. 12.64; two real solutions
38. 0; one real solution
39. 0; one real solution
40. 18.25; two real solutions
41b. Sample answer: No the parabola has a maximum at about 66, meaning only 66% of the store's customers would ever have the store's loyalty card.

## Follow-Up
Students have explored solving quadratic equations.

Ask:
- How do you know what method to use when solving a quadratic equation? Sample answer: If the equation has terms that you know are easily factorable, you could solve by factoring. If the equation has more complex terms, you could solve by using the Quadratic Formula or by graphing. You could also use one method to solve and a second method to check your answer.

### Go Online!   eBook

**Interactive Student Guide**
Use the *Interactive Student Guide* to deepen conceptual understanding.
· Solving by Using the Quadratic Formula

# Lesson 9-6 | Solving Quadratic Equations by Using the Quadratic Formula

## Levels of Complexity Chart

The levels of the exercises progress from 1 to 3, with Level 1 indicating the lowest level of complexity.

| Exercises | 16–41 | 42–49, 59–64 | 50–58 |
|---|---|---|---|
| Level 3 | | | ● |
| Level 2 | | ● | |
| Level 1 | ● | | |

## Assess

**Ticket Out the Door** Make several copies each of five quadratic equations. Give one equation to each student. As students leave the room, ask each of them to tell you the discriminant of the equation and the number of real solutions.

## Additional Answers

**42.** No; Sample answer: Hannah was traveling at about 61 mph, so she was not speeding.

**50c.** Sample answer: Multiply three factors written from the roots.
$(x-1)(x-2)(x-3) = 0$
$x^3 - 6x^2 + 11x - 6 = 0$
The equation is not quadratic because its degree is 3 not 2.

**57.** Sample answer: If the discriminant is positive, the Quadratic Formula will result in two real solutions because you are adding and subtracting the square root of a positive number in the numerator of the expression. If the discriminant is zero, there will be one real solution because you are adding and subtracting the square root of zero. If the discriminant is negative, there will be no real solutions because you are adding and subtracting the square root of a negative number in the numerator of the expression.

## Go Online!

### Quizzes

Students can use *Self-Check Quizzes* to check their understanding of this lesson. You can also give *Quiz 3*, which covers the content in Lessons 9-6 and 9-7.

---

**B** ▶ **42. TRAFFIC** The equation $d = 0.05v^2 + 1.1v$ models the distance $d$ in feet it takes a car traveling at a speed of $v$ miles per hour to come to a complete stop. If Hannah's car stopped after 250 feet on a highway with a speed limit of 65 miles per hour, was she speeding? Explain your reasoning. **See margin.**

Without graphing, determine the number of x-intercepts of the graph of the related function for each equation.

**43.** $4.25x + 3 = -3x^2$  **0**    **44.** $x^2 + \frac{2}{25} = \frac{3}{5}x$  **2**    **45.** $0.25x^2 + x = -1$  **1**

Solve each equation by using the Quadratic Formula. Round to the nearest tenth.

**46.** $-2x^2 - 7x = -1.5$    **47.** $2.3x^2 - 1.4x = 6.8$    **48.** $x^2 - 2x = 5$  **3.4, −1.4**
**46. −3.7, 0.2**    **47. −1.4, 2.1**

**49. POSTER** Bartolo is making a poster for the dance. He wants to cover three fourths of the area with text.

**49c.** about 0.7 in. on the sides, 2.8 in. on the top, and 2.1 in. on the bottom

**a.** Write an equation for the area of the section with text. $(20 - 2x)(25 - 7x) = 375$

**b.** Solve the equation by using the Quadratic Formula. about 12.9, 0.7

**c.** What should be the margins of the poster?

Poster dimensions: 20 in. wide, 25 in. tall, with $4x$ in. top margin, $x$ in. side margins, $3x$ in. bottom margin. "Spring Fling Friday at 8 p.m."

**C** ▶ **50. MULTIPLE REPRESENTATIONS** In this problem, you will investigate writing a quadratic equation with given roots. If $p$ is a root of $0 = ax^2 + bx + c$, then $(x - p)$ is a factor of $ax^2 + bx + c$.

**a. Tabular** Copy and complete the first two columns of the table.

**b. Algebraic** Multiply the factors to write each equation with integral coefficients. Use the equations to complete the last column of the table. Write each equation.

**c. Analytical** How could you write an equation with three roots? Test your conjecture by writing an equation with roots 1, 2, and 3. Is the equation quadratic? Explain. **See margin.**

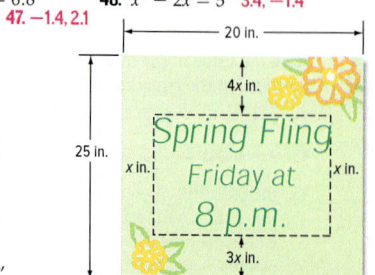

| Roots | Factors | Equation |
|---|---|---|
| 2, 5 | $(x-2), (x-5)$ | $(x-2)(x-5) = 0$ $x^2 - 7x + 10 = 0$ |
| 1, 9 | $(x-1), (x-9)$ | $x^2 - 10x + 9 = 0$ |
| −1, 3 | $(x+1), (x-3)$ | $x^2 - 2x - 3 = 0$ |
| 0, 6 | $x, (x-6)$ | $x^2 - 6x = 0$ |
| $\frac{1}{2}$, 7 | $\left(x - \frac{1}{2}\right), (x-7)$ | $2x^2 - 15x + 7 = 0$ |
| $-\frac{2}{3}$, 4 | $\left(x + \frac{2}{3}\right), (x-4)$ | $3x^2 - 10x - 8 = 0$ |

A.REI.4a, A.REI.4b, A.CED.1

### H.O.T. Problems  Use Higher-Order Thinking Skills

**51. CHALLENGE** Find all values of $k$ such that $2x^2 - 3x + 5k = 0$ has two solutions. $k < \frac{9}{40}$

**52.** **SENSE-MAKING** Use factoring techniques to determine the number of real zeros of $f(x) = x^2 - 8x + 16$. Compare this method to using the discriminant. **52. See Ch. 9 Answer Appendix.**

**CHALLENGE** Determine whether there are *two*, *one*, or *no* real solutions of each equation.

**53.** The graph of the related quadratic function does not have an x-intercept. **none**

**54.** The graph of the related quadratic function touches but does not cross the x-axis. **one**

**55.** The graph of the related quadratic function intersects the x-axis twice. **two**

**56.** Both $a$ and $b$ are greater than 0 and $c$ is less than 0 in a quadratic equation. **two**

**57.** **WRITING IN MATH** Why can the discriminant be used to confirm the number of real solutions of a quadratic equation? **See margin.**

**58. WRITING IN MATH** Describe the advantages and disadvantages of each method of solving quadratic equations. Why are the methods equivalent? Which method do you prefer, and why? **See Ch. 9 Answer Appendix.**

### Standards for Mathematical Practice

| Emphasis On | Exercises |
|---|---|
| 1 Make sense of problems and persevere in solving them. | 43–45, 51–56, 58, 60 |
| 4 Model with mathematics. | 15, 28, 41, 42, 49, 64 |
| 6 Attend to precision. | 16–27, 46–48, 57–59, 61–64 |
| 7 Look for and make use of structure. | 35–40, 57, 63 |

612 | Lesson 9-6 | Solving Quadratic Equations by Using the Quadratic Formula

## Preparing for Assessment

**59. MULTI-STEP** Gabrielle is using the Quadratic Formula to solve the equation $8x^2 - 3x = 9$. Her work so far is shown below.
**MP 3, 6** A.REI.4b

$$\frac{-b \pm \sqrt{b^2 - 4ac}}{2a}$$

$$\frac{-(-3) \pm \sqrt{(-3)^2 - 4(8)(9)}}{2(8)}$$

$$\frac{3 \pm \sqrt{-9 - 288}}{16}$$

a. What errors, if any, did Gabrielle make when using the Quadratic Formula to solve this equation? Select all that apply.  **D, F**

- [ ] A   There are no errors.
- [ ] B   She made a sign error when substituting for $a$.
- [ ] C   She made a sign error when substituting for $b$.
- [ ] D   She made a sign error when substituting for $c$.
- [ ] E   She mixed up the values of $a$, $b$, and $c$ when substituting.
- [ ] F   She squared the value of $b$ incorrectly.

b. Which are the correct solutions of $8x^2 - 3x = 9$? Select all that apply.  **B, E**

- [ ] A   0.9
- [ ] B   1.3
- [ ] C   0
- [ ] D   −1.3
- [ ] E   −0.9
- [ ] F   no solution

59c. Sample answer: She could write out $a = 8$, $b = -3$, $c = -9$. She could double-check her negative signs. She could rewrite the equation so that it equals zero, so she realizes that $c = -9$.

c. What are two strategies Gabrielle could have used to reduce errors in this calculation?

**60.** Which equation has no real solutions? **MP 1** A.REI.4b  **C**

- [ ] A   $2x^2 - 4x = 5$
- [ ] B   $2x^2 + 4x = 5$
- [ ] C   $-2x^2 + 4x = 5$
- [ ] D   $-2x^2 + 4x = -5$

**61.** Examine the quadratic equation $3x^2 - 4x + 2 = 0$.
**MP 6** A.REI.4b

a. What do you know about the discriminant of $3x^2 - 4x + 2$?  **A**

- [ ] A   It is a negative integer.
- [ ] B   It is a positive integer.
- [ ] C   It is an irrational number.
- [ ] D   It is a rational number.

b. What is one solution of $3x^2 - 4x + 2$?  **D**

- [ ] A   about 0.2
- [ ] B   about 0.3
- [ ] C   about 1.7
- [ ] D   There is no solution.

**62.** Which equation has solutions, rounded to the nearest tenth, of −2.1 and 2.4? **MP 6** A.REI.4b  **A**

- [ ] A   $3x^2 - x - 15 = 0$
- [ ] B   $2x^2 - x - 15 = 0$
- [ ] C   $3x^2 - 4x + 2 = 0$
- [ ] D   $2x^2 - 4x + 2 = 0$

**63.** For what value of $c$ does the quadratic equation $-2x^2 + 12x - c = 0$ have exactly one real solution? **MP 6, 7** A.REI.4b  **B**

- [ ] A   $c = 3$
- [ ] B   $c = 18$
- [ ] C   $c = -18$
- [ ] D   $c = 72$

**64.** A piece of scaffolding falls to the ground from a height of 50 feet. The situation is modeled by the function $h = -16t^2 + 50$, where $h$ is the height in feet and $t$ is the time in seconds. About how many seconds is the best estimate for the time it will take for the scaffolding to hit the ground? **MP 4, 6** A.REI.4b

1.8 seconds

---

### Differentiated Instruction  OL  BL

**Extension** Since students are much more likely to retain a concept that they have researched and explained, ask them to further research the derivation of the Quadratic Formula and to write short paragraphs explaining the derivation.

---

Lesson 9-6 | Solving Quadratic Equations by Using the Quadratic Formula

## Preparing for Assessment

Exercises 59–64 require students to use the skills they will need on standardized assessments. Each exercise is dual-coded with content standards and mathematical practice standards.

| \multicolumn{3}{c}{Dual Coding} |
|---|---|---|
| Items | Content Standards | MP Mathematical Practices |
| 59 | A.REI.4b | 3, 6 |
| 60 | A.REI.4b | 1 |
| 61 | A.REI.4b | 6 |
| 62 | A.REI.4b | 6 |
| 63 | A.REI.4b | 6, 7 |
| 64 | A.REI.4b | 4, 6 |

## Diagnose Student Errors

Survey student responses for each item. Class trends may indicate common errors and misconceptions.

**60.**

| A | Did not write equation in standard form before using discriminant |
|---|---|
| B | Did not write equation in standard form before using discriminant |
| C | CORRECT |
| D | Used incorrect sign when calculating discriminant |

**61a.**

| A | Forgot to multiply $ac$ by 4 |
|---|---|
| B | CORRECT |
| C | Multiplied by 16 instead of dividing |
| D | Forgot to divide by $2a$ |

**62.**

| A | CORRECT |
|---|---|
| B | Did not factor $2x^2 - x - 15$ as $(2x + 5)(x - 3)$ |
| C | Identified equation that has no solutions |
| D | Identified equation that has exactly one solution |

**63.**

| A | Added $b + b$ instead of squaring $b$ when finding the discriminant |
|---|---|
| B | CORRECT |
| C | Did not take into account that $c$ is subtracted, not added, when finding the discriminant |
| D | Did not multiply $ac$ by 4 when finding the discriminant |

connectED.mcgraw-hill.com   613

# LESSON 9-7
## Solving Systems of Linear and Quadratic Equations

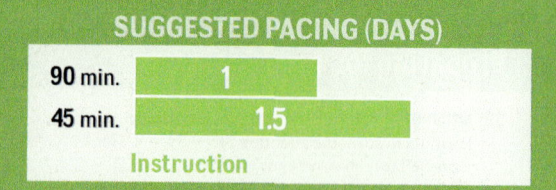

**SUGGESTED PACING (DAYS)**
- 90 min. — 1
- 45 min. — 1.5

Instruction

## Track Your Progress

### Objectives
1. Solve systems of linear and quadratic equations by graphing.
2. Solve systems of linear and quadratic equations by using algebraic methods.

### Mathematical Background
Students have already solved systems of equations that consist of two linear equations in two variables. In this lesson, students extend their knowledge to systems that consist of one linear equation and one quadratic equation. Some of the methods students learned earlier can be adapted to linear and quadratic systems. In particular, students will explore how to solve by graphing and how to solve by substitution.

### THEN
**A.REI.4b** Solve quadratic equations by inspection (e.g., for $x^2 = 49$), taking square roots, completing the square, the quadratic formula, and factoring, as appropriate to the initial form of the equation. Recognize when the quadratic formula gives complex solutions and write them as $a \pm bi$ for real numbers $a$ and $b$.

### NOW
**A.CED.2** Create equations in two or more variables to represent relationships between quantities; graph equations on coordinate axes with labels and scales.

**A.REI.7** Solve a simple system consisting of a linear equation and a quadratic equation in two variables algebraically and graphically.

### NEXT
**F.IF.6** Calculate and interpret the average rate of change of a function (presented symbolically or as a table) over a specified interval. Estimate the rate of change from a graph.

**F.LE.1** Distinguish between situations that can be modeled with linear functions and with exponential functions.

---

**Go Online!** All of these resources and more are available at connectED.mcgraw-hill.com

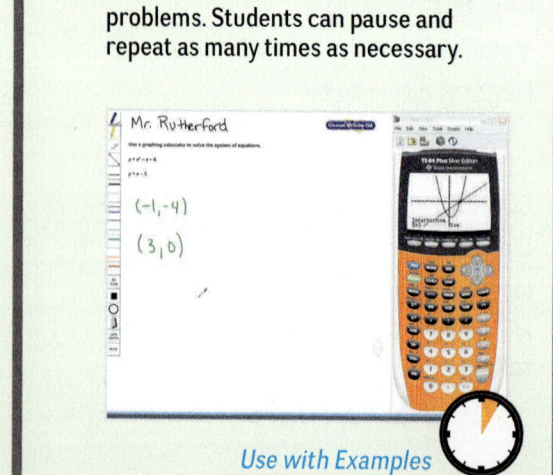

**Graphing Calculator Personal Tutors** let students hear real teachers solve problems. Students can pause and repeat as many times as necessary.

*Use with Examples*

Use **The Geometer's Sketchpad** to explore the graphs of systems of linear and quadratic equations.

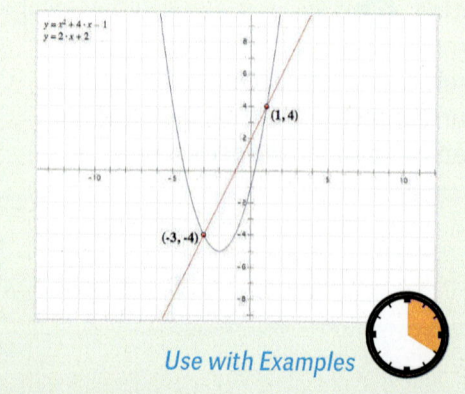

*Use with Examples*

The **Chapter Project** allows students to create and customize a project as a non-traditional method of assessment.

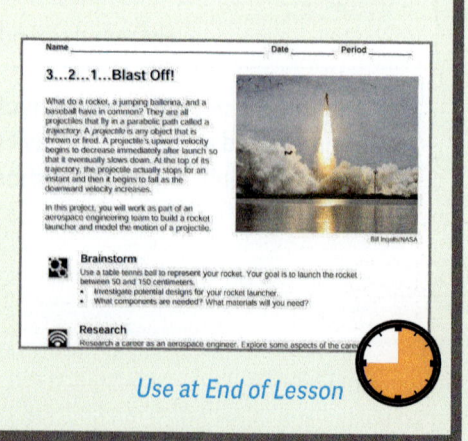

*Use at End of Lesson*

---

### OER Using Open Educational Resources
**Video Sharing** Have students work in groups to record and upload examples on **SchoolTube** about solving systems of linear and quadratic equations. Have them watch videos of others for ideas. If you are unable to access **SchoolTube**, try **KidsTube**, **MathATube**, **YouTube**, or **TeacherTube**. *Use as homework*

# Go Online!
connectED.mcgraw-hill.com  Worksheets

# Differentiate Your Resources

**Extra Practice** Additional practice or homework; Skills Practice is best for approaching-level students and Practice is best for on-level and beyond-level students

## Skills Practice
AL  OL  ELL

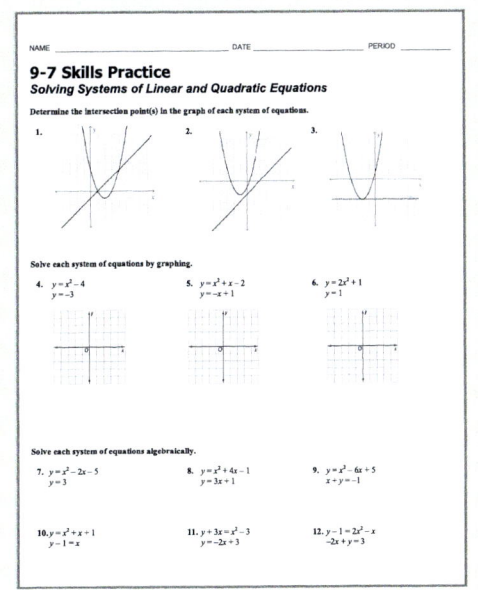

## Practice
AL  OL  BL  ELL

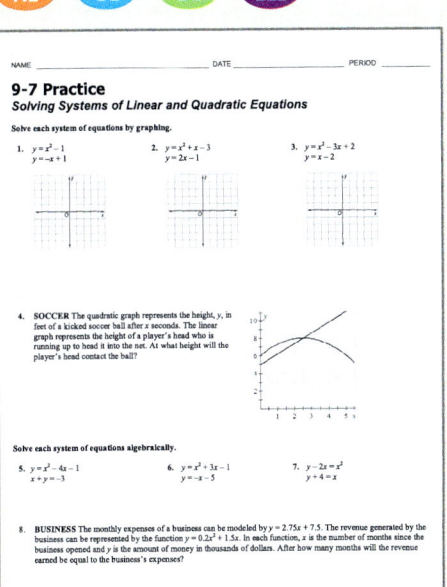

## Word Problem Practice
AL  OL  BL  ELL

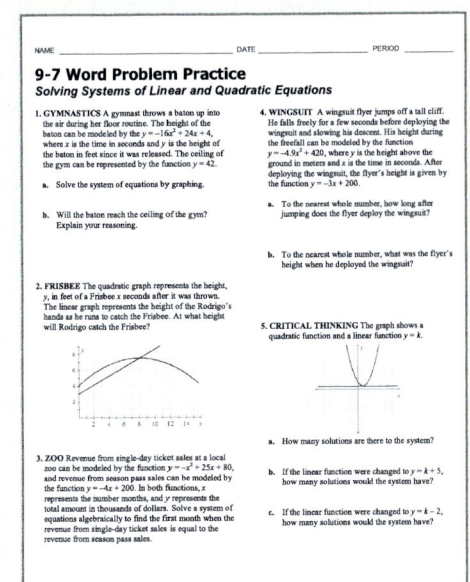

**Intervention** Reteaching and vocabulary activities that can be used with struggling or absent students and as ELL support

## Study Guide and Intervention
AL  OL  ELL

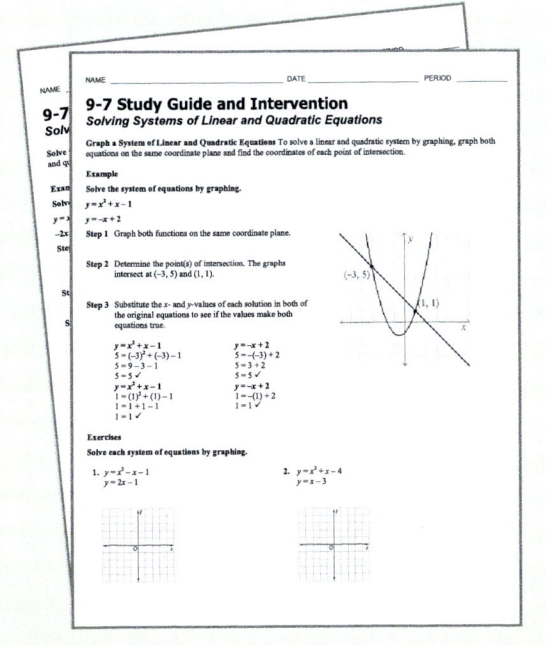

## Study Notebook
AL  OL  BL  ELL

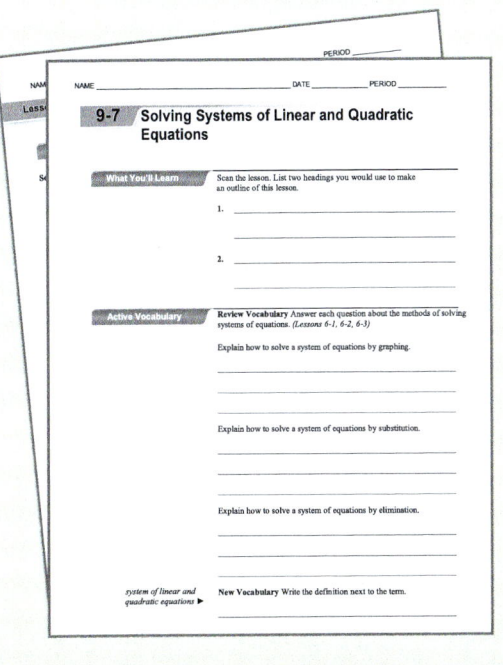

**Extension** Activities that can be used to extend lesson concepts

## Enrichment
OL  BL  ELL

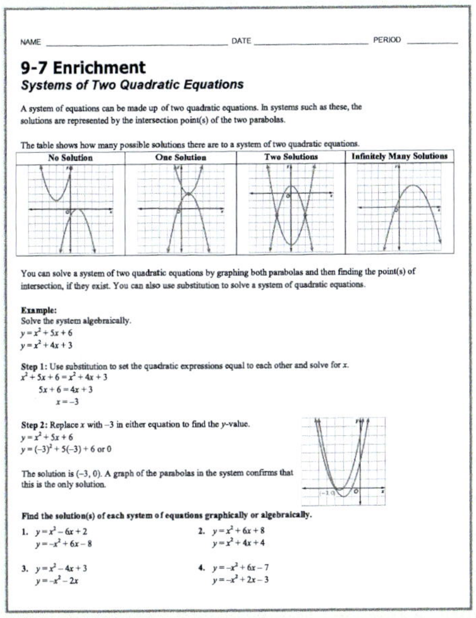

connectED.mcgraw-hill.com  614B

Lesson 9-7 | Solving Systems of Linear and Quadratic Equations

# Launch

Have students read the Why? section of the lesson.

- **Why does it make sense that the height of a kicked football is modeled by a quadratic equation?** The football rises, reaches a maximum height, and then it descends. The graph of a quadratic equation can have these same characteristics.

- **How do the graphs of quadratic equations and linear equations differ?** The graph of a quadratic equation is a parabola. The graph of a linear equation is a straight line.

## LESSON 7
# Solving Systems of Linear and Quadratic Equations

**Then** • You graphed and solved linear and quadratic functions.

**Now**
1. Solve systems of linear and quadratic equations by graphing.
2. Solve systems of linear and quadratic equations by using algebraic methods.

**Why?** The height of a kicked football can be modeled by a quadratic equation. The height of a player running up a hill to catch the ball during a practice drill can be modeled by a linear equation. The intersection of the graphs of the equations tells the time and height when the player should catch the ball.

**MP Mathematical Practices**
2 Reason abstractly and quantitatively.
4 Model with mathematics.

**Content Standards**
A.CED.2 Create equations in two or more variables to represent relationships between quantities; graph equations on coordinate axes with labels and scales.
A.REI.7 Solve a simple system consisting of a linear equation and a quadratic equation in two variables algebraically and graphically.

**1 Graph a System of Linear and Quadratic Equations** Like solving systems of linear equations, you can solve systems of linear and quadratic equations by graphing the equations on the same coordinate plane. A system with one linear equation and one quadratic equation can have two solutions, zero solutions, or one solution. The number of solutions depends upon the number of intersections of the two graphs.

**Key Concept** Number of Solutions

| Number of Solutions | Description | Graph |
|---|---|---|
| one solution | two equations with graphs that intersect at one point | |
| two solutions | two equations with graphs that intersect at two points | |
| no solution | two equations with graphs that do not intersect | |

## MP Mathematical Practices Strategies

**Reason abstractly and quantitatively.**
Help students make sense of quantities and their relationships in problem situations. For example, ask:

- **What can you determine about a system of linear and quadratic equations from its graph?** the number of solutions of the system
- **How are the solutions of a linear and quadratic system represented?** as ordered pairs
- **How is solving a linear and quadratic system similar to solving a system with two linear equations?** You can solve both types of systems by graphing or by using substitution.
- **How is solving a linear and quadratic system different from solving a system with two linear equations algebraically?** When you solve a linear and quadratic system by substitution, you need to solve a quadratic equation in one variable.

You can solve a system of linear and quadratic equations by using the following steps. The points of intersection of the graphs are the solutions to the system.

**Key Concept** Solving a Linear and Quadratic System by Graphing

**Step 1** Graph the equations on the same coordinate plane.
**Step 2** Find the coordinates of each point of intersection.
**Step 3** Check the solutions. Substitute the x- and y-values of each solution in the original equation.

A.REI.7

**Example 1** Solve a System by Graphing

Solve the system of equations by graphing.
$y = x^2 + 4x - 1$
$y = 2x + 2$

**Step 1** Graph $y = x^2 + 4x - 1$ and $y = 2x + 2$ on the same coordinate plane.

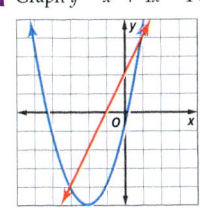

**Study Tip**
**MP Sense-Making** The order in which you graph the linear and quadratic equations does not matter.

**Step 2** Find the coordinates of each point of intersection. The graphs intersect at $(-3, -4)$ and $(1, 4)$.

**Step 3** To check your solutions, substitute each solution into both of the original equations to see if the values make both equations true.

| $y = x^2 + 4x - 1$ | Original equation | $y = 2x + 2$ |
|---|---|---|
| $-4 \stackrel{?}{=} (-3)^2 + 4(-3) - 1$ | $x = -3, y = -4$ | $-4 \stackrel{?}{=} 2(-3) + 2$ |
| $-4 \stackrel{?}{=} 9 - 12 - 1$ | Simplify. | $-4 \stackrel{?}{=} -6 + 2$ |
| $-4 = -4$ ✓ | Simplify. | $-4 = -4$ ✓ |

| $y = x^2 + 4x - 1$ | Original equation | $y = 2x + 2$ |
|---|---|---|
| $4 \stackrel{?}{=} 1^2 + 4(1) - 1$ | $x = 1, y = 4$ | $4 \stackrel{?}{=} 2(1) + 2$ |
| $4 \stackrel{?}{=} 1 + 4 - 1$ | Simplify. | $4 \stackrel{?}{=} 2 + 2$ |
| $4 = 4$ ✓ | Simplify. | $4 = 4$ ✓ |

**Guided Practice**

1. $y = x^2 - x - 5$  See margin
   $y = x + 3$

**Additional Answer (Guided Practice)**

1. $(-2, 1)$ and $(4, 7)$

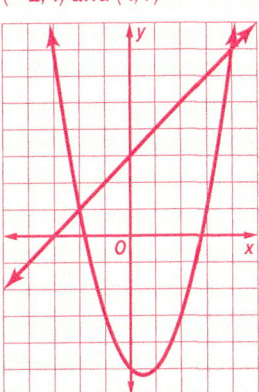

---

Lesson 9-7 | Solving Systems of Linear and Quadratic Equations

# Teach

Ask the scaffolded questions for each example to build conceptual understanding for students at all levels.

## 1 Graph a System of Linear and Quadratic Equations

**Example 1 Solve a System by Graphing**

**AL** How many times do the graphs intersect? **2**

**OL** At what points do the graphs intersect? **$(-3, -4)$ and $(1, 4)$**

**BL** How could you use a graphing calculator to solve the system of equations? **Enter each equation and graph. Then use the intersect feature from the CALC menu twice to find each point of intersection.**

**Need Another Example?**
Solve the system of equations by graphing.
$y = x^2 - 4x + 1$
$y = -x - 1$
**$(2, -3)$ and $(1, -2)$**

**MP Teaching the Mathematical Practices**

**Reasoning** Mathematically proficient students make sense of the quantities and their relationships in problem situations, including determining the number of solutions in a system of equations. For example, the system $y = 4$ and $y = -x^2 - 2$ has no solutions because the graph of $y = 4$ lies entirely above the x-axis while the graph of $y = -x^2 - 2$ lies entirely below the x-axis. This means the graphs of the equations do not intersect.

connectED.mcgraw-hill.com  615

Lesson 9-7 | Solving Systems of Linear and Quadratic Equations

## 2 Solve a Linear and Quadratic System Algebraically

### Example 2 Solve a System Algebraically

**AL** Why can you solve the system by setting both of the expressions in the variable *x* equal to each other? Since both expressions are equal to *y*, they are equal to each other.

**OL** What is another way you could check your solution? Substitute each solution into each equation to check that it results in a true statement.

**BL** Is it possible to solve the system by using elimination? Explain. Yes; subtract one equation from the other to eliminate the variable *y*. Solve the resulting quadratic equation for *x* and then substitute to find the values of *y*.

**Need Another Example?**
Solve the system of equations algebraically.
$y = x^2 - 4x + 5$
$y = 2x$
(1, 2) and (5, 10)

---

**Watch Out!**

**Solutions** Some students may find the two values of *x*, 3 and −2, and write the solution of the system as (3, −2). Remind students that each *x*-value has a corresponding *y*-value. This results in two ordered pairs.

---

**2 Solve a Linear and Quadratic System Algebraically** Linear and quadratic systems can also be solved algebraically. Just as you can use substitution to solve systems of linear equations, you can use this method to solve systems of linear and quadratic equations.

> **Key Concept** Solving a Linear and Quadratic System Algebraically
> 
> **Step 1** Solve each equation for *y*.
> **Step 2** Substitute one expression for *y* in the other equation.
> **Step 3** Solve for *x*.
> **Step 4** Substitute the *x*-value(s) in either of the original equations.
> **Step 5** Solve for *y*.
> **Step 6** Graph the equations to check the solution(s).

A.REI.7

### Example 2  Solve a System Algebraically

Solve the system of equations algebraically.
$y = x^2 - 2x - 3$
$x + y = 3$

**Step 1** Solve each equation for *y*.
The first equation is already solved for *y*. Solve the second equation for *y*.

| | |
|---|---|
| $x + y = 3$ | Second equation |
| $-x + x + y = -x + 3$ | Subtract *x* from to each side. |
| $y = -x + 3$ | Simplify. |

**Step 2** Substitute the quadratic expression for *y* in the linear equation.
$x^2 - 2x - 3 = -x + 3$   Substitute $x^2 - 2x - 3$ for *y*

**Step 3** Solve for *x*.

| | |
|---|---|
| $x^2 - 2x - 3 = -x + 3$ | Equation from Step 2 |
| $x^2 - x - 3 = 3$ | Add *x* to each side. |
| $x^2 - x - 6 = 0$ | Subtract 3 from each side. |
| $(x - 3)(x + 2) = 0$ | Factor. |
| $x - 3 = 0$ or $x + 2 = 0$ | Zero Product Property |
| $x = 3$ or $x = -2$ | Solve for *x*. |

**Step 4** Substitute the *x*-value(s) in either of the original equations.

| $x + y = 3$ | $x + y = 3$ | Original equation |
|---|---|---|
| $3 + y = 3$ | $-2 + y = 3$ | $x = 3$  $x = -2$. |

**Study Tip**
**MP Structure** After you have found the values of *x*, you can substitute them in either equation to find the values of *y*. Most of the time, it will be easier to use the linear function instead of the quadratic function.

**Step 5** Solve for *y*.

| $3 + y = 3$ | $-2 + y = 3$ |
|---|---|
| $3 + y - 3 = 3 - 3$ | $-2 + y + 2 = 3 + 2$ |
| $y = 0$ | $y = 5$ |

The solutions of the system are (3, 0) and (−2, 5).

616 | Lesson 9-7 | Solving Systems of Linear and Quadratic Equations

**Study Tip**

**MP Sense-Making** After you solve algebraically, you can check by graphing. You can also check by substituting the x- and y-values into both equations in the system to make sure the values result in true equations.

**Step 6** Graph the equations to check the solutions.

The graphs of the equations intersect at (3, 0) and (−2, 5), so (3, 0) and (−2, 5) are the solutions of this system.

▶ **Guided Practice**

2. $y = x^2 - 5x - 12$  (−2, 2) and (10, 38)
   $2y - 16 = 6x$

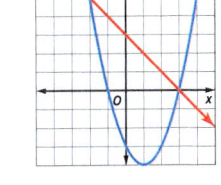

You can use systems of linear and quadratic equations to model revenue and look for trends in sales data.

A.REI.7

**Real-World Example 3** Apply a System of Linear and Quadratic Equations

**SALES** Camping equipment sales at a sporting goods store can be modeled by the function $y = -x^2 + 12x + 25$ and gift card sales can be modeled by the function $y = 5x + 7$, where $x$ represents the number of months past January and $y$ represents the revenue in thousands of dollars. Solve a system of equations algebraically to find the month in which the revenue from camping equipment sales is equal to the revenue from gift card sales.

Substitute one expression for $y$ in the other equation. Then solve for $x$.

$$-x^2 + 12x + 25 = 5x + 7 \quad \text{Substitute.}$$
$$12x + 25 = x^2 + 5x + 7 \quad \text{Add } x^2 \text{ to each side.}$$
$$25 = x^2 - 7x + 7 \quad \text{Subtract } 12x \text{ from each side.}$$
$$0 = x^2 - 7x - 18 \quad \text{Subtract 25 from each side.}$$
$$0 = (x - 9)(x + 2) \quad \text{Factor.}$$
$$x - 9 = 0 \quad x + 2 = 0 \quad \text{Zero Product Property}$$
$$x = 9 \quad x = -2 \quad \text{Solve for } x.$$

A solution of the system occurs when $x = 9$, so the revenue from camping equipment sales and from gift card sales are equal 9 months past January, or in October.

Because the functions only model data for months past January, $x = -2$ is not a viable solution as it represents a month prior to January.

Graph the equations to check the solutions.

The graphs intersect at $x = 9$, so the solution is correct.

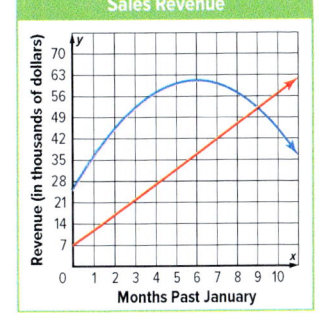

▶ **Guided Practice** April

3. **AMUSEMENT PARKS** Revenue from single-day ticket sales at a local amusement park can be modeled by the function $y = -x^2 + 11x + 42$ and revenue from season pass sales can be modeled by the function $y = -5x + 81$, where $x$ represents the number of months past January and $y$ represents the total revenue in tens of thousands of dollars. Solve a system of equations algebraically to find the first month when the revenue from single-day ticket sales is equal to the revenue from season pass sales.

---

**Lesson 9-7** | Solving Systems of Linear and Quadratic Equations

**Example 3 Apply a System Linear and Quadratic Equations**

**AL** What does $x = 9$ represent in the problem? Explain. October; $x$ represents the number of months past January, so an $x$-value of 9 represents October

**OL** What other method could you use to solve the resulting quadratic equation? the Quadratic Formula

**BL** Why do you think the quadratic revenue model has a negative coefficient for the $x^2$-term? Camping equipment sales increase as the summer months approach and then decrease, so the parabola should open down.

**Need Another Example?**

**SALES** Car wash sales can be modeled by the function $y = -x^2 + 8x - 4$ and prepaid car wash card sales can be modeled by the function $y = 2x + 5$, where $x$ represents the number of months since May and $y$ represents the total revenue in thousands of dollars. Solve a system of equations algebraically to find the first month after May in which revenue from car wash sales is equal to the revenue from prepaid card sales. August

**Watch Out!**

**Preventing Errors** Encourage students to consider the context of real-world situations in order to determine whether their solutions make sense. Also remind students to interpret their solutions in the context of the situation. In Example 3, the solution represents a month and not just a value.

connectED.mcgraw-hill.com  **617**

**Lesson 9-7 | Solving Systems of Linear and Quadratic Equations**

**Example 4 Solve a System of Linear and Quadratic Equations Algebraically**

**AL** If a linear and quadratic system has no solution, what can conclude about the graphs of the equations in the system? **The graphs do not intersect.**

**OL** How can you use the discriminant to determine the number of solutions of a quadratic equation? **If the value of $b^2 - 4ac$ is negative, the equation has no solutions. If the value is 0, the equation has one solution. If the value is positive, the equation has two solutions.**

**BL** The equation $x^2 - x + 1 = 0$ cannot be solved by factoring. Does this mean the equation has no real solutions? Explain. **No, the solutions might not be whole numbers. You can find the real solutions using another method.**

**Need Another Example?**
Solve the system of equations algebraically.
$y = -x^2 + 3x - 1$
$y = 4x + 3$ **no solution**

**Additional Answers**

1. $(3, 0), (0, -3)$

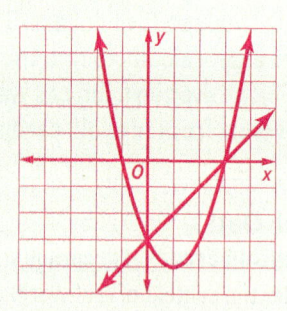

2. $(-1, -3), (1, -3)$

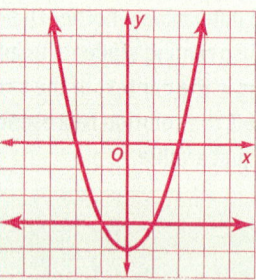

3. $(0, 1), (3, -2)$

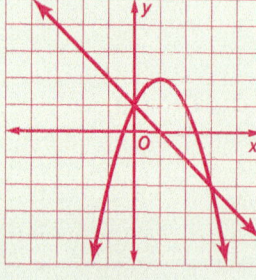

4. $(-2, -3), (1, 3)$

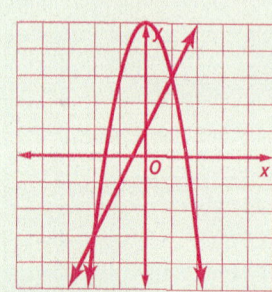

---

When you solve a linear and quadratic system algebraically, you may arrive at a an equation in one variable that has no real solutions. This means the original system has no real solutions.

A.REI.7

**Example 4** Solve a System of Linear and Quadratic Equations Algebraically

Solve the system of equations algebraically.
$y = x + 1$
$y = x^2 + 2$

The equations are already solved for $y$.
Substitute one expression for $y$ in the other equation. Then solve for $x$.

$x^2 + 2 = x + 1$   Substitute.
$x^2 - x + 2 = 1$   Subtract $x$ from each side.
$x^2 - x + 1 = 0$   Subtract 1 from each side.

Find the discriminant, $b^2 - 4ac$, of the quadratic equation $x^2 - x + 1 = 0$.
$b^2 - 4ac = (-1)^2 - 4(1)(1)$   $a = 1, b = -1$, and $c = 1$
$= 1 - 4$   Simplify.
$= -3$   Subtract.

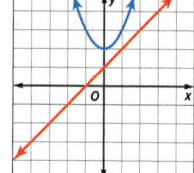

The discriminant is negative, so $x^2 - x + 1 = 0$ has no real solutions. Therefore, the system does not have a solution.
Graph the equations to check your solution.
The graphs do not intersect, so the system has no solution.

**Study Tip**
**Structure** If the discriminant of a quadratic equation is negative, then the quadratic equation has no real solutions.

▶ **Guided Practice**

4. $y = -x^2 + x - 1$   **no solution**
   $-3 + y = x$

**Check Your Understanding**   ◯ = Step-by-Step Solutions begin on page R11.

**Example 1** Solve each system of equations by graphing. **1–4. See margin.**
A.REI.7

1. $y = x^2 - 2x - 3$         2. $y = x^2 - 4$
   $y = x - 3$                     $y = -3$

3. $y = -x^2 + 2x + 1$       4. $y = -2x^2 + 5$
   $y = -x + 1$                    $y = 2x + 1$

**Examples 2–4** 5. **TICKET PRICES** The revenue for a school play can be modeled by $y = -x^2 + 8x$, where $x$ is the ticket price in dollars. The cost to produce the school play can be modeled by $y = -20 + 9x$. Determine the ticket price that will allow the school play to break even. (Hint: Breaking even means that the revenue equals the cost.) **$4**

Solve each system of equations algebraically.

6. $y = -x + 3$   **(1, 2), (−2, 5)**      7. $y = x^2 - 5x + 5$   **(5, 5)**
   $y = x^2 + 1$                                 $y = 5x - 20$

8. $-x^2 - x + 19 = y$   **(9, −71), (−11, −91)**   9. $y = 3x^2 + 21x - 5$   $\left(\frac{1}{3}, \frac{7}{3}\right), (-4, -41)$
   $x = y + 80$                                              $-10x + y = -1$

---

**Differentiated Instruction** OL BL

**IF** students are successful with solving linear and quadratic systems algebraically,

**THEN** have them work in groups to solve systems that involve two quadratic equations. Students may choose to solve such systems graphically or algebraically, or both. For example, ask students to solve the system $y = x^2 + x + 7$ and $y = 2x^2 + 2x + 1$. **(2, 13) and (−3, 13)**

618 | Lesson 9-7 | Solving Systems of Linear and Quadratic Equations

## Practice and Problem Solving

Extra Practice is found on page R9.

**Example 1**
A.REI.7

Solve each system of equations by graphing. 11–13. See Ch. 9 Answer Appendix.

10. $y = x^2 - 2x - 2$   See margin.
    $y = 4x - 11$

11. $y = x - 5$
    $y = x^2 - 6x + 5$

12. $y = x + 1$
    $y = x^2 + 2x + 4$

13. $y = 2x$
    $y = -x^2 + 2x + 4$

**Examples 2–4**
A.REI.7

14b. 5 cm or 2 cm; the fish catches the fly at a height of 5 cm if it catches the fly 3 cm from the left edge of the screen; the fish catches the fly at a height of 2 cm if it catches the fly 6 cm from the left edge of the screen.

14. **VIDEO GAMES** Jamar is using a coordinate plane to program a video game in which a fish jumps out of the water to catch a fly. The path of the fish is modeled by $y = -x^2 + 8x - 10$ and the path of the fly is modeled by $y = -x + 8$, where $y$ is the height in centimeters above the water and $x$ is the distance in centimeters from the left edge of the screen.

a. How far is the fish from left edge of the screen when it catches the fly? Is there more than one possible distance? Explain.

b. What is the height of the fish when it catches the fly? Is there more than one possible height? Explain. **14a.** 3 cm or 6 cm; there are two solutions to the system of equations which represent two points of intersection of the paths.

Solve each system algebraically.

15. $y = 6x^2$  (−1, 6), (0.5, 1.5)
    $3x + y = 3$

16. $y = x^2 + 1$  (1, 2), (−9, 82)
    $8x + y = 10$

17. $y = x^2$  (−3, 9), (3, 9)
    $y - 9 = 0$

18. $y = x^2 - 8x + 19$  (5, 4)
    $x - 0.5y = 3$

19. $y = x^2 + 11$  (−1, 12), (−11, 132)
    $y = -12x$

20. $y = 5x - 20$  (5, 5)
    $y = x^2 - 5x + 5$

**B** 21. **CITY PLANNING** A city planner is using the first quadrant of a coordinate plane to lay out new streets in a suburban development. Franklin Road is modeled by $y = -5x^2 + 32x + 2$ and Jefferson Street is modeled by $y = 28x + 1$.

a. At what point do Franklin Road and Jefferson Street intersect? (1, 29)

b. The city planner changes the equation for Jefferson Street to $y = 28x + 4$. How does this affect the intersection of the streets? Explain.
The streets do not intersect since the system has no real solutions.

22. **PARACHUTING** A parachutist jumps off a tall building. She falls freely for a few seconds before releasing her parachute. Her height during the free fall can be modeled by $y = -4.9x^2 + x + 360$, where $y$ is the height in meters and $x$ is the number of seconds after jumping. Her height after she releases the parachute can be modeled by $y = -4x + 142$.

a. How long after jumping did she release her parachute? 7.2 s

b. What was her height when she released her parachute? about 113 m

23. **MP STRUCTURE** The graph shows a quadratic function and a linear function $y = k$.

a. If the linear function were changed to $y = k + 4$, how many solutions would the system have? 2

b. If the linear function were changed to $y = k - 1$, how many solutions would the system have? 0

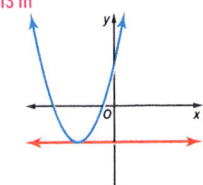

## Differentiated Homework Options

| Levels | AL Basic | OL Core | BL Advanced |
|---|---|---|---|
| Exercises | 10–20, 29–40 | 11–23 odd, 29–40 | 24–33 (optional: 34–40) |
| 2-Day Option | 11–19 odd, 34–40 | 10–20, 34–40 | |
| | 10–20 even, 29–33 | 21, 22, 29–33 | |

You can use ALEKS to provide additional remediation support with personalized instruction and practice.

---

Lesson 9-7 | Solving Systems of Linear and Quadratic Equations

# Practice

**Formative Assessment** Use Exercises 1–9 to assess students' understanding of the concepts in the lesson.

The Practice and Problem Solving exercises assess the content taught in the lesson. The Preparing for Assessment page is meant to be used as preparation for end-of-course assessments.

### MP Teaching the Mathematical Practices

**Sense-Making** Mathematically proficient students can understand a variety of approaches to solving a problem and can identify correspondences between different approaches. In Exercise 14, have students discuss the advantages of solving the problem algebraically versus graphically. Ask students which method leads to a more accurate solution and which method makes it easier to interpret the solution.

## Extra Practice

See page R9 for extra exercises for students who are approaching level or for on-level students who need additional reinforcement.

### Levels of Complexity Chart

The levels of the exercises progress from 1 to 3, with Level 1 indicating the lowest level of complexity.

| Exercises | 10–20 | 21–23, 34–40 | 24–33 |
|---|---|---|---|
| C Level 3 | | | ● |
| B Level 2 | | ● | |
| Level 1 | ● | | |

## Additional Answers

10. (3, 1)

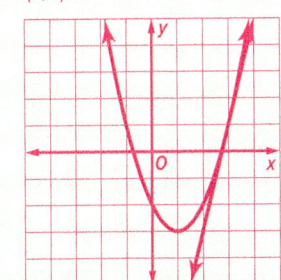

connectED.mcgraw-hill.com 619

**Lesson 9-7 | Solving Systems of Linear and Quadratic Equations**

## e Follow-Up

Students have explored two different ways to find the solution to a system of linear and quadratic equations. Ask: When is it best to solve a system by graphing, and when is it best to solve it by using substitution? Sample answer: Graphing is best when the equations can be graphed easily or when an approximate solution is sufficient. Solving by substitution is best when a precise solution is required.

## Assess

**Ticket Out the Door** Make several copies of cards labeled one solution, no solution, or two solutions. Give one card to each student and have them sketch a graph of a system of linear and quadratic equations with the given number of solutions.

### Additional Answer

30. Yes; since $y = c$ represents a horizontal line, it must intersect the parabola at its vertex.

---

24. **SENSE-MAKING** Solve $x^2 + 3x - 5 = 2x + 1$ by using a system of linear and quadratic equations. Describe your solution method. −3 and 2; Sample answer: Set each side of the equation equal to $y$ and solve the resulting system of equation by graphing.

Each graph shows the graph of a linear equation that is part of a system of equations. Write a quadratic equation that could be part of the system so that the system has the given number of solutions.

25. one solution

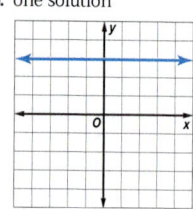

26. two solutions

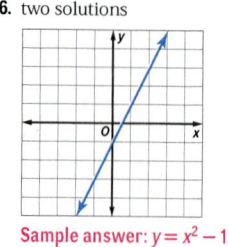

27. no solution

Sample answer: $y = x^2 + 3$   Sample answer: $y = x^2 - 1$   Sample answer: $y = -x^2$

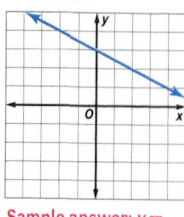

A.CED.2, A.REI.7

**H.O.T. Problems** Use Higher-Order Thinking Skills

28. **CHALLENGE** How many solutions does the system of equations have? Explain.
    $y = -2x^2 + 4x - 1$
    $-5x + y = 5$
    no solution; solving the system results in the quadratic equation $2x^2 + x + 6 = 0$, for which the discriminant is negative.

29. **TOOLS** Use a graphing calculator to solve the system involving two quadratic equations.
    $y = x^2 + 3x - 5$   (−2.5, −6.25), (1, −1)
    $y = -x^2$

30. **CRITIQUE ARGUMENTS** Sarah says that if a system of linear and quadratic equations involves a linear equation of the form $y = c$, where $c$ is a constant, and there is only one solution, then the graph of the linear equation intersects that of the quadratic equation at the vertex. Is she correct? Explain your answer. See margin.

31. **ERROR ANALYSIS** Tomas solved the system of equations and found no solution. What error did he make? Explain.
    $y = 0$
    $y - x^2 = 0$   The solution of the system is (0, 0), which is not the same as no solution.

32. **WRITING IN MATH** Explain why a linear and quadratic system of equations can have no more than two solutions. The graph of a linear equation is a straight line and the graph of a quadratic equation is a parabola. These graphs can intersect in at most two points.

33. **TOOLS** Use a graphing calculator to determine the solutions of each system of equations.
    a. $y = x^2 - 2x + 1$
       $y = 2x - 3$    (2, 1)
    b. $y = x^2 - 2x + 5$
       $y = 2x$    no solution
    c. $y = x^2 - 2x + 1$
       $y = 2x + 5$    (−0.83, 3.34), (4.83, 14.66)

---

### Standards for Mathematical Practice

| Emphasis On | Exercises |
|---|---|
| 1 Make sense of problems and persevere in solving them. | 10–13, 15–20, 24, 37 |
| 2 Reason abstractly and quantitatively. | 21–27, 36, 38 |
| 3 Construct viable arguments and critique the reasoning of others. | 28, 30, 31, 37 |
| 5 Use appropriate tools strategically. | 29, 33 |
| 7 Look for and make use of structure. | 23, 32, 35, 40 |

---

**Go Online!**   eBook

**Interactive Student Guide**

Use the *Interactive Student Guide* to deepen conceptual understanding.
· Solving Linear-Quadratic Systems

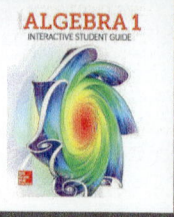

Lesson 9-7 | Solving Systems of Linear and Quadratic Equations

## Preparing for Assessment

34. Which ordered pair is a solution of the following system? **MP 7 A.REI.7  A**

    $y = x^2 + 7x + 10$
    $y = x + 1$

    - ◯ A  $(-3, -2)$
    - ◯ B  $(3, 40)$
    - ◯ C  $(3, 4)$
    - ◯ D  $(-4, -3)$

35. Which of the following are solutions of the given system? Select all that apply. **MP 7 A.REI.7  A, B**

    $y = x^2 + 7x + 12$
    $y = 2x + 8$

    - ☐ A  $(-4, 0)$
    - ☐ B  $(-1, 6)$
    - ☐ C  $(1, 10)$
    - ☐ D  $(4, 16)$
    - ☐ E  $(-1, -4)$
    - ☐ F  $(12, 8)$

36. In the system shown below, $k$ is a constant. If the system has exactly one solution, what is the value of $k$? **MP 2 A.REI.7  C**

    $y = x^2 - 3x + 7$
    $y = x + k$

    - ◯ A  $-4$
    - ◯ B  $2$
    - ◯ C  $3$
    - ◯ D  $7$

37. The quadratic equation $y = x^2 + 1$ is part of a linear and quadratic system of equations. If the system has two solutions, select all of the linear equations that could complete the system. **MP 1, 3 A.REI.7  B, D, F**

    - ☐ A  $y = 1$
    - ☐ B  $y = 3$
    - ☐ C  $y = x$
    - ☐ D  $y = x + 2$
    - ☐ E  $y = x - 2$
    - ☐ F  $y = -x + 3$

38. The graph below shows the system $y = x - 1$ and $y = -x^2 - x$. Based on the graph, which of the following is a solution of the system, assuming coordinates are rounded to the nearest tenth? **MP 1, 6  A.CED.2, A.REI.7  D**

    - ◯ A  $(0.4, -2.4)$
    - ◯ B  $(-1.0, -3.4)$
    - ◯ C  $(-0.5, 0.25)$
    - ◯ D  $(-2.4, -3.4)$

39. **MULTI-STEP** An animator is using a coordinate plane to create a cartoon about outer space. The path of a comet is represented by $y = x^2 - 3x + 4$. The path of a rocket is represented by $y - x = 1$. **MP 4  A.CED.2, A.REI.7**

    a. Graph the equations in the system on the same coordinate plane. **See margin for graph.**

    b. Is it possible that the rocket and the comet will collide? If so, what are the coordinates at which this might happen? If not, why not? **Yes, there are two places where they might collide: (3, 4) and (1, 2)**

    c. The animator wants to change the path of the rocket so that there is exactly one point at which the rocket and the comet could collide. How can the animator change the equation for the path of the rocket? **Sample answer: y = 1.75**

40. Which system has $(1, -1)$ as a solution? **MP 7 A.REI.7  C**

    - ◯ A  $y = x^2 - 3x + 1$
          $y = x + 2$
    - ◯ B  $y = x^2 + 3x - 1$
          $y = x - 2$
    - ◯ C  $y = x^2 - 2$
          $y = x - 2$
    - ◯ D  $y = x^2 - 2$
          $y = -x + 2$

## Preparing for Assessment

Exercises 34–40 require students to use the skills they will need on assessments. Each exercise is dual-coded with content standards and mathematical practice standards.

| Items | Content Standards | MP Mathematical Practices |
|---|---|---|
| 34 | A.REI.7 | 7 |
| 35 | A.REI.7 | 7 |
| 36 | A.REI.7 | 2 |
| 37 | A.REI.7 | 1, 3 |
| 38 | A.CED.2, A.REI.7 | 1, 6 |
| 39 | A.CED.2, A.REI.7 | 4 |
| 40 | A.REI.7 | 7 |

## Diagnose Student Errors

Survey student responses for each item. Class trends may indicate common errors and misconceptions.

34.

| A | CORRECT |
|---|---|
| B | When using Zero Product Property, used $x = 3$ as solution to $(x + 3)(x + 3) = 0$ |
| C | When using Zero Product Property, took $x = 3$ as solution to $(x + 3)(x + 3) = 0$ |
| D | Solved for $x$ and used it as the $y$-coordinate of the solution |

38.

| A | Found the incorrect $x$-coordinate |
|---|---|
| B | Found the incorrect $x$-coordinate |
| C | Found the vertex of the parabola |
| D | CORRECT |

## Additional Answer

39a.

## Go Online!

The most up-to-date resources available for your program can be found at connectED.mcgraw-hill.com.

# LESSON 9-8
# Analyzing Functions with Successive Differences

**SUGGESTED PACING (DAYS)**

| | | |
|---|---|---|
| 90 min. | 0.5 | 0.75 |
| 45 min. | 1 | 1.25 |
| | Instruction | Explore Lab |

## Track Your Progress

### Objectives
1. Identify linear, quadratic, and exponential functions from given data.
2. Write equations that model data.

### Mathematical Background
Linear, quadratic, and exponential functions can be used to model data. For linear functions, the first differences of the $y$-values are all equal. For quadratic functions, the first differences are not all equal, but the second differences are. For exponential functions, the ratios of successive $y$-values are all equal.

### THEN
**F.IF.6** Calculate and interpret the average rate of change of a function (presented symbolically or as a table) over a specified interval.

### NOW
**F.LE.1** Distinguish between situations that can be modeled with linear functions and with exponential functions.

**F.LE.2** Construct linear and exponential functions, including arithmetic and geometric sequences, given a graph, a description of a relationship, or two input-output pairs (include reading these from a table).

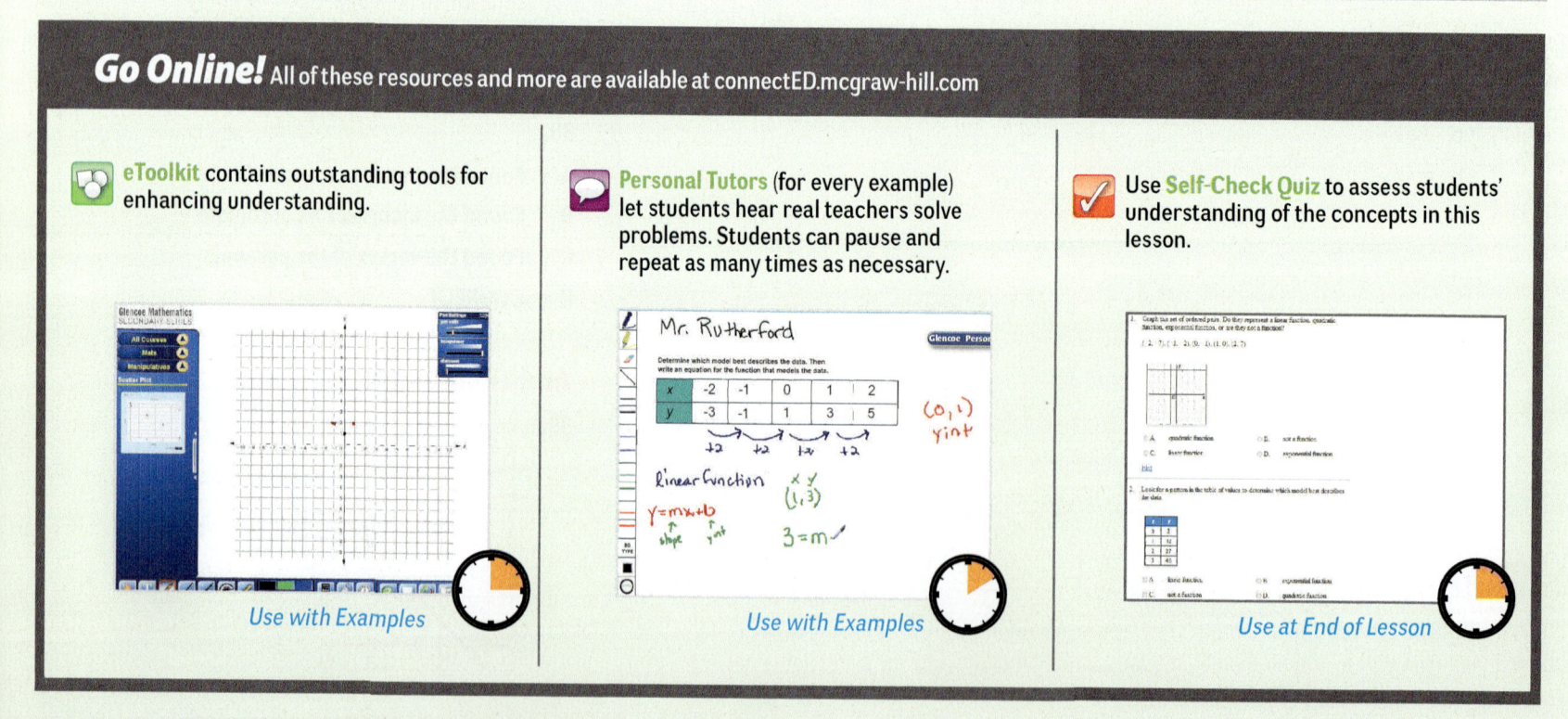

### NEXT
**F.BF.1b** Combine standard function types using arithmetic operations.

## Go Online! All of these resources and more are available at connectED.mcgraw-hill.com

**eToolkit** contains outstanding tools for enhancing understanding.

*Use with Examples*

**Personal Tutors** (for every example) let students hear real teachers solve problems. Students can pause and repeat as many times as necessary.

*Use with Examples*

Use **Self-Check Quiz** to assess students' understanding of the concepts in this lesson.

*Use at End of Lesson*

## OER Using Open Educational Resources
**Submitting Assignments** Communicate with students and parents or have students submit their assignments using **DropBox**. Dropbox is a free cloud service that allows you to set up a folder for each student. Assignments can be submitted or distributed and can be peer-reviewed. *Use as organization tool*

622A | Lesson 9-8 | Analyzing Functions with Successive Differences

# Go Online!
connectED.mcgraw-hill.com — Worksheets

# Differentiate Your Resources

**Extra Practice** Additional practice or homework; Skills Practice is best for approaching-level students and Practice is best for on-level and beyond-level students

## Skills Practice
AL  OL  ELL

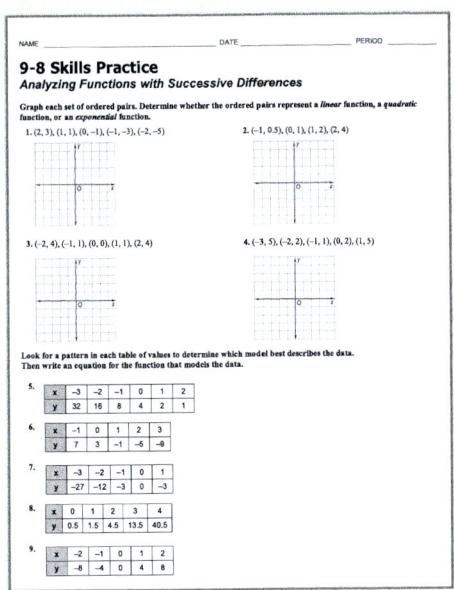

## Practice
AL  OL  BL  ELL

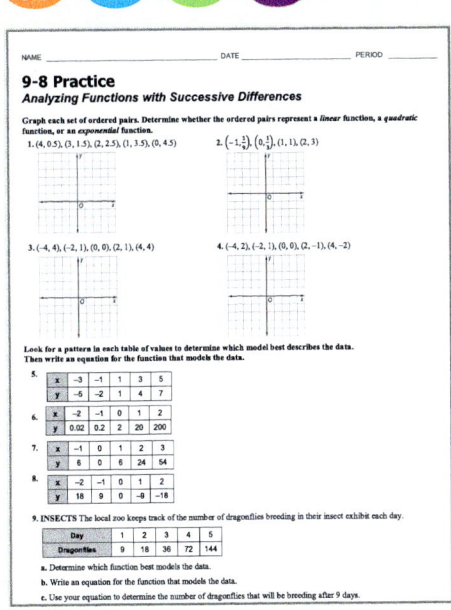

## Word Problem Practice
AL  OL  BL  ELL

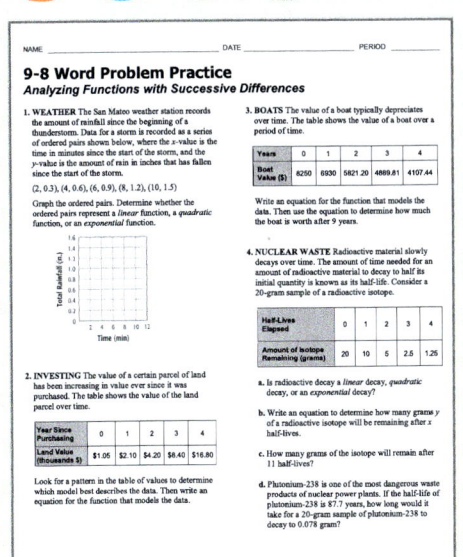

**Intervention** Reteaching and vocabulary activities that can be used with struggling or absent students and as ELL support

## Study Guide and Intervention
AL  OL  ELL

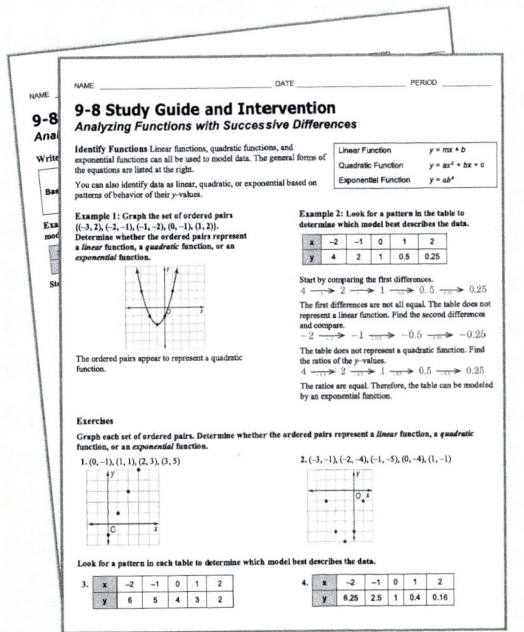

## Study Notebook
AL  OL  BL  ELL

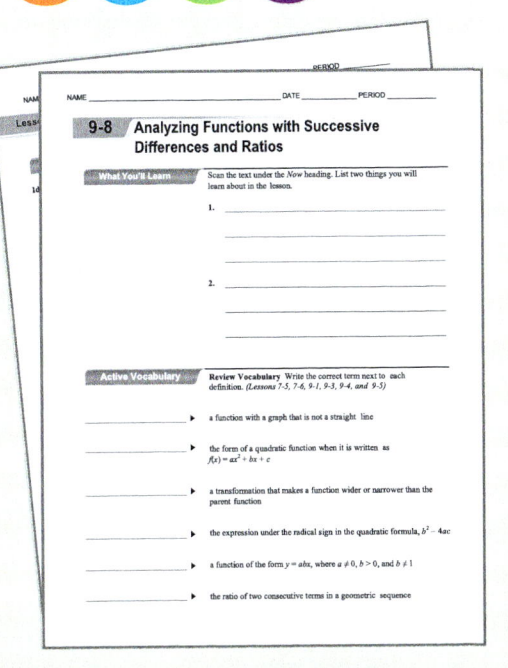

**Extension** Activities that can be used to extend lesson concepts

## Enrichment
OL  BL  ELL

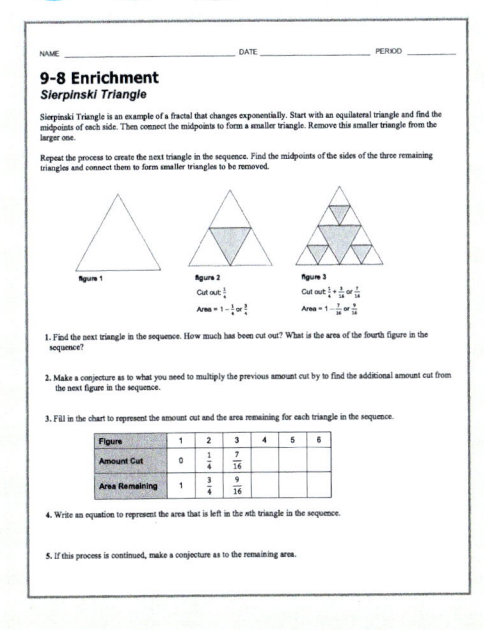

connectED.mcgraw-hill.com          622B

Lesson 9-8 | Analyzing Functions with Successive Differences

# Launch

Have students read the Why? section of the lesson. Ask:

- If the cost of 1–4 pounds of candy sold is given in the table, is this a geometric sequence? **no**

| Pounds | 1 | 2 | 3 | 4 |
|---|---|---|---|---|
| Price ($) | 4 | 8 | 12 | 16 |

- What is the common difference for this sequence? **4**

- What equation would represent this function? **$y = 4x$, where $x$ = number of pounds and $y$ = price**

# Teach

Ask the scaffolded questions for each example to build conceptual understanding for students at all levels.

## 1 Identify Functions

**Example 1 Choose a Model Using Graphs**

- **AL** What does the graph of a linear function look like? quadratic? exponential? **line; parabola, curve**

- **OL** How many points do you need to graph to see the function begin to take shape? **at least 3**

- **BL** Can you examine the coordinates and determine the function type without graphing? Explain. **See students' work.**

# Go Online!

**Interactive Whiteboard**
Use the *eLesson* or *Lesson Presentation* to present this lesson.

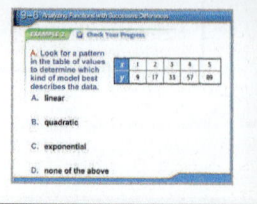

---

## LESSON 8
# Analyzing Functions with Successive Differences

**Then**
- You graphed linear, quadratic, and exponential functions.

**Now**
1. Identify linear, quadratic, and exponential functions from given data.
2. Write equations that model data.

**Why?**
Every year, the golf team sells candy to raise money for charity. By knowing what type of function models the sales of the candy, they can determine the best price for the candy.

**MP Mathematical Practices**
7 Look for and make use of structure.

**Content Standards**
F.IF.6 Calculate and interpret the average rate of change of a function (presented symbolically or as a table) over a specified interval. Estimate the rate of change from a graph.
F.LE.1 Distinguish between situations that can be modeled with linear functions and with exponential functions.
F.LE.2: Construct linear and exponential functions, including arithmetic and geometric sequences, given a graph, a description of a relationship, or two input-output pairs (include reading these from a table).
F.LE.3: Observe using graphs and tables that a quantity increasing exponentially eventually exceeds a quantity increasing linearly, quadratically, or (more generally) as a polynomial function.

**1 Identify Functions** You can use linear functions, quadratic functions, and exponential functions to model data. The general forms of the equations and a graph of each function type are listed below.

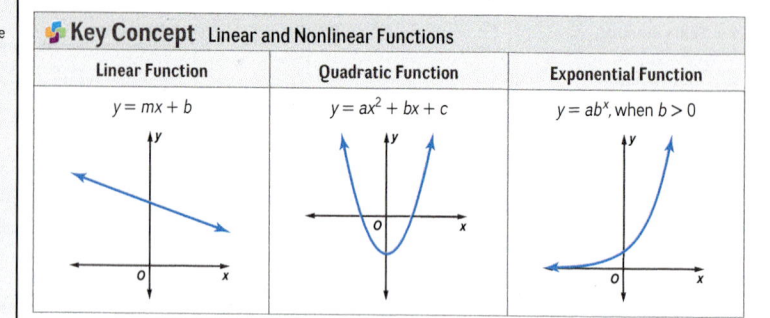

**Key Concept** Linear and Nonlinear Functions

| Linear Function | Quadratic Function | Exponential Function |
|---|---|---|
| $y = mx + b$ | $y = ax^2 + bx + c$ | $y = ab^x$, when $b > 0$ |

F.LE.1c

**Example 1    Choose a Model Using Graphs**

Graph each set of ordered pairs. Determine whether the ordered pairs represent a *linear* function, a *quadratic* function, or an *exponential* function.

a. $\{(-2, 5), (-1, 2), (0, 1), (1, 2), (2, 5)\}$

The ordered pairs appear to represent a quadratic function.

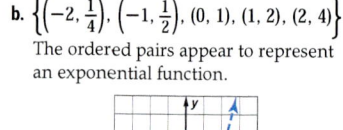

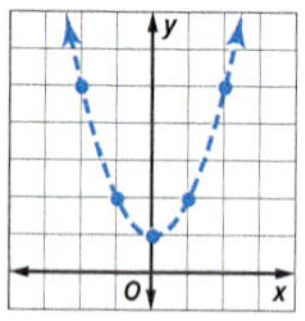

b. $\left\{\left(-2, \frac{1}{4}\right), \left(-1, \frac{1}{2}\right), (0, 1), (1, 2), (2, 4)\right\}$

The ordered pairs appear to represent an exponential function.

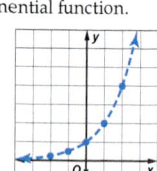

**Guided Practice**

1A. $(-2, -3), (-1, -1), (0, 1), (1, 3)$ **linear**

1B. $(-1, 0.25), (0, 1), (1, 4), (2, 16)$ **exponential**

---

**MP  Mathematical Practices Strategies**

**Look for and express regularity in repeated reasoning.**
Help students generalize and maintain oversight of the process of analyzing functions with successive differences. For example, ask:

- What is the key difference between exponential and quadratic functions?
  **For exponential functions, the ratio of successive *y*-values is equal; for quadratic functions, the differences of successive first differences are all equal.**

- If quadratic functions have equal second differences, what do you expect to be true about higher degree polynomial functions, such as a cubic function?
  **Sample answer: They will eventually have equal differences. For example, a cubic would have equal third differences.**

- How do you express arithmetic sequences as functions? geometric sequences?
  **as linear functions; as exponential functions**

622 | Lesson 9-8 | Analyzing Functions with Successive Differences

Another way to determine which model best describes data is to use patterns. The differences of successive $y$-values are called *first differences*. The differences of successive first differences are called *second differences*.

- If the differences of successive $y$-values are all equal, the data represent a linear function.
- If the second differences are all equal, but the first differences are not equal, the data represent a quadratic function.
- If the ratios of successive $y$-values are all equal and $r \neq 1$, the data represent an exponential function.

F.IF.6

**Watch Out!**

*x*-Values Before you check for successive differences or ratios, make sure the *x*-values are increasing by the same amount.

### Example 2   Choose a Model Using Differences or Ratios

Look for a pattern in each table of values to determine which kind of model best describes the data.

a.
| x | −2 | −1 | 0 | 1 | 2 |
|---|---|---|---|---|---|
| y | −8 | −3 | 2 | 7 | 12 |

First differences: −8  −3  2  7  12
                      5   5  5  5

Since the first differences are all equal, the table of values represents a linear function.

b.
| x | −1 | 0 | 1 | 2 | 3 |
|---|---|---|---|---|---|
| y | 8 | 4 | 2 | 1 | 0.5 |

First differences: 8  4  2  1  0.5
                   −4 −2 −1 −0.5

The first differences are not all equal. So, the table of values does not represent a linear function. Find the second differences and compare.

First differences:  −4  −2  −1  −0.5
Second differences:    2    1    0.5

The second differences are not all equal. So, the table of values does not represent a quadratic function. Find the ratios of the $y$-values and compare.

Ratios: 8  4  2  1  0.5
       $\frac{4}{8}=\frac{1}{2}$  $\frac{2}{4}=\frac{1}{2}$  $\frac{1}{2}$  $\frac{0.5}{1}=\frac{1}{2}$

The ratios of successive $y$-values are equal. Therefore, the table of values can be modeled by an exponential function.

**Guided Practice**

2A.
| x | −3 | −2 | −1 | 0 | 1 |
|---|---|---|---|---|---|
| y | −3 | −7 | −9 | −9 | −7 |

quadratic

2B.
| x | −2 | −1 | 0 | 1 | 2 |
|---|---|---|---|---|---|
| y | −18 | −13 | −8 | −3 | 2 |

linear

**Go Online!**

Review with your graphing calculator. Ask your teacher to assign the Graphing Calculator Easy File™ 5-Minute Check to you in ConnectED.

**2 Write Equations** Once you find the model that best describes the data, you can write an equation for the function. For a quadratic function in this lesson, the equation will have the form $y = ax^2$.

---

## Lesson 9-8 | Analyzing Functions with Successive Differences

**Need Another Example?**

Graph each set of ordered pairs. Determine whether the ordered pairs represent a *linear*, *quadratic*, or *exponential* function.

a. (1, 2), (2, 5), (3, 6), (4, 5), (5, 2) quadratic

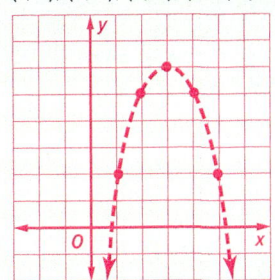

b. $(-1, 6), (0, 2), \left(1, \frac{2}{3}\right), \left(2, \frac{2}{9}\right)$ exponential

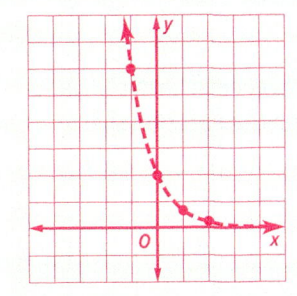

**Example 2  Choose a Model Using Differences or Ratios**

**AL** What is the difference between successive terms in an arithmetic sequence called? the common difference What is the ratio between successive terms in a geometric sequence called? the common ratio

**OL** How is the common difference of an arithmetic sequence related to first and second differences? If there is a common difference of the successive function values, the function is linear. If there is a common difference of the first differences, the function is quadratic.

**BL** What type of function would you expect to have all third differences equal? cubic

**Need Another Example?**

Look for a pattern in each table of values to determine which kind of model best describes the data.

a. linear
| x | −2 | −1 | 0 | 1 | 2 |
|---|---|---|---|---|---|
| y | −1 | 1 | 3 | 5 | 7 |

b. exponential
| x | −2 | −1 | 0 | 1 | 2 |
|---|---|---|---|---|---|
| y | 36 | 12 | 4 | $\frac{4}{3}$ | $\frac{4}{9}$ |

connectED.mcgraw-hill.com   623

**Lesson 9-8 | Analyzing Functions with Successive Differences**

## 2 Write Equations

### Example 3 Write an Equation

**AL** What is the first step to finding an equation for the data? choosing the function type

**OL** How is the common difference of an arithmetic sequence related to first and second differences? If there is a common difference of the successive function values, the function is linear. If there is a common difference of the first differences, the function is quadratic.

**BL** If you were to graph the function of the set of first differences, what type of function would it be? second differences? linear, constant

**Need Another Example?**
Determine which model best describes the data. Then write an equation for the function that models the data.

| x | 0  | 1  | 2   | 3    | 4     |
|---|----|----|-----|------|-------|
| y | −1 | −8 | −64 | −512 | −4096 |

exponential, $y = -(8)^x$

### Example 4 Write an Equation for a Real-World Situation

**AL** How is the constant ratio related to the equation of the function? It is the base of the exponential function.

**OL** Why is (0, 5) an easy choice for finding the value of b? $2^0 = 1$, so it is just one step to solve.

**BL** Do you think this sequence will continue indefinitely? Explain. Sample answer: No, the membership is not likely to continue doubling.

**Need Another Example?**
**Karate** The table shows the number of children enrolled in a beginner's karate class for four consecutive years. Determine which model best represents the data. Then write a function that models that data.

| x | 0 | 1  | 2  | 3  | 4  |
|---|---|----|----|----|----|
| y | 8 | 11 | 14 | 17 | 20 |

linear, $y = 3x + 8$

## Go Online!

**eSolutions Manual**
Create worksheets, answer keys, and solutions handouts for your assignments.

---

**Example 3** Write an Equation  F.LE.2

Determine which kind of model best describes the data. Then write an equation for the function that models the data.

| x | −4 | −3 | −2 | −1 | 0 |
|---|----|----|----|----|---|
| y | 32 | 18 | 8  | 2  | 0 |

**Step 1** Determine which model fits the data.

32, 18, 8, 2, 0
  −14, −10, −6, −2
     4,   4,  4

First differences are not equal.
Second differences are equal.
A quadratic function models the data.

**Step 2** Write an equation for the function that models the data.

The equation has the form $y = ax^2$. Find the value of $a$ by choosing one of the ordered pairs from the table of values. Let's use (−1, 2).

$y = ax^2$    Equation for quadratic function
$2 = a(-1)^2$    $x = -1$ and $y = 2$
$2 = a$    An equation that models the data is $y = 2x^2$.

**Watch Out!**
Finding a In Example 3, the point (0, 0) cannot be used to find the value of a. You will have to divide each side by 0, giving you an undefined value for a.

**Guided Practice**

3A. linear; $y = -4x + 3$

| x | −2 | −1 | 0 | 1  | 2  |
|---|----|----|---|----|----|
| y | 11 | 7  | 3 | −1 | −5 |

3B. exponential; $y = 3(2)^x$

| x | −3    | −2   | −1  | 0 | 1 |
|---|-------|------|-----|---|---|
| y | 0.375 | 0.75 | 1.5 | 3 | 6 |

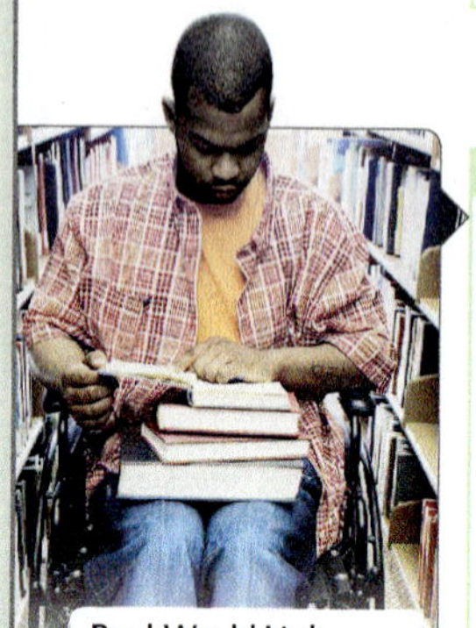

---

**Real-World Example 4** Write an Equation for a Real-World Situation  F.LE.2

**BOOK CLUB** The table shows the number of book club members for four consecutive years. Determine which model best represents the data. Then write a function that models the data.

**Understand** Find a model for the data, and then write a function.

| Time (years) | 0 | 1  | 2  | 3  | 4  |
|--------------|---|----|----|----|----|
| Members      | 5 | 10 | 20 | 40 | 80 |

**Plan** Find a pattern using successive differences or ratios. Then use the general form of the equation to write a function.

**Solve** The constant ratio is 2. This is the value of the base. An exponential function of the form $y = ab^x$ models the data.

$y = ab^x$    Equation for exponential function
$5 = a(2)^0$    $x = 0, y = 5$ and $b = 2$
$5 = a$    The equation that models the data is $y = 5(2)^x$.

**Check** You used (0, 5) to write the function. Verify that every other ordered pair satisfies the equation.

**Real-World Link**
A poll by the National Education Association found that 87% of all teens polled found reading relaxing, 85% viewed reading as rewarding, and 79% found reading exciting.
Source: American Demographics

**Guided Practice**

4. **ADVERTISING** The table shows the cost of placing an ad in a newspaper. Determine a model that best represents the data and write a function that models the data. linear; $C = 2.1n + 4$

| No. of Lines    | 5     | 6     | 7     | 8     |
|-----------------|-------|-------|-------|-------|
| Total Cost ($)  | 14.50 | 16.60 | 18.70 | 20.80 |

---

## Differentiated Instruction ELL

**Verbal/Linguistic Learners** Have students write a list of tips to help someone determine which model best describes a set of data and to write an equation for the function.

## Teaching Tip

**Checking the Equation** Encourage students to verify that the ordered pairs satisfy the written equation.

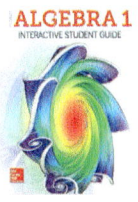

**Lesson 9-8 | Analyzing Functions with Successive Differences**

## Check Your Understanding

= Step-by-Step Solutions begin on page R11.

**Go Online!** for a Self-Check Quiz

**Example 1** Graph each set of ordered pairs. Determine whether the ordered pairs represent a *linear*
F.LE.1c function, a *quadratic* function, or an *exponential* function.

1. (−2, 8), (−1, 5), (0, 2), (1, −1)
2. (−3, 7), (−2, 3), (−1, 1), (0, 1), (1, 3)
3. (−3, 8), (−2, 4), (−1, 2), (0, 1), (1, 0.5)
4. (0, 2), (1, 2.5), (2, 3), (3, 3.5)

1–4. See Ch. 9 Answer Appendix.

**Example 2** Look for a pattern in each table of values to determine which kind of model best describes
F.IF.6 the data.

5. quadratic

| x | 0 | 1 | 2 | 3 | 4 |
|---|---|---|---|---|---|
| y | 5 | 8 | 17 | 32 | 53 |

6. linear

| x | −3 | −2 | −1 | 0 |
|---|---|---|---|---|
| y | −6.75 | −7.5 | −8.25 | −9 |

7. exponential

| x | −1 | 0 | 1 | 2 | 3 |
|---|---|---|---|---|---|
| y | 3 | 6 | 12 | 24 | 48 |

8. quadratic

| x | 3 | 4 | 5 | 6 | 7 |
|---|---|---|---|---|---|
| y | −1.5 | 0 | 2.5 | 6 | 10.5 |

**Example 3** Determine which kind of model best describes the data. Then write an equation for the
F.LE.2 function that models the data.

9. 

| x | −1 | 0 | 1 | 2 | 3 |
|---|---|---|---|---|---|
| y | 1 | 3 | 9 | 27 | 81 |

exponential; $y = 3(3)^x$

10. 

| x | −5 | −4 | −3 | −2 | −1 |
|---|---|---|---|---|---|
| y | 125 | 80 | 45 | 20 | 5 |

quadratic; $y = 5x^2$

11. 

| x | −3 | −2 | −1 | 0 | 1 |
|---|---|---|---|---|---|
| y | 1 | 1.5 | 2 | 2.5 | 3 |

linear; $y = \frac{1}{2}x + \frac{5}{2}$

12. 

| x | −1 | 0 | 1 | 2 |
|---|---|---|---|---|
| y | −1.25 | −1 | −0.75 | −0.5 |

linear; $y = \frac{1}{4}x − 1$

**Example 4** 13. **PLANTS** The table shows the height of
F.LE.2 a plant for four consecutive weeks.
Determine which kind of function
best models the height. Then write a
function that models the data. linear; $y = 0.5x + 3$

| Week | 0 | 1 | 2 | 3 | 4 |
|---|---|---|---|---|---|
| Height (in.) | 3 | 3.5 | 4 | 4.5 | 5 |

## Practice and Problem Solving

Extra Practice is on page R9.

**Example 1** Graph each set of ordered pairs. Determine whether the ordered pairs represent a *linear*
F.LE.1c function, a *quadratic* function, or an *exponential* function.

14–19. See Ch. 9 Answer Appendix.

14. (−1, 1), (0, −2), (1, −3), (2, −2), (3, 1)
15. (1, 2.75), (2, 2.5), (3, 2.25), (4, 2)
16. (−3, 0.25), (−2, 0.5), (−1, 1), (0, 2)
17. (−3, −11), (−2, −5), (−1, −3), (0, −5)
18. (−2, 6), (−1, 1), (0, −4), (1, −9)
19. (−1, 8), (0, 2), (1, 0.5), (2, 0.125)

**Examples 2–3** Look for a pattern in each table of values to determine which kind of model best describes
F.IF.6 the data. Then write an equation for the function that models the data.
F.LE.2

20. 

| x | −3 | −2 | −1 | 0 |
|---|---|---|---|---|
| y | −8.8 | −8.6 | −8.4 | −8.2 |

linear; $y = 0.2x − 8.2$

21. 

| x | −2 | −1 | 0 | 1 | 2 |
|---|---|---|---|---|---|
| y | 10 | 2.5 | 0 | 2.5 | 10 |

quadratic; $y = 2.5x^2$

22. 

| x | −1 | 0 | 1 | 2 | 3 |
|---|---|---|---|---|---|
| y | 0.75 | 3 | 12 | 48 | 192 |

exponential; $y = 3(4)^x$

23. 

| x | −2 | −1 | 0 | 1 | 2 |
|---|---|---|---|---|---|
| y | 0.008 | 0.04 | 0.2 | 1 | 5 |

exponential; $y = 0.2(5)^x$

24. quadratic; $y = 4.2x^2$

| x | 0 | 1 | 2 | 3 | 4 |
|---|---|---|---|---|---|
| y | 0 | 4.2 | 16.8 | 37.8 | 67.2 |

25. 

| x | −3 | −2 | −1 | 0 | 1 |
|---|---|---|---|---|---|
| y | 14.75 | 9.75 | 4.75 | −0.25 | −5.25 |

linear; $y = −5x − 0.25$

## Differentiated Homework Options

| Levels | AL Basic | OL Core | BL Advanced |
|---|---|---|---|
| Exercises | 14–26, 32–41 | 15–25 odd, 26–30, 32–41 | 27–35, (optional: 36–41) |
| 2-Day Option | 15–25 odd, 36–41 | 14–26, 36–41 | |
| | 14–26 even, 32–35 | 27–30, 32–35 | |

You can use ALEKS to provide additional remediation support with personalized instruction and practice.

## Practice

**Formative Assessment** Use Exercises 1–13 to assess students' understanding of the concepts in this lesson.

The Practice and Problem Solving exercises assess the content taught in the lesson. The Preparing for Assessment page is meant to be used as preparation for end-of-course assessments.

### Exercise Alert

**Grid Paper** For Exercises 1–4, 14–19, and 27, students will need grid paper.

### Extra Practice

See page R9 for extra exercises for students who are approaching level or for on-level students who need additional reinforcement.

### Go Online!  eBook

**Interactive Student Guide**

Use the *Interactive Student Guide* to deepen conceptual understanding.

· Analyzing Functions with Successive Differences

# Lesson 9-8 | Analyzing Functions with Successive Differences

## Levels of Complexity Chart

The levels of the exercises progress from 1 to 3, with Level 1 indicating the lowest level of complexity.

| Exercises | 14–26 | 27, 28, 36–41 | 29–35 |
|---|---|---|---|
| Level 3 | | | ● |
| Level 2 | | ● | |
| Level 1 | ● | | |

## MP Teaching the Mathematical Practices

**Construct Arguments** Mathematically proficient students reason inductively. In Exercise 32, tell students to look for a pattern.

## Assess

**Yesterdays News** Ask students to write how yesterday's lesson on quadratic equations helped them with today's lesson.

## Additional Answers

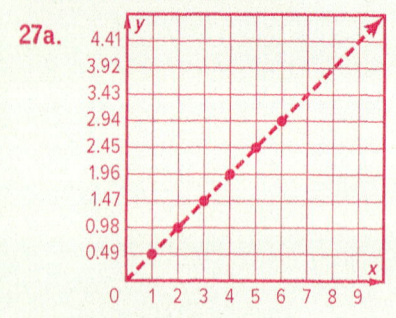

27a.

[graph showing linear relationship with points at 0.49, 0.98, 1.47, 1.96, 2.45, 2.94, 3.43, 3.92, 4.41 along x-values 1 through 9]

34. If one linear term is $ax$, the next term is $a(x + 1)$, and the difference between the terms is $a(x + 1) - ax = ax + a - ax$ or $a$. If one exponential term is $a^x$, the next term is $a^{x+1}$, and the ratio of the terms is $\frac{a^{x+1}}{a^x}$ or $a$.

35. Sample answer: The data can be graphed to determine which function best models the data. You can also find the differences in ratios of the $y$-values. If the first differences are constant, the data can be modeled by a linear function. If the second differences are constant but the first differences are not, the data can be modeled by a quadratic function. If the ratios are constant, then the data can be modeled by an exponential function.

---

**Example 4**
**F.LE.2**

26. **WEBSITES** A company tracked the number of visitors to its website over four days. Determine which kind of model best represents the number of visitors to the website with respect to time. Then write a function that models the data. **quadratic; $y = 0.9x^2$**

| Day | 0 | 1 | 2 | 3 | 4 |
|---|---|---|---|---|---|
| Visitors (in thousands) | 0 | 0.9 | 3.6 | 8.1 | 14.4 |

**B** 27. **FROZEN YOGURT** The cost of a build-your-own cup of frozen yogurt depends on the weight of the contents. The table shows the cost for up to 6 ounces.

| Ounces | 1 | 2 | 3 | 4 | 5 | 6 |
|---|---|---|---|---|---|---|
| Cost ($) | 0.49 | 0.98 | 1.47 | 1.96 | 2.45 | 2.94 |

a. Graph the data and determine which kind of function best models the data. **See margin for graph; linear.**
b. Write an equation for the function that models the data. **$y = 0.49x$**
c. Use your equation to determine how much 10 ounces of yogurt would cost. **$4.90**

28. **DEPRECIATION** The value of a car depreciates over time. The table shows the value of a car over a period of time.

| Year | 0 | 1 | 2 | 3 | 4 |
|---|---|---|---|---|---|
| Value ($) | 18,500 | 15,910 | 13,682.60 | 11,767.04 | 10,119.65 |

a. Determine which kind of function best models the data. **exponential**
b. Write an equation for the function that models the data. **$y = 18,500(0.86)^t$**
c. Use your equation to determine how much the car is worth after 7 years. **$6436.66**

**C** 29. **BACTERIA** A scientist estimates that a bacteria culture with an initial population of 12 will triple every hour.
a. Make a table to show the bacteria population for the first 4 hours. **See Ch.9 Answer Appendix.**
b. Which kind of model best represents the data? **exponential**
c. Write a function that models the data. **$b = 12(3)^t$**
d. How many bacteria will there be after 8 hours? **78,732**

30. **PRINTING** A printing company charges the fees shown to print flyers. Write a function that models the total cost of the flyers, and determine how much 30 flyers would cost. **$c = 0.15t + 25$; $29.50**

> **Quick 2 U Printing**
> Set Up Fee $25
> 15¢ each flyer

32. Linear functions have a constant first difference and quadratic functions have a constant second difference, so cubic equations would have a constant third difference. **F.IF.6, F.LE.1, F.LE.2, F.LE.3**

**H.O.T. Problems** Use Higher-Order Thinking Skills

31. **MP REASONING** Write a function that has constant second differences, first differences that are not constant, a $y$-intercept of $-5$, and contains the point (2, 3). **Sample answer: $y = 2x^2 - 5$**

32. **MP CONSTRUCT ARGUMENTS** What type of function will have constant third differences but not constant second differences? Explain.

33. **MP STRUCTURE** Write a linear function that has a constant first difference of 4. **$y = 4x + 1$**

34. **PROOF** Write a paragraph proof to show that linear functions grow by equal differences over equal intervals, and exponential functions grow by equal factors over equal intervals. (*Hint:* Let $y = ax$ represent a linear function and let $y = a^x$ represent an exponential function.) **See margin.**

35. **Ⓔ WRITING IN MATH** How can you determine whether a given set of data should be modeled by a *linear* function, a *quadratic* function, or an *exponential* function? **See margin.**

## MP Standards for Mathematical Practice

| Emphasis On | Exercises |
|---|---|
| 2 Reason abstractly and quantitatively. | 31 |
| 3 Construct viable arguments and critique the reasoning of others. | 32, 34, 35 |
| 4 Model with mathematics. | 13, 26–30 |
| 7 Look for and make use of structure. | 14–25, 33, 36, 39–41 |
| 8 Look for and express regularity in repeated reasoning. | 36–38 |

Lesson 9-8 | Analyzing Functions with Successive Differences

## Preparing for Assessment

**36a. MULTI-STEP** Which function best models the data in the table of values? (MP) 7, 8 F.IF.6, F.LE.1, F.LE.2 **D**

| x | 0 | 1 | 2 | 3 | 4 |
|---|---|---|---|---|---|
| y | 3 | 0 | −1 | 0 | 3 |

- A $y = x - 3$
- B $y = -\left(\frac{1}{2}\right)^x$
- C $y = x^2 + 4x + 3$
- D $y = x^2 - 4x + 3$

**b.** Copy and complete the table below so that the data is best modeled by a linear equation. Write this linear equation.

$y = -3x + 3$

| x | 0 | 1 | 2 | 3 | 4 |
|---|---|---|---|---|---|
| y | 3 | 0 | −3 | −6 | −9 |

**c.** Copy and complete the table below so that the data is best modeled by an exponential equation. Write this exponential equation.

$y = 4^x$

| x | 0 | 1 | 2 | 3 | 4 |
|---|---|---|---|---|---|
| y | 1 | 4 | 16 | 64 | 256 |

**37.** Which tables of values are best modeled by an exponential equation? Select all that apply. (MP) 8 F.LE.1, F.LE.2, F.IF.6 **B, D, E**

- A 
| x | −2 | −1 | 0 | 1 | 2 |
|---|---|---|---|---|---|
| y | −4 | −1 | 0 | 1 | 4 |

- B 
| x | −2 | −1 | 0 | 1 | 2 |
|---|---|---|---|---|---|
| y | 0.5 | 1.5 | 4.5 | 13.5 | 40.5 |

- C 
| x | −2 | −1 | 0 | 1 | 2 |
|---|---|---|---|---|---|
| y | 1 | 0 | 1 | 4 | 9 |

- D 
| x | −2 | −1 | 0 | 1 | 2 |
|---|---|---|---|---|---|
| y | 2 | 4 | 8 | 16 | 32 |

- E 
| x | −2 | −1 | 0 | 1 | 2 |
|---|---|---|---|---|---|
| y | 20 | 10 | 5 | 2.5 | 1.25 |

- F 
| x | −2 | −1 | 0 | 1 | 2 |
|---|---|---|---|---|---|
| y | −1 | 0 | −1 | −4 | −9 |

**38.** Jake graphed the sets of ordered pairs in the table below.

| x | −2 | −1 | 0 | 1 | 2 |
|---|---|---|---|---|---|
| y | 10 | 6 | 2 | −2 | −6 |

Which equation best models the data in the table? (MP) 8 F.LE.1, F.LE.2 **A**

- A $y = -4x + 2$
- B $y = -4x - 2$
- C $y = -2x^2 + 2$
- D $y = -2x^2 - 2$

**39.** For what value of $c$ does the equation $y = x^2 + 6x + c$ model the data in the table? (MP) 7 F.LE.1, F.LE.2 **11**

| x | −4 | −3 | −2 | −1 | 0 |
|---|---|---|---|---|---|
| y | 3 | 2 | 3 | 6 | 11 |

**40.** What type of model best fits the data? (MP) 3, 7 F.LE.1 **D**

| x | −1 | 0 | 1 | 2 | 3 |
|---|---|---|---|---|---|
| y | 5 | 6 | 9 | 13 | 17 |

- A linear
- B quadratic
- C exponential
- D none of the above

**41.** What type of model best fits the data? (MP) 3, 7 F.LE.1 **B**

| x | −2 | −1 | 0 | 1 | 2 |
|---|---|---|---|---|---|
| y | 4 | 3 | 4 | 7 | 12 |

- A linear
- B quadratic
- C exponential
- D none of the above

## Preparing for Assessment

Exercises 36–41 require students to use the skills they will need on standardized assessments. Each exercise is dual-coded with content standards and mathematical practice standards.

### Dual Coding

| Items | Content Standards | (MP) Mathematical Practices |
|---|---|---|
| 36 | F.IF.6, F.LE.1, F.LE.2 | 7, 8 |
| 37 | F.IF.6, F.LE.1, F.LE.2 | 8 |
| 38 | F.LE.1, F.LE.2 | 8 |
| 39 | F.LE.1, F.LE.2 | 7 |
| 40, 41 | F.LE.1 | 3, 7 |

## Diagnose Student Errors

Survey student responses for each item. Class trends may indicate common errors and misconceptions.

**36a.**

| A | Identified linear instead of quadratic equation |
|---|---|
| B | Identified exponential instead of quadratic equation |
| C | Used wrong sign for x-term |
| D | CORRECT |

**38.**

| A | CORRECT |
|---|---|
| B | Used wrong sign for y-intercept |
| C | Identified quadratic instead of linear equation |
| D | Identified quadratic instead of linear equation |

## Differentiated Instruction (OL) (BL)

**Extension** Ask students to make a table of values for $y = \sqrt{x}$ and then graph it. Be sure students do not include any negative values for x, because the square root of a negative number is not a real number.

| x | 0 | 1 | 4 | 9 | 16 | 25 |
|---|---|---|---|---|---|---|
| y | 0 | 1 | 2 | 3 | 4 | 5 |

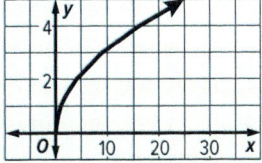

## Go Online!

### Quizzes

Students can use *Self-Check Quizzes* to check their understanding of this lesson. You can also give the *Chapter Quiz*, which covers the content in Lessons 9-1 through 9-8.

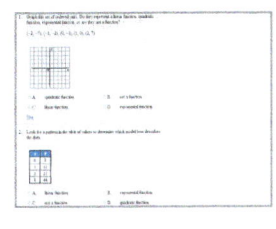

connectED.mcgraw-hill.com 627

# Extend 9-8

## Launch

**Objective** Use a graphing calculator to find an appropriate regression equation for a set of data.

### Materials for Each Group
- TI-83/84 Plus or other graphing calculator

### Teaching Tip
Remind students that they learned about regression and median fit lines in Chapter 4. Before students begin, make sure they turn their Diagnostic setting on. To do this from the home screen, press 2nd [CATALOG], scroll down and click **DiagnosticOn**, then press ENTER ENTER.

## Teach

**Working in Cooperative Groups** Put students in groups of two or three, mixing abilities. Have groups complete the Activities and Exercises 1–4.
**ELL**

- In Step 1 of Activity 1, make sure students clear previous lists before entering the data. Students should enter years after 2010 in **L1** and the number of flights in **L2**.

- In Step 2 of Activity 1, point out that the $R^2$ value, 0.9998751467 is the coefficient of determination. Generally, the closer this is to 1, the better the model.

- For Step 3 of Activity 1, tell students that they must copy the quadratic regression exactly to the Y = LIST in order to get the proper graph.

**Practice** Have students complete Exercise 5.

## Go Online!

### Graphing Calculators
Students can use the Graphing Calculator Personal Tutors to review the use of the graphing calculator to represent functions. They can also use the Other Calculator Keystrokes, which cover lab content for students with calculators other than the TI-84 Plus.

---

# EXTEND 9-8
## Graphing Technology Lab
# Curve Fitting

If there is a constant increase or decrease in data values, there is a linear trend. If the values are increasing or decreasing more and more rapidly, there may be a quadratic or exponential trend.

With a graphing calculator, you can find the appropriate regression equation and an $R^2$ value. $R^2$ is the **coefficient of determination**. The closer $R^2$ is to 1, the better the model fits the data.

**Mathematical Practices**
5 Use tools strategically

**Content Standards**
**F.LE.2** Construct linear and exponential functions, including arithmetic and geometric sequences, given a graph, a description of a relationship, or two input-output pairs (include reading these from a table).

**S.ID.6a** Fit a function to the data; use functions fitted to data to solve problems in the context of the data. Use given functions or choose a function suggested by the context. Emphasize linear, quadratic, and exponential models.

### Activity 1

**CHARTER AIRLINE** The table shows the average monthly number of flights made each year by a charter airline that was founded in 2010. Work cooperatively to predict the number of flights in 2030 and when the airline will meet its goal of 200 flights per month.

| Year | 2010 | 2011 | 2012 | 2013 | 2014 | 2015 | 2016 | 2017 |
|---|---|---|---|---|---|---|---|---|
| Flights | 17 | 20 | 24 | 28 | 33 | 38 | 44 | 50 |

**Step 1** Make a scatter plot.

Enter the number of years since 2010 in **L1** and the number of flights in **L2**. Graph the scatter plot. It appears that the data may have either a quadratic or exponential trend.

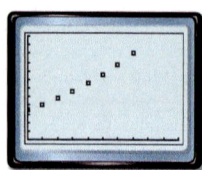

[0, 10] scl: 1 by [0, 60] scl: 5

**Step 2** Find the regression equation.

Check both trends by examining their regression equations. To acquire the exponential or quadratic equation select **ExpReg** or **QuadReg** on the **STAT** menu. To choose, fit both and use the one with the $R^2$ value closest to 1. In this case the quadratic is a better fit.

```
QuadReg
y=ax²+bx+c
a=.25
b=2.988095238
c=16.91666667
R²=.9998751467
```

**Step 3** Graph the quadratic regression equation.

Copy the equation to the **Y=** list and graph.

**Step 4** Predict using the equation.

Use the graph to predict the number of flights in 2030. There will be approximately 177 flights per month if this trend continues.

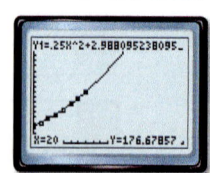

**Step 5** Find a solution using the model.

Use the graph to determine when the airline will reach its goal of 200 flights a month. Graph the equation $y = 200$ and find the intersection. If this trend continues the airline will reach their goal in 2032.

[0, 25] scl: 1 by [0, 200] scl: 5

*(continued on the next page)*

# Extend 9-8

## Activity 2

**FROGS** The table shows the frog population in a small pond. Work cooperatively to predict what the frog population will be in year 11 and when the pond will reach its capacity of 500 frogs.

| Year | 0 | 1 | 2 | 3 | 4 | 5 | 6 | 7 | 8 |
|---|---|---|---|---|---|---|---|---|---|
| Population | 80 | 90 | 102 | 115 | 130 | 147 | 166 | 188 | 212 |

**Step 1** Make a scatter plot. Enter the years 1–8 in **L1** and the population in **L2**. Graph the scatter plot. The data appear to have either a quadratic or exponential trend.

**Step 2** Find the regression equation. Check both trends by examining their regression equations. To acquire the exponential or quadratic equation select **ExpReg** or **QuadReg** on the **STAT** menu. To choose, fit both and use the one with the $R^2$ value closest to 1.

**Step 3** Graph the exponential regression equation. Copy the equation to the **Y=** list and graph.

**Step 4** Predict using the equation.

Use the graph to predict the number of frogs in year 11. There will be approximately 306 frogs in the pond if this trend continues.

**Step 5** Find a solution using the model. The pond cannot maintain more than 500 frogs. Use the graph to determine when the pond will reach capacity. Graph the equation $y = 500$ and find the intersection. If this trend continues the pond will reach capacity in year 15.

## Exercises

Plot each set of data points. Determine whether to use a *linear*, *quadratic* or *exponential* regression equation. State the coefficient of determination.

1. quadratic; $R^2 \approx 0.969$

| x | y |
|---|---|
| 1 | 30 |
| 2 | 40 |
| 3 | 50 |
| 4 | 55 |
| 5 | 50 |
| 6 | 40 |

2. quadratic; $R^2 \approx 0.964$

| x | y |
|---|---|
| 0.0 | 12.1 |
| 0.1 | 9.6 |
| 0.2 | 6.3 |
| 0.3 | 5.5 |
| 0.4 | 4.8 |
| 0.5 | 1.9 |

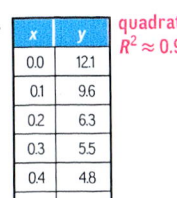

3. quadratic; $R^2 \approx 0.980$

| x | y |
|---|---|
| 0 | 1.1 |
| 2 | 3.3 |
| 4 | 2.9 |
| 6 | 5.6 |
| 8 | 11.9 |
| 10 | 19.8 |

4. quadratic; $R^2 \approx 0.840$

| x | y |
|---|---|
| 1 | 1.67 |
| 5 | 2.59 |
| 9 | 4.37 |
| 13 | 6.12 |
| 17 | 5.48 |
| 21 | 3.12 |

5c. $y = (306)(0.96^x)$; $R^2 = 0.957$; $D = \{x \mid x > 0\}$; $R = \{y \mid y > 0\}$

5. **BAKING** Alyssa baked a cake and is waiting for it to cool so she can ice it. The table shows the temperature of the cake every 5 minutes after Alyssa took it out of the oven.

   a. Make a scatter plot of the data. **See margin.**

   b. Which regression equation has an $R^2$ value closest to 1? Is this the equation that best fits the context of the problem? Explain your reasoning. **See margin.**

   c. Find an appropriate regression equation, and state the coefficient of determination. What is the domain and range?

   d. Alyssa will ice the cake when it reaches room temperature (70°F). Use the regression equation to predict when she can ice her cake. **36 min**

| Time (min) | Temperature (°F) |
|---|---|
| 0 | 350 |
| 5 | 244 |
| 10 | 178 |
| 15 | 137 |
| 20 | 112 |
| 25 | 96 |
| 30 | 89 |

## Assess

### Formative Assessment

In Step 4 of Activity 1, the quadratic regression estimated that during the 7th month the most passengers flew, about 59 million. Ask students to explain why the estimated number of passengers that flew in the 7th month may not be completely accurate. **Sample answer: The quadratic regression equation is a best fit to data points that do not fall on an actual graphed function. There will be differences between actual data points and points that fall on the regression function.**

### From Concrete to Abstract

**Ask:**
How do you determine whether to use a linear, quadratic, or exponential regression equation for your data? **Sample answer: Make a scatter plot of your data points. If it looks close to a straight line, use a linear regression equation. If the data points follow a curve, fit a quadratic regression equation and an exponential regression equation to your points. The model with the coefficient of determination closest to 1 is the model to use.**

### Additional Answers

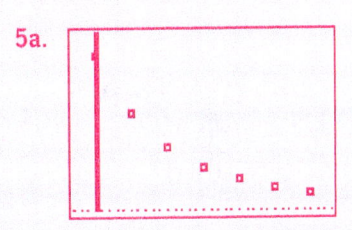

5a.

[−5, 35] scl:1 by [0, 400] scl:1

5b. The quadratic equation has an $R^2$ value closest to 1. However in a quadratic trend, the cake would cool and then heat up again. The exponential regression best fits the context of the problem.

## Go Online!

The most up-to-date resources available for your program can be found at connectED.mcgraw-hill.com.

# LESSON 9-9
# Combining Functions

**SUGGESTED PACING (DAYS)**

90 min. | 1
45 min. | 1.5

Instruction

## Track Your Progress

### Objective
1. Combine functions by using addition and subtraction.
2. Combine functions by using multiplication.

### Mathematical Background
Students have had experience working with linear functions, exponential functions, and quadratic functions. In modeling situations it is often useful to combine various function types through arithmetic operations. This lesson focuses on addition, subtraction, and multiplication. It is also possible to combine functions using division. Students will explore this in future mathematics courses when they study rational expressions and learn to divide polynomials.

### THEN
**F.LE.1** Distinguish between situations that can be modeled with linear functions and with exponential functions.

**F.IF.6** Calculate and interpret the average rate of change of a function (presented symbolically or as a table) over a specified interval. Estimate the rate of change from a graph.

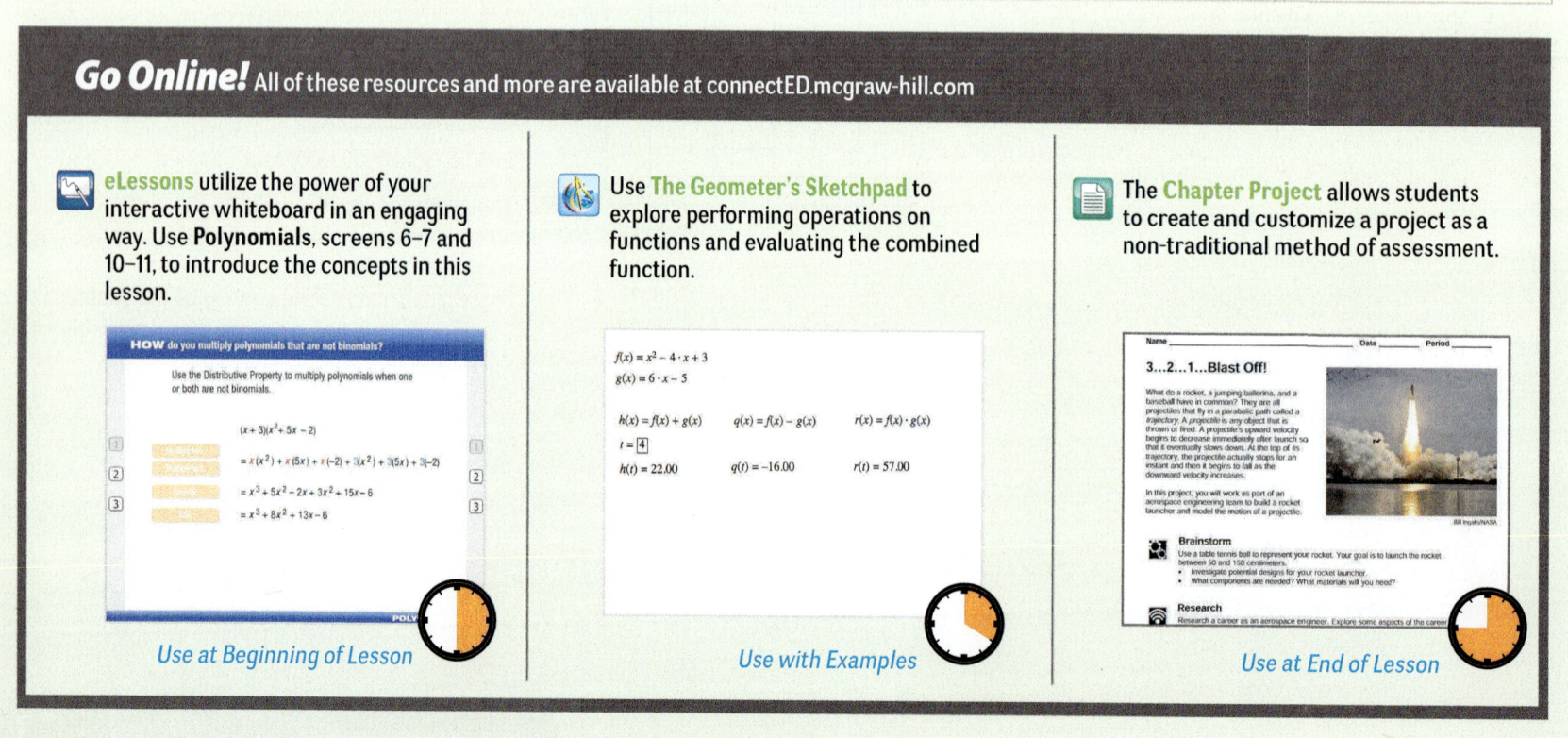

### NOW
**F.BF.1b** Combine standard function types using arithmetic operations.

### NEXT
**A.APR.1** Understand that polynomials form a system analogous to the integers, namely, they are closed under the operations of addition, subtraction, and multiplication; add, subtract, and multiply polynomials.

**F.BF.1c** Compose functions.

---

**Go Online!** All of these resources and more are available at connectED.mcgraw-hill.com

**eLessons** utilize the power of your interactive whiteboard in an engaging way. Use **Polynomials**, screens 6–7 and 10–11, to introduce the concepts in this lesson.

*Use at Beginning of Lesson*

Use **The Geometer's Sketchpad** to explore performing operations on functions and evaluating the combined function.

*Use with Examples*

The **Chapter Project** allows students to create and customize a project as a non-traditional method of assessment.

*Use at End of Lesson*

---

## OER Using Open Educational Resources
**Video Sharing** Have students work in groups to record and upload examples on **SchoolTube** about combining functions. Have them watch videos of others for ideas. If you are unable to access SchoolTube, try **KidsTube**, **MathATube**, **YouTube**, or **TeacherTube**. *Use as homework*

# Go Online!
connectED.mcgraw-hill.com — Worksheets

# Differentiate Your Resources

**Extra Practice** Additional practice or homework; Skills Practice is best for approaching-level students and Practice is best for on-level and beyond-level students

## Skills Practice
AL  OL  ELL

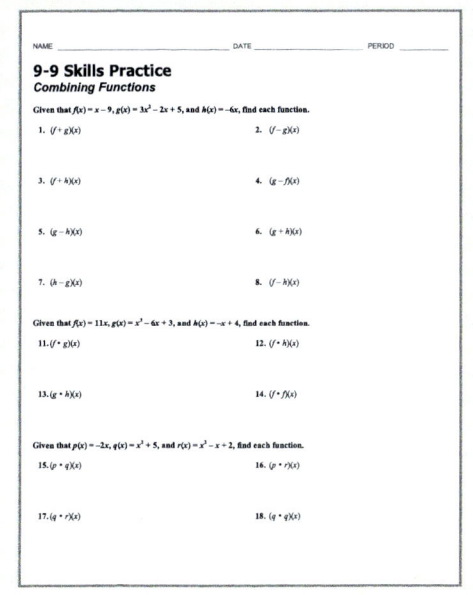

## Practice
AL  OL  BL  ELL

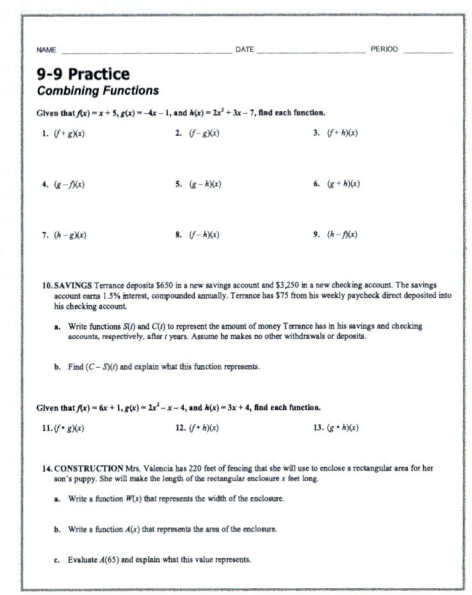

## Word Problem Practice
AL  OL  BL  ELL

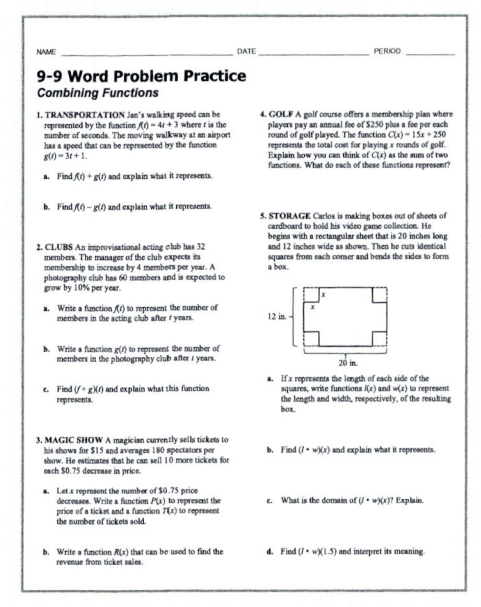

**Intervention** Reteaching and vocabulary activities that can be used with struggling or absent students and as ELL support

**Extension** Activities that can be used to extend lesson concepts

## Study Guide and Intervention
AL  OL  ELL

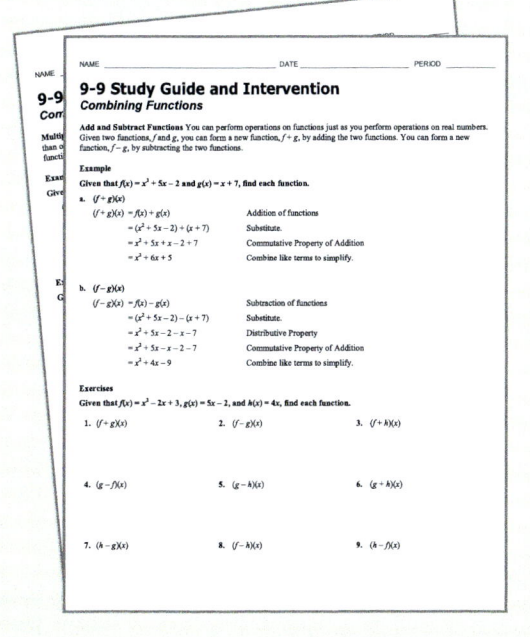

## Study Notebook
AL  OL  BL  ELL

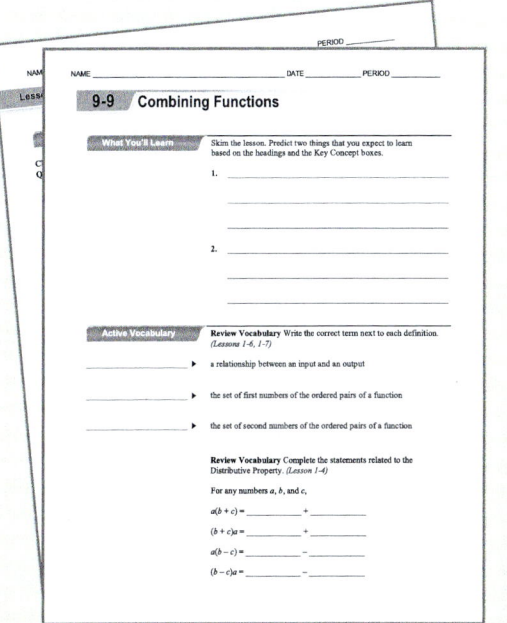

## Enrichment
OL  BL  ELL

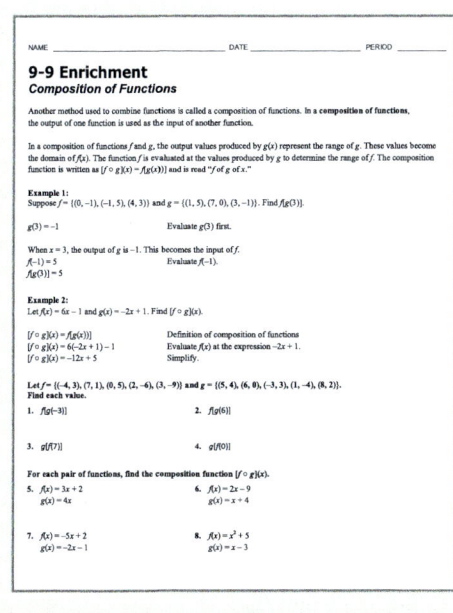

connectED.mcgraw-hill.com    630B

Lesson 9-9 | Combining Functions

# Launch

Have students read the Why? section of the lesson. Ask:

- What do you think happens to the temperature of the meteorite once it lands on Earth? **The temperature decreases until it is the same as the local air temperature.**

- Why do you think the local air temperature is modeled by a constant function? **During the time the meteorite is cooling, the local air temperature is not changing significantly.**

# Teach

Ask the scaffolded questions for each example to build conceptual understanding for students at all levels.

## 1 Add and Subtract Functions

**Example 1 Add and Subtract Functions**

**AL** What type of function is $f(x)$? What type of function is $g(x)$? **quadratic; linear**

**OL** How can you check that you found $(f + g)(x)$ correctly? **Sample answer: Evaluate $(f + g)(x)$ at a specific value of $x$. The result should be the same as when you evaluate $f(x)$ and $g(x)$ at this value of $x$ and add the results.**

**BL** Is $(f + g)(x)$ equivalent to $(g + f)(x)$? Is $(f - g)(x)$ equivalent to $(g - f)(x)$? Explain. **$(f + g)(x)$ is equivalent to $(g + f)(x)$ because addition is commutative. Since subtraction is not commutative, $(f - g)(x)$ is not necessarily equivalent to $(g - f)(x)$.**

**Need Another Example?**
Given that $f(x) = 3x + 7$ and $g(x) = -4x^2 + x - 2$, find each function.
a. $(f + g)(x)$ **$-4x^2 + 4x + 5$**
b. $(f - g)(x)$ **$4x^2 + 2x + 9$**

## Go Online!

**Interactive Whiteboard**
Use the eLesson or Lesson Presentation to present this lesson.

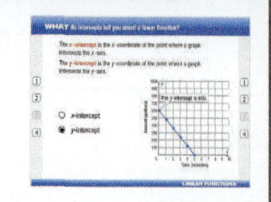

---

# LESSON 9
# Combining Functions

| :: Then | :: Now | :: Why? |
|---|---|---|
| ● You wrote and graphed linear, exponential, and quadratic functions. | 1 Combine functions by using addition and subtraction.<br>2 Combine functions by using multiplication. | ● On February 15, 2013, a meteor exploded over Chelyabinsk, Russia, sending a 1400-pound meteorite to Earth's surface. You can write a function that models the temperature of the meteorite by combining an exponential decay function and a constant function that represents the local air temperature. |

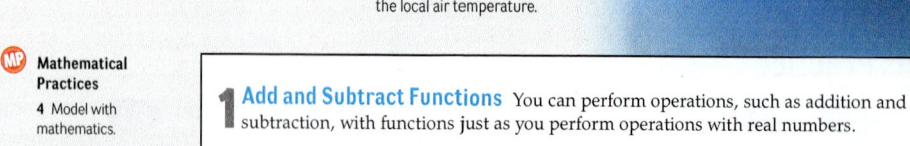

**MP Mathematical Practices**
4 Model with mathematics.
7 Look for and make use of structure.

**Content Standards**
F.BF.1b Combine standard function types using arithmetic operations.

### 1 Add and Subtract Functions
You can perform operations, such as addition and subtraction, with functions just as you perform operations with real numbers.

**Key Concept** Adding and Subtracting Functions

Words: Given two functions, $f(x)$ and $g(x)$, you can form new functions, $(f + g)(x)$, by adding the two functions, and $(f - g)(x)$, by subtracting the two functions.

Symbols: $(f + g)(x) = f(x) + g(x)$
$(f - g)(x) = f(x) - g(x)$

F.BF.1b

**Example 1** Add and Subtract Functions

Given that $f(x) = x^2 + 3x - 7$ and $g(x) = x - 5$, find each function.

a. $(f + g)(x)$
$(f + g)(x) = f(x) + g(x)$    Addition of functions
$= (x^2 + 3x - 7) + (x - 5)$    Substitution
$= x^2 + 3x + x - 7 - 5$    Combine like terms
$= x^2 + 4x - 12$    Simplify.

b. $(f - g)(x)$
$(f - g)(x) = f(x) - g(x)$    Subtraction of functions
$= (x^2 + 3x - 7) - (x - 5)$    Substitution
$= x^2 + 3x - 7 - x + 5$    Distributive Property
$= x^2 + 3x - x - 7 + 5$    Combine like terms.
$= x^2 + 2x - 2$    Simplify.

▶ **Guided Practice**

Given that $f(x) = 4^x + 1$ and $g(x) = 5x^2 + x - 1$, find each function.
1A. $(f + g)(x)$    $4^x + 5x^2 + x$
1B. $(f - g)(x)$    $4^x - 5x^2 - x + 2$

---

**MP Mathematical Practices Strategies**

**Look for and make use of structure.**
Help students make connections between the structure of a pair of functions and the structure of their sum, difference, or product. For example, ask:

- What is the slope-intercept form of a linear function? **$y = mx + b$**
- What is the standard form of a quadratic function? **$y = ax^2 + bx + c$**
- What is the general form of an exponential function? **$y = ab^x$**
- How do you simplify when you add two linear expressions or two quadratic expressions? **combine like terms**
- What do you know about the product two non-constant linear expressions? **It will be a quadratic expression since the product of the x-terms results in an x-squared term.**

630 | Lesson 9-9 | Combining Functions

**Lesson 9-9 | Combining Functions**

### Real-World Example 2  Build a New Function by Adding

F.BF.1b

**FINANCIAL LITERACY** Hugo graduates from college with $28,500 in student loan debt. He decides to defer his payments while he is in graduate school. During that time, he still accrues interest on his student loans at an annual rate of 5.65%. While he is in graduate school, Hugo's parents also lend him $400 per month for rent. His parents decide not to charge him interest on this loan.

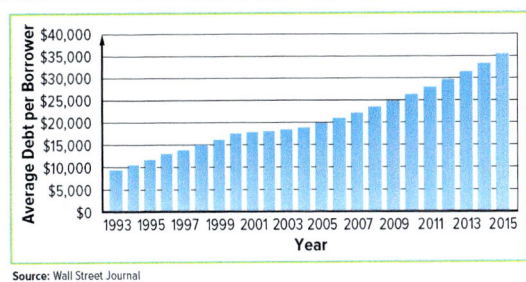

*Source: Wall Street Journal*

**a.** Write an exponential function $f(t)$ to represent the amount of money Hugo owes on his student loans $t$ years after interest begins to accrue.

$f(t) = a(1 + r)^t$    Equation for exponential growth
$\phantom{f(t)} = 28{,}500(1 + 0.0565)^t$    Substitution with $a = 28{,}500$ and $r = 5.65\%$ or $0.0565$
$\phantom{f(t)} = 28{,}500(1.0565)^t$    Simplify.

**b.** Write a function $g(t)$ to represent the amount of money Hugo owes his parents after $t$ years.

Since $t$ is the time in years, first find the amount of money Hugo borrows from his parents each year.

$12(400) = 4800$

Hugo borrows $4800 from his parents each year.

So, $g(t) = 4800t$ represents the total amount Hugo owes his parents after $t$ years.

**c.** Find $C(t) = f(t) + g(t)$ and explain what this function represents.

$C(t) = f(t) + g(t)$    Addition of functions
$\phantom{C(t)} = 28{,}500(1.0565)^t + 4800t$    Substitution

$C(t)$ represents the total amount of money Hugo has to repay after $t$ years.

**d.** If Hugo spends 3 years in graduate school, find the total amount of money he will have to repay.

Because Hugo spends 3 years in graduate school, $t = 3$.

$C(t) = 28{,}500(1.0565)^t + 4800t$
$C(3) = 28{,}500(1.0565)^3 + 4800(3)$
$\phantom{C(3)} = 33{,}608.83 + 14{,}400$
$\phantom{C(3)} = \$48{,}008.83$

**Watch Out!**
**Percents** To convert a percent to a decimal, divide the percent by 100 or just move the decimal point 2 places to the left. 5.65% = 0.0565

### Guided Practice

**BANKING** Malia deposits $350 in a new savings account and $2100 in a new checking account. The savings account pays 2% interest, compounded annually. Malia's weekly paycheck of $148 is deposited directly into her checking account and earns no interest.

**2A.** Write functions $S(t)$ and $C(t)$ to represent the amount of money Malia has in her savings account and in her checking account, respectively, after $t$ years. Assume she makes no withdrawals from or additional deposits into the accounts.

**2B.** Find $M(t) = S(t) + C(t)$ and explain what this function represents. Then find $M(2)$ and explain what this represents.

**2A.** $S(t) = 350(1.02)^t$; $C(t) = 7696t + 2100$
**2B.** $M(t) = 350(1.02)^t + 7696t + 2100$; this represents the total amount of money in the bank after $t$ years. $M(2) = 17{,}856.14$; this represents the total amount of money in the bank after 2 years.

## Teaching Tip

**Use Tools** A function like $(f + g)(t) = 28{,}500(1.0565)^t + 4800t$ may look unfamiliar to students because it contains an exponential term and a linear term. It can be helpful for students to explore the function on a graphing calculator or other graphing tool. Have students graph the function and notice its shape. Ask them to use the graph to find the value of the function when $t = 3$. Students can also create a table of values for the function and scroll down to find $(f + g)(3)$.

---

**Watch Out!**
**Preventing Errors** Remind students that when they write and simplify expressions for $(f + g)(x)$, $(f - g)(x)$, or $(f \cdot g)(x)$, they should only combine like terms. Some students may be tempted to combine terms such as $2^x$ and $x^2$. Be sure students understand that like terms have the same variable raised to the same power.

**MP Teaching the Mathematical Practices**

**Modeling** Mathematically proficient students apply mathematics to solve problems in everyday life, society, and the workplace. They recognize that real-world constraints may affect the domain of a function. Have students describe a reasonable domain for each of the functions they write for this lesson's modeling problems.

### Example 2  Build a New Function by Adding

**AL** In the equation for exponential growth, $f(t) = a(1 + r)^t$, what does each variable represent? *$a$ is the initial amount; $r$ is the rate of growth, written as a decimal; $t$ is time*

**OL** What can you say about the value of $(f + g)(t)$ as $t$ increases? *The value of $(f + g)(t)$ also increases.*

**BL** Suppose Hugo wanted to know when the total amount he has to repay will reach $40,000. How could he use $(f + g)(t)$ and technology to determine this? *He could use a graphing calculator or other tool to solve $(f + g)(t) = 40{,}000$.*

### Need Another Example?

**DEPRECIATION** Jamar buys a new car for $19,500. He expects the value of the car to decrease by 20% per year. He also buys a motorcycle for $4800 and he expects the value of the motorcycle to decrease by $350 per year.

**a.** Write a function $f(t)$ to represent the value of the car after $t$ years. *$f(t) = 19{,}500(0.8)^t$*

**b.** Write a function $g(t)$ to represent the value of the motorcycle after $t$ years. *$g(t) = 4800 - 350t$*

**c.** Find $(f + g)(t)$ and explain what it represents. *$(f + g)(t) = 19{,}500(0.8)^t + 4800 - 350t$; the function represents the total value of both vehicles after $t$ years.*

**d.** If Jamar keeps both vehicles for 5 years, find their total value at the end of this time. *$9439.76*

connectED.mcgraw-hill.com    **631**

# Lesson 9-9 | Combining Functions

## 2 Multiply Functions

### Example 3 Multiply Functions

**AL** What does the Distributive Property state?
$a(b + c) = ab + ac$

**OL** Why does it make sense that $(f \cdot g)(x)$ is a polynomial of degree 3? $f(x)$ is quadratic and $g(x)$ is linear. When you multiply the expressions, you get a term with $x$ raised to the third power.

**BL** Does the order of the two functions matter when you multiply them? Why or why not? No; multiplication is commutative.

**Need Another Example?**
Given that $f(x) = 5x + 1$ and $g(x) = -x^2 + 3x - 2$, find $(f \cdot g)(x)$. $(f \cdot g)(x) = -5x^3 + 14x^2 - 7x - 2$

### Example 4 Build a New Function by Multiplying

**AL** When the price of a large sub is $8.50, how can you determine the number of $0.25 price increases? Subtract the original price from the new price to get $1.50, then divide the result by $0.25 to get 6.

**OL** How are the functions $P(x)$ and $T(x)$ similar? How are they different? Sample answer: They are both linear functions; the graph of $P(x)$ has a positive slope, while the graph of $T(x)$ has a negative slope.

**BL** If the owner of the shop wants to increase revenue from sales of large subs, should the shop charge $8.50 for a large sub? Why or why not? No; the revenue from sales at this price is $2040, but the revenue at the original price was 7(360) = $2520.

**Need Another Example?**
**NEWS** A news website charges $12 per month for a subscription. At this price, the site has 1450 subscribers. The site's business manager predicts that every decrease of $0.50 in the subscription price will result in 240 new subscribers.

a. Let $x$ represent the number of $0.50 price decreases. Write a function $P(x)$ to represent the price of a monthly subscription. $P(x) = 12 - 0.5x$
b. Write a function $T(X)$ to represent the number of subscribers to the site. $T(X) = 1450 + 240x$
c. Write a function $R(X)$ that can be used to find the monthly revenue from subscriptions to the site. $R(X) = -120x^2 + 2155x + 17{,}400$
d. If the business manager decides to lower the price of a subscription to $11, find the monthly revenue from subscriptions to the site. $21,230

---

**2 Multiply Functions** Just as you can add and subtract functions, you can also multiply them. When you multiply functions that have more than one term, use the Distributive Property to multiply each term in the first function by each term in the second function.

### Key Concept  Multiplying Functions

**Words**   Given two functions, $f(x)$ and $g(x)$, you can form a new function, $(f \cdot g)(x)$, by multiplying the two functions.

**Symbols**  $(f \cdot g)(x) = f(x) \cdot g(x)$

F.BF.1b

**Study Tip**
**Structure** Be sure to distribute all terms of one function to all terms of the other function. When a function has more than two terms, you may find it helpful to draw arrows from one term to the others.

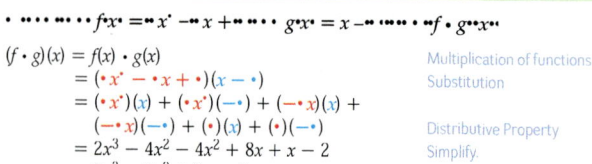

### Example 3  Multiply Functions

If $f(x) = \cdots x^2 - \cdots x + \cdots$ and $g(x) = x - \cdots$, find $f \cdot g(x)$.

$(f \cdot g)(x) = f(x) \cdot g(x)$  Multiplication of functions
$= (\cdot x^2 - \cdot x + \cdot)(x - \cdot)$  Substitution
$= (\cdot x^2)(x) + (\cdot x^2)(-\cdot) + (-\cdot x)(x) +$
$\quad (-\cdot x)(-\cdot) + (\cdot)(x) + (\cdot)(-\cdot)$  Distributive Property
$= 2x^3 - 4x^2 - 4x^2 + 8x + x - 2$  Simplify.
$= 2x^3 - 8x^2 + 9x - 2$  Simplify.

### Guided Practice

If $f(x) = \cdots x - \cdots$, $g(x) = \cdots x^2 + \cdots x - \cdots$, $h(x) = \cdots x + x \cdots$

3A. $(f \cdot g)(x)$  $12x^3 + 7x^2 - 13x + 2$
3B. $(f \cdot h)(x)$  $3x(8^x) + 3x^2 - 2(8^x) - 2x$
3C. $(h \cdot g)(x)$  $4x^2(8^x) + 5x(8^x) - 8^x + 4x^3 + 5x^2 - x$

Some real-world situations are best modeled by the product of functions.

### Real-World Example 4  Build a New Function by Multiplying

F.BF.1b

**Real World Link**
The submarine sandwich, or sub, may have gotten its name because its shape is similar to that of a submarine ship. However, this style of sandwich goes by many different names, including hoagie, hero, and grinder.

**BUSINESS** ...

a. ... $P(x) = \cdots$
The price of a large sub is $7 plus $0.25 times the number of price increases.
$P(x) = 7 + 0.25x$

b. ... $T(x) = \cdots$
The number of large subs sold is 360 minus 20 times the number of price increases.
$T(x) = 360 - 20x$

c. ... $R(x) = \cdots$
The revenue from sales of large subs is equal to the price times the number of large subs sold.

$R(x) = P(x) \cdot T(x)$  Multiplication of functions
$= (7 + 0.25x)(360 - 20x)$  Substitution
$= -5x^2 - 50x + 2520$  Multiply

---

### Differentiated Instruction  AL  OL

**IF** students have difficulty writing equations to represent real-world situations,

**THEN** have them work in small groups and experiment with specific values before they write equations. For instance, in Example 4, ask students to explore some different scenarios, such as a single $0.25 price increase. In this case, the price of a sub is $7.25 and the number sold is 360 − 20 = 340. The revenue from these sales is 7.25(340) = $2465. Have students calculate the revenue for different price increases. Then ask them to write the equations and evaluate for the same values to see how they compare.

632 | Lesson 9-9 | Combining Functions

## Lesson 9-9 | Combining Functions

### Study Tip

**MP Modeling** When you write a function to model a situation, you will often need to evaluate the function for a specific value. Before you do this, it may be helpful to recall what the variable represents.

**d.** If the owner decides to charge $8.50 for a large sub, find the revenue from sales of large subs.

Since $x$ represents the number of $0.25 price increases, and the price increased from $7.00 to $8.50, $x = 6$. Substitute 6 for $x$ in $R(x)$.

$R(6) = -5(6)^2 - 50(6) + 2520$
$= -180 - 300 + 2520$
$= \$2040$

#### Guided Practice

**THEATERS** A theater currently sells tickets for $18 and they sell an average of 250 tickets per show. The box office manager estimates that they can sell 25 more tickets for every $1.50 decrease in the price.

**4A.** $P(x) = 18 - 1.5x$; $T(x) = 250 + 25x$; $R(x) = -37.5x^2 + 75x + 4500$

**4A.** Let $x$ represent the number of $1.50 price decreases. Write a function $P(x)$ to represent the price of a ticket, a function $T(x)$ to represent the number of tickets sold, and a function $R(x)$ that can be used to find the revenue from ticket sales.

**4B.** $4387.50

**4B.** If the manager decides to sell tickets at $13.50, find the revenue from ticket sales.

## Practice

**Formative Assessment** Use Exercises 1–11 to assess students' understanding of the concepts in the lesson.

The Practice and Problem Solving exercises assess the content taught in the lesson.

### Additional Answer

**7c.** $V(t) = 20(1.045)^t + 2t + 42$; this represents the total value of the coins after $t$ years.

---

### Check Your Understanding
○ = Step-by-Step Solutions begin on page R11.
**Go Online!** for a Self-Check Quiz

**Example 1**
F.BF.1b

Given that $f(x) = x^2 - 5x - 9$, $g(x) = 4x + 1$, and $h(x) = 3x$, find each function.

**1.** $(f + g)(x)$  $x^2 - x - 8$    **2.** $(f - g)(x)$  $x^2 - 9x - 10$    **3.** $(f + h)(x)$  $x^2 - 2x - 9$

**4.** $(g - f)(x)$  $-x^2 + 9x + 10$    **5.** $(g - h)(x)$  $x + 1$    **6.** $(g + h)(x)$  $7x + 1$

**Example 2**
F.BF.1b

**7. COINS** A coin collector buys a rare nickel and a rare quarter. She pays $20 for the nickel and $42 for the quarter. The value of the nickel increases by 4.5% per year. The value of the quarter increases by $2 per year.

**a.** Write a function $f(t)$ to represent the value of the nickel after $t$ years.  $f(t) = 20(1.045)^t$
**b.** Write a function $g(t)$ to represent the value of the quarter after $t$ years.  $g(t) = 42 + 2t$
**c.** Find $V(t) = f(t) + g(t)$ and explain what this function represents. See margin.
**d.** How much will the coins we worth after 5 years?  $76.92

**Example 3**
F.BF.1b

Given that $f(x) = 3x + 10$, $g(x) = x^2 - 6x - 2$, and $h(x) = 2x - 5$, find each function.

**8.** $(g \cdot h)(x)$  $2x^3 - 17x^2 + 26x + 10$    **9.** $(f \cdot g)(x)$  $3x^3 - 8x^2 - 66x - 20$    **10.** $(f \cdot h)(x)$  $6x^2 + 5x - 50$

**Example 4**
F.BF.1b

**11. FARMERS' MARKETS** A vendor at a farmers' market sells jars of strawberry jam for $4 per jar. At this price, he sells an average of 160 jars of jam per day. The vendor predicts that he will sell 10 fewer jars of jam for each $0.15 increase in the price.

**a.** Let $x$ represent the number of $0.15 price increases. Write a function $P(x)$ to represent the price of a jar of strawberry jam.  $P(x) = 4 + 0.15x$
**b.** Write a function $T(x)$ to represent the number of jars of jam sold per day.  $T(x) = 160 - 10x$
**c.** Write a function $R(x)$ that can be used to find the revenue from sales of strawberry jam.  $R(x) = -1.5x^2 - 16x + 640$
**d.** If the vendor decides to charge $4.60 for a jar of jam, find the revenue from sales of strawberry jam.  $552

---

### Differentiated Homework Options

| Levels | AL Basic | OL Core | BL Advanced |
|---|---|---|---|
| Exercises | 12–25, 49–51, 53–59 | 13–39 odd, 40, 49–51, 53–59 | 41–52, (optional: 53–59) |
| 2-Day Option | 13–25 odd, 53–39 | 12–25, 53–59 | |
| | 12–24 even, 49–51 | 26–31, 33–38, 40–43, 49–51 | |

You can use ALEKS to provide additional remediation support with personalized instruction and practice.

## Go Online!    eBook

### Interactive Student Guide
Use the *Interactive Student Guide* to deepen conceptual understanding.
· Function Intersections
· Combining Functions

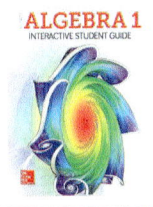

connectED.mcgraw-hill.com  633

**Lesson 9-9 | Combining Functions**

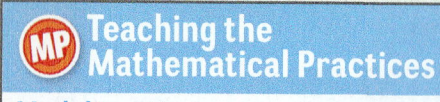

## Teaching the Mathematical Practices

**Modeling** Mathematically proficient students routinely interpret their results in the context of the problem. In Exercise 25, students must understand that the number of subscribers to the blog must be a whole number. This will affect how students round their answer in part d of the problem.

## Extra Practice

See page R9 for extra exercises for students who are approaching level or for on-level students who need additional reinforcement.

## Additional Answers

12. $2^x - x^2 + 11$
13. $2x^2 - 6x$
14. $-2^x - x^2 + 7$
15. $2^x + x^2 - 7$
16. $2^x + 3x^2 - 6x - 7$
17. $-4x^2 + 6x + 18$
19. $-5x^3 + 18x^2 - 11x - 8$
20. $3x^3 - 3x^2 - 9x - 3$
21. $-15x^3 - 15x^2$
22. $-15x^2 + 9x + 24$
23. $25x^3 - 40x^2$
24. $-5x^4 + 10x^3 + 5x^2$

---

**Practice and Problem Solving**  Extra Practice is found on page R9.

**Example 1**
F.BF.1b

Given that $f(x) = -x^2 + 9$, $g(x) = 2^x + 2$, and $h(x) = 3x^2 - 6x - 9$, find each function.

12. $(f + g)(x)$
13. $(f + h)(x)$
14. $(f - g)(x)$   12-17 See margin.
15. $(g - f)(x)$
16. $(g + h)(x)$
17. $(f - h)(x)$

**Example 4**
F.BF.1b

18. **MUSEUMS** An art museum has an admission price of $15. An average of 620 people visit the museum each day. The museum's director predicts that each $0.50 decrease in the price of admission will result in 40 additional visitors per day.

   a. Let $x$ represent the number of $0.50 price decreases. Write a function $A(x)$ to represent the museum's admission price. $A(x) = 15 - 0.5x$
   b. Write a function $V(x)$ to represent the number of visitors per day. $V(x) = 620 + 40x$
   c. Write a function $R(x)$ that can be used to find the museum's daily revenue. $R(x) = -20x^2 + 290x + 9300$
   d. If the director decides to change the museum's admission price to $13, what daily revenue can the museum expect? $10,140

**Example 3**
F.BF.1b

Given that $f(x) = -x^2 + 2x + 1$, $g(x) = 5x - 8$, $h(x) = -3x - 3$, and $k(x) = 5x^2$, find each function.

19. $(f \cdot g)(x)$
20. $(f \cdot h)(x)$
21. $(h \cdot k)(x)$   19-24 See margin.
22. $(g \cdot h)(x)$
23. $(g \cdot k)(x)$
24. $(f \cdot k)(x)$

**Example 2**
F.BF.1b

25c. $(f + g)(t) = 590(1.15)^t + 360t + 385$; this represents the total number of subscribers to Aurelio's blogs after $t$ years.

25. **BLOGS** Aurelio writes a sports blog and a photography blog. The sports blog has 385 subscribers and Aurelio expects the number of subscribers to increase at a rate of 30 subscribers per month. The photography blog has 590 subscribers and Aurelio expects the number of subscribers to increase by 15% per year.

   a. Write a function $f(t)$ to represent the number of subscribers to the sports blog after $t$ years. $f(t) = 385 + 360t$
   b. Write a function $g(t)$ to represent the number of subscribers to the photography blog after $t$ years. $g(t) = 590(1.15)^t$
   c. Find $(f + g)(t)$ and explain what this function represents.
   d. Find the total number of subscribers Aurelio expects to have after 6 years. 3910

Use the table of values to find each of the following.

| x | f(x) | g(x) |
|---|---|---|
| -2 | 2 | 3 |
| -1 | 0 | -5 |
| 0 | -4 | 1 |
| 1 | 8 | 5 |
| 2 | -7 | -3 |

26. $(f + g)(-1)$  $-5$
27. $(f \cdot g)(2)$  $21$
28. $(f - g)(0)$  $-5$
29. $(f \cdot g)(0)$  $-4$
30. $(g - f)(1)$  $-3$

31. **MANUFACTURING** A company makes boxes in the shape of a rectangular prism with an open top. They start with a rectangular piece of cardboard that is 10 inches long and 8 inches wide. Then they cut identical squares from each corner, as shown, and fold up the sides.

   a. Let $x$ represent the side length of the squares that are cut from the corners of the cardboard. Write a function $f(x)$ to represent the length of the resulting box. $f(x) = 10 - 2x$

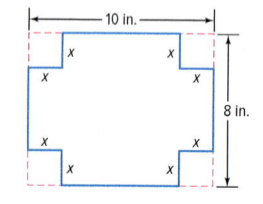

---

### Levels of Complexity Chart

The levels of the exercises progress from 1 to 3, with Level 1 indicating the lowest level of complexity.

| Exercises | 12–25 | 26–40, 53–59 | 41–52 |
|---|---|---|---|
| Level 3 | | | ● |
| Level 2 | | ● | |
| Level 1 | ● | | |

**Lesson 9-9 | Combining Functions**

**31d.** D = {0 < x < 4}; to create a square, x must be greater than 0; also, x must be less than half the width of the cardboard.

b. Write a function g(x) to represent the width of the resulting box. $g(x) = 8 - 2x$
c. Find $(f \cdot g)(x)$ and explain what this function represents. $4x^2 - 36x + 80$; this represents the area of the base of the box.
d. What is the domain of $(f \cdot g)(x)$? Explain.

**32.** PERSEVERANCE Suppose $f(x) = x^2 - 2x + 14$ and $(f + g)(x) = 4x^2 - 3x - 5$. Find $g(x)$. $g(x) = 3x^2 - x - 19$

Use the graph of f(x) and g(x) to find each of the following.

**33.** $(f + g)(1)$  4
**34.** $(f - g)(1)$  4
**35.** $(f + g)(-2)$  $-2$
**36.** $(f \cdot g)(0)$  $-3$
**37.** $(f \cdot g)(-1)$  $-4$
**38.** $(f \cdot g)(-3)$  0

**39.** Let $f(x) = 2x^2 + x + 3$ and let $g(x) = x^2 + x + 2$.

**39a.** $(f - g)(x) = x^2 + 1$; $(g - f)(x) = -x^2 - 1$; the functions are opposites since $(f - g)(x) = -(g - f)(x)$.

a. Find $(f - g)(x)$ and $(g - f)(x)$. Explain how these two functions are related.
b. Graph $(f - g)(x)$ and $(g - f)(x)$. See margin.
c. How are the graphs of $(f - g)(x)$ and $(g - f)(x)$ related?
d. Do you think the relationship you noticed between the graphs of $(f - g)(x)$ and $(g - f)(x)$ will hold true given any functions f(x) and g(x)? Explain.

**40.** TABLETS A telecommunications company offers a plan in which you purchase a tablet for $149 and then pay a monthly fee for online access. The function $C(x) = 29x + 149$ models the total cost after x months. Explain how you can think of $C(x)$ as the sum of two functions. What do each of these functions represent? See margin.

Given that $f(x) = 3^x$ and $g(x) = 2(3^x)$, find each function.

**41.** $(f + g)(x)$  $3^{x+1}$
**42.** $(f - g)(x)$  $-3^x$
**43.** $(f \cdot g)(x)$  $2(3^{2x})$

**39c.** The graphs are reflection images of each other after a reflection across the x-axis.

Describe each statement as *sometimes*, *always*, or *never* true.

**44.** If f(x) and g(x) are both linear functions, then the domain of $(f + g)(x)$ is all real numbers. always
**45.** If f(x) and g(x) are both quadratic functions, then $(f - g)(x)$ is a linear function. sometimes
**46.** If $f(0) = 1$ and $g(0) = 1$, then the graph of $(f + g)(x)$ passes through the origin. never
**47.** The function $(f \cdot g)(x)$ is a quadratic function. sometimes

**39d.** Yes; the functions are opposites of each other since $(f - g)(x) = -(g - f)(x)$. This means one graph is a reflection of the other across the x-axis.

F.BF.1b

**H.O.T. Problems** Use Higher-Order Thinking Skills

48-50. See margin.

**48.** OPEN-ENDED Write two quadratic functions, f(x) and g(x), so that $(f + g)(x) = x^2 - 2x + 1$.
**49.** CRITIQUE ARGUMENTS Jeremy said that if f(x) and g(x) are both linear functions, then $(f \cdot g)(x)$ cannot be a linear function. Is he correct? Explain your answer.
**50.** SENSE-MAKING If f(x) is a linear function and g(x) is a quadratic function, what type of function is $(f + g)(x)$? Explain.
**51.** ERROR ANALYSIS Mikayla was asked to find $(f - g)(x)$ given that $f(x) = 3x^2 + 7x - 1$ and $g(x) = 2x^2 - 3x - 4$. She wrote $(f - g)(x) = x^2 + 4x - 5$.
  a. What error did Mikayla make? Mikayla did not distribute the minus sign to every term of g(x).
  b. Find the correct function $(f - g)(x)$. $(f - g)(x) = x^2 + 10x + 3$

52. See margin.

**52.** WRITING IN MATH Describe an example of a real-world situation in which you might write two functions, f(x) and g(x), and then subtract the functions to find $(f - g)(x)$. Explain what f(x), g(x), and $(f - g)(x)$ would represent in the real-world situation.

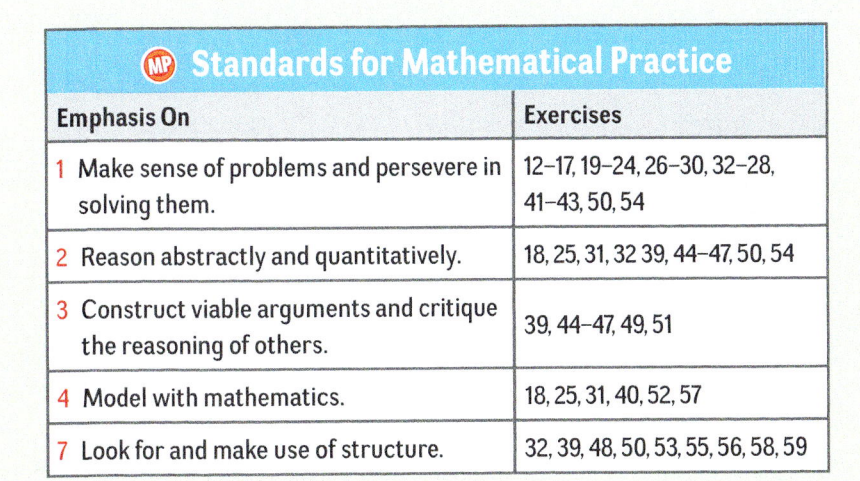

## Assess

**Ticket Out the Door** Make several copies of five different pairs of linear or quadratic functions. Give one pair of functions to each student. As students leave the room, ask them to classify the sum, difference, or product of the functions as a linear function, quadratic function, or third-degree polynomial.

## Additional Answers

**39b.**

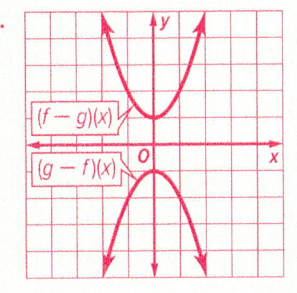

**39c.** The graphs are reflection images of each other after a reflection across the x-axis.

**40.** $C(X) = (f + g)(x)$, where $f(x) = 29x$ and $g(x) = 149$; f(x) represents the total cost of online access after x months and g(x) represents the fixed cost of the tablet.

**48.** Sample answer: $f(x) = 2x^2 - x$ and $g(x) = -x^2 - x + 1$

**49.** No; $(f \cdot g)(x)$ can be a linear function if one or both of f(x) or g(x) is a constant function.

**50.** Since f(x) is linear, it has the form $f(x) = ax + b$. Since g(x) is quadratic, it has the form $cx^2 + dx + e$, where $c \neq 0$. This means $(f + g)(x) = cx^2 + (a + d)x + (b + e)$, so $(f + g)(x)$ is quadratic.

**52.** Sample answer: For a business that makes and sells candles, f(x) might represent the revenue from selling x candles and g(x) might represent the cost of making x candles; the function $(f - g)(x)$ represents the profit from selling x candles.

## Standards for Mathematical Practice

| Emphasis On | Exercises |
| --- | --- |
| 1 Make sense of problems and persevere in solving them. | 12–17, 19–24, 26–30, 32–28, 41–43, 50, 54 |
| 2 Reason abstractly and quantitatively. | 18, 25, 31, 32 39, 44–47, 50, 54 |
| 3 Construct viable arguments and critique the reasoning of others. | 39, 44–47, 49, 51 |
| 4 Model with mathematics. | 18, 25, 31, 40, 52, 57 |
| 7 Look for and make use of structure. | 32, 39, 48, 50, 53, 55, 56, 58, 59 |

connectED.mcgraw-hill.com  635

Lesson 9-9 | Combining Functions

# Preparing for Assessment

Exercises 53–59 require students to use the skills they will need on assessments. Each exercise is dual-coded with content standards and mathematical practice standards.

## Dual Coding

| Exercises | Content Standards | MP Mathematical Practices |
|---|---|---|
| 53 | F.BF.1b | 7 |
| 54 | F.BF.1b | 1 |
| 55 | F.BF.1b | 7 |
| 56 | F.BF.1b | 7 |
| 57 | F.BF.1b | 4 |
| 58 | F.BF.1b | 1, 7 |
| 59 | F.BF.1b | 7 |

## Diagnose Student Errors

Survey student responses for each item. Class trends may indicate common errors and misconceptions.

**53.**

| A | Added the functions instead of multiplying |
|---|---|
| B | CORRECT |
| C | Omitted the term $-10x^2$ when distributing |
| D | Subtracted the functions instead of multiplying |

**54.**

| A | Found $(g-f)(-2)$ |
|---|---|
| B | Found $(f-g)(2)$ |
| C | Found $f(-2)$ |
| D | CORRECT |

## Go Online!

### Quizzes

Students can use *Self-Check Quizzes* to check their understanding of this lesson. You can also give the *Chapter Quiz*, which covers the content in Lessons 9-1 through 9-9.

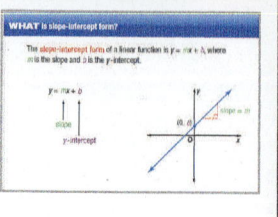

---

# Preparing for Assessment

**53.** Given that $f(x) = 2x^2 - 4x$ and $g(x) = 6x - 5$, which of the following is $(f \cdot g)(x)$?  MP 7  F.BF.1b  **B**

- A  $(f \cdot g)(x) = 2x^2 + 2x - 5$
- B  $(f \cdot g)(x) = 12x^3 - 34x^2 + 20x$
- C  $(f \cdot g)(x) = 12x^3 - 24x^2 + 20x$
- D  $(f \cdot g)(x) = 2x^2 - 10x + 5$

**54.** The graphs of $f(x)$ and $g(x)$ are shown. What is $(f - g)(-2)$?  MP 1  F.BF.1b  **D**

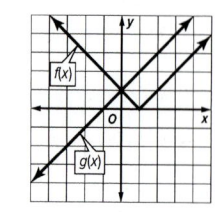

- A  $-4$
- B  $-2$
- C  $3$
- D  $4$

**55.** Given that $f(x) = 3x + 4$, $g(x) = -x^2 + 2x - 1$, and $h(x) = 10$, which of the following are quadratic functions?  MP 7  F.BF.1b  **A, B, E, F**

- A  $(f + g)(x)$
- B  $(f - g)(x)$
- C  $(f \cdot g)(x)$
- D  $(f + h)(x)$
- E  $(g \cdot h)(x)$
- F  $(g - h)(x)$

**56.** What is the value of $(f + g)(3)$ if $f(x) = 2^x - 1$ and $g(x) = 3^x$?  MP 7  F.BF.1b

34

---

**57.** MULTI-STEP  In 2010, the population of Oakville was 82,400 and increasing at a rate of 2.5% per year. In 2010, the population of Elmwood was 75,600 and decreasing by an average of 300 residents each month.  MP 4  F.BF.1b

a. Write a function $f(x)$ to represent the population of Oakville $x$ years after 2010.
   $f(x) = 82{,}400(1.025)^x$
b. Write a function $g(x)$ to represent the population of Elmwood $x$ years after 2010.
   $g(x) = 75{,}600 - 300x$
c. Find $(f - g)(x)$ and explain what this function represents.
d. Find $(f - g)(6)$ and explain what it represents.
   21,759; there were approximately 21,759 more residents of Oakville than Elmwood in 2016.
   57c. $(f - g)(x) = 82{,}400(1.025)^x + 300x - 75{,}600$; this represents how many more residents lived in Oakville than in Elmwood.

**58.** Based on the table of values, what is $(f \cdot g)(2)$?  MP 1, 7  F.BF.1b  **B**

| x | f(x) | g(x) |
|---|---|---|
| -2 | 2 | -8 |
| -1 | -3 | 3 |
| 0 | 2 | -5 |
| 1 | 0 | 2 |
| 2 | 4 | -3 |

- A  $-16$
- B  $-12$
- C  $1$
- D  $7$

**59.** For which pair of functions does $(f - g)(x) = x$?  MP 7  F.BF.1b  **C**

- A  $f(x) = 2x^2 + 2x - 5$; $g(x) = -2x^2 - x + 5$
- B  $f(x) = -4x^2 - x + 1$; $g(x) = -4x^2 + x + 1$
- C  $f(x) = -x^2 + 3x - 2$; $g(x) = -x^2 + 2x - 2$
- D  $f(x) = x^2 + 10x + 4$; $g(x) = -x^2 + 9x + 4$

---

**58.**

| A | Found $(f \cdot g)(-2)$ |
|---|---|
| B | CORRECT |
| C | Added the function values instead of multiplying |
| D | Subtracted the function values instead of multiplying |

**59.**

| A | Added the functions instead of subtracting |
|---|---|
| B | Incorrectly subtracted the x-terms |
| C | CORRECT |
| D | Incorrectly subtracted the $x^2$-terms |

# CHAPTER 9
# Study Guide and Review

*Go Online!* for Vocabulary Review Games and key vocabulary in 13 languages

## Study Guide

### Key Concepts

**Graphing Quadratic Functions** (Lesson 9-1)
- A quadratic function can be described by an equation of the form $y = ax^2 + bx + c$, where $a \neq 0$.
- The axis of symmetry for the graph of $y = ax^2 + bx + c$, where $a \neq 0$, is $x = -\frac{b}{2a}$.

**Transformations of Quadratic Functions** (Lesson 9-2)
- $f(x) = x^2 + k$ translates the graph up or down.
- $f(x) = ax^2$ compresses or expands the graph vertically.

**Solving Quadratic Equations** (Lessons 9-3 through 9-6)
- Quadratic equations can be solved by graphing. The solutions are the $x$-intercepts or zeros of the related quadratic function.
- Some quadratic equations of the form $ax^2 + bx + c = 0$ can be solved by factoring and then using the Zero Product Property.

**Solving Systems of Linear and Quadratic Equations** (Lesson 9-7)
- To solve a system graphically, determine the point(s) of intersection of the graphs.

**Analyzing Functions with Successive Differences** (Lesson 9-8)
- If the differences of successive $y$-values are all equal, the data represent a linear function.
- If the second differences are all equal, but the first differences are not, the data represent a quadratic function.

**Combining Functions** (Lesson 9-9)
- To add or subtract functions, combine like terms.
- To multiply functions, apply the distributive property and combine like terms.

**FOLDABLES Study Organizer**

Use your Foldable to review the chapter. Working with a partner can be helpful. Ask for clarification of concepts as needed.

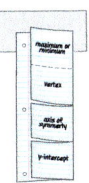

### Key Vocabulary

| | |
|---|---|
| axis of symmetry (p. 559) | Quadratic Formula (p. 606) |
| completing the square (p. 596) | quadratic function (p. 559) |
| discriminant (p. 610) | standard form (p. 559) |
| double root (p. 581) | vertex (p. 559) |
| maximum (p. 559) | vertex form (p. 575) |
| minimum (p. 559) | |
| parabola (p. 559) | |

### Vocabulary Check

State whether each sentence is *true* or *false*. If *false*, replace the underlined term to make a true sentence.

1. The <u>axis of symmetry</u> of a quadratic function can be found by using the equation $x = -\frac{b}{2a}$.  **true**

2. The <u>vertex</u> is the maximum or minimum point of a parabola.  **true**

3. The graph of a quadratic function is a(n) <u>straight line</u>.
   **false; parabola**

4. The graph of a quadratic function has a(n) <u>maximum</u> if the coefficient of the $x^2$-term is positive.  **false; minimum**

5. A quadratic equation with a graph that has two $x$-intercepts has <u>one</u> real root.  **false; two**

6. The expression $b^2 - 4ac$ is called the <u>discriminant</u>.  **true**

7. The solutions of a quadratic equation are called <u>roots</u>.
   **true**

8. The graph of the parent function is <u>translated down</u> to form the graph of $f(x) = x^2 + 5$.
   **false; translated up 5 units**

### Concept Check

9. Explain the relationship between the solutions of a quadratic equation and the graph of the related quadratic function.  **See margin.**

10. Explain how functions can be combined to form new functions.  **Functions can be combined using basic operations, such as addition, subtraction, or multiplication.**

---

## Chapter 9 Study Guide and Review

### FOLDABLES Study Organizer

A completed Foldable for this chapter should include the Key Concepts related to quadratic functions and equations.

### Key Vocabulary ELL

The page reference after each word denotes where that term was first introduced. If students have difficulty answering questions 1–8, remind them that they can use these page references to refresh their memories about the vocabulary terms.

Have students work with a partner to complete the Vocabulary Check. Encourage them to reference and compare their notes from Chapter 9.

You can use the detailed reports in ALEKS to automatically monitor students' progress and pinpoint remediation needs prior to the chapter test.

### Additional Answer

9. The solutions of a quadratic equation are the $x$-intercepts of the graph of the related quadratic function.

---

## Answering the Essential Question

Before answering the Essential Question, have students review their answers to the *Building on the Essential Question* exercises found throughout the chapter.

- Why do we use different methods to solve math problems? (p. 556)
- How can the graph of a quadratic function help you solve the corresponding quadratic equation? (p. 585)
- How is the symmetry of the graph of a quadratic function reflected in the solutions found by completing the square? (p. 600)
- How do you know which method to use when solving a quadratic equation? (p. 611)
- When is it best to solve a system by graphing and when is it best to solve it by using substitution? (p. 620)

## Go Online!

### Vocabulary Review

Students can use the *Vocabulary Review Games* to check their understanding of the vocabulary terms in this chapter. Students should refer to the *Student-Built Glossary* they have created as they went through the chapter to review important terms. You can also give a *Vocabulary* Test over the content of this chapter.

connectED.mcgraw-hill.com   637

# Chapter 9 Study Guide and Review

## Lesson-by-Lesson Review

**Intervention** If the given examples are not sufficient to review the topics covered by the questions, remind students that the lesson references tell them where to review that topic in their textbook.

**Two-Day Option** Have students complete the Lesson-by-Lesson Review. Then you can use McGraw-Hill eAssessment to customize another review worksheet that practices all the objectives of this chapter or only the objectives on which your students need more help.

## Additional Answers

11a. minimum
11b. 0
11c. $D = \{-\infty < x < \infty\}$; $R = \{y \mid y \geq 0\}$
12a. maximum
12b. 2.25
12c. $D = \{-\infty < x < \infty\}$; $R = \{y \mid y \leq 2.25\}$
13a. minimum
13b. −4
13c. $D = \{-\infty < x < \infty\}$; $R = \{y \mid y \geq -4\}$
14a. maximum
14b. 2
14c. $D = \{-\infty < x < \infty\}$; $R = \{y \mid y \leq 2\}$
16. translated up 8 units
17. translated down 3 units
18. vertical stretch and translated right 1 unit
19. vertical stretch and translated down 18 units
20. vertical compression
21. vertical compression
23. reflected across the *x*-axis, vertically stretched and translated up 100 units

---

## CHAPTER 9
# Study Guide and Review *Continued*

### Lesson-by-Lesson Review

#### 9-1 Graphing Quadratic Functions
F.IF.4, F.IF.7a

Consider each equation.
a. Determine whether the function has a *maximum* or *minimum* value.
b. State the maximum or minimum value.
c. What are the domain and range of the function?

11. $y = x^2 - 4x + 4$ **11–14. See margin.**
12. $y = -x^2 + 3x$
13. $y = x^2 - 2x - 3$
14. $y = -x^2 + 2$.

15. **ROCKET** A toy rocket is launched with an upward velocity of 32 feet per second. The equation $h = -16t^2 + 32t$ gives the height of the ball $t$ seconds after it is launched.
  a. Determine whether the function has a *maximum* or *minimum* value. **maximum**
  b. State the maximum or minimum value. **16**
  c. State a reasonable domain and range for this situation.
  $D = \{t \mid 0 \leq t \leq 2\}$; $R = \{h \mid 0 \leq h \leq 16\}$

**Example 1**
Consider $f(x) = x^2 + 6x + 5$.
a. Determine whether the function has a *maximum* or *minimum* value.

For $f(x) = x^2 + 6x + 5$, $a = 1$, $b = 6$, and $c = 5$. Because $a$ is positive, the graph opens up, so the function has a minimum value.

b. State the *maximum* or *minimum* value of the function.

The minimum value is the *y*-coordinate of the vertex. The *x*-coordinate of the vertex is $\frac{-b}{2a}$ or $\frac{-6}{2(1)}$ or −3.

$f(x) = x^2 + 6x + 5$   Original function
$f(-3) = (-3)^2 + 6(-3) + 5$   $x = -3$
$f(-3) = -4$   Simplify.

The minimum value is −4.

c. State the domain and range of the function.

The domain is all real numbers. The range is all real numbers greater than or equal to the minimum value, or $\{y \mid y \geq -4\}$.

#### 9-2 Transformations of Quadratic Functions
F.IF.7a, F.BF.3

Describe the transformations in each function as it relates to the graph of $f(x) = x^2$.
**16–21. See margin.**
16. $f(x) = x^2 + 8$
17. $f(x) = x^2 - 3$
18. $f(x) = 2(x - 1)^2$
19. $f(x) = 4x^2 - 18$
20. $f(x) = \frac{1}{3}x^2$
21. $f(x) = \frac{1}{4}x^2$

**Example 2**
Describe the transformations in $f(x) = x^2 - 2$ as it relates to the graph of $f(x) = x^2$.

The graph of $f(x) = x^2 + k$ represents a translation up or down of the parent graph.

Since $k = -2$, the translation is down.

So, the graph is translated down 2 units from the parent function.

# Chapter 9 Study Guide and Review

**22.** Write an equation for the function shown in the graph. $y = 2x^2 - 3$

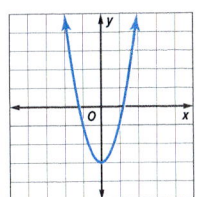

**23. PHYSICS** A ball is dropped off a cliff that is 100 feet high. The function $h = -16t^2 + 100$ models the height $h$ of the ball after $t$ seconds. Compare the graph of this function to the graph of $h = t^2$. **See margin.**

### Example 3
Write an equation for the function shown in the graph.

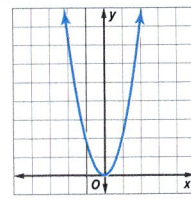

Since the graph opens upward, the leading coefficient must be positive. The parabola has not been translated up or down, so $c = 0$. Since the graph is stretched vertically, it must be of the form of $f(x) = ax^2$ where $a > 1$. The equation for the function is $y = 2x^2$.

## 9-3 Solving Quadratic Equations by Graphing
A.REI.4b, F.IF.7a

Solve each equation by graphing. If integral roots cannot be found, estimate the roots to the nearest tenth.

**24.** $x^2 - 3x - 4 = 0$    $-1, 4$
**25.** $-x^2 + 6x - 9 = 0$    $3$
**26.** $x^2 - x - 12 = 0$    $-3, 4$
**27.** $x^2 + 4x - 3 = 0$    $-4.6, 0.6$
**28.** $x^2 - 10x = -21$    $3, 7$
**29.** $6x^2 - 13x = 15$    $-0.8, 3$

**30. NUMBER THEORY** Find two numbers that have a sum of 2 and a product of $-15$.    $-3$ and $5$

### Example 4
Solve $x^2 - x - 6 = 0$ by graphing.

Graph the related function $f(x) = x^2 - x - 6$.

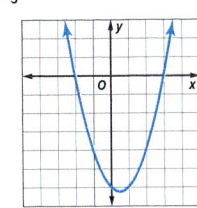

The x-intercepts of the graph appear to be at $-2$ and $3$, so the solutions are $-2$ and $3$.

## 9-4 Solving Quadratic Equations by Factoring
A.SSE.3a, A.REI.4b

Solve each equation. Check your solutions.

**31.** $x^2 + 6x - 55 = 0$    $-11, 5$
**32.** $(g + 8)^2 = 49$    $-1, -15$
**33.** $y^2 - 4y = 32$    $-4, 8$
**34.** $3k^2 - 8k = 3$    $-\frac{1}{3}, 3$
**35.** $2n^2 + 4n = 16$    $-4, 2$
**36.** $4w^2 + 9 = 12w$    $\frac{3}{2}$

### Example 5
Solve $x^2 - 2x = 120$ by factoring.

Write the equation in standard form and factor.

$x^2 - 2x = 120$    Original equation
$x^2 - 2x - 120 = 0$    Subtract 120 from each side.
$(x - 12)(x + 10) = 0$    Factor.
$x - 12 = 0$   or   $x + 10 = 0$    Zero Product Property
$x = 12$        $x = -10$    Solve each equation

The solutions are 12 and $-10$.

## Go Online!

**Anticipation Guide**

Students should complete the *Chapter 9 Anticipation Guide*, and discuss how their responses have changed now that they have completed Chapter 9.

# Chapter 9 Study Guide and Review

**Before the Test**
Have students complete the Study Notebook Tie it Together activity to review topics and skills presented in the chapter.

## CHAPTER 9
## Study Guide and Review Continued

### 9-5 Solving Quadratic Equations by Completing the Square
A.SSE.3b, A.REI.4, F.IF.8a

Solve each equation by completing the square. Round to the nearest tenth if necessary.

37. $x^2 + 6x + 9 = 16$  **1, −7**
38. $-a^2 - 10a + 25 = 25$  **0, −10**
39. $y^2 - 8y + 16 = 36$  **10, −2**
40. $y^2 - 6y + 2 = 0$  **5.6, 0.4**
41. $n^2 - 7n = 5$  **−0.7, 7.7**
42. $-3x^2 + 4 = 0$  **−1.2, 1.2**
43. $a^2 - 4a + 9 = 0$  **no solution**
44. $2a^2 - 4a + 1 = 0$  **1.7, 0.3**
45. **NUMBER THEORY** Find two numbers that have a sum of −2 and a product of −48.  **−8, 6**

**Example 6**
Solve $x^2 - 16x + 32 = 0$ by completing the square. Round to the nearest tenth if necessary.

Isolate the $x^2$- and $x$-terms. Then complete the square and solve.

$x^2 - 16x + 32 = 0$   Original equation
$x^2 - 16x = -32$   Isolate the $x^2$- and $x$-terms.
$x^2 - 16x + 64 = -32 + 64$   Complete the square.
$(x - 8)^2 = 32$   Factor.
$x - 8 = \pm\sqrt{32}$   Take the square root.
$x = 8 \pm \sqrt{32}$   Add 8 to each side.
$x = 8 \pm 4\sqrt{2}$   Simplify.

The solutions are about 2.3 and 13.7.

### 9-6 Solving Quadratic Equations by Using the Quadratic Formula
A.REI.4b, A.CED.1

Solve each equation by using the Quadratic Formula. Round to the nearest tenth if necessary.

46. $x^2 - 8x = 20$  **−2, 10**
47. $21x^2 + 5x - 7 = 0$  **−0.7, 0.5**
48. $d^2 - 5d + 6 = 0$  **2, 3**
49. $2f^2 + 7f - 15 = 0$  **−5, 1.5**
50. $2h^2 + 8h + 3 = 3$  **−4, 0**
51. $4x^2 + 4x = 15$  **−2.5, 1.5**
52. **GEOMETRY** The area of a square can be quadrupled by increasing the side length and width by 4 inches. What is the side length?  **4 in.**

State the discriminant for each equation. Then determine the number of real solutions of the equation.

53. $a^2 - 4a + 5 = 0$  **−4; no real solutions**
54. $-6x^2 + 2x + 3 = 0$  **76; two real solutions**

**Example 7**
Solve $x^2 + 10x + 9 = 0$ by using the Quadratic Formula.

$x = \dfrac{-b \pm \sqrt{b^2 - 4ac}}{2a}$   Quadratic Formula

$= \dfrac{-10 \pm \sqrt{10^2 - 4(1)(9)}}{2(1)}$   $a = 1, b = 10, c = 9$

$= \dfrac{-10 \pm \sqrt{64}}{2}$   Simplify.

$x = \dfrac{-10 + 8}{2}$ or $x = \dfrac{-10 - 8}{2}$   Separate the solutions.

$= -1$   $= -9$   Simplify.

## Go Online!
### eAssessment
Customize and create multiple versions of chapter tests and answer keys that align to your standards. Tests can be delivered on paper or online.

## 9-7 Solving Systems of Linear and Quadratic Equations

A.REI.7, A.CED.2

Solve each system of equations.

55. $y = x^2 - 4x + 4$  (1, 1), (5, 9)
    $y = 2x - 1$

56. $y = x^2 + 5x - 3$  (−9, 33), (2, 11)
    $y = 15 - 2x$

57. $y = 2x^2 - 4x + 1$  $\left(\frac{1}{2}, -\frac{1}{2}\right)$, (4, 17)
    $y - 5x = -3$

58. $y = 4x^2 + 4x - 3$  (−2, 5), $\left(\frac{3}{4}, 2\frac{1}{4}\right)$
    $x + y = 3$

59. $y = x^2 - 4x + 9$  (3, 0)
    $y = 2x$

60. $y = x^2 + 9$  (1, 10), (9, 90)
    $y = 10x$

61. $y = x^2 + 9$  no solution
    $y = x$

### Example 8

Solve the system of equations $\begin{cases} y = x^2 + 2x \\ y = x + 2 \end{cases}$.

The equations are solved for y. Substitute and solve for x.

$x^2 + 2x = x + 2$    Substitute.
$x^2 + x = 2$    Subtract x from each side.
$x^2 + x - 2 = 0$    Subtract 2 from each side.
$(x + 2)(x - 1) = 0$    Factor.
$x + 2 = 0$  or  $x - 1 = 0$    Zero Product Property
$x = -2$  or  $x = 1$    Solve for x.

Substitute to find the values of y.

$y = x + 2$      $y = x + 2$
$y = (-2) + 2$   $y = 1 + 2$
$y = 0$          $y = 3$

The solutions are (−2, 0) and (1, 3).

## 9-8 Analyzing Functions with Successive Differences

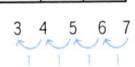

F.IF.6, F.LE.1, F.LE.2, F.LE.3

Look for a pattern in each table of values to determine which kind of model best describes the data. Then write an equation for the function that models the data.

62. 
| x | 0 | 1 | 2 | 3 | 4 |
|---|---|---|---|---|---|
| y | 0 | 3 | 12 | 27 | 48 |

quadratic; $y = 3x^2$

63. 
| x | 0 | 1 | 2 | 3 | 4 |
|---|---|---|---|---|---|
| y | 1 | 2 | 4 | 8 | 16 |

exponential; $y = 2^x$

64. 
| x | 0 | 1 | 2 | 3 | 4 |
|---|---|---|---|---|---|
| y | 0 | −1 | −4 | −9 | −16 |

quadratic; $y = -x^2$

65. 
| x | 0 | 1 | 2 | 3 | 4 |
|---|---|---|---|---|---|
| y | 8 | 5 | 2 | −1 | −4 |

Linear; $y = 8 - 3x$

### Example 9

Determine the model that best describes the data. Then write an equation for the function that models the data.

| x | 0 | 1 | 2 | 3 | 4 |
|---|---|---|---|---|---|
| y | 3 | 4 | 5 | 6 | 7 |

**Step 1** First differences:   3   4   5   6   7
                                1   1   1   1

A linear function models the data.

**Step 2** The slope is 1 and the y-intercept is 3, so the equation is $y = x + 3$.

# Chapter 9 Study Guide and Review

## CHAPTER 9
## Study Guide and Review Continued

### 9-9 Combining Functions
F.BF.1b

Consider the following functions.
$f(x) = 4x^2 - 8x + 2$
$g(x) = -5x^2 + x + 1$
$h(x) = 3 - x^2$

Determine each of the following.

66. $(f + g)(x)$  $-x^2 - 7x + 3$
67. $(g + h)(x)$  $-6x^2 + x + 4$
68. $(f - h)(x)$  $5x^2 - 8x - 1$
69. $(h - g)(x)$  $4x^2 - x + 2$
70. $(f \cdot h)(x)$  $-4x^4 + 8x^3 + 10x^2 - 24x + 6$
71. $(g \cdot h)(x)$  $5x^4 - x^3 - 16x^2 + 3x + 3$

**Example 10**

Let $f(x) = -2x^2 - x + 5$ and $g(x) = x^2 - 1$. Determine $(f \cdot g)(x)$ and $(f - g)(x)$.

| | |
|---|---|
| $(f \cdot g)(x)$ | Original expression |
| $= (-2x^2 - x + 5)(x^2 - 1)$ | Substitute. |
| $= -2x^4 + 2x^2 - x^3 + x + 5x^2 - 5$ | Distribute. |
| $= -2x^4 - x^3 + 7x^2 + x - 5$ | Combine like terms. |
| $(f - g)(x)$ | Original expression |
| $= (-2x^2 - x + 5) - (x^2 - 1)$ | Substitute. |
| $= -2x^2 - x + 5 - x^2 + 1$ | Distribute. |
| $= -3x^2 - x + 6$ | Combine like terms. |

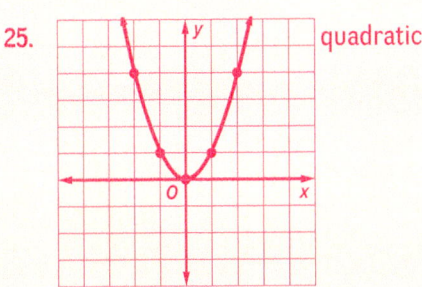

# CHAPTER 9
## Practice Test

Go Online! for another Chapter Test

Use a table of values to graph the following functions. State the domain and range.

1. $y = x^2 + 2x + 5$  **1–2. See Ch. 9 Answer Appendix.**
2. $y = 2x^2 - 3x + 1$

Consider $y = x^2 - 7x + 6$.

3. Determine whether the function has a *maximum* or *minimum* value. **minimum**
4. State the maximum or minimum value. **−6.25**
5. What are the domain and range?
   **D = {all real numbers}; R = {y | y ≥ −6.25}**

Describe how the graph of each function is related to the graph of $f(x) = x^2$.

6. $g(x) = x^2 - 5$ **translated down 5 units**
7. $g(x) = -3x^2$ **reflected across the x-axis, stretched vertically**
8. $h(x) = \frac{1}{2}x^2 + 4$ **compressed vertically and translated 4 units up**
9. **MULTIPLE CHOICE** Which is an equation for the function shown in the graph? **D**

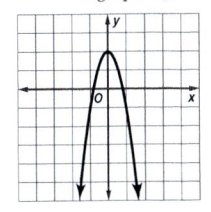

   A  $y = -3x^2$
   B  $y = 3x^2 + 1$
   C  $y = x^2 + 2$
   D  $y = -3x^2 + 2$

Solve each equation by graphing. If integral roots cannot be found, estimate the roots to the nearest tenth.

10. $x^2 + 7x + 10 = 0$ **−5, −2**
11. $x^2 - 5 = -3x$ **−4.2, 1.2**

Solve each equation by factoring.

12. $x^2 + 18x + 81 = 0$ **−9**
13. $x^2 - 18x = 40$ **−2, 20**
14. $5x^2 - 31x + 6 = 0$ **$\frac{1}{5}$, 6**

Solve each equation by completing the square. Round to the nearest tenth if necessary.

15. $x^2 + 2x + 5 = 0$ **no real solution**
16. $x^2 + 5x - 8 = 12$ **−7.6, 2.6**
17. $2x^2 - 36 = -6x$ **−6, 3**

Solve each equation by using the Quadratic Formula. Round to the nearest tenth if necessary.

18. $x^2 - x - 30 = 0$ **−5, 6**
19. $x^2 - 10x = -15$ **1.8, 8.2**
20. $2x^2 + x - 15 = 0$ **2.5, −3**

21. **BASEBALL** Elias hits a baseball into the air. The equation $h = -16t^2 + 60t + 3$ models the height $h$ in feet of the ball after $t$ seconds. How long is the ball in the air? **about 3.8 seconds**

Solve each system of equations algebraically.

22. $y = x^2 - 7x + 3$ **(1, −3), (4, −9)**
    $y = -2x - 1$
23. $y = 3x^2 - 8x - 1$ **$(-\frac{1}{3}, 2)$, (2, −5)**
    $3x + y = 1$
24. $y = 1 - x^2$ **(−2, −3)**
    $y = 4x + 5$
25. Graph {(−2, 4), (−1, 1), (0, 0), (1, 1), (2, 4)}. Determine whether the ordered pairs represent a *linear function*, a *quadratic function*, or an *exponential function*. **See margin.**

26. **CAR CLUB** The table shows the number of car club members for four consecutive years after it began.

| Time (years) | 0 | 1 | 2 | 3 | 4 |
|---|---|---|---|---|---|
| Members | 10 | 20 | 40 | 80 | 160 |

a. Determine which model best represents the data. **exponential**
b. Write a function that models the data. $y = (10)2^x$
c. Predict the number of car club members after 6 years. **640**

Consider the functions $f(x) = 2x^2 - 4x + 1$ and $g(x) = -6x^2 + 9x - 8$.

27. Determine $(f + g)(x)$. **$-4x^2 + 5x - 7$**
28. Determine $(f - g)(x)$. **$8x^2 - 13x + 9$**
29. Determine $(f \cdot g)(x)$. **$-12x^4 + 42x^3 - 58x^2 + 41x - 8$**

# Chapter 9 Preparing for Assessment

## Launch

**Objective** Apply concepts and skills from this chapter in a real-world setting.

## Teach

Ask:

- How are the substitutions used to complete Part A different? In exercise 1, we substitute 20 for $x$ to find $y$ and in exercise 2, we substitute 15 for $y$ to find $x$.

- What assumptions are made when determining whether the player hit a home run in Part B? Sample answer: The ball is not intercepted and makes it over the fence.

- What do your answers in Part C tell you about the graph of the function? Sample answer: It has a vertex at (195, 80.1) and $y$-intercept of 4.

- Why is there only one reasonable solution in Part D? Sample answer: Since distance cannot be negative, the negative solution cannot be used.

The Performance Task focuses on the following content standards and standards for mathematical practice.

### Dual Coding

| Items | Content Standards | Mathematical Practices |
|---|---|---|
| 1 | F.IF.2 | 1, 7 |
| 2 | A.REI.4b | 1, 7 |
| 3 | A.CED.2 | 4 |
| 4 | A.REI.4b | 1, 4, 7 |
| 5 | F.IF.4, F.IF.7a | 2 |
| 6 | F.IF.4, F.IF.7a | 2 |
| 7 | F.IF.7a | 2 |
| 8 | A.REI.7 | 1, 4 |
| 9 | F.LE.1 | 3 |

## Go Online! eBook

**Interactive Student Guide** Refer to *Interactive Student Guide* for an additional Performance Task.

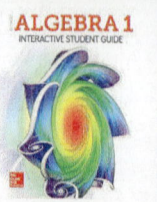

---

# CHAPTER 9
# Preparing for Assessment

## Performance Task

Provide a clear solution to each part of the task. Be sure to show all of your work, include all relevant drawings, and justify your answers.

**BASEBALL** A professional baseball player hits a baseball. Its height above the field can be modeled by the equation $y = -0.002x^2 + 0.78x + 4$, where $y$ is height of the baseball, in feet, and $x$ is the horizontal distance the baseball is from home plate, in feet.

### Part A

1. What is the height of the baseball when it is 20 feet from home plate? **18.8 ft**

2. At what distance(s) is the baseball from home plate when its height is 15 feet? Round to the nearest tenth. **14.7 ft and 375.4 ft**

### Part B

We can use the quadratic equation to find how far the baseball will travel before it hits the ground. If the baseball is hit towards left center field, where the fence is 370 feet from home plate, we can determine whether the baseball will travel far enough to be a home run.

3. **Modeling** Write an equation that could be used to determine if the baseball will travel far enough to be home run. $-0.002x^2 + 0.78x + 4 = 0$

4. **Reasoning** Solve the equation and round to the nearest tenth, if necessary. Based on the solution, does the baseball travel far enough to be home run? Explain. **Yes; sample answer: If there is no one to intercept the baseball, it will travel about 395.1 feet which is farther than the 375 feet it needs to go to be a home run.**

### Part C

The graph of the related quadratic function reveals characteristics of the baseball's trajectory.

5. What is the maximum height of the baseball? **80.1 ft**

6. How far away from home plate does the baseball reach its maximum height? **195 ft**

7. How far above the ground was the baseball at the moment it was hit by the player? **4 ft**

### Part D

Another player hits a foul ball. Its height $y$ can be modeled by $y = -0.02x^2 + x + 4$ where $x$ is the distance, in feet, from home plate. A fan runs up the stands to catch the foul ball. His height above the field $y$, in feet, is approximately modeled by $y = 0.7x - 21$, where $x$ is his distance from home plate, in feet.

8. At what distance and height will the fan catch the baseball? **43.6 ft from home plate, 9.5 ft above the field**

### Part E

9. **Construct an Argument** Is it more reasonable to use a linear or quadratic model for the trajectory of a baseball after it is hit by the batter? **Quadratic; sample answer: The parabolic trajectory is more in line with vertical motion affected by gravity than a linear model.**

---

### Levels of Complexity Chart

The levels of the exercises progress from 1 to 3, with Level 1 indicating the lowest level of complexity.

| Parts | Level 1 | Level 2 | Level 3 |
|---|---|---|---|
| A | ● | | |
| B | | ● | |
| C | ● | | |
| D | | ● | |
| E | | | ● |

### Differentiated Instruction

**Extension** Write a similar trajectory function for your favorite baseball player. Then test its validity by plugging in some possible ordered pairs of distance from home plate paired with height of the ball.

## Test-Taking Strategy

### Example

Read the problem. Identify what you need to know. Then use the information in the problem to solve.

Find the exact roots of the quadratic equation $-2x^2 + 6x + 5 = 0$.

A $\dfrac{3 \pm \sqrt{17}}{4}$  
B $\dfrac{4 \pm \sqrt{17}}{3}$  
C $\dfrac{3 \pm \sqrt{19}}{2}$  
D $\dfrac{3 \pm \sqrt{19}}{4}$

**Step 1** What quantities are given in the problem?
A quadratic equation in the form $ax^2 + bx + c = 0$, where $a = -2$, $b = 6$, and $c = 5$.

**Step 2** What quantities do you need to find?
The roots, which are the values of $x$ that make the equation true.

**Step 3** Is there a formula that relates these quantities? If so, write it.
Yes, the Quadratic Formula can be used. $x = \dfrac{-b \pm \sqrt{b^2 - 4ac}}{2a}$

**Step 4** Substitute the known quantities to solve for the unknown quantity.

$$x = \dfrac{-6 \pm \sqrt{(6)^2 - 4(-2)(5)}}{2(-2)} = \dfrac{-6 \pm \sqrt{76}}{-4}$$
$$= \dfrac{3 \pm \sqrt{19}}{2}$$

The correct answer is C.

**Test-Taking Tip**

**Using a Formula** A formula is an equation that shows a relationship among certain quantities. The two most important things to remember when using a formula are to substitute the correct values for the appropriate variables, and be careful to simplify or solve correctly. If you do these two things, the formula will guide you to the correct answer.

### Apply the Strategy

Read each problem. Identify what you need to know. Then use the information in the problem to solve.

1. Find the exact roots of the quadratic equation $x^2 + 5x - 12 = 0$. **A**

   A $\dfrac{-5 \pm \sqrt{73}}{2}$  
   B $\dfrac{4 \pm \sqrt{61}}{3}$  
   C $\dfrac{-3 \pm \sqrt{73}}{4}$  
   D $\dfrac{-1 \pm \sqrt{61}}{2}$

Answer the questions below.

1. What quantities are given in the problem? $ax^2 + bx + c = 0$, where $a = 1$, $b = 5$, and $c = -12$
2. What quantities do you need to find? the roots, or solutions, of the equation
3. Is there a formula that relates these quantities? If so, write it. yes, the Quadratic Formula; $x = \dfrac{-b \pm \sqrt{b^2 - 4ac}}{2a}$
4. Substitute the known quantities to solve for the unknown quantity. What is the correct answer? **A**

---

## Chapter 9 Preparing for Assessment

### Test-Taking Strategy
### Using a Formula

**Step 1** Identify what is given in the problem.

**Step 2** Identify what quantities need to be found in the problem.

**Step 3** Determine whether there is a formula that can be used. If so, write the formula.

**Step 4** Substitute the known values into the formula, then simplify or solve.

### Need Another Example?

Find the exact roots of the quadratic equation $2x^2 + 5x - 5 = 0$.

A $\dfrac{5 \pm \sqrt{65}}{4}$  
B $\dfrac{5 \pm \sqrt{15}}{4}$  
C $\dfrac{-5 \pm \sqrt{65}}{4}$  
D $\dfrac{-5 \pm \sqrt{15}}{4}$

1. What quantities are given in the problem? a quadratic equation in the form $ax^2 + bx + c = 0$, where $a = 2$, $b = 5$, and $c = -5$
2. What quantities do you need to find? the roots
3. Is there a formula that related these quantities? If so, write it. the Quadratic Formula; $x = \dfrac{-b \pm \sqrt{b^2 - 4ac}}{2a}$
4. Substitute the known quantities to solve for the unknown quantity. What is the correct answer? C

### Go Online!

The most up-to-date resources available for your program can be found at **connectED.mcgraw-hill.com**.

connectED.mcgraw-hill.com 645

# Chapter 9 Preparing for Assessment

## Diagnose Student Errors

Survey student responses for each item. Class trends may indicate common errors and misconceptions.

| 1. | A | Mistakenly thought data was linear |
|---|---|---|
| | B | Mistakenly thought data was linear |
| | C | Made a sign error |
| | D | CORRECT |

| 4. | A | CORRECT |
|---|---|---|
| | B | Confused *x*-intercept with *y*-intercept |
| | C | Mistook minimum value as maximum value |
| | D | Used the *y*-value of vertex instead of *x*-value |
| | E | CORRECT |

| 5. | A | Forgot to add half of −4 squared to each side |
|---|---|---|
| | B | CORRECT |
| | C | Forgot to add half of 4 squared to each side |
| | D | Forgot to add half of 4 squared to each side |

| 7. | A | CORRECT |
|---|---|---|
| | B | Forgot to reverse the inequality sign |
| | C | Made a sign error when solving |
| | D | Made a sign error when solving |

| 8. | A | Added 9 and factored incorrectly |
|---|---|---|
| | B | Factored incorrectly |
| | C | Incorrectly added 9 to each side |
| | D | CORRECT |

| 9. | A | Incorrectly added 8 to each side |
|---|---|---|
| | B | Forgot to find both solutions |
| | C | Reversed the signs of the solutions |
| | D | CORRECT |

| 10. | A | Chose the wrong starting number for set |
|---|---|---|
| | B | Chose the wrong starting number for set |
| | C | Chose numbers on the wrong side of −2 |
| | D | CORRECT |

| 12. | A | Forgot to find the cube root |
|---|---|---|
| | B | Used exponent incorrectly |
| | C | CORRECT |
| | D | Reversed meaning of fractional exponent |

| 13. | A | Confused vertical with horizontal |
|---|---|---|
| | B | Does not understand reflections |
| | C | CORRECT |
| | D | Used (2, 6) as vertex instead of (2, −6) |

## Go Online!

### Standardized Test Practice

Students can take self-checking tests in standardized format to plan and prepare for standardized assessments.

---

# CHAPTER 9
# Preparing for Assessment
## Cumulative Review

Read each question. Then fill in the correct answer on the answer document provided by your teacher or on a sheet of paper.

**1.** Celia graphed the set of ordered pairs in the table below.

| x | 2 | 3 | 4 | 5 | 6 |
|---|---|---|---|---|---|
| y | 3 | 0 | −1 | 0 | 3 |

Which equation best models the data in the table? A.CED.2 **D**

- A  $y = 2x - 1$
- B  $y = -2x + 7$
- C  $y = x^2 + 8x + 15$
- D  $y = x^2 - 8x + 15$

**2.** The graph of $f(x) = x^2$ is reflected across the *x*-axis and translated to the right 5 units. What is the value of *b* when the equation of the transformed graph is written in standard form $f(x) = ax^2 + bx + c$? F.BF.3

**10**

**3.** MULTI-STEP Maria starts with $200 in her savings account and deposits $25 each month. Stefan starts with $275 in his savings account and deposits $20 each month. A.CED.3, A.REI.6

**a.** Write an algebraic equation to represent the amount, *a*, in in dollars, that Maria and Stefan each have in their respective savings accounts after *t* months.

Maria: $a = 200 + 25t$

Stefan: $a = 275 + 20t$

**b.** In what month do Maria and Stefan have the same amount in their savings accounts?

**15**

**c.** MP What mathematical practice did you use to solve this problem? See students' work.

**4.** Which statements describe the function graphed below? Select all that apply. F.IF.7a **A, E**

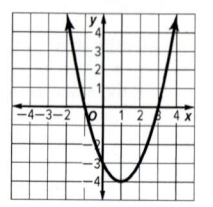

- ☐ A  The range is {*y* | *y* ≥ −4}.
- ☐ B  The *y*-intercept is 3.
- ☐ C  The maximum value is −4.
- ☐ D  The equation of the axis of symmetry is $x = -4$.
- ☐ E  The *x*-intercepts are −1 and 3.

**5.** Which equation has no real solutions? A.REI.4b **B**

- A  $x^2 - 4x = -4$
- B  $x^2 - 4x = -8$
- C  $x^2 + 4x = -2$
- D  $x^2 + 4x = 2$

**6.** Let $f(x) = 3x^2 - 5x + 1$ and $g(x) = x^2 - x - 6$. Write an expression for $(g - f)(x)$ in standard form. F.BF.1b

$-2x^2 + 4x - 7$

**7.** Which inequality has the solution set graphed below? A.REI.3 **A**

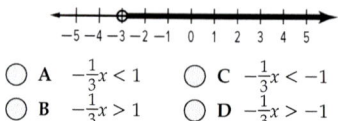

- A  $-\frac{1}{3}x < 1$
- B  $-\frac{1}{3}x > 1$
- C  $-\frac{1}{3}x < -1$
- D  $-\frac{1}{3}x > -1$

646 | Chapter 9 | Preparing for Assessment

## Chapter 9 Preparing for Assessment

**Formative Assessment**
You can use these pages to benchmark student progress.

📄 Standardized Test Practice

**Test Item Formats**
In the Cumulative Review, students will encounter different formats for assessment questions to prepare them for standardized tests.

| Question Type | Exercises |
| --- | --- |
| Multiple-Choice | 1, 5, 7–10, 12, 13 |
| Multiple Correct Answers | 4 |
| Type Entry: Short Response | 2, 3, 6, 11 |
| Type Entry: Extended Response | 3 |

**Answer Sheet Practice**
Have students simulate taking a standardized test by recording their answers on a practice recording sheet.

**Homework Option**
**Get Ready for Chapter 10** Assign students the exercises on p. 650 as homework to assess whether they possess the prerequisite skills needed for the next chapter.

---

8. Karl solved a quadratic equation by graphing the related function as shown below.

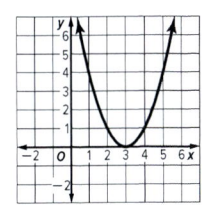

Which equation did he solve? F.IF.7a **D**
- A  $x^2 + 6x = 9$
- B  $x^2 + 6x = -9$
- C  $x^2 - 6x = 9$
- D  $x^2 - 6x = -9$

9. What are the solutions of the quadratic equation $5x^2 - 2x = 8$? Round to the nearest tenth if necessary. A.REI.4b **D**
- A  no solution
- B  $-1.1$
- C  1.1 or $-1.5$
- D  $-1.1$ or 1.5

10. For what domain values of the function $y = -2x + 1$ is the range of the function $\{y \mid y < 5\}$? F.IF.1 **D**
- A  $\{x \mid x < 2\}$
- B  $\{x \mid x > 2\}$
- C  $\{x \mid x < -2\}$
- D  $\{x \mid x > -2\}$

**Test-Taking Tip**
Question 8 First, use the graph to find the zero of the function. Then, use what you know about the relationship between the zeros of a quadratic function and their corresponding linear factors to identify the correct equation.

11. If $b = 12$, what value of $c$ makes $x^2 + bx + c$ a perfect square trinomial? A.SSE.3b
    **36**

12. Which expression has a value of 25? N.RN.2 **C**
- A  $5^{\frac{2}{3}}$
- B  $5^{\frac{3}{2}}$
- C  $125^{\frac{2}{3}}$
- D  $125^{\frac{3}{2}}$

13. The graph of the quadratic function $h(x)$ is shown below.

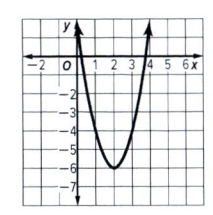

Which of the following statements best describes the function? F.BF.3 **C**
- A  $h(x)$ is a vertical compression of the graph of $f(x) = x^2$.
- B  $h(x)$ is a reflection across the x-axis of the graph of $f(x) = x^2$.
- C  In vertex form, the equation of the function is $f(x) = 2(x - 2)^2 - 6$.
- D  In standard form, the equation of the function is $f(x) = 2x^2 - 8x + 14$.

### Need Extra Help?

| If you missed Question... | 1 | 2 | 3 | 4 | 5 | 6 | 7 | 8 | 9 | 10 | 11 | 12 | 13 |
| --- | --- | --- | --- | --- | --- | --- | --- | --- | --- | --- | --- | --- | --- |
| Go to Lesson... | 9-8 | 9-2 | 6-2 | 9-1 | 9-5 | 9-9 | 5-2 | 9-3 | 9-6 | 3-1 | 9-5 | 7-3 | 9-2 |

---

## LEARNSMART

Use LearnSmart as part of your test-preparation plan to measure student topic retention. You can create a student assignment in LearnSmart to additional practice on these topics.

- Solve Quadratic Equations in One Variable
- Solve a System of Equations Consisting of a Linear and a Quadratic Equation in Two Variables
- Use Properties of Rational and Irrational Numbers
- Create and Interpret Quadratic Models
- Build New Functions From Existing Functions
- Construct and Compare Linear, Quadratic, and Exponential Models and Solve Problems

**Go Online!**

**eAssessment**

Customize and create multiple versions of chapter tests and answer keys that align to your standards. Tests can be delivered on paper or online.

## Lesson 9-1 (Guided Practice)

**1.**

| x | y |
|---|---|
| −2 | 7 |
| −1 | 4 |
| 0 | 3 |
| 1 | 4 |
| 2 | 7 |

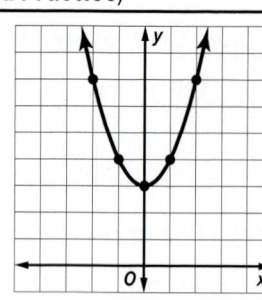

$D = \{-\infty < x < \infty\}$; $R = \{y \mid y \geq 3\}$

## Lesson 9-1

**1.**

| x | −3 | −2 | −1 | 0 | 1 | 2 |
|---|---|---|---|---|---|---|
| y | 0 | −6 | −8 | −6 | 0 | 10 |

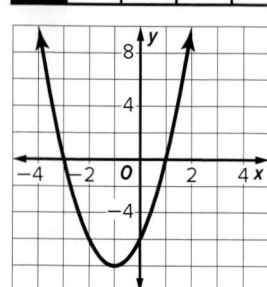

$D = \{-\infty < x < \infty\}$; $R = \{y \mid y \geq -8\}$

**2.**

| x | −3 | −2 | −1 | 0 | 1 | 2 |
|---|---|---|---|---|---|---|
| y | 2 | −1 | −2 | −1 | 2 | 7 |

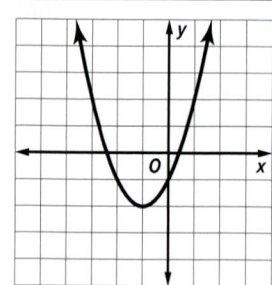

$D = \{-\infty < x < \infty\}$; $R = \{y \mid y \geq -2\}$

**3.**

| x | y |
|---|---|
| −1 | 4 |
| 0 | −3 |
| 1 | −8 |
| 2 | −11 |
| 3 | −12 |
| 4 | −11 |
| 5 | −8 |
| 6 | −3 |
| 7 | 4 |

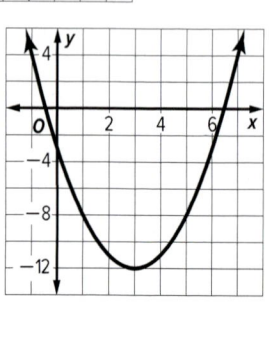

$D = \{-\infty < x < \infty\}$; $R = \{y \mid y \geq -12\}$

**4.**

| x | −2 | −1 | 0 | 1 | 2 | 3 |
|---|---|---|---|---|---|---|
| y | 19 | 4 | −5 | −8 | −5 | 4 |

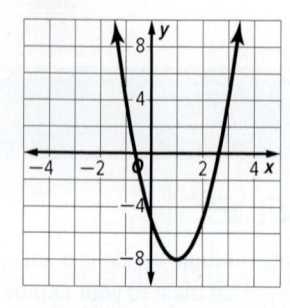

$D = \{-\infty < x < \infty\}$; $R = \{y \mid y \geq -8\}$

**22.**

| x | −4 | −3 | −2 | −1 | 0 |
|---|---|---|---|---|---|
| y | 6 | 3 | 2 | 3 | 6 |

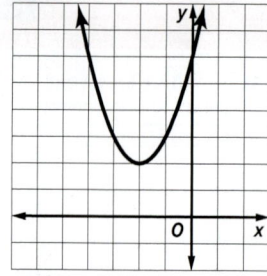

$D = \{-\infty < x < \infty\}$; $R = \{y \mid y \geq 2\}$

**23.**

| x | −3 | −2 | −1 | 0 | 1 |
|---|---|---|---|---|---|
| y | 13 | 7 | 5 | 7 | 13 |

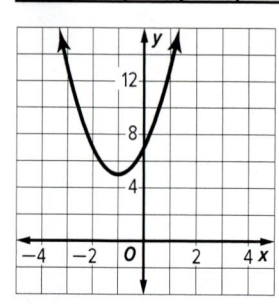

$D = \{-\infty < x < \infty\}$; $R = \{y \mid y \geq 5\}$

**24.**

| x | 4 | 3 | 2 | 1 | 0 |
|---|---|---|---|---|---|
| y | −5 | −11 | −13 | −11 | −5 |

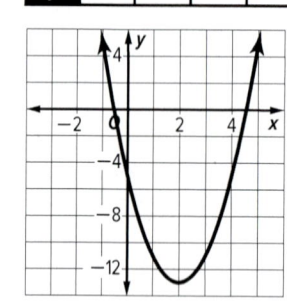

$D = \{-\infty < x < \infty\}$; $R = \{y \mid y \geq -13\}$

**25.**

| x | 0 | −1 | −2 | −3 | −4 |
|---|---|---|---|---|---|
| y | 5 | −4 | −7 | −4 | 5 |

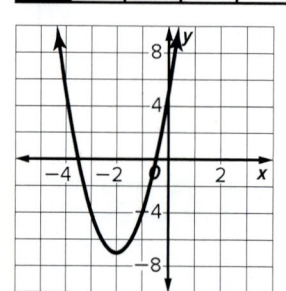

$D = \{-\infty < x < \infty\}$; $R = \{y \mid y \geq -7\}$

**26.**

| x | 3 | 2 | 1 | 0 | −1 |
|---|---|---|---|---|---|
| y | 7 | −2 | −5 | −2 | 7 |

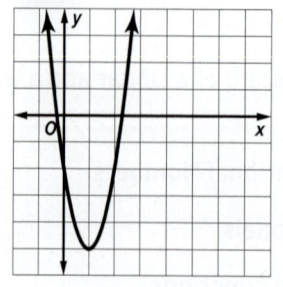

$D = \{-\infty < x < \infty\}$; $R = \{y \mid y \geq -5\}$

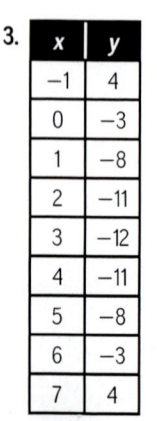

647A | Chapter 9 | Answer Appendix

**27.**

| x | 3 | 2 | 1 | 0 | −1 |
|---|---|---|---|---|---|
| y | 2 | −1 | −2 | −1 | 2 |

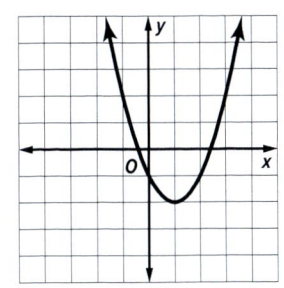

$D = \{-\infty < x < \infty\}$;
$R = \{y \mid y \geq -2\}$

**52.**

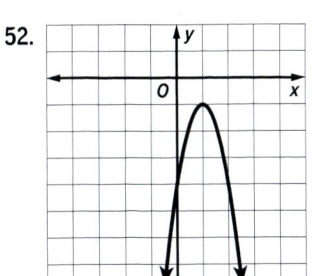

**53.**

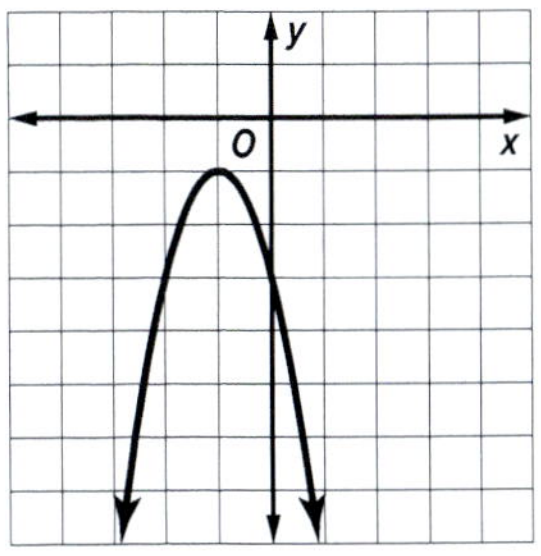

**54.**

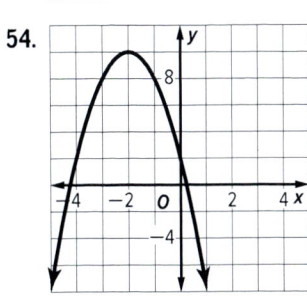

**55.**

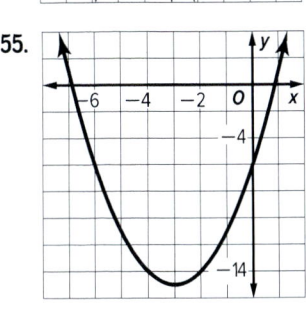

**56.**

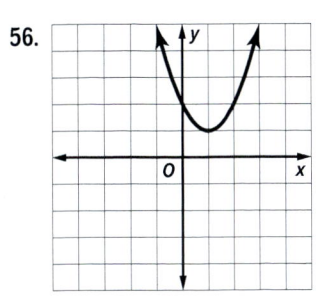

**57.**

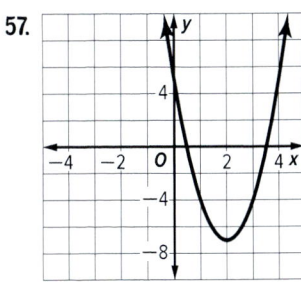

**59.** 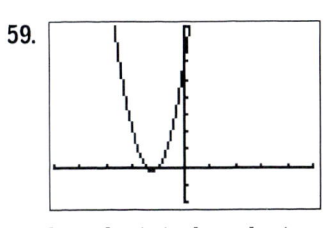 [−5, 5] scl:1 by [−5, 5] scl:1

**60.** [−5, 5] scl: 1 by [−2, 18] scl: 2

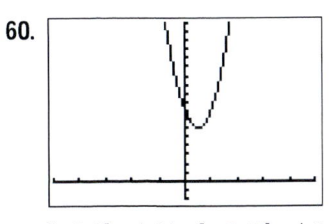

**61.** [−5, 5] scl: 1 by [−20, 2] scl: 2

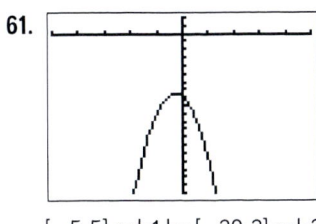

**62.** [−5, 5] scl: 1 by [−20, 2] scl: 2

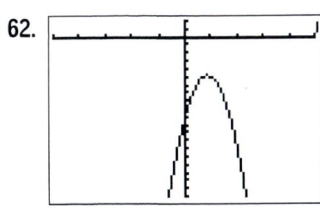

**74.** Sample answer: suppose $a = 1$, $b = 2$, and $c = 1$

| x | $f(x) = 2^x + 1$ | $g(x) = x^2 + 1$ | $h(x) = x + 1$ |
|---|---|---|---|
| −10 | 1.00098 | 101 | −9 |
| −8 | 1.00391 | 65 | −7 |
| −6 | 1.01563 | 37 | −5 |
| −4 | 1.0625 | 17 | −3 |
| −2 | 1.25 | 5 | −1 |
| 0 | 2 | 1 | 1 |
| 2 | 5 | 5 | 3 |
| 4 | 17 | 17 | 5 |
| 6 | 65 | 37 | 7 |
| 8 | 257 | 65 | 9 |
| 10 | 1205 | 101 | 11 |

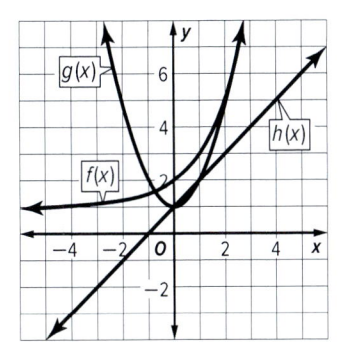

Intercepts: $f(x)$ and $g(x)$ have no x-intercepts, $h(x)$ has one at −1 because $c = 1$. $g(x)$ and $h(x)$ have one y-intercept at 1 and $f(x)$ has one y-intercept at 2. The graphs are all shifted up 1 unit from the graph of the parent functions because $c = 1$.

Increasing/decreasing: $f(x)$ and $h(x)$ are increasing on the entire domain. $g(x)$ is increasing to the right of the vertex and decreasing to the left.

Positive/negative: The function values for $f(x)$ and $g(x)$ are all positive. The function values of $h(x)$ are negative for $x < -1$ and positive for $x > -1$.

Maxima/minima: $f(x)$ and $h(x)$ have no maxima or minima. $g(x)$ has a minimum at $(0, 1)$.

Symmetry: $f(x)$ and $g(x)$ have no symmetry. $g(x)$ is symmetric about the y-axis.

End behavior: For $f(x)$ and $h(x)$, as x increases, y increases and as x decreases, y decreases. For $g(x)$, as x increases, y increases and as x decreases, y increases.

The exponential function $f(x)$ eventually exceeds the others.

### Extend 9-1

**1.**

| x | −4 | −3 | −2 | −1 | 0 | 1 | 2 | 3 | 4 |
|---|---|---|---|---|---|---|---|---|---|
| y | 16 | 9 | 4 | 1 | 0 | 1 | 4 | 9 | 16 |

**2.** The function is increasing for $x > 0$ and decreasing for $x < 0$.

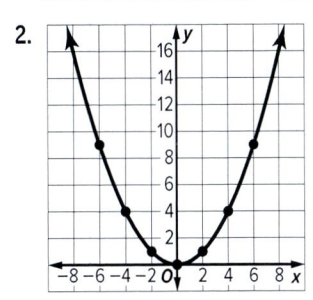

3.

| x | −4 | −3 | −2 | −1 | 0 | 1 | 2 | 3 | 4 |
|---|---|---|---|---|---|---|---|---|---|
| y | 16 | 9 | 4 | 1 | 0 | 1 | 4 | 9 | 16 |
| Rate of Change | -- | −7 | −5 | −3 | −1 | 1 | 3 | 5 | 7 |

Sample answer: The rate of change as the function decreases is the opposite of the rate of change as the function increases.

4.

| x | −4 | −3 | −2 | −1 | 0 | 1 | 2 | 3 | 4 |
|---|---|---|---|---|---|---|---|---|---|
| y | −156 | −44 | 36 | 84 | 100 | 84 | 36 | −44 | −156 |
| Rate of Change | -- | 112 | 80 | 48 | 16 | −16 | −48 | −80 | −112 |

Sample answer: The rate of change as the function increases has the same absolute value as when the function decreases.

## Lesson 9-2

**31a.** See students' graphs.

**31b.** For $f(x) = x^2$, when $b > 1$, the graph is compressed horizontally.

**31c.** For $f(x) = x^2$, when $0 < b < 1$, the graph is stretched horizontally.

**31d.** For $f(x) = x^2$, when $b < 0$, the square term offsets the negative. For example, when $b = -3$, $(-3x)^2 = 9x^2$. This is the same transformation as when $b = 3$. Therefore, the sign of $b$ has no effect on the transormation of this function.

**31e.** See students' graphs.

**34.**

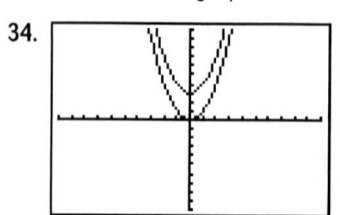

[−10, 10] scl: 1 by [−10, 10] scl: 1
Both graphs have the same shape, but the graph of $y = x^2 + 3$ is translated up 3 units from the graph of $y = x^2$.

**35.**

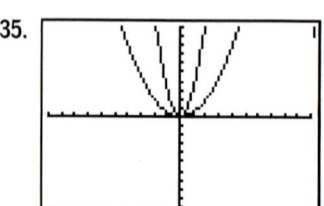

[−10, 10] scl: 1 by [−10, 10] scl: 1
The graph of $y = 3x^2$ is narrower than the graph of $y = \frac{1}{2}x^2$.

**36.**

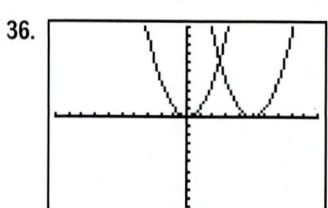

[−10, 10] scl: 1 by [−10, 10] scl: 1
Both graphs have the same shape, but the graph of $y = (x - 5)^2$ is translated right 5 units from the graph of $y = x^2$.

**37.**

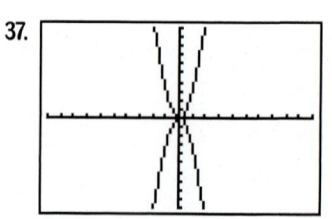

[−10, 10] scl: 1 by [−10, 10] scl: 1
Both graphs have the same shape, but the graph of $y = -3x^2$ opens down while the graph of $y = 3x^2$ opens up.

**38.**

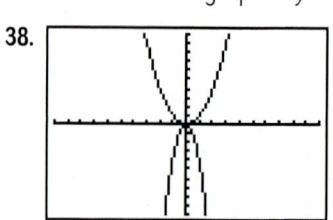

[−10, 10] scl: 1 by [−10, 10] scl: 1
The graph of $y = -4x^2$ opens down and is narrower than the graph of $y = x^2$.

**39.**

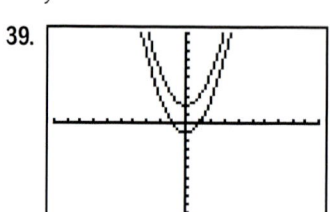

[−10, 10] scl: 1 by [−10, 10] scl: 1
Both graphs have the same shape, but the graph of $y = x^2 + 2$ is translated up 2 units from the graph of $y = x^2$ and the graph of $y = x^2 - 1$ is translated down 1 unit from the graph of $y = x^2$.

**40.**

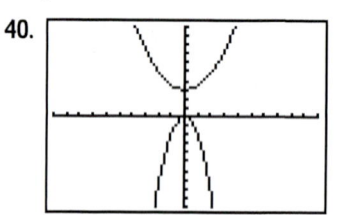

[−10, 10] scl: 1 by [−10, 10] scl: 1
The graph of $y = -2x^2$ opens down and is narrower than the graph of $y = \frac{1}{2}x^2 + 3$; also the graph of $y = \frac{1}{2}x^2 + 3$ is translated up 3 units from the graph of $y = x^2$.

**41.**

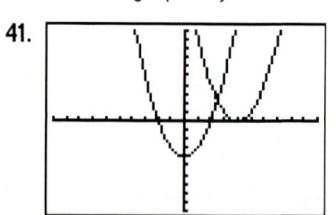

[−10, 10] scl: 1 by [−10, 10] scl: 1
Both graphs have the same shape, but the graph of $y = x^2 - 4$ is translated down 4 units from the graph of $y = x^2$ and the graph of $y = (x - 4)^2$ is translated right 4 units from the graph of $y = x^2$.

## Lesson 9-3 (Guided Practice)

**1A.**

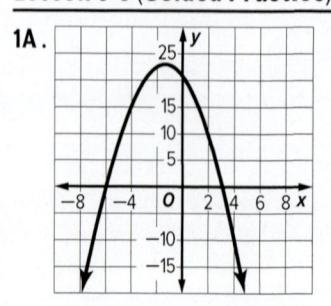

**1B.**

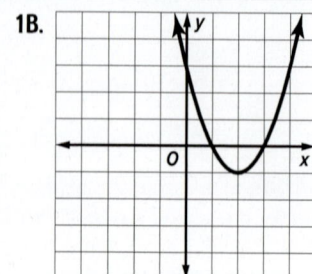

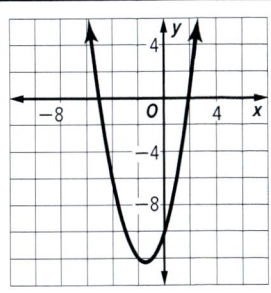

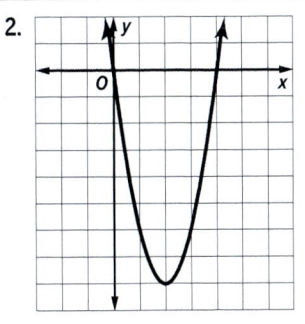

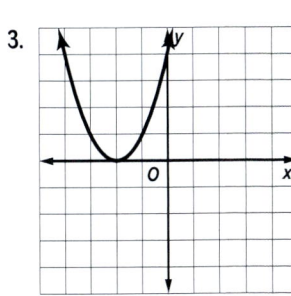

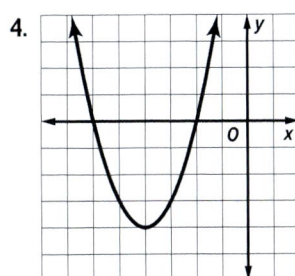

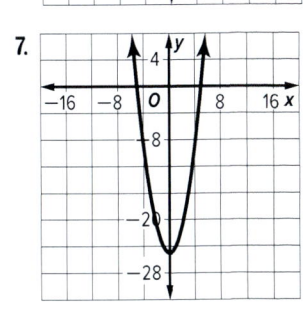

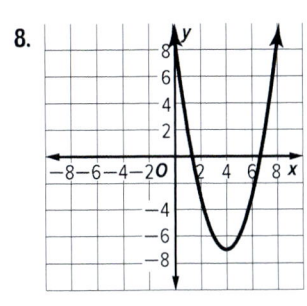

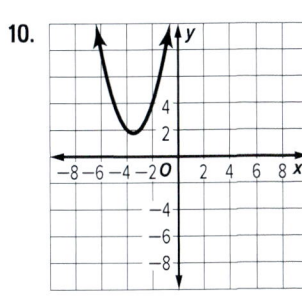

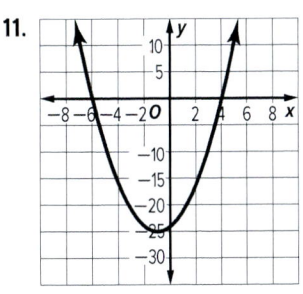

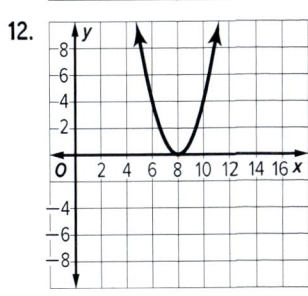

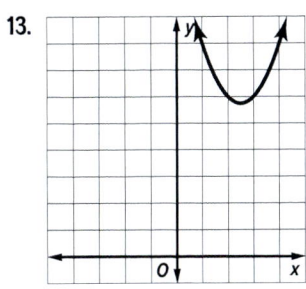

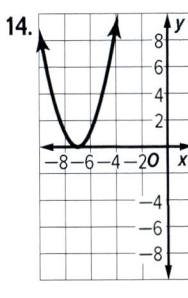

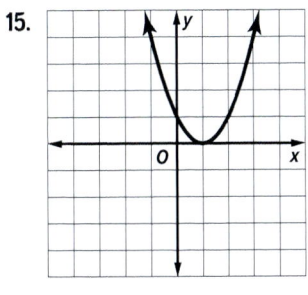

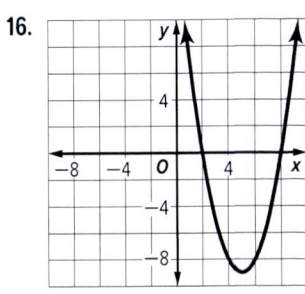

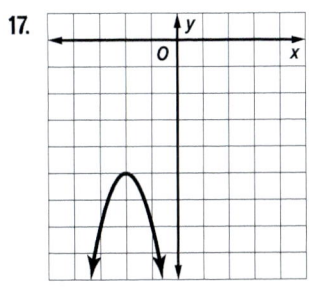

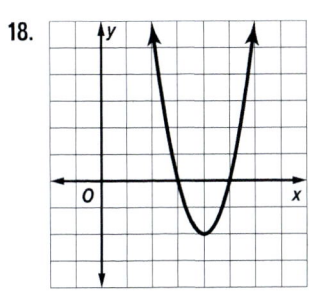

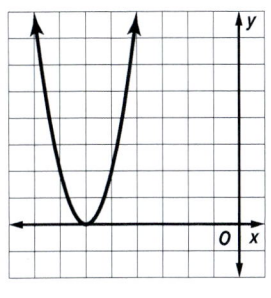

## Lesson 9-3

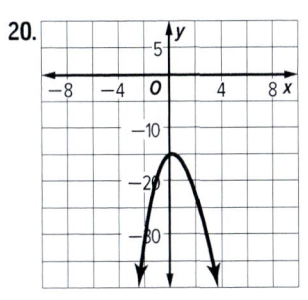

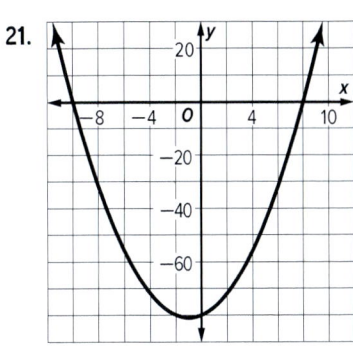

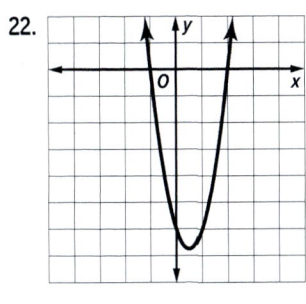

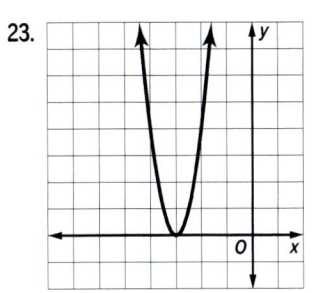

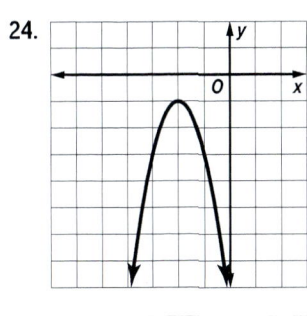

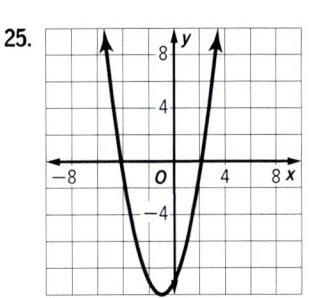

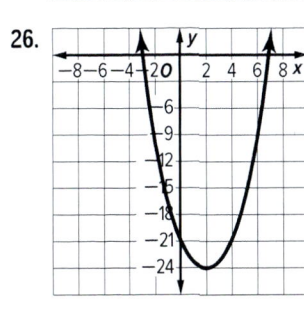

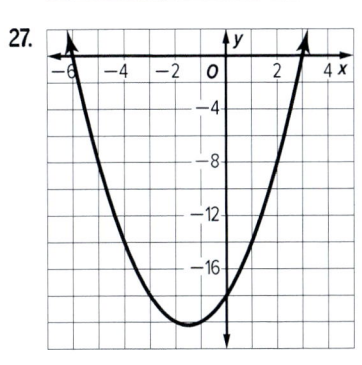

1–27. (Graphs of parabolas)

## Extend 9-5

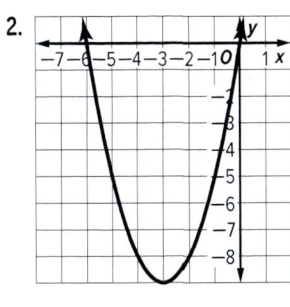

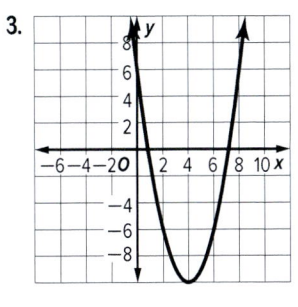

2. $y = (x + 3)^2 - 9$; $x = -3$; min. at $(-3, -9)$; $-6, 0$

3. $y = (x - 4)^2 - 10$; $x = 4$; min. at $(4, -10)$; $0.84, 7.16$

**4.**

$y = (x + 1)^2 - 13$; $x = -1$; min. at $(-1, -13)$; $-4.61, 2.61$

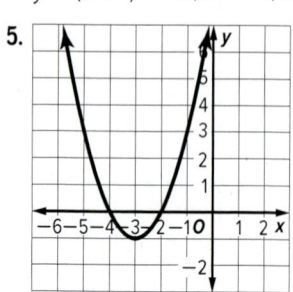

**5.**

$y = (x + 3)^2 - 1$; $x = -3$; min. at $(-3, -1)$; $-4, -2$

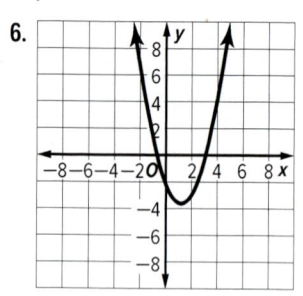

**6.**

$y = (x - 2)^2 - 1$; $x = 2$; min. at $(2, -1)$; $1, 3$

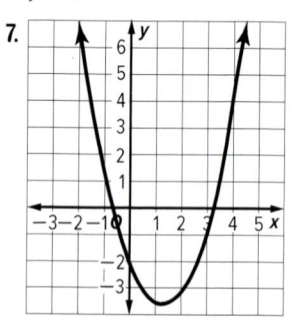

**7.**

$y = (x - 1.2)^2 - 3.64$; $x = 1.2$; min. at $(1.2, -3.64)$; $-0.71, 3.12$

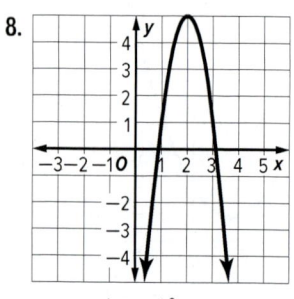

**8.**

$y = -4(x - 2)^2 + 5$; $x = 2$; max. at $(2, 5)$; $0.88, 3.11$

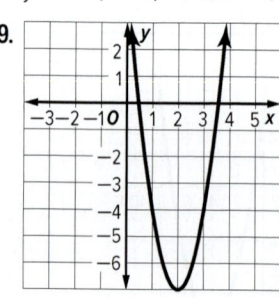

**9.**

$y = 3(x - 2)^2 - 7$; $x = 2$; min. at $(2, -7)$; $0.5, 3.5$

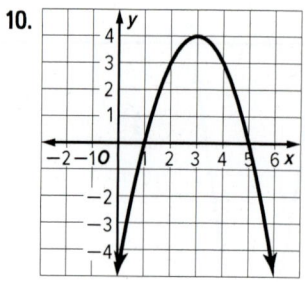

**10.**

$y = -(x - 3)^2 + 4$; $x = 3$; max. at $(3, 4)$; $1, 5$

### Mid-Chapter Quiz

**1.**

| x | y |
|---|---|
| −3 | 1 |
| −2 | −1 |
| −1 | −1 |
| 0 | 1 |
| 1 | 5 |
| 2 | 11 |

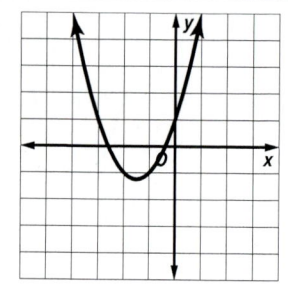

$D = \{-\infty < x < \infty\}$;
$R = \{y \mid y \geq -1.25\}$

**2.**

| x | y |
|---|---|
| −3 | 33 |
| −2 | 19 |
| −1 | 9 |
| 0 | 3 |
| 1 | 1 |
| 2 | 3 |
| 3 | 9 |

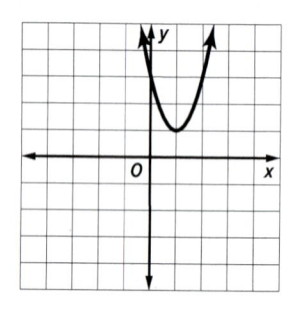

$D = \{-\infty < x < \infty\}$;
$R = \{y \mid y \geq 1\}$

**3.**

| x | y |
|---|---|
| −3 | −3 |
| −2 | −1 |
| −1 | −1 |
| 0 | −3 |
| 1 | −7 |
| 2 | −13 |

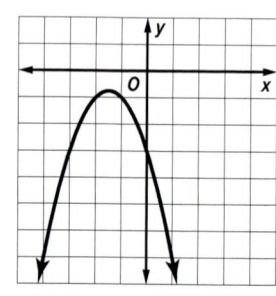

$D = \{-\infty < x < \infty\}$;
$R = \{y \mid y \leq -0.75\}$

**4.**

| x | y |
|---|---|
| −3 | −23 |
| −2 | −9 |
| −1 | −1 |
| 0 | 1 |
| 1 | −3 |
| 2 | −13 |

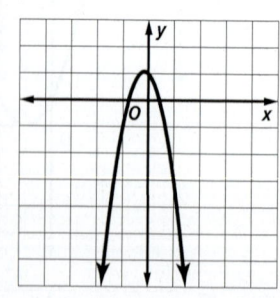

$D = \{-\infty < x < \infty\}$;
$R = \{y \mid y \leq 1\frac{1}{12}\}$

647E | Chapter 9 | Answer Appendix

**7.** [graph of upward parabola]

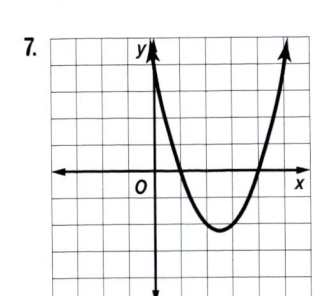

## Lesson 9-6

**52.** Sample answer: The polynomial can be factored to get $f(x) = (x - 4)^2$, so the only real zero is 4. The discriminant is 0, so there is 1 real zero. The discriminant tells us how many real zeros there are. Factoring tells us what they are.

**58.** Sample answer: Factoring is easy if the polynomial is factorable and complicated if it is not. Not all equations are factorable. Graphing only gives approximate answers, but it is easy to see the number of solutions. Using square roots is easy when there is no x-term. Completing the square can be used for any quadratic equation and exact solutions can be found, but the leading coefficient has to be 1 and the $x^2$- and x-term must be isolated. It is also easier if the coefficient of the x-term is even; if not, the calculations become harder when dealing with fractions. The Quadratic Formula will work for any quadratic equation and exact solutions can be found. This method can be time consuming, especially if an equation is easily factored. See students' preferences.

## Lesson 9-7

**11.** (5, 0), (2, -3)

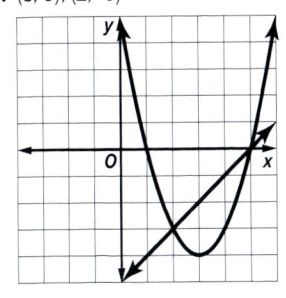

**12.** no solution

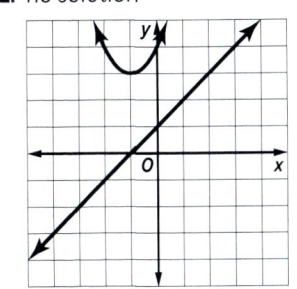

**13.** (-2, -4), (2, 4)

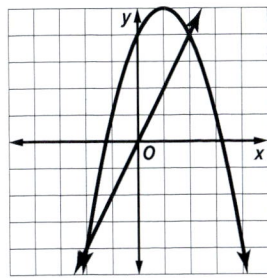

## Lesson 9-8

**1.** linear

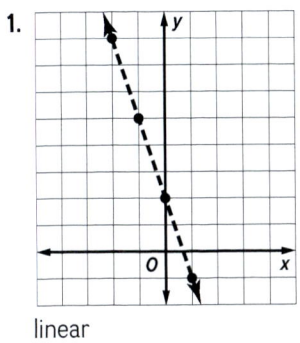

**2.** quadratic

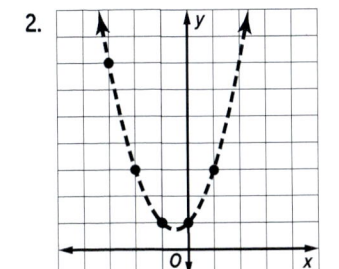

**3.** exponential

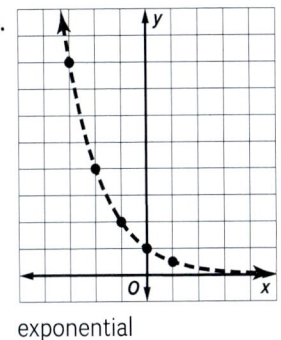

**4.** linear

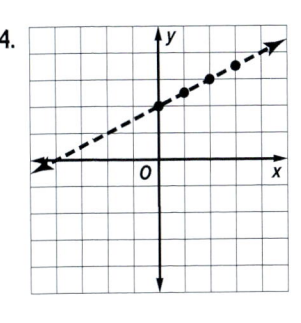

**14.** quadratic

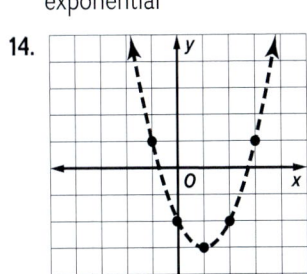

**15.** linear

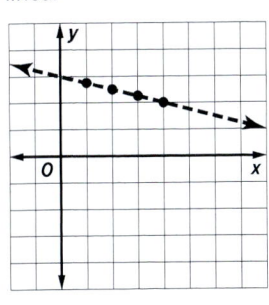

**16.** exponential

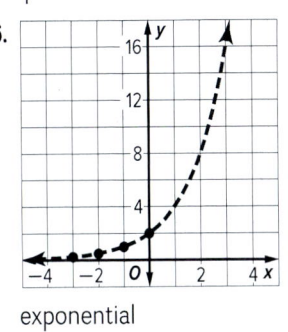

**17.** quadratic

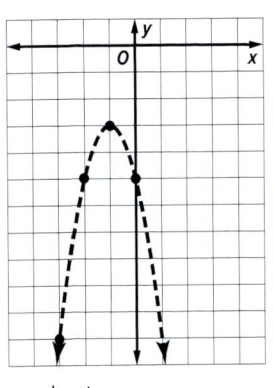

**18.** linear

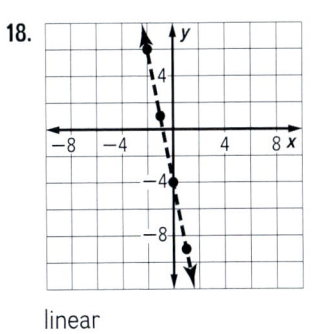

**19.** exponential

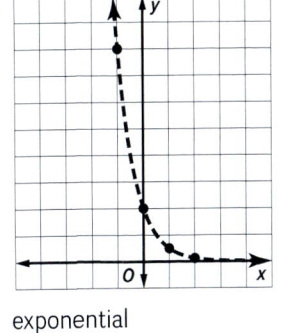

**29a.**

| Time (hour) | 0 | 1 | 2 | 3 | 4 |
|---|---|---|---|---|---|
| Amount of Bacteria | 12 | 36 | 108 | 324 | 972 |

### Practice Test

**1.**

| x | -3 | -2 | -1 | 0 | 1 | 2 |
|---|---|---|---|---|---|---|
| y | 8 | 5 | 4 | 5 | 8 | 13 |

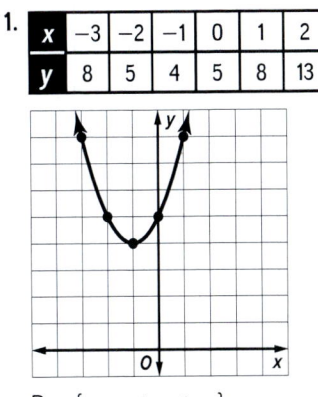

$D = \{-\infty < x < \infty\}$;
$R = \{y \mid y \geq 4\}$

**2.**

| x | -2 | -1 | 0 | 1 | 2 | 3 |
|---|---|---|---|---|---|---|
| y | 15 | 6 | 1 | 0 | 3 | 10 |

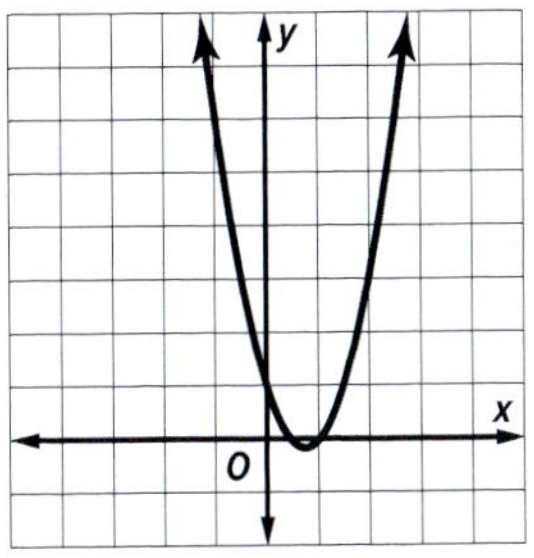

$D = \{-\infty < x < \infty\}$;
$R = \{y \mid y \geq -0.125\}$

# CHAPTER 10
# Statistics

**SUGGESTED PACING (DAYS)**

| 90 min. | 3 | 1 |
| 45 min. | 7 | 2 |

Instruction | Review & Assess

## Track Your Progress

This chapter focuses on content from the Interpreting Categorical and Quantitative Data and Quantities domains.

### THEN

**A.REI.4b** Solve quadratic equations by inspection (e.g., for $x^2 = 49$), taking square roots, completing the square, the quadratic formula, and factoring, as appropriate to the initial form of the equation. Recognize when the quadratic formula gives complex solutions and write them as $a \pm bi$ for real numbers $a$ and $b$.

**F.IF.7a** Graph linear and quadratic functions and show intercepts, maxima, and minima.

**F.BF.3** Identify the effect on the graph of replacing $f(x)$ by $f(x) + k$, $k f(x)$, $f(kx)$, and $f(x + k)$ for specific values of $k$ (both positive and negative); find the value of $k$ given the graphs. Experiment with cases and illustrate an explanation of the effects on the graph using technology.

### NOW

**S.ID.1** Represent data with plots on the real number line (dot plots, histograms, and box plots).

**S.ID.2.** Use statistics appropriate to the shape of the data distribution to compare center (median, mean) and spread (interquartile range, standard deviation) of two or more different data sets.

**S.ID.3** Interpret differences in shape, center, and spread in the context of the data sets, accounting for possible effects of extreme data points (outliers).

**S.ID.5** Summarize categorical data for two categories in two-way frequency tables. Interpret relative frequencies in the context of the data (including joint, marginal, and conditional relative frequencies). Recognize possible associations and trends in the data.

### NEXT

**S.ID.4** Use the mean and standard deviation of a data set to fit it to a normal distribution and to estimate population percentages. Recognize that there are data sets for which such a procedure is not appropriate. Use calculators, spreadsheets, and tables to estimate areas under the normal curve.

## Standards for Mathematical Practice

All of the Standards for Mathematical Practice will be covered in this chapter. The MP icon notes specific areas of coverage.

**MP** **Teaching the Mathematical Practices**
Help students develop the mathematical practices by asking questions like these.

**Questioning Strategies** As students approach problems in this chapter, help them develop mathematical practices by asking:

**Sense-Making**
- What is the difference between percent and a percentile?
- Is the mean of a data set the most accurate representation of the set?

**Reasoning**
- How are the upper and lower quartiles found?
- Is this a discrete or continuous graphical representation? Is a histogram a suitable way to represent the data?
- What can be inferred if two data sets have a similar mean, but one has a greater variability?

**Tools**
- How might changing the scale of the $y$-axis affect the results?
- What happens when the frequency of data is changed so that intervals are no longer equal? Is a histogram a good representation to use in this case?

**Structure**
- What would happen to the results if the standard deviation were changed to a mean absolute deviation?
- How would the graph change if the scale of the horizontal axis changed?

### Go Online!

**StudySync: SMP Modeling Videos**

These demonstrate how to apply the Standards for Mathematical Practice to collaborate, discuss, and solve real-world math problems.

648A | Chapter 10 | Statistics

# Go Online!
connectED.mcgraw-hill.com

LearnSmart | The Geometer's Sketchpad | Vocabulary | Personal Tutor | Tools | Calculator Resources | Self-Check Practice | Animations

## Customize Your Chapter
Use the *Plan & Present, Assignment Tracker,* and *Assessment* tools in ConnectED to introduce lesson concepts, assign personalized practice, and diagnose areas of student need.

**Differentiated Instruction** Throughout the program, look for the icons to find specialized content designed for your students.

- AL Approaching Level
- OL On Level
- BL Beyond Level
- ELL English Language Learners

## Personalize

### Differentiated Resources

| FOR EVERY CHAPTER | AL | OL | BL | ELL |
|---|---|---|---|---|
| ✓ Chapter Readiness Quizzes | ● | ● | ◐ | ● |
| ✓ Chapter Tests | ● | ● | ● | ● |
| ✓ Standardized Test Practice | ● | ● | ● | ● |
| abc Vocabulary Review Games | ● | ● | ● | ● |
| Anticipation Guide (English/Spanish) | ● | ● | ● | ● |
| Student-Built Glossary | ● | ● | ● | ● |
| Chapter Project | ● | ● | ● | ● |
| **FOR EVERY LESSON** | **AL** | **OL** | **BL** | **ELL** |
| Personal Tutors (English/Spanish) | ● | ● | ● | ● |
| Graphing Calculator Personal Tutors | ● | ● | ● | ● |
| ▷ Step-by-Step Solutions | ● | ● | ◐ | ● |
| ✓ Self-Check Quizzes | ● | ● | ● | ● |
| 5-Minute Check | ● | ● | ● | ● |
| Study Notebook | ● | ● | ● | ● |
| Study Guide and Intervention | ● | ● | ◐ | ● |
| Skills Practice | ● | ◐ | ● | ● |
| Practice | ◐ | ● | ● | ● |
| Word Problem Practice | ● | ● | ● | ◐ |
| Enrichment | ◐ | ● | ● | ● |
| ✦ Extra Examples | ● | ◐ | ◐ | ◐ |
| Lesson Presentations | ● | ● | ● | ● |

◐ Aligned to this group     ● Designed for this group

## Engage

### Featured IWB Resources

**eLessons** engage students and help build conceptual understanding of big ideas. *Use with Lessons 10-1 and 10-2*

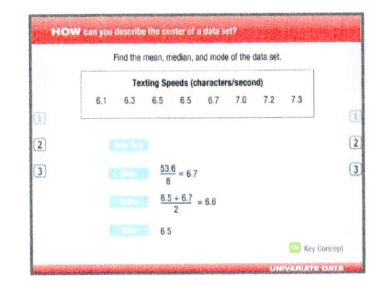

**Graphing Calculator Personal Tutors** presents an experienced educator explaining step-by-step solutions to problems. *Use with Lessons 10-2 through 10-4*

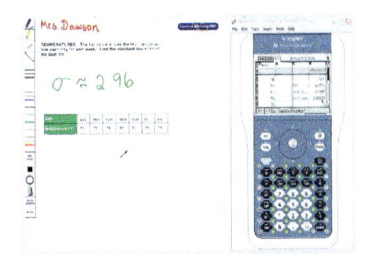

**Chapter Project** provide students with a tangible way to apply all learning in the chapter to a real-world project. *Use with Lessons 10-1 through 10-6*

**Time Management** How long will it take to use these resources? Look for the clock in each lesson interleaf.

connectED.mcgraw-hill.com

Chapter 10 | Statistics

# Introduce the Chapter

## Mathematical Background
Statistics are useful in identifying various sampling techniques, recognizing biased samples, and counting outcomes.

## Essential Question
At the end of this chapter, students should be able to answer the Essential Question.

How are statistics used in the real world?
**Sample answer: Statistics are used in charts, graphs, articles, reports, websites, newspapers, magazines, and television news.**

## Apply Math to the Real World
**FOOD SERVICE** In this activity, students use what they already know about using means and medians to evaluate data. They will also use a bar graph to illustrate data. Have students complete this activity individually or in small groups. MP 1, 2, 4, 7

## Go Online!

### Chapter Project
**I Have the Best Idea for a Show** Students can use what they have learned about distributions of data to complete a project. This chapter project addresses business literacy, as well as several specific skills identified as being essential to success by the Framework for 21st Century Learning. MP 1, 2, 3, 4

---

# CHAPTER 10
# Statistics

**THEN**
You calculated simple probability.

**NOW**
In this chapter, you will:
- Determine which measure of center best describes a set of data.
- Represent data using dot plots, histograms, bar graphs, and box plots and analyze their shapes.
- Summarize data in two-way frequency tables.
- Describe the effects linear transformations have on measures of center and spread.

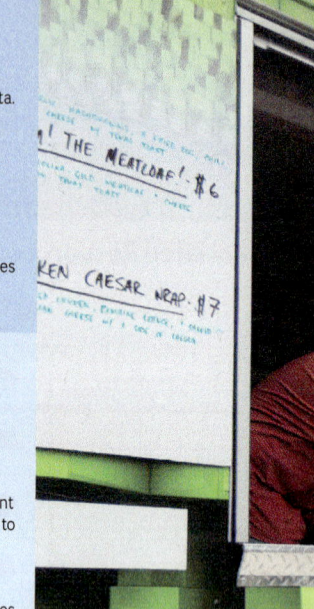

**WHY**
**FOOD SERVICE** The food service industry includes everything from food trucks to five-star restaurants. Regardless of the venue, customer satisfaction is important. A restaurant owner can use surveys and analysis to improve the customer experience.

**1. Use Tools** A food truck owner surveys a random sampling of customers who rate their experiences from 1 (poor) to 10 (excellent).

**Food Truck Customer Survey**

| 8 | 9 | 9 | 10 | 7 |
|---|---|---|----|---|
| 9 | 4 | 8 | 9  | 10 |

**2. Apply Reasoning** What can you infer from this data set? What is the average rating?

**3. Create a Graph** Create a histogram using the data from the survey.

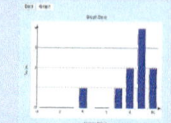

**4. Discuss** What rating did the food truck receive the most? What does this tell you about customer satisfaction?

---

## ALEKS

**Your Student Success Tool** ALEKS is an adaptive, personalized learning environment that identifies precisely what each student knows and is ready to learn—ensuring student success at all levels.

- **Formative Assessment:** Dynamic, detailed reports monitor students' progress toward standards mastery.
- **Automatic Differentiation:** Strengthen prerequisite skills and target individual learning gaps.
- **Personalized Instruction:** Supplement in-class instruction with personalized assessment and learning opportunities.

Chapter 10 | Statistics

## Go Online to Guide Your Learning

### Explore & Explain

**Normal Distribution**
The Normal Distribution mat in eToolkit is a useful tool for enhancing your understanding of the distribution of data discussed in Lesson 10-4.

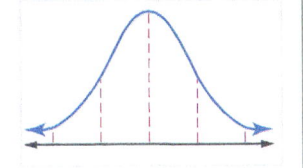

**Box-and-Whisker Plots**
Use two of the Box-and-Whisker Plot tools from the eToolkit to compare sets of data, as explored in Lesson 10-5.

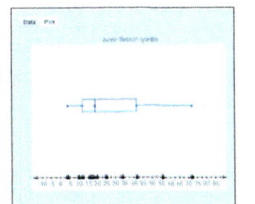

**eBook**
**Interactive Student Guide**
Before starting the chapter, answer the **Chapter Focus** preview questions. Check your answers as you complete each lesson. At the end of the chapter, try the **Performance Task**.

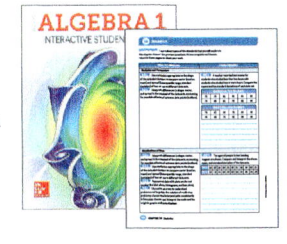

### Organize

**Foldables**
Get organized! Create this Foldable to help you organize your Chapter 10 notes about statistics.

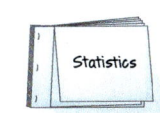

### Collaborate

**Chapter Project**
In the **I Have the Best Idea for a Show** project, you will use what you have learned about distributions of data to complete a project that addresses business literacy.

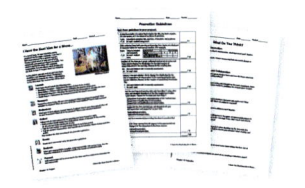

### Focus

**LEARNSMART**
Need help studying? Complete the **Descriptive Statistics** domain in LearnSmart to review for the chapter test.

**ALEKS**
You can use the **Data Analysis and Probability** topic in ALEKS to find out what you know about statistics and what you are ready to learn.*

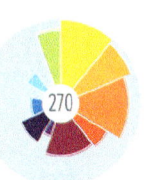

*Ask your teacher if this is part of your program.

## Dinah Zike's FOLDABLES

**Focus** Students write notes about statistics for each lesson in this chapter.

**Teach** Have students make and label their Foldables as illustrated. For each lesson, have students record definitions and examples on the appropriate sheets.

**When to Use It** Encourage students to add to their Foldables as they work through the chapter and to use them to review for the chapter test.

## Go Online!

**Extending Vocabulary**
Looking for more interesting assessments? Learn strategies for teaching and assessing vocabulary with pocketbooks and notebook Foldables.
MP 2, 3, 4, 5

connectED.mcgraw-hill.com 649

# Chapter 10 | Statistics

## Get Ready for the Chapter

**RtI Response to Intervention**
Use the Concept Check results and the Intervention Planner chart to help you determine your Response to Intervention.

### Intervention Planner

**TIER 1 On Level OL**

**IF** students miss 25% of the exercises or less,

**THEN** choose a resource:

*Go Online!*
- ✓ Self-Check Quiz

**TIER 2 Approaching Level AL**

**IF** students miss 50% of the exercises,

**THEN** choose a resource:

*Go Online!*
- ➕ Extra Examples
- 💬 Personal Tutors
- 📄 Homework Help

*Quick Review Math Handbook*

**TIER 3 Intensive Intervention**

**IF** students miss 75% of the exercises,

**THEN** Use *Math Triumphs, Alg. 1, Ch. 3*

*Go Online!*
- ➕ Extra Examples
- 💬 Personal Tutors
- 📄 Homework Help
- 🔤 Review Vocabulary

### Additional Answers
1. Total all the cubes, set that as the denominator; find the total of a specific color, set as the numerator.
2. the number of ways an event can occur divided by all of the possible outcomes
3. the Distributive Property
4. the magnitude of real number irrespective of its sign
5. The range remains the same.

## Get Ready for the Chapter

*Go Online!* for Vocabulary Review Games and key vocabulary in 13 languages.

### Connecting Concepts

**Concept Check**
Review the concepts used in this chapter by answering the questions below. 1-8. See margin.

1. Describe how to determine the probability of selecting a specific color cube from a bag with 1 green cube, 6 red cubes, and 4 yellow cubes.
2. Define probability using the terms "event" and "outcomes."
3. What is the equation $a(b + c) = ab + ac$ an example of?
4. What is an absolute value?
5. If all the values in a data set are increased by the same amount, what happens to the range?
6. Describe how you would determine the volume of a cylinder.
7. Given the linear equation $y = 3x + 1$, illustrated here, how were the data points determined so that the line could be graphed?
8. What do you call the sets of data points you create to graph a linear equation, as was done for the equation and graph in question 7?

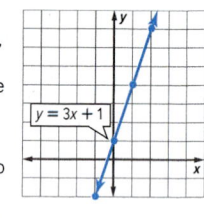

**Performance Task Preview**
You can use the concepts and skills in this chapter to solve problems in a real-world setting. Understanding statistics will help you finish the Performance Task at the end of the chapter.

**In this Performance Task you will:**
- construct viable arguments
- attend to precision
- use appropriate tools strategically
- reason abstractly and quantitatively
- look for and make use of structure

### New Vocabulary

| English | | Español |
|---|---|---|
| measures of central tendency | p. 651 | medidas de tendencia central |
| mean | p. 651 | media |
| median | p. 651 | mediana |
| mode | p. 651 | moda |
| frequency table | p. 659 | tabla de frecuencias |
| bar graph | p. 659 | gráfico de barra |
| cumulative frequency | p. 659 | frecuencia acumulativa |
| histogram | p. 659 | histograma |
| range | p. 665 | rango |
| quartile | p. 665 | cuartile |
| measures of position | p. 665 | medidas de la posición |
| lower quartile | p. 665 | cuartil inferior |
| upper quartile | p. 665 | cuartil superior |
| interquartile range | p. 666 | amplitud intercuartílica |
| outlier | p. 666 | valores atípico |
| standard deviation | p. 667 | desviación típica |
| variance | p. 667 | varianza |
| distribution | p. 673 | distribución |
| linear transformation | p. 680 | transformación lineal |
| relative frequency | p. 690 | frecuencia relativa |

### Review Vocabulary

**bivariate data** *datos bivariate* data with two variables

**box-and-whisker plot** *diagrama de caja patillas* a diagram that divides a set of data into four parts using the median and quartiles; a box is drawn around the quartile values, and whiskers extend from each quartile to the extreme data points

## Key Vocabulary ELL

Introduce the key vocabulary in the chapter using the routine below.

**Define** A random sample is a sample that is chosen without any preference, and is representative of the entire population.

**Example** Thirty students are in a class. Three of the 30 students are chosen at random to answer a survey about a new school policy.

**Ask** Are the 3 students representative of a random sample? Explain why or why not? **No; for it to be a random sample, the sample must be representative of the entire school population, not just one class of 30 students.**

6. determine the area of the base of the cylinder by finding $\pi r^2$ and then multiply by the height of the cylinder
7. Substitute a value for $x$ and solve for $y$ or vice versa to generate data points.
8. A table of ordered pairs

650 | Chapter 10 | Statistics

# LESSON 10-1
# Measures of Center

**SUGGESTED PACING (DAYS)**

90 min. **0.5**
45 min. **1**
Instruction

## Track Your Progress

### Objectives
1. Represent sets of data by using measures of center.
2. Represent sets of data by using percentiles.

### Mathematical Background
A measure of center is a value that describes a set of data by identifying the central position of the data set. The mean, median, and mode are common measures of center and can give you insight into what a typical value in the data set might look like.

### THEN

**8.SP.A1** Construct and interpret scatter plots for bivariate measurement data to investigate patterns of association between two quantities. Describe patterns such as clustering, outliers, positive or negative association, linear association, and nonlinear association.

**8.SP.A4** Understand that patterns of association can also be seen in bivariate categorical data by displaying frequencies and relative frequencies in a two-way table. Construct and interpret a two-way table summarizing data on two categorical variables collected from the same subjects. Use relative frequencies calculated for rows or columns to describe possible association between the two variables.

### NOW

**Prep for S.ID.2** Use statistics appropriate to the shape of the data distribution to compare center (median, mean) and spread (interquartile range, standard deviation) of two or more different data sets.

### NEXT

**S.ID.1** Represent data with plots on the real number line (dot plots, histograms, and box plots).

**S.ID.3** Interpret differences in shape, center, and spread in the context of the data sets, accounting for possible effects of extreme data points (outliers).

## Go Online! All of these resources and more are available at connectED.mcgraw-hill.com

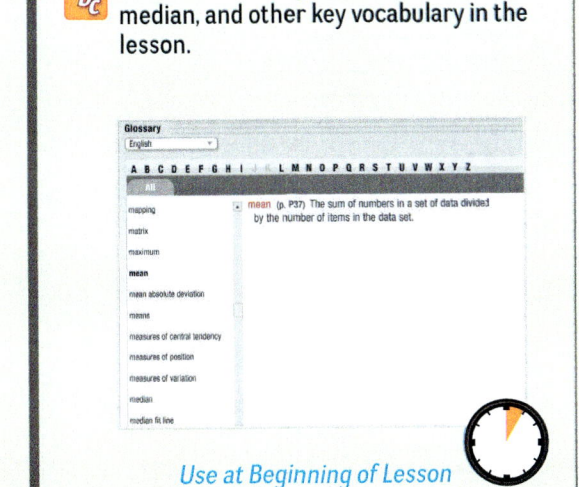

Use the **eGlossary** to define mean, median, and other key vocabulary in the lesson.

*Use at Beginning of Lesson*

**eLessons** utilize the power of your interactive whiteboard in an engaging way. Use **Univariate Data**, screen 5, to introduce the concepts in this lesson.

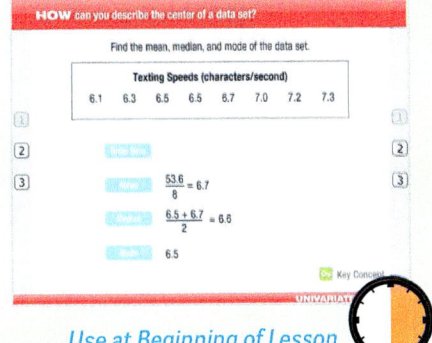

*Use at Beginning of Lesson*

**Personal Tutors** let students hear real teachers solve problems. Students can pause and repeat as many times as necessary.

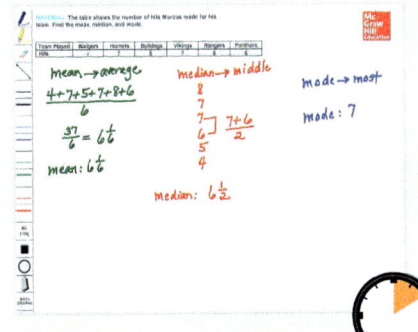

*Use with Examples*

### OER Using Open Educational Resources
**Video Sharing** Have students work in groups to record and upload videos on **MathATube** about mean, median, and mode. If you are unable to access **MathATube**, try **KidsTube**, **YouTube**, **SchoolTube**, or **Teacher Tube**. *Use as homework.*

connectED.mcgraw-hill.com  651A

# Go Online!
connectED.mcgraw-hill.com
Worksheets

# Differentiate Your Resources

**Extra Practice** Additional practice or homework; Skills Practice is best for approaching-level students and Practice is best for on-level and beyond-level students

## Skills Practice
AL  OL  ELL

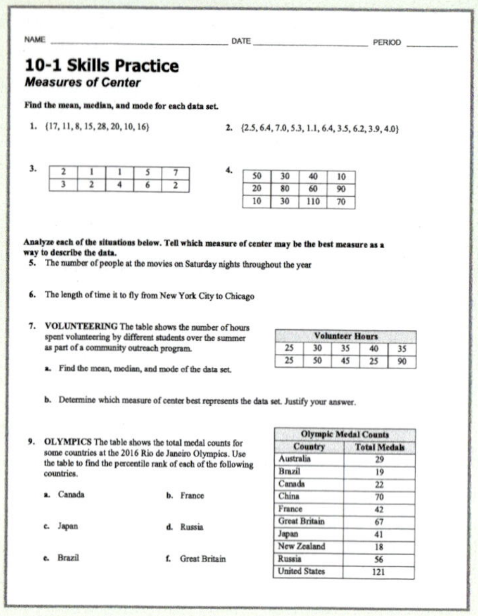

## Practice
AL  OL  BL  ELL

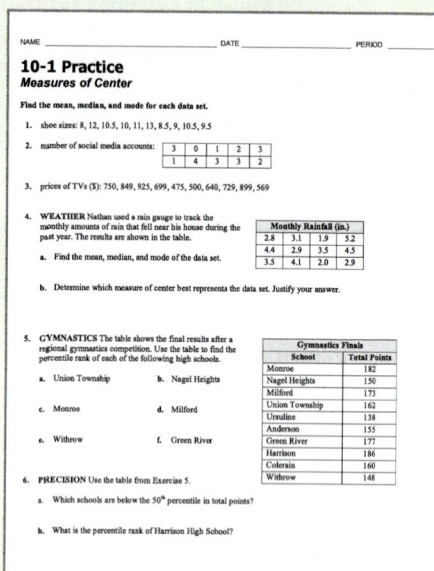

## Word Problem Practice
AL  OL  BL  ELL

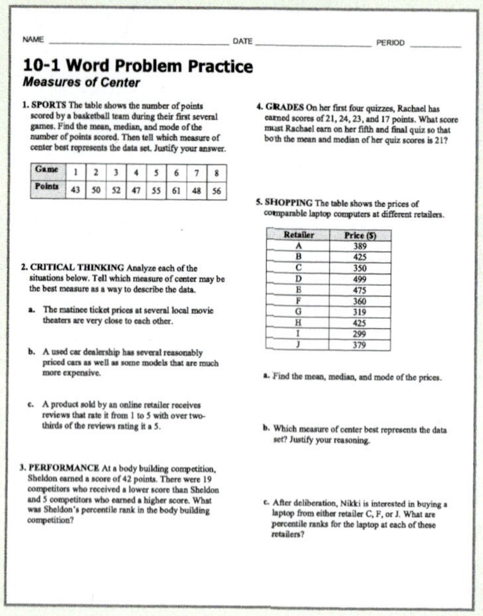

**Intervention** Reteaching and vocabulary activities that can be used with struggling or absent students and as ELL support

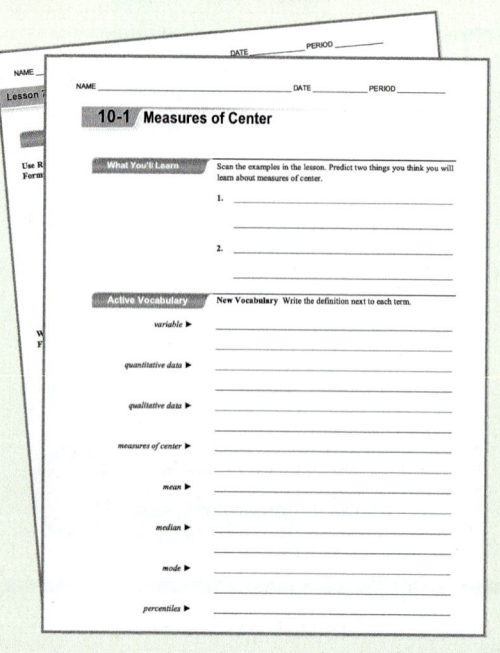

**Extension** Activities that can be used to extend lesson concepts

## Study Guide and Intervention
AL  OL  ELL

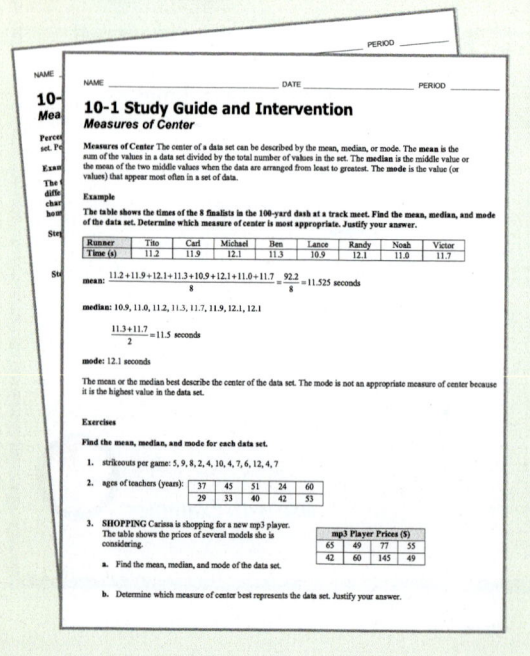

## Study Notebook
AL  OL  BL  ELL

## Enrichment
OL  BL  ELL

651B | Lesson 10-1 | Measures of Center

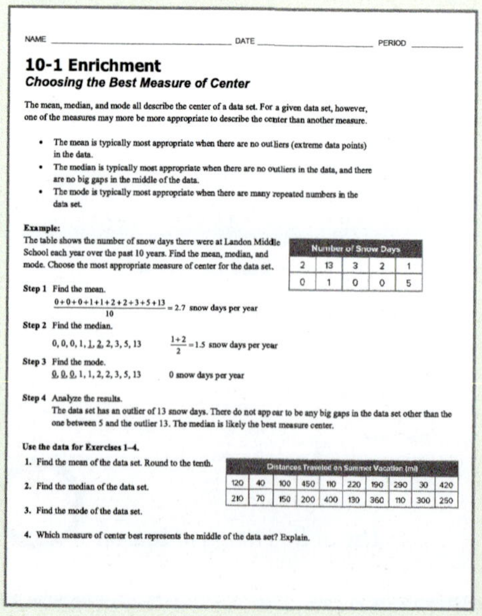

# LESSON 1
## Measures of Center

**Then**
- You collected data and were introduced to the concepts of mean, median, and mode.

**Now**
1. Represent sets of data by using measures of center.
2. Represent sets of data by using percentiles.

**Why?**
Each year, many students are given standardized achievement tests. Most of these test scores are reported as percentiles, and they may also include graphics with sliding scales to clarify scores that may otherwise be difficult to interpret.

**New Vocabulary**
variable
quantitative data
qualitative data
measures of center
measures of central tendency
mean
median
mode
percentiles

**Mathematical Practices**
3 Construct viable arguments and critique the reasoning of others.
4 Model with mathematics.
7 Look for and make use of structure.

**Content Standards**
Prep for S.ID.2 Use statistics appropriate to the shape of the data distribution to compare center (median, mean) and spread (interquartile range, standard deviation) of two or more different data sets.

### 1 Measures of Center
A **variable** is any characteristic, number, or quantity that can be counted or measured. A variable is an item of data. Data that can be measured are called **quantitative data**. Data that can be organized into different categories are called categorical or **qualitative data**. Quantitative data in one variable are often summarized using a single number to represent what is average or typical. Measures of what is average are also called **measures of center** or **central tendency**. The most common measures of center are mean, median, and mode.

**Key Concept** Measures of Center
- The **mean** is the sum of the values in a data set divided by the total number of values in the set.
- The **median** is the middle value or the mean of the two middle values in a set of data when the data are arranged in numerical order.
- The **mode** is the value or values that appear most often in a set of data. A set of data can have no mode, one mode, or more than one mode.

S.ID.2

**Real-World Example 1** Measures of Center

**BASEBALL** The table shows the number of hits Julius made for his baseball team. Find the mean, median, and mode.

| Team Played | Hits |
|---|---|
| Badgers | 3 |
| Hornets | 6 |
| Bulldogs | 4 |
| Vikings | 0 |
| Rangers | 3 |
| Panthers | 7 |

**Mean:** To find the mean, find the sum of all the hits and divide by the number of games in which he made these hits.

Julius's team played 6 other teams.

Mean = $\frac{3 + 6 + 4 + 0 + 3 + 7}{6} = \frac{23}{6} \approx 3.83$ or about 4 hits

**Median:** To find the median, order the numbers from least to greatest and find the middle value or values.

0, 3, 3, 5, 6, 7

median = $\frac{3 + 5}{2}$, or 4 hits

Because there is an even number of values, find the mean of the middle two.

**Mode:** From the arrangement of data values, you can see that the value that occurs most often is 3, so the mode of the data set is 3 hits.

Julius's mean and median number of hits for these baseball games was 4, and his mode was 3 hits.

---

### Mathematical Practices Strategies

**Make sense of problems and persevere in solving them.**
Help students recognize the importance of the mean, median, and mode when analyzing sets of data. They can be used to describe correspondences between data values and search for regularity or trends. For example, ask:

- Why is the mean often equal to a value that is not a data point in the set? **The mean is an average of the data values.**

- How do you find the median if there is an even number of values in the data set? **The median is the average of the two middle terms of the data set after they are ordered consecutively from least to greatest.**

- Can the mean ever be greater than the greatest value in the data set? **No. It will always be between the least and greatest value. It can never be less than the least value or greater than the greatest value.**

---

**Lesson 10-1** | Measures of Center

## Launch

Have students read the Why? section of the lesson. Ask:

- What is an example of a standardized test subject that you may have taken? **Sample answer: math, language arts**

- How are standardized test results used in your school district? **Sample answer: They are used to compare student achievement in different schools across the district.**

## Teach

Ask the scaffolded questions for each example to build conceptual understanding for students at all levels.

### 1 Measures of Center

**Example 1** Measures of Center

**AL** To find the mode, do you have to arrange the data in numerical order? **No, as long as you can determine the mode by inspection.**

**OL** How could you modify this data set so that there would be two modes? **Sample answer: If Julius had one more hit against the Hornets, he would have 7 hits against each team, the Hornets and the Panthers. So the modes of the data set would be 3 and 7.**

**BL** What does the phrase "measure of center" mean? **Sample answer: If the data are all arranged in numerical order, you can break up the data set into parts, with the center being the values in the center of the order. The measures of center are values that may be useful if you want to analyze the data. The values of the mean, median, and mode are the normal measures of center.**

### Go Online!

**Personal Tutor**
Personal Tutors (for every example) let students hear real teachers solve problems. Students can pause and repeat as many times as necessary.

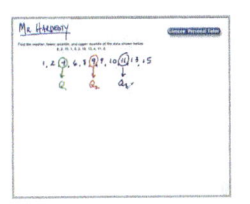

connectED.mcgraw-hill.com 651

Lesson 10-1 | Measures of Center

## Need Another Example?

**Swimming** The table shows the number of laps Ekta swam each day. Find the mean, median, and mode.

| Day | Number of Laps |
|---|---|
| Saturday | 6 |
| Sunday | 8 |
| Monday | 8 |
| Tuesday | 6 |
| Wednesday | 4 |
| Thursday | 9 |
| Friday | 5 |

≈ 6.6 laps; 6 laps; 6 and 8 laps

## Additional Answer (Guided Practice)

2. Mariana: mean: 10.96; median: 10.25; mode: 10, 10.25

   Rachael: mean: 10.96; median: 10.5; mode: 10.5

   Sample answer: Because the means are the same and Rachael has a greater median and mode, she should win the tournament.

## Example 2 Use Measures of Center to Analyze Data

**AL** Why is it not sufficient enough to say that because the means of the salaries are equivalent, the stores pay their employees equally? The mean alone is not a good enough measure of center, and in this case, the medians and modes of these data are not the same.

**OL** Why is the data point 20.50 considered an outlier in this case? Since the salaries at Game Place in this data set are between $9.50 and $10.80, a salary of $20.50 will skew the data and, in turn, increase the mean.

**BL** Would your solution change if an additional salary of $9.50 was recorded at Big Win Games? Explain. Sample answer: The mean would decrease to $11.07, the median would decrease to $10.95, and the mode would not change. However, these changes would not affect the solution and Big Win Games would still pay their employees better.

## Need Another Example?

**BASKETBALL** The table shows the total number of points scored in several men's and women's NCAA Championship Basketball Games. Compare and contrast the measures of center of the men's and women's scores. Based on the statistics, which group scores more points in NCAA Championship Basketball Games?

### Guided Practice

1. **TIPS** Gloria is a server at a popular restaurant. The table shows the total tips she received each day she worked. Find the mean, median, and mode.
   mean 66; median 68; no mode

| Day | Tips ($) |
|---|---|
| Monday | 47 |
| Tuesday | 52 |
| Wednesday | 68 |
| Friday | 90 |
| Sunday | 73 |

In Example 1, the mean and median are close together, so they both represent the average of Julius's number of hits well. Notice that the median is greater than the mean. This indicates that the games with fewer hits than the median are more spread out than the games with more hits than the median. The mode is greater than most of the games.

**Study Tip**
Median If there are an even number of values in the set of data, you will have to find the average of the two middle values to find the median. To do this, divide the sum of the two middle values by two.

### Example 2 Use Measures of Center to Analyze Data

**SALARY** Compare and contrast the measures of center of the employee salaries for the two stores. Based on the statistics, which store pays its employees better?

| Hourly Salaries ($) | |
|---|---|
| Big Win Games | Game Place |
| 10.80, 11, 11.50, 10, 10.90, 13.90, 10.80, 11.20 | 9.50, 9.50, 10.40, 10.40, 9.50, 10.80, 20.50, 9.50 |

**Big Win Games**
Mean:
$$\frac{10.8 + 11 + 11.5 + 10 + 10.9 + 13.9 + 10.8 + 11.2}{8}$$
≈ $11.26
Median: $\frac{10.9 + 11}{2}$ or $10.95
Mode: $10.80

**Game Place**
Mean:
$$\frac{9.5 + 9.5 + 10.4 + 10.4 + 9.5 + 10.8 + 20.5 + 9.5}{8}$$
= $9.95
Median: $\frac{9.5 + 10.4}{2}$ or $9.95
Mode: $9.50

The mean salary for both stores is equal, but the median and mode for Big Win Games is greater. At Game Place, $20.50 is an outlier causing the mean to increase. So, the mean is not an accurate representation of the salary of most employees at that store.

Because the median and mode are greater, Big Win Games pays their employees better.

### Guided Practice

2. **TOURNAMENT** Mariana and Rachael are in a fishing tournament. Compare and contrast the measures of center of the fish they caught. Based on the statistics, who should win the tournament? See margin.

| Fish Length (in.) | |
|---|---|
| Mariana | Rachael |
| 12.5, 10.25, 10, 11.75, 12, 10.25, 10 | 10.5, 11.5, 8, 10.5, 8.5, 18.5, 9.25 |

You may find that certain measures of center do not give you the information you need to fully analyze a situation. In that case, you will need to determine which additional statistics would be useful to have.

### Example 3 Determine Best Measures of Center

Analyze each situation. Which measure of center would best describe the data? Explain.

a. Researching the employee salary at a specific company

   The median would be best. Because a few employees may have a significantly greater salary, the mean may not accurately describe the salary of most employees.

| Year | Men's Score | Women's Score |
|---|---|---|
| 2006 | 130 | 153 |
| 2007 | 159 | 105 |
| 2008 | 143 | 112 |
| 2009 | 161 | 130 |
| 2010 | 120 | 100 |
| 2011 | 94 | 146 |
| 2012 | 126 | 141 |
| 2013 | 158 | 153 |
| 2014 | 114 | 137 |
| 2015 | 131 | 116 |
| 2016 | 150 | 133 |
| 2017 | 136 | 122 |

Men: mean: 135.2; median: 133.5; mode: none

Women: mean: 129; median: 131.5; mode: 153

Sample answer: Because the mean and median of the men's scores is greater, the men score more points.

b. The mean attendance for each game in a season would be best. While the attendance changes with each game, it is unlikely that there is an outlier.

▶ Guided Practice

3A. A professional basketball player negotiating his salary. **mode or median**

3B. Planning a food budget by analyzing the monthly costs for the previous year **mean**

## 2 Percentiles

A **percentile** is a measure that is often used to report test data, such as standardized test scores. Percentiles measure rank from the bottom and tell us what percent of the scores were below a given score. The lowest score is the 1st percentile and the highest score is the 99th percentile. There is no 0 or 100th percentile rank.

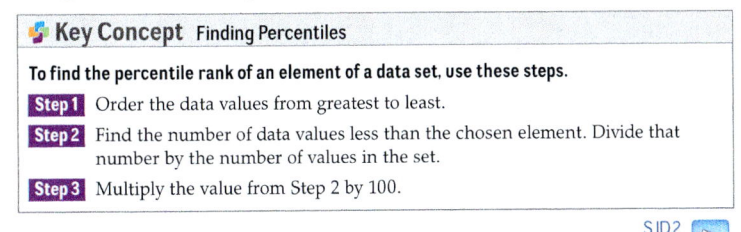

### Key Concept  Finding Percentiles

To find the percentile rank of an element of a data set, use these steps.

**Step 1** Order the data values from greatest to least.
**Step 2** Find the number of data values less than the chosen element. Divide that number by the number of values in the set.
**Step 3** Multiply the value from Step 2 by 100.

S.ID.2

**Study Tip**
Percent vs. Percentile
*Percent* and *percentile* have different meanings. For example, a score at the 40th percentile means 40% of the scores are either the same as the score of the 40th percentile or less than that rank. It does not mean a person scored 40% of the possible points.

### Real-World Example 4  Find the Percentile Rank of a Data Value

TALENT ...

**Step 1** Order the scores from greatest to least.

| 29 | 28 | 27 | 26 | 25 | 22 | 21 | 20 | 18 | 17 |
| 16 | 15 | 14 | 12 | 11 | 10 | 9 | 6 | 5 | 4 |

| Name | Score | Name | Score |
|---|---|---|---|
| Arnold | 17 | Ishi | 27 |
| Benito | 9 | James | 20 |
| Brooke | 25 | Kat | 16 |
| Carmen | 21 | Malik | 10 |
| Daniel | 14 | Natalie | 26 |
| Delia | 29 | Pearl | 4 |
| Fernando | 15 | Twyla | 6 |
| Heather | 12 | Victor | 28 |
| Horatio | 5 | Warren | 22 |
| Ingrid | 11 | Yolanda | 18 |

**Steps 2 and 3** Find Victor's percentile rank.

Victor had a score of 28. There are 18 scores below his score. To find his percentile rank, use the following formula.

$$\frac{\text{number of scores below 28}}{\text{total number of scores}} \cdot 100 = \frac{18}{20} \cdot 100 = 90$$

So, Victor scored at the 90th percentile in the contest.

▶ Guided Practice

4. Find Fernando's percentile rank. **40th percentile**

1. mean: 4.6; median: 4; mode: none; The mean and median are close enough in value that either one of them can be used.

### Check Your Understanding

⬤ = Step-by-Step Solutions begin on page R11.

**Example 1**
S.ID.2

1. **CONCESSIONS** The table shows the number of students working at the concession stand each hour. Find the mean, median, and mode of the number of students. Which measure represents the data best? Explain.

| Hour | 1 | 2 | 3 | 4 | 5 |
|---|---|---|---|---|---|
| Number of Students | 3 | 8 | 6 | 4 | 2 |

---

**Need Another Example?**
Use the data table for the example.
The highest score is Delia's. What is her percentile rank? **95th percentile**

## Teaching Tip
You may want to discuss the differences between *percent* and *percentile*. In Example 4, if a student scores in the 65th percentile on a test, he or she has a test score that is greater than or equal to 65% of his or her classmates' scores. It does not mean that he or she scored 65% on the test.

---

## 2 Percentiles

### Example 3  Determine Best Measures of Center

**AL** Why may the mode not be the best measure of center in part **b**? It is highly unlikely that more than one game would have the same number of attendees.

**OL** Why may the mean not be the best measure when researching salaries in part **a**? Sample answer: Some employee salaries may be so far above the salaries of the majority of employees that the mean will be much higher than the median, which is a closer value to the majority of the data.

**BL** How do you interpret a set of data such that about 70 percent of the values are very close to the value of the mean? The data as described are probably a symmetric data set, so the median and mean are close together. Either measure of center may be good to use.

**Need Another Example?**
Analyze each of the situations below. Tell which measure of center may be the best to use to describe the data.

a. As a wrestling coach, you are listing the weights of all the wrestlers in each weight category. The mean or the median would be best because most wrestlers in that category would weigh close to those values.

b. You download 5 songs every weekday from the Internet, and 4 times as many downloads on weekend days. The mode would be the best because it represents most of the days of the week.

### Example 4  Find the Percentile Rank of a Data Value

**AL** What is Pearl's percentile rank? **1st percentile**

**OL** Why is arranging the scores in Step 1 necessary? The percentile rank is a statement of all the scores at or below the rank, so scores need to be in order so you can determine which scores are lower than the rank.

**BL** Why doesn't the 100th percentile rank exist? The quotient of the number of scores below the top score and the total number of scores can never be 100. For example, if there are 100 scores, 99 ÷ 100 = 0.99; for 1000 scores, 999 ÷ 1000 = 0.999, and so on.

connectED.mcgraw-hill.com  653

# Lesson 10-1 | Measures of Center

## Follow-Up

Students have explored how to find mean, median, mode, and percentile rank. Ask: **How can measures of center and percentile rank be used to analyze real-world data sets?** Sample answer: Analyze the context for which the real-world data is compiled. Then find the centers and percentile rank of data values as needed with the data.

## Practice

**Formative Assessment** Use Exercises 1–6 to assess students' understanding of the measures of center and the percentile rank, and how to calculate them.

### Teaching the Mathematical Practices

**Precision** Mathematically proficient students understand and use definitions. For Exercise 5, ask students to identify similarities and differences between percentile rank and percent score.

### Levels of Complexity Chart

The levels of the exercises progress from 1 to 3, with Level 1 indicating the lowest level of complexity.

| Exercises | 7–9 | 10–12, 22–27 | 13–21 |
|---|---|---|---|
| Level 3 | | | ● |
| Level 2 | | ● | |
| Level 1 | ● | | |

### Additional Answers

**2** For Lee's class, the mean is 229 pages, and the median is 114 pages. For Romina's class, the mean is 215.6 pages, and the median is 219.5 pages. The mean and median are further apart in Lee's class, indicating a greater range in the data. The mean in both classes are close, but the median in Romina's class is much less. So, Lee's class is reading longer books.

**5a.** 60th percentile

### Go Online! eBook

**Interactive Student Guide**
Use the *Interactive Student Guide* to deepen conceptual understanding.
- Statistics and Parameters

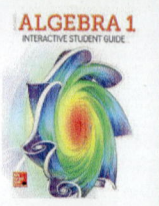

---

**Example 2** S.ID.2
**2. READING** Lee and Romina have different English teachers. Both teachers have 5 books that are required reading for their class. The page counts for the books are shown in the table. Compare and contrast the measures of center for the books in each class. Based on the statistics, which class is reading longer books? **See margin.**

| Book Page Counts | |
|---|---|
| Lee's Class | Romina's Class |
| 103, 114, 708, 98, 122 | 237, 178, 225, 206, 232 |

**Example 3** S.ID.2 Analyze each situation. Which measure of center best describes the data? Explain.
**3.** The miles per gallon ratings for cars from all major car manufacturers **mean**
**4.** The property value of homes in a suburban town **median**

**Example 4** S.ID.2
**5. OLYMPICS** The table shows the total medal counts for some countries at the 2014 Sochi Olympics.
a. Find Canada's percentile rank. **See margin.**
b. Are there any countries at the 50th percentile mark? If so, which ones?

**5b.** Yes; the nation with 5 scores below it will be in the 50th percentile. Netherlands's rank is 50th percentile.

| Country | Total | Country | Total |
|---|---|---|---|
| Belarus | 6 | Switzerland | 11 |
| France | 15 | Netherlands | 24 |
| Austria | 17 | United States | 28 |
| Norway | 26 | Germany | 19 |
| Russia | 33 | Canada | 25 |

Source: Business Insider

**6. OLYMPICS** Use the table for Exercise 5.
a. Find the percentile rank for the United States. **80th percentile**
b. Which countries are below the 50th percentile in medal counts? **Germany, Austria, France, Switzerland, Belarus**
c. Which countries are above the 70th percentile? **United States, Russia**

### Practice and Problem Solving

Extra Practice is on page R10.

**Example 1** S.ID.2
**7. COMPUTER TABLETS** The table shows the prices of comparable computer tablets at ten different locations.
a. Find the mean, median, and mode for the data set. **11a. mean = $342.37, median = $329.98, mode = $259.95**
b. You want to buy one of the tablets. Which is the better measure of center to consider? Justify your reasoning. **See margin.**

| Tablet Prices ($) |
|---|
| 289.95, 259.95, 310, 349.95, 459.95, 399.95, 259.95, 300, 389, 405 |

**Examples 2, 3** S.ID.2
**8. POPULATIONS** The table shows the populations of the six largest cities in North Carolina and Colorado, according to a recent census. Compare and contrast the measures of center for the two states. Based on the statistics, which state has more populated cities? **See margin.**

| North Carolina | | Colorado | |
|---|---|---|---|
| City | Population (thousands) | City | Population (thousands) |
| Charlotte | 731 | Aurora | 325 |
| Durham | 228 | Colorado Springs | 416 |
| Fayetteville | 201 | Denver | 600 |
| Greensboro | 270 | Fort Collins | 144 |
| Raleigh | 404 | Lakewood | 143 |
| Winston-Salem | 230 | Thornton | 119 |

**Example 4** S.ID.2
**9. TEST SCORES** The table below shows the results of a recent Algebra test in a class of 18 students.

| Name | Abby | Brian | Cassie | Dan | Emma | Fritz | Gabby | Huang | Imogen |
|---|---|---|---|---|---|---|---|---|---|
| Score | 84 | 96 | 80 | 72 | 80 | 80 | 84 | 92 | 84 |
| Name | Jenny | Kelly | Larry | Marcus | Nathan | Owen | Patti | Rachel | Yolanda |
| Score | 100 | 84 | 80 | 88 | 80 | 76 | 80 | 84 | 92 |

Find Marcus's percentile rank. **72nd percentile**

### Differentiated Homework Options

| Levels | AL Approaching Level | OL On Level | BL Beyond Level |
|---|---|---|---|
| Exercises | 7–9, 15–17, 19–27 | 7–9 odd, 10–17, 19–27 | 13–21, 22–27 (optional) |
| 2-Day Option | 7, 9, 22–27 | 7–9 | |
| | 8, 15–17, 19–21 | 10–17, 19–27 | |

You can use ALEKS to provide additional remediation support with personalized instruction and practice.

**7b.** Sample answer: The mean and the median are close to the same values. Either measure would be good, but choosing the tablet closest to the median because it is less than the mean might be best.

654 | Lesson 10-1 | Measures of Center

Lesson 10-1 | Measures of Center

**10.** Give a real-world example of data for which the given measure of center best represents the data.
   a. mean  Sample answer: Attendance at basketball games during each regular season home game.
   b. median  Sample answer: Researching prices of houses in a community.
   c. mode  Sample answer: Ratings from 1 to 5 stars for a local restaurant.

**11.** Darla is training for a 5K race. Her practice times (m:s) are 20:45, 21:30, 21:15, 22:32, and 21:40.
   a. Find her mean practice time in minutes and seconds. (*Hint:* Change the practice times to seconds first.)  **21:32.4**
   b. Darla ran the actual race 30 seconds faster than her median practice time. What was her race time?  **21 minutes**

**13a.** John: 9.65, Raul: 9.475, Laird, Boris, Han: 9.525; John won.
**13b.** John: 9.633, Raul: 9.483, Laird, Boris, Han: 9.517; John still won.

**12. HEIGHT** The heights of the students in class are given in the table.
   a. In what percentile is the tallest person in the class?  **90th percentile**
   b. In what percentile is the third shortest person in the class?  **20th percentile**
   c. Is there a person in the 50th percentile? If so, who?  **Yes, Jin**
   d. Is there someone in the 100th percentile? If so, who?  **No**

| Student | Height (inches) |
|---|---|
| Charlene | 61 |
| Anthony | 64 |
| Bart | 65 |
| Renata | 59 |
| Pedro | 63 |
| Georgio | 58 |
| Jin | 62 |
| Essie | 63 |
| Caroline | 60 |
| Antoine | 59 |

**13.** The table shows the scores from six judges at a gymnastics meet. The final score is calculated by dropping the highest and lowest scores and then finding the mean of the 4 remaining scores.
   a. Find the final score for each gymnast. Who won the meet?
   b. What is the mean score for each gymnast if all six judges are considered? Who would win?
   c. Compare your result for parts **a** and **b**. What can you conclude about the scoring for the meet?

**13c.** Sample answer: For this meet, it didn't matter which method of scoring they used because the overall standings turned out the same.

| | Judge 1 | Judge 2 | Judge 3 | Judge 4 | Judge 5 | Judge 6 |
|---|---|---|---|---|---|---|
| John | 9.6 | 9.6 | 9.8 | 9.4 | 9.6 | 9.8 |
| Raul | 9.5 | 9.3 | 9.8 | 9.4 | 9.2 | 9.7 |
| Laird | 9.6 | 9.8 | 9.4 | 9.2 | 9.7 | 9.4 |
| Boris | 9.4 | 9.2 | 9.6 | 9.8 | 9.5 | 9.6 |
| Han | 9.7 | 9.3 | 9.7 | 9.4 | 9.3 | 9.7 |

**14.** A value that divides a set of data into ten parts of equal size is called a *decile*. The first decile contains data up to but not including the 10th percentile. The second decile contains data from the 10th percentile up to but not including the 20th decile, and so on.
   a. Use the table for Example 4. Which contestants' scores fall into the sixth decile?  **Arnold and Yolanda**
   b. In which decile are Heather and Daniel?  **4th decile**

### H.O.T. Problems  Use Higher-Order Thinking Skills

**18.** Sample answer: 60, 55, 50, 45, 40, 35, 30, 25, 20, 15, 10, 5

**20.** Sample answer: The number of students volunteering during the hours of a carwash: 1, 2, 2, 3, 3, 5, 5, 5, 8; mode = 5, median = 3, mean = 3.8

**15. ARGUMENTS** A student in another class says that a set of data cannot have the same mean and mode. Do you agree or disagree? Explain. Include an example or counterexample.  **15–17. See margin.**

**16. WRITING IN MATH** How can you describe a data value based on its position in the data set?

**17. REASONING** A set of data can only have one mean and one median. How many values can a set of data have for the mode? Explain your reasoning.

**18. CHALLENGE** Write an example of a data set with an equal number of scores at the 25th percentile, 50th percentile, and 75th percentile.

**19. WRITING IN MATH** Compare and contrast percentile rank and percent score.  **See margin.**

**20. OPEN ENDED** Write an example of a data set whose mode is greater than the mean and median.

**21. MODELING** Create a data set of test scores with the median lower than the mean. Describe what would happen to the mean if an unusually high test score were added to the data set.  **See margin.**

## Extra Practice
See page R10 for extra exercises for students who are approaching level or for on-level students who need additional reinforcement.

## Assess

### Ticket Out the Door
Make several copies of each of the exercises on percentile rank, and mean, median, and mode on note cards. Give one card to each student. As the students leave the room, ask them to describe how to determine the calculated values for the measures on their cards.

### Additional Answers

**8** For North Carolina, the mean population in thousands is 344 and the median is 250. For Colorado, the mean population in thousands is 291 and the median is 234.5. Both the mean and median are greater for North Carolina, so its cities are more populated.

**15.** Disagree; Sample counterexample: The data set 3, 3, 3, 3, 3, 3, has a mean of = 3 and mode of = 3.

**16.** Sample answer: Using percentile rank; The position in the data set tells how many data values are below that position.

**17.** Sample answer: A data set can have zero, one, or more than one mode. The mode is not dependent on where the data value is positioned in the set or on the distribution of the data values. The mode only depends on how many of a certain data value there are.

**19.** The percentile rank shows where a score ranks among the other scores. The percent score is a comparison to the highest score possible.

**21.** Sample answer: 40, 40, 50, 50, 93, 95; mean = 61.3, median = 50; If an unusually high test score such as 99 is added to the data set, it will not change the median, but the mean will increase by over 5 points.

## Standards for Mathematical Practice

| Emphasis On | Exercises |
|---|---|
| 1 Make sense of problems and persevere in solving them. | 14, 20–22 |
| 2 Reason abstractly and quantitatively. | 17 |
| 3 Construct viable arguments and critique the reasoning of others. | 15 |
| 4 Model with mathematics. | 5–8, 12, 21 |
| 6 Attend to precision. | 16, 19, 21 |

## Go Online!

### Self-Check Quiz
Students can use *Self-Check Quizzes* to check their understanding of this lesson.

Lesson 10-1 | Measures of Center

# Preparing for Assessment

Exercises 22–27 require students to use the skills they will need on standardized assessments. Each exercise is dual-coded with content standards and mathematical practice standards.

| | Dual Coding | |
|---|---|---|
| Items | Content Standards | MP Mathematical Practices |
| 22 | S.ID.2 | 1, 4, 6, 7 |
| 23 | S.ID.2 | 4, 6 |
| 24 | S.ID.2 | 1, 4, 7 |
| 25 | S.ID.2 | 2, 6 |
| 26 | S.ID.2 | 3 |
| 27 | S.ID.2 | 1, 3, 4, 6 |

## Diagnose Student Errors

Survey student responses for each item. Class trends may indicate common errors and misconceptions.

**22b.**

| | |
|---|---|
| A | Did not put the data in numerical order first. |
| B | Chose the 5th number in the list instead of the average of the 5th and 6th numbers |
| C | CORRECT |
| D | Chose the 6th number in the list instead of the average of the 5th and 6th numbers |

**23.**

| | |
|---|---|
| A | Did not include two values of 80 in their calculations |
| B | CORRECT |
| C | Did not notice that there are more values of 80 than other values |
| D | Calculated the median incorrectly |

## Preparing for Assessment

**22. MULTI-STEP** The top ten state populations, in millions, according to the 2010 Census, are shown in the table below. MP 1, 4, 6, 7  S.ID.2

| State | Population (Millions) |
|---|---|
| California | 37.3 |
| Florida | 18.8 |
| Georgia | 9.7 |
| Illinois | 12.8 |
| Michigan | 9.9 |
| New York | 19.4 |
| North Carolina | 9.5 |
| Ohio | 11.5 |
| Pennsylvania | 12.7 |
| Texas | 25.1 |

a. What is the mean population for these states?

16.67 million

b. Which is the median population for these states? **C**
- A 9.9 million
- B 12.7 million
- C 12.75 million
- D 12.8 million

c. Find the percentile rank of North Carolina among these ten states.

1st percentile

d. Find the percentile rank of Georgia.

10th percentile

e. Is there any state at the 50th percentile? If so, list it.

Illinois

f. Which states are below the 20th percentile? Select all that apply.  **B, D**
- A Ohio
- B Georgia
- C Illinois
- D North Carolina

**23.** Amit scored 98, 80, 92, 79, 84, 88, and 80 for one semester. Find the mean, median, and mode for his test scores. MP 4, 6  S.ID.2  **B**
- A mean = 86.83, median = 84, mode = 80
- B mean = 85.86, median = 84, mode = 80
- C mean = 85.86, median = 84, no mode
- D mean = 85.86, median = 86, mode = 80

**24.** You are researching the number of free throws the basketball team makes each game over the course of the regular season. Which measure of center may be used best to describe the data you collect? MP 1, 4, 7  S.ID.2  **A**
- A mean
- B median
- C mode

**25.** Write an example of a data set consisting of five numbers in which the median is equal to the mean. MP 2, 6  S.ID.2

Sample answer: 20, 30, 40, 50, 60

**26.** Determine whether the following statement is *true* or *false*. Justify your answer. MP 3  S.ID.2

*The percentile rank of the team with the highest score is at the 100th percentile.* See margin.

**27.** Which measure of center may best describe the number of participants in the school activities? Explain. MP 1, 3, 6, 4  S.ID.2  See margin.

| Club | Number of Members |
|---|---|
| Student Council | 48 |
| Dance Committee | 15 |
| Student Government | 60 |
| Chemistry Club | 7 |
| Principal's Assistants | 2 |
| Band | 120 |
| Library Assistants | 3 |
| Culinary Club | 18 |
| Honors Society | 90 |
| Community Service | 99 |

## Additional Answers

**26.** False; if there were 10 teams being considered, the highest percentile would be the 90th percentile, because there are 9 teams with lower rank.

**27.** Sample answer: mean; The mean is 46.2 people, and the median is 33 people. The mean is closer to the number of participants in the activities.

# LESSON 10-2
# Representing Data

**SUGGESTED PACING (DAYS)**
- 90 min. — 0.5
- 45 min. — 1

Instruction

## Track Your Progress

### Objectives

1. Represent sets of data by using dot plots.
2. Determine whether a discrete or continuous graphical representation is appropriate, and then create the bar graph or histogram.

### Mathematical Background

Representing data using a graphic is an important tool for organizing data. Dot plots, bar graphs, and histograms are used for different types of data. Using the appropriate display will help you understand and analyze the data to make informed decisions.

#### THEN
**S.ID.2** Use statistics appropriate to the shape of the data distribution to compare center (median, mean) and spread (interquartile range, standard deviation) of two or more different data sets.

#### NOW
**S.ID.1** Represent data with plots on the real number line (dot plots, histograms, and box plots)

**N.Q.1** Use units as a way to understand problems and to guide the solution of multi-step problems; choose and interpret units consistently in formulas; choose and interpret the scale and the origin in graphs and data displays.

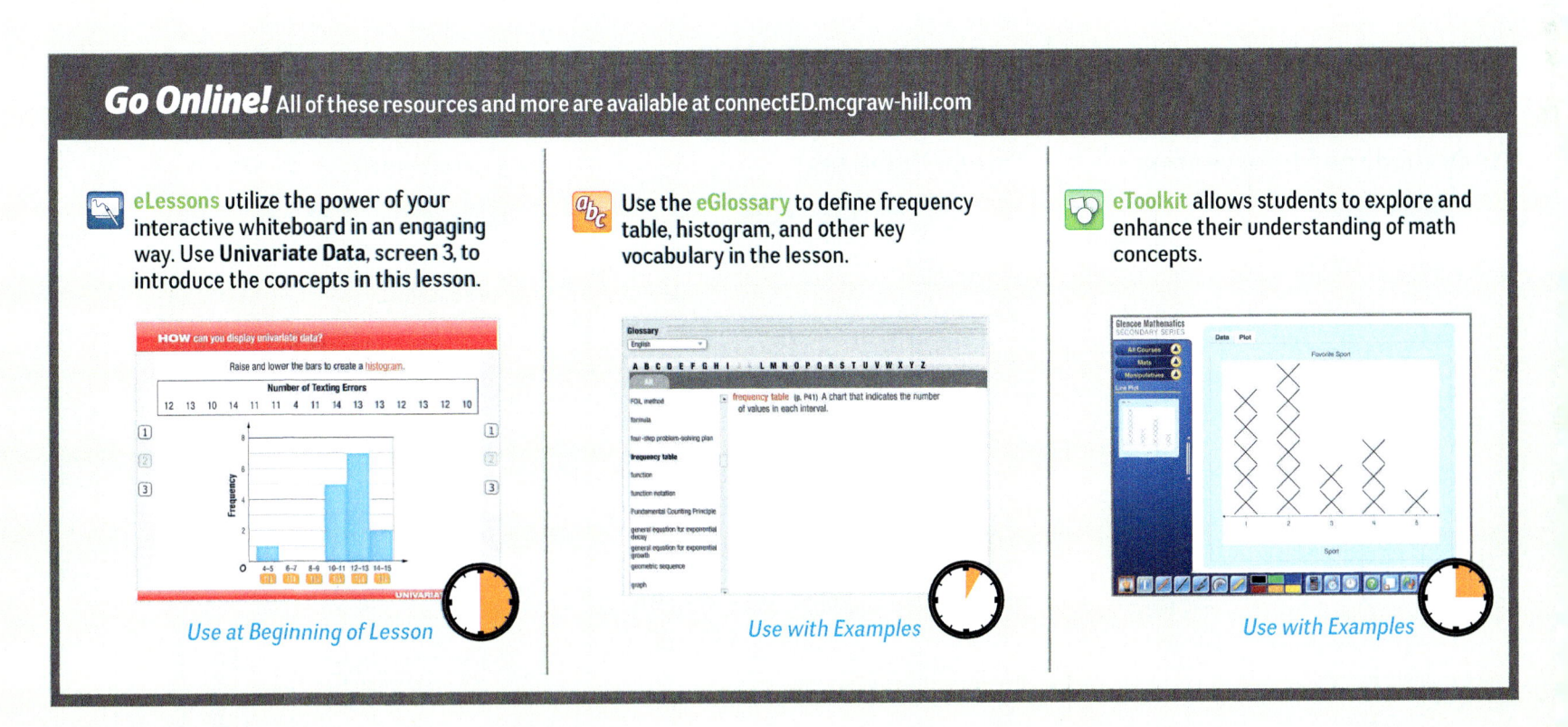

#### NEXT
**S.ID.3** Interpret differences in shape, center, and spread in the context of the data sets, accounting for possible effects of extreme data points (outliers).

## Go Online! All of these resources and more are available at connectED.mcgraw-hill.com

**eLessons** utilize the power of your interactive whiteboard in an engaging way. Use **Univariate Data**, screen 3, to introduce the concepts in this lesson.

*Use at Beginning of Lesson*

Use the **eGlossary** to define frequency table, histogram, and other key vocabulary in the lesson.

*Use with Examples*

**eToolkit** allows students to explore and enhance their understanding of math concepts.

*Use with Examples*

### OER Using Open Educational Resources

**Digital Resource** Statkey is an online tool that can create graphs and displays. Have student create graphs online for different data sets.

connectED.mcgraw-hill.com  **657A**

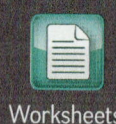

# Go Online!
connectED.mcgraw-hill.com
Worksheets

# Differentiate Your Resources

**Extra Practice** Additional practice or homework; Skills Practice is best for approaching-level students and Practice is best for on-level and beyond-level students

## Skills Practice
AL  OL  ELL

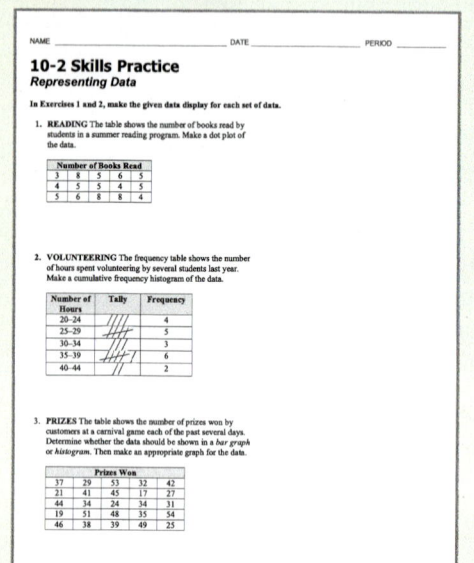

## Practice
AL  OL  BL  ELL

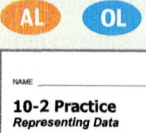

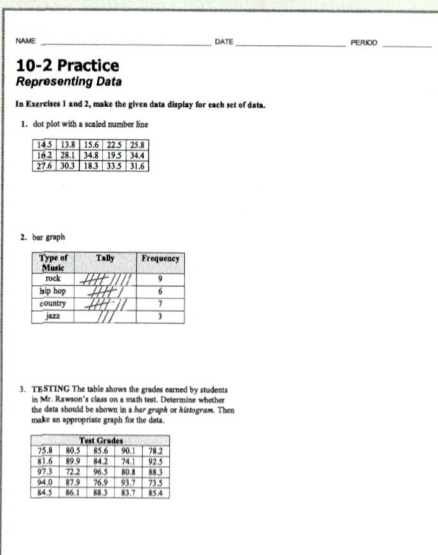

## Word Problem Practice
AL  OL  BL  ELL

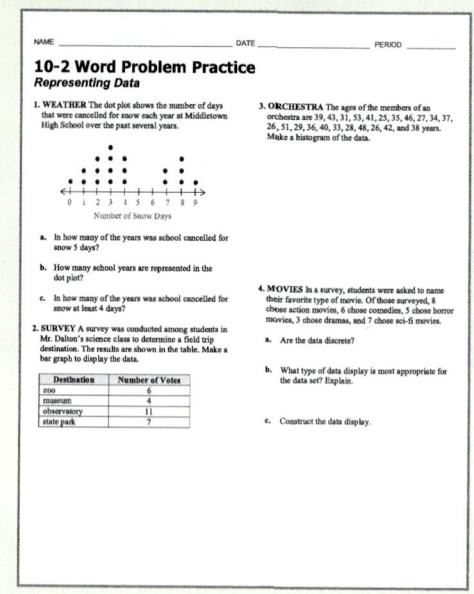

**Intervention** Reteaching and vocabulary activities that can be used with struggling or absent students and as ELL support

**Extension** Activities that can be used to extend lesson concepts

## Study Guide and Intervention
AL  OL  ELL

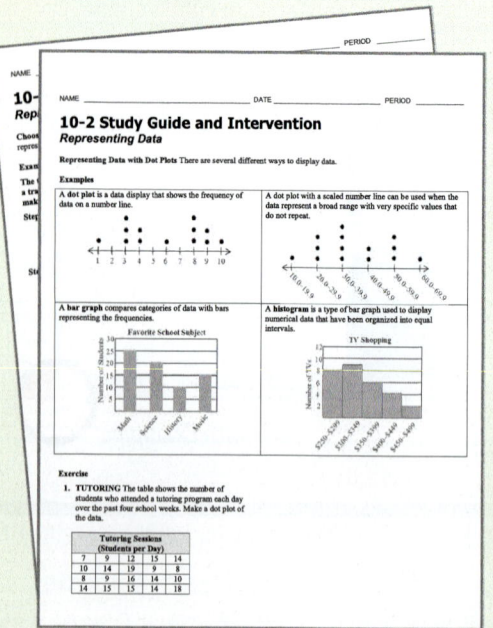

## Study Notebook
AL  OL  BL  ELL

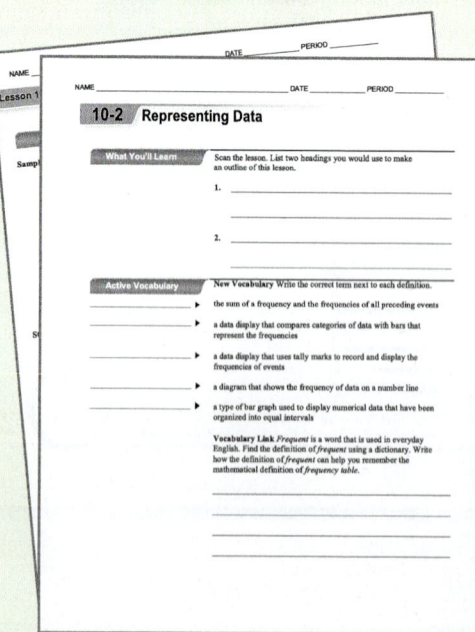

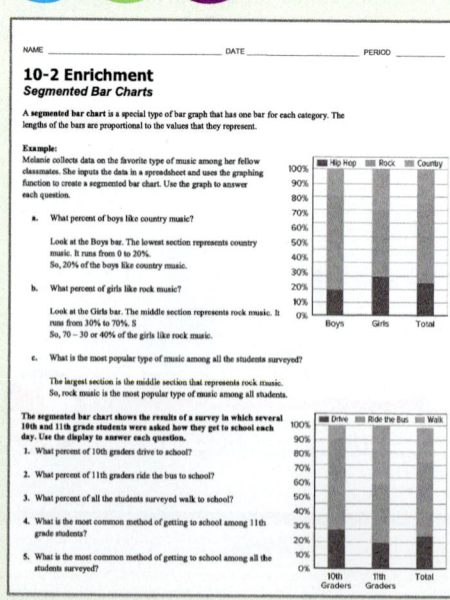

## Enrichment
OL  BL  ELL

657B | Lesson 10-2 | Representing Data

# LESSON 2
## Representing Data

**Then**
- You used measures of center to compare data sets.

**Now**
1. Represent data by using dot plots.
2. Determine whether a discrete or continuous graphical representation is appropriate, and then create the bar graph or histogram.

**Why?**
The final scores for golfers in a tournament can be divided into equal intervals. A histogram is a visual way to see the frequency of each interval.

**New Vocabulary**
dot plot
frequency table
bar graph
cumulative frequency
histogram

**Mathematical Practices**
1 Make sense of problems and persevere in solving them.
4 Model with mathematics.

**Content Standards**
S.ID.1 Represent data with plots on the real number line (dot plots, histograms, and box plots).
N.Q.1 Use units as a way to understand problems and to guide the solution of multi-step problems; choose and interpret units consistently in formulas; choose and interpret the scale and the origin in graphs and data displays.

### 1 Representing Data with Dot Plots
A **dot plot** is a diagram that shows the frequency of data on a number line. Dot plots are also called *line plots*. When data are represented as a dot plot, the gaps and clusters of the data become more apparent.

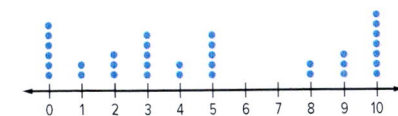

**Key Concept** Making a Dot Plot

**Step 1** Write the data in order from least to greatest.
**Step 2** Draw a number line that starts at the least data point and ends at the greatest data point. Choose an appropriate scale.
**Step 3** Plot the dots on the number line. Stack the points when there is more than one data point with the same number.
**Step 4** If appropriate, include a label for the number line and title for the dot plot.

**Example 1** Make a Dot Plot

Represent the data as a dot plot.

11, 12, 14, 15, 12, 13, 15, 13, 9, 15, 12, 13, 15, 15, 11

**Step 1** Write the data points in order from least to greatest.

9, 11, 11, 12, 12, 12, 13, 13, 13, 14, 15, 15, 15, 15, 15

**Step 2** Make a number line that starts at the least data point and ends at the greatest data point. Choose an appropriate scale.

The data are whole numbers ranging from 9 to 15. So, make a number line starting at 9 with intervals of 1.

**Step 3** Plot dots on the number line. Stack the dots when there is more than one of the same value.

---

## Mathematical Practices Strategies

**Make sense of problems.**
Help students analyze each data set to determine the best representation for the data.

- Give an example of a data set that would be best represented by a dot plot.
  *the number of RBIs from each player on the baseball team*

- What is the difference between discrete and continuous data?
  *discrete data are counted while continuous data are measured*

- When would you use a histogram?
  *to represent continuous data*

**Need Another Example?**
Represent the data as a dot plot.
20, 22, 30, 23, 20, 22, 20, 22, 23, 25, 22, 26, 20, 22, 25

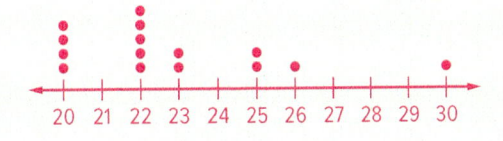

---

**Lesson 10-2 | Representing Data**

## Launch

Have students read the Why? section of the lesson. Ask:

- How else can you present the data besides a histogram? *a tally chart or a bar graph*

- What is the benefit of representing the data as a histogram instead of a list? *You can see the frequency of each final score.*

- Could you use a dot plot to represent the data? How does this differ from a histogram? *Yes; a dot plot would show the number of players with each final score while a histogram would show the number of players within a frequency interval.*

## Teach

Ask the scaffolded questions for each example to build conceptual understanding for students at all levels.

### 1 Representing Data with Dot Plots

**Example 1 Make a Dot Plot**

**AL** In Step 1, why should we order the numbers? *It helps us find duplicates and prepares us for the dot plot.*

**OL** How is a dot plot better than a list of numbers? *You can see which number occurs the most at a glance.*

**BL** If there is a number or interval that has no data points, is there any need to include it in a dot plot? *Yes, even if a number or interval has no data points, it should be included in the dot plot so you can see clearly the overall distribution of data points, including the fact that this number or interval has none.*

---

### Go Online!

**Interactive Whiteboard**
Use the *eLesson* or *Lesson Presentation* to present this lesson.

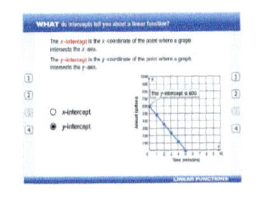

connectED.mcgraw-hill.com  657

Lesson 10-2 | Representing Data

### Example 2 Make a Dot Plot with a Scaled Number Line

**AL** How can you order the data from least to greatest without needing to physically cross out numbers on the original list? Look for the least number in the list, count how many times it appears, and write it down that number of times. Then look for the next greater number and repeat.

**OL** How can you choose appropriate intervals? The range is 22.6 to 96.4, which is close to 20 and 100. To use equal intervals, divide that range into groups of 10 without overlapping the intervals.

**BL** Is there more than one way to choose appropriate intervals? Yes; the intervals must be of equal size and not overlap. If the intervals are too small or too large, the overall shape of the data may not help interpret its meaning.

**Need Another Example?**
**TEST SCORES** The test scores for the 1st period math class are listed below. Represent the data as a dot plot.
75, 80, 90, 92, 65, 70, 75, 95, 90, 92, 99, 65, 75, 88

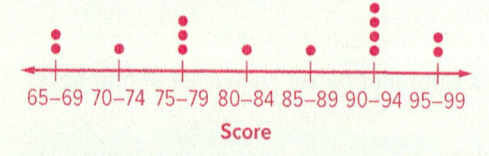

**1st Period Math Test Scores**

65–69  70–74  75–79  80–84  85–89  90–94  95–99
Score

## 2 Representing Data with Bar Graphs or Histograms

### Example 3 Make a Bar Graph

**AL** What does the height of the bars in this bar graph represent? The number of students who chose each sport as their favorite.

**OL** Would a dot plot be a good alternate way to display this data? Why or why not? No. A dot plot displays data on a number line, not in categories.

**BL** Name one advantage and one disadvantage of using a bar graph to display data in categories. An advantage is that you can easily compare the sizes of the categories by observing the sizes of the bars. A disadvantage is that, depending on the scale, you may not be able to tell the exact number in each category.

---

**Step 4** If appropriate, include a label for the number line and title for the dot plot.
Since no information is given regarding what these data represent, no title is needed for this dot plot.

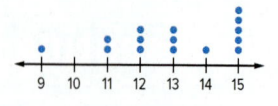

9  10  11  12  13  14  15

> **Guided Practice**

**1A.** 42, 40, 40, 45, 50, 42, 50, 46, 50, 40, 45, 40, 43, 45  See margin.

**1B.** 100, 101, 106, 105, 100, 102, 102, 101, 101, 100, 100, 108, 100, 101  See margin.

S.ID.1, N.Q.1

**Real-World Example 2** Make a Dot Plot with a Scaled Number Line

**INTERNET USAGE** The table shows Internet users of Middle Eastern countries as a percentage of their total population. Represent the data as a dot plot.

**Step 1** Write the data points in order from least to greatest.
22.6, 28.1, 33.0, 57.2, 64.6, 65.9, 74.7, 78.6, 78.7, 80.4, 86.1, 91.9, 93.2, 96.4

**Step 2** Make a number line. Choose an appropriate scale.
The data range from 22.6 to 96.4. The data represent a broad range with specific values, so it is unlikely that any data point is represented more than once. To represent the data in a meaningful way, divide the range into equal intervals.

**Step 3** Plot the dots on the number line. Stack the points when there is more than one data point in the same interval.

**Step 4** Include a label for the number line and title for the dot plot.

| Country | Internet Users (% of Population) |
|---|---|
| Bahrain | 96.4 |
| Iran | 57.2 |
| Iraq | 33.0 |
| Israel | 74.7 |
| Jordan | 86.1 |
| Kuwait | 78.7 |
| Lebanon | 80.4 |
| Oman | 78.6 |
| Palestine | 64.6 |
| Qatar | 91.9 |
| Saudi Arabia | 65.9 |
| Syria | 28.1 |
| United Arab Emirates | 93.2 |
| Yemen | 22.6 |

**Study Tip**
**MP Modeling** Choose an appropriate scale for the data set when making the number line. Having too many or too few intervals will not cluster the data in a meaningful way.

**Internet Usage by Middle Eastern Countries**

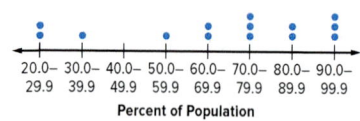

20.0–29.9  30.0–39.9  40.0–49.9  50.0–59.9  60.0–69.9  70.0–79.9  80.0–89.9  90.0–99.9
Percent of Population

> **Guided Practice**

**2A. SALARIES** The table shows salaries for employees at Julio's company. Represent the data as a dot plot.  See margin.

| 38,150 | 40,500 | 42,750 | 43,685 | 57,890 | 37,550 |
| 41,235 | 78,990 | 66,000 | 44,435 | 45,775 | 39,800 |

### Additional Answers (Guided Practice)

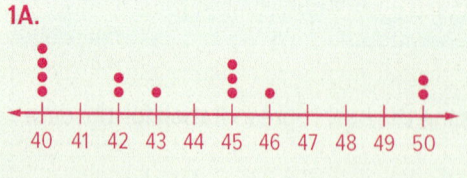

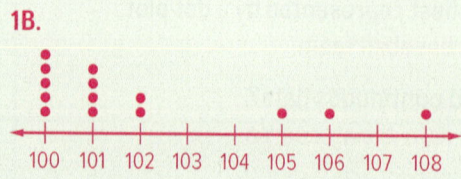

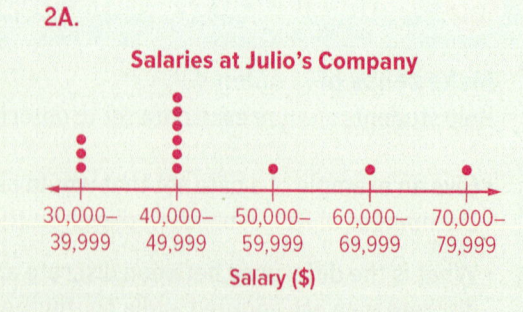

**1A.**

40  41  42  43  44  45  46  47  48  49  50

**1B.**

100  101  102  103  104  105  106  107  108

**2A.**

**Salaries at Julio's Company**

30,000–39,999   40,000–49,999   50,000–59,999   60,000–69,999   70,000–79,999
Salary ($)

658 | Lesson 10-2 | Representing Data

Lesson 10-2 | Representing Data

**2B. MILES** The table shows the number of miles people in Manda's fitness class drive from home to the gym. Represent the data as a dot plot. **See margin.**

| 11 | 21 | 14 | 9 | 15 | 16 | 25 |
| 26 | 5 | 22 | 13 | 22 | 15 | 8 |
| 22 | 19 | 16 | 10 | 4 | 19 | 17 |

## 2 Representing Data with Bar Graphs or Histograms

A **frequency table** uses tally marks to record and display frequencies of events. A **bar graph** compares categories of data with bars representing the frequencies. Bar graphs are used when the data are discrete, which means that the data belong in specific categories and there are no "in between" values. To indicate this, there is space between the bars.

S.ID1

**Study Tip**

**MP Modeling** A bar graph must have an appropriate scale with equal intervals on the *y*-axis that does not misrepresent the data.

### Example 3 Make a Bar Graph

Make a bar graph to display the data gathered from a survey of students about their favorite sport.

| Sport | Tally | Frequency |
|---|---|---|
| basketball | ||||  ||||  ||||  | 15 |
| football | ||||  ||||  ||||  ||||  ||||  | 25 |
| soccer | ||||  ||||  ||||  ||| | 18 |
| baseball | ||||  ||||  ||||  ||||  | | 21 |

**Step 1** Draw a horizontal axis and a vertical axis. Label the axes as shown. Add a title.

**Step 2** Draw a bar to represent each sport. The vertical scale is the number of students who chose each sport. The horizontal scale identifies the sport.

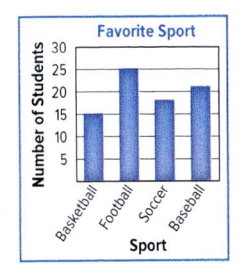

▶ **Guided Practice**

3. Make a bar graph to display the data regarding members of the orchestra.

| Instrument | brass | percussion | strings | woodwinds |
|---|---|---|---|---|
| Frequency | 16 | 3 | 31 | 17 |

**See margin.**

The **cumulative frequency** for each event is the sum of its frequency and the frequencies of all preceding events. A **histogram** is a type of bar graph used to display numerical data that have been organized into equal intervals. Each interval is represented by an interval called a *bin*. A histogram represents continuous data, so the bins have no spaces between them.

S.ID1

### Example 4 Make a Histogram and a Cumulative Frequency Histogram

Make histograms of the frequency and the cumulative frequency.

| Age at Inauguration | 40–44 | 45–49 | 50–54 | 55–59 | 60–64 | 65–69 |
|---|---|---|---|---|---|---|
| U.S. Presidents | 2 | 7 | 13 | 12 | 7 | 3 |

Find the cumulative frequency for each interval.

| Age | < 45 | < 50 | < 55 | < 60 | < 65 | < 70 |
|---|---|---|---|---|---|---|
| Presidents | 2 | 2 + 7 = 9 | 9 + 13 = 22 | 22 + 12 = 34 | 34 + 7 = 41 | 41 + 3 = 44 |

**Teaching Tip**

**Bar Graph vs. Histogram** In Example 3, the frequency of categorical data is represented with bars while in Example 4, the frequency of numerical data is represented by the bars.

Cumulative Frequency Graph
Ages of Volunteers at Shelter

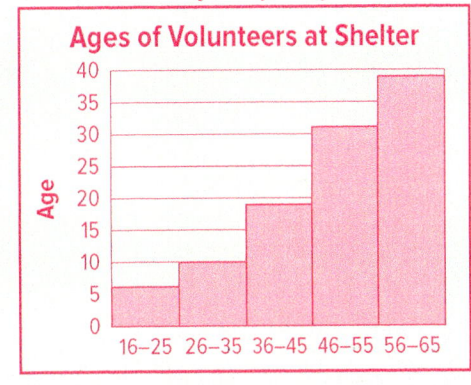

## Additional Answer (Guided Practice)

**2B.** Distance From Gym

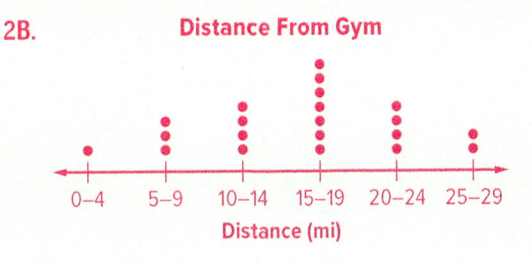

### Need Another Example?

**Music** The table shows the results of a student survey on favorite music. Make a bar graph to display the data.

| Music | Frequency |
|---|---|
| Country | 13 |
| Hip-Hop | 35 |
| Metal | 20 |
| Rock | 32 |

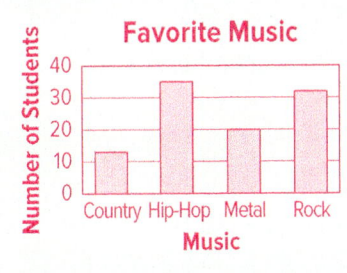

Favorite Music

### Example 4 Make a Histogram and a Cumulative Frequency Histogram

**AL** What does the height of each bar in the histogram represent? The height of each bar in the cumulative frequency histogram? **The number of presidents in each age range at the time of their inauguration. The number of presidents younger than the given age at the time of their inauguration.**

**OL** Would a line graph be a good alternate way to display this data? Why or why not? **No; a line graph is good for showing change in data over time, not the frequency of data in different intervals.**

**BL** From a cumulative frequency histogram, how can you get an idea of the number of items within each interval? **The increase in size from one bar to the next represents the number of items in that interval. For example, from <45 to <50, the bar increases from 2 to 9, for a change of 7. Therefore, there are 7 presidents in the 45–49 interval.**

### Need Another Example?

**Volunteers** The table shows the ages of the volunteers at an animal shelter. Make histograms of the frequencies and the cumulative frequencies.

| Ages | Frequency |
|---|---|
| 16–25 | 6 |
| 26–35 | 4 |
| 36–45 | 9 |
| 46–55 | 12 |
| 56–65 | 8 |

Frequency Graph

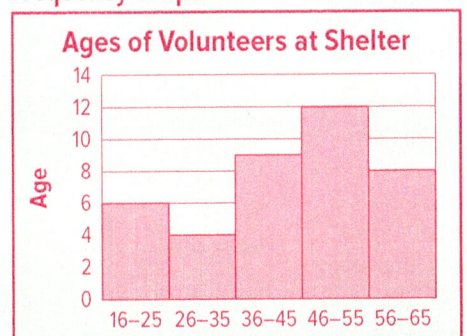

Ages of Volunteers at Shelter

connectED.mcgraw-hill.com 659

Lesson 10-2 | Representing Data

## Example 5 Determine an Appropriate Graph

**part a** Determine an Appropriate Graph for Discrete Data

**AL** Why is the data discrete? The data can be organized by category.

**OL** Could you use a histogram to represent the data? Why or why not? No; the data is discrete and organized by sporting event so a histogram cannot be used.

**BL** Which representation of data, the table or bar graph, is best suited to find the event with fewest total medals? The table because three events have total medals below 200 so the bar graph is not clear.

### Need Another Example?

The table shows the total number of pieces of fudge sold over the weekend from a store. Make a graph of the data to show the total amount of each type of fudge bought over the weekend.

| Type | Friday | Saturday | Sunday |
|---|---|---|---|
| chocolate | 120 | 145 | 111 |
| chocolate walnut | 76 | 90 | 100 |
| chocolate peanut butter | 67 | 82 | 90 |
| maple walnut | 90 | 65 | 83 |
| peanut butter | 85 | 77 | 78 |

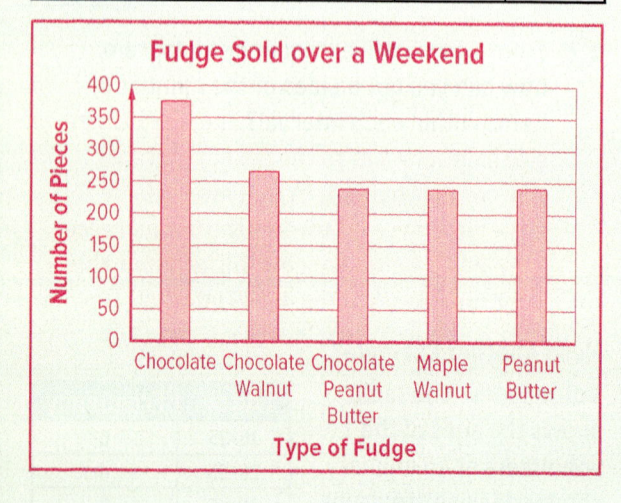

**Fudge Sold over a Weekend** (bar graph: Chocolate ~375, Chocolate Walnut ~265, Chocolate Peanut Butter ~240, Maple Walnut ~240, Peanut Butter ~240; y-axis: Number of Pieces 0–400; x-axis: Type of Fudge)

### Go Online!

**Personal Tutor**

Personal Tutor (for every example) let student hear real teacher solve problems. Students can pause and repeat as many times as necessary.

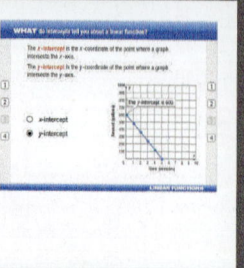

---

Make each histogram like a bar graph but with no space between the bars.

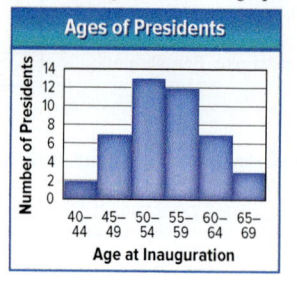

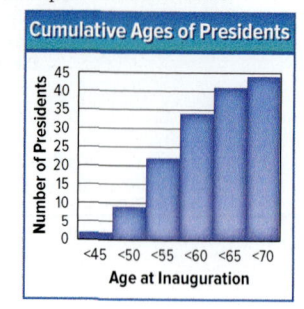

**Ages of Presidents** (histogram: Number of Presidents vs. Age at Inauguration; intervals 40–44, 45–49, 50–54, 55–59, 60–64, 65–69)

**Cumulative Ages of Presidents** (histogram: Number of Presidents vs. Age at Inauguration; intervals <45, <50, <55, <60, <65, <70)

**Study Tip**
Notice that the bars of each histogram are of equal width. This equal width provides a visual representation of the equal intervals.

### Guided Practice

4. Make histograms of the frequency and the cumulative frequency. See margin.

| Gas Price per Gallon | $1.80–1.84 | $1.85–1.89 | $1.90–1.94 | $1.95–1.99 | $2.00–2.04 | $2.05–2.09 | $2.10–2.14 |
|---|---|---|---|---|---|---|---|
| Frequency | 2 | 7 | 6 | 3 | 3 | 2 | 1 |

Given a set of data, you will need to decide the best way to graphically display it. Use a bar graph for discrete data and a histogram for continuous data.

### Real-World Example 5 Determine an Appropriate Graph

Determine whether the data are *discrete* or *continuous*. Then, make a graph of the data to show the total medals in each sport.

**a. OLYMPICS** The table shows the total number of Olympic medals won by U.S. athletes from the first Summer Olympics in 1896 through 2012.

| Event | Gold | Silver | Bronze |
|---|---|---|---|
| boxing | 49 | 23 | 39 |
| diving | 48 | 41 | 43 |
| swimming | 230 | 164 | 126 |
| track & field | 319 | 247 | 193 |
| wrestling | 52 | 43 | 34 |

**Step 1** Determine whether the data should be represented as a bar graph or histogram.

These data represent discrete, categorical data, so use a bar graph.

**Step 2** Determine appropriate categories, and tally the data.

Each sport will represent a category of the bar graph.

Complete the table to find the total number of medals in each sport.

| Event | Total |
|---|---|
| boxing | 111 |
| diving | 132 |
| swimming | 520 |
| track & field | 759 |
| wrestling | 129 |

---

### Additional Answer (Guided Practice)

4. **Frequency Graph**

**Gas Prices** (histogram with bars at heights ~2, 7, 6, 3, 3, 2, 1; x-axis: Price Per Gallon ($) 1.80–1.84, 1.85–1.89, 1.90–1.94, 1.95–1.99, 2.00–2.04, 2.05–2.09, 2.10–2.14)

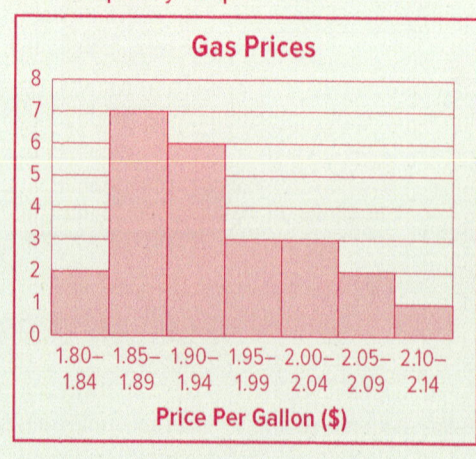

**Cumulative Frequency Graph**

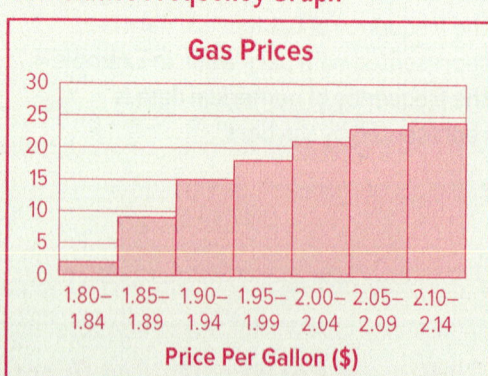

**Gas Prices** (cumulative histogram; y-axis 0–30; x-axis: Price Per Gallon ($) 1.80–1.84, 1.85–1.89, 1.90–1.94, 1.95–1.99, 2.00–2.04, 2.05–2.09, 2.10–2.14)

660 | Lesson 10-2 | Representing Data

## Lesson 10-2 | Representing Data

**Steps 3 and 4** Draw a bar to represent each category. Label the axes, and include a title for the graph.

Label the *x*-axis as *Event* and the *y*-axis as *Medals*. Then title the graph.

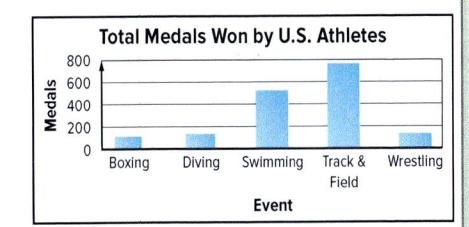

**Total Medals Won by U.S. Athletes**
(bar graph: Boxing, Diving, Swimming, Track & Field, Wrestling)

**b. MARATHON** The results of the top finishers of the 2015 New York City Marathon, wheelchair division, are given in the table.

**Step 1** Determine whether the data should be represented as a bar graph or histogram.

Because racers can finish with any time, the data are continuous and you can use a histogram.

| Place | Time (h:m:s) | Place | Time (h:m:s) |
|---|---|---|---|
| 1 | 1:30:54 | 6 | 1:35:37 |
| 2 | 1:30:55 | 7 | 1:35:38 |
| 3 | 1:34:05 | 8 | 1:36:45 |
| 4 | 1:35:19 | 9 | 1:36:59 |
| 5 | 1:35:21 | 10 | 1:38:39 |

**Step 2** Determine appropriate intervals and tally the data.

Since the data are spread over several minutes, group the data by the minute. Then, tally each interval.

Use a table to determine the frequency of each minute.

| Time (h:m:s) | Frequency |
|---|---|
| 1:30:00 – 1:30:59 | 2 |
| 1:31:00 – 1:31:59 | 0 |
| 1:32:00 – 1:32:59 | 0 |
| 1:33:00 – 1:33:59 | 0 |
| 1:34:00 – 1:34:59 | 1 |
| 1:35:00 – 1:35:59 | 4 |
| 1:36:00 – 1:36:49 | 2 |
| 1:37:00 – 1:37:59 | 0 |
| 1:38:00 – 1:38:59 | 1 |

**Steps 3 and 4** Draw a bar to represent each interval.

Label the axes, and include a title for the graph.

**Marathon Results** (histogram)

### Real-World Link
The first New York City Marathon was held in 1970. The official wheelchair division was introduced in 2000.

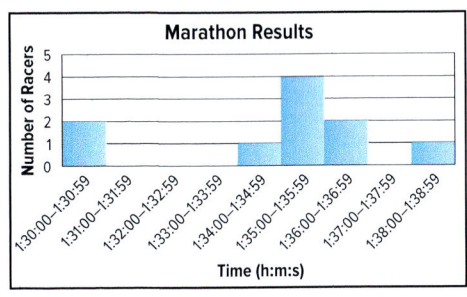

### Study Tip
**MP Modeling** The appropriate graph to represent example B can either be a dot plot or a bar graph since each type shows which interval has the most runners.

### Guided Practice

**5. HEIGHTS** The table shows the heights of players on the Cedar Ridge High School basketball team. Determine whether the data are *discrete* or *continuous*. Then make an appropriate graph. **See margin.**

| 67 | 70 | 69 | 73 | 75 | 72 | 68 |
| 70 | 70 | 71 | 68 | 73 | 69 | 71 |

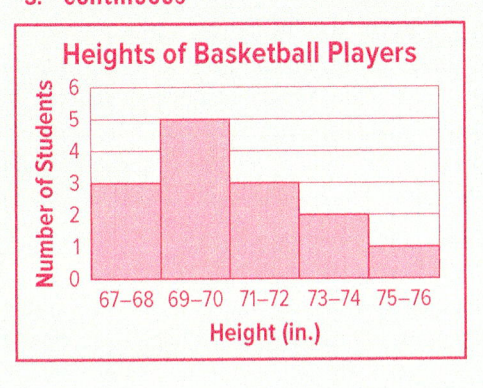

### part b Determine an Appropriate Graph for Continuous Data

**AL** Why is the racing data continuous? the racers can finish the race with any time.

**OL** What conclusions can you reach from the histogram? The most people crossed the finish line after 1:35 and before 1:36. Most of the times were between 1:34 and 1:40.

**BL** Could the data be represented in another way? Explain. Yes; the data could also be represented by a dot plot.

### Need Another Example?
The table shows the different temperatures of sick high school students that came into the nurse's office in a week. Determine whether the data are *discrete* or *continuous*. Then make a graph.

| Temperature (°F) | | |
|---|---|---|
| 98.5 | 98.4 | 99.7 |
| 99.2 | 100.7 | 100.9 |
| 101.1 | 102.3 | 98.7 |
| 100.7 | 101.2 | 99.2 |
| 99.1 | 100.3 | 99.9 |

continuous

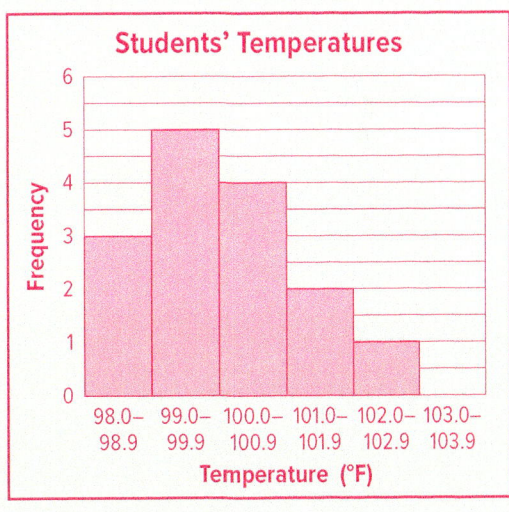

**Students' Temperatures** (histogram)

## Additional Answer (Guided Practice)

5. continuous

**Heights of Basketball Players** (histogram: 67–68, 69–70, 71–72, 73–74, 75–76; Height (in.))

Lesson 10-2 | Representing Data

# Practice

**Formative Assessment** Use Exercises 1–5 to assess students' understanding of the concepts in this lesson.

The Practice and Problem Solving exercises assess the content taught in the lesson. The Preparing for Assessment page is meant to be used as preparation for end-of-course assessments.

## Extra Practice

See page R10 for extra exercises for students who are approaching level or for on-level students who need additional reinforcement.

### Levels of Complexity Chart

The levels of the exercises progress from 1 to 3, with Level 1 indicating the lowest level of complexity.

| Exercises | 6–11 | 12–15, 19–25 | 16–18 |
|---|---|---|---|
| Level 3 | | | ● |
| Level 2 | | ● | |
| Level 1 | ● | | |

## Additional Answers

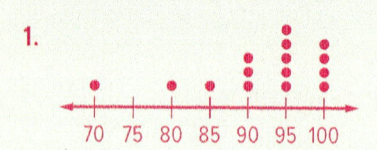

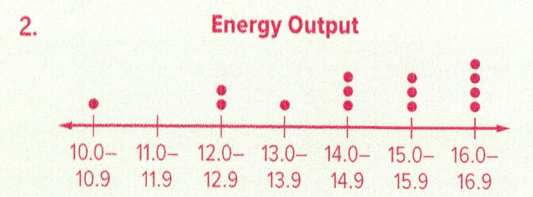

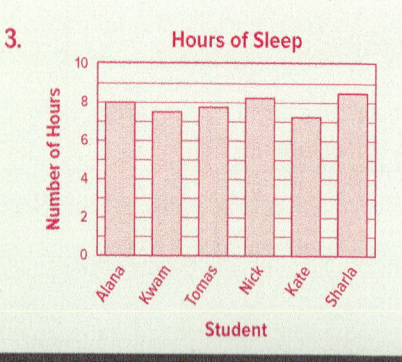

1. [dot plot: 70 75 80 85 90 95 100]

2. Energy Output [dot plot: 10.0–10.9, 11.0–11.9, 12.0–12.9, 13.0–13.9, 14.0–14.9, 15.0–15.9, 16.0–16.9]

3. Hours of Sleep [bar graph: Alana, Kwam, Tomas, Nick, Kate, Sharla]

## Go Online!

**eSolutions Manual** Create worksheets, answer keys, and solutions handouts for your assignments.

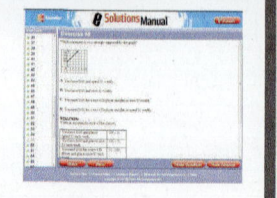

---

**Check Your Understanding** ● = Step-by-Step Solutions begin on page R11.

**Go Online!** for a Self-Check Quiz

**Example 1** S.ID.1
1. Represent the data as a dot plot.
100, 80, 95, 90, 100, 95, 70, 95, 90, 90, 95, 85, 100, 100, 95  See margin.

**Example 2** S.ID.1, N.Q.1
2. **SOLAR POWER** Kellen tracked the energy output, in kilowatt hours, of the solar panels on his house for two weeks in June. Represent the data as a dot plot. See margin.

| Energy Output in June | | | | | |
|---|---|---|---|---|---|
| 16.8 | 14.4 | 15.2 | 16.6 | 14.0 | 16.9 | 12.8 |
| 13.8 | 12.3 | 15.9 | 16.4 | 15.6 | 14.2 | 10.2 |

**Example 3** S.ID.1
3. **SURVEYS** Alana surveyed several students to find how many hours of sleep they typically get each night. The results are shown in the table. Make a bar graph of the data. See margin.

| Hours of Sleep | | | | | |
|---|---|---|---|---|---|
| Alana | 8 | Kwam | 7.5 | Tomas | 7.75 |
| Nick | 8.25 | Kate | 7.25 | Sharla | 8.5 |

**Example 4** S.ID.1
4. **PLAYS** The frequency table at the right shows the ages of people attending a high school play.
   a. Make a histogram to display the data.
   b. Make a cumulative frequency histogram showing the number of people attending who were less than 20-, 40-, 60-, or 80-years of age.  a–b. See Ch. 10 Answer Appendix.

| Age | Tally | Frequency |
|---|---|---|
| 0–19 | ||||| ||||| ||||| ||||| ||||| ||||| ||||| || | 47 |
| 20–39 | ||||| ||||| ||||| ||||| ||||| ||||| ||||| ||| | 43 |
| 40–59 | ||||| ||||| ||||| ||||| ||||| ||||| | | 31 |
| 60–79 | ||||| ||| | 8 |

**Example 5** S.ID.1
5. **PARKS** The table shows the areas of national parks in Alaska. Determine whether the data are *discrete* or *continuous*. Then graph the data.

| Area (km²) of National Parks in Alaska | | | |
|---|---|---|---|
| 13,050.5 | 2711.3 | 33,682.6 | 7084.9 |
| 14,870.3 | 10,601.7 | 19,185.8 | 30,448.1 |

5–9. See Ch. 10 Answer Appendix.

### Practice and Problem Solving
Extra Practice is found on page R10.

**Example 1** S.ID.1
6. Represent the data as a dot plot. 8, 6, 0, 2, 7, 1, 8, 1, 4, 8, 0, 1, 2, 8, 4, 7, 1, 5, 9, 1

**Example 2** S.ID.1, N.Q.1
7. **MOUNTAINS** The table gives the elevation of the highest mountain peaks in the United States in feet. Represent the data as a dot plot.

| Elevation (ft) | | | | |
|---|---|---|---|---|
| 20,308 | 17,402 | 16,391 | 15,325 | 14,829 |
| 18,009 | 16,421 | 16,237 | 14,951 | 14,573 |

**Example 3** S.ID.1
8. **VIDEO GAMES** The table shows the number of active video game players by country. Make a bar graph of the data.

| Country | Players (millions) | Country | Players (millions) |
|---|---|---|---|
| Australia | 9.5 | Poland | 11.8 |
| Brazil | 40.2 | Spain | 17 |
| France | 25.3 | Turkey | 21.8 |
| Germany | 38.5 | United Kingdom | 33.6 |
| Italy | 18.6 | United States | 157 |

**Example 4** S.ID.1
9. **PHOTO SHARING** The table shows the users of a photo-sharing app by age group. Use a histogram to graph the data.

| Age | 18–24 | 25–34 | 35–44 | 45–54 | 55–64 | 65+ |
|---|---|---|---|---|---|---|
| Users (%) | 45 | 26 | 13 | 10 | 6 | 1 |

### Differentiated Homework Options

| Levels | AL Basic | OL Core | BL Advanced |
|---|---|---|---|
| Exercises | 6–11, 16–18, 19–25 | 6–11 odd, 12–25 | 16–25 |
| 2-Day Option | 7–11 odd, 19–25 | 6–11 | |
| | 6–10 even, 16–18 | 12–25 | |

You can use ALEKS to provide additional remediation support with personalized instruction and practice.

662 | Lesson 10-2 | Representing Data

Lesson 10-2 | Representing Data

Example 5
S.ID.1

**10. AGE** The table shows ages of students who are taking a college calculus course. Determine whether the data is *discrete* or *continuous*, and then make an appropriate graph for the data. **See margin.**

| 18 | 18 | 20 | 19 | 19 | 20 | 18 | 20 | 20 |
|----|----|----|----|----|----|----|----|----|
| 19 | 19 | 18 | 18 | 20 | 18 | 20 | 21 | 18 |

**11. GRADES** The table shows all the grades received on a biology test. Determine whether the data is *discrete* or *continuous*, and then make an appropriate graph for the data. **See margin.**

| 96.5 | 94.3 | 75 | 75.7 | 82 | 89 | 82.6 | 96.8 | 91 | 94.5 | 78 | 68.7 |
|------|------|----|------|----|----|------|------|----|------|----|------|
| 68 | 76 | 82.6 | 89.7 | 82.3 | 75.1 | 96.3 | 91.5 | 89.8 | 75.7 | 91.7 | 94.9 |

**12.** Shelby wrote down the day of the month on which her classmates were born. Use the data to make a graph. **See margin.**

| 2 | 4 | 7 | 11 | 1 | 18 | 12 | 3 | 9 | 28 |
|---|---|---|----|---|----|----|---|---|----|
| 4 | 17 | 10 | 2 | 15 | 30 | 20 | 25 | 26 | 8 |
| 6 | 19 | 23 | 28 | 16 | 31 | 24 | 12 | 6 | 31 |

**13. SURVEY** Harper took a survey of the families in her neighborhood to find out how many mobile devices they have. Determine whether her data are *discrete* or *continuous*, and then make an appropriate graph for the data.
**See Ch. 10 Answer Appendix.**

| Family | Mobile Devices | Family | Mobile Devices |
|--------|---------------|--------|----------------|
| Anderson | 6 | Patel | 4 |
| Clark | 3 | Perez | 2 |
| Davis | 3 | Roberts | 4 |
| Garcia | 3 | Turner | 1 |
| Li | 4 | Ward | 2 |

**14. DRIVING** Freddy is taking the exam to get his driver's license in 20 days. Before the exam, he needs to drive a total of 9 hours with his parents. He records his daily driving time, in hours, in a journal. Determine whether the data is *discrete* or *continuous*, and then make an appropriate graph for the data. **See Ch. 10 Answer Appendix.**

| .35 | .55 | .9 | .25 | .35 | .4 | 1 | .75 | .25 | .55 |
|-----|-----|----|-----|-----|----|----|-----|-----|-----|
| .25 | .45 | .75 | .35 | .75 | .55 | .9 | .35 | .55 | 1 |

**15.** A survey of students' favorite movie genres was conducted at school. Make a bar graph to display the data.
**See Ch. 10 Answer Appendix.**

| Comedy | Drama | Crime | Thriller | Action |
|--------|-------|-------|----------|--------|
| 58 | 23 | 47 | 38 | 32 |

**H.O.T. Problems** Use Higher-Order Thinking Skills

**16. SENSE MAKING** The dot plot shows the number of hours students study per week. 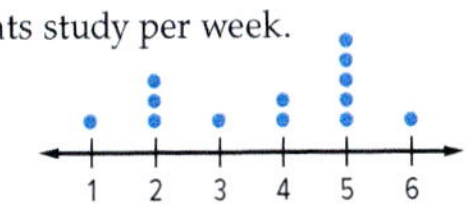
a. How many students studied 5 hours per week? **5**
b. What is the total number of students in this class? **13**
c. How many of the students study 4 or more hours per week? **8**

**17. SENSE MAKING** Henry recorded the daily temperatures in his town and displayed the data in a dot plot. 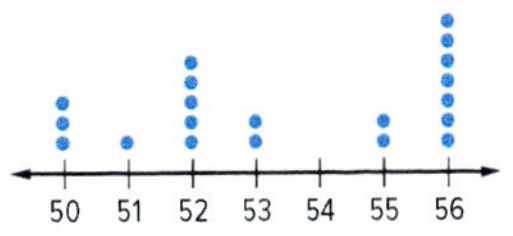
a. What is the most frequent temperature? **56 °F**
b. For how many days did Henry record the temperature? **20**
c. How many days reached a temperature of 53 °F or higher? **11**

**18. WRITING IN MATH** Describe one difference and one similarity between a histogram and a bar graph. **See Ch. 10 Answer Appendix.**

---

## Teaching the Mathematical Practices

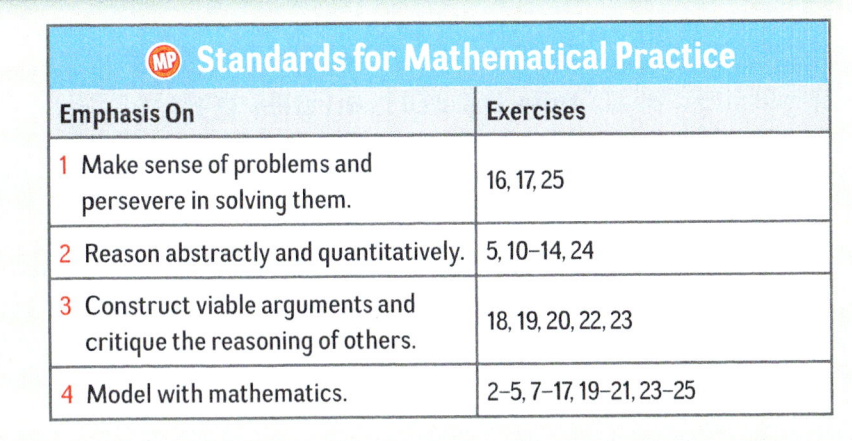

**Sense-Making** Mathematically proficient students are able to analyze data that is visually represented to them by breaking them down into pieces. Suggest students think about what each dot in a dot plot represents.

## Assess

**Ticket Out the Door** On small pieces of paper, write down a situation for which there is numerical data. Give one to each student. Ask students to tell whether the data for that situation is discrete or continuous, and whether the appropriate graph for the data is a histogram or a bar graph.

### Additional Answers

**10.** continuous

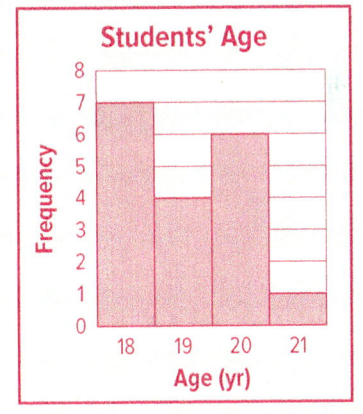

*Students' Age* histogram

**11.** continuous

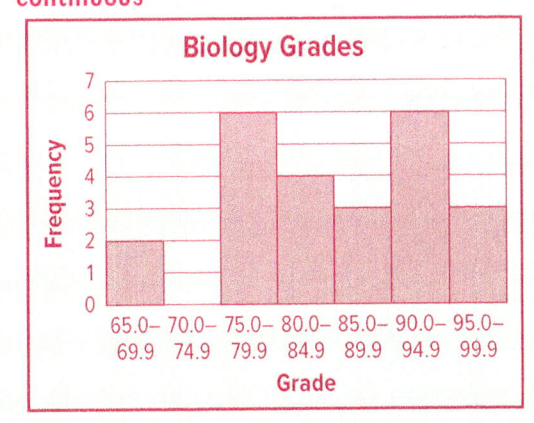

*Biology Grades* histogram

**12.**

*Birthday Frequency* histogram

---

### Standards for Mathematical Practice

| Emphasis On | Exercises |
|-------------|-----------|
| 1 Make sense of problems and persevere in solving them. | 16, 17, 25 |
| 2 Reason abstractly and quantitatively. | 5, 10–14, 24 |
| 3 Construct viable arguments and critique the reasoning of others. | 18, 19, 20, 22, 23 |
| 4 Model with mathematics. | 2–5, 7–17, 19–21, 23–25 |

connectED.mcgraw-hill.com  663

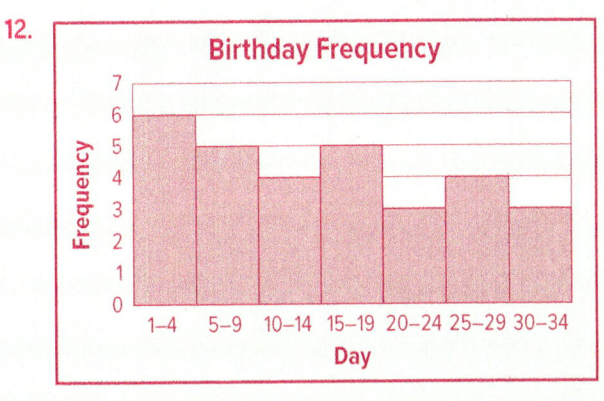

## Lesson 10-3 | Measures of Spread

### Example 2 Effect of Outliers

**AL** What are the scores that make up the data set? 88, 79, 94, 90, 45, 71, 82, 88

**OL** What is an outlier, and how do you find it? An outlier is a data point either extremely high or extremely low when compared with the other data in a data set. To find an outlier, determine the interquartile range and multiply by 1.5. Subtract this value from $Q_1$ to find the value that would be a lower outlier. Add this value to $Q_3$ to find the upper outlier. Any values beyond the lower and upper would also be outliers.

**BL** Why does an outlier seem to affect the mean more than the median? The mean is the average of all of the numbers, where the median is either the middle number in a set or the average of the two middle numbers.

### Need Another Example?

**Text Messages** The table shows the number of text messages Tumu received.

| Text Messages |    |    |    |    |    |
|---|---|---|---|---|---|
| 15 | 28 | 20 | 17 | 24 | 14 |
| 11 | 24 | 15 | 16 | 19 | 40 |
| 24 | 11 | 21 | 18 | 10 | 15 |

a. Identify any outliers in the data. 40
b. Find the mean and median of the data set with and without the outlier. Describe the effect.

| Data Set | Mean | Median |
|---|---|---|
| with outlier | 19 | 17.5 |
| without outlier | ≈17.8 | 17 |

Removal of the outlier causes the mean and median to decrease, but the mean is affected more.

---

1. 35.2 oz; 35.7 oz; 35.9 oz; 36.2 oz; 36.5 oz; The minimum number of ounces the manager measured was 35.2, the maximum was 36.5. He measured less than 35.7 ounces 25% of the time, less than 35.9 ounces 50% of the time, and less than 36.2 ounces 75% of the time.

**Study Tip**

**Interquartile Range** When the interquartile range is a small value, the data in the set are close together. A large interquartile range means that the data are spread out.

### Guided Practice

**1. SODA** The manager of a convenience store measured the amount of soda dispensed into 36-ounce cups of cola and gathered this data: 36.1, 35.8, 35.2, 36.5, 36.0, 36.2, 35.7, 35.8, 35.9, 36.4, 35.6. Find the minimum, lower quartile, median, upper quartile, and maximum. Then interpret this five-number summary.

The difference between the upper and lower quartiles is called the **interquartile range**. The interquartile range, or IQR, contains about 50% of the values.

14, 16, **17**, 17, 20, 21, 22, 26, **28**, 29, 35
       ↑                              ↑
       $Q_1$                          $Q_3$
       |← IQR = $Q_3$ − $Q_1$ or 11 →|

Before deciding on which measure of center best describes a data set, check for outliers. An **outlier** is an extremely high or extremely low value when compared with the rest of the values in the set. To check for outliers, look for data values that are beyond the upper or lower quartiles by more than 1.5 times the interquartile range.

S.ID.3

### Real-World Example 2    Effect of Outliers

**TEST SCORES** Students taking a test received the following scores: 88, 79, 94, 90, 45, 71, 82, and 88.

a. Identify any outliers in the data.

First determine the median and upper and lower quartiles of the data.

45,    71,    79,    82,    88,    88,    90,    94
              ↑             ↑             ↑

$Q_1 = \frac{71 + 79}{2}$ or 75    $Q_2 = \frac{82 + 88}{2}$ or 85    $Q_3 = \frac{88 + 90}{2}$ or 89

Find the interquartile range.

IQR = $Q_3 - Q_1 = 89 - 75$ or 14

Use the interquartile range to find the values beyond which any outliers would lie.

$Q_1 - 1.5(\text{IQR})$   and   $Q_3 + 1.5(\text{IQR})$   Values beyond which outliers lie
75 − 1.5(14)                   89 + 1.5(14)              $Q_1 = 75$, $Q_3 = 89$, and IQR = 14
54                              110                       Simplify.

There are no scores greater than 110, but there is one score less than 54. The score of 45 can be considered an outlier for this data set.

b. Find the mean and median of the data set with and without the outlier. Describe what happens.

| Data Set | Mean | Median |
|---|---|---|
| with outlier | $\frac{88 + 79 + 94 + 90 + 45 + 71 + 82 + 88}{8}$ or about 79.6 | 85 |
| without outlier | $\frac{88 + 79 + 94 + 90 + 71 + 82 + 88}{7}$ or about 84.6 | 88 |

Removal of the outlier causes the mean and median to increase, but notice that the mean is affected more by the removal of the outlier than the median.

---

### Differentiated Instruction  OL  BL

**IF** students demonstrate an understanding of outliers and how to find them using interquartile range,

**THEN** have students describe the meaning of an outlier in context. For example, when looking at data about temperatures during a given month, what would an outlier of a very cold or very warm temperature indicate? Encourage students to think of other examples and explain the meaning of a possible outlier.

Lesson 10-3 | Measures of Spread

▶ **Guided Practice**

2. **SOCIAL NETWORKS** A student surveyed friends to find the amount of time in minutes that they spend on social networking Websites each day. The results were: 25, 35, 45, 30, 65, 50, 25, 100, 45, 35, 5, 105, 110, 190, 40, 30, 80. Find the mean and median and identify any outliers. If there is an outlier, find the mean and median without the outlier. State which measure is affected more by the removal of the outlier.

2. ≈59.71 min, 45 min; 190 min; ≈51.6 min, 42.5 min; mean

**2 Statistical Analysis** In a set of data, the **standard deviation** shows how the data deviate from the mean. A low standard deviation indicates that the data tend to be very close to the mean, while a high standard deviation indicates that the data are spread out over a larger range of values.

The standard deviation is represented by the lowercase Greek letter sigma, $\sigma$.
The **variance** $\sigma^2$ of the data is the square of the standard deviation.

**Study Tip**
**Symbols** The mean of a sample and the mean of a population are calculated the same way. $\bar{x}$ refers to the mean of a sample and $\mu$ refers to the mean of a population. In this text, $\bar{x}$ will refer to both.

🔑 **Key Concept** Standard Deviation

**Step 1** Find the mean, $\bar{x}$.
**Step 2** Find the square of the difference between each data value $x_n$ and the mean, $(\bar{x} - x_n)^2$.
**Step 3** Find the sum of all of the values in Step 2.
**Step 4** Divide the sum by the number of values in the set of data $n$. This value is the variance.
**Step 5** Take the square root of the variance.

Formula $\sigma = \sqrt{\dfrac{(\bar{x} - x_1)^2 + (\bar{x} - x_2)^2 + \ldots + (\bar{x} - x_n)^2}{n}}$

S.ID.2

**Real-World Example 3** Variance and Standard Deviation

**ELECTRONICS** Ed surveys his classmates to find out how many electronic devices each person has in their home. Find and interpret the standard deviation of the data set.

{9, 10, 11, 6, 9, 11, 9, 8, 11, 8, 7, 9, 11, 11, 5}

**Step 1** Find the mean.
$$\bar{x} = \frac{9 + 10 + 11 + 6 + 9 + 11 + 9 + 8 + 11 + 8 + 7 + 9 + 11 + 11 + 5}{15} \text{ or } 9$$

**Step 2** Find the square of the differences, $(\bar{x} - x_n)^2$.

$(9 - 9)^2 = 0$   $(9 - 10)^2 = 1$   $(9 - 11)^2 = 4$   $(9 - 6)^2 = 9$   $(9 - 9)^2 = 0$
$(9 - 11)^2 = 4$   $(9 - 9)^2 = 0$   $(9 - 8)^2 = 1$   $(9 - 11)^2 = 4$   $(9 - 8)^2 = 1$
$(9 - 7)^2 = 4$   $(9 - 9)^2 = 0$   $(9 - 11)^2 = 4$   $(9 - 11)^2 = 4$   $(9 - 5)^2 = 16$

**Step 3** Find the sum.
$0 + 1 + 4 + 9 + 0 + 4 + 0 + 1 + 4 + 1 + 4 + 0 + 4 + 4 + 16 = 52$

**Step 4** Find the variance.
$\sigma^2 = \dfrac{(\bar{x} - x_1)^2 + (\bar{x} - x_2)^2 + \ldots + (\bar{x} - x_n)^2}{n}$   Formula for variance
$= \dfrac{52}{15}$ or about 3.47   The sum is 52 and $n = 15$.

**Step 5** Find the standard deviation.
$\sigma = \sqrt{\sigma^2}$   Square root of the variance
$\approx \sqrt{3.47}$ or about 1.86

**Go Online!**
Finding standard deviations and other statistics is made easier on a graphing calculator. Watch the **Personal Tutor** with a partner to see how it's done. **ELL**

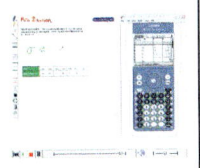

**2 Statistical Analysis**

**Example 3 Variance and Standard Deviation**

🔴 **AL** How many classmates responded? **15**

🔵 **OL** Explain how to find the square of the differences. **Subtract 9 from each data value, then square each difference.**

🟢 **BL** What does the standard deviation of 1.86 tell you about the data? **Sample answer: On average, most students have about the same number of electronic devices in their homes, 9. There is not a lot of deviation from that average.**

**Need Another Example?**
**Scores** Leo tracked his homework scores for the past week: {100, 0, 100, 50, 0}. Find and interpret the standard deviation of the data set. ≈ 44.7; a standard deviation very close to the mean suggests that the data deviate quite a bit. Most of Leo's scores are far away from the mean of 50.

connectED.mcgraw-hill.com  667

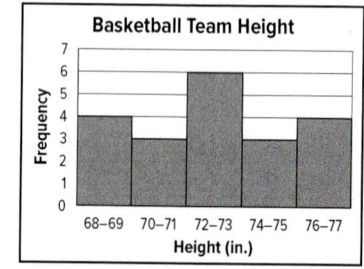

Lesson 10-2 | Representing Data

# Preparing for Assessment

Exercises 19–25 require students to use the skills they will need on standardized assessments. Each exercise is dual-coded with content and mathematical practice standards.

## Dual Coding

| Items | Content Standards | MP Mathematical Practices |
|---|---|---|
| 19 | S.ID.1, N.Q.1 | 1, 3, 4 |
| 20 | N.Q.1, S.ID.1 | 1, 3, 4 |
| 21 | N.Q.1, S.ID.1 | 1, 4 |
| 22 | S.ID.1 | 3 |
| 23 | N.Q.1, S.ID.1 | 3, 4 |
| 24 | N.Q.1, S.ID.1 | 2, 4 |
| 25 | S.ID.1 | 1, 4 |

## Diagnose Student Errors

Survey student responses for each item. Class trends may indicate common errors and misconceptions.

**19.**

| | |
|---|---|
| A | May have misinterpreted the smallest value in the cumulative frequency column as the least frequency |
| B | CORRECT |
| C | CORRECT |
| D | CORRECT |

**24.**

| | |
|---|---|
| A | Chose an option that has overlapping intervals |
| B | Chose an option with different sized intervals |
| C | CORRECT |
| D | Chose an option that does not include every data value in the intervals |

## Go Online!

**Quizzes**

Students can use *Self-Check Quizzes* to check their understanding of this lesson. You can also give the *Chapter Quiz*, which covers the content in Lessons 10-1 and 10-2.

664 | Lesson 10-2 | Representing Data

---

# Preparing for Assessment

**19.** •••••••••••••••••••••••••••••••••••••
•••••••••••••••••••••••••••••••
••••••• MP 1, 3, 4 N.Q.1, S.ID.1 **B, C, D**

| Time (h) | Frequency | Cumulative Frequency |
|---|---|---|
| 0–4 | 7 | 7 |
| 5–9 | 4 | 11 |
| 10–14 | 11 | 22 |
| 15–19 | 17 | 39 |
| 20–24 | 15 | 54 |
| 25–29 | 6 | 60 |

☐ A ••••••••••••••••••••••••
••••••••••••••
☐ B ••••••••••••••••••••••••
••••••••••
☐ C ••••••••••••••••••••••••
•••••••••
☐ D ••••••••••••••••••••••••
•••••••

**20.** •••••••••••••••••••••

| Number of Books | | | | |
|---|---|---|---|---|
| 4 | 4 | 3 | 8 | 0 |
| 1 | 9 | 8 | 4 | 3 |
| 5 | 6 | 7 | 8 | 0 |
| 1 | 2 | 3 | 9 | 10 |
| 3 | 3 | 4 | 7 | 6 |

a. ••••••••••••••
•••••••
b. ••••••••••••••••••••••
••••• a-b. See margin.

**21.** ••••••••••••••••••••••••••
••••••••••••••••••••
MP 1, 4 N.Q.1, S.ID.1 **10; 3; 4; 6; 2; 5, 1 and 7**

**22.** ••••••••••••••••••••••••
•••••••••••••••••
MP 3 S.ID.1 See margin.

**23.** •••••••••••••••••••••••••••
•••••••••••• MP 3, 4 N.Q.1, S.ID.1 See margin.

| Allowance | |
|---|---|
| How Spent | Amount ($) |
| savings | 15 |
| downloaded music | 8 |
| snacks | 5 |
| T-shirt | 12 |

**24.** ••••••••••••••••••••••••••••••••
•••••••••••••••••• MP 2, 4 N.Q.1, S.ID.1 **C**

| 42 | 40 | 40 | 35 | 50 |
|---|---|---|---|---|
| 32 | 50 | 36 | 50 | 40 |
| 45 | 70 | 43 | 45 | 32 |
| 40 | 35 | 61 | 48 | 35 |

Source: *The World Almanac*

○ A •••••••••
○ B •••••••••
○ C •••••••••
○ D •••••••••

**25.** **MULTI-STEP** •••••••••••••••••••••
•••••••••••••••• MP 1, 4 S.ID.1

Basketball Team Height histogram with Frequency on y-axis (0–7) and Height (in.) on x-axis: 68–69, 70–71, 72–73, 74–75, 76–77

a. ••••••••••••••
••••• **7**
b. •••••••••••••••••
**72–73 in.**
c. •••••••••••••••••••
**35%**

## Additional Answers

**20a.** Books Read histogram: Frequency (0–12) vs Books (0–2: 5, 3–5: 10, 6–8: 7, 9–11: 3)

**20b.** Sample answer: I chose a histogram to display the frequency of number of books read because the data is continuous.

**22.** Because each charity represents a category to which the organization donated, a bar graph would be a better representation.

**23.** The data is categorical, so a bar graph should be used to display the data.

# LESSON 10-3
## Measures of Spread

**SUGGESTED PACING (DAYS)**
- 90 min. 0.5
- 45 min. 1

Instruction

## Track Your Progress

### Objectives
1. Identify and interpret factors affecting variation.
2. Analyze data sets using statistics.

### Mathematical Background
Measures of variation show the spread of data. Range describes the spread of all values in the data. Quartiles and interquartile range describe the spread in the middle half of the data. Variance and standard deviation describe the spread around the mean. Two data sets can have the same range and mean, but the spread around the mean can be quite different.

### THEN
**S.ID.1** Represent data with plots on the real number line (dot plots, histograms, and box plots).

**S.ID.2** Use statistics appropriate to the shape of the data distribution to compare center (median, mean) and spread (interquartile range, standard deviation) of two or more different data sets.

### NOW
**S.ID.2** Use statistics appropriate to the shape of the data distribution to compare center (median, mean) and spread (interquartile range, standard deviation) of two or more different data sets.

### NEXT
**S.ID.3** Interpret differences in shape, center, and spread in the context of the data sets, accounting for possible effects of extreme data points (outliers).

**S.ID.5** Summarize categorical data for two categories in two-way frequency tables. Interpret relative frequencies in the context of the data. Recognize possible associations and trends in the data.

---

### Go Online! All of these resources and more are available at connectED.mcgraw-hill.com

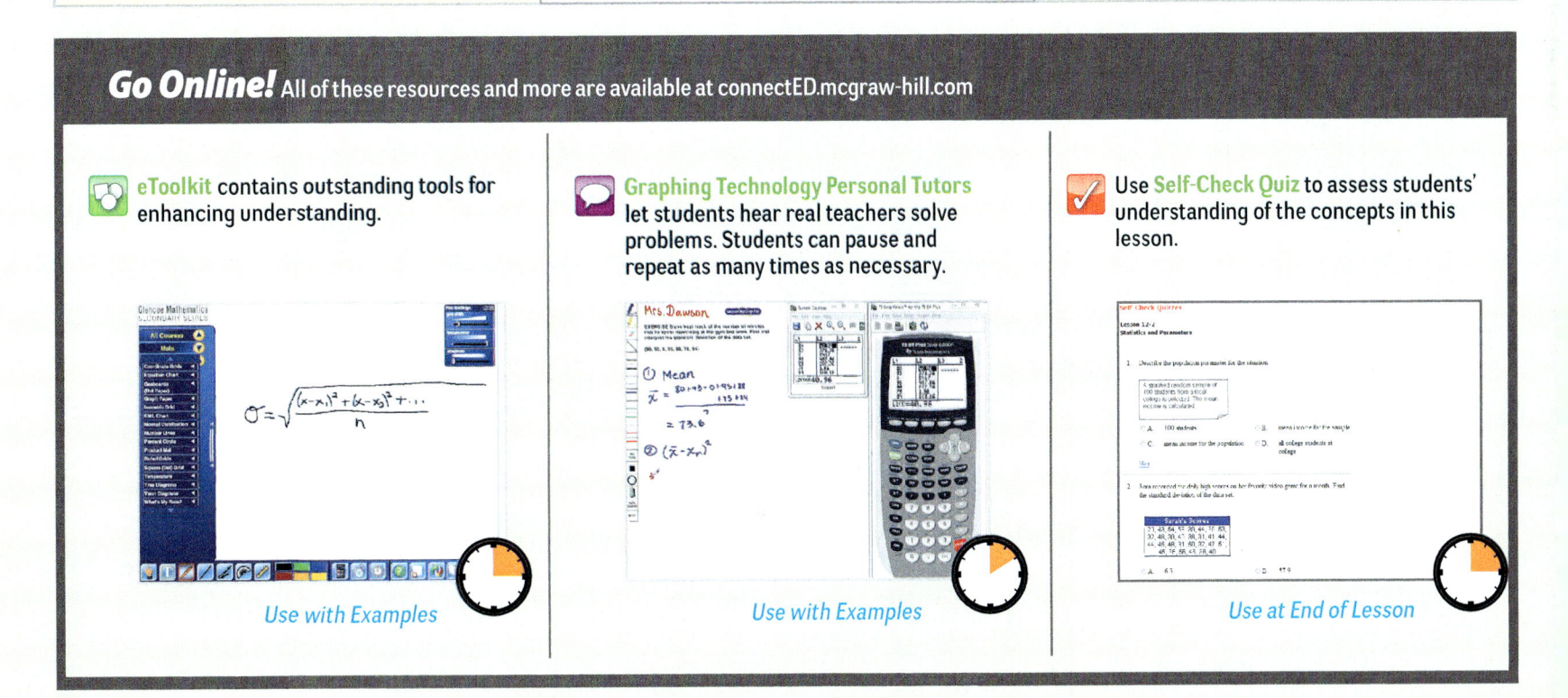

**eToolkit** contains outstanding tools for enhancing understanding.

*Use with Examples*

**Graphing Technology Personal Tutors** let students hear real teachers solve problems. Students can pause and repeat as many times as necessary.

*Use with Examples*

Use **Self-Check Quiz** to assess students' understanding of the concepts in this lesson.

*Use at End of Lesson*

---

### OER Using Open Educational Resources
**Lesson Sharing** Go to **Do the Math Stats** for ideas about discussion topics and ways for the students to critique statistics in the media. Ask students to bring in articles with statistics to critique in class. Have students work in pairs or small groups to analyze the validity of the article and its parameters. Ask students to turn in written critiques. *Use as group work*

# Go Online!
connectED.mcgraw-hill.com — Worksheets

# Differentiate Your Resources

**Extra Practice** Additional practice or homework; Skills Practice is best for approaching-level students and Practice is best for on-level and beyond-level students

## Skills Practice
AL  OL  ELL

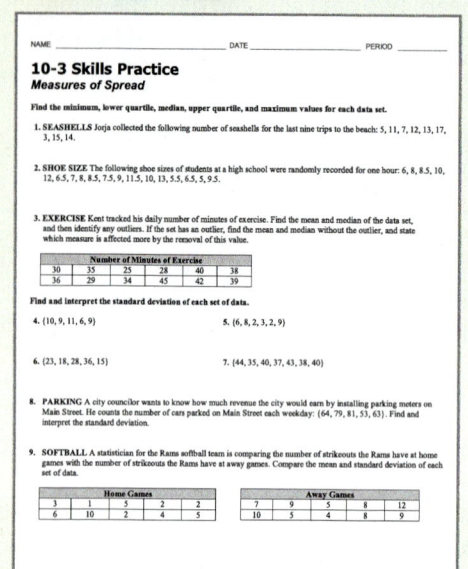

## Practice
AL  OL  BL  ELL

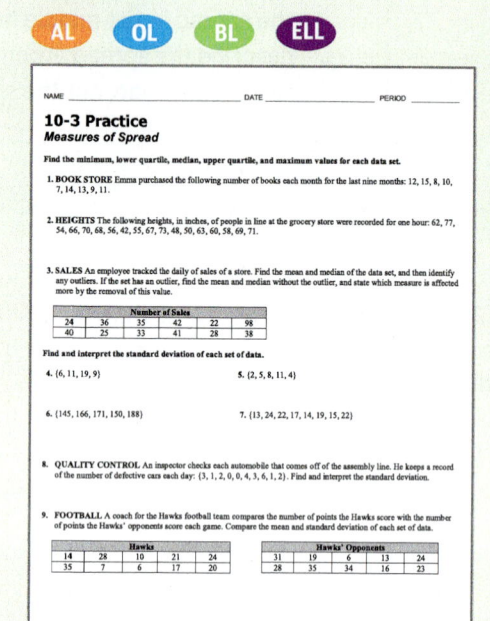

## Word Problem Practice
AL  OL  BL  ELL

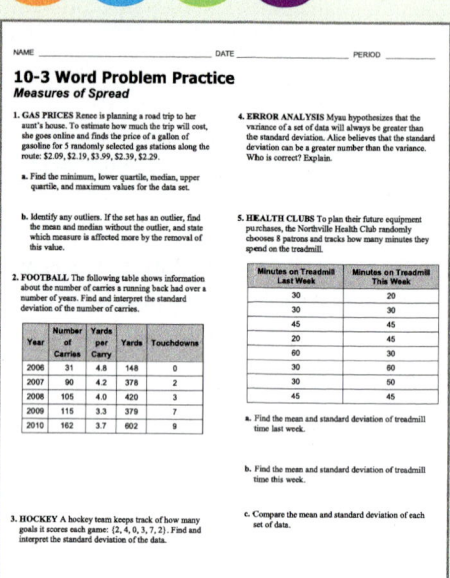

**Intervention** Reteaching and vocabulary activities that can be used with struggling or absent students and as ELL support

**Extension** Activities that can be used to extend lesson concepts

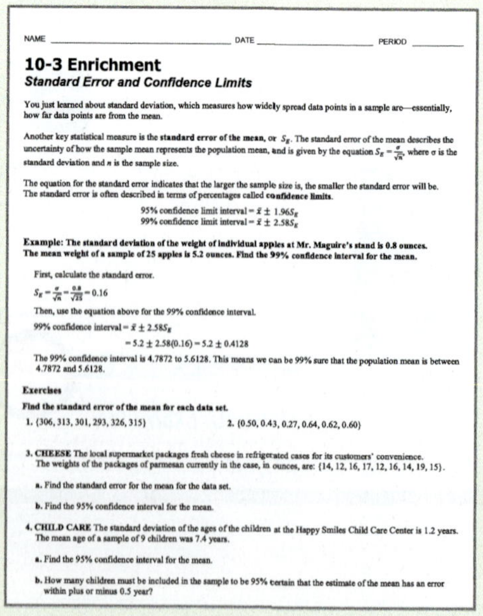

## Study Guide and Intervention
AL  OL  ELL

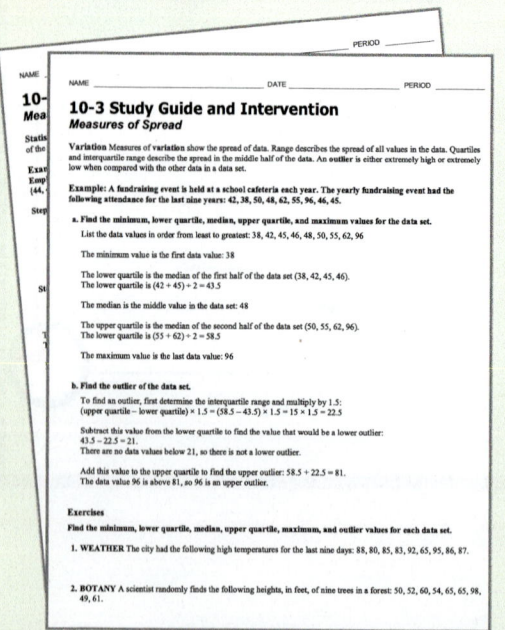

## Study Notebook
AL  OL  BL  ELL

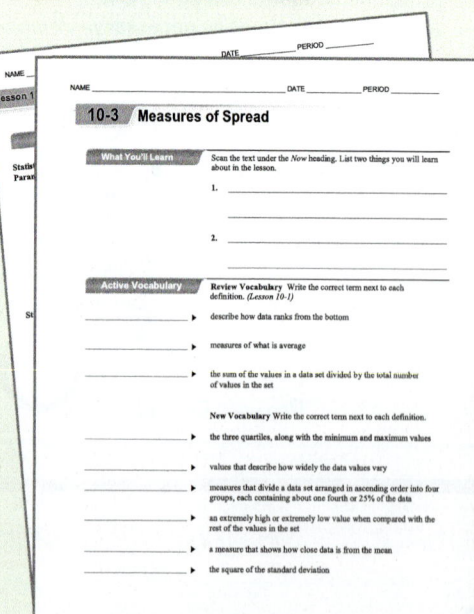

## Enrichment
OL  BL  ELL

665B | Lesson 10-3 | Measures of Spread

## LESSON 3
# Measures of Spread

**Then**
- You analyzed data collection techniques.

**Now**
1. Calculate measures of spread.
2. Analyze data sets using statistics.

**Why?**
At the start of every class period for one week, each of Mr. Day's algebra students randomly draws 9 pennies from a jar of 1000 pennies and note the dates on the coins. What measures of central tendency can each student find for the set of coins that they have?

How else can the data be compared?

### New Vocabulary
measures of spread or variation
range
quartiles
measure of position
lower quartile
upper quartile
five-number summary
interquartile range
outlier
standard deviation
variance

### Mathematical Practices
2 Reason abstractly and quantitatively.
6 Attend to precision.

### Content Standards
**S.ID.2** Use statistics appropriate to the shape of the data distribution to compare center (median, mean) and spread (interquartile range, standard deviation) of two or more different data sets.

## 1 Variation
Two very different data sets can have the same mean, so statisticians also use **measures of spread** or **variation** to describe how widely the data values vary. One such measure that you studied in earlier courses is the **range**, which is the difference between the greatest and least values in a set of data.

Values in a data set can be described based on the position of a value relative to other values in a set. **Quartiles** are common **measures of position** that divide a data set arranged in ascending order into four groups, each containing about one fourth or 25% of the data. The median marks the second quartile $Q_2$ and separates the data into upper and lower halves. The first or **lower quartile** $Q_1$ is the median of the lower half, while the third or **upper quartile** $Q_3$ is the median of the upper half.

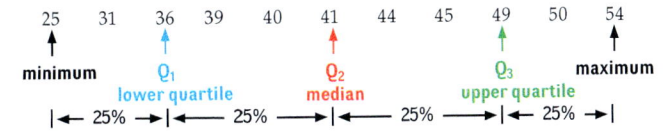

```
  25   31   36   39   40   41   44   45   49   50   54
   ↑         ↑              ↑              ↑         ↑
minimum      Q₁             Q₂             Q₃      maximum
         lower quartile   median      upper quartile
   |←── 25% ──→|←── 25% ──→|←── 25% ──→|←── 25% ──→|
```

The three quartiles, along with the minimum and maximum values, are called a **five-number summary** of a data set. Note that when calculating quartiles, if the number of values in a set of data are odd, the median is not included in either half of the data when calculating $Q_1$ or $Q_3$.

### Real-World Example 1  Five-Number Summary

**FUNDRAISER** The number of boxes of donuts Aang sold for a fundraiser each day for the last 11 days were 22, 16, 35, 26, 14, 17, 28, 29, 21, 17, and 20. Find the minimum, lower quartile, median, upper quartile, and maximum of the data set. Then interpret this five-number summary.

Order the data from least to greatest. Use the list to determine the quartiles.

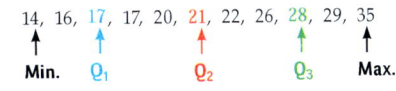

```
  14, 16, 17, 17, 20, 21, 22, 26, 28, 29, 35
   ↑       ↑           ↑           ↑       ↑
  Min.    Q₁          Q₂          Q₃      Max.
```

The minimum is 14, the lower quartile is 17, the median is 21, the upper quartile is 28, and the maximum is 35. Over the last 11 days, Aang sold a minimum of 14 boxes and a maximum of 35 boxes. He sold fewer than 17 boxes 25% of the time, fewer than 21 boxes 50% of the time, and fewer than 28 boxes 75% of the time.

---

## MP Mathematical Practices Strategies

### Attend to precision.
Help students to communicate their ideas and use clear definitions as they learn new terms presented in this lesson.

- In this problem, what are the two quartiles used to find the interquartile range? $Q_3$ and $Q_1$; students should state the values for those quartiles.

- For a set of data, how can you tell if an outlier is likely? How can you confirm this? If a number is greatly higher or lower than the other values in the set of data, then it may be an outlier. For a number that seems to be lower, it is an outlier if it is less than $Q_1 - 1.5$ times the interquartile range. For a number that seems to be higher, it is an outlier if it is greater than $Q_3 + 1.5$ times the interquartile range.

- In this problem, what does the standard of deviation tell you? If the standard of deviation is low, it tells you that the data tend to be close to the mean. If the standard deviation is high, then the data are more spread out.

---

**Lesson 10-3 | Measures of Spread**

## Launch
Have students read the Why? section of the lesson. Ask:

- Will the measures of central tendency for all sets of coins be the same? Explain. **Sample answer: It is possible that some of the measures are the same, if one or more students chose coins that have the same dates, but it is not likely they will all be the same.**

- How else can the data be compared? **Sample answer: The ranges of the data can be compared. The number of pennies in each year range can be graphed in a histogram.**

## Teach
Ask the scaffolded questions for each example to build conceptual understanding for students at all levels.

### 1 Variation

**Example 1  Five-Number Summary**

**AL** How many values are in the data set? **11**

**OL** What two five-number summary values can you find without putting the data in order? **minimum and maximum**

**BL** Suppose there were 12 values in the data set. How would you find $Q_2$? **Find the mean of values 6 and 7, when the items are in order.** How many items would be in each quarter of the data? **3**

### Need Another Example?
**FUNDRAISER** The senior class held a bake sale to raise money. They tracked the number of items sold each day. Find the minimum, lower quartile, median, upper quartile, and maximum of the data. Then interpret this five-number summary.
27, 25, 44, 13, 29, 44, 52, 28, 41
**13; 26; 29; 44; 52 ; The minimum number of items sold was 13 and the maximum was 52. They sold fewer than 26 items 25% of the time, fewer than 29 items 50% of the time, and fewer than 44 items 75% of the time.**

### Go Online!

**Interactive Whiteboard**
Use the *eLesson* or *Lesson Presentation* to present this lesson.

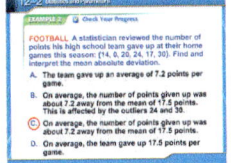

connectED.mcgraw-hill.com  **665**

## Lesson 10-3 | Measures of Spread

### Example 4 Compare Two Sets of Data

**AL** At which course does Miguel have higher scores, on average? **Memorial Park**

**OL** The two data sets have a similar mean, but the second set has a greater standard deviation. What conclusions can you make? **There is greater variability in the data values in the second set of data.**

**BL** Miguel plays at Redstone again and scores an 89. What happens to the mean and standard deviation of his scores at that course? **The mean and standard deviation both increase.**

### Need Another Example?

**Baseball** Kyle can throw a baseball left-handed or right-handed. Below are the speeds in miles per hour of 16 throws from each hand. Compare the means and standard deviations.

| Left-Handed |    |    |    |
|---|---|---|---|
| 68 | 71 | 70 | 69 |
| 67 | 67 | 73 | 71 |
| 74 | 68 | 68 | 71 |
| 72 | 70 | 66 | 70 |

| Right-Handed |    |    |    |
|---|---|---|---|
| 71 | 78 | 77 | 70 |
| 81 | 72 | 74 | 80 |
| 70 | 69 | 79 | 83 |
| 81 | 68 | 83 | 82 |

Sample answer: The left-handed throws had a mean of about 69.7 miles per hour with a standard deviation of about 2.2. The right-handed throws had a mean of about 76.1 miles per hour with a standard deviation of about 5.3. While the right-handed throws had a higher average speed, there was also greater variability in the speeds of the throws.

## Practice

**Formative Assessment** Use Exercises 1–4 to assess students' understanding of the concepts in this lesson.

The Practice and Problem Solving exercises assess the content taught in the lesson. The Preparing for Assessment page is meant to be used as preparation for end-of-course assessments.

### Extra Practice

See page R10 for extra exercises for students who are approaching level or for on-level students who need additional reinforcement.

---

A standard deviation of 1.86 is small compared to the mean of 9. This suggests that most of the data values are relatively close to the mean.

**Guided Practice**

3. 67.8; Sample answer: the standard deviation is very small compared to the mean, so most of the data values are very close to the mean of 2007 Calories.

3. **NUTRITION** Caleb tracked his Calorie intake for a week. Find and interpret the standard deviation of his Calorie intake.
   1950, 2000, 2100, 2000, 1900, 2100, 2000

The mean and standard deviation can be used to compare data.

**Real-World Example 4** Compare Two Sets of Data

**GOLF**

| Redstone |    |    |    |    |
|---|---|---|---|---|
| 81 | 78 | 79 | 82 | 80 |
| 80 | 79 | 83 | 81 | 80 |

| Memorial Park |    |    |    |    |
|---|---|---|---|---|
| 84 | 79 | 86 | 78 | 77 |
| 88 | 85 | 79 | 87 | 86 |

Use a graphing calculator to find the mean and standard deviation. Clear all lists. Then press STAT ENTER, and enter each data value into L1. To view the statistics, press STAT ▶ 1 ENTER.

**Study Tip**
Symbols The standard deviation of a sample $s$ and the standard deviation of a population $\sigma$ are calculated in different ways. In this text, you will calculate the standard deviation of a population.

Miguel's mean score at Redstone is 80.3 with a standard deviation of about 1.4. His mean score at Memorial Park is 82.9 with a standard deviation of about 4.0. Therefore, he tends to score lower at Redstone. The greater standard deviation at Memorial Park indicates that there is greater variability to his scores at that course.

**Guided Practice**

4. Sample answer: lineup A had a mean time of 4.354 with a standard deviation of 0.16 while lineup B had a mean time of 4.42 with a standard deviation of 0.16. Therefore, since the times for both lineups have the same variability and lineup A has a lower mean time, Anna should use lineup A.

4. **SWIMMING** Anna is considering two different lineups for her 4 × 100 relay team. Below are the times in minutes recorded for each lineup. Compare the means and standard deviations of each set of data.

| Lineup A |    |    |    |    |
|---|---|---|---|---|
| 4.25 | 4.31 | 4.19 | 4.40 | 4.23 |
| 4.18 | 4.71 | 4.56 | 4.32 | 4.39 |

| Lineup B |    |    |    |    |
|---|---|---|---|---|
| 4.47 | 4.68 | 4.25 | 4.41 | 4.49 |
| 4.18 | 4.27 | 4.69 | 4.32 | 4.44 |

### Check Your Understanding

= Step-by-Step Solutions begin on page R11.

**Example 1** 1. **SMARTPHONES** A student surveyed the prices of smartphones at local stores and collected this data: $199.99, $99.99, $249.99, $399.99, $439.99, $349.99. Find the five-number summary.

1–4. See margin.

### Differentiated Homework Options

| Levels | **AL** Basic | **OL** Core | **BL** Advanced |
|---|---|---|---|
| Exercises | 5–11, 15–16, 19–20, 21–27 | 5–11 (odd), 12–14, 15–16, 19–20, 21–27 | 15–20, 21–27 |
| 2-Day Option | 5–11 (odd), 21–27 | 5–11, 21–27 | |
|  | 6–10 (even), 15–16, 19–20 | 12–16, 19–20 | |

You can use ALEKS to provide additional remediation support with personalized instruction and practice.

**Lesson 10-3 | Measures of Spread**

**Example 2**
S.ID.2

2. **PRICING** A department store manager recorded these prices for eight similar pairs of pants with different brands: $69.99, $41.99, $29.99, $24.99, $29.99, $33.99, $46.99, and $36.99. Find the mean and median of the data set, and identify any outliers. If the set has an outlier, find the mean and median without the outlier, and state which measure is affected more by the removal of this value.

**Example 3**
S.ID.2

3. **PART-TIME JOBS** Ms. Johnson asks all of the members of the girls' tennis team to find the number of hours each week they work at part-time jobs: {10, 12, 0, 6, 9, 15, 12, 10, 11, 20}. Find and interpret the standard deviation of the data set.

**Example 4**
S.ID.2

4. **MODELING** Mr. Jones recorded the number of pull-ups done by his students. Compare the means and standard deviations of each group.
Boys: {5, 16, 3, 8, 4, 12, 2, 15, 0, 1, 9, 3}   Girls: {2, 4, 0, 3, 5, 4, 6, 1, 3, 8, 3, 4}

### Practice and Problem Solving

Extra Practice is on page R10.

**Example 1**
S.ID.2

Find the minimum, lower quartile, median, upper quartile, and maximum values for each data set. 5–6. See margin.

5. **ART MUSEUM** The city art museum's annual gala event had the following attendance for the last nine years: 68, 99, 73, 65, 67, 62, 80, 81, 83.

6. **AMUSEMENT RIDE** The following ages of riders on a roller coaster were recorded for the last hour of the day: 45, 17, 16, 22, 25, 19, 20, 21, 32, 37, 19, 21, 24, 20, 18, 22, 23, 19.

**Example 2**
S.ID.2

Find the mean and median of the data set, and then identify any outliers. If the set has an outlier, find the mean and median without the outlier, and state which measure is affected more by the removal of this value.

7. **MOVIES** A math teacher asked a group of his students to count the number of movies they owned. See margin.

**Number of Movies**
| 26 | 39 | 5 | 82 | 12 | 14 |
|----|----|---|----|----|----|
| 0  | 3  | 15| 19 | 41 | 6  |

8. **SWIMMING** The owner of a public swimming pool tracked the daily attendance. See margin.

**Daily Attendance**
| 86  | 45 | 91 | 104 | 95 | 88  |
|-----|----|----|-----|----|-----|
| 127 | 85 | 79 | 102 | 98 | 103 |

**Example 3**
S.ID.2

9. **MODELING** Samantha earns $8.50 per hour for babysitting. She takes a survey of her friends to see what they charge per hour. The results are {$8.00, $8.50, $9.00, $7.50, $15.00, $8.25, $8.75}. Find and interpret the standard deviation of the data. See margin.

10. **ARCHERY** Carla participates in competitive archery. Each competition allows a maximum of 90 points. Carla's results for the last 8 competitions are {76, 78, 81, 75, 80, 80, 76, 77}. Find and interpret the standard deviation of the data. See margin.

**Example 4**
S.ID.2

11. **BASKETBALL** The coach of the Wildcats basketball team is comparing the number of fouls called against his team with the number called against their rivals, the Trojans. He records the number of fouls called against each team for each game of the season. Compare the means and standard deviations of each set of data. See margin.

**Wildcats**
| 15 | 12 | 13 | 9  |
|----|----|----|----|
| 11 | 12 | 14 | 12 |
| 8  | 16 | 9  | 9  |
| 11 | 13 | 12 | 14 |

**Trojans**
| 9  | 10 | 14 | 13 |
|----|----|----|----|
| 7  | 8  | 10 | 10 |
| 9  | 7  | 11 | 9  |
| 12 | 11 | 13 | 8  |

B 12. **BATTING AVERAGES** The batting averages for the last 10 seasons for a baseball team are 0.267, 0.305, 0.304, 0.201, 0.284, 0.302, 0.311, 0.289, 0.300, and 0.292. 12a–b. See margin.
  a. From reading the data, do you notice an obvious outlier? Explain.
  b. Find the range and the five-number summary for the set of data and interpret the summary.

### Levels of Complexity Chart

The levels of the exercises progress from 1 to 3, with Level 1 indicating the lowest level of complexity.

| Exercises | 5–11 | 12–14, 21–27 | 15–20 |
|-----------|------|--------------|-------|
| Level 3   |      |              | ●     |
| Level 2   |      | ●            |       |
| Level 1   | ●    |              |       |

### Additional Answers

1. $99.99; $199.99; $299.99; $399.99; $439.99

2. ≈$39.37; $35.49; $69.99; $34.99; $33.99; mean

3. 4.98; Sample answer: The standard deviation is relatively high due to outliers 0 and 20.

4. Sample answer: The boys had a mean of 6.5 pull-ups with a standard deviation of about 5.2 pull-ups. The girls had a mean of about 3.6 pull-ups with a standard deviation of about 2.1 pull-ups. The boys had a higher average, but there was much more variability in the data set. The girls had a lower average number of pull-ups, but their data values were more consistent.

5. 62 people; 66 people; 73 people; 82 people; 99 people

6. 16 years old; 19 years old; 21 years old; 24 years old; 45 years old

7. ≈21.83; 14.5; 82; ≈16.36; 14; mean

8. ≈91.92; 45; ≈96.18; mean

9. 2.4; Sample answer: With a mean of about $9.29, the standard deviation of about $2.38 suggests that there is a good amount of deviation to the data. This deviation is mostly caused by the outlier of $15.00. If this outlier were removed, the new mean of the data would be about $8.33 with a standard deviation of about $0.49.

10. 2.1; Sample answer: With a mean of 77.875, the standard deviation of about 2.1 suggests that there is a very little deviation to the data. Therefore, you can conclude that Carla's archery scores are pretty consistent.

11. The Wildcats had a mean of about 11.9 fouls per game with a standard deviation of about 2.2 fouls. The Trojans had a mean of about 10.1 fouls per game with a standard deviation of about 2.1 fouls. The data sets had nearly the same variability while the Wildcats had a mean of almost 2 fouls more than the Trojans. The Wildcats coach can conclude that his team consistently has about 2 more fouls than the Trojans.

12a. Sample answer: Yes, 0.201 is much lower than the other values. It may be an outlier.

12b. range: 0.11; min: 0.201; $Q_1$: 0.284, median: 0.296, $Q_3$: 0.304, max: 0.311. Over the last 10 years, the minimum batting average was 0.201 and the maximum batting average was 0.311. 25% of the time the batting averages were below 0.284; 50% of the time they were below 0.296; 75% of the time they were below 0.304.

### Go Online!   eBook

**Interactive Student Guide**
Use the *Interactive Student Guide* to deepen conceptual understanding.
· Statistics and Parameters

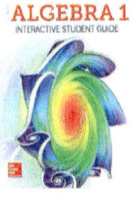

## Lesson 10-3 | Measures of Spread

**MP Teaching the Mathematical Practices**

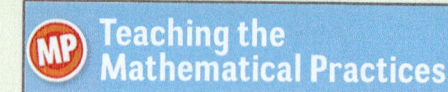

**Construct Arguments** Mathematically proficient students are able to compare the effectiveness of two plausible arguments. In Exercise 15, have students make up a data set and test Jennifer's and Megan's statements.

## Assess

**Yesterday's News** Have students write a sentence on how yesterday's lesson on summarizing and analyzing survey results helped with today's lesson on sample statistics and population parameters.

### Additional Answers

**13a.** Sample answer: Movie A had a mean of about 7.2 with a standard deviation of about 0.81. Movie B had a mean of about 6.8 with a standard deviation of about 2.86. While both movies had a mean rating of close to 7, the ratings for Movie B had a wider range than those for Movie A.

**13b.** Movie A: Sample answer: All the reviews are between 6 and 8. The ratings were consistent amongst the group. Movie B: Sample answer: The ratings range from 2 to 10 with half of the students rating the movie between 8 and 10.

**14b.** Sample answer: first 15 race finishers; population: all racers.

**14c.** quantitative; No, since the sample is the top 15 runners in the race, it is not random. So, it would not be accurate to apply the mean and standard deviation of the running times to the population.

**15.** Both; When an outlier is removed, the spread and standard deviation will decrease. When more values that are equal to the mean of a data set are added, the outliers will have less influence.

**16.** Sometimes; if the samples are truly random, they would rarely contain identical elements and the mean and standard deviation would differ. If the sample produces identical elements, the mean and standard deviation would be the same.

**17.** Sample answer: Poll voters to determine if a particular presidential candidate is favored to win the election. Use a stratified random sample to call 100 people throughout the country.

### Go Online!

**eSolutions Manual**
Create worksheets, answer keys, and solutions handouts for your assignments.

---

c. Identify any outlier(s) and find the mean and median without the outlier(s). State which measure was more affected by the outlier. **0.2949; 0.300; mean**

d. What is the lowest batting average that would result in no outlier? **0.255**

**13. MOVIE RATINGS** Two movies were rated by the same group of students. Ratings were from 1 to 10, with 10 being the best. **13a–b. See margin.**

a. Compare the means and standard deviations of each set of data.

b. Provide an argument for why movie A would be preferred, then an argument for movie B.

| Movie A |   |   |   |
|---|---|---|---|
| 7 | 8 | 7 | 6 |
| 8 | 6 | 7 | 8 |
| 6 | 8 | 8 | 6 |
| 7 | 7 | 8 | 8 |

| Movie B |   |   |   |
|---|---|---|---|
| 9 | 5 | 10 | 6 |
| 3 | 10 | 9 | 4 |
| 8 | 3 | 9 | 9 |
| 2 | 8 | 10 | 3 |

**14. RUNNING** The results of a 5-kilometer race are published in a local paper. Over a hundred people participated, but only the times of the top 15 finishers are listed.

**15th Annual 5K Road Race**

| Place | Time (min:s) | Place | Time (min:s) | Place | Time (min:s) |
|---|---|---|---|---|---|
| 1 | 17:51 | 6 | 19:03 | 11 | 19:50 |
| 2 | 18:01 | 7 | 19:06 | 12 | 20:07 |
| 3 | 18:17 | 8 | 19:27 | 13 | 20:11 |
| 4 | 18:22 | 9 | 19:49 | 14 | 20:13 |
| 5 | 18:26 | 10 | 19:49 | 15 | 20:13 |

a. Find the mean and standard deviation of the top 15 running times. (*Hint*: Convert each time to seconds.) ≈**19.25 min, ≈50.05 seconds**

b. Identify the sample and population. **See margin.**

c. Analyze the sample. Classify the data as *quantitative* or *qualitative*. Can a statistical analysis of the sample be applied to the population? Explain your reasoning. **See margin.**

S.ID.2

**H.O.T. Problems** Use Higher-Order Thinking Skills

**15. ERROR ANALYSIS** Jennifer and Megan are determining one way to decrease the size of the standard deviation of a set of data. Is either of them correct? Explain. **See margin.**

| Jennifer | Megan |
|---|---|
| Remove the outliers from the data set. | Add data values to the data set that are equal to the mean. |

**16. CONSTRUCT ARGUMENTS** Determine whether the statement *Two random samples taken from the same population will have the same mean and standard deviation* is *sometimes*, *always*, or *never* true. Explain. **See margin.**

**17. CHALLENGE** Describe a situation in which identifying an outlier and revising the mean to take the outlier into account results in a better descriptor for the data. **See margin.**

**18. CHALLENGE** Write a set of data with a standard deviation that is equal to the variance. **See margin.**

**WRITING IN MATH** Compare and contrast each of the following. **19.–20. See margin.**

19. range and interquartile range

20. mean and standard deviation

### Standards for Mathematical Practice

| Emphasis On | Exercises |
|---|---|
| 1 Make sense of problems and persevere in solving them. | 23 |
| 2 Construct viable arguments and critique the reasoning of others. | 3–4, 9–12, 14–20 |
| 4 Model with mathematics. | 2, 7–8, 13, 22, 23, 25 |
| 6 Attend to precision. | 1, 5–6, 21, 24, 26, 27 |

**18.** Sample answer: Any set of data with identical terms. For example: 5, 5, 5, 5, 5, 5, 5.

**19.** Both values are helpful in identifying the measure of spread. The range of a data set is the difference between the highest and lowest values in a set of data. The interquartile range is the difference between $Q_3$, the upper quartile, and $Q_1$, the lower quartile. This value is helpful in identifying outliers.

**20.** Mean is a measure of center and standard of deviation indicates variance from the mean. Mean is the average of the data. The standard deviation shows how the data deviate from the mean. A low standard of deviation indicates that the data is very close to the mean. A high standard of deviation tells you that the data are spread out over a larger range of values.

## Lesson 10-3 | Measures of Spread

### Preparing for Assessment

**21.** Identify the outlier in the following data set. **MP** 2, 6  S.ID.2  **D**

{11, 13, 7, 9, 14, 18, 10, 13, 9}

- ○ A  7
- ○ B  9
- ○ C  18
- ○ D  none

**22.** The worldwide grosses for the top 10 highest grossing films are shown below. **MP** 1, 4, 6  S.ID.2

| Rank | Gross (million $) | Rank | Gross (million $) |
|---|---|---|---|
| 1 | 2788 | 6 | 1516 |
| 2 | 2187 | 7 | 1405 |
| 3 | 2060 | 8 | 1342 |
| 4 | 1670 | 9 | 1277 |
| 5 | 1520 | 10 | 1215 |

a. Find the minimum, lower quartile, median, upper quartile, and maximum of the data set. **1215; 1342; 1518; 2060; 2788**

b. Interpret the five-number summary. **See margin.**

**23.** Which is the standard deviation of the following data set? **MP** 1  S.ID.2  **A**

{0, 8, 14, 1, 12, 6, 6, 10}

- ○ A  4.6
- ○ B  7
- ○ C  7.1
- ○ D  21.4

**24.** Brandon records the temperature in Celsius at various times during a chemistry lab. His data is shown below. Find the mean and standard deviation. **MP** 1, 4  S.ID.2  **mean −17.0; standard deviation 0.94**

{−16.0, −18.3, −15.8, −16.6, −18.2, −17.4, −16.5}

**25.** Find the upper and lower quartiles for the following data. **MP** 1  S.ID.2  **C**

{37, 62, 10, 13, 54, 44, 47, 28, 30}

- ○ A  $Q_1 = 13$; $Q_3 = 54$
- ○ B  $Q_1 = 20.5$; $Q_3 = 50.5$
- ○ C  $Q_1 = 28$; $Q_3 = 47$
- ○ D  $Q_1 = 36$; $Q_3 = 37.5$

**26.** The members of a science club in New Orleans recorded the monthly rainfall in inches for one year. Their findings were:

5.87, 5.47, 5.24, 5.02, 4.62, 6.83, 6.2, 6.15, 5.55, 3.05, 5.09, 5.07

Select all the statements that are true. **MP** 6  S.ID.2, S.ID.3  **A, B, D**

- ☐ A  The data has an outlier, 3.05 inches.
- ☐ B  A standard deviation of about 0.9114 is small compared to the mean of 5.3467. This suggests that most of the data values are relatively close to the mean.
- ☐ C  The interquartile range for this data is 0.961 inches.
- ☐ D  The median for the data is 5.355 inches.
- ☐ E  The measure of central tendency affected most by the outlier is the median.

**27. MULTI-STEP** The weekly pay for 10 employees at a small sandwich shop is $54, $278, $70, $159, $482, $49, $205, $70, $386, and $63. **MP** 6  S.ID.2, S.ID.3

a. Find the range and the interquartile range of the data. **$433; $215**

b. Find the mean, median, and mode of the data. **$181.60, $114.50, $70**

c. Identify any outliers. **There is no outlier.**

d. What is the standard deviation, and what does this indicate about the data?
**27d. $142.34; a standard deviation of $142.34 is large compared to the mean of $181.60. This suggests that most of the data values are not relatively close to the mean.**

---

### Differentiated Instruction  OL  BL

**Extension** Have students compute the range and the standard deviation for a few different data sets. Ask them to compare the values and see if they can make any generalizations about the relationship between range and standard deviation. For large data sets, the range is usually from 4 to 6 times as large as the standard deviation. This provides a rule of thumb that is sometimes used to find a quick estimate of standard deviation without going through the complete calculation.

### Additional Answer

**22b.** For the top ten highest grossing films, the maximum is $2788 million and the minimum is $1215 million. About 25% of the films grossed less than $1342 million, 50% grossed less than $1518 million and 75% grossed less than $2060 million.

---

## Preparing for Assessment

Exercises 21–27 require students to use the skills they will need on standardized assessments. Each exercise is dual-coded with content standards and mathematical practice standards

### Dual Coding

| Exercises | Content Standards | Mathematical Practices |
|---|---|---|
| 21 | S.ID.2 | 2, 6 |
| 22 | S.ID.2 | 1, 4, 6 |
| 23 | S.ID.2 | 1 |
| 24 | S.ID.2 | 1, 4 |
| 25 | S.ID.2 | 1 |
| 26 | S.ID.2, S.ID.3 | 6 |
| 27 | S.ID.2, S.ID.3 | 6 |

### Diagnose Student Errors

Survey student responses for each item. Class trends may indicate common errors and misconceptions.

**21.**

| A | Found the minimum |
|---|---|
| B | Found the mode |
| C | Found the maximum |
| D | CORRECT |

**23.**

| A | CORRECT |
|---|---|
| B | Found the median |
| C | Found the mean |
| D | Found the variance |

**25.**

| A | Did not find the mean of the quartiles correctly |
|---|---|
| B | Did not find the mean of the quartiles correctly |
| C | CORRECT |
| D | Did not arrange data in ascending order |

### Go Online!

**Quizzes**
Students can use Self-Check Quizzes to check their understanding of this lesson.

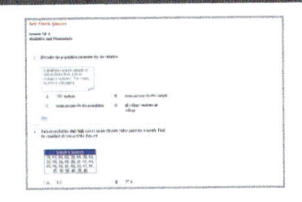

# Chapter 10 Mid-Chapter Quiz

**RtI Response to Intervention**
Use the Intervention Planner to help you determine your Response to Intervention.

## Intervention Planner

### TIER 1 — On Level (OL)

**IF** students miss 25% of the exercises or less,

**THEN** choose a resource:
- SE Lessons 10-1, 10-2, and 10-3

**Go Online!**
- Skills Practice
- Chapter Project
- Self-Check Quizzes

### TIER 2 — Strategic Intervention (AL)
Approaching grade level

**IF** students miss 50% of the exercises,

**THEN** choose a resource:
- Quick Review Math Handbook

**Go Online!**
- Study Guide and Intervention
- Extra Examples
- Personal Tutors
- Homework Help

### TIER 3 — Intensive Intervention
2 or more grades below level

**IF** students miss 75% of the exercises,

**THEN** choose a resource:
- Use *Math Triumphs, Alg. 1*

**Go Online!**
- Extra Examples
- Personal Tutors
- Homework Help
- Review Vocabulary

## Go Online!

**eAssessment**
You can use the premade Mid-Chapter Test to assess students' progress in the first half of the chapter. Customize and create multiple versions of your Mid-Chapter Quiz and answer keys that align to your standards. Tests can be delivered on paper or online.

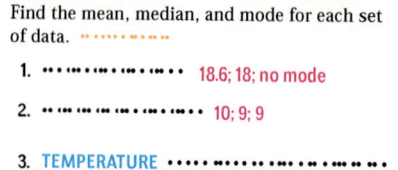

---

# CHAPTER 10
## Mid-Chapter Quiz
### Lessons 10-1 through 10-3

Find the mean, median, and mode for each set of data.

1. •••••••••• ••• •••• 18.6; 18; no mode
2. •••• ••• ••• 10; 9; 9
3. **TEMPERATURE** •••••••••••••••••••••••
•••••••••••••••••••••••••
•••••••••••••••••••••••••
••••••••• See Ch. 10 Answer Appendix.

| Day | Sunnydale Temperature (°F) | Sun Valley Temperature (°F) |
|---|---|---|
| Sunday | 95 | 80 |
| Monday | 88 | 86 |
| Tuesday | 86 | 91 |
| Wednesday | 90 | 102 |
| Thursday | 90 | 103 |
| Friday | 93 | 91 |
| Saturday | 89 | 85 |

4. **COMPETITION** •••••••••••••••••••
•••••••••••••••••   •••••••••

| Name | Score | Name | Score |
|---|---|---|---|
| Adam | 87 | Emilio | 79 |
| Bella | 91 | Fran | 82 |
| Camila | 90 | Greg | 94 |
| Devonte | 88 | Holly | 81 |

a. •••••••••••••••• Devonte
b. •••••••••••••••
•••••••• Holly and Emilio
c. •••••••••••••••• 75th percentile

5. **CARNIVAL** ••••••••
••••••••••••••
••••••••••••
a. ••••••••••••

| Age | Frequency |
|---|---|
| 0–19 | 66 |
| 20–39 | 49 |
| 40–59 | 54 |
| 60–79 | 16 |

b. •••••••••••••••••••••••
••••••••••••••••••
a–b. See Ch. 10 Answer Appendix.

6. **MUSIC** •••••••••••••••••••••
•••••••••••••••••••••
••••••••••• See Ch. 10 Answer Appendix.

| Favorite Instrument | |
|---|---|
| **Instrument** | **Number of Students** |
| drums | 8 |
| guitar | 12 |
| piano | 5 |
| trumpet | 7 |

Find the range, median, lower quartile, and upper quartile for each set of data.

7. ••••••••••••••••••• 19; 24; 19; 31
8. •••••••••••••••••••• 9; 75.5; 71; 77

9. **PLAY AREA** ••••••••••••••••
•••••••••••••••••••••••••
••••••••••••••••••
a. •••••••••••••••••••••••
•••••• See Ch. 10 Answer Appendix.
b. (MP) •••••••••••••••••••
•••••••• See students' work.

10. **GIFTS** ••••••••••••••••••••
••••••••••••••••••
•••••• A

A ••••• C •••••
B ••••• D •••••

11. •••••••••••••••••••••
••••••••••••••••••
•••••••••••••• See Ch. 10 Answer Appendix.

---

## Foldables Study Organizer

**Dinah Zike's FOLDABLES**

Before students complete the Mid-Chapter Quiz, encourage them to review the information for Lessons 10-1 through 10-3 in their Foldables. Encourage students to review the vocabulary terms in each lesson with a partner. They should seek clarification of any concepts, as needed.

**ALEKS** can be used as a formative assessment tool to target learning gaps for those who are struggling, while providing enhanced learning for those who have mastered the concepts.

# LESSON 10-4
# Distributions of Data

**SUGGESTED PACING (DAYS)**
- 90 min. — 0.5
- 45 min. — 1

Instruction

## Track Your Progress

### Objectives
1. Describe the shape of a distribution.
2. Use the shapes of distributions to select appropriate statistics.

### Mathematical Background
A distribution of data shows the frequency of each possible data value. The shape of a distribution can be determined by looking at its histogram or box-and-whisker plot. When describing a distribution, use the mean and standard deviation if the graph is symmetric and the five-number summary if the distribution is skewed.

### THEN
**S.ID.1** Represent data with plots on the real number line (dot plots, histograms, and box plots).

**S.ID.2** Use statistics appropriate to the shape of the data distribution to compare center (median, mean) and spread (interquartile range, standard deviation) of two or more different data sets.

### NOW
**S.ID.1** Represent data with plots on the real number line (dot plots, histograms, and box plots).

**S.ID.3** Interpret differences in shape, center, and spread in the context of the data sets, accounting for possible effects of extreme data points (outliers).

### NEXT
**S.ID.4** Use the mean and standard deviation of a data set to fit it to a normal distribution and to estimate population percentages. Recognize that there are data sets for which such a procedure is not appropriate. Use calculators, spreadsheets, and tables to estimate areas under the normal curve.

---

**Go Online!** All of these resources and more are available at connectED.mcgraw-hill.com

**eToolkit** contains outstanding tools for enhancing understanding.
*Use with Examples*

**Graphing Technology Personal Tutors** let students hear real teachers solve problems. Students can pause and repeat as many times as necessary.
*Use with Examples*

Use **Self-Check Quiz** to assess students' understanding of the concepts in this lesson.
*Use at End of Lesson*

---

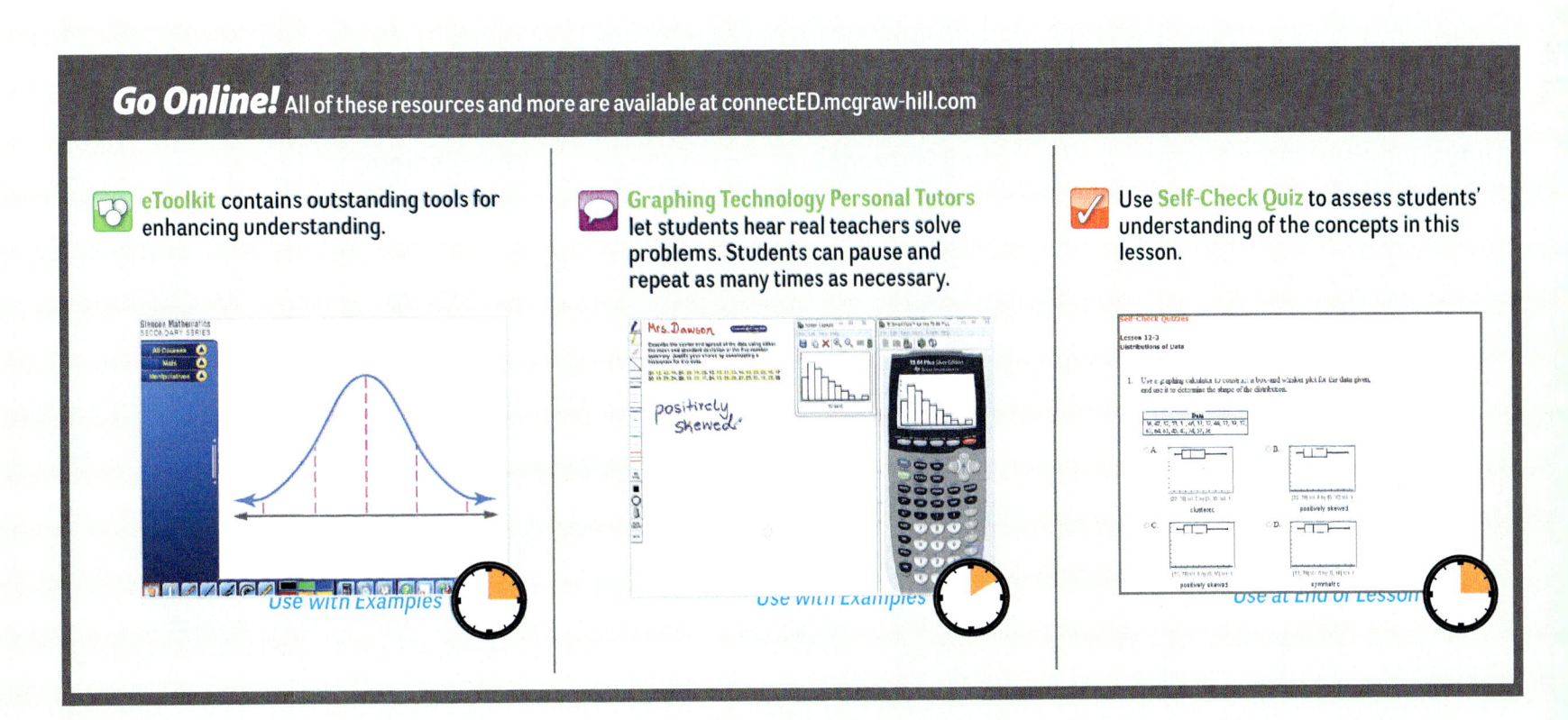

**OER Using Open Educational Resources**
**Surveys** Have students create a poll on **Polldaddy** to collect data. Then students can describe the distribution created by that data and use the shape of the distribution to select appropriate statistics. *Use as homework*

connectED.mcgraw-hill.com  673A

# Go Online!
connectED.mcgraw-hill.com — Worksheets

# Differentiate Your Resources

**Extra Practice** Additional practice or homework; Skills Practice is best for approaching-level students and Practice is best for on-level and beyond-level students

## Skills Practice
AL  OL  ELL

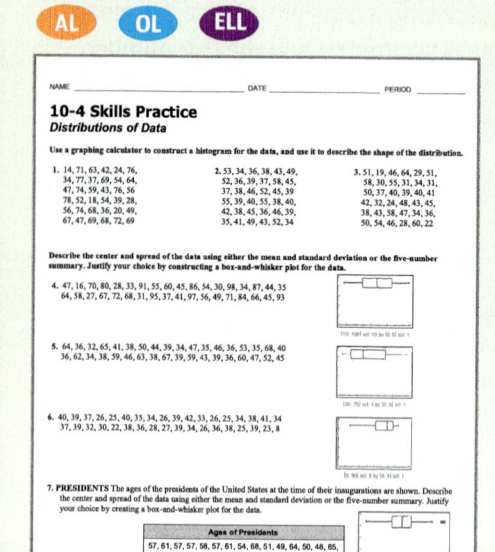

## Practice
AL  OL  BL  ELL

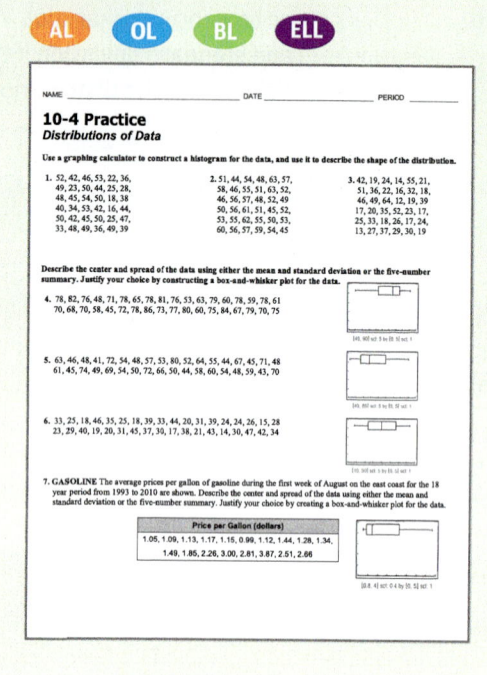

## Word Problem Practice
AL  OL  BL  ELL

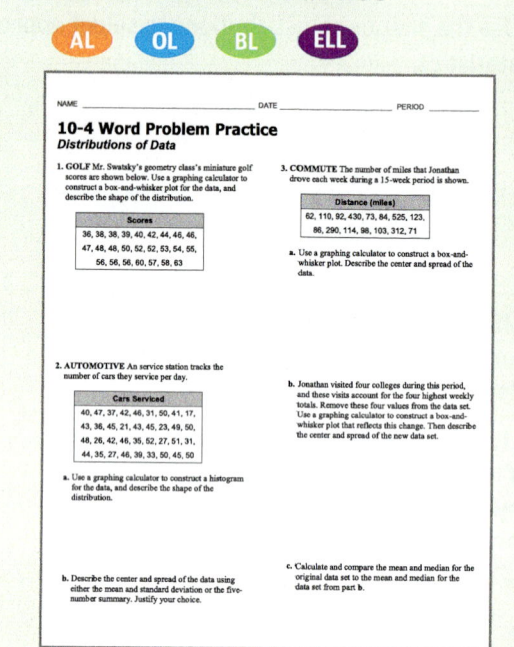

**Intervention** Reteaching and vocabulary activities that can be used with struggling or absent students and as ELL support

## Study Guide and Intervention
AL  OL  ELL

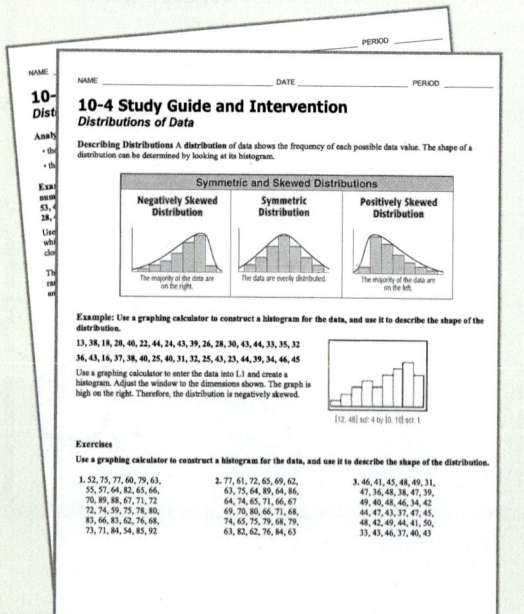

## Study Notebook
AL  OL  BL  ELL

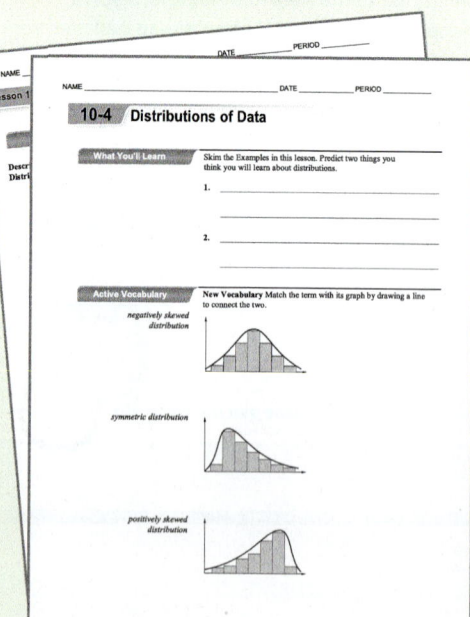

**Extension** Activities that can be used to extend lesson concepts

## Enrichment
OL  BL  ELL

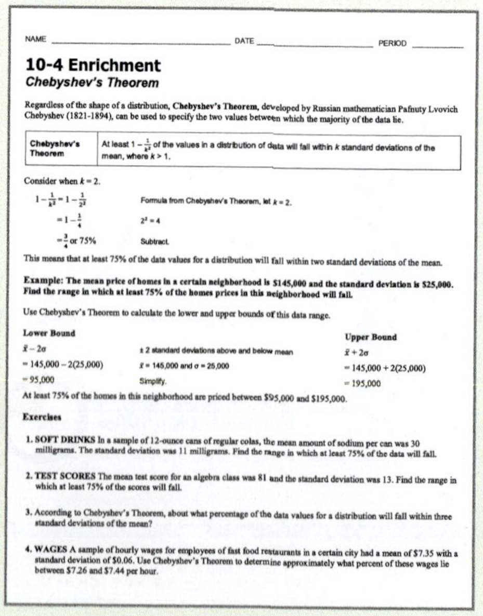

673B | Lesson 10-4 | Distributions of Data

# LESSON 4
## Distributions of Data

**Then**
- You calculated measures of central tendency and variation.

**Now**
1. Describe the shape of a distribution.
2. Use the shapes of distributions to select appropriate statistics.

**Why?**
Over many years, the number of bluebirds in the United States declined dramatically. In order to increase the bluebird population, volunteers, in some states, supply bluebird nestboxes and monitor bluebird trails. The population of bluebirds in each state can be displayed in a distribution of data.

**New Vocabulary**
distribution
negatively skewed distribution
symmetric distribution
positively skewed distribution

**Mathematical Practices**
5 Use appropriate tools strategically.

**Content Standards**
S.ID.1 Represent data with plots on the real number line (dot plots, histograms, and box plots).
S.ID.3 Interpret differences in shape, center, and spread in the context of the data sets, accounting for possible effects of extreme data points (outliers).

### 1 Describing Distributions
A **distribution** of data shows the observed or theoretical frequency of each possible data value. Recall that a histogram is a type of bar graph used to display data that have been organized into equal intervals. A histogram is useful when viewing the overall distribution of the data within a set over its range. You can see the shape of the distribution by drawing a curve over the histogram.

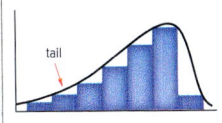

**Key Concept** Symmetric and Skewed Distributions

| Negatively Skewed Distribution | Symmetric Distribution | Positively Skewed Distribution |
|---|---|---|
| The majority of the data are on the right. | The data are evenly distributed. | The majority of the data are on the left. |

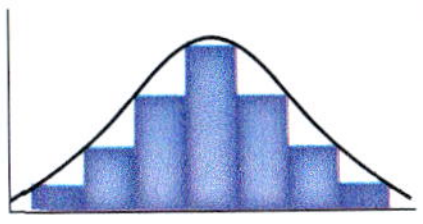

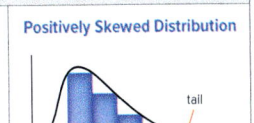

**Example 1** Distribution Using a Histogram

Use a graphing calculator to construct a histogram for the data, and use it to describe the shape of the distribution.

25, 22, 31, 25, 26, 35, 18, 39, 22, 32, 34, 26, 42, 23, 40, 36, 18, 30
26, 30, 37, 23, 19, 33, 24, 29, 39, 21, 43, 25, 34, 24, 26, 30, 21, 22

First, press STAT ENTER and enter each data value. Then, press 2nd [STAT PLOT] ENTER ENTER and choose histogram. Press ZOOM [ZoomStat] to adjust the window.

The graph is high on the left and has a tail on the right. Therefore, the distribution is positively skewed.

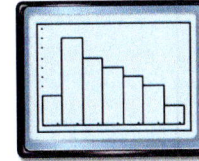

[17, 45] scl: 4 by [0, 10] scl: 1

**Guided Practice**

1. Use a graphing calculator to construct a histogram for the data, and use it to describe the shape of the distribution. **See Ch. 10 Answer Appendix.**

8, 11, 15, 25, 21, 26, 20, 12, 32, 20, 31, 14, 19, 27, 22, 21, 14, 8
6, 23, 18, 16, 28, 25, 16, 20, 29, 24, 17, 35, 20, 27, 10, 16, 22, 12

---

**Mathematical Practices Strategies**

**Use appropriate tools strategically.**
Help students use graphing calculators to graph data sets, to describe the shapes of the graphs, and to select appropriate statistics to describe the data sets.

- How does choosing the appropriate window setting and bin width affect the shape of the data? **A window that is too small may not display all of the data, while a window that is too large may inaccurately skew the data. Bins that are too large may group too many data points together and bins that are too small may not cluster data in a meaningful way.**

- How can you choose appropriate statistics to represent a data set? **If the distribution is relatively symmetric, use the mean and standard deviation. If the distribution is skewed or has outliers, use the five-number summary.**

---

## Lesson 10-4 | Distributions of Data

### Launch
Have students read the Why? section of the lesson. Ask:

- In what ways can the population of bluebirds in each state be displayed? **Sample answer: histogram, frequency table, stem and leaf plot, or dot plot**

- Which of these displays would best represent the data? why? **Sample answer: a histogram will show a quick view of the overall distribution of population throughout the country.**

- If you graphed the bluebird population over time, what would you hope to see? **The number of bluebirds increasing over time.**

### Teach
Ask the scaffolded questions for each example to build conceptual understanding for students at all levels.

### 1 Describing Distributions

**Example 1 Distribution Using a Histogram**

**AL** Describe how the graph looks. **Sample answer: high on the left and goes lower and lower as it moves to the right**

**OL** What type of distribution is this? **positively skewed**

**BL** What values would need to be added to the distribution to make it appear symmetric? **5 values between 17–20, 4 between 13–16, 3 between 9–12, 2 between 5–8, and 1 between 1–4**

(continued on next page)

---

**Go Online!**

**Interactive Whiteboard**
Use the *eLesson* or *Lesson Presentation* to present this lesson.

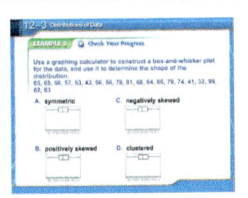

connectED.mcgraw-hill.com 673

# Lesson 10-4 | Distributions of Data

**Need Another Example?**

Use a graphing calculator to construct a histogram for the data, and use it to describe the shape of the distribution.

9, 18, 22, 12, 24, 25, 19, 25, 2, 5, 28, 12, 22, 19, 28, 15, 23, 6, 8, 27, 17, 14, 22, 21, 13, 24, 21, 9, 25, 16, 24, 16, 25, 27, 21, 10

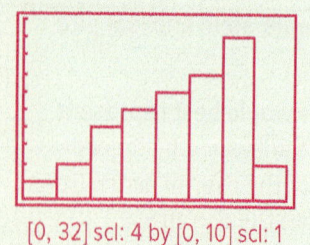

[0, 32] scl: 4 by [0, 10] scl: 1

**negatively skewed**

---

**Example 2  Distribution Using a Box-and-Whisker Plot**

**AL** What does the middle of a box plot represent? *the median of the data*

**OL** Where does the outlier appear on the box plot? Does it affect the shape of the distribution? *Sample answer: The outlier appears as a dot on the far right of the graph. It does not affect the shape of the distribution.*

**BL** What would a histogram with this data look like? *The bars would be evenly distributed with the tallest bar in the middle.*

**Need Another Example?**

Use a graphing calculator to construct a box-and-whisker plot for the data, and use it to determine the shape of the distribution.

9, 18, 22, 12, 24, 25, 19, 25, 2, 5, 28, 12, 22, 19, 28, 15, 23, 6, 8, 27, 17, 14, 22, 21, 13, 24, 21, 9, 25, 16, 24, 16, 25, 27, 21, 10

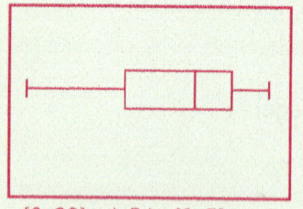

[0, 30] scl: 5 by [0, 5] scl: 1

**negatively skewed**

---

A box-and-whisker plot can also be used to identify the shape of a distribution. Box-and-whisker plots are sometimes called box plots. Recall from Lesson 10-3 that a box-and-whisker plot displays the spread of a data set by dividing it into four quartiles. The data from Example 1 are displayed below.

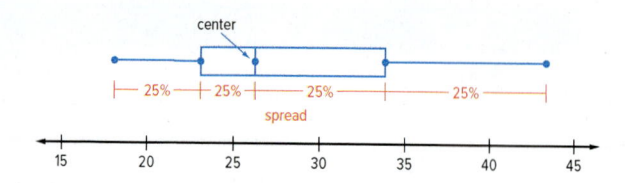

Notice that the left whisker is shorter than the right whisker, and that the line representing the median is closer to the left whisker. This represents a peak on the left and a tail to the right.

**Go Online!**

You can use the box-and-whisker plot Virtual Manipulative in the eToolkit to create box-and-whisker plots from given data.

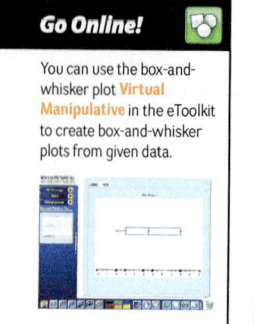

**Key Concept**  Symmetric and Skewed Box-and-Whisker Plots

| Negatively Skewed | Symmetric | Positively Skewed |
|---|---|---|
| The left whisker is longer than the right. The median is closer to the shorter whisker. | The whiskers are the same length. The median is in the center of the data. | The right whisker is longer than the left. The median is closer to the shorter whisker. |

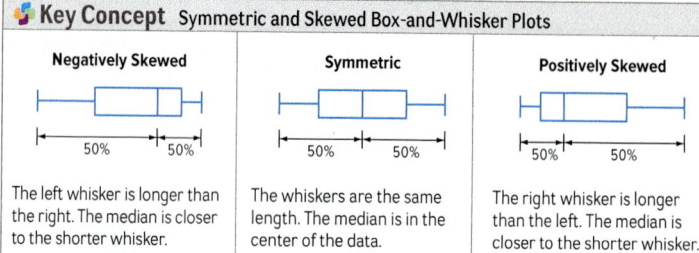

**Study Tip**

**Outliers** In Example 2, notice that the outlier does not affect the shape of the distribution.

S.ID.1, S.ID.3

**Example 2**  Distribution Using a Box-and-Whisker Plot

Use a graphing calculator to construct a box-and-whisker plot for the data, and use it to determine the shape of the distribution.

9, 17, 15, 10, 16, 2, 17, 19, 10, 18, 14, 8, 20, 20, 3, 21, 12, 11
5, 26, 15, 28, 12, 5, 27, 26, 15, 53, 12, 7, 22, 11, 8, 16, 22, 15

Enter the data as L1. Press [2nd] [STAT PLOT] [ENTER] [ENTER] and choose ▯. Adjust the window to the dimensions shown.

The lengths of the whiskers are approximately equal, and the median is in the middle of the data. This indicates that the data are equally distributed to the left and right of the median. Thus, the distribution is symmetric.

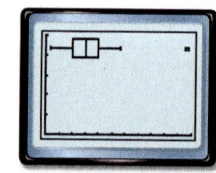

[0, 55] scl: 5 by [0, 5] scl: 1

**Guided Practice**

2. Use a graphing calculator to construct a box-and-whisker plot for the data, and use it to describe the shape of the distribution.

40, 50, 35, 48, 43, 31, 52, 42, 54, 38, 50, 46, 49, 43, 40, 50, 32, 53
51, 43, 47, 41, 49, 50, 34, 54, 51, 44, 54, 39, 47, 35, 51, 44, 48, 37

2.

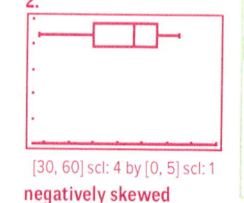

[30, 60] scl: 4 by [0, 5] scl: 1
**negatively skewed**

---

**Differentiated Instruction** AL OL BL

**Interpersonal Learners**  Have students work in pairs to think of ways to help them remember how to recognize the different shapes of the distributions from histograms and box-and-whisker plots.

**Lesson 10-4 | Distributions of Data**

**2 Analyzing Distributions** You have learned that data can be described using statistics. The mean and median describe the center. The standard deviation and quartiles describe the spread. You can use the shape of the distribution to choose the most appropriate statistics that describe the center and spread of a set of data.

When a distribution is symmetric, the mean accurately reflects the center of the data. However, when a distribution is skewed, this statistic is not as reliable.

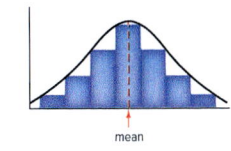

In Lesson 10-3, you discovered that outliers can have a strong effect on the mean of a data set, while the median is less affected. So, when a distribution is skewed, the mean lies away from the majority of the data toward the tail. The median is less affected and stays near the majority of the data.

**Negatively Skewed Distribution**   **Positively Skewed Distribution**

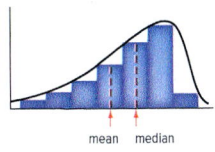

mean  median                median  mean

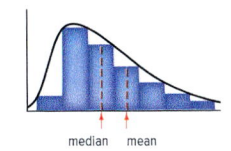

When choosing appropriate statistics to represent a set of data, first determine the shape of the distribution.
- If the distribution is relatively symmetric, the mean and standard deviation can be used.
- If the distribution is skewed or has outliers, use the five-number summary.

S.ID.1, S.ID.3

**Example 3  Choose Appropriate Statistics**

Describe the center and spread of the data using either the mean and standard deviation or the five-number summary. Justify your choice by constructing a histogram for the data.

21, 28, 16, 30, 25, 34, 21, 47, 18, 36, 24, 28, 30, 15, 33, 24, 32, 22 27, 38, 23, 29, 15, 27, 33, 19, 34, 29, 23, 26, 19, 30, 25, 13, 20, 25

Use a graphing calculator to create a histogram. The graph is high in the middle and low on the left and right. Therefore, the distribution is symmetric.

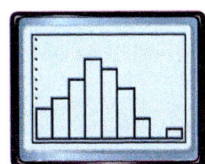

[12, 48] scl: 4 by [0, 10] scl: 1

The distribution is relatively symmetric, so use the mean and standard deviation to describe the center and spread. Press STAT ▶ ENTER ENTER.

The mean $\bar{x}$ is about 26.1 with standard deviation $\sigma$ of about 7.1.

**Technology Tip**

**Tools** On a graphing calculator, each bar is called a *bin*. The width of each bin can be adjusted by pressing WINDOW and changing Xscl. View the histogram using different bin widths and compare the results to determine the appropriate bin width.

---

**2 Analyzing Distributions**

**Example 3  Choose Appropriate Statistics**

**AL** Describe the shape of the histogram.
symmetric

**OL** Which statistics best represent the data?
mean and standard deviation

**BL** Why would the mean and standard deviation not be the best statistics to use when the data is skewed? When the data is skewed, the mean lies away from the majority of the data and is less representative of the data than the median.

**Need Another Example?**

Describe the center and spread of the data using either the mean and standard deviation or the five-number summary. Justify your choice by constructing a histogram for the data.
78, 68, 72, 71, 79, 67, 71, 78, 70 80, 76, 82, 82, 70, 84, 72, 71, 85 67, 86, 74, 86, 73, 72, 77, 87, 70 66, 88, 75, 72, 76, 71, 90, 69, 94

Sample answer: The distribution is skewed, so use the five-number summary. The range is 94 − 66 or 28. The median is 74.5, and half of the data are between 71 and 82.

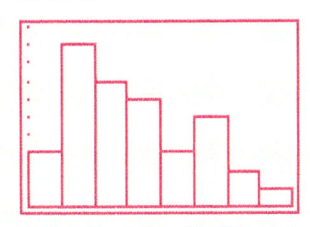

[64, 96] scl: 4 by [0, 10] scl: 1

---

**Differentiated Instruction** AL OL ELL

**Interpersonal Learners** Have students work in pairs to generate a set of data, create a histogram using the data, and describe the shape of the distribution. Then have the students describe the center and spread of the data using either the mean and standard deviation or the five-number summary.

**MP Teaching the Mathematical Practices**

**Tools** Mathematically proficient students recognize both the insight to be gained and the limitations of mathematical tools. Discuss the pros and cons of creating statistical graphs with graphing calculators. Allow students to work together with these tools.

Lesson 10-4 | Distributions of Data

## Example 4 Choose Appropriate Statistics

**AL** Describe the shape of the box plot. **positively skewed**

**OL** Which statistics best represent the data? **five-number summary**

**BL** What data would need to be added to the data set in order to change median to 14? **four data values that are at least 14**

### Need Another Example?

**Bowling** The averages for the bowlers on five teams are shown below. Describe the center and spread of the data using either the mean and standard deviation or the five-number summary. Justify your choice by constructing a box-and-whisker plot for the data.

| Bowling Average |     |     |     |     |
| --- | --- | --- | --- | --- |
| 142 | 180 | 161 | 131 | 201 |
| 179 | 152 | 177 | 196 | 148 |
| 198 | 123 | 203 | 170 | 187 |
| 159 | 193 | 176 | 137 | 183 |

Sample answer: The distribution is skewed, so use the five-number summary. The range is 203 − 123 or 80. The median bowling average is 176.5, and half of the bowlers have an average between 150 and 190.

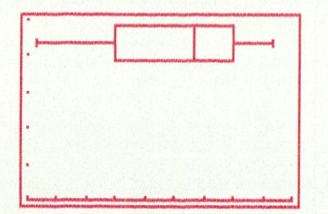

[120, 210] scl: 10 by [0, 5] scl: 1

3. Sample answer: The distribution is skewed, so use the five-number summary. The range is 32 − 2 or 30. The median is 24, and half of the data are between 18 and 26.

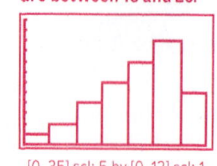

[0, 35] scl: 5 by [0, 12] scl: 1

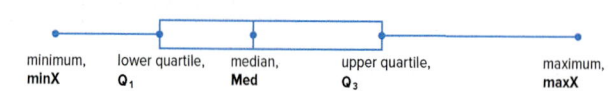

**Real-World Link**
Teens can volunteer in many different capacities at certain hospitals. Volunteers might help in both clinical and non-clinical settings. Training is required.

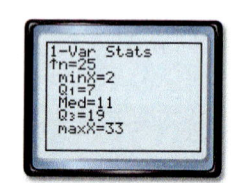

4. Sample answer: The distribution is symmetric, so use the mean and standard deviation to describe the center and spread. The mean amount raised was $36.70 with a standard deviation of about $18.58.

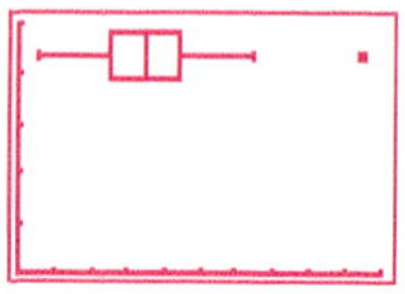

[0, 100] scl: 10 by [0, 5] scl: 1

### Guided Practice

3. Describe the center and spread of the data using either the mean and standard deviation or the five-number summary. Justify your choice by creating a histogram for the data.

19, 2, 25, 14, 24, 20, 27, 30, 14, 25, 19, 32, 21, 31, 25, 16, 24, 22
29, 6, 26, 32, 17, 26, 24, 26, 32, 10, 28, 19, 26, 24, 11, 23, 19, 8

A box-and-whisker plot is helpful when viewing a skewed distribution since it is constructed using the five-number summary.

minimum, minX — lower quartile, $Q_1$ — median, Med — upper quartile, $Q_3$ — maximum, maxX

S.ID.3

### Real-World Example 4 Choose Appropriate Statistics

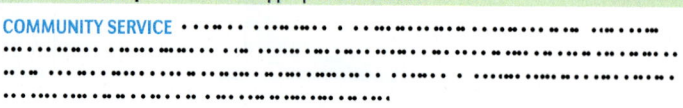

**COMMUNITY SERVICE**

| Community Service Hours | | | | | | | | | |
| --- | --- | --- | --- | --- | --- | --- | --- | --- | --- |
| 6 | 13 | 8 | 7 | 19 | 12 | 2 | 19 | 11 | 22 | 7 | 33 | 13 |
| 3 | 8 | 10 | 5 | 25 | 16 | 6 | 14 | 7 | 20 | 10 | 30 | |

Use a graphing calculator to create a box-and-whisker plot. The right whisker is longer than the left and the median is closer to the left whisker. Therefore, the distribution is positively skewed.

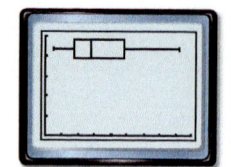

[0, 36] scl: 4 by [0, 5] scl: 1

The distribution is positively skewed, so use the five-number summary. The range is 33 − 2 or 31. The median number of hours completed is 11, and half of the students completed between 7 and 19 hours.

1-Var Stats
n=25
minX=2
$Q_1$=7
Med=11
$Q_3$=19
maxX=33

### Guided Practice

4. **FUNDRAISER** The money raised per student in Mr. Bulanda's fifth period class is shown. Describe the center and spread of the data using either the mean and standard deviation or the five-number summary. Justify your choice by creating a box-and-whisker plot for the data.

| Money Raised per Student (dollars) | | | | | | | | |
| --- | --- | --- | --- | --- | --- | --- | --- | --- |
| 41 | 27 | 52 | 18 | 42 | 32 | 16 | 95 | 27 | 65 |
| 36 | 45 | 5 | 34 | 50 | 15 | 62 | 38 | 57 | 20 |
| 38 | 21 | 33 | 58 | 25 | 42 | 31 | 8 | 40 | 28 |

## Go Online!

The most up-to-date resources available for your program can be found at connectED.mcgraw-hill.com.

Lesson 10-4 | Distributions of Data

## Check Your Understanding

= Step-by-Step Solutions begin on page R11.   **Go Online!** for a Self-Check Quiz

**Examples 1–2**
S.ID.1, S.ID.3
Use a graphing calculator to construct a histogram and a box-and-whisker plot for the data. Then describe the shape of the distribution.  1–2. See margin.

1. [dot plot]

2. [dot plot]

**Example 3**
S.ID.1, S.ID.3
Describe the center and spread of the data using either the mean and standard deviation or the five-number summary. Justify your choice by constructing a histogram for the data. See margin.

3. [dot plot]

**Example 4**
S.ID.1, S.ID.3
4. **PRESENTATIONS** [dot plot]
See Chapter 10 Answer Appendix.

**Presentations**
20, 18, 15, 17, 18, 10, 15
10, 18, 19, 17, 19, 12, 6
19, 15, 21, 10, 9, 18

## Practice and Problem Solving

Extra Practice is on page R10.

**Examples 1–2**
S.ID.1, S.ID.3
Use a graphing calculator to construct a histogram and a box-and-whisker plot for the data. Then describe the shape of the distribution. 5–6. See Chapter 10 Answer Appendix.

5. [dot plot]

6. [dot plot]

**Example 3**
S.ID.1, S.ID.3
Describe the center and spread of the data using either the mean and standard deviation or the five-number summary. Justify your choice by constructing a histogram for the data. 7–8. See Chapter 10 Answer Appendix.

7. [dot plot]

8. [dot plot]

**Example 4**
S.ID.1, S.ID.3
9. **WEATHER** [dot plot]
See Chapter 10 Answer Appendix.

| Temperature (°F) | | | | | | | | | | | | |
|---|---|---|---|---|---|---|---|---|---|---|---|---|
| 48 | 50 | 55 | 53 | 57 | 53 | 44 | 61 | 57 | 49 | 51 | 58 | 46 | 54 | 57 |
| 50 | 55 | 47 | 57 | 48 | 58 | 53 | 49 | 56 | 59 | 52 | 48 | 55 | 53 | 51 |

### Differentiated Homework Options

| Levels | AL Basic | OL Core | BL Advanced |
|---|---|---|---|
| Exercises | 5–9, 15–23 | 5–11, 15–23 | 10–23 |
| 2-Day Option | 5–9 odd, 18–23 | 5–9, 18–23 | |
| | 6, 8, 15–17 | 10, 11, 15–17 | |

You can use ALEKS to provide additional remediation support with personalized instruction and practice.

## Practice

**Formative Assessment** Use Exercises 1–4 to assess students' understanding of the concepts in this lesson.

The Practice and Problem Solving exercises assess the content taught in the lesson. The Preparing for Assessment page is meant to be used as preparation for end-of-course assessments.

## Additional Answers

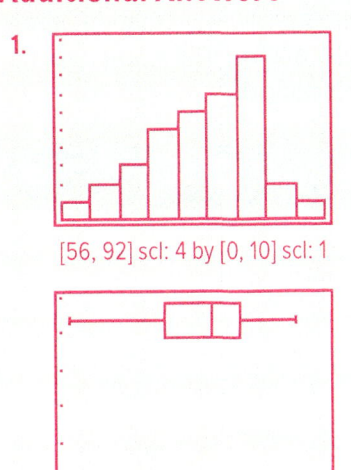

1. [histogram]

[56, 92] scl: 4 by [0, 10] scl: 1

[box-and-whisker plot]

[56, 92] scl: 4 by [0, 5] scl: 1

negatively skewed

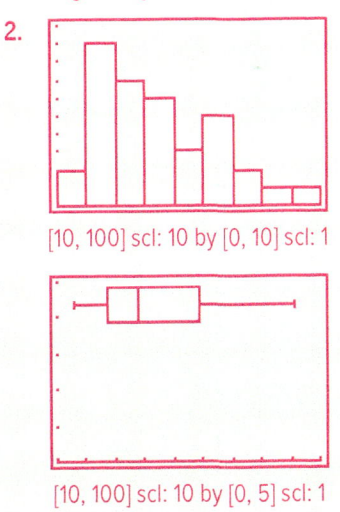

2. [histogram]

[10, 100] scl: 10 by [0, 10] scl: 1

[box-and-whisker plot]

[10, 100] scl: 10 by [0, 5] scl: 1

positively skewed

3. Sample answer: The distribution is skewed, so use the five-number summary. The range is 92 − 52 or 40. The median is 65, and half of the data are between 59.5 and 74.

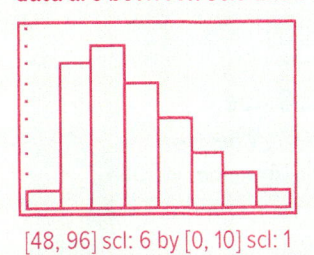

[histogram]

[48, 96] scl: 6 by [0, 10] scl: 1

connectED.mcgraw-hill.com  677

Lesson 10-4 | Distributions of Data

| Teaching the Mathematical Practices |
|---|
| **Construct Arguments** Mathematically proficient students use previously established results in constructing arguments. In Exercise 15, have students use what they have learned about distributions to discuss how to analyze a bimodal distribution. |

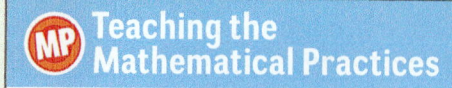

## Extra Practice

See page R10 for extra exercises for students who are approaching level or for on-level students who need additional reinforcement.

### Levels of Complexity Chart

The levels of the exercises progress from 1 to 3, with Level 1 indicating the lowest level of complexity.

| Exercises | 5–9 | 10, 11, 18–23 | 12–17 |
|---|---|---|---|
| Level 3 | | | ● |
| Level 2 | | ● | |
| Level 1 | ● | | |

## Assess

**Ticket Out the Door** Have students explain the difference between *positively skewed*, *negatively skewed*, and *symmetric* sets of data, and give an example of each.

### Additional Answers

15. Sample answer: A bimodal distribution is a distribution of data that is characterized by having data divided into two clusters, thus producing two modes, and having two peaks. The distribution can be described by summarizing the center and spread of each cluster of data.

16. Sample answer: The average high temperature over the course of a year for a city may have a symmetrical distribution. The attendance at a baseball stadium over the course of a season for a team may be skewed.

## Go Online!   eBook

**Interactive Student Guide**
Use the *Interactive Student Guide* to deepen conceptual understanding.
· Distributions of Data

---

**B** 10. **TRACK** While training for the 100-meter dash, Sarah pulled a muscle. After being cleared for practice, she continued to train. Sarah's median time was about 12.34 seconds, but her average time dropped to about 12.53 seconds. Sarah's 100-meter dash times are shown.

a. Use a graphing calculator to create a box-and-whisker plot. Describe the center and spread of the data. **a–c. See Ch. 10 Answer Appendix.**

b. Sarah's slowest time prior to her injury was 12.50 seconds. Use a graphing calculator to create a box-and-whisker plot that *does not* include the times that she ran after her injury. Then describe the center and spread of the new data set.

c. What effect does removing the times recorded after Sarah pulled a muscle have on the shape of the distribution and on how you should describe the center and spread?

| 100-meter dash (seconds) |||||
|---|---|---|---|---|
| 12.20 | 12.35 | 13.60 | 12.24 | 12.72 |
| 12.18 | 12.06 | 12.41 | 12.28 | 13.06 |
| 12.87 | 12.04 | 12.38 | 12.20 | 13.12 |
| 12.30 | 13.27 | 12.93 | 12.16 | 12.02 |
| 12.50 | 12.14 | 11.97 | 12.24 | 13.09 |
| 12.46 | 12.33 | 13.57 | 11.96 | 13.34 |

11. **MENU** The prices for entrees at a restaurant are shown.

a. Use a graphing calculator to create a box-and-whisker plot. Describe the center and spread of the data.

b. The owner of the restaurant decides to eliminate all entrees that cost more than $15. Use a graphing calculator to create a box-and-whisker plot that reflects this change. Then describe the center and spread of the new data set. **a–b. See Ch. 10 Answer Appendix.**

| Entree Prices ($) |||||
|---|---|---|---|---|
| 9.00 | 11.25 | 16.50 | 9.50 | 13.00 |
| 18.50 | 7.75 | 11.50 | 13.75 | 9.75 |
| 8.00 | 16.50 | 12.50 | 10.25 | 17.75 |
| 13.00 | 10.75 | 16.75 | 8.50 | 11.50 |

S.ID.1, S.ID.3

**H.O.T. Problems**   Use Higher-Order Thinking Skills

**C** **REASONING** Identify the box-and-whisker plot that corresponds to each of the following histograms.

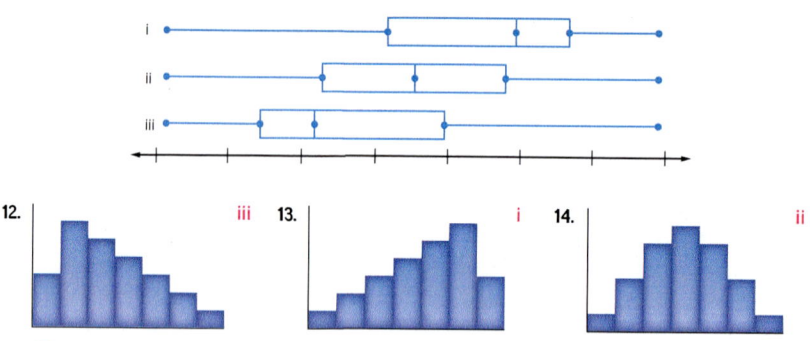

12. iii   13. i   14. ii

15. **CONSTRUCT ARGUMENTS** Research and write a definition for a *bimodal distribution*. How can the measures of center and spread of a bimodal distribution be described? **15–16. See margin.**

16. **OPEN-ENDED** Give an example of a set of real-world data with a distribution that is symmetric and one with a distribution that is not symmetric.

17. **WRITING IN MATH** Explain why the mean and standard deviation are used to describe the center and spread of a symmetrical distribution and the five-number summary is used to describe the center and spread of a skewed distribution. **See Ch. 10 Answer Appendix.**

### Standards for Mathematical Practice

| Emphasis On | Exercises |
|---|---|
| 1 Make sense of problems and persevere in solving them. | 16, 18–23 |
| 2 Reason abstractly and quantitatively. | 12–15, 19 |
| 3 Construct viable arguments and critique the reasoning of others. | 15, 17 |
| 4 Model with mathematics. | 4, 9–11, 19, 21–22 |
| 5 Use appropriate tools strategically. | 1–3, 5–8 |

Lesson 10-4 | Distributions of Data

## Preparing for Assessment

**18.** The histogram at the right represents a data set.
What is the shape of the data? **MP** 1 S.ID.1, S.ID.3 **C**

- ○ A  negatively skewed
- ○ B  relatively symmetric
- ○ C  positively skewed
- ○ D  There is not enough information to determine the shape.

**19.** The following data represents the weights, in pounds, of a particular species of fish caught at random in a pond of a fish farm.

21, 19, 28, 23, 20, 18, 18, 24, 22, 27, 24, 26, 22, 26, 25, 28

What are the most appropriate statistics to use to describe the distribution of the weights of the fish, and why? **MP** 1, 2, 4 S.ID.3 **A**

- ○ A  mean and standard deviation because the distribution is relatively symmetric
- ○ B  mean and standard deviation because the distribution is skewed
- ○ C  five-number summary because the distribution is relatively symmetric
- ○ D  five-number summary because the distribution is skewed

**20.** The box-and-whisker plot below represents a data set.

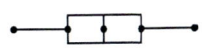

Which of the following statements are true? **MP** 1 S.ID.1, S.ID.3 **B, D**

- ☐ A  The distribution is negatively skewed.
- ☐ B  The distribution is relatively symmetric.
- ☐ C  The distribution is positively skewed.
- ☐ D  The mean and median are approximately the same.
- ☐ E  The median is a more appropriate measure of center than the mean.

**21.** MULTI-STEP  The following data represents the salaries of the employees of a company, in dollars. **MP** 1, 4 S.ID.3

| Employee Salaries ($) | | | | |
|---|---|---|---|---|
| 31,000 | 48,000 | 37,000 | 43,000 | 52,000 |
| 48,000 | 36,000 | 56,000 | 59,000 | 58,000 |
| 34,000 | 256,000 | 63,000 | 45,000 | 37,000 |

**a.** What is the median of the data?

$48,000

**b.** What is the mean of the data?

$60,200

**c.** Explain why the median is a more appropriate statistic than the mean to represent the data.

$256,000 is an outlier

**22.** The data represent the heights of students in a class, in centimeters.

| Heights (cm) | | | | |
|---|---|---|---|---|
| 151 | 158 | 209 | 153 | 197 |
| 150 | 178 | 155 | 209 | 217 |
| 148 | 164 | 158 | 162 | 158 |
| 195 | 168 | 178 | 186 | 190 |

What are the most appropriate statistics to use to describe the distribution of the heights of the students, and why? **MP** 1, 2, 4 S.ID.3 **D**

- ○ A  mean and standard deviation because the distribution is relatively symmetric
- ○ B  mean and standard deviation because the distribution is skewed
- ○ C  five-number summary because the distribution is relatively symmetric
- ○ D  five-number summary because the distribution is skewed

**23.** The histogram at the right represents a data set.
What is the shape of the data? **MP** 1 S.ID.1, S.ID.3 **A**

- ○ A  negatively skewed
- ○ B  relatively symmetric
- ○ C  positively skewed
- ○ D  There is not enough information to determine the shape.

---

## Preparing for Assessment

Exercises 18–23 require students to use the skills they will need on standardized assessments. Each exercise is dual-coded with content standards and mathematical practice standards.

### Dual Coding

| Items | Content Standards | Mathematical Practices |
|---|---|---|
| 18 | S.ID.1, S.ID.3 | 1 |
| 19 | S.ID.3 | 1, 2, 4 |
| 20 | S.ID.1, S.ID.3 | 1 |
| 21 | S.ID.3 | 1, 4 |
| 22 | S.ID.3 | 1, 2, 4 |
| 23 | S.ID.1, S.ID.3 | 1 |

### Diagnose Student Errors

Survey student responses for each item. Class trends may indicate common errors and misconceptions.

**18.**

| | |
|---|---|
| A | Confused negative and positively skewed |
| B | Did not understand the meaning of *relatively symmetric* |
| C | CORRECT |
| D | Did not realize that there is enough information |

**19.**

| | |
|---|---|
| A | CORRECT |
| B | Did not realize that the distribution is relatively symmetric |
| C | Did not realize which statistics are more appropriate for a relatively symmetric distribution |
| D | Did not realize that the distribution is relatively symmetric |

**20.**

| | |
|---|---|
| A | Did not realize that the distribution is relatively symmetric |
| B | CORRECT |
| C | Did not realize that the distribution is relatively symmetric |
| D | CORRECT |
| E | The median and mean are approximately equal, so one is not necessarily more appropriate than the other |

---

### Differentiated Instruction  OL  BL

**Extension**  Have students generate a set of data that is neither skewed nor symmetrical. Discuss situations that may result in data that resembles such distributions.

---

**22.**

| | |
|---|---|
| A | Did not realize that the distribution is skewed |
| B | Did not realize which statistics are more appropriate for a skewed distribution |
| C | Did not realize that the distribution is skewed |
| D | CORRECT |

**23.**

| | |
|---|---|
| A | CORRECT |
| B | Did not understand the meaning of *relatively symmetric* |
| C | Confused negative and positively skewed |
| D | Did not realize that there is enough information |

### Go Online!

**Quizzes**

Students can use *Self-Check Quizzes* to check their understanding of this lesson. You can also give the *Chapter Quiz*, which covers the content in Lessons 10-3 and 10-4.

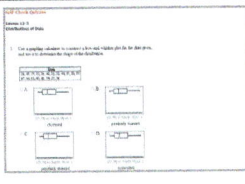

# LESSON 10-5
# Comparing Sets of Data

**SUGGESTED PACING (DAYS)**

90 min. 0.5
45 min. 1
Instruction

## Track Your Progress

### Objectives

1. Determine the effect that transformations of data have on measures of central tendency and variation.
2. Compare data using measures of central tendency and variation.

### Mathematical Background

If a real number, $k$, is added to every value in a set of data, the mean, median, and mode of the data set can be found by adding $k$ to the mean, median, and mode of the original data set. The range and standard deviation will be the same. If every value in a set of data is multiplied by a constant $k$, $k > 0$, then the mean, median, mode, range, and standard deviation of the new data set can be found by multiplying each original statistic by $k$.

### THEN

**S.ID.1** Represent data with plots on the real number line (dot plots, histograms, and box plots).

### NOW

**S.ID.2** Use statistics appropriate to the shape of the data distribution to compare center (median, mean) and spread (interquartile range, standard deviation) of two or more different data sets.

**S.ID.3** Interpret differences in shape, center, and spread in the context of the data sets, accounting for possible effects of extreme data points (outliers).

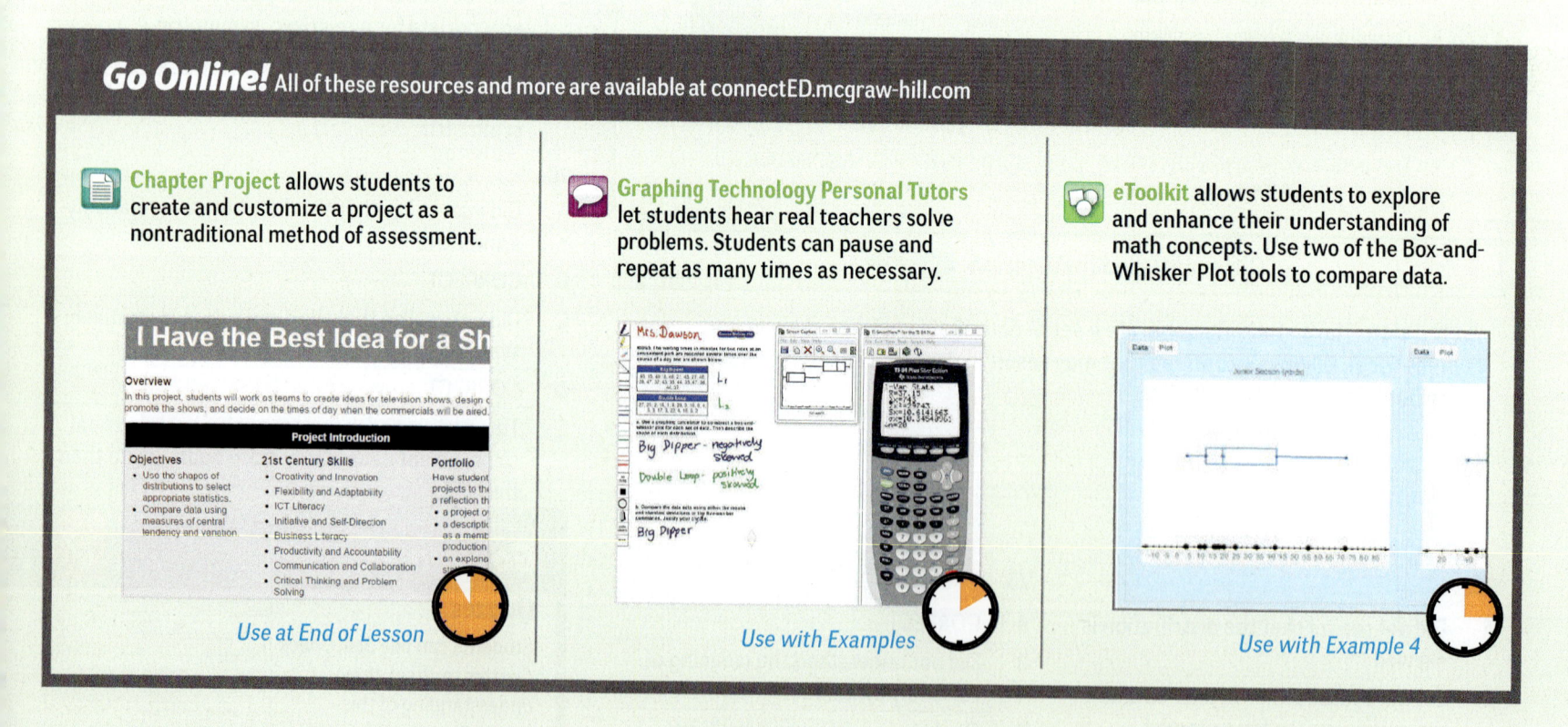

### NEXT

**S.ID.4** Use the mean and standard deviation of a data set to fit it to a normal distribution and to estimate population percentages. Recognize that there are data sets for which such a procedure is not appropriate. Use calculators, spreadsheets, and tables to estimate areas under the normal curve.

---

## Go Online! All of these resources and more are available at connectED.mcgraw-hill.com

**Chapter Project** allows students to create and customize a project as a nontraditional method of assessment.

*Use at End of Lesson*

**Graphing Technology Personal Tutors** let students hear real teachers solve problems. Students can pause and repeat as many times as necessary.

*Use with Examples*

**eToolkit** allows students to explore and enhance their understanding of math concepts. Use two of the Box-and-Whisker Plot tools to compare data.

*Use with Example 4*

---

## OER Using Open Educational Resources

**Game** Have students play the game about mean, median, mode, and range on **Questionaut** to review skills before teaching this lesson. *Use as homework or flipped learning*

# Go Online!
connectED.mcgraw-hill.com — Worksheets

# Differentiate Your Resources

**Extra Practice** Additional practice or homework; Skills Practice is best for approaching-level students and Practice is best for on-level and beyond-level students

## Skills Practice
AL  OL  ELL

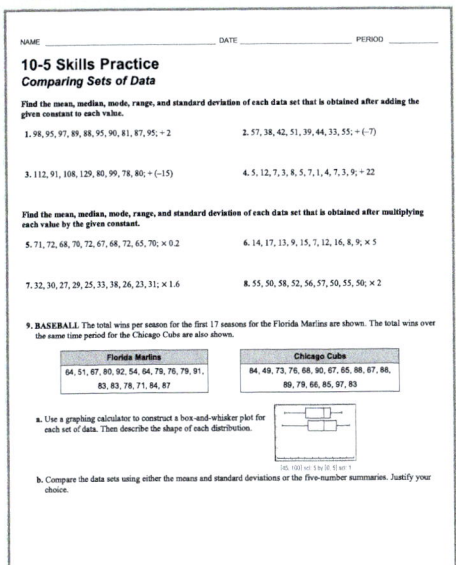

## Practice
AL  OL  BL  ELL

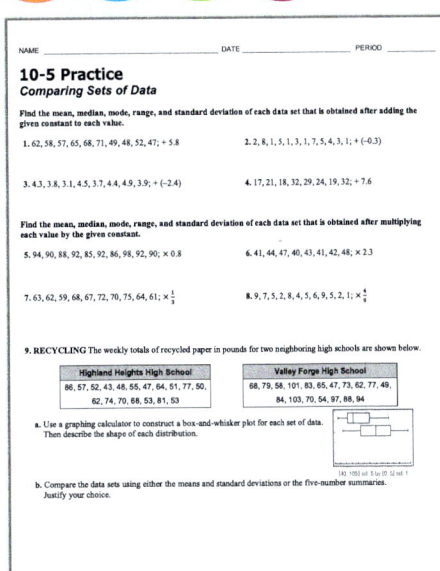

## Word Problem Practice
AL  OL  BL  ELL

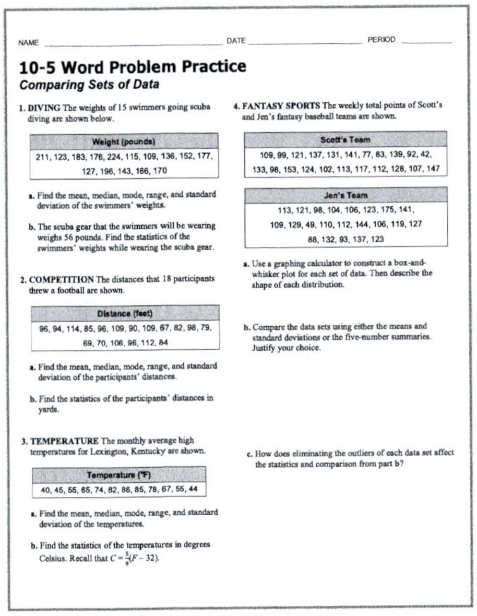

**Intervention** Reteaching and vocabulary activities that can be used with struggling or absent students and as ELL support

## Study Guide and Intervention
AL  OL  ELL

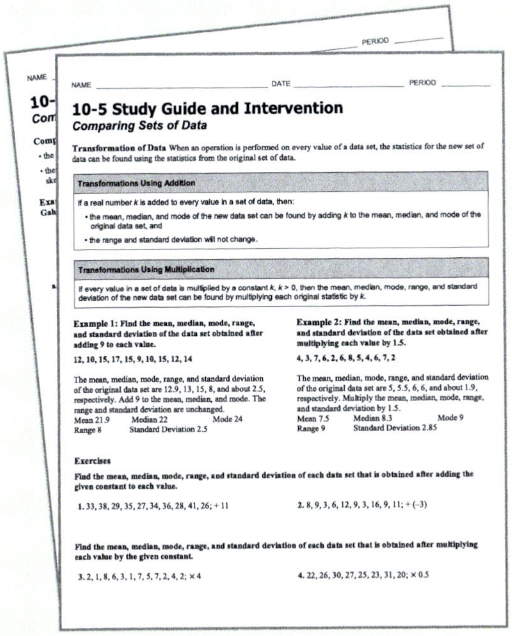

## Study Notebook
AL  OL  BL  ELL

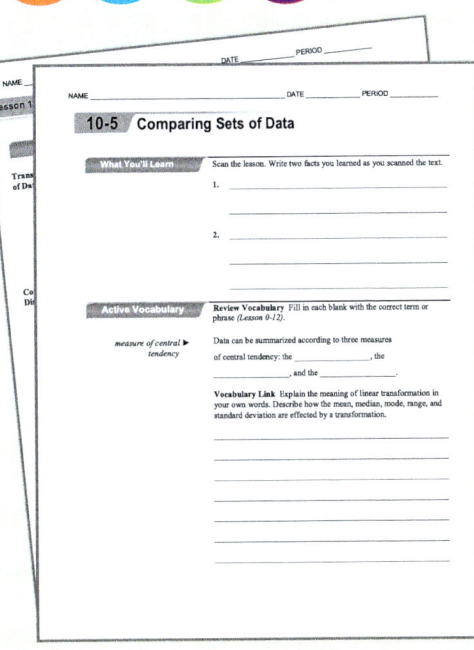

**Extension** Activities that can be used to extend lesson concepts

## Enrichment
OL  BL  ELL

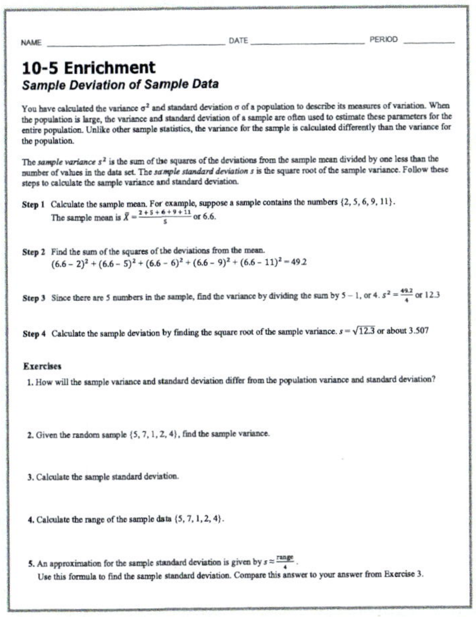

connectED.mcgraw-hill.com  680B

Lesson 10-5 | Comparing Sets of Data

# Launch

Have students read the Why? section of the lesson. Ask:

- **How can Tom calculate the effect of each bonus?** Sample answer: For the extra $5 per day, Tom can add $5 to each daily total. For the 10% increase, Tom can multiply each daily total by 1.10.

- **When is it to Tom's benefit that he accepts the extra $5 over the 10% increase and vice versa? Explain.** Sample answer: On days when Tom makes less than $50, he should take the extra $5 because 10% of any value less than $50 will be less than $5. On days when Tom makes more than $50, he should take the 10% increase because 10% of any value greater than $50 will be greater than $5. When Tom makes $50, the bonuses are equal.

# Teach

Ask the scaffolded questions for each example to build conceptual understanding for students at all levels.

## 1 Transformations of Data

---

## LESSON 5
# Comparing Sets of Data

**::Then** — You calculated measures of central tendency and variation.

**::Now**
1. Determine the effect that transformations of data have on measures of central tendency and variation.
2. Compare data using measures of central tendency and variation.

**::Why?** — Tom gets paid hourly to do landscaping work. Because he is such a good employee, Tom is planning to ask his boss for a bonus. Tom's initial pay for a month is shown. He is trying to decide whether he should ask for an extra $5 per day or a 10% increase in his daily wages.

**Tom's Pay ($)**

| 44 | 52 | 50 |
| 40 | 48 | 46 |
| 44 | 52 | 54 |
| 58 | 42 | 52 |
| 54 | 50 | 52 |
| 42 | 52 | 46 |
| 56 | 48 | 44 |
| 50 | 42 |    |

**New Vocabulary**
linear transformation

**Mathematical Practices**
2 Reason abstractly and quantitatively.

**Content Standards**
**S.ID.2** Use statistics appropriate to the shape of the data distribution to compare center (median, mean) and spread (interquartile range, standard deviation) of two or more different data sets.
**S.ID.3** Interpret differences in shape, center, and spread in the context of the data sets, accounting for possible effects of extreme data points (outliers).

### 1 Transformations of Data

......... linear transformation ......... **linear transformation** .........

$y = \text{ } + x$

**Tom's Earnings Before Extra $5**

40   45   50   52   55
           median  mode
Range = 58 − 40 or 18
Mean 48.6      Standard Deviation 4.9

**Tom's Earnings With Extra $5**

45   50   55   57   60
           median  mode
Range = 63 − 45 or 18
Mean 53.6      Standard Deviation 4.9

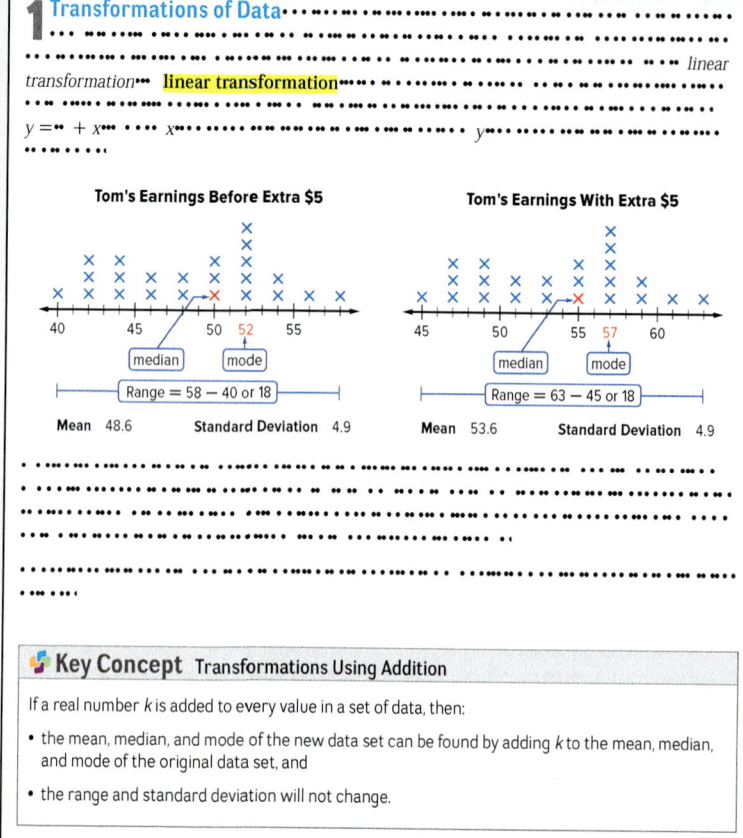

### Key Concept  Transformations Using Addition

If a real number k is added to every value in a set of data, then:

- the mean, median, and mode of the new data set can be found by adding k to the mean, median, and mode of the original data set, and
- the range and standard deviation will not change.

---

### MP  Mathematical Practices Strategies

**Make sense of problems and persevere in solving them.**
Help students understand the effect of transformations on data sets and to compare data sets.

- **What was the effect of adding a fixed number to each value in the data set?** The mean, median, and mode can be found by adding the fixed number to those statistics for the original data set. The range and standard deviation are unchanged.

- **What was the effect of multiplying each value in the data set by a fixed positive number?** The mean, median, mode, range, and standard deviation of the new data can be found by multiplying each of those statistics for the original data set by the fixed number.

---

## Go Online!

**Interactive Whiteboard**
Use the eLesson or Lesson Presentation to present this lesson.

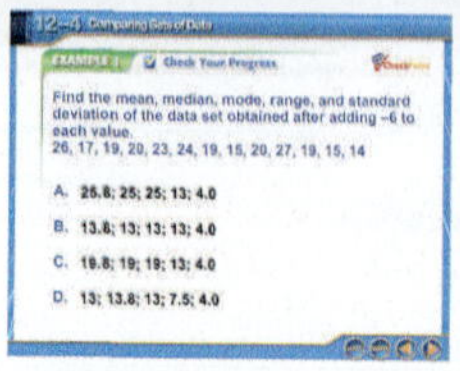

680 | Lesson 10-5 | Comparing Sets of Data

Lesson 10-5 | Comparing Sets of Data

S.ID.2

### Example 1 Transformation Using Addition

Find the mean, median, mode, range, and standard deviation of the data set obtained after adding 7 to each value.

13, 5, 8, 12, 7, 4, 5, 8, 14, 11, 13, 8

**Method 1** Find the mean, median, mode, range, and standard deviation of the original data set.

Mean  9   Mode  8   Standard Deviation  3.3
Median  8   Range  10

Add 7 to the mean, median, and mode. The range and standard deviation are unchanged.

Mean  16   Mode  15   Standard Deviation  3.3
Median  15   Range  10

**Method 2** Add 7 to each data value.

20, 12, 15, 19, 14, 11, 12, 15, 21, 18, 20, 15

Find the mean, median, mode, range, and standard deviation of the new data set.

Mean  16   Mode  15   Standard Deviation  3.3
Median  15   Range  10

#### Guided Practice

1. Find the mean, median, mode, range, and standard deviation of the data set obtained after adding −4 to each value.  30, 32.5, 22, 38, 10.1

   27, 41, 15, 36, 26, 40, 53, 38, 37, 24, 45, 26

To see the effect that a daily increase of 10% has on the data set, we can multiply each value by 1.10 and recalculate the measures of center and variation.

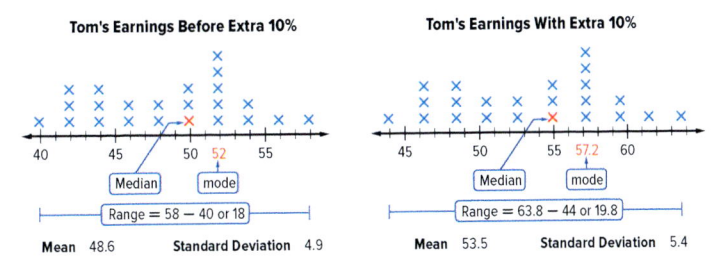

**Tom's Earnings Before Extra 10%**

40  45  50 **52** 55
       Median  mode
Range = 58 − 40 or 18
Mean  48.6   Standard Deviation  4.9

**Tom's Earnings With Extra 10%**

45  50  55 **57.2** 60
       Median  mode
Range = 63.8 − 44 or 19.8
Mean  53.5   Standard Deviation  5.4

Notice that each value did not increase by the same amount, but did increase by a factor of 1.10. Thus, the mean, median, and mode increased by a factor of 1.10. Since each value was increased by a constant percent and not by a constant amount, the range and standard deviation both changed, also increasing by a factor of 1.10.

> **Key Concept** Transformations Using Multiplication
>
> If every value in a set of data is multiplied by a constant $k$, $k > 0$, then the mean, median, mode, range, and standard deviation of the new data set can be found by multiplying each original statistic by $k$.

**Technology Tip**
**1-Var Stats** To quickly calculate the mean $\bar{x}$, median Med, standard deviation $\sigma$, and range of a data set, enter the data as L1 in a graphing calculator, and then press STAT ▶ ENTER ENTER. Subtract minX from maxX to find the range.

connectED.mcgraw-hill.com  681

---

### Differentiated Instruction AL OL BL ELL

**Interpersonal Learners** Have students work in pairs to perform transformations on data sets. Have each student create a data set and then determine an operation that should be done to each value.

The students should exchange data sets, and then find the mean, median, mode, range, and standard deviation of the original data set and of the new data set after the operation is performed. Have students describe the effect of the transformation on the data set that they received.

---

### Example 1 Transformation Using Addition

**AL** What happens to the sum of a set of 10 data values if each value is increased by 7? The sum increases by 70.

**OL** Why is the range of a set of data unaffected when the data are all increased by the same value? The least value and greatest value are increased by the same amount, so the difference between the two remains the same.

**BL** Each value in a data set with $n$ values and a mean of $x$ is increased by $y$. What is the mean of the new data set? $x + y$

**Need Another Example?**

Find the mean, median, mode, range, and standard deviation of the data set obtained after adding 12 to each value. 73, 78, 61, 54, 88, 90, 63, 78, 80, 61, 86, 78
mean: 86.2; median: 90; mode:90; range: 36; standard deviation: 11.3

### Example 2 Transformation Using Multiplication

**AL** How do you determine the mean in Example 2? Add all the data values and divide by 12.

**OL** Is it necessary to multiply every data value by 3? Explain. No; the mean, median, mode, range, and standard deviation can be multiplied by 3 instead.

**BL** Why is the range of a set of data affected when the data are all multiplied by the same value? Explain verbally and algebraically. The least value and greatest value are multiplied by the same amount, so the difference between the two will also be multiplied by the same amount. $3x − 3y = 3(x − y)$

**Need Another Example?**

Find the mean, median, mode, range, and standard deviation of the data set obtained after multiplying each value by 2.5. 4, 2, 3, 1, 4, 6, 2, 3, 7, 5, 1, 4
mean: 8.75; median: 8.75; mode: 10; range: 15; standard deviation: 4.5

connectED.mcgraw-hill.com  681

Lesson 10-5 | Comparing Sets of Data

## 2 Comparing Distributions

### Example 3 Compare Data Using Histograms

**AL** Describe the shape of each histogram. **They are both symmetric.**

**OL** Which statistics best represent the data? **mean and standard deviation**

**BL** What conclusions can you make when two symmetrical distributions of data are graphed on the same scale and one distribution appears to be narrower than the other? **The narrower distribution has a smaller standard deviation.**

### Need Another Example?

**Games** Brittany and Justin are playing a computer game. Their high scores for each game are shown below.

| Brittany's Scores |
|---|
| 29, 43, 54, 58, 39, 44, 39, 53, 32, 48, 39, 49, 38, 31, 41, 44, 44, 45, 48, 31 |

| Justin's Scores |
|---|
| 48, 26, 28, 53, 39, 28, 30, 58, 45, 37, 30, 31, 40, 32, 30, 44, 33, 35, 43, 35 |

a. Use a graphing calculator to create a histogram for each set of data. Then describe the shape of each distribution.

Brittany's Scores

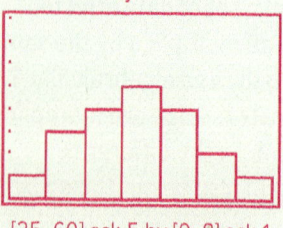

[25, 60] scl: 5 by [0, 8] scl: 1

Justin's Scores

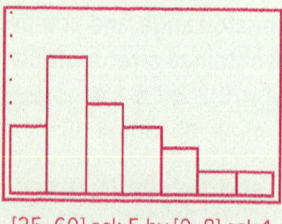

[25, 60] scl: 5 by [0, 8] scl: 1

**Brittany, symmetric; Justin, positively skewed**

b. Compare the distributions using either the means and standard deviations or the five-number summaries. Justify your choice.

---

### Go Online!

In *Personal Tutor* videos for this lesson, teachers describe how to compare sets of data. Watch with a partner, then try describing how to solve a problem for them. Have them ask questions to help your understanding. **ELL**

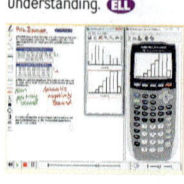

### Technology Tip

**Histograms** To create a histogram for a set of data in L2, press [2nd] [STAT PLOT] [ENTER] [ENTER], choose ▙▟▖, and enter L2 for Xlist.

### Additional Answer (Need Another Example?)

**3b.** Sample answer: One distribution is symmetric and the other is skewed, so use the five-number summaries. Both distributions have a maximum of 58, but Brittany's minimum score is 29 compared to Justin's minimum scores of 26. The median for Brittany's scores is 43.5 and the upper quartile for Justin's scores is 43.5. This means that 50% of Brittany's scores are between 43.5 and 58, while only 25% of Justin's scores fall within this range. Therefore, we can conclude that overall, Brittany's scores are higher than Justin's scores.

---

Since the medians for both bonuses are equal and the means are approximately equal, Tom should ask for the bonus that he thinks he has the best chance of receiving.

S.ID2

**Example 2** Transformation Using Multiplication

Find the mean, median, mode, range, and standard deviation of the data set obtained after multiplying each value by 3.

21, 12, 15, 18, 16, 10, 12, 19, 17, 18, 12, 22

Find the mean, median, mode, range, and standard deviation of the original data set.

Mean   16           Mode   12           Standard Deviation   3.7
Median   16.5       Range   12

Multiply the mean, median, mode, range, and standard deviation by 3.

Mean   48           Mode   36           Standard Deviation   11.1
Median   49.5       Range   36

### Guided Practice

2. Find the mean, median, mode, range, and standard deviation of the data set obtained after multiplying each value by 0.8.  **44, 44.4, 48, 13.6, 4.2**

63, 47, 54, 60, 55, 46, 51, 60, 58, 50, 56, 60

### 2 Comparing Distributions
Recall that when choosing appropriate statistics to represent data, you should first analyze the shape of the distribution. The same is true when comparing distributions.

- Use the mean and standard deviation to compare two symmetric distributions.
- Use the five-number summaries to compare two skewed distributions or a symmetric distribution and a skewed distribution.

S.ID2, S.ID.3

**Real-World Example 3**   Compare Data Using Histograms

**QUIZ SCORES** Robert and Elaine's quiz scores for the first semester of Algebra 1 are shown below.

| Robert's Quiz Scores |
|---|
| 85, 95, 70, 87, 78, 82, 84, 84, 85, 99, 88, 74, 75, 89, 79, 80, 92, 91, 96, 81 |

| Elaine's Quiz Scores |
|---|
| 89, 76, 87, 86, 92, 77, 78, 83, 83, 82, 81, 82, 84, 85, 85, 86, 89, 93, 77, 85 |

a. Use a graphing calculator to construct a histogram for each set of data. Then describe the shape of each distribution.

Enter Robert's quiz scores as L1 and Elaine's quiz scores as L2.

Robert's Quiz Scores

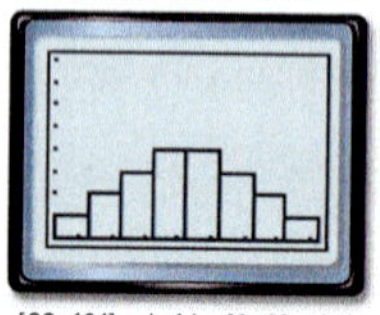

Elaine's Quiz Scores

[69, 101] scl: 4 by [0, 8] scl: 1        [69, 101] scl: 4 by [0, 8] scl: 1

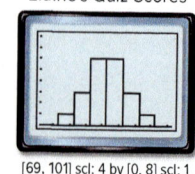

Both distributions are high in the middle and low on the left and right. Therefore, both distributions are symmetric.

---

### Teaching Tip

**Sense-Making** If every value in a set of data is multiplied by a negative constant $k$, then the mean, median, and mode of the new data set can be found by multiplying the mean, median, and mode of the original data set by $k$. The range and standard deviation of the new data set can be found by multiplying the range and standard deviation of the original data set by $|k|$. Students will explore this in Exercise 22.

# Lesson 10-5 | Comparing Sets of Data

**b. Compare the data sets using either the means and standard deviations or the five-number summaries. Justify your choice.**

Both distributions are symmetric, so use the means and standard deviations to describe the centers and spreads.

**Robert's Quiz Scores**
1-Var Stats
x̄=84.7
Σx=1694
Σx²=144594
Sx=7.650937335
σx=7.457211275
↓n=20

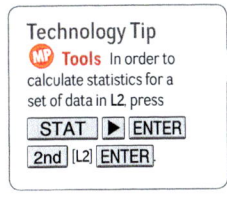

**Elaine's Quiz Scores**
1-Var Stats
x̄=84
Σx=1680
Σx²=141552
Sx=4.768316485
σx=4.647580015
↓n=20

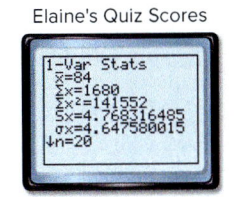

The means for the students' quiz scores are approximately equal, but Robert's quiz scores have a much higher standard deviation than Elaine's quiz scores. This means that Elaine's quiz scores are generally closer to her mean than Robert's quiz scores are to his mean.

### Guided Practice

**COMMUTE** The students in two of Mr. Martin's classes found the average number of minutes that they each spent traveling to school each day.

**3A.** Use a graphing calculator to construct a histogram for each set of data. Then describe the shape of each distribution.

**3B.** Compare the data sets using either the means and standard deviations or the five-number summaries. Justify your choice. *See margin.*

| 2nd Period (minutes) |
|---|
| 8, 4, 18, 7, 13, 26, 12, 6, 20, 5, 9, 24, 8, 16, 31, 13, 17, 10, 8, 22, 12, 25, 13, 11, 18, 12, 16, 22, 25, 33 |

| 7th Period (minutes) |
|---|
| 21, 4, 20, 13, 22, 6, 10, 23, 13, 25, 14, 16, 19, 21, 19, 8, 20, 18, 9, 14, 21, 17, 19, 22, 4, 19, 21, 26 |

Box-and-whisker plots are useful for comparisons of data because they can be displayed on the same screen.

S.ID.2, S.ID.3

### Real-World Example 4  Compare Data Using Box-and-Whisker Plots

**FOOTBALL** Kurt's total rushing yards per game for his junior and senior seasons are shown.

| Junior Season (yards) |||||| 
|---|---|---|---|---|---|
| 16 | 20 | 72 | 4 | 25 | 18 |
| 34 | 10 | 42 | 17 | 56 | 12 |

| Senior Season (yards) ||||||
|---|---|---|---|---|---|
| 77 | 54 | 109 | 60 | 156 | 72 |
| 39 | 83 | 73 | 101 | 46 | 80 |

**a. Use a graphing calculator to construct a box-and-whisker plot for each set of data. Then describe the shape of each distribution.**

Enter Kurt's rushing yards from his junior season as **L1** and his rushing yards from his senior season as **L2**. Graph both box-and-whisker plots on the same screen by graphing **L1** as **Plot1** and **L2** as **Plot2**.

For Kurt's junior season, the right whisker is longer than the left, and the median is closer to the left whisker. The distribution is positively skewed.

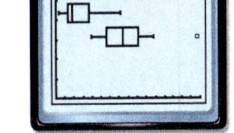

[0, 160] scl: 10 by [0, 5] scl: 1

For Kurt's senior season, the lengths of the whiskers are approximately equal, and the median is in the middle of the data. The distribution is symmetric.

---

**Technology Tip**

**Tools** In order to calculate statistics for a set of data in L2, press STAT ▶ ENTER 2nd [L2] ENTER.

**3A.**

2nd Period

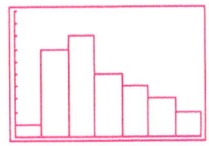

[0, 35] scl: 5 by [0, 10] scl: 1

7th Period

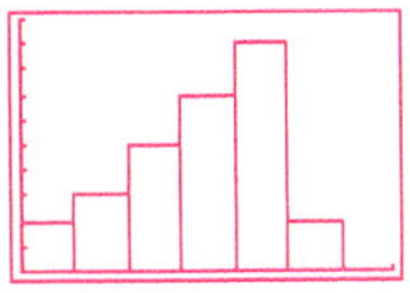

[0, 35] scl: 5 by [0, 10] scl: 1

2nd period, positively skewed; 7th period, negatively skewed

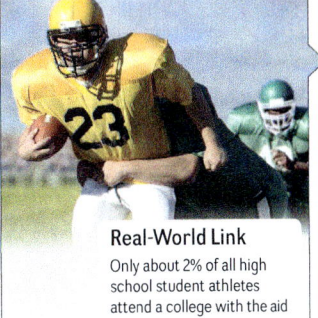

**Real-World Link**
Only about 2% of all high school student athletes attend a college with the aid of athletic scholarships.
**Source:** NCAA

---

## Additional Answer (Guided Practice)

**3B.** Sample answer: The distributions are skewed, so use the five-number summaries. Both classes have a minimum of 4, but 2nd period has a median of 13 and 7th period has a lower quartile of 13. This means that the lower 50% of the data from 2nd period spans the same range as the lower 25% of the data from 7th period. The upper 50% of the data from 2nd period spans from 13 to 33, while the upper 75% of data from 7th period spans from 13 to 26. Therefore, we can conclude that while the median for 7th period is significantly higher than the median for 2nd period, the data from 2nd period is dispersed over a wider range than the data from 7th period.

---

### Example 4  Compare Data Using Box-and-Whisker Plots

**AL** Describe the shapes of the box plots. The Junior Season box plot is positively skewed and the Senior Season box plot is symmetric.

**OL** Which statistics best represent the data? five-number summary

**BL** Describe the appearance of two box plots when one set of data is more spread out than the other. The data set that is more spread out will have a wider box plot.

### Need Another Example?

**Fishing** Steve and Mateo went fishing for the weekend. The weights of the fish they each caught are shown below.

| Steve's Fish (pounds) |
|---|
| 1.6, 2.1, 2.6, 1.3, 2.7, 3.2, 1.4, 2.3, 3.5, 1.9, 2.2, 2.7, 3.5, 1.4, 3.7, 3.4, 1.8, 2.5, 3.0 |

| Mateo's Fish (pounds) |
|---|
| 1.1, 3.2, 2.3, 3.7, 1.7, 2.7, 2.1, 4.0, 1.0, 2.9, 2.9, 1.2, 3.3, 2.3, 4.5, 2.4, 3.9 |

**a.** Use a graphing calculator to create a box-and-whisker plot for each data set. Then describe the shape of the distribution for each data set.

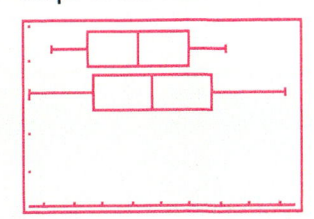

[1, 4.6] scl: 0.4 by [0, 5] scl: 1

both symmetric

**b.** Compare the distributions using either the means and standard deviations or the five-number summaries. Justify your choice. The distributions are symmetric, so use the means and standard deviations. The mean weight for Steve's fish is about 2.5 pounds with standard deviation of about 0.8 pound. The mean weight for Mateo's fish is about 2.7 pounds with standard deviation of about 1 pound. While the mean weight for Mateo's fish is greater, the weights of Mateo's fish also have more variability. This means the weights for Steve's fish are generally closer to his mean than the weights for Mateo's fish.

Lesson 10-5 | Comparing Sets of Data

# Practice

**Formative Assessment** Use Exercises 1–6 to assess students' understanding of the concepts in this lesson.

The Practice and Problem Solving exercises assess the content taught in the lesson. The Preparing for Assessment page is meant to be used as preparation for end-of-course assessments.

## Extra Practice

See page R10 for extra exercises for students who are approaching level or for on-level students who need additional reinforcement.

## Additional Answer (Guided Practice)

**4B.** Sample answer: One distribution is symmetric and the other is skewed, so use the five-number summaries. The maximum for Vanessa's junior season is 21, while the median for her senior season is 22. This means that in half of her games during her senior season, Vanessa scored more points than in any of the games during her junior season. Therefore, we can conclude that overall, Vanessa scored more points during her senior season.

## Additional Answers

**5a.**

Kyle's Distances

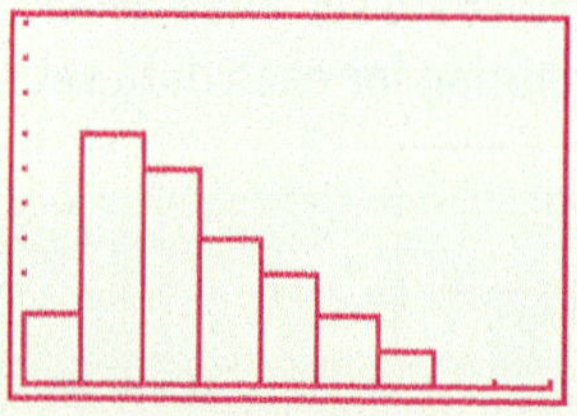

[17, 19.25] scl: 0.25 by [0, 10] scl: 1

Mark's Distances

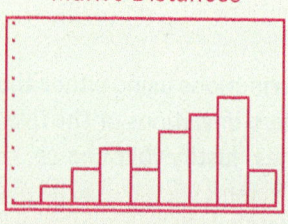

[17, 19.25] scl: 0.25 by [0, 10] scl: 1

Kyle, positively skewed; Mark, negatively skewed

## Go Online!

**eBook**

**Interactive Student Guide**
Use the *Interactive Student Guide* to deepen conceptual understanding.
· Comparing Sets of Data

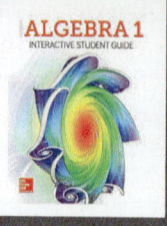

---

**Study Tip**

**Box-and-Whisker Plots** Recall that a box-and-whisker plot, also called a box plot, displays the spread of a data set by dividing it into four quartiles. Each quartile accounts for 25% of the data.

**4A.**

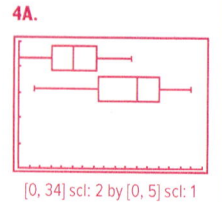

[0, 34] scl: 2 by [0, 5] scl: 1

junior season, symmetric; senior season, negatively skewed

**b.** Compare the data sets using either the means and standard deviations or the five-number summaries. Justify your choice.

One distribution is symmetric and the other is skewed, so use the five-number summaries to compare the data.

The upper quartile for Kurt's junior season was 38, while the minimum for his senior season was 39. This means that Kurt rushed for more yards in every game during his senior season than 75% of the games during his junior season.

The maximum for Kurt's junior season was 72, while his median for his senior season was 75. This means that in half of his games during his senior year, he rushed for more yards than in any game during his junior season. Overall, we can conclude that Kurt rushed for many more yards during his senior season than during his junior season.

**Guided Practice**

**BASKETBALL** The points Vanessa scored per game during her junior and senior seasons are shown.

**4A.** Use a graphing calculator to construct a histogram for each set of data. Then describe the shape of each distribution.

**4B.** Compare the data sets using either the means and standard deviations or the five-number summaries. Justify your choice. **See margin.**

| Junior Season (points) | Senior Season (points) |
|---|---|
| 10, 12, 6, 10, 13, 8, 12, 3, 21, 14, 7, 0, 15, 6, 16, 8, 17, 3, 17, 2 | 10, 32, 3, 22, 20, 30, 26, 24, 5, 22, 28, 32, 26, 21, 6, 20, 24, 18, 12, 25 |

---

**Check Your Understanding** ● = Step-by-Step Solutions begin on page R11.

**Go Online!** for a Self-Check Quiz

**Example 1** Find the mean, median, mode, range, and standard deviation of each data set that is obtained after adding the given constant to each value.
S.ID.2

1. 10, 13, 9, 8, 15, 8, 13, 12, 7, 8, 11, 12; + (−7)
3.5, 3.5, 1, 8, 2.4

2. 38, 36, 37, 42, 31, 44, 37, 45, 29, 42, 30, 42; + 23
60.8, 60.5, 65, 16, 5.5

**Example 2** Find the mean, median, mode, range, and standard deviation of each data set that is obtained after multiplying each value by the given constant.
S.ID.2

● 3. 6, 10, 3, 7, 4, 9, 3, 8, 5, 11, 2, 1; × 3
17.3, 16.5, 9, 30, 9.4

4. 42, 39, 45, 44, 37, 42, 38, 37, 41, 49, 42, 36; × 0.5
20.5, 20.8, 21, 6.5, 1.8

**Example 3** 5. **TRACK** Mark and Kyle's long jump distances are shown.
S.ID.2, S.ID.3

| Kyle's Distances (ft) | Mark's Distances (ft) |
|---|---|
| 17.2, 18.28, 18.56, 17.28, 17.36, 18.08, 17.43, 17.71, 17.46, 18.26, 17.51, 17.58, 17.41, 18.21, 17.34, 17.63, 17.55, 17.26, 17.18, 17.78, 17.51, 17.83, 17.92, 18.04, 17.91 | 18.88, 19.24, 17.63, 18.69, 17.74, 19.18, 17.92, 18.96, 18.19, 18.21, 18.46, 17.47, 18.49, 17.86, 18.93, 18.73, 18.34, 18.67, 18.56, 18.79, 18.47, 18.84, 18.87, 17.94, 18.7 |

**a.** Use a graphing calculator to construct a histogram for each set of data. Then describe the shape of each distribution. **a–b. See margin.**

**b.** Compare the data sets using either the means and standard deviations or the five-number summaries. Justify your choice.

---

### Differentiated Homework Options

| Levels | AL Basic | OL Core | BL Advanced |
|---|---|---|---|
| Exercises | 7–18, 22–30 | 7–19 odd, 23–29 | 19–25<br>26–29 (optional) |
| 2-Day Option | 7–17 odd, 26–29 | 7–18, 26–29 | |
| | 8–18 even, 22–25 | 19, 20, 22–25 | |

You can use ALEKS to provide additional remediation support with personalized instruction and practice.

**5b.** Sample answer: The distributions are skewed, so use the five-number summaries. Kyle's upper quartile is 17.98, while Mark's lower quartile is 18.065. This means that 75% of Mark's distances are greater than 75% of Kyle's distances. Therefore, we can conclude that overall, Mark's distances are longer than Kyle's.

**Lesson 10-5 | Comparing Sets of Data**

**Example 4**
S.ID.2, S.ID.3

**6. TIPS** Miguel and Stephanie are servers at a restaurant. The tips that they earned to the nearest dollar over the past 15 workdays are shown.

| Miguel's Tips ($) |
|---|
| 14, 68, 52, 21, 63, 32, 43, 35, 70, 37, 42, 16, 47, 38, 48 |

| Stephanie's Tips ($) |
|---|
| 34, 52, 43, 39, 41, 50, 46, 36, 37, 47, 39, 49, 44, 36, 50 |

a. Use a graphing calculator to construct a box-and-whisker plot for each set of data. Then describe the shape of each distribution. **a–b. See margin.**
b. Compare the data sets using either the means and standard deviations or the five-number summaries. Justify your choice.

**Practice and Problem Solving**

Extra Practice is on page R10.

**Example 1**
S.ID.2

Find the mean, median, mode, range, and standard deviation of each data set that is obtained after adding the given constant to each value.

7. 52, 53, 49, 61, 57, 52, 48, 60, 50, 47; + 8
   60.9, 60, 60, 14, 4.7
8. 101, 99, 97, 88, 92, 100, 97, 89, 94, 90; + (−13)
   81.7, 82.5, 84, 13, 4.5
9. 27, 21, 34, 42, 20, 19, 18, 26, 25, 33; + (−4)
   22.5, 21.5, no mode, 24, 7.4
10. 72, 56, 71, 63, 68, 59, 77, 74, 76, 66; + 16
    84.2, 85.5, no mode, 21, 6.8

**Example 2**
S.ID.2

Find the mean, median, mode, range, and standard deviation of each data set that is obtained after multiplying each value by the given constant.

11. 11, 7, 3, 13, 16, 8, 3, 11, 17, 3; × 4
    36.8, 38, 12, 56, 20.0
12. 64, 42, 58, 40, 61, 67, 58, 52, 51, 49; × 0.2
    10.8, 11, 11.6, 5.4, 1.7
13. 33, 37, 38, 29, 35, 37, 27, 40, 28, 31; × 0.8
    26.8, 27.2, 29.6, 10.4, 3.5
14. 1, 5, 4, 2, 1, 3, 6, 2, 5, 1; × 6.5
    19.5, 16.3, 6.5, 32.5, 11.6

**Example 3**
S.ID.2, S.ID.3

**15. BOOKS** The page counts for the books that the students chose are shown.

| 1st Period |
|---|
| 388, 439, 206, 438, 413, 253, 311, 427, 258, 511, 283, 578, 291, 358, 297, 303, 325, 506, 331, 482, 343, 372, 456, 267, 484, 227 |

| 6th Period |
|---|
| 357, 294, 506, 392, 296, 467, 308, 319, 485, 333, 352, 405, 359, 451, 378, 490, 379, 401, 409, 421, 341, 438, 297, 440, 500, 312, 502 |

a. Use a graphing calculator to construct a histogram for each set of data. Then describe the shape of each distribution. **a–b. See margin.**
b. Compare the data sets using either the means and standard deviations or the five-number summaries. Justify your choice.

**16. MINIATURE GOLF** A sample of scores for the Blue and Red courses at Top Golf are shown.

| Blue Course |
|---|
| 46, 25, 62, 45, 30, 43, 40, 46, 33, 53, 35, 38, 39, 40, 52, 42, 44, 48, 50, 35, 32, 55, 28, 58 |

| Red Course |
|---|
| 53, 49, 26, 61, 40, 50, 42, 35, 45, 48, 31, 48, 33, 50, 35, 55, 38, 50, 42, 53, 44, 54, 48, 58 |

a. Use a graphing calculator to construct a histogram for each set of data. Then describe the shape of each distribution. **a–b. See Ch. 12 Answer Appendix.**
b. Compare the data sets using either the means and standard deviations or the five-number summaries. Justify your choice.

**Example 4**
S.ID.2, S.ID.3

**17 BRAINTEASERS** The amount of time (in minutes) that it took Leon and Cassie to complete puzzles is shown.

| Leon's Times |
|---|
| 4.5, 1.8, 3.2, 5.1, 2.0, 2.6, 4.8, 2.4, 2.2, 2.8, 1.8, 2.2, 3.9, 2.3, 3.3, 2.4 |

| Cassie's Times |
|---|
| 2.3, 5.8, 4.8, 3.3, 5.2, 4.6, 3.6, 5.7, 3.8, 4.2, 5.0, 4.3, 5.5, 4.9, 2.4, 5.2 |

a. Use a graphing calculator to construct a box-and-whisker plot for each set of data. Then describe the shape of each distribution. **a–b. See Ch. 12 Answer Appendix.**
b. Compare the data sets using either the means and standard deviations or the five-number summaries. Justify your choice.

**15b.** Sample answer: One distribution is symmetric and the other is skewed, so use the five-number summaries. The lower quartile for 1st period is 291 pages, while the minimum for 6th period is 294 pages. This means that the lower 25% of data for 1st period is lower than any data from 6th period. The range for 1st period is 578 − 206 or 372 pages. The range for 6th period is 506 − 294 or 212 pages. The median for 1st period is about 351 pages, while the median for 6th period is 392 pages. This means that, while the median for 6th period is greater, 1st period's pages have a greater range and include greater values than 6th period.

**Levels of Complexity Chart**

The levels of the exercises progress from 1 to 3, with Level 1 indicating the lowest level of complexity.

| Exercises | 7–18 | 19–21, 26–29 | 22–25 |
|---|---|---|---|
| Level 3 |  |  | ● |
| Level 2 |  | ● |  |
| Level 1 | ● |  |  |

**Additional Answers**

**6a.**

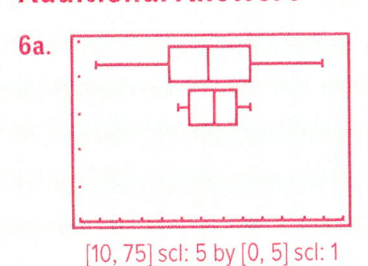

[10, 75] scl: 5 by [0, 5] scl: 1

both symmetric

**6b.** Sample answer: The distributions are symmetric, so use the means and standard deviations. The mean for Miguel's tips is about $41.73 with standard deviation of about $16.64. The mean for Stephanie's tips is about $42.87 with standard deviation of about $5.73. While the means only differ by $1.14, the standard deviations show that Stephanie's tips are generally closer to her mean than Miguel's tips.

**15a.**

1st Period

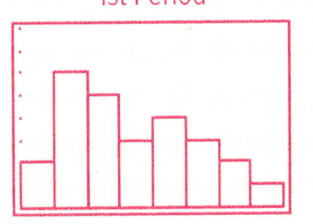

[200, 600] scl: 50 by [0, 8] scl: 1

6th Period

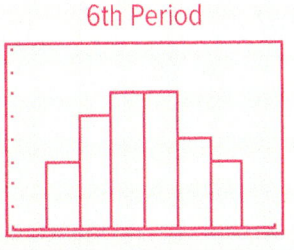

[200, 600] scl: 50 by [0, 8] scl: 1

1st period, positively skewed; 6th period, symmetric

**Go Online!**

**eSolutions Manual**
Create worksheets, answer keys, and solutions handouts for your assignments.

## Lesson 10-5 | Comparing Sets of Data

### Watch Out!

**Transformations** In Exercises 19 and 20, the data given has already been transformed. Students will need to subtract $5 from each of Rhonda's daily earnings, and will need to subtract 7% sales tax from each item Lorenzo purchased.

### MP Teaching the Mathematical Practices

**Regularity** Mathematically proficient students notice if calculations are repeated, and look both for general methods and for shortcuts. In Exercise 24, have students look at the transformations they have done to make a conclusion.

## Assess

**Ticket Out the Door** Have students summarize how the mean, median, mode, range, and standard deviation of a new data set can be found after a transformation is performed on the original data set.

## Additional Answers

**18b.** Sample answer: One distribution is symmetric and the other is skewed, so use the five-number summaries. The maximum for the boys is $131, while the upper quartile for the girls is $135.50. This means that 25% of the data from the girls is greater than all of the data from the boys. When listed from least to greatest, each statistic for the girls is greater than its corresponding statistic for the boys. We can conclude that in general, the girls spent more money on the dance than the boys.

**22.** Sample answer: The mean, median, and mode of the new data set can be found by multiplying each original statistic by $k$. The range and the standard deviation can be found by multiplying each original statistic by $|k|$.

**23.** Sample answer: Histograms show the frequency of values occurring within set intervals. This makes the shape of the distribution easy to recognize. However, no specific values of the data set can be identified from looking at a histogram, and the overall spread of the data can be difficult to determine. The box-and-whisker plots show the data divided into four sections. This aids when comparing the spread of one set of data to another. However, the box-and-whisker plots are limited because they cannot display the data any more specifically than showing it divided into four sections.

**24.** Sample answer: The mean, median, and mode of the new data set can be found by adding $k$ to each original statistic, and then multiplying each resulting value by $m$. Since the range and the

---

**18. DANCE** The total amount of money that a sample of students spent to attend the homecoming dance is shown.

| Boys (dollars) |
|---|
| 114, 98, 131, 83, 91, 64, 94, 77, 96, 105, 72, 108, 87, 112, 58, 126 |

| Girls (dollars) |
|---|
| 124, 74, 105, 133, 85, 162, 90, 109, 94, 102, 98, 171, 138, 89, 154, 76 |

a. Use a graphing calculator to construct a box-and-whisker plot for each set of data. Then describe the shape of each distribution. **See Ch. 12 Answer Appendix.**

b. Compare the data sets using either the means and standard deviations or the five-number summaries. Justify your choice. **See margin.**

**19. LANDSCAPING** Refer to the beginning of the lesson. Rhonda, another employee that works with Tom, earned the following over the past month.

a. Find the mean, median, mode, range, and standard deviation of Rhonda's earnings. **52.96, 53, 53, 19, 6.08**

b. A $5 bonus had been added to each of Rhonda's daily earnings. Find the mean, median, mode, range, and standard deviation of Rhonda's earnings before the $5 bonus. **47.96, 48, 48, 19, 6.08**

| Rhonda's Pay ($) | | |
|---|---|---|
| 45 | 55 | 53 |
| 47 | 53 | 54 |
| 44 | 56 | 59 |
| 63 | 47 | 53 |
| 60 | 57 | 62 |
| 44 | 50 | 45 |
| 60 | 53 | 49 |
| 62 | 47 | |

**20. SHOPPING** The items Lorenzo purchased are shown.

a. Find the mean, median, mode, range, and standard deviation of the prices. **18.95, 16.05, no mode, 40.66, 11.62**

b. A 7% sales tax was added to the price of each item. Find the mean, median, mode, range, and standard deviation of the items without the sales tax. **17.71, 15, no mode, 38, 10.86**

| | |
|---|---|
| Baseball hat | $14.98 |
| Jeans | $24.61 |
| T-shirt | $12.84 |
| T-shirt | $16.05 |
| Backpack | $42.80 |
| Folders | $2.14 |
| Sweatshirt | $19.26 |

**21. MP MODELING** A salesperson has 15 SUVs priced between $33,000 and $37,000 and 5 luxury cars priced between $44,000 and $48,000 on a car lot. The average price for all of the vehicles is $39,250. The sales manager decides to reduce the prices of all 15 SUVs on the car lot by $2000 per vehicle. What is the new average price for the SUVs and luxury cars? **$37,750**

S.ID.2 S.ID.3

### H.O.T. Problems   Use Higher-Order Thinking Skills

**22. MP REASONING** If every value in a set of data is multiplied by a constant $k$, $k < 0$, then how is the mean, median, mode, range, and standard deviation of the new data set found? **See margin.**

**23. WRITING IN MATH** Compare and contrast the benefits of displaying data using histograms and box-and-whisker plots. **23–24. See margin.**

**24. MP REGULARITY** If $k$ is added to every value in a set of data, and then each resulting value is multiplied by a constant $m$, $m > 0$, how can the mean, median, mode, range, and standard deviation of the new data set be found? Explain your reasoning.

**25. WRITING IN MATH** Explain why the mean and standard deviation are used to compare the center and spread of two symmetrical distributions and the five-number summary is used to compare the center and spread of two skewed distributions or a symmetric distribution and a skewed distribution. **See margin.**

### MP Standards for Mathematical Practice

| Emphasis On | Exercises |
|---|---|
| 1 Make sense of problems and persevere in solving them. | 1–4, 7–14, 26–29 |
| 2 Reason abstractly and quantitatively. | 22, 27–29 |
| 3 Construct viable arguments and critique the reasoning of others. | 23, 25, 27–29 |
| 4 Model with mathematics. | 5–6, 15–21, 26–29 |

---

standard deviation are not affected when a constant is added to a set of data, they can be found by multiplying each original value by the constant $m$.

**25.** Sample answer: When two distributions are symmetric, the first thing to determine is how close the averages are and how spread out each set of data is. The mean and standard deviation are the best values to use for this comparison. When distributions are skewed, we also want to determine the direction and the degree to which it is skewed. The mean and standard deviation cannot provide any information in this regard, but we can get this information by comparing the range, quartiles, and medians found in the five-number summaries. So, if one or both sets of data are skewed, it is best to compare their five-number summaries.

Lesson 10-5 | Comparing Sets of Data

## Preparing for Assessment

**26.**

32, 18, 12, 33, 20, 17, 23, 42, 28, 27, 12, 9, 21, 16, 32, 37

MP 1, 4   S.ID.2

a. 22
b. 27
c. 23.7
d. 28.7
e. 33
f. 33

27b. Their means and standard deviations are more appropriate because both distributions are relatively symmetric.

**27. MULTI-STEP**

MP 1, 2, 4   S.ID.2, S.ID.3

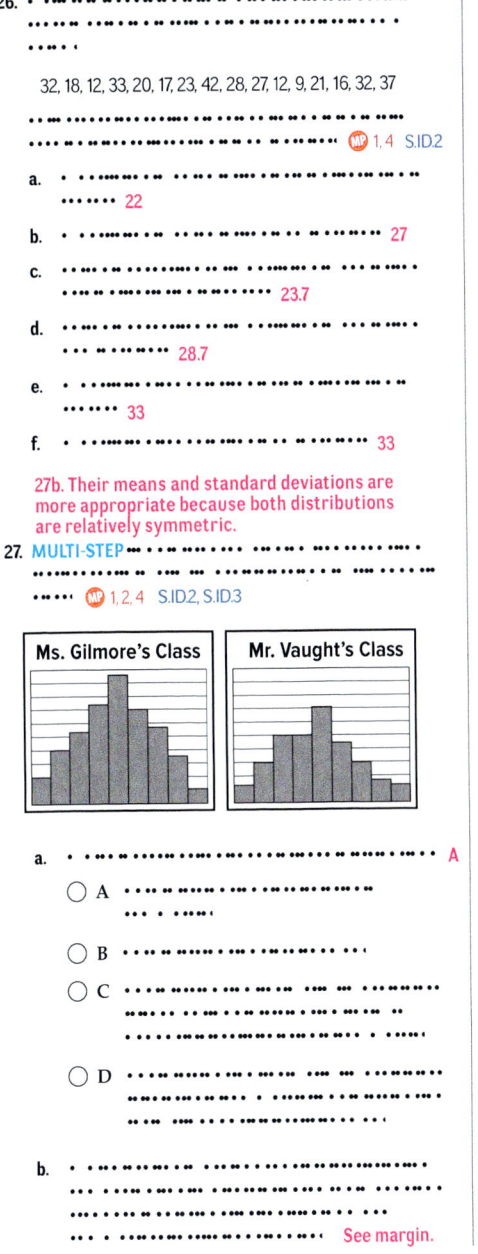

Ms. Gilmore's Class    Mr. Vaught's Class

a.
○ A
○ B
○ C
○ D

A

b.
See margin.

**28. MULTI-STEP**

MP 1, 2, 4   S.ID.2, S.ID.3

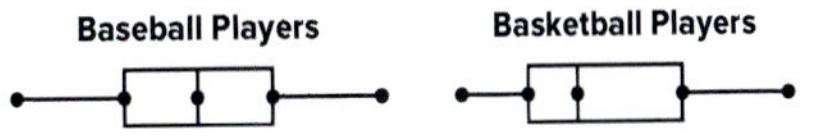

Baseball Players    Basketball Players

a.
○ A
○ B
○ C
○ D

C

b.
Their five-number summaries are more appropriate because one or more of the distributions is skewed.

**29. MULTI-STEP**

MP 1, 2, 4   S.ID.2, S.ID.3

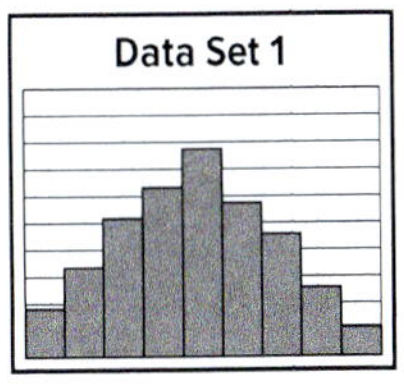

Data Set 1    Data Set 2

a.
○ A
○ B
○ C
○ D

C

b.
29b. Their five-number summaries are more appropriate because one of the distributions is skewed.

## Preparing for Assessment

Exercises 26–29 require students to use the skills they will need on standardized assessments. Each exercise is dual-coded with content standards and mathematical practice standards.

| Dual Coding |||
|---|---|---|
| Items | Content | MP Mathematical Practices |
| 26 | S.ID.2 | 1, 4 |
| 27 | S.ID.2, S.ID.3 | 1, 2, 4 |
| 28 | S.ID.2 | 1, 2, 4 |
| 29 | S.ID.2 | 1, 2, 4 |

### Diagnose Student Errors

Survey student responses for each item. Class trends may indicate common errors and misconceptions.

**27a.**

| A | CORRECT |
|---|---|
| B | Did not understand the meaning of *skewed* |
| C | Did not understand the meaning of *skewed* |
| D | Did not understand the meaning of *skewed* |

**28a.**

| A | Did not understand the meaning of *relatively symmetric* |
|---|---|
| B | Did not understand the meaning of *skewed* |
| C | CORRECT |
| D | Did not realize that meaning of the terms |

**29a.**

| A | Did not understand the meaning of *relatively symmetric* |
|---|---|
| B | Did not understand the meaning of *relatively symmetric* |
| C | CORRECT |
| D | Did not understand the meaning of *relatively symmetric* |

## Differentiated Instruction  OL  BL

**Extension** Have students research the Internet to find data about two cities in the United States. This could include population, median household incomes, or other data. Ask students to make box-and-whisker plots of each data set and compare them. Their analyses should include a comparison using either the means and standard deviations or the five-number summaries.

## Additional Answer

**27b.** Their means and standard deviations are more appropriate because both distributions are relatively symmetric.

## Go Online!

### Quizzes

Students can use *Self-Check Quizzes* to check their understanding of this lesson. You can also give the *Chapter Quiz*, which covers the content in Lessons 10-4 and 10-5.

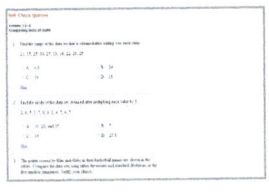

connectED.mcgraw-hill.com    687

# LESSON 10-6
## Summarizing Categorical Data

**SUGGESTED PACING (DAYS)**
90 min. — 1
45 min. — 1.5
Instruction

## Track Your Progress

### Objectives
1. Summarize data in two-way frequency tables.
2. Summarize data in two-way relative frequency tables.

### Mathematical Background
Arranging categorical frequency data in two-way frequency tables makes it easier to compare and contrast the data categories. Turning this data into percentages makes it even easier to compare. This is called a relative frequency table. For even more detailed comparison of data, conditional relative frequency can be calculated within data categories.

### THEN
**SID.2** Use statistics appropriate to the shape of the data distribution to compare center (median, mean) and spread (interquartile range, standard deviation) of two or more different data sets.

**SID.3** Interpret differences in shape, center, and spread in the context of the data sets, accounting for possible effects of extreme data points (outliers).

### NOW
**SID.5** Summarize categorical data for two categories in two-way frequency tables. Interpret relative frequencies in the context of the data (including joint, marginal, and conditional relative frequencies). Recognize possible associations and trends in the data.

### NEXT
**SID.6** Represent data on two quantitative variables on a scatter plot, and describe how the variables are related.

## Go Online! All of these resources and more are available at connectED.mcgraw-hill.com

Use the **eGlossary** to define relative frequency and other key vocabulary in the lesson.
*Use at Beginning of Lesson*

**eToolkit** allows students to explore and enhance their understanding of math concepts.
*Use with Examples*

The **Chapter Project** allows students to create and customize a project as a non-traditional method of assessment.
*Use at End of Lesson*

## Using Open Educational Resources
**Apps** Have students access **Google Apps for Education** to collaborate on tips for making two-way frequency tables. As an educator it also offers spreadsheets, calendars, and surveys. You can also try **WhoTeaches**, or **TeachAde**. *Use as planning tool*

# Differentiate Your Resources

**Extra Practice** Additional practice or homework; Skills Practice is best for approaching-level students and Practice is best for on-level and beyond-level students

## Skills Practice
AL OL ELL

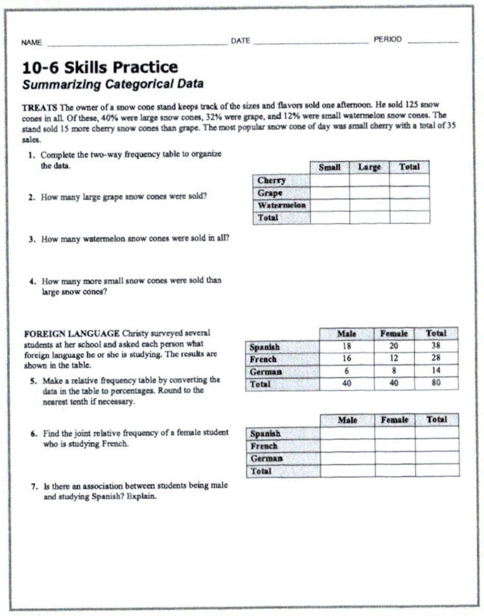

## Practice
AL OL BL ELL

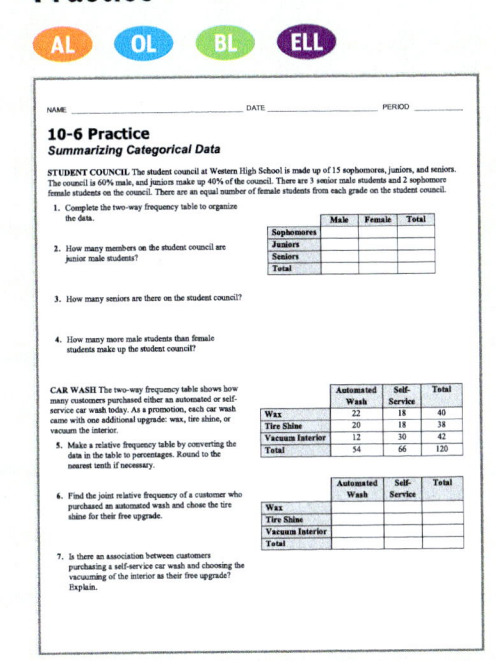

## Word Problem Practice
AL OL BL ELL

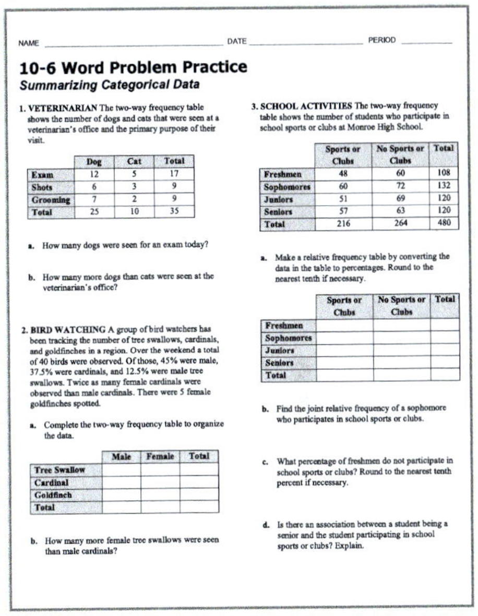

---

**Intervention** Reteaching and vocabulary activities that can be used with struggling or absent students and as ELL support

**Extension** Activities that can be used to extend lesson concepts

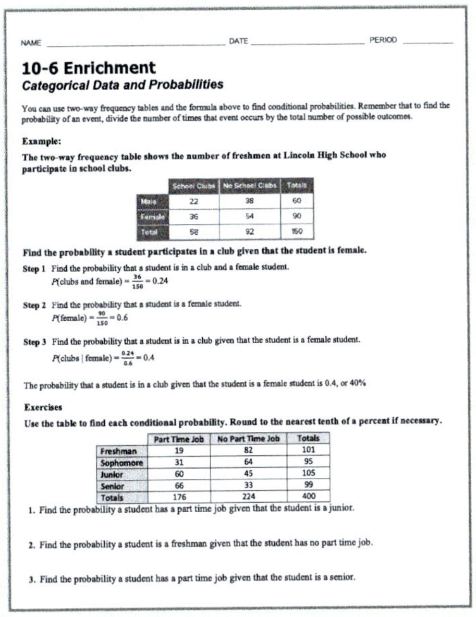

## Study Guide and Intervention
AL OL ELL

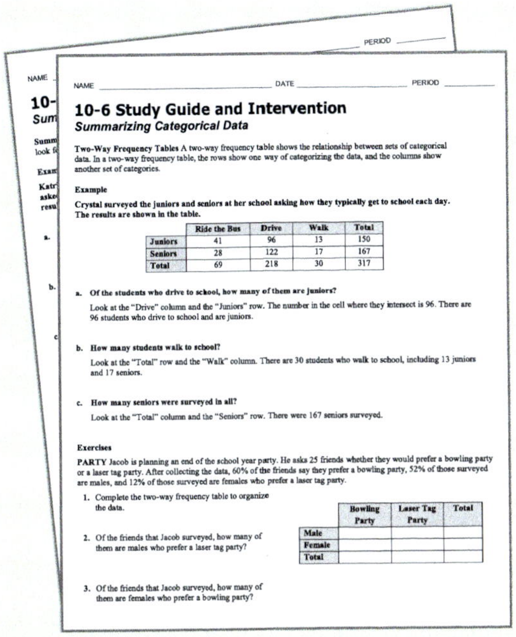

## Study Notebook
AL OL BL ELL

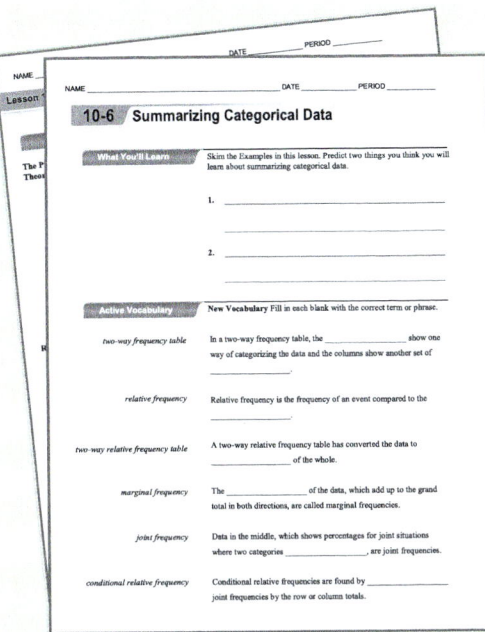

## Enrichment
OL BL ELL

Lesson 10-6 | Summarizing Categorical Data

# Launch

Have the students read the Why? section of the lesson.

Ask:
- Which age group seems to want skateboarding facilities more? **ages 12–25**
- Which activities does the older age group like more?
  **walking and running**
- Who seems to use the park more already, or has answered the survey in greater numbers? **ages 12–25**
- How might this be useful for the city to know?
  **If the younger group will use it more, then perhaps the funding should be spent the way that they would enjoy the most.**

# Teach

Ask the scaffolded questions for each example to build conceptual understanding for students at all levels.

## 1 Two-Way Frequency Tables

### Example 1 Interpret a Two-Way Frequency Table

**AL** Where in the chart do you find the total of all students? **the bottom right corner**

**OL** What do the categories along the top of the chart mean? **students' grade level**
What do the categories along the left of the chart mean? **students' preferences in game types**

**BL** Does it make sense that the bottom row has the same total as the rightmost column of data? Explain. **Yes. The total number of students in each class should equal the total number for game preferences.**

**Go Online!**

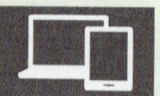

The most up-to-date resources available for your program can be found at connectED.mcgraw-hill.com.

688 | Lesson 10-6 | Summarizing Categorical Data

---

## LESSON 6
# Summarizing Categorical Data

**Then**
- You measured central tendency and spread to describe data sets.

**Now**
1. Summarize data in two-way frequency tables.
2. Summarize data in two-way relative frequency tables.

**Why?**
- To make plans for a community park, the city may survey residents about how they use the park and what features they would like.

| Age | Skate-boarding | Walking/Running | Total |
|---|---|---|---|
| 12–25 | 678 | 82 | 760 |
| 26–99 | 64 | 435 | 499 |
| Total | 742 | 517 | 1259 |

A two-way frequency table can help city officials quantify the residents' preferences and behavior patterns.

**New Vocabulary**
two-way frequency table
relative frequency
two-way relative frequency table
marginal frequency
joint frequency
conditional relative frequency
association

**Mathematical Practices**
4 Model with mathematics.
7 Look for and make use of structure.
8 Look for and express regularity in repeated reasoning.

**Content Standards**
S.ID.5 Summarize categorical data for two categories in two-way frequency tables. Interpret relative frequencies in the context of the data (including joint, marginal, and conditional relative frequencies). Recognize possible associations and trends in the data.

**1 Two-Way Frequency Tables** A two-way frequency table or contingency table is used to show the frequencies of data from a survey or experiment classified according to two categories, with rows indicating one category and columns indicating the other. The subcategories are the column and row headers of the table that represent the two different types of categories. Using a two-way frequency table allows people to improve their community planning, supply ordering, and other functions where they need to know the preferences of a variety of people.

S.ID.5

**Real-World Example 1**   Interpret a Two-Way Frequency Table

**STUDENT CENTER** Organizers at Lincoln High School are planning a new student center. They surveyed freshmen and sophomores about what types of games they would like to have, video games, board games, or table sports. Students were asked to select only one type of game. The results are shown in the table.

| Student Center Preferences | Freshmen | Sophomores | Total |
|---|---|---|---|
| Video Games | 20 | 170 | 190 |
| Board Games | 55 | 35 | 90 |
| Table Sports | 30 | 190 | 220 |
| Total | 105 | 395 | 500 |

a. Of the students who chose table sports, how many of them are freshmen?
   Find the value that is in both the *Freshmen* column and *Table Sports* row. There are 30 students who chose table sports and are freshmen.

b. How many students chose video games?
   Find the value that is in both the *Total* column and *Video Games* row. There are 190 students who chose video games.

c. How many students are freshmen?
   Find the value that is in both the *Freshmen* column and *Total* row. There are 105 students who are freshmen.

---

## MP Mathematical Practices Strategies

**Make sense of problems and persevere in solving them.**
Help students understand how to follow the logic through the charts to fill in numbers when they are missing.

- Where do the totals appear in the charts? **the right column and bottom rows**
- Where is the *total* total? **the bottom right cells**
- How would you find a value in the center of the chart? **Subtract another number in the same row or column from the margin totals. These values are called joint frequency data values.**
- Why are the data in the center of the table called joint frequency data? **They describe a joint situation of circumstances.**
- Why would the values in the right column and bottom row be called margin frequency data? **They are the totals at the margins of the table.**

**Lesson 10-6** | Summarizing Categorical Data

d. **How many students were surveyed in total?**
   Find the value that is in both the *Total* row and column. There are a total of 500 students who were surveyed.

e. **One teacher suggested that the budget for the student center should be split evenly between the freshmen and sophomores. Another teacher disagrees. What argument can be made?**
   The table shows that only 105 out of 500, or about 20% of students, are freshmen. Therefore, it might make more sense for 80% of the money to be spent on the preferences of the sophomores.

### Guided Practice

**CRAFTS** The table shows craft options chosen by freshmen and sophomores at a camp event.

| Craft | Freshmen | Sophomores | Total |
|---|---|---|---|
| Sand Art | 20 | 17 | 37 |
| Key Chain | 15 | 18 | 33 |
| Picture Frame | 12 | 6 | 18 |
| Total | 47 | 41 | 88 |

**1A.** Which craft was most popular overall? **sand art**
**1B.** Which craft was most popular with the sophomores? **key chain**
**1C.** How many freshmen attended the camp event? **47**
**1D.** How many attendees did crafts overall? **88**

You can use a two-way frequency table and information that is known about some parts of a situation to deduce the rest of the data for the table.

S.ID.5

**Real-World Example 2** Complete a Two-Way Frequency Table

**ACTIVITIES** A small town conducted a survey to help decide on which community initiatives they should spend annual funding. The town has 2400 residents who responded to the survey. According to the survey, 45% of respondents are cyclists, of whom 80% enjoy spending time at the lake. Of the residents who are not cyclists, 75% enjoy spending time at the lake. Make a two-way frequency table to display the data.

How many residents reported that they enjoy spending time at the lake?

**Step 1** Make a table and label the two sets of categories. Then, complete the table with the values from the given information.

Based on the only information given, only the total can be added to the table. The remaining values of the table must be calculated.

**Survey Results**

| Resident | Enjoy Lake | Do Not Enjoy Lake | Total |
|---|---|---|---|
| Cyclists | | | |
| Non-Cyclists | | | |
| Total | | | 2400 |

**Step 2** Find the values for the remaining cells of the table.
45% of the 2400 respondents are cyclists.
$(0.45)2400 = 1080$   There are 1080 total cyclists.

**Study Tip**

**Appropriate Tools** You can use spreadsheet software to make a two-way frequency table. Use formulas to calculate missing values.

---

### Need Another Example?

The table shows the animals seen by a veterinary office in one week.

| Animal | Sick | Regular Checkups | Total |
|---|---|---|---|
| Cats | 2 | 27 | 29 |
| Dogs | 6 | 40 | 46 |
| Other | 5 | 14 | 19 |
| Total | 13 | 81 | 94 |

a. How many pets were sick or injured this week? **13**
b. How many pets other than cats or dogs visited the office this week? **19**
c. How many cats had regular checkups? **27**
d. If the veterinarian takes approximately the same amount of time to examine a sick animal as a regular checkup, how should she plan to schedule her time? **She should plan to spend about 14% of her time with sick animals and 86% of her time on regular checkups.**

### Example 2 Use Data to Complete a Two-Way Frequency Table

**AL** How do you find the total number of cyclists? **Multiply by the percent given.**

**OL** How do you find the total number of people who do not enjoy cycling? **Subtract the cyclists from the total.**

**BL** How do you check that the totals work out? **Add them both vertically and horizontally.**

### Need Another Example?

The junior and senior classes were surveyed to determine their preference for the theme of the winter dance, either *Fire and Ice* or *Winter Wonderland*. Of the 250 students who responded, 42% were juniors. Of the seniors, 60% preferred *Fire and Ice* as the theme. *Winter Wonderland* was selected by 80% of the juniors. Complete a two-way frequency table.

| Theme | Juniors | Seniors | Total |
|---|---|---|---|
| Winter Wonderland | 84 | 58 | 142 |
| Fire and Ice | 21 | 87 | 108 |
| Total | 105 | 145 | 250 |

connectED.mcgraw-hill.com   689

## Lesson 10-6 | Summarizing Categorical Data

> **MP Teaching the Mathematical Practices**
>
> **Tools** Mathematically proficient students make sound decisions about when tools might be helpful, recognizing both the insight to be gained and their limitations. Discuss how and when using a two-way frequency table would be useful for analyzing and organizing data.

Of the cyclists, 80% enjoy spending time at the lake.
(0.80)1080 = 864     There are 864 cyclists who enjoy the lake.

To find the number of cyclists who do not enjoy the lake, find the difference of the total number of cyclists and the number of cyclists who enjoy the lake.
1080 − 864 = 216     There are 216 cyclists who do not enjoy the lake.

To find the total number of non-cyclists, find the difference of the total respondents and the cyclists.
2400 − 1080 = 1320     There are 1320 non-cyclists.

Of the residents who are non-cyclists, 75% enjoy spending time at the lake.
(0.75)1320 = 990     There are 990 non-cyclists who enjoy the lake.

To find the number of non-cyclists who do not enjoy the lake, find the difference of the total number of non-cyclists and the number of non-cyclists who enjoy the lake.
1320 − 990 = 330     There are 330 non-cyclists who do not enjoy the lake.

Find the totals for each column by adding cyclists and non-cyclists.
Enjoy Lake: 864 + 990 = 1854
Do Not Enjoy Lake: 216 + 330 = 546

| Survey Results | | | |
|---|---|---|---|
| Resident | Enjoy Lake | Do Not Enjoy Lake | Total |
| Cyclists | 864 | 216 | 1080 |
| Non-Cyclists | 990 | 330 | 1320 |
| Total | 1854 | 546 | 2400 |

> **Study Tip**
>
> **MP Structure** When completing a frequency table, the *Total* row and the *Total* column must add up to the same value.

Check that the totals of each column have a sum equal to the total number of respondents.

### Guided Practice

2. **BIRDS** A bird-watching club has counted all of the birds of a two certain species that they found in a park one afternoon. There are two color variations: plain wings and red-banded wings. They can tell the males and females apart by other color markings. The club found 18 males and a total of 42 birds. Of the females, 75% had plain wings, and 50% of the males had plain wings. Complete the two-way frequency table.

| Gender | Red-Banded Wings | Plain Wings | Total |
|---|---|---|---|
| Males | 15 | 9 | 18 |
| Females | 9 | 18 | 24 |
| Total | 6 | 27 | 42 |

**2 Two-Way Relative Frequency Tables** A **relative frequency** is the ratio of the number of observations in a category to the total number of observations. To create a **two-way relative frequency table**, divide each of the values by the total number of observations and replace them with their corresponding decimals or percents.

Relative frequency tables express the categories of information as percentages. Then the data can be compared, to help decide if there is an association between two events.

690 | Lesson 10-6 | Summarizing Categorical Data

Some parts of a relative frequency table have specific names. The subtotals of each subcategory, are called **marginal frequencies**.

Data which shows percentages for joint situations where two categories intersect, are called **joint frequencies**.

S.ID.5

### Real-World Example 3    Two-Way Relative Frequency Table

**GYM** A manager of a gym conducted a survey to determine whether the gym members prefer rock climbing, weight lifting, or neither. She also asked members whether they regularly bring guests to the gym. The frequency table below shows the surveyed responses of the surveyed members.

| Activity | Bring Guests | Do Not Bring Guests | Total |
|---|---|---|---|
| Weight Lifting | 86 | 2 | 88 |
| Cooking Class | 24 | 40 | 64 |
| Neither | 16 | 32 | 48 |
| Total | 126 | 74 | 200 |

**a.** Make a relative frequency table by converting the data in the table to percentages.

Divide each value in the table by the total number of members and multiply by 100% to find each frequency. For example:

Percentage of total members who bring guests: $\frac{126}{200} \cdot 100\% = 63\%$

Percentage of total members who neither prefer rock climbing nor weight lifting: $\frac{48}{200} \cdot 100\% = 24\%$

| Activity | Bring Guests | Do Not Bring Guests | Total |
|---|---|---|---|
| Rock Climbing | 43% | 1% | 44% |
| Weight Lifting | 12% | 20% | 32% |
| Neither | 8% | 16% | 24% |
| Total | 63% | 37% | 100% |

**b.** Find the joint frequency of rock climbing and bringing guests.

Find the value of both the *Rock Climbing* row and the *Bring Guests* column, which is 43%.

**Reading Math**

*MP* **Finding Patterns**
In a relative frequency table, the sum of *Total* row and column should be equal 100%. You can use this fact to check that your math is correct.

### Guided Practice

**PETS** The table shows data about pet owners having dogs and cats.

| Pet Owners | Owns a Dog | Does Not Own a Dog | Total |
|---|---|---|---|
| Owns a Cat | 6  15% | 11  27.5% | 17; 42.5% |
| Does Not Own a Cat | 9  22.5% | 14  35% | 23; 57.5% |
| Total | 15; 37.5% | 25; 62.5% | 40; 100% |

**3A.** Fill in the missing numbers for the frequency table. Then convert the frequency table to a relative frequency table by changing the numbers to percentages of the whole.

**3B.** What is the joint frequency data for someone owning neither a dog nor a cat? 35%

---

**Lesson 10-6** | Summarizing Categorical Data

### Example 3 Two-Way Relative Frequency Table

**AL** What number do you divide by to find the percentages for the two-way relative frequency tables? the grand total

**OL** Are the percentages expected to add to 100%? yes, except for slight possible differences due to rounding

**BL** What is the benefit of changing the numbers to percentages? You can compare the data on the same basis.

### Need Another Example?

**SCHEDULES** The general manager at Sam's Sports Shop is trying to determine the best way to schedule the managers and staff based on their shift preferences. The frequency table shows the responses of the employees.

| Employee | Day | Evening | Total |
|---|---|---|---|
| Management | 5 | 3 | 8 |
| Staff | 10 | 22 | 32 |
| Total | 15 | 25 | 40 |

**a.** Make a relative frequency table using the data.

| Employee | Day | Evening | Total |
|---|---|---|---|
| Management | 12.5% | 7.5% | 20% |
| Staff | 25% | 55% | 80% |
| Total | 37.5% | 62.5 | 100% |

**b.** What is the joint frequency data for a staff member who would prefer to work the day shift? 25%

---

**Watch Out!**

**Finding Errors** Have students check that the marginal frequencies add to 100%. If they add to a number less than or greater than 100%, have students determine whether the error is due to reasonable rounding or a mathematical error.

---

connectED.mcgraw-hill.com  691

## Lesson 10-6 | Summarizing Categorical Data

### Example 4  Conditional Relative Frequency

**AL** Is there a difference between the junior varsity and varsity squads for color preference? How do you know?  **Yes, because their conditional relative frequencies are very different.**

**OL** Is it possible to put all of the conditional relative frequencies into the same table format?  **No, because the frequencies could be conditional in the horizontal or in the vertical direction.**

**BL** What is the benefit of using conditional relative frequencies instead of just relative frequencies?  **You can compare the data on a better, and equivalent basis.**

### Need Another Example?

**Transportation** The two-way relative frequency table shows the how freshmen and sophomores of a certain high school typically arrive to school each day.

| Transportation | Freshmen | Sophomores | Total |
|---|---|---|---|
| Bus | 27% | 17% | 44% |
| Drive | 10% | 21% | 31% |
| Walk or Bike | 15% | 10% | 25% |
| Total | 52% | 48% | 100% |

a. What percentage of students who drive to school are sophomores?  ≈ 68%
b. What percentage of students walk or bike to school are freshmen?  60%
c. What percentage of sophomores ride the bus to school?  ≈ 35%
d. What percentage of freshmen drive to school?  ≈ 19%

---

You can refine the comparisons between data points even more by examining the conditional relative frequency. A **conditional relative frequency** is the ratio of the joint frequency to the marginal frequency. Because each two-way frequency table has two categories, each two-way relative frequency table can provide two different conditional relative frequency tables.

Conditional frequencies can be used to compare data in a meaningful way and to eliminate bias that might appear because of skewed data sets. Bias can result when one group has a larger population than another.

S.ID.5

**Real-World Example 4**   Conditional Relative Frequency

**DANCE**  The dance squad is planning to buy new uniforms. The coach asked members of the varsity and junior varsity squads which color of uniform they would prefer. The results are recorded in a relative frequency table.

| Color | Junior Varsity | Varsity | Total |
|---|---|---|---|
| Black | 22% | 16% | 38% |
| Blue | 11% | 19% | 30% |
| Red | 28% | 4% | 32% |
| Total | 61% | 39% | 100% |

a. What percentage of members who prefer blue uniforms are on the varsity squad?

$\frac{19}{30} \cdot 100\% \approx 63.3\%$

b. What percentage of the junior varsity squad prefers red uniforms?

$\frac{28}{61} \cdot 100\% \approx 45.9\%$

c. What percentage of members who prefer black uniforms are on the junior varsity squad?

$\frac{22}{38} \cdot 100\% \approx 57.9\%$

d. What percentage of the varsity squad prefers blue uniforms?

$\frac{19}{39} \cdot 100\% \approx 48.7\%$

**Study Tip**

**Precision**  When rounding percentages in your calculations, check that sums of the percentages are close to the expected totals.

### Guided Practice

Use the relative frequency table in Example 4 to answer the following question.

**4A.** What percentage of members who prefer red uniforms are on the varsity squad? **87.5%**

**4B.** What percentage of members who prefer blue uniforms are on the junior varsity squad? **36.7%**

**4C.** What percentage of the varsity squad prefers black uniforms? **41.0%**

**4D.** What percentage of the junior varsity squad prefers blue uniforms? **18.0%**

If the conditional frequency of one group is very different from that of another group, then there is an **association** between the two categories of data.

If the conditional frequencies of two groups are similar, then there may not be an association between the two data categories, and the combinations are simply formed by random chance.

# Lesson 10-6 | Summarizing Categorical Data

**Study Tip**

**MP Structure** Note that there are different ways to compare data when looking for associations. You can calculate each joint data point as a percentage of the marginal data point to the right of the row, or at the bottom of the column.

**Real-World Example 5** Associations Between Data in Relative Frequency Tables

**LIBRARY** ⋯

| Distance (mi) | Weekend Book Club | After School Tutoring | Total |
|---|---|---|---|
| < 1 | 7% | 36% | 43% |
| 1–3 | 30% | 1% | 31% |
| > 3 | 16% | 10% | 26% |
| Total | 53% | 47% | 100% |

a. ⋯

$\frac{30}{53} \cdot 100\% \approx 57\%$

b. ⋯

$\frac{36}{47} \cdot 100\% \approx 77\%$

c. ⋯

Yes, there seems to be a stronger association between living very close to the library and after school tutoring than there is between weekend book club and living farther from the library.

### Guided Practice

**5A. GARDENING** Ronald noticed that several homes in his neighborhood have flowerbeds and gardens. He also noticed that some neighbors seemed to participate more in community activities. He collected the following information. Is there an association between having flowerbeds and participating in community events?

| Home Landscape | Participate in at Least Half of Community Events | Participate in Less than Half of Community Events | Total |
|---|---|---|---|
| Flowerbeds/Gardens | 30 | 10 | 40 |
| Plain Lawn | 20 | 80 | 100 |
| Total | 50 | 90 | 140 |

**5A.** The conditional frequency of garden-keepers attending events is 75%. The conditional frequency of non-garden-keepers attending events is 20%. Therefore, there is an association.

## Check Your Understanding

= Step-by-Step Solutions begin on page R11.

**Go Online!** for a Self-Check Quiz

**Example 1** **EVENTS** ⋯
**S.ID.5**

| Role | Normans | Saxons | Total |
|---|---|---|---|
| Peasants | 42 | 36 | 78 |
| Nobles | 29 | 14 | 43 |
| Total | 71 | 50 | 121 |

1. How many Norman nobles were there? **29**
2. How many Saxons were there in total? **50**
3. How many peasants were there in total? **78**

---

**Example 5** Associations Between Data and Relative Frequency Tables

**AL** What is an association in the data? When there seems to be a trend because of how the data changes in one category compared to another, there is an association.

**OL** How do you know there is an association in this data set? When the joint frequency data is divided by the marginal frequency data for the categories, there is a clear difference.

**BL** What other conclusions or association can you make from the data? Students who live close to the library are uninterested in the book club. Students who live between 1 and 3 miles from the library are uninterested in tutoring.

### Need Another Example?

**Transportation** Refer to the conditional frequency table in Need Another Example 4.

a. What percentage of the sophomores drive to school? $\approx 44\%$
b. What percentage of the freshmen ride the bus to school? $\approx 52\%$
c. Is there an association between students' grade level and driving to school or riding the bus? Yes, they are both associated, though the association between freshmen and riding the bus is stronger.

---

## MP Teaching the Mathematical Practices

**Modeling** Mathematically proficient students are able to identify important qualities in practical situations and map their relationships in two-way tables. They can analyze relationships mathematically to draw conclusions. Encourage students to think about their results in the context of the situation and to determine whether their results make sense using their own experiences.

Lesson 10-6 | Summarizing Categorical Data

# Practice

**Formative Assessment** Use Exercises 1–14 to assess students' understanding of the concepts in this lesson. The Preparing for Assessment page is meant to be used as preparation for end-of-course assessment.

### Teaching the Mathematical Practices

**Sense-Making** Mathematically proficient students can explain the correspondences between a verbal description and a two-way frequency table. In Exercises 10-12, have students identify relevant information before creating the table. Then, read the information once again after the table is complete.

### Levels of Complexity Chart

The levels of the exercises progress from 1 to 3, with Level 1 indicating the lowest level of complexity.

| Exercises | 15–35 | 36, 37, 41–43 | 38–40 |
|---|---|---|---|
| Level 3 | | | ● |
| Level 2 | | ● | |
| Level 1 | ● | | |

## Additional Answers

**5.**

| Role | Normans | Saxons | Total |
|---|---|---|---|
| Peasants | 35% | 30% | 64% |
| Nobles | 24% | 12% | 36% |
| Total | 59% | 41% | 100% |

**9.**

| Gender | Retriever | Small Dog | Total |
|---|---|---|---|
| Female | 24 | 20 | 44 |
| Male | 24 | 12 | 36 |
| Total | 48 | 32 | 80 |

**10.**

| Nationality | Full Suit | Spring Skiwear | Total |
|---|---|---|---|
| Norwegian | 21% | 29% | 50% |
| Swiss | 12% | 21% | 33% |
| American | 4% | 13% | 17% |
| Total | 37% | 63% | 100% |

### Go Online!  eBook

**Interactive Student Guide**
Use the *Interactive Student Guide* to deepen conceptual understanding.
• Two-Way Frequency Tables

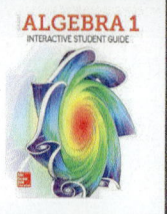

---

**Example 3**
S.ID.5

4. How many reenactors were at the event?  **121**
5. Make a two-way relative frequency table. Round to the nearest percent.  **See margin.**
6. Which numbers are joint frequencies?  **35%, 30%, 24%, 12%**
7. Which numbers are marginal frequencies?  **64%, 36%, 59%, 41%**
8. What is the joint frequency of Norman nobles?  **24%**

**Example 2**
S.ID.5

9. At a dog show, there were 80 dogs competing. Of which the competitors, 60% were retrievers, while the rest were small dogs. Thirty percent of the dogs were male retrievers and 25% were female lap dogs. Make a two-way frequency table.  **See margin.**

**Example 4**
S.ID.5

SKIING At an international ski race, Giorgio and Susie recorded the nationality and skiwear for 24 skiers. Twelve skiers were Norwegian, 8 were Swiss, and the rest were American. Seventy-five percent of the Americans were wearing spring skiwear. Five of the Norwegians were wearing full ski suits. In all, 15 people were wearing spring skiwear.

10. Make a relative frequency table. Round to the nearest percent.  **See margin.**
11. What percent of the skiers wearing full ski suits were American?  **11%**
12. What percent of the skiers wearing spring skiwear were Swiss?  **33%**

**Example 5**
S.ID.5

13. Explain whether there is an association between nationality and skiwear at this event.
14. **REASONING** How reliable is the association? Explain your answer.

**13.** It seems that there is an association; Americans are less likely to wear a full ski suit.

**14.** The association might not be that solid or reliable because the American data set is very small. Adding additional people to the data set could change the percentages significantly.

### Practice and Problem Solving
Extra Practice is found on page R10.

**Example 1**
S.ID.5

BEACH Jordan made a survey of the activities of people at the beach and recorded the data in a two-way frequency table. Use the table for Exercises 15–20.

| Activities | Swimming Suits | Shorts and T-shirts | Dresses | Total |
|---|---|---|---|---|
| Swimming | 18 | 0 | 0 | 18 |
| Sitting or Lying on the Sand | 21 | 3 | 3 | 27 |
| Walking | 5 | 9 | 5 | 19 |
| Playing Sports | 2 | 8 | 1 | 11 |
| Total | 46 | 20 | 9 | 75 |

15. How many people were swimming in total?  **18**
16. How many people were walking on the beach while wearing a dress?  **5**
17. How many people were walking on the beach with clothes other than their swimming suits?  **14**
18. How many people were swimming in shorts and T-shirts?  **0**
19. How many people were playing sports?  **11**
20. How many people were at the beach?  **75**

**Example 2**
S.ID.5

MEALS Mary manages a banquet center, and she is compiling data on a meal options that were served. Of 24 vegetarian meals, 18 people left some food on the plate. Of the 18 chicken meals, 9 had leftovers. Thirty people ordered the beef option and 21 had nothing left of their plates.

21. Make a two-way frequency table.  **See margin.**
22. Describe two ways to find the total number of diners.
23. Make a two-way relative frequency table of the data. Round to the nearest percent.  **See margin.**

**22.** Sample answer: Add all vegetarian, chicken, and beef meals; add the total number of people who ate everything on their plate and the total number of people who left food behind.

### Differentiated Homework Options

| Levels | AL Basic | OL Core | BL Advanced |
|---|---|---|---|
| Exercises | 15–35, 38–43 | 15–35 odd, 36–43 | 21–40<br>41–43 (optional) |
| 2-Day Option | 15–35 odd, 41–43 | 15–35 | |
| | 16–34 even, 38–40 | 36–43 | |

**21.**

| Meal | Ate Everything | Left Food | Total |
|---|---|---|---|
| Vegetarian | 6 | 18 | 24 |
| Chicken | 18 | 9 | 27 |
| Beef | 21 | 9 | 30 |
| Total | 45 | 36 | 81 |

**Lesson 10-6 | Summarizing Categorical Data**

24. What percentage is the joint data point for ate everything and beef? **26%**
25. Did more people overall eat everything or have leftovers? **They are equal.**
26. What percent of people who had the vegetarian meal ate everything from their plate? **23%**
27. Explain whether there an association between the type of meal a diner chose and whether they ate all of their food.
28. **REASONING** Discuss some possible reasons that the vegetarian meals had more leftovers than the meat meals. What actions might the manager consider?

Example 3
S.ID.5

29. **STRUCTURE** Elizabeth and her extended family went to the zoo. Complete the two-way frequency table about the frozen yogurt they bought.

| Flavor | Waffle Cone | Regular Cone | Total |
|---|---|---|---|
| Strawberry | 2 | 2 | 4 |
| Vanilla | 1 | 1 | 2 |
| Chocolate | 3 | 5 | 8 |
| Total | 6 | 8 | 14 |

**27.** Yes, it appears that there is an association, because the conditional relative frequency of vegetarians eating everything on their plates is much lower than the conditional relative frequency of diners with other meals eating everything on their plates.

Example 4
S.ID.5

30. What is the conditional relative frequency of chocolate in waffle cones compared to total chocolate cones? **38%**
31. What is the marginal frequency for total strawberry cones bought? **4**
32. What is the relative frequency for regular cones? **57%**
33. What is the frequency for waffle cones? **6**
34. What is the joint frequency for vanilla regular cones? **1**

Example 5
S.ID.5

35. **STRUCTURE** A manager takes count of the level of preparedness of 24 employees on their first day on a construction job. Four have their paperwork ready but forgot to bring safety gear. Six have forgotten both safety gear and their paperwork.

a. Complete the two-way frequency table.

| Safety Gear | Paperwork Ready | Paperwork Not Ready | Total |
|---|---|---|---|
| Have Safety Gear | 8 | 6 | 14 |
| Do Not Have Safety Gear | 4 | 6 | 10 |
| Total | 12 | 12 | 24 |

**28.** The association might mean that the vegetarian meal is not very tasty. The manager could interview the diners who had the vegetarian meal, especially the 77% who left food on their plates.

b. What percentage of new employees in total had their paperwork ready? **50%**
c. What percentage of new employees were completely prepared? **33%**
d. What percentage of new employees had all of their safety gear ready? **58%**
e. Complete the two-way relative frequency table.

| Safety Gear | Paperwork Ready | Paperwork Not Ready | Total |
|---|---|---|---|
| Have Safety Gear | 33% | 25% | 58% |
| Do Not Have Safety Gear | 17% | 25% | 42% |
| Total | 50% | 50% | 100% |

f. What is the conditional relative frequency for having paperwork but missing safety glasses, compared to the total people who had their paperwork ready? **34%**
g. What is the percentage of people who did not have their paperwork ready? **50%**

---

**Teaching the Mathematical Practices**

**Structure** Mathematically proficient students can see complicated things, such as a two-way frequency table, being composed of single objects. As students are completing the two-way frequency table in Exercise 35, students will need to calculator each part of the table using the information provided.

**Study Tip**
Remind students that it is possible that their *Total* column and row may not equal exactly 100% due to rounding.

---

23.

| Meal | Ate Everything | Left Food | Total |
|---|---|---|---|
| Vegetarian | 7% | 22% | 30% |
| Chicken | 22% | 11% | 33% |
| Beef | 26% | 11% | 37% |
| Total | 56% | 44% | 100% |

**Lesson 10-6** | Summarizing Categorical Data

---

### Teaching the Mathematical Practices

**Construct Arguments** Mathematically proficient students can justify their conclusions and communicate them to others. In Exercise 38, encourage students to discuss situations where one type of frequency table would be better than the other. They can use these situations to build an argument.

---

## Assess

**Crystal Ball** Ask students to summarize what they learned about two-way frequency tables and how they might be able to apply the technique to their own real-world situations.

---

**B** 36. **REASONING** A manager of an Indian restaurant decided to keep track of customer orders for a week to determine which dishes are ordered. Complete the two-way frequency table.

| Curry Flavor | With Lentils and Rice | With Plain Rice | With Naan Bread | Total |
|---|---|---|---|---|
| Green Curry | 285 | 304 | 256 | 845 |
| Red Curry | 156 | 180 | 170 | 506 |
| Yellow Curry | 346 | 511 | 427 | 1284 |
| Total | 787 | 995 | 853 | 2635 |

a. How many dishes were served overall? **2635**
b. What is the conditional relative frequency of green curry with naan bread compared to green curry dishes served overall? **30%**
c. What is the relative frequency of red curry with lentils and rice? **6%**
d. What is the marginal frequency of dishes served with white rice? **38%**
e. What is the marginal frequency for dishes served with lentils and rice? **30%**
f. What is the conditional relative frequency of yellow curry with plain rice compared to plain rice dishes served overall? **51%**

37. **REGULARITY** Ms. Ramos is a preschool teacher, and she collected data about the allergies of students in her class. This would allow her to remember what foods are safe in her classroom. Copy and complete the two-way frequency table, and copy and complete the two-way relative frequency table.

**Two-Way Frequency Table**

| Allergy | Eat Gluten | Should Not Eat Gluten | Total |
|---|---|---|---|
| Peanuts | 4 | 3 | 7 |
| Soy | 2 | 1 | 3 |
| Milk | 1 | 2 | 3 |
| Total | 7 | 6 | 13 |

40. Sample answer: Fill in any values that are given in the problem, then try to complete rows or columns based on given information by subtracting.

**Two-Way Relative Frequency Table**

| Allergy | Eat Gluten | Should Not Eat Gluten | Total |
|---|---|---|---|
| Peanuts | 31% | 23% | 54% |
| Soy | 15% | 8% | 23% |
| Milk | 8% | 15% | 23% |
| Total | 54% | 46% | 100% |

38. Relative frequency tables have all of the numbers converted to percentages. This makes relative two-way frequency tables easier to see comparisons than in two-way frequency tables that have raw data.

39. Relative frequencies give the percentage of each joint data point within the data set. However, the data set may be skewed, as one category in the data set may have only a fraction of the data points that another category has.

S.ID.5

---

**H.O.T. Problems** Use Higher-Order Thinking Skills

**C** 38. **CONSTRUCT ARGUMENTS** Discuss the difference between two-way frequency tables and two-way relative frequency tables. Which type of frequency table makes comparison easier?

39. **OPEN-ENDED** Discuss how conditional frequencies help to compare data points better than simple relative frequencies. Give an example.

40. **WRITING IN MATH** Describe the processes that you use to fill in two-way frequency tables, starting from minimal data.

---

### Standards for Mathematical Practice

| Emphasis On | Exercises |
|---|---|
| 1 Make sense of problems and persevere in solving them. | 40, 42 |
| 2 Reason abstractly and quantitatively. | 17, 28, 36, 39, 42 |
| 3 Construct viable arguments and critique the reasoning of others. | 38 |
| 7 Look for and make use of structure. | 21, 29, 35, 41, 42 |
| 8 Look for and express regularity in repeated reasoning. | 37, 41, 43 |

## Lesson 10-6 | Summarizing Categorical Data

### Preparing for Assessment

**41. PETS** A pet store has three new litters of guinea pigs. The animal caretaker records the weight in grams of each new guinea pig and their colors and markings. MP 4, 7, 8  S.ID.5

| Weight (g) | Brown | Black and White | Total |
|---|---|---|---|
| < 80 | 4 | 5 | 9 |
| > 80 | 2 | 1 | 3 |
| Total | 6 | 6 | 12 |

a. How many brown guinea pigs are greater than 80 g? **A**
- ○ A  2
- ○ B  3
- ○ C  4
- ○ D  5

b. What is the relative frequency for black and white guinea pigs that weigh less than 80 g? **C**
- ○ A  8%
- ○ B  33%
- ○ C  42%
- ○ D  50%

c. What percent of the black and white guinea pigs weigh greater than 80 g? **B**
- ○ A  8%
- ○ B  17%
- ○ C  42%
- ○ D  83%

43d. tomatoes–79%; peas–40%; squash–40%. Yes, there is an association. A much greater percentage of tomatoes were damaged in the storm.

**42. ACTIVITIES** James surveyed 150 classmates about their involvement in band and sports. He found that 48% are in band and 25% of those in band also play sports. Eighteen percent of the surveyed students are neither in band nor play sports. Make a two-way frequency table to display the data.
MP 1, 2, 4, 7  S.ID.5  See margin.

**43. MULTI-STEP** After heavy storm, Maria evaluates the vegetables in her garden. Damaged vegetables must be harvested right away, while undamaged vegetables can stay in the garden to further mature.
MP 4, 7, 8  S.ID.5

| Vegetables | Damaged | Undamaged | Total |
|---|---|---|---|
| Tomatoes | 46 | 12 | |
| Peas | 200 | 300 | |
| Squash | 6 | 9 | |
| Total | | | |

a. Which of the following statements are true? **B, D, E, G**
- ☐ A  The total number of peas is 600.
- ☐ B  The marginal frequency for squash is 15.
- ☐ C  The joint frequency of damaged peas is 500.
- ☐ D  The marginal frequency for vegetables that are undamaged is 321.
- ☐ E  The total number of vegetables is 573.
- ☐ F  The joint frequency of undamaged tomatoes is 58.
- ☐ G  The marginal frequency of vegetables that are damaged is 252.

b. What percentage of the tomatoes are damaged? **79%**

c. What percentage of the vegetables are damaged? **44%**

d. Calculate the conditional relative frequencies of each damaged vegetable. Is there an association between the type of vegetable and its likelihood to have been damaged in the storm?

e. What math operation explains how to find the marginal frequency data that would go in the last column, third cell down? **C**
- ○ A  add the two cells above it
- ○ B  add the two cells below it
- ○ C  add the two cells to the left of it
- ○ D  subtract the cell in the middle from the cell at the left

**42.**

| Band Involvement | Sports | Not in Sports | Total |
|---|---|---|---|
| Band | 18 | 54 | 72 |
| Not in Band | 51 | 27 | 78 |
| Total | 69 | 81 | 150 |

---

### Preparing for Assessment

Exercises 41–43 require students to use the skills they will need on standardized assessments. Each exercise is dual-coded with content and standards and mathematical practice standards.

#### Dual Coding

| Items | Content | MP Mathematical Practices |
|---|---|---|
| 41 | S.ID.5 | 4, 7, 8 |
| 42 | S.ID.5 | 1, 2, 4, 7 |
| 43 | S.ID.5 | 4, 7, 8 |

### Diagnose student errors

Survey student responses for each item. Class trends may indicate common errors and misconceptions.

**41a.**

| A | CORRECT |
|---|---|
| B | Chose marginal cell instead of joint cell |
| C | Found brown guinea pigs less than 80 g |
| D | Chose incorrect joint cell |

**41b.**

| A | Found relatively frequency for guinea pigs more than 80 g |
|---|---|
| B | Found the relative frequency for brown guinea pigs |
| C | CORRECT |
| D | Found the relative frequency for total black and white guinea pigs |

**41c.**

| A | Divided by the total number of guinea pigs |
|---|---|
| B | CORRECT |
| C | Divided by the total number of guinea pigs |
| D | Found the percentage weighing less than 80 g |

### Go Online!

The most up-to-date resources available for your program can be found at **connectED.mcgraw-hill.com**.

# Chapter 10 Study Guide and Review

## FOLDABLES Study Organizer

A completed Foldable for this chapter should include the Key Concepts related to statistics and probability.

## Key Vocabulary ELL

The page reference after each word denotes where that term was first introduced. If students have difficulty answering questions 1–4, remind them that they can use these page references to refresh their memories about the vocabulary terms.

Have students work in pairs to discuss each term in the Key Vocabulary list as they complete the Vocabulary Check. Encourage students to review using their notes and the text.

You can use the detailed reports in ALEKS to automatically monitor students' progress and pinpoint remediation needs prior to the chapter test.

5. A bar graph compares categories of data, while a histogram displays data that have been organized into equal intervals.

6. Draw a histogram. If data are evenly distributed, then there is a symmetric distribution. If the majority of the data are on the right, then there is a negatively skewed distribution. If the majority of the data are on the left, then there is a positively skewed distribution.

7. The mean is the average of the data set, the median is the middle numebr when written in order, and mode occurs the most often.

8. Sample answer: The mean best represents the average grade on the test. The median is best used when comparing incomes. The mode is the best used when comparing the number of points scored during the season.

## Go Online!

### Vocabulary Review

Students can use the *Vocabulary Review Games* to check their understanding of the vocabulary terms in this chapter. Students should refer to the *Student-Built Glossary* they have created as they went through the chapter to review important terms. You can also give a *Vocabulary Test* over the content of this chapter.

---

# CHAPTER 10
# Study Guide and Review

Go Online! for Vocabulary Review Games and key vocabulary in 13 languages

## Study Guide

### Key Concepts

**Measures of Center** (Lesson 10-1)
- The mean, median, and mode are measures of center, which summarize data with a single number.
- A percentile gives the percent of the data values in a set that are below a certain data value.

**Representing Data** (Lesson 10-2)
- Bar graphs, histograms, dot plots, and box plots can be used to represent data.

**Measures of Spread** (Lesson 10-3)
- A low standard deviation indicates that the data tend to be very close to the mean, while a high standard deviation indicates that the data are spread out over a larger range.

**Distributions of Data and Comparing Sets of Data** (Lessons 10-4 and 10-5)
- In a negatively skewed distribution, the majority of the data are on the right. In a positively skewed distribution, the majority of the data are on the left. In a symmetric distribution, the data are evenly distributed.
- Two symmetric distributions can be compared using mean and standard deviation. If one of the distributions is skewed, compare the five-number summary.

**Summarizing Categorical Data** (Lesson 10-6)
- A two-way frequency table is used to show the frequencies of data classified according to two categories, with rows indicating one category and columns indicating the other.
- Relative frequencies are found by dividing each value in a two-way table by the total number. Conditional relative frequencies are found by dividing each joint frequency by the marginal frequency.

### Key Vocabulary

bar graph (p. 659)
cumulative frequency (p. 659)
distribution (p. 673)
dot plot (p. 657)
frequency table (p. 659)
histogram (p. 659)
joint frequencies (p. 691)
linear transformation (p. 680)
marginal frequencies (p. 691)
mean (p. 651)
measures of center (p. 651)
median (p. 651)
mode (p. 651)
percentile (p. 653)
qualitative data (p. 651)
quantitative data (p. 651)
relative frequency (p. 690)
skewed distribution (p. 673)
standard deviation (p. 667)
symmetric distribution (p. 673)
two-way frequency table (p. 688)

### Vocabulary Check

Choose the term that best completes each sentence.

1. The (**mean**, median, mode) of a data set is the average data value.  **mean**

2. A (symmetric, **skewed**) distribution is one in which there are more data values on one side than the other.  **skewed**

3. Data that can be measured are (qualitative, **quantitative**) data.  **quantitative**

4. The totals of each subcategory in a two-way table are the (joint, **marginal**) frequencies.  **marginal**

### Concept Check

5. Explain the difference between a bar graph and a histogram.  **5–8. See Margin.**

6. Explain how to determine the shape of a data distribution.

7. Explain how to find the mean, median, and mode.

8. Create examples for when the mean, median and mode are the best measures of central tendency.

## FOLDABLES Study Organizer

Use your Foldable to review the chapter. Working with a partner can be helpful. Ask for clarification of concepts as needed.

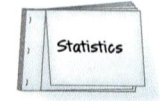

## Answering the Essential Question

Before answering the Essential Question, have students review their answers to the *Building on the Essential Question* exercises found throughout the chapter.

- How are statistics used in the real world? (p. 648)
- How can measures of center and percentile rank be used to analyze real-world data sets? (p. 654)

Chapter 10 Study Guide and Review

## Lesson-by-Lesson Review ▶

### 10-1 Measures of Center
Preparation for S.ID.2

Find the mean, median, and mode of each data set.

9. {6, 20, 55, 20, 7, 18}  21; 19; 20

10. {10, 81, 66, 15}  43; 40.5; no mode

11. **TESTS** Aliyah's first four science test scores are 91, 72, 97, and 82. Find the mean, median, and mode of the test scores. Which statistic best describes Aliyah's science tests? See margin.

12. **BOWLING** The scores for a bowling team are shown in the table.

    | Team Scores |       |        |       |
    |-------------|-------|--------|-------|
    | Bowler      | Score | Bowler | Score |
    | Lamar       | 112   | Paige  | 251   |
    | Mike        | 136   | Quinn  | 74    |
    | Nicole      | 177   | Roman  | 68    |
    | Opal        | 65    | Sergio | 103   |

    a. Which bowlers are above the 50th percentile?
    b. Which bowler is in the 38th percentile? Sergio

    12a. Mike, Nicole, Paige

**Example 1**

Find the mean, median, and mode of the set {34, 5, 1, 22, 18, 29, 1}.

To find the mean, find the sum and divide by the number of values.

$$\frac{34 + 5 + 1 + 22 + 18 + 29 + 1}{7} = \frac{110}{7} \approx 15.7$$

To find the median, order the numbers from least to greatest and find the middle value.

1, 1, 5, ⑱, 22, 29, 34

To find the mode, find the number that occurs most often. The value 1 occurs most often.

The mean is about 15.7, the median is 18, and the mode is 1.

### 10-2 Representing Data
S.ID.1, N.Q.1

13. **MOVIES** The table shows the results of a survey in which students were asked to choose which of four genres of movies they like best. Make a bar graph of the data. See margin.

    | Genre       | Frequency |
    |-------------|-----------|
    | Action      | 25        |
    | Romance     | 16        |
    | Comedy      | 30        |
    | Documentary | 2         |

14. **CARS** The table shows the number of each type of vehicle that passes an intersection during an hour. Make a bar graph of the data. See margin.

    | Type  | Frequency |
    |-------|-----------|
    | Sedan | 54        |
    | Truck | 21        |
    | Bus   | 4         |
    | SUV   | 16        |

**Example 2**

Make a bar graph to display the data collected regarding students' favorite color.

Step 1: Decide whether the data is continuous or discrete. Bar graphs are used for discrete data and histograms are used for continuous data.

| Color  | Frequency |
|--------|-----------|
| Red    | 12        |
| Yellow | 8         |
| Green  | 15        |
| Blue   | 9         |

Step 2: Draw a horizontal axis and a vertical axis. Label the axes as shown. Add a title.

Step 3: Draw a bar to represent each category.

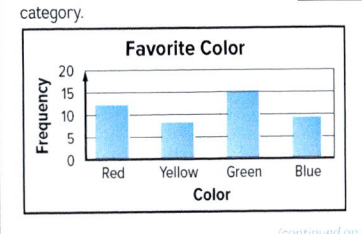

*(continued on the next page)*

## Lesson-by-Lesson Review

**Intervention** If the given examples are not sufficient to review the topics covered by the questions, remind students that the lesson references tell them where to review that topic in their textbook.

**Two-Day Option** Have students complete the Lesson-by-Lesson Review. Then you can use McGraw-Hill eAssessment to customize another review worksheet that practices all the objectives of this chapter or only the objectives on which your students need more help.

## Additional Answers

11. 85.5; 86.6; no mode; Because the data seems to be spread out evenly, both the mean and median are good measures of center.

13.

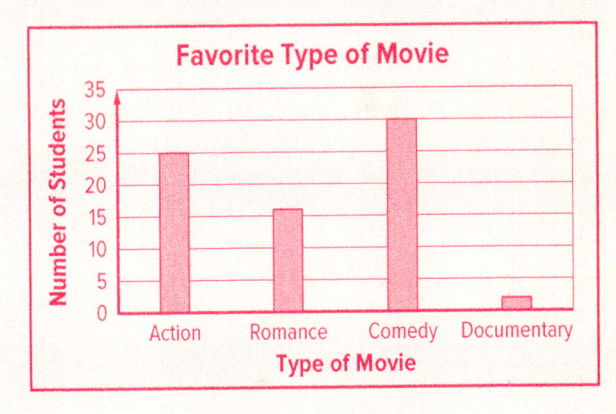

14.

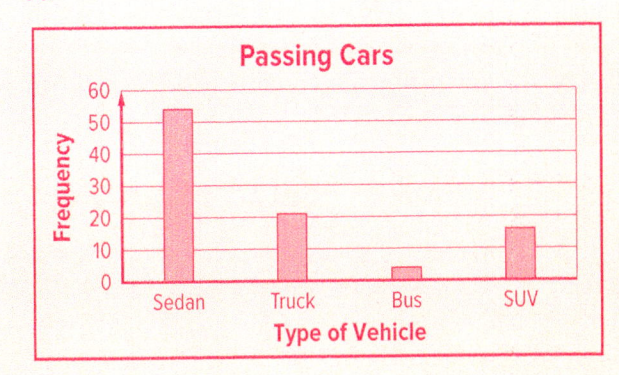

### Go Online! 🖥️📱

The most up-to-date resources available for your program can be found at connectED.mcgraw-hill.com.

connectED.mcgraw-hill.com  699

# Chapter 10 Study Guide and Review

## Additional Answers

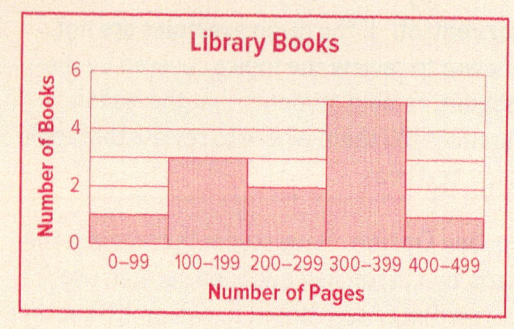

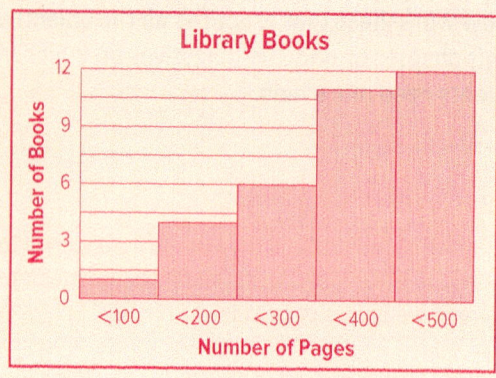

**15.**

*Library Books* histogram — Number of Books vs Number of Pages (0–99, 100–199, 200–299, 300–399, 400–499): bars at 1, 3, 2, 5, 1.

*Library Books* cumulative frequency histogram — Number of Books vs Number of Pages (<100, <200, <300, <400, <500): bars at 1, 4, 6, 11, 12.

**16.**

*Ages of People at a Restaurant* histogram — Number of People vs Age (0–19, 20–39, 40–59, 60–79): bars at 4, 12, 18, 3.

*Ages of People at a Restaurant* cumulative frequency histogram — Number of People vs Age (<20, <40, <60, <80): bars at 4, 16, 34, 37.

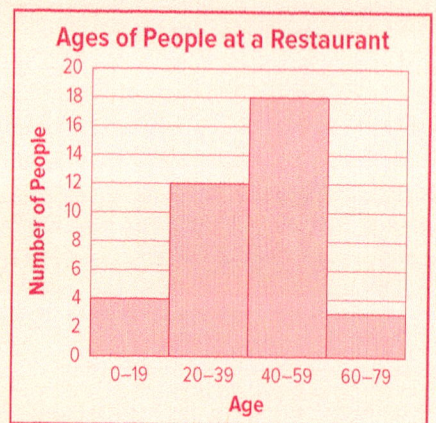

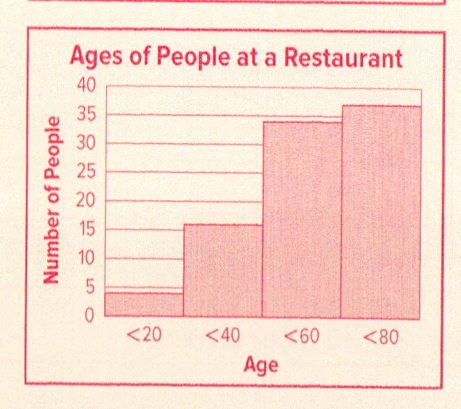

**17.** 3.4; 1.02; Sample answer: The standard deviation is relatively small compared to the mean. So, Ben mows about the same number of lawns each week.

**18.** 33; 15.23; Sample answer: The standard deviation is great compared to the mean. Therefore, there is a wide range in the number of candy bars sold.

**19.** The day customers had a mean of about 9.6 times per month with a standard deviation of about 3.5. The night customers had a mean of about 11.7 times per month with a standard deviation of about 2.5. The night customers had a higher average and their data values were more consistent.

---

# CHAPTER 10
# Study Guide and Review *Continued*

**Make a histogram of the frequency and the cumulative frequency.**

**15. BOOKS** The frequency table shows the number of pages in the books Amelia borrows from a library. *See margin.*

| Pages | Books |
|---|---|
| 0–99 | 1 |
| 100–199 | 3 |
| 200–299 | 2 |
| 300–399 | 5 |
| 400–499 | 1 |

**16. DINING** The frequency table shows the ages of people at a restaurant. *See margin.*

| Age | Frequency |
|---|---|
| 0–19 | 4 |
| 20–39 | 12 |
| 40–59 | 18 |
| 60–79 | 3 |

### Example 3
Make a histogram of the frequency and the cumulative frequency.

| Height (ft) | 0–0.9 | 1–1.9 | 2–2.9 |
|---|---|---|---|
| Plants | 5 | 12 | 7 |

Find the cumulative frequency for each interval.

| Height (ft) | <1 | <2 | <3 |
|---|---|---|---|
| Plants | 5 | 17 | 24 |

Make each histogram like a bar graph but without spaces between the bars.

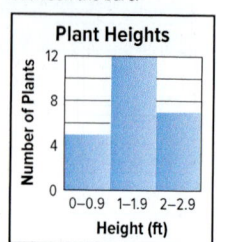

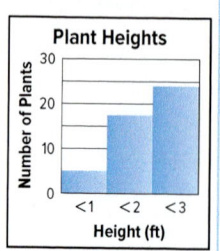

*Plant Heights* histograms shown.

## 10-3 Measures of Spread
S.ID.2

**17. SHOVELING** Ben mows lawns to earn money. The number of lawns he cuts for five weeks is {2, 4, 3, 5, 3}. Find and interpret the mean and standard deviation. *See margin.*

**18. CANDY BARS** Luci is keeping track of the number of candy bars each member of the drill team sold. The results are {20, 25, 30, 50, 40, 60, 20, 10, 42}. Find and interpret the mean and standard deviation. *See margin.*

**19. FOOD** A fast food company polls a random sample of its day and night customers to find how many times a month they eat out. Compare the means and standard deviations of each data set. *See margin.*

| Day Customers | Night Customers |
|---|---|
| 10, 3, 12, 15, 7, 8, 4, 12, 9, 14, 12, 9 | 15, 12, 13, 9, 11, 12, 14, 12, 8, 16, 9, 9 |

### Example 4
**GIFTS** Joshua is collecting money from his family for a Mother's Day gift. He keeps track of how much each person has donated: {10, 5, 20, 15, 10}. Find and interpret the mean absolute deviation.

**Step 1** Find the mean: $\bar{x} = \frac{10 + 5 + 20 + 15 + 10}{5}$ or 12

**Step 2** Find the squares of the differences.
$(12 - 10)^2 = 4 \quad (12 - 5)^2 = 49$
$(12 - 20)^2 = 64 \quad (12 - 15)^2 = 9$
$(12 - 10)^2 = 4$

**Step 3** Find the sum.
$4 + 49 + 64 + 9 + 4 = 130$

**Step 4** Find the variance.
$\sigma^2 = \frac{130}{5}$ or 26

**Step 5** Find the standard deviation.
$\sigma = \sqrt{\sigma^2} = \sqrt{26}$ or about 5.10

The standard deviation is fairly great compared to the mean. This suggests that many of the data values are not close to the mean.

---

**20.** [8, 80] scl: 10 by [−2, 10] scl: 1
negatively skewed

**21.** [0, 90] scl: 10 by [−2, 8] scl: 1
negatively skewed

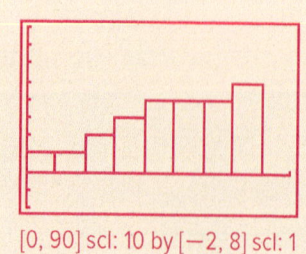

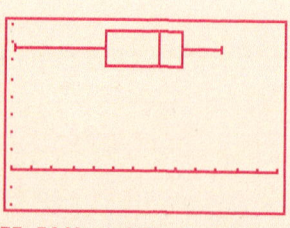

**22.** [175, 500] scl: 25 by [−2, 8] scl: 1

Sample answer: The data are negatively skewed, so use the five-number summary. The range is 252 gallons. The median is 354.5 gallons. Half of the data are between 288 and 383 gallons.

## Chapter 10 Study Guide and Review

### 10-4 Distributions of Data

S.ID.3

Use a graphing calculator to construct a histogram for the data. Then describe the shape of the distribution.

20. 55, 62, 32, 56, 31, 59, 19, 61, 8, 48, 41, 69, 32, 63, 48, 60, 43, 66, 71, 70, 49, 56, 21, 67   20–22. See margin.

21. 4, 19, 62, 28, 26, 59, 33, 39, 36, 72, 46, 48, 49, 44, 72, 76, 55, 53, 55, 62, 66, 69, 71, 74

22. **MILK** A grocery store manager tracked the amount of milk in gallons sold each day. Describe the center and spread of the data using either the mean and standard deviation or the five-number summary. Justify your choice by constructing a box-and-whisker plot for the data.

| Gallons of Milk Sold per Day |     |     |     |     |     |
|---|---|---|---|---|---|
| 383 | 296 | 354 | 288 | 195 | 372 |
| 421 | 367 | 411 | 355 | 296 | 321 |
| 403 | 357 | 432 | 229 | 180 | 266 |

23. 22.5, 21.5, no mode, 24, 7.4

#### Example 5

**DRIVING TESTS** Several driving test results are shown. Describe the center and spread of the data using either the mean and standard deviation or the five-number summary. Justify your choice by constructing a box-and-whisker plot for the data.

| Driving Test Scores |     |     |     |     |     |
|---|---|---|---|---|---|
| 80 | 95 | 100 | 95 | 95 | 100 |
| 100 | 90 | 75 | 60 | 90 | 80 |

Use a graphing calculator to create a box-and-whisker plot.

The left whisker is longer than the right and the median is closer to the right whisker. Therefore, the distribution is negatively skewed.

Use the five-number summary. The range is 40. The median score is 92.5, and half of the drivers scored between 80 and 97.5.

[56, 104] scl: 10 by [−2, 12] scl: 1

24. 84.2, 85.5, no mode, 21, 6.8

### 10-5 Comparing Sets of Data

S.ID.2, S.ID.3

Find the mean, median, mode, range, and standard deviation of each data set that is obtained after adding the given constant to each value.

23. 27, 21, 34, 42, 20, 19, 18, 26, 25, 33; +(−4)

24. 72, 56, 71, 63, 68, 59, 77, 74, 76, 66; +16

25. **SCHOOL** Principal Andrews tracked the number of disciplinary actions given by Ms. Miller and Ms. Anderson to their students each week.  See margin.

| Ms. Miller |
|---|
| 9, 16, 12, 11, 12, 9, 10, 14, 13, 10, 9, 10, 11, 9, 12, 10, 11, 12 |

| Ms. Anderson |
|---|
| 7, 1, 0, 4, 2, 1, 6, 2, 2, 1, 4, 3, 0, 7, 0, 2, 5, 0 |

a. Use a graphing calculator to construct a histogram for each set of data. Then describe the shape of each distribution.

b. Compare the data sets using either the means and standard deviations or the five-number summaries. Justify your choice.

#### Example 6

Find the mean, median, mode, range, and standard deviation of the data set obtained after adding 6 to each value.

12, 15, 11, 12, 14, 16, 15, 12, 10, 13

Find the mean, median, mode, range, and standard deviation of the original data set.

Mean 13   Mode 12   Standard Deviation 1.8

Median 12.5   Range 6

Add 6 to the mean, median, and mode. The range and standard deviation are unchanged.

Mean 19   Mode 18   Standard Deviation 1.8

Median 18.5   Range 6

---

**25a.** Ms. Miller:

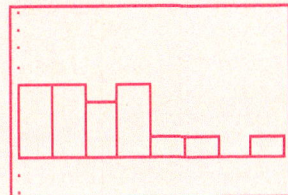

[9, 17] scl: 1 by [−2, 8] scl: 1

Ms. Anderson:

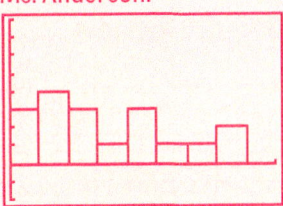

[0, 9] scl: 1 by [−2, 8] scl: 1

both positively skewed

**25b.** Sample answer: Use the five-number summaries. The range for Ms. Miller is 7. The median is 11. Half of the data are between 10 and 12. The range for Ms. Anderson is 7. The median is 2. Half of the data are between 1 and 4. All of the data in Ms. Anderson's distribution are less than all of the data in Ms. Miller's distribution. Therefore, Ms. Miller will more than likely hand out more disciplinary actions than Ms. Anderson.

---

## Go Online!

### Anticipation Guide

Students should complete the *Anticipation Guide*, and discuss how their responses have changed now that they have completed Chapter 10.

# Chapter 10 Study Guide and Review

## Before the Test
Have students complete the Study Notebook Tie it Together activity to review topics and skills presented in the chapter.

## Additional Answers

**26.**

| Class | Drive | Walk | Bus | Total |
|---|---|---|---|---|
| Junior | 12 | 42 | 29 | 83 |
| Senior | 34 | 49 | 9 | 92 |
| Total | 46 | 91 | 38 | 175 |

**27a.**

| Gender | Yes | No | Total |
|---|---|---|---|
| Boys | 8 | 20 | 28 |
| Girls | 19 | 15 | 34 |
| Total | 27 | 35 | 62 |

**27b.**

| Gender | Yes | No | Total |
|---|---|---|---|
| Boys | 12.9% | 32.3% | 45.2% |
| Girls | 30.6% | 24.2% | 54.8% |
| Total | 43.5% | 56.5% | 100% |

**27c.**

| Gender | Yes | No | Total |
|---|---|---|---|
| Boys | 28.6% | 71.4% | 100% |
| Girls | 55.9% | 44.1% | 100% |

| Gender | Yes | No |
|---|---|---|
| Boys | 29.6% | 57.1% |
| Girls | 70.4% | 42.9% |
| Total | 100% | 100% |

---

# CHAPTER 10
## Study Guide and Review Continued

### 10-6 Summarizing Categorical Data     S.ID.5

**26. TRANSPORTATION** A school conducts a survey to determine how students get to school. Of the 92 seniors, 34 responded that they drive to school and 9 responded that they ride the school bus. Of the 83 juniors, 12 responded that they drive to school and 42 responded that they walk to school. Create a two-way frequency table of the data. **See margin.**

**27. CLUBS** Hiroshi sent out a survey asking whether anyone would be interested in starting an environmental club. Of the 28 boys that responded, 8 said yes. Of the 34 girls that responded, 15 said no. **27a–c. See margin.**
  a. Create a two-way frequency table.
  b. Convert the two-way frequency table into a relative frequency table.
  c. Create two conditional relative frequency tables: one for the interest in the club and one for gender.

#### Example 7

**ELECTIONS** Of the eighty-four 18–24 year olds in a survey, 62 responded that they would vote in the next election. Of the seventy-seven 25–34 year olds in the survey, 54 responded that they would not vote in the next election. Create a two-way frequency table of the data.

**Step 1** Find the values for every combination of subcategories.
**Step 2** Place every combination in the corresponding cell.
**Step 3** Find the totals of each subcategory.
**Step 4** Find and record the sum of the set of marginal frequencies. These two sums should be equal.

| Age Group | Yes | No | Totals |
|---|---|---|---|
| 18-24 | 62 | 22 | 84 |
| 25-34 | 23 | 54 | 77 |
| Totals | 85 | 76 | 161 |

---

## Go Online!
### eAssessment
Customize and create multiple versions of chapter tests and answer keys that align to standards. Tests can be delivered on paper or online.

# CHAPTER 10
## Practice Test

**Go Online!** for another Chapter Test

Find the mean, median, and mode for each set of data.

1. {99, 88, 88, 92, 100}  **93.4; 92; 88**
2. {30, 22, 38, 41, 33, 41, 30, 24}  **32.375; 31.5; 30 and 41**
3. **TESTS** Kevin's scores on the first four science tests are 88, 92, 82, and 94. What score must he earn on the fifth test so that the mean will be 90? **94**
4. **FOOD** The table shows the results of a survey in which students were asked to choose their favorite food. Make a bar graph of the data.
   **See Chapter 10 Answer Appendix.**

| Favorite Foods |                    |
|----------------|--------------------|
| Food           | Number of Students |
| pizza          | 15                 |
| chicken nuggets| 10                 |
| cheesy potatoes| 8                  |
| ice cream      | 5                  |

5. **SALES** Nate is keeping track of how much people spent at the school store in one day. Find and interpret the mean and standard deviation for the data: 1, 1, 2, 3, 4, 5, 12. **See margin.**

6. **MULTIPLE CHOICE** Use a graphing calculator to construct a histogram for the data, and use it to describe the shape of the distribution. **A**

   16, 18, 14, 31, 19, 18, 10, 29,
   12, 12, 28, 19, 17, 26, 15, 20

   A  positively skewed
   B  negatively skewed
   C  symmetric
   D  none of the above

7. Use a graphing calculator to construct a histogram for the data, and use it to describe the shape of the distribution. **See Chapter 10 Answer Appendix.**

   19, 36, 26, 36, 40, 31, 30, 33, 23, 38, 23, 46

Find the mean, median, mode, range, and standard deviation of each data set that is obtained after multiplying each value by the given constant.

8. 9, 17, 31, 21, 17, 25, 13, 9, 12, 9; × 3  **48.9; 45; 51, 66, 21.3**
9. 16, 14, 23, 41, 38, 29, 18, 13, 16; × 0.25  **5.78; 4.5, 4, 7, 2.48**

Find the mean, median, mode, range, and standard deviation of each data set that is obtained after adding the given constant to each value.

10. 6, 9, 0, 15, 9, 14, 11, 13, 9, 5, 8, 6; + (−3)  **5.75; 6; 6; 15; 4.04**
11. 19, 22, 10, 17, 26, 24, 12, 22, 18, 17; + 8  **26.7; 26.5; 30; 16; 4.80**
12. **MULTIPLE CHOICE** Which pair of box-and-whisker plots depicts two positively skewed sets of data in which 75% of one set of data is greater than 75% of the other set of data? **A**

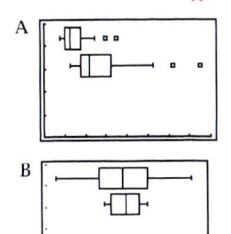

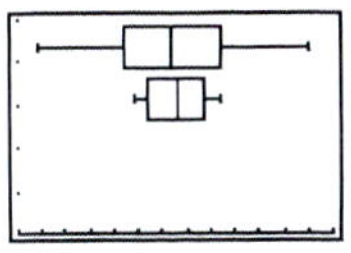

    A
    B
    C 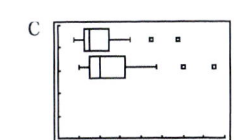
    D

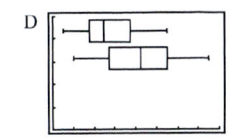

13. Two school districts send out a survey to residents asking for their opinions on a potential merger. Of the 225 residents in District 1 who responded to the survey, 178 said they are in favor of the merger. Of the 306 residents in District 2, 145 said they are against the merger. **a–c. See Ch. 10 Answer Appendix.**

    a. Create a two-way frequency table.
    b. Convert the two-way frequency table into a relative frequency table.
    c. Create a conditional relative frequency table for the districts and for the responses.

---

## Go Online!

### Chapter Tests

You can use premade leveled *Chapter Tests* to differentiate assessment for your students. Students can also take self-checking *Chapter Tests* to plan and prepare for chapter assessments.

*MC* = multiple-choice questions
*FR* = free-response questions

| Form | Type | Level |
|------|------|-------|
| 1    | MC   | AL    |
| 2A   | MC   | OL    |
| 2B   | MC   | OL    |
| 2C   | FR   | OL    |
| 2D   | FR   | OL    |
| 3    | FR   | BL    |
| Vocabulary Test |||
| Extended-Response Test |||

---

## Chapter 10 Practice Test

### RtI Response to Intervention

Use the Intervention Planner to help you determine your Response to Intervention.

**Intervention Planner**

**TIER 1 — On Level OL**

**IF** students miss 25% of the exercises or less,

**THEN** choose a resource:

SE 10-1, 10-2, 10-3, 10-4, 10-5, and 10-6

**Go Online!**
- Skills Practice
- Chapter Project
- Self-Check Quizzes

**TIER 2 — Strategic Intervention AL**
Approaching grade level

**IF** students miss 50% of the exercises,

**THEN** choose a resource:

*Quick Review Math Handbook*

**Go Online!**
- Study Guide and Intervention
- Extra Examples
- Personal Tutors
- Homework Help

**TIER 3 — Intensive Intervention**
2 or more grades below level

**IF** students miss 75% of the exercises,

**THEN** choose a resource:

Use *Math Triumphs, Alg. 1*

**Go Online!**
- Extra Examples
- Personal Tutors
- Homework Help
- Review Vocabulary

### Additional Answer

5. 4; 3.55; Sample answer: These values are significantly influenced by the outlier 12. Because the standard deviation is great compared to the mean, the range of the data is great.

# Chapter 10 Performance Task

## Launch

**Objective** Apply concepts and skills from this chapter in a real-world setting.

## Teach

Ask:
- **Part A:** Why was a histogram used to represent the data instead of a bar graph? Sample answer: The data is quantitative, not categorical.

- **Part B:** Do you think that Tamiko gave the coach enough information for the coach to consider her for the state diving meet? Explain. No; Sample answer: To fairly compare Tamiko to the other divers, she should give her coach all of the values of the five-number summary for her diving data.

- **Part C:** Would you describe Amani's diving scores as consistent? Why or why not? Sample answer: Her scores do not seem consistent. The range for the data is greater than Isabelle's data and the data is not symmetric.

- **Part D:** What are the categories that will be used in the two-way frequency table? Sample answer: There are two categories regarding snacks: *yes* and *no*. There are two categories regarding age: $> 45$ years old and $\leq 45$ years old.

The Performance Task focuses on the following content standards and standards for mathematical practice.

| Dual Coding ||| 
|---|---|---|
| Items | Content Standards | MP Mathematical Practices |
| 1 | S.ID.1 | 1, 4, 5 |
| 2 | S.ID.2 | 4, 6 |
| 3 | S.ID.2 | 4, 6 |
| 4 | S.ID.2 | 1, 2, 3, 4 |
| 5 | S.ID.2 | 1, 2, 3, 4 |
| 6 | S.ID.5 | 1, 4, 7, 8 |

---

# CHAPTER 10
# Preparing for Assessment

## Performance Task

Provide a clear solution to each part of the task. Be sure to show all of your work. Include all relevant drawings and justify your answers.

**DIVING TEAM** Kate Rodriguez is the coach of a diving team. She has been asked to choose one of the members of the team to compete at a state diving meet. Coach Rodriguez is considering two of the team's divers. To help her make the decision, she collects the divers' scores from their last 20 diving meets. The scores are shown in the table.

### Part A

1. Make a histogram for each data set. See margin.
2. For each data set, describe the shape of the distribution and explain what the plots tell you about each diver. See margin.
3. Compare the data sets using either the means and standard deviations or the five-number summaries. See margin.

| Diver | Scores |
|---|---|
| Amani | 120.0, 135.2, 102.3, 120.6, 95.2, 99.3, 125.5, 136.1, 115.3, 124.6, 136.0, 75.6, 102.3, 126.0, 104.7, 90.3, 130.6, 126.9, 92.4, 80.3 |
| Isabelle | 97.5, 115.3, 104.6, 103.9, 108.3, 97.5, 117.3, 92.6, 125.3, 106.5, 108.3, 103.3, 121.1, 124.0, 91.3, 96.6, 116.3, 111.5, 103.3, 109.3 |

### Part B

4. A third member of the team, Tamiko, tells Coach Rodriguez that she should also be considered for the state diving meet. She tells the coach that over her last 10 diving meets her median score is 110 and her interquartile range is about the same as Isabelle's interquartile range. Give a set of 10 scores that could represent Tamiko's data and justify your answer. See Ch. 10 Answer Appendix.

### Part C

5. Coach Rodriguez wants to choose the diver with the best scores, but she also wants to choose a diver who is consistent. Before she makes her decision, she discovers that Isabelle's scores were incorrectly reported, and each score should be 10 points greater than what is shown.

   Taking this new information into consideration, who should Coach Rodriguez send to the competition? See Ch. 10 Answer Appendix.

### Part D

6. During the competition, a survey was conducted to find out more about the crowd watching and what they might purchase. Of the 300 people surveyed, 114 of the people were greater than 45 years old. And of the 92 that said they would not purchase a snack while at the competition, 45 of them were 45 or younger.

   Construct a two-way frequency table to show the results of the survey and describe the categories and subcategories. See Ch. 10 Answer Appendix.

---

### Levels of Complexity Chart

The levels of the exercises progress from 1 to 3, with Level 1 indicating the lowest level of complexity.

| Parts | Level 1 | Level 2 | Level 3 |
|---|---|---|---|
| A | ● | | |
| B | | ● | |
| C | | | ● |
| D | | | ● |

# Chapter 10 Preparing for Assessment

## Test-Taking Strategy

**Example**

Read the problem. Identify what you need to know. Then use the information in the problem to solve. Show your work.

Determine which data point, if any, is an outlier in the following data set.
{0.85, 0.18, 1.04, 0.26, 2.09, 0.98, 0.40, 0.51, 0.79}

A  0.18
B  1.04
C  2.09
D  none

**Step 1** What do you need to find?
Any outliers in the data set

**Step 2** How can you solve the problem?
I will start by arranging the data in ascending order and finding the interquartile range IQR. Then, I can use the IQR to find the values beyond which any outliers would lie.

**Step 3** What is the correct answer?

0.18  0.26  0.40  0.51  0.79  0.85  0.98  1.04  2.09
         ↑               ↑              ↑
     $Q_1 = 0.33$    $Q_2 = 0.79$   $Q_3 = 1.01$

$IQR = Q_3 - Q_1 = 0.68$

$Q_1 - 1.5(IQR)$  and  $Q_3 + 1.5(IQR)$

$0.33 - 1.5(0.68) = -0.69$    $1.01 + 1.5(0.68) = 2.03$

Because 2.09 is greater than 2.03, the answer is C.

**Test-Taking Tip**

**Strategy for Organizing Data** Sometimes you may be given a set of data that you need to analyze in order to solve problems. When you are given a problem statement containing data, consider:
- making a list of the data
- using a table to organize the data
- using a data display (such as a bar graph, Venn diagram, circle graph, line graph, or box-and-whisker plot) to organize the data

### Apply the Strategy

Read the problem. Identify what you need to know. Then use the information in the problem to solve. Show your work.

Determine which data point, if any, is an outlier in the following data set.
{22, 1.2, 12.5, 11.5, 12, 11, 15.5, 14, 10, 6.5, 23}

A  1.2
B  12
C  23
D  none

Answer the questions below.

a. What do you need to find? Any outliers in the data set
b. How can you solve the problem? Find the IQR and find the values beyond which any outliers would lie.
c. What is the correct answer? A

2. For Amani, the distribution is negatively skewed. The mean of the data is less than the median. For Isabelle, the distribution is symmetric. The mean of the data is close to the median.

3. Because one distribution is skewed, compare using the five-number summaries. For Amani, the range is 136.1 − 75.6 = 60.5. The median is 117.7, and half the data are between 97.25 and 126.45. For Isabelle, the range is 125.3 − 91.3 = 34. The median is 107.4, and half the data are between 100.4 and 115.8. This shows that Amani has a greater median score, but Isabelle's scores are clustered more closely around the median.

## Test-Taking Strategy

**Step 1** Identify what the problem is asking.

**Step 2** Determine a method to solve the problem.

**Step 3** Evaluate using that method.

### Need Another Example?

Determine which data point, if any, is an outlier in the following data set. {87, 93, 84, 87, 100, 88, 70}  **D**

A  71
B  87
C  100
D  none

a. What do you need to find?
Any outliers in the data set

b. How can you solve the problem?
Find the IQR and find the values beyond which any outliers would lie.

c. What is the correct answer?  **D**

### Additional Answers

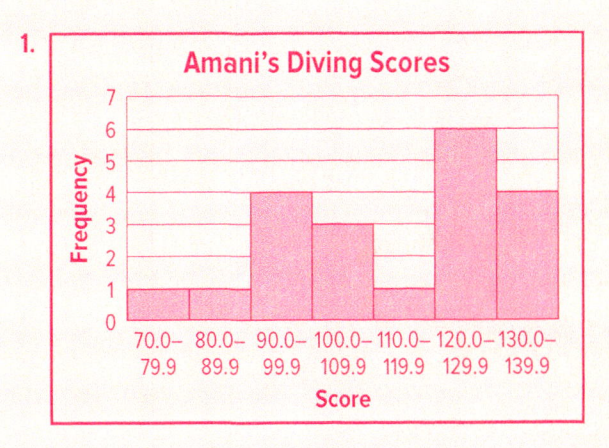

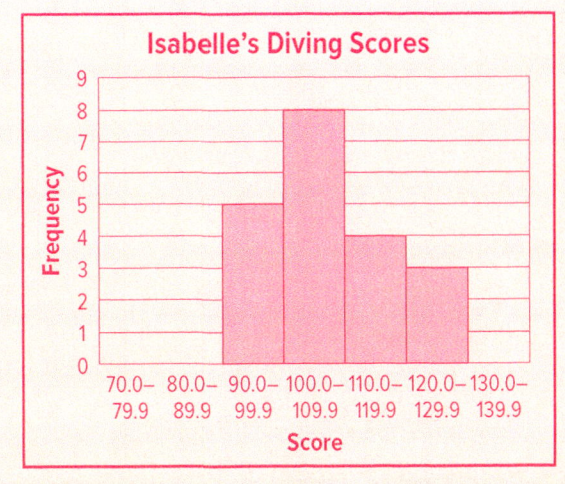

1.

**Amani's Diving Scores**

(histogram with Score on x-axis: 70.0–79.9, 80.0–89.9, 90.0–99.9, 100.0–109.9, 110.0–119.9, 120.0–129.9, 130.0–139.9; Frequency on y-axis 0–7)

**Isabelle's Diving Scores**

(histogram with Score on x-axis: 70.0–79.9, 80.0–89.9, 90.0–99.9, 100.0–109.9, 110.0–119.9, 120.0–129.9, 130.0–139.9; Frequency on y-axis 0–9)

### Go Online!

The most up-to-date resources available for your program can be found at connectED.mcgraw-hill.com.

# Chapter 10 Preparing for Assessment

## Diagnose Student Errors
Survey student responses for each item. Class trends may indicate common errors and misconceptions.

| 1. | A | Did not distribute properly |
|---|---|---|
|   | B | Did not combine terms correctly |
|   | C | Made a calculation error |
|   | D | CORRECT |
| 2. | A | Thought an *x*-intercept was the solution |
|   | B | Thought a *y*-intercept was the solution |
|   | C | CORRECT |
|   | D | Thought an *x*-intercept was the solution |
| 4. | A | Thought −5 shifted to the left |
|   | B | CORRECT |
|   | C | Confused vertical and horizontal translation |
|   | D | Confused reflection and translation |
| 5. | A | Confused the coefficients |
|   | B | CORRECT |
|   | C | Confused coefficients and total costs |
|   | D | Confused total costs |
| 6. | A | CORRECT |
|   | B | CORRECT |
|   | C | CORRECT |
|   | D | Did not realize range stays the same |
|   | E | Did not realize standard deviation stays the same |
| 7. | A | Substituted 19 for *x* |
|   | B | CORRECT |
|   | C | Divided by 3 instead of −3 |
|   | D | Substituted −19 for *x* |
| 8. | A | CORRECT |
|   | B | CORRECT |
|   | C | Mistook graph for a translation |
|   | D | Mistook graph for a translation |
|   | E | Mistook graph for a translation |
|   | F | Mistook graph for a translation |
| 10. | A | Does not understand symmetry |
|   | B | CORRECT |
|   | C | CORRECT |
|   | D | Does not understand symmetry |
|   | E | Does not understand symmetry |
| 12. | A | Confused vertical compression and stretch |
|   | B | CORRECT |
|   | C | Confused equation with a translation |
|   | D | Confused equation with a translation |
| 13. | A | Thought numbers under square roots could be added |
|   | B | CORRECT |
|   | C | Simplified incorrectly |
|   | D | Thought numbers under square roots could be added |

## Go Online!

### Standardized Test Practice
Students can take self-checking tests in standardized format to plan and prepare for standardized assessment.

---

# CHAPTER 10
# Preparing for Assessment
## Cumulative Review

Read each question. Then fill in the correct answer on the answer document provided by your teacher or on a sheet of paper.

1. Solve the following equation.
   $2x - 3(4 - x) = 5x + 3$  A.REI.3  **D**
   - A  −2
   - B  1
   - C  9
   - D  no solution

2. Charlene graphed the linear system below.

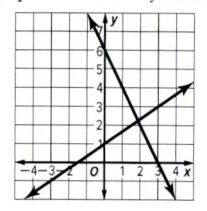

   Which is the best estimate of the solution? A.REI.6  **C**
   - A  (−1.5, 0)
   - B  (0, 1)
   - C  (2, 2)
   - D  (3, 0)

3. There were 24 cars that were either red or black in the parking lot. There were 7 times as many black cars as red cars. A.REI.6
   a. How many black cars were in the parking lot?
      **21** black cars
   b. What mathematical practice did you use to solve this problem? See students' work.

   **Test-Taking Tip**
   Question 1 If solving an equation results in a true statement, such as 4 = 4, then the solution is all real numbers. If a false statement results, such as 4 = 5, then there is no solution.

4. How does the graph of $g(x) = f(x - 5)$ compare to the graph of the linear parent function $f(x) = x$?  F.BF.3  **B**
   - A  $f(x)$ is translated left 5 units.
   - B  $f(x)$ is translated right 5 units.
   - C  $f(x)$ is translated up 5 units.
   - D  $f(x)$ is reflected across the *x*-axis.

5. The table shows the total cost for two different trips to a movie theater.

   | Adult Tickets | Child Tickets | Total Cost |
   |---|---|---|
   | 2 | 3 | $48 |
   | 4 | 1 | $56 |

   Let *a* represent the cost of an adult ticket and *c* represent the cost of a child ticket. Which system of equations best represents this situation? A.CED.3  **B**
   - A  $2a + 4c = 48$
     $3a + c = 56$
   - B  $2a + 3c = 48$
     $4a + c = 56$
   - C  $2a + 4c = 56$
     $4a + c = 48$
   - D  $2a + 3c = 56$
     $4a + c = 48$

6. Each value in a data set is increased by 4. Select the statistics of the new data set that can be found by adding 4 to the statistic of the original data set. S.ID.2  **A, B, C**
   - ☐ A  mean
   - ☐ B  median
   - ☐ C  mode
   - ☐ D  range
   - ☐ E  standard deviation

11. Students may have taken the square root of 256 without first factoring out 4. Students may have divided by 2 rather than taken the square root of 64.

706 | Chapter 10 | Preparing for Assessment

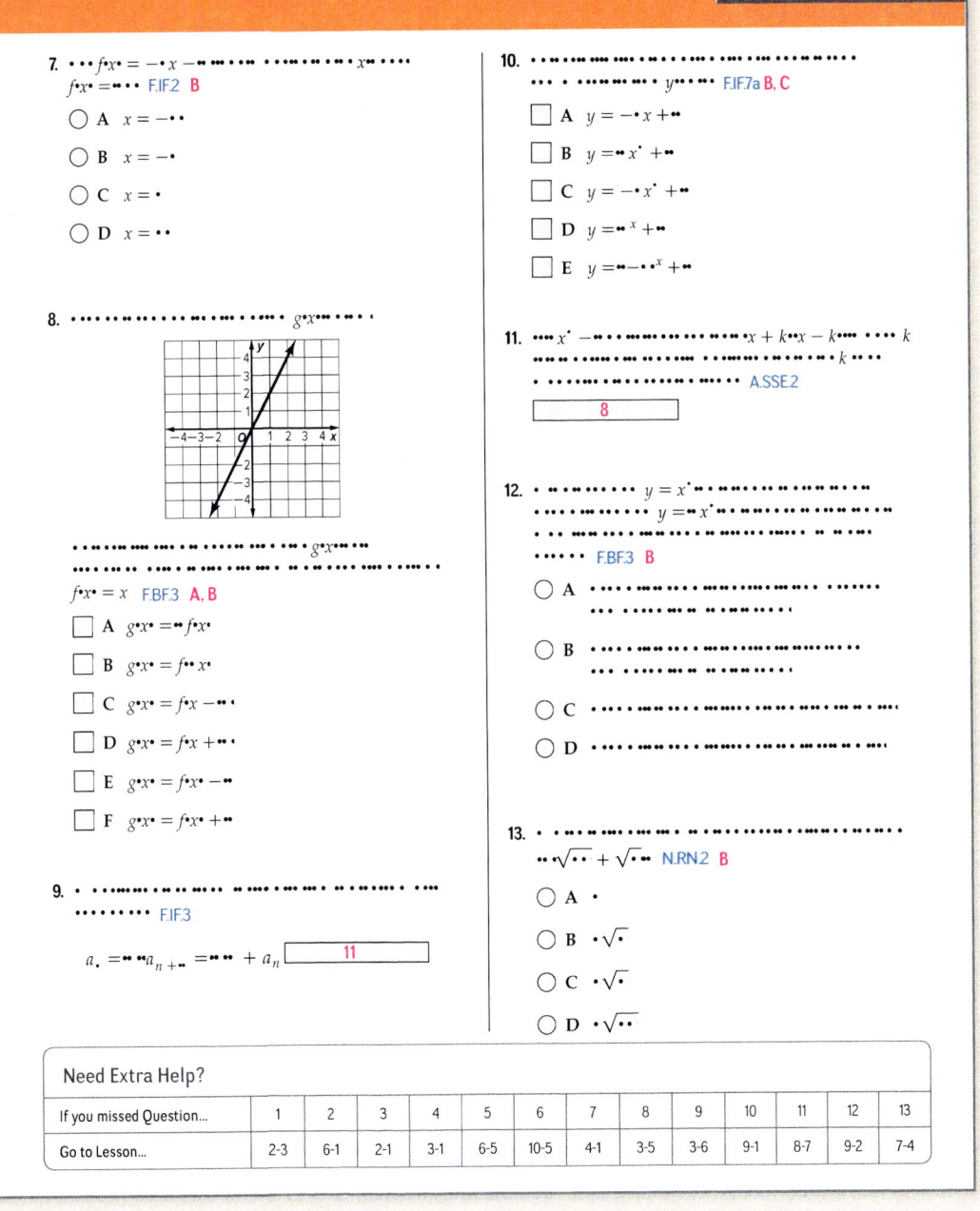

Chapter 10 Preparing for Assessment

**Formative Assessment**

You can use these pages to benchmark student progress.

Standardized Test Practice

**Test Item Formats**

In the Cumulative Review, students will encounter different formats for assessment questions to prepare them for standardized tests.

| Question Type | Exercises |
| --- | --- |
| Multiple-Choice | 1–2, 4–5, 7, 12–13 |
| Multiple Correct Answers | 6, 8, 10 |
| Type Entry: Short Response | 3, 9, 11 |
| Type Entry: Extended Response | 3 |

**Answer Sheet Practice**

Have students simulate taking a standardized test by recording their answers on a practice recording sheet.

**Need Extra Help?**

| If you missed Question… | 1 | 2 | 3 | 4 | 5 | 6 | 7 | 8 | 9 | 10 | 11 | 12 | 13 |
| --- | --- | --- | --- | --- | --- | --- | --- | --- | --- | --- | --- | --- | --- |
| Go to Lesson… | 2-3 | 6-1 | 2-1 | 3-1 | 6-5 | 10-5 | 4-1 | 3-5 | 3-6 | 9-1 | 8-7 | 9-2 | 7-4 |

# LEARNSMART

Use LearnSmart as part of your test-preparation plan to measure student topic retention. You can create a student assignment in LearnSmart for additional practice on these topics.

- Interpreting Categorical and Quantitative Data
- Summarize, Represent, and Interpret Data on Two Categorical and Quantitative Variables

**Go Online!**

**eAssessment**

Customize and create multiple versions of chapter tests and answer keys that align to your standards. Tests can be delivered on paper or online.

connectED.mcgraw-hill.com 707

**Lesson 10-2**

4a. Ages of People Attending the Play

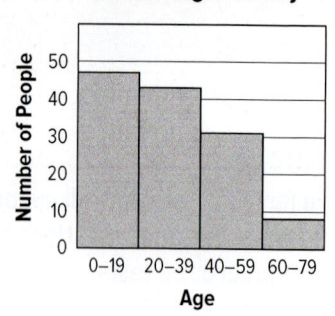

4b. Ages of People Attending the Play

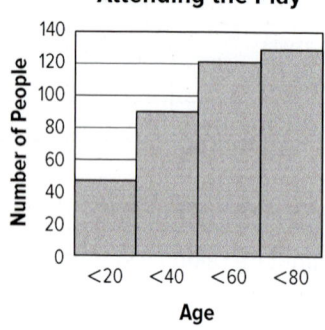

5. continuous;

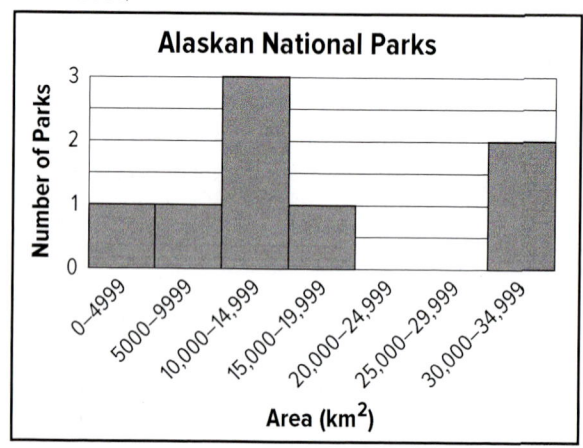

Alaskan National Parks

6. [dot plot on number line 0–9]

7. Elevation of Mountains in US

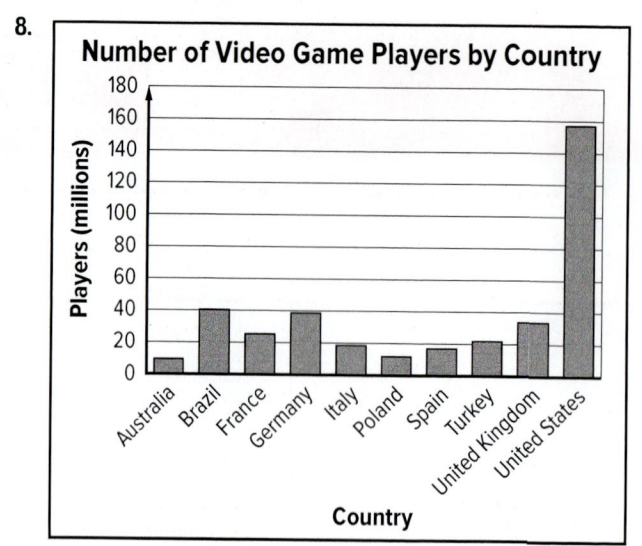

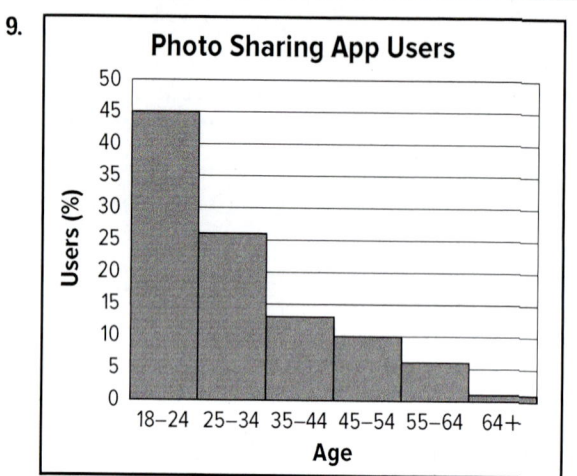

8. Number of Video Game Players by Country

9. Photo Sharing App Users

13. discrete;

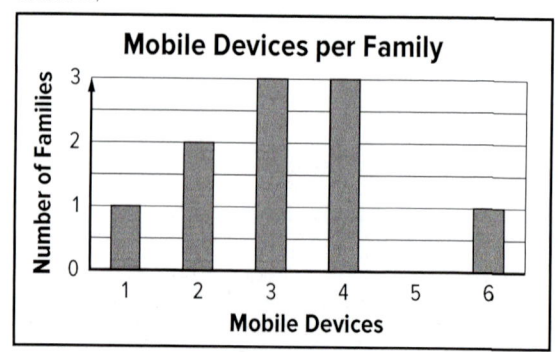

Mobile Devices per Family

14. continuous;

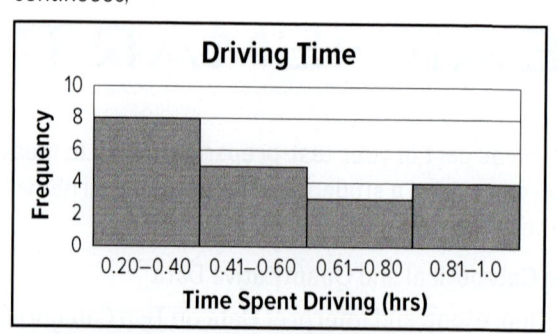

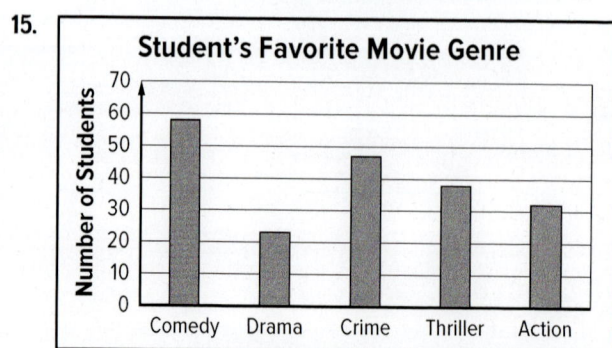

Driving Time

15. Student's Favorite Movie Genre

707A | Chapter 10 | Answer Appendix

18. Sample answer: Both use bars. In a bar graph the bars do not touch to represent categorical or discrete data. In a histogram the bars touch but do not overlap to represent continuous data.

## Mid-Chapter Quiz

5a. **Ages of People Attending the Carnival** (histogram with bars at ages 0-19: ~65, 20-39: ~50, 40-59: ~55, 60-79: ~17)

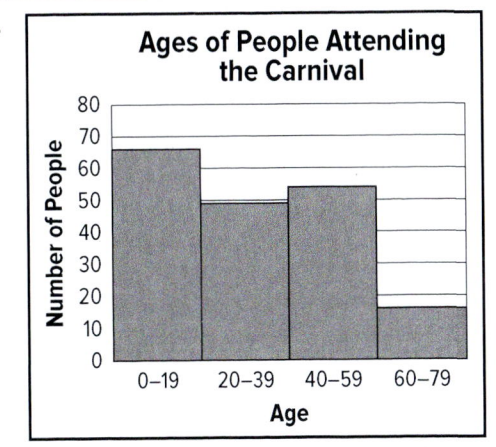

5b. **Ages of People Attending the Carnival** (cumulative histogram <20: ~35, <40: ~115, <60: ~170, <80: ~188)

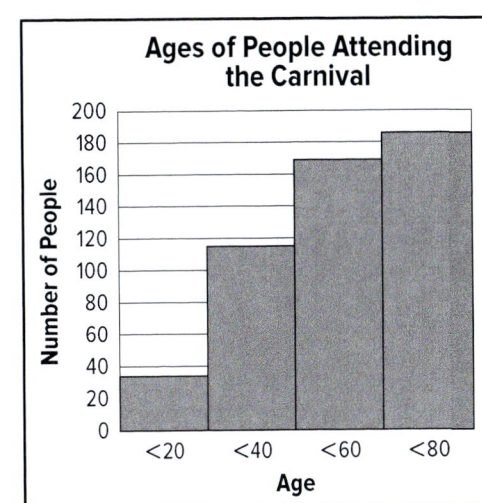

6. **Favorite Instrument** (bar graph: Drums 8, Guitar 12, Piano 5, Trumpet 7)

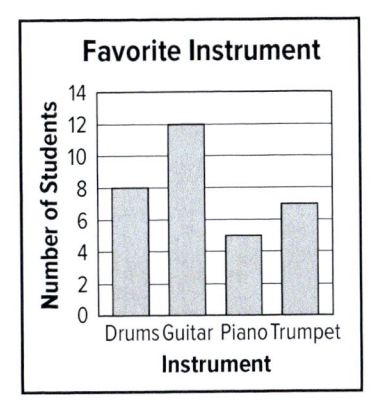

9. 1.66; Sample answer: With a mean of about 3.08, the standard deviation of about 1.66 suggests that there is some deviation to the data. However, if you remove the outlier, the mean becomes about 2.64 with a standard deviation of about 0.77. Therefore, almost all of the children at the play area are very similar in age.

11. The distribution has a mean of 7.88 and a standard deviation of about 2.88.

## Lesson 10-4 Guided Practice

1. [6, 38] scl: 4 by [0, 10] scl: 1
symmetric

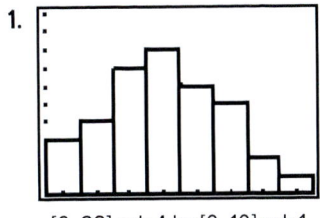

## Lesson 10-4

4. Sample answer: The distribution is skewed, so use the five-number summary. The range is 21 − 6 or 15. The median time is 17, and half of the times are between 11 and 18.5 minutes.

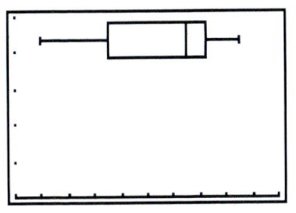

[4, 24] scl: 2 by [0, 5] scl: 1

5. [24, 78] scl: 6 by [0, 10] scl: 1
symmetric

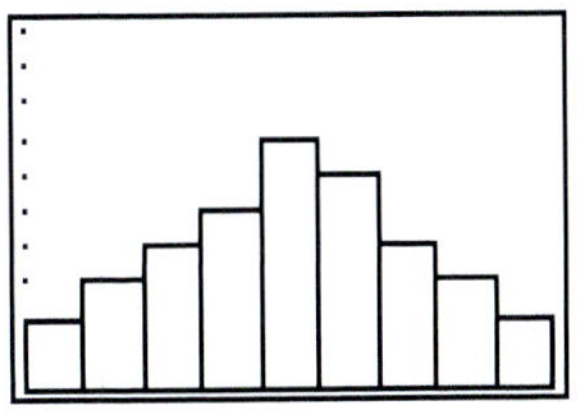

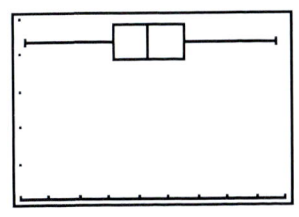

[24, 78] scl: 6 by [0, 5] scl: 1

6. [24, 78] scl: 6 by [0, 10] scl: 1

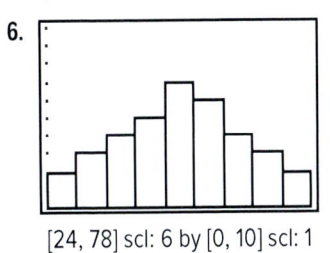

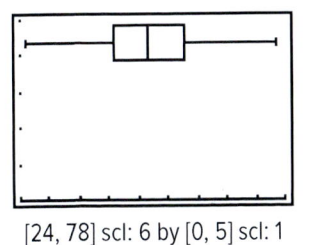

[24, 78] scl: 6 by [0, 5] scl: 1
negatively skewed

7. Sample answer: The distribution is skewed, so use the five-number summary. The range is 53 − 12 or 41. The median is 39.5, and half of the data are between 28 and 48.

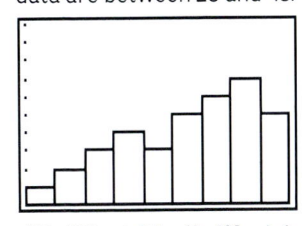

[10, 55] scl: 5 by [0, 10] scl: 1

8. Sample answer: The distribution is symmetric, so use the mean and standard deviation. The mean is 82 with a standard deviation of about 7.4.

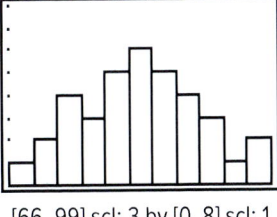

[66, 99] scl: 3 by [0, 8] scl: 1

9. Sample answer: The distribution is symmetric, so use the mean and standard deviation to describe the center and spread. The mean temperature is 52.8° with a standard deviation of about 4.22°.

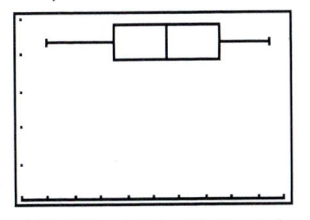

[42, 62] scl: 2 by [0, 5] scl: 1

**10a.**

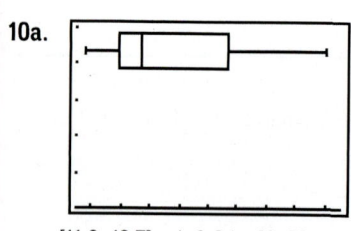

[11.9, 13.7] scl: 0.2 by [0, 5] scl: 1

Sample answer: The distribution is skewed, so use the five-number summary. The range is 13.6 − 11.96 or 1.64. The median time is 12.34, and half of the times are between 12.18 and 12.93 seconds.

**10b.**

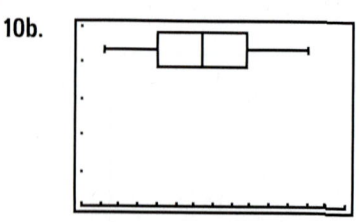

[11.9, 12.6] scl: 0.05 by [0, 5] scl: 1

Sample answer: The distribution is symmetric, so use the mean and standard deviation. The mean is about 12.22 seconds with a standard deviation of about 0.15 second.

**10c.** Sample answer: Removing the times causes the shape of the distribution to go from being skewed to being symmetric. Therefore, the center and spread should be described using the mean and standard deviation.

**11a.**

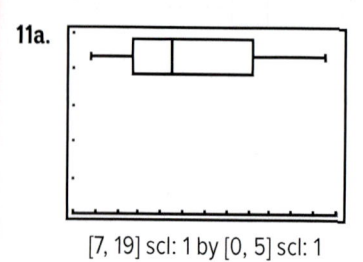

[7, 19] scl: 1 by [0, 5] scl: 1

Sample answer: The distribution is skewed, so use the five-number summary. The range is $18.50 − $7.75 or $10.75. The median price is $11.50, and half of the prices are between $9.63 and $15.13.

**11b.**

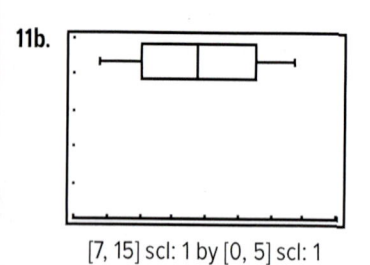

[7, 15] scl: 1 by [0, 5] scl: 1

Sample answer: The distribution is symmetric, so use the mean and standard deviation. The mean is about $10.67 with a standard deviation of about $1.84.

**17.** Sample answer: In a symmetrical distribution, the majority of the data are located near the center of the distribution. The mean of the distribution is also located near the center of the distribution. Therefore, the mean and standard deviation should be used to describe the data. In a skewed distribution, the majority of the data lies either on the right or left side of the distribution. Since the distribution has a tail or may have outliers, the mean is pulled away from the majority of the data. The median is less affected. Therefore, the five-number summary should be used to describe the data.

## Lesson 10-5

**16a.** Blue Course / Red Course

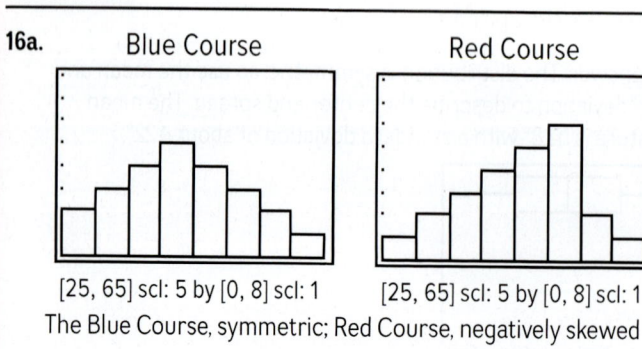

[25, 65] scl: 5 by [0, 8] scl: 1    [25, 65] scl: 5 by [0, 8] scl: 1

The Blue Course, symmetric; Red Course, negatively skewed

**16b.** Sample answer: One distribution is symmetric and the other is skewed, so use the five-number summaries. The minimum and maximum for the Blue Course are 25 and 62. The minimum and maximum for the Course are 26 and 61. Therefore, the ranges are approximately equal. The upper quartile for the Electronic Superstore is 49, while the median for the Red Course is 48. Since these two values are approximately equal, this means that about 50% of the data for the Red Course is greater than 75% of the data from the Blue Course. Overall, while both courses have similar ranges, the Red Course has higher scores.

**17a.**

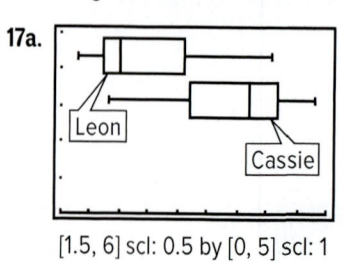

Leon, positively skewed; Cassie, negatively skewed

[1.5, 6] scl: 0.5 by [0, 5] scl: 1

**17b.** Sample answer: The distributions are skewed, so use the five-number summaries. The lower quartile for Leon's times is 2.2 minutes, while the minimum for Cassie's times is 2.3 minutes. This means that 25% of Leon's times are less than all of Cassie's times. The upper quartile for Leon's times is 3.6 minutes, while the lower quartile for Cassie's times is 3.7 minutes. This means that 75% of Leon's times are less than 75% of Cassie's time. Overall, we can conclude that Leon completed the brainteasers faster than Cassie.

**18a.**

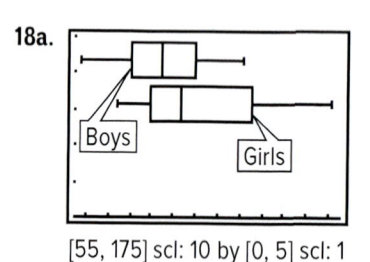

boys, symmetric; girls, positively skewed

[55, 175] scl: 10 by [0, 5] scl: 1

## Practice Test

**4.** Favorite Food (bar graph: Pizza 15, Chicken Nuggets 10, Cheesy Potatoes 8, Ice Cream 5)

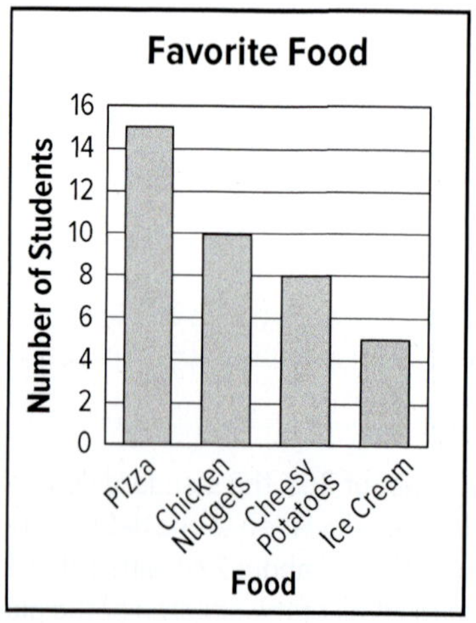

**7.**

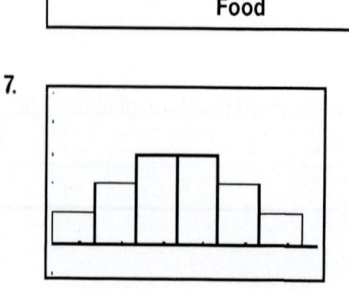

[14, 52] scl: 6 by [−1, 5] scl: 1

## Lesson 10-6

13a.
| District | Favor | Against | Total |
|---|---|---|---|
| District #1 | 178 | 47 | 225 |
| District #2 | 161 | 145 | 306 |
| Total | 339 | 192 | 531 |

13b.
| District | Favor | Against | Total |
|---|---|---|---|
| District #1 | 33.5% | 8.9% | 42.4% |
| District #2 | 30.3% | 27.3% | 57.6% |
| Total | 63.8% | 36.2% | 100% |

13c.
| District | Favor | Against | Total |
|---|---|---|---|
| District #1 | 79.1% | 20.9% | 100% |
| District #2 | 52.6% | 47.4% | 100% |

| District | Favor | Against |
|---|---|---|
| District #1 | 52.5% | 24.5% |
| District #2 | 47.5% | 75.5% |
| Total | 100% | 100% |

**Note**

## Performance Task

4. 100, 102, 102.5, 103, 110, 110, 117, 117.5, 118, 120. The median is 110 and the interquartile range is 15, which is close to 15.4 (the interquartile range for Isabelle).

5. Coach Rodriguez should send Isabelle. The change in the scores increases her median to 117.4, but the interquartile ranges stays at 15.4. Now Isabelle's median is about the same as that for Amani, and greater than that for Tamiko, but Isabelle is a more consistent diver than Amani.

6.
| Snacks | > 45 years old | ≤ 45 years old | Total |
|---|---|---|---|
| Yes | 67 | 141 | 208 |
| No | 47 | 45 | 92 |
| Total | 114 | 186 | 300 |

The main categories were age and buying snacks. The age subcategories broke down who was over 45 and who was 45 or younger. And the purchasing subcategories were broken down as those that would buy and those that would not.

# Student Handbook

This **Student Handbook** can help you answer these questions.

## What if I Need More Practice?

**Extra Practice**     R1
The **Extra Practice** section provides additional problems for each lesson so you have ample opportunity to practice new skills.
See Teacher Edition Volume 1 for Chapters 1–5; see Volume 2 for Chapters 6–10.

## What if I Need to Check a Homework Answer?

**Selected Answers and Solutions**     R11
The answers to odd-numbered problems are included in **Selected Answers and Solutions**.
See Teacher Edition Volume 1 for Chapters 1–5; see Volume 2 for Chapters 6–10.

## What if I Forget a Vocabulary Word?

**Glossary/Glosario**     R82
The **English-Spanish Glossary** provides definitions and page numbers of important or difficult words used throughout the textbook.

## What if I Need to Find Something Quickly?

**Index**     R100
The **Index** alphabetically lists the subjects covered throughout the entire textbook and the pages on which each subject can be found.

## What if I Forget a Formula?

**Formulas and Measures, Symbols and Properties**     Inside Back Cover
Inside the back cover of your math book is a list of **Formulas and Symbols** that are used in the book.

# Extra Practice

## CHAPTER 6 — Systems of Linear Equations and Inequalities

Use the graph below to determine whether each system is consistent or inconsistent and if it is independent or dependent. (Lesson 6-1) **1. consistent and independent**

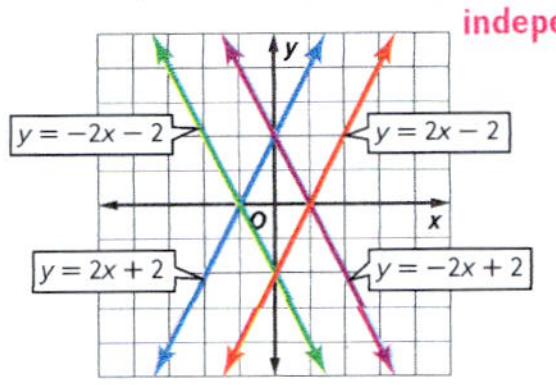

1. $y = 2x + 2$
   $y = -2x - 2$

2. $y = -2x + 2$
   $y = -2x - 2$ **inconsistent**

3. **DANCES** Mario and Tanesha are inflating balloons for the school dance. Mario has 12 balloons inflated and is inflating additional balloons at a rate of 3 balloons per minute. Tanesha has 16 balloons inflated and is inflating additional balloons at a rate of 2 balloons per minute. (Lesson 6-1)

   a. Write a system of equations to represent the situation.
   b. Graph each equation.
   c. How long will it take Mario to have more balloons filled than Tanesha?
   **See Extra Practice Answer Appendix.**

Use substitution to solve each system of equations. (Lesson 6-2)

4. $x = -y + 3$ **(−1, 4)**
   $3y + 2x = 10$

5. $-x + 2y = 6$ **no solution**
   $4y - 2x = 11$

6. $y - 7x = 2$ $\left(-\frac{7}{33}, \frac{17}{33}\right)$
   $2y + 3 = 5y$

7. $-2y = \frac{x+3}{3}$ **infinitely many**
   $\frac{3}{2} = \frac{1}{2}x - y$

8. **FRUIT** Sarah and Toni each bought fruit for a fundraiser. If Toni spent $4.30 and Sarah spent $2.80, how much does each type of fruit cost? (Lesson 6-2) **apple, $0.30; orange, $0.50**

| Girl | Apples | Oranges |
|---|---|---|
| Toni | 6 | 5 |
| Sarah | 6 | 2 |

Use elimination to solve each system of equations. (Lesson 6-3)

9. $2m + 3n = 16$ **(−12, 40/3)**
   $-3m - 3n = -4$

10. $-5k + 4j = 8$ **(0, 2)**
    $-5k - 6j = -12$

11. The difference of three times a number and a second number is two. The sum of the two numbers is fourteen. What are the two numbers? (Lesson 6-3) **4 and 10**

Use elimination to solve each system of equations. (Lesson 6-4)

12. $1.6x + 2.2y = 5.4$ **(2, 1)**
    $-3.2x + 4y = -2.4$

13. $2x + 5y = -8$ $\left(-\frac{2}{3}, -\frac{4}{3}\right)$
    $4x - 2y = 0$

14. **CARNIVALS** Scott and Isaac went to the school carnival. Use the table shown to determine how many tickets a ride and game each cost. (Lesson 6-4) **ride, 3 tickets; game, 1 ticket**

| Rider | Rides | Games | Tickets |
|---|---|---|---|
| Scott | 5 | 4 | 19 |
| Isaac | 7 | 2 | 23 |

15. **SAVINGS** Caleb made $105 mowing lawns and walking dogs, charging the rates shown. If he mowed half as many lawns as dogs walked, how many lawns did he mow and how many dogs did he walk? (Lesson 6-5) **lawns, 3; dogs, 6**

$7.50 per dog     $20 per lawn

Determine the best method to solve each system of equations. Then solve the system. (Lesson 6-5)

16. $4x + 2y = 12$ **elim (+); (−4, 14)**
    $-y - 4x = 2$

17. $y + 3x = 11$ **subst; (2, 5)**
    $-2x + 3y = 11$

18. **BABYSITTING** Kelsey and Emma babysit after school to earn extra money. Kelsey made $52 by charging $10 per hour and $4 per child. Emma made $67.50 by charging $15 per hour and $2.50 per child. (Lesson 6-5)

   a. Write a system of equations to represent the situation. **10x + 4y = 52, 15x + 2.5y = 67.5**
   b. How many hours and how many children did each babysit? **4 hours, 3 children**

Solve each system of inequalities by graphing. (Lesson 6-6) **19, 20. See Extra Practice Answer Appendix.**

19. $0.5x - y \geq 3$
    $x + y < 3$

20. $y < 2x - 1$
    $y > 4(1 + 0.5x)$

---

## CHAPTER 7 — Exponents and Exponential Functions

Simplify each expression. (Lesson 7-1)

1. $(2xy^2)^3(2x^2y^3z)$  **$16x^5y^9z$**
2. $(2ab^2c^3)(3ad^2)^2$  **$18a^3b^2c^3d^4$**

**GEOMETRY** Express the area of each triangle as a monomial. (Lesson 7-1)

3. 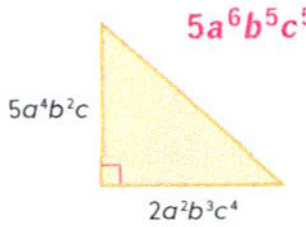 **$5a^6b^5c^5$**  (triangle with sides $5a^2b^3c$, $2a^2b^2c^4$)

4. 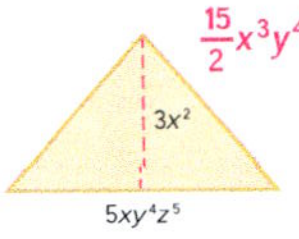 **$\frac{15}{2}x^3y^4z^5$**  (triangle with $3x^2$, $5xy^2z^5$)

Simplify each expression. Assume that no denominator equals zero. (Lesson 7-2)

5. $\dfrac{3m^6n^3p^4}{5m^6n^4p^{-4}}$  **$\dfrac{3p^2}{5mnq^5}$**

6. $\left(\dfrac{x^2yz^2}{2xy^{-3}}\right)^3$  **$\dfrac{x^3}{8^6y^6}$**

7. $\dfrac{2a^{-1}b^{-2}c^{-3}}{7a^2b^3c^4d^{-5}}$  **$\dfrac{2d^5}{7a^3b^5c^7}$**

8. $\left(\dfrac{4x^{-2}y^4z^5}{5x^5y^{-3}z^{-2}}\right)^{\frac{1}{2}}$  **$\dfrac{5x^7}{6y^7z^7}$**

Write each expression in radical form, or write each radical in exponential form. (Lesson 7-3)

9. $13^{\frac{1}{3}}$  **$\sqrt[3]{13}$**

10. $(7k)^{\frac{1}{2}}$  **$\sqrt{7k}$**

11. $\sqrt{17a}$  **$(17a)^{\frac{1}{2}}$**

12. $3\sqrt{2xyz^2}$  **$3(2xyz^2)^{\frac{1}{2}}$**

Simplify. (Lesson 7-3)

13. $\sqrt[4]{\dfrac{81}{625}}$  **$\dfrac{3}{5}$**

14. $\sqrt[5]{0.00001}$  **0.1**

15. $4096^{\frac{1}{3}}$  **16**

16. $\left(\dfrac{125}{343}\right)^{\frac{2}{3}}$  **$\dfrac{625}{2401}$**

Solve each equation. (Lesson 7-4)

17. $81^x = \dfrac{1}{3} = \dfrac{1}{4}$  **$-\dfrac{1}{4}$**

18. $3^{4x} = 3^{x+1}$  **$\dfrac{1}{3}$**

Simplify each expression. (Lesson 7-4)

19. $\dfrac{6}{\sqrt{3}+2} \cdot \dfrac{12 - 6\sqrt{3}}{}$  **$12 - 6\sqrt{3}$**

20. $(3\sqrt{6} + 2\sqrt{4})(5\sqrt{2} - 4\sqrt{3})$  **$14\sqrt{3} - 16\sqrt{2}$**

Graph each equation. Find the y-intercept, and state the domain and range. (Lesson 7-5)

21. $f(x) = -3^x - 1$  **21–23. See Extra Practice Answer Appendix.**

22. $f(x) = \left(\dfrac{1}{2}\right)^x + 3$

23. Determine whether the set of data shown below displays exponential behavior. Write yes or no. Explain why or why not. (Lesson 7-5)

| x | −1 | 0 | 1 | 2 | 3 | 4 |
|---|---|---|---|---|---|---|
| y | 3 | 1 | $\dfrac{1}{3}$ | $\dfrac{1}{9}$ | $\dfrac{1}{27}$ | $\dfrac{1}{81}$ |

24. Write a function $g(x)$ to represent the graph of $f(x) = 9^x$ translated up 2 units and left 4 units. (Lesson 7-6)  **$g(x) = 9^{(x+4)} + 2$**

25. Describe the transformations in the graph of $g(x) = 1.5^{-x} - 4$ as it relates to the graph of $f(x) = 1.5^x$. (Lesson 7-6)  **reflected across the y-axis and translated down 4 units**

26. **POPULATION** A neighborhood had 4518 residents in 2006. The number of residents has been declining by 3.5% each year. How many residents were there in 2012? (Lesson 7-7)  **3648 residents**

27. **MONEY** Sarah put $3000 in an investment that gets 6.2% compounded quarterly for 8 years. What will her investment be worth at the end of the 8 years? (Lesson 7-7)  **$4902.71**

28. **SOCCER** The Westside Soccer League has 186 players. They expect a 7.5% increase in players for at least the next 4 years. How many players will they have at that point? (Lesson 7-7)  **248 players**

29. Bank A offers a savings account with 4.5% interest compounded annually. Bank B offers a savings account with a monthly compounded interest rate of 0.39%. Which is the better plan? Explain. (Lesson 7-8)  **Bank B; the effective monthly interest rate for Bank A is about 0.37%, which is less than the monthly interest rate at Bank B.**

Determine whether each sequence is arithmetic, geometric, or neither. Explain. (Lesson 7-9)

30. $-\dfrac{1}{2}, -\dfrac{3}{4}, 0, \dfrac{1}{4}, \dfrac{1}{2}, \ldots$

31. $100, 90, 85, 75, 60, \ldots$

**30, 31. See Extra Practice Answer Appendix.**

Find the next three terms in each geometric sequence. (Lesson 7-9)

32. $48, -96, 192, -384, 768, \ldots$  **−1536, 3072, −6144**

33. $150, 75, 37.5, 18.75, 9.375, \ldots$  **4.6875, 2.3438, 1.1719**

Find the first five terms of each sequence. (Lesson 7-10)

34. $a_1 = 5, a_n = 3.5a_{n-1} + 1, n \geq 2$

35. $a_1 = 12, a_n = -\dfrac{1}{2}a_{n-1} + \dfrac{5}{2}, n \geq 2$

**34, 35. See Extra Practice Answer Appendix.**

Write a recursive formula for each sequence. (Lesson 7-10)  **36, 37. See Extra Practice Answer Appendix.**

36. $7, 16, 43, 124, \ldots$

37. $729, 243, 81, 27, \ldots$

# Extra Practice

## CHAPTER 8  Polynomials

**Find each sum or difference.** (Lesson 8-1)

1. $(7g^3 + 2g^2 - 12) - (-2g^3 - 4g)$  
2. $(-3h^2 + 3h - 6) + (5h^2 - 3h - 10)$  $2h^2 - 16$

**Simplify each expression.** (Lesson 8-2)

3. $\frac{1}{2}n^3p^2(5np^3 - 3n^2p^2 + 8n)$  
4. $6j^2(-3j + 3k^2) - 2k^2(2j + 10j^2)$  $3\frac{5}{2}n^4p^5 - \frac{3}{2}n^5p^4 + 4n^4p^2$  $-18j^3 - 2j^2k^2 - 4jk^2$

**Solve each equation.** (Lesson 8-2)

5. $-4(b + 3) + b(b - 3) = -b(6 - b) + 2(b - 3)$  $-2$  
6. $3(a - 3) + a(a - 1) + 12 = a(a - 2) + 3(a - 2) + 4$  $-5$

**Find each product.** (Lesson 8-3)

7. $(-3t - 16)(5t + 2)$   $-15t^2 - 86t - 32$  
8. $(4p + \frac{1}{2})(\frac{1}{2}p + 4)$   $8.\ 2p^2 + \frac{65}{4}p + 2$

9. **SIDEWALKS** Reynoldsville is repairing sidewalks. If the sidewalk is the same width around a city block, write an expression for the area of the block and the sidewalk. $(80 + 2x)(100 + 2x); 4x^2 + 360x + 8000$

**Find each product.** (Lesson 8-4)

10. $(\frac{1}{2}m + 3)^2$  
11. $(2n - 6)(2n + 6)$  
12. $(5a - 4)^2$  
13. $(x - 2y)(x + 2y)$

**10–13. See Extra Practice Answer Appendix.**

**Use the Distributive Property to factor each polynomial.** (Lesson 8-5)

14. $4m^3j^2 + 16m^2n^3 - 8m^3n^4$  $4m^2n^2(m + 4n - 2mn^2)$  
15. $12j^4k^4 + 36j^3k^2 - 3j^2k^5$  $3j^2k^2(4j^2k^2 + 12j - k^3)$

**Factor each polynomial.** (Lesson 8-5)

16. $x^2 - 4x + 3xy - 12y$  $(x + 3y)(x - 4)$  
17. $4a - 10ab + 6b - 15b^2$  $(2a + 3b)(2 - 5b)$

18. **HEIGHT** The height in feet of a ball bounced off the ground after $t$ seconds is modeled by the expression $-16t^2 + 28t$. (Lesson 8-5)
  a. Write the factored form of the expression for the height of the ball. $-4t(t - 7)$
  b. What is the height of the ball after 1.5 seconds? **6 ft**

**Factor each polynomial.** (Lesson 8-6)

19. $t^2 + 2t - 15$  $(t - 3)(t + 5)$  
20. $d^2 - 3d - 28$  $(d - 7)(d + 4)$  
21. $m^2 + 5m - 14$  $(m - 2)(m + 7)$  
22. $x^2 - 4x - 45$  $(x - 9)(x + 5)$

23. **GEOMETRY** Find an expression for the perimeter of a rectangle with the area $6x^2 + x - 15$. (Lesson 8-6)  $10x + 4$

**24–27. See Extra Practice Answer Appendix.**
Factor each polynomial, if possible. If the polynomial cannot be factored using integers, write *prime*. (Lesson 8-6)

24. $6x^2 + 21x - 90$  
25. $3x^2 - 11x - 42$  
26. $6x^2 - 13x - 5$  
27. $5y^2 - 3y + 11$

**28–33. See Extra Practice Answer Appendix.**
Factor each polynomial. (Lesson 8-7)

28. $\frac{1}{2}t^2 - 16^2$  
29. $25d^3 - 49d$  
30. $196z^2u^3 - 144u^3$  
31. $169a^4b^6 - 121c^8$  
32. $4g^2 - 1296ft^2$  
33. $18a^3 + 27a^2 - 50a - 75$

34. The area of a square is represented by $16x^2 - 40x + 25$. Find the length of each side. (Lesson 8-7) $|4x - 5|$

35. The volume of a rectangular prism is represented by $6x^3 + x^2 - 2x$. Find the possible dimensions of the prism if the dimensions are represented by polynomials with integer coefficients. (Lesson 8-7) $x, 2x - 1, 3x + 2$

36. **FRUIT** An apple fell 25 feet from a tree. The expression $25 - 16t^2$ models the height in feet of the apple after $t$ seconds. (Lesson 8-7)
  a. Write the factored form of the expression for the height of the apple. $(5 - 4t)(5 + 4t)$
  b. What is the height of the apple after 0.5 second? **21 ft**

**Determine whether each trinomial is a perfect square trinomial. Write *yes* or *no*. If so, factor it.** (Lesson 8-7) **37–40. See Extra Practice Answer Appendix.**

37. $64x^2 - 32x + 4$  
38. $4a^2 - 12a + 16$  
39. $12y^2 - 36y + 27$  
40. $75b^3 - 60ab^2 + 12a^2b$

---

## CHAPTER 9  Quadratic Functions and Equations

**Find the vertex, the equation of the axis of symmetry, and the $y$-intercept of the graph of each equation.** (Lesson 9-1) **1–4. See Extra Practice Answer Appendix.**

1. $y = 4x^2 + 8x - 5$  
2. $y = -2x^2 + 8x + 5$  
3. $y = x^2 - 8x + 9$  
4. $y = 4x^2 + 16x - 6$

5. **KICKBALL** A kickball is kicked into the air. The equation $h = -16t^2 + 60t$ gives the height $h$ of the ball after $t$ seconds. (Lesson 9-1)
  a. What is the height of the ball after one second? **44 ft**
  b. When will the ball reach its maximum height? **1.875 s**
  c. When will the ball hit the ground? **3.75 s**

**Describe the transformations in each function as it relates to the graph of $f(x) = x^2$.** (Lesson 9-2)

6. $g(x) = -x^2 - 4$  
7. $h(x) = 7x^2 + 2$

**6, 7. See Extra Practice Answer Appendix.**

**Match each equation to its graph.** (Lesson 9-2)

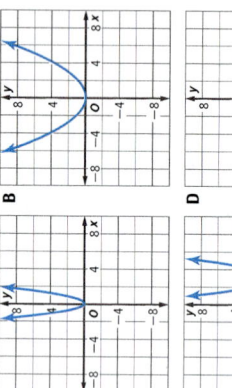

A [graph]   B [graph]   C [graph]   D [graph]

8. $y = 3x^2$  **A**  
9. $y = \frac{1}{4}x^2$  **B**  
10. $y = -(x + 4)^2$  **D**  
11. $y = 2(x - 3)^2$  **C**

**Solve each equation by graphing.** (Lesson 9-3)  **12, 13. See Extra Practice Answer Appendix for graphs.**

12. $-2x^2 - 2x + 4 = 0$  
13. $x^2 - 2x - 3$  $-3, 1$

14. **MARBLES** Jason shot a marble straight up using a slingshot. The equation $h = -16t^2 + 42t + 5.5$ models the height $h$, in feet, of the marble after $t$ seconds. After how long will the marble hit the ground? (Lesson 9-3) **about 2.7 seconds**

15. Solve each equation. Check your solutions. (Lesson 9-4)
  a. $x^2 = 16$  $x = \pm 4$
  b. $x^2 = 144$  $x = \pm 12$
  c. $x^2 = 9$  $x = \pm 3$

16. Solve $(x - 2)^2 = 16$. (Lesson 9-4) $x = 6$ or $x = -2$  
17. Solve $(2x + 8)(4x - 20) = 0$. (Lesson 9-4) $x = -4; x = 5$  
18. Solve $x^2 + 5x + 6 = 0$. (Lesson 9-4) $x = -2; x = -3$

19. **BIOLOGY** The number of cells in a Petri dish can be modeled by the quadratic equation $n = 6t^2 - 4.5t + 74$, where $t$ is the number of hours the cells have been in the dish. When will there be 200 cells in the Petri dish? (Lesson 9-5) **after 7 hours**

**Solve each equation by completing the square. Round to the nearest tenth if necessary.** (Lesson 9-5)

20. $x^2 + 4x - 8 = 5$  $-6.1, 2.1$  
21. $3x^2 + 5x = 18$  $-3.4, 1.8$

22. Find the value of $x$ in the figure if the area is 36 square inches. (Lesson 9-5) **4**

[triangle figure with $(x + 2)$ in. and $(x + 8)$ in.]

**Solve each equation by using the Quadratic Formula. Round to the nearest tenth if necessary.** (Lesson 9-6)

23. $3x^2 + 10x = 15$  $-4.5, 1.1$  
24. $\frac{1}{2}x^2 - 8x + 6 = 0$  $0.8, 15.2$

**State the value of the discriminant. Then determine the number of real solutions of the equation.** (Lesson 9-6)

25. $4x^2 - 12x = -9$  **0, one real solution**  
26. $3x^2 + 8 = 9x$  **−15, no real solutions**  
27. Solve the system of equations algebraically. (Lesson 9-7)
$y = x^2 + 2x + 1$
$y - x = 1$  **(0, 1) and (−1, 0)**

**Look for a pattern in each table of values to determine which kind of model best describes the data.** (Lesson 9-8)

28. 
| x | 2 | 3 | 4 | 5 | 6 |
|---|---|---|---|---|---|
| y | $\frac{9}{4}$ | $\frac{27}{8}$ | $\frac{81}{16}$ | $\frac{243}{32}$ | $\frac{729}{64}$ |

**exponential**

29.
| x | −2 | −1 | 0 | 1 | 2 |
|---|---|---|---|---|---|
| y | −13 | −625 | 0 | 575 | 11 |

**quadratic**

30. Given that $f(x) = x^2 + 3x - 4$ and $g(x) = x - 4$, find each function. (Lesson 9-9)
  a. $(f + g)(x)$  $x^2 + 4x - 8$
  b. $(f - g)(x)$  $x^2 + 2x$

31. Given that $f(x) = 2x^2 - x + 3$ and $g(x) = x + 1$, find $(f \cdot g)(x)$. (Lesson 9-9) $2x^3 + x^2 + 2x + 3$

connectED.mcgraw-hill.com

# CHAPTER 10  Statistics

1. 
| 66 | 66 | 60 | 59 | 69 | 71 | 62 | 63 | 64 |
|---|---|---|---|---|---|---|---|---|
| 67 | 64 | 65 | 66 | 67 | 68 | 62 | 69 | 70 |

The table represents the height, in inches, of the girls on the basketball team. (Lesson 10-1)

   a. What is the mean height? **65.44 in.**
   b. What is the median height? **66 in.**
   c. What is the mode? **66 in.**

2. Which measure of center would best describe the data, *mean, median,* or *mode*? (Lesson 10-1)

   a. Jamal scores mostly As on his math tests. **mode**
   b. A teacher wants to explore how many minutes students study per day, with answers ranging from 15 minutes to 200 minutes and the most common answer being 100 minutes. **mode**

3. Kendra scored a 95% on her math test. Out of the 25 students in the class, 22 students scored less than Kendra. In what percentile did she score? (Lesson 10-1) **88th percentile**

4. The table represents favorite ice cream flavors in Ms. Isbel's fourth grade class.

| Chocolate | 7 |
|---|---|
| Vanilla | 9 |
| Strawberry | 5 |
| Cookies and Cream | 6 |

   Represent the data in a bar graph. (Lesson 10-2) **See Extra Practice Answer Appendix.**

5. Which type of graph would best represent the scenario? (Lesson 10-2)

   a. the height ranges of NBA basketball players **histogram**
   b. the types of music to which the students in high school listen **bar graph**

6. **GAS** The price of gasoline was recorded at 7 gas stations on the same day. They were: $2.63, $2.59, $2.70, $2.58, $2.83, $2.65, $2.71. Find the minimum, lower quartile, median, upper quartile, and maximum for the data set. (Lesson 10-3) **$2.58; $2.59; $2.65; $2.71; $2.83**

7. **BOWLING** Tina's results for 8 bowling games are shown below.
   {110, 123, 147, 119, 153, 142, 113, 143}
   Find and interpret the standard deviation of the data. (Lesson 10-3) **15.7; Compared to the mean of 131.25, the standard deviation is small, so Tina's scores are relatively close together.**

   For Exercises 8 and 9, use these data.
   {12, 18, 21, 18, 19, 18, 16, 23, 20, 15, 17, 18}

8. Describe the center and spread of the data using either the mean and standard deviation or the five-number summary. Justify your choice by constructing a histogram. (Lesson 10-4) **See Extra Practice Answer Appendix.**

9. Find the mean, median, mode, range, and standard deviation of the data after multiplying each value by 3. (Lesson 10-5) **53.75; 54; 54; 33; 8.17**

10. Use the table to complete the following exercises. (Lesson 10-6)

| Favorite Subject | 7th Grade | 8th Grade | Total |
|---|---|---|---|
| Math | 92 | 47 | 139 |
| Science | 79 | 51 | 130 |
| English | 68 | 25 | 93 |
| Band | 13 | 59 | 72 |
| Total | 252 | 182 | 434 |

   a. How many seventh graders chose math as their favorite subject? **92**
   b. How many students are in the eighth grade? **182**
   c. How many more students prefer math than prefer science? **9**

11. The PTA president is organizing a celebration and asked members whether they would prefer a brunch, lunch, or dinner celebration and whether they would bring an appetizer or dessert. Use the table to complete the following exercises. (Lesson 10-6)

| Preference | Appetizer | Dessert | Total |
|---|---|---|---|
| Brunch | 8 | 4 | 12 |
| Lunch | 7 | 2 | 9 |
| Dinner | 6 | 11 | 17 |
| Total | 21 | 17 | 38 |

   a. Make a relative frequency table of the data. **See Extra Practice Answer Appendix.**
   b. What is the joint relative frequency of bringing an appetizer and preferring dinner? **15.8%**

## Chapter 6

**3a.** Mario: $y = 12 + 3x$, Tanesha: $y = 16 + 2x$

**3b.**

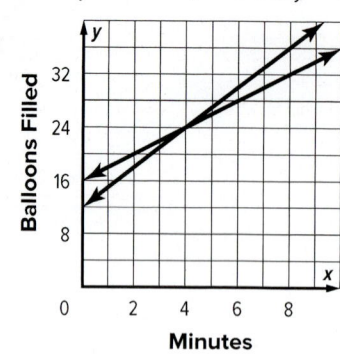

**3c.** (4, 24); Mario will have more filled balloons after 4 minutes.

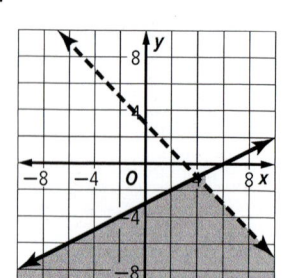

**19.**   **20.** no solution

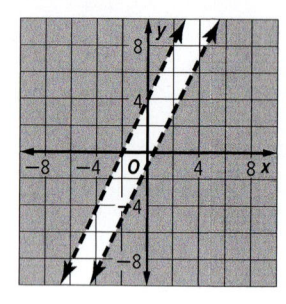

## Chapter 7

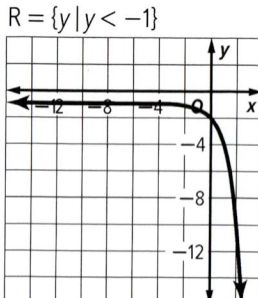

**21.** $-2$; D = {all real numbers}; R = {$y\,|\,y < -1$}

**22.** 4; D = {all real numbers}; R = {$y\,|\,y > 3$}

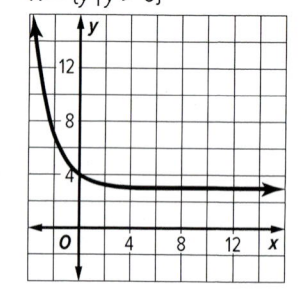

**23.** Yes; the domain values are at regular intervals and the range values have a common factor of $\frac{1}{3}$.

**30.** Arithmetic; the difference between terms is $\frac{1}{2}$.

**31.** Neither; there is no common difference or common factor.

**34.** 5, 18.5, 65.75, 231.125, 809.9375

**35.** $-3.5, 4.25, 0.375, 2.3125$

**36.** $a_1 = 7, a_n = 3a_{n-1} - 5, n \geq 2$

**37.** $a_1 = 729, a_n = \frac{1}{3}a_{n-1}, n \geq 2$

## Chapter 8

**10.** $\frac{1}{4}m^2 + 3m + 9$

**11.** $4n^2 - 36$

**12.** $25a^2 - 40a + 16$

**13.** $x^2 - 4y^2$

**24.** $3(2x - 5)(x + 6)$

**25.** $(3x + 7)(x - 6)$

**26.** $(2x - 5)(3x + 1)$

**27.** prime

**28.** $\frac{1}{2}(t - 18)(t + 18)$

**29.** $d(5d - 7)(5d + 7)$

**30.** $4u^3(7t - 6)(7t + 6)$

**31.** $(13a^2b^3 - 11c^4)(13a^2b^3 + 11c^4)$

**32.** $4(g + 18h)(g - 18h)$

**33.** $(3a + 5)(3a - 5)(2a + 3)$

**37.** yes; $4(4x - 1)(4x - 1)$

**38.** no

**39.** yes; $3(2x - 3)(2x - 3)$

**40.** yes; $3b(5b - 2a)(5b - 2a)$

## Chapter 9

**1.** $(-1, -9)$; $x = -1$; $-5$

**2.** $(2, 13)$; $x = 2$; $5$

**3.** $(4, -7)$; $x = 4$; $9$

**4.** $(-2, -22)$; $x = -2$; $-6$

**6.** reflected across the x-axis, translated down

**7.** stretched vertically, translated up

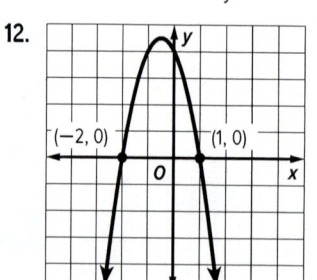

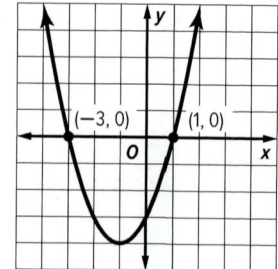

**12.**   **13.**

## Chapter 10

**4.**

**Favorite Flavors**

(Bar graph: Chocolate 7, Vanilla 9, Strawberry 5, Cookies and Cream 6)

**8.** Sample answer: The distribution is symmetric, so use the mean and standard deviation. The mean is about 17.9 with standard deviation of about 2.8.

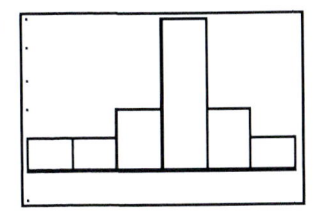

[12, 24] scl: 2 by [−1, 5] scl: 1

**11a.**

|  | Appetizer | Dessert | Total |
|---|---|---|---|
| **Brunch** | 21% | 10% | 31% |
| **Lunch** | 19% | 5% | 24% |
| **Prefer dinner** | 15% | 30% | 45% |
| **Total** | 55% | 45% | 100% |

## CHAPTER 6
## Systems of Linear Equations and Inequalities

### Get Ready
1. Yes, these expressions are the same.
3. Rewrite the expression as $\cdot \frac{7}{11}$.
5. Subtract 4y from each side to isolate the term with x on one side of the equation.   7. $y = \frac{x-8}{2}$

### Lesson 6-1
1. consistent and independent   3. inconsistent
5. consistent and independent
7. 1 solution, (−4, 0)

9a. Alberto: $y = 20x + 35$; Ashanti: $y = 10x + 85$
9b.
9c. (5, 135); Alberto will have read more after 5 days.   11. consistent and independent
13. Because these two graphs intersect at one point, there is exactly one solution. Therefore, the system is consistent and independent.
15. consistent and independent
17. 1 solution; (−1, −2)   19. infinitely many
21. 1 solution; (5, −1)   23. no solution
25a. Akira: $y = 30x + 22$; Jen: $y = 20x + 53$
25b.
25c. (3.1, 115); After about 3 days Akira will have sold more tickets.
27. Graph the two equations on the same coordinate plane. The graphs appear to intersect at (−4, −2). Check by substituting into the equations.

$y = \frac{1}{2}x$
$-2 = \frac{1}{2}(-4)$
$-2 = -2$ Yes

$y = x + 2$
$-2 = -4 + 2$
$-2 = -2$ Yes

So, the solution is (−4, −2).
29. 1 solution, (7, −3)

# Selected Answers and Solutions

**31.** 1 solution, (5, 3)

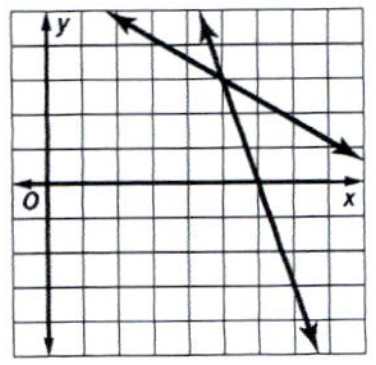

**33.** infinitely many

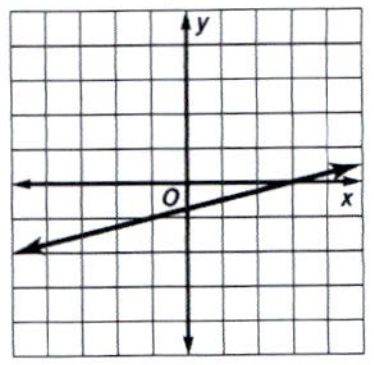

**35.** no solution

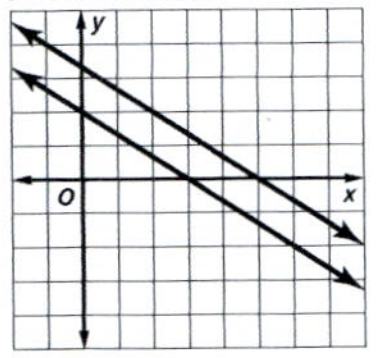

**37.** no solution

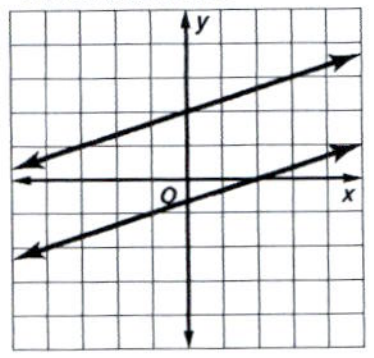

**39a.** Sample answers: At this rate, website A will have no visitors by 2069. Both websites will have the same number of visitors sometime during 2021. Website B will have 1 billion visitors by 2039.
**39b.** Sample answer: First, I found the average change in visitors per year for both websites: $\frac{476 - 512}{4}$ or $-9$, $\frac{251 - 131}{4}$ or 30. Next, I wrote equations to represent the situation: $y = -9x + 512$ and $y = 30x + 131$. Then, I graphed the equations and saw that they intersected between 9 and 10. So the number of visitors to the websites is the same sometime during the ninth year after 2012, or during 2021. I also saw from the graph when website A would reach 0 and website B would reach 1 billion.
**39c.** I assumed that these trends happened at a constant rate throughout the years, instead of the number of visitors going up and down over the years.

**41.** no solution

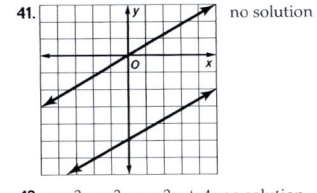

**43.** $y = 3x - 3$, $y = 3x + 4$; no solution
**45.** $y = -x + 2$, $y = 2x - 1$; (1, 1)

**47.** First solve each equation for $y$.
$2x + 3y = 5$    First equation
$3y = -2x + 5$    Subtract $2x$.
$y = -\frac{2}{3}x + \frac{5}{3}$    Divide by 3.

$3x + 4y = 6$    Second equation
$4y = -3x + 6$    Subtract $3x$.
$y = -\frac{3}{4}x + \frac{3}{2}$    Divide by 4.

$4x + 5y = 7$    Third equation
$5y = -4x + 7$    Subtract $4x$.
$y = -\frac{4}{5}x + \frac{7}{5}$    Divide by 5.

Now graph all three new equations.

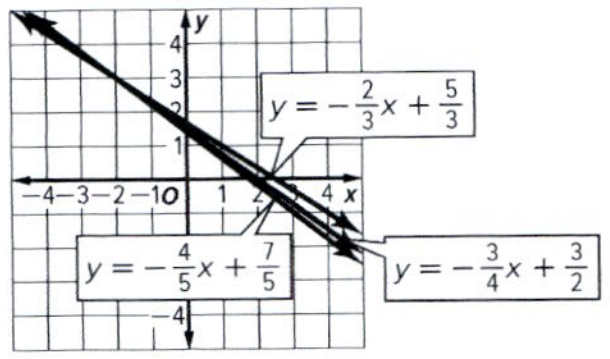

The graphs intersect at $(-2, 3)$, so this is the solution. **49.** Always; if the equations are linear and have more than one common solution, they must be consistent and dependent, which means that they have an infinite number of solutions in common. **51.** Sample answers: $y = 5x + 3$; $y = -5x - 3$; $2y = 10x - 6$ **53.** A **55.** B **57a.** $y = 7200 + 100x$; $y = 8850 - 50x$ **57b.** 2028 **57c.** 8300 people **57d.** Sample answer: Create a system of equations, graph the equations, and find the point of intersection. **57e.** Sample answer: I assumed that the populations continue to change at the same rate. **59.** C

## Lesson 6-2

**1.** (5, 10)   **3.** (2, 0)   **5.** infinitely many
**7a.** $x = m\angle X$, $y = m\angle Y$; $x + y = 180$, $x = 24 + y$
**7b.** $x = 102°$, $y = 78°$

**9.** Step 1: One equation is already solved for $y$.
$y = 4x + 5$
$2x + y = 17$
Step 2: Substitute $4x + 5$ for $y$ in the second equation.
$2x + y = 17$
$2x + 4x + 5 = 17$
$6x + 5 = 17$
$6x = 12$
$x = 2$
Step 3: Substitute 2 for $x$ in either equation to find $y$.
$y = 4x + 5$
$y = 4(2) + 5$
$y = 8 + 5$
$y = 13$
The solution is (2, 13).

**11.** $(-3, -11)$   **13.** $(-1, 0)$   **15.** infinitely many   **17.** (2, 3)   **19.** no solution   **21.** (2, 0)   **23a.** Let $d$ = demand for nurses; $s$ = supply of nurses; $t$ = number years; $d = 40,521t + 2,000,000$, $s = 5600t + 1,890,000$
**23b.** during 1996

---

**5.** 6, 18   **7.** $(-3, 4)$   **9.** $(-3, 1)$   **11.** $(4, -2)$
**13.** $(8, -7)$   **15.** $(4, 7)$   **17.** $(4, 1.5)$   **19.** 5, 17

**21.** 
| Three times a number | minus | another number | is | $-3$ |
|---|---|---|---|---|
| $3x$ | $-$ | $y$ | $=$ | $-3$ |

| The first number | plus | the second number | is | 11 |
|---|---|---|---|---|
| $x$ | $+$ | $y$ | $=$ | 11 |

Steps 1 and 2: Write the equations vertically and add.
$3x - y = -3$
$x + y = 11$
$\overline{4x \phantom{+ y} = 8}$
$x = 2$
Step 3: Substitute 2 for $x$ in either equation to find $y$.
$x + y = 11$
$2 + y = 11$
$y = 9$
The numbers are 2 and 9.
**23.** adult, $17.95; children, $13.95   **25.** $(2, -1)$
**27.** $\left(-\frac{5}{6}, 3\right)$   **29.** $\left(2\frac{7}{9}, 13\frac{1}{3}\right)$
**31a.** $x + y = 66$; $x = 30 + y$
**31b.** (48, 18)   **31c.** There are 48 teams that are not from the United States and 18 teams that are from the United States.
**31d.**

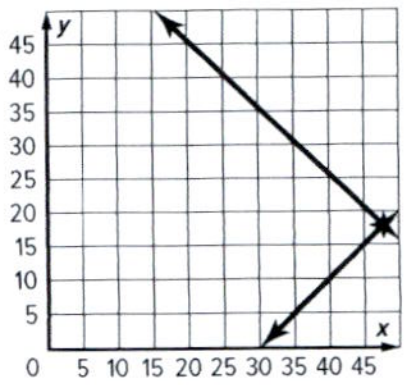

**33. a.** Sample answer: If you choose 4 pennies and 5 paper clips, the score will be $4(3) + 5$ or 17.
**b.** The total number of objects is 9.
$p + c = 9$
Pennies are worth 3 points each and paper clips are worth 1 point each for a total of 15 points.
$3p + c = 15$
Solve:
$p + c = 9$
$(-)3p + c = 15$
$\overline{-2p = -6}$
$p = 3$
Substitute 3 for $p$ in either equation to find $c$.
$p + c = 9$
$3 + c = 9$
$c = 6$
So, $p = 3$ and $c = 6$.

---

**25. a.** Men: $8:21:00 = 8(60) + 21 = 501$ and $8:18:37 = 8(60) + 18 = 498$, then round up because the number of seconds is greater than 30. So, $8:18:37$ rounds to 499. Women: $9:26:16 = 9(60) + 26 = 566$, then because the number of seconds is less than 30, $9:26:16$ rounds to 566 and $9:15:54 = 9(60) + 15 = 555$, then round up because the number of seconds is greater than 30. So, $9:15:54$ rounds to 556.
**b.** The $y$-intercept is (0, 501). Find the rate of change.
$m = \frac{501 - 499}{0 - 12}$
$= \frac{2}{-12}$
$= -\frac{1}{6}$
So, the equation is $y = -\frac{1}{6}x + 501$.
The $y$-intercept is (0, 566). Find the rate of change.
$m = \frac{566 - 556}{0 - 12}$
$= \frac{10}{-12}$
$= -\frac{5}{6}$
So, the equation is $y = -\frac{5}{6}x + 566$.
**c.** 2097; the graphs intersect around (97, 485)
**27.** Neither; Guillermo substituted incorrectly for $b$. Cara solved correctly for $b$ but misinterpreted the pounds of apples bought.   **29.** Sample answer: The solutions found by each of these methods should be the same. However, it may be necessary to estimate using a graph. So, when a precise solution is needed, you should use substitution.   **31.** An equation containing a variable with a coefficient of 1 can easily be solved for the variable. That expression can then be substituted into the second equation for the variable.   **33.** C
**35.** D   **37a.** $x + y = 3(x - y)$; $x = y + 5$   **37b.** 5, 10
**37c.** The systems can be solved by substituting either $y + 5$ for $x$ or $x - 5$ for $y$.   **39.** C

## Lesson 6-3

**1.** (2, 3)

**3.** Step 1: The like terms are already aligned.
$7f + 3g = -6$
$7f - 2g = -31$
Step 2: Subtract the equations.
$\phantom{(-)}7f + 3g = -6$
$\underline{(-)7f - 2g = -31}$
$\phantom{(-)7f + }5g = 25$
$\phantom{(-)7f + }g = 5$
Step 3: Substitute 5 for $g$ in either equation to find $f$.
$7f + 3g = -6$
$7f + 3(5) = -6$
$7f + 15 = -6$
$7f = -21$
$f = -3$
The solution is $(-3, 5)$.

# Selected Answers and Solutions

c.

| p | c = 9 − p | 3p + c |
|---|---|---|
| 0 | 9 | 3(0) + 9 = 9 |
| 1 | 8 | 3(1) + 8 = 11 |
| 2 | 7 | 3(2) + 7 = 13 |
| 3 | 6 | 3(3) + 6 = 15 |
| 4 | 5 | 3(4) + 5 = 17 |
| 5 | 4 | 3(5) + 4 = 19 |

d. Yes; because the pennies are 3 points each, 3 of them makes 9 points. Add the 6 points from 6 paper clips and you get 15 points.

35. The result of the statement is false, so there is no solution. 37. Sample answer: $-x + y = 5$; I used the solution to create another equation with the coefficient of the $x$-term being the opposite of its corresponding coefficient.

39. Sample answer: It would be most beneficial when one variable has either the same coefficient or opposite coefficients in each of the equations.

41. D  43. E  45. B

## Lesson 6-4

1. (3, 2)

3. Eliminate $y$:
$$(4x + 2y = -14)(-3) \quad -12x - 6y = 42$$
$$(5x + 3y = -17)(2) \quad \underline{10x + 6y = -34}$$
$$-2x = 8$$
$$x = -4$$

Now, substitute −4 for $x$ in either equation to find the value of $y$.
$4x + 2y = -14$
$4(-4) + 2y = -14$
$-16 + 2y = -14$
$2y = 2$
$y = 1$

The solution is (−4, 1).

5. 6 mph  7. (−1, 3)  9. (−3, 4)  11. (−2, 3)
13. (3, 5)  15. (1, −5)  17. (0, 1)

19. Four times a minus five times another equals 21.
  number       number
$4x \quad - \quad 5y \quad = \quad 21$

Three times the sum of the two is 36
  numbers
$3 \quad (x + y) \quad = \quad 36$

$(4x - 5y = 21)(3) \quad 12x - 15y = 63$
$(3(x + y) = 36)(5) \quad \underline{15x + 15y = 180}$
$27x = 243$
$x = 9$

Now substitute 9 for $x$ in either equation to find $y$.
$4x - 5y = 21$
$4(9) - 5y = 21$

$36 - 5y = 21$
$-5y = -15 \quad$ The two numbers are 9 and 3.
$y = 3$

21. (2.5, 3.25)  23. $(3, \frac{1}{2})$

25a. Michelle should bake 7 tubes of cookies in 56 minutes and Julie should bake 3 tubes of cookies in 36 minutes.  25b. Sample answer: Let $m$ = the number of tubes Michelle bakes. Let $j$ = the number of tubes Julie bakes. The number of tubes Michelle bakes plus the number of cookies Julie bakes should equal 264. Each batch Michelle makes produces 29 cookies. Each batch Julie makes produces 24 cookies. Therefore, $29m + 24j = 264$ relates the number of cookies baked to the number of tubes each girl uses. We also know that it takes Michelle's tubes 8 minutes to bake, and Julie's tubes take 12 minutes to bake. The girls have a little over 90 minutes to bake their cookies. So, $8m + 12j = 90$. Solve the system.

$29m + 24j = 264 \quad$ First equation
$(-)16m + 24j = 180 \quad$ Multiply second equation by 2 and subtract.
$13m = 84 \quad j$ is eliminated.
$m \approx 6.46 \quad$ Divide each side by 13.

It does not make sense for Michelle to bake part of a tube of cookie dough, so $m = 7$. Substituting 7 into the first equation, we determine that Julie bakes ≈2.54 or 3 tubes of cookies.  25c. Sample answer: I assumed that Michelle and Julie cannot bake their cookies simultaneously. I assumed that one of the cookies that Michelle makes per tube is slightly smaller than the others. (You could have assumed that she ate the leftover cookie dough.) I assumed that the girls were able to bake an entire tube of cookie dough in one batch. I assumed that the amount of time needed to transfer baked cookies from the oven and unbaked cookies to the oven is negligible.

27. a. Let $x$ be the cost of each trip to the batting cage and let $y$ be the cost of each miniature golf game. For the first group, the equation is $16x + 3y = 44.81$. For the second group, the equation is $22x + 5y = 67.73$.

b. Solve.
$(16x + 3y = 44.81)(5) \quad 80x + 15y = 224.05$
$(22x + 5y = 67.73)(-3) \quad \underline{-66x - 15y = -203.19}$
$14x = 20.86$
$x = 1.49$

Now, substitute 1.49 for $x$ in either equation to find $y$.
$16x + 3y = 44.81$
$16(1.49) + 3y = 44.81$
$23.84 + 3y = 44.81$
$3y = 20.97$
$y = 6.99$

A trip to the batting cage costs $1.49 and a game of miniature golf costs $6.99.

29. One of the equations will be a multiple of the other.  31. Sample answer: $2x + 3y = 6, 4x + 9y = 5$

33. Sample answer: It is more helpful to use substitution when one of the variables has a coefficient of 1 or if a coefficient can be reduced to 1 without turning other coefficients into fractions. Otherwise, elimination is more helpful because it will avoid the use of fractions when solving the system.  35. A  37a. $x$ = number of cartons of cookies; $y$ = number of cartons of candy bars; $9x + 12y = 1845$ and $8x + 15y = 2127.50$

37b. cookies: $\frac{\$55}{\text{carton}}$; candy: $\frac{\$112.50}{\text{carton}}$

37c. $5.50, $2.25

## Lesson 6-5

1. elim (×); (2, −5)  3. elim (+); $\left(-\frac{1}{3}, 1\right)$
5a. $y = -0.6x + 46.6$, $y = x - 11$
5b. The profit from pizza sales is equal to $46.60 minus the profit from sub sales. The number of pizzas sold is equal to the number of subs sold minus 11.
5c. (36, 25); The debate team sold 36 subs and 25 pizzas.
7. subst.; (2, −2)  9. elim. (−); $\left(1, -\frac{1}{2}\right)$

11. $-5x + 4y = 7$
$\underline{-5x - 3y = -14}$

Because there are no coefficients of 1, elimination is the best method.
$(-5x + 4y = 7)(-1) \quad 5x - 4y = -7$
$-5x - 3y = -14 \quad \underline{-5x - 3y = -14}$
$-7y = -21$
$y = 3$

Now substitute 3 for $y$ in either equation to find $x$.
$-5x + 4y = 7$
$-5x + 4(3) = 7$
$-5x + 12 = 7$
$-5x = -5$
$x = 1$

The solution is (1, 3).

13. $g + b = 40$ and $g = 3b - 4$; 29 girls, 11 boys
15. 880 books; If they sell this number, then their income and expenses both equal $35,200.
17. $y = -2x + 3$, $y = x - 3$; (2, −1)

19. a. Let $x$ be the cost per pound of the aluminum cans and $y$ be the cost per pound of the newspapers. For Mara, the equation is $9x + 26y = 3.77$. For Ling, the equation is $9x + 114y = 4.65$.

b. Elimination is the best method for solving these equations.
$(9x + 26y = 3.77)(-1) \quad -9x - 26y = -3.77$
$9x + 114y = 4.65 \quad \underline{9x + 114y = 4.65}$
$88y = 0.88$
$y = 0.01$

Now substitute 0.01 for $y$ in either equation to find $x$.
$9x + 26y = 3.77$
$9x + 26(0.01) = 3.77$
$9x + 0.26 = 3.77$
$9x = 3.51$
$x = 0.39$

The aluminum cans are $0.39 per pound. This solution is reasonable.

21a. $1.15  21b. $9.15  23. Sample answer: $x + y = 12$ and $3x + 2y = 29$, where $x$ represents the cost of a student ticket for the basketball game and $y$ represents the cost of an adult ticket; substitution could be used to solve the system; (5, 7) means the cost of a student ticket is $5 and the cost of an adult ticket is $7.

25. Graphing: (2, 5)

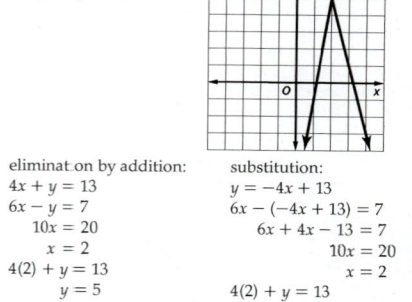

elimination by addition:
$4x + y = 13$
$\underline{6x - y = 7}$
$10x = 20$
$x = 2$
$4(2) + y = 13$
$y = 5$

substitution:
$y = -4x + 13$
$6x - (-4x + 13) = 7$
$6x + 4x - 13 = 7$
$10x = 20$
$x = 2$
$4(2) + y = 13$
$y = 5$

27. The third system; this system is the only one that is not a system of linear equations.  29. 12  31. C
33. C, D, G  35. 12, 17

## Lesson 6-6

1.

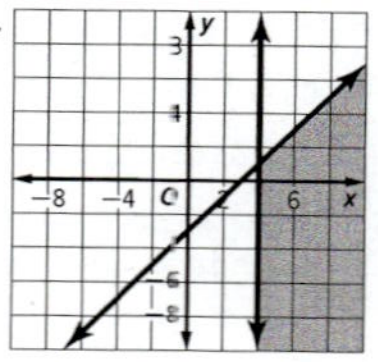

3.

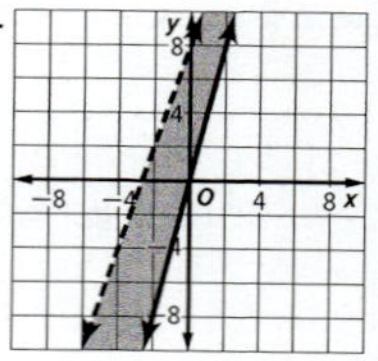

5.

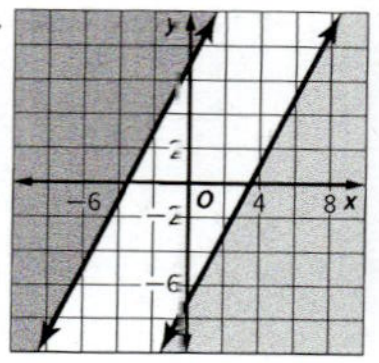

no solution

7.

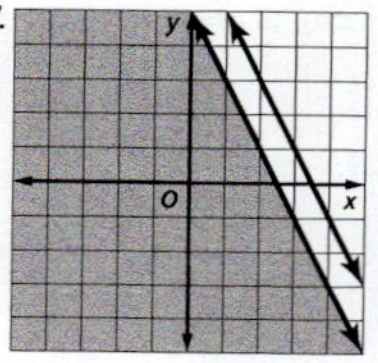

9a. Sample answer: Let $h$ = the height of the driver in inches and $w$ = the weight of the driver in pounds; $h < 79$ and $w < 295$

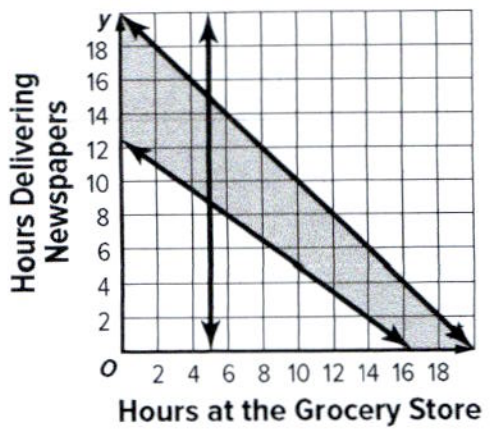

# Selected Answers and Solutions

**9a.** [Driving Requirements graph]

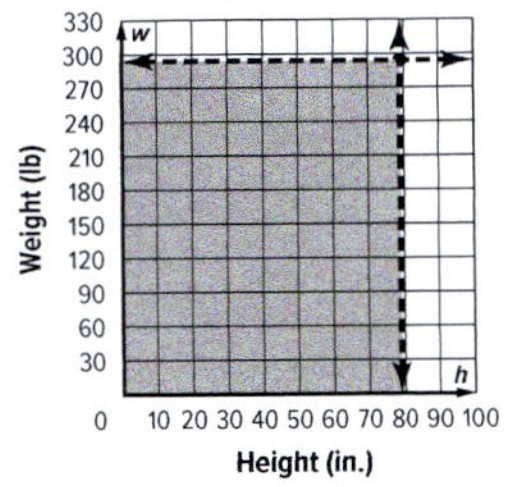

**9b.** Sample answer: 72 in. and 220 lb

**9c.** Yes, the point falls in the overlapping region.

**11.** Graph both inequalities on the same coordinate plane. $y \geq 0$ has a solid line. $y \leq x - 5$ has a solid line. The solution is the intersection of the shading.

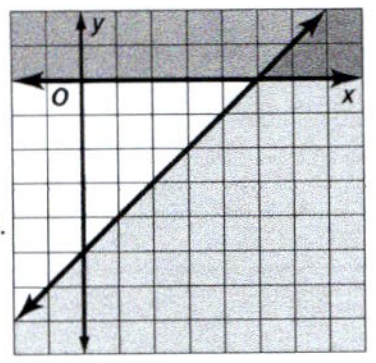

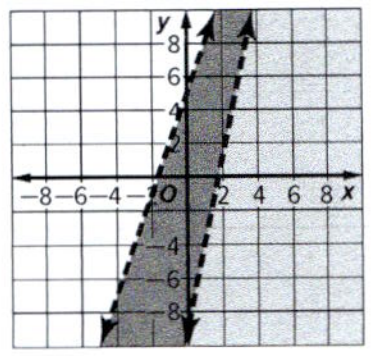

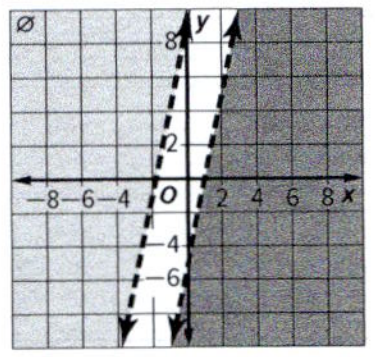

**13.** [graph]  **15.** [graph]  **17.** [graph] Ø no solution

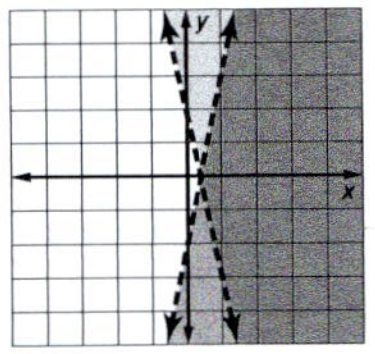

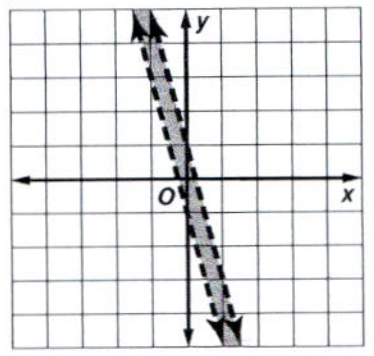

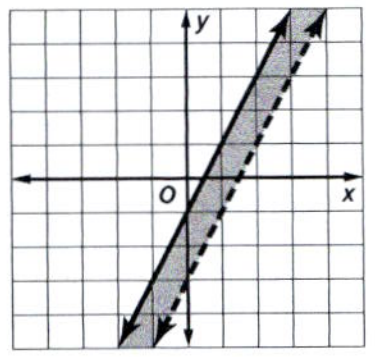

**19.** [graph]  **21.** [graph] no solution  **23.** [graph]

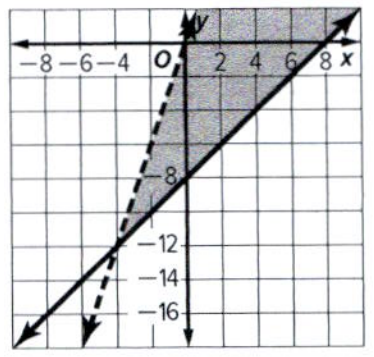

**25a.** Sample answer: Let $f$ = square footage and let $p$ = price; $1000 \leq f \leq 17{,}000$ and $10{,}000 \leq p \leq 150{,}000$ [Ice Rink Resurfacers graph]

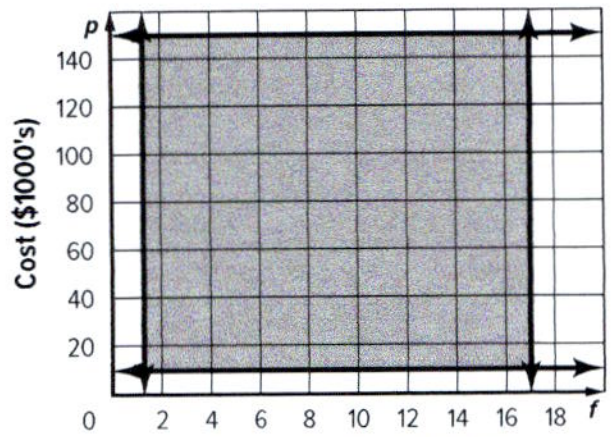

**25b.** Sample answer: an ice resurfacer for a rink of 5000 ft² and a price of $20,000

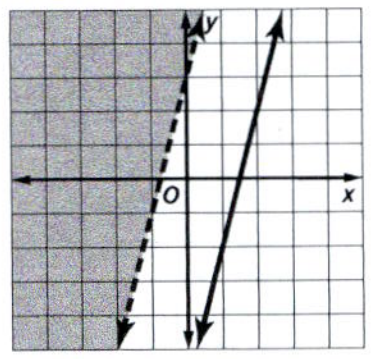

**25c.** Yes; the point satisfies each inequality.

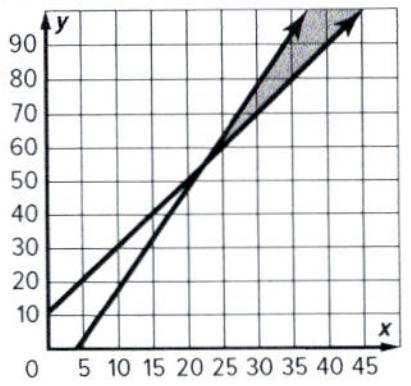

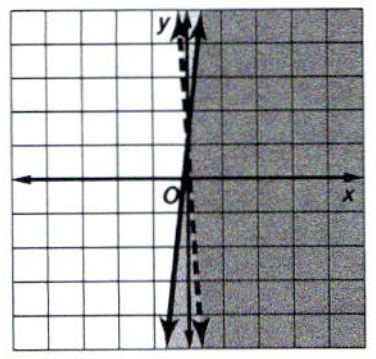

**27.** [graph]  **29.** [graph]  **31.** [graph]  **33.** [graph]  **35.** [graph]

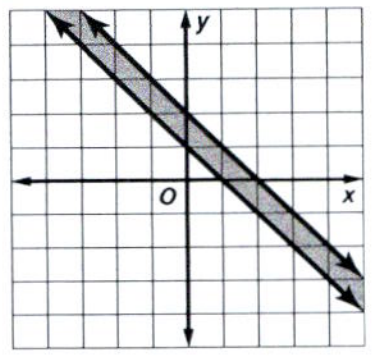

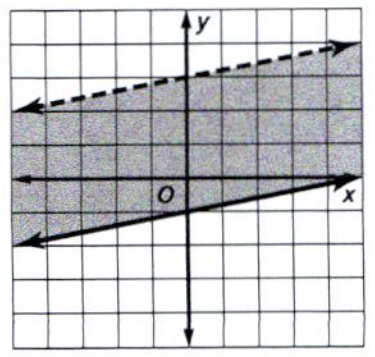

**37. a.** Let $x$ be the number of hours she works for a photographer and $y$ be the number of hours she works coaching.
$x + y \leq 20$
$15x + 10y \geq 90$

**b.** Graph both inequalities on the same grid. $x + y \leq 20$ and $15x + 10y \geq 90$ have solid lines. The solution is the intersection of the shading. [Earnings graph]

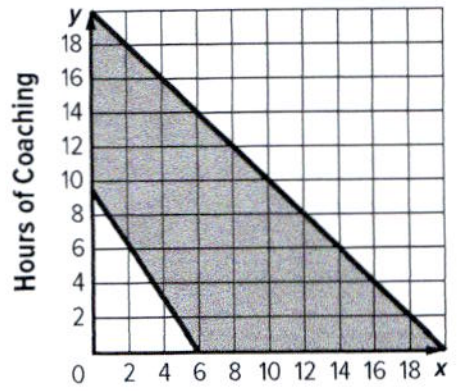

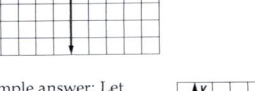

**c.** Two ordered pairs that are in the shaded area are (6, 10) and (8, 10). This means she could work for the photographer for 6 hours and coach for 10 or work for the photographer for 8 hours and coach for 10.

**d.** (2, 2) is not a solution because it does not fall in the shaded region. She would not earn enough money.

**39.** Sometimes; sample answer: $y > 3$, $y < -3$ will have no solution, but $y > -3$, $y < 3$ will have solutions.

**41.** Sample answer: $3x - y < -4$

**43.** Sample answer: The yellow region represents the beats per minute below the target heart rate. The blue region represents the beats per minute above the target heart rate. The green region represents the beats per minute within the target heart rate. Shading in different colors clearly shows the overlapping solution set of the system of inequalities. **45.** A  **47.** C

## Chapter 6 Study Guide and Review

**1.** true  **3.** false; dependent  **5.** true  **7.** false; system of inequalities  **9.** No, multiplication and division may also be used depending on the coefficients in the system of equations.

**11.** one; (3, 2)  **13.** one; (0, 2)

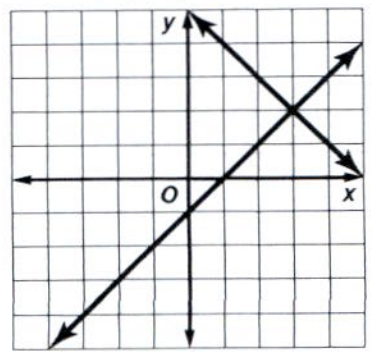

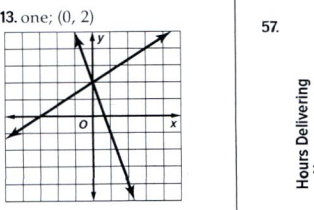

**15.** [graph] no solution

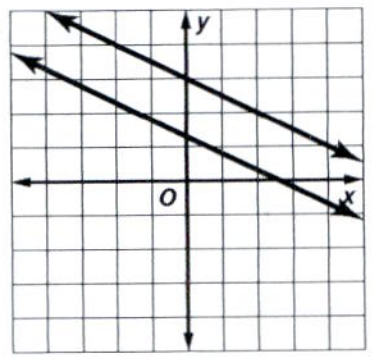

**17.** Sample answer: Let $x$ be one number and $y$ the other number; $x + y = 14$; $x - y = 4$; 9 and 5

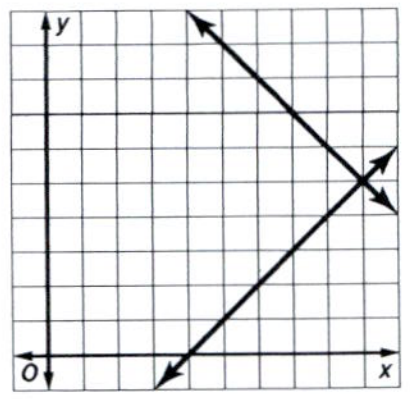

**19.** (2, −10)  **21.** (2, −6)  **23.** (−3, 4)  **25.** (9, 4)
**27.** (4, −2)  **29.** $\left(\frac{1}{2}, 6\right)$  **31.** (−3, 5)  **33.** Sample answer: Let $f$ be the first type of card and let $c$ be the second type of card; $f + c = 24$, $f + 3c = 50$; 11 $1 cards and 13 $3 cards.  **35.** (5, 7)  **37.** (2, 5)  **39.** (6, −1)
**41.** (1, −2)  **43.** Subs; (2, −6)  **45.** Subs; (24, −4)
**47.** Elim (−); (−2, 1)  **49.** Elim (×); (2, 5)  **51.** Sample answer: Let $d$ represent the dimes and let $q$ represent the quarters; $d + q = 25$, $0.10d + 0.25q = 4$; 15 dimes, 10 quarters

**53.** [graph]  **55.** [graph]

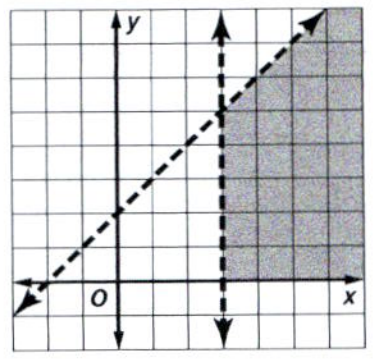

**57.** [Jobs graph: Hours Delivering Newspapers vs Hours at the Grocery Store]

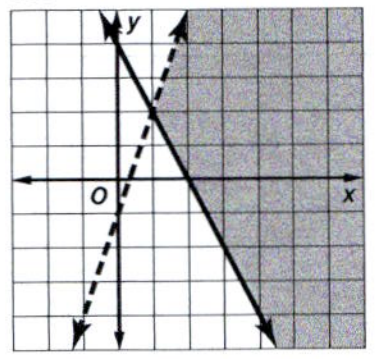

# CHAPTER 7
## Exponents and Exponential Functions

### Get Ready
1. Using exponents simplifies expressions. 3. The area would be in square inches. 5. The units would be feet cubed. 7. Yes, these expressions are different.

### Lesson 7-1
1. Yes; constants are monomials. 3. No; there is a variable in the denominator. 5. Yes; this is a product of a number and variables. 7. $k^4$

9. $2q^2(9q^4) = (2 \cdot 9)(q^2 \cdot q^4)$
$= 18q^{2+4}$
$= 18q^6$

11. $3^8$ or 6561  13. $16a^8b^{18}c^2$  15. $81p^{20}y^{24}$
17. $800x^8y^{12}z^4$  19. $-18g^7h^3j^{10}$  21. Yes; constants are monomials. 23. No; there is addition and more than one term. 25. Yes; this can be written as the product of a number and a variable.

27. $(q^2)(2q^4) = 2(q^2 \cdot q^4)$
$= 2q^{2+4}$
$= 2q^6$

29. $9a^8x^{12}$  31. $7b^{14}c^8a^6$  33. $j^{20}k^{28}$  35. $2^8$ or 256
37. $4096p^{12}t^6$  39. $20c^5d^5$  41. $16a^{21}$  43. $512g^{27}h^{18}$
45. $294p^{27}r^{19}$  47. $30a^3b^7c^6$  49. $0.25x^6$  51. $\frac{27}{64}c^3$
53. $-9x^3y^9$  55. $2,985,984t^{28}w^{32}$  57a. $0.12c$
57b. $280  59. $15x^7$

61a. $V = \pi r^2 h$
$= \pi(2pr^3)^2(4pr^3)$
$= \pi(2^2)(p^3)^2(4pr^3)$
$= \pi(4)(p^6)(4pr^3)$
$= \pi(4 \cdot 4)(p^6 \cdot p^3)$
$= (16\pi)(p^{6+3})$
$= 16\pi p^9$

b. 

| Radius | Height | Volume |
|---|---|---|
| $4p$ | $p^7$ | $16\pi p^9$ |
| $4p^2$ | $p^5$ | $16\pi p^9$ |
| $4p^3$ | $p^3$ | $16\pi p^9$ |
| $4p^4$ | $p$ | $16\pi p^9$ |
| $2p$ | $4p^7$ | $16\pi p^9$ |

c. If the height of the container is doubled, the volume of the container is doubled. So, the volume is $32\pi p^9$.

63a.
| Power | $3^3$ | $3^2$ | $3^1$ | $3^0$ | $3^{-1}$ | $3^{-2}$ | $3^{-3}$ | $3^{-4}$ |
|---|---|---|---|---|---|---|---|---|
| Value | 81 | 27 | 9 | 3 | $\frac{1}{3}$ | $\frac{1}{9}$ | $\frac{1}{27}$ | $\frac{1}{81}$ |

63b. 1 and $\frac{1}{5}$  63c. $\frac{1}{a^n}$  63d. Any nonzero number raised to the zero power is 1.

65a.
| Equation | Related Expression | Power of $x$ | Linear or Nonlinear |
|---|---|---|---|
| $y = x$ | $x$ | 1 | linear |
| $y = x^2$ | $x^2$ | 2 | nonlinear |
| $y = x^3$ | $x^3$ | 3 | nonlinear |

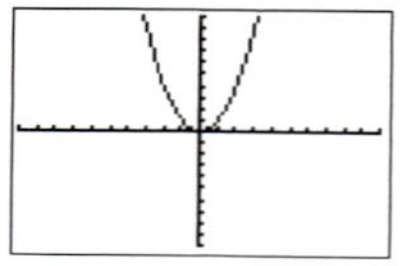

65b. [graphs] [−10, 10] scl: 1 by [−10, 10] scl: 1

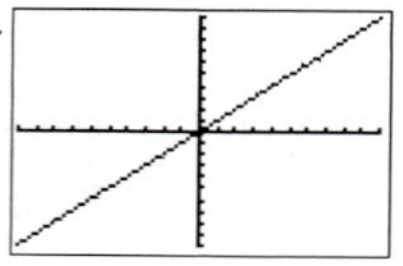

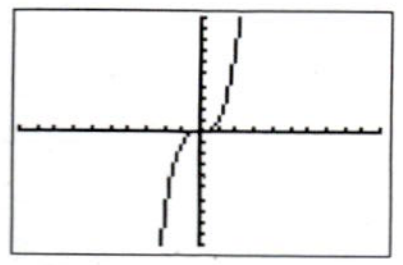

65c. See chart for 65a. 65d. If the power of $x$ is 1, the equation or its related expression is linear. Otherwise, it is nonlinear. 67. Sample answer: The area of a circle or $A = \pi r^2$, where $r$ is the radius, can be used to find the area of any circle. The area of a rectangle or $A = w \times \ell$, where $w$ is the width and $\ell$ is the length, can be used to find the area of any rectangle. 69. D 71. E 73a. B
73b. $30g^6b^3$

### Lesson 7-2
1. $t^3u^3$

3. $\dfrac{m^6p^3}{m^5p^3} \cdot \left(\dfrac{m^6}{m^5}\right)\left(\dfrac{r^3}{r^2}\right)\left(\dfrac{p^3}{p^3}\right)$
$= m^{6-5}r^{5-2}p^{3-3}$
$= m^1r^3p^0$
$= mr^3$

5. $ghm$  7. $xyz$  9. $\dfrac{4a^6b^{10}}{9}$  11. $\dfrac{32c^{15}d^{25}}{3125g^{10}}$  13. 1
15. $\dfrac{g^2h^4}{f^3}$  17. $\dfrac{a^5c^{13}}{3b^9}$  19. $m^2p$  21. $\dfrac{r^4p^2}{4m^3t^4}$  23. $\dfrac{9x^2y^8}{25z^4}$
25. $\dfrac{p^6t^{21}}{1000}$

29. $\dfrac{(2r^3t^6)^4}{(5u^9)^4} \dfrac{2^4(r^3)^4(t^6)^4}{5^4(u^9)^4}$
$= \dfrac{16r^{12}t^{24}}{625u^{36}}$

31. $\dfrac{p^4t^2}{t^3}$  33. $\dfrac{r^5}{t^3}$  35. $\dfrac{-f}{4}$  37. $k^2mp^2$  39. $\dfrac{3t^7}{u^6v^2}$
41. $\dfrac{r^3}{t^2z^{10}}$  43. $10^6, 10^9$, about $10^3$ or 1000 times as many users as servers  45. $\dfrac{w^9}{3}$

47. $1600k^{13}$  49. $\dfrac{5q}{r^6p^3}$  51. $\dfrac{4g^{12}}{h^4}$  53. $\dfrac{4x^8y^4}{z^6}$
55. $\dfrac{16z^2}{y^8}$  57. 1000

59a. The probability is $\dfrac{1}{6}$ multiplied $d$ times, or $\left(\dfrac{1}{6}\right)^d$.

b. $\left(\dfrac{1}{6}\right)^d = (6^{-1})^d$
$= 6^{-d}$

61. Sometimes; sample answer: The equation is true when $x = 1$, $y = 2$, and $z = 3$, but it is false when $x = 2$, $y = 2$, and $z = 3$.

63. $\dfrac{1}{x^n} = \dfrac{x^0}{x^n} = x^{0-n} = x^{-n}$

65. The Quotient of Powers Property is used when dividing two powers with the same base. The exponents are subtracted. The Power of a Quotient Property is used to find the power of a quotient. You find the power of the numerator and the power of the denominator. 67. C 69. D 71. C

### Lesson 7-3
1. $\sqrt{12}$  3. $33\frac{1}{2}$  5. 8  7. 7  9. 49

11. $216^{\frac{1}{3}} = (\sqrt[3]{216})^4 = (\sqrt[3]{6 \cdot 6 \cdot 6})^4 = 6^4$ or 1296

13. 4  15. 5.5  17. $\sqrt{15}$  19. $4\sqrt{k}$  21. $26^{\frac{1}{2}}$
23. $2(ab)^{\frac{1}{2}}$  25. 2  27. 6  29. 0.1  31. 11  33. 15
35. $\dfrac{1}{3}$  37. 4  39. 243  41. 625  43. $\dfrac{27}{1000}$  45. 5

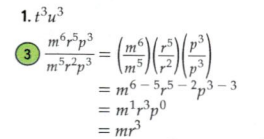

47. $\dfrac{3}{2}$  49. $\dfrac{3}{2}$  51. 8  53. 8

55. $4^{3x} = 512$
$(2^2)^{3x} = 2^9$
$2^{6x} = 2^9$
$6x = 9$
$x = \dfrac{3}{2}$

57. 4 ft  59. $\sqrt[3]{17}$  61. $7\sqrt{3}b$  63. $29^{\frac{1}{3}}$  65. $2a^{\frac{1}{3}}$

67. 0.3  69. $a$  71. 16  73. $\dfrac{1}{3}$  75. $\dfrac{1}{27}$  77. $\dfrac{1}{\sqrt{k}}$
79. 12  81. $-5$  83. $-\dfrac{3}{2}$  85a. 440 Hz

85b. A below middle C, the 37th note

87. $r = 0.62V^{\frac{1}{3}}$    Original equation
$3.65 = 0.62V^{\frac{1}{3}}$    $r = \dfrac{7.3}{2}$ or 3.65
$\dfrac{3.65}{0.62} = V^{\frac{1}{3}}$    Divide each side by 0.62.
$\left(\dfrac{3.65}{0.62}\right)^3 = \left(\dfrac{3.65}{0.62}\right)^3$    Power Property of Equality
$\left(\dfrac{3.65}{0.62}\right)^3 \approx V$    Simplify.
$204.0 \approx V$

$r = 0.62V^{\frac{1}{3}}$    Original equation
$3.8 = 0.62V^{\frac{1}{3}}$    $r = \dfrac{7.6}{2}$ or 3.8
$\dfrac{3.8}{0.62} = V^{\frac{1}{3}}$    Divide each side by 0.62.
$\left(\dfrac{3.8}{0.62}\right)^3 = \left(\dfrac{3.8}{0.62}\right)^3$    Power Property of Equality
$\left(\dfrac{3.8}{0.62}\right)^3 \approx V$    Simplify.
$230.2 \approx V$

So the volume of a size 3 ball is 204.0 to 230.2 in³.

$r = 0.62V^{\frac{1}{3}}$    Original equation
$4.0 = 0.62V^{\frac{1}{3}}$    $r = \dfrac{8.0}{2}$ or 4.0
$\dfrac{4.0}{0.62} = V^{\frac{1}{3}}$    Divide each side by 0.62.
$\left(\dfrac{4.0}{0.62}\right)^3 = \left(\dfrac{4.0}{0.62}\right)^3$    Power Property of Equality
$\left(\dfrac{4.0}{0.62}\right)^3 \approx V$    Simplify.
$268.5 \approx V$

$r = 0.62V^{\frac{1}{3}}$    Original equation
$4.15 = 0.62V^{\frac{1}{3}}$    $r = \dfrac{8.3}{2}$ or 4.15
$\dfrac{4.15}{0.62} = V^{\frac{1}{3}}$    Divide each side by 0.62.
$\left(\dfrac{4.15}{0.62}\right)^3 = \left(\dfrac{4.15}{0.62}\right)^3$    Power Property of Equality
$\left(\dfrac{4.15}{0.62}\right)^3 \approx V$    Simplify.
$299.9 \approx V$

So the volume of a size 4 ball is 268.5 to 299.9 in³.

$r = 0.62V^{\frac{1}{3}}$    Original equation
$4.3 = 0.62V^{\frac{1}{3}}$    $r = \dfrac{8.6}{2}$ or 4.3
$\dfrac{4.3}{0.62} = V^{\frac{1}{3}}$    Divide each side by 0.62.
$\left(\dfrac{4.3}{0.62}\right)^3 = \left(\dfrac{4.3}{0.62}\right)^3$    Power Property of Equality
$\left(\dfrac{4.3}{0.62}\right)^3 \approx V$    Simplify.
$333.5 \approx V$

$r = 0.62V^{\frac{1}{3}}$    Original equation
$4.5 = 0.62V^{\frac{1}{3}}$    $r = \dfrac{9.0}{2}$ or 4.5
$\dfrac{4.5}{0.62} = V^{\frac{1}{3}}$    Divide each side by 0.62.
$\left(\dfrac{4.5}{0.62}\right)^3 = \left(\dfrac{4.5}{0.62}\right)^3$    Power Property of Equality
$\left(\dfrac{4.5}{0.62}\right)^3 \approx V$    Simplify.
$382.4 \approx V$

So the volume of a size 5 ball is 333.6 to 382.4 in³.

89. Sample answer: $2^{\frac{1}{2}}$ and $4^{\frac{1}{4}}$  91. $-1$, 0, 1
93. Sample answer: 2 is the principal fourth root of 16 because 2 is positive and $2^4 = 16$.  95. C  97a. 17.95 s
97b. 8.88 $\dfrac{m}{s^2}$  99. B  101. A

## Lesson 7-4

**1.** $2\sqrt{6}$  **3.** 10  **5.** $3\sqrt{6}$  **7.** $2x^2y^3\sqrt{15y}$
**9.** $3b^2c|\sqrt{11ab}$  **11.** $\dfrac{9-3\sqrt{5}}{4}$  **13.** $\dfrac{2+2\sqrt{10}}{-9}$
**15.** $\dfrac{24+4\sqrt{7}}{29}$
**17.** $3\sqrt{5}+6\sqrt{5} = (3+6)\sqrt{5}$
$= 9\sqrt{5}$
**19.** $-5\sqrt{7}$  **21.** $8\sqrt{5}$  **23.** $5\sqrt{2}+2\sqrt{3}$  **25.** $72\sqrt{3}$
**27.** $\sqrt{21}+3\sqrt{6}$  **29.** $14.5+3\sqrt{15}$ units²
**31.** $2\sqrt{14}$  **33.** 80  **35.** $45q^2\sqrt{q}$  **37.** $5r\sqrt{3qr}$
**39.** $4|g|h^2\sqrt{66}$  **41.** $4c^3d^2\sqrt{2}$
**43.** $\sqrt{\dfrac{32}{t^4}} = \dfrac{\sqrt{32}}{\sqrt{t^4}}$
$= \dfrac{\sqrt{16 \cdot 2}}{t^2}$
$= \dfrac{\sqrt{16} \cdot \sqrt{2}}{t^2}$
$= \dfrac{4\sqrt{2}}{t^2}$
**45.** $\dfrac{35-7\sqrt{3}}{22}$  **47.** $\dfrac{6\sqrt{3}+9\sqrt{2}}{2}$  **49.** $\dfrac{5\sqrt{6}-5\sqrt{3}}{3}$
**51. a.** $v = \sqrt{64h}$
$= \sqrt{64h}$
$= \sqrt{8 \cdot 8h}$
$= \sqrt{8^2 \cdot \sqrt{h}}$
$= 8\sqrt{h}$
**b.** $v = 8\sqrt{h}$
$= 8\sqrt{134}$
$\approx 92.6$ ft/s
**53.** $11\sqrt{5}$  **55.** $5\sqrt{10}$  **57.** $3\sqrt{5}+6-\sqrt{30}-2\sqrt{6}$
**59.** $5\sqrt{5}+5\sqrt{2}$
**61.** $10\sqrt{7}+2\sqrt{5}$ units; 12 units²  **63a.** $v = \dfrac{\sqrt{6k}}{3}$
**63b.** 73.5 joules  **65.** $4-2\sqrt{3}$
**67.** $\dfrac{-4\sqrt{5}}{5}$  **69.** $\sqrt{2}$  **71.** $14-6\sqrt{5}$
**73a.** $v_0 = \sqrt{v^2-64h}$
$= \sqrt{(120)^2-64(225)}$
$= \sqrt{0}$
$= 0$ ft/s
**b.** Sample answer: In the formula, we are taking the square root of the difference, not the square root of each term.
**75.** Cross multiply and then divide. Rationalize the denominator to find that $x = \dfrac{5\sqrt{3}+9}{2}$.
**77.** Sample answer: $1+\sqrt{2}$ and $1-\sqrt{2}$; $(1+\sqrt{2}) \cdot (1-\sqrt{2}) = 1-2 = -1$
**79.** Irrational; irrational; no rational number could be added to or multiplied by an irrational number so that the result is rational.  **81.** Sample answer: You can use the FOIL method. You multiply the first terms within the parentheses. Then you multiply the outer terms within the parentheses. Then you would multiply the inner terms within the parentheses. Then you would multiply the last terms within the parentheses. Combine any like terms and simplify any radicals. For example:
$(\sqrt{2}+\sqrt{3})(\sqrt{5}+\sqrt{7}) = \sqrt{10}+\sqrt{14}+\sqrt{15}+\sqrt{21}$
**83.** B  **85.** A  **87.** C  **89a.** 2.8%  **89b.** $P = \dfrac{A}{(1+r)^t}$
**89c.** $2678.02

## Lesson 7-5

**1.**

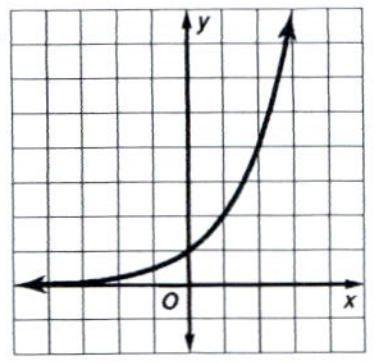

[graph]

1; D = {all real numbers};
R = {y | y > 0}; y = 0

**3.**

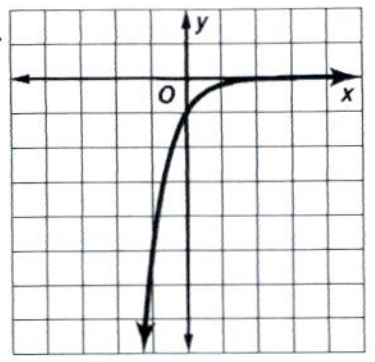

[graph]

−1; D = {all real numbers};
R = {y | y < 0}; y = 0

**5.**

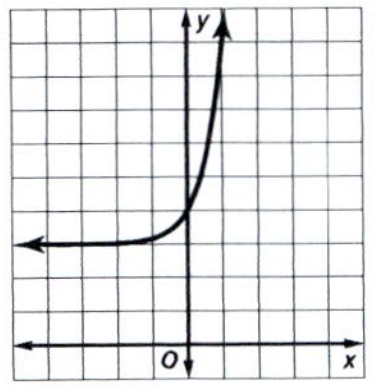

[graph]

4; D = {all real numbers};
R = {y | y > 3}; y = 3

**7a.** D = {t | t ≥ 0}; R = {f(t) | f(t) ≥ 100}, the number of fruit flies is greater than or equal to 100.  **7b.** about 198 fruit flies  **9.** Yes; the domain values are at regular intervals, and the range values have a common factor of 4.

**11.**

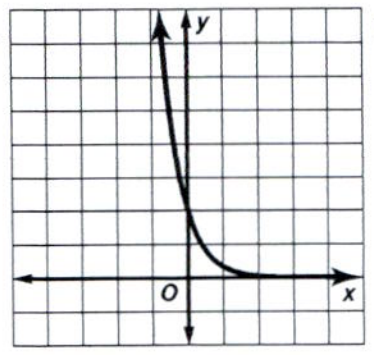

[graph]

2; D = {all real numbers};
R = {y | y > 0}; y = 0

**13.**

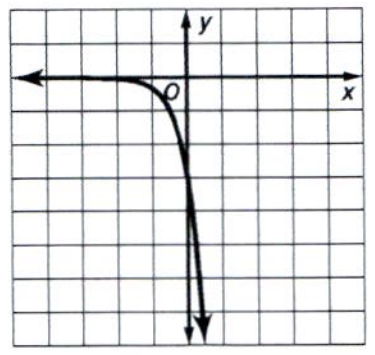

[graph]

−3; D = {all real numbers};
R = {y | y < 0}; y = 0

**15.**

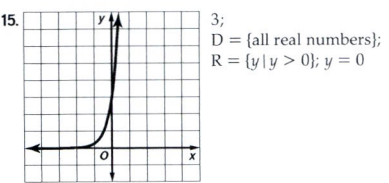

[graph]

3; D = {all real numbers};
R = {y | y > 0}; y = 0

**17.**

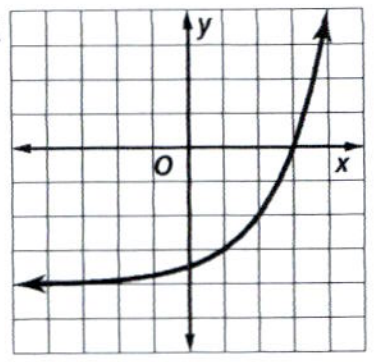

[graph]

The y-intercept is −3.5;
D = all real numbers;
R = {y | y > −4}; y = −4

**19.**

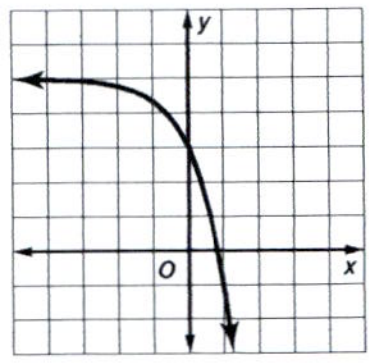

[graph]

The y-intercept is 3;
D = all real numbers;
R = {y | y < 5}; y = 5

**21.** No; the domain values are at regular intervals of 4.
$2 \times (-2) = 4$
$-4 \times (-2) = 8$
$8 \times (-2) = -16$
$-16 \times (-2) = 32$
The range values differ by the common factor of −2. The range values do not have a positive common factor.
**23.** Yes; the domain values are at regular intervals, and the range values have a common factor of 2.
**25.** This enlargement is about 506% bigger than the original.  **27.** exponential  **29.** linear  **31.** neither
**33.** about 75  **35.** $y = 0$  **37.** $y = -3$  **39.** $y = 1$
**41.** $f(x) = 3(2)^x$
**43.** Sample answer: The number of teams competing in a basketball tournament can be represented by $y = 2^x$, where the number of teams competing is $y$ and the number of rounds is $x$.

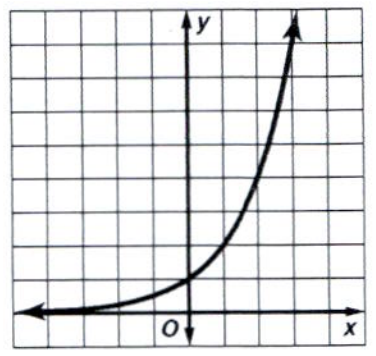

[graph]

The y-intercept of the graph is 1. The graph increases quickly for $x > 0$. With an exponential model, each team that joins the tournament will play all of the other teams. If the scenario were modeled with a linear function, each team that joined would play a fixed number of teams.
**45.** Sample answer: First, look for a pattern by making sure that the domain values are at regular intervals and the range values differ by a common factor.
**47.** D  **49a.** A  **49b.** D

## Lesson 7-6

**1.** translated down 1 unit  **3.** translated left 1 unit
**5.** $g(x) = 2^{x-4}$  **7.** stretched vertically  **9.** reflected across y-axis and compressed horizontally  **11.** reflected across x-axis and stretched vertically  **13.** reflected across y-axis, compressed vertically, and translated up 1 unit
**15.** For $g(x) = \left(\dfrac{5}{4}\right)^x - 2$, the parent function is $f(x) = \left(\dfrac{5}{4}\right)^x$, $a = 4$, so $f(x)$ is compressed horizontally because $a > 1$ and multiplied before $x$ is evaluated. $f(x)$ is translated down 2 units because $k = -2$.
**17.** translated right 2 units  **19.** stretched horizontally
**21.** stretched vertically  **23.** compressed vertically
**25.** reflected across x-axis  **27.** compressed vertically and translated right 1 unit  **29.** reflected across y-axis and compressed horizontally, and translated left 3 units
**33.** compressed horizontally and translated down 1 unit  **35.** reflected across x-axis, compressed vertically, and translated left 4 units and down 5 units
**37.** $f(x+n)$ indicates a horizontal translation of $n$ units. All parent functions $f(x)$ pass through the point $(0, 1)$. $n = -4$ because $g(x)$ passes through $(-4, 1)$.
**39.** 0.5  **41.** −3  **43.** 3

**45.** [graph]

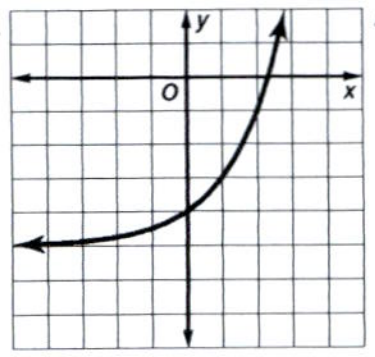

**47.** [graph]

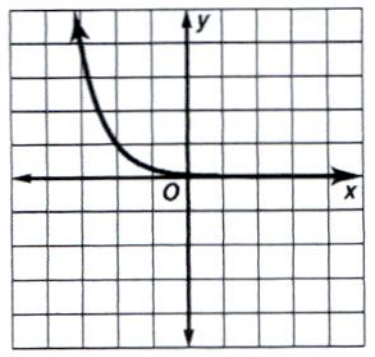

**49.** [graph]

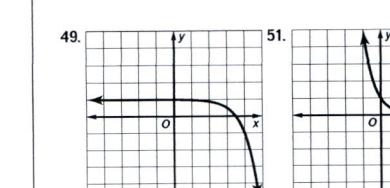

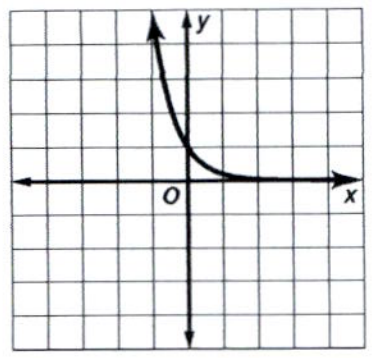

**51.** [graph]

# Selected Answers and Solutions

**53.** $m(x) = f(x) - 2$  **55.** $j(x) = -2f(x + 2)$  **57.** $n(x) = 5f(x)$  **59.** $j(x), g(x), p(x)$  **61a.** They have the same asymptote and end behavior as $x \to \infty$. For $y = \frac{2^x}{5}$, the function is decreasing as $x \to \infty$. The function $y = -\frac{2^x}{5}$ is increasing as $x \to \infty$. The y-intercepts are 1 and $-1$, respectively.
**61.** They have the same asymptote and y-intercept. For $y = \frac{2^x}{5}$, the function is decreasing as $x \to \infty$. The function $y = \frac{2^{-x}}{5}$ is increasing as $x \to \infty$.
**63a.** The graph is compressed vertically.
**63b.** The graph becomes more compressed and farther away from the y-axis.
**63c.** The graph is stretched vertically.
**63d.** The graph becomes more stretched and closer to the y-axis.
**65.** For any point $(x, y)$ of the function, the reflected function will have $(-x, y)$.
**67.** Case 1: As x decreases, y still approaches 0, but as x increases, y goes to negative infinity instead of positive infinity. Case 2: As x decreases, y goes to negative infinity instead of positive infinity, but as x increases, y still approaches 0.
**69.** Sample answer: $y = -\left(\frac{1}{4}\right)^x - 2$
**71.** C  **73.** B, C
**75.** $g(x) = -7\left(\frac{1}{2}\right)^{x+1}$
**77a.** −2; reflects across the x-axis and vertically stretches the graph
**77b.** −1; translates the graph left 1 unit
**77c.** 4; translates the graph up 4 units

## Lesson 7-7

**1.** $y = 0.5(3)^x$  **3.** $y = 4(3)^x$  **5.** initial: 240; rate of change: 65% per week; if this growth rate continues, then the number of likes will continue to increase, exceeding 100,000 in 12 weeks.  **7a.** $y = 2200(0.98)^t$  **7b.** about 1625  **9.** $y = 0.25(2)^x$  **11.** $y = 10(0.5)^x$  **13.** $y = -0.5(4)^x$  **15.** $y = 5(2)^x$
**17.** Initial: 82 grams of food; rate of change: decreasing by 35% per minute; if this decay rate continues, there will be less than a gram of food after 11 minutes.
**19.** $y = a(1 + r)^t$
$= 300(1 + 0.05)^5$
$= 300(1.05)^5$
$\approx 382.88$ or about $382.88
**21.** $3964.93
**23.** Sample answer: No; she will have about $199.94 in the account in 4 years.
**25.** Sample answer: No; the car is worth about $5774.61.
**27a.** Write an equation to represent the loss of value: $I = 194.375(1 - 0.0425)^t$
**27b.** Solve the equation: about $81,549
**29a.** $w(t) = 19,000(0.995)^t$  **29b.** $p(t) = 300t$
**29c.** $C(t) = 300t + 19,000(0.995)^t$; The function represents the number of gallons of water in the pool at any time after the hose is turned on.  **29d.** about 7.3 h
**31.** about 9.2 yr  **33.** Sample answer: Exponential models can grow without bound, which is usually not the case of the situation that is being modeled. For instance, a population cannot grow without bound due to space and food constraints. Therefore, when using a model, the situation that is being modeled should be carefully considered when used to make decisions.
**35.** A  **37.** $240  **39.** A, B, D  **41.** A

## Lesson 7-8

**1a.** The interest rate is 3.1% and the initial investment is $1, so
$A(t) = a(1 + r)^t$
$A(t) = 1(1 + 0.031)^t$
$A(t) = 1.031^t$
To write an equivalent function with an exponent of 4t:
$A(t) = 1.031^{\left(\frac{1}{4} \cdot 4\right)t}$
$A(t) = (1.031^{\frac{1}{4}})^{4t}$
$A(t) = (1.0077)^{4t}$
**1b.** The second equation shows 1.0077, or 1 + 0.0077, as the base of the exponential expression. So the effective quarterly interest rate is 0.0077, or 0.77%.
**1c.** Oak Hill Financial; The effective quarterly rate of 0.77% is greater than the 0.7% quarterly rate at First City Bank.

**3.** World Mutual; The effective monthly interest rate is about 1.19%, which is lower than the monthly rate for Super City Card.
**5.** Write an equivalent function with an exponent of 12t:
$P(t) = 10,200(1.08)^t$
$P(t) = 10,200(1.08)^{\left(\frac{1}{12} \cdot 12\right)t}$
$P(t) = 10,200(1.08^{\frac{1}{12}})^{12t}$
$P(t) \approx 10,200(1.0064)^{12t}$
The base of the exponential expression is 1.0064 or 1 + 0.0064, so the effective monthly growth rate is 0.0064 or about 0.64%.
**7a.** about 25.5%  **7b.** about 19.77 crowns  **9.** No; in the expression 0.987$^{12t}$, the base of the expression represents a decrease of 1.3%, so the effective monthly decrease is about 1.3%, not 98.7%.  **11.** Sample answer: Any amount can be used for the initial investment, and $1 is convenient. Equivalent expressions used to compare rates of increase or decrease require transformations of the exponential expression, not the coefficient outside the parentheses. The initial amount does not have an effect on the rates you want to compare.  **13.** B, C, E  **15.** B
**17.** 1.1%

## Lesson 7-9

**1.** Geometric; the common ratio is $\frac{1}{5}$.
**3.** Arithmetic; the common difference is 3.
**5.** 160, 320, 640  **7.** $-\frac{1}{16}, -\frac{1}{64}, -\frac{1}{256}$
**9.** $a_n = -6 \cdot (4)^{n-1}$; −1536  **11.** $a_n = 72 \cdot \left(\frac{2}{3}\right)^{n-1}$; $\frac{4096}{2187}$
**13.**

Experiment

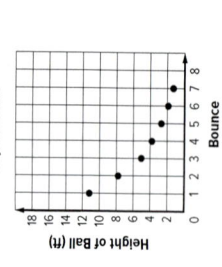

[graph: Height of Ball (ft) vs Bounce, points descending from 16 to near 0]

**15.** Arithmetic; the common difference is 10.
**17.** Geometric; the common ratio is $\frac{1}{2}$.
**19.** Neither; there is no common ratio or difference.
**21.** Step 1: Find the common ratio.
$36 \times \frac{1}{3} = 12$
$12 \times \frac{1}{3} = 4$
The common ratio is $\frac{1}{3}$.
Step 2: Multiply each term by the common ratio to find the next three terms.
$4\left(\frac{1}{3}\right) = \frac{4}{3}$
$\frac{4}{3}\left(\frac{1}{3}\right) = \frac{4}{9}$

$\left(\frac{4}{9}\right)\left(\frac{1}{3}\right) = \frac{4}{27}$

So, the next three terms are $\frac{4}{3}, \frac{4}{9}$, and $\frac{4}{27}$.
**23.** $\frac{25}{4}, \frac{25}{16}, \frac{25}{64}$  **25.** $-2, \frac{1}{4}, -\frac{1}{32}$  **27.** 134,217,728
**29.** −1,572,864  **31.** 19,683  **33a.** the second option She should choose whichever option would earn her the most money over the summer. For the first option, $30 a week for 9 weeks would yield $270.
**33b.** She should choose the second option, totaling $511 over 9 weeks. So, although this option starts out slow, it ends up being the most profitable.
**35.** Divide the 3rd term by the 2nd term to find the common ratio. The common ratio is $\frac{1}{3}$. Substitute 2 for n and $\frac{1}{3}$ for r to find the first term.
$a_n = a_1 r^{n-1}$
$a_2 = a_1 \left(\frac{1}{3}\right)^{2-1}$
$3 = a_1 \left(\frac{1}{3}\right)$
$9 = a_1$
The first term is 9. Find the 4th term.
$a_n = a_1 r^{n-1}$
$a_4 = 9\left(\frac{1}{3}\right)^{4-1}$
$a_4 = 9\left(\frac{1}{3}\right)^3$
$a_4 = \frac{1}{3}$
The fourth term is $\frac{1}{3}$.

**37a.**

| Richter Number (x) | Increase in Magnitude (y) | Rate of Change (slope) |
|---|---|---|
| 1 | 1 | — |
| 2 | 10 | 9 |
| 3 | 100 | 90 |
| 4 | 1,000 | 900 |
| 5 | 10,000 | 9000 |

**37b.**

[graph: Magnitude vs Richter Number, exponential curve]

**37c.** The graph appears to be exponential. The rate of change between any two points does not match any others.  **37d.** $1 \cdot (10)^{x-1} = y$  **39.** Neither; Haro calculated the exponent incorrectly. Matthew did not calculate $(-2)^8$ correctly.  **41.** Sample answer: When graphed, the terms of a geometric sequence lie

on a curve that can be represented by an exponential function. They are different in that the domain of a geometric sequence is the set of natural numbers, while the domain of an exponential function is all real numbers. Thus, geometric sequences are discrete, while exponential functions are continuous. **43.** C
**45.** B  **47.** A, D, E  **49a.** 8295  **49b.** 13,054

**Lesson 7-10**
**1.** 16, 13, 10, 7, 4  **3.** $a_1 = 1$, $a_n = a_{n-1} + 5$, $n \geq 2$
**5a.** $a_1 = 10$, $a_n = 0.6a_{n-1}$, $n \geq 2$
**5b.** $a_n = 10(0.6)^{n-1}$
**7.** $a_n = 5n + 8$ is an explicit formula for an arithmetic sequence with $d = 5$ and $a_1 = 5(1) + 8$ or 13. Therefore, $a_1 = 13$, $a_n = a_{n-1} + 5$, $n \geq 2$.
**9.** $a_n = 22(4)^{n-1}$  **11.** 48, −16, 16, 0, 8
**13.** 12, 15, 24, 51, 132  **15.** $\frac{1}{2}$, $\frac{2}{3}$, $\frac{7}{5}$, $\frac{5}{8}$, $\frac{13}{2}$
**17.** $a_1 = 27$, $a_n = a_{n-1} + 14$, $n \geq 2$
**19.** $a_1 = 100$, $a_n = 0.8a_{n-1}$, $n \geq 2$
**21.** $a_1 = 81$, $a_n = \frac{1}{3}a_{n-1}$, $n \geq 2$
**23.** $a_1 = 3$, $a_n = 4a_{n-1}$, $n \geq 2$  **25.** $a_n = 38\left(\frac{1}{2}\right)^{n-1}$
**27. a.** Barbara was the first to receive the message, so the first term is 1. She then forwarded it to 5 of her friends, so the second term is 5. Each of her 5 friends forwarded the message to 5 more friends, so the third term is 5 · 5 or 25. This pattern continues. The fourth term is 25 · 5 or 125, and the fifth term is 125 · 5 or 625. Therefore, the first five terms are 1, 5, 25, 125, and 625.
**b.** There is a common ratio of 5. The sequence is geometric.
$a_n = r \cdot a_{n-1}$
$a_n = 5a_{n-1}$
The first term $a_1$ is 1, and $n \geq 2$. So, $a_1 = 1$, $a_n = 5a_{n-1}$, $n \geq 2$.
**c.** We found that $a_5 = 625$.
$a_6 = 5a_{6-1}$
  $= 5a_5$
  $= 5(625)$ or 3125
$a_7 = 5a_{7-1}$
  $= 5a_6$
  $= 5(3125)$ or 15,625
$a_8 = 5a_{8-1}$
  $= 5a_7$
  $= 5(15,625)$ or 78,125
**29a.** $a_1 = 10$, $a_n = 1.1a_{n-2}$, $n \geq 2$  **29b.** 16.1 ft
**31.** Both; sample answer: The sequence can be written as the recursive formula $a_1 = 2$, $a_n = (-1)a_{n-1}$, $n \geq 2$. The sequence can also be written as the explicit formula $a_n = 2(-1)^{n-1}$.
**33.** False; sample answer: A recursive formula for the sequence 1, 2, 3, ... can be written as $a_1 = 1$, $a_n = a_{n-1} + 1$, $n \geq 2$ or as $a_1 = 1$, $a_2 = 2$, $a_n = a_{n-2} + 2$, $n \geq 3$.

**35.** Sample answer: In an explicit formula, the $n$th term $a_n$ is given as a function of $n$. In a recursive formula, the $n$th term $a_n$ is found by performing operations to one or more of the terms that precede it.
**37.** C  **39a.** B  **39b.** 1024

**Chapter 7  Study Guide and Review**
**1.** cube root  **3.** exponential function  **5.** exponential equation  **7.** If the domain values are at equal intervals, find the common factors among each consecutive pair of range values. If the common factors are the same, the data are exponential.
**9.** $x^9$  **11.** $20a^6b^6$  **13.** $64t^{18}y^6$  **15.** $8x^{15}$
**17.** $45\pi x^4$  **19.** $\frac{27x^4y^9}{8z^3}$  **21.** $\frac{c^6}{a^3}$  **23.** $x^6$  **25.** $\frac{6}{yx^3}$
**27.** 7  **29.** 5  **31.** 64  **33.** 2401  **35.** 5
**37.** $6|xy|^3\sqrt{y}$  **39.** $3\sqrt{2}$  **41.** $21 − 8\sqrt{5}$  **43.** $\frac{5\sqrt{2}}{|a|}$
**45.** $−6 − 3\sqrt{5}$  **47.** $−2\sqrt{6} + 11\sqrt{3}$  **49.** $5\sqrt{2} + 3\sqrt{6}$
**51.** $24\sqrt{10} + 8\sqrt{2} + 6\sqrt{15} + 2\sqrt{3}$
**53.** y-intercept 1; D = {all real numbers};
R = {y | y > 0}; y = 0

[graph]

**55.** y-intercept 3; D = {all real numbers};
R = {y | y > 2}; y = 2

[graph]

**57.** about 568  **59.** reflected across the x-axis, compressed vertically, and translated down 2 units
**61.** $g(x) = -2(4)^x$  **63.** $y = 4 \cdot (3)^x$  **65.** $3053.00
**67.** For First Bank, the effective yearly rate is about 1.8% which is greater than the yearly rate of 1.5% at Main Street Bank.  **69.** about 0.6%  **71.** 81, 243, 729
**73.** $a_n = -1(-1)^{n-1}$
**75.** $a_n = 256\left(\frac{1}{2}\right)^{n-1}$  **77.** 11, 7, 3, −1, −5
**79.** $a_1 = 2$, $a_n = a_{n-1} + 5$, $n \geq 2$
**81.** $a_1 = 2$, $a_n = 2a_{n-1} + 1$, $n \geq 2$

---

## CHAPTER 8
## Quadratic Expressions and Equations

**Chapter 8  Get Ready**
**1.** A factor in front of parentheses multiplies across all terms inside the parentheses. $a(b + c) = ab + ac$
**3.** Multiply length by width; area = $a(b + 3c)$.  **5.** −2, 6
**7.** Yes; $-18y^7$

**Lesson 8-1**
**1.** yes; 3; trinomial  **3.** yes; 2; monomial
**5.** yes; 5; binomial  **7.** $2x^5 + 3x - 12$; 2
**9.** $-5z^4 - 2z^2 + 4z$; −5  **11.** $4x^3 + 5$
**13.** $(4 + 2a^2 - 2a) - (3a^2 - 8a + 7) = (2a^2 - 2a + 4) - (3a^2 - 8a + 7) = (2a^2 - 2a + 4) + (-2a^2 - (-8a) + (-7))$
$(4-7) = -a^2 + 6a - 3$
**15.** $-8z^3 - 3z^2 - 2z + 13$  **17.** $4y^2 + 3y + 3$
**19a.** $D(n) = 6n + 14$  **19b.** 116,000 students
**19c.** 301,000 students  **21.** yes; 0; monomial
**23.** No; the exponent is a variable.  **25.** yes; 4; binomial  **27.** $7y^3 + 8y$; 7
**29.** $-y^3 - 3y^2 + 3y + 2$; −1  **33.** $-b^6 - 9b^2 + 10b$; −1
**31.** $-r^3 + r + 2$; −1  **33.** $-b^6 - 9b^2 + 10b$; −1
**35.** $(2x + 3x^2) - (7 - 8x^2) = (2x + 3x^2) + (-7 + 8x^2) = [3x^2 + 8x^2] + 2x + (-7) = 11x^2 + 2x - 7$
**37.** $2z^2 + z - 11$  **39.** $-2b^2 + 2a + 9$
**41.** $7x^2 - 2xy - 7y$  **43.** $3x^2 - rxt - 8r^2x - 6rx^2$
**45.** quadratic trinomial  **47.** quartic binomial
**49.** quintic polynomial  **51a.** $s = 0.55t^2 - 0.05t + 3.7$
**51b.** 3030 students
**53a.** the area of the rectangle
**53b.** the perimeter of the rectangle
**55.** $10a^2 - 8a + 16$
**57.** $7n^3 - 7n^2 - n - 6$
**59. a.** Words: $25 plus $0.35 per mile
Expression: $25 + 0.35m$
The expression is $25 + 0.35m$.
**b.** $25 + 0.35m = 25 + 0.35(145)$
  $= 25 + 50.75$
  $= 75.75$
The cost is $75.75.
**c.** $247  **d.** $714
**61.** Neither; neither of them found the additive inverse correctly. All terms should have been multiplied by −1.  **63.** $6n + 9$  **65.** Sample answer: To add polynomials in a horizontal format, combine like terms. For the vertical format, write the polynomials in

standard form, align like terms in columns, and combine like terms. To subtract polynomials in a horizontal format you find the additive inverse of the polynomial you are subtracting, and then combine like terms. For the vertical format, you write the polynomials in standard form, align like terms in columns, and subtract by adding the additive inverse.  **67.** C  **69.** $x^2 + 2x + 9$  **71.** $14x + 7$

**Lesson 8-2**
**1.** $-15a^3 + 10w^2 - 20w$
**3.** $32k^2mt^4 + 8k^3m^3 + 20k^2m^2$
**5.** $2ab(7a^4b^2 + a^5b - 2a) = 2ab(7a^4b^2) + 2ab(a^5b) + 2ab(-2a)$
  $= 14a^5b^3 + 2a^6b^2 + (-4a^2b)$
  $= 14a^5b^3 + 2a^6b^2 - 4a^2b$
**7.** $4t^3 + 15t^2 - 8t + 4$
**9.** $-5d^4c^2 + 8d^2c^2 - 4d^3c + dc^4$  **11.** 30  **13.** $\frac{20}{9}$
**21.** $10f^5 - 30f^4 + 4f^3 + 2f^2 + 25f$
**21.** $10b^5 - 30b^4 + 4b^3 + 2b^2 + 4b^2$
**23.** $8b^5u^3 - 40b^4u^5 + 8t^3u$
**25.** $-8a^3 + 20a^2 + 4a - 12$  **27.** $-9g^3 + 21g^3 + 12$
**29.** $8n^4p^2 + 12n^2p^2 + 20n^2 - 8np^3 + 12p^2$
**31.** $7(t^2 + 5t - 9) + t = t(7t - 2) + 13$
  $7t^2 + 35t - 63 + t = 7t^2 - 2t + 13$
  $7t^2 + 36t - 63 = 7t^2 - 2t + 13$
  $36t - 63 = -2t + 13$
  $38t = 76$
  $t = 2$
**33.** $\frac{43}{6}$  **35.** $\frac{30}{43}$  **37.** $20np^4 + 6n^3p^3 - 8np^2$
**39.** $-q^2w^3 - 35q^3w^4 + 8q^2w^2 - 27qw$
**41a.** 100.5 − 0.5h  **41b.** $94.50
**43. a.** $A = \ell w$
  $= (1.5x + 24)x$
  $= 1.5x^2 + 24x$
**b.** $x(x-9) = x^2 - 9x$
**c.** $2(2.5x) = 2(2.5)(36)$
  $2(x + 6) = 2(36 + 6)$
   $= 2(42)$
   $= 84$ ft
Perimeter = 180 + 84 or 264 ft
**45.** Ted; Pearl used the Distributive Property incorrectly.
**47.** $8x^2y^{-2} + 24x^{-10}y^8 - 16x^{-3}$
**49.** Sample answer: $3n$, $4n + 1$; $12n^2 + 3n$  **51.** B
**53a.** A  **53b.** C
**53c.** 20 m²
**53d.** $30x^3 + 45x^2 + 90x = 240 + 180 + 180 = 600$ m²
**53e.** 600 m² × $10/m² = $6000

## Selected Answers and Solutions

### Lesson 8-3

**1.** $x^2 + 7x + 10$  **3.** $b^2 - 4b - 21$  **5.** $16h^2 - 26h + 3$  **7.** $4x^2 + 72x + 320$  **9.** $16y^4 + 28y^3 - 4y^2 - 21y - 6$  **11.** $10n^4 + 11n^3 - 52n^2 - 12n + 48$  **13.** $2g^2 + 15g - 50$  **15.** $(4x + 1)(6x + 3) = 4x(6x) + 4x(3) + 1(6x) + 1(3)$
$= 24x^2 + 12x + 6x + 3$
$= 24x^2 + 18x + 3$

**17.** $24d^2 - 62d + 35$  **19.** $49m^2 - 84n + 36$  **21.** $25r^2 - 49$  **23.** $33.3z^2 - 17y^2 + 37y - 22$  **25.** $2y^3$  **27.** $m^4 + 2m^3 - 34m^2 + 43m - 12$  **29.** $6b^5 - 3b^4 - 35b^3 - 10b^2 + 43b + 63$  **31.** $2m^3 + 5m^2 - 4$  **33.** $4\pi x^2 + 12\pi x + 9\pi - 3x^2 - 5x - 2$  **35a.** $A = \ell w$
$= (3y + 4)(6y - 5)$
$= 3y(6y) + 3y(-5) + 4(6y) + 4(-5)$
$= 18y^2 - 15y + 24y - 20$
$= 18y^2 + 9y - 20$

**b.** $3y + 4 = 31$
$3y = 27$
$y = 9$
So, the width is $6y - 5 = 6(9) - 5 = 54 - 5 = 49$
$A = \ell w$
$= (31)(49)$
$= 1519$ ft$^2$

**37.** $a^2 - 4ab + 4b^2$  **39.** $x^2 - 10xy + 25y^2$  **41.** $125p^3 + 150p^2h + 60ph^2 + 8h^3$  **43a.** $x > 4$; If $x = 4$, the width of the rectangular sandbox would be zero and if $x < 4$ the width of the rectangular sandbox would be negative.  **43b.** square  **43c.** 4 ft$^2$  **45.** Always; by grouping two adjacent terms, a trinomial can be written as a binomial, the sum of two quantities, and apply the FOIL method. For example, $(2x + 3)(x^2 + 5x + 7) = (2x + 3)[x^2 + (5x + 7)] = 2x(x^2) + 2x(5x + 7) + 3(x^2) + 3(5x + 7)$. Then use the Distributive Property and simplify.  **47.** Sample answer: $x - 1, x^2 - x - 1, (x - 1)(x^2 - x - 1) = x^3 - 2x^2 + 1$  **49.** The Distributive Property can be used with a vertical or horizontal format by distributing, multiplying, and combining like terms. The FOIL method is used with a horizontal format. You multiply the first, outer, inner, and last terms of the binomials and then combine like terms. A rectangular method can also be used by writing the terms of the polynomials along the top and left side of a rectangle and then multiplying the terms and combining like terms.  **51a.** $x^2$, $(x + 3)^2$ or $x^2 + 6x + 9$  **51b.** 1  **51c.** 1 unit$^2$  16 units$^2$  **53.** D  **55.** $8x^3 + 6x^2 - 3x - 2$  **57a.** $2x^3 - 11x^2 + 14x - 3$  **57b.** $6x^3 + 16x^2 + 15x - 3$  **59.** $8x^2 + 32x + 32$

### Lesson 8-4

**1.** $x^2 + 10x + 25$
**3.** $(2x + 7y)^2 = (2x)^2 + 2(2x)(7y) + (7y)^2$
$= 4x^2 + 28xy + 49y^2$

**5.** $g^2 - 8gh + 16h^2$  **7a.** $D^2 + 2Dy + y^2$  **7b.** 75%  **9.** $x^2 - 25$  **11.** $81t^2 - 36$  **13.** $b^2 - 12b + 36$  **15.** $x^2 + 12x + 36$  **17.** $81 - 36y + 4y^2$  **19.** $25t^2 - 20t + 4$  **21a.** $(T + t)^2 = T^2 + 2Tt + t^2$  **21b.** TT: 25%; Tt: 50%; tt: 25%  **23.** $(b + 7)(b - 7) = b^2 - (7)^2 = b^2 - 49$  **25.** $16 - x^2$  **27.** $9a^4 - 49b^2$  **29.** $64 - 160a + 100a^2$  **31.** $9x^2 - 144$  **33.** $9g^2 - 30qr + 25r^2$  **35.** $g^2 + 10gh + 25h^2$  **37.** $9g^8 - b^2$  **39.** $64a^4 - 81b^6$  **41.** $\frac{4}{25}y^2 - \frac{16}{5}y + 16$  **43.** $4m^3 + 16m^2 - 9m - 36$  **45.** $2x^2 + 2x + 5$  **47.** $6x + 3$  **49.** $c^3 + 3c^2d + 3cd^2 + d^3$  **51.** $f^3 + f^2g - fg^2 - g^3$  **53.** $n^3 - n^2p - np^2 + p^3$  **55. a.** $A = 3.14(r + 9)^2$
$= 3.14(r^2 + 18r + 81)$
$= 3.14r^2 + 56.52r + 254.34$
$\approx (3.14r^2 + 56.52r + 254.34)$ ft$^2$

**b.** $38^2 - 3.14r^2 + 56.52r + 254.34$
$= 1444 - 3.14r^2 - 56.52r - 254.34$
$\approx (1189.66 - 3.14r^2 - 56.52r)$ ft$^2$

**57.** Sample answer: $(2c + d)(2c - d)$; The product of these binomials is a difference of two squares and does not have a middle term. The other three do.  **59.** 81  **61.** Sample answer: To find the square of a sum, apply the FOIL method or apply the pattern. The square of the sum of two quantities is the first quantity squared plus two times the product of the two quantities plus the second quantity squared. The square of the difference of two quantities is the first quantity squared minus two times the product of the two quantities plus the second quantity squared. The product of the sum and difference of two quantities is the square of the first quantity minus the square of the second quantity.  **63a.** $x^2 + 16x + 16$  **63b.** $25x^2 - 9$  **63c.** $-21x^2 + 16x + 7$  **65.** C  **67.** C  **69.** $(x^4 + 4x^2 + 4)$ yd$^2$

### Lesson 8-5

**1.** $3(7b - 5a)$  **3.** $gh(10gh + 9h - g)$
**5.** $np + 2n + 8p + 16 = (np + 2n) + (8p + 16)$
$= n(p + 2) + 8(p + 2)$
$= (n + 8)(p + 2)$

**7.** $(b + 5)(3c - 2)$  **9.** $(3k + 2)(m - 7)$  **11.** $(5p^3 - 3q)(2q - p)$  **13a.** $4t(3 - 4t)$  **13b.** 1.04 ft  **13c.** 2 ft  **15.** $8(2t - 5y)$

**17.** $2k(k + 2)$  **19.** $2ab(2ab + a - 5b)$
**21.** $fg - 5g + 4f - 20 = (fg - 5g) + (4f - 20)$
$= g(f - 5) + 4(f - 5)$
$= (g + 4)(f - 5)$

**23.** $(h + 5)(j - 2)$  **25.** $(9q - 10)(5p - 3)$  **27.** $(3d - 5)(t - 7)$  **29.** $(3t - 5)(7h - 1)$  **31.** $(r - 5)(5b + 2)$  **33.** $g(15f + g + 15)$  **35.** $3cd(9d - 6cd + 1)$  **37.** $2(8u - 15)(3t + 2)$  **39.** $(5p + 2r)(4p + 3)$  **41.** $(3k^2 - 2)(3m - 5)$  **43.** $(6f + 1)(8g - 3)$  **45a.** $ab$  **45b.** $(a + 6)(b + 6)$  **45c.** $6(a + b + 6)$  **47a.** $8(55 - 2t)$  **47b.** 2800 ft, 2400 ft  **47c.** 3025 ft  **49.** $72 - 16t^2 = 8(9) + 8t(-2t)$
$8(3)[9 - 2(3)] = 24(3)$
$= 8t(9 - 2t)$
$= 24(3)$
$= 72$ ft

The height of the arrow after 3 seconds is 72 feet.

**51a.** 3 and $-2$

**51b.**
| | $+3x$ | $-6$ |
|---|---|---|
| $x^2$ | $+3x$ | |
| $-2x$ | | $-6$ |

**51c.**
| | $x$ | $+3$ |
|---|---|---|
| $x$ | $x^2$ | $+3x$ |
| $-2$ | $-2x$ | $-6$ |

$(x + 3)(x - 2)$

**51d.** Sample answer: Place $x^2$ in the top left-hand corner and place $-40$ in the lower right-hand corner. Then determine which two factors have a product of $-40$ and a sum of $-3$. Then place these factors in the box. Then find the factor of each row and column. The factors will be listed on the very top and far left of the box.

**53.** Sample answer:
$4yz^2 + 24z + 5yz + 30 = (4yz^2 + 24z) + (5yz + 30)$
$= (4z + 5)(yz + 6)$

$4yz^2 + 24z + 5yz + 30 = 4yz^2 + 5yz + 24z + 30$
$= (4yz^2 + 5yz) + (24z + 30)$
$= (yz + 6)(4z + 5)$

**55.** $ac^2, ac^2; (b^2c + a^2b + a^2c^2)$  **57.** $(5x^2 + 2)$ and $(2x + 3)$  **59.** A  **61.** D  **63a.** $3y^2 + 2)(2y + 7)$  **63b.** $(4a + 3y)(3a - 5b)$  **63c.** $(w - 2)(4w^2 + 3z)$  **63d.** $3mn(5m - 9n)$  **63e.** $(4y + 5z)(2w + 3x)$  **65.** $x^2 + 2x - 8$

### Lesson 8-6

**1.** $(x + 2)(x + 12)$  **3.** $(n + 7)(n - 3)$  **5.** $2(2x - 3)(x - 6)$  **7.** $(3x + 2)(x + 5)$  **9.** prime  **11.** $(x + 3)(x + 14)$  **13.** $(a - 4)(a + 12)$  **15.** $(h + 4)(h + 11)$  **17.** $(x + 2)(x - 12)$  **19.** Find two numbers with a sum of 19 and a product of 48. $(2x + 3)(x + 8)$  **21.** $2(2x + 1)(x + 5)$  **23.** $(2x + 3)(x - 3)$  **25.** prime  **27.** $3(4x + 3)(x + 5)$  **29.** $(5x + 8)(x + 3)$  **31.** $(q + 2r)(q + 9r)$  **33.** $(x - y)(x - 5y)$  **35.** $4x + 48$  **37.** $-(2x + 5)(3x + 4)$  **39.** $-(x - 4)(5x + 2)$  **41.** prime  **43a.** $a^2$ and $b^2$  **43b.** $a^2 - b^2$  **43c.** width: $a - b$; length: $a + b$  **43d.** $(a - b)(a + b)$  **43e.** $(a - b)(a + b)$; the figure with area $a^2 - b^2$ and the rectangle with area $(a - b)(a + b)$ have the same area, so $a^2 - b^2 = (a - b)(a + b)$.  **45.** $-15, -9, 9, 15$  **47.** 7, 12, 15, 16  **49.** Sometimes; Sample answer: The trinomial $x^2 + 10x + 9 = (x + 1)(x + 9)$ and $10 > 9$. The trinomial $x^2 + 7x + 10 = (x + 2)(x + 5)$ and $7 < 10$.

**51.** $(4y - 5)^2 + 3(4y - 5) - 70 = [(4y - 5) + 10]\cdot[(4y - 5) - 7]$
$= (4y + 5)(4y - 12) = 4(y + 5)(y - 3)$

**53.** 4 ft  **55.** B  **57.** A  **59a.** 524; 56, 100  **59b.** D  **59c.** $(9x - 13)$ ft and $(3x + 1)$ ft  **59d.** 115 ft by 167 ft; 16 ft by 32 ft  **61.** $x - 1$

### Lesson 8-7

**1.** yes; $(q + 11)(q - 11)$  **3.** no  **5.** yes; $(4m + k^2)(4m - k^2)$  **7.** yes; $(5x + 6)^2$  **9.** yes; $(y^2 + 1)^2$  **11.** $(u + 3)(u - 3)(u^2 + 9)$
**13.** $20b^4 - 45n^4 = 5(4r^4 - 9n^4)$
$= 5((2r^2)^2 - (3n^2)^2)$
$= 5(2r^2 + 3n^2)(2r^2 - 3n^2)$

**15.** $(c + 1)(c - 1)(2c + 3)$  **17.** $(t + 4)(t - 4)(3t + 2)$  **19a.** $(4n + 1)^2 - 5^2$  **19b.** $(4n + 6)$ by $(4n - 4)$  **21.** $2m(2m - 7)(3m + 5)$  **23.** $3(2x - 7)^2$  **25.** $3p(2p + 1)(2p - 1)$  **27.** $(a + 7)(a - 7)$  **29.** $3(m^2 + 81)$  **31.** $2(a + 4)(a - 4)(6a + 1)$  **33.** $(24 + x)$ by $(24 - x)$; 96 ft  **35.** $(x + 2y)(x - 2y)(x^2 + 2)$  **37.** $(r - 6)(r + 6)(2r - 1)$  **39.** $2cd(c^2 + d^2)(2c - 5)$

## Selected Answers and Solutions

**41.** $x^4 + 6x^3 - 36x^2 - 216x = x^3(x+6) - 36x(x+6)$
$= (x^3 - 36x)(x+6) = x(x^2 - 36)(x+6) = x(x+6)(x-6)(x+6) = x(x+6)^2(x-6)$
**43.** $(y-2)(y+2)(y^2+4)(y^2+16)$ **45a.** $h(h-6)^2$; $h$, $(h-6)$, $(h-6)$ **45b.** Yes; because length cannot be negative, the value of $h$ has to be greater than 6 so that $(h-6)$ is greater than 0. So the sides of the box are each greater than 6 inches.
**47.** Lorenzo; sample answer: Checking Elizabeth's answer gives us $16x^2 - 25y^2$. The exponent on $x$ in the final product should be 4.
**49.** Sample answer: $x^2 - 3x + \frac{9}{4} = 0$; $\left\{\frac{3}{2}\right\}$
**51.** When the difference of squares pattern is multiplied together using the FOIL method, the outer and inner terms are opposites of each other. When these terms are added together, the sum is zero.
**53.** $4x^2 + 10x + 4$ because it is the only expression that is not a perfect square trinomial. **55.** A
**57.** $|4x + 5|$ **59.** B **61.** Sample answer: $3x$, $(2x - 7)$, $(2x + 7)$

### Chapter 8  Study Guide and Review

**1.** false; sample answer: $x^2 + 5x + 7$ **3.** true
**5.** true **7.** true **9.** false; difference of squares
**11.** difference of squares
**13.** prime polynomial **15.** $-x^4 + 1$
**17.** $3x^5 + x^3 - 2x^2 + 6x - 2$ **19.** $-a^2 + 9a - 6$
**21.** $4x^2 + 4x + 8$ **23.** 1 **25.** $3x^3 + 3x^2 - 21x$
**27.** $18a^2 + 3a - 10$ **29.** $10x^2 + 29x + 10$
**31.** $x^2 - 25$ **33.** $25x^2 + 40x + 16$ **35.** $4r^2 + 20rt + 25t^2$
**37.** $3x^2 - 21$ **39.** $7xy(2x - 3 + 5y)$ **41.** $(a+b)(a-4c)$
**43.** $(3a + 5b)(8m - 3n)$ **45.** $(3r^2 + p)(r-4)$
**47.** $3f^4g^2(6g^3 - f^2 + 3g)$
**49.** $(x-5)(x-3)$ **51.** $(x-6)(x+1)$ **53.** $(x+10)(x-5)$
**55.** $(x+4)(x+8)$ **57.** $(x+10)(x+1)$
**59.** $2(2x-1)(3x+7)$ **61.** $3(x-5)(x+3)$
**63.** $(4x-3)(5x+4)$ **65.** $(3x+2)(x-5)$ **67.** $3x+7$
**69.** $(8+5x)(8-5x)$ **71.** $3(x+1)(x-1)$
**73.** $(3x-5)(3x+5)$ **75.** prime **77.** $(2-7a)^2$
**79.** $x^2(x+4)(x-4)$ **81.** $-3(x+2)^2$

---

## CHAPTER 9
# Quadratic Functions and Equations

### Chapter 9  Get Ready

**1.** A table can organize data so that you can determine points to use for creating a graph.
**3.** Plot each point given by the $(x, y)$ pairs, then draw a straight line through all the points.

**5.**
| x | y |
|---|---|
| 0 | 3 |
| 1 | 2 |
| 2 | 1 |

**7.** Sample answer: Find two integers that have a sum of $b$ and a product of $c$.

### Lesson 9-1

**1.**
| x | y |
|---|---|
| -3 | 4 |
| -2 | -6 |
| -1 | -8 |
| 0 | -6 |
| 1 | 0 |
| 2 | 10 |

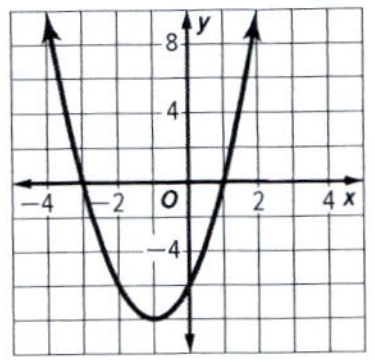

$D = \{-\infty < x < \infty\}$; $R = \{y \mid y \geq -8\}$

**3.**
| x | y |
|---|---|
| -1 | 4 |
| 0 | -3 |
| 1 | -8 |
| 2 | -11 |
| 3 | -12 |
| 4 | -11 |
| 5 | -8 |
| 6 | -3 |
| 7 | 4 |

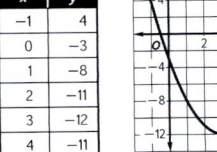

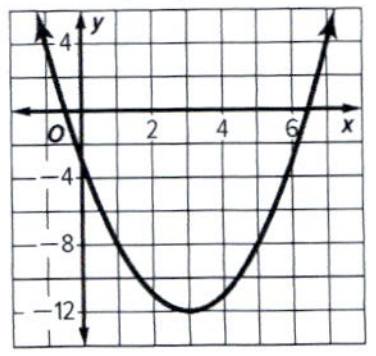

$D = \{-\infty < x < \infty\}$; $R = \{y \mid y \geq -12\}$

**5.** vertex $(2, 0)$, axis of symmetry $x = 2$, y-intercept 4, zero $(2, 0)$ **7.** vertex $(-2, -1)$, axis of symmetry $x = -2$, y-intercept 3, zeros $(-1, 0)$, $(-3, 0)$ **9.** vertex $(1, 2)$, axis of symmetry $x = 1$, y-intercept $-1$ **11.** vertex $(2, 1)$, axis of symmetry $x = 2$, y-intercept 5

**13. a.** Because the $a$ value is 3, the graph opens downward and has a maximum.
**b.** In this equation $a = -1$, $b = 4$, and $c = -3$.
$x = \frac{-b}{2a}$
$= \frac{-4}{2(-1)}$
$= 2$
To find the vertex, use the value you found for the x-coordinate of the vertex. To find the y-coordinate, substitute the value for $x$ in the original equation.

$y = -x^2 + 4x - 3$
$= -(2)^2 + 4(2) - 3$
$= -4 + 8 - 3$
$= 1$
The maximum is at $(2, 1)$.
**c.** The domain of the function is $D = \{-\infty < x < \infty\}$. The range is $R = \{y \mid y \leq 1\}$.

**15a.** maximum **15b.** 6 **15c.** $D = $ {all real numbers}; $R = \{y \mid y \leq 6\}$

**17.**

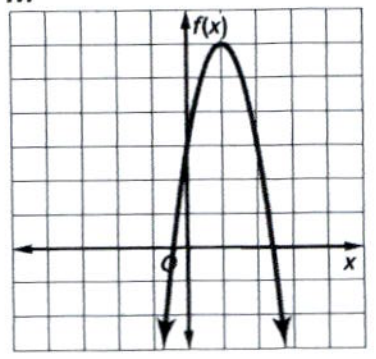

**19.**

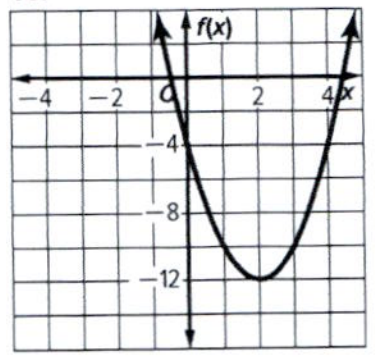

**21a.**

**21b.** 5 ft **21c.** 9 ft

**23.**
| x | y |
|---|---|
| -3 | 13 |
| -2 | 7 |
| -1 | 5 |
| 0 | 7 |
| 1 | 13 |

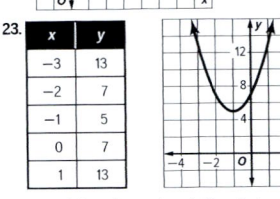

$D = $ {all real numbers}; $R = \{y \mid y \geq 5\}$

**25.**
| x | y |
|---|---|
| 0 | 5 |
| -1 | -4 |
| -2 | -7 |
| -3 | -4 |
| -4 | 5 |

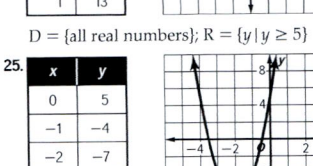

$D = \{-\infty < x < \infty\}$; $R = \{y \mid y \geq -7\}$

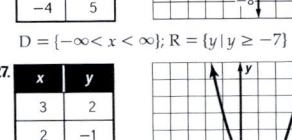

**27.**
| x | y |
|---|---|
| 3 | 2 |
| 2 | -1 |
| 1 | -2 |
| 0 | -1 |
| -1 | 2 |

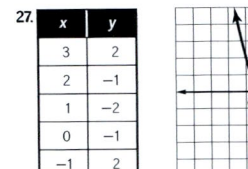

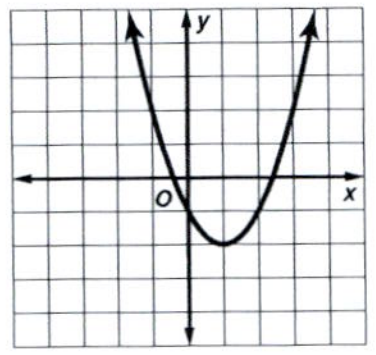

$D = \{-\infty < x < \infty\}$; $R = \{y \mid y \geq -2\}$

connectED.mcgraw-hill.com

# Selected Answers and Solutions

**29.** vertex (2, −4), axis of symmetry $x = 2$, $y$-intercept 0, zeros (0, 0) and (4, 0) **31.** vertex (1, 1), axis of symmetry $x = 1$, $y$-intercept 4, no real zeros **33.** vertex (0, 0), axis of symmetry $x = 0$, $y$-intercept 0, zero (0, 0) **35.** In this equation $a = 2$, $b = 12$, and $c = 10$.

$$x = \frac{-b}{2a} = \frac{-12}{2(2)}$$

The equation for the axis of symmetry is $x = -3$.

To find the vertex, use the value you found for the axis of symmetry as the $x$-coordinate of the vertex. To find the $y$-coordinate, substitute the value for $x$ in the original equation.

$y = 2x^2 + 12x + 10$
$= 2(-3)^2 + 12(-3) + 10$
$= -8$

The vertex is at (−3, −8).
The $y$-intercept occurs at (0, c). So, in this case, the $y$-intercept is 10.
**37.** vertex (−3, 4), axis of symmetry $x = -3$, $y$-intercept −5 **39.** vertex (2, −14), axis of symmetry $x = 2$, $y$-intercept 14 **41.** vertex (1, −15), axis of symmetry $x = 1$, $y$-intercept −18
**43a.** maximum **43b.** 9
**43c.** D = $\{-\infty < x < \infty\}$, R = $\{y | y \leq 9\}$
**45a.** minimum **45b.** −48
**45c.** D = $\{-\infty < x < \infty\}$, R = $\{y | y \geq -48\}$
**47a.** maximum **47b.** 33
**47c.** D = $\{-\infty < x < \infty\}$, R = $\{y | y \leq 33\}$
**49a.** maximum **49b.** 4
**49c.** D = $\{-\infty < x < \infty\}$, R = $\{y | y \leq 4\}$
**51a.** maximum **51b.** 3
**51c.** D = $\{-\infty < x < \infty\}$, R = $\{y | y \leq 3\}$
**53.**

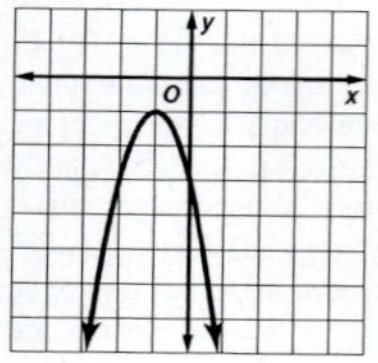

**55.**

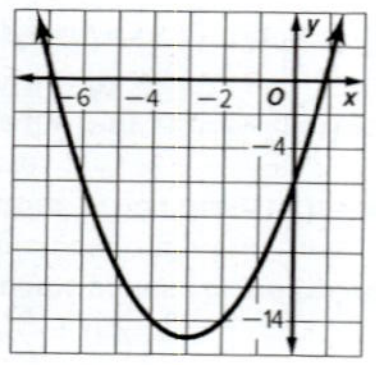

**57.**

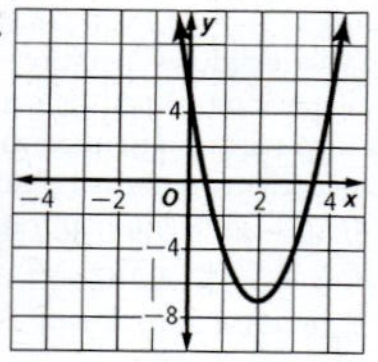

**59.** (−1.25, −0.25)

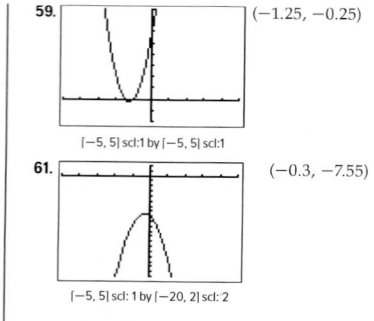

**61.** (−0.3, −7.55)

**63a.**

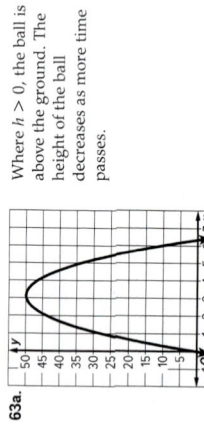

Where $h > 0$, the ball is above the ground. The height of the ball decreases as more time passes.

**63b.** 0 m **63c.** ≈50.0 m **63d.** ≈6.4 seconds
**63e.** D = $\{t | 0 \leq t \leq 6.4\}$; R = $\{h | 0 \leq h \leq 50.0\}$
**65a.** $h = -16t^2 + 90t$
$= -16(1)^2 + 90(1)$
$= 74$ ft
**b.** $126 = -16t^2 - 90t$
$0 = -16t^2 - 90t - 126$
$0 = (t - 3)(-16t + 42)$
$t = 3$ and $t = 2.625$
**c.** $h = -16t^2 + 90t$
$0 = -16t^2 + 90t$
$0 = -16t(t - 5)$
$t = 0$ and $t = 5.625$
These represent the time that the ball leaves the ground initially and the time it returns to the ground.

**67a.**

| Equation | Related Function | Zeros | y-Values |
|---|---|---|---|
| $x^2 - x = 12$ | $y = x^2 - x - 12$ | −3, 4 | −3.8, −6.4; −6, 8 |
| $x^2 + 8x = 9$ | $y = x^2 + 8x - 9$ | −9, 1 | −9, 11; −9.1; −9, 11 |
| $x^2 = 14x - 24$ | $y = x^2 - 14x + 24$ | 2, 12 | 2, 11; −9, 12; −9, 11 |
| $x^2 + 16x = -28$ | $y = x^2 + 16x + 28$ | −14, −2 | −14.13, −11; −2, −11.13 |

**b.** Use a graphing calculator to graph.

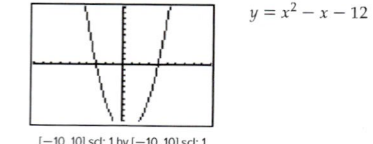

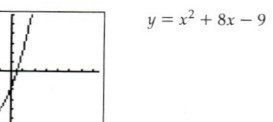

$y = x^2 - x - 12$
[−10, 10] scl:1 by [−10, 10] scl:1

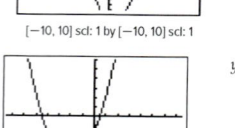

$y = x^2 + 8x - 9$
[−10, 10] scl:1 by [−30, 30] scl:5

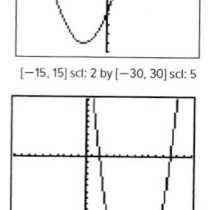

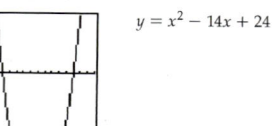

$y = x^2 - 14x + 24$
[−15, 15] scl:1 by [−10, 10] scl:1

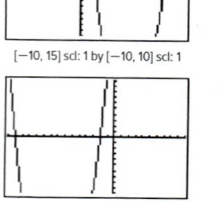

$y = x^2 + 16x + 28$
[−15, 10] scl:1 by [−10, 10] scl:1

**c.** Use the table function on the calculator to identify the zeros. The zeros are the values where $y$ is 0. Also list the $y$-values that are to the left and right of the zeros.
**d.** The function values have opposite signs just before and just after the zeros.
**69.** Chase; the lines of symmetry are $x = 2$ and $x = 1.5$. **71.** (−1, 9); Sample answer: I graphed the points given, and sketched the parabola that goes through them. I counted the spaces over and up from the vertex and did the same on the opposite side of the line $x = 2$. **73.** Sample answer: The function $y = -x^2 - 4$ has a vertex at (0, −4), but it is a maximum. **75.** B **77.** 3 **79a.** D **79b.** A **79c.** C **79d.** B, D, E, F

## Lesson 9-2

**1.** translated down 11 units **3.** reflected across the $x$-axis, translated up 8 units **5.** reflected across the $x$-axis, translated left 3 units and stretched vertically **7.** Because $a = -0.204$, the parent function is reflected across the $x$-axis and compressed vertically. $h = 6.2$ and $k = 13.8$, so the parent function is translated right 6.2 units and up 13.8 units.
**9.** $f(x) = 4(x - 1)^2 - 1$, $f(x) = 4x^2 - 8x + 3$
**11.** The function can be written as $f(x) = ax^2 + k$ where $a = -1$ and $k = -7$. Because $-7 < 0$ and $-1 < 0$, the graph of $y = -x^2 - 7$ translates the graph of $y = x^2$ down 7 units and reflects it across the $x$-axis.
**13.** compressed vertically, translated up 6 units **15.** stretched vertically, translated up 3 units **17.** translated left 1 unit and up 2.6 units and stretched vertically **19a.** $c(x)$ is reflected across the $x$-axis, stretched vertically, and translated right 2.5 units and up 105 units. $p(x)$ is reflected across the $x$-axis, stretched vertically, and translated right 2.8 units and up 126.5 units. **19b.** Paulo's **19c.** Paulo's
**21.** $f(x) = \frac{4}{9}(x+1)^2 - 4$, $f(x) = \frac{4}{9}x^2 + \frac{8}{9}x - \frac{32}{9}$
**23.** $f(x) = \frac{1}{3}(x+5)^2 - 9$, $f(x) = \frac{1}{3}x^2 + \frac{10}{3}x - \frac{2}{3}$
**25a.** The two equations are $h = -16t^2 + 300$ and $h = -16t^2 + 700$.
**b.** $0 = -16t^2 + 300$, $t \approx 4.3$; $0 = -16t^2 + 700$, $t \approx 6.6$; $6.6 - 4.3 \approx 2.3$ seconds
**27a.** The graph of $g(x)$ is the graph of $f(x)$ translated 200 yards right and 20 yards up, compressed vertically, and reflected across the $x$-axis **27b.** $h(x) = 0.0005(x - 230)^2 + 20$ **29.** Translate the graph of $f(x)$ up 7 units and to the right 2 units. **31a.** For $f(x) = x^2$ when $b > 1$, the graph is compressed horizontally. **31b.** For $f(x) = x^2$, when $0 < b < 1$, the graph is stretched horizontally.
**31d.** For $f(x) = x^2$, when $b < 0$, the squared term offsets the negative. For example, when $b = -3$, $(-3x)^2 = 9x^2$. This is the same transformation as when $b = 3$. Therefore, the sign of $b$ has no effect on the transformation of this function.
**33.** Graph 1: $y = -\frac{1}{3}x^2$, Graph 2: $y = -2x^2$. The graph of $y = 3x^2$ is narrower than the graph of $y = \frac{1}{3}x^2$.
**35.** The graph of $y = -3x^2$ opens down while the graph of $y = 3x^2$ opens up. **37.** Both graphs have the same shape, but the graph of $y = -3x^2$ opens down while the graph of $y = 3x^2$ opens up. **39.** Both graphs have the same shape, but the graph of $y = x^2 + 2$ is translated up 2 units from the graph of $y = x^2$ and the graph of $y = x^2 - 1$ is translated down 1 unit from the graph of $y = x^2$.
**41.** Both graphs have the same shape, but the graph of $y = x^2 - 4$ is translated down 4 units from the graph of $y = x^2$ and the graph of $y = (x - 4)^2$ is translated right 4 units from the graph of $y = x^2$.
**43.** $y = x^2 - 1$ **45.** Sample answer: $f(x) = -\frac{1}{5}x^2$
**47.** C **49.** −4 **51a.** 0.009 **51b.** −7.05 **51c.** −0.071
**51d.** B, E, F

## Lesson 9-3

**1.** 2, −5

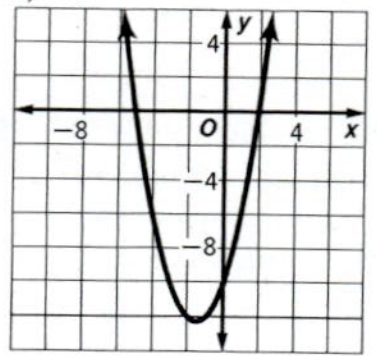

**3.** −2

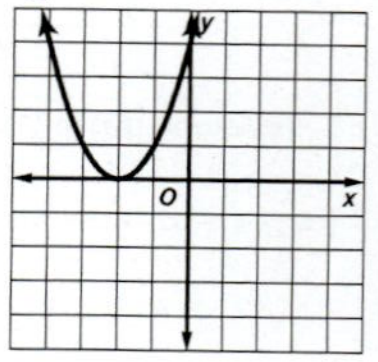

**5.** −8, 1

# Selected Answers and Solutions

**7.** $5, -5$  **9.** about 8.4 seconds  **11.** Step 1: Graph the related function of $f(x) = x^2 + 2x - 24 = 0$.

Step 2: The x-intercepts appear to be at $-6$ and $4$, so the solutions are $-6$ and $4$.

**13.** ∅  **15.** 1  **17.** ∅  **19.** $-6$  **21.** $8, -10$  **23.** $-3$  **25.** $2.2, -4.2$  **27.** $3, -6$  **29.** about 7.6 seconds  **31.** $1; -2$  **33.** $2; -4, -8$  **35.** $-3, 4$

**37a.** $h = -16t^2 + 30t + 10$
$0 = -16t^2 + 30t + 10$
Graph the equation and find the x-intercepts.

The positive x-intercept appears to be at 2.2, so she is in the air 2.2 seconds.

**b.** From the graph, she appears to hit a height of 15 feet at 0.2 second and 1.7 seconds.

**c.** $x = \frac{-b}{2a}$
$= \frac{-30}{2(-16)}$
$h = (-16)(0.9375)^2 + 30(0.9375) + 10$
$\approx 24$ feet

Her maximum height is about 24 feet, so she gets the bonus points.

**39.** $-2, 1, 4$  **41.** Sample answer: A tennis ball being hit in the air; an equation is $h = -16t^2 + 25t + 2$. The ball is in the air for about 1.6 seconds.  **43.** 1.5 and $-1.5$; Sample answer: Make a table of values for $x$ from $-2.0$ to 2.0. Use increments of 0.1.  **45a.** B  **45b.** B, E  **47a.** 2  **47b.** 1  **49c.** C, E  **49.** A, C, E

## Lesson 9-4

**1** Take the square root of each side of the equation. The square root of $x^2$ is $x$. To find the square root of 88, multiply 4 by 22 and take the square root of 4 to get $\pm 2\sqrt{22}$.
$x = \pm 2\sqrt{22} = \pm 9.38$  **3.** $x = 3, x = -5$
**5.** 0.6 second  **7.** $-10, 0$  **9.** $0, \frac{3}{4}$
**11.** $1, 9$  **13.** $\frac{1}{5}, \frac{7}{5}$  **15.** $f(x) = -\frac{1}{4}x^2 + \frac{1}{4}x + 3$  **17.** $\pm 3$
**19.** $\pm 10$  **21.** $\pm 6\sqrt{2}$

**23.** $-5 \pm 2\sqrt{5}$

**25.** $-2, 6$  **27.** $0, 3$

**29.** $-2, -\frac{1}{2}$

**31.** $-3, 0$  **33.** $8, 10$

**35.** $-6, 11$

**37** $16b^2 + 24b + 20 = 15$
$16b^2 + 24b + 5 = 0$
$(4b+5)(4b+1) = 0$
$4b + 5 = 0$ or $4b + 1 = 0$
$b = -\frac{5}{4}$   $b = -\frac{1}{4}$

**39.** $-\frac{3}{4}, -\frac{2}{3}$

**41.** $f(x) = \frac{3}{4}x^2 - 3x - 9$

**43.** 20 feet  **45.** $-9, 1$

**47.** $f(x) = \frac{1}{4}x^2 - 2x + 3$

**49** $V = \ell w h$
$96 = (h)(h-2)(h+8)$
$96 = h^3 + 6h^2 - 16h$
$0 = h^3 + 6h^2 - 16h - 96$
$0 = (h^3 + 6h^2) + (-16h - 96)$
$0 = h^2(h+6) - 16(h+6)$
$0 = (h^2 - 16)(h+6)$
$0 = (h+4)(h-4)(h+6)$
$h + 4 = 0$ or $h - 4 = 0$ or $h + 6 = 0$
$h = -4$   $h = 4$   $h = -6$

Since height cannot be negative, $h = 4$ is the only viable solution. Substitute 4 for $h$ in the length and width. The ballot box is 4 in. high by 12 in. long by 2 in. wide.

**51.** The quadratic is a perfect square, so I can factor it as $(x+4)^2 = 0$. Then, I can use the Square Root Property to solve. So, $x = -4$.

**53.** Because the solutions are $-\frac{b}{a}$ and $\frac{b}{a}, a \neq 0$ and $b$ is any real number.

**55.** $-\frac{5}{2}, 3$

**57.** $x^2 - 7x = 0$

**59a.** $h = -16t^2 + 29t + 6$

**59b.** 2 seconds

**61.** $-3, 3$  **63.** A, D  **65.** 0

**67a.** $2x^2 - 16x - 31 = \frac{1}{2}(2x)(x + 14)$

**67b.** 45 inches  **67c.** No; the negative value cannot be used for the solutions since length cannot be negative and $2(-1) = -2$.

## Lesson 9-5

**1** Step 1: Find $\frac{1}{2}$ of $-18 = -9$.
Step 2: Square the result in step 1: $(-9)^2 = 81$
Step 3: Add the result of step 2 to $x^2 - 18x$:
$x^2 - 18x + 81$
Thus, $c = 81$.

**3.** $\frac{81}{4}$  **5.** $-5.2, 1.2$  **7.** $-2.4, 0.1$  **9.** 8 ft by 18 ft

**11.** $y = (x+9)^2 - 45$

**13.** $\frac{169}{4}$  **15.** $\frac{361}{4}$  **17.** $\frac{25}{4}$

**19** $x^2 + 6x - 16 = 0$
$x^2 + 6x = 16$
$x^2 + 6x + 9 = 16 + 9$
$(x+3)^2 = 25$
$x + 3 = \pm 5$
$x = -3 \pm 5$

The solutions are 2 and $-8$.

**21.** $-1, 9$  **23.** $-0.2, 11.2$

**25.** ∅  **27.** on the 30th and 40th days after purchase

**29.** 5.3  **31.** $y = (x-4)^2 - 26$  **33.** $-1, 2$  **35.** 0.2, 0.9

**37a.** Earth  **37b.** Earth: 4.9 seconds; Mars: 8.0 seconds

**37c.** Yes. The acceleration due to gravity is much greater on Earth than it is on Mars. So the time to reach the ground should be much less.

**39.** $y = 3(x+4)^2 - 3$  **41a.** 97, $-72, -24, 0, 57$

**41b.** 2, 2, 0, 1, 2  **41c.** If $b^2 - 4ac$ is negative, the equation has no real solutions. If $b^2 - 4ac$ is zero, the equation has one real solution. If $b^2 - 4ac$ is positive, the equation has 2 real solutions.

**41d.** 0 because $b^2 - 4ac$ is negative. The equation has no real solutions because taking the square root of a negative number does not produce a real number.

**43** None; sample answer: If you add $\left(\frac{b}{2}\right)^2$ to each side of the equation and each side of the inequality, you get $x^2 + bx + \left(\frac{b}{2}\right)^2 = c + \left(\frac{b}{2}\right)^2$ and $c + \left(\frac{b}{2}\right)^2 < 0$. Because the left side of the last equation is a perfect square, it cannot equal the negative number $c + \left(\frac{b}{2}\right)^2$. So, there are no real solutions.

**45.** $y = -(x-1)^2 + 2$; The function is of the form $y = a(x-h)^2 + k$. Using the coordinates given in the graph, the vertex is (1, 2). So, the function $y = a(x-1)^2 + 2$. Substitute the coordinates of one of the points to get $-7 = a(-2-1)^2 + 2$, or $-7 = 9a + 2$. Solve for $a$ to get $a = -1$. So the function is $y = -(x-1)^2 + 2$.

**47.** B  **49.** C  **51.** 4, 6  **53.** C

### Lesson 9-6

**1.** −3, 5   **3.** 6.4, 1.6   **5.** 0.6, 2.5   **7.** −6, $\frac{1}{2}$
**9.** ±$\frac{5}{3}$   **11.** −3; no real solutions
**13.** 0; one real solution   **15.** The discriminant is −14.71, so the equation has no real solutions. Thus, the jaguarundi will not reach a height of 10 feet.

**17.** $x^2 + 16 = 0$
For this equation, $a = 1$, $b = 0$, and $c = 16$.
$x = \frac{-b \pm \sqrt{b^2 - 4ac}}{2a}$
$= \frac{0 \pm \sqrt{(0)^2 - 4(1)(16)}}{2(1)}$
$= \frac{\sqrt{-64}}{2}$
So, there is no real solution. The solution can be written ∅.

**19.** 2.2, −0.6   **21.** −3, $\frac{6}{5}$   **23.** 0.5, −2
**25.** 0.5, −1.2   **27.** 3   **29.** −1.2, 5.2   **31.** −2, 5
**33.** −6.2, −0.8   **35.** −0.07; no real solution   **37.** 12.64; two real solutions   **39.** 0; one real solution
**41a.** 2036   **41b.** Sample answer: No; the parabola has a maximum at about 66, meaning only 66% of the store's customers would ever have the store's loyalty card.   **43.** 0   **45.** 1   **47.** −1.4, 2.1

**49a.** $(20 - 2x)(25 - 7x) = 375$
**b.** $500 - 50x - 140x + 14x^2 = 375$
$14x^2 - 190x + 125 = 0$
$x = \frac{-b \pm \sqrt{b^2 - 4ac}}{2a}$
$= \frac{190 \pm \sqrt{(-190)^2 - 4(14)(125)}}{2(14)}$
$= \frac{190 \pm \sqrt{29{,}100}}{28}$
≈ 0.7 and 12.9
**c.** The margins should be 0.7 inches on the sides and 4(0.7) or 2.8 inches on the top and 3(0.7) or 2.1 inches on the bottom.

**51.** $k < \frac{9}{40}$   **53.** none   **55.** two   **57.** Sample answer: If the discriminant is positive, the Quadratic Formula will result in two real solutions because you are adding and subtracting the square root of a positive number in the numerator of the expression. If the discriminant is zero, there will be one real solution because you are adding and subtracting the square root of zero. If the discriminant is negative, there will be no real solutions because you are adding and subtracting the square root of a negative number in the numerator of the expression.   **59a.** D, F   **59b.** B, E   **59c.** Sample answer: She could rewrite the equation so that it equals zero. She could write out the values of $a$, $b$, and $c$.   **61a.** A   **61b.** D
**63.** B

### Lesson 9-7

**1.** (3, 0), (0, −3)

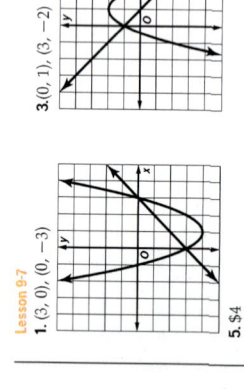

**3.** (0, 1), (0, −3)

**5.** ∅4

**7.** Substitute one expression for $y$ in the other equation and solve for $x$.
$x^2 - 5x + 5 = 5x - 20$
$x^2 - 10x + 25 = 0$
$(x - 5)^2 = 0$
$x - 5 = 0$
$x = 5$
Substitute 5 for $x$ in either equation.
$y = 5(5) - 20$
$y = 5$
The solution is (5, 5).

**11.** (5, 0), (2, −3)   **13.** (−2, −4), (2, 4)

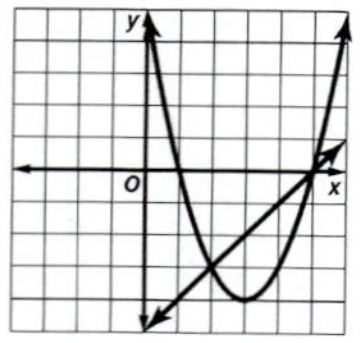

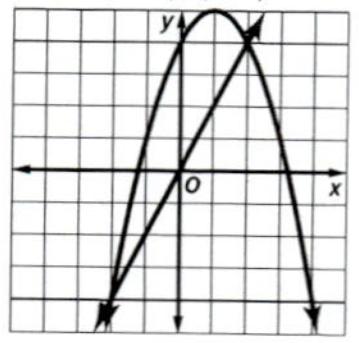

**15.** (−1, 6), (0.5, 1.5)

**17.** Substitute the first expression for $y$ in the second equation and solve for $x$.
$x^2 - 9 = 0$
$(x - 3)(x + 3) = 0$
$x - 3 = 0$ or $x + 3 = 0$
$x = 3$ or $x = -3$
Substitute the $x$-values in either equation.
$y = (3)^2$    $y = (-3)^2$
$y = 9$         $y = 9$
The solutions are (3, 9) and (−3, 9).

**19.** (−1, 12), (−11, 132)   **21a.** (1, 29)   **21b.** The streets do not intersect because the system has no real solutions.   **23a.** 2   **23b.** 0   **25.** Sample answer: $y = x^2 + 3$
**27.** $y = -x^2$   **29.** (−2.5, −6.25), (1, −1)   **31.** The solution of the system is (0, 0) which is not the same as no solution.   **33a.** (2, 1)   **33b.** no solution   **33c.** (−0.83, 3.34), (4.83, 14.66)   **35.** A, B   **37.** B, D, F

**39a.**

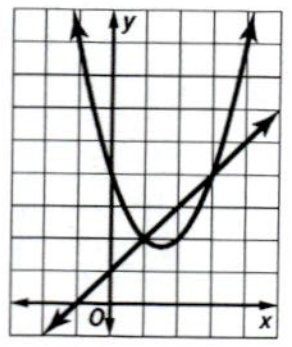

### Lesson 9-8

**1.**

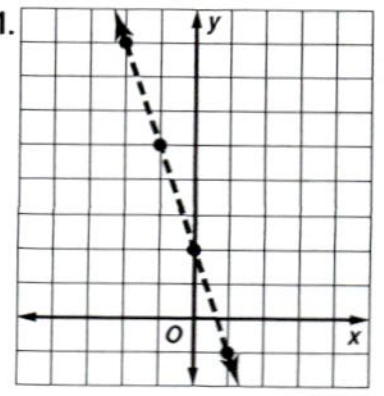

linear

**3.**

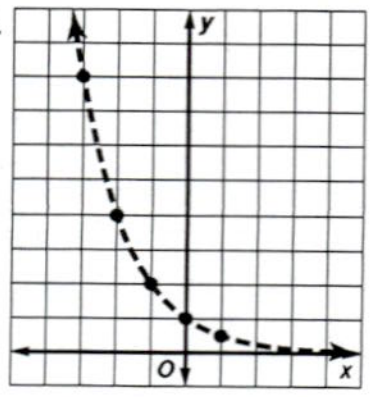

exponential

**5.** quadratic   **7.** exponential
**9.** exponential; $y = 3(3)^x$   **11.** linear; $y = \frac{1}{2}x + \frac{5}{2}$
**13.** linear; $y = 0.5x + 3$

**15.**

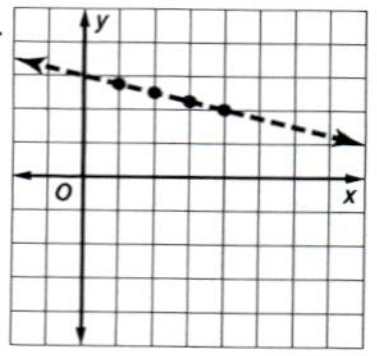

linear

**17.**

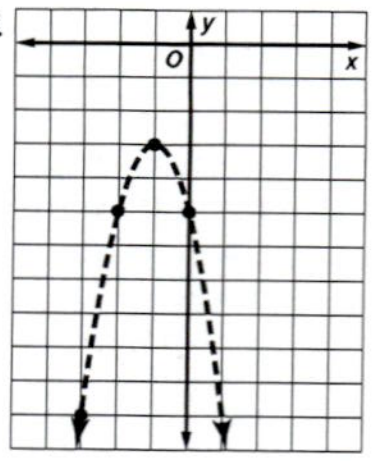

quadratic

**19.**

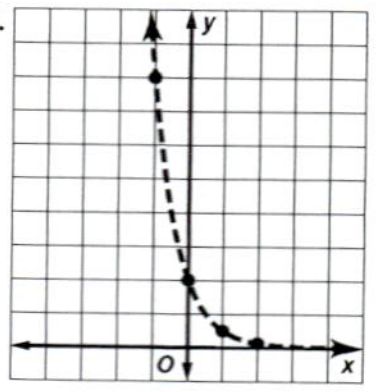

exponential

**21.** Look for a pattern in the $y$-values. Start with comparing first differences.
10   2.5   2.5   10
  −7.5  −2.5  2.5   7.5
        5    5    5

The first differences are not all equal. So, the table of values does not represent a linear function. Find the second differences and compare.
−7.5  −2.5  2.5   7.5
    5    5    5

The second differences are all equal, so the table of values represents a quadratic function.
Write an equation for the function that models the data.

The equation has the form $y = ax^2$. Find the value of $a$ by choosing one of the ordered pairs from the table of values. Let's use (2, 10).
$y = ax^2$
$10 = a(2)^2$
$10 = 4a$
$\frac{5}{2} = a$
$2.5 = a$
An equation that models the data is $y = 2.5x^2$.

**23.** exponential; $y = 0.25(5)^x$
**25.** linear; $y = -5x - 0.25$

**27a.** Graph the ordered pairs on a coordinate plane.

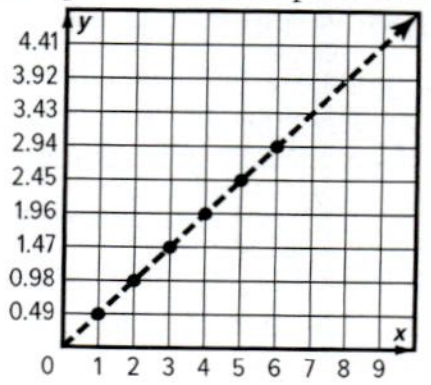

The graph appears to be linear.
**b.** Look at the first differences of the $y$-values.
0.49  0.98  1.47  1.96  2.45  2.94
   0.49  0.49  0.49  0.49  0.49
The common difference is 0.49.
**c.** $y = 0.49x$
The equation is $y = 0.49x$.
$y = 0.49(10)$
= $4.90

| Time (hour) | 0 | 1 | 2 | 3 | 4 |
|---|---|---|---|---|---|
| Amount of Bacteria | 12 | 36 | 108 | 324 | 972 |

**29b.** exponential   **29c.** $b = 12(3)^t$   **29d.** 78,732
**31.** Sample answer: $y = 2x^2 - 5$   **33.** $y = 4x + 1$
**35.** Sample answer: The data can be graphed to determine which function best models the data. You can also find the differences in ratios of the $y$-values. If the first differences are constant, the data can be modeled by a linear function. If the second differences are constant but the first differences are not, the data can be modeled by a quadratic function. If the ratios are constant, then the data can be modeled by an exponential function.   **37.** B, D, E   **39.** 11   **41.** B

### Lesson 9-9

**1.** $x^2 + x - 8$   **3.** $x^2 - 2x - 9$
**5.** $x + 1$
**7a.** $f(t) = 20(1.045)^t$
**b.** $g(t) = 42 + 2t$
**c.** $v(t) = 20(1.045)^t + 42 + 2t$. It represents the total value of the coins after $t$ years.
**d.** $76.92

**39b.** Yes there are two places where they might collide: (3, 4) and (1, 2).   **39c.** Sample answer: $y = 1.75$

**9.** $(f \cdot g)(x) = (3x + 10)(x^2 - 6x - 20)$
$= 3x^3 - 18x^2 - 6x^2 + 10x^2 - 60x - 20$
$= 3x^3 - 8x^2 - 66x - 20$
**11a.** $P(x) = 4 + 0.15x$  **b.** $T(x) = 160 - 10x$
**c.** $R(x) = -1.5x^2 - 16x + 640$  **d.** $552
**13.** $2x^2 - 6x$  **15.** $2^x + x^2 - 7$
**17.** $-4x^2 - 6x + 18$
**19.** $-5x^3 + 18x^2 - 11x - 8$
**21.** $-15x^3 - 15x^2$
**23.** $25x^5 - 40x^2$
**25. a.** $f(t) = 385 + 360t$
**b.** $g(t) = 590(1.15)^t$
**c.** $(f + g)(t) = 590(1.15)^t + 360t + 385$; this represents the total number of subscribers to Aurelio's blogs after $t$ years.
**d.** $590(1.15)^6 + 360(6) + 385$
$1364.7 + 2160 + 385 \approx 3910$
**27.** $21$  **29.** $-4$
**31a.** $f(x) = 10 - 2x$  **31b.** $g(x) = 8 - 2x$
**31c.** $4x^2 - 36x + 80$; this represents the area of the base.
**31d.** $D = \{0 < x < 4\}$; to create a square, $x$ must be greater than 0; also, $x$ must be less than half the width of the cardboard.
**33.** 4  **35.** $-2$  **37.** $-4$
**39a.** $(f - g)(x) = x^2 + 1$; $(g - f)(x) = -x^2 - 1$; the functions are opposites since $(f - g)(x) = -(g - f)(x)$
**39b.**

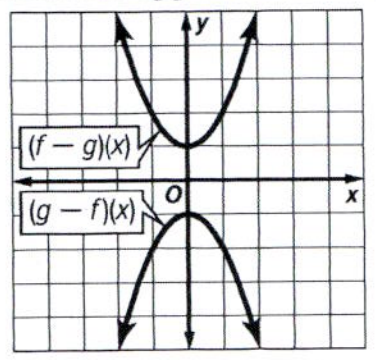

**39c.** The graph of $(g - f)(x)$ is the graph of $(f - g)(x)$ reflected across the x-axis.
**39d.** Yes; the functions are opposites of each other. This means one graph is a reflection of the other across the x-axis.
**41.** $(f + g)(x) = f(x) + g(x)$
$= 3^x + 2(3^x)$
$= 3(3^x)$
$= (3^1)(3^x)$
$= 3^{x+1}$
**43.** $2(3)^{2x}$
**45.** sometimes  **47.** sometimes  **49.** No; $(f \cdot g)(x)$ can be a linear function if one or both of $f(x)$ or $g(x)$ is a constant function.
**51a.** Mikayla did not distribute the minus sign to every term in $g(x)$.  **51b.** $x^2 + 10x + 3$  **51b.** $x^2(1.025)^x$  **55.** A, B, E, F
**57a.** $f(x) = 82,400(1.025)^x$  **57b.** $g(x) = 75,600 - 300x$
**57c.** $82,400(1.025)^x + 300x - 75,600$; this represents how many more residents live in Oakville than in Elmwood.  **57d.** 21,759; there were 21,759 more residents of Oakville than Elmwood in 2016.  **59.** C

---

## Chapter 9  Study Guide and Review

**1.** true  **3.** false; parabola  **5.** false; two  **7.** true
**9.** The solutions of a quadratic equation are the x-intercepts of the graph of the related quadratic function.
**11a.** minimum  **11b.** 0  **11c.** $D = \{-\infty < x < \infty\}$; $R = \{y | y \geq 0\}$  **13a.** minimum  **13b.** $-4$
**13c.** $D = \{-\infty < x < \infty\}$; $R = \{y | y \geq -4\}$
**15a.** maximum  **15b.** 16  **15c.** $D = \{t | 0 \leq t \leq 2\}$; $R = \{h | 0 \leq h \leq 16\}$  **17.** translated down 3 units
**19.** vertical stretch and translated down 18 units
**21.** vertical compression  **23.** reflected across the x-axis, vertically stretched, and translated up 100 units  **25.** 3
**27.** $-4, 6, 0.6$  **29.** $-0.8, 3$
**31.** $-11, 5$  **33.** $-4, -8$  **35.** $-4, 2$  **37.** $1, -7$
**39.** $10, -2$  **41.** $-0.7, 7.7$  **43.** no solution
**45.** $-8, 6$  **47.** $-0.7, 0.5$
**49.** $-5, 1.5$  **51.** $-2.5, 1.5$
**53.** $-4$; no solution  **55.** $(1, 1), (5, 9)$  **57.** $(0.5, -0.5), (4, 17)$
**61.** no solution  **63.** exponential; $y = 2^x$
**65.** linear; $y = 8 - 3x$  **67.** $-6x^2 + x + 4$
**69.** $4x^2 - x + 2$  **71.** $5x^4 - x^3 - 16x^2 + 3x + 3$

---

## CHAPTER 10
## Statistics

### Get Ready

**1.** Total all the cubes, set that as the denominator; find the total of a specific color, set as the numerator.
**3.** the Distributive Property  **5.** The range remains the same.  **7.** Substitute a value for $x$ and solve for $y$ or vice versa to generate data points.

### Lesson 10-1

**1.** mean: 4.6; median: 4; mode: none; The mean and median are close enough in value that either of them can represent the data.  **3.** mean
**5. a.** Canada has more medals than 6 of the countries, and there are 10 countries.
$\frac{6}{10} = 60$th percentile
**b.** Yes, the nation with 5 medals below it will be in the 50th percentile. The Netherlands has more medals than 5 countries.
Netherlands = $\frac{5}{10}$ = 50th percentile
**7a.** mean = $342.37; median $329.98; mode = $259.95
**7b.** The mean and median are close to the same value. Either measure would be good, but choosing a tablet closest to the median might be best because it is less than the mean.
**9.** $75^{th}$ percentile
**11. a.** Change all the scores to seconds by multiplying the minutes by 60 and then adding the seconds.
20:45 = 20 × 60 + 45 = 1245. Similarly, the other scores in seconds are 1290, 1275, 1352, 1300.
Mean = $\frac{1245 + 1290 + 1275 + 1352 + 1300}{5} = \frac{6462}{5} =$ 1292.4. Convert 1292.4 seconds back to minutes by dividing by 60 to get 21.54 minutes. Then convert 0.54 minutes to seconds to multiplying by 60. 0.54 × 60 = 32.4 seconds. So, her mean practice time in minutes and seconds is 21:32.4.
**b.** Arrange the practice times (in seconds) in order from least to greatest to find the median. 1245, 1275, 1290, 1300, 1352. The middle value is 1290 seconds. Convert it to minutes as seconds as shown in Part a, or 1290 ÷ 60 = 21.5, or 21 minutes 30 seconds. So, the median is 21:30. Since her time is 30 seconds less, the actual race time is exactly 21 minutes.
**13a.** John: 9.65, Raul: 9.475, Laird, Boris, Han: 9.525; John won.
**13b.** John: 9.633, Raul: 9.483, Laird, Boris, Han: 9.517; John still won.
**13c.** Sample answer: For this meet, it didn't matter which method of scoring they used because the overall standings turned out the same.
**15.** The student is not correct. A counterexample is the set 3, 3, 3, 3, 3, 3, with mean = 3 and mode = 3.
**17.** Sample answer: A data set can have zero, one, or more than one mode. This is true because the mode is not dependent on where the data value is positioned in the set or on the distribution of the data values. The mode only depends on how many of a certain data value there are. If no values appear more than once, there will be no mode. If multiple values appear more than once, there will be more than one mode.
**19.** The percentile rank tells where a score ranks among the other scores. The percent is a comparison of a score to the highest score possible.
**21.** Sample answer: 40, 40, 50, 50, 93, 95. The mean is 61.3, and the median is 50. If an unusually high test score like 99 is added to the data set, it will not change the median, but the mean will increase by over 5 points.
**23.** B  **25.** Sample answer: 20, 30, 40, 50, 60  **27.** Sample answer: Mean; The mean is 46.2 people, and the median is 33 people. The mean is closer to the number of participants in the activities.

### Lesson 10-2

**1.**

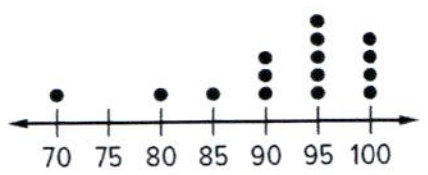

**3.**

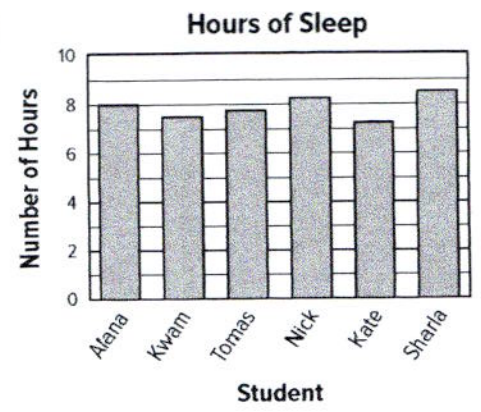

**5.**

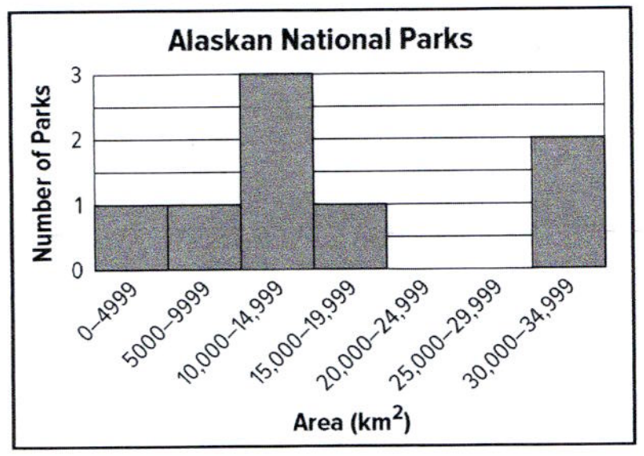

# Selected Answers and Solutions

**7.**

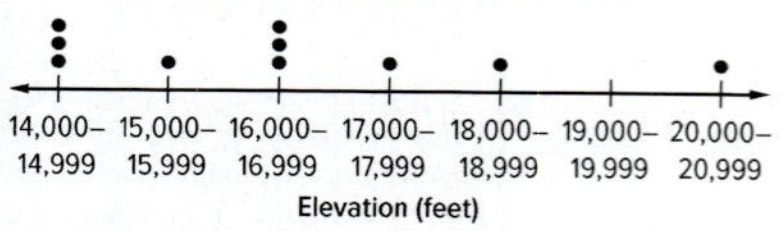

**Elevation of Mountains in United States**

(dot plot with elevation ranges 14,000–14,999 through 20,000–20,999 feet)

**9.**

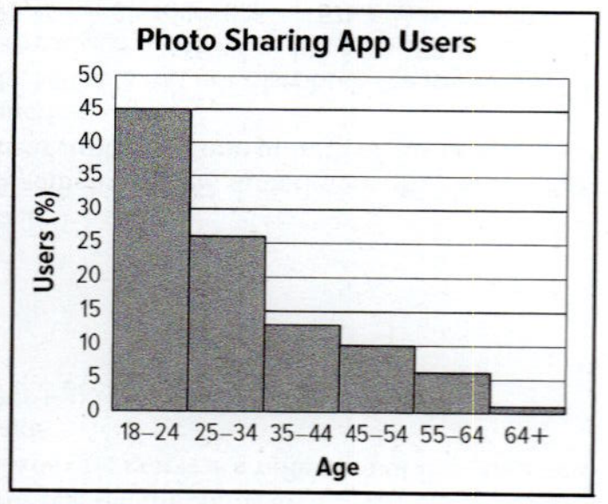

**Photo Sharing App Users**

(histogram: Users (%) vs Age ranges 18–24, 25–34, 35–44, 45–54, 55–64, 64+)

**11.** continuous;

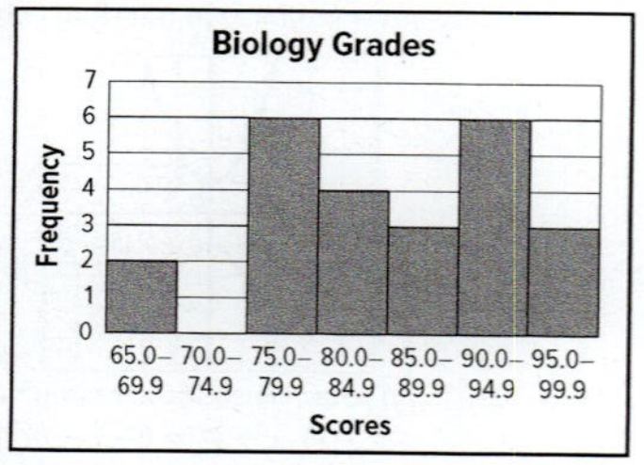

**Biology Grades**

(histogram: Frequency vs Scores 65.0–69.9, 70.0–74.9, 75.0–79.9, 80.0–84.9, 85.0–89.9, 90.0–94.9, 95.0–99.9)

**13.** discrete;

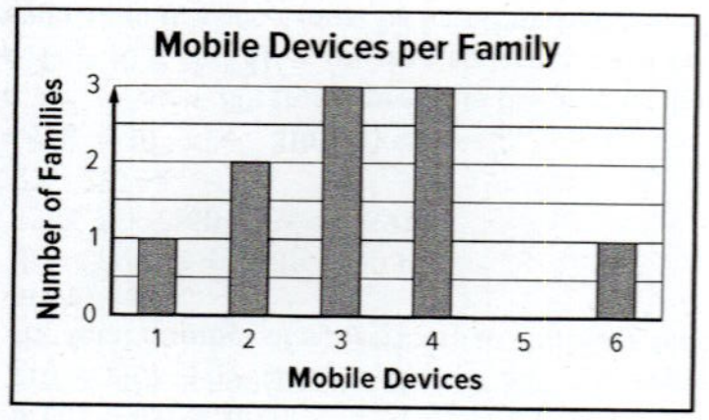

**Mobile Devices per Family**

(bar graph: Number of Families vs Mobile Devices 1–6)

**15.**

**Students' Favorite Movie Genre**

(bar graph: Number of Students vs Comedy, Drama, Crime, Thriller, Action)

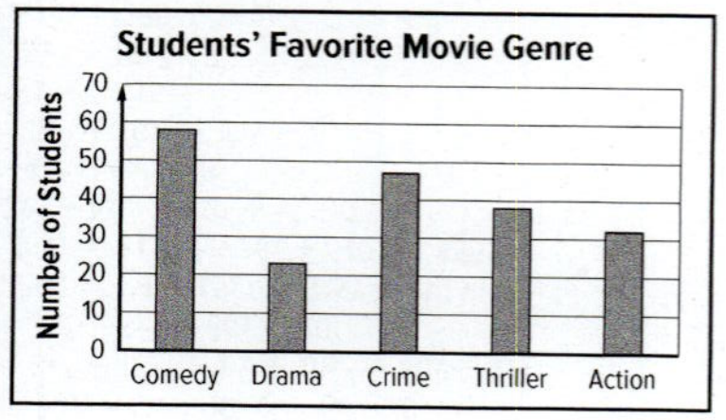

**17a.** 56°F  **17b.** 20  **17c.** 11  **19.** B, C, D  **21.** 10; 3; 4; 6; 2; 5; 1 and 7  **23.** The data is categorical, so a bar graph should be used to display the data.  **25a.** 7 in.  **25b.** 72–73 in.  **25c.** 35

## Lesson 10-3

**1.** $99.99; $199.99; $299.99; $399.99; $439.99

**3.** Find the mean.

$$\overline{x} = \frac{10 + 12 + 0 + 6 + 9 + 15 + 12 + 10 + 11 + 20}{10}$$

or 10.5

Find the sum of the square of the differences,
$(\overline{x} - x_1)^2 + (\overline{x} - x_2)^2 + \ldots + (\overline{x} - x_n)^2$.

$(10 - 10.5)^2 = 0.25$
$(12 - 10.5)^2 = 2.25$
$(0 - 10.5)^2 = 110.25$
$(6 - 10.5)^2 = 20.25$
$(9 - 10.5)^2 = 2.25$
$(15 - 10.5)^2 = 20.25$
$(12 - 10.5)^2 = 2.25$
$(10 - 10.5)^2 = 0.25$
$(11 - 10.5)^2 = 0.25$
$+ (20 - 10.5)^2 = 90.25$
$\phantom{+ (20 - 10.5)^2 =} 248.50$

Find the variance.

$$\sigma^2 = \frac{(\overline{x} - x_1)^2 + (\overline{x} - x_2)^2 + \ldots + (\overline{x} - x_n)^2}{n}$$

$= \frac{248.50}{10}$ or 24.85

Find the standard deviation.

$\sigma = \sqrt{\sigma^2}$

$= \sqrt{24.85}$ or about 4.98

Sample answer: The mean is 10.5 and the standard deviation is about 4.98. The standard deviation is relatively high due to outliers 0 and 20.

**5.** 62 people; 66 people; 73 people; 82 people; 99 people
**7.** ≈21.83; 14.5; 82; ≈16.36; 14; mean  **9.** 2.4; Sample answer: With a mean of about $9.29, the standard deviation of about $2.38 suggests that there is a good amount of deviation to the data. This deviation is mostly caused by the outlier of $15.00. If this outlier were removed, the new mean of the data would be about $8.33 with a standard deviation of about $0.49.  **11.** The Wildcats had a mean of about 11.9 fouls per game with a standard deviation of about 2.2 fouls. The Trojans had a mean of about 10.1 fouls per game with a standard deviation of about 2.1 fouls. The data sets had nearly the same variability while the Wildcats had a mean of almost 2 fouls more than the Trojans. The Wildcats coach can conclude that his team consistently has about 2 more fouls than the Trojans.

**13. a.** Movie A:

Find the mean.

$$\overline{x} = \frac{7 + 8 + 7 + 6 + 8 + 6 + 7 + 8 + 6 + 8 + 6 + 7 + 8 + 7 + 8 + 8}{16}$$

or about 7.2

Find the sum of the square of the differences,
$(\overline{x} - x_1)^2$.

$(7 - 7.2)^2 = 0.04$
$(8 - 7.2)^2 = 0.64$
$(7 - 7.2)^2 = 0.04$
$(6 - 7.2)^2 = 1.44$
$(8 - 7.2)^2 = 0.64$
$(6 - 7.2)^2 = 1.44$
$(7 - 7.2)^2 = 0.04$
$(8 - 7.2)^2 = 0.64$
$(6 - 7.2)^2 = 1.44$
$(8 - 7.2)^2 = 0.64$
$(6 - 7.2)^2 = 1.44$
$(7 - 7.2)^2 = 0.04$
$(8 - 7.2)^2 = 0.64$
$(7 - 7.2)^2 = 0.04$
$(8 - 7.2)^2 = 0.64$
$+ (8 - 7.2)^2 = 0.64$
$\phantom{+ (8 - 7.2)^2 =} 10.44$

Find the variance.

$$\sigma^2 = \frac{(\overline{x} - x_1)^2 + (\overline{x} - x_2)^2 + \ldots + (\overline{x} - x_n)^2}{n}$$

$= \frac{10.44}{16}$ or 0.6525

Find the standard deviation.

$\sigma = \sqrt{\sigma^2}$

$= \sqrt{0.6525}$ or about 0.81

Movie B:

Find the mean.

$$\overline{x} = \frac{9 + 3 + 8 + 2 + 5 + 10 + 3 + 8 + 10 + 9 + 9 + 10 + 6 + 4 + 9 + 3}{16}$$

or about 6.8

Find the sum of the square of the differences,
$(\overline{x} - x_n)^2$.

$(9 - 6.8)^2 = 4.84$
$(3 - 6.8)^2 = 14.44$
$(8 - 6.8)^2 = 1.44$
$(2 - 6.8)^2 = 23.04$
$(5 - 6.8)^2 = 3.24$
$(10 - 6.8)^2 = 10.24$
$(3 - 6.8)^2 = 14.44$
$(8 - 6.8)^2 = 1.44$
$(10 - 6.8)^2 = 10.24$
$(9 - 6.8)^2 = 4.84$
$(9 - 6.8)^2 = 4.84$
$(10 - 6.8)^2 = 10.24$
$(6 - 6.8)^2 = 0.64$
$(4 - 6.8)^2 = 7.84$
$(9 - 6.8)^2 = 4.84$
$+ (3 - 6.8)^2 = 14.44$
$\phantom{+ (3 - 6.8)^2 =} 131.04$

Find the variance.

$$\sigma^2 = \frac{(\overline{x} - x_1)^2 + (\overline{x} - x_2)^2 + \ldots + (\overline{x} - x_n)^2}{n}$$

$= \frac{131.04}{16}$ or 8.19

Find the standard deviation.

$\sigma = \sqrt{\sigma^2}$

$= \sqrt{8.19}$ or about 2.86

Sample answer: Movie A had a mean of about 7.2 with a standard deviation of about 0.81. Movie B had a mean of about 6.8 with a standard deviation of about 2.86. While both movies had a mean rating of close to 7, the ratings for Movie B had a wider range than those for Movie A.

**b.** Movie A: Sample answer: All the reviews are between 6 and 8. The ratings were consistent amongst the group.
Movie B: Sample answer: The ratings range from 2 to 10 with half of the students rating the movie between 8 and 10.

**15.** Both; when an outlier is removed, the spread and standard deviation will decrease. When more values that are equal to the mean of a data set are added, the outliers will have less influence.

**17.** Sample answer: Poll voters to determine if a particular presidential candidate is favored to win the election. Use a stratified random sample to call 100 people throughout the country.

**19.** Both values are helpful in identifying the measure of spread. The range of a data set is the difference between the highest and lowest values in a set of data. The interquartile range is the difference between Q3, the upper quartile, and Q1, the lower quartile. This value is helpful in identifying outliers.

**21.** D  **23.** A  **25.** C  **27a.** $433; $215
**27b.** $181.6, $114.50, $70  **27c.** There is no outlier.
**27d.** $142.34; a standard deviation of $142.34 is large compared to the mean of $181.60. This suggests that most of the data values are not relatively close to the mean.

## Lesson 10-4

**1.**

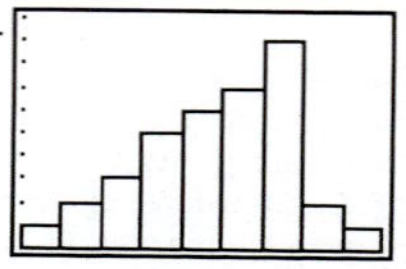

(histogram and box plot)

[56, 92] scl: 4 by [0, 10] scl: 1

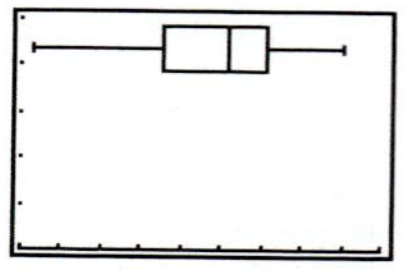

[56, 92] scl: 4 by [0, 5] scl: 1

negatively skewed

connectED.mcgraw-hill.com  R75

## Selected Answers and Solutions

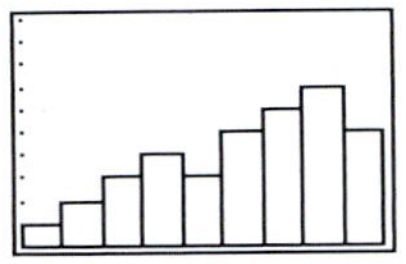

**3.** Sample answer: The distribution is skewed, so use the five-number summary. The range is 92 − 52 or 40. The median is 65, and half of the data are between 59.5 and 74.

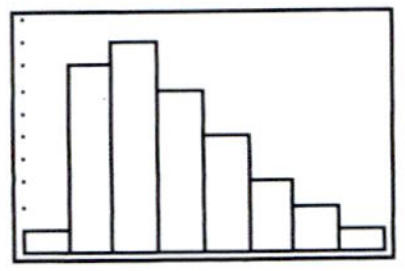

**5.**

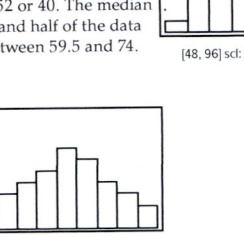

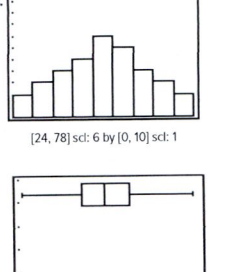

[24, 78] scl: 6 by [0, 10] scl: 1

[24, 78] scl: 6 by [0, 5] scl: 1

The distribution is negatively skewed, so use the five-number summary.

```
1-Var Stats
n=36
minX=12
Q1=28.5
Med=48
Q3=48
maxX=53
```

The range is 53 − 12 or 41. The median of the numbers is 39.5, and half the numbers are between 28 and 48.

**9.** Sample answer: The distribution is symmetric, so use the mean and standard deviation to describe the center and spread. The mean temperature is 52.8° with standard deviation of about 4.22°.

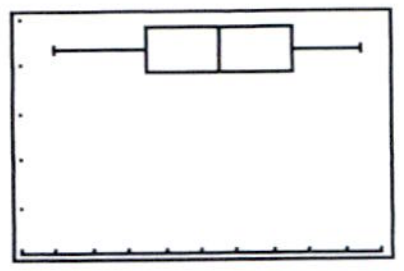

**11.**

[48, 96] scl: 6 by [0, 10] scl: 1

[42, 62] scl: 2 by [0, 5] scl: 1

**a.** Enter the list of prices as L1 and create a box-and-whisker plot. Adjust the window to the dimensions shown. The right whisker is longer than the left. The median is closer to the shorter whisker. Thus, the distribution is positively skewed.

```
1-Var Stats
n=20
minX=7.75
Q1=9.625
Med=11.25
Q3=15.125
maxX=18.5
```

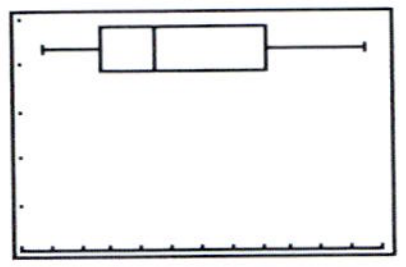

[7, 19] scl: 1 by [0, 5] scl: 1

The distribution is positively skewed, so use the five-number summary.

The range of prices is $18.50 − $7.75 or $10.75. The median price is $11.50, and half of the prices are between $9.63 and $15.13.

**b.** Delete all the entries in L1 that are greater than $15 and create another box-and-whisker plot. The lengths of the whiskers are approximately equal, and the median is in the middle of the data. Thus, the distribution is symmetric.

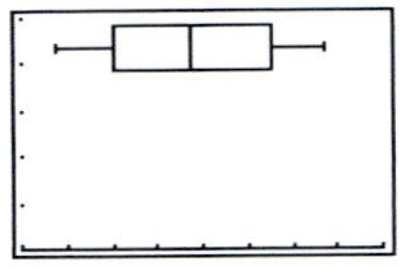

[7, 15] scl: 1 by [0, 5] scl: 1

The distribution is symmetric, so use the mean and standard deviation to describe the center and the spread of the new data.

```
1-Var Stats
x̄=10.66666667
Σx=160
Σx²=1757.625
Sx=1.907847204
σx=1.843155507
n=15
```

The mean is about $10.67 with standard deviation of about $1.84.

**13.** i **15.** Sample answer: A bimodal distribution is a distribution of data that is characterized by having data divided into two clusters, thus producing two modes, and having two peaks. The distribution can be described by summarizing the center and spread of each cluster of data.

**17.** Sample answer: In a symmetrical distribution, the majority of the data is located near the center of the distribution. The mean of the distribution is also located near the center of the distribution. Therefore, the mean and standard deviation should be used to describe the data. In a skewed distribution, the majority of the data lies either on the right or left side of the distribution. Because the distribution has a tail or may have outliers, the mean is pulled away from the majority of the data. The median is less affected. Therefore, the five-number summary should be used to describe the data.

**19.** A **21a.** $48,000 **21b.** $60,200 **21c.** $256,000 is an outlier. **23.** A

### Lesson 10-5

**1.** 3.5, 3.5, 1, 8, 2.4

**3.** Find the mean, median, mode, range, and standard deviation of the original set.
Mean: 5.75
Median: 5.5
Mode: 3
Range: 10
Standard deviation: 3.14
Multiply the mean, median, mode, range, and standard deviation by 3.
Mean: 3 × 5.75 ≈ 17.3
Median: 3 × 5.5 = 16.5
Mode: 3 × 3 = 9
Range: 3 × 10 = 30
Standard deviation: 3 × 3.14 ≈ 9.4

**5a.** Kyle's Distances

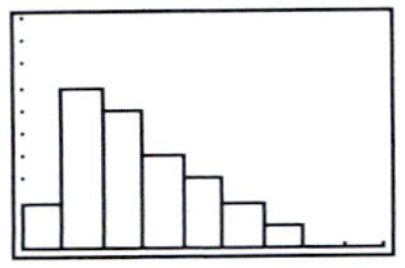

Mark's Distances

[7, 19.25] scl: 0.25 by [0, 10] scl: 1

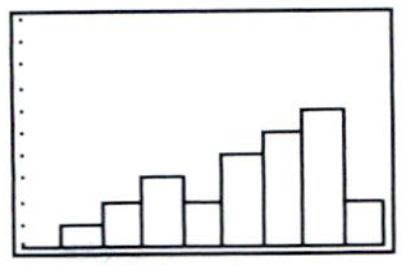

[7, 19.25] scl: 0.25 by [0, 10] scl: 1

Kyle, positively skewed; Mark, negatively skewed

**5b.** Sample answer: The distributions are skewed, so use the five-number summaries. Kyle's upper quartile is 17.98, while Mark's lower quartile is 18.065. This means that 75% of Mark's distances are greater than 75% of Kyle's distances. Therefore, we can conclude that overall, Mark's distances are higher than Kyle's distances.

**7.** 60, 9, 60, 14, 4.7

**9.** 22.5, 21.5, no mode, 24, 7.4

**11.** 36.8, 38, 12, 56, 20.0

**13.** 26.8, 27.2, 29.6, 10.4, 3.5

**15a.** 1st Period

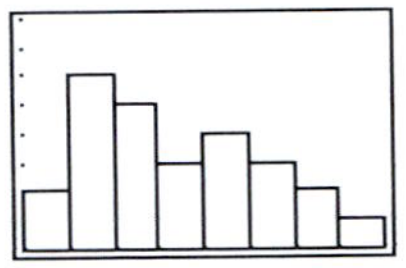

6th Period

[200, 600] scl: 50 by [0, 8] scl: 1

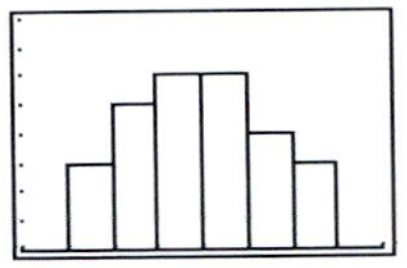

[200, 600] scl: 50 by [0, 8] scl: 1

1st period, positively skewed; 6th period, symmetric

**15b.** Sample answer: One distribution is symmetric and the other is skewed, so use the five-number summaries. The lower quartile for 1st period is 291 pages, while the minimum for 6th period is 294 pages. This means that the lower 25% of data for 1st period is lower than any data from 6th period. The range for 1st period is 578 − 206 or 372 pages. The range for 6th period is 506 − 294 or 212 pages. The median for 1st period is about 351 pages, while the median for 6th period is 392 pages. This means, that while the median for 6th period is greater, 1st period's pages have a greater range and include greater values than 6th period.

**17. a.** Enter Leon's times as L1 and Cassie's times as L2. Then use STAT PLOT to create a histogram for each list.

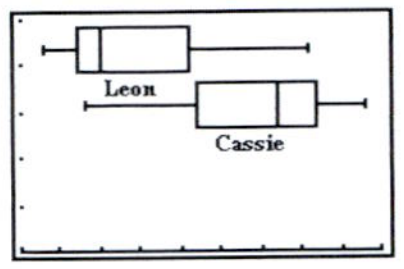

[1.5, 6] scl: 0.5 by [0, 5] scl: 1

For Leon's times, the right whisker is longer than the left, and the median is closer to the left whisker. The distribution is positively skewed. For Cassie's times, the left whisker is longer than the right and the median is closer to the right whisker. The distribution is negatively skewed.

# Selected Answers and Solutions

**b.** One distribution is positively skewed and the other is negatively skewed, so use the five-number summaries to compare the data.

```
Leon's Times          Cassie's Times
1-Var Stats           1-Var Stats
n=16                  n=16
min X=1.8             min X=2.3
Q1=2.2                Q1=3.1
Med=2.5               Med=4.7
Q3=3.6                Q3=5.2
max X=5.1             max X=5.8
```

The lower quartile for Leon's times is 2.2 minutes, while the minimum for Cassie's times is 2.3 minutes. This means that 25% of Leon's times are less than all of Cassie's times. The upper quartile for Leon's times is 3.6 minutes, while the lower quartile for Cassie's times is 3.7 minutes. This means that 75% of Leon's times are less than 75% of Cassie's time. Overall, we can conclude that Leon completed the brainteasers faster than Cassie.

**19a.** 52.96, 53, 53, 19, 6.08
**19b.** 47.96, 48, 48, 19, 6.08

**21** Find the total cost of all 20 SUVs and luxury cars.
20 × $39,250 = $785,000
Next subtract the price reduction of the 15 SUVs.
$785,000 − (15 × 2000) = $755,000
Finally, divide this new total by the 20 cars.
$755,000 ÷ 20 = $37,750
The new average price for all the vehicles is $37,750.

**23.** Sample answer: Histograms show the frequency of values occurring within set intervals. This makes the shape of the distribution easy to recognize. However, no specific values of the data set can be identified from looking at a histogram, and the overall spread of the data can be difficult to determine. The box-and-whisker plots show the data divided into four sections. This aids with comparing the spread of one set of data to another. However, the box-and-whisker plots are limited because they cannot display the data any more specifically than showing it divided into four sections. **25.** Sample answer: When two distributions are symmetric, the first thing to determine is how close the averages are and how spread out each set of data is. The mean and standard deviation are the best values to use for this comparison. When distributions are skewed, we also want to determine the direction and the degree to which it is skewed. The mean and standard deviation cannot provide any information in this regard, but we can get this information by comparing the range, quartiles, and medians found in the five-number summaries. So, if one or both sets of data are skewed, it is best to compare their five-number summaries. **27a.** A **27b.** Their means and standard deviations are relatively symmetric. **29a.** C **29b.** Their five-number summaries are more appropriate because one of the distributions is skewed.

## Lesson 10-6

**1** The total number of Norman nobles is found in column 2, row 2. **29 3.** 78

**5.**

| Role | Normans | Saxons | Total |
|---|---|---|---|
| Peasants | 34% | 30% | 64% |
| Nobles | 25% | 11% | 36% |
| Total | 59% | 41% | 100% |

**7** Marginal frequencies are found by adding the totals for each column and row. 34% + 25% = 59%; 30% + 11% = 41%; 34% + 30% = 64%; 25% + 11% = 36%; The marginal frequencies are 64%, 36%, 59%, and 41%.

**9.**

| Gender | Retriever | Small Dog | Total |
|---|---|---|---|
| Female | 24 | 20 | 44 |
| Male | 24 | 12 | 36 |
| Total | 48 | 32 | 80 |

**11.** 11%
**13.** It seems that there is an association; Americans are less likely to wear a full ski suit.
**15.** 18 **17.** 14 **19.** 11

**21.**

| Meal | Ate Everything | Left Food | Total |
|---|---|---|---|
| Vegetarian | 6 | 18 | 24 |
| Chicken | 18 | 9 | 27 |
| Beef | 21 | 9 | 30 |
| Total | 45 | 36 | 81 |

**23.**

| Meal | Ate Everything | Left Food | Total |
|---|---|---|---|
| Vegetarian | 7% | 22% | 30% |
| Chicken | 22% | 11% | 33% |
| Beef | 26% | 11% | 37% |
| Total | 56% | 44% | 100% |

**25.** They are equal. **27.** Yes, it appears that there is an association, because the conditional relative frequency of vegetarians eating everything on their plates is much lower than the conditional relative frequency of diners with other meals eating everything on their plates.

**29.**

| Flavor | Waffle Cone | Regular Cone | Total |
|---|---|---|---|
| Strawberry | 2 | 2 | 4 |
| Vanilla | 1 | 1 | 2 |
| Chocolate | 3 | 5 | 8 |
| Total | 6 | 8 | 14 |

**31.** 4% **33.** 6%

**35a.**

| Safety Gear | Paperwork Ready | Paperwork Not Ready | Total |
|---|---|---|---|
| Have Safety Gear | 8 | 6 | 14 |
| Do Not Have Safety Gear | 4 | 6 | 10 |
| Total | 12 | 12 | 24 |

**35b.** 50% **35c.** 33% **35d.** 58%

**35e.**

| Safety Gear | Paperwork Ready | Paperwork Not Ready | Total |
|---|---|---|---|
| Have Safety Gear | 33% | 25% | 58% |
| Do Not Have Safety Gear | 17% | 25% | 42% |
| Total | 50% | 50% | 100% |

**35f.** 34% **35g.** 50%

**37.**

| Allergy | Eat Gluten | Should Not Eat Gluten | Total |
|---|---|---|---|
| Peanuts | 4 | 3 | 7 |
| Soy | 2 | 1 | 3 |
| Milk | 1 | 2 | 3 |
| Total | 7 | 6 | 13 |

**39** Relative frequencies give the percentage of each joint data point within the data set. However, the data set may be skewed, as one category in the data set may have only a fraction of the data points that another category has.

| Allergy | Eat Gluten | Should Not Eat Gluten | Total |
|---|---|---|---|
| Peanuts | 31% | 23% | 54% |
| Soy | 15% | 8% | 23% |
| Milk | 8% | 15% | 23% |
| Total | 54% | 46% | 100% |

**41a.** A **41b.** C **41c.** B **43a.** B, D, E, G **43b.** 79% **43c.** 44% **43d.** tomatoes: 79%; peas: 40%; squash: 40%; Yes, there is an association. A much larger percentage of tomatoes were damaged in the storm **43e.** C

## Chapter 10 Study Guide and Review

**1.** mean **3.** quantitative **5.** A bar graph compares categories of data, while a histogram displays data that have been organized into equal intervals. The mean is the average of the data set, the median is the middle number when written in order, and mode occurs the most often. **9.** 21; 19; 20 **11.** 85.5; 86.6; no mode; Because the data set seems to be spread evenly, both the mean and the median are good measures of center.

**13.**

Favorite Type of Movie (bar graph)

**15.**

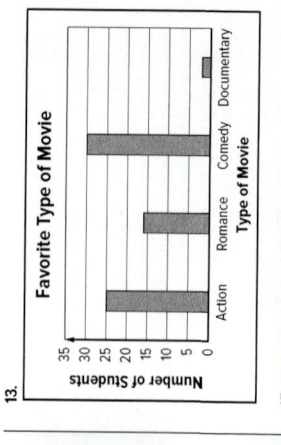

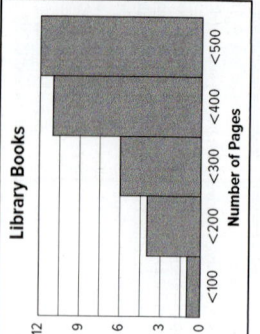

Library Books (histogram)

Library Books (histogram)

**17.** 3.4; 1.02; Sample answer: The standard deviation is relatively small compared to the mean. So, Ben mows about the same number of lawns each week. **19.** The day customers had a mean of about 9.6 times per month with a standard deviation of about 3.5. The night customers had a mean of about 11.7 times per month with a standard deviation of about 2.5. The night customers had a higher average and their data values were more consistent.
**21.** negatively skewed;

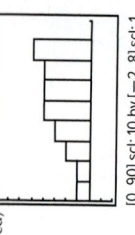

[0, 90] scl: 10 by [−2, 8] scl: 1

**23.** 22.5, 21.5, no mode, 24, 7.4
**25a.** Ms. Miller:

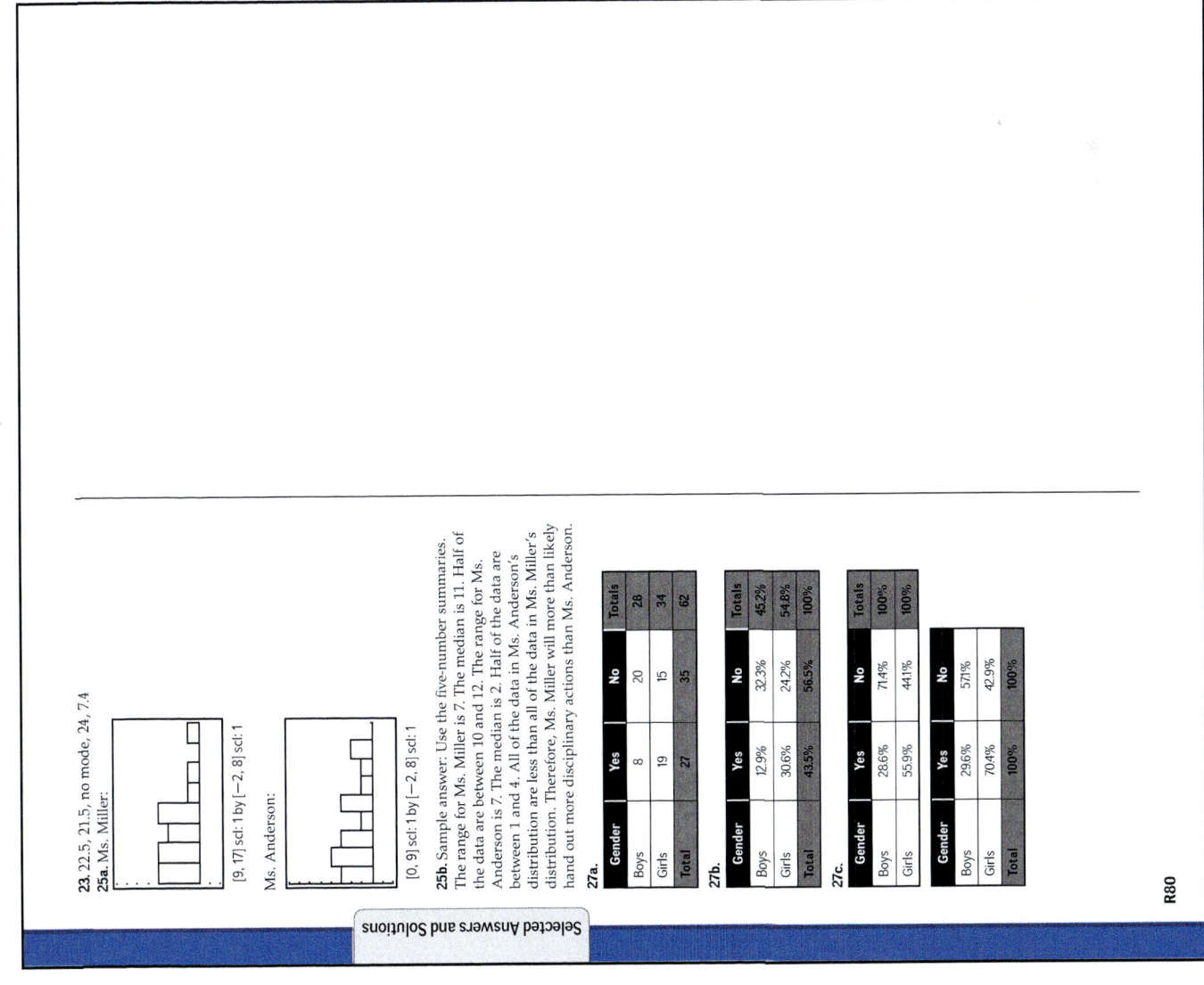

[9, 17] scl: 1 by [−2, 8] scl: 1

Ms. Anderson:

[0, 9] scl: 1 by [−2, 8] scl: 1

**25b.** Sample answer: Use the five-number summaries. The range for Ms. Miller is 7. The median is 11. Half of the data are between 10 and 12. The range for Ms. Anderson is 7. The median is 2. Half of the data are between 1 and 4. All of the data in Ms. Anderson's distribution are less than all of the data in Ms. Miller's distribution. Therefore, Ms. Miller will more than likely hand out more disciplinary actions than Ms. Anderson.

**27a.**

| Gender | Yes | No | Totals |
|---|---|---|---|
| Boys | 8 | 20 | 28 |
| Girls | 19 | 15 | 34 |
| Total | 27 | 35 | 62 |

**27b.**

| Gender | Yes | No | Totals |
|---|---|---|---|
| Boys | 12.9% | 32.3% | 45.2% |
| Girls | 30.6% | 24.2% | 54.8% |
| Total | 43.5% | 56.5% | 100% |

**27c.**

| Gender | Yes | No | Totals |
|---|---|---|---|
| Boys | 28.6% | 71.4% | 100% |
| Girls | 55.9% | 44.1% | 100% |

| Gender | Yes | No |
|---|---|---|
| Boys | 29.6% | 57.1% |
| Girls | 70.4% | 42.9% |
| Total | 100% | 100% |

# Glossary/Glosario

**Multilingual eGlossary**
Go to connectED.mcgraw-hill.com for a glossary of terms in these additional languages:

| | | |
|---|---|---|
| Arabic | Haitian Creole | Tagalog |
| Bengali | Hmong | Urdu |
| Brazilian Portuguese | Korean | Vietnamese |
| Chinese | Russian | |
| English | Spanish | |

## English

### A

**absolute error** (p. 313) The absolute error of a measurement is equal to one half the unit of measure.

**absolute value** (p. P11) The distance a number is from zero on the number line.

**absolute value function** (p. 205) A function written as $f(x) = |x|$, in which $f(x) \geq 0$ for all values of $x$.

**accuracy** (p. 34) The degree to which a measured value comes close to the actual or desired value.

**additive identity** (p. 16) For any number $a$, $a + 0 = 0 + a = a$.

**additive inverses** (p. P11) Two integers, $x$ and $-x$, are called additive inverses. The sum of any number and its additive inverse is zero.

**algebraic expression** (p. 5) An expression consisting of one or more numbers and variables along with one or more arithmetic operations.

**area** (p. P26) The measure of the surface enclosed by a geometric figure.

**arithmetic sequence** (p. 190) A numerical pattern that increases or decreases at a constant rate or value. The difference between successive terms of the sequence is constant.

**association** (p. 247) See correlation.

**asymptote** (p. 430) A line that a graph approaches.

**augmented matrix** (p. 374) A coefficient matrix with an extra column containing the constant terms.

### B

**axis of symmetry** (p. 559) The vertical line containing the vertex of a parabola.

**bar graph** (p. 659) A graphic form using bars to make comparisons of statistics.

**base** (p. 5) In an expression of the form $x^n$, the base is $x$.

**best-fit line** (p. 259) The line that most closely approximates the data in a scatter plot.

**binomial** (p. 491) The sum of two monomials.

**bivariate data** (p. 247) Data with two variables.

**boundary** (p. 321) A line or curve that separates the coordinate plane into regions.

### C

**causation** (p. 254) A relationship in which one event causes the other.

**center** (p. P24) The given point from which all points on the circle are the same distance.

**circle** (p. P24) The set of all points in a plane that are the same distance from a given point called the center.

**circumference** (p. P24) The distance around a circle.

**closed** (p. 30) A set is closed under an operation if for any numbers in the set, the result of the operation is also in the set.

**closed half-plane** (p. 321) The solution of a linear inequality that includes the boundary line.

**coefficient** (p. 26) The numerical factor of a term.

**coefficient of determination** (p. 628) A measure of how well a regression line fits a data set, denoted by $r^2$.

**common difference** (p. 190) The difference between successive terms in an arithmetic sequence.

**common ratio** (p. 462) The ratio of successive terms of a geometric sequence.

**complements** (p. P33) One of two parts of a probability making a whole.

**completing the square** (p. 596) To add a constant term to a binomial of the form $x^2 + bx$ so that the resulting trinomial is a perfect square.

## Español

### A

**error absoluto** El error absoluto de una medida es igual a un medio de la unidad de medida.

**valor absoluto** Es la distancia que dista de cero en una recta numérica.

**función del valor absoluto** Una función que se escribe $f(x) = |x|$, donde $f(x) \geq 0$, para todos los valores de $x$.

**exactitud** El grado de cercanía entre un valor medido y el valor real o deseado.

**identidad de la adición** Para cualquier número $a$, $a + 0 = 0 + a = a$.

**inverso aditivo** Dos enteros $x$ y $-x$ reciben el nobre de inversos aditivos. La suma de cualquier número y su inverso aditivo es cero.

**expresión algebraica** Una expresión que consiste en uno o más números y variables, junto con una o más operaciones aritméticas.

**área** La medida de la superficie incluida por una figura geométrica.

**sucesión aritmética** Un patrón numérico que aumenta o disminuye a una tasa o valor constante. La diferencia entre términos consecutivos de la sucesión es siempre la misma.

**asociación** Véase correlación.

**asíntota** Una línea a que un gráfico acerca.

**matriz aumentada** una matriz del coeficiente con una columna adicional que contiene los términos de la constante.

### B

**eje de simetría** La recta vertical que pasa por el vértice de una parábola.

**gráfico de barra** Forma gráfica usando barras para comparar estadísticas.

**base** En una expresión de la forma $x^n$, la base es $x$.

**recta de ajuste óptimo** La recta que mejor aproxima los datos de una gráfica de dispersión.

**binomio** La suma de dos monomios.

**datos bivariate** Datos con dos variables.

**frontera** Recta o curva que divide el plano de coordenadas en regiones.

### C

**causalidad** Una relación en la que un suceso causa el otro.

**centro** Punto dado del cual equidistan todos los puntos de un círculo.

**círculo** Conjunto de todos los puntos del plano que están a la misma distancia de un punto dado del plano llamado centro.

**circunferencia** Longitud del contorno de un círculo.

**cerrado** Un conjunto es cerrado bajo una operación si para cualquier número en el conjunto, el resultado de la operación es también en el conjunto.

**mitad-plano cerrado** La solución de una desigualdad linear que incluye la línea de límite.

**coeficiente** Factor numérico de un término.

**coeficiente de determinación** Una medida de la proximidad o el ajuste de una recta de regresión a un conjunto de datos, expresado como $r^2$.

**diferencia común** Diferencia entre términos consecutivos de una sucesión aritmética.

**azar común** El razón de términos sucesivos de una secuencia geométrica.

**complementos** Una de dos partes de una probabilidad que forma un todo.

**completar el cuadrado** Adición de un término constante a un binomio de la forma $x^2 + bx$, para que el trinomio resultante sea un cuadrado perfecto.

# Glossary/Glosario

**compound inequality** (p. 309) Two or more inequalities that are connected by the words *and* or *or*.

**compound interest** (p. 451) A special application of exponential growth.

**conditional relative frequency** (p. 692) In a two-way frequency table, conditional relative frequencies are found by dividing joint frequencies by row or column totals, which are marginal frequencies.

**conjugates** (p. 421) Binomials of the form $a\sqrt{b} + c\sqrt{d}$ and $a\sqrt{b} - c\sqrt{d}$.

**consecutive integers** (p. 96) Integers in counting order.

**consistent** (p. 339) A system of equations that has at least one ordered pair that satisfies both equations.

**constant** (pp. 143, 395) A monomial that is a real number.

**constant functions** (pp. 172, 142) A linear function of the form $y = b$.

**constraint** (p. 227) A condition that a solution must satisfy.

**continuous function** (p. 50) A function that can be graphed with a line or a smooth curve.

**coordinate** (p. P8) The number that corresponds to a point on a number line.

**coordinate plane** (p. 42) The plane containing the $x$- and $y$-axes.

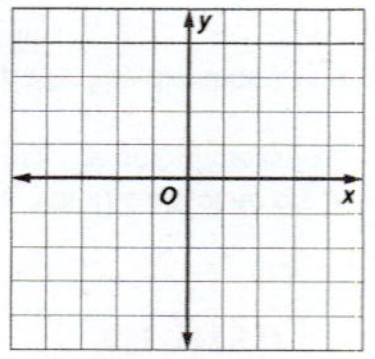

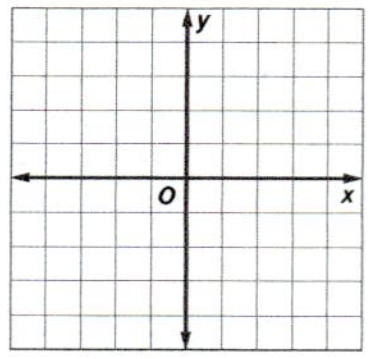

**coordinate system** (p. 42) The grid formed by the intersection of two number lines: the horizontal axis and the vertical axis.

**correlation** (p. 247, 254) A relationship between a set of data with two variables.

**correlation coefficient** (p. 259) A value that shows how close data points are to a line.

**cube root** (p. 411) If $a^3 = b$, then $a$ is the cube root of $b$.

**cumulative frequency** (p. 659) The sum of frequencies of all preceding events.

---

**desigualdad compuesta** Dos o más desigualdades que están unidas por las palabras *y* u *o*.

**interés compuesto** Aplicación especial de crecimiento exponencial.

**frecuencia relativa condicionada** En una tabla de frecuencias de doble entrada, las frecuencias relativas condicionadas se calculan dividiendo las frecuencias conjuntas entre los totales de las filas o las columnas, que son las frecuencias marginales.

**conjugados** Binomios de la forma $a\sqrt{b} + c\sqrt{d}$ y $a\sqrt{b} - c\sqrt{d}$.

**enteros consecutivos** Enteros en el orden de contar.

**consistente** Sistema de ecuaciones para el cual existe al menos un par ordenado que satisface ambas ecuaciones.

**constante** Monomio que es un número real.

**función constante** Función lineal de la forma $f(x) = b$.

**restricción** Una condición que una solución debe satisfacer.

**función continua** Función cuya gráfica puede ser una recta o una curva suave.

**coordenada** Número que corresponde a un punto en una recta numérica.

**plano de coordenadas** Plano que contiene los ejes $x$ y $y$.

**sistema de coordenadas** Cuadriculado formado por la intersección de dos rectas numéricas: los ejes $x$ y $y$.

**correlación** Es la relación que hay en un conjunto de datos con dos variables.

**coeficiente de correlación** Un valor que demostraciones cómo los puntos de referencias cercanos están a una línea.

**raíz cúbica** Si $a^3 = b$, entonces $a$ es la raíz cúbica de $b$.

**frecuencia acumulada** La suma de las frecuencias de todos los sucesos inferiores.

---

## D

**debt-to-income ratio** (p. 33) The ratio of how much a person owes per month to how much the person earns per month; used to determine if a person qualifies for a loan.

**decreasing** (p. 59) The graph of a function goes down on a portion of its domain when viewed from left to right.

**defining a variable** (p. P5) Choosing a variable to represent one of the unspecified numbers in a problem and using it to write expressions for the other unspecified numbers in the problem.

**degree of a monomial** (p. 491) The sum of the exponents of all its variables.

**degree of a polynomial** (p. 491) The greatest degree of any term in the polynomial.

**dependent** (p. 339) A system of equations that has an infinite number of solutions.

**dependent variable** (p. 44) The variable in a relation with a value that depends on the value of the independent variable.

**diameter** (p. P24) The distance across a circle through its center.

**difference of two squares** (p. 539) Two perfect squares separated by a subtraction sign.
$a^2 - b^2 = (a + b)(a - b)$ or
$a^2 - b^2 = (a - b)(a + b)$.

**dilation** (pp. 182, 572) A transformation that stretches or compresses the graph of a function.

**dimensional analysis** (p. 124) The process of carrying units throughout a computation.

**dimensions** (p. 374) The number of rows $m$ and the number of columns $n$ of a matrix written as $m \times n$.

**discrete function** (p. 50) A function of points that are not connected.

**discriminant** (p. 610) In the Quadratic Formula, the expression under the radical sign, $b^2 - 4ac$.

**distribution** (p. 673) A graph or table that shows the theoretical frequency of each possible data value.

**domain** (p. 42) The set of the first numbers of the ordered pairs in a relation.

**dot plot** (p. 657) A diagram that shows the frequency of data on a number line; also called a line plot.

---

**relación deuda-ingresos** La razón de la cantidad de dinero que una persona debe por mes a la cantidad de dinero que gana por mes; se usa para determinar si una persona califica para un préstamo.

**decreciente** El gráfico de una función va abajo en una porción de su dominio cuando está visto de izquierda a derecha.

**definir una variable** Consiste en escoger una variable para representar uno de los números desconocidos en un problema y luego usarla para escribir expresiones para otros números desconocidos en el problema.

**grado de un monomio** Suma de los exponentes de todas sus variables.

**grado de un polinomio** El grado mayor de cualquier término del polinomio.

**dependiente** Sistema de ecuaciones que posee un número infinito de soluciones.

**variable dependiente** La variable de una relación cuyo valor depende del valor de la variable independiente.

**diámetro** La distancia a través de un círculo a través de su centro.

**diferencia de cuadrados** Dos cuadrados perfectos separados por el signo de sustracción.
$a^2 - b^2 = (a + b)(a - b)$ or
$a^2 - b^2 = (a - b)(a + b)$.

**homotecia** Transformación que estira ó comprime la gráfica de una función.

**análisis dimensional** Proceso de tomar en cuenta las unidades de medida al hacer cálculos.

**dimensión** El número de filas, de $m$, y del número de la columna, $n$, de una matriz escrita como $m \times n$.

**función discreta** Función de puntos desconectados.

**discriminante** En la fórmula cuadrática, la expresión debajo del signo radical, $b^2 - 4ac$.

**distribución** Un gráfico o una tabla que muestra la frecuencia teórica de cada valor de datos posible.

**dominio** Conjunto de los primeros números de los pares ordenados de una relación.

**diagrama de puntos** Un diagrama que muestra la frecuencia de los datos en una recta numérica.

connectED.mcgraw-hill.com

**double root** (p. 581) The roots of a quadratic function that are the same number.

**elements** (p. 374) Each entry in a matrix.

**elimination** (p. 354) The use of addition or subtraction to eliminate one variable and solve a system of equations.

**end behavior** (p. 59) Describes how the values of a function behave at each end of the graph.

**equally likely** (p. P33) The outcomes of an experiment are equally likely if there are $n$ outcomes and the probability of each is $\frac{1}{n}$.

**equation** A mathematical sentence that contains an equal sign, $=$.

**equivalent equations** (p. 87) Equations that have the same solution.

**equivalent expressions** (p. 16) Expressions that denote the same value for all values of the variable(s).

**evaluate** (p. 10) To find the value of an expression.

**exponent** (p. 5) In an expression of the form $x^n$, the exponent is $n$. It indicates the number of times $x$ is used as a factor.

**exponential decay** (p. 430) When an initial amount decreases by the same percent over a given period of time.

**exponential decay function** (p. 430) A function of the form $y = ab^x$ where $a > 0$ and $0 < b < 1$.

**exponential equation** (p. 413) An equation in which the variables occur as exponents.

**exponential function** (p. 430) A function that can be described by an equation of the form $y = a^x$, where $a > 0$ and $a \neq 1$.

**exponential growth** (p. 430) When an initial amount increases by the same percent over a given period of time.

**exponential growth function** (p. 430) A function of the form $y = ab^x$ where $a > 0$ and $b > 1$.

**extrema** (p. 59) The greatest or least value of a function over an interval.

**extremes** (p. 116) In the ratio $\frac{a}{b} = \frac{c}{d}$, $a$ and $d$ are the extremes.

---

**raíces dobles** Las raíces de una función cuadrática que son el mismo número.

**elemento** Cada entrada de una matriz.

**eliminación** El uso de la adición o la sustracción para eliminar una variable y resolver así un sistema de ecuaciones.

**comportamiento extremo** Describe como los valores de una función se comportan en el cada fin del gráfico.

**igualmente probablemente** Los resultados de un experimento son igualmente probables si hay resultados de $n$ y la probabilidad de cada uno es $\frac{1}{n}$.

**ecuación** Enunciado matemático que contiene el signo de igualdad, $=$.

**ecuaciones equivalentes** Ecuaciones que poseen la misma solución.

**expresiones equivalentes** Expresiones que denotan el mismo valor para todos los valores de la(s) variable(s).

**evaluar** Calcular el valor de una expresión.

**exponente** En una expresión de la forma $x^n$, el exponente es $n$. Éste indica cuántas veces se usa $x$ como factor.

**desintegración exponencial** La cantidad inicial disminuye según el mismo porcentaje a lo largo de un período de tiempo dado.

**función de decaimiento exponencial** Una función con la forma $y = ab^x$, donde $a > 0$ y $0 < b < 1$.

**ecuación exponencial** Ecuación en que las variables aparecen en los exponentes.

**función exponencial** Función que puede describirse mediante una ecuación de la forma $y = a^x$, donde $a > 0$ y $a \neq 1$.

**crecimiento exponencial** La cantidad inicial aumenta según el mismo porcentaje a lo largo de un período de tiempo dado.

**función de crecimiento exponencial** Una función con la forma $y = ab^x$, donde $a > 0$ y $b > 1$.

**extremo** El valor mayor o el valor menor de una función en un intervalo.

**extremos** En la razón $\frac{a}{b} = \frac{c}{d}$, $a$ y $d$ son los extremos.

---

**F**

**factoring** (p. 520) To express a polynomial as the product of monomials and polynomials.

**factoring by grouping** (p. 521) The use of the Distributive Property to factor some polynomials having four or more terms.

**factors** (p. 5) In an algebraic expression, the quantities being multiplied are called factors.

**family of graphs** (p. 151) Graphs and equations of graphs that have at least one characteristic in common.

**five-number summary** (p. 665) The three quartiles and the minimum and maximum values of a data set.

**FOIL method** (p. 507) To multiply two binomials, find the sum of the products of the First terms, the Outer terms, the Inner terms, and the Last terms.

**formula** (p. 80) An equation that states a rule for the relationship between certain quantities.

**four-step problem-solving plan** (p. P5)
Step 1 Understand the problem.
Step 2 Plan the solution.
Step 3 Solve the problem.
Step 4 Check the solution.

**frequency table** (p. 659) A chart that indicates the number of values in each interval.

**function** (p. 49) A relation in which each element of the domain is paired with exactly one element of the range.

**function notation** (p. 52) A way to name a function that is defined by an equation. In function notation, the equation $y = 3x - 8$ is written as $f(x) = 3x - 8$.

**Fundamental Counting Principle** (p. P34) If an event $M$ can occur in $m$ ways and is followed by an event $N$ that can occur in $n$ ways, then the event $M$ followed by the event $N$ can occur in $m \times n$ ways.

---

**factorización** La escritura de un polinomio como producto de monomios y polinomios.

**factorización por agrupamiento** Uso de la Propiedad distributiva para factorizar polinomios que poseen cuatro o más términos.

**factores** En una expresión algebraica, los factores son las cantidades que se multiplican.

**familia de gráficas** Gráficas y ecuaciones de gráficas que tienen al menos una característica común.

**resumen de cinco números** Los tres cuartiles, el valor mínimo y el valor máximo de un conjunto de datos.

**método FOIL** Para multiplicar dos binomios, busca la suma de los productos de los primeros (First) términos, los términos exteriores (Outer), los términos interiores (Inner) y los últimos términos (Last).

**fórmula** Ecuación que establece una relación entre ciertas cantidades.

**solución de problemas en cinco pasos**
Paso 1 Comprender el problema.
Paso 2 Planifica la solución.
Paso 3 Resuelve el problema.
Paso 4 Examina la solución.

**Tabla de frecuencias** Tabla que indica el número de valores en cada intervalo.

**función** Una relación en que a cada elemento del dominio le corresponde un único elemento del rango.

**notación funcional** Una manera de nombrar una función definida por una ecuación. En notación funcional, la ecuación $y = 3x - 8$ se escribe $f(x) = 3x - 8$.

**Principio fundamental de contar** Si un evento $M$ puede ocurrir de $m$ maneras y lo sigue un evento $N$ que puede ocurrir de $n$ maneras, entonces el evento $M$ seguido del evento $N$ puede ocurrir de $m \times n$ maneras.

---

**G**

**geometric sequence** (p. 462) A sequence in which each term after the first is found by multiplying the previous term by a constant $r$, called the common ratio.

**graph** (p. P8) To draw, or plot, the points named by certain numbers or ordered pairs on a number line or coordinate plane.

---

**secuencia geométrica** Una secuencia en la cual cada término después de la primera sea encontrada multiplicando el término anterior por un $r$ constante, llamado el razón común.

**graficar** Marcar los puntos que denotan ciertos números en una recta numérica o ciertos pares ordenados en un plano de coordenadas.

# Glossary/Glosario

**greatest integer function** (p. 197) A step function, written as $f(x) = [\![x]\!]$, where $f(x)$ is the greatest integer less than or equal to $x$.

**La función más grande del número entero** Una función del paso, escrita como $f(x) = [\![x]\!]$, donde está el número entero $f(x)$ es el número más grande menos que o igual a $x$.

## H

**half-plane** (p. 321) The region of the graph of an inequality on one side of a boundary.

**semiplano** Región de la gráfica de una desigualdad en un lado de la frontera.

**histogram** (p. 659) A graphical display that uses bars to display numerical data that have been organized into equal intervals.

**histograma** Una exhibición gráfica que utiliza barras para exhibir los datos numéricos que se han organizado en intervalos iguales.

## I

**identity** (pp. 35, 102) An equation that is true for every value of the variable.

**identidad** Ecuación que es verdad para cada valor de la variable.

**identity matrix** (p. 375) A square matrix that, when multiplied by another matrix, equals that same matrix. If $A$ is any $n \times n$ matrix and $I$ is the $n \times n$ identity matrix, then $A \cdot I = A$ and $I \cdot A = A$.

**matriz de la identidad** Una matriz cuadrada que, cuando es multiplicada por otra matriz, iguala que la misma matriz. Si $A$ es alguna de matriz $n \times n$ e $I$ es la matriz de la identidad de $n \times n$, entonces $A \cdot I = A$ e $I \cdot A = A$.

**inconsistent** (p. 339) A system of equations with no ordered pair that satisfy both equations.

**inconsistente** Un sistema de ecuaciones para el cual no existe par ordenado alguno que satisfaga ambas ecuaciones.

**increasing** (p. 59) The graph of a function goes up on a portion of its domain when viewed from left to right.

**creciente** El gráfico de una función va arriba en una porción de su dominio cuando está visto de izquierda a derecha.

**independent** (p. 339) A system of equations with exactly one solution.

**independiente** Un sistema de ecuaciones que posee una única solución.

**independent variable** (p. 44) The variable in a function with a value that is subject to choice.

**variable independiente** La variable de una función sujeta a elección.

**inequality** (p. 289) An open sentence that contains the symbol $<$, $\leq$, $>$, or $\geq$.

**desigualdad** Enunciado abierto que contiene uno o más de los símbolos $<$, $\leq$, $>$, o $\geq$.

**integers** (p. P7) The set {..., −2, −1, 0, 1, 2, ...}.

**enteros** El conjunto {..., −2, −1, 0, 1, 2, ...}.

**intercepts** (p. 58) Points where the graph intersects an axis.

**puntos de corte** Los puntos en los que la gráfica se interseca con un eje.

**interquartile range** (p. 666) The range of the middle half of a set of data. It is the difference between the upper quartile and the lower quartile.

**amplitud intercuartílica** Amplitud de la mitad central de un conjunto de datos. Es la diferencia entre el cuartil superior y el inferior.

**intersection** (p. 309) The graph of a compound inequality containing *and*; the solution is the set of elements common to both inequalities.

**intersección** Gráfica de una desigualdad compuesta que contiene la palabra *y*; la solución es el conjunto de soluciones de ambas desigualdades.

**interval** (p. 179) The points between two points or numbers between two values.

**intervalo** Los puntos que están entre dos puntos o los números que están entre dos valores.

**inverse function** (p. 268) A function that undoes the action of another function; can be written as $f^{-1}(x)$ and read as "the inverse of $f$ of $x$."

**función inversa** Una función que cancela la acción de otra función; se puede escribir $f^{-1}(x)$ y se lee "el inverso de $f$ de $x$."

**inverse relation** (p. 267) The set of ordered pairs obtained by exchanging the $x$- and $y$-coordinates of each ordered pair in a relation.

**relación inversa** El conjunto de pares ordenados que se obtiene al intercambiar las coordenadas $x$ e $y$ de cada par ordenado de una relación.

**irrational numbers** (p. P7) Numbers that cannot be expressed as terminating or repeating decimals.

**números irracionales** Números que no pueden escribirse como decimales terminales o periódicos.

## J

**joint frequency** (p. 691) In a two-way frequency table, joint frequencies display the frequency of two conditions happening together.

**frecuencia conjunta** En una tabla de frecuencias de doble entrada, las frecuencias conjuntas muestran la frecuencia de dos condiciones que ocurren juntas.

## L

**leading coefficient** (p. 492) The coefficient of the term with the highest degree in a polynomial.

**coeficiente inicial** El coeficiente del término con el grado más alto (el primer coeficiente inicial) en un polinomio.

**like terms** (p. 25) Terms that contain the same variables, with corresponding variables having the same exponent.

**términos semejantes** Expresiones que tienen las mismas variables, con las variables correspondientes elevadas a los mismos exponentes.

**line of fit** (p. 248) A line that describes the trend of the data in a scatter plot.

**recta de ajuste** Recta que describe la tendencia de los datos en una gráfica de dispersión.

**line symmetry** (p. 59) A graph possesses line symmetry if one half of the graph is a reflection of the other half across the line.

**simetría axial** Una gráfica tiene simetría axial si una mitad de la gráfica es una reflexión de la mitad que está del otro lado del eje.

**linear equation** (p. 143) An equation in the form $Ax + By = C$, with a graph that is a straight line.

**ecuación lineal** Ecuación de la forma $Ax + By = C$, cuya gráfica es una recta.

**linear extrapolation** (p. 227) The use of a linear equation to predict values that are outside the range of data.

**extrapolación lineal** Uso de una ecuación lineal para predecir valores fuera de la amplitud de los datos.

**linear function** (p. 141) A function with ordered pairs that satisfy a linear equation.

**función lineal** Función cuyos pares ordenados satisfacen una ecuación lineal.

**linear interpolation** (p. 249) The use of a linear equation to predict values that are inside the data range.

**interpolación lineal** Uso de una ecuación lineal para predecir valores dentro de la amplitud de los datos.

**linear regression** (p. 259) An algorithm to find a precise line of fit for a set of data.

**regresión lineal** Un algoritmo para encontrar una línea exacta del ajuste para un sistema de datos.

**linear transformation** (p. 680) One or more operations performed on a set of data that can be written as a linear function.

**transformación lineal** Una o más operaciones que se hacen en un conjunto de datos y que se pueden escribir como una función lineal.

**literal equation** (p. 123) A formula or equation with several variables.

**ecuación literal** Un fórmula o ecuación con varias variables.

**lower quartile** (p. 665) Divides the lower half of the data into two equal parts.

**cuartil inferior** Éste divide en dos partes iguales la mitad inferior de un conjunto de datos.

connectED.mcgraw-hill.com    R89

# Glossary/Glosario

## M

**mapping** (p. 42) Illustrates how each element of the domain is paired with an element in the range.

| X | Y |
|---|---|
| 1 | 2 |
| −2 | 4 |
| 0 | −3 |

**aplicaciones** Ilustra la correspondencia entre cada elemento del dominio con un elemento del rango.

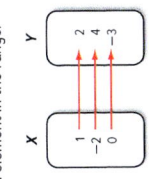

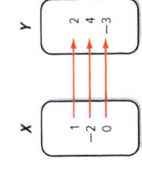

| X | Y |
|---|---|
| 1 | 2 |
| −2 | 4 |
| 0 | −3 |

**marginal frequency** (p. 691) In a two-way frequency table, marginal frequencies are totals for each row or column.

**frecuencia marginal** En una tabla de frecuencias de doble entrada, las frecuencias marginales son los totales de cada fila o columna.

**matrix** (p. 374) Any rectangular arrangement of numbers in rows and columns.

**matriz** Disposición rectangular de números colocados en filas y columnas.

**maximum** (p. 559) The highest point on the graph of a curve.

**máximo** El punto más alto en la gráfica de una curva.

**mean** (p. 651) The sum of numbers in a set of data divided by the number of items in the data set.

**media** La suma de los números en un conjunto de datos dividida entre el numero total de articulos.

**measure of spread** (p. 665) A measure of the variation in a data set.

**medida de dispersión** Una medida de la variación en un conjunto de datos.

**measures of central tendency** (p. 651) Numbers or pieces of data that can represent the whole set of data.

**medidas de tendencia central** Números o fragmentos que pueden representar el conjunto de datos total de datos.

**measures of position** (p. 665) Measures that compare the position of a value relative to other values in a set.

**medidas de la posición** Las medidas que comparar la posición de un valor relativo a otros valores de un conjunto.

**measures of variation** (p. 665) Used to describe the distribution of statistical data.

**medidas de variación** Números que se usan para describir la distribución o separación de un conjunto de datos.

**median** (p. 651) The middle number in a set of data when the data are arranged in numerical order. If the data set has an even number, the median is the mean of the two middle numbers.

**mediana** El número central de conjunto de datos, una vezque los datos han sido ordenados numéricamente. Si hay un número par de datos, la mediana es el promedio de los datos centrales.

**median fit line** (p. 262) A type of best fit line that is calculated using the medians of the coordinates of the data points.

**línea apta del punto medio** Tipo de mejor cupo la linea se calcula que usando los puntos medios de los coordenadas de los puntos de referencias.

**metric** (p. 33) A rule for assigning a number to some characteristic or attribute.

**métrico** Una regla para asignar un número a alguna característica o atribuye.

**minimum** (p. 559) The lowest point on the graph of a curve.

**mínimo** El punto más bajo en la gráfica de una curva.

**mode** (p. 651) The number(s) that appear most often in a set of data.

**moda** El número(s) que aparece más frecuencia en un conjunto de datos.

**monomial** (p. 395) A number, a variable, or a product of a number and one or more variables.

**monomio** Número, variable o producto de un número por una o más variables.

**multiplicative identity** (p. 17) For any number $a$, $a \cdot 1 = 1 \cdot a = a$.

**identidad de la multiplicación** Para cualquier número $a \cdot 1 = 1 \cdot a = a$.

**multiplicative inverses** (pp. P18, 17) Two numbers with a product of 1.

**inversos multiplicativos** Dos números cuyo producto es igual a 1.

**multi-step equation** (p. 95) Equations with more than one operation.

**multi-step equation** Eventos que no pueden ocurrir simultáneamente.

## N

**nth root** (p. 411) If $a^n = b$ for a positive integer $n$, then $a$ is an nth root of $b$.

**raíz enésima** Si $a^n = b$ para cualquier entero positivo $n$, entonces a se llama una raíz enésima de $b$.

**natural numbers** (p. P7) The set {1, 2, 3, ...}.

**números naturales** El conjunto {1, 2, 3, ...}.

**negative** (p. 59) A function is negative on a portion of its domain where its graph lies below the x-axis.

**negativo** Una función es negativa en una porción de su dominio donde su gráfico está debajo del eje-x.

**negative correlation** (p. 247) In a scatter plot, as x increases, y decreases.

**correlación negativa** En una gráfica de dispersión, a medida que x aumenta, y disminuye.

**negative exponent** (p. 404) For any real number $a \neq 0$ and any integer $n$, $a^{-n} = \frac{1}{a^n}$ and $\frac{1}{a^{-n}} = a^n$.

**exponente negativo** Para números reales, si $a \neq 0$, y cualquier número entero $n$, entonces $a^{-n} = \frac{1}{a^n}$ and $\frac{1}{a^{-n}} = a^n$.

**negative number** (p. P7) Any value less than zero.

**número negativo** Cualquier valor menor que cero.

**nonlinear function** (p. 52) A function with a graph that is not a straight line.

**función no lineal** Una función con un gráfica que no es una línea recta.

**number theory** (p. 96) The study of numbers and the relationships between them.

**teoría del número** El estudio de números y de las relaciones entre ellas.

## O

**odds** (p. P35) The ratio of the probability of the success of an event to the probability of its complement.

**probabilidades** El cociente de la probabilidad del éxito de un acontecimiento a la probabilidad de su complemento.

**open half-plane** (p. 321) The solution of a linear inequality that does not include the boundary line.

**abra el mitad plano** La solución de una desigualdad linear que no incluya la línea de límite.

**opposites** (p. P11) Two numbers with the same absolute value but different signs.

**opuestos** Dos números que tienen el mismo valor absoluto, pero que tienen distintos signos.

**order of magnitude** (p. 405) The order of magnitude of a quantity is the number rounded to the nearest power of 10.

**orden de magnitud de una cantidad** Un número redondeado a la potencia más cercana de 10.

**order of operations** (p. 10)
1. Evaluate expressions inside grouping symbols.
2. Evaluate all powers.
3. Do all multiplications and/or divisions from left to right.
4. Do all additions and/or subtractions from left to right.

**orden de las operaciones**
1. Evalúa las expresiones dentro de los símbolos de agrupamiento.
2. Evalúa todas las potencias.
3. Multiplica o divide de izquierda a derecha.
4. Suma o resta de izquierda a derecha.

**ordered pair** (p. 42) A set of numbers or coordinates used to locate any point on a coordinate plane, written in the form (x, y).

**par ordenado** Un par de números que se usa para ubicar cualquier punto de un plano de coordenadas y que se escribe en la forma (x, y).

# Glossary/Glosario

**origin** (p. 42) The point where the two axes intersect at their zero points.

**outliers** (p. 666) Data that are more than 1.5 times the interquartile range beyond the quartiles.

**origen** Punto donde se intersecan los dos ejes en sus puntos cero.

**valores atípicos** Datos que distan de los cuartiles más de 1.5 veces la amplitude intercuartílica.

## P

**parabola** (p. 559) The graph of a quadratic function.

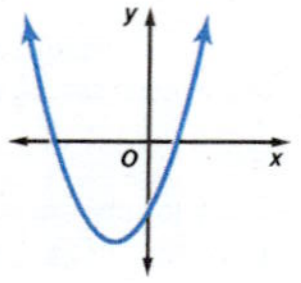

**parábola** La gráfica de una función cuadrática.

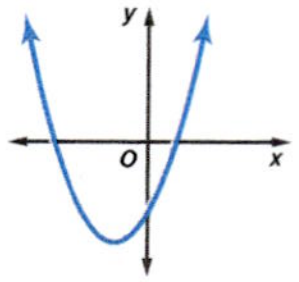

**parallel lines** (p. 239) Lines in the same plane that do not intersect and either have the same slope or are vertical lines.

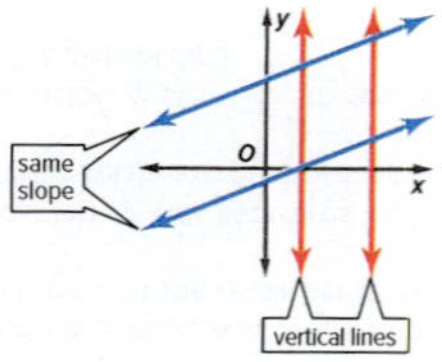

**rectas paralelas** Rectas en el mismo plano que no se intersecan y que tienen pendientes iguales, o las mismas rectas verticales.

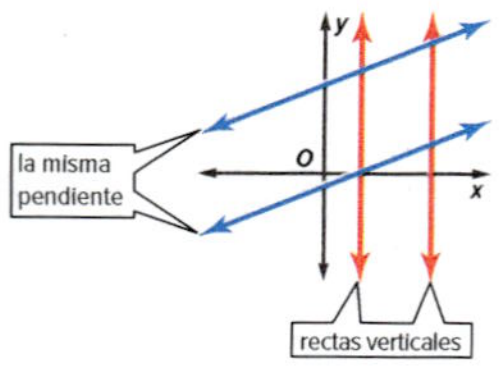

**parent function** (p. 151) The simplest of functions in a family.

**función básica** La función más fundamental de una familia de funciones.

**percent** (p. P20) A ratio that compares a number to 100.

**porcentaje** Razón que compara un número con 100.

**percent of change** (p. 457) When an increase or decrease is expressed as a percent.

**porcentaje de cambio** Cuando un aumento o disminución se escribe como un tanto por ciento.

**percent proportion** (p. P20)
$$\frac{part}{whole} = \frac{percent}{100} \text{ or } \frac{a}{b} = \frac{p}{100}$$

**proporción porcentual**
$$\frac{parte}{todo} = \frac{por\ ciento}{100} \text{ or } \frac{a}{b} = \frac{p}{100}$$

**percentile** (p. 653) A measure that tells what percent of a data set are below a reported value in the set.

**percentil** Una medida que indica qué porcentaje de un conjunto de datos está por debajo de un valor informado del conjunto.

**perfect square** (p. P7) A number with a square root that is a rational number.

**cuadrado perfecto** Número cuya raíz cuadrada es un número racional.

**perfect square trinomial** (p. 540) A trinomial that is the square of a binomial.
$(a + b)^2 = (a + b)(a + b) = a^2 + 2ab + b^2$ or
$(a - b)^2 = (a - b)(a - b) = a^2 - 2ab + b^2$

**trinomio cuadrado perfecto** Un trinomio que es el cuadrado de un binomio.
$(a + b)^2 = (a + b)(a + b) = a^2 + 2ab + b^2$ or
$(a - b)^2 = (a - b)(a - b) = a^2 - 2ab + b^2$

**perimeter** (p. P23) The distance around a geometric figure.

**perímetro** Longitud alrededor una figura geométrica.

**perpendicular lines** (p. 240) Lines that intersect to form a right angle.

**recta perpendicular** Recta que se intersecta formando un ángulo recto.

**piecewise-defined function** (p. 198) A function that is written using two or more expressions.

**función definida por partes** Función que se escribe usando dos o más expresiones.

**piecewise-linear function** (p. 197) A function written using two or more linear expressions.

**función lineal por partes** Función que se escribe usando dos o más expresiones lineal.

**point-slope form** (p. 233) An equation of the form $y - y_1 = m(x - x_1)$, where $m$ is the slope and $(x_1, y_1)$ is a given point on a nonvertical line.

**forma punto-pendiente** Ecuación de la forma $y - y_1 = m(x - x_1)$, donde $m$ es la pendiente y $(x_1, y_1)$ es un punto dado de una recta no vertical.

**polynomial** (p. 491) A monomial or sum of monomials.

**polinomio** Un monomio o la suma de monomios.

**positive** (p. 59) A function is positive on a portion of its domain where its graph lies above the x-axis.

**positiva** Una función es positiva en una porción de su dominio donde su gráfico está encima del eje-x.

**positive correlation** (p. 247) In a scatter plot, as $x$ increases, $y$ increases.

**correlación positiva** En una gráfica de dispersión, a medida que $x$ aumenta, $y$ aumenta.

**positive number** (p. P7) Any value that is greater than zero.

**números positivos** Cualquier valor mayor que cero.

**power** (p. P7) An expression of the form $x^n$, read $x$ to the $n$th power.

**potencia** Una expresión de la forma $x^n$, se lee $x$ a la enésima potencia.

**prime polynomial** (p. 535) A polynomial that cannot be written as a product of two polynomials with integral coefficients.

**polinomio primo** Polinomio que no puede escribirse como producto de dos polinomios con coeficientes enteros.

**principal square root** (p. P7) The nonnegative square root of a number.

**raíz cuadrada principal** La raíz cuadrada no negativa de un número.

**probability** (p. P33) The ratio of the number of favorable equally likely outcomes to the number of possible equally likely outcomes.

**probabilidad** La razón del número de maneras en que puede ocurrir el evento al número de resultados posibles.

**product** (p. 5) In an algebraic expression, the result of quantities being multiplied is called the product.

**producto** En una expresión algebraica, se llama producto al resultado de las cantidades que se multiplican.

**proof** (p. 21) A proof is a logical argument in which each statement is supported by a true statement.

**demostración** Una demostración es un argumento lógico en el que cada enunciado se reafirma con un enunciado verdadero.

**proportion** (p. 15) An equation of the form $\frac{a}{b} = \frac{c}{d}$, where $b, d \neq 0$, stating that two ratios are equivalent.

**proporción** Ecuación de la forma $\frac{a}{b} = \frac{c}{d}$, donde $b, d \neq 0$, que afirma la equivalencia de dos razones.

**Pythagorean Theorem** If $a$ and $b$ are the measures of the legs of a right triangle and $c$ is the measure of the hypotenuse, then $c^2 = a^2 + b^2$.

**Teorema de Pitágoras** Si $a$ y $b$ son las longitudes de los catetos de un triángulo rectángulo y si $c$ es la longitud de la hipotenusa, entonces $c^2 = a^2 + b^2$.

connectED.mcgraw-hill.com     R93

# Glossary/Glosario

## Q

**quadratic equation** (p. 571) An equation of the form $ax^2 + bx + c = 0$, where $a \neq 0$.

**ecuación cuadrática** Ecuación de la forma $ax^2 + bx + c = 0$, donde $a \neq 0$.

**quadratic expression** (p. 507) An expression in one variable with a degree of 2 written in the form $ax^2 + bx + c$.

**expresión cuadrática** Una expresión en una variable con un grado de 2, escritos en la forma $ax^2 + bx + c$.

**Quadratic Formula** (p. 606) The solutions of a quadratic equation in the form $ax^2 + bx + c = 0$, where $a \neq 0$, are given by the formula
$$x = \frac{-b \pm \sqrt{b^2 - 4ac}}{2a}$$

**Fórmula cuadrática** Las soluciones de una ecuación cuadrática de la forma $ax^2 + bx + c = 0$, donde $a \neq 0$, vienen dadas por la fórmula
$$x = \frac{-b \pm \sqrt{b^2 - 4ac}}{2a}$$

**quadratic function** (p. 559) An equation of the form $y = ax^2 + bx + c$, where $a \neq 0$.

**función cuadrática** Función de la forma $y = ax^2 + bx + c$, donde $a \neq 0$.

**qualitative data** (p. 651) Data that can be categorized, but not measured.

**datos cualitativos** Los datos que se pueden categorizar, pero no se pueden medir.

**quantitative data** (p. 651) Data that can be measured.

**datos cuantitativos** Los datos que se pueden medir.

**quartile** (p. 665) The values that divide a set of data into four equal parts.

**cuartil** Valores que dividen en conjunto de datos en cuarto partes iguales.

## R

**radical expression** (p. 419) An expression that contains a square root.

**expresión radical** Expresión que contiene una raíz cuadrada.

**range** (pp. 42, 665) 1. The set of second numbers of the ordered pairs in a relation. 2. The difference between the greatest and least data values.

**rango** 1. Conjunto de los segundos números de los pares ordenados de una relación. 2. La diferencia entre los valores de datos más grande o menos.

**rate** (p. 117) The ratio of two measurements having different units of measure.

**tasa** Razón de dos medidas que tienen distintas unidades de medida.

**rate of change** (p. 160) How a quantity is changing with respect to a change in another quantity.

**tasa de cambio** Cómo cambia una cantidad con respecto a un cambio en otra cantidad.

**ratio** (p. 115) A comparison of two numbers by division.

**razón** Comparación de dos números mediante división.

**rational exponent** (p. 410) For any positive real number $b$ and any integers $m$ and $n > 1$, $b^{\frac{m}{n}} = (\sqrt[n]{b})^m = \sqrt[n]{b^m}$. $\frac{m}{n}$ is a rational exponent.

**exponent racional** Para cualquier número real no nulo $b$ cualquier entero $m$ y $n > 1$, $b^{\frac{m}{n}} = (\sqrt[n]{b})^m = \sqrt[n]{b^m}$. $\frac{m}{n}$ es un exponent racional.

**rational numbers** (p. P7) The set of numbers expressed in the form of a fraction $\frac{a}{b}$, where $a$ and $b$ are integers and $b \neq 0$.

**números racionales** Conjunto de los números que pueden escribirse en forma de fracción $\frac{a}{b}$, donde $a$ y $b$ son enteros y $b \neq 0$.

**rationalizing the denominator** (p. 421) A method used to eliminate radicals from the denominator of a fraction.

**racionalizar el denominador** Método que se usa para eliminar radicales del denominador de una fracción.

**real numbers** (p. P7) The set of rational numbers and the set of irrational numbers together.

**números reales** El conjunto de los números racionales junto con el conjunto de los números irracionales.

**reciprocal** (pp. P18, 17) The multiplicative inverse of a number.

**recíproco** Inverso multiplicativo de un número.

**recursive formula** (p. 469) Each term is formulated from one or more previous terms.

**fórmula recursiva** Cada término proviene de uno o más términos anteriores.

**reflection** (p. 184, 574) A transformation where a figure, line, or curve, is flipped across a line.

**reflexión** Transformación en que cadapunto de una figura se aplica a través de una recta de simetría a su imagen correspondiente.

**relation** (p. 42) A set of ordered pairs.

**relación** Conjunto de pares ordenados.

**relative error** (p. 313) The ratio of the absolute error to the expected measure.

**error relativo** La razón del error absoluto a la medida esperada.

**relative frequency** (p. 690) The number of times an event occurred compared to the whole.

**frecuencia relativa** Número de veces que aparece un resultado en un experimento probabilístico.

**relative maximum** (p. 59) A point on graph is a relative maximum if no other nearby points have a greater $y$-coordinate.

**máximo relativo** Un punto de una gráfica es un máximo relativo si no hay puntos cercanos que tengan una coordenada $y$ mayor.

**relative minimum** (p. 59) A point on graph is a relative minimum if no other nearby points have a lesser $y$-coordinate.

**mínimo relativo** Un punto de una gráfica es un mínimo relativo si no hay puntos cercanos que tengan una coordenada $y$ menor.

**residual** (p. 260) The difference between an observed $y$-value and its predicted $y$-value on a regression line.

**residual** Diferencia entre el valor observado de $y$ y el valor redicho de $y$ en la recta de regresión.

**root** (p. 151) The solutions of a quadratic equation.

**raíces** Las soluciones de una ecuación cuadrática.

**row reduction** (p. 375) The process of performing elementary row operations on an augmented matrix to solve a system.

**reducción de la fila** El proceso de realizar operaciones elementales de la fila en una matriz aumentada para solucionar un sistema.

## S

**sample space** (p. P33) The list of all possible outcomes.

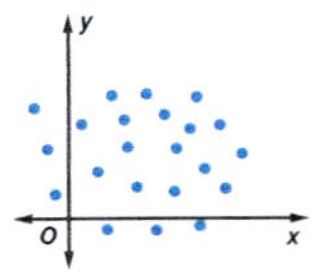

**espacio muestral** Lista de todos los resultados posibles.

**scale** (p. 118) The relationship between the measurements on a drawing or model and the measurements of the real object.

**escala** Relación entre las medidas de un dibujo o modelo y las medidas de la figura verdadera.

**scale model** (p. 118) A model used to represent an object that is too large or too small to be built at actual size.

**modelo a escala** Modelo que se usa para representar un figura que es demasiado grande o pequeña como para ser construida de tamaño natural.

**scatter plot** (p. 247) A scatter plot shows the relationship between a set of data with two variables, graphed as ordered pairs on a coordinate plane.

**gráfica de dispersión** Es un diagrama que muestra la relación entre un conjunto de datos con dos variables, graficados como pares ordenados en un plano coordenadas.

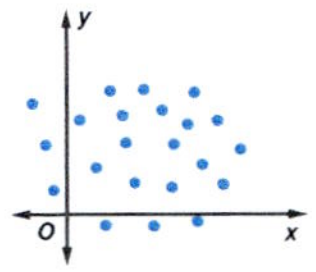

# Glossary/Glosario

**sequence** (p. 190) A set of numbers in a specific order.

**set-builder notation** (p. 152) A concise way of writing a solution set. For example, {t | t < 17} represents the set of all numbers t such that t is less than 17.

**simplest form** (p. 25) An expression is in simplest form when it is replaced by an equivalent expression having no like terms or parentheses.

**slope** (p. 162) The ratio of the change in the y-coordinates (rise) to the corresponding change in the x-coordinates (run) as you move from one point to another along a line.

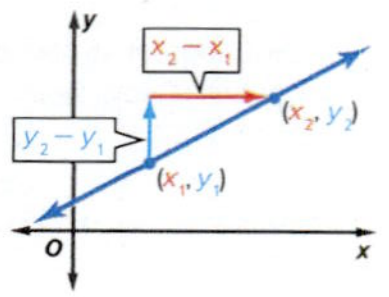

**slope-intercept form** (p. 171) An equation of the form $y = mx + b$, where $m$ is the slope and $b$ is the y-intercept.

**solve an equation** (p. 87) The process of finding all values of the variable that make the equation a true statement.

**square root** (p. P7) One of two equal factors of a number.

**standard deviation** (p. 667) The square root of the variance.

**standard form of a linear equation** (pp. 143, 232) The standard form of a linear equation is $Ax + By = C$, where $A \geq 0$, $A$ and $B$ are not both zero, and $A$, $B$, and $C$ are integers with a greatest common factor of 1.

**standard form of a polynomial** (p. 492) A polynomial that is written with the terms in order from greatest degree to least degree.

**step function** (p. 197) A function with a graph that is a series of horizontal line segments.

**substitution** (p. 348) Use algebraic methods to find an exact solution of a system of equations.

**surface area** (p. P31) The sum of the areas of all the surfaces of a three-dimensional figure.

**system of equations** (p. 339) A set of equations with the same variables.

---

**sucesión** Conjunto de números en un orden específico.

**notación de construcción de conjuntos** Manera concisa de escribir un conjunto solución. Por ejemplo, {t | t < 17} representa el conjunto de todos los números t que son menores o iguales que 17.

**forma reducida** Una expresión está reducida cuando se puede sustituir por una expresión equivalente que no tiene ni términos semejantes ni paréntesis.

**pendiente** Razón del cambio en la coordenada y (elevación) al cambio correspondiente en la coordenada x (desplazamiento) a medida que uno se mueve de un punto a otro en una recta.

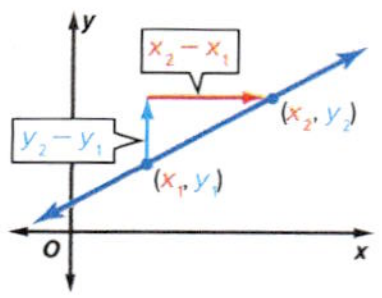

**forma pendiente-intersección** Ecuación de la forma $y = mx + b$, donde $m$ es la pendiente y $b$ es la intersección y.

**resolver una ecuación** Proceso en que se hallan todos los valores de la variable que hacen verdadera la ecuación.

**raíz cuadrada** Uno de dos factores iguales de un número.

**desviación típica** Calculada como la raíz cuadrada de la varianza.

**forma estándar** La forma estándar de una ecuación lineal es $Ax + By = C$, donde $A \geq 0$, ni $A$ ni $B$ son ambos cero, y $A$, $B$, y $C$ son enteros cuyo máximo común divisor es 1.

**forma de estándar de un polinomio** Un polinomio que se escribe con los términos en orden del grado más grande a menos grado.

**función escalonada** Función cuya gráfica es una serie de segmentos de recta.

**sustitución** Usa métodos algebraicos para hallar una solución exacta a un sistema de ecuaciones.

**área de superficie** Suma de las áreas de todas las superficies (caras) de una figura tridimensional.

**sistema de ecuaciones** Conjunto de ecuaciones con las mismas variables.

---

**system of inequalities** (p. 376) A set of two or more inequalities with the same variables.

**term** (p. 5) A number, a variable, or a product or quotient of numbers and variables.

**terms of a sequence** (p. 190) The numbers in a sequence.

**transformation** (pp. 181, 571) A movement of a graph on the coordinate plane.

**translation** (pp. 181, 571) A transformation where a figure is slid from one position to another without being turned.

**tree diagram** (p. P34) A diagram used to show the total number of possible outcomes.

**trinomials** (p. 491) The sum of three monomials.

**two-way frequency table** (p. 688) A table used to organize the frequency of data in two categories.

**two-way relative frequency table** (p. 690) A table used to organize the relative frequency of data in two categories; frequencies may be relative to row totals, column totals, or the total count.

**union** (p. 310) The graph of a compound inequality containing or; the solution is a solution of either inequality, not necessarily both.

**unit analysis** (p. 124) The process of including units of measurement when computing.

**unit rate** (p. 117) A ratio of two quantities, the second of which is one unit.

**upper quartile** (p. 665) The median of the upper half of a set of data.

**variable** (pp. 5, 651) 1. Symbols used to represent unspecified numbers or values. 2. a characteristic of a group of people or objects that can assume different values.

**variance** (p. 667) The mean of the squares of the deviations from the arithmetic mean.

---

**sistema de desigualdades** Conjunto de dos o más desigualdades con las mismas variables.

**término** Número, variable o producto, o cociente de números y variables.

**términos** Los números de una sucesión.

**transformación** Desplazamiento de una gráfica en un plano de coordenadas.

**translación** Transformación en que una figura se desliza sin girar, de una posición a otra.

**diagrama de árbol** Diagrama que se usa para mostrar el número total de resultados posibles.

**trinomios** Suma de tres monomios.

**tabla de frecuencias de doble entrada** Una tabla que se usa para organizar las frecuencias de los datos en dos categorías.

**tabla de frecuencias relativas de doble entrada** Una tabla que se usa para organizar las frecuencias relativas de los datos en dos categorías; las frecuencias pueden ser relativas a los totales de las filas, a los totales de las columnas o al total general.

**unión** Gráfica de una desigualdad compuesta que contiene la palabra o; la solución es el conjunto de soluciones de por lo menos una de las desigualdades, no necesariamente ambas.

**análisis de la unidad** Proceso de incluir unidades de medida al computar.

**tasa unitaria** Tasa reducida que tiene denominador igual a 1.

**cuartil superior** Mediana de la mitad superior de un conjunto de datos.

**variable** 1. Símbolos que se usan para representar números o valores no especificados. 2. una característica de un grupo de personas u objetos que pueden asumir valores diferentes.

**varianza** Media de los cuadrados de las desviaciones de la media aritmética.

connectED.mcgraw-hill.com

## Glossary/Glosario

**variation** (p. 665) A measure of how spread out or scattered a data set is.

**vertex** (pp. 205, 559) The maximum or minimum point of a parabola.

**vertex form** (p. 576) A quadratic function in the form $f(x) = a(x - h)^2 + k$.

**vertical line test** (p. 51) If any vertical line passes through no more than one point of the graph of a relation, then the relation is a function.

**volume** (p. P29) The measure of space occupied by a solid region.

### W

**whole numbers** (p. P7) The set {0, 1, 2, 3, …}.

### X

**x-axis** (p. 42) The horizontal number line on a coordinate plane.

**x-coordinate** (p. 42) The first number in an ordered pair.

**x-intercept** (pp. 58, 144) The x-coordinate of a point where a graph crosses the x-axis.

### Y

**y-axis** (p. 42) The vertical number line on a coordinate plane.

**y-coordinate** (p. 42) The second number in an ordered pair.

---

**variación** Una medida de la dispersión de los datos de un conjunto de datos.

**vértice** Punto máximo o mínimo de una parábola.

**forma de vértice** Una función cuadrática de la forma $f(x) = a(x - h)^2 + k$.

**prueba de la recta vertical** Si cualquier recta vertical pasa por un solo punto de la gráfica de una relación, entonces la relación es una función.

**volumen** Medida del espacio que ocupa un sólido.

**números enteros** El conjunto {0, 1, 2, 3, …}.

**eje x** Recta numérica horizontal que forma parte de un plano de coordenadas.

**coordenada x** El primer número de un par ordenado.

**intersección x** La coordenada x de un punto donde la gráfica corta el eje x.

**eje y** Recta numérica vertical que forma parte de un plano de coordenadas.

**coordenada y** El segundo número de un par ordenado.

---

**y-intercept** (pp. 58, 144) The y-coordinate of a point where a graph crosses the y-axis.

### Z

**zero exponent** (p. 403) For any nonzero number a, $a^0 = 1$. Any nonzero number raised to the zero power is equal to 1.

**zeros** (p. 151) The x-intercepts of the graph of a function; the values of x for which $f(x) = 0$.

---

**intersección y** La coordenada y de un punto donde la gráfica corta al eje de y.

**exponente cero** Para cualquier número distinto a cero a, $a^0 = 1$. Cualquier número distinto a cero levantado al potente cero es igual a 1.

**cero** Las intersecciones x de la gráfica de una función; los puntos x para los que $f(x) = 0$.

# Index

21st Century Skills, 1, 75, 137, 221, 285, 335, 391, 485, 555, 647

## A

**ALEKS,** viii, 2, 7, 13, 20, 27, 32, 45, 54, 61, 65, 76, 82, 90, 98, 104, 110, 114, 119, 125, 128, 138, 139, 147, 154, 165, 169, 175, 185, 194, 200, 208, 212, 222, 228, 236, 243, 246, 250, 256, 262, 271, 276, 286, 292, 299, 305, 308, 312, 318, 324, 328, 336, 342, 351, 357, 364, 367, 371, 378, 392, 399, 407, 414, 418, 424, 433, 452, 465, 472, 475, 486, 494, 501, 509, 515, 518, 523, 536, 546, 556, 565, 576, 584, 592, 600, 605, 611, 619, 625, 633, 637, 648, 654, 662, 668, 672, 677, 684, 698

**Animations.** See Go Online!

**Anticipation Guide,** 68, 132, 214, 279, 329, 477, 549, 639, 701

**Apply Math to the Real World,** 2, 76, 138, 222, 286, 336, 392, 486, 556, 648

**Assessment**
  Crystal Ball, P12, 21, 55, 120, 252, 293, 319, 366, 426, 434, 502, 696
  Diagnostic Assessment, 4, 9, 15, 22, 29, 32, 39, 41, 48, 56, 64, 72–75, 78, 84, 93, 100, 106, 113, 121, 127, 134–137, 140, 150, 156, 168, 169, 178, 189, 196, 203, 211, 220–221, 224, 231, 238–239, 245, 246, 253, 258, 266, 274, 284–285, 288, 294, 301, 307, 308, 314, 320, 326, 332–335, 338, 345, 353, 360, 366, 367, 373, 380, 388–391, 394, 401, 409, 417, 418, 427, 435, 447, 454, 467, 474, 488, 490, 497, 503, 511, 517, 526, 538, 545, 554–555, 558, 569, 579, 586, 595, 602, 605, 613, 621, 627, 636, 646–647, 650, 656, 664, 671, 672, 679, 687, 697, 704, 707
  Extending the Concept, 86, 347, 505, 528, 530
  Formative Assessment, P6, P10, P12, P16, P19, P22, P25, P28, P30, P32, P36, 7, 13, 19, 26, 30, 38, 45, 53, 57, 57, 61, 75, 82, 86, 86, 90, 94, 94, 98, 104, 109, 118, 124, 137, 141, 147, 154, 157, 157, 159, 159, 165, 170, 170, 175, 179, 179, 185, 194, 200, 204, 204, 208, 221, 228, 235, 242, 250, 256, 262, 271, 275, 275, 285, 292, 295, 295, 299, 305, 312, 315, 315, 324, 327, 327, 335, 342, 347, 347, 350, 371, 375, 375, 381, 381, 406, 428, 428, 436, 456, 468, 490, 494, 505, 514, 519, 528, 530, 570, 604, 629
  From Concrete to Abstract, 57, 86, 94, 141, 157, 159, 170, 204, 275, 295, 315, 327, 347, 375, 381, 456, 468, 490, 505, 519, 528, 530, 570, 604, 629
  Name the Math, P6, P22, P30, 14, 28, 92, 105, 177, 195, 313, 353, 380, 466, 510
  Summative Assessment, 9, 15, 22, 29, 32, 39, 41, 48, 56, 64, 71, 74–75, 84, 93, 100, 106, 113, 121, 127, 133, 136–137, 150, 156, 168, 169, 178, 189, 196, 203, 211, 217, 220–221, 231, 238, 245, 246, 253, 258, 266, 274, 281, 284–285, 294, 301, 307, 308, 314, 320, 326, 331, 334–335, 345, 353, 360, 366, 367, 373, 380, 387, 390–391, 401, 409, 417, 418, 427, 435, 447, 454, 467, 474,

## A

**Absolute error,** 313
**Absolute value,** P11
  adding, P11, P15
  equations, 107–112, 128, 131
  functions, 205–210, 212
  inequalities, 316–319, 328, 330
  radical expressions, 420
**Accuracy,** 34–36, 65
**Activities.** See Algebra Labs; Graphing Technology Labs
**Addition**
  associative property, 18, 26, 65
  commutative property, 18, 26, 65
  data transformations using, 681
  of decimals, P15
  elimination using, 354–355, 357–359, 368, 384
  equation models for, 85
  of fractions, P14
  of functions, 630–631, 637, 642
  of integers, P11
  of polynomials, 489–490, 492, 546, 548
  of radical expressions, 422, 424
  of rational and irrational numbers, 429
  of rational numbers, P14–P15, 31, 429
  solving equations using, 87
  solving inequalities by, 289–290, 329
  squares of sums and differences, 512–513
**Addition Property,** 26
  of equality, 87–88, 128
  of inequality, 289
**Additive identity,** 16
**Additive inverse,** P11, 16, 493, 521
**Ahmes,** 81
**Algebraic expressions**
  evaluating, 5–8, 11–12
  simplifying, 25–26
  from verbal expressions, 5–8, 26
**Algebra Labs**
  Adding and Subtracting Polynomials, 489–490
  Analyzing Linear Graphs, 141–142

Average Rate of Change of Exponential Functions, 468
Drawing Inverses, 275
Exponential Growth Patterns, 436–437
Factoring Trinomials, 529–530
Factoring Using the Distributive Property, 519
Finding the Maximum or Minimum Value, 603–604
Linear Growth Patterns, 179–180
Multiplying Polynomials, 504–505
Operations with Rational Numbers, 30–31, 428–429
Proving the Elimination Method, 527–528
Rate of Change of a Linear Function, 159
Rate of Change of a Quadratic Function, 570
Rational Numbers, 30–31
Reading Compound Statements, 315
Solving Equations, 85–86
Solving Inequalities, 295
Solving Multi-Step Equations, 94
Using Matrices to Solve Systems of Equations, 374–375
**Algebra tiles**
  adding and subtracting polynomials, 489
  algebraic expressions, 21
  factoring trinomials, 529–530
  multiplying polynomials, 504–505
  to solve equations, 85–86, 94, 102
  to solve inequalities, 295
**Animations,** 88, 96, 297, 492, 514
**Area,** P26
  of a circle, P27, 397
  estimating, P27
  of a parallelogram, P26–P27
  of a rectangle, P26, 25, 423
  of a rhombus, 423
  of a square, P26
  surface area of a cube, 398
  surface area of a cylinder, P31, 19
  surface area of a rectangular prism, P29
  of a trapezoid, 132
  of a triangle, P26–P27, 210, 399, 424, 591–592
**Arithmetic sequences,** 190–195
  identifying, 191, 462–463
  as linear functions, 193, 215
  nth term of, 191–192, 212
  writing recursive formulas for, 470–471
**Association,** 247
**Associative Property,** 18, 26, 65
**Asymptote,** 430–431, 438–439
**Augmented matrix,** 374
**Average rate of change,** 468
**Averages.** See Means
**Axis of symmetry,** 559–561, 603–604, 637

## B

**Bar graph,** 659, 699
**Bases,** 5, 395–396
**Best-fit lines,** 259–262
**Binomials,** 491. See also Polynomials
  additive inverse, 521
  conjugates, 421
  factoring, 519, 540
  multiplication of, 423, 504–505, 506–510
  squares of sums and differences, 512–513
**Bivariate data,** 247
**Blood pressure equation,** 606
**Body mass index,** 123
**Boundaries,** 321–322
**Box-and-whisker plot,** 650, 674, 676, 683, 701
**BrainPOP® Animation,** 96

## C

**Calculator.** See Graphing calculators
**Careers.** See Real-World Careers
**Causation,** 254–257, 276, 279
**Celsius,** 82, 134, 270

**Center**, of a circle, P24

**Challenge.** *See* Higher-Order Thinking Skills

**Chapter 0**
adding and subtracting rational numbers, P13–P16
area, P26–P28
multiplying and dividing rational numbers, P17–P19
operations with integers, P11–P12
percent proportion, P20–P22
perimeter, P23–P25
plan for problem solving, P5–P6
real numbers, P7–P10
representing data, P41–P44
simple probability and odds, P33–P36
surface area, P31–P32
volume, P29–P30

**Check Your Understanding**, 12, 19, 27, 36–37, 45, 53, 61, 81, 90, 97, 104, 109, 118, 124, 147, 165, 174–175, 185, 194, 200, 208, 228, 235, 242, 250, 256, 263, 271, 292, 299, 304, 311, 318, 324, 342, 351, 357, 363, 371, 378, 398, 406, 414, 424, 433, 443, 451–452, 461, 465, 472, 494, 500, 509, 514, 523, 536, 543, 565, 577, 584, 592–593, 599, 611, 618, 625, 633, 654, 662, 668–669, 677, 684, 693

**Circle**
area, P27, 397
center, P24
circumference, P24
diameter, P24
radius, P24

**Circumference**, P24

**Closed half-planes**, 321

**Closure Property**, 21

**Coefficients**, 26
correlation, 259
of determination, 628
of exponents, 396
in inequalities, 302
leading, 492

**Common denominator, least**, P13–P14

**Common difference (d)**, 190, 192–193

**Common ratio**, 462–464

**Commutative Property**, 18, 26, 65

**Complement**, P33

**Complete graphs**, 157–158

**Completing the square**, 596–601, 640

**Complex numbers**, 581

**Compound inequalities**, 309, 328, 330

**Compound interest**, 451, 458–460, 468

**Compound statements**, 315

**Concept Check**, 65, 78, 128, 140, 212, 224, 276, 288, 328, 338, 394, 488, 546, 558, 637, 650, 698

**Concept Summary**
absolute value functions of the form $g(x) = a|x - h| + k$, 200
consecutive integers, 96–97
factoring methods, 542
operations with radical expressions, 424
parallel and perpendicular lines, 242
percentiles, 653
phrases for inequalities, 291
piecewise and step functions, 200
possible solutions, 339
properties of numbers, 26
representations of a function, 51
solving quadratic equations, 610
solving systems of equations, 368
transformations of quadratic functions, 576
writing equations, 234

**Conditional relative frequencies**, 692, 698

**Conjugate**, 421

**ConnectED.** *See* Go Online!

**Consecutive integers**, 96–97

**Consistent systems of equations**, 339

**Constant functions**, 142, 172

**Constants**
adding to $f(x)$, 181, 438
definition of, 143, 395
in dilation, 183–184
in linear equations, 143
multiplying $f(x)$ by, 183, 440, 442
multiplying $x$ by, 441–442

**Constraints**, 227, 323

**Construct Arguments**, 21

**Continuous function**, 50

**Contraction.** *See* Dilation

**Coordinate of a point**, P8

**Coordinate system**, 42

**Correlation coefficient**, 259

**Correlations**, 247
causation and, 254–257, 276, 279
coefficient, 259
lines of fit, 248–249, 276, 278
negative, 247
none, 247
positive, 247

**Critique Arguments**, 63, 105, 167, 188, 210, 265, 365, 464, 620, 635

**Cross products**, 116–117

**Cube, surface area of**, 398

**Cube root**, 411

**Cumulative frequency**, 659, 700

**Cumulative Review**, 74–75, 136–137, 220–221, 284–285, 334–335, 390–391, 484–485, 554–555, 646–647, 706–707

**Curve fitting**, 628–329

**Cylinders**
surface area, P31, 19
volume, P29, 7

**D**

**Data.** *See also* Graphs; Statistics
associations between, 693
bivariate, 247
categorical, 690–693, 698, 702
comparing sets of, 668, 682–684, 701
correlation and causation in, 254–256, 276, 279
discrete *vs.* continuous, 661
distributions of, 673–676, 701
five-number summary, 665–666, 676
frequency table, 659
interquartile range, 666
means of, 116, 651–653, 666, 675, 681–684, 699
measures of spread, 665–668
medians of, 701, 651–653, 666, 675, 681–682
modeling real-world, 173–174, 248
modeling with linear functions, 248
modes of, 651, 681–682, 699
outliers, 666, 675
qualitative, 651
quantitative, 651
quartiles, 665
range of, 42, 50, 141, 561, 665–666, 681–682

481, 484–485, 490, 497, 503, 511, 517, 526, 538, 545, 551, 554–555, 569, 579, 586, 595, 602, 605, 613, 621, 627, 636, 643, 646–647, 656, 664, 671, 672, 679, 687, 697, 703, 706–707

Ticket Out the Door, P10, P19, P25, P36, 8, 40, 47, 62, 83, 126, 149, 167, 188, 202, 210, 237, 244, 273, 300, 325, 345, 359, 360, 400, 416, 446, 453, 473, 496, 516, 544, 568, 578, 594, 601, 612, 620, 635, 655, 663, 678, 686

Yesterday's News, P16, P32, 99, 112, 155, 230, 265, 306, 373, 408, 525, 585, 626, 670

## B

**Big Ideas.** *See* Essential Questions

**Brain POPs®.** *See* Go Online!

## C

**Chapter Projects.** *See* Go Online!

**Chapter Tests.** *See* Go Online!

**Cooperative Groups**, 30, 57, 85, 94, 141, 157, 159, 170, 179, 204, 275, 295, 315, 327, 346, 374, 381, 428, 436, 456, 468, 489–490, 504–505, 519, 527, 529, 570, 587, 603, 628

## D

**Diagnostic Assessment.** *See* Assessment

**Differentiated Homework Options**, 7, 13, 20, 27, 45, 54, 82, 90, 98, 104, 110, 119, 125, 147, 154, 165, 175, 185, 194, 200, 208, 228, 236, 243, 250, 256, 262, 271, 292, 299, 305, 312, 318, 324, 342, 351, 364, 371, 378, 399, 407, 414, 424, 433, 443, 452, 465, 472, 494, 501, 509, 515, 523, 536, 565, 576, 584, 592, 611, 619, 633, 654, 662, 668, 677, 684, 694

**Differentiated Instruction**, 2B, 138B. *See also* Learning Styles
Advanced, 297, 350, 397, 560
Advanced High, 297, 350, 560
Approaching Level, 4, 5, 6, 11, 18, 25, 52, 59, 76B, 81, 82, 96, 97, 98, 102, 104, 108, 110, 117, 119, 123, 125, 147, 152, 154, 163, 174, 193, 227, 234, 241, 248, 261, 269, 290, 304, 311, 317, 323, 349, 363, 370, 377, 404, 412, 421, 423, 454, 471, 495, 499, 513, 522, 533, 535, 541, 563, 573, 581, 598, 610, 632, 674, 675, 681
Beginning, 260, 297, 350, 397, 560
Beyond Level, 5, 9, 15, 22, 28, 46, 48, 52, 64, 76B, 81, 82, 84, 93, 97, 98, 100, 104, 106, 110, 118, 121, 125, 127, 146, 147, 150, 154, 164, 168, 176, 184, 191, 193, 201, 209, 231, 234, 238, 245, 255, 264, 294, 298, 301, 304, 307, 311, 314, 320, 326, 353, 358, 360, 366, 373, 380, 396, 401, 412, 427, 432, 435, 467, 474, 497, 503, 507, 511, 517, 526, 538, 541, 545, 569, 579, 602, 610, 613, 618, 627, 666, 671, 674, 679, 681, 687

English Language Learners, 4, 6, 76B, 17, 24, 52, 59, 78, 97, 140, 162, 174, 224, 228, 241, 260, 290, 297, 304, 311, 338, 350, 363, 370, 397, 404, 412, 423, 463, 513, 541, 549, 560, 587, 598, 610, 624, 637, 675, 681
Intermediate, 260, 297, 350, 397
On Level, 4, 5, 6, 9, 11, 15, 22, 25, 46, 52, 59, 64, 81, 82, 76B, 93, 96, 97, 98, 100, 104, 108, 110, 117, 118, 119, 123, 125, 127, 146, 147, 152, 154, 163, 164, 168, 176, 184, 191, 193, 201, 209, 227, 231, 234, 238, 245, 248, 255, 261, 264, 269, 294, 298, 301, 304, 307, 314, 317, 320, 323, 326, 349, 353, 358, 360, 366, 370, 373, 377, 380, 396, 401, 404, 412, 421, 423, 427, 432, 435, 454, 467, 471, 497, 499, 503, 507, 511, 517, 522, 526, 533, 538, 541, 545, 563, 569, 573, 579, 581, 598, 610, 613, 618, 627, 632, 666, 671, 674, 675, 679, 681, 687

Differentiated Resources, 2B, 5B, 10B, 16B, 23B, 33B, 42B, 49B, 58A, 76B, 79B, 85B, 94B, 101B, 107B, 122B, 138B, 141B, 151B, 158B, 170B, 181B, 190B, 197B, 205B, 222B, 225B, 232B, 239B, 247B, 254B, 259B, 267B, 286B, 289B, 295B, 302B, 309B, 315B, 321B, 336B, 348B, 354B, 361B, 376B, 392B, 395B, 402B, 410B, 419B, 448B, 458B, 469B, 486B, 489B, 504B, 512B, 539B, 556B, 588B, 606B, 548B, 688B

# E

**eAssessment.** *See* Go Online!

**English Language Learners,** 4, 6, 17, 24, 52, 59, 76B, 78, 97, 140, 162, 174, 224, 228, 241, 260, 290, 297, 304, 311, 338, 350, 363, 370, 397, 404, 412, 423, 463, 513, 541, 549, 560, 587, 598, 610, 624, 637, 675, 681

**eSolutions Manual.** *See* Go Online!

**Essential Questions,** 2, 34, 76, 138, 145, 155, 222, 233, 286, 336, 392, 439, 486, 540, 556, 597, 648, 652
Answering, 65, 128, 212, 276, 328, 475, 546, 637, 698

**eToolkit.** *See* Go Online!

regression in, 259–265, 279
relative frequency, 690–693, 698
representing, 657–661, 699
residuals, 260–261
as set of ordered pairs, 50
standard deviation, 667–668, 681–684
summarizing, 691–693, 698, 702
transformations of, 680–684
two-way frequency tables, 688–693, 698, 702

**Decay, exponential,** 430–431, 450–451, 461

**Defining a variable,** P5

**Degree**
of a monomial, 491
of a polynomial, 491

**Denominators**
least common, P13–P14
rationalizing, 421

**Dependent systems of equations,** 339, 349

**Dependent variables,** 44

**Depreciation,** 461

**Descriptive modeling,** 33–40, 65

**Diagrams**
tree, P34
Venn, 30, 428

**Diameter,** P24

**Differences.** *See also* Subtraction
common, 190, 192–193
first, 623
second, 623
squares of sums and, 512–513
successive, 622–624, 641
of two squares, 539

**Dilation**
of absolute value functions, 207
of exponential functions, 438, 440–441
horizontal, 573
of linear functions, 182–184
of quadratic functions, 572–573
vertical, 573

**Dimensional analysis,** 124, 128, 132

**Dimensions of a matrix,** 374

**Discrete functions,** 50

**Discriminant,** 610, 618

**Distribution,** 673–676
analyzing, 675–676

comparing data sets, 682–684
negatively skewed, 673, 675, 698
positively skewed, 673, 675–676
representing, 674–676, 682–683
symmetric, 673, 675, 698

**Distributive Property,** 23–25, 26, 65, 303, 506–508, 519, 520, 521, 546, 548

**Division**
equation models for, 86
exponents, 402–408
integers, P12
monomials, 402–408
radical expressions, 420, 424
rational numbers, P18
solving inequalities by, 298, 329
square roots, 420
by a variable, 522, 590

**Division Property**
of equality, 88, 128
of inequality, 298

**Domains,** 42, 50, 141

**Dot plots (line plots),** 657–661

**Double roots,** 581

# E

**Einstein, Albert,** 397

**Elements of a matrix,** 374

**Elimination,** 354, 382
proving the method, 527–528
using addition, 354–355, 357–359, 368, 384
using multiplication, 354–359, 361–365, 382, 384
using subtraction, 356–359, 368, 384

**Ellipsis,** 190

**Empty set,** 303–304

**End behavior,** 59, 65

**Equality**
addition property of, 87–88, 128
division property of, 88, 129
multiplication property of, 88, 128
power property of, 413
subtraction property of, 88, 128

**Equally likely,** P33

**Equations.** *See also* Functions
absolute value, 107–112, 128, 131
addition, 85, 87
compound interest, 449, 458

equivalent, 87, 146
exponential, 413
exponential decay, 450–451
exponential growth, 448
in function notation, 52
as functions, 51
graphing by using intercepts, 145–146
grouping symbols in, 11, 102
identities, 102
linear (*See* Linear equations)
literal, 123, 128, 132
multi-step, 94, 95–99, 130
one-step, 87–92, 129
point–slope form, 233–235, 276, 277
proportions, 115–116, 128, 131
quadratic (*See* Quadratic equations)
roots of, 151
slope–intercept form, 171–177, 214, 225–230, 234–235, 276, 277
solving with polynomial expressions, 500
standard form of a linear equation, 143–144, 212, 232–233, 276, 277
subtraction, 85
systems of (*See* Systems of equations)
translating into sentences, 81
translating sentences into, 79–80
in unit conversions, 134
using slope and a point, 225
using two points, 225–226
with the variable on each side, 101–105, 130
writing, 79–83, 103, 109, 128–129, 173, 225–227, 623–624

**Equivalent equations,** 87, 146

**Error**
absolute, 313
greatest possible, 40
margin of, 107
relative, 313

**Error analysis.** *See* Higher-Order Thinking Skills

**Evaluate,** 10–12, 17

**Exclusive statements,** 315

**Explicit formulas,** 471

**Exponential decay,** 430–431, 450–451, 461

**Exponential equations,** 413

**Exponential functions,** 430–434. *See also* Functions

compared with quadratic and linear functions, 563
compound interest, 449–450, 458–460, 468
decay, 430–431, 450–451
end behavior, 431, 432
exponential growth, 430, 435–436, 448–449, 475
geometric sequences as, 462–466
graphing, 430–432, 438–440, 451, 456–457, 468
growth, 430–431, 435–436, 448–449
identifying, 622–623
inequalities, 457
rate of change, 432, 458–461, 468
reflecting, 441–443
regression, 629
solving, 456–457
translating, 438–440, 443
using to solve a problem, 341–432
writing, 448–451, 624

**Exponential growth,** 430–431, 435–436, 448–449, 475

**Exponents,** 5, 395
coefficients and powers of 1, 396
division properties of, 402–408
multiplication properties of, 395–400, 475
negative, 404–406
in order of magnitude, 405–406
radical form, 410
rational, 410–416, 475
zero, 403–404

**Expressions.** *See also* Algebraic expressions
absolute value, 107, 420
algebraic, 5–8, 11–12, 25–26
evaluating, 10–12, 23–24
linear, 395
monomial, 395–398
numerical, 10–14
quadratic, 507
radical, 419–426, 475
in simplest form, 25
simplifying, 24–26
verbal, 5–8, 26, 507

**Extrapolation,** 227, 249, 261

**Extrema,** 59–60, 603–604

**Extremes,** in proportions, 116

### F

**Factored form,** 520
**Factoring,** 520–525
differences of squares, 539, 546, 591
into factored form, 520
greatest common factor, 520, 534, 542
by grouping, 521, 546
perfect square trinomials, 513, 540, 546, 591
prime polynomials, 535, 542, 550
quadratic trinomials, 531–535, 546, 549, 591
signs of factors, 533
solving equations by, 522–523
solving quadratic equations by, 588–592, 637, 639
trinomials, 529–530
using the distributive property, 506–508, 519, 521, 546, 548, 591

**Factors,** 5, 520, 534, 542
**Fahrenheit scale,** 82, 134, 270
**Families**
of functions, 592
of graphs, 151
**Fibonacci (Leonardo Pisano),** 369
**Financial literacy,** 7, 14, 19, 33–34, 38, 82, 98, 109, 148, 153, 175, 186, 227, 257, 292, 302, 318, 371, 379, 399, 426, 468, 631
compound interest, 449–450, 458–460, 468
investment plans, 468
**First differences,** 623
**Five-number summary,** 665–666, 676, 682–684, 698
**FOIL method,** 423, 507–508, 531
**Foldables® Study Organizer**
equations of linear functions, 223, 276
exponents and exponential functions, 393, 475
expressions and functions, 3, 65
linear equations, 77, 128
linear functions, 139, 212
linear inequalities, 287, 328
quadratic expressions, 487, 546
quadratic functions, 557, 637

statistics, 649, 698
systems of linear equations and inequalities, 337, 382
**Formulas,** 80
area, P26
circle, P27, 397
parallelogram, P26–P27
rectangle, P26, 25, 423
rhombus, 423
square, P26
body mass index, 123
Celsius to Fahrenheit conversion, 82, 270
circumference of a circle, P24
compound interest, 449, 458
distance between wave crests, 122
electrical resistance, 123
explicit, 471
horsepower, 395
literal equations as, 123, 128, 132
$n$th term of a geometric sequence, 471
perimeter, P23
parallelogram, P23–P24
rectangle, P23
of a regular hexagon, 28
square, P23
triangle, P23, 20, 28, 80
quadratic, 606–610, 640
recursive, 469–473
slope, 163
surface area
cube, 398
cylinder, P31, 19
rectangular prism, P29
translating sentences into, 80
volume, P29
cylinder, P29, 7
pyramid, 14
rectangular prism, P29
sphere, 12

**Four-step problem-solving plan,** P5–P6
**Fractions.** *See also* Rational numbers
adding, P14–P15, 31
dividing, P18
multiplying, P17–P18, 30–31
percents as, P20
repeating decimals as, P8
subtracting, P14–P16
writing as decimals, P13

## F

**Foldables,** xxii–xxiii, P2, 3, 32, 65, 77, 114, 128, 139, 169, 212, 223, 246, 276, 287, 308, 328, 337, 367, 393, 418, 475, 487, 518, 546, 557, 605, 637, 649, 672, 698. *See also* Go Online!

**Formative Assessment.** *See* Assessment

# G

Geometer's Sketchpad. *See* Go Online!

**Go Online!**
  Animations, xiv, 33A, 49A, 76B, 85A, 101A, 138B, 141A, 286B, 315A, 336B, 376A, 458A, 486B, 556B, 580A, 657A
  Anticipation Guide, 68, 132, 214, 279, 329, 477, 549, 639, 701
  Brain POPs®, 2B, 16A, 23A, 222B, 295A, 392B, 402A, 489A
  Chapter Projects, 2, 5A, 58A, 76, 79A, 122A, 138, 190A, 222, 239A, 247A, 254A, 286, 289A, 336, 361A, 368A, 392, 448A, 462A, 469A, 486, 556, 648, 648B
  Chapter Tests, 71, 133, 217, 281, 331, 387, 481, 551, 643, 703
  eAssessment, 32, 70, 75, 114, 132, 137, 169, 215, 221, 246, 280, 285, 308, 330, 335, 367, 391, 418, 478, 485, 518, 550, 605, 640, 647, 672, 702, 707
  eLessons, 2B, 42A, 49A, 76B, 85A, 94A, 101A, 138B, 141, 141A, 151A, 159A, 170A, 181A, 197A, 205A, 222B, 247A, 254A, 259A, 267A, 275, 286B, 289A, 295A, 302A, 309A, 321A, 336B, 339A, 348A, 354A, 361A, 368, 392A, 430A, 438A, 448A, 458A, 469A, 486B, 489A, 488A, 504A, 511A, 512A, 519A, 529A, 539A, 556B, 559A, 571A, 580A, 596A, 606A, 614A, 630A, 648B, 657A, 688A
  eSolutions Manual, P6, P9, P12, P15, P18, P21, P24, P27, P30, P32, P35, 8, 14, 20, 28, 40, 46, 54, 62, 83, 91, 98, 104, 111, 119, 125, 149, 155, 167, 175, 188, 195, 201, 209, 229, 236, 243, 251, 264, 272, 293, 300, 306, 313, 319, 344, 352, 358, 365, 372, 400, 407, 415, 426, 434, 446, 453, 466, 473, 495, 501, 510, 515, 524, 566, 577, 585, 594, 600, 620 626, 663, 670, 685
  eToolkit, vi, 23A, 42A, 49A, 86, 94, 94A, 107A, 151A, 190A, 254A, 289A, 295, 309A, 348A, 410A, 419A, 438A, 489A, 490, 504A, 512A, 519, 527, 529A, 588A, 596A, 606A, 614A, 622A, 630A, 665A, 673A, 688A,
  Foldables, xxii–xxiii, P2, 3, 32, 65, 77, 114, 128, 139, 169, 212, 223, 246, 276, 287, 308, 328, 337, 367, 393, 418, 475, 487, 518, 546, 557, 605, 637, 649, 672, 688
  Geometer's Sketchpad, vi, xvi, 2B, 10A, 16A, 23A, 42A, 76B, 94A, 101A, 138B, 139, 159A, 170A, 181A, 222B, 225A, 232A, 239A, 247A, 259A, 286B, 295A, 302A, 309A, 321A, 336B, 339A, 376A, 392B, 395A, 402A, 430A, 486B, 512A, 556B, 571A, 580A
  Graphing Calculators, xv, xviii, 57, 158, 170, 327, 346, 381, 586, 628, 648B, 665A, 673A
  Graphing Tools, xii, 139, 141A, 159A, 170A, 181A, 259A, 321A, 339A, 430A, 458A, 559A, 571A, 651A, 657A
  Interactive Student Guide, vii, 7, 12, 19, 27, 45, 53, 61, 72, 82, 97, 109, 118, 124, 134, 139, 147, 154, 165, 185, 194, 200, 208, 218, 228, 250, 263, 271, 282, 292, 299, 305, 318, 324, 332, 342, 351, 357, 363, 379, 398, 406, 414, 433, 442, 443, 452, 465, 472, 482, 494, 500, 509, 514, 523, 536, 541, 543, 565, 576, 584, 592, 588, 611, 625, 633, 644, 669, 678, 684
  Interactive Whiteboard, 5, 10, 16, 23, 33, 42, 49, 58, 79, 87, 95, 101, 107, 115, 122, 143, 151, 160, 171, 181, 190, 197, 225, 232, 239, 247, 259, 267, 289, 296, 302, 309, 316, 321, 339, 348, 354, 361, 368, 376, 395, 402, 410, 419, 430, 438, 448, 462, 469, 491, 448, 506, 512, 520, 531,

Frequency table, 659
Functions, 49–56
  absolute value, 205–210, 212
  adding and subtracting, 630–631, 637, 642
  constant, 142, 172
  continuous, 50
  decreasing, 59–60
  dilation of, 182–184, 207, 438, 440–441, 572–573
  discrete, 50
  end behavior, 59–60
  exponential (*See* Exponential functions)
  exponential decay, 430–431, 450–451, 461
  exponential growth, 430–431, 435–436, 448–449
  extrema, 59–60
  in graphing calculators, 57
  greatest integer, 197, 200, 204, 212
  identifying, 49–51, 53, 622–623
  identity, 441
  increasing, 59–60
  intercepts, 58–59, 65, 144–146
  inverse, 268–270, 280
  linear (*See* Linear functions)
  multiplying, 632–633, 637, 642
  negative, 59–60, 65
  nonlinear, 52, 58
  notation, 52
  parent, 151, 183, 571
  piecewise-defined, 198–202
  piecewise-linear, 197, 204, 216
  positive, 59–60, 65
  quadratic (*See* Quadratic functions)
  rate of change in, 432, 458–461, 468, 570
  reflection of, 184–185, 206, 441–443, 574
  relative maximum, 59–60
  relative minimum, 59–60
  representations of, 51
  step, 197–198, 200, 216
  successive differences, 622–624, 641
  translation of, 181–182, 205–208, 216, 438–440
  values of, 52
  zeros of linear, 151–155, 212, 213
Fundamental Counting Principle, P34–P35

# G

GCF (greatest common factor), 520, 534, 542
Geometric sequences, 462–466
  finding terms of, 463
  identifying, 462–463
  *n*th term of, 463–464
  writing recursive formulas for, 470–471
Geometry
  circles, P24, P26, 397
  cubes, 398
  cylinders, 7, 19
  hexagons, 28
  parallelograms, P23–P24, P26–P27
  pyramids, 14
  rectangles, 25, 423
  rectangular prisms, P29, P31
  rhombuses, 423
  spheres, 12
  squares, P23, P26, 235
  trapezoids, 132, 243
  triangles, 20, 28, 80, 236, 364, 399, 424, 591–592
Get Ready for the Chapter, 65, 78, 128, 140, 212, 224, 276, 288, 328, 338, 394, 488, 546, 558, 637, 650, 698
Go Online!. *See also* Foldables® Study Organizer
  ALEKS, 3, 77, 139, 223, 287, 337, 393, 487, 557, 649
  Algebra Lab, 242
  Animations, 88, 96, 297, 492, 514
  Box-and-Whisker Plots, 649
  Chapter Project, 3, 139, 223, 287, 337, 393, 487, 557, 649
  eLessons, 223
  eStudent Edition, 43, 59, 80, 109, 310, 370, 500, 507, 540, 542, 608
  eToolkit, 102, 144, 162, 192, 287, 303, 322, 337, 463, 487, 557, 689
  Geometer's Sketchpad®, 3, 77, 139, 207, 223, 337, 393, 404, 487, 557
  Graphing Calculators, 11, 192, 533, 623
  Graphing Tools, 139, 287, 341, 393, 431, 560, 572, 649
  Interactive Student Guide, 3, 77, 139, 223, 287, 337, 393, 487, 557, 649
  LearnSmart, 3, 77, 139, 223, 287, 337, 393, 487, 557, 649
  Personal Tutors, 6, 261, 350, 582, 667, 682

Self-Check Quizzes, 7, 12, 19, 27, 36, 45, 52, 53, 61, 81, 90, 97, 104, 109, 118, 147, 152, 154, 174, 185, 194, 200, 208, 235, 242, 256, 263, 268, 271, 292, 299, 304, 311, 318, 324, 342, 351, 357, 363, 371, 378, 398, 406, 412, 414, 421, 424, 443, 451, 461, 465, 472, 493, 500, 509, 514, 523, 543, 565, 577, 584, 592, 599, 611, 618, 625, 633, 654, 662, 668, 677, 684, 693
Spreadsheet Activity, 450
Standardized Test Practice, 75, 137, 285, 335, 391, 555, 647
Tools, P2, 3, 26, 77
Virtual Manipulatives, 102, 172, 248, 303, 674
Vocabulary, P2, 78, 128, 224, 276, 288, 382, 475, 488, 546, 637
Watch icons, 18
Worksheets, 116, 363, 522
**Graphing calculators**
  bin widths, 675
  box-and-whisker plots, 676, 683
  curve fitting, 628–629
  exponential equations, 456–457
  factoring polynomials, 533
  functions, 57
  histograms, 675, 683
  inequalities, 327, 381, 587
  linear functions, 57, 157, 170, 204
  median-fit lines, 262, 276, 279
  *n*th roots, 411
  rational exponents, 412
  Trace feature, 567
**Graphing Technology Labs**
  curve fitting, 628–629
  exponential equations and inequalities, 456–457
  graphing inequalities, 327
  linear functions, 157–158
  parallel and perpendicular lines, 239–241
  piecewise-linear functions, 204
  quadratic inequalities, 587
  representing functions, 57
  slope-intercept form, 170
  systems of equations, 346–347
  systems of inequalities, 381
**Graphs.** *See also* Data; Statistics
  absolute value functions, 205–207, 216
  analyzing, 44, 141–142
  bar, 659, 699
  best-fit line and residuals, 260

R104 | Index

box-and-whisker plots, 650, 674, 676, 683, 701
complete, 157–158
correlation and causation, 254–255, 279
dilation, 183, 207, 573
drawing, 50, 145–146
end behavior, 59
exponential decay, 451, 461
exponential functions, 430–432, 438–440, 451, 456–457, 468
extrema, 59–60
families, 151
functions, 50–52
geometric sequence, 464
greatest integer function, 197, 204
histograms, 659–660, 673, 675, 682, 700–701
inequalities in two variables, 321–325, 327, 328, 330
intercepts, 58–59, 65, 144–145
intersection of two, 309
inverse linear functions, 269
inverse relations, 268, 275
linear equations, 151, 157–158, 171–174, 179–180, 213
linear functions, 141–142, 144–146, 157–158
linear regression and residuals, 260–261, 276
line symmetry in, 59, 141–142
median-fit line, 262, 276, 279
parabolas, 559, 587
percent rate of change, 458–460
piecewise-defined functions, 198–199, 204, 216
quadratic functions, 559–560, 562–568, 598, 638–639
quadratic inequalities, 587
rate of change and slope, 159, 160–164
real numbers, P8
relations as, 42–48
scatter plots, 247–249, 254, 276, 278, 628
slope of a line, 162–164, 171–177, 225–226
step functions, 198
systems of equations, 340–341, 368, 383, 614–618
systems of inequalities, 376–377, 381, 382
transformations of exponential functions, 438–442

transformations of linear functions, 181–185, 215
translations of quadratic functions, 571–576
union of two, 310
zeros of a linear function and, 152–153, 212, 213
**Greatest common factor (GCF),** 520, 534, 542
**Greatest integer function,** 197, 200, 204, 212
**Greatest possible error,** 40
**Grouping symbols,** 11, 102
braces, 11
brackets, 11
parentheses, 11, 102
**Growth rate,** 179–180, 436–437, 450
**Guided Practice,** 5, 6, 10, 11, 12, 17, 18, 19, 24, 26, 34, 35, 36, 44, 50, 51, 52, 58, 60, 61, 79, 80, 87, 88, 89, 95, 97, 101, 102, 107, 108, 109, 115, 116, 117, 118, 122, 123, 124, 143, 144, 145, 146, 152, 153, 160, 161, 162, 163, 164, 171, 172, 173, 174, 182, 183, 185, 191, 192, 197, 198, 199, 205, 207, 225, 226, 227, 233, 234, 235, 239, 240, 241, 247, 249, 255, 262, 268, 270, 289, 290, 291, 297, 298, 302, 303, 310, 311, 316, 317, 322, 323, 340, 341, 348, 349, 350, 355, 356, 357, 361, 362, 363, 369, 370, 376, 377, 395, 396, 397, 398, 402, 403, 404, 405, 406, 410, 412, 413, 419, 420, 421, 422, 423, 430, 431, 432, 439, 440, 442, 443, 448, 449, 450, 451, 459, 460, 461, 463, 464, 470, 471, 491, 492, 493, 498, 499, 500, 507, 508, 512, 513, 514, 520, 521, 522, 523, 532, 533, 534, 535, 539, 541, 542, 559, 560, 561, 562, 563, 564, 572, 573, 574, 575, 576, 581, 582, 583, 589, 590, 591, 592, 597, 598, 607, 608, 610., 614, 615, 617, 618, 622, 623, 624, 630, 631, 632, 633, 652, 653, 658, 659, 660, 661, 666, 667, 668, 673, 674, 676, 681, 682, 683, 684, 689, 690, 691, 692, 693

## H

**Half-pipes,** 438
**Half-planes,** 321
**Harriot, Thomas,** 298
**Hexagons,** perimeter of, 28
**Higher-Order Thinking Skills (H.O.T)**
Challenge, 8, 47, 55, 83, 92, 99, 105, 112, 120, 126, 155, 167, 177, 188, 195, 252, 265, 271, 293, 300, 306, 313, 344, 359, 365, 372, 379, 408, 416, 426, 466, 473, 496, 502, 510, 516, 525, 537, 568, 570, 578, 585, 601, 612, 620, 655, 670
Construct Arguments, 537
Error Analysis, 8, 14, 55, 112, 126, 155, 202, 210, 230, 237, 244, 300, 313, 319, 344, 352, 416, 473, 496, 502, 525, 537, 544, 568, 594, 620, 635, 655, 670
Multiple Representations, 8, 21, 28, 83, 105, 120, 149, 167, 244, 273, 293, 306, 313, 319, 325, 344, 359, 400, 408, 416, 473, 496, 502, 510, 516, 525, 537, 544, 568, 578, 600, 612
Open Ended, 8, 14, 21, 40, 47, 55, 63, 83, 92, 99, 105, 112, 126, 149, 155, 167, 177, 188, 195, 202, 230, 237, 257, 265, 273, 293, 300, 313, 325, 344, 352, 359, 365, 372, 379, 400, 408, 416, 418, 426, 434, 502, 525, 578, 635, 678
Proof, 626
Which One Doesn't Belong?, 21, 92, 237, 252, 306, 344, 372, 516, 601
Write a Question, 55, 372
Writing in Math, 8, 14, 21, 28, 40, 47, 55, 63, 83, 86, 92, 99, 105, 112, 120, 126, 149, 155, 167, 177, 188, 195, 202, 210, 230, 237, 244, 252, 265, 273, 293, 300, 306, 313, 319, 325, 344, 347, 352, 359, 365, 372, 379, 400, 408, 416, 426, 434, 453, 466, 473, 496, 502, 505, 510, 516, 519, 525, 530, 537, 568, 578, 585, 594, 612, 620, 626, 655, 663, 670, 678, 686
**Histograms,** 659–660, 673, 675, 682, 700–701
**Horizontal lines,** 142
**Horizontal translations,** 571–572, 575–576
**Horsepower,** 395
**H.O.T.** *See* Higher-Order Thinking Skills
**Hypotenuse,** 80, 517

## I

**Identity**
additive, 16
multiplicative, 17
**Identity function,** 102, 349, 441
**Identity matrix,** 375
**Identity properties,** 16, 17, 26
**Inclusive statements,** 315

539, 559, 571, 580, 598, 596, 606, 614, 622, 630, 657, 665, 673, 680
Personal Tutors, x, xix, P5, P7, P11, P13, P17, P20, P23, P26, P29, P31, P33, 5A, 10A, 16A, 33A, 58A, 79A, 85A, 107A, 122A, 151A, 190A, 197A, 205A, 225A, 232A, 239A, 259A, 267A, 302A, 315A, 348A, 354A, 361A, 368A, 376A, 395A, 402A, 410A, 419A, 438A, 448A, 462A, 469A, 498A, 504A, 519A, 529A, 539A, 559A, 588A, 596A, 606A, 614A, 622A, 630A, 648B, 651, 651A, 559, 665A, 673A, 688A
Self-Check Quizzes, xiii, P10, P16, P19, P22, P25, P28, P36, 5A, 9, 10A, 15, 22, 29, 33A, 48, 56, 58A, 64, 79A, 84, 93, 100, 106, 107A, 113, 121, 122A, 127, 150, 156, 168, 178, 196, 197A, 203, 205A, 211, 225A, 231, 232A, 238, 245, 253, 266, 267A, 274, 294, 301, 307, 314, 315A, 320, 326, 345, 353, 354A, 360, 366, 373, 380, 395A, 409, 410A, 417, 419A, 427, 435, 447, 454, 462A, 467, 474, 497, 498A, 503, 511, 517, 519A, 526, 538, 539A, 545, 569, 579, 586, 588A, 595, 601, 612, 622A, 627, 636, 655, 664, 665A, 671, 673A, 679, 687
Standardized Test Practice, 74, 136, 220, 284, 334, 390, 484, 646, 706
Time Management, 2B, 76B, 138B, 222B, 286B, 336B, 392B, 486B, 556B, 648B
Videos
  Choosing Foldables, 393
  Creating VKV Flashcards, 77
  Extending Vocabulary, 649
  Flashcards for Spanish English Cognates, 139
  Getting Started with Notebooking, 223
  Independent and Notebook Foldables, 337, 487
  Introducing VKV Flashcards, 3
  Notebooking Tips, 287
  StudySync: SMP Modeling Videos, 2A, 76A, 138A, 222A, 286A, 336A, 392A, 486A, 556A, 648A
  Vocabulary Learning, 557
Vocabulary, 4, 65, 78, 128, 140, 212, 224, 276, 288, 328, 338, 394, 475, 488, 546, 558, 637, 650, 698

**Graphing Calculators.** *See* Go Online!

## I

**Interactive Classroom.** *See* Go Online! Interactive Whiteboard

**Interactive Student Guide.** *See* Go Online!

**Interactive Whiteboard.** *See* Go Online!

**Interpersonal Learners.** *See* Learning Styles

**Intervention,** 4, 5B, 10B, 16B, 23B, 33B, 42B, 49B, 58B, 79B, 85B, 94B, 101B, 107B, 122B, 141B, 151B, 158B, 170B, 181B, 190B, 197B, 205B, 225B, 232B, 239B, 247B, 254B, 259B, 267B, 289B, 295B, 302B, 309B, 315B, 321B, 339B, 348B, 354B, 361B, 368B, 376B, 395B, 402B, 410B, 419B, 430B, 438B, 448B, 458B, 462B, 469B, 489B, 498B, 504B, 512B, 519B, 529B, 539B, 559B, 571B, 580B, 588B, 596A, 606B, 614B, 622B, 630B, 651B, 657B, 665B, 673B, 688B

# K

**Key Vocabulary,** 4, 65, 78, 128, 140, 212, 224, 276, 288, 328, 338, 394, 475, 488, 546, 558, 637, 650, 698

**Kinesthetic Learners.** *See* Learning Styles

**Inconsistent systems of equations,** 339

**Independent systems of equations,** 339

**Independent variables,** 44

**Inequalities,** 289
- *and* in, 309–310
- absolute value, 316–319, 328, 330
- compound, 309–311, 328, 330
- empty set as solution of, 303–304
- exponential, 457
- graphing inequalities in two variables, 321–325, 327, 328, 330
- modeling, 295
- multi-step, 302–306, 328, 329
- negative coefficients in, 302
- *or* in, 310–311
- quadratic, 587
- set-builder notation, 290–291
- solving by addition, 289–290, 329
- solving by division, 298, 329
- solving by multiplication, 296–297, 329
- solving by subtraction, 290–291, 329
- solving using distributive property, 303
- systems of, 376–379, 381, 382, 386
- verbal phrases for, 291

**Integers,** P7
- adding, P11
- consecutive, 96–97
- greatest integer function, 197, 204, 212
- subtracting, P11

**Interactive Student Guide,** 3

**Intercepts,** 58–59, 65, 144–146

**Interdisciplinary connections**
- archeology, 244
- architecture, 208, 526
- astronomy, 406, 408, 418
- aviation, 201, 227, 263
- biology, 82, 114, 176, 418, 432, 433, 434, 452, 523, 611, 626, 629
- chemistry, 317, 318, 413, 453
- economics, 351
- engineering, 501
- environmental studies, 12, 154, 237, 282, 452, 677
- genetics, 514, 515
- geography, 119, 619, 662
- geology, 98, 252, 466
- government, 80
- health, 144
- history, 191, 208
- marine science, 176, 207
- music, 350, 371, 372, 415, 452
- physical science, 589
- physics, 72, 111, 123, 125, 213, 270, 343, 400, 415, 425, 565
- physiology, 47

**Interest, compound,** 449–450, 458–460, 468

**Interpolation,** 249, 261

**Interquartile range (IQR),** 666

**Intersection of two graphs,** 309

**Intervals,** 179

**Inverse functions,** 268–270, 280

**Inverse relations,** 267–268, 275, 276, 297

**Inverses**
- additive, P11, 16, 493, 521
- multiplicative, P18, 17

**IQR (interquartile range),** 666

**Irrational numbers,** P7, 428–429

# J

**Joint frequencies,** 691

**Justify Arguments,** 40, 626

# K

**Kelvin scale,** 134

**Key Concepts**
- absolute value equations, 108, 128
- absolute value functions, 205, 212
- absolute value inequalities, 328
- adding a constant to $f(x)$, 181, 438
- adding and subtracting functions, 630–631
- addition properties, 16
- addition property of equality, 87–88, 289
- arithmetic sequences, 190, 212, 215
- associations in two-way data, 692
- associative property, 16
- $b^{1/n}$, 410, 411
- $b^{m/n}$, 412
- causation, 255
- commutative property, 16
- comparing the mean and the median, 652
- completing the square, 596
- compound interest equation, 449
- correlation and causation, 254, 276
- data distributions, 698
- data representation, 698
- data transformations using multiplication, 681
- dilations, 572
- dimensional analysis, 128
- distributive property, 23–25
- division property of equality, 88
- division property of inequality, 298
- elementary row operations, 374
- exponential decay equation, 450
- exponential functions, 430
- factoring by grouping, 521
- factoring differences of squares, 539
- factoring perfect square trinomials, 540
- factoring $x^2 + bx + c$, 531
- finding inverse functions, 269
- FOIL method, 507
- functions, 49–52
- functions with successive differences, 622–624
- Fundamental Counting Principle, P35
- general equation for exponential growth, 448
- graphing exponential functions, 431
- graphing linear equations, 212
- graphing linear inequalities, 321
- graphing quadratic functions, 562, 637
- greatest integer function, 197
- grouping a constant with $f(x)$, 182
- horizontal translations, 572
- inequalities in two variables, 328
- inverse relations, 267, 276
- linear and nonlinear functions, 622
- linear equations, 151
- literal equations, 128
- maximum and minimum values, 561
- means-extremes property of proportion, 116
- measures of center, 651, 698
- measures of spread, 698
- methods of factoring, 591
- multiple transformations, 443
- multiplication properties, 16
- multiplication property of equality, 88
- multiplication property of inequality, 296

R106 | Index

multiplying functions, 632
multiplying f(x) by −1, 441
multiplying f(x) by a constant, 183, 440
multiplying f(x) by a negative constant, 442
multiplying x by −1, 442
multiplying x by a negative constant, 442
multiplying x by a positive constant, 441
multi-step and compound inequalities, 328
negative exponent property, 404, 475
nth root, 411
nth term of a geometric sequence, 464
nth term of an arithmetic sequence, 191, 212
number of solutions, 614
order of operations, 10, 65
parallel and perpendicular lines, 276
perfect squares, P9
point–slope form, 233, 276
positive, negative, increasing, decreasing, extrema, and end behavior, 59
power of a product, 397, 475
power of a quotient, 403, 475
power property of equality, 413
product of a sum and a difference, 514
product of powers, 396, 475
product property of square roots, 419
properties of equality, 16
quadratic equation solutions, 580
quadratic formula, 607
quadratic functions, 559, 562
quotient of powers, 402, 475
quotient property of square roots, 420
rate of change, 160
rate of change and slope, 212
ratios and proportions, 128
reflection, 574
regression and median-fit lines, 276
scatter plots, 247, 276
simplifying monomial expressions, 398
slope, 163, 164
slope–intercept form, 171, 212, 276
solving a linear and quadratic system by graphing, 615, 637
solving by elimination, 354, 361

solving by elimination using multiplication, 361
solving by substitution, 348
solving equations, 128
solving one-step inequalities, 328
solving quadratic equations, 637
special functions, 212
square of a sum, 512–513
square root property, 588, 590
standard deviation, 667
standard form of a linear equation, 143–144, 232–233, 276
steps for solving equations, 103
subtraction property of equality, 88
subtraction property of inequality, 290
summarizing data, 698
symmetric and skewed box-and-whisker plots, 674
systems of inequalities, 382
transformations of data using addition, 680
transformations of quadratic functions, 637
translating verbal to algebraic expressions, 6
two-way frequency tables, 688
using a linear function to model data, 248
using the discriminant, 612
vertex form, 598
vertical translations, 571
writing equations, 128
writing recursive formulas, 470
zero exponent property, 404, 475
zero product property, 522
zeros of linear functions, 212
**Key Vocabulary,** 65, 128, 212, 276, 328, 382, 475, 546, 637, 598

### L

**Labs.** *See* Algebra Labs; Graphing Technology Lab
**Leading coefficient,** 492
**Least common denominator (LCD),** P13–P14
**Least-squares regression line,** 259–262
**Lesson-by-Lesson Review,** 66–70, 129–132, 213–216, 277–280, 329–330, 383–386, 477–480, 547–550, 638–642, 699–702

**Like terms,** 25, 489–490
**Line of fit,** 248–249, 276, 278
**Linear equations**
changing inputs to, 182
constant, 172
dilation of, 182–184
graphing calculators, 157–158, 170
graphing using intercepts, 146
graphing using tables, 146
graphs, 151, 157–158, 171–174, 179–180, 213
identifying, 143–144
linear growth patterns, 179–180
median-fit, 262, 276, 279
modeling real-world data, 173–174
models of, 152–153
parent function, 151
point–slope form, 233–235, 276, 277
rate of change, 159, 160–167, 212, 214
reflection, 184–185
slope–intercept form, 170, 171–177, 214, 225–230, 234–235, 276, 277
standard form of, 143–144, 212, 232–233, 276, 277
systems of, 370–372, 385, 527–528, 614–618, 627–637
writing forms of, 234–235
zeros of, 151–155, 212, 213
**Linear expressions,** 395
**Linear extrapolation,** 227, 261
**Linear functions,** 58, 141, 151
absolute value functions, 205–210, 212, 216
analyzing graphs of, 141–142
arithmetic sequences as, 193–195, 212, 215
in data transformation, 680–682
end behavior, 59, 65
identifying, 622–623
inverse functions, 268–270, 276, 280
inverse relations, 267–268, 275, 276
quadratic and exponential functions compared with, 563
transformations of, 181–188, 215, 680–682
**Linear interpolation,** 249
**Linear regression,** 259–262, 276, 279
**Linear transformation,** 680–682

## L

**Learning Styles**
Interpersonal Learners, 269, 340, 412, 463, 471, 495, 674, 675, 681
Kinesthetic Learners, 241, 598
Logical Learners, 396, 432
Verbal/Linguistic Learners, 624
Visual Learners, 358
Visual/Spatial Learners, 499, 541

**LearnSmart,** viii, 139, 221, 285, 335, 391, 485, 647, 707

**Lesson Objectives,** 2A, 5A, 10A, 16A, 23A, 33A, 49A, 58A, 72, 79A, 85A, 94A, 101A, 107A, 122A, 141A, 151A, 159A, 170A, 181A, 190A, 197A, 205A, 225A, 232A, 239A, 247A, 254A, 259A, 267A, 289A, 295A, 302A, 309A, 315A, 321A, 339A, 348A, 354A, 361A, 368A, 376A, 395A, 402A, 410A, 419A, 430A, 438A, 448A, 458A, 462A, 469A, 489A, 498A, 504A, 512A, 519A, 529A, 539A, 559A, 571A, 580A, 588A, 596A, 606A, 614A, 622A, 630A, 651A, 657A, 665A, 673A, 688A

**Levels of Complexity,** 8, 13, 20, 27, 38, 46, 53, 61, 72, 83, 90, 98, 104, 110, 118, 125, 134, 148, 154, 165, 175, 186, 195, 201, 209, 218, 229, 236, 243, 250, 256, 263, 271, 282, 292, 300, 305, 312, 319, 325, 332, 343, 351, 357, 365, 371, 379, 399, 407, 414, 424, 433, 444, 452, 466, 472, 482, 494, 501, 509, 515, 523, 537, 543, 565, 576, 584, 593, 600, 612, 619, 626, 634, 644, 654, 662, 669, 678, 685, 694, 704

**Logical Learners.** *See* Learning Styles

# M

**Manipulatives,** xi
   Algebra template, 159, 468
   Algebra Tiles, 21, 28, 85, 94, 295, 489, 504, 519, 527, 529
   Equation mats, 85, 94, 295, 504, 519, 527, 529

**Materials,** 57, 85, 94, 157, 159, 170, 204, 275, 295, 327, 346, 381, 456, 468, 489, 504, 519, 527, 529, 570, 587, 628

**Mathematical Background,** 2, 5A, 10A, 16A, 23A, 33A, 42A, 49A, 58A, 76, 79A, 85A, 94A, 101A, 107A, 122A, 138, 141A, 151A, 159A, 170A, 181A, 190A, 197A, 205A, 222, 225A, 232A, 239A, 247A, 254A, 259A, 267A, 286, 289A, 295A, 302A, 309A, 315A, 321A, 336, 339A, 348A, 354A, 361A, 368A, 376A, 392, 395A, 402A, 410A, 419A, 430A, 438A, 448A, 458A, 462A, 469A, 486, 489A, 498A, 504A, 512A, 519A, 529A, 539A, 556, 559A, 571A, 580A, 588A, 596A, 606A, 614A, 622A, 630A, 648, 657A, 665A, 673A, 688A

**Mathematical Practices,** xxi, 2A, 40, 47
   Challenge, 271
   Construct Arguments, 17, 20, 21, 49, 76A, 107, 115, 122, 138A, 166, 222A, 230, 247, 254, 257, 319, 336A, 344, 416, 419, 426, 473, 537, 556A, 577, 670, 676
   Critique Arguments, 49, 105, 237, 244, 254, 265, 364, 419, 466, 496, 524
   Justify Arguments, 597, 626, 654
   Modeling, 7, 14, 82, 101, 107, 115, 122, 138A, 151, 154, 194, 198, 222A, 228, 236, 286A, 289, 292, 324, 343, 364, 371, 379, 415, 433, 472, 501, 556A, 585, 600, 631, 634, 648A
   Model with Mathematics, 259
   Multiple Representations, 43
   Organize Ideas, 39, 543
   Perseverance, 12, 24, 28, 42, 46, 54, 58, 87, 126, 195, 209, 243, 309, 313, 352, 355, 361, 400, 430, 434, 448, 502, 544, 559, 571, 651, 655, 680, 688
   Precision, 21, 35, 46, 76A, 91, 111, 119, 124, 125, 222A, 239, 242, 268, 298, 299, 336A, 379, 392A, 452, 486A, 491, 556A, 580, 607, 609, 665
   Reasoning, 27, 34, 53, 76A, 79, 82, 87, 90, 98, 138A, 147, 149, 160, 161, 175, 181, 182, 222A, 225, 229, 235, 247, 252, 273, 286A, 292, 306, 336A, 348, 351, 357, 369, 372, 392A, 395, 464, 469, 486A, 498, 506, 512, 520, 523, 531, 543, 588, 596, 598, 614, 615, 622, 648A, 678, 696
   Regularity, 86, 97, 99, 143, 149, 171, 177, 190, 192, 205, 210, 232, 316, 318, 376, 395, 397, 408, 468, 506, 512, 513, 520, 531, 533, 596, 622, 685, 695
   Repeated Reasoning, 286A
   Sense-Making, 7, 12, 24, 38, 42, 54, 58, 61, 76A, 87, 101, 104, 110, 115, 138A, 157, 165, 222A, 309, 312, 336A, 339, 358, 361, 363, 392A, 407, 430, 439, 448, 494, 514, 524, 541, 556A, 559, 566, 571, 573, 619, 648A, 651, 663, 680, 688, 694
   Standards for, 8, 14, 21, 23, 28, 55, 62, 76A, 83, 92, 99, 105, 112, 120, 126, 138A, 149, 155, 167, 177, 188, 195, 202, 210, 222A, 230, 244, 252, 265, 273, 286A, 293, 300, 306, 313, 319, 325, 336A, 345, 351, 359, 365, 372, 379, 392A,

**Lines**
   best-fit, 259–262
   horizontal, 142, 172, 240
   line of fit, 248–249, 276, 278
   median-fit, 262, 276, 279
   parallel, 239–244, 276, 278
   perpendicular, 240–244, 278
   slope, 162–164, 172, 225–226
   vertical, 142, 172, 240

**Line symmetry,** 59, 141–142

**Literal equations,** 123, 128, 132

**Lower quartile,** 665

## M

**Make a conjecture,** 86

**Mapping,** 42–43, 51

**Marginal frequencies,** 691

**Margin of error,** 107

**Mathematical Practices**
   Analyze Relationships, 626
   Arguments, 601, 654, 655
   Attend to Precision, 225
   Construct Arguments, 17, 21, 169, 180, 202, 225, 230, 236, 257, 319, 344, 416, 426, 437, 473, 568, 579, 670, 678
   Critique Arguments, 63, 105, 167, 188, 210, 265, 365, 466, 620, 635
   Justify Arguments, 40, 626
   Make Use of Structure, 232
   Modeling, 36, 45, 46, 82, 154, 169, 194, 201, 228, 230, 232, 236, 244, 273, 292, 324, 342, 343, 371, 378, 415, 433, 501, 543, 544, 585, 599, 600, 611, 633, 642, 654, 655, 669, 686
   Multiple Representations, 8, 21, 28, 83, 105, 120, 149, 167, 244, 273, 293, 306, 313, 319, 325, 344, 359, 400, 408, 416, 473, 496, 502, 510, 516, 525, 537, 544, 568, 578, 600, 612
   Organize Ideas, 236, 626
   Perseverance, 14, 24, 28, 55, 126, 195, 209, 313, 349, 352, 355, 362, 367, 400, 434, 502, 544, 635
   Precision, 20, 38, 47, 91, 111, 118, 119, 124, 125, 144, 226, 242, 243, 282, 299, 379, 425, 452, 453, 510, 654
   Problem Solving, 232
   Reason Abstractly, 33, 35, 232, 651, 652, 653
   Reasoning, 8, 12, 14, 19, 21, 27, 28, 53, 55, 63, 83, 88, 90, 92, 98, 99, 105, 112, 120, 126, 147, 148, 149, 161, 167, 169, 175, 176, 177, 187, 188, 195, 229, 237, 252, 264, 273, 290, 293, 300, 306, 319, 322, 325, 351, 352, 357, 359, 365, 369, 372, 379, 400, 403, 408, 426, 434, 450, 453, 465, 496, 502, 510, 521, 523, 525, 537, 543, 592, 593, 594, 620, 635, 644, 678, 686, 694, 695, 696
   Regularity, 97, 99, 149, 177, 191, 200, 206, 209, 318, 397, 408, 513, 696
   Repeated Reasoning, 597
   Sense-Making, 7, 24, 43, 54, 61, 62, 104, 110, 146, 158, 165, 210, 244, 252, 257, 358, 363, 388, 407, 493, 494, 524, 542, 568, 573, 594, 612, 615, 616, 655, 663, 674
   Structure, 13, 92, 112, 120, 163, 169, 182, 199, 230, 244, 250, 251, 282, 300, 304, 305, 312, 359, 421, 426, 495, 508, 509, 510, 515, 537, 544, 565, 585, 616, 619, 695
   Tools, 57, 103, 142, 155, 166, 241, 272, 325, 399, 414, 499, 620, 675

**Math History Link**
   Ahmes, 81
   Einstein, Albert, 397
   Harriot, Thomas, 298
   Pisano, Leonardo, 369
   Rees, Mina, 191

**Matrices** (singular, *matrix*), 374–375. *See also* Systems of equations

**Maximum values,** 559, 561–562, 603–604

**Mean absolute deviation,** 698, 700

**Means**
   in comparing data sets, 682–684
   in data transformation, 681–682
   distribution and, 675
   as measures of center, 651–652, 699
   outliers and, 666
   in proportions, 116

**Means–Extremes Property of Proportion,** 116

**Measurement**
   area, P26–P27, P31, 19, 25, 132, 210, 397–399, 423–424, 591–592
   circumference, P24
   dimensional analysis, 124, 128, 132
   greatest possible error, 40
   perimeter, P23–P24, 20, 28, 80
   surface area, P31, 19, 398
   temperature, 82, 134, 270
   volume, P29, 7, 12, 14
   126, 147, 148, 149, 161, 167, 169, 175, 176, 177, 187, 188, 195, 229, 237, 252, 264, 273, 290, 293, 300, 306, 319, 322, 325, 351, 352, 357, 359, 365, 369, 372, 379, 400, 403, 408, 426, 434, 450, 453, 465, 496, 502, 510, 521, 523, 525, 537, 543, 592, 593, 594, 620, 635, 644, 678, 686, 694, 695, 696

**Measures of center** (measures of central tendency), 651–653, 681–682, 698–699
   in data transformation, 681–682
   mean, 116, 651–653, 666, 675, 681–684, 799
   median, 651–653, 666, 675, 681–682
   mode, 651, 681–682, 699

**Measures of position,** 665–666

**Measures of spread** (variation), 665–668, 700
   interquartile range, 666
   range, 42, 50, 141, 561, 665–666, 681–682
   standard deviation, 667–668, 681–684

**Median,** 651–653, 666, 675, 681, 682, 699

**Median-fit line,** 262, 276, 279

**Metric,** 33, 65

**Mid-Chapter Quiz,** 32, 39, 169, 246, 308, 367, 418, 490, 605, 672

**Midpoint,** 109

**Minimum values,** 559, 561–562, 603–604

**Mode,** 651, 681–682, 699

**Modeling,** 36, 45, 46, 82, 154, 169, 194, 201, 228, 230, 232, 236, 244, 273, 292, 324, 342, 343, 371, 378, 415, 433, 501, 543, 544, 585, 599, 600, 611, 6333, 644, 654, 655, 669, 686

**Monomials**
   degree of, 491
   dividing, 402–408
   identifying, 395
   multiplying, 396–397
   simplifying, 398

**Multiple Representations,** 8, 21, 28, 83, 105, 120, 149, 167, 244, 273, 293, 306, 313, 319, 325, 344, 359, 400, 408, 416, 573, 496, 502, 510, 516, 525, 537, 544, 568, 578, 600, 612

**Multiplication**
   associative property, 18, 26, 65
   binomials, 423, 504–505, 506–510
   commutative property, 18, 26, 65
   of conjugates, 421
   cross products, 116–117
   data transformations using, 681–862
   elimination using, 354–359, 361–365, 382, 384
   equation models for, 86

exponents, 395–400, 475
FOIL method, 423, 507–508, 531
of functions, 632–633, 637, 642
of $f(x)$ by a constant, 181–183
of integers, P12
of monomials, 395–397
of a polynomial by a monomial, 498–499
of polynomials, 504–505, 506–510, 546
product of a sum and a difference, 514
product property of square roots, 419
radical expressions, 419–420, 421, 424
radicands, 423, 424
of rational and irrational numbers, 428
rational numbers, P17–P18, 30–31
solving inequalities by, 296–297, 329
square roots, 419–420
of $x$ by a constant, 441–442
zero product property, 522

**Multiplication Property**
of equality, 88, 128
of inequality, 296

**Multiplicative identity,** 17
**Multiplicative inverses,** P18, 17
**Multiplicative Property of Zero,** 17
**Multi-step equations,** 94, 95–99, 130
**Multi-step inequalities,** 302–306, 328, 329
**Multi-step problems,** 9, 15, 21, 22, 29, 64, 84, 93, 94, 99, 100, 127, 150, 156, 168, 177, 178, 189, 202, 209, 230, 245, 253, 258, 294, 301, 307, 314, 318, 320, 325, 326, 345, 353, 364, 373, 380, 401, 409, 413, 435, 444, 445, 447, 454, 466, 467, 474, 497, 503, 511, 515, 517, 526, 538, 545, 553, 567, 569, 586, 594, 595, 621, 627, 646, 664, 697

## N

**Natural numbers,** P7
**Negative correlation,** 247
**Negative Exponent Property,** 404, 475
**Negative exponents,** 404–406
**Negative numbers,** P7, 302
**Nonlinear expressions,** 395

**Nonlinear functions,** 52, 58
**Non-terminating decimals,** 428
**Notation**
function, 52
inverse linear functions, 269
set-builder, 290–291
**Note-taking.** *See* Foldables® Study Organizer
*n*th **roots,** 411
*n*th **terms,** recursive formulas and, 469–473
**Number line**
absolute value, 107–109
arithmetic sequences, 190–193
coordinates on, P8
inequalities, 290, 297–298, 303–304, 309–310, 316–317
number sets graphed on, P8
in operations with integers, P11, P15
scaled, 658
**Numbers**
complex, 581
graphing a set of, P8
integers, P7, 96–97, 200
irrational, P7, 428–429
natural, P7
negative, P7
positive, P7
rational, P7, P13–P16, 30–31, 428
real, P7–P10, 30, 303
rounding, 35, 131
whole, P7
**Number theory,** 96–97, 104, 105, 364
**Numerical expressions,** 10–14

## O

**Odds,** P35
**Ohm's Law,** 123
**Open Ended.** *See* Higher-Order Thinking Skills
**Open half-planes,** 321
**Opposite reciprocals,** 240–241
**Opposites,** P11
**Ordered pair,** 42, 43
**Order of magnitude,** 405–406
**Order of operations,** 10–14, 398
**Outliers,** 666, 675

## P

**Parabolas,** 559–561, 587, 603–604
**Parallel lines,** 239–244, 276, 278, 377
**Parallelograms**
area of, P26–P27
perimeter of, P23–P24
**Parent functions,** 151, 183, 571
**Percentiles,** 653, 698
**Percent proportion,** P20–P21
**Percent rate of change,** 458–460
**Percents,** P20–P21
**Perfect squares,** P7, P9
**Perfect square trinomials,** 513, 540, 546
**Performance Task,** 72, 134, 218, 332, 388, 482, 552, 644, 704
**Performance Task Preview,** 4, 78, 140, 224, 288, 338, 394, 488, 558, 650
**Perimeters,** P23
parallelogram, P23–P24
rectangle, P23
regular hexagon, 28
square, P23
triangle, P23, 20, 28, 80
**Perpendicular lines,** 240–244, 278
**Perseverance,** 14, 24, 28, 55, 126, 195, 209, 313, 349, 352, 355, 362, 367, 400, 434, 502, 544, 635
**Piecewise-defined functions,** 198–202
**Piecewise-linear functions,** 197, 204, 216
**Pisano, Leonardo (Fibonacci),** 369
**Point–slope form,** 233–235, 276, 277
**Polynomials,** 491
adding, 489–490, 492–493, 546, 548
degree of, 491–492
factoring, 520–525, 531–535, 539
factoring quadratic trinomials, 531–535
identifying, 491
modeling, 489
multiplication by a monomial, 498–499
multiplying, 504–505, 506–510, 546
perfect square trinomial, 513
prime, 535
product of a sum and a difference, 514
solving, 499

400, 408, 416, 426, 434, 446, 453, 466, 473, 486A, 496, 502, 510, 516, 525, 531, 537, 544, 556A, 568, 577, 585, 594, 601, 612, 620, 626, 635, 648A, 655, 663, 670, 678, 686, 696

Strategies, 5, 10, 16, 33, 42, 49, 58, 79, 87, 95, 143, 151, 160, 171, 181, 190, 197, 205, 225, 232, 239, 247, 254, 259, 267, 289, 298, 302, 309, 316, 321, 339, 348, 354, 361, 368, 376, 395, 402, 410, 419, 438, 448, 469, 491, 498, 506, 539, 559, 571, 580, 588, 596, 606, 614, 622, 630, 651, 657, 665, 673, 680, 688, 693

Structure, 92, 95, 101, 107, 112, 115, 120, 122, 138A, 197, 207, 234, 251, 267, 286A, 302, 305, 312, 321, 354, 359, 368, 392A, 402, 410, 421, 438, 462, 463, 486A, 495, 509, 510, 516, 536, 539, 540, 556A, 565, 589, 630, 648A, 657

Tools and Techniques, 34, 57, 103, 142, 155, 166, 240, 272, 324, 336A, 392A, 399, 411, 414, 456, 486A, 499, 528, 530, 556A, 583, 606, 652, 653, 673, 675

## O

**Open Educational Resources**
Activities, 580A
Animations, 122A
Apps, 10A, 49A, 107A, 141A, 181A, 410A, 519A, 588A, 688A
Assessment, 33A, 267A
Blogs, 58A, 94A
Communicating, 151A, 469A
Digital Resources, 657A
Extra Help, 376A
Games, 16A, 85A, 159A, 225A, 361A, 395A, 448A, 504A, 529A, 539A
Homework Help, 430A
Instructional Aids, 289A
Interactives, 571A
Lesson Planning, 462A
Lesson Sharing, 302A, 354A, 512A, 665A
Mnemonic Devices, 368A
Practice Problems, 33A, 247A, 267A, 295A, 489A
Professional Development, 458A
Projects, 101A
Publishing, 5A, 170A, 190A, 254A, 438A
Simulations, 259A
Submitting Assignments, 622A
Surveys, 673A
Tutorials, 23A, 239A, 309A, 315A, 339A, 402A, 419A, 498A, 596A
Video Sharing, 79A, 197A, 205A, 232A, 321A, 348A, 606A, 614A, 630A, 651A
Vocabulary Resources, 559A
Word Cloud, 42A

## P

**Pacing,** i, 2A, 10A, 16A, 23A, 33A, 42A, 58A, 94A, 138A, 151A, 159A, 170A, 181A, 190A, 197A, 205A, 222A, 232A, 239A, 247A, 254A, 259A, 267A, 286A, 295A, 302A, 315A, 321A, 336A, 339A, 348A, 354A, 361A, 368A, 376A, 392A, 395A, 402A, 410A, 430A, 438A, 448A, 458A, 462A, 469A, 486A, 489A, 498A, 504A, 512A, 519A, 529A, 539A, 556A, 559A, 571A, 580A, 588A,

connectED.mcgraw-hill.com  R109

596A, 606A, 614A, 622A, 630A, 648A, 651A, 657A, 665A, 673A, 688A

**Personal Tutors.** *See* Go Online!

**Planning Guide.** *See* Pacing

# Q

**Quizzes.** *See* Go Online!

solving equations with, 500
squares of sums and differences, 512–513
standard form of a, 492
subtracting, 490, 493, 546, 548
**Positive correlations,** 247
**Positive numbers,** P7
**Posttest,** P37–P38
**Power of a Power Property,** 397, 412, 475
**Power of a Product Property,** 397, 475
**Power of a Quotient Property,** 403
**Power Property of Equality,** 413
**Powers,** 5, 395–397. *See also* Exponents
**Practice Tests,** 71, 133, 217, 281, 331, 387, 481, 551, 643, 703
**Precision,** 20, 38, 47, 91, 111, 118, 119, 124, 125, 144, 226, 242, 243, 282, 299, 379, 425, 452, 453, 510, 654
**Prediction,** from slope-intercept form, 180, 227
**Preparing for Algebra.** *See* Chapter 0
**Preparing for Assessment,** 9, 15, 22, 29, 41, 48, 56, 64, 72–75, 84, 93, 100, 106, 113, 121, 127, 134–137, 150, 156, 168, 178, 189, 196, 203, 211, 218, 220–221, 231, 238–239, 245, 253, 258, 266, 274, 282, 284–285, 294, 301, 307, 314, 320, 326, 332–335, 345, 353, 360, 366, 373, 380, 388–391, 401, 409, 413, 427, 435, 447, 454, 467, 474, 497, 503, 511, 517, 526, 538, 545, 552, 554–555, 569, 579, 586, 595, 602, 613, 621, 627, 636, 644, 646–647, 656, 664, 671, 679, 687, 697, 704–707
**Pretest,** P3–P4
**Prime polynomials,** 535, 542
**Principal square root,** P7
**Probability,** P33–P35
in rolling dice, 407, 408
**Problem Solving.** *See also* Problem-Solving Tips; Real-World Examples
consecutive integers, 96–97
estimating, P9–P10, P18, P27, 583
four-step problem-solving plan, P5–P6
four-step, 80
writing problems, 81
**Problem-Solving Tips**
checking your answers, 355
deciding whether an answer is reasonable, 227
guess and check, 532
make a model, 25
make an organized list, 432
precision, 227
**Product of Powers Property,** 396, 403, 475
**Product Property of Square Roots,** 419
**Products,** 5. *See also* Multiplication
**Proofs,** 21
**Properties**
addition, 16
addition property of equality, 87–88, 128, 289
addition property of inequality, 289
additive identity, 16
additive inverse, 16
associative, 18, 26, 65
closure, 16
commutative, 18, 26, 65
distributive, 23–25, 26, 65, 303, 506–508, 519, 548
division properties of exponents, 402–408
division property of equality, 88, 128
division property of inequality, 298
of equality, 16, 65
identity, 16, 17, 26
means–extremes property of proportion, 116, 128
multiplication properties of exponents, 395–400, 475
multiplication property of equality, 88, 128
multiplication property of inequality, 296
multiplicative identity, 17
multiplicative property of zero, 17, 26
negative exponent, 404, 475
power of a power, 397, 412, 475
power of a product, 397, 475
power of a quotient, 403, 475
power property of equality, 413
product of powers, 396, 403, 475
product property of square roots, 419
quotient of powers rule, 402, 475
quotient property of square roots, 420
reflexive, 16, 65
square root, 588, 590
substitution, 16, 26, 65
subtraction property of equality, 88, 128
subtraction property of inequality, 88, 290–291
symmetric, 16, 23, 65
transitive, 16, 65
zero exponent, 404, 475
zero product, 522
**Proportions,** 115–120
equations, 115–116, 128, 131
extremes in, 116
means–extremes property of, 116, 128
percent proportion, P20–P21
solving, 117, 131
**Pyramids,** 14
**Pythagorean theorem,** 517

# Q

**Quadratic equations**
differences of squares, 539, 546
double root, 581
factoring, 531–535, 546, 549
FOIL method and, 507, 531
solutions of, 580–582
solving
algebraically, 616–618
by completing the square, 596–601, 640
by estimating, 583
by factoring, 588–592, 637, 639
by graphing, 580–585, 615, 637, 639
using different methods, 609–610
using square root method, 588–589
using the quadratic formula, 606–610, 640
systems of, 614–618, 637
using algebra tiles, 529–530
**Quadratic expressions,** 507
**Quadratic Formula,** 606–610, 640
**Quadratic functions,** 559, 637
characteristics of, 559–562, 575
graphing, 559–560, 562–568, 638
identifying equations from graphs, 575–576, 582, 622–623
linear and exponential functions compared with, 563

R110 | Index

rate of change of, 570
solutions of, 580
standard form of, 559
transformations of, 571–578, 638–639
vertex form of, 576, 598, 603–604
writing equations for, 623–624

**Quadratic inequalities,** 587

**Qualitative data,** 651

**Quantitative data,** 651

**Quartiles,** 665

**Quotient of Powers Rule,** 402, 475

**Quotient Property of Square Roots,** 420

**Quotients.** *See* Division

## R

**Radical expressions,** 419–426, 475. *See also* Square roots
adding, 422, 424
conjugates, 421
dividing, 420, 424
multiplying, 419–420, 421, 424
rationalizing the denominator, 421
simplifying, 419, 420–421
subtracting, 422

**Radical form,** 410

**Radicands,** 419
adding, 422, 424
multiplying, 423, 424
subtracting, 422, 424

**Radius,** P24

**Rate of change**
in exponential functions, 432, 468
in linear functions, 159, 160–167, 173–174, 212, 214
in quadratic functions, 570

**Rates,** 117

**Rational exponents,** 410–416, 475

**Rationalizing the denominator,** 421

**Rational numbers,** P7, 30. *See also* Percents
adding, P14–P15, 31
comparing, P13
dividing, P18
multiplying, P17–P18, 30–31

operations with irrational and, 428–429
ordering, P13
percents as, P20
repeating decimals as, P8
subtracting, P14–P16

**Ratios,** 115–120
common, 462–464
debt-to-income, 33–34
equivalent, 115, 131
means-extremes property of proportion, 116
proportions and, 115–116
rate of change, 117–118, 160–167
slope of a line, 162–164, 225–226

**Reading Math**
absolute values, 108
checking two-way frequency tables, 691
fractions in the radicand, 420
intercepts, 144
reading problems, 219
set-builder notation, 290
subscripts, 163
totals for two-way frequency tables, 691

**Real numbers,** P7–P10, 30, 303

**Real-World Careers**
astronomer, 403
in Chile, 270
entertainment manager, 153
financial advisor, 449
ground crew, 227
retail buyer, 117
sound engineering technician, 350
sports marketer, 6
trucking, 471
urban planner, 535
veterinarian, 303
wheelchair marathoner, 660

**Real-World Examples,** 12, 18, 33, 44, 50, 58, 59, 60, 61, 80, 89, 96, 118, 123, 152, 153, 160, 161, 181, 193, 198, 199, 207, 227, 232–233, 240, 247, 248, 259–260, 261, 291, 297, 302, 310, 317, 323, 341, 350, 356, 363, 377, 406, 413, 431, 439, 449, 450, 464, 493, 508, 513, 523, 564, 583, 589, 591, 598, 604, 614, 617, 624, 631, 632, 651, 676, 688–689

**Real-World Links,** 12, 18, 24, 34, 44, 50, 60, 80, 89, 96, 108, 119, 174, 193, 198, 207, 240, 248, 262, 291, 317, 323, 341, 350, 356, 377, 405, 406, 413, 431, 664, 493, 499, 508, 523, 564, 583, 598, 624, 676, 683

**Reason Abstractly,** 33, 35, 232, 651, 652, 653

**Reasoning,** 8, 12, 14, 19, 21, 27, 28, 39, 53, 55, 63, 83, 88, 90, 92, 98, 99, 105, 112, 120, 126, 147, 148, 149, 161, 167, 169, 175, 176, 177, 187, 188, 195, 229, 237, 252, 264, 273, 290, 293, 300, 306, 319, 322, 325, 351, 352, 357, 359, 365, 369, 372, 379, 400, 403, 408, 426, 434, 450, 453, 465, 496, 502, 510, 521, 523, 525, 537, 543, 592, 593, 594, 620, 635, 644, 678, 686, 694, 695, 696

**Reciprocals,** P18, 17

**Rectangles**
area, P26, 25, 423
perimeter of, P23

**Rectangular prisms**
surface area, P29
volume, P29

**Recursive formulas,** 469–473

**Rees, Mina,** 191

**Reflections**
absolute value functions, 206
exponential functions, 441–443
linear functions, 184–185
quadratic functions, 574

**Reflexive Property,** 16, 65

**Regression,** 259–262, 276, 279
best-fit line, 259–262
linear, 259
median-fit line, 262, 276, 279
quadratic, 628–629
residuals, 260–261

**Regularity,** 97, 99, 149, 177, 191, 200, 206, 209, 318, 397, 408, 513, 696

**Relations,** 42–48, 65
dependent variables, 44
domains of, 42, 50
independent variables, 44
range of, 42, 50
representing, 43–44

**Relative error,** 313

**Relative frequency,** 690–693, 698

**Relative maximum,** 59–60

**Relative minimum,** 59–60

**Repeating decimals,** P8

**Residuals,** 260–261

## R

**Reading and Writing Mathematics,** 5, 10, 16, 23, 33, 42, 49, 58, 79, 87, 95, 101, 107, 115, 122, 143, 151, 160, 171, 181, 190, 197, 205, 225, 232, 239, 247, 254, 259, 267, 289, 296, 302, 309, 316, 321, 339, 348, 354, 361, 368, 376, 395, 402, 410, 419, 430, 438, 448, 462, 469, 491, 498, 506, 512, 520, 531, 539, 559, 571, 580, 588, 596, 606, 614, 622, 630, 651, 657, 665, 673, 680, 688

**Response to Intervention (RTI),** 4, 32, 71, 78, 114, 133, 140, 169, 217, 224, 246, 281, 288, 308, 331, 338, 367, 387, 394, 418, 481, 488, 518, 551, 558, 605, 643, 650, 672, 703

# S

**Scaffolding Questions,** 5, 6, 10, 11, 12, 17, 18, 19, 23, 24, 25, 26, 33, 34, 35, 36, 43, 44, 49, 50, 51, 52, 58, 59, 60, 79, 80, 81, 87, 88, 89, 95, 96, 97, 101, 102, 103, 107, 108, 109, 115, 116, 117, 118, 122, 123, 124, 143, 144, 145, 146, 152, 153, 160, 161, 162, 163, 164, 171, 172, 173, 174, 181, 182, 183, 184, 185, 191, 192, 193, 197, 198, 199, 205, 206, 207, 225, 226, 227, 232, 233, 234, 235, 239, 240, 241, 247, 248, 249, 254, 255, 259, 260, 261, 262, 267, 268, 269, 270, 289, 290, 291, 297, 298, 302, 303, 304, 309, 310, 311, 316, 317, 321, 322, 323, 339, 340, 341, 348, 349, 350, 355, 356, 361, 362, 369, 370, 376, 377, 378, 395, 396, 397, 398, 402, 403, 404, 405, 406, 410, 411, 412, 413, 419, 420, 421, 422, 423, 430, 431, 432, 438, 439, 440, 441, 442, 448, 449, 450, 451, 462, 463, 464, 469, 470, 471, 491, 492, 493, 498, 499, 500, 506, 507, 508, 512, 513, 514, 520, 521, 522, 531, 532, 533, 534, 535, 539, 540, 541, 542, 559, 560, 561, 562, 563, 564, 571, 572, 573, 574, 575, 576, 580, 581, 582, 583, 588, 589, 590, 591, 596, 597, 598, 606, 608, 609, 610, 614, 615, 616, 617, 618, 622, 623, 624, 630, 631, 632, 651, 652, 653, 657, 658, 660, 661, 665, 666, 667, 668, 673, 674, 675, 676, 680, 681, 682, 683, 688, 689, 691, 692, 693

**Self-Check Quizzes.** *See* Go Online!

**Sketchpad.** *See* Go Online! Geometer's Sketchpad

**Skills Trace,** 2A, 5A, 10A, 16A, 23A, 33A, 42A, 49A, 58A, 76A, 79A, 85A, 94A, 101A, 107A, 122A, 138A, 141A, 151A, 159A, 170A, 181A, 190A, 197A, 205A, 222A, 225A, 239A, 247A, 254A, 259A, 267A, 286A, 289A, 295A, 302A, 309A, 315A, 321A, 336A, 339A, 348A, 354A, 361A, 368A, 376A, 392A, 395A, 402A, 410A, 419A, 430A, 438A, 448A, 458A, 462A, 469A, 486A, 489A, 498A, 504A, 512A, 519A, 529A, 539A, 556A, 559A, 571A, 580A, 588A, 596A, 606A, 614A, 622A, 630A, 648A, 651A, 657A, 665A, 673A, 688A

**Standardized Test Practice.** *See* Go Online!

**Summative Assessment.** *See* Assessment

**Review.** *See* Go Online!; Guided Practice; Preparing for Assessment; Review Vocabulary; Study Guide and Review

**Review Vocabulary**
absolute value, 488
additive inverse, 4
algebraic expression, 78
base, 394
bivariate data, 650
box-and-whisker plot, 650
coefficient, 224
coordinate system, 78
Distributive Property, 394
domain, 338, 558, 562
equivalent equations, 288
exponent, 394
FOIL method, 423
function, 78, 224
intersection, 338
leading coefficient, 558
linear equation, 288
multiplicative inverse, 4
opposite reciprocal, 240
origin, 140
parallel lines, 340
perfect square, 488
perimeter, 4
proportion, 338
range, 558, 562
ratio, 224
reciprocal, 89
solution set, 288
standard form of a linear equation, 234
x-axis, 140
y-axis, 140

**Rhombus,** 423

**Rise,** 162–164. *See also* Slope of a line

**Root of an equation,** 151, 588

**Roots.** *See also* Square roots
cube, 411
double, 581
estimating, P9–P10
negative sign and, 420
nth, 411
simplifying, P9, 419–420
square, P7

**Row reduction,** 375

**Run,** 162–164. *See also* Slope of a line

# S

**Samples,** 667–668

**Sample space,** P33

**Scale,** 118

**Scale model,** 118

**Scatter plots,** 247–252
correlations, 247, 254–256
curve fitting, 628–629
interpolation and extrapolation, 249
lines of fit, 248–249, 260–261, 276, 278
median-fit line, 262

**Science on a Sphere (SOS)®,** 12

**Second differences,** 623

**Sense-Making,** 7, 24, 43, 54, 61, 62, 104, 110, 146, 158, 165, 210, 244, 252, 257, 358, 363, 388, 407, 493, 494, 524, 542, 568, 573, 612, 615, 616, 655, 663, 674

**Sequence,** 190–195, 469–470

**Set-builder notation,** 290–291

**Sets**
closed, 30
hierarchy of, 428

**Sierpinski's triangle,** 466

**Simplifying**
algebraic expressions, 24–26
monomial expressions, 398
radical expressions, 419, 420–421
radicals, 422
radicands, 421
roots, P9, 419–420

**Slope–intercept form,** 170–177
of constant functions, 172
graphing calculators and, 170
graphing equations, 171–172
for line of fit, 248–249, 278
modeling real-world data, 173–174
for parallel lines, 239, 276, 278
for perpendicular lines, 241, 278
prediction from, 180, 227
using, 227
writing equations in, 171, 225–227, 234–235, 277

**Slope of a line,** 162–164
coordinates from, 164
in equations, 225–226
formula for, 163
horizontal lines, 172
negative, 163–164
of parallel lines, 239, 276, 278
of perpendicular lines, 240, 278
positive, 163–164
rate of change and, 173–174, 212, 214
in squares, 235
undefined, 164
vertical lines, 172
zero, 163–164

**Solving**
consecutive integer problems, 87–92
dimensional analysis, 124, 132
equations
with absolute value, 107–112, 131
addition or subtraction, 85, 87–88
algebra tiles, 85–86, 94, 102
elimination, 354, 361
factoring, 522–523, 588–592, 637, 639
grouping symbols in, 11, 102
identities, 102
matrices, 374–375
multiplication or division, 86, 88–89
multi-step, 94, 95–99, 130
one-step, 87–92, 129
polynomials, 499–500
for a specific variable, 122–123
steps in, 128
with the variable on each side, 101–105, 122, 130
exponential equations, 413, 456–457
finding intercepts, 144–145
formulas, 123
four-step problem-solving plan, P5–P6
inequalities, 289–290, 295, 329
addition, 289–290, 329
division, 298, 329
multiplication, 296–297, 329
subtraction, 290–291, 329
using algebra tiles, 295
using the distributive property, 303
proportions, 117, 131
quadratic equations
algebraically, 616–618
by completing the square, 596–602, 640
by estimating, 583
by factoring, 588, 592, 637, 639
by graphing, 580, 585, 615, 637, 639
quadratic formula, 606–610, 640
square root method, 588–589
using different methods, 609–610

R112 | Index

radical expressions, 421
systems of equations
　algebraically, 616
　choose a method to, 368–369, 382
　by graphing, 340–341, 368, 383, 615
　no real solutions, 618
　possible number of solutions, 339–340, 349
　by substitution, 348–352, 368, 382, 383
　using elimination, 354–359, 361–365, 368, 382, 384, 527–528
　using matrices, 374–375

**Special functions,** 212
　absolute value functions, 205–210, 212, 216
　greatest integer function, 197, 204, 212

**SPF (sun protection factor),** 410

**Spheres,** 12

**Square**
　area of, P26
　completing the, 596–601, 640
　of a difference, 513
　difference of two squares, 539
　perfect squares, P7, P9
　perimeter of, P23
　of a sum, 512–513

**Square Root Property,** 588

**Square roots,** 410–411, 419–421
　dividing, 420
　estimating, P9–P10
　multiplying, 419–420
　perfect squares and, P7, P9
　product property of, 419
　quotient property of, 420
　in rationalizing the denominator, 421
　reading, 420
　simplifying, P9, 419–420
　in solving quadratic equations, 588–589

**Standard deviation**
　in comparing data sets, 682–684
　in data transformation, 681–682
　in statistical analysis, 667–668

**Standard form**
　of a linear equation, 143–144, 212, 232–233, 276, 277
　of a polynomial, 492
　of a quadratic function, 559

**Standard viewing window,** 157

**Statistics.** *See also* Data; Graphs; Measures of center; Measures of spread; Scatter plots
　in comparing data sets, 682–684
　correlations, 247–249, 254–257, 276, 279
　curve fitting, 628–629
　in data transformation, 680–682
　distribution and, 675–676, 682–684
　five-number summary, 665–666, 676
　histograms, 659–660, 673, 675, 700–701
　interquartile range, 666
　linear extrapolation, 249, 261
　linear interpolation, 249
　lines of fit, 248–249, 252–256, 276, 278
　mean, 116, 651–653, 666, 675, 681–684, 699
　mean absolute deviation, 698, 700
　median, 651–653, 666, 675, 681–682, 699
　mode, 651, 681–682, 699
　outliers, 666
　quartiles, 665
　range, 42, 50, 141, 561, 665, 666, 681–682
　standard deviation, 667–668, 681–684
　variance, 667

**Step functions,** 197–198, 200, 216

**Structure,** 13, 92, 112, 120, 163, 169, 182, 199, 230, 244, 250, 251, 282, 300, 304, 305, 312, 359, 421, 426, 495, 508, 509, 510, 515, 536, 544, 565, 585, 616, 619, 695

**Study Guide and Review,** 65–70, 128–132, 212–216, 276–280, 328–330, 382–386, 475–481, 546–550, 637–642, 698–702

**Study Tips**
　additive inverse of a polynomial, 493
　algorithms, 597
　box-and-whisker plots, 684
　characteristics of a function, 561
　checking answers, 521, 542, 616
　coefficients, 355
　coefficients and powers of 1, 396
　compare and contrast, 26
　congruent marks, P24
　constant functions, 60
　construct arguments, 17
　correlation and causation, 255
　counting possible outcomes, P34
　cross products, 117
　defining $n$, 470
　defining variables, 145
　dependent systems, 349
　direction, 206
　discriminant of a square trinomial, 610
　distribution of data sets, 652, 653
　domain of $n$, 470
　draw a diagram, P10
　eliminating choices, 173
　end behavior, 59
　equivalent equations, 146
　estimating, P18
　exponential growth functions, 450
　exponents, 397
　families of functions, 592
　finding the midpoint, 109
　formula sheets, 499
　function notation, 52
　geometric sequences, 463, 471
　graphing perpendicular lines, 241
　graphing systems of equations, 615
　graphing the inverse of a line, 268
　greatest common factor, 534
　greatest integer and piecewise-defined functions, 200
　guess and check, 532
　horizontal translations, 439
　identity function, 441
　inequalities, 304, 317, 322, 323
　intercepts, 145
　interpolation and extrapolation, 145
　interquartile range, 666
　intersections and unions, 311
　irrational numbers, 608
　linear or nonlinear function, 162
　location of zeros, 583
　means of samples and populations, 667
　measures of center, 651
　mental math, P14, P27
　modeling, 116
　modeling with functions, 633
　multiplying functions, 632
　negative constant in trinomial, 540
　negative discriminants, 618
　negative sign in rise and run, 172
　negative sign placement, 405
　$n$th roots, 411
　$n$th terms, 192
　number lines, P15
　odds, P35

# T

**Teaching the Mathematical Practices.** *See* Mathematical Practices

**Teaching Tips**
  Accuracy, 39
  Algebra Tiles, 504
  Alternate Method, P32, 375, 513
  Analyze Relationships, 186, 444
  Apply Math to the Real World, 34, 36, 233
  Bar Graph *vs.* Histogram, 659
  Building on Prior Knowledge, 192
  Checking the Equation, 624
  Check Solutions, 292, 298, 407
  Coefficients, 90
  Common Misconceptions, 27
  Correlation and Causation, 254
  Correlation Coefficient, 261
  Cross Products, 117
  Definitions, 30
  Dependent Variables, 173
  Draw a Diagram, P30
  Eliminating the Variable, 103
  Equal Factors Over Equal Intervals, 436
  Estimation, P6
  FOIL, 507
  Function Notation, 153
  Geometry Concept, 242
  Graphing Calculators, 57, 157, 170, 204, 261, 327, 346, 381, 456, 468, 532, 587, 628
  Horizontal Lines, 163
  Horizontal Translations, 439
  Interactive Whiteboard, 340
  Interpreting Solutions, 615
  Intersection and Union Symbols, 310
  Intervention, 420
  Inverse Functions, 269
  Irrational numbers, 428
  Isolating Variables, 91
  Mental Math, 24
  Modeling, 596, 652, 653
  Monomials, 405
  Multiple Representations, 182, 439
  Multiplication Facts, 500
  Multiply by a Negative Monomial, 500
  Negative Fractions, P19
  Notation, 411
  Open-Ended, 34
  Order of Operations, 411
  Pacing, 166
  Pencils on Coordinate Planes, 141
  Percent and Percentile, 653
  Perseverance, 540
  Polynomials, 527, 529
  Power of *r*, 464
  Probability and Odds, P35
  Problem Solving, 369
  Properties, 403
  Quadratic Formula, 609
  Reasoning, P21, 153, 251, 398, 521, 597
  Recall Definitions, 315

outliers, 674
parent function, 183
patterns, 514
percent proportion, P21
perfect squares, P9
pi, P24
piecewise-defined functions, 199
points equidistant from axis of symmetry, 563
polynomial multiplication, 508
positive rate of change, 161
power rules, 403
precision, 124, 226
probability, P34
products and quotients, P12
rational exponents on a calculator, 412
reason abstractly, 33, 35, 232
reasoning, 88, 255
reflection, 574
regularity, 191, 907
rounding percentages, 692
sense-making and perseverance, 24, 43
simplifying radicals, 422
slope, 226, 234
slopes in squares, 235
solving an equation, 102
solving an equation by factoring, 534
solving by elimination, 362
solving by subtraction, 357
solving equations mentally, 589
solving for a specific variable, 123
solving inequalities, 290
standard deviation of sample and population, 668
stretching quadratic functions, 573
substituting values of *x* to find *y*, 616
substitution, 349, 370
subtracting inside absolute value symbols, 206
systems of equations, 340, 369
systems of inequalities, 377
testing after correct factors found, 532
translating exponential functions, 439
translating linear functions, 182
using your text, 5, 17
vertex form, 598
vertical line test, 51
vertical translations, 439
when the *x*-coefficient is a fraction, 146

writing equations, 226
writing inequalities, 291
writing verbal models, 232
*y*-intercept, 561
zero and undefined slopes, 164
zero exponent, 404

**Substitution,** 348–352, 368, 382, 383
**Substitution Property,** 16, 26, 65
**Subtraction.** *See also* Differences
  of decimals, P16
  elimination using, 356–359, 368, 384
  equation models for, 85
  of fractions, P14
  of functions, 630–631, 637, 642
  integers, P11
  of polynomials, 490, 493, 546, 548
  of rational numbers, P14–P16
  solving equations using, 88
  solving inequalities using, 290–291, 329
  squares of sums and differences, 512–513
**Subtraction Property**
  of equality, 88, 128
  of inequality, 290–291
**Successive differences,** 622–624, 641
**Sunscreens,** 410
**Surface area,** P31
  cube, 398
  cylinder, P31, 19
  rectangular prism, P31
**Symmetric Property,** 16, 23, 65
**System of linear and quadratic equations,** 614–618, 637
**Systems of equations,** 339
  consistent, 339, 382
  dependent, 339, 349, 382
  graphing calculators, 346–347
  inconsistent, 339
  independent, 339
  linear, 370–372, 385, 527–528, 614–618, 641
  quadratic, 614–618, 641
  solve, 641
    algebraically, 616
    choose a method to, 368–369, 382
    by elimination, 354–359, 361–365, 368, 382, 384, 527–528
    by graphing, 340–341, 368, 383, 615
    by matrices, 374–375

no real solutions, 618
possible number of solutions, 339–340, 349
substitution, 348–352, 368, 382, 383
**Systems of inequalities,** 376–379, 381, 382, 386

# T

**Technology.** *See* Graphing Technology Labs; Technology Tips
**Technology Tips**
  1-Var stats, 681
  bin widths, 675
  calculating statistics, 683
  histograms, 682
**Temperature conversions,** 82, 134, 270
**Terms,** 5
  like terms, 25, 489–490
  *n*th term of a geometric sequence, 471
  *n*th term of arithmetic sequence, 190–192, 212
  *n*th terms, recursive formulas and, 469–473
**Test-Taking Strategies,** 73, 135, 219, 283, 333, 389, 553, 645, 705
**Test-Taking Tips,** 390
  assigning a variable, 135
  consecutive numbers and consecutive odd (or even) numbers, 136
  guess and check strategy, 389
  look for a pattern, 283
  reading math problems, 219
  solve multi-step problems, 553
  strategies for organizing data, 705
  tools, 103
  using a formula, 645
  write and solve an inequality, 333, 334
**Transformations,** 181, 571. *See also* Dilation; Reflection, Translation
  of data, 680–684
  of exponential functions, 438–443
  of linear functions, 181–188, 215
  multiple, 443
  of quadratic functions, 571–578, 637, 638–639
  of recursive and explicit formulas, 471

R114 | Index

**Transitive Property,** 16, 65
**Translation,** 181, 571
   of absolute value functions, 205–208, 216
   of exponential functions, 436–438, 441
   horizontal, 571–573, 575–576
   of a linear function, 181–182
   of quadratic equations, 571–572, 637
   of recursive and explicit formulas, 471
   vertical, 571, 573, 575–576
**Trapezoids,** 132
**Tree diagrams,** P34
**Triangles**
   area, P26–P27, 210, 399, 424, 591–592
   hypotenuse, 80, 517
   perimeter, P23, 20, 28
**Trinomials,** 491 *See also* Polynomials
   factoring, 529–530, 531–535, 540–541, 546, 549
   perfect square, 513, 540, 546, 610
**Two-way frequency tables,** 688–693, 698, 702
**Two-way relative frequency tables,** 690–692

### U

**Union of two graphs,** 310
**Unit analysis,** 124
**Unit conversion,** 134
**Unit rates,** 117
**Upper quartile,** 665

### V

**Variables,** 5, 651
   defining, P5, 145
   dependent, 44
   division by, 522, 590
   independent, 44
   solving for specific, 123
**Variance,** 667
**Venn diagrams,** 30, 428
**Verbal expressions,** 5–8, 26, 507
**Vertex,** 205, 559–561
**Vertex form,** 576, 598, 603–604

**Vertical dilation,** 571, 573, 575–576
**Vertical line test,** 51, 65
**Vertical lines,** 142
**Vertical translation,** 571, 573, 575–576
**Vocabulary Check,** 65, 128, 212, 276, 328, 382, 475, 546, 637, 698
**Volume,** P29
   cylinder, P29, 7
   pyramid, 14
   rectangular prism, P29
   sphere, 12

### W

**Watch Out!**
   check roots found by graphing, 581
   completing the square, 597
   converting percents to decimals, 631
   dividing by a variable, 522, 590
   maximum and minimum values, 562
   modeling, 36
   multiplying radicands, 423
   negative common ratio, 464
   negative signs in inequalities, 298
   notation, 269
   order, 163
   solutions of quadratic equations, 609
   solutions of system of equations, 617
   square of a difference, 513
   successive differences, 623
   transformations, 574
   writing equations, 624
**Which One Doesn't Belong?.** *See* Higher-Order Thinking Skills
**Whole numbers,** P7
**Writing in Math.** *See* Higher-Order Thinking Skills

### X

$x$-**coordinate,** 42
$x$-**intercept,** 141, 144–145, 151

### Y

$y$-**coordinate,** 42
$y$-**intercept,** 58, 141, 144–145, 225–226, 561

### Z

**Zero,** multiplicative property of, 17, 26
**Zero exponent,** 403–404
**Zero Exponent Property,** 404, 475
**Zero pairs,** 85, 94, 489–490
**Zero Product Property,** 590
**Zeros of linear functions,** 151–155, 212, 213
**Zeros of quadratic functions,** 560

---

Row Operations, 374
Self-Adhesive Notes, 295
Sense-Making, 46, 80, 161, 183, 228, 290, 541, 575, 590, 598, 610, 682
Set Building Notation, 290
Slope, 179, 234
Structure, 182, 440, 589, 593, 652
Symmetry and Graphing, 563
Translating, 80
Use Appropriate Tools Strategically, 631
Word Problems, P22
Zero Pairs, 85, 94, 489
Zero to the Zero Power, 404

**Teach with Tech**
   Blogs, P23
   Checking Answers, 25
   Graphing Calculators, 275
   Interactive Whiteboards, P5, P7, P14, P17, P20, P26, 5
   Interpreting Graphs, 145
   Like Terms, 25
   Number Lines, P8
   Portable Media Players, P34
   Video Recordings, P31
   Web Pages, P29
   Wikis, P11

## V

**Verbal/Linguistic Learners.** *See* Go Online!

**Vertical Alignment,** P5, P7, P11, P13, P17, P20, P23, P26, P29, P31, P33

**Visual Learners.** *See* Go Online!

**Visual/Spatial Learners.** *See* Go Online!

**Vocabulary.** *See* Go Online!

## W

**Watch Out!**
   Combining Like Terms, 631
   Common Errors, P30, 362, 509
   Common Misconceptions, P36
   Error Analysis, 14, 37, 54, 112, 126, 154, 166, 230, 237, 300, 313, 344, 351, 416, 473, 502, 537, 540, 544, 567, 600
   Error-hunting, 691
   Factoring, 543
   Preventing Errors, P11, P15, P18, P28, P32, 8, 13, 19, 21, 26, 52, 83, 88, 91, 116, 124, 164, 172, 192, 234, 242, 299, 322, 349, 356, 371, 403, 404, 405, 425, 532, 535, 541, 563, 608, 610, 694
   Reasoning, 542
   Sense-Making, 617
   Solutions, 616
   Standard Form, 144
   Student Misconceptions, P25, 12, 228, 249, 304, 310, 397, 431, 449, 536
   Transformations, 685

## Symbols

| | | | |
|---|---|---|---|
| ≠ | is not equal to | AB | measure of $\overline{AB}$ |
| ≈ | is approximately equal to | ∠ | angle |
| ~ | is similar to | △ | triangle |
| >, ≥ | is greater than, is greater than or equal to | ° | degree |
| <, ≤ | is less than, is less than or equal to | π | pi |
| −a | opposite or additive inverse of a | sin x | sine of x |
| \|a\| | absolute value of a | cos x | cosine of x |
| $\sqrt{a}$ | principal square root of a | tan x | tangent of x |
| a : b | ratio of a to b | ! | factorial |
| (x, y) | ordered pair | P(a) | probability of a |
| f(x) | f of x, the value of f at x | P(n, r) | permutation of n objects taken r at a time |
| $\overline{AB}$ | line segment AB | C(n, r) | combination of n objects taken r at a time |

## Algebraic Properties and Key Concepts

| | |
|---|---|
| **Identity** | For any number a, $a + 0 = 0 + a = a$ and $a \cdot 1 = 1 \cdot a = a$. |
| **Substitution (=)** | If $a = b$, then a may be replaced by b. |
| **Reflexive (=)** | $a = a$ |
| **Symmetric (=)** | If $a = b$, then $b = a$. |
| **Transitive (=)** | If $a = b$ and $b = c$, then $a = c$. |
| **Commutative** | For any numbers a and b, $a + b = b + a$ and $a \cdot b = b \cdot a$. |
| **Associative** | For any numbers a, b, and c, $(a + b) + c = a + (b + c)$ and $(a \cdot b) \cdot c = a \cdot (b \cdot c)$. |
| **Distributive** | For any numbers a, b, and c, $a(b + c) = ab + ac$ and $a(b - c) = ab - ac$. |
| **Additive Inverse** | For any number a, there is exactly one number $-a$ such that $a + (-a) = 0$. |
| **Multiplicative Inverse** | For any number $\frac{a}{b}$, where a, $b \neq 0$, there is exactly one number $\frac{b}{a}$ such that $\frac{a}{b} \cdot \frac{b}{a} = 1$. |
| **Multiplicative (0)** | For any number a, $a \cdot 0 = 0 \cdot a = 0$. |
| **Addition (=)** | For any numbers a, b, and c, if $a = b$, then $a + c = b + c$. |
| **Subtraction (=)** | For any numbers a, b, and c, if $a = b$, then $a - c = b - c$. |
| **Multiplication and Division (=)** | For any numbers a, b, and c, with $c \neq 0$, if $a = b$, then $ac = bc$ and $\frac{a}{c} = \frac{b}{c}$. |
| **Addition (>)*** | For any numbers a, b, and c, if $a > b$, then $a + c > b + c$. |
| **Subtraction (>)*** | For any numbers a, b, and c, if $a > b$, then $a - c > b - c$. |
| **Multiplication and Division (>)*** | For any numbers a, b, and c, <br> 1. if $a > b$ and $c > 0$, then $ac > bc$ and $\frac{a}{c} > \frac{b}{c}$. <br> 2. if $a > b$ and $c < 0$, then $ac < bc$ and $\frac{a}{c} < \frac{b}{c}$. |
| **Zero Product** | For any real numbers a and b, if $ab = 0$, then $a = 0$, $b = 0$, or both a and b equal 0. |
| **Square of a Sum** | $(a + b)^2 = (a + b)(a + b) = a^2 + 2ab + b^2$ |
| **Square of a Difference** | $(a - b)^2 = (a - b)(a - b) = a^2 - 2ab + b^2$ |
| **Product of a Sum and a Difference** | $(a + b)(a - b) = (a - b)(a + b) = a^2 - b^2$ |

* These properties are also true for <, ≥, and ≤.

## Formulas

| | |
|---|---|
| Slope | $m = \dfrac{y_2 - y_1}{x_2 - x_1}$ |
| Distance on a coordinate plane | $d = \sqrt{(x_2 - x_1)^2 + (y_2 - y_1)^2}$ |
| Midpoint on a coordinate plane | $M = \left(\dfrac{x_1 + x_2}{2}, \dfrac{y_1 + y_2}{2}\right)$ |
| Pythagorean Theorem | $a^2 + b^2 = c^2$ |
| Quadratic Formula | $x = \dfrac{-b \pm \sqrt{b^2 - 4ac}}{2a}$ |
| Perimeter of a rectangle | $P = 2\ell + 2w$ or $P = 2(\ell + w)$ |
| Circumference of a circle | $C = 2\pi r$ or $C = \pi d$ |

### Area

| | | | |
|---|---|---|---|
| rectangle | $A = \ell w$ | trapezoid | $A = \frac{1}{2}h(b_1 + b_2)$ |
| parallelogram | $A = bh$ | circle | $A = \pi r^2$ |
| triangle | $A = \frac{1}{2}bh$ | | |

### Surface Area

| | | | |
|---|---|---|---|
| cube | $S = 6s^2$ | regular pyramid | $S = \frac{1}{2}P\ell + B$ |
| prism | $S = Ph + 2B$ | cone | $S = \pi r\ell + \pi r^2$ |
| cylinder | $S = 2\pi rh + 2\pi r^2$ | | |

### Volume

| | | | |
|---|---|---|---|
| cube | $V = s^3$ | regular pyramid | $V = \frac{1}{3}Bh$ |
| prism | $V = Bh$ | cone | $V = \frac{1}{3}\pi r^2 h$ |
| cylinder | $V = \pi r^2 h$ | | |

## Measures

| Metric | Customary |
|---|---|
| **Length** | |
| 1 kilometer (km) = 1000 meters (m) | 1 mile (mi) = 1760 yards (yd) |
| 1 meter = 100 centimeters (cm) | 1 mile = 5280 feet (ft) |
| 1 centimeter = 10 millimeters (mm) | 1 yard = 3 feet |
| | 1 foot = 12 inches (in.) |
| | 1 yard = 36 inches |
| **Volume and Capacity** | |
| 1 liter (L) = 1000 milliliters (mL) | 1 gallon (gal) = 4 quarts (qt) |
| 1 kiloliter (kL) = 1000 liters | 1 gallon = 128 fluid ounces (fl oz) |
| | 1 quart = 2 pints (pt) |
| | 1 pint = 2 cups (c) |
| | 1 cup = 8 fluid ounces |
| **Weight and Mass** | |
| 1 kilogram (kg) = 1000 grams (g) | 1 ton (T) = 2000 pounds (lb) |
| 1 gram = 1000 milligrams (mg) | 1 pound = 16 ounces (oz) |
| 1 metric ton (t) = 1000 kilograms | |